ATLAS of PRENATAL RAT BRAIN DEVELOPMENT

Joseph Altman
Shirley A. Bayer

With the assistance of Clark Gedney, Julie Henderson, Kunda Das, and Mark O'Neil of the Department of Biological Sciences and a staff of Computer Engineering students at Purdue University

CRC Press
Boca Raton Ann Arbor London Tokyo

Library of Congress Cataloging-in-Publication Data

Catalog information is available from the Library of Congress.

Printed in the United States of America 2 3 4 5 6 7 8 9 0
Printed on acid-free paper

TABLE OF CONTENTS

ACKNOWLEDGMENTS

Completion of this Atlas would have been impossible without the cooperation of many individuals with diverse talents. We thank Libbey Craft and Sarah Frazer for their excellent work in sectioning and staining our collection of developing rat brains. We relied on Dr. Clark Gedney's computer science expertise to produce the 3-D images of the coronally sectioned brains. Clark Gedney was ever ready to give advice and a helping hand when we got into difficulties with overworked computers and idiosyncratic computer programs, and he spent endless hours in training the students who prepared some of the 3-D images. We wish to single out among them Fred Marshall, Manish H. Naik, Matthew Carpenter, and Husni Fahmi, each of whom has succeeded in reconstructing one or two 3-D images. Mark O'Neil was most helpful in putting the finishing touches to all the brain reconstructions. The meticulous task of checking the alphanumeric labels and preparing the Chronological Index was ably carried out by Kunda Das. Above all, we wish to thank Julie Henderson who took part in every facet of this work and succeeded in transferring to hard-disk cartridges the completed, camera-ready copy of this Atlas. Among Julie Henderson's tasks, all of them performed with great enthusiasm and skill, was the printing and scanning of the photomicrographs, the professional labeling of the figures, and the placement of Plates, Figures and captions in their proper places for easy use of the Atlas.

We thank Dr. Louis Sherman for helping us out with Departmental funds as this work was in progress. Paul Petralia of CRC Press advised and encouraged us each step along the way, and Chris Richardson and Gary Bennett of CRC Press helped us with their technical expertise.

INTRODUCTION

This atlas of the developing rat brain contains 10 embryonic (E) ages: Ell, E12, E13, E14, E15, E16, E17, E18, E20, and E22. Each of these ages is represented by two specimens, one cut in the sagittal plane, the other in the coronal plane. The specimens were selected from a large collection of developing rat brains prepared in our laboratory that range in age (at half-daily or daily intervals) from E10 to birth (E22-E23), and (at greater intervals) through the postnatal period into adulthood. We excluded from this atlas the E10 and E10.5 specimens because the closure of the neural tube is not completed in the rat until E11. We also excluded the E11.5 and E12.5 specimens because these brains were not sufficiently different from the brains of the E11 and E12 embryos. Finally, we omitted the brains of the E19 and E21 fetuses because the differences in these brains with respect to the adjacent ages were not pronounced enough to justify making the Atlas more bulky.

Our original purpose for preparing this large collection of prenatal and postnatal rat brains was for use as normative material to support our ongoing research on the dynamics of brain development in the rat. This research is based primarily on a still larger collection of experimental specimens of the developing rat brain prepared with the techniques of short-survival, sequential-survival, and long-survival [^{3}H]thymidine autoradiography. This experimental material has enabled us to determine the *site of origin*, *time of origin*, *pattern of migration*, and *mode of differentiation* of neurons throughout the rat nervous system during the pre- and postnatal periods.

Although the Atlas does not contain any experimental material, we have made an effort to apply much of what we have learned from those specimens to this Atlas. Specifically, we have made use of the experimentally determined timetables of neuron origin in different brain regions for the identification of patches of the germinal matrix (the neuroepithelium and secondary germinal zones) as putative sources of neurons in different brain regions at different stages of embryonic development. We also used our experimental materials for the identification of migrating streams of cells as well as differentiating fields of neurons prior to their settling in their final locations. The following example illustrates the utility of this approach.

There are two prominent embryonic structures in the basal telencephalon during the embryonic period, usually referred to noncommittally as the medial and lateral ventricular ridges or the medial and lateral ganglionic eminences, or more specifically, as the medial and lateral basal telencephalic neuroepithelia. Examination of these two structures through successive stages of embryonic development revealed that the more posterior medial ganglionic eminence is prominent at E14 and E15 but shrinks from E16 onward. In contrast, the more anterior lateral ganglionic eminence becomes prominent on E16 and continues to expand until birth. We coupled this observation with quantitative data on the time of neuron origin in the globus pallidus and the striatum (caudate-putamen complex). These data showed that the bulk of globus pallidus neurons are generated between days E13 and E16, whereas the bulk of striatal neurons are generated between days E16 and E22. On the basis of this evidence, and additional topographic indications, we identify in this Atlas the medial ganglionic eminence of the putative pallidal neuroepithelium and the pallidal differentiating field, and the lateral ganglionic eminence as the putative striatal neuroepithelium and the striatal differentiating field. Using similar evidence, we also identify several additional basal telencephalic germinal matrices, such as the septal, the rhinencephalic, and the amygdaloid neuroepithelia, and the neuroepithelia of the nucleus accumbens and the bed nucleus of the stria terminalis.

This Atlas is the most comprehensive treatment of embryonic brain development that has yet appeared. In addition to an extended coverage of 10 embryonic ages, it contains some innovations. Most important of these are

(1) the use of a rational alphanumeric notation system to label different developing and maturing brain regions,

(2) the use of computer graphics to delineate different brain regions in each illustrated section, and

(3) the use of 3-dimensional reconstructions of the developing brain at each age to help the reader visualize the plane of each illustrated coronal section. Such a 3-D reconstruction is particularly helpful for the visualization of the brains of young embryos in which, due to the folding of the embryo, a "coronal" cut at the forebrain level transects the rest of the nervous system at other angles.

MATERIALS AND METHODS

Dams. Laboratory-bred Wistar rats were used. Adult female rats were put into the home cages of individually housed male rats at 4:00 pm and were removed at 7:00 am the next morning. At that time, vaginal smears were taken from each female and those that had sperm were placed into separate maternity cages. The dams were provided with food and water *ad lib* throughout gestation. Two or more dams were deeply anesthestized with sodium pentobarbital (35 mg/kg) at successive gestational ages between 8:30 and 9:30 am and the embryos were removed by cesarean section and killed by immersion in Bouin's fixative.

Dating of the Embryos. The date when sperm were found in the vagina of a mated female was designated as the first day of gestation, or embryonic day 1 (E1). Although some embryologists designate the first day of gestation as day 0, we followed those who use day 1. Logically, an embryo may be called, if the date of fertilization is precisely known, as 1 minute, 1 hour or 1 day old, it cannot be called 0 minute, 0 hour or 0 day old. Because coitus typically occurs within hours if a female rat is placed in a male's home cage, we assume that fertilization antedated the inspection for sperms by several hours. In fact, since fertilization may have occurred soon after 4:00 pm on the previous day and the embryos were not killed until 8:30-9:30 on the next morning, up to 17 hours may have elapsed between conception and sacrifice. We could have designated the prospective embryos at the time the vaginal smear were taken as roughly 0.5 days old but, as we noted earlier, we could not reliably distinguish the brains of 11 rats from 11.5 rats (the latter were killed between 4:30-5:00 pm) or E12 rats from E12.5 rats. Readers used to the designation of day 0 as the first day of gestation should, therefore, add 1 day to the age of the embryos as presented in this Atlas.

Preparation of the Specimens. Embryos aged E11-E16 were immersed *de toto* into Bouin's fluid. From E17 on, the embryos were decapitated before fixation, and from E19 on, the calvarium was removed after the heads had been fixed for approximately 20 minutes to aid better penetration of the fixative into the brain. The embryos or brains remained in Bouin's fixative for 24 hours, then were transferred to 10% neutral formalin (pH 7.4) until the time of embedding. Embryos E14 or younger were embedded in methacrylate (JB-4, Polysciences); the brains of embryos E15 or older were embedded in paraffin (Tissue Prep, Fisher) according to standard histological procedures. The methacrylate-embedded specimens were cut at 3 µm serially with every 10th section is a series saved on slides. The paraffin embedded specimens were cut at 10 µm with every section saved on slides. The methacrylate sections were stained with toluidine blue, and the paraffin sections were stained with Harris' hematoxylin.

Photomicrography and Magnifications. Photographs of the stained sections were taken either with a Zeiss photomicroscope (younger embryos) or with a Wild-Leitz photomacroscope (older embryos and fetuses) using 35 mm Kodak Technical Pan film. Ilford multigrade paper with the highest contrast setting, and an Agfa-Gevaert automatic processor, were used in the preparation of the prints. The horizontal and vertical scales used for the Atlas figures are based on the net magnification during photography and printing without any correction for shrinkage during tissue preparation. We assume that shrinkage is variable across ages because of the changing ratio between ventricular and parenchymal space, and changing ratio of gray and white matter, as development proceeds. The horizontal and vertical scales are not designed as stereotactic coordinates because that requires external reference points (such as bony landmarks) which are absent in the embryos. In sagittal sections the horizontal 0 point has been set at the anterior pole of the brain and the vertical 0 point at the highest level of the brain, as determined by the particular orientation of the sectioned specimen at a given age. In coronal sections the vertical 0 point has, likewise, been set at the highest point, whereas the horizontal 0 point was set at the midline.

Preparation of the Drawings. The photographs from all specimens were taken at regular intervals (30 µm to 150 µm) throughout the brain in both the sagittal and coronal planes. These were printed on 4" X 5" paper, and sections which contained notable features were selected for illustration as Plates. This procedure was preferable to showing sections at regular intervals, because some interesting but small structures (such as the suprachiasmatic nucleus or interpeduncular nucleus) would often appear between intervals and would therefore not be illustrated. Those photographs selected as Plates were reprinted at higher magnification. A Rapidograph pen was used to outline the entire brain, the ventricular lumen, the superficial edge of the neuroepithelium, and any other obvious feature in the brain parenchyma. The outlined photographs were transilluminated, and the inked outlines were retraced with a fine pen or pencil onto either translucent mylar sheets or transparent acetate film. The outline images were scanned in as high contrast line art using a Silverscan II interfaced to a Macintosh

Quadra 950 computer. The final drawings shown in the Atlas were done using Adobe Illustrator software package (versions 3 and 5) after the scanned image files were imported as templates.

The Need for 3-D Reconstructions. Most novice users of brain atlases find it difficult to visualize the three-dimensional relationship of different brain regions when they examine two-dimensional photomicrographs or drawings of serial brain sections. This task becomes even more difficult in the case of the developing embryonic brain for two reasons. First, the structure of the embryonic brain is not static (like the adult brain) but undergoes a series of topological transformations in the course of its early development. This includes folding and unfolding along its axis, the evagination and invagination of different parts, and the differential growth of particular regions at different stages of development. Second, the nominal use of a standard "coronal" plane of sectioning (where top is dorsal and bottom is ventral) applies only to a selected region, usually the cerebral neocortex. This plane transects other regions of the brain, particularly those that undergo folding changes, in different planes, ranging from the oblique, to the horizontal, to the inverted coronal (where top is ventral and bottom is dorsal). To help the reader, every coronal Plate and Figure is supplied with a 3-D model of the brain from which the section was taken, with the Plate being reproduced on the face of the cutaway model. Paging through the Plates of a given embryonic age, the viewer will see at what level and in what orientation the sections were taken in succession from anterior (rostral) to posterior (caudal).

Construction of the 3-D Images. This involved several steps. First, all of the photographed coronal sections were printed at the same magnification along with the sagittal plates from another specimen of the same age. Second, the total length of the coronal brain was determined by counting all of the coronal sections, using the slides as a record of the total number of sections. The first section was the most anterior extent of the forebrain, and the last section was the most posterior extent of what was judged to be the beginning of the cervical spinal cord. To determine the placement of each photographed section in the Z axis (specimen length), the section number was multiplied by the section thickness and the magnification factor of the photographic prints. The total length of the coronal brain was compared to that of the sagittal brain at the same age, and the levels of various coronal sections with good landmarks (such as the section with the most posterior extent of the mammillary body) were indicated in the sagittal photographs along with the angle at which the coronal brain was cut in the Y axis (vertical dimension). Once the plane of several coronal sections has been successfully matched with the sagittal brain, all of the coronal sections were realigned using the curvature of the sagittal brain as a guide to determine the offset between sections in the Y axis. Theoretically, this procedure placed the coronal sections in juxtaposition to each other as they were before sectioning. Next, the outline of the perimeter of the brain was drawn onto the photograph with a fine Rapidograph pen, and the outline was traced onto a mylar sheet along with the reference marks on each photograph. The outline tracings were scanned as high contrast line art using a Silverscan II interfaced to a Macintosh Quadra 950 computer. A transparent sheet with a master set of reference marks in each tracing was aligned with those on the scanner. In this way, all of the contours were scanned into the computer already aligned in the X and Y planes. The scanned image files were saved in PICT format and were imported into Presenter Professional (VIDI Inc., Monrovia, CA). Spline curves were drawn using each contour as a template. The surface was built in ribbon-like segments from section to section by creating spline meshes between the contours as they were spaced apart in the Z axis. The 3-dimensional model in each Plate was cut away at the level where the illustrated section is situated in the brain. A photograph of that section was scanned into the computer and the section was cut out of the scan and placed into a new file using Adobe Photoshop software package. That file was imported into Presenter Professional and was projected onto an autosurface of the section at the face of the model. In this way, the rendered image contained a photograph of the section in exactly the correct perspective and at exactly the correct size as if the brain has been actually cut and rotated in the same way as the model.

The completed model can be rotated and viewed in any desired plane. The views that were judged to be most informative were rendered at high resolution using the Presenter Professional software package. For each Plate, the best view was judged to be of the upper left anterior part of the brain, looking downward to the posterior right side. Additional views of the entire model are shown in the first page of each age group.

NOMENCLATURE AND ABBREVIATIONS

The Preparation of Brain Atlases. Brain atlases are typically composed of three elements: (1) photographic plates that show how the brain appears in histologically prepared sections; (2) graphic reconstructions of the photographic plates that delineate or highlight important features seen in the plates; and (3) a nomenclature that identifies discrete structures by their conventional names. The guiding principle in the preparation of brain atlases has always been to supplement the landmarks of the nervous system that are discernible with the naked eye or light microscopy with details gathered by neuroanatomists using descriptive as well as experimental techniques. Not infrequently the structural details are qualified by functional considerations based on physiological, neuropsychological and clinical evidence.

Nomenclature. The nomenclature designating different brain structures by a distinctive term has evolved over several centuries, with a few of the names traceable to Antiquity. The origin of some of the terms (e.g., thalamus) is obscure, others are obvious (e.g., "dentate" nucleus of the cerebellum or "dentate" gyrus of the hippocampus). Topographic qualifiers like inferior and superior (as in the inferior olive and superior olive) are easy to follow, as are dorsal and ventral (as in the dorsal horn and ventral horn of the spinal cord), or dorsomedial and ventromedial (as in the dorsomedial and ventromedial nuclei of the hypothalamus).

Unfortunately, there is disagreement among neuroanatomists in the delineation and naming of several brain regions, and in some cases different terms are used to designate the same brain region or the same name is applied to different brain structures. As in other scientific disciplines where classification is important, there are "lumpers" and "splitters" among neuroanatomists. Lumpers use a single name for a structure that has discrete parts, while splitters use separate names for barely distinguishable components of a seemingly homogeneous structure.

Abbreviations. Coupled with the problem of nomenclature (which, of course, we cannot resolve here) are some conventions that tend to make the use of brain atlases difficult. This is the highly variable and sometimes inconsistent use of abbreviations to identify brain regions in the photographs or graphic reconstructions. Consider the example (Table 1) from two recently published atlases of the rat brain. Note the different abbreviations used for some selected components of a single brain region, the amygdala. For comparison we include in Table 1 our simplified and more rational alphanumerical coding that is used in this Atlas.

THE ALPHANUMERICAL CODING OF BRAIN STRUCTURES

The alphanumerical coding used in this Atlas was specially designed for the developing embryonic nervous system; its possible extension to the adult nervous system remains to be established. The design of this code was influenced by two widely used library methods for cataloging books and periodicals, the Dewey Decimal Classification and the Library of Congress Classification. These systems use a number or letter to designate a major subject matter (e.g., philosophy, religion, literature, history, science) and add numbers or letters to specify subdisciplines within a particular field (e.g., physics, chemistry, biology, etc. within the broader field of science).

Our alphanumerical coding consists of a *single initial letter* and *one or two attached numerals*, and this alphanumeric base may be followed by an *optional suffix* of one or two lower-case letters. The single initial letter specifies the *primary* neuroanatomic unit or division. The attached numeral or numerals provides the *secondary* division. The optional suffix makes possible the addition of a *tertiary* division.

THE INITIAL ALPHABETIC CODE FOR PRIMARY BRAIN DIVISIONS

Initial Capital Letter. In the form of a *capital letter*, the initial code designates a primary division or unit of the nervous system. In most instances, as a mnemonic aid, the initial capital letter corresponds to the first letter of the structure's common name, e.g.,

A = ***A***mygdala
B = ***B***asal ganglia
C = ***C***erebral cortex
M = ***M***idbrain
T = ***T***halamus

Because some brain structures share the same initial letter (cerebral cortex and cerebellum, hypothalamus and hippocampus) another available capital letter is used for the

TABLE 1

An Illustration of the Abbreviations used to Designate Some Parts of the Amygdala in Two Recent Atlases of the Rat Brain. Compare These With Our Alphanumeric Coding System

	ABBREVIATIONS		
Name of Structure	*Paxinos and Watson* [1]	*Swanson* [2]	*This Atlas*
Amygdala	Amg	AMY	A3
Amygdala - Hippocampal area	AHi	(PA)	A5
Basolateral amygdaloid nucleus	BL	BLA	A8
Basomedial amygdaloid nucleus	BM	BMA	A7
Central amygdaloid nucleus	Ce	CEA	A6
Central amygdaloid nucleus, lateral	CeL	CEAl	A6*l*
Central amygdaloid nucleus, medial	CeM	CEAm	A6m
Cortical amygdaloid nucleus	-	CoA	A9
Cortical amygdaloid nucleus, anterior	ACo	CoAa	A9a
Cortical amygdaloid nucleus, posterior	-	CoAp	A9p
Cortical amygdaloid nucleus, posterolateral	PLCo	CoApl	A9pl*
Cortical amygdaloid nucleus, posteromedial	PMCo	CoApm	A9pm*
Medial amygdaloid nucleus	M3	MEA	A14

[1] G. Paxinos and C. Watson. *The Rat Brain in Stereotaxic Coordinates*. (Second edition) San Diego: Academic Press,1986.

[2] L. W. Swanson. *Brain Maps: Structure of the Rat Brain*. Amsterdam: Elsevier, 1992.

*These are potential codes that have not been used in this developmental atlas

second structure, if possible one that provides a mnemonic aid, e.g.,

K = hippo***K***ampus
L = cerebe***L***lum

or else, an arbitrary capital letter has been chosen, e.g.,

X = pontine area
Y = medulla
Z = spinal cord

Initial Lower-Case Letter. Trans-regional components of the central nervous system (for instance, fiber tracts and brain ventricles) and components of the peripheral nervous system are designated by an initial lower case letter, e.g.,

f = ***f***iber tracts
g = ***g***lands and paraneural structures
n = ***n***erves and nuclei, cranial
v = ***v***entricles

THE NUMERICAL CODE FOR SECONDARY BRAIN DIVISIONS

"Higher Order" and "Lower Order" Secondary Brain Divisions. In the case of secondary brain divisions, we have tried to distinguish "higher order" and "lower order" brain regions. The term higher order secondary brain division refers to distinct, usually large brain regions that could very well have been treated as primary divisions but, for some reason or another, are usually considered to be components of a primary division. Examples are the striatum and pallidum that are considered to be part of the basal ganglia (B), or the inferior colliculus and superior colliculus that are considered to be part of the midbrain (M). To the category of lower order secondary brain division are assigned smaller structures that are typically called "nuclei" in ganglionic regions or "layers" in cortical areas. Where possible, the alphanumeric code makes a distinction between the "higher order" and "lower order" secondary brain divisions, as will be described below.

Reserved Numbers. Because of the great importance in the developing nervous system of the germinal matrices where neurons are generated, the numeral ***1*** in the *first decimal place* has been reserved for the primary periventricular neuroepithelium and the numeral ***2*** in the *first decimal place* has been reserved for the secondary germinal matrices of the primary divisions of the brain, e.g.,

A***1*** = amygdaloid neuroepithelium
H***1*** = hypothalamic neuroepithelium
P***1*** = preoptic neuroepithelium
T***1*** = thalamic neuroepithelium

and

C***2*** = neocortical subventricular zone
L***2*** = cerebellar external germinal layer
T2 = thalamic subventricular zone

Where the primary brain division has higher order secondary divisions with distinct neuroepithelia, the numeral **1** in the first decimal place is supplemented by a numeral in the *second decimal place* to distinguish the different neuroepithelia. Thus, in the basal ganglia (B) we have:

B1 = striatal neuroepithelium
B***1***1 = pallidal neuroepithelium
B***2***1 = strionuclear neuroepithelium*
B***3***1 = accumbens neuroepithelium

In the same way, the numeral **2** in the first decimal place is supplemented by a numeral in the *second decimal place* to distinguish the secondary germinal matrices (and sometimes migratory streams) of the higher order divisions of the brain. Thus in the case of the large pontine area (X) we have:

X2 = pontine subventricular zone
X***1***2 = precerebellar migration
X***2***2 = cochlear external germinal layer

Finally, the numeral ***3*** in the *first decimal place* has been reserved as a comprehensive designator for either the differentiating or

* neuroepithelium of the bed nucleus of stria terminalis

mature principal brain region preceded by a capital letter, e.g.,

A*3* =	amygdala (or differentiating field)
H*3* =	hypothalamus (or differentiating field)
P*3* =	preoptic area (or differentiating field)
T*3* =	thalamus (or differentiating field)

Where the differentiating primary brain division has several higher order subdivisions, the numeral **3** in the first decimal place is supplemented by a numeral in the *second decimal place* to distinguish these subdivisions, e.g. in the case of the basal ganglia:

B3 =	striatum (or differentiating field)
B*1*3 =	pallidum (or differentiating field)
B*2*3 =	bed nucleus of stria terminalis (or differentiating field)
B*3*3 =	nucleus accumbens (or differentiating field)

or, in the case of the midbrain:

M3 =	superior colliculus (or differentiating field)
M***1***3 =	tegmentum (or differentiating field)
M*2*3 =	pretectum (or differentiating field)
M*3*3 =	inferior colliculus (or differentiating field)

Consecutive Numbers. The numerals 4, 5, 6... and 14, 15, 16... have been consecutively assigned to "lower order" secondary brain division on a first come, first served basis. An example in the septum (S) is:

S***4*** =	indesium griseum
S***5*** =	lateral septum
S***6*** =	medial septum
S***14*** =	bed nucleus of anterior commissure
S***15*** =	triangular septal nucleus
S***16*** =	taenia tecta

Where the numeral in the second decimal place has been reserved to desginate a higher order structure, an attempt has been made to keep the lower order brain structures belonging to that division with the same designator. Thus, we have in the midbrain:

M13 =	tegmentum (or differentiating field)
M14 =	substantia nigra
M15 =	interpeduncular nucleus

M23 =	pretectum (or differentiating field)
M24 =	anterior pretectal nucleus
M25 =	nucleus of the posterior commissure

M33 =	inferior colliculus (or differentiating field)
M34 =	external nucleus of inferior colliculus
M35 =	central nucleus of inferior colliculus

As an aid to the novice reader, the numeral code for principal structures (***3***) is always presented in coronal sections on the right half of the brain and are printed in bold font both in the figures and the captions (e.g., **A3**, **B3**, **B13**, **B23**).

OPTIONAL SUFFIXES FOR TERTIARY BRAIN DIVISIONS

Many brain regions may be adequately identified by an alphanumeric code consisting of an initial capital letter and the attached single or double numeral. Other brain structures have several tertiary divisions and a suffix consisting of one or two lower case letters is used to distinguish these subdivisions. Examples in the rhinencephalon (R) are:

R6	= anterior olfactory nucleus
R6***d***	= anterior olfactory nucleus, dorsal
R6***l***	= anterior olfactory nucleus, lateral
R6***m***	= anterior olfactory nucleus, medial
R6***p***	= anterior olfactory nucleus, posterior

THE ALPHANUMERIC CODING OF TRANSREGIONAL, NON-NEURONAL, AND PERIPHERALLY-RELATED STRUCTURES

A different coding principle was adopted for the identification of structures, such as fiber tracts and ventricles, that are not topographically limited to a single brain division, and for structures connected with or forming part of the peripheral nervous system, such as the cranial nerves and cranial nerve nuclei. As described earlier, lower case letters designate fiber tracts (f), glands and paraneural structures (g), cranial nerves and nuclei (n), and ventricles (v). The numerals attached to these codes have no reserved numbers or decimal positions but have been used consecutively, roughly in an order from anterior to posterior. Examples are:

f***1***	= lateral olfactory tract
f2	= medial olfactory tract
f***3***	= corpus callosum
f***11***	= external capsule
f***12***	= cingulum bundle
f***13***	= fimbria

In the case of the cranial nerves and nuclei, the lower case letter ***n*** is followed by the traditional number of the nerve, e.g.,

n***1***	= olfactory nerve
n***2***	= optic nerve
n***7***	= facial nerve

The cranial nerve ganglia, nuclei or other components of the nerve are identified by a suffix, e.g.,

n5	= trigeminal nerve
n5***g***	= trigeminal ganglion
n5***n***	= trigeminal motor nucleus
n5***p***	= principal (sensory) trigeminal nucleus

In a similar fashion, the code for the ventricles is the lower case ***v*** and the different ventricles are distinguished by the corresponding number or letter as a suffix, e.g.,

v***a***	= aqueduct
v***l***	= lateral ventricle
v***3***	= third ventricle

Because of the great developmental significance of subdivisions of the different regions or patches of the periventricular primary neuroepithelium, one or two lower-case letters as suffixes identify the changing subdivisions of the ventricles, e.g.,

v***la***	= lateral ventricle, anterior horn
v***lfa***	= lateral ventricle, amygdaloid fork
v***lfp***	= lateral ventricle, pallidal fork
v***lfs***	= lateral ventricle, septal fork

SOME PRINCIPLES OF BRAIN DEVELOPMENT

An atlas is not the appropriate forum for a lengthy discussion of the principles of brain development, much of which is hypothetical and controversial. The few principles that we shall deal with here are those that will aid the user of this Atlas to appreciate the nature of the complex transformations that take place in the structure of the rat brain from the day of the closure of the neural tube (E11) to the day before birth (E22).

The Neural Tube and the Brain Vesicles. The transformation of the superficial neural plate (neuroectoderm) into the neural tube through invagination and subsequent fusion at its dorsal edges is described and illustrated in most textbooks of biology or embryology. The neural tube extends along the length of the trunk of the embryo and, as development proceeds, becomes transformed into the spinal cord. A modified region of the neural tube are the brain vesicles in the head, or cephalic, region of the embryo. Most commonly, three primary brain vesicles are distinguished: the prosencephalic (forebrain), mesencephalic (midbrain), and rhombencephalic (hindbrain). The prosencephalic vesicle rapidly becomes partitional into the telencephalic vesicle (which gives rise to the cerebral cortex, basal ganglia and some other structures) and the diencephalic vesicle (which gives rise to the thalamus and the hypothalamus). Some authors also partition the rhombencephalon into the metencephalon (the primordium of the cerebellum and the pons) and the myelencephalon (the primordium of the medulla). Since from the beginning of the closure of the brain vesicles, the anterior portion of the prosencephalon (the primordium of the telencephalon) and the posterior portion of the prosencephalon (the primordium of the diencephalon) are easily distinguished, we shall partition from the outset the cephalic extension of the neural tube into four vesicles: the *telencephalic*, *diencephalic*, *mesencephalic*, and *rhombencephalic*. This accords well with the division of the lumen of the cephalic vesicles into four parts: the lateral ventricle (telencephalon), the third ventricle (diencephalon), the aqueduct (mesencephalon), and the fourth ventricle (rhombencephalon).

The Alar Plate and the Basal Plate. As important as the division of the different cephalic vesicles is the division of the developing neuraxis into an upper and a lower half, the *alar plate* (or alar lamina) and the *basal plate* (or basal lamina). In the neural tube caudally, the alar plate is the source of neurons of the dorsal horn, and the basal plate is the source of neurons of the ventral horn. This suggested the hypothesis that the alar plate is the primordium of the afferent or sensory components of the central nervous system, and the basal plate is the primordium of the efferent or motor components of the central nervous system. This hypothesis can easily be extended to the mesencephalon and the diencephalon (Fig. 1). In the mesencephalon, the alar plate is the source of neurons of the superior and inferior colliculi (which receive optic and auditory afferents, respectively) and the basal plate is the primordium of the tegmentum (a major motor outflow region). In the diencephalon the alar plate is the source of neurons of the thalamic nuclei (which are relay stations in the somesthetic, auditory and visual pathways) and the basal plate is the primordium of the hypothalamus (which contains many of the efferent centers of the autonomic, endocrine and other motor systems). In the spinal cord and medulla, a furrow, the sulcus limitans, is said to divide the alar and basal plates. However, the course of this sulcus is not easy to follow into the brain.

Figure 1 (p. xiv) illustrates the two possible divisions of the rostral neuraxis into an alar and a basal plate. In these unfolded representations of the early-embryonic brain, hatched lines descriptively divide the neuraxis into a dorsal and a ventral half, and dotted lines hypothetically divide the neuraxis into an alar plate and a basal plate. Caudally, the two lines overlap, but rostrally they separate. According to one view (Theory 1), the alar plate gives rise to one ventral telencephalic region, the rhinencephalon, which together with the olfactory bulb and the primary olfactory cortex is obviously an afferent system. This theory assumes that the other major ventral telencephalic regions (septum, basal ganglia, amygdala) are basal plate derivatives. It is this theory that has served as the basis for the delineations presented in this Atlas. However, an equally plausible view (Theory 2) is that the basal plate extends only to the rostral end of the diencephalon while the entire telencephalon is of alar plate origin. A more-detailed derivation of some of the major brain structures from the neuroepithelium of the alar plate and the basal plate caudally, and from the neuroepithelium of the dorsal and ventral portions of the telencephalon rostrally, is presented in Figure 2 (pp. xvi-xvii).

FIGURE 1

Derivatives of the Alar Plate and the Basal Plate

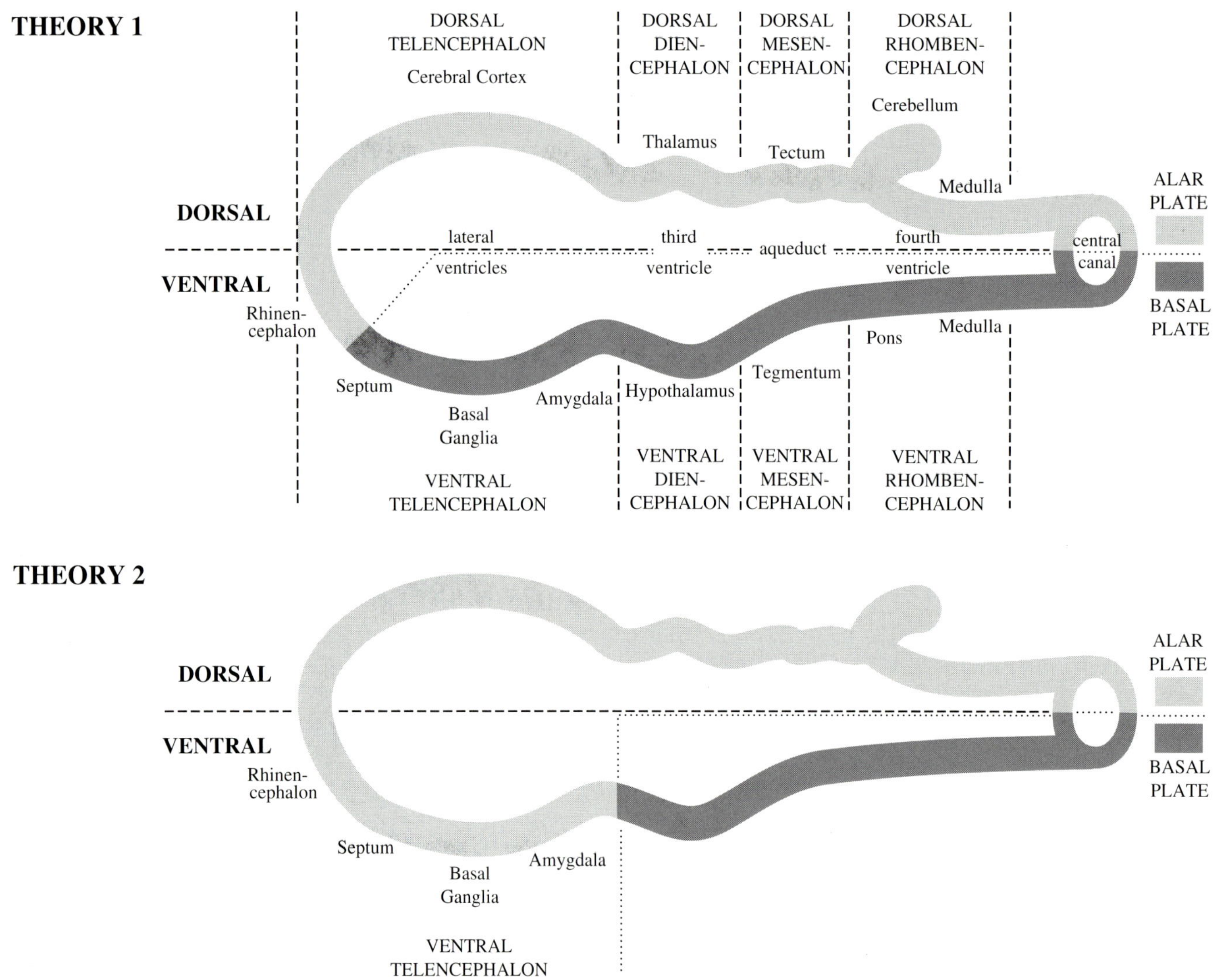

The Primary and Secondary Germinal Matrices. The ultimate source (and in most, but not all cases, the immediate source) of all neurons in the mammalian brain is a proliferative matrix that surrounds the ventricular lumen. The early embryologists called this germinal matrix the ependymal layer, a term that has been largely abandoned because the true ependymal layer is composed of specialized cells that line the ventricles in the mature nervous system and are not involved in the generation of neurons. Another widely used term is the ventricular zone, but this term fails to highlight the importance of this germinal matrix which, during early development, is the sole constituent (rather that a mere zone) of the developing nervous system. The term *neuroepithelium* (some call it neural epithelium) implies a dedicated epithelial tissue, directly derived from the *neuroectoderm* of the neural plate. The neuroepithelium is the ultimate source of all *neural* elements of the brain and spinal cord, i.e., neurons, neuroglia, tanycytes, ependymal cells, and the cells of some paraneural structures.

During early stages of embryonic development, the mammalian nervous system is composed almost exclusively of the neuroepithelium and for an extended period (several days in the rat) the growth of the nervous system constitutes little more than an increase in the stock of neural precursor cells through mitotic division. Importantly, this explosive increase in the undifferentiated neuroepithelium does not mean an absence of regional specialization. Thymidine-autoradiographic and related experimental evidence indicates that the neuroepithelium is a heterogeneous, or mosaic, system in which aggregates of cells (recognized as neuroepithelial lobules and sublobules, evaginations and invaginations, or patches marked by different thickness or rates of cell proliferation) are dedicated to produce specialized brain structures and particular cell types. It is in this sense that we designate one neuroepithelial lobule, or ridge, for instance, as the striatal neuroepithelium because, according to available evidence, it is destined to produce neurons for the striatum. Similarly, we designate a sublobule, or protuberance, in the thalamic neuroepithelium as the reticular neuroepithelium because it is destined to produce neurons for the reticular nucleus of the thalamus.

[^{3}H]thymidine-autoradiographic evidence has also established that in addition to the periventricular primary neuroepithelium, there are also secondary germinal matrices that produce neurons in certain brain regions. The secondary germinal matrices that are close to the ventricle are known as subependymal layers or subventricular zones, and those farther removed from the ventricle, usually in a subpial position, are known as external granular layers or external germinal layers. The subventricular zone of the cerebral cortex and the external germinal layer of the cerebellum are illustrative of these secondary matrices. Typically, secondary germinal matrices begin to expand in a given brain region after the primary neuroepithelium has begun to shrink. The progeny of cells proliferating in secondary germinal matrices typically give rise to late developing short-axoned neurons, or microneurons, known in many brain regions as granule cells.

The Translocation and Active Migration of Differentiating Neurons. The precursors of neurons (neuroblasts) produced in the primary and secondary germinal matrices are considered to have become differentiated into young neurons when they leave these proliferative sites and lose their potential to divide. These differentiating neurons fail to incorporate [^{3}H]thymidine into their chromosomal DNA and are thus easily identified with thymidine autoradiography. Upon leaving the germinal matrices, the young neurons either settle near their site of production, these are the "passively" translocated neurons; or they migrate to distant sites, these are the "actively" migrating neurons. The passively translocated neurons that settle near their site of production may become anchored in that location, in which case they remain permanently near the ventricular lumen (or where the lumen was at their time of production); an example of this is the early-generated neurons of the cranial nerve nuclei that are all situated near the midline. Alternatively, the translocated neurons may become displaced from their original position by new waves of later-produced neurons leaving the neuroepithelium (Fig. 3; p. xviii). Even more drastic transformations take place over time in the location of brain structures that are composed of actively migrating neurons. Experimental evidence indicates that neuronal migration may be a relatively simple radially-directed process, or it may be one where neurons migrate short or long distances through dense brain tissue, over the surface of the brain, or in rare cases across the midline from one side to the other, in various directions. Figure 4 (p. xix) illustrates the complex migration of neurons from a single location (the precerebellar pontine neuroepithelium) in several directions through the medulla (intramural migration) and over the wall of the medulla and pons (extramural migration). The migration of neurons in different brain regions, like their production, is governed by a strict timetable.

FIGURE 2

Derivatives of the Telencephalic, Diencephalic,

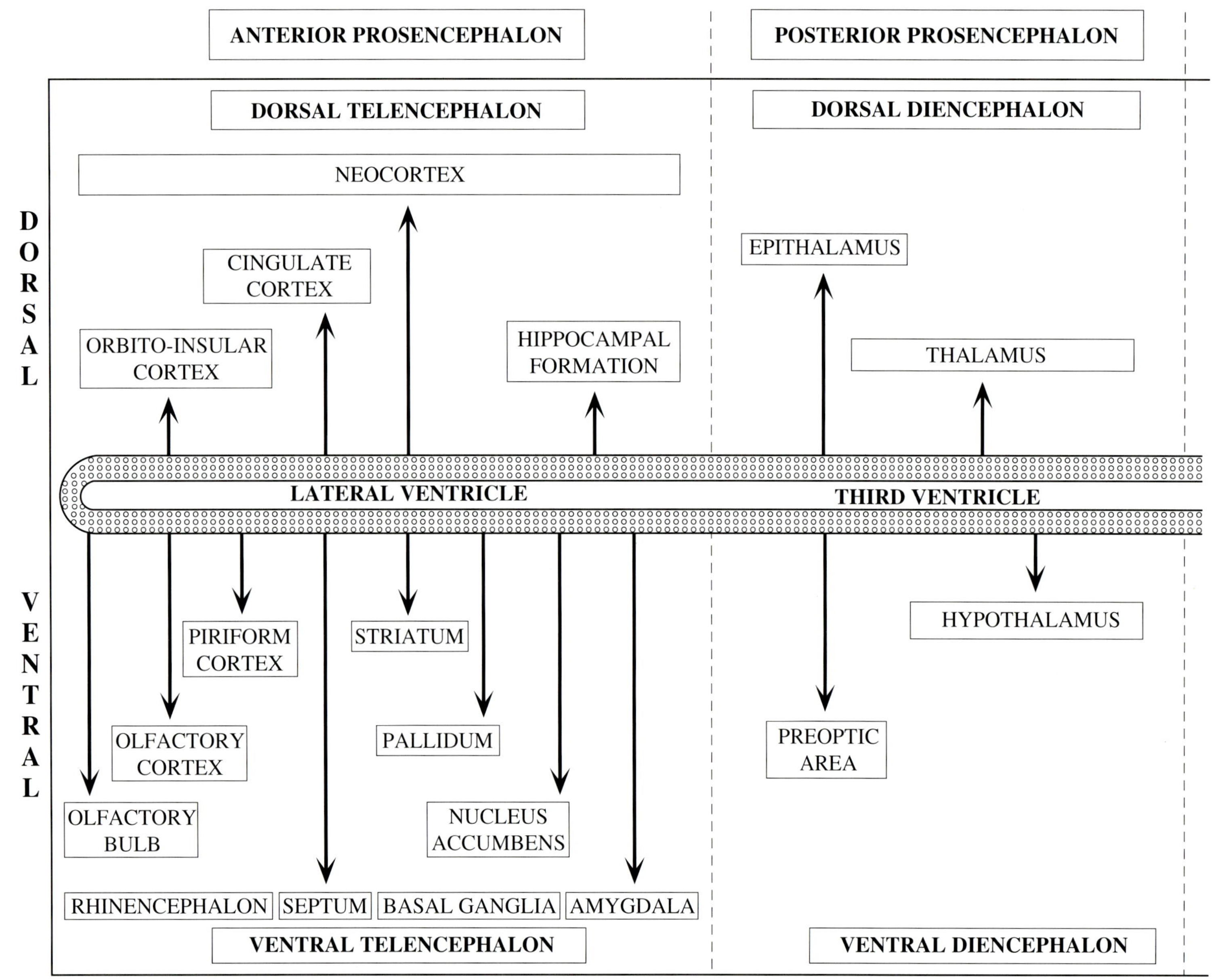

FIGURE 2 (contd.)

Mesencephalic and Rhombencephalic Neuroepithelium

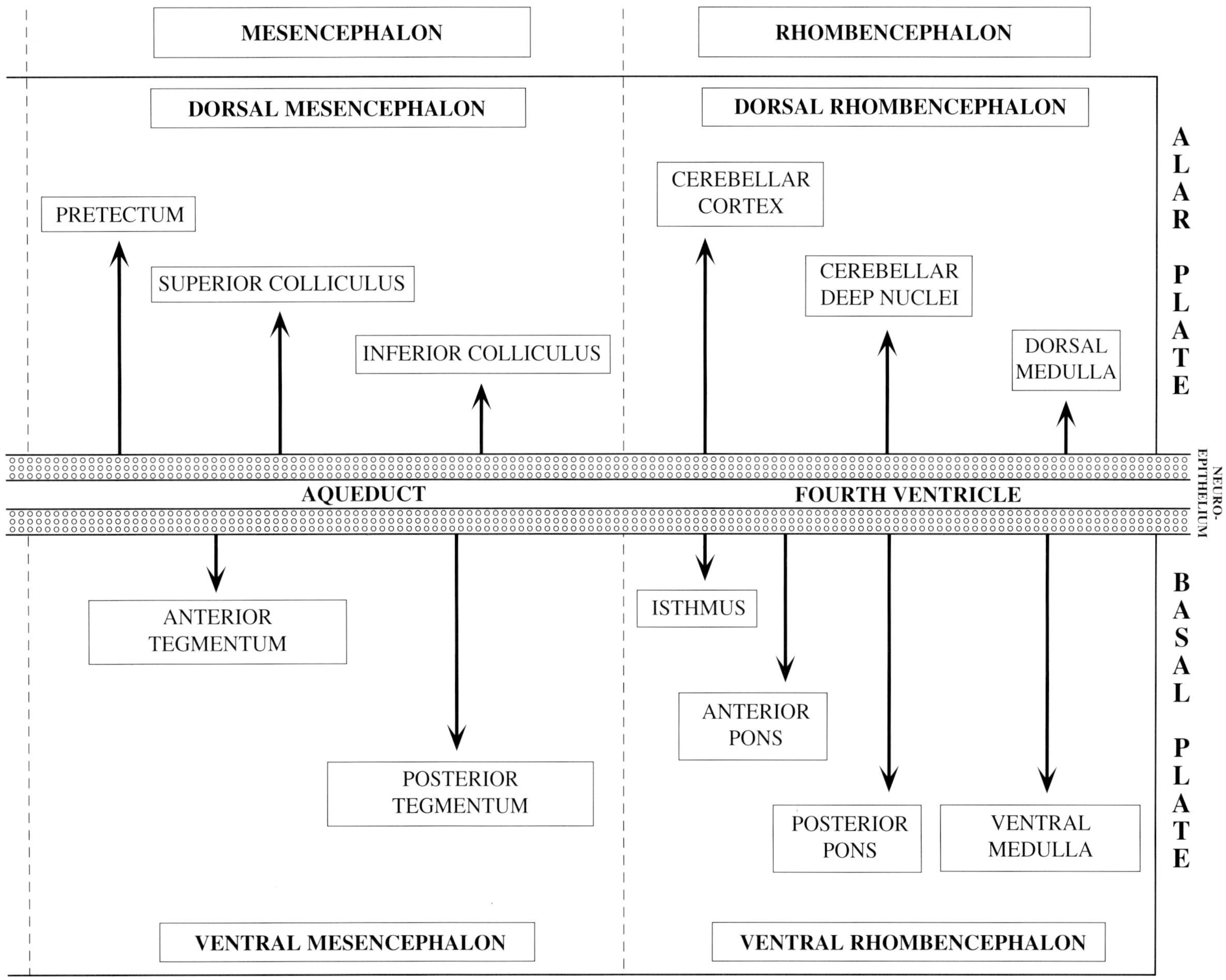

FIGURE 3

Changing Location of Passively Translocated Successively Produced Neuronal Structures

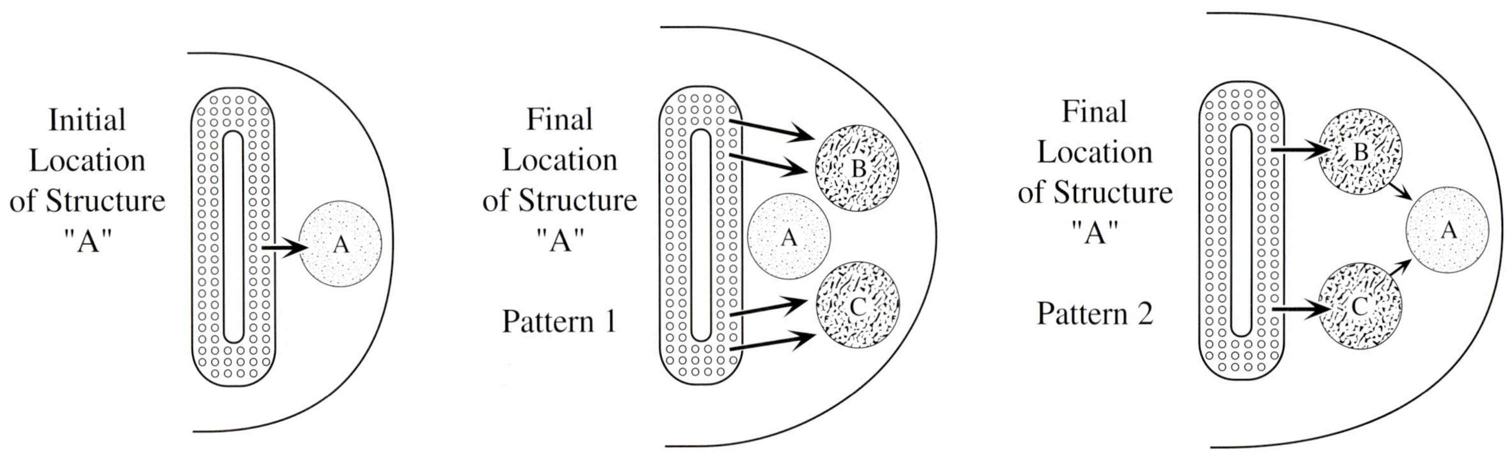

Figure 4. Abbreviations. **L3**, cerebellum (cut); **M33**, inferior colliculus (cut); va, aqueduct; v4, fourth ventricle; **X3**, pons; X5, pontine gray nucleus; X6, pontine reticulotegmental nucleus; X11a, anterior precerebellar neuroepithelium; X11p, posterior precerebellar neuroepithelium; X12a, anterior precerebellar extramural migration; X12i, precerebellar intramural migration; X12p, posterior precerebellar extramural migration; **Y3**, medulla; Y8, inferior olive; Y24, external cuneate nucleus; Y25, lateral reticular nucleus.

FIGURE 4

Complex Migratory Routes of Neurons from the Precerebellar Neuroepithelia to Target Structures

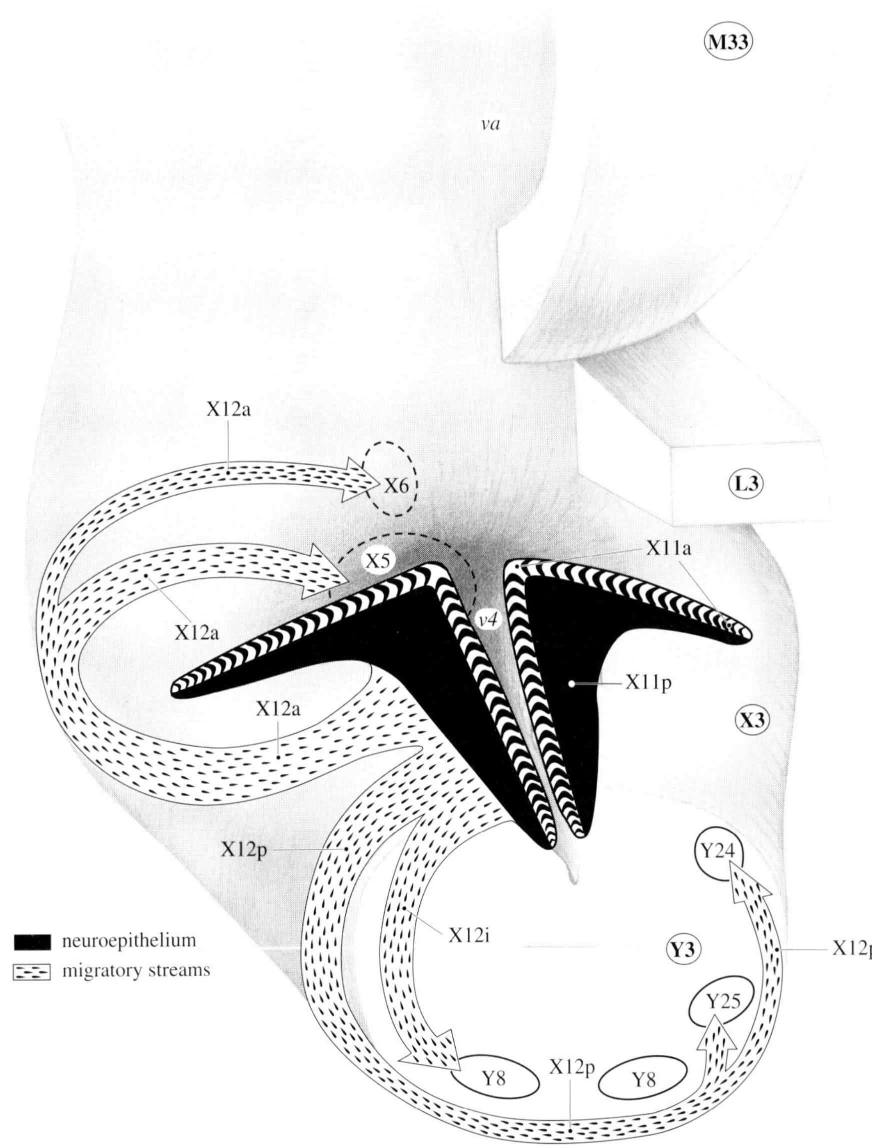

SOME PRINCIPLES OF BRAIN DEVELOPMENT

Stages in Prenatal Brain Development. Examination of the structural features of the developing rat brain between days E11 and E22 suggest three stages of mammalian brain development (Figs. 5-12).

Stage I is the *Initial Stage of Prenatal Brain Development* (Figs. 5, 8). During this stage, the rapid growth of brain development consists principally in the expansion of the primary proliferative matrix, the neuroepithelium (in black). In the rat, this phase begins on day E11 and ends during day E14. By day E14, differentiating brain regions (in gray) with postmitotic neurons are obvious in the caudal and ventral portion of the developing central nervous system.

Stage II is the *Intermediate Stage of Prenatal Brain Development.* This stage covers the period between days E15 and E18 (Figs. 6, 9-10). The neuroepithelium remains prominent and active during this period, particularly dorsally and rostrally, and the expansion of the secondary germinal matrices (black dots) adds to the continued increase in the stock of neuroblasts. But this is coupled with the progressive enlargement of the brain parenchyma by differentiating neurons (shades of gray), and by the end of the period all major brain systems are identifiable.

Stage III is the *Final Stage of Prenatal Brain Development.* In the rat, this stage covers the span between E20 and E22 (Figs. 7, 11-12). This stage is marked by the gradual dissolution of the neuroepithelium and the rapid growth of the brain parenchyma. By the end of this stage (i.e., at birth) most of the larger brain structures are identifiable in their final locations. Neurogenesis continues in the rat after birth, primarily as a result of the continuing generation of neuroblasts in secondary germinal matrices in selected brain regions (particularly in the cerebellar cortex and the hippocampal dentate gyrus). However, the major event in postnatal brain development is the accelerated differentiation of neurons and the myelination of nerves. Postnatal brain development is not dealt with in this Atlas.

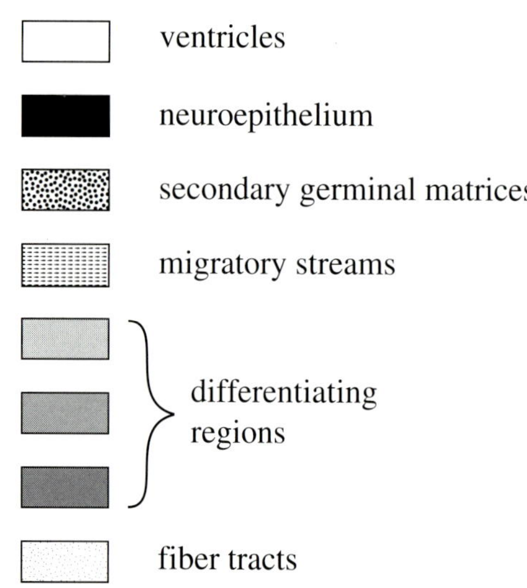

FIGURE 5

Stage I. Initial Stage of Prenatal Rat Brain Development

Sagittal

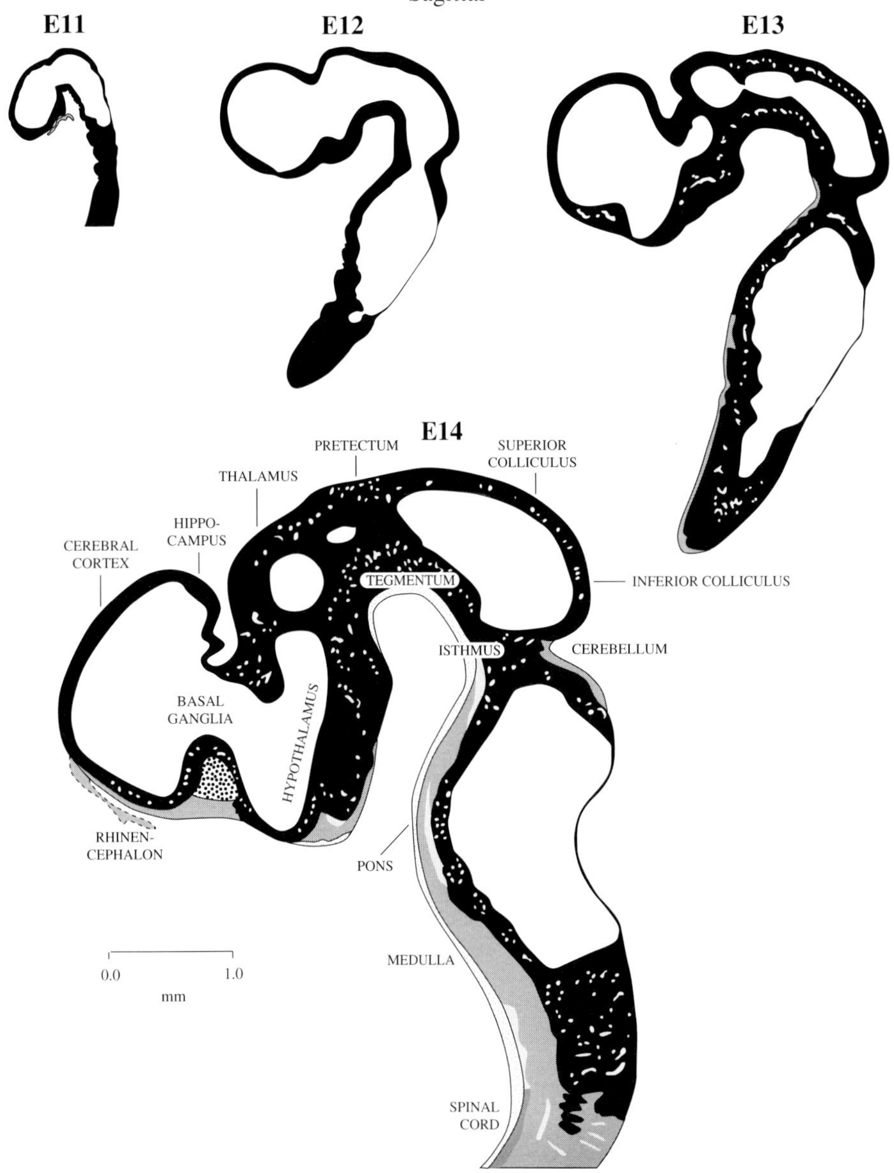

FIGURE 6

Stage II. Intermediate Stage of Prenatal Rat Brain Development

Sagittal

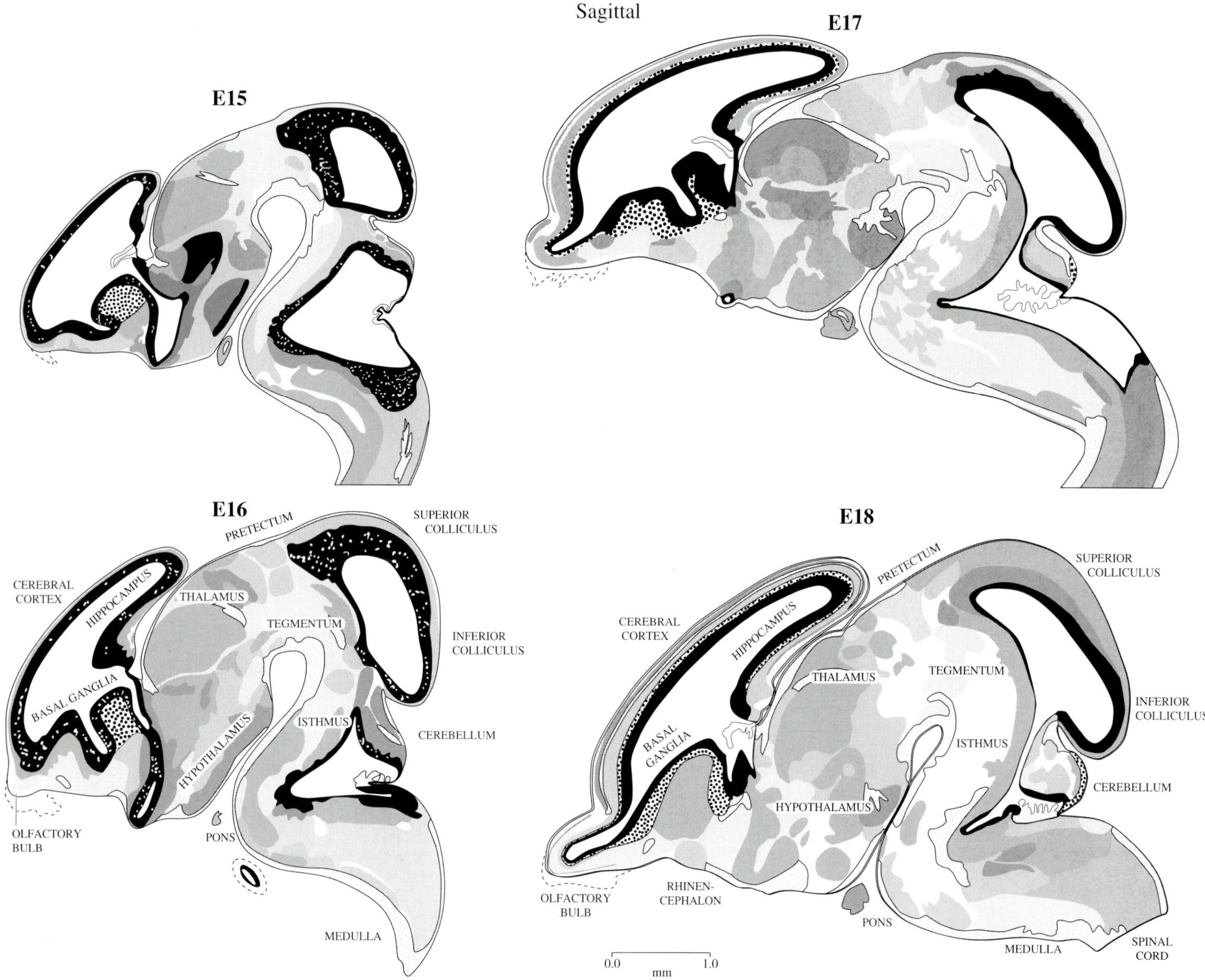

FIGURE 7

Stage III. Final Stage of Prenatal Rat Brain Development
Sagittal

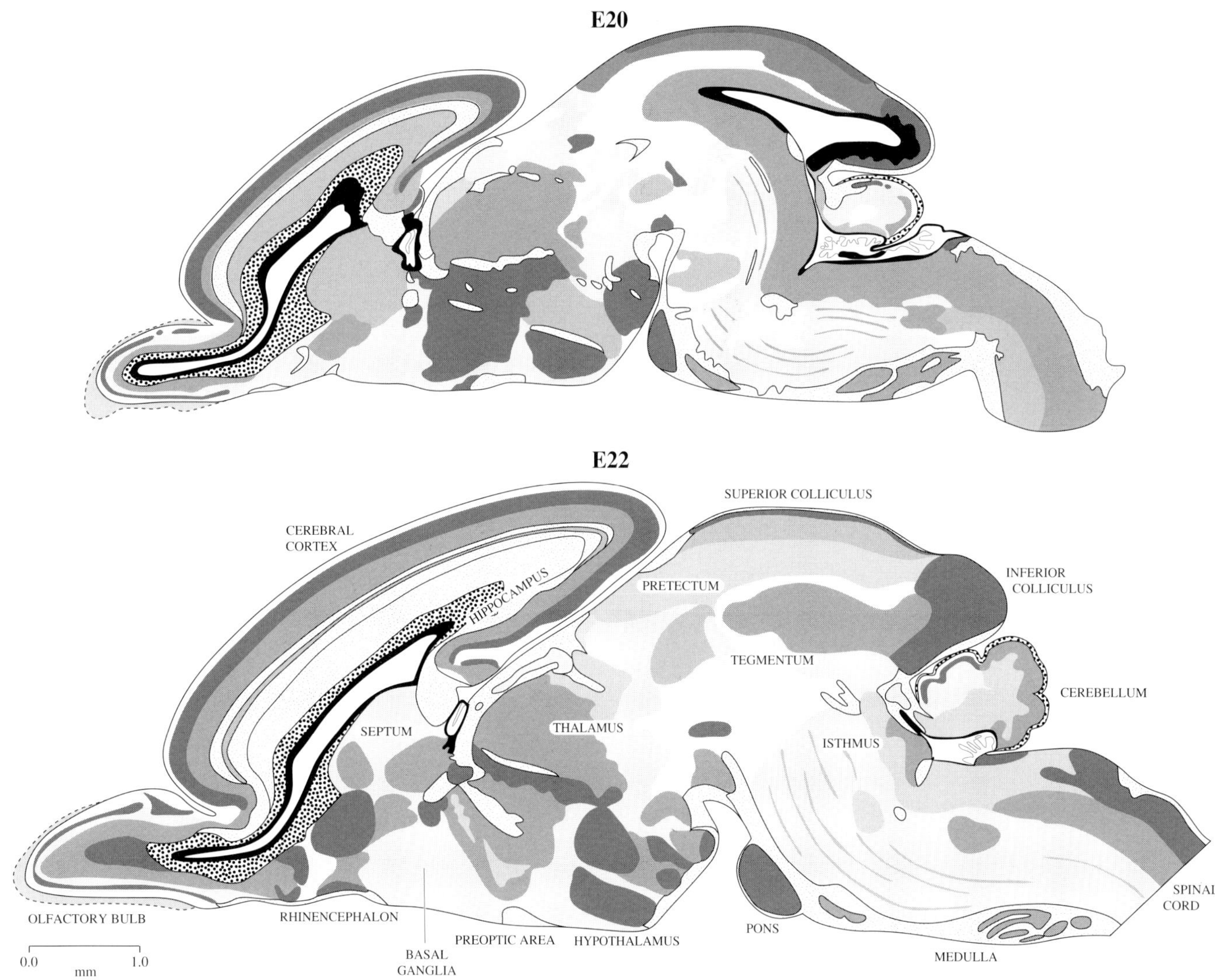

FIGURE 8

Stage I. Initial Stage of Prenatal Rat Brain Development
Coronal

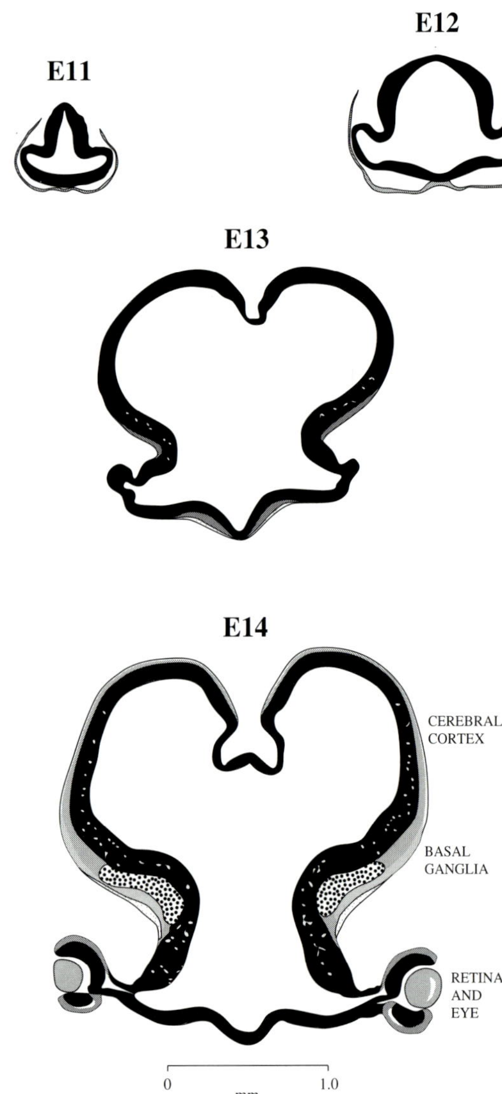

FIGURE 9

Stage II. Intermediate Stage of Prenatal Rat Brain Development

Coronal

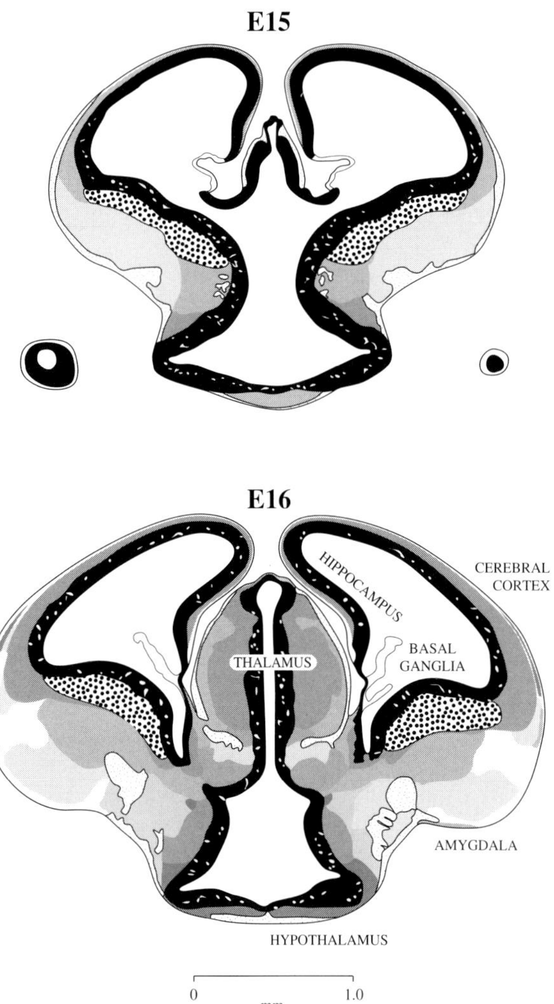

FIGURE 10

Stage II. Intermediate Stage of Prenatal Rat Brain Development (continued)
Coronal

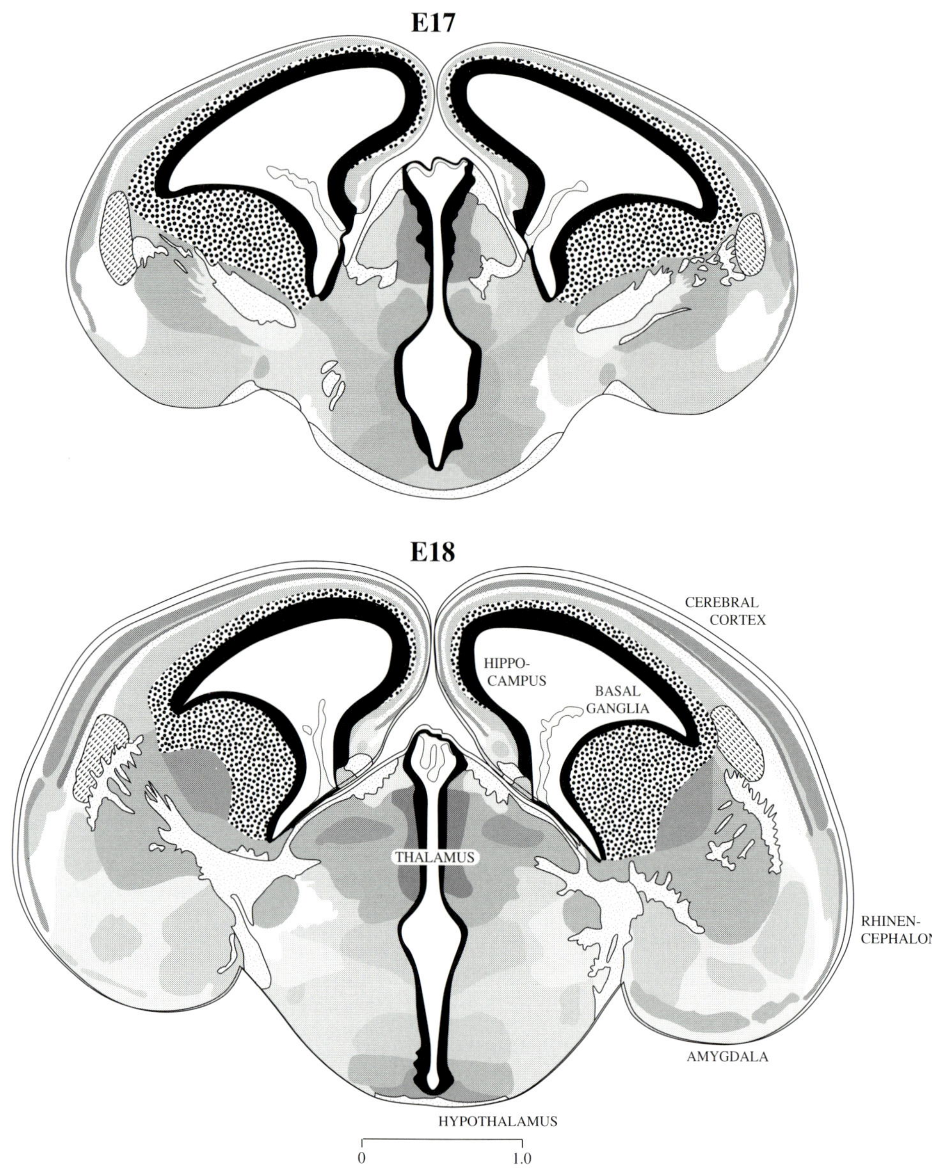

FIGURE 11

Stage III. Final Stage of Prenatal Rat Brain Development
Coronal

E20

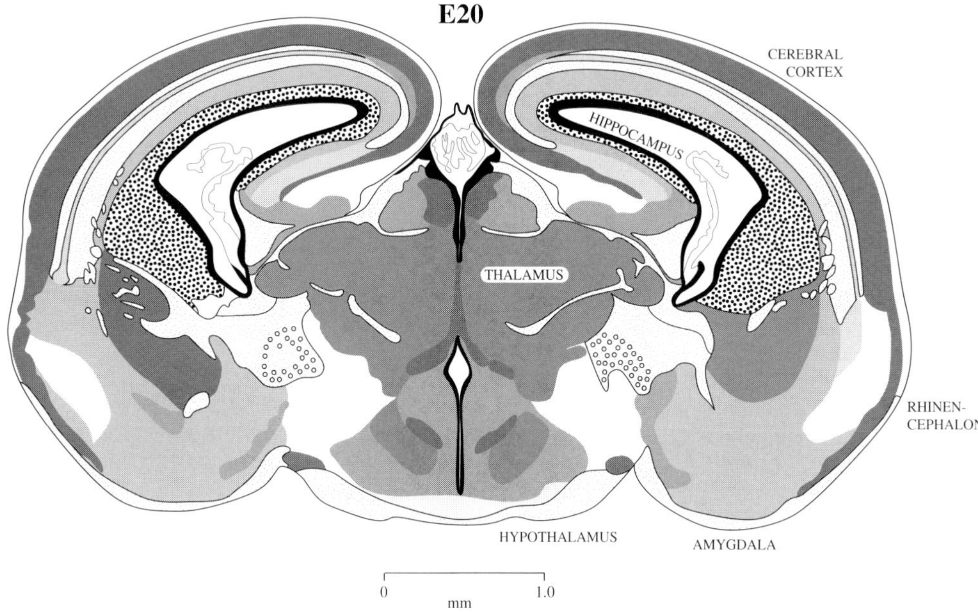

FIGURE 12

Stage III. Final Stage of Prenatal Rat Brain Development (continued)
Coronal

E22

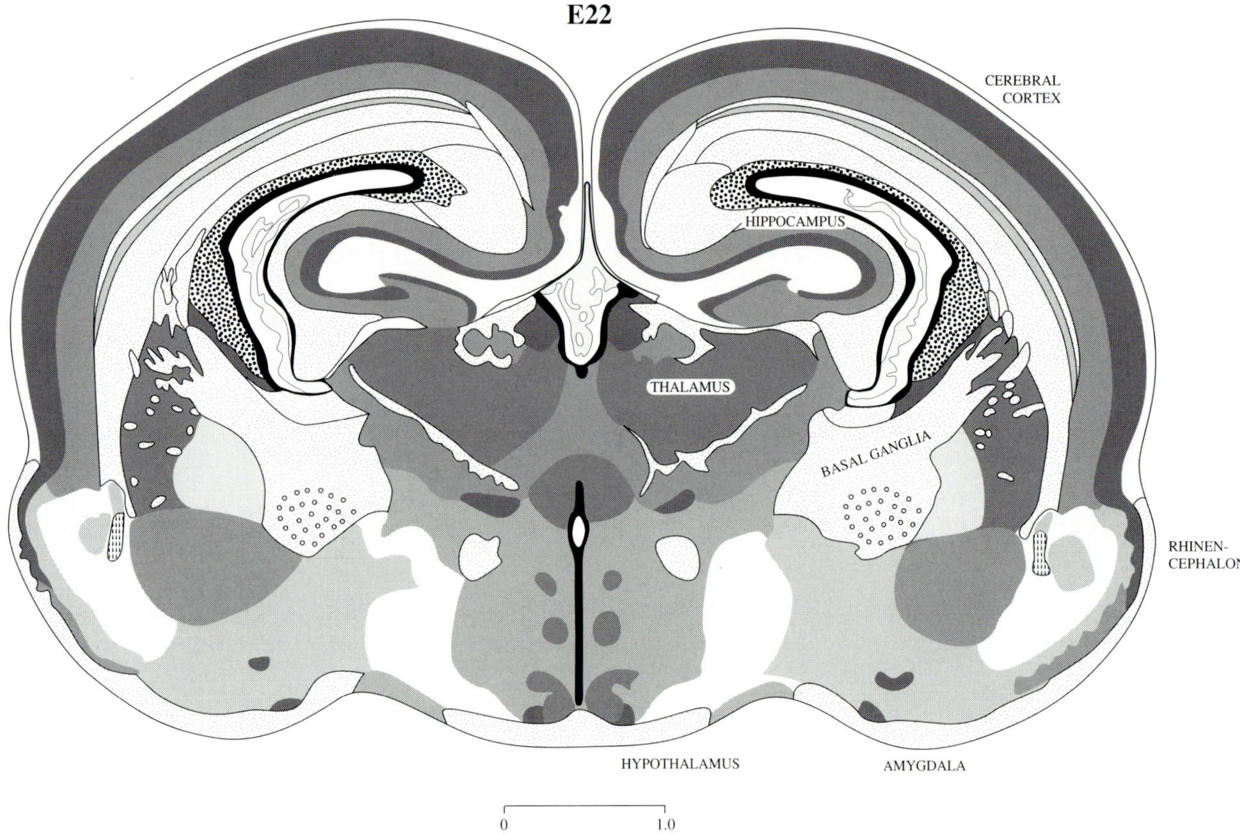

BACKGROUND READINGS

There is vast literature on the embryonic development of the mammalian central nervous system, including the rat brain. Instead of presenting here a lengthy Bibliography, we have opted to cite a selected list of our own publications on rat brain development that has been the immediate source of the structural identifications made in this Atlas. Beginning with a few recent review articles, we list these publications under the same major brain divisions that we use throughout the Atlas. The interested reader will find in these references the rationale for some of the identifications made as well as an extensive bibliography.

REVIEW ARTICLES

Altman, J. 1992. The early stages of nervous system development: Neurogenesis and neuronal migration. In: A. Björklund, T. Hökfelt and M. Tohyama (eds.), *Handbook of Chemical Neuroanatomy, Volume 10. Ontogeny of Transmitters and Peptides in the CNS*, pp. 1-31. Amsterdam: Elsevier.

Bayer, S. A. and J. Altman. 1995. Neurogenesis and neuronal migration. In: G. Paxinos (ed.), *The Rat Nervous System*, pp. 1041-1078. (Second edition) Orlando: Academic Press.

Bayer, S. A. and J. Altman. 1995. Principles of neurogenesis, neuronal migration, and neural circuit formation. In: G. Paxinos (ed.), *The Rat Nervous System*, pp. 1079-1098. (Second edition) Orlando: Academic Press.

AMYGDALA

Bayer, S. A. 1980. Quantitative [^{3}H]thymidine radiographic analyses of neurogenesis in the rat amygdala. *Journal of Comparative Neurology* 194:845-875.

BASAL GANGLIA

Bayer, S. A. 1984. Neurogenesis in the rat neostriatum. *International Journal of Developmental Neuroscience* 2:163-175.

Bayer, S. A. 1985. Neurogenesis of the magnocellular basal telencephalic nuclei in the rat. *International Journal of Developmental Neuroscience* 3:229-243.

Bayer, S. A. 1987. Neurogenetic and morphogenetic heterogeneity in the bed nucleus of the stria terminalis. *Journal of Comparative Neurology* 265:47-64.

CEREBELLUM

Altman, J. and S. A. Bayer. 1978. Prenatal development of the cerebellar system in the rat: I. Cytogenesis and histogenesis of the deep nuclei and the cortex of the cerebellum. *Journal of Comparative Neurology* 179:23-48.

Altman, J. 1982. Morphological development of the rat cerebellum and some of its mechanisms. In: S. L. Palay and V. Chan-Palay (eds.), *The Cerebellum: New Vistas*, pp. 8-49. Berlin: Springer.

Altman, J. and S. A. Bayer. 1985. Embryonic development of the rat cerebellum. I. Delineation of the cerebellar primordium and early cell movements. *Journal of Comparative Neurology* 231:1-26

Altman, J. and S. A. Bayer. 1985. Embryonic development of the rat cerebellum. II. Translocation and regional distribution of the deep neurons. *Journal of Comparative Neurology* 231:27-41.

Altman, J. and S. A. Bayer. 1985. Embryonic development of the rat cerebellum. III. Regional differences in the time of origin, migration and settling of Purkinje cells. *Journal of Comparative Neurology* 231:42-65.

CEREBRAL CORTEX

Altman, J. and S. A. Bayer. 1990. Vertical compartmentation and cellular transformations in the germinal matrices of the embryonic rat cerebral cortex. *Experimental Neurology* 107:23-35.

Altman, J. and S. A. Bayer. 1990. Horizontal compartmentation in the germinal matrices and intermediate zone of the embryonic rat cerebral cortex. *Experimental Neurology* 107:36-47.

Bayer, S. A. 1990. Development of the lateral and medial limbic cortices in the rat in relation to cortical phylogeny. *Experimental Neurology* 107:118-131.

Bayer, S. A. and J. A. Altman. 1990. Development of layer I and the subplate in the rat neocortex. *Experimental Neurology* 107:48-62.

Bayer, S. A. and J. Altman. 1991. *Neocortical Development*. New York: Raven Press.

Bayer, S. A., J. Altman, X. Dai and L. Humphreys. 1991. Planar differences in nuclear area and orientation in the subventricular and intermediate zones of the rat embryonic neocortex. *Journal of Comparative Neurology* 307:487-498.

Bayer, S. A., J. Altman, R. J. Russo, X. Dai and J. A. Simmons. 1991. Cell migration in the rat embryonic neocortex. *Journal of Comparative Neurology* 307:499-516.

CRANIAL NERVES AND NUCLEI

Altman, J. and S. A. Bayer. 1982. *Development of the Cranial Nerve Ganglia and the Nuclei in the Rat*. (Advances in Anatomy, Embryology and Cell Biology, Volume 74, pp. 1-90.) Berlin: Springer.

HIPPOCAMPUS

Bayer, S. A. 1980. Development of the hippocampal region in the rat. I. Neurogenesis examined with [^{3}H]thymidine autoradiography. *Journal of Comparative Neurology* 190:87-114.

Bayer, S. A. 1980. Development of the hippocampal region in the rat. II. Morphogenesis during embryonic and early postnatal life. *Journal of Comparative Neurology* 190:115-134.

Altman, J. and S. A. Bayer. 1990. Mosaic organization of the hippocampal neuroepithelium and the multiple germinal sources of dentate granule cells. *Journal of Comparative Neurology* 301:325-342.

BACKGROUND READINGS

Altman, J. and S. A. Bayer. 1990. The prolonged sojourn of developing pyramidal cells in the intermediate zone of the hippocampus and their settling in the stratum pyramidale. *Journal of Comparative Neurology* 301:343-364.

Altman, J. and S. A. Bayer. 1990. The migration and distribution of two populations of hippocampal granule cell precursors during the perinatal and postnatal periods. *Journal of Comparative Neurology* 301:365-381.

HYPOTHALAMUS

Altman, J. and S. A. Bayer. 1978. Development of the diencephalon in the rat: I. Autoradiographic study of the time of origin and settling patterns of neurons of the hypothalamus. *Journal of Comparative Neurology* 182:945-972.

Altman, J. and S. A. Bayer. 1978. Development of the diencephalon in the rat: II. Correlation of the embryonic development of the hypothalamus with the time of origin of its neurons. *Journal of Comparative Neurology* 182:973-994.

Altman, J. and S. A. Bayer. 1978. Development of the diencephalon in the rat: III. Ontogeny of the specialized ventricular linings of the hypothalamic third ventricle. *Journal of Comparative Neurology* 182:995-1016.

Altman, J. and S. A. Bayer. 1986. *The Development of the Rat Hypothalamus.* (Advances in Anatomy, Embryology and Cell Biology, Volume 100, pp. 1-178.) Berlin: Springer.

MEDULLA

Altman, J. and S. A. Bayer. 1980. Development of the brain stem in the rat. I. Thymidine-radiographic study of the time of origin of neurons of the lower medulla. *Journal of Comparative Neurology* 194:1-35.

Altman, J. and S. A. Bayer. 1980. Development of the brain stem in the rat. II. Thymidine-radiographic study of the time of origin of neurons of the upper medulla, excluding the vestibular and auditory nuclei. *Journal of Comparative Neurology* 194:37-56.

Altman, J. and S. A. Bayer. 1980. Development of the brain stem in the rat. III. Thymidine-radiographic study of the time of origin of neurons of the vestibular and auditory nuclei of the upper medulla. *Journal of Comparative Neurology* 194:877-904.

MIDBRAIN

Altman, J. and S. A. Bayer. 1981. Time of origin of neurons of the rat inferior colliculus and the relations between cytogenesis and tonotopic order in the auditory pathway. *Experimental Brain Research* 42:411-423.

Altman, J. and S. A. Bayer. 1981. Time of origin of neurons of the rat superior colliculus in relation to other components of the visual and visuomotor pathways. *Experimental Brain Research* 42:424-434.

Altman, J. and S. A. Bayer. 1981. Development of the brain stem in the rat. V. Thymidine-radiographic study of the time of origin of neurons in the midbrain tegmentum. *Journal of Comparative Neurology* 198:677-716.

OLFACTORY BULB

Bayer, S. A. 1983. [^{3}H]thymidine radiographic studies of neurogenesis in the rat olfactory bulb. *Experimental Brain Research* 50:329-340.

PONTINE AREA

Altman, J. and S. A. Bayer. 1978. Prenatal development of the cerebellar system in the rat: II. Cytogenesis and histogenesis of the inferior olive, pontine gray, and the precerebellar reticular nuclei. *Journal of Comparative Neurology* 179:49-76.

Altman, J. and S. A. Bayer. 1980. Development of the brain stem in the rat. IV. Thymidine-radiographic study of the time of origin of neurons in the pontine region. *Journal of Comparative Neurology* 194:905-929.

Altman, J. and S. A. Bayer. 1987. Development of the precerebellar nuclei in the rat. I. The precerebellar neuroepithelium of the rhombencephalon. *Journal of Comparative Neurology* 257:477-489.

Altman, J. and S. A. Bayer. 1987. Development of the precerebellar nuclei in the rat. II. The intramural olivary migratory stream and the neurogenetic organization of the inferior olive. *Journal of Comparative Neurology* 257:490-512.

Altman, J. and S. A. Bayer. 1987. Development of the precerebellar nuclei in the rat. III. The posterior precerebellar extramural migratory stream and the lateral reticular and external cuneate nuclei. *Journal of Comparative Neurology* 257:513-528.

Altman, J. and S. A. Bayer. 1987. Development of the precerebellar nuclei in the rat. IV. The anterior precerebellar extramural migratory stream and the nucleus reticularis tegmenti pontis and the basal pontine gray. *Journal of Comparative Neurology* 257:529-552.

PREOPTIC AREA

Bayer, S. A. and J. Altman. 1987. Development of the preoptic area: time and site of origin, migratory routes, and settling patterns of its neurons. *Journal of Comparative Neurology* 265:65-95.

RHINENCEPHALON

Bayer, S. A. 1985. Neurogenesis in the olfactory tubercle and islands of Calleja in the rat. *International Journal of Developmental Neuroscience* 3:135-147.

Bayer, S. A. 1986. Neurogenesis in the anterior olfactory nucleus and its associated transition areas in the rat brain. *International Journal of Developmental Neuroscience* 4:225-249.

Bayer, S. A. 1986. Neurogenesis in the rat primary olfactory cortex. *International Journal of Developmental Neuroscience* 4:251-271.

Bayer, S. A. and J. Altman. 1991. Development of the endopiriform nucleus and the claustrum in the rat brain. *Neuroscience* 45(2):391-412.

SEPTAL AREA

Bayer, S. A. 1979. The development of the septal region in the rat. I. Neurogenesis examined with [^{3}H]thymidine autoradiography. *Journal of Comparative Neurology* 183:89-106.

Bayer, S. A. 1979. The development of the septal region in the rat. II. Morphogenesis in normal and x-irradiation embryos. *Journal of Comparative Neurology* 183:107-120.

SPINAL CORD

Altman, J. and S. A. Bayer. 1984. *The Development of the Rat Spinal Cord.* (Advances in Anatomy, Embryology and Cell Biology, Volume 85, pp. 1-116.) Berlin: Springer.

THALAMUS

Altman, J. and S. A. Bayer. 1979. Development of the diencephalon in the rat. IV. Quantitative study of the time of origin of neurons and the internuclear chronological gradients in the thalamus. *Journal of Comparative Neurology* 188:455-472.

Altman, J. and S. A. Bayer. 1979. Development of the diencephalon in the rat. V. Thymidine-radiographic observations on internuclear and intranuclear gradients in the thalamus. *Journal of Comparative Neurology* 188:473-500.

Altman, J. and S. A. Bayer. 1979. Development of the diencephalon in the rat. VI. Re-evaluation of the embryonic development of the thalamus on the basis of thymidine-radiographic datings. *Journal of Comparative Neurology* 188:501-524.

Altman, J. and S. A. Bayer. 1988. Development of the rat thalamus: I. Mosaic organization of the thalamic neuroepithelium. *Journal of Comparative Neurology* 275:346-377.

Altman, J. and S. A. Bayer. 1988. Development of the rat thalamus: II. Time and site of origin and settling pattern of neurons derived from the anterior lobule of the thalamic neuroepithelium. *Journal of Comparative Neurology* 275:378-405.

Altman, J. and S. A. Bayer. 1988. Development of the rat thalamus: III. Time and site of origin and settling pattern of neurons of the reticular nucleus. *Journal of Comparative Neurology* 275:406-428.

Altman, J. and S. A. Bayer. 1989. Development of the rat thalamus. IV. The intermediate lobule of the thalamic neuroepithelium, and the time and site of origin and settling pattern of neurons of the ventral nuclear complex. *Journal of Comparative Neurology* 284:534-566.

Altman, J. and S. A. Bayer. 1989. Development of the rat thalamus. V. The posterior lobule of the thalamic neuroepithelium, and the time and site of origin and settling pattern of neurons of the medial geniculate body. *Journal of Comparative Neurology* 284:567-580.

Altman, J. and S. A. Bayer. 1989. Development of the rat thalamus. VI. The posterior lobule of the thalamic neuroepithelium, and the time and site of origin and settling pattern of neurons of the lateral geniculate and lateral posterior nuclei. *Journal of Comparative Neurology* 284:581-601.

INDEX OF THE ALPHANUMERIC CODE

AMYGDALA (A)

A1	amygdaloid neuroepithelium
A2	amygdaloid subventricular zone
A3	amygdala (or differentiating field)
A4	anterior amygdaloid area
A5	amygdala-hippocampal area
A6	central amygdaloid nucleus
A6*l*	central amygdaloid nucleus, lateral
A6m	central amygdaloid nucleus, medial
A7	basomedial amygdaloid nucleus
A8	basolateral amygdaloid nucleus
A9	cortical amygdaloid nucleus
A9a	cortical amygdaloid nucleus, anterior
A9p	cortical amygdaloid nucleus, posterior
A10	lateral amygdaloid nucleus
A14	medial amygdaloid nucleus
A15	amygdala-piriform transition area
A16	intercalated masses

BASAL GANGLIA (B)

	STRIATUM
B1	striatal neuroepithelium
B2	striatal subventricular zone
B3	striatum (or differentiating field)
	PALLIDUM
B11	pallidal neuroepithelium
B12	pallidal subventricular zone
B13	pallidum (or differentiating field)
B14	entopeduncular nucleus
B15	substantia innominata
	BED NUCLEUS OF STRIA TERMINALIS
B21	strionuclear neuroepithelium
B22	strionuclear subventricular zone
B23	bed nucleus of stria terminalis (or differentiating field)
B24	bed nucleus of stria terminalis, dorsal division
B25	bed nucleus of stria terminalis, ventral division
B26	bed nucleus of stria terminalis, medial division
B27	bed nucleus of stria terminalis, lateroventral division
B28	bed nucleus of stria terminalis, intermediate division
	NUCLEUS ACCUMBENS
B31	accumbens neuroepithelium
B32	accumbens subventricular zone
B33	nucleus accumbens (or differentiating field)

CEREBELLUM (see under L)

CEREBRAL CORTEX (C)

	NEOCORTEX
C1	neocortical neuroepithelium
C2	neocortical subventricular zone
C3	neocortex (or differentiating field)
C4	molecular layer (or primordial plexiform layer)
C5	intermediate zone
C6	cortical plate
C6i	cortical plate, infragranular part
C6s	cortical plate, supragranular part
C7	white matter
C8	subplate
C9	lateral migratory stream
C9h	lateral migratory stream, head
C9r	lateral migratory stream, reservoir
	CINGULATE CORTEX
C11	cingulate cortex neuroepithelium
C12	cingulate cortex subventricular zone
C13	cingulate cortex (or differentiating field)
	ORBITAL AND INSULAR CORTEX
C23	orbital and insular area (or differentiating field)
C24	insular cortex
C25	orbital cortex
C26	claustrum

CRANIAL NERVES AND NUCLEI (see under n)

FIBER TRACTS (f)

f1	lateral olfactory tract
f2	medial olfactory tract
f3	corpus callosum
f4	anterior commissure
f5	medial forebrain bundle
f6	fornix
f7	hippocampal commissure
f8	internal capsule
f9	cerebral peduncle
f10	external medullary lamina
f11	external capsule
f12	cingulum bundle
f13	fimbria
f14	stria terminalis
f15	stria medullaris
f16	medial lemniscus
f17	mammillothalamic tract
f18	habenulopeduncular tract
f19	posterior commissure

f20 supramammillary decussation
f21 brachium of superior colliculus
f22 commissure of superior colliculus
f23 superior cerebellar peduncle
f24 middle cerebellar peduncle
f25 inferior cerebellar peduncle
f26 ventral tegmental decussation
f27 medial longitudinal fasciculus
f28 lateral lemniscus
f29 brachium of inferior colliculus
f30 spinal tract of trigeminal
f31 cuneate fasciculus

GLANDS AND PARANEURAL STRUCTURES (g)

g1 choroid plexus of lateral ventricle
g2 subfornical organ
g3 choroid plexus of third ventricle
g4 choroid plexus of fourth ventricle
g5 pineal gland (primordium)
g6 pituitary gland (primordium)
g7 subcommissural organ
g8 area postrema
g9 lamina terminalis
g10 velum medullare

HIPPOCAMPAL FORMATION (see under K)

HYPOTHALAMUS (H)

H1 hypothalamic neuroepithelium
H1a anterior hypothalamic neuroepithelium
H1f Forelian neuroepithelium
H1i intermediate hypothalamic neuroepithelium
H1*l* lateral hypothalamic neuroepithelium
H1m mammillary neuroepithelium
H1p posterior hypothalamic neuroepithelium
H1pm premammillary neuroepithelium
H1v ventral hypothalamic neuroepithelium
H3 hypothalamus (or differentiating field)
H3a anterior hypothalamus (or differentiating field)
H3f fields of Forel (or differentiating field)
H3i intermediate hypothalamus (or differentiating field)
H3*l* lateral hypothalamus (or differentiating field)
H3m mammillary body (or differentiating field)
H3p posterior hypothalamus (or differentiating field)
H3v ventral hypothalamus (or differentiating field)
H4 supraoptic nucleus
H5 suprachiasmatic nucleus
H6 paraventricular nucleus
H7 zona incerta
H8 dorsomedial nucleus
H9 ventromedial nucleus
H10 anterobasal nucleus
H14 supramammillary nucleus
H15 premammillary nucleus
H16 medial mammillary nucleus
H17 lateral mammillary nucleus
H19 subthalamic nucleus
H20 arcuate nucleus
H24 anterodorsal nucleus
H25 anteroventral nucleus
H26 posterior hypothalamic area
H27 tuberomammillary nucleus

HIPPOCAMPAL FORMATION (K)

AMMON'S HORN

K1 hippocampal (Ammon's horn) neuroepithelium
K3 hippocampus (or differentiating field)
K3d dorsal hippocampus (or differentiating field)
K3v ventral hippocampus (or differentiating field)
K4 CA1 field
K5 CA3 field
K6 alveus

SUBICULUM

K11 subicular neuroepithelium
K13 subicular area (or differentiating field)

DENTATE GYRUS

K21 dentate gyrus neuroepithelium
K23 dentate gyrus (or differentiating field)
K24 dentate granular layer
K25 dentate polymorph cell layer

CEREBELLUM (L)

L1 cerebellar neuroepithelium
L1h cerebellar hemisphere neuroepithelium
L1v cerebellar vermis neuroepithelium
L2 external germinal layer
L3 cerebellum (or differentiating field)
L3h cerebellar hemisphere (or differentiating field)
L3v cerebellar vermis (or differentiating field)
L4 molecular layer

L5 Purkinje cell layer
L6 medullary layer
L7 lateral deep nucleus (dentate)
L7mi lateral deep nucleus, migration
L8 medial deep nucleus (fastigial)
L8mi medial deep nucleus, migration
L9 cerebellar germinal trigone
L10 cerebellar commissure

MEDULLA (see under Y)

MIDBRAIN (M)

SUPERIOR COLLICULUS

M1 superior tectal neuroepithelium
M3 superior colliculus (or differentiating field)
M4 periaqueductal gray
M4d dorsal periaqueductal gray
M4v ventral periaqueductal gray
M5 superior colliculus, superficial layers
M6 superior colliculus, deep layers
M7 superior colliculus, stratum opticum

M8 red nucleus
M9 midbrain reticular formation

TEGMENTUM

M11 tegmental neuroepithelium
M11a anterior tegmental neuroepithelium
M11p posterior tegmental neuroepithelium
M13 tegmentum (or differentiating field)
M13a tegmentum, anterior (or differentiating field)
M13p tegmentum, posterior (or differentiating field)
M14 substantia nigra
M14c substantia nigra, pars compacta
M14r substantia nigra, pars reticulata
M15 interpeduncular nucleus
M15c interpeduncular nucleus, caudal
M15r interpeduncular nucleus, rostral
M16 dorsal raphe (mesencephalic)
M17 ventral tegmental area
M18 parabigeminal nucleus

PRETECTUM

M21 pretectal neuroepithelium
M23 pretectum (or differentiating field)
M24 anterior pretectal nucleus
M25 nucleus of the posterior commissure
M26 olivary pretectal nucleus
M27 nucleus of the optic tract
M28 posterior pretectal nucleus

INFERIOR COLLICULUS

M31 inferior tectal neuroepithelium
M33 inferior colliculus (or differentiating field)
M34 external nucleus of inferior colliculus
M35 central nucleus of inferior colliculus

NERVES AND NUCLEI, CRANIAL (n)

n1 olfactory nerve
n2 optic nerve
n2c optic chiasma
n2t optic tract
n3n oculomotor nucleus
n4n trochlear nucleus
n5 trigeminal nerve
n5b trigeminal boundary cap
n5g trigeminal ganglion
n5m mesencephalic trigeminal nucleus
n5n trigeminal motor nucleus
n5p principal (sensory) trigeminal nucleus
n5s spinal trigeminal nucleus
n7 facial nerve
n7n facial motor nucleus
n8 vestibulocochlear nerve
n8b vestibulocochlear boundary cap
n8g vestibular ganglion
n9g ganglion of glossopharyngeal nerve
n10g ganglion of the vagus
n10n dorsal motor nucleus of vagus
n12n hypoglossal nucleus

OLFACTORY BULB (O)

OLFACTORY BULB

O1 olfactory bulb neuroepithelium
O2 olfactory bulb subventricular zone
O3 olfactory bulb (or differentiating field)
O4 glomerular layer
O5 external plexiform layer
O6 mitral cell layer
O7 internal plexiform layer
O8 granular layer

ACCESSORY OLFACTORY BULB

O13 accessory olfactory bulb (or differentiating field)
O14 glomerular layer (accessory olfactory bulb)
O15 external plexiform layer (accessory olfactory bulb)

O16	mitral cell layer (accessory olfactory bulb)
O17	internal plexiform layer (accessory olfactory bulb)
O18	granular layer (accessory olfactory bulb)

PONTINE AREA (see under X)

PREOPTIC AREA AND OPTIC EPITHELIA (P)

	PREOPTIC AREA
P1	preoptic neuroepithelium
P3	preoptic area (or differentiating field)
P4	medial preoptic region
P5	lateral preoptic region
P6	superior preoptic region
P7	magnocellular preoptic nucleus
	EYE
P11	retinal neuroepithelium
P12	eye stalk epithelium

RHINENCEPHALON (R)

R1	rhinencephalic neuroepithelium
R1a	rhinencephalic neuroepithelium, anterior
R1p	rhinencephalic neuroepithelium, posterior
R2	rhinencephalic subventricular zone
R3	rhinencephalon (or differentiating field)
R3a	rhinencephalon, anterior (or differentiating field)
R3p	rhinencephalon, posterior (or differentiating field)
R4	piriform cortex
R5	endopiriform nucleus
R6	anterior olfactory nucleus
R6d	anterior olfactory nucleus, dorsal
R6*l*	anterior olfactory nucleus, lateral
R6m	anterior olfactory nucleus, medial
R6p	anterior olfactory nucleus, posterior
R7	rhinal fissure
R8	entorhinal cortex
R9	olfactory tubercle
R14	nucleus of the lateral olfactory tract
R15	islands of Calleja

SEPTAL AREA (S)

S1	septal neuroepithelium
S2	septal subventricular zone
S3	septum (or differentiating field)
S4	indesium griseum
S5	lateral septum
S6	medial septum
S7	diagonal band (or differentiating field)
S8	diagonal band (vertical limb)
S9	diagonal band (horizontal limb)
S14	bed nucleus of anterior commissure
S15	triangular septal nucleus
S16	taenia tecta
S17	septofimbrial nucleus

SPINAL CORD (see under Z)

THALAMUS (T)

T1	thalamic neuroepithelium
T1a	anterior thalamic neuroepithelium
T1e	epithalamic neuroepithelium
T1i	intermediate thalamic neuroepithelium
T1p	posterior thalamic neuroepithelium
T1r	reticular (or ventral) thalamic neuroepithelium
T2	thalamic subventricular zone
T3	thalamus (or differentiating field)
T3a	anterior thalamus (or differentiating field)
T3e	epithalamus (or differentiating field)
T3i	intermediate thalamus (or differentiating field)
T3p	posterior thalamus (or differentiating field)
T3v	ventral thalamus (or differentiating field)
T4	medial habenular nucleus
T5	lateral habenular nucleus
T6	paraventricular nucleus
T7	paratenial nucleus
T8	anteromedial nucleus
T9	anteroventral nucleus
T10	anterodorsal nucleus
T14	centromedian nucleus
T15	centrolateral nucleus
T16	mediodorsal nucleus
T17	rhomboid nucleus
T18	reuniens nucleus
T19	parafascicular nucleus
T20	laterodorsal nucleus
T24	lateral posterior nucleus
T25	reticular nucleus
T26	ventral medial nucleus
T27	ventrolateral nuclear complex
T28	ventrobasal nuclear complex
T29	dorsal lateral geniculate nucleus
T30	ventral lateral geniculate nucleus

T34	medial geniculate nucleus
T35	posterior nuclear complex

VENTRICLES (v)

va	aqueduct
vai	aqueduct, inferior tectal
vapr	aqueduct, pretectal
vas	aqueduct, superior tectal
vat	aqueduct, tegmental
vl	lateral ventricle
vla	lateral ventricle, anterior horn
vlc	lateral ventricle, central
vld	lateral ventricle, dorsal
vlfa	lateral ventricle, amygdaloid fork
vlfp	lateral ventricle, pallidal fork
vlfs	lateral ventricle, septal fork
vlm	foramen of Monro
vlo	olfactory ventricle
vlp	lateral ventricle, posterior horn
vs	spinal canal
v3	third ventricle
v3e	third ventricle, epithalamic
v3h	third ventricle, hypothalamic
v3i	third ventricle, infundibular recess
v3m	third ventricle, mammillary recess
v3o	third ventricle, optic recess
v3pi	third ventricle, pineal recess
v3pr	third ventricle, preoptic recess
v3t	third ventricle, thalamic
v4	fourth ventricle
v4a	fourth ventricle, anterior
v4i	fourth ventricle, isthmal
v4ir	fourth ventricle, isthmal recess
v4l	fourth ventricle, lateral recess
v4p	fourth ventricle, posterior

PONTINE AREA (X)

PONS

X1	pontine neuroepithelium
X1a	anterior pontine neuroepithelium
X1p	posterior pontine neuroepithelium
X2	pontine subventricular zone
X2a	anterior pontine subventricular zone
X2p	posterior pontine subventricular zone
X3	pons (or differentiating field)
X3a	anterior pons (or differentiating field)
X3p	posterior pons (or differentiating field)
X4	dorsal raphe nucleus (pontine)
X5	pontine gray nucleus
X6	pontine reticulotegmental nucleus
X7	pontine reticular formation
X8	superior central raphe nucleus
X9	dorsal tegmental nucleus
X10	ventral tegmental nucleus

PRECEREBELLAR AND VESTIBULAR DIVISIONS

X11a	anterior precerebellar neuroepithelium
X11p	posterior precerebellar neuroepithelium
X12	precerebellar migration
X12a	anterior precerebellar extramural migration
X12i	precerebellar intramural migration
X12p	posterior precerebellar extramural migration
X14	spinal vestibular nucleus
X15	lateral vestibular nucleus
X16	superior vestibular nucleus
X17	medial vestibular nucleus

AUDITORY DIVISION

X21	cochlear nuclei neuroepithelium
X22	cochlear external germinal layer
X23	cochlear nuclei (or differentiating field)
X23d	dorsal cochlear nucleus (or differentiating field)
X23v	ventral cochlear nucleus (or differentiating field)
X24	superior olivary nucleus
X25	trapezoid body

ISTHMAL DIVISION

X31	isthmal neuroepithelium
X31d	isthmal neuroepithelium, dorsal
X31v	isthmal neuroepithelium, ventral
X33	isthmus (or differentiating field)
X34	locus coeruleus
X35	parabrachial nucleus
X36	nucleus of lateral lemniscus
X36d	nucleus of lateral lemniscus, dorsal
X36v	nucleus of lateral lemniscus, ventral

MEDULLA (Y)

Y1	medullary neuroepithelium
Y1d	medullary neuroepithelium, dorsal
Y1v	medullary neuroepithelium, ventral
Y3	medulla (or differentiating field)
Y4	medullary glial palisade
Y5	medullary raphe
Y6	medullary reticular formation
Y7	prepositus nucleus
Y8	inferior olive
Y14	gracile nucleus
Y15	cuneate nucleus
Y16	solitary nucleus
Y17	ambiguus nucleus

Y18	magnocellular reticular nucleus
Y19	gigantocellular reticular nucleus
Y20	parvicellular reticular nucleus
Y24	external cuneate nucleus
Y25	lateral reticular nucleus
Y26	medial reticular nucleus

SPINAL CORD (Z)

Z1	spinal cord neuroepithelium
Z1d	spinal cord neuroepithelium, dorsal
Z1v	spinal cord neuroepithelium, ventral
Z3	spinal cord (or differentiating field)
Z3d	spinal cord (or differentiating field), dorsal
Z3v	spinal cord (or differentiating field), ventral
Z4	ventral horn
Z5	dorsal horn
Z6	dorsal funiculus
Z7	ventral funiculus
Z8	dorsal root ganglion

GUIDE TO THE CHRONOLOGICAL INDEX

GUIDE TO INDEX

A

B

C

D

GUIDE TO INDEX

CHRONOLOGICAL INDEX

CHRONOLOGICAL INDEX

			E11	E12	E13	E14	E15	E16	E17	E18	E20	E22
AMYGDALA (A)												
amygdaloid neuroepithelium	A1	S	8	58	104-108	164	216-218	276	342	406	462-464	
		C			122-126	178	236-238	300-302	368	430-432	492-494	566-568
amygdaloid subventricular zone	A2	S								406		528
		C							368	432	492-494	564-568
amygdala (or differentiating field)	A3	S				164	216-218	276	340-342	406	462-466	528-532
		C			122-126	178	236-238	292-302	360-368	422-432	486-494	554-568
anterior amygdaloid area	A4	S						276		406	462-466	528-532
		C					236-238	292-296	360-362	422	486-488	554
amygdala-hippocampal area	A5	S									464-466	530
		C									494	564-568
central amygdaloid nucleus lateral, medial	A6, *l*, m	S						276			464-466	530-532
		C						300-302	366-368	428-430	488-492	556-566
basomedial amygdaloid nucleus	A7	S						276		406	462-464	532
		C					238	298-302	364-366	424-426	490-492	556-564
basolateral amygdaloid nucleus	A8	S										530
		C						298-300	364-366	426-430	490-492	556-564
cortical amygdaloid nucleus anterior, posterior	A9, a, p	S						276	342	406	462-466	528-532
		C					238	296-302	364-368	424-432	490-494	556-568
lateral amygdaloid nucleus	A10	S										
		C						292-302	364-368	424-430	490-494	558-566
medial amygdaloid nucleus	A14	S										528
		C					238	298-302	364-368	424-428	488-492	556-564
intercalated masses	A16	S									466	530-532
		C										558-560
BASAL GANGLIA (B)												
striatal neuroepithelium	B1	S				160-162	214-218	270-276	336-342	402-406	458-466	526
		C		64-66	114-120	172-176	226-234	284-300	354-368	416-432	478-490	546-562
striatal subventricular zone	B2	S						270-276	338-342	402-406	460-466	526
		C					228-234	286-300	354-368	416-432	478-490	544-562

S = Sagittal C = Coronal

BASAL GANGLIA (contd.)		E11	E12	E13	E14	E15	E16	E17	E18	E20	E22
striatum (or differentiating field) **B3**	S				160	216-218	272-276	340-342	404-406	462-466	524-532
	C				172-176	226-234	284-298	354-368	416-432	478-490	544-562
pallidal neuroepithelium **B11**	S	4 - 8	56-58	104-106	154-164	210-216	266-276	332-340	398-406	458	
	C	12-18	64-70	114-120	172-178	228-234	288-298	358-362	422-426	486-490	
pallidal subventricular zone **B12**	S				156-158	210-212	266-276	336-340	400-404	460	
	C				174-176	228-234	290-298	358-362	424-426		
pallidum (or differentiating field) **B13**	S				156-164	210-216	266-274	338-340	402-406	462-466	526-532
	C			118-120	172-178	230-234	288-296	358-364	422-430	484-490	552-556
entopeduncular nucleus **B14**	S						270	342		462	
	C							364	426-428	488-492	556-558, 562
substantia innominata **B15**	S									458-462	522-528
	C							360-362	420-422	482-486	550-554
strionuclear neuroepithelium **B21**	S				158	210-214	266-270	334			
	C			124-126	178	236-238	296-298	364	422-426	484	552
strionuclear subventricular zone **B22**	S										
	C								422-426	484	
bed nucleus of stria terminalis (or differentiating field) **B23-B28**	S					210-214	266-270	332-336		452-458	520-524
	C					238	296-298		422-428	482-488	550-554
accumbens neuroepithelium **B31**	S							338			
	C								416-420	476-480	544
accumbens subventricular zone **B32**	S										
	C								416-420	476-480	544-546
nucleus accumbens (or differentiating field) **B33**	S									460-464	522-526
	C								416-420	476-482	544-548
CEREBELLUM (see under L)											
CEREBRAL CORTEX (C)											
neocortical neuroepithelium **C1**	S	4 - 6	54-58	102-108	158-164	210-218	262, 266-276	328-342	396-406	456-466	522-532
	C	12-20	62-70	112-128	168-180	222-242	280-304	352-372	414-438	478-494	546-566
neocortical subventricular zone **C2**	S							328-342	396-406	456-466	520-532
	C							352-372	414-438	476-496	542-568

S = Sagittal C = Coronal

CHRONOLOGICAL INDEX

CEREBRAL CORTEX (contd.)		E11	E12	E13	E14	E15	E16	E17	E18	E20	E22
neocortex (or differentiating field) **C3**	S					210-218	262, 266-276	328-342	392-406	454-466	520-532
	C				168-180	222-242	280-306	348-374	414-440	474-498	538-572
molecular layer (or primordial plexiform layer) **C4**	S						262	328-342	392-406	454-466	520-532
	C			114-116	168-180	224-242	280-304	348-374	414-440	474-500	538-570
intermediate zone **C5**	S						262, 266-276	330-342	396-406	456-466	
	C					224-242	280-304	350-374	414-440	476-498	
cortical plate inferior, superior **C6, i, s**	S							330-342	392-406	454-466	520-532
	C						284-302	350-374	414-440	474-498	538-570
white matter **C7**	S							330-342	392, 396-406	454-466	520-532
	C					226-238	286-296	350-372	414-440	476-498	540-570
subplate **C8**	S									458-466	520-532
	C								414-438	476-494	540-568
lateral migratory stream head, reservoir **C9, h, r**	S									462-466	526-532
	C							356-366	414-430	476-488	544-562
cingulate cortex neuroepithelium **C11**	S	4	50-52	98-100	152-156	208-214	266-268		396-398		
	C	14-18	62-70	112-126	168-180	224-234	280-292	352-362	416-434	478-480	544
cingulate cortex subventricular zone **C12**	S										
	C									478-480	544
cingulate cortex (or differentiating field) **C13**	S				152-154	206-214	266		396-400		518
	C				168-180	224-234	280-292	350-362	414-434	476-494	540-566
orbital and insular area (or differentiating field) **C23**	S								396-402	454-466	518-532
	C					226-232	284-296	352-364	416-420	474-486	538-554
insular cortex **C24**	S										
	C					226-232		354	414-420	476-486	542-554
orbital cortex **C25**	S										
	C					226-228					540-542
claustrum **C26**	S										526-532
	C						286-296	354-364	414-428	476-484	542-552
CRANIAL NERVES AND NUCLEI (see under n)											

S = Sagittal C = Coronal

CHRONOLOGICAL INDEX

		E11	E12	E13	E14	E15	E16	E17	E18	E20	E22
FIBER TRACTS (f)											
lateral olfactory tract **f1**	S							340-342	404-406	462-466	526-532
	C						284-294	352-364	412-426	476-486	540-554
medial olfactory tract **f2**	S							328-330			
	C								414		
corpus callosum **f3**	S									454	518-526
	C									480	540-566
anterior commissure **f4**	S							328-322, 340	392-406	452-456 460-466	518-524 528-532
	C							356-360	416-422	476-486	544-552
medial forebrain bundle **f5**	S							340-342		460	
	C				174-178	234-238	294-298		414		548
fornix **f6**	S								394-396, 400	452-454, 458	518-524
	C						286-288	356-358, 372	420-422	482, 486-488 494-496	548-566
hippocampal commissure **f7**	S									452	518-520
	C									482-484	550-554
internal capsule **f8**	S						270	338-342	400-406	460-462	526-532
	C						294-300	362-366	424-430	478-492	546-562
cerebral peduncle **f9**	S							328-340	392-404	462	
	C					242-244		366-376	428-440	496	562-578
external medullary lamina **f10**	S										
	C										556-564
external capsule **f11**	S								404-406		
	C						288-294	354-364	416-430	478	552
cingulum bundle **f12**	S										
	C										544-564
fimbria **f13**	S								402-404	454-464	522-530
	C							360-362, 374	424-432	486-492	554-566
stria terminalis **f14**	S					210	266-268			460-464	524, 530
	C					236-238	294-296	364	424	486-490	552-562

S = Sagittal C = Coronal

CHRONOLOGICAL INDEX

FIBER TRACTS (contd.)		E11	E12	E13	E14	E15	E16	E17	E18	E20	E22
stria medullaris **f15**	S					210		332-336	396-398	454-456	518-524
	C					234-238	294-298	364	428-434	486	554, 558-562
medial lemniscus **f16**	S										
	C									494	560, 566-568
mammillothalamic tract **f17**	S							334-336	396-398		
	C					238-242	302-308	372-374	432-434	496	564, 568-570
habenulopeduncular tract **f18**	S					210	266	332-338	396-398	454-456	520-522
	C					244	302-308	372-374	432-436	494-496	564-568
posterior commissure **f19**	S				152-154	206-208	262-264	328-330	392-398	452	518, 522
	C				190-194	246-248	304-310	372-376	438-440	494-496	566-570
supramammillary decussation **f20**	S										
	C							374			570
brachium of superior colliculus **f21**	S										
	C										574-576
commissure of superior colliculus **f22**	S								392-394	452	
	C								442-446	498-500	
superior cerebellar peduncle **f23**	S						266-274		398-404	458-460	
	C						318-320		436-438	506-508	572, 578-580
middle cerebellar peduncle **f24**	S									464	528
	C									498-500	572-576
inferior cerebellar peduncle **f25**	S						266-272	342			
	C						318-320	376-380			578, 582-584
medial longitudinal fasciculus **f27**	S							328	394		
	C					242			440		
lateral lemniscus **f28**	S									462	
	C								436, 440		576, 580
brachium of inferior colliculus **f29**	S										
	C								446		
spinal tract of trigeminal **f30**	S										
	C										578-588
cuneate fasciculus **f31**	S									458-460	518-524
	C										

S = Sagittal C = Coronal

		E11	E12	E13	E14	E15	E16	E17	E18	E20	E22
HIPPOCAMPAL FORMATION (see under K)											
HYPOTHALAMUS (H)											
hypothalamic neuroepithelium **H1, a-v**	S	4 - 8	50-58	98-108	152-162	206-216	262-266	328-330	392-394	452	518
	C	18-28	68-80	120-138	176-188	232-246	292-308	362-374	422-432	486-496	554-566
hypothalamus (or differentiating field) **H3, a-v**	S				156-160	206-214	262-272	328-338	392-404	452-460	518-526
	C			120-138	178-188	232-246	294-308	362-374	422-434	486-496	554-570
supraoptic nucleus **H4**	S										526
	C						300		424-426	490	556
suprachiasmatic nucleus **H5**	S								394-396	460	520
	C								420	484	556
paraventricular nucleus **H6**	S										
	C						300-302	364-368	424-428	488-492	556-562
zona incerta **H7**	S					212	266-270	332-340	396-402	458-460	518, 524
	C					240	298-302	366-370	428-434	488-492	556-566
dorsomedial nucleus **H8**	S								396-398	454-456	520-522
	C						302	366-370	430-434	494	564-566
ventromedial nucleus **H9**	S								396-398	454-456	520-522
	C						302-304	366-370	430-434	494	560-566
anterobasal nucleus **H10**	S										518
	C					234-236	298-300	362-364	422	490	558-560
supramammillary nucleus **H14**	S										
	C				186-188	246	308	372-374	434	496	568-570
premammillary nucleus **H15**	S									452-454	518-520
	C						306				
medial mammillary nucleus **H16**	S						266	332	396	452-456	518-522
	C						308	372-376	434	496	570
lateral mammillary nucleus **H17**	S					210-212		334-336	398	458	524
	C				186		308	372-374		496	568-570
subthalamic nucleus **H19**	S							340	402-404	460	526
	C					244-246	302-308	370-372	432-434	494	566-568

S = Sagittal C = Coronal

CHRONOLOGICAL INDEX

HYPOTHALAMUS (contd.)		E11	E12	E13	E14	E15	E16	E17	E18	E20	E22
arcuate nucleus **H20**	S						266	332	394-396	454	520
	C					238-242	302-306	366-370	426-432	492, 496	564-570
anterodorsal nucleus **H24**	S										
	C					238	298-300	362-364	422-428	486-492	556-560
anteroventral nucleus **H25**	S										
	C					238	298-300	362-364	422-428	486-492	556-560
posterior hypothalamic area **H26**	S									456	
	C									494	566-568
tuberomammillary nucleus **H27**	S									460	
	C							370		496	568
HIPPOCAMPAL FORMATION (K)											
hippocampal (Ammon's horn) neuroepithelium **K1**	S		54-58	102-108	156-164	210-218	266-276	332-342	396-406	464	
	C			116-118	170-176	226-238	286-304	358-370	424-436	484-490	
hippocampus (or differentiating field) dorsal, ventral **K3, d, v**	S					214-218	266-274	332-342	396-406	454-466	520-532
	C				180	228-238	286-304	358-370	424-436	484-494	554-568
CA1 field **K4**	S									466	
	C								424-434	488-496	554-568
CA3 field **K5**	S									460	
	C										554-568
alveus **K6**	S										532
	C										554-568
subicular neuroepithelium **K11**	S	6	50-58	98-108	156-164	210-218	266-276	332-342	396-406	458	
	C	20	68-70	114-120	170-176	226-240	286-304	358-372	424-436	482-494	
subicular area (or differentiating field) **K13**	S					216-218	268-276	330-342	396-406	454-466	520-532
	C				174-176	226-238	286-300	358-372	424-436	482-496	556-570
dentate gyrus neuroepithelium **K21**	S						266-274	332-342	398-404		
	C						290-302	364-368	430-432		
dentate gyrus (or differentiating field) **K23**	S								400-404	458-466	522-532
	C							364-370	428-432	486-496	554-568
dentate granular layer **K24**	S										
	C									488-494	554-568

S = Sagittal C = Coronal

HIPPOCAMPAL FORMATION (contd.)		E11	E12	E13	E14	E15	E16	E17	E18	E20	E22
dentate polymorph cell layer **K25**	S										
	C										558-564
CEREBELLUM (L)											
cerebellar neuroepithelium hemisphere, vermis **L1, h, v**	S	4 - 8	50-58	98-106	152-164	206-218	262-274	328-342	392-406		
	C	36-38	84-88	132-142	190-200	248-256	312-322	376-386	436-442		
external germinal layer **L2**	S							328-342	392-406	452-466	518-530
	C							376-384	436-444	506-512	578-586
cerebellum (or differentiating field) hemisphere, vermis **L3, h, v**	S				156-164	208-218	262-274	328-342	392-406	452-466	518-530
	C			132-142	190-200	248-256	312-322	376-386	436-444	504-512	578-586
molecular layer **L4**	S									452	528
	C										578-584
Purkinje cell layer **L5**	S							340	398-406	452-466	518-528
	C							384	438-444	506-512	578-586
medullary layer **L6**	S										
	C										578-584
lateral deep nucleus (dentate) **L7**	S									464	526
	C									506-508	580-582
lateral deep nucleus, migration **L7mi**	S										
	C					248	312-316				
medial deep nucleus (fastigial) **L8**	S										518-520
	C										582
medial deep nucleus, migration **L8mi**	S						264-266		396-398		
	C						318-320	380-384	442-444		
cerebellar germinal trigone **L9**	S							332-336	400-406	462-464	526-528
	C							378-384	438-440	510	582
cerebellar commissure **L10**	S										
	C								444		
MEDULLA (see under Y)											

S = Sagittal C = Coronal

CHRONOLOGICAL INDEX

		E11	E12	E13	E14	E15	E16	E17	E18	E20	E22
MIDBRAIN (M)											
superior tectal neuroepithelium **M1**	S	4 - 8	50-58	98-104	152-158	206-212	262-270	328-338	392-400		
	C	30-32	88-92	142-146	196-200	250-256	310-322	378-386	442-446		574
superior colliculus (or differentiating field) **M3**	S					206-212	262-270	328-340	392-404	452-462	518-526
	C				196	250-254	310-322	376-386	440-446	498-510	570-578
periaqueductal gray dorsal, ventral **M4, d, v**	S						267	336-340	392-400	454-460	518-524
	C						306-314	374-380	436-446	496-508	568-580
superior colliculus, superficial layers **M5**	S							334-340	392-404	452-460	518-524
	C							376-384	440-446	496-510	570-578
superior colliculus, deep layers **M6**	S							334-340	392-404	452-460	518-524
	C						310-322	376-386	440-446	498-508	570-578
superior colliculus, stratum opticum **M7**	S										522-526
	C										572-576
red nucleus **M8**	S									454-458	520-524
	C					242-246	304-308	372-374	438-440	498-500	570-572
midbrain reticular formation **M9**	S								398-404	460	522-526
	C					248-252	310-316	376-386	436-446	498-504	570-576
tegmental neuroepithelium anterior, posterior **M11, a, p**	S	4 - 8	50-56	98-104	152-158	206-212	262-268	338	392-400		518
	C	30-32	82-92	140-144	190-198	248-254	306-316	382-384	444-446		
tegmentum (or differentiating field) **M13, a, p**	S				158-162	206-212	262-274	328-340	392-404	452-462	518-526
	C			140-144	190-198	248-254	304-316	374-384	436-446	496-506	568-576
substantia nigra compacta, reticulata **M14, c, r**	S										
	C					248	308	374-376	436-440	498-500	570-574
interpeduncular nucleus caudal, rostral **M15, c, r**	S									456-460	518-526
	C					248	310-312	374-376	436-440	496-500	572-574
dorsal raphe (mesencephalic) **M16**	S										520
	C					248	308	378	438	496-504	572-576
ventral tegmental area **M17**	S										
	C						310	376	438	496-500	570-574
parabigeminal nucleus **M18**	S										
	C							378-380	438	500-504	576

S = Sagittal C = Coronal

MIDBRAIN (contd.)		E11	E12	E13	E14	E15	E16	E17	E18	E20	E22
pretectal neuroepithelium	S	4 - 8	50-56	98-104	152-158	206-208	262-264	328-330	392-394		
M21	C	26-28	80-86	134-140	190-194	246-248	306-308		438		
pretectum (or differentiating field)	S					206-212	262-268	328-340	392-402	452-462	518-526
M23	C				190-194	246-248	306-308	372-374	436-438	494-496	566-568
anterior pretectal nucleus	S								398	454	
M24	C						306	372	436		566
nucleus of the posterior commissure	S								398	454	
M25	C								438	494-496	568
olivary pretectal nucleus	S										
M26	C						308	374			568
nucleus of the optic tract	S										
M27	C										568
posterior pretectal nucleus	S									454, 458	
M28	C								438	496	568
inferior tectal neuroepithelium	S	4 - 6	50-58	98-106	152-160	206-216	262-274	328-340	392-402	454-458	518
M31	C		90-92	144-146	196-200	252-256	316-322	384-386	446	508-512	578-580
inferior colliculus (or differentiating field)	S					206-216	262-274	328-340	392-406	452-464	518-526
M33	C					252-254	316-322	382-386	446	506-512	576-582
external nucleus of inferior colliculus	S									460	
M34	C									508, 512	576
central nucleus of inferior colliculus	S									460	
M35	C							382		506, 512	578-582
NERVES AND NUCLEI, CRANIAL (n)											
olfactory nerve	S				156-158	210-212	266-268	332-336	396-400	454-456	518-526
n1	C				168-170	224	280-284	346-350	410-412		536
optic nerve, chiasma, and tract	S						262-264	328-334	392-398, 406	452-460	518-526
n2, c, t	C						296	360-366	418-426	486-494	554-564
oculomotor nucleus	S										
n3n	C						310	376	438-442	498	572-574
trochlear nucleus	S										
n4n	C									502	

S = Sagittal C = Coronal

NERVES AND NUCLEI, CRANIAL (contd.)		E11	E12	E13	E14	E15	E16	E17	E18	E20	E22
trigeminal nerve, ganglion, boundary cap, and nuclei **n5, b-s**	S			108	164	218	262, 270-276	340-342	400-406	460-462	524-526
	C	34-40	78-82	126-130	178-184	236-246	300-322	364-386	432-444	500-512	576-588
facial nerve and nucleus **n7, n**	S						266	338	398	456-460	524-526
	C				182-186	242-246	310-312	374-378	432	504	578-580
vestibulocochlear nerve, boundary cap, and ganglia **n8, b, g**	S		56-58	106-108	162-164	216-218		342	404-406		
	C	40-42	74-78	122-126	182-184	242	312	376	432		
ganglion of glossopharyngeal nerve **n9g**	S								403		
	C										
ganglion of the vagus **n10g**	S								403		
	C										
dorsal motor nucleus of vagus **n10n**	S								399		
	C										586
hypoglossal nucleus **n12n**	S						266	334	392-398	452-454	518-524
	C					242-244	318-322	382-386	434-440	512	586-588
OLFACTORY BULB (O)											
olfactory bulb neuroepithelium **O1**	S					210-216	266-268	332-336	398-400	454-456	520-524
	C				168-170	222-224	282	348-350	410-412	470-474	536-540
olfactory bulb subventricular zone **O2**	S							336	398-400	454-458	520-524
	C							346-350	410-412	470-474	536-540
olfactory bulb (or differentiating field) **O3**	S					210-216	266-272	330-338	396-402	452-460	520-524
	C					222-224	280-282	346-350	410-412	470-474	536-540
glomerular layer **O4**	S									454-460	520-524
	C									470-474	536-540
external plexiform layer **O5**	S								396-400	452-460	520-526
	C					222		346-350	410-412	470-474	536-540
mitral cell layer **O6**	S							330-336	396-402	454-460	520-524
	C							346-350	410-412	470-474	536-540
internal plexiform layer **O7**	S									454-458	520-524
	C									470-474	536-538
granular layer **O8**	S									454-458	520-524
	C							346-350	410-412	472-474	536-540

S = Sagittal C = Coronal

OLFACTORY BULB (contd.)		E11	E12	E13	E14	E15	E16	E17	E18	E20	E22
accessory olfactory bulb (or differentiating field) **O13**	S									454-458	522-524
	C								412	472	538-540
glomerular layer (accessory olfactory bulb) **O14**	S										524
	C									472	538
external plexiform layer (accessory olfactory bulb) **O15**	S									458	522-524
	C								412	472	538-540
mitral cell layer (accessory olfactory bulb) **O16**	S								402		524
	C								412	472	538-540
internal plexiform layer (accessory olfactory bulb) **O17**	S										524
	C								412	472	538-540
granular layer (accessory olfactory bulb) **O18**	S										524
	C									472	538
PONTINE AREA (see under X)											
PREOPTIC AREA AND OPTIC EPITHELIA (P)											
preoptic neuroepithelium **P1**	S	4	50-56	98-108	152-164	206-216	262-268	328-334	392-394	452	
	C	16, 20	68	118-120	174-176	230-234	292-294	360	418-420	482	552
preoptic area (or differentiating field) **P3**	S					206-212	262-268	330-338	394-400	452-460	518-526
	C				174	230-234	292-294	360	418-420	482-484	552
medial preoptic region **P4**	S								392-396		520
	C					230	292	360	418-420	482-484	552
lateral preoptic region **P5**	S								398-400		526
	C					232-234	292-294	360	418-420	482-484	552
superior preoptic region **P6**	S								394-398		520
	C							360	418-422	482-484	552
retinal neuroepithelium **P11**	S										
	C	16-24	68-72								
eye stalk epithelium **P12**	S										
	C		72	120	174-176	234	294-296			486-488	

S = Sagittal C = Coronal

CHRONOLOGICAL INDEX

		E11	E12	E13	E14	E15	E16	E17	E18	E20	E22
RHINENCEPHALON (R)											
rhinencephalic neuroepithelium anterior, posterior **R1, a, p**	S	4 - 8	50-58	100-108	156-164	208-210 214-218	268-276	332-340	396-400		522
	C	12-14	62-66	112-116	168-172	222-226	282-290	348-358	414	474	540-542
rhinencephalic subventricular zone **R2**	S							332-340	396-402	452-458	522-524
	C							348-358	414	474	540-542
rhinencephalon (or differentiating field) **R3, a, p**	S				158-164	208-218	266-276	330-342	398-406	456-466	522-526
	C				168-172	222-238	280-300	348-370	414-430	474-496	540-570
piriform cortex **R4**	S						274-276		406	460-466	526-532
	C					226-238	282-300	352-370	416-430	476-494	542-566
endopiriform nucleus **R5**	S							342	404	466	526-532
	C							364-368	414-428	478-488	542-566
anterior olfactory nucleus **R6, d-p**	S						274				524
	C							348-350		474	540-542
rhinal fissure **R7**	S									462-464	526-532
	C						288-292	350-352		476-496	542-570
entorhinal cortex **R8**	S										528-532
	C							370-372		496	568-570
olfactory tubercle **R9**	S								406	458-466	522-528
	C							358	414-420	478-482	542-552
nucleus of the lateral olfactory tract **R14**	S									464	528
	C									488	554
islands of Calleja **R15**	S									464	528
	C										546
SEPTAL AREA (S)											
septal neuroepithelium **S1**	S	4	52-54	98-102	152-154	206-214	262-272	328-336	392-400	458	524
	C	12	62	112-114	168-170	226-228	282-290	352-358	414-422	476-480	546-548
septal subventricular zone **S2**	S							332-334	398-400	454	524
	C							352-358	414-422	476-480	544-548
septum (or differentiating field) **S3**	S				152-154	206-214	262-270	328-334	392-398	452-456	518-523
	C				168-170	226-228	280-290	352-358	414-422	476-482	544-550

S = Sagittal C = Coronal

SEPTAL AREA (contd.)		E11	E12	E13	E14	E15	E16	E17	E18	E20	E22
indesium griseum	S										518
S4	C								416-422	480-484	546-552
lateral septum	S								398	452-454	520-522
S5	C							356	416-422	478-482	546-550
medial septum	S							328	392-396	452	520
S6	C					226-228	284-290	354-358	416-422	480-482	548-550
diagonal band (or differentiating field) **S7-**	S							330-336	394-400	454-456	520
S9	C					228	286-290	352-358	416-420	478-482	546-552
triangular septal nucleus	S										
S15	C									482-484	552
taenia tecta	S										518-522
S16	C							354	414	476-478	542-546
septofimbrial nucleus	S										
S17	C							358		484	550-552
SPINAL CORD (see under Z)											
THALAMUS (T)											
thalamic neuroepithelium	S	4 - 8	50-58	98-104	152-158	208	264	328-330	392-394	452	
T1, a-r	C	22-26	72-80	126-140	178-190	234-246	292-304	362-372	424-434	486-492	554, 562, 566
thalamic subventricular zone	S										
T2	C							368-370	432-434		
thalamus (or differentiating field)	S					210-212	266-270	332-340	394-404	452-462	518-528
T3, a-v	C				182-190	236-246	294-308	362-374	424-438	486-496	554-572
medial habenular nucleus	S										518-520
T4	C								432-434	488-492	556-564
lateral habenular nucleus	S							332	396-398	454-456	520
T5	C					240	296-302	366-368	428-434	490-492	556-564
paraventricular nucleus	S										
T6	C								424-428	486-490	554-564
anteromedial nucleus	S										
T8	C								426	486-488	554-556

S = Sagittal C = Coronal

CHRONOLOGICAL INDEX

THALAMUS (contd.)		E11	E12	E13	E14	E15	E16	E17	E18	E20	E22
anteroventral nucleus	S										
T9	C								426	488	554-556
anterodorsal nucleus	S								400		
T10	C								424-426	488	556
centromedian nucleus	S										520
T14	C								430-434	492	558-562
centrolateral nucleus	S										
T15	C									490	556-558
mediodorsal nucleus	S										520
T16	C									490-492	558-560
rhomboid nucleus	S										
T17	C								426-430	488-492	556-562
reuniens nucleus	S										
T18	C									490-492	556-564
parafascicular nucleus	S							332		454	520
T19	C							372			564-568
laterodorsal nucleus	S								400		
T20	C							368	428	490-492	558-564
lateral posterior nucleus	S										
T24	C						302-304	370	432-434	495	564-566
reticular nucleus	S					210	266-270	332-340	396-402	458-460	520-524
T25	C				182-186	238-242	296-302	366-370	426-434	488-490	556-564
ventral medial nucleus	S										
T26	C							370	430	492-494	558-566
ventrolateral nuclear complex	S								402		
T27	C					240	298-300	366-368	428-434	490-492	556-564
ventrobasal nuclear complex	S									462	
T28	C					242	302-304	370	432-434	494-496	562-566
dorsal lateral geniculate nucleus	S								402	460	
T29	C					242-246	302-304	370-374	430-434	492-494	558-566
ventral lateral geniculate nucleus	S								402		
T30	C						302-304	372-374	428-434	490-494	560-566

S = Sagittal C = Coronal

THALAMUS (contd.)		E11	E12	E13	E14	E15	E16	E17	E18	E20	E22
medial geniculate nucleus	S										
T34	C				188-190	242-246	304-308	372-374	436-442	496	568-574
posterior nuclear complex	S								400	456-460	
T35	C								436		566-568
PONTINE AREA (X)											
pontine neuroepithelium anterior, posterior	S	4 - 8	50-58	98-108	152-164	206-218	262-272	328-340	392-394		
X1, a, p	C	34-42	74-86	120-144	182-194	244-250	312-316	376-378	434, 438-442		
pontine subventricular zone	S										
X2a, p	C						310				
pons (or differentiating field)	S				156-162	206-218	262-274	328-342	392-404	452-462	518-526
X3, a, p	C			120-144	182-194	242-250	306-316	372-380	430-442	498-508	572-584
dorsal raphe nucleus	S							332	398		
X4	C							378	440-442		
pontine gray nucleus	S								396	454-456	518-524
X5	C								430	498	572-574
pontine reticulotegmental nucleus	S								398-400	452	518, 526
X6	C									500	576
pontine reticular formation	S							336-338	400-402	460	
X7	C					248	308-314	372-376	440-444	502	576-578
superior central raphe nucleus	S								394-398	452, 456	
X8	C						308-314	372-380	432-442	500-504	576
dorsal tegmental nucleus	S							332-336	394-398	454-456	520-522
X9	C						314	378-380	442-444	504-506	578
ventral tegmental nucleus	S							332-338	394-398	454-456	
X10	C						312-314	376	436-440	502-506	578
precerebellar neuroepithelium	S						262-268	336-340	396-402	456-458	
X11a, p	C						316-318	378-380	438-440		
precerebellar migrations	S							330-340	392-402	458-460	518-520
X12a-p	C					242	306, 314-318	378-384	430-440	498-500	574
spinal vestibular nucleus	S								396-402	458	524
X14	C						312-316	378		508	582-584

S = Sagittal C = Coronal

PONTINE AREA (contd.)		E11	E12	E13	E14	E15	E16	E17	E18	E20	E22
lateral vestibular nucleus **X15**	S							340	400-402	458-460	
	C					244	312-316	376	434-436	506	580
superior vestibular nucleus **X16**	S							340	400	460	
	C					244	312	376-378	434-436	506	580
medial vestibular nucleus **X17**	S							336-338	392-398	452-456	522-524
	C					244	312-316	376	436	506-508	580-584
cochlear nuclei neuroepithelium **X21**	S		58	106	164		274	342	404-406	462-464	526-528
	C					244-246	312-314	376	436	508	
cochlear external germinal layer **X22**	S										528-530
	C									508	578-580
cochlear nuclei (or differentiating field) **X23, d, v**	S						274	342	404-406	462-464	526-530
	C					246	312-314	376	434-436	504-508	578-580
superior olivary nucleus **X24**	S							336-338	400	456-460	522-524
	C						310	372-374	432	500-502	576-578
trapezoid body **X25**	S										
	C									502	576-578
isthmal neuroepithelium **X31, d, v**	S	4 - 8	50-58	98-102	152-156	206-210	262-264		392		
	C	34	82-88	140-144	188-198	248-254	314-316	380	444		
isthmus (or differentiating field) **X33**	S				156	208-212	262-270	330-340	392-402	452-460	522-526
	C			140-144	188-196	248-254	312-316	376-382	438-444	502-506	576-578
locus coeruleus **X34**	S						266-268			458-460	524
	C					250-252	316-318	380-382	442-444	506-508	578-580
parabrachial nucleus **X35**	S										526
	C						308-314	372-380	432-440	504	578
nucleus of lateral lemniscus dorsal, ventral **X36, d, v**	S									462	526
	C					250	314	378-380	442-444	502-504	576
MEDULLA (Y)											
medullary neuroepithelium dorsal, ventral **Y1, d, v**	S	4 - 6	50-54	98-104	152-158	206-214	262-266	328-338	392-394		
	C	34-44	72-82	120-130	182-188	244-250	318-322	384-386	440-444		
medulla (or differentiating field) **Y3**	S				156-158	206-214	262-270	328-340	392-402	452-462	518-526
	C			120-126	182-186	242-250	312-322	376-386	434-446	504-512	580-588

S = Sagittal C = Coronal

		E11	E12	E13	E14	E15	E16	E17	E18	E20	E22
medullary glial palisade	S				152-154						
Y4	C										
medullary raphe	S						262-264	328-332	392-396		
Y5	C							378-386	434-444	508	582-586
medullary reticular formation	S							336-340	398-402	460	
Y6	C						312-320	376-386	434-444	508-512	582-588
prepositus nucleus	S							334-336	392-398	452-454	518-524
Y7	C					242-244	318-322	380-384	434-440	508-512	582-588
inferior olive	S						262	328-338	392-394	452-458	518-524
Y8	C					242	314-318	376-386	434-438	508-512	582-588
gracile nucleus	S						262	328	396	452-456	518-520
Y14	C						320-322	382-386	442-444	512	588
cuneate nucleus	S							336-338	398-400	458	520-524
Y15	C						320-322	384-386	442-444		588
solitary nucleus	S							338	398	458	522
Y16	C					246	318-322	380-382	436-438	508-512	582-586
ambiguus nucleus	S										
Y17	C									506-512	582
magnocellular reticular nucleus	S										
Y18	C						312		434	512	580
gigantocellular reticular nucleus	S										518
Y19	C										
parvicellular reticular nucleus	S										524
Y20	C										
external cuneate nucleus	S										
Y24	C								438-440	510	586-588
lateral reticular nucleus	S										
Y25	C								438	508-510	582-588
medial reticular nucleus	S										
Y26	C										580-582

S = Sagittal C = Coronal

		E11	E12	E13	E14	E15	E16	E17	E18	E20	E22
SPINAL CORD (Z)											
spinal cord neuroepithelium dorsal, ventral **Z1, d, v**	S	4 - 6	54	98	152	208					
	C	32-44	62-76	112-118	182-188						
spinal cord (or differentiating field) **Z3, d, v**	S					206-212	262-264	328-340	392-400	454	
	C			114-118	182-188	244-246	320-322				
ventral horn **Z4**	S				156	210	262-264	328-338	394-396	452, 456-458	518-520, 526
	C										
dorsal horn **Z5**	S				156	210	262-264	328-338	394-398	452, 456-460	520, 526
	C										
dorsal funiculus **Z6**	S							328-330	398		
	C										
ventral funiculus **Z7**	S							328-330	398		
	C										
dorsal root ganglion **Z8**	S										
	C		62-66, 72-74								

S = Sagittal C = Coronal

THE E11 BRAIN

COMPUTERIZED 3-D RECONSTRUCTION: 3 VIEWS

RIGHT, POSTERIOR VIEW

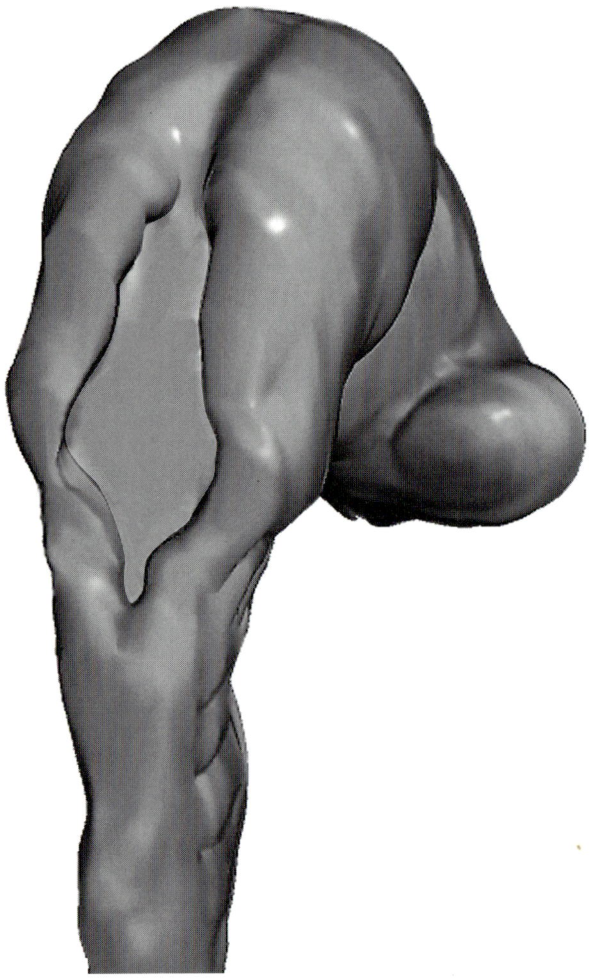

LEFT, ANTERIOR VIEW

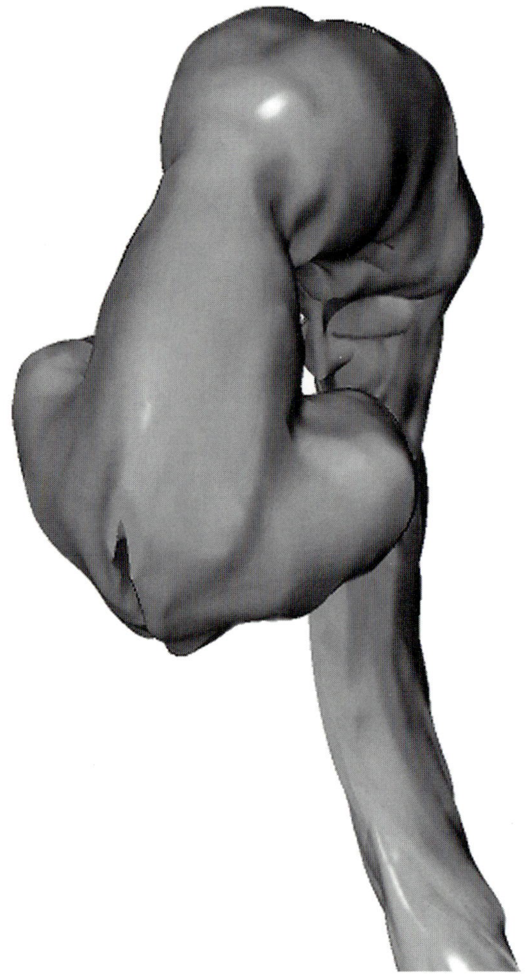

FRONTAL VIEW

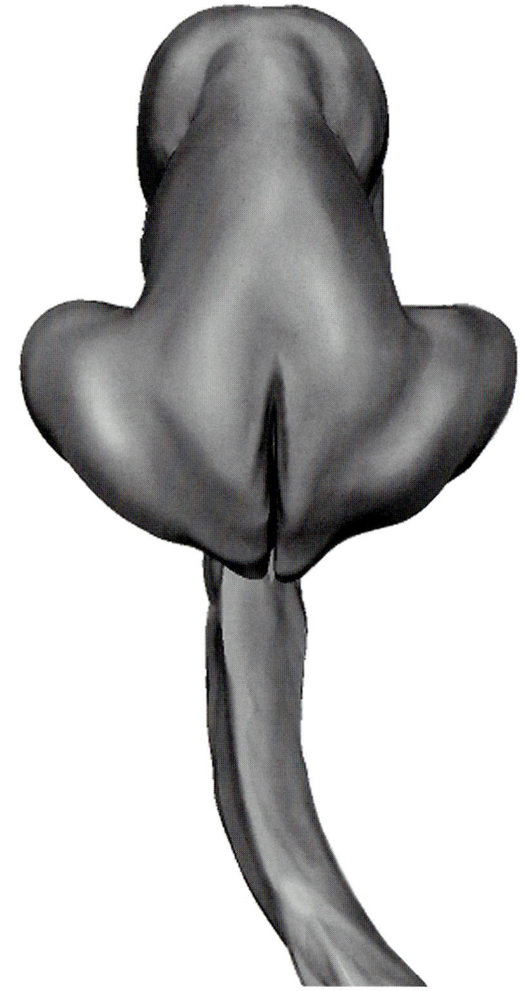

SAGITTAL

FIGURES and PLATES 1-3

SAGITTAL FIGURE 1

BASAL GANGLIA (B)

B11 pallidal neuroepithelium

CEREBRAL CORTEX (C)

C1 neocortical neuroepithelium
C11 cingulate cortex neuroepithelium

GLANDS AND PARANEURAL STRUCTURES (g)

g6 pituitary gland (primordium)
g10 velum medullare

HYPOTHALAMUS (H)

H1m mammillary neuroepithelium
H1v ventral hypothalamic neuroepithelium

CEREBELLUM (L)

L1 cerebellar neuroepithelium

MIDBRAIN (M)

M1 superior tectal neuroepithelium
M11a anterior tegmental neuroepithelium
M11p posterior tegmental neuroepithelium
M21 pretectal neuroepithelium
M31 inferior tectal neuroepithelium

PREOPTIC AREA AND OPTIC EPITHELIA (P)

P1 preoptic neuroepithelium

RHINENCEPHALON (R)

R1 rhinencephalic neuroepithelium

SEPTAL AREA (S)

S1 septal neuroepithelium

THALAMUS (T)

T1 thalamic neuroepithelium

VENTRICLES (v)

vai aqueduct, inferior tectal
vapr aqueduct, pretectal
vas aqueduct, superior tectal
vat aqueduct, tegmental
vla lateral ventricle, anterior horn
vld lateral ventricle, dorsal
v3i third ventricle, infundibular recess
v3m third ventricle, mammillary recess
v3t third ventricle, thalamic
v4 fourth ventricle
v4i fourth ventricle, isthmal

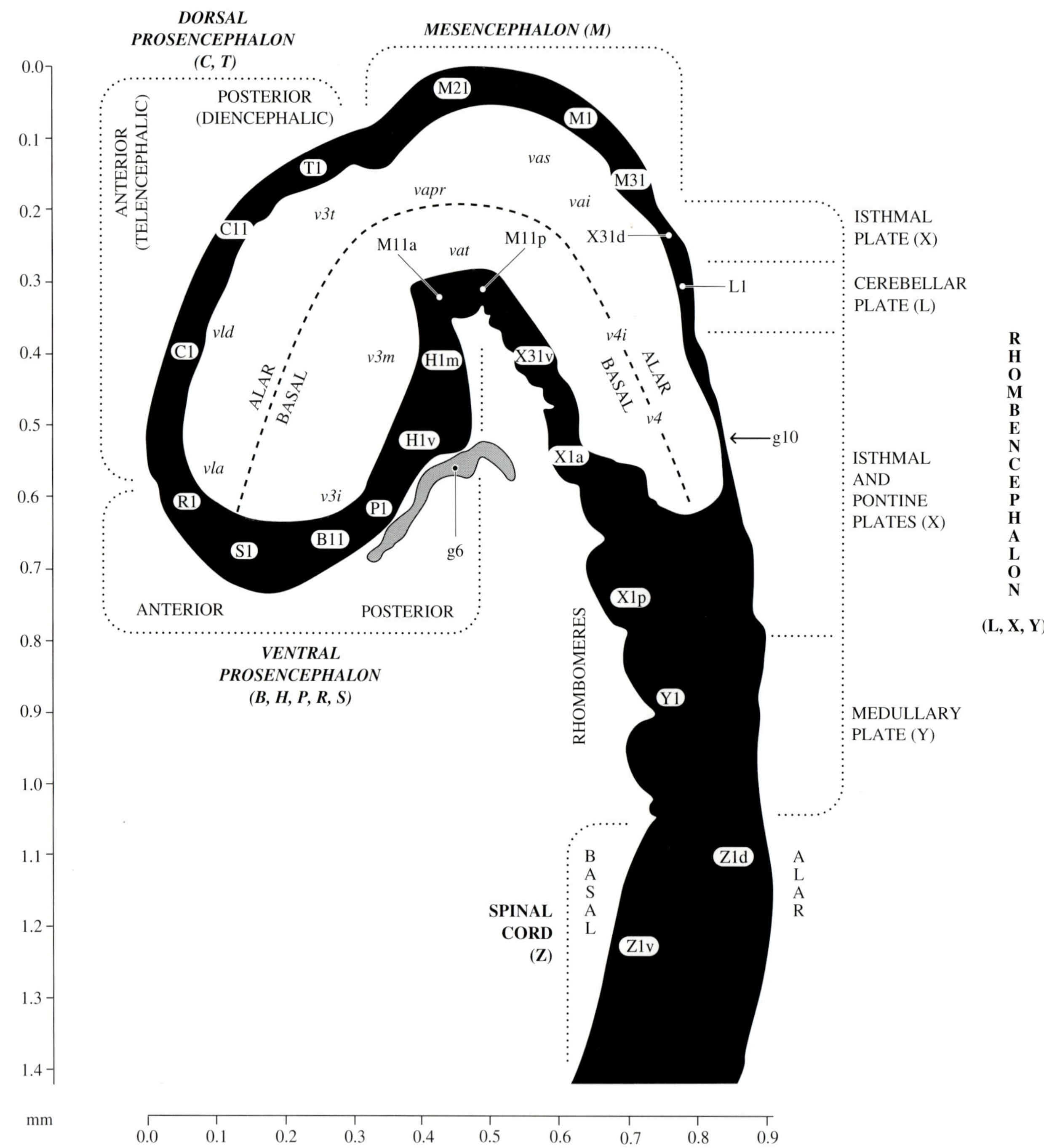

SAGITTAL PLATE 1

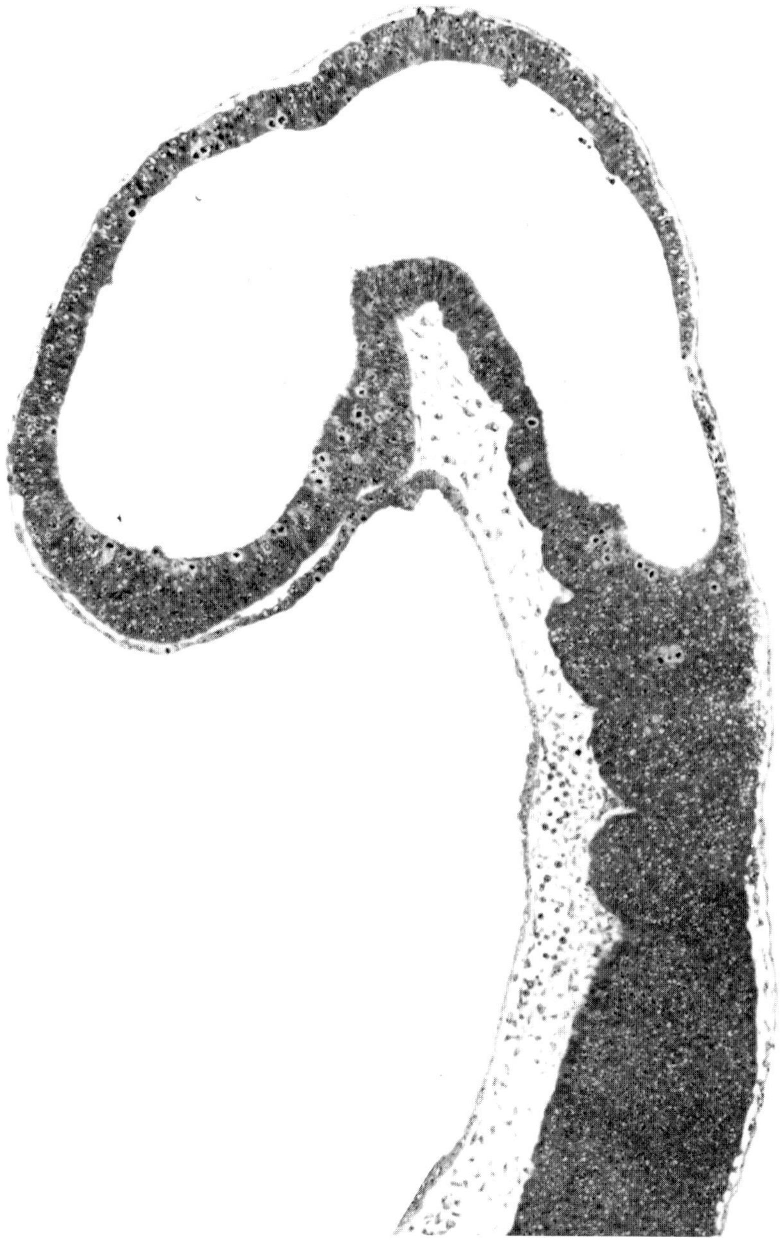

PONTINE AREA (X)

X1a	anterior pontine neuroepithelium
X1p	posterior pontine neuroepithelium
X31d	isthmal neuroepithelium, dorsal
X31v	isthmal neuroepithelium, ventral

MEDULLA (Y)

Y1	medullary neuroepithelium

SPINAL CORD (Z)

Z1d	spinal cord neuroepithelium, dorsal
Z1v	spinal cord neuroepithelium, ventral

E11

6

SAGITTAL FIGURE 2

BASAL GANGLIA (B)

B11 pallidal neuroepithelium

CEREBRAL CORTEX (C)

C1 neocortical neuroepithelium

GLANDS AND PARANEURAL STRUCTURES (g)

g6 pituitary gland (primordium)
g10 velum medullare

HYPOTHALAMUS (H)

H1m mammillary neuroepithelium
H1p posterior hypothalamic neuroepithelium
H1v ventral hypothalamic neuroepithelium

HIPPOCAMPAL FORMATION (K)

K11 subicular neuroepithelium

CEREBELLUM (L)

L1 cerebellar neuroepithelium

MIDBRAIN (M)

M1 superior tectal neuroepithelium
M11 tegmental neuroepithelium
M21 pretectal neuroepithelium
M31 inferior tectal neuroepithelium

RHINENCEPHALON (R)

R1 rhinencephalic neuroepithelium

THALAMUS (T)

T1 thalamic neuroepithelium

VENTRICLES (v)

vai aqueduct, inferior tectal
vapr aqueduct, pretectal
vas aqueduct, superior tectal
vat aqueduct, tegmental
vla lateral ventricle, anterior horn
vld lateral ventricle, dorsal
v3h third ventricle, hypothalamic
v3m third ventricle, mammillary recess
v3t third ventricle, thalamic
v4 fourth ventricle
v4i fourth ventricle, isthmal

PONTINE AREA (X)

X1 pontine neuroepithelium
X31d isthmal neuroepithelium, dorsal
X31v isthmal neuroepithelium, ventral

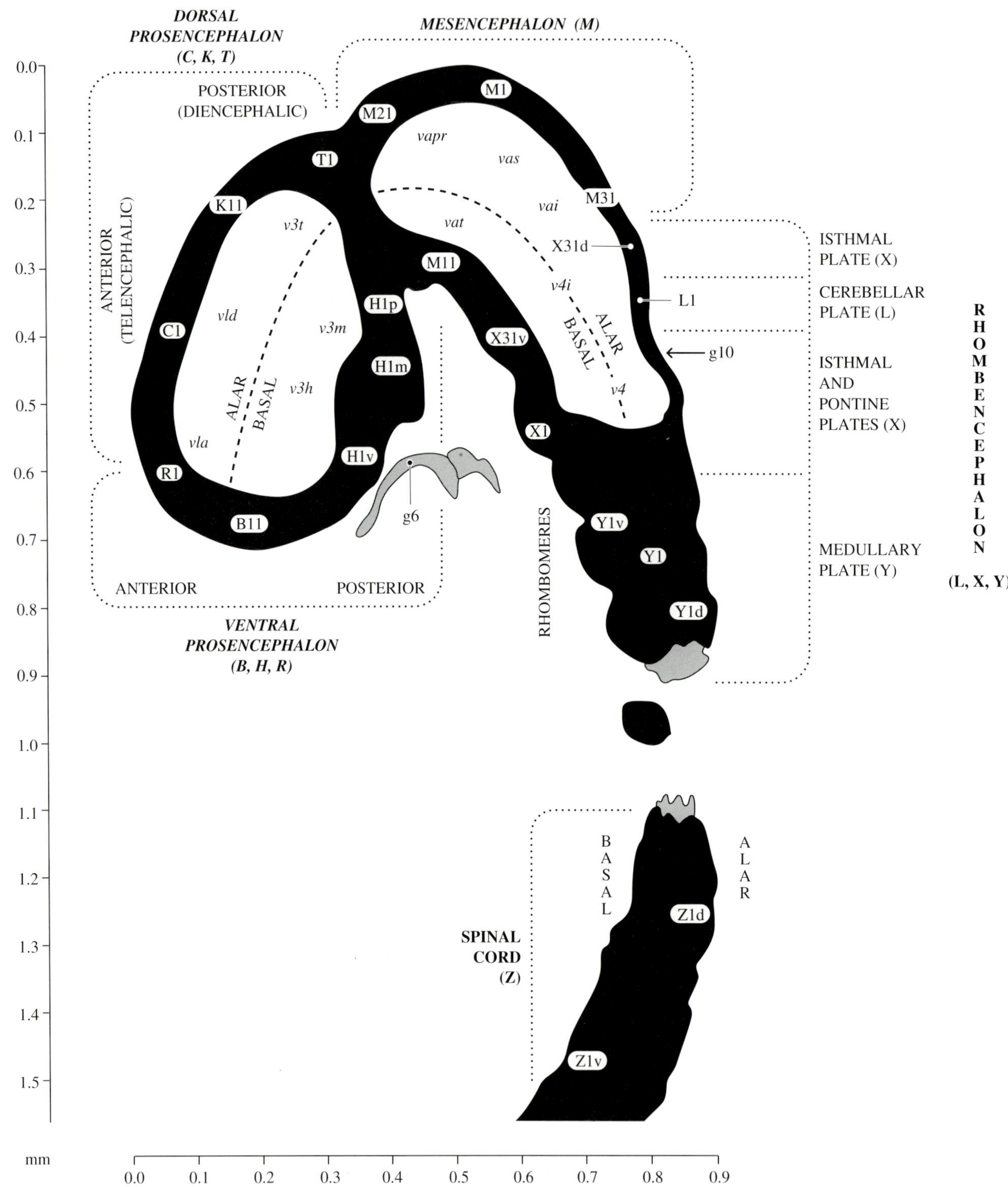

MEDULLA (Y)

Y1	medullary neuroepithelium
Y1d	medullary neuroepithelium, dorsal
Y1v	medullary neuroepithelium, ventral

SPINAL CORD (Z)

Z1d	spinal cord neuroepithelium, dorsal
Z1v	spinal cord neuroepithelium, ventral

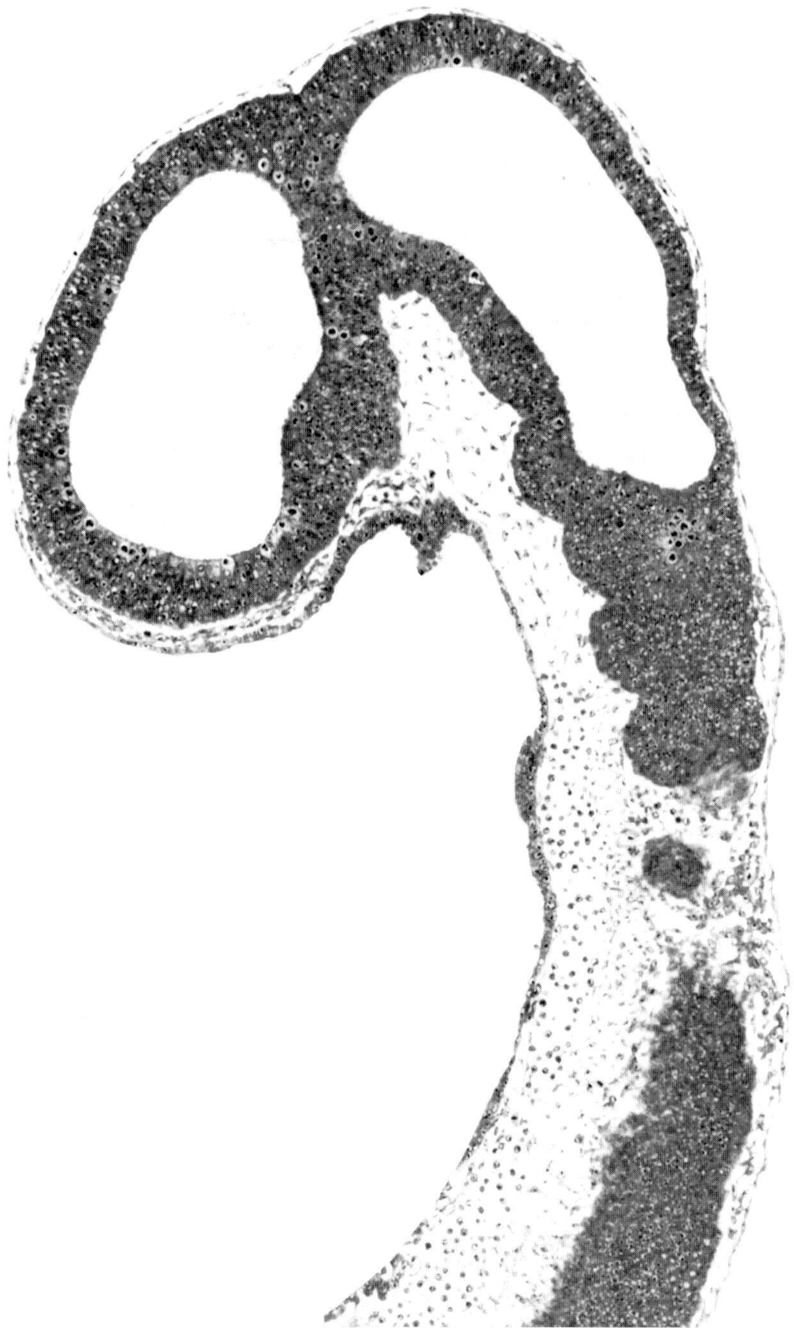

SAGITTAL FIGURE 3

AMYGDALA (A)

A1	amygdaloid neuroepithelium

BASAL GANGLIA (B)

B11	pallidal neuroepithelium

HYPOTHALAMUS (H)

H1i	intermediate hypothalamic neuroepithelium
H1p	posterior hypothalamic neuroepithelium

CEREBELLUM (L)

L1	cerebellar neuroepithelium

MIDBRAIN (M)

M1	superior tectal neuroepithelium
M11a	anterior tegmental neuroepithelium
M11p	posterior tegmental neuroepithelium
M21	pretectal neuroepithelium

RHINENCEPHALON (R)

R1	rhinencephalic neuroepithelium

THALAMUS (T)

T1	thalamic neuroepithelium

VENTRICLES (v)

vla	lateral ventricle, anterior horn
vlp	lateral ventricle, posterior horn

PONTINE AREA (X)

X1	pontine neuroepithelium
X31d	isthmal neuroepithelium, dorsal
X31v	isthmal neuroepithelium, ventral

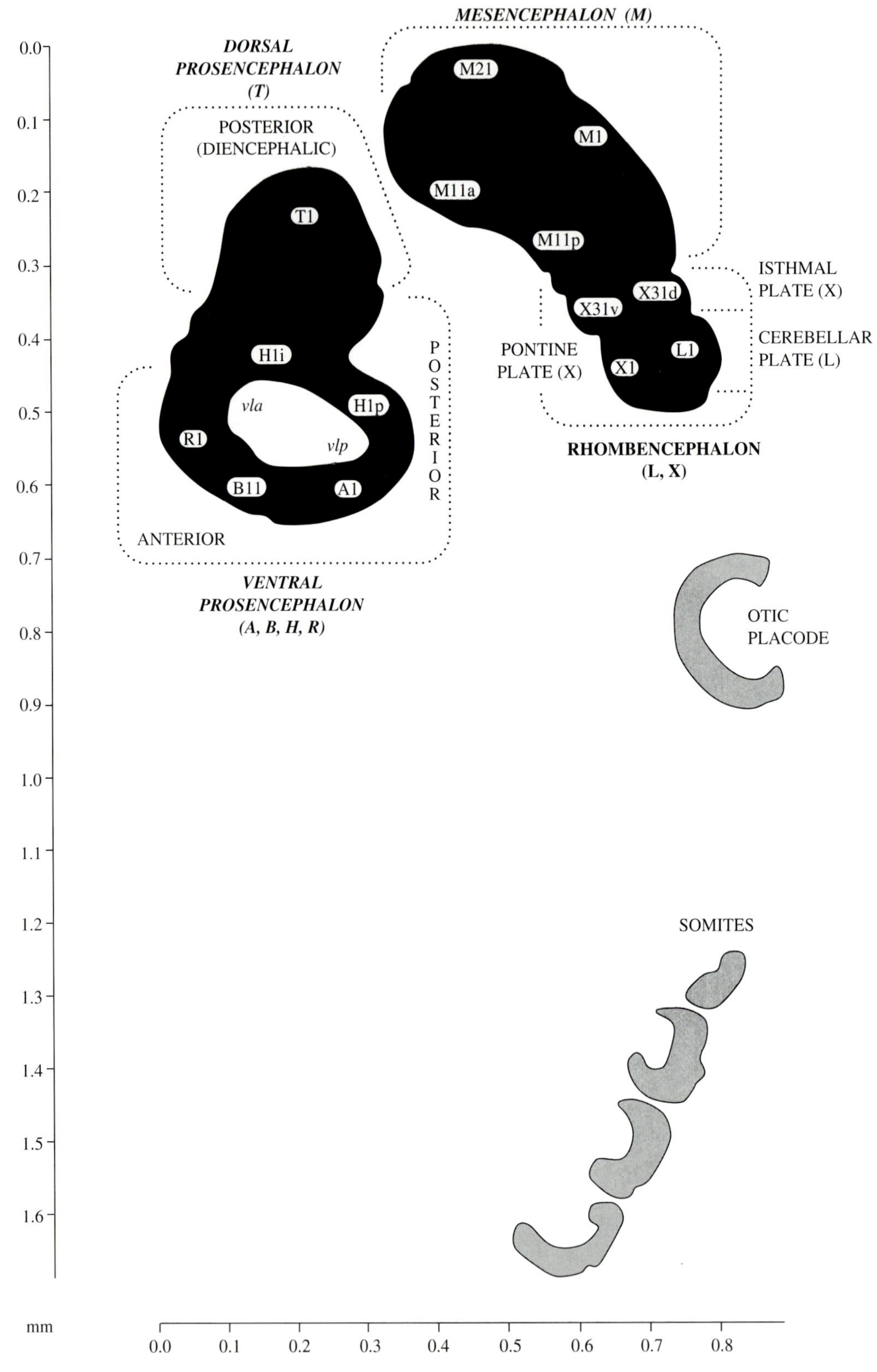

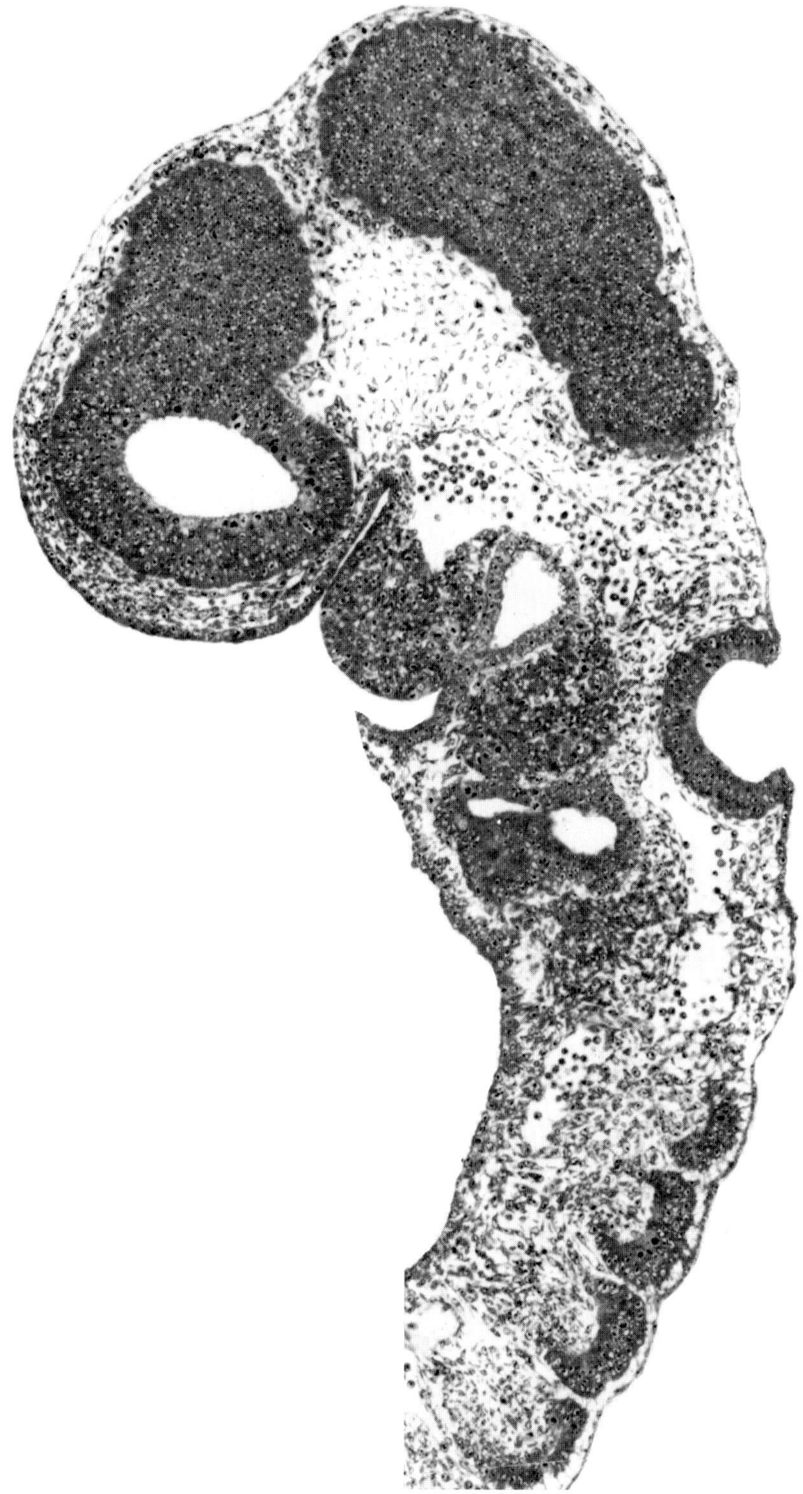

CORONAL

FIGURES and PLATES 1-17

CORONAL FIGURE 1

BASAL GANGLIA (B)

B11 pallidal neuroepithelium

CEREBRAL CORTEX (C)

C1 neocortical neuroepithelium

RHINENCEPHALON (R)

R1 rhinencephalic neuroepithelium

SEPTAL AREA (S)

S1 septal neuroepithelium

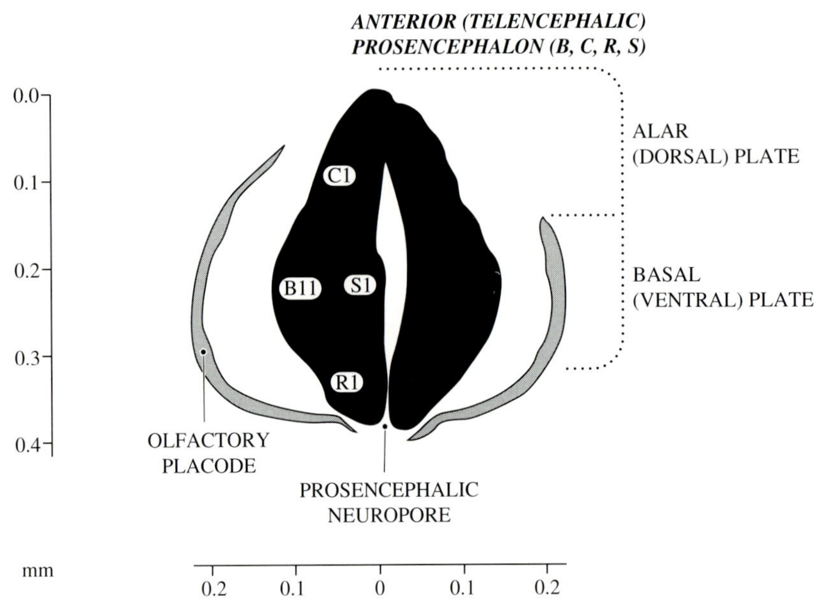

CORONAL FIGURE 2

BASAL GANGLIA (B)

B11 pallidal neuroepithelium

CEREBRAL CORTEX (C)

C1 neocortical neuroepithelium
C11 cingulate cortex neuroepithelium

RHINENCEPHALON (R)

R1 rhinencephalic neuroepithelium

VENTRICLES (v)

vla lateral ventricle, anterior horn
vld lateral ventricle, dorsal

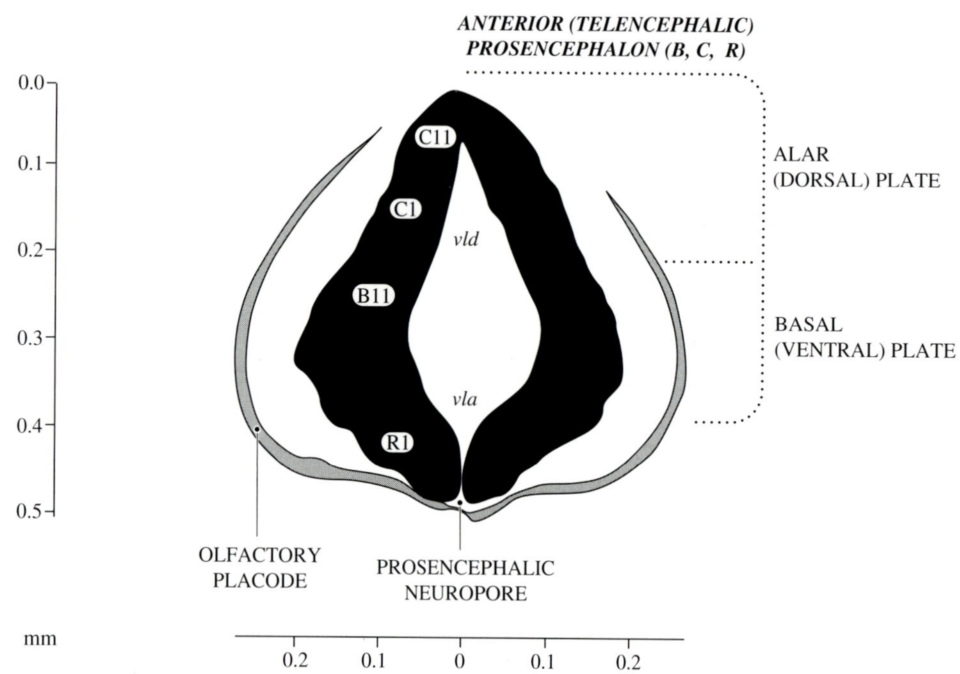

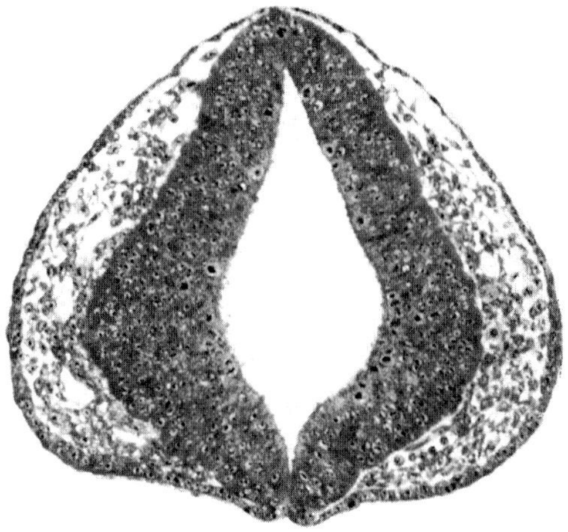

CORONAL FIGURE 3

BASAL GANGLIA (B)

B11 pallidal neuroepithelium

CEREBRAL CORTEX (C)

C1 neocortical neuroepithelium
C11 cingulate cortex neuroepithelium

PREOPTIC AREA AND OPTIC EPITHELIA (P)

P1 preoptic neuroepithelium
P11 retinal neuroepithelium

VENTRICLES (v)

vla lateral ventricle, anterior horn
vld lateral ventricle, dorsal
v3o third ventricle, optic recess
v3pr third ventricle, preoptic recess

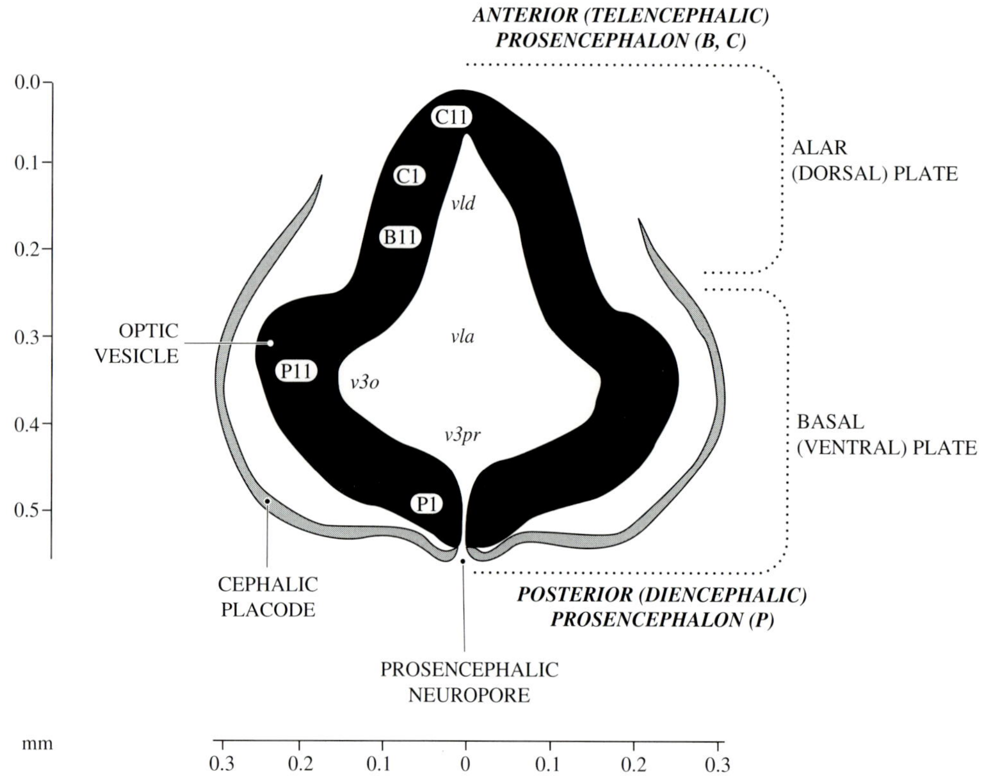

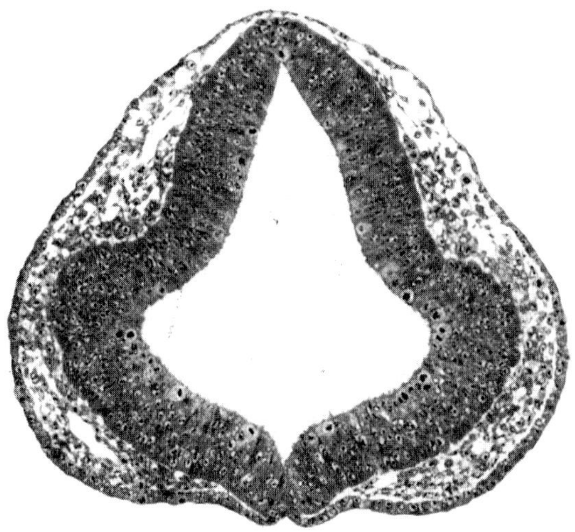

CORONAL FIGURE 4

BASAL GANGLIA (B)

B11 pallidal neuroepithelium

CEREBRAL CORTEX (C)

C1 neocortical neuroepithelium
C11 cingulate cortex neuroepithelium

HYPOTHALAMUS (H)

H1a anterior hypothalamic neuroepithelium
H1v ventral hypothalamic neuroepithelium

PREOPTIC AREA AND OPTIC EPITHELIA (P)

P11 retinal neuroepithelium

VENTRICLES (v)

vla lateral ventricle, anterior horn
vld lateral ventricle, dorsal
v3o third ventricle, optic recess
v3pr third ventricle, preoptic recess

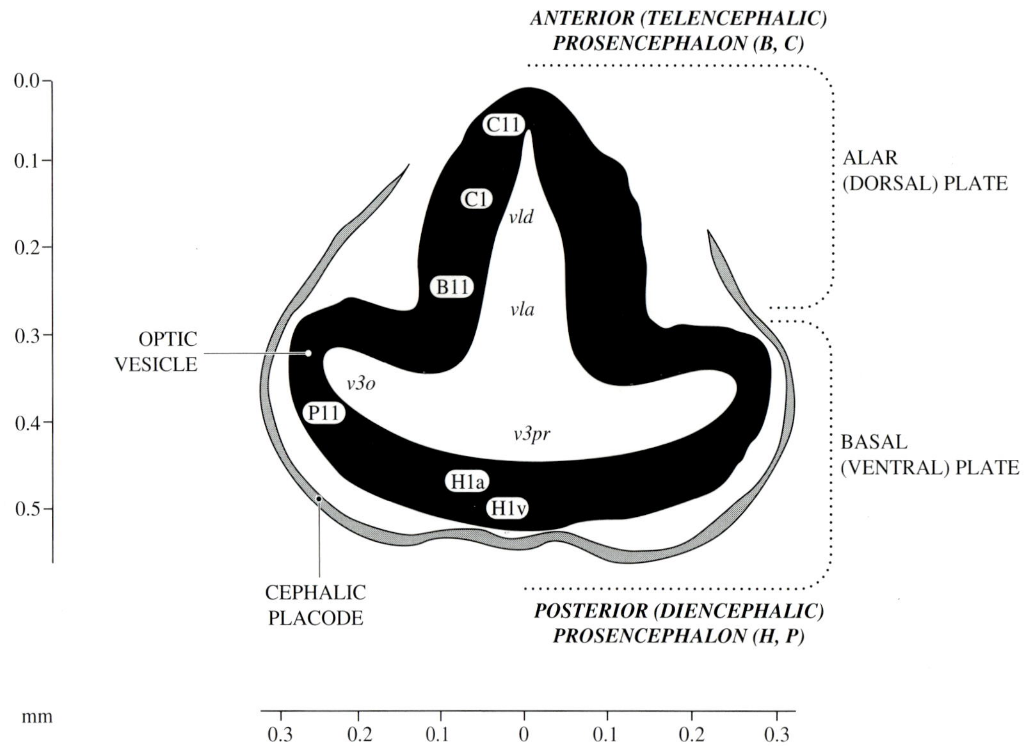

CORONAL FIGURE 5

CEREBRAL CORTEX (C)

C1 neocortical neuroepithelium

HYPOTHALAMUS (H)

H1a anterior hypothalamic neuroepithelium
H1v ventral hypothalamic neuroepithelium

HIPPOCAMPAL FORMATION (K)

K11 subicular neuroepithelium

PREOPTIC AREA AND OPTIC EPITHELIA (P)

P1 preoptic neuroepithelium
P11 retinal neuroepithelium

VENTRICLES (v)

vl lateral ventricle
v3h third ventricle, hypothalamic
v3o third ventricle, optic recess

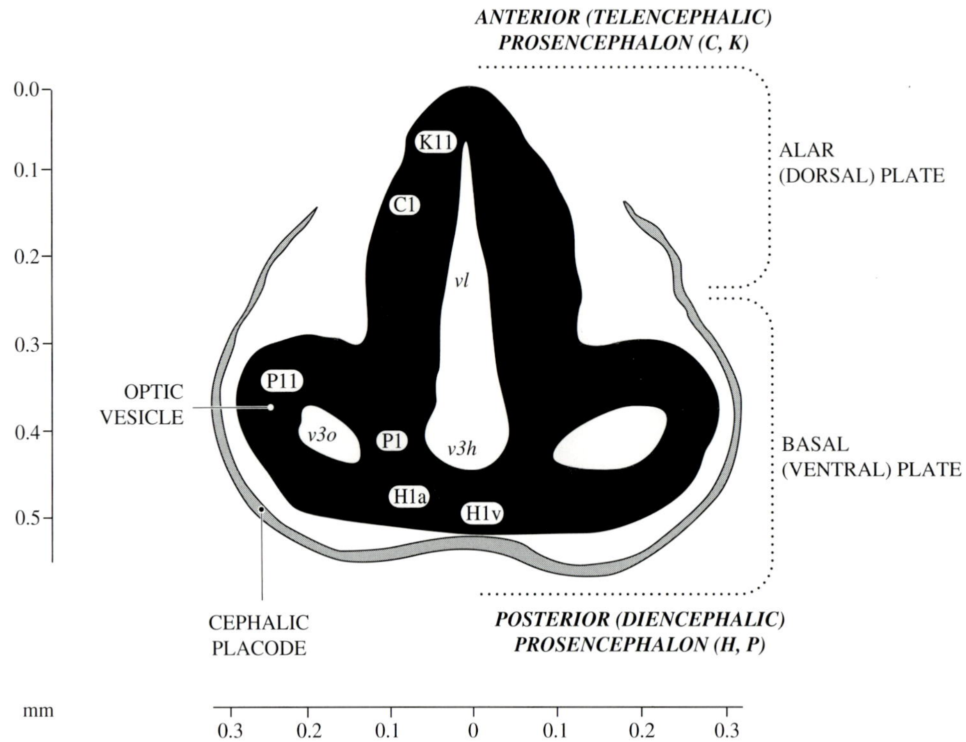

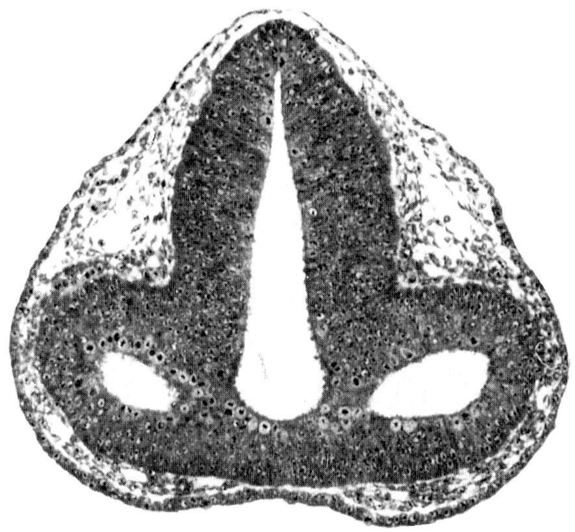

CORONAL FIGURE 6

HYPOTHALAMUS (H)

H1i intermediate hypothalamic neuroepithelium
H1v ventral hypothalamic neuroepithelium

PREOPTIC AREA AND OPTIC EPITHELIA (P)

P11 retinal neuroepithelium

THALAMUS (T)

T1a anterior thalamic neuroepithelium
T1e epithalamic neuroepithelium

VENTRICLES (v)

v3h third ventricle, hypothalamic
v3t third ventricle, thalamic

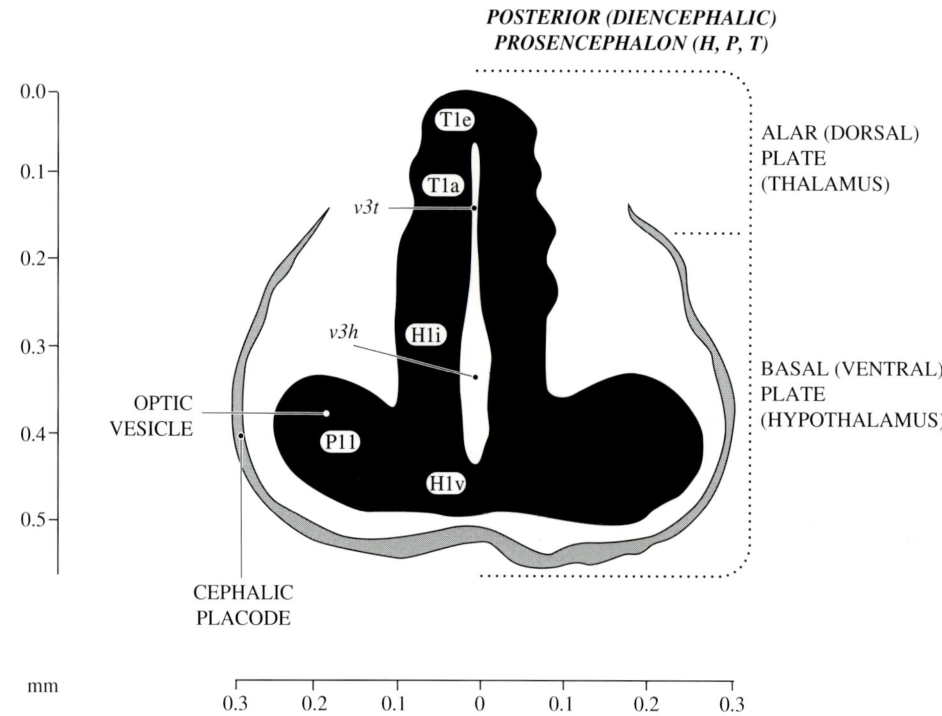

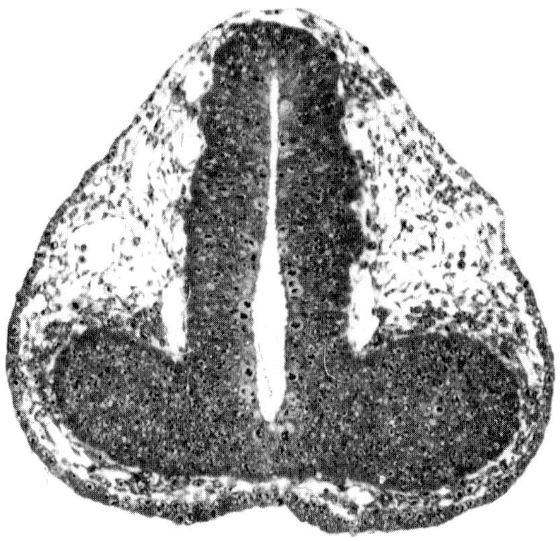

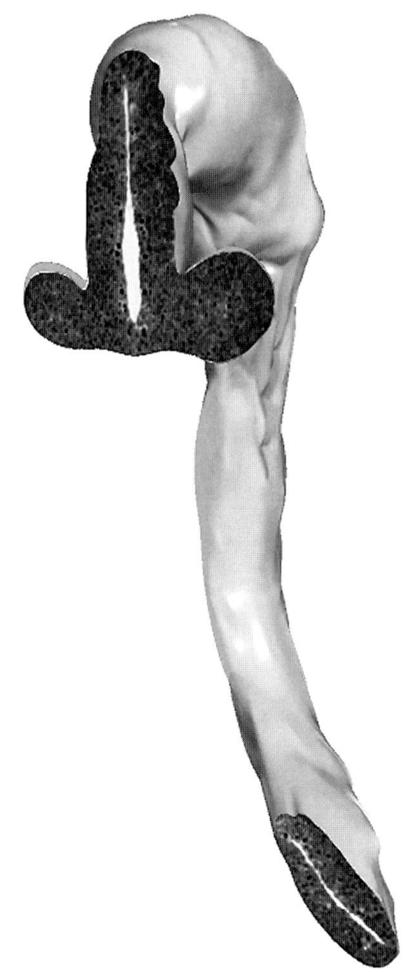

CORONAL FIGURE 7

HYPOTHALAMUS (H)

H1i	intermediate hypothalamic neuroepithelium
H1v	ventral hypothalamic neuroepithelium

PREOPTIC AREA AND OPTIC EPITHELIA (P)

P11	retinal neuroepithelium

THALAMUS (T)

T1e	epithalamic neuroepithelium
T1i	intermediate thalamic neuroepithelium

VENTRICLES (v)

v3h	third ventricle, hypothalamic
v3t	third ventricle, thalamic

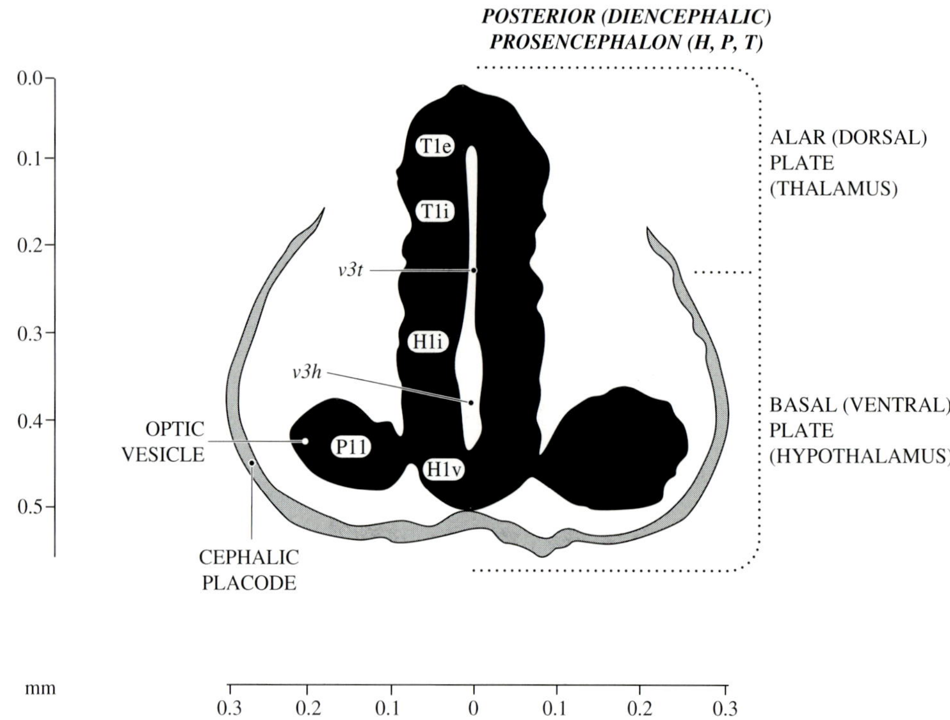

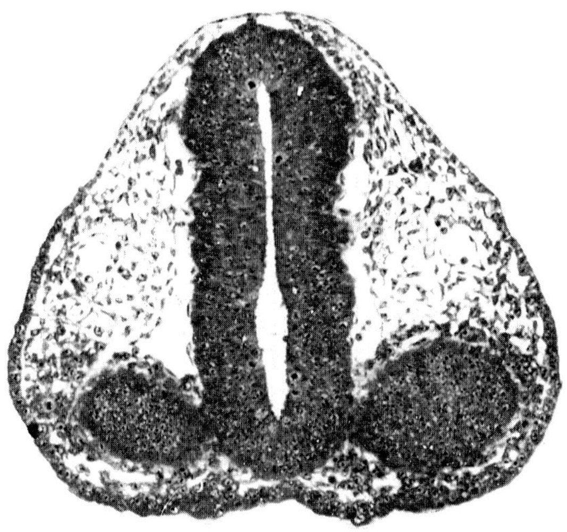

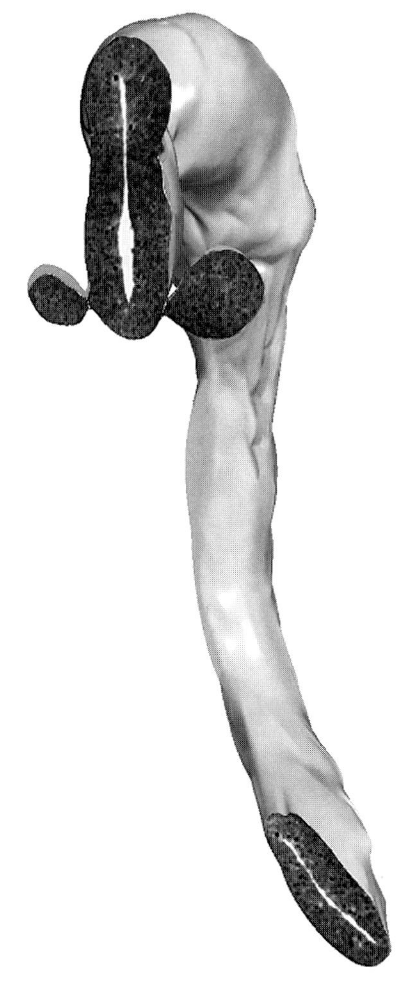

GLANDS AND PARANEURAL STRUCTURES (g)

g6 pituitary gland (primordium)

HYPOTHALAMUS (H)

H1i intermediate hypothalamic neuroepithelium
H1v ventral hypothalamic neuroepithelium

MIDBRAIN (M)

M21 pretectal neuroepithelium

THALAMUS (T)

T1p posterior thalamic neuroepithelium

VENTRICLES (v)

vapr aqueduct, pretectal
v3h third ventricle, hypothalamic
v3t third ventricle, thalamic

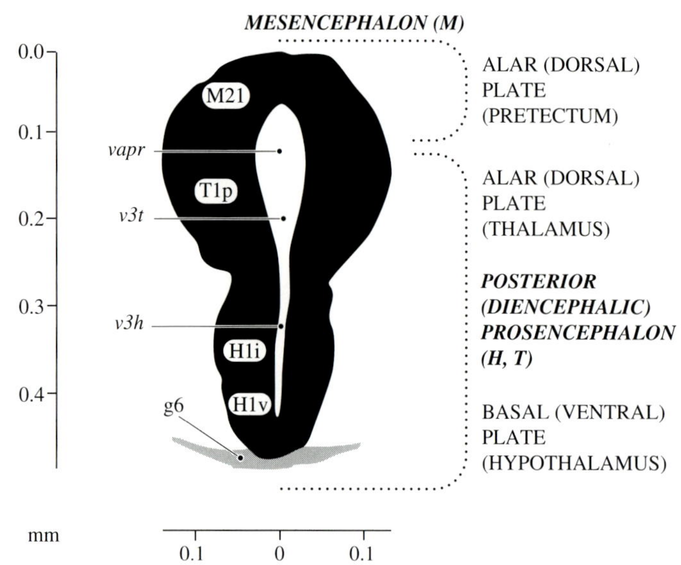

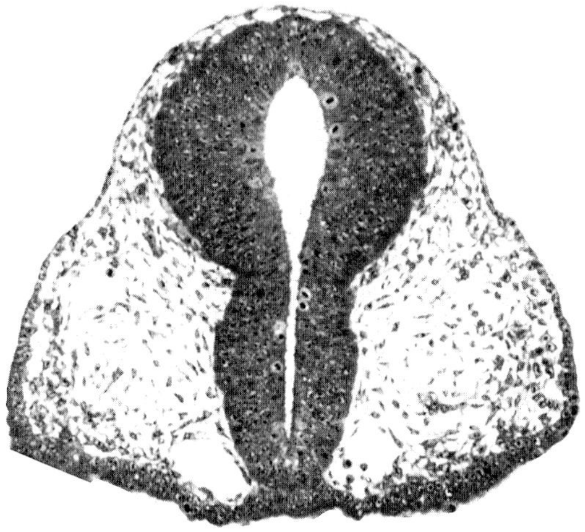

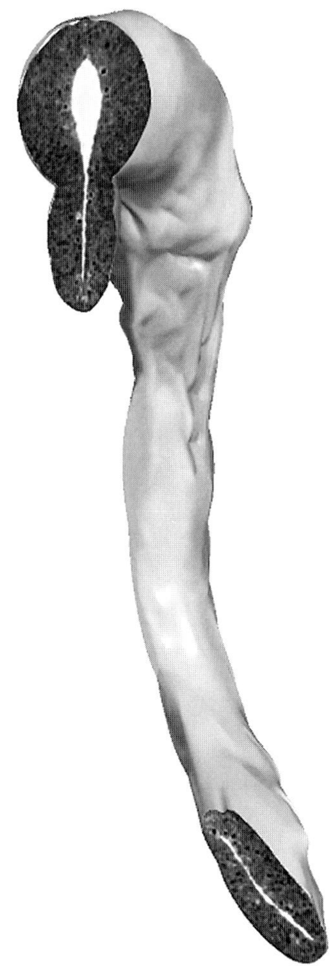

CORONAL FIGURE 9

HYPOTHALAMUS (H)

H1*l*	lateral hypothalamic neuroepithelium
H1m	mammillary neuroepithelium
H1v	ventral hypothalamic neuroepithelium

MIDBRAIN (M)

M21	pretectal neuroepithelium

VENTRICLES (v)

vapr	aqueduct, pretectal
v3h	third ventricle, hypothalamic

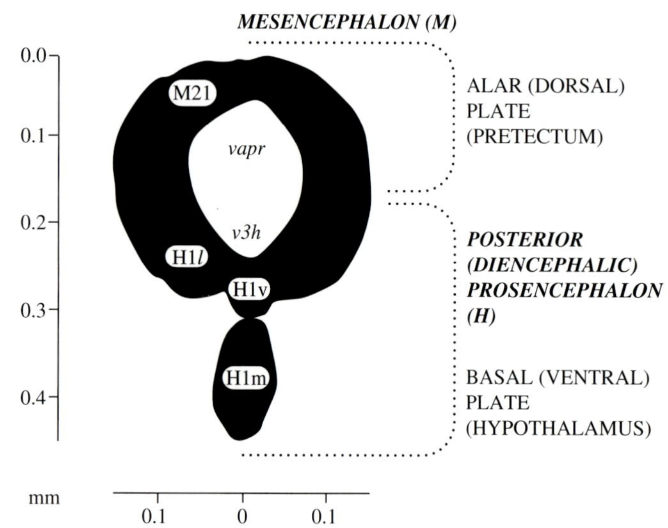

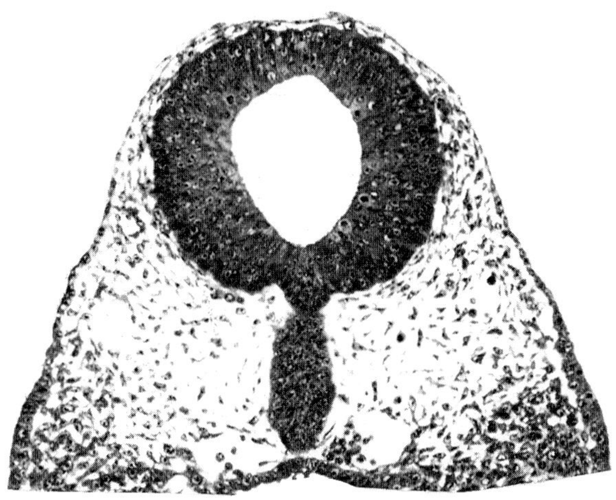

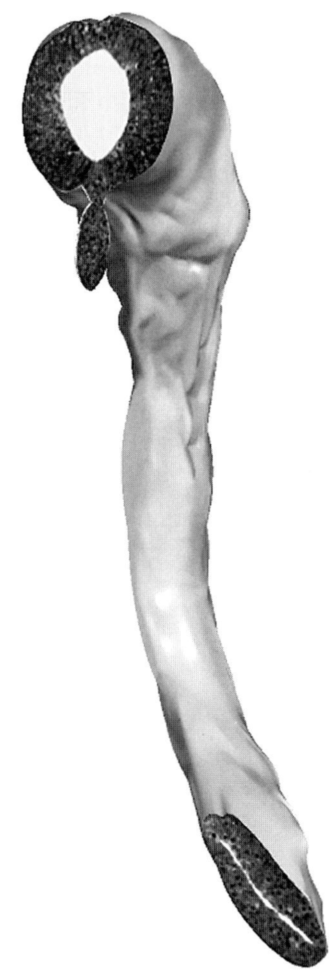

CORONAL FIGURE 10

MIDBRAIN (M)

M1 superior tectal neuroepithelium
M11 tegmental neuroepithelium

VENTRICLES (v)

vas aqueduct, superior tectal
vat aqueduct, tegmental

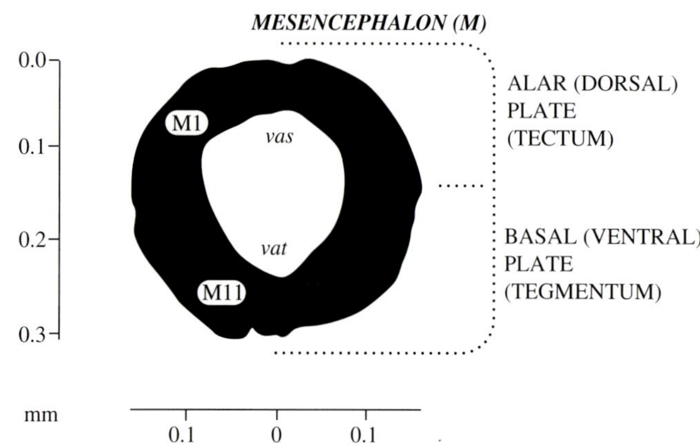

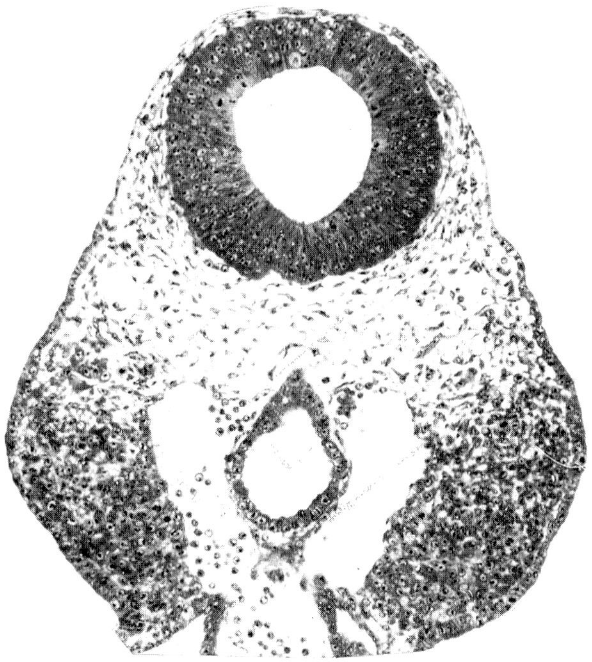

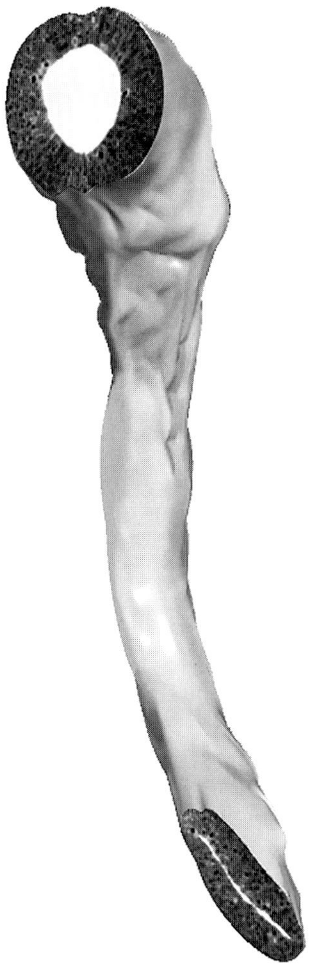

CORONAL FIGURE 11

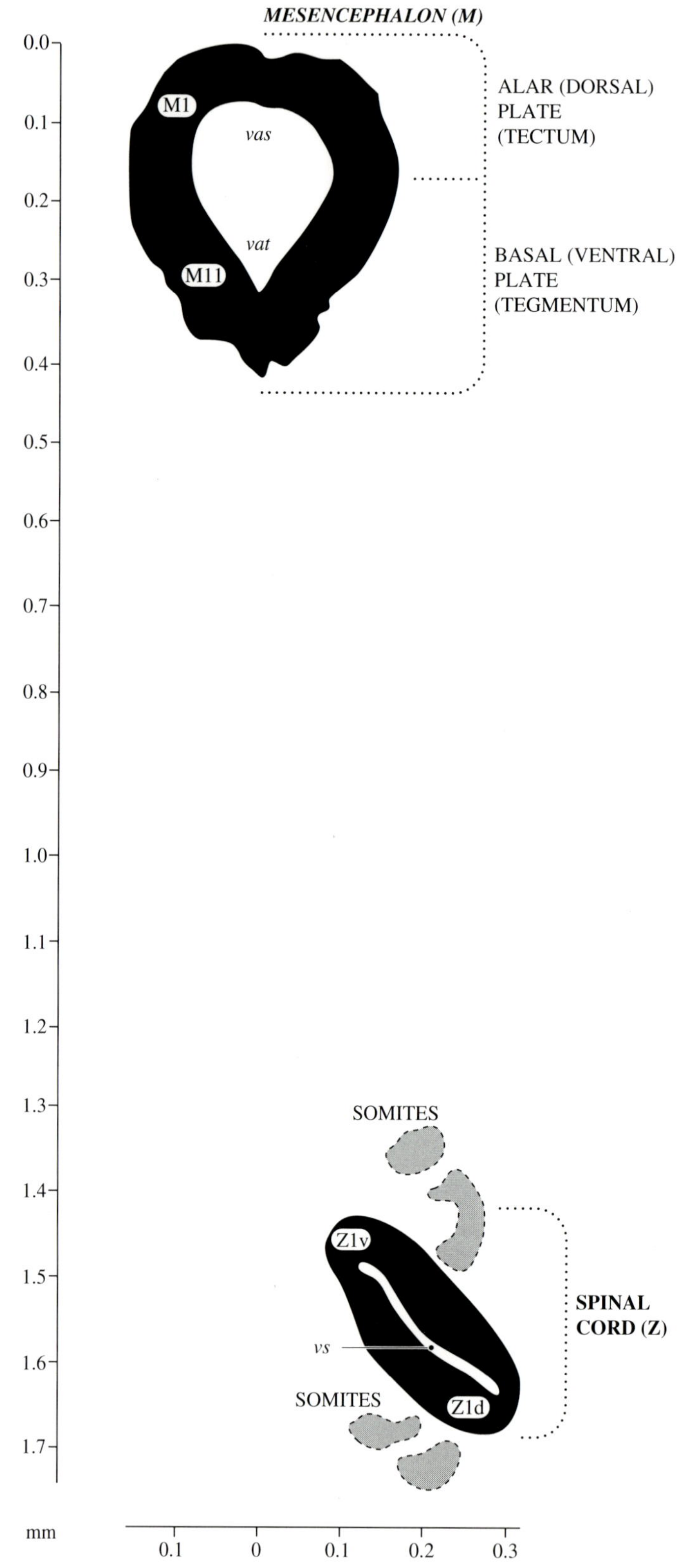

MIDBRAIN (M)

M1 superior tectal neuroepithelium
M11 tegmental neuroepithelium

VENTRICLES (v)

vas aqueduct, superior tectal
vat aqueduct, tegmental
vs spinal canal

SPINAL CORD (Z)

Z1d spinal cord neuroepithelium, dorsal
Z1v spinal cord neuroepithelium, ventral

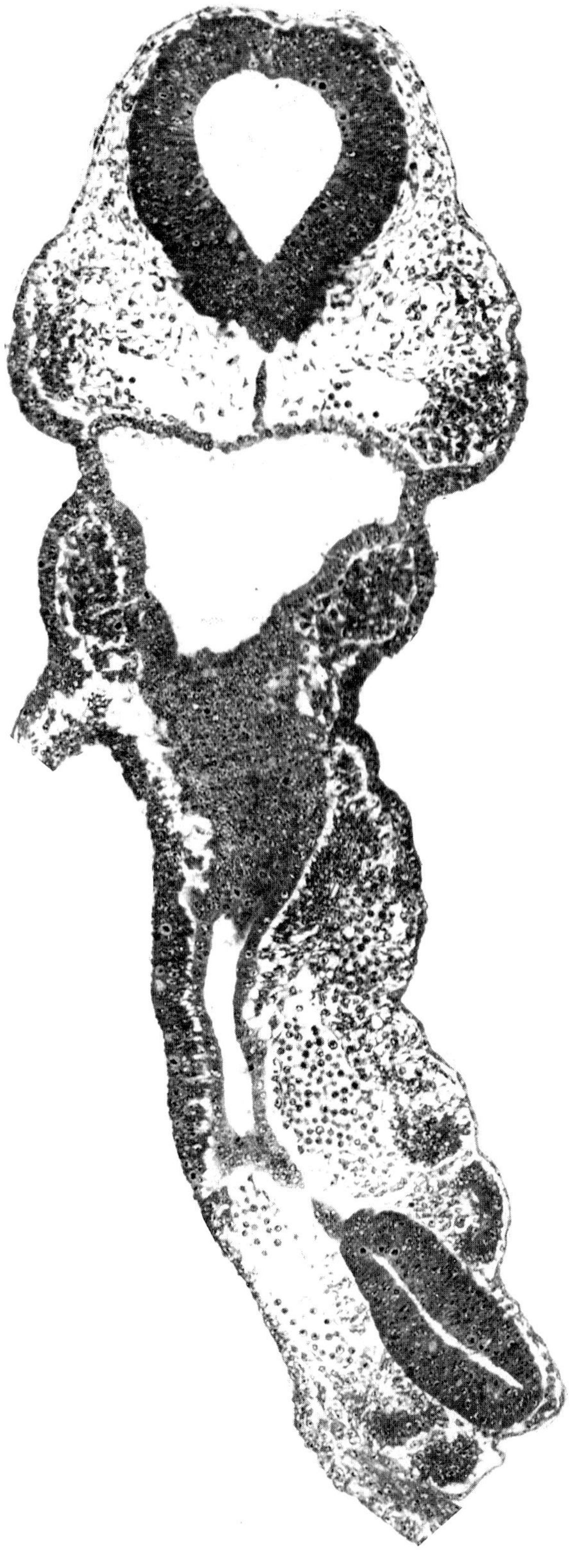

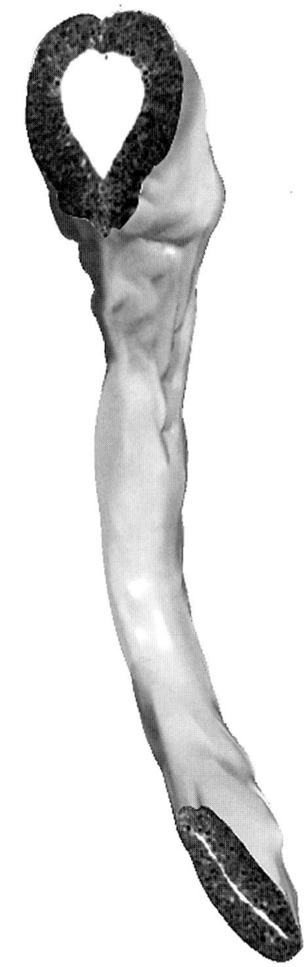

NERVES AND NUCLEI, CRANIAL (n)

n5g trigeminal ganglion

VENTRICLES (v)

v4i fourth ventricle, isthmal
v4p fourth ventricle, posterior

PONTINE AREA (X)

X1 pontine neuroepithelium
X31 isthmal neuroepithelium

MEDULLA (Y)

Y1 medullary neuroepithelium

SPINAL CORD (Z)

Z1v spinal cord neuroepithelium, ventral

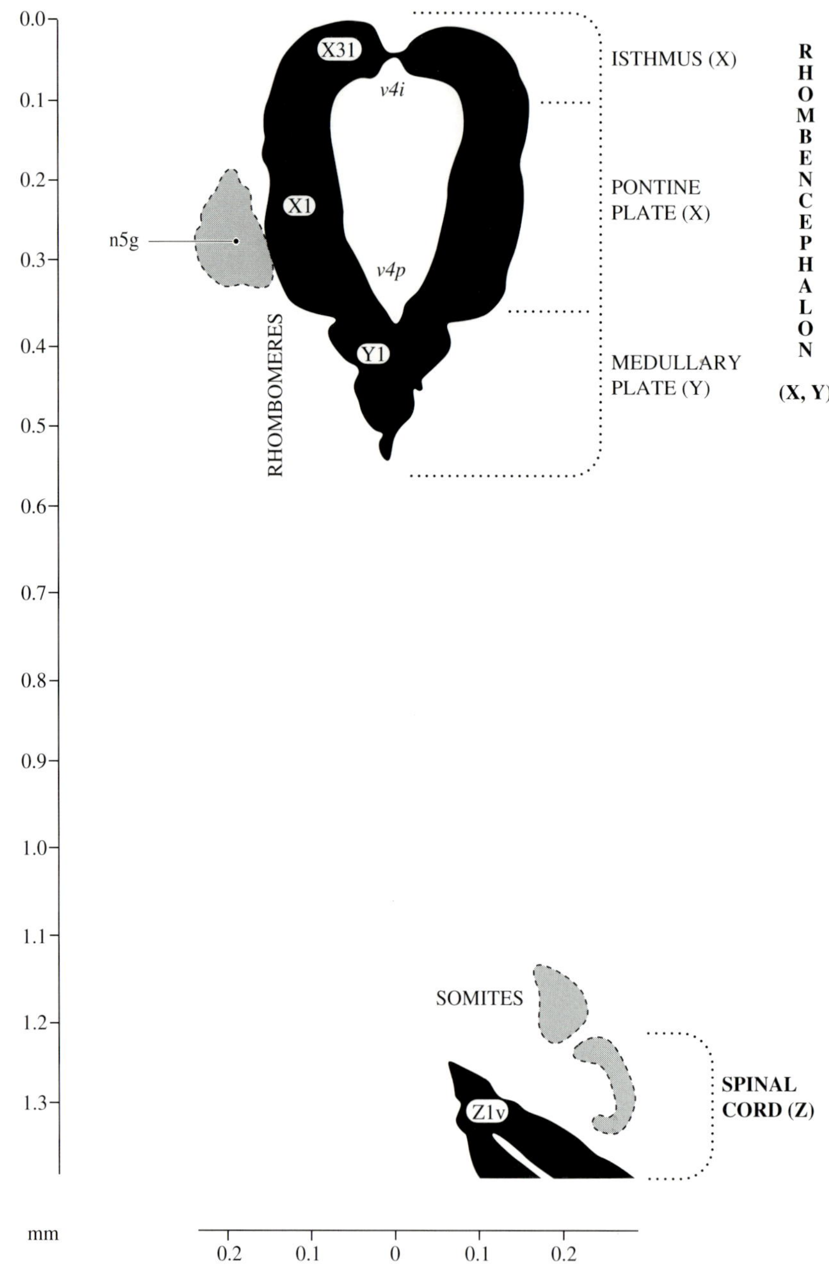

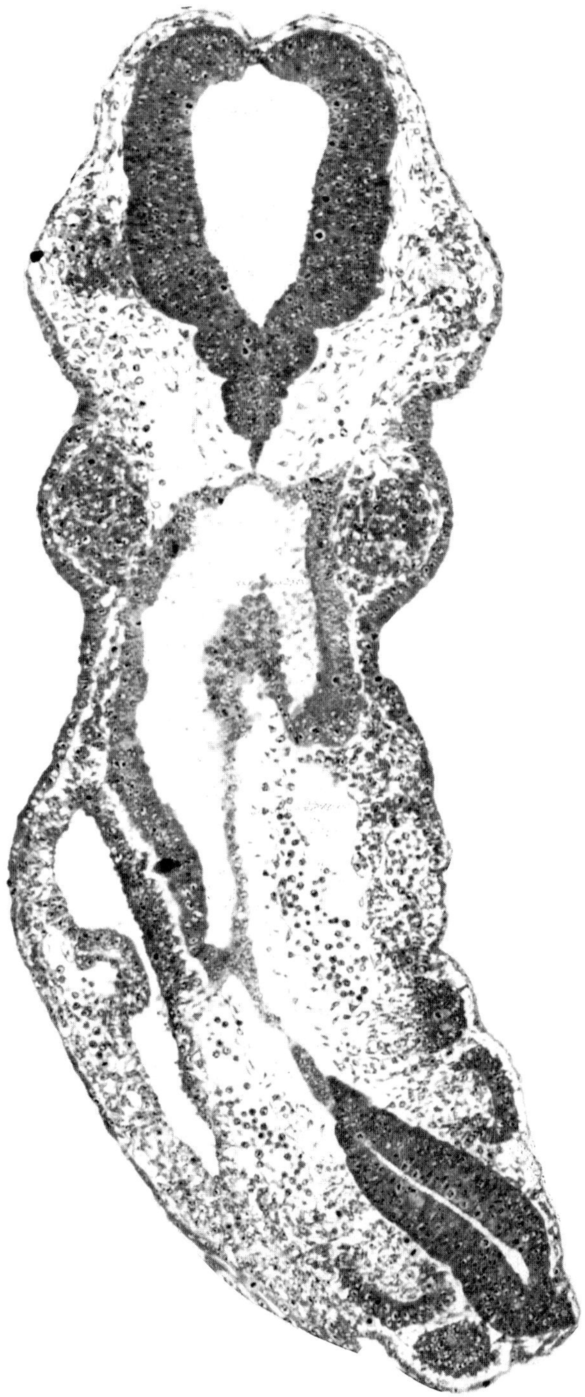

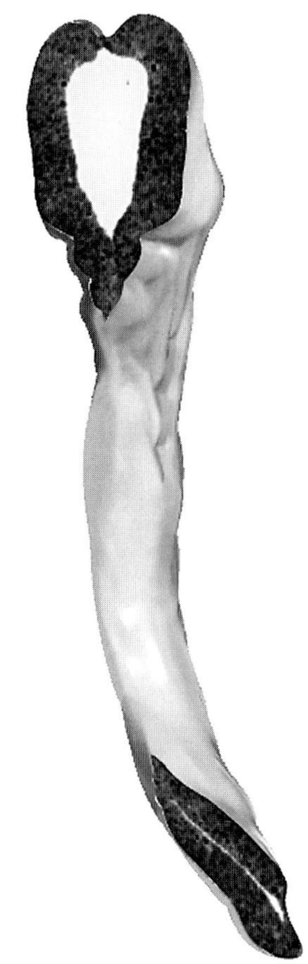

CORONAL FIGURE 13

GLANDS AND PARANEURAL STRUCTURES (g)

g10 velum medullare

CEREBELLUM (L)

L1 cerebellar neuroepithelium

NERVES AND NUCLEI, CRANIAL (n)

n5g trigeminal ganglion

VENTRICLES (v)

vs spinal canal
v4a fourth ventricle, anterior
v4p fourth ventricle, posterior

PONTINE AREA (X)

X1 pontine neuroepithelium

MEDULLA (Y)

Y1 medullary neuroepithelium

SPINAL CORD (Z)

Z1 spinal cord neuroepithelium

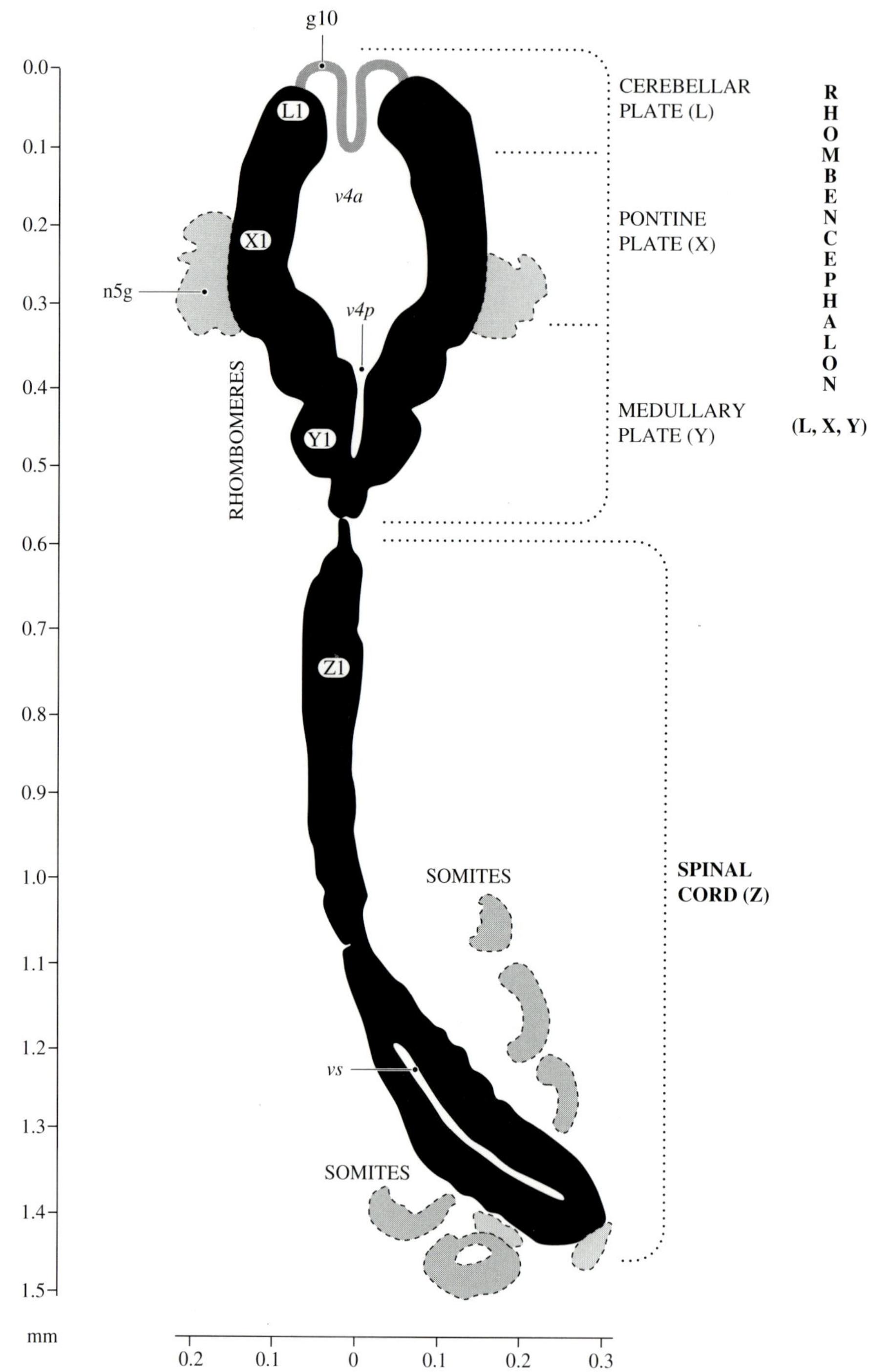

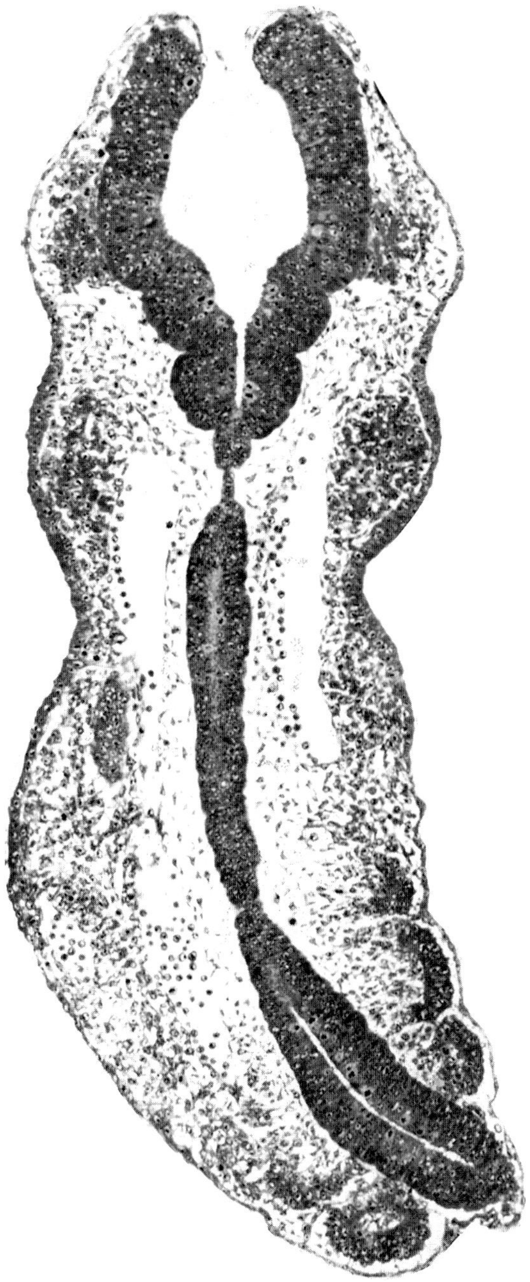

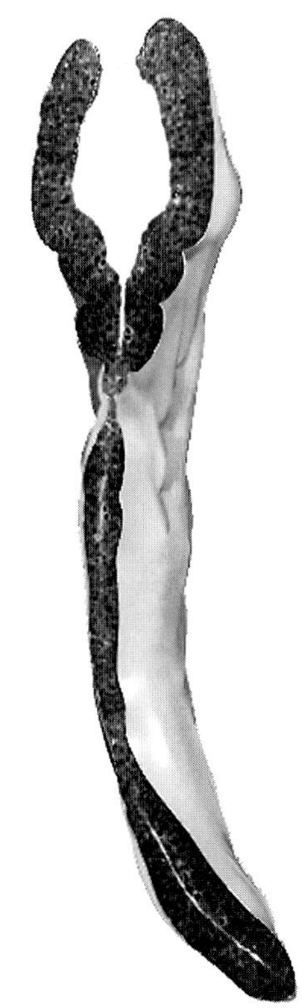

CORONAL FIGURE 14

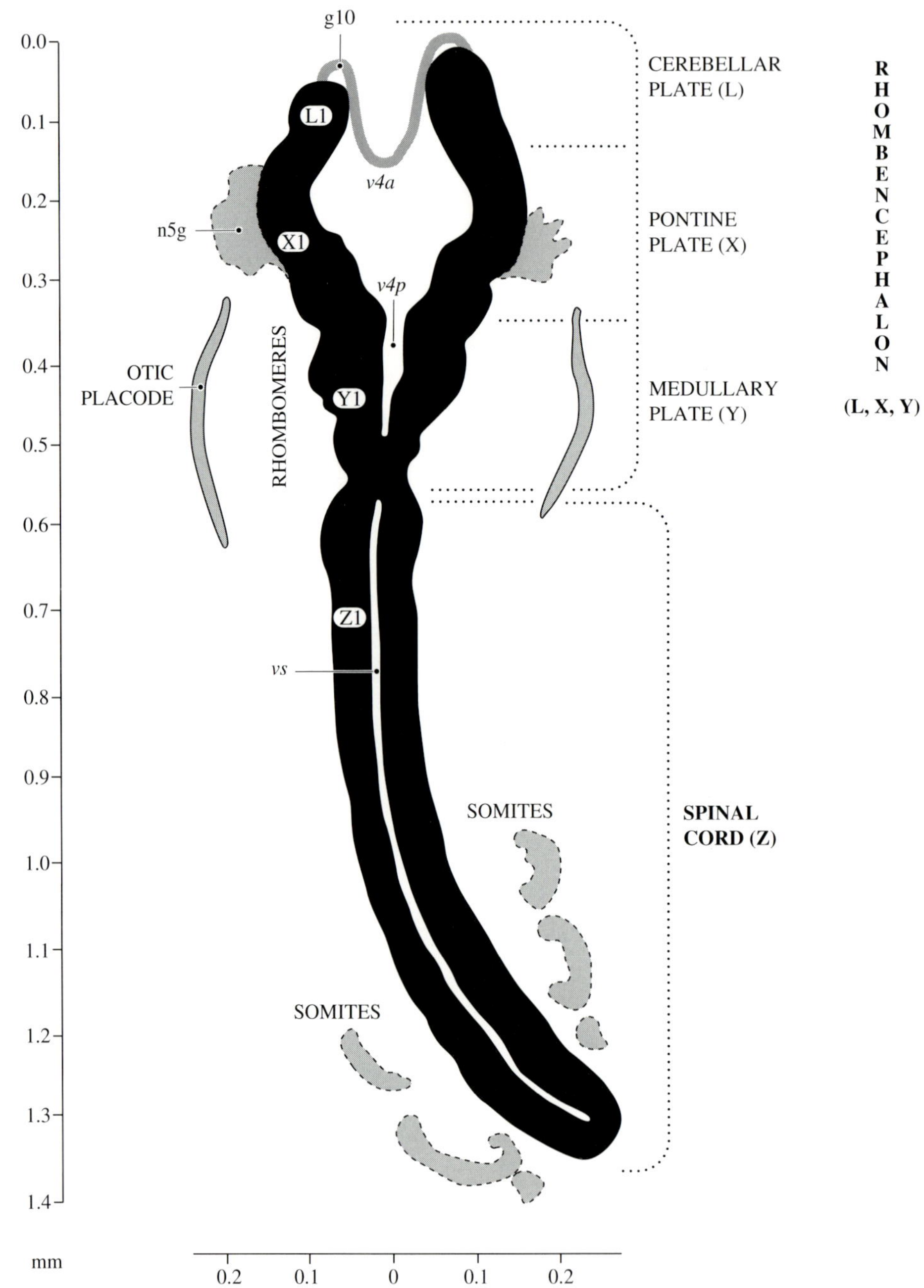

GLANDS AND PARANEURAL STRUCTURES (g)

g10 velum medullare

CEREBELLUM (L)

L1 cerebellar neuroepithelium

NERVES AND NUCLEI, CRANIAL (n)

n5g trigeminal ganglion

VENTRICLES (v)

vs spinal canal
v4a fourth ventricle, anterior
v4p fourth ventricle, posterior

PONTINE AREA (X)

X1 pontine neuroepithelium

MEDULLA (Y)

Y1 medullary neuroepithelium

SPINAL CORD (Z)

Z1 spinal cord neuroepithelium

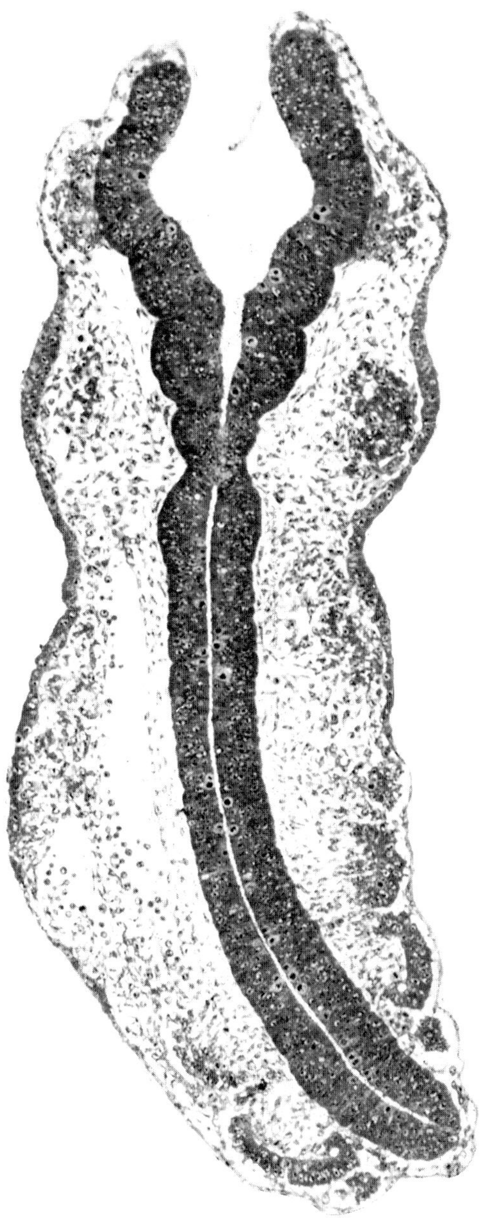

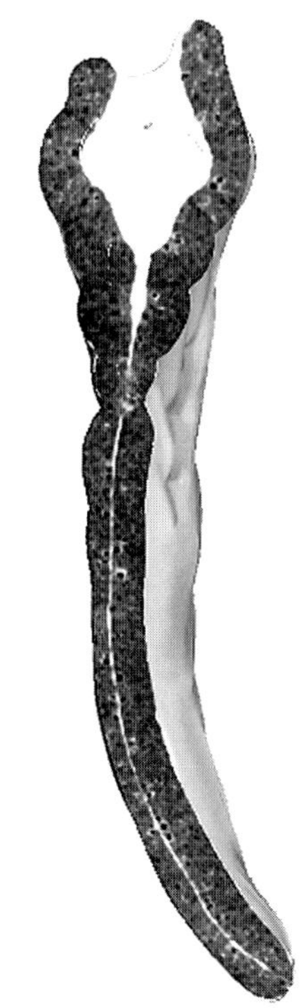

CORONAL FIGURE 15

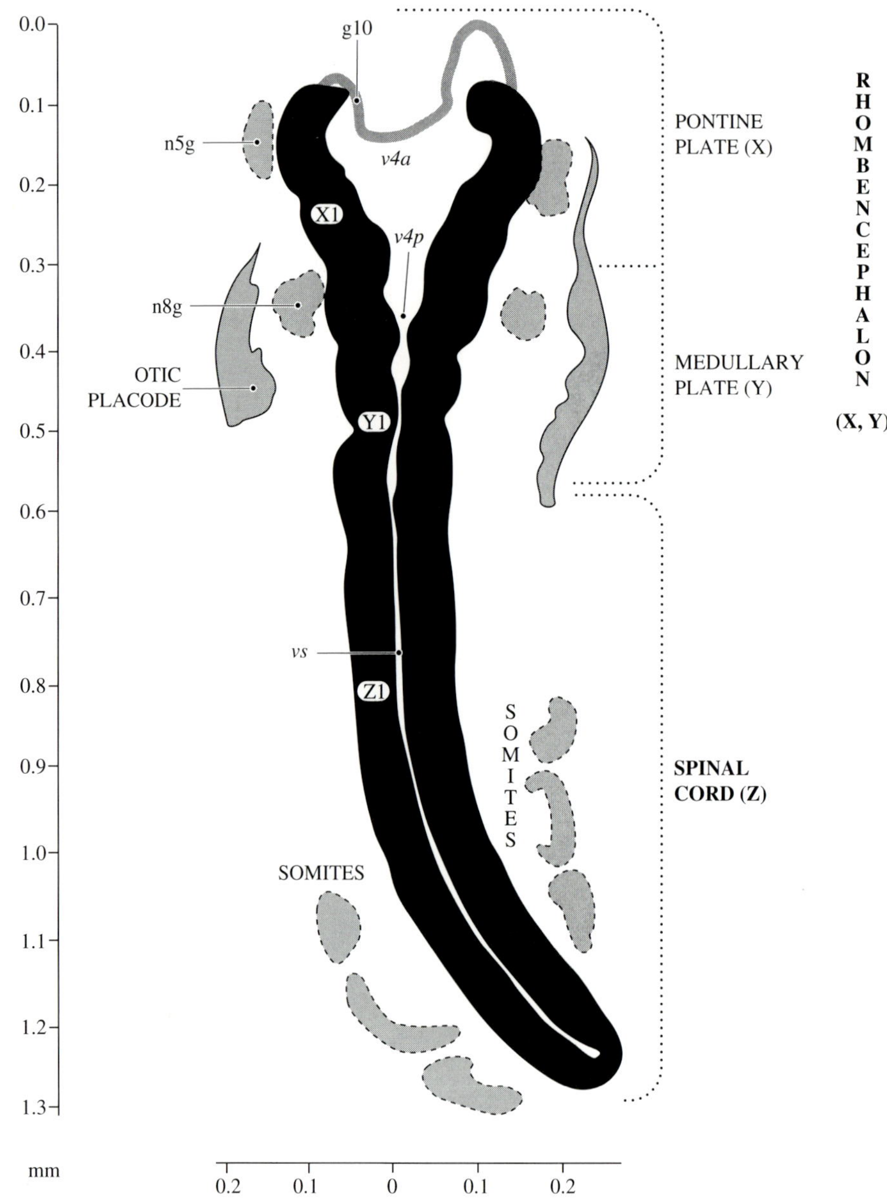

GLANDS AND PARANEURAL STRUCTURES (g)

g10 velum medullare

NERVES AND NUCLEI, CRANIAL (n)

n5g trigeminal ganglion
n8g vestibular ganglion

VENTRICLES (v)

vs spinal canal
v4a fourth ventricle, anterior
v4p fourth ventricle, posterior

PONTINE AREA (X)

X1 pontine neuroepithelium

MEDULLA (Y)

Y1 medullary neuroepithelium

SPINAL CORD (Z)

Z1 spinal cord neuroepithelium

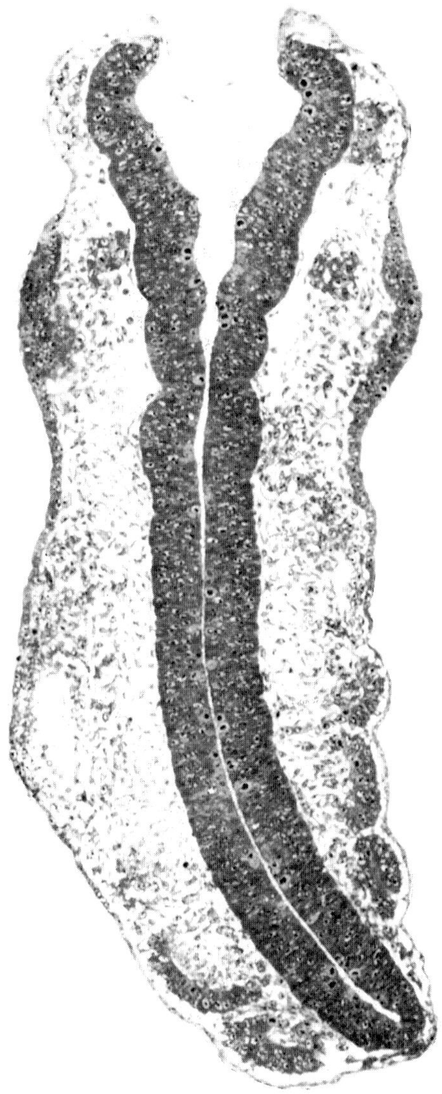

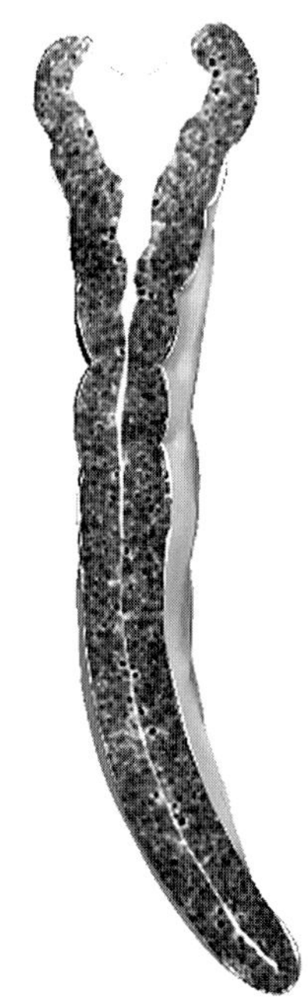

CORONAL FIGURE 16

GLANDS AND PARANEURAL STRUCTURES (g)

g10 velum medullare

NERVES AND NUCLEI, CRANIAL (n)

n8g vestibular ganglion

VENTRICLES (v)

vs spinal canal
v4p fourth ventricle, posterior

PONTINE AREA (X)

X1 pontine neuroepithelium

MEDULLA (Y)

Y1 medullary neuroepithelium

SPINAL CORD (Z)

Z1 spinal cord neuroepithelium

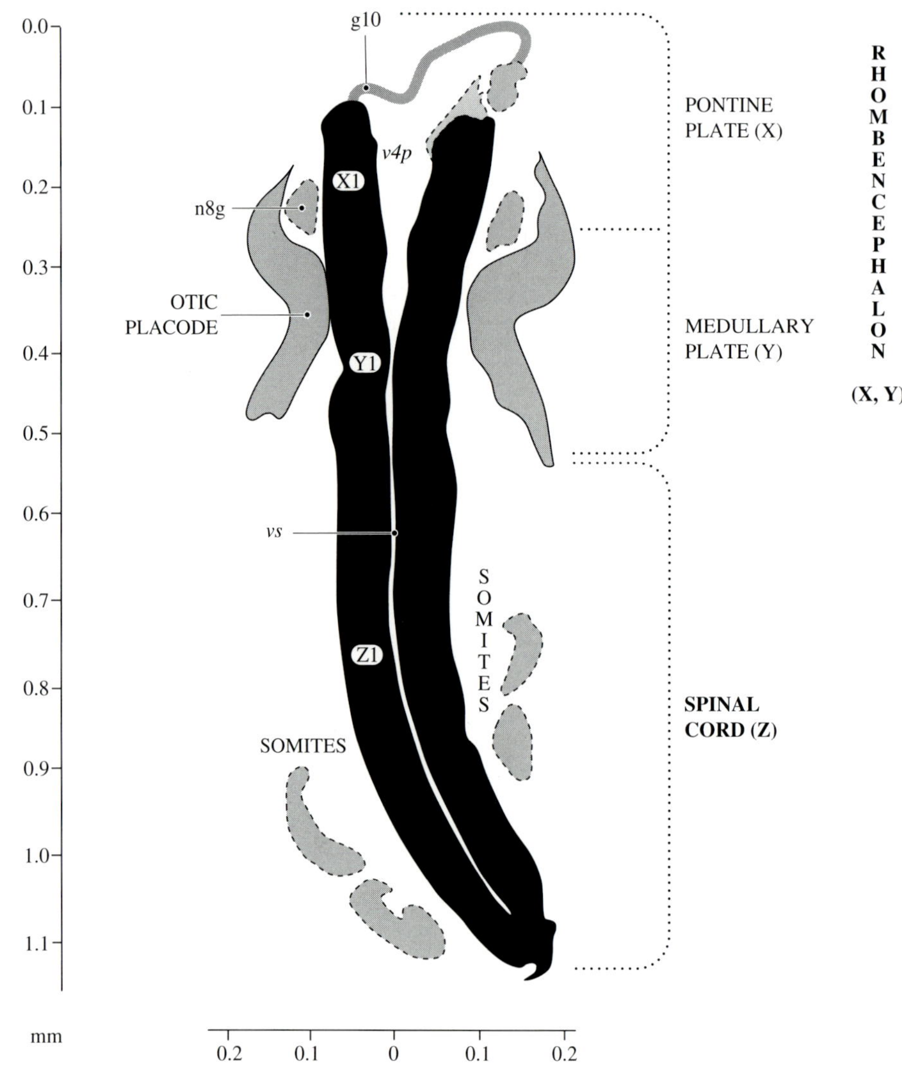

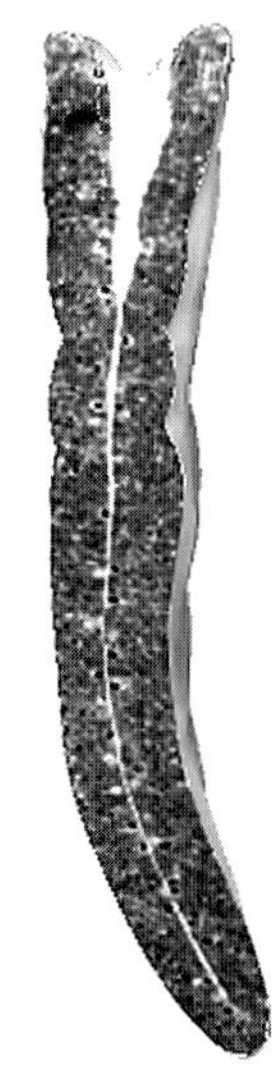

CORONAL FIGURE 17

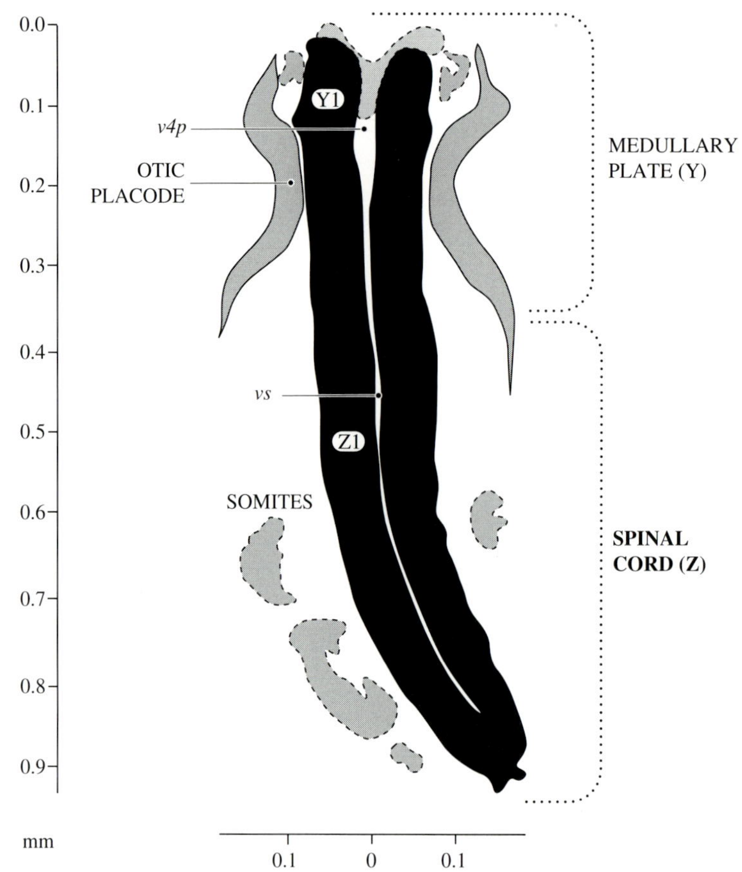

VENTRICLES (v)

vs	spinal canal
v4p	fourth ventricle, posterior

MEDULLA (Y)

Y1	medullary neuroepithelium

SPINAL CORD (Z)

Z1	spinal cord neuroepithelium

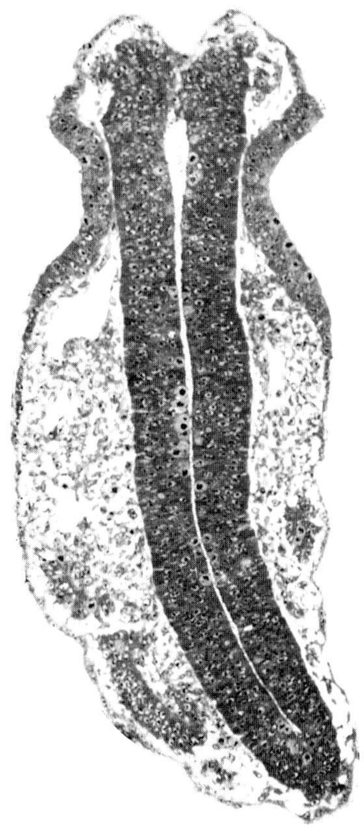

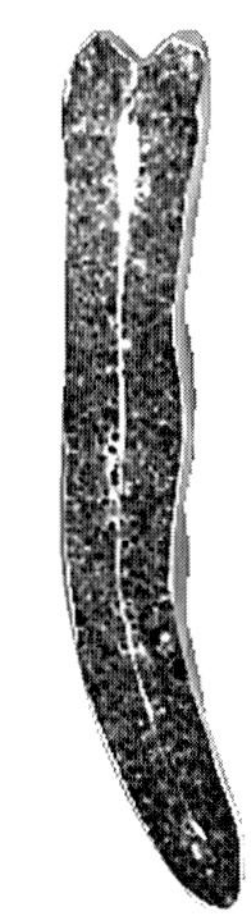

THE E12 BRAIN

RIGHT, POSTERIOR VIEW

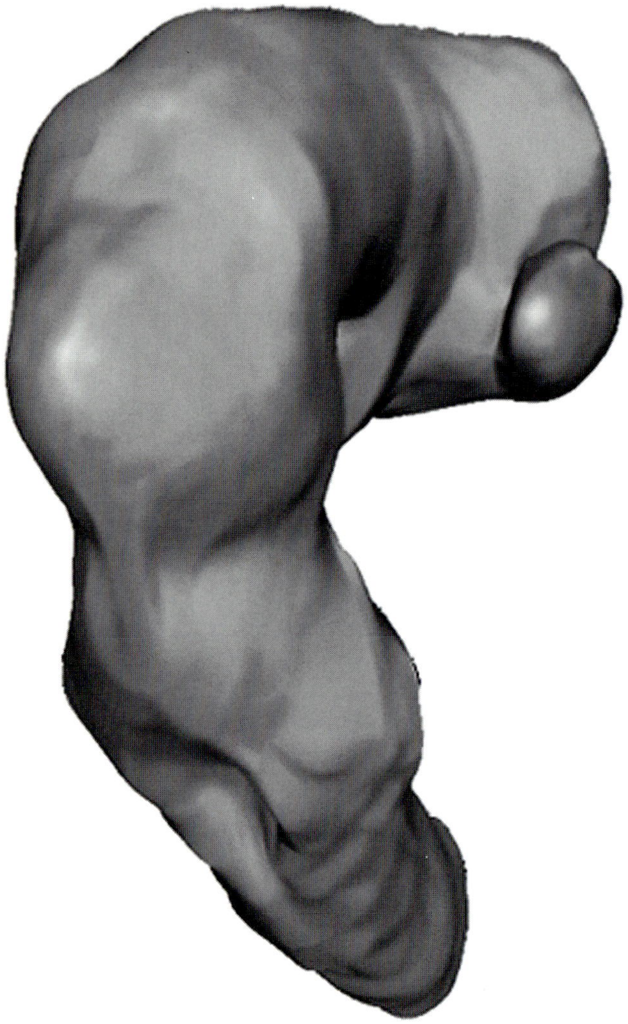

LEFT, ANTERIOR VIEW

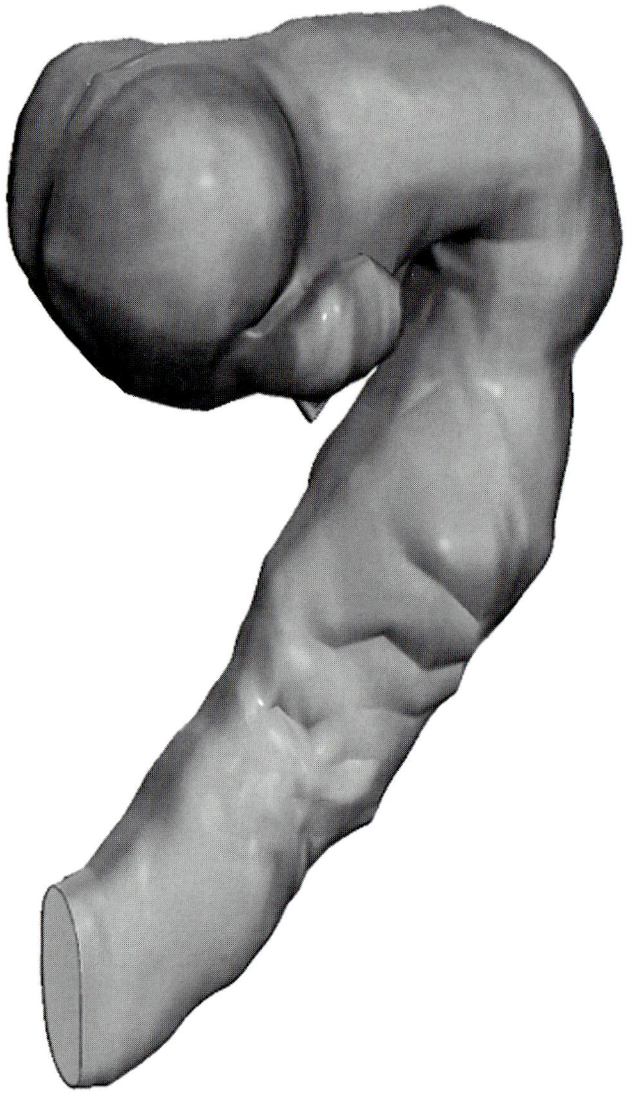

FRONTAL VIEW

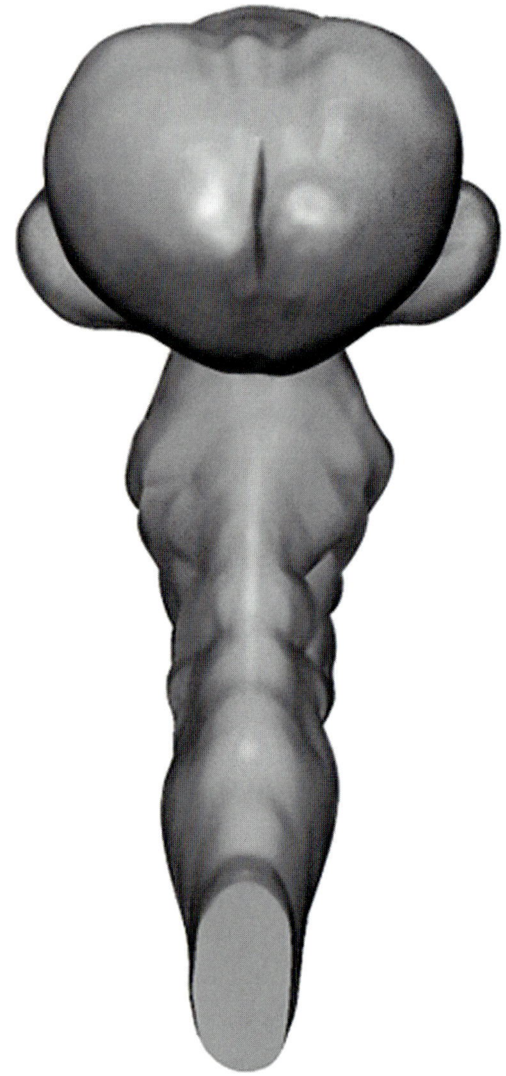

SAGITTAL

FIGURES and PLATES 1-5

SAGITTAL FIGURE 1

CEREBRAL CORTEX (C)

C11 cingulate cortex neuroepithelium

GLANDS AND PARANEURAL STRUCTURES (g)

g6 pituitary gland (primordium)
g10 velum medullare

HYPOTHALAMUS (H)

H1a anterior hypothalamic neuroepithelium
H1p posterior hypothalamic neuroepithelium

HIPPOCAMPAL FORMATION (K)

K11 subicular neuroepithelium

CEREBELLUM (L)

L1 cerebellar neuroepithelium

MIDBRAIN (M)

M1 superior tectal neuroepithelium
M11a anterior tegmental neuroepithelium
M11p posterior tegmental neuroepithelium
M21 pretectal neuroepithelium
M31 inferior tectal neuroepithelium

PREOPTIC AREA AND OPTIC EPITHELIA (P)

P1 preoptic neuroepithelium

RHINENCEPHALON (R)

R1 rhinencephalic neuroepithelium

THALAMUS (T)

T1 thalamic neuroepithelium

VENTRICLES (v)

vai aqueduct, inferior tectal
vapr aqueduct, pretectal
vas aqueduct, superior tectal
vat aqueduct, tegmental
vla lateral ventricle, anterior horn
vld lateral ventricle, dorsal
vlm foramen of Monro
v3h third ventricle, hypothalamic
v3i third ventricle, infundibular recess
v3pr third ventricle, preoptic recess
v3t third ventricle, thalamic
v4a fourth ventricle, anterior
v4i fourth ventricle, isthmal
v4p fourth ventricle, posterior

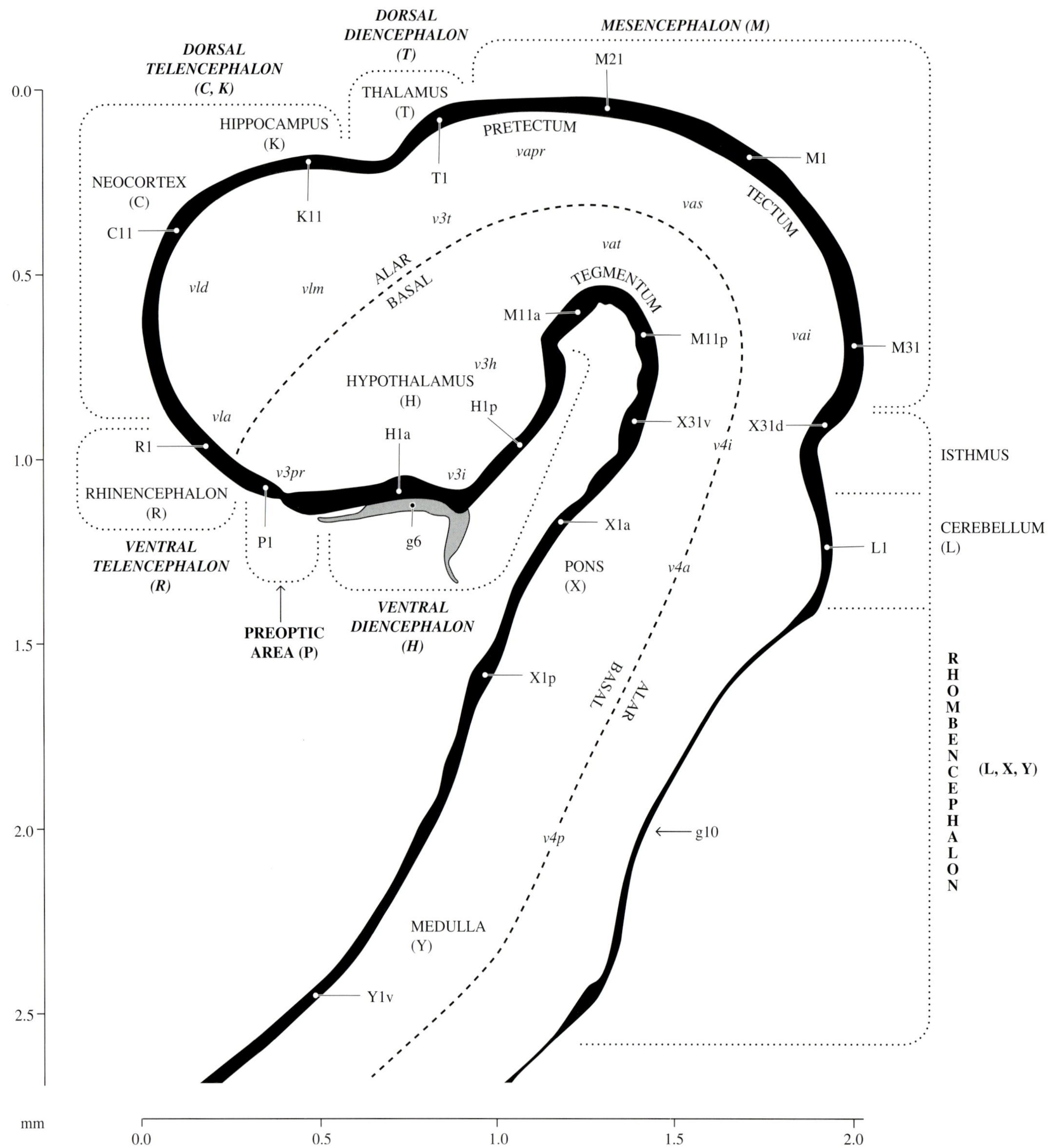

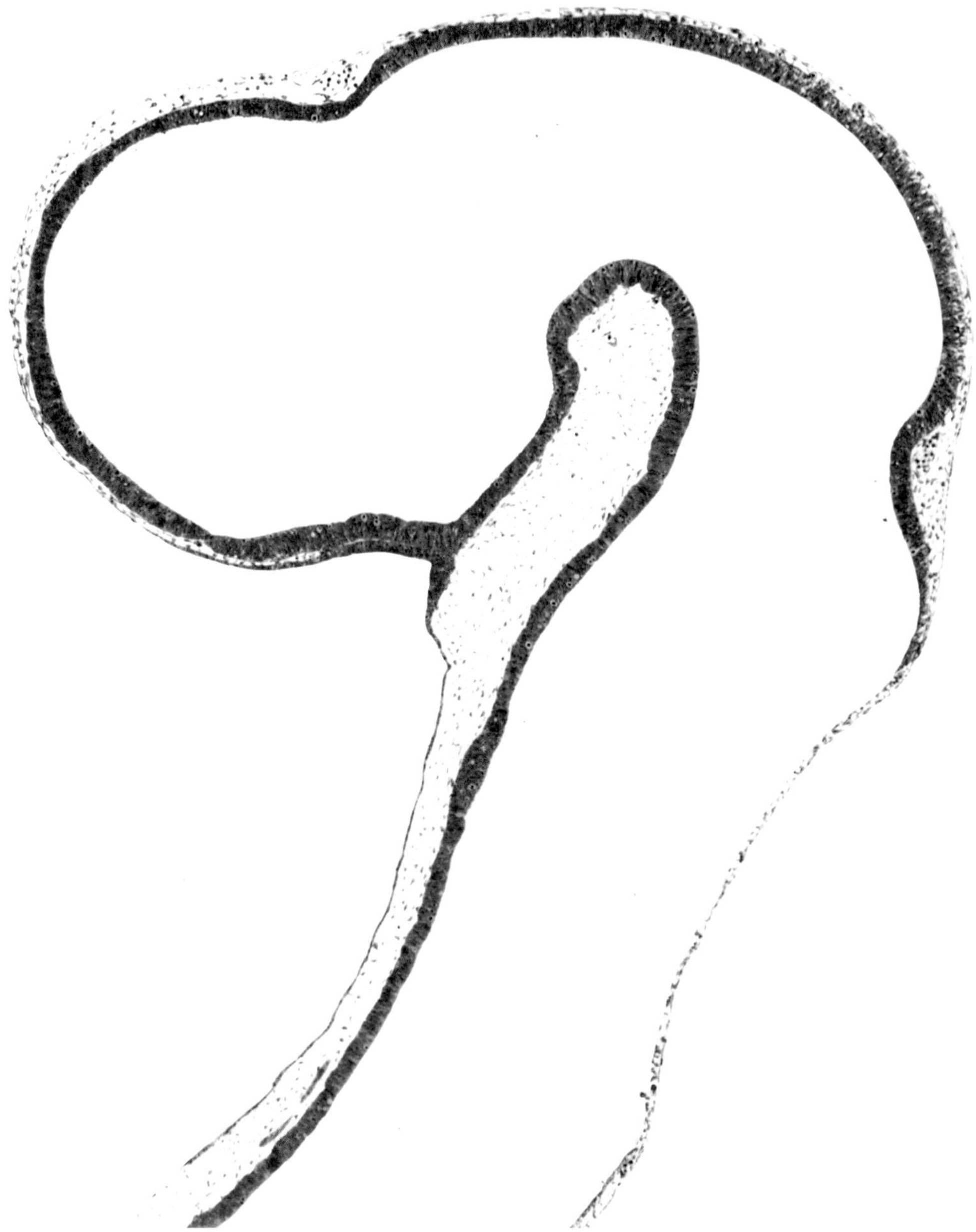

PONTINE AREA (X)

X1a anterior pontine neuroepithelium
X1p posterior pontine neuroepithelium
X31d isthmal neuroepithelium, dorsal
X31v isthmal neuroepithelium, ventral

MEDULLA (Y)

Y1v medullary neuroepithelium, ventral

SAGITTAL FIGURE 2

CEREBRAL CORTEX (C)

C11 cingulate cortex neuroepithelium

GLANDS AND PARANEURAL STRUCTURES (g)

g6 pituitary gland (primordium)
g10 velum medullare

HYPOTHALAMUS (H)

H1a anterior hypothalamic neuroepithelium
H1m mammillary neuroepithelium
H1v ventral hypothalamic neuroepithelium

HIPPOCAMPAL FORMATION (K)

K11 subicular neuroepithelium

CEREBELLUM (L)

L1 cerebellar neuroepithelium

MIDBRAIN (M)

M1 superior tectal neuroepithelium
M11a anterior tegmental neuroepithelium
M11p posterior tegmental neuroepithelium
M21 pretectal neuroepithelium
M31 inferior tectal neuroepithelium

PREOPTIC AREA AND OPTIC EPITHELIA (P)

P1 preoptic neuroepithelium

RHINENCEPHALON (R)

R1 rhinencephalic neuroepithelium

SEPTAL AREA (S)

S1 septal neuroepithelium

THALAMUS (T)

T1 thalamic neuroepithelium

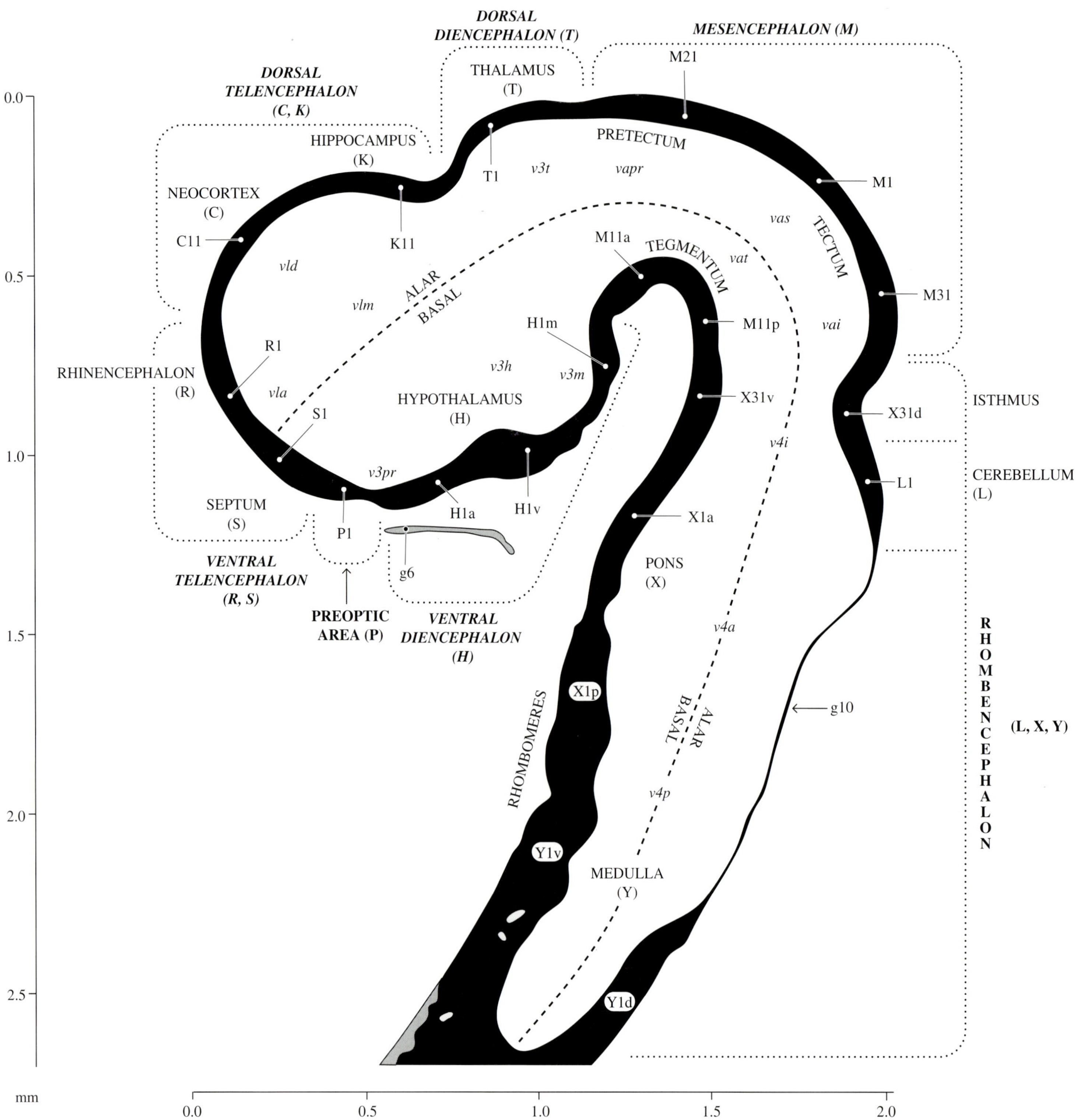

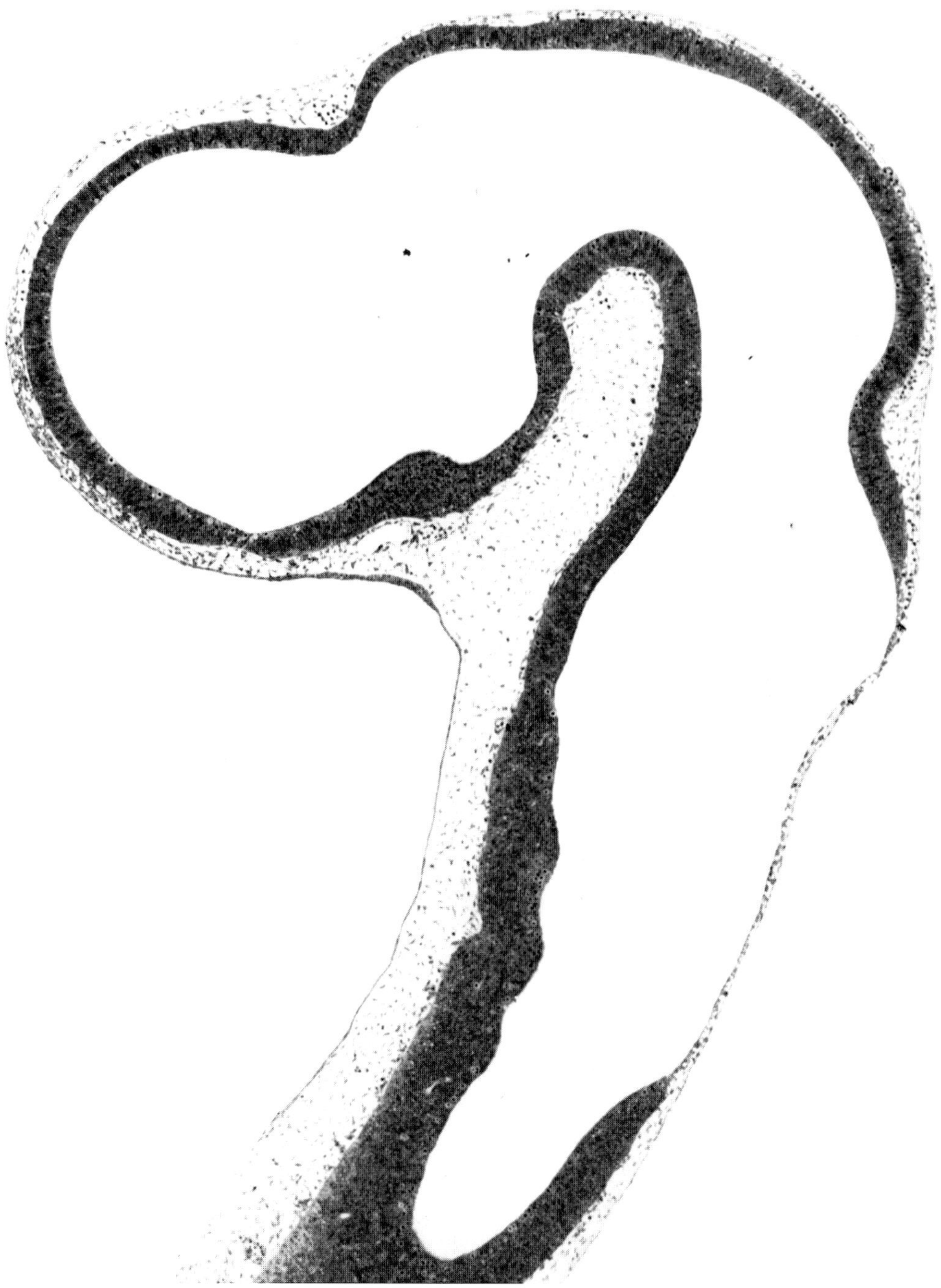

VENTRICLES (v)

vai	aqueduct, inferior tectal
vapr	aqueduct, pretectal
vas	aqueduct, superior tectal
vat	aqueduct, tegmental
vla	lateral ventricle, anterior horn
vld	lateral ventricle, dorsal
vlm	foramen of Monro
v3h	third ventricle, hypothalamic
v3m	third ventricle, mammillary recess
v3pr	third ventricle, preoptic recess
v3t	third ventricle, thalamic
v4a	fourth ventricle, anterior
v4i	fourth ventricle, isthmal
v4p	fourth ventricle, posterior

PONTINE AREA (X)

X1a	anterior pontine neuroepithelium
X1p	posterior pontine neuroepithelium
X31d	isthmal neuroepithelium, dorsal
X31v	isthmal neuroepithelium, ventral

MEDULLA (Y)

Y1d	medullary neuroepithelium, dorsal
Y1v	medullary neuroepithelium, ventral

SAGITTAL FIGURE 3

CEREBRAL CORTEX (C)

C1 neocortical neuroepithelium

GLANDS AND PARANEURAL STRUCTURES (g)

g10 velum medullare

HYPOTHALAMUS (H)

H1a anterior hypothalamic neuroepithelium
H1v ventral hypothalamic neuroepithelium

HIPPOCAMPAL FORMATION (K)

K1 hippocampal (Ammon's horn) neuroepithelium
K11 subicular neuroepithelium

CEREBELLUM (L)

L1 cerebellar neuroepithelium

MIDBRAIN (M)

M1 superior tectal neuroepithelium
M11a anterior tegmental neuroepithelium
M11p posterior tegmental neuroepithelium
M21 pretectal neuroepithelium
M31 inferior tectal neuroepithelium

PREOPTIC AREA AND OPTIC EPITHELIA (P)

P1 preoptic neuroepithelium

RHINENCEPHALON (R)

R1 rhinencephalic neuroepithelium

SEPTAL AREA (S)

S1 septal neuroepithelium

THALAMUS (T)

T1 thalamic neuroepithelium

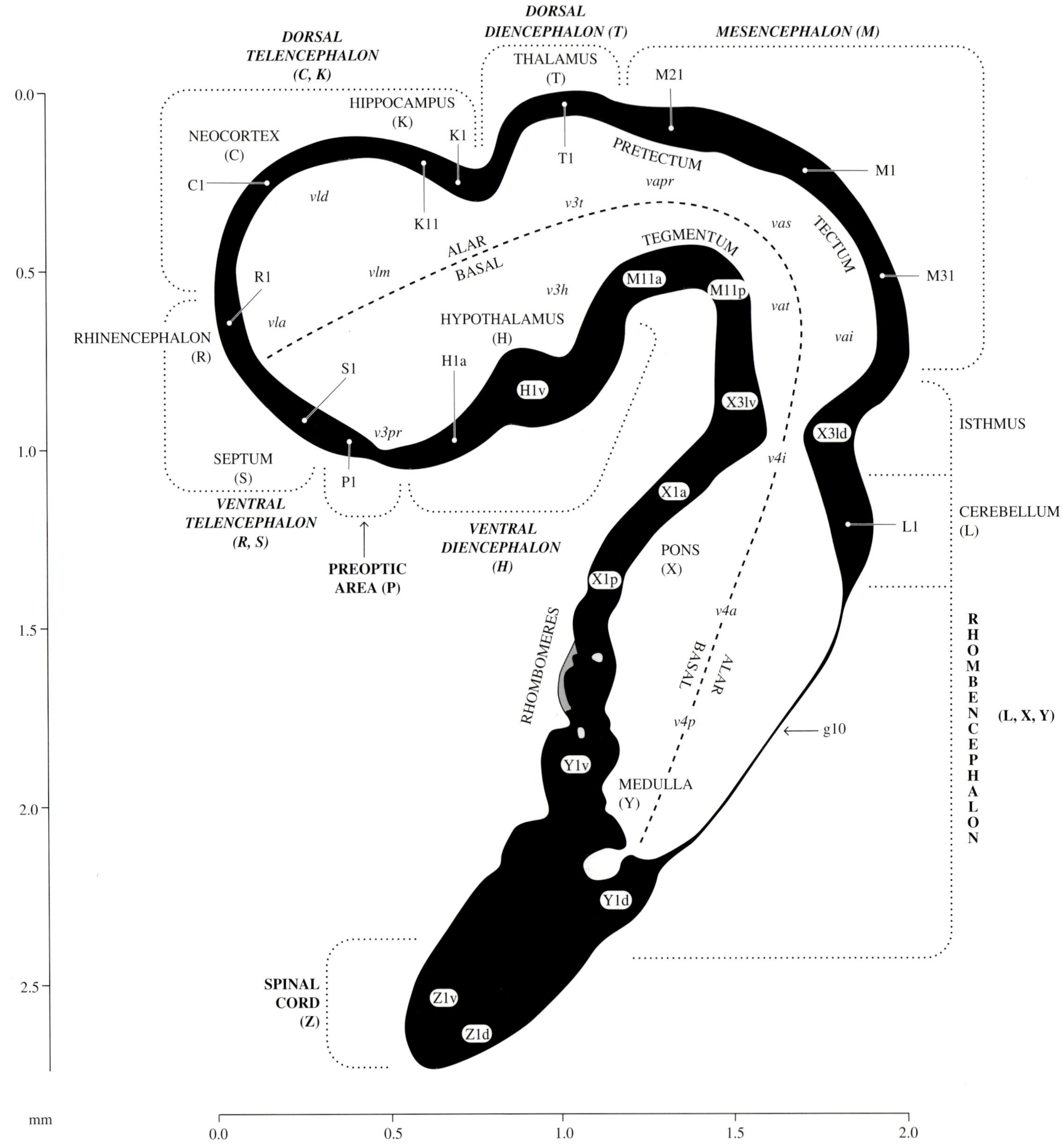

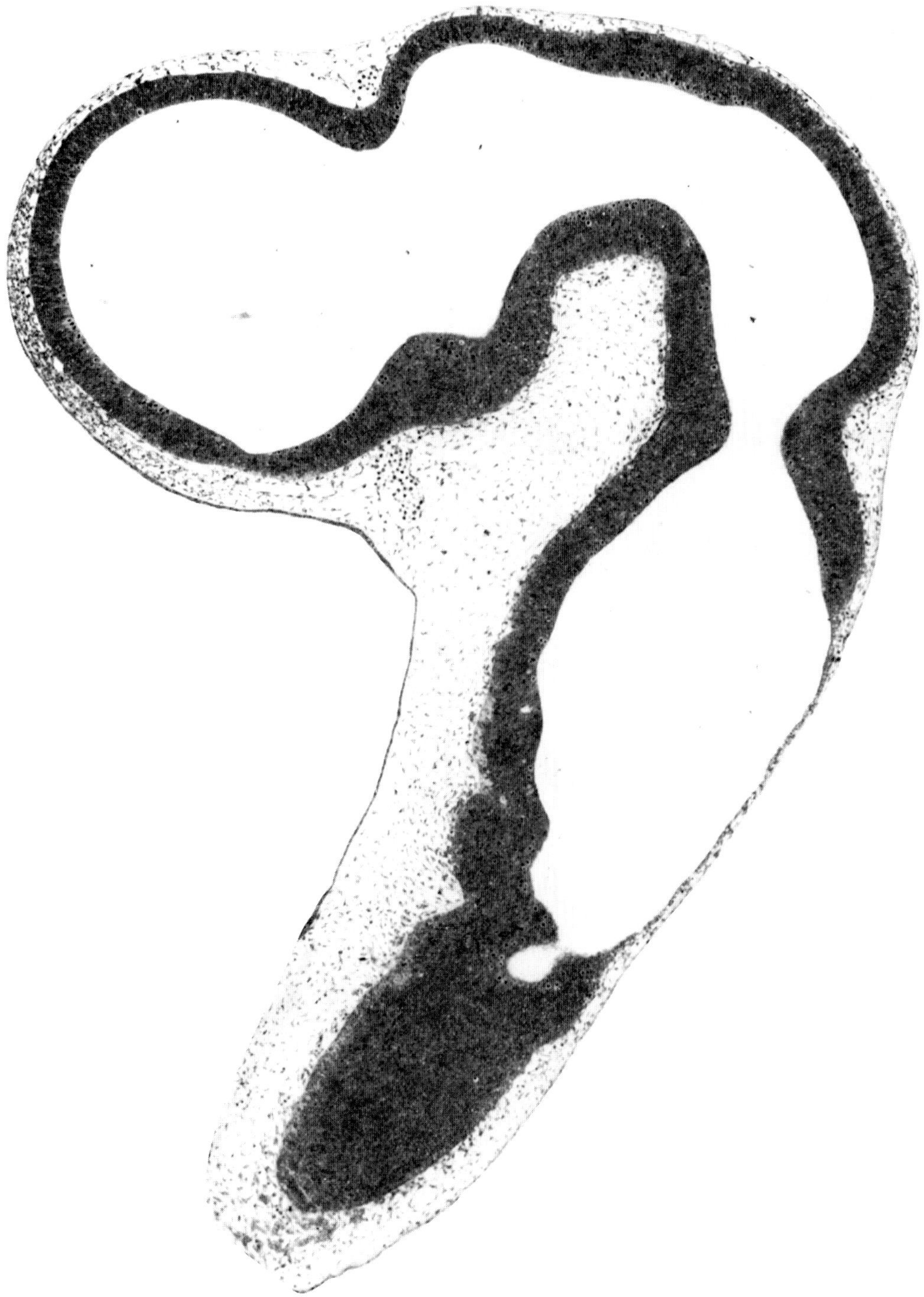

VENTRICLES (v)

vai	aqueduct, inferior tectal
vapr	aqueduct, pretectal
vas	aqueduct, superior tectal
vat	aqueduct, tegmental
vla	lateral ventricle, anterior horn
vld	lateral ventricle, dorsal
vlm	foramen of Monro
v3h	third ventricle, hypothalamic
v3pr	third ventricle, preoptic recess
v3t	third ventricle, thalamic
v4a	fourth ventricle, anterior
v4i	fourth ventricle, isthmal
v4p	fourth ventricle, posterior

PONTINE AREA (X)

X1a	anterior pontine neuroepithelium
X1p	posterior pontine neuroepithelium
X31d	isthmal neuroepithelium, dorsal
X31v	isthmal neuroepithelium, ventral

MEDULLA (Y)

Y1d	medullary neuroepithelium, dorsal
Y1v	medullary neuroepithelium, ventral

SPINAL CORD (Z)

Z1d	spinal cord neuroepithelium, dorsal
Z1v	spinal cord neuroepithelium, ventral

SAGITTAL FIGURE 4

BASAL GANGLIA (B)

B11 pallidal neuroepithelium

CEREBRAL CORTEX (C)

C1 neocortical neuroepithelium

HYPOTHALAMUS (H)

H1a anterior hypothalamic neuroepithelium
H1f Forelian neuroepithelium
H1i intermediate hypothalamic neuroepithelium
H1p posterior hypothalamic neuroepithelium

HIPPOCAMPAL FORMATION (K)

K1 hippocampal (Ammon's horn) neuroepithelium
K11 subicular neuroepithelium

CEREBELLUM (L)

L1 cerebellar neuroepithelium

MIDBRAIN (M)

M1 superior tectal neuroepithelium
M11a anterior tegmental neuroepithelium
M11p posterior tegmental neuroepithelium
M21 pretectal neuroepithelium
M31 inferior tectal neuroepithelium

NERVES AND NUCLEI, CRANIAL (n)

n8b vestibulocochlear boundary cap
n8g vestibular ganglion

PREOPTIC AREA AND OPTIC EPITHELIA (P)

P1 preoptic neuroepithelium

RHINENCEPHALON (R)

R1 rhinencephalic neuroepithelium

THALAMUS (T)

T1i intermediate thalamic neuroepithelium
T1p posterior thalamic neuroepithelium

VENTRICLES (v)

vla lateral ventricle, anterior horn
vlc lateral ventricle, central
vld lateral ventricle, dorsal
vlm foramen of Monro
v3pr third ventricle, preoptic recess
v3t third ventricle, thalamic
v4l fourth ventricle, lateral recess

PONTINE AREA (X)

X1a anterior pontine neuroepithelium
X1p posterior pontine neuroepithelium
X31d isthmal neuroepithelium, dorsal
X31v isthmal neuroepithelium, ventral

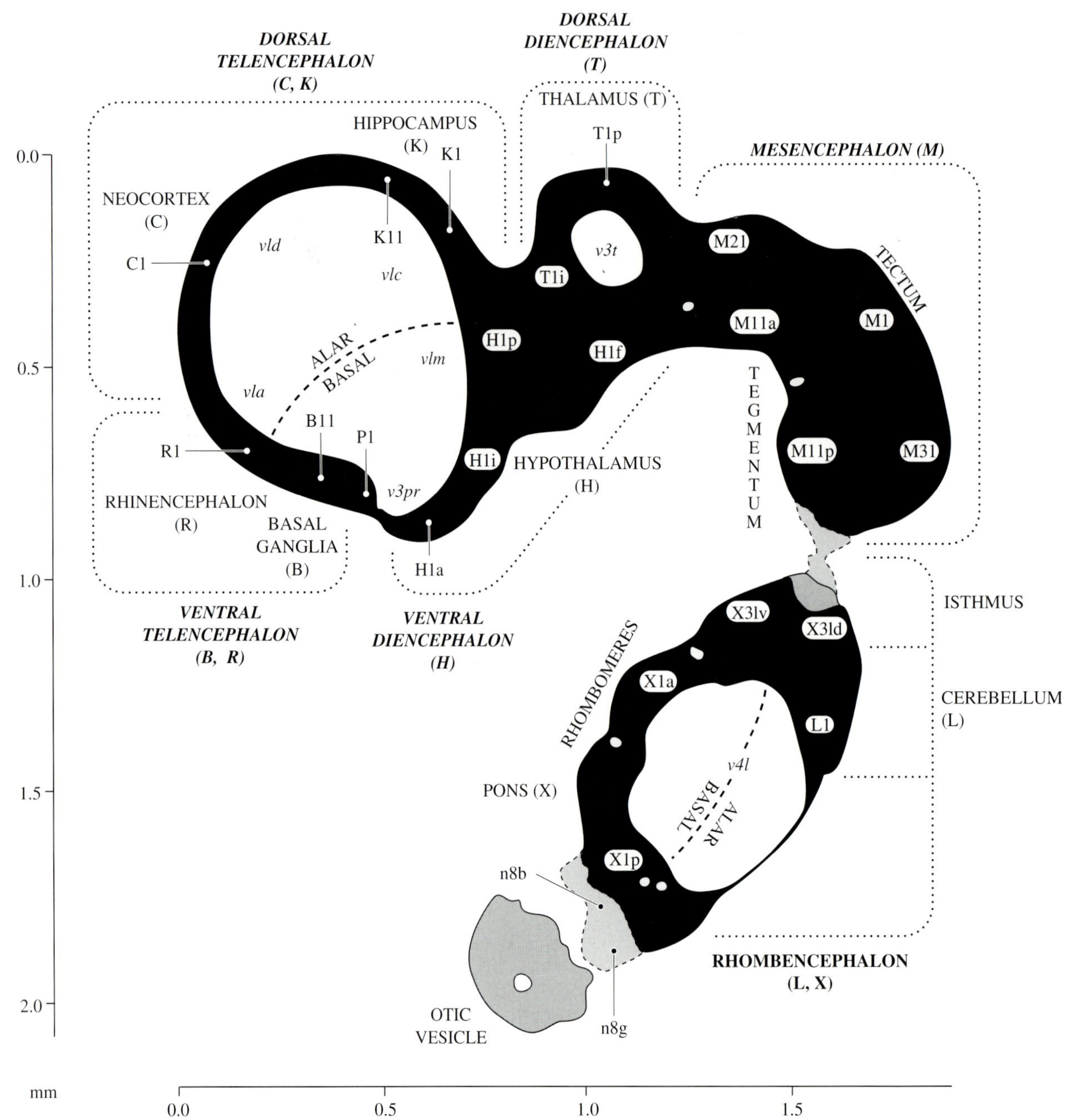

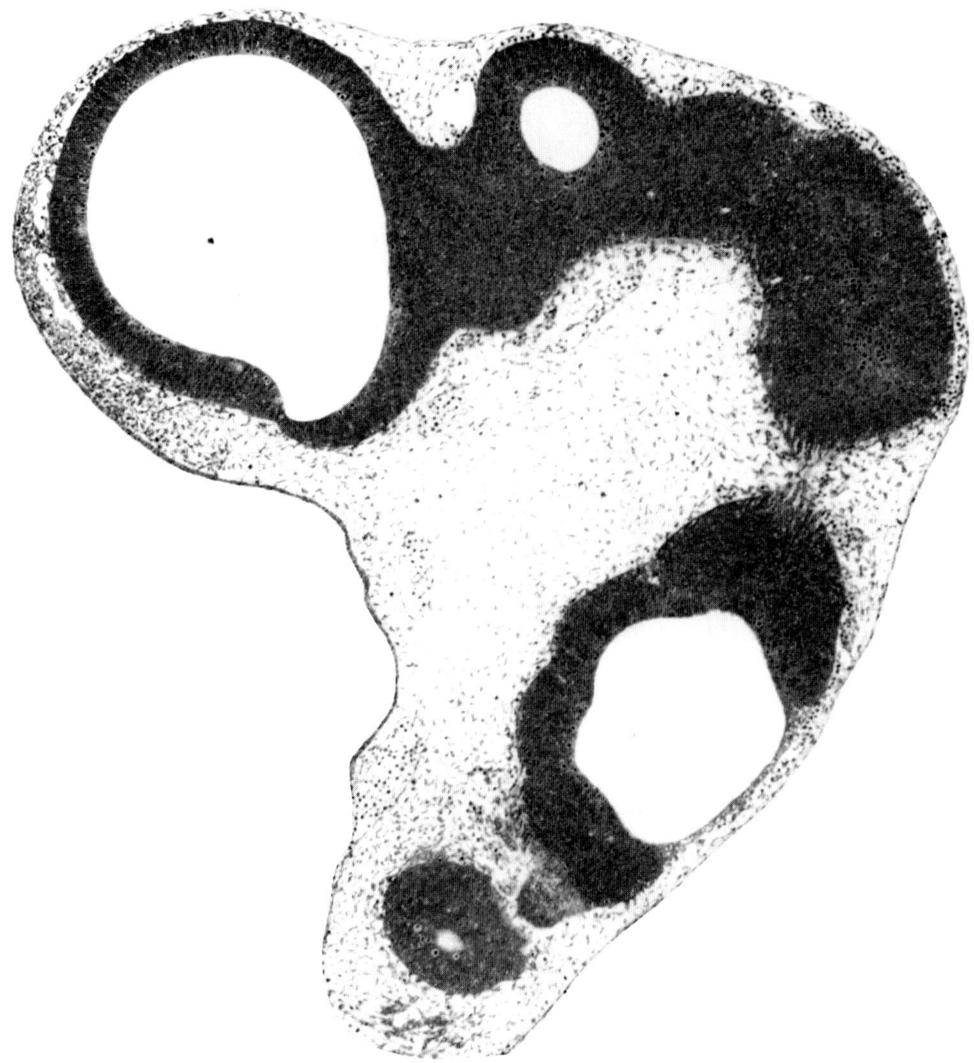

AMYGDALA (A)

A1 amygdaloid neuroepithelium

BASAL GANGLIA (B)

B11 pallidal neuroepithelium

CEREBRAL CORTEX (C)

C1 neocortical neuroepithelium

HYPOTHALAMUS (H)

H1*l* lateral hypothalamic neuroepithelium
H1p posterior hypothalamic neuroepithelium

HIPPOCAMPAL FORMATION (K)

K1 hippocampal (Ammon's horn) neuroepithelium
K11 subicular neuroepithelium

CEREBELLUM (L)

L1 cerebellar neuroepithelium

MIDBRAIN (M)

M1 superior tectal neuroepithelium
M31 inferior tectal neuroepithelium

NERVES AND NUCLEI, CRANIAL (n)

n8b vestibulocochlear boundary cap
n8g vestibular ganglion

RHINENCEPHALON (R)

R1 rhinencephalic neuroepithelium

THALAMUS (T)

T1p posterior thalamic neuroepithelium

VENTRICLES (v)

vla lateral ventricle, anterior horn
vlc lateral ventricle, central
vld lateral ventricle, dorsal
vlp lateral ventricle, posterior horn
v3h third ventricle, hypothalamic
v4l fourth ventricle, lateral recess

PONTINE AREA (X)

X1a anterior pontine neuroepithelium
X1p posterior pontine neuroepithelium
X21 cochlear nuclei neuroepithelium
X31 isthmal neuroepithelium

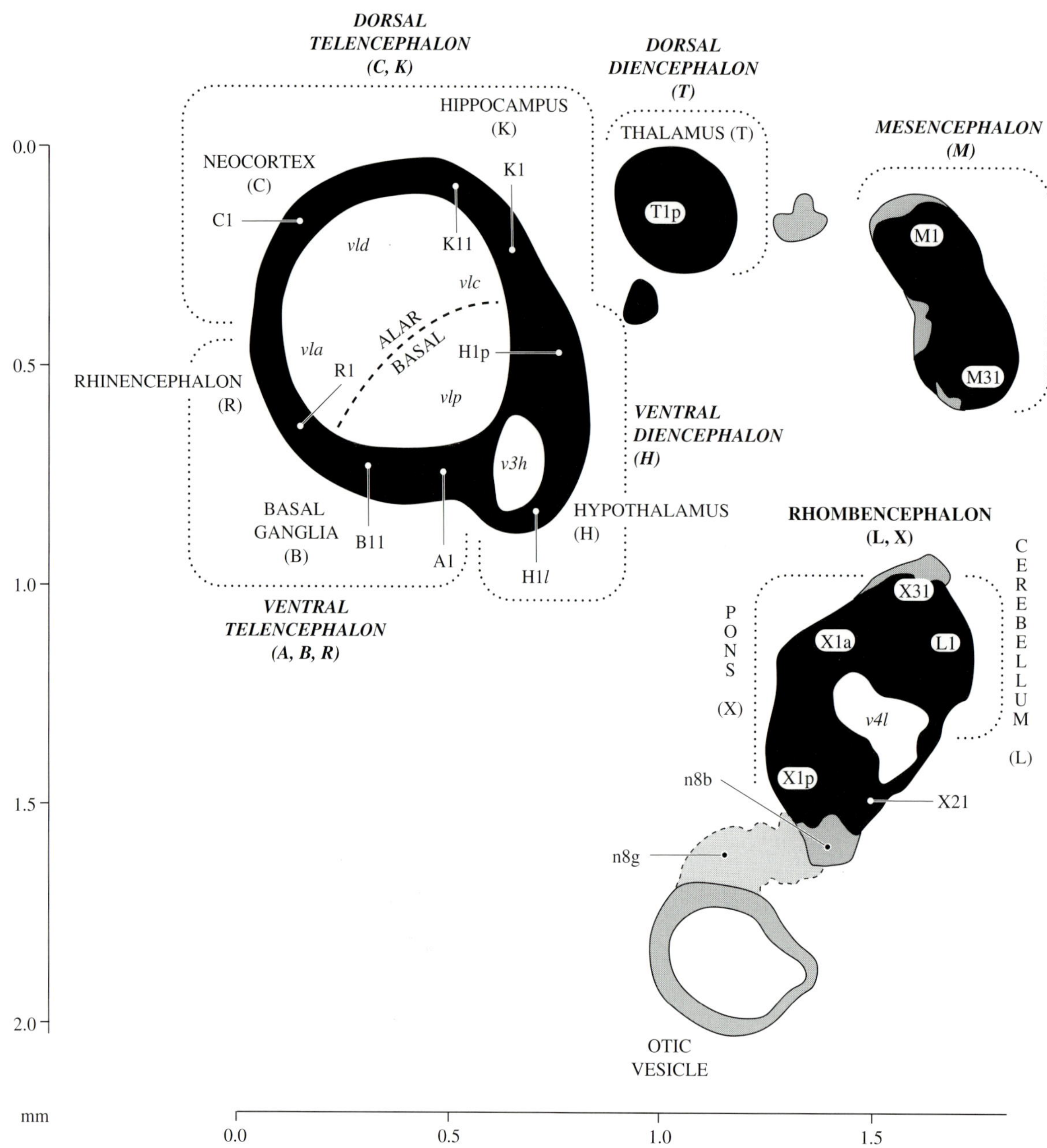

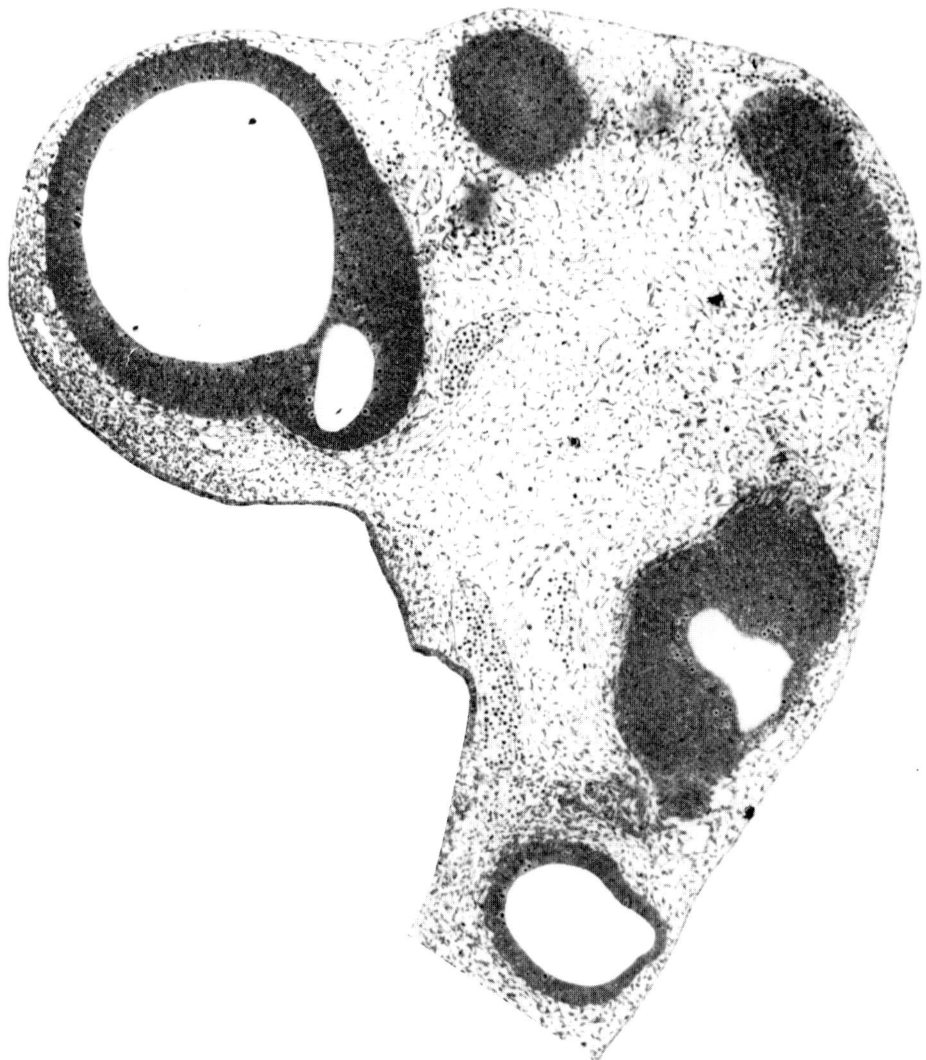

CORONAL

FIGURES and PLATES 1-16

CORONAL FIGURE 1

CEREBRAL CORTEX (C)

C1 neocortical neuroepithelium
C11 cingulate cortex neuroepithelium

RHINENCEPHALON (R)

R1 rhinencephalic neuroepithelium

SEPTAL AREA (S)

S1 septal neuroepithelium

VENTRICLES (v)

vs spinal canal

SPINAL CORD (Z)

Z1d spinal cord neuroepithelium, dorsal
Z1v spinal cord neuroepithelium, ventral
Z8 dorsal root ganglion

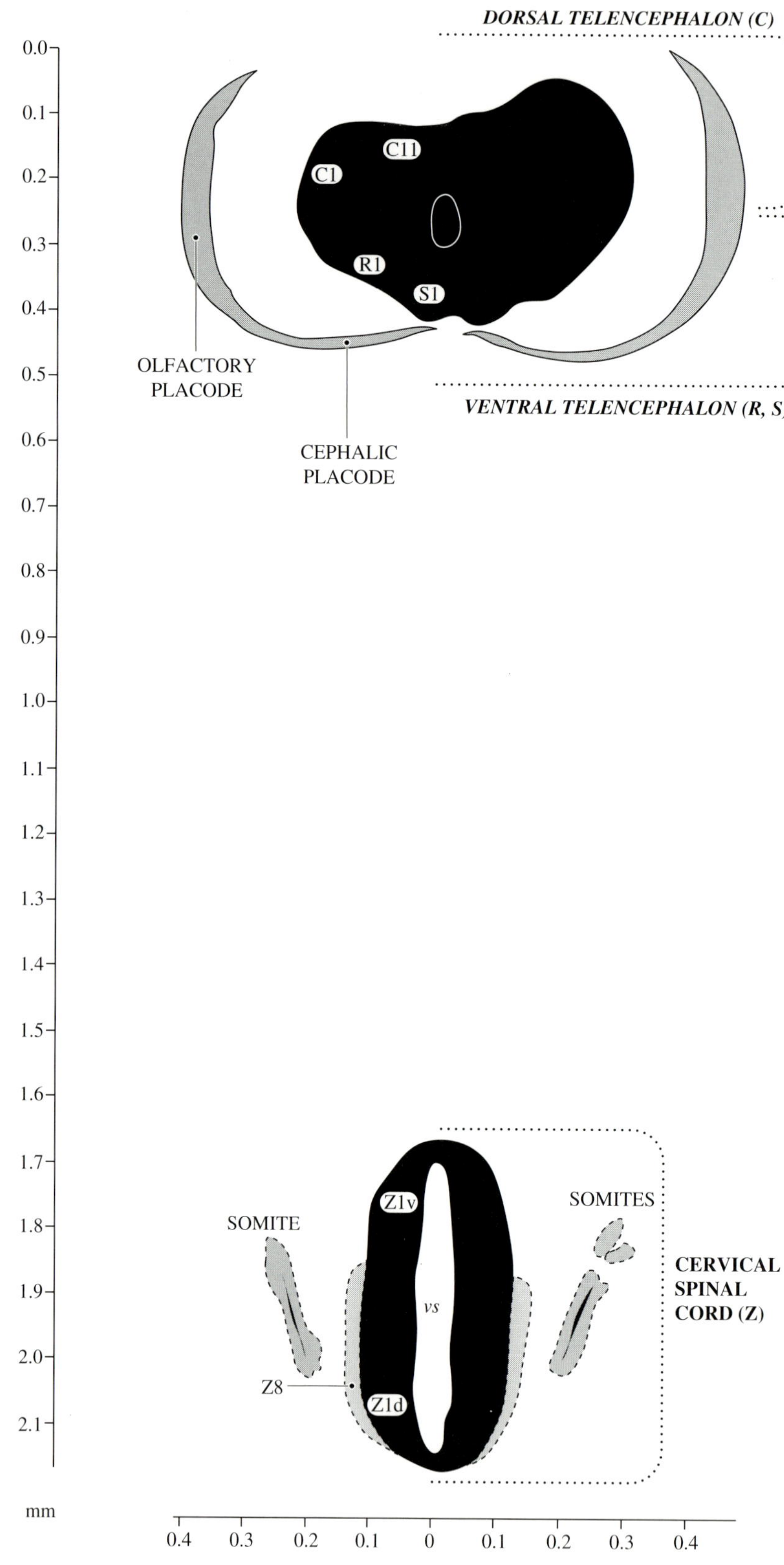

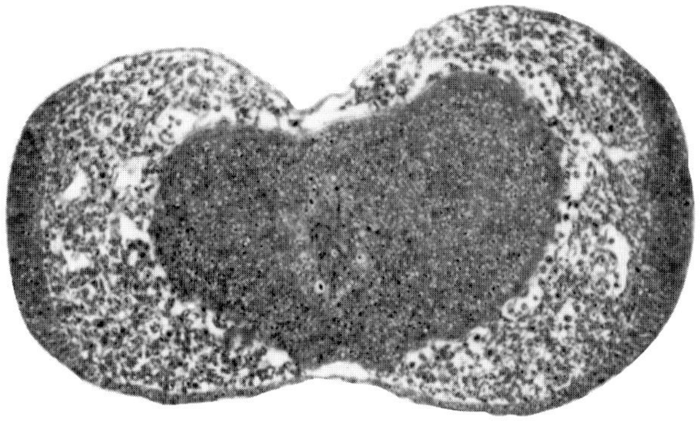

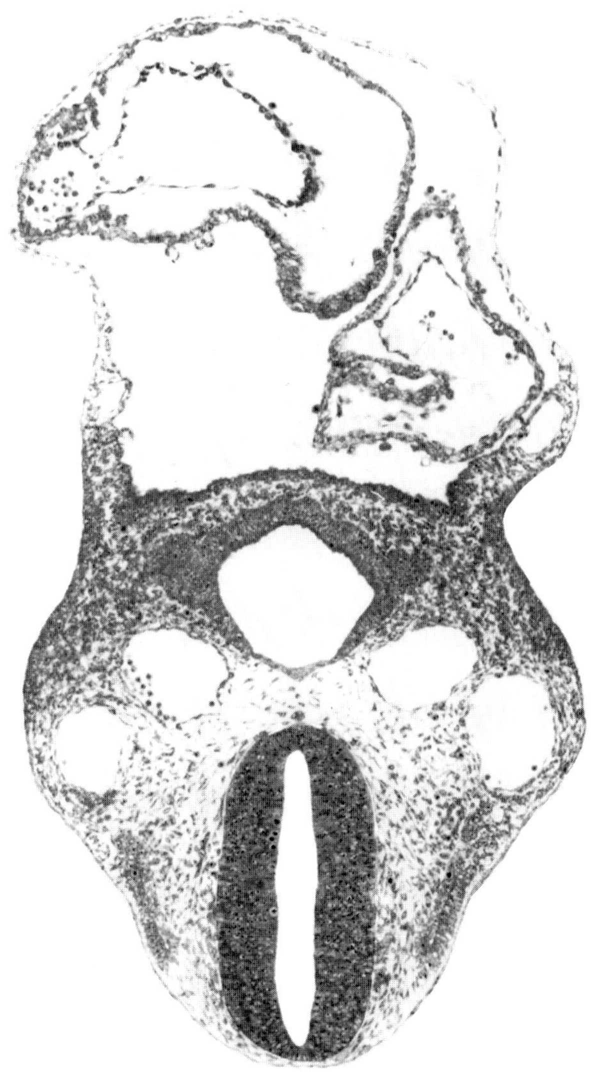

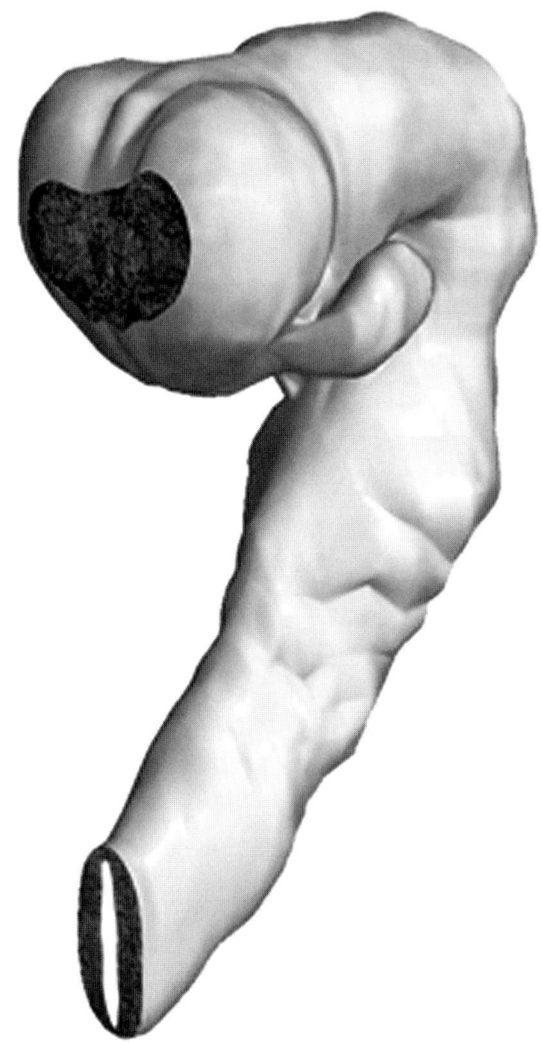

CORONAL FIGURE 2

BASAL GANGLIA (B)

B1 striatal neuroepithelium
B11 pallidal neuroepithelium

CEREBRAL CORTEX (C)

C1 neocortical neuroepithelium
C11 cingulate cortex neuroepithelium

GLANDS AND PARANEURAL STRUCTURES (g)

g9 lamina terminalis

RHINENCEPHALON (R)

R1 rhinencephalic neuroepithelium

VENTRICLES (v)

vla lateral ventricle, anterior horn
vld lateral ventricle, dorsal
vs spinal canal

SPINAL CORD (Z)

Z1d spinal cord neuroepithelium, dorsal
Z1v spinal cord neuroepithelium, ventral
Z8 dorsal root ganglion

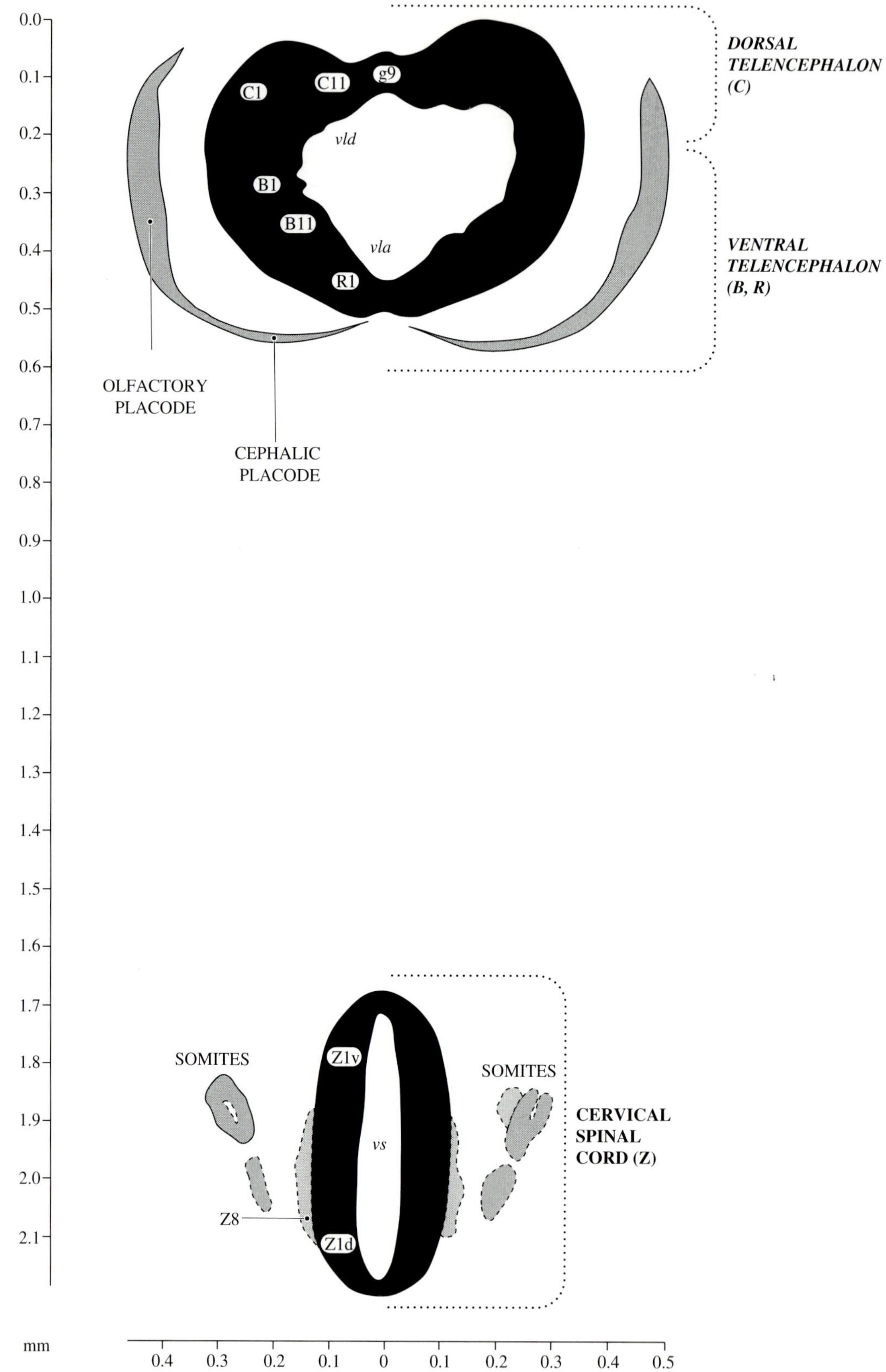

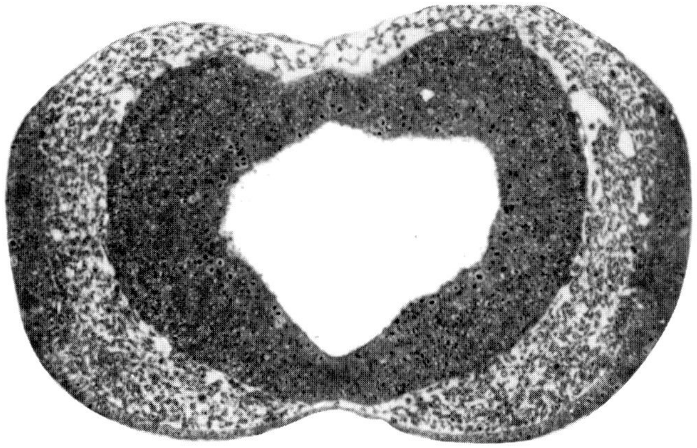

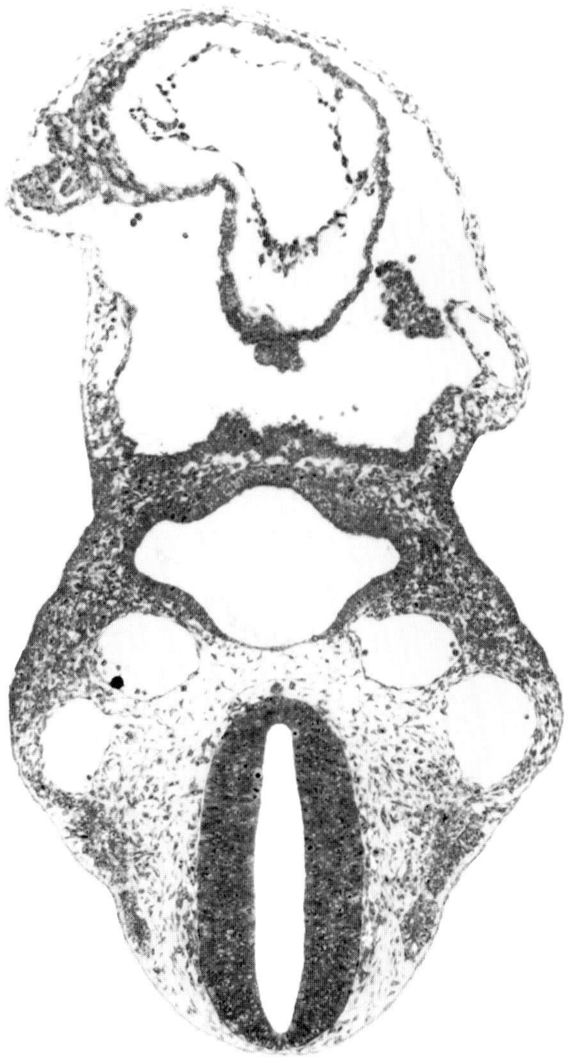

CORONAL FIGURE 3

BASAL GANGLIA (B)

B1 striatal neuroepithelium
B11 pallidal neuroepithelium

CEREBRAL CORTEX (C)

C1 neocortical neuroepithelium
C11 cingulate cortex neuroepithelium

GLANDS AND PARANEURAL STRUCTURES (g)

g9 lamina terminalis

RHINENCEPHALON (R)

R1 rhinencephalic neuroepithelium

VENTRICLES (v)

vla lateral ventricle, anterior horn
vld lateral ventricle, dorsal
vlm foramen of Monro
vs spinal canal

SPINAL CORD (Z)

Z1d spinal cord neuroepithelium, dorsal
Z1v spinal cord neuroepithelium, ventral
Z8 dorsal root ganglion

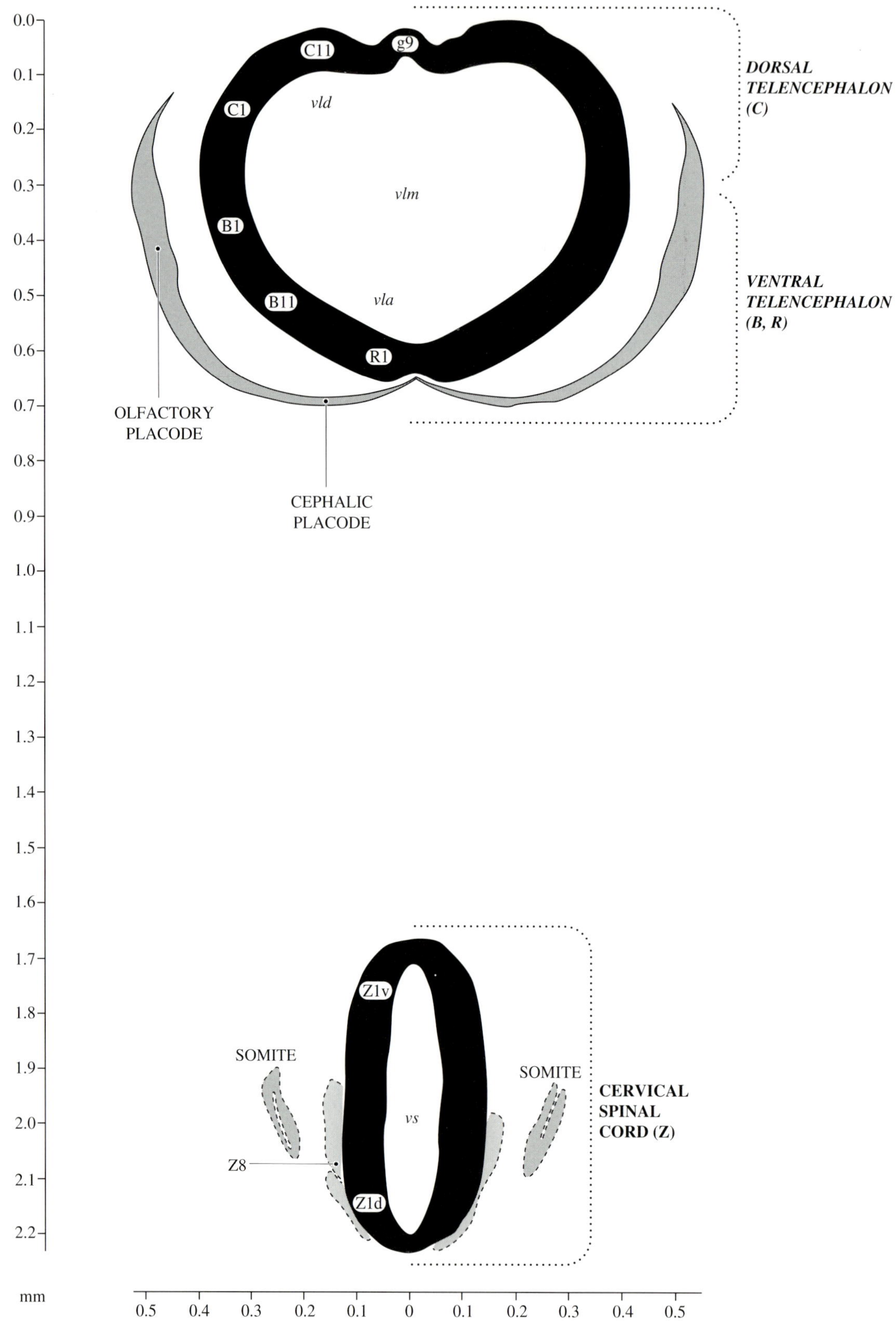

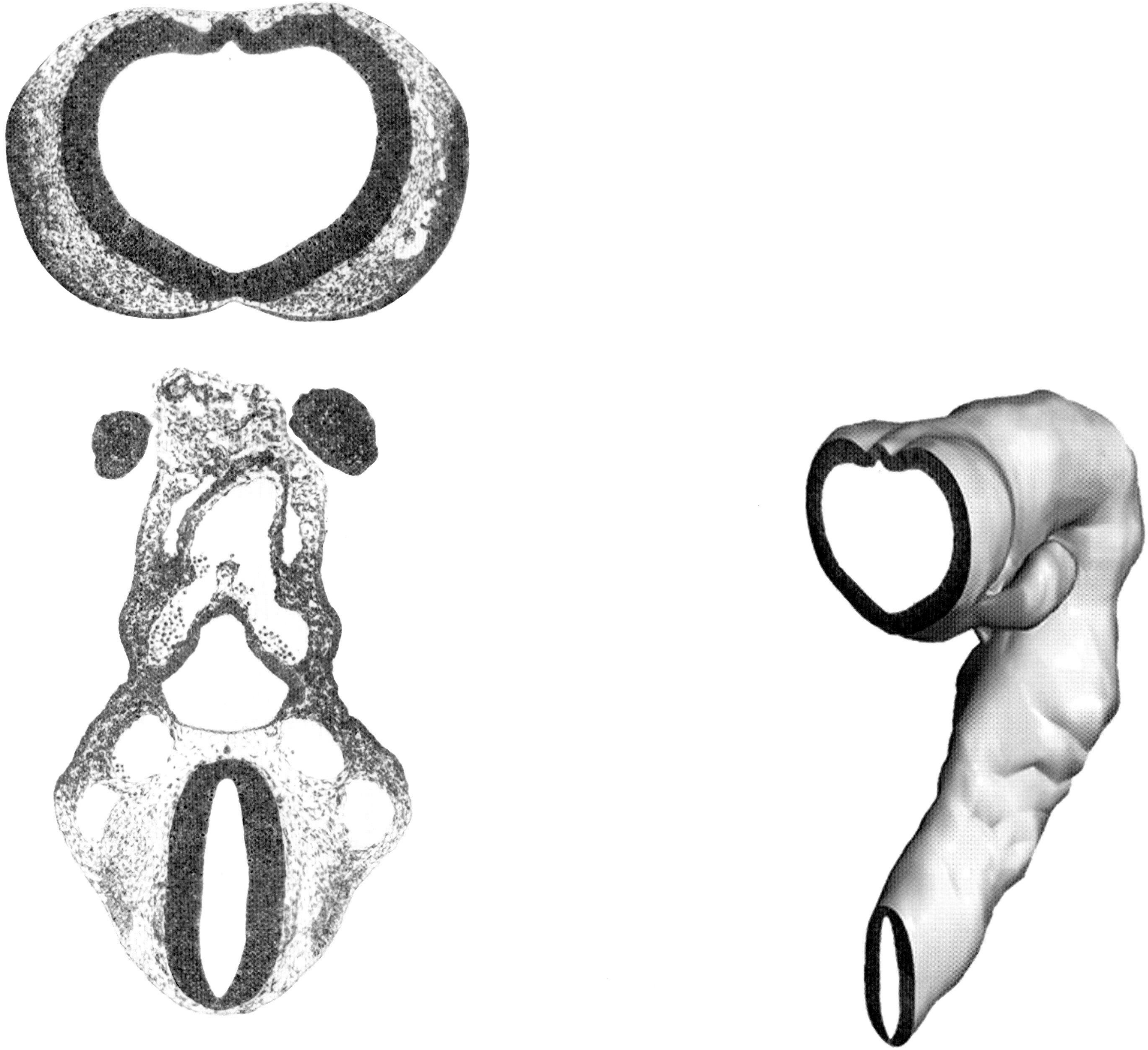

CORONAL FIGURE 4

BASAL GANGLIA (B)

B11 pallidal neuroepithelium

CEREBRAL CORTEX (C)

C1 neocortical neuroepithelium
C11 cingulate cortex neuroepithelium

GLANDS AND PARANEURAL STRUCTURES (g)

g9 lamina terminalis

HYPOTHALAMUS (H)

H1a anterior hypothalamic neuroepithelium

HIPPOCAMPAL FORMATION (K)

K11 subicular neuroepithelium

PREOPTIC AREA AND OPTIC EPITHELIA (P)

P1 preoptic neuroepithelium
P11 retinal neuroepithelium

VENTRICLES (v)

vld lateral ventricle, dorsal
vlm foramen of Monro
vs spinal canal
v3h third ventricle, hypothalamic

SPINAL CORD (Z)

Z1d spinal cord neuroepithelium, dorsal
Z1v spinal cord neuroepithelium, ventral

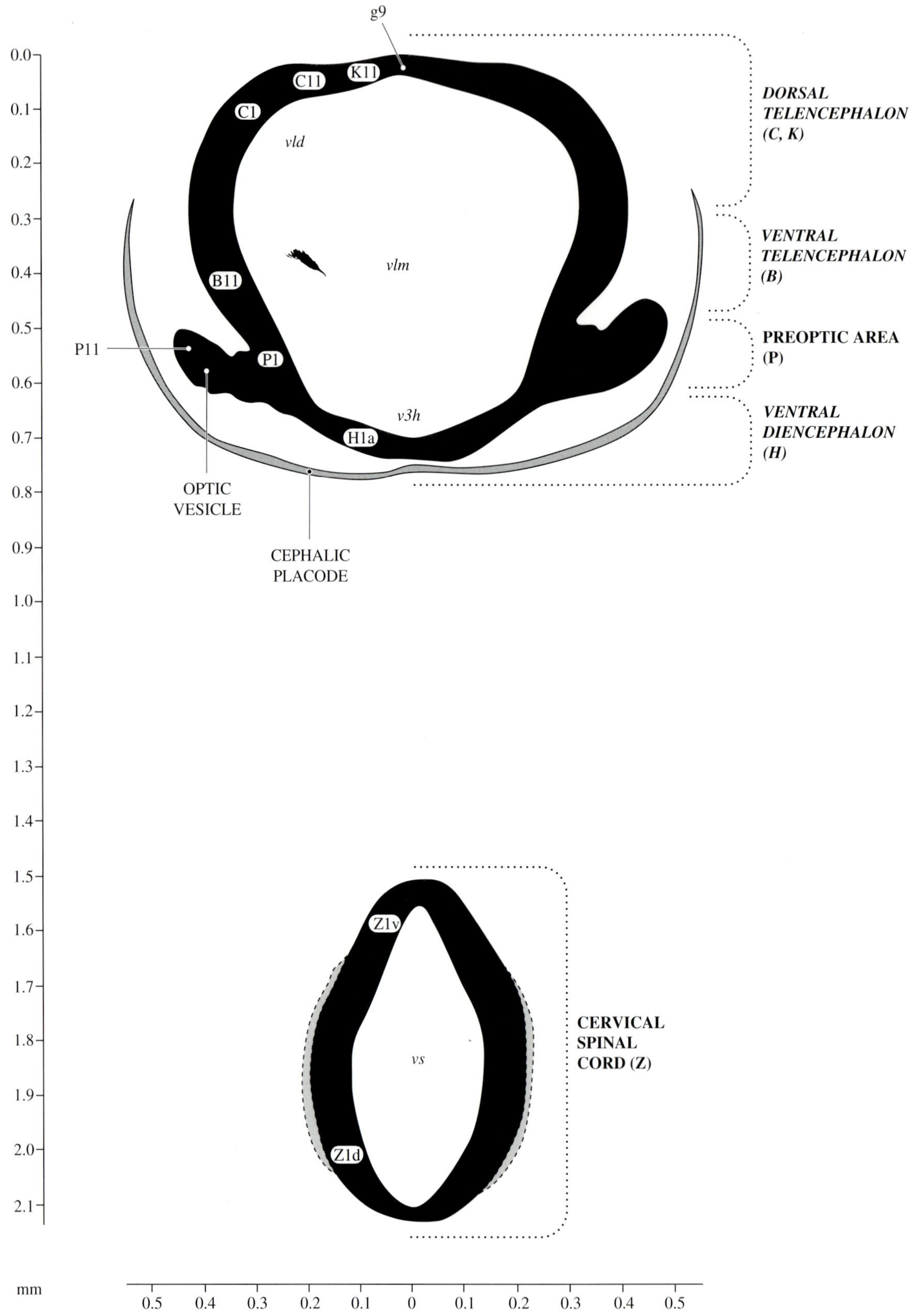

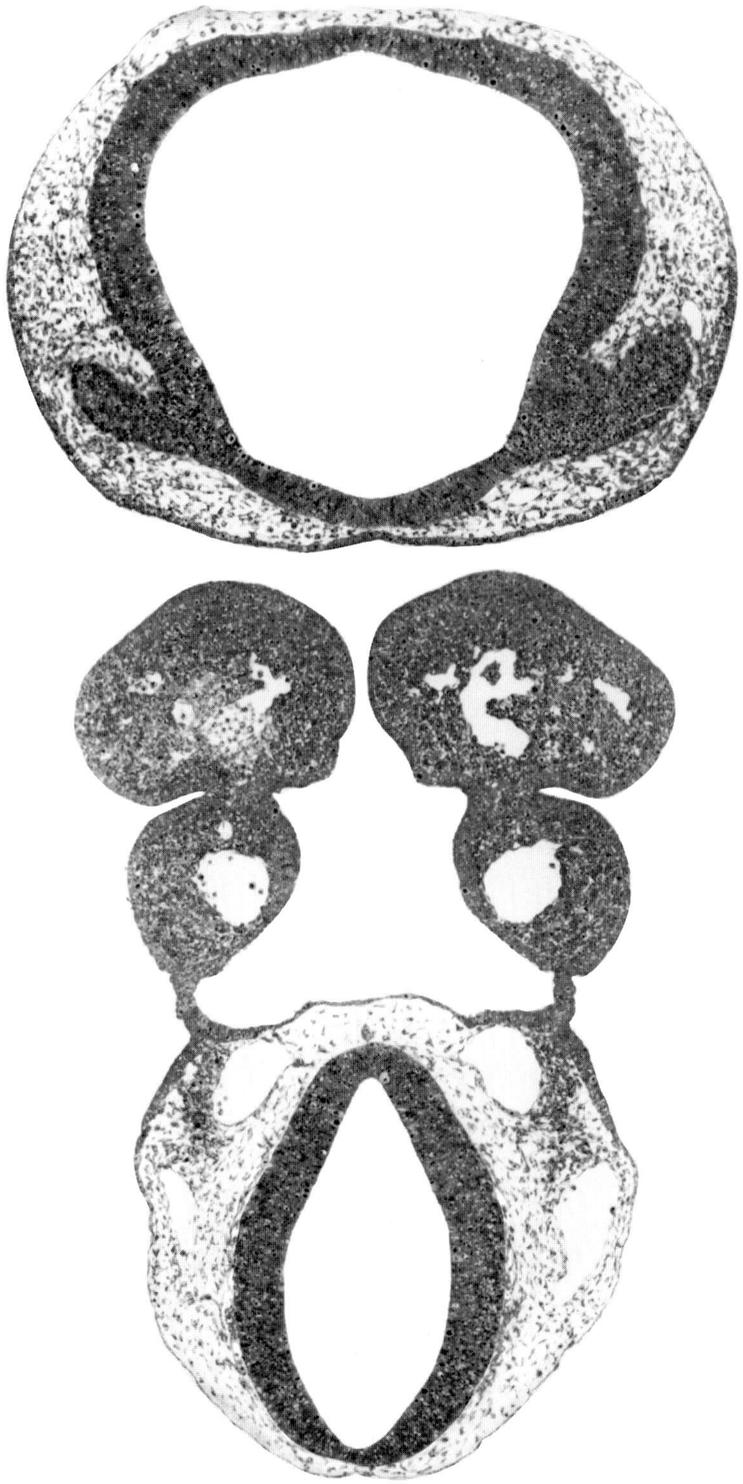

CORONAL FIGURE 5

BASAL GANGLIA (B)

B11 pallidal neuroepithelium

CEREBRAL CORTEX (C)

C1 cortical neuroepithelium
C11 cingulate cortex neuroepithelium

GLANDS AND PARANEURAL STRUCTURES (g)

g6 pituitary gland (primordium)

HYPOTHALAMUS (H)

H1a anterior hypothalamic neuroepithelium

HIPPOCAMPAL FORMATION (K)

K11 subicular neuroepithelium

PREOPTIC AREA AND OPTIC EPITHELIA (P)

P11 retinal neuroepithelium

VENTRICLES (v)

vld lateral ventricle, dorsal
vlm foramen of Monro
vs spinal canal
v3h third ventricle, hypothalamic
v3o third ventricle, optic recess

SPINAL CORD (Z)

Zld spinal cord neuroepithelium, dorsal
Z1v spinal cord neuroepithelium, ventral

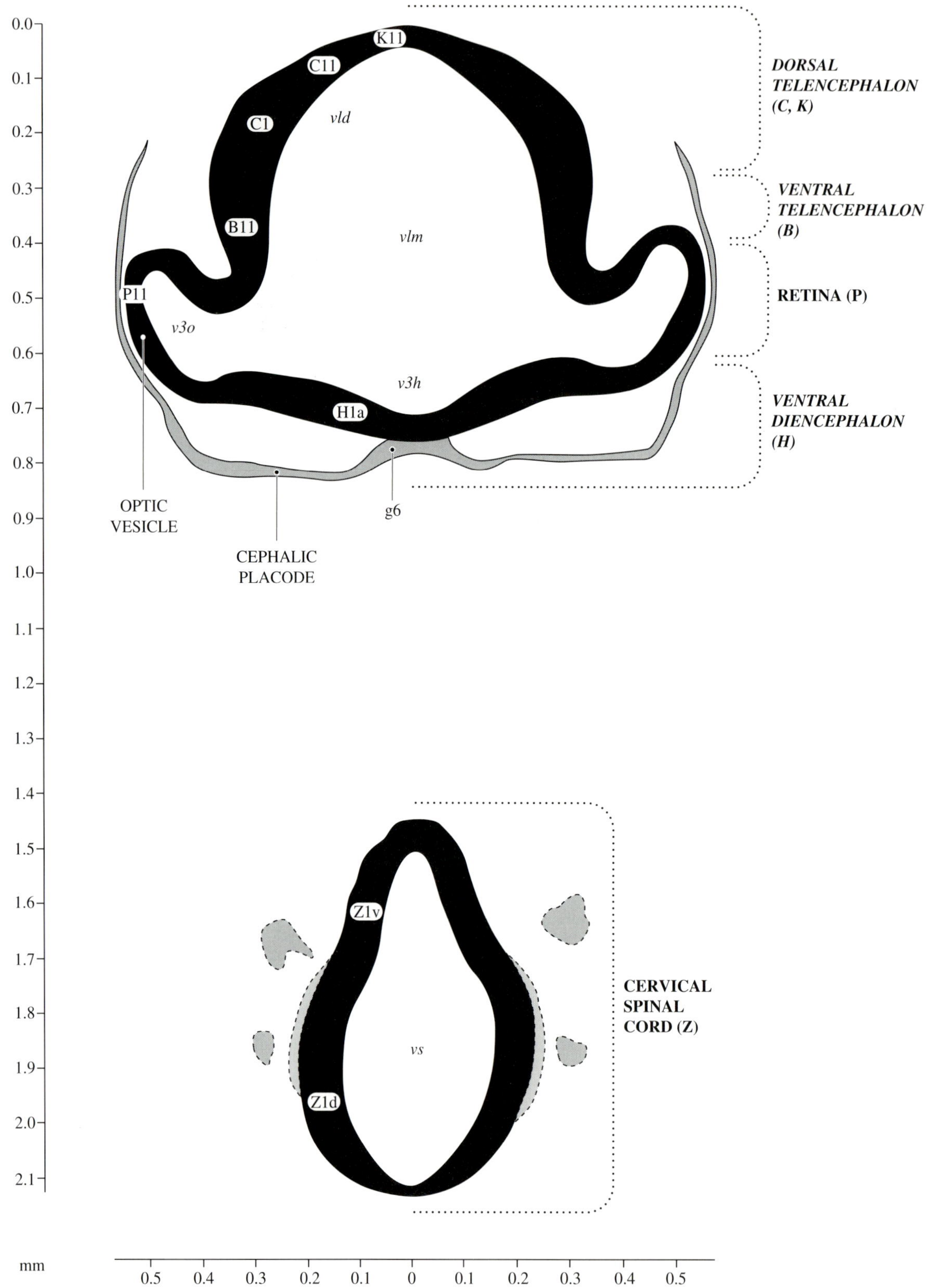

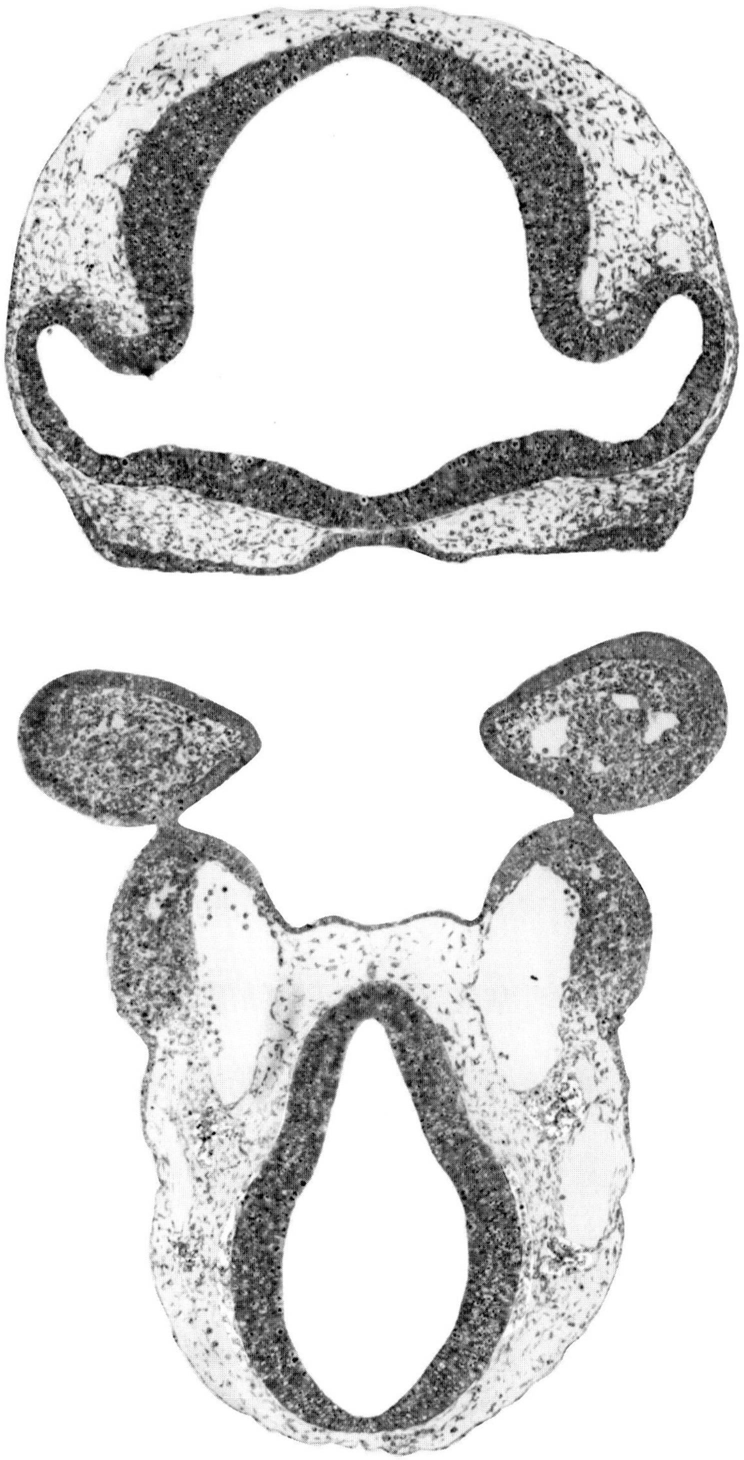

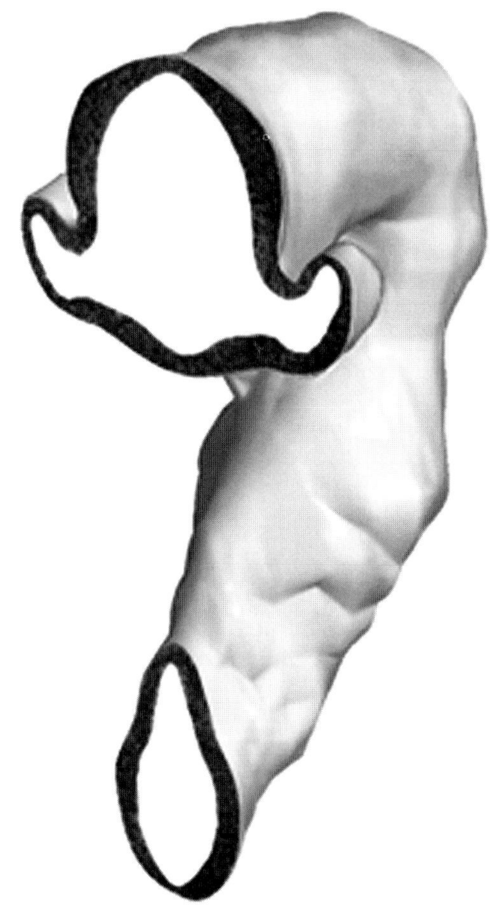

CORONAL FIGURE 6

GLANDS AND PARANEURAL STRUCTURES (g)

g6 pituitary gland (primordium)

HYPOTHALAMUS (H)

H1a anterior hypothalamic neuroepithelium
H1f Forelian neuroepithelium
H1v ventral hypothalamic neuroepithelium

PREOPTIC AREA AND OPTIC EPITHELIA (P)

P11 retinal neuroepithelium
P12 eye stalk epithelium

THALAMUS (T)

T1a anterior thalamic neuroepithelium
T1e epithalamic neuroepithelium

VENTRICLES (v)

vs spinal canal
v3h third ventricle, hypothalamic
v3o third ventricle, optic recess
v3t third ventricle, thalamic
v4p fourth ventricle, posterior

MEDULLA (Y)

Y1 medullary neuroepithelium

SPINAL CORD (Z)

Z1d spinal cord neuroepithelium, dorsal
Z8 dorsal root ganglion

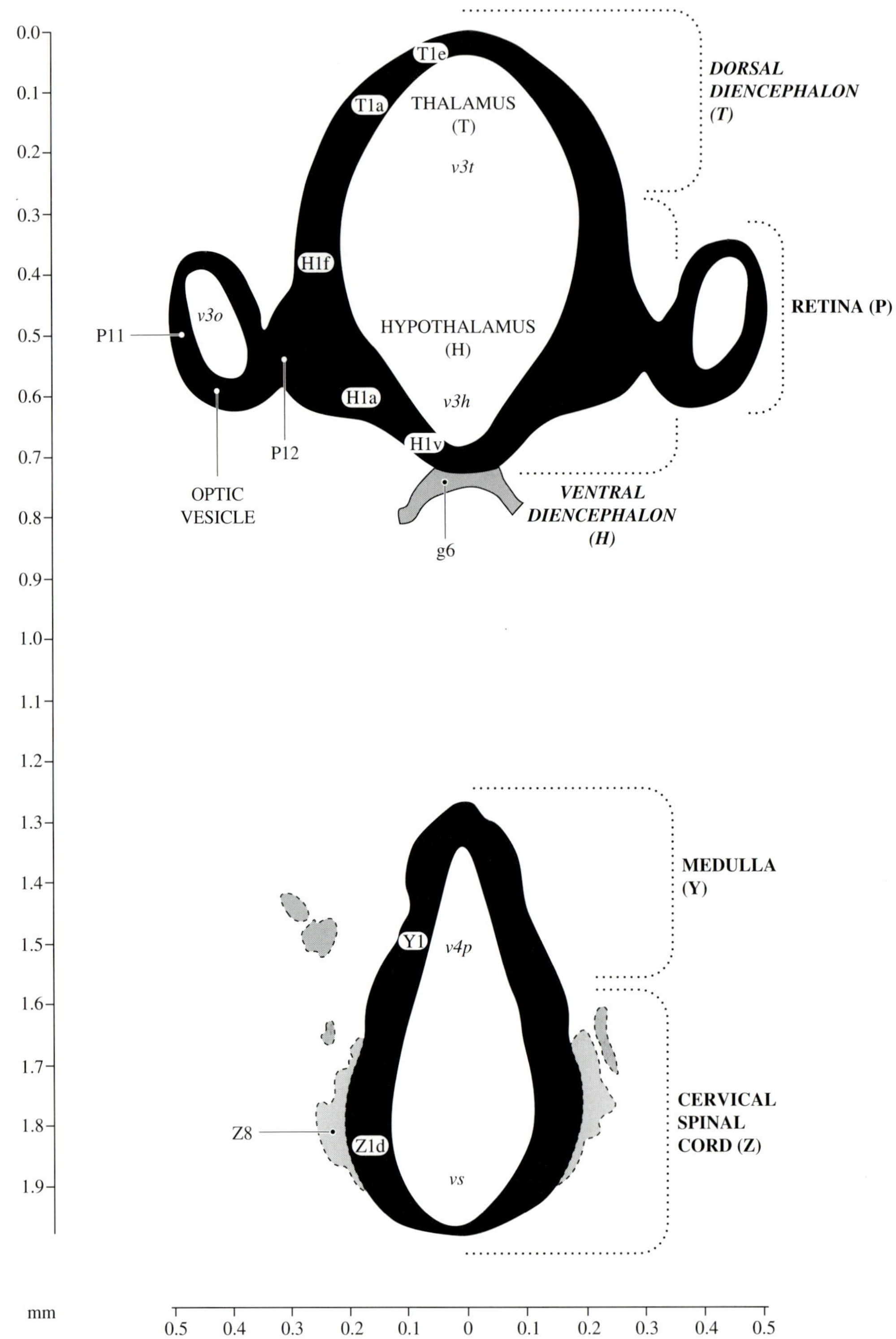

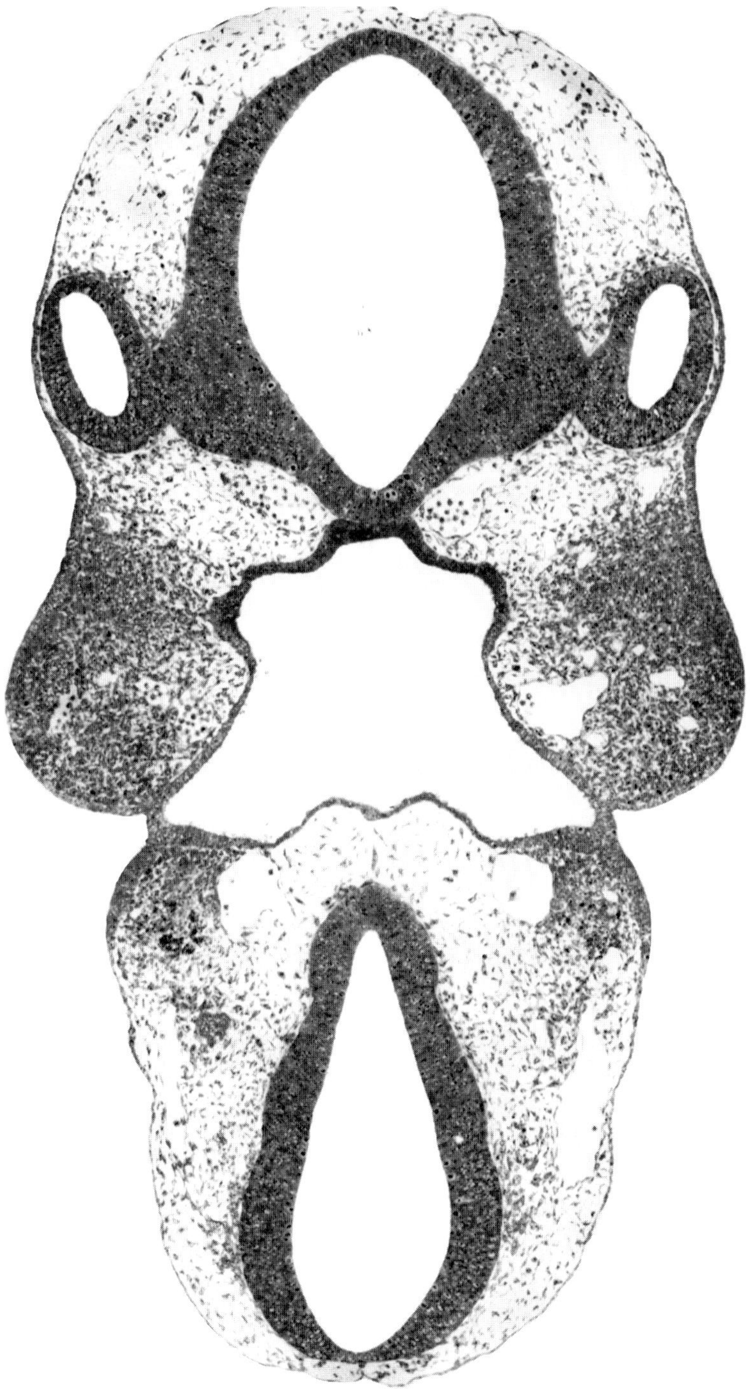

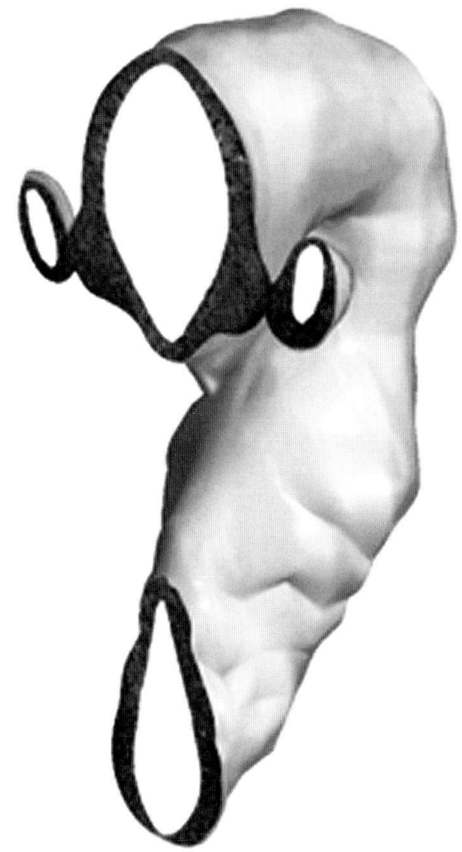

CORONAL FIGURE 7

GLANDS AND PARANEURAL STRUCTURES (g)

g6 pituitary gland (primordium)

HYPOTHALAMUS (H)

H1f Forelian neuroepithelium
H1i intermediate hypothalamic neuroepithelium
H1v ventral hypothalamic neuroepithelium

NERVES AND NUCLEI, CRANIAL (n)

n8g vestibular ganglion

THALAMUS (T)

T1e epithalamic neuroepithelium
T1i intermediate thalamic neuroepithelium

VENTRICLES (v)

vs spinal canal
v3h third ventricle, hypothalamic
v3t third ventricle, thalamic
v4p fourth ventricle, posterior

PONTINE AREA (X)

X1 pontine neuroepithelium

MEDULLA (Y)

Y1 medullary neuroepithelium

SPINAL CORD (Z)

Z1d spinal cord neuroepithelium, dorsal
Z8 dorsal root ganglion

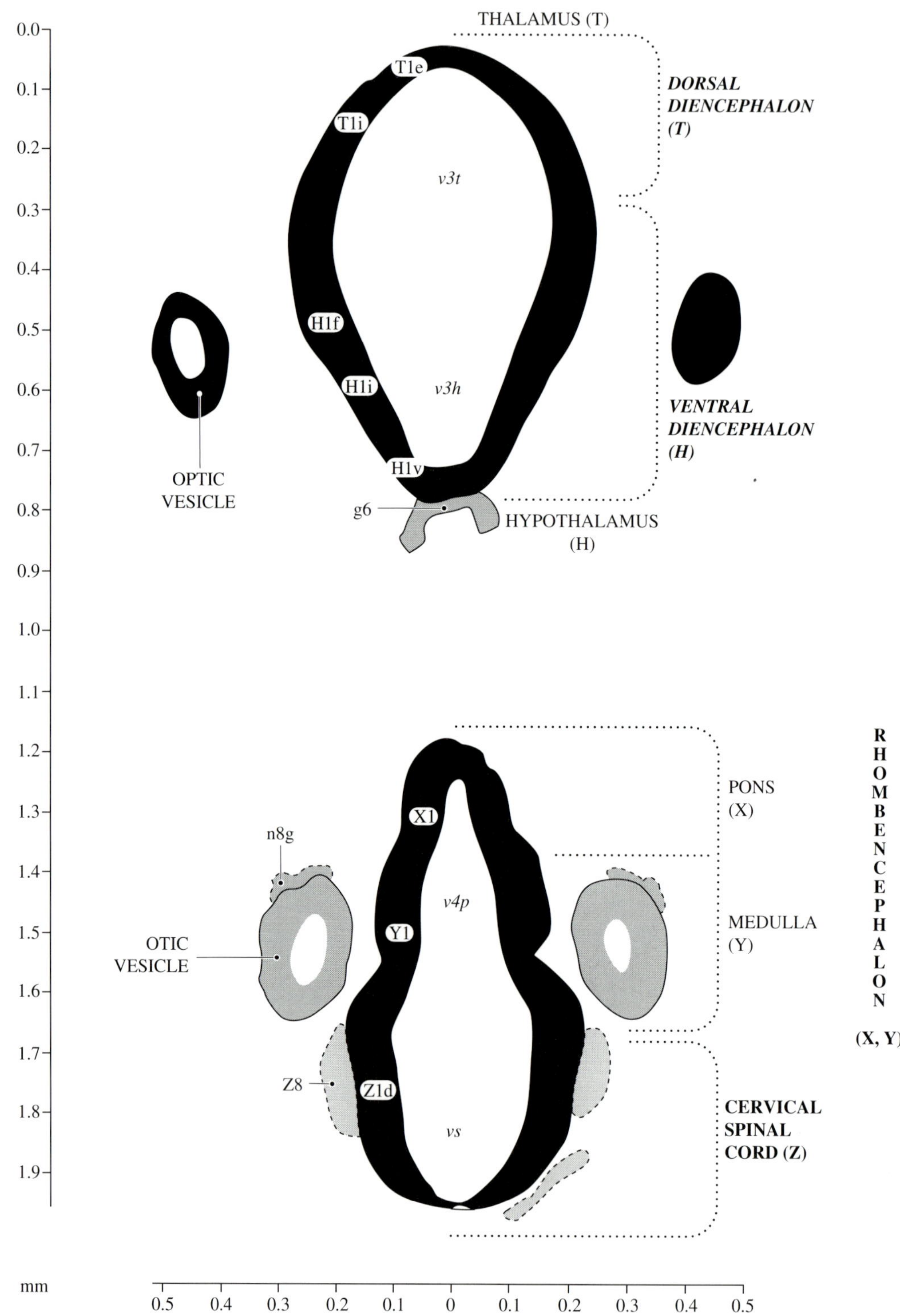

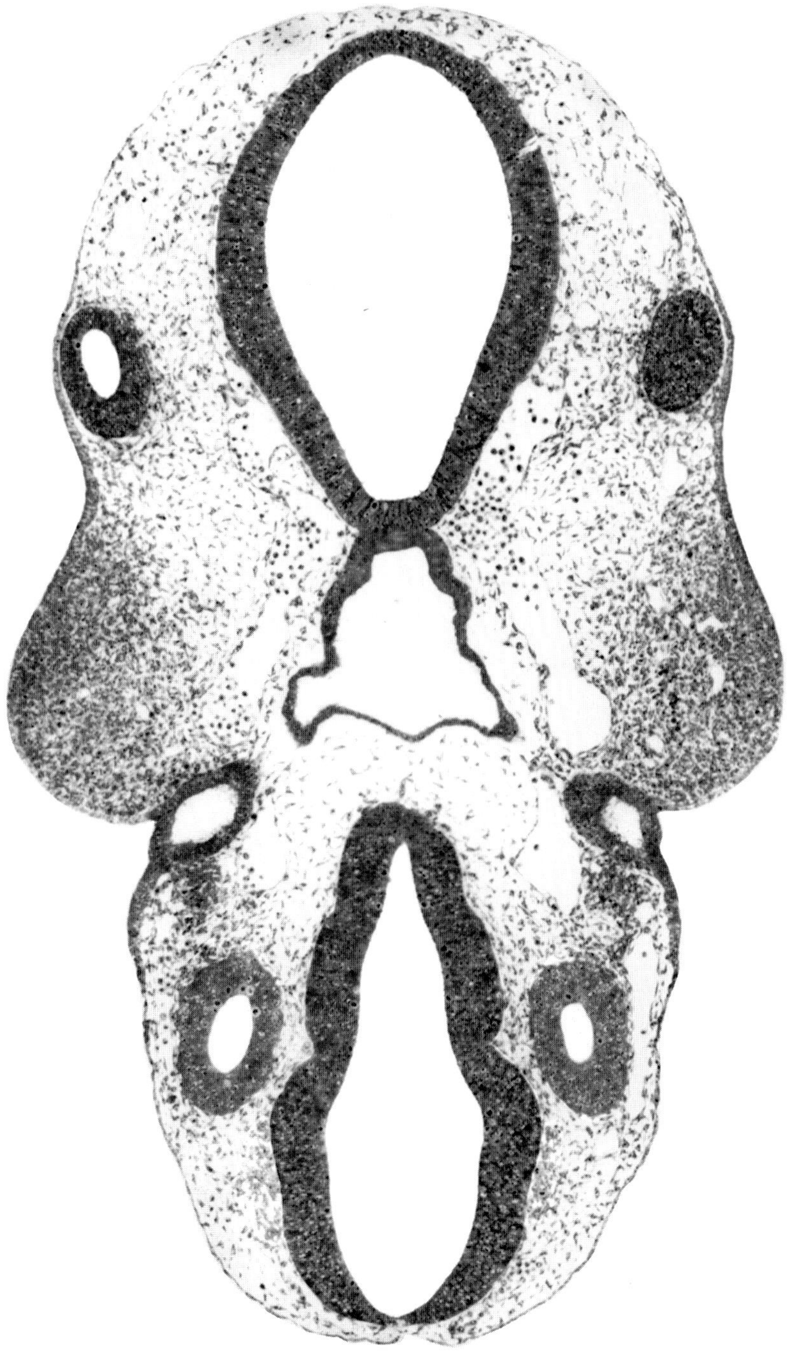

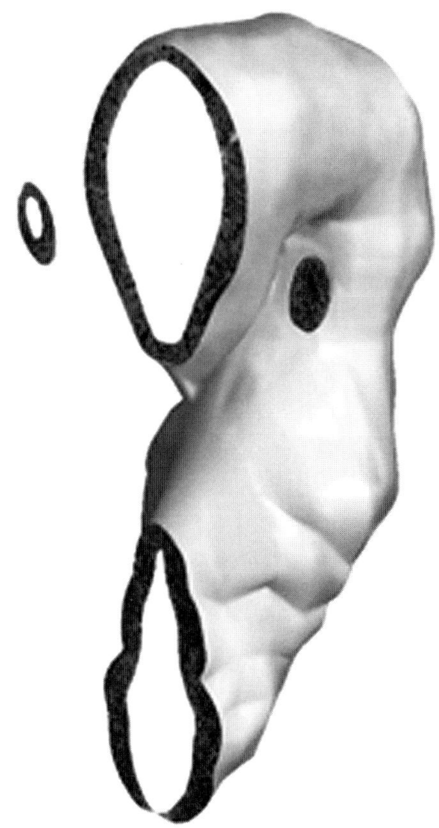

CORONAL FIGURE 8

HYPOTHALAMUS (H)

H1f Forelian neuroepithelium
H1i intermediate hypothalamic neuroepithelium
H1v ventral hypothalamic neuroepithelium

NERVES AND NUCLEI, CRANIAL (n)

n8g vestibular ganglion

THALAMUS (T)

T1e epithalamic neuroepithelium
T1i intermediate thalamic neuroepithelium

VENTRICLES (v)

vs spinal canal
v3h third ventricle, hypothalamic
v3t third ventricle, thalamic
v4p fourth ventricle, posterior

PONTINE AREA (X)

X1 pontine neuroepithelium

MEDULLA (Y)

Y1 medullary neuroepithelium

SPINAL CORD (Z)

Z1d spinal cord neuroepithelium, dorsal

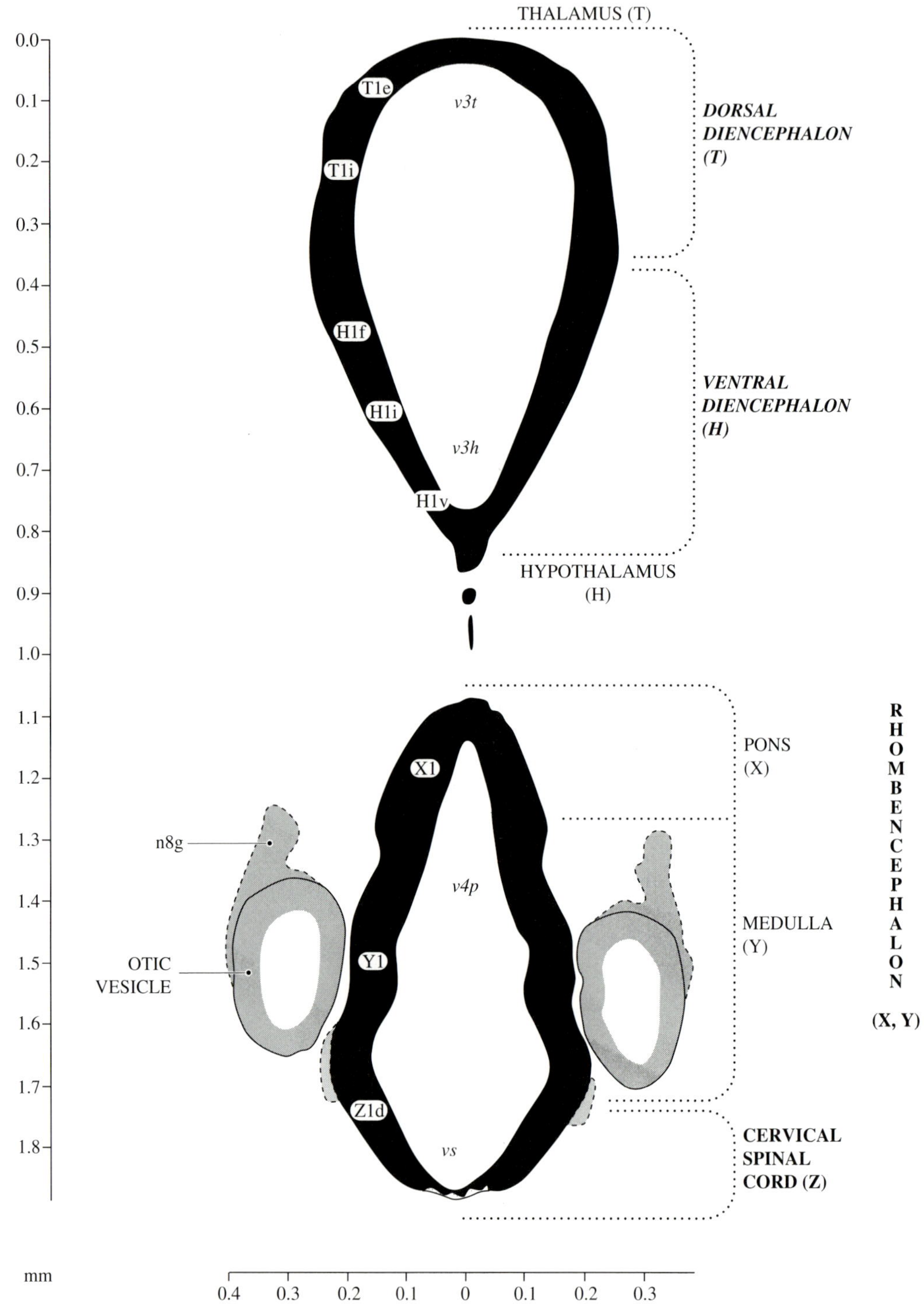

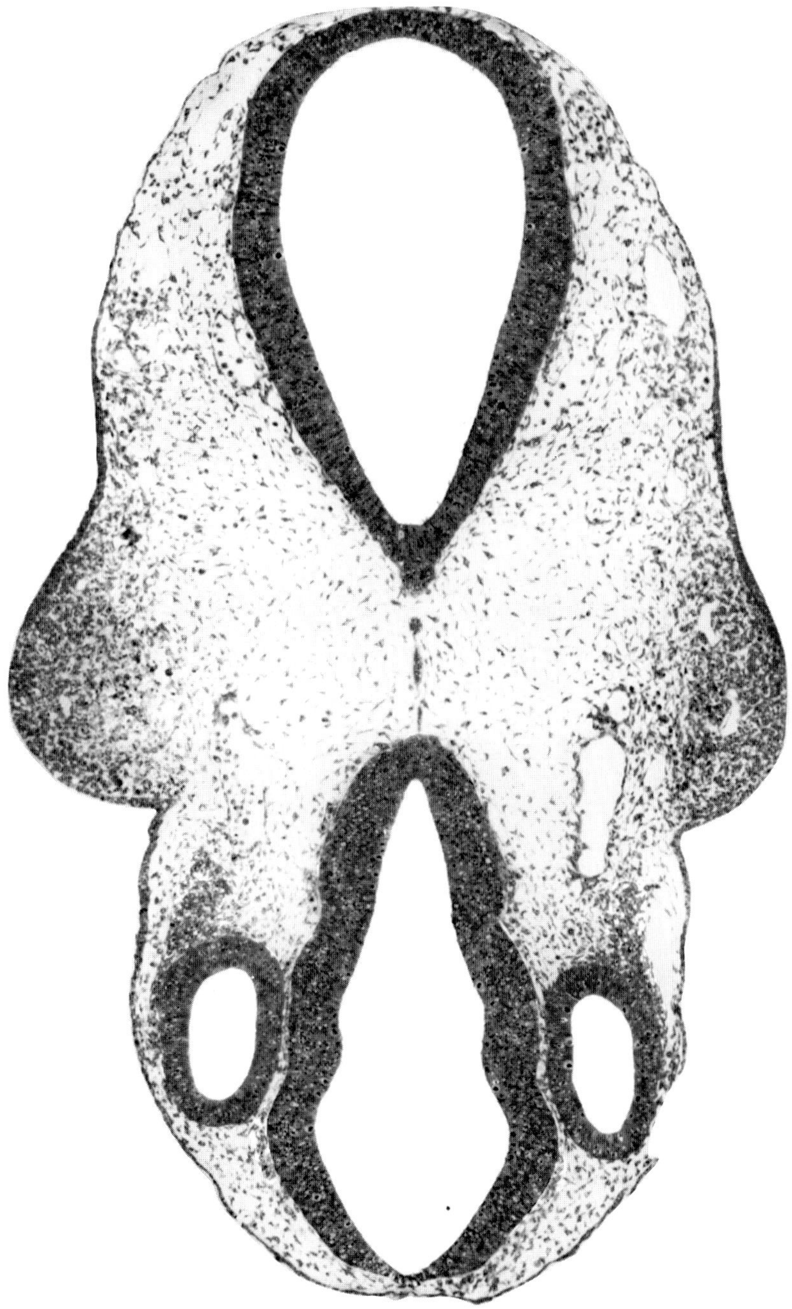

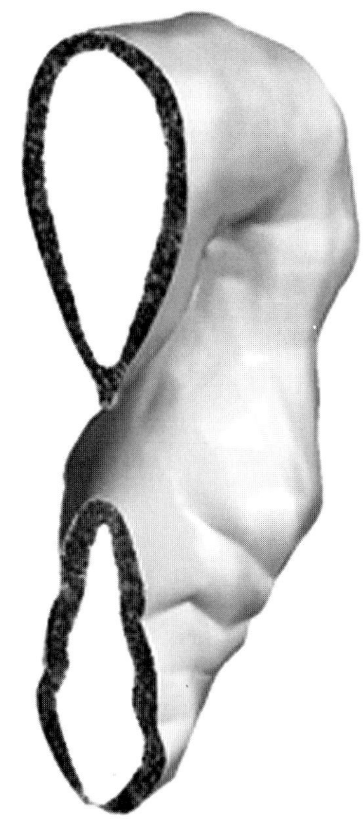

CORONAL FIGURE 9

GLANDS AND PARANEURAL STRUCTURES (g)

g10 velum medullare

HYPOTHALAMUS (H)

H1i intermediate hypothalamic neuroepithelium
H1v ventral hypothalamic neuroepithelium

NERVES AND NUCLEI, CRANIAL (n)

n5g trigeminal ganglion
n8g vestibular ganglion

THALAMUS (T)

T1p posterior thalamic neuroepithelium

VENTRICLES (v)

v3h third ventricle, hypothalamic
v3t third ventricle, thalamic
v4a fourth ventricle, anterior
v4p fourth ventricle, posterior

PONTINE AREA (X)

X1 pontine neuroepithelium

MEDULLA (Y)

Y1 medullary neuroepithelium

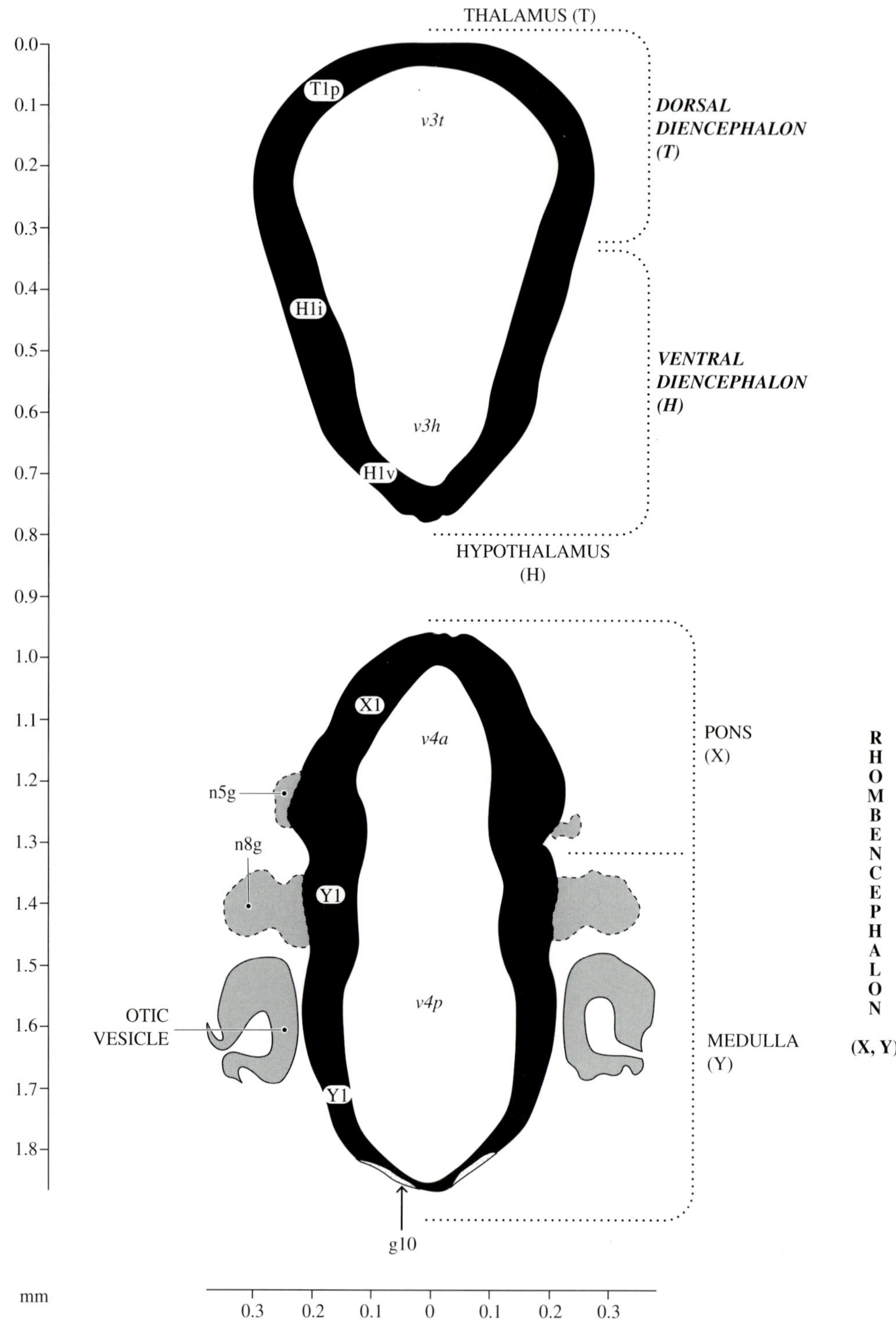

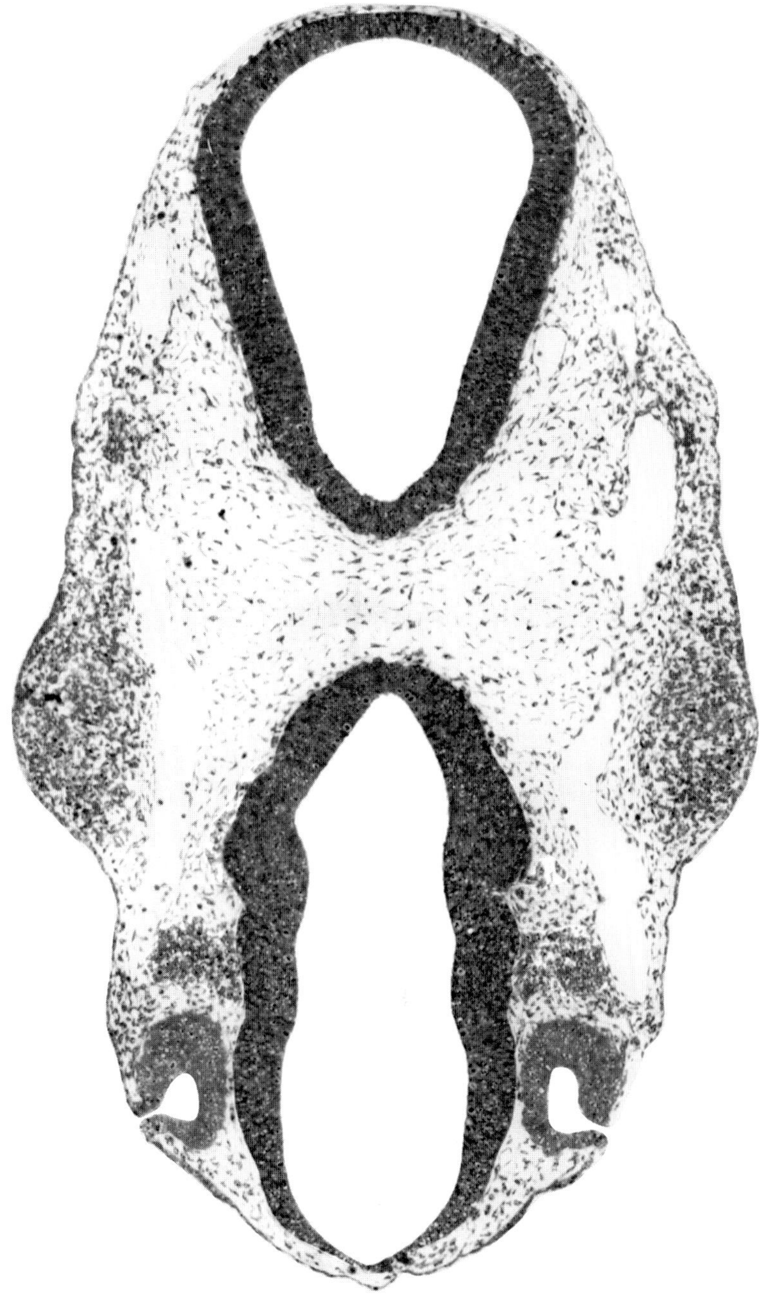

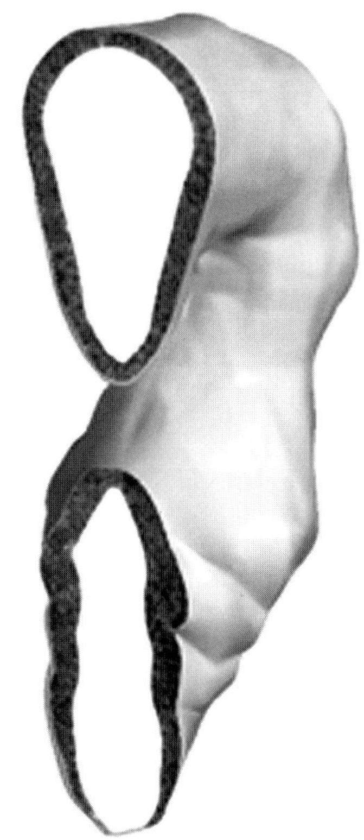

CORONAL FIGURE 10

GLANDS AND PARANEURAL STRUCTURES (g)

g10 velum medullare

HYPOTHALAMUS (H)

H1m mammillary neuroepithelium
H1p posterior hypothalamic neuroepithelium

MIDBRAIN (M)

M21 pretectal neuroepithelium

NERVES AND NUCLEI, CRANIAL (n)

n5g trigeminal ganglion

THALAMUS (T)

T1p posterior thalamic neuroepithelium

VENTRICLES (v)

vapr aqueduct, pretectal
v3m third ventricle, mammillary recess
v3t third ventricle, thalamic
v4a fourth ventricle, anterior
v4p fourth ventricle, posterior

PONTINE AREA (X)

X1 pontine neuroepithelium

MEDULLA (Y)

Y1 medullary neuroepithelium

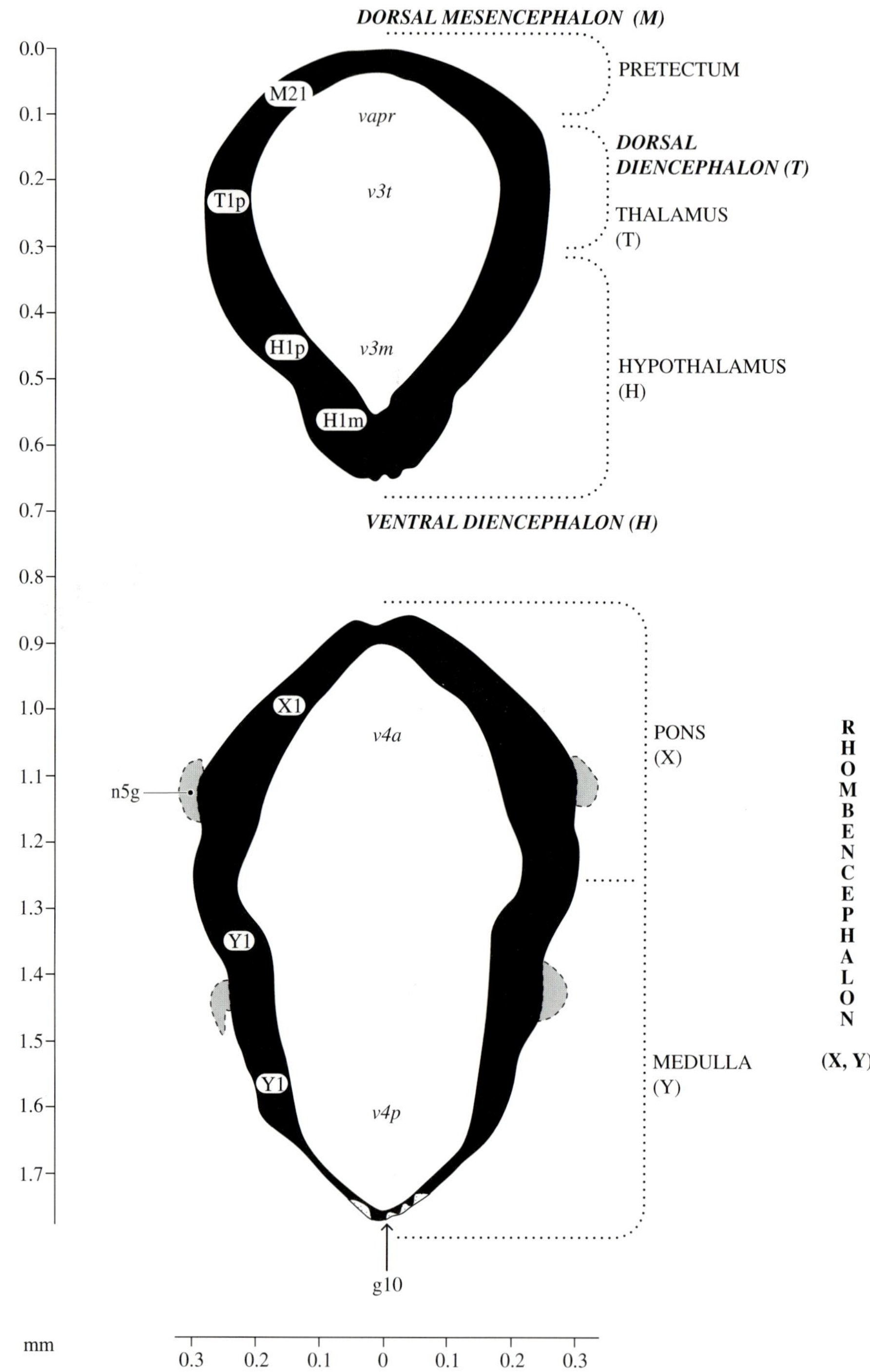

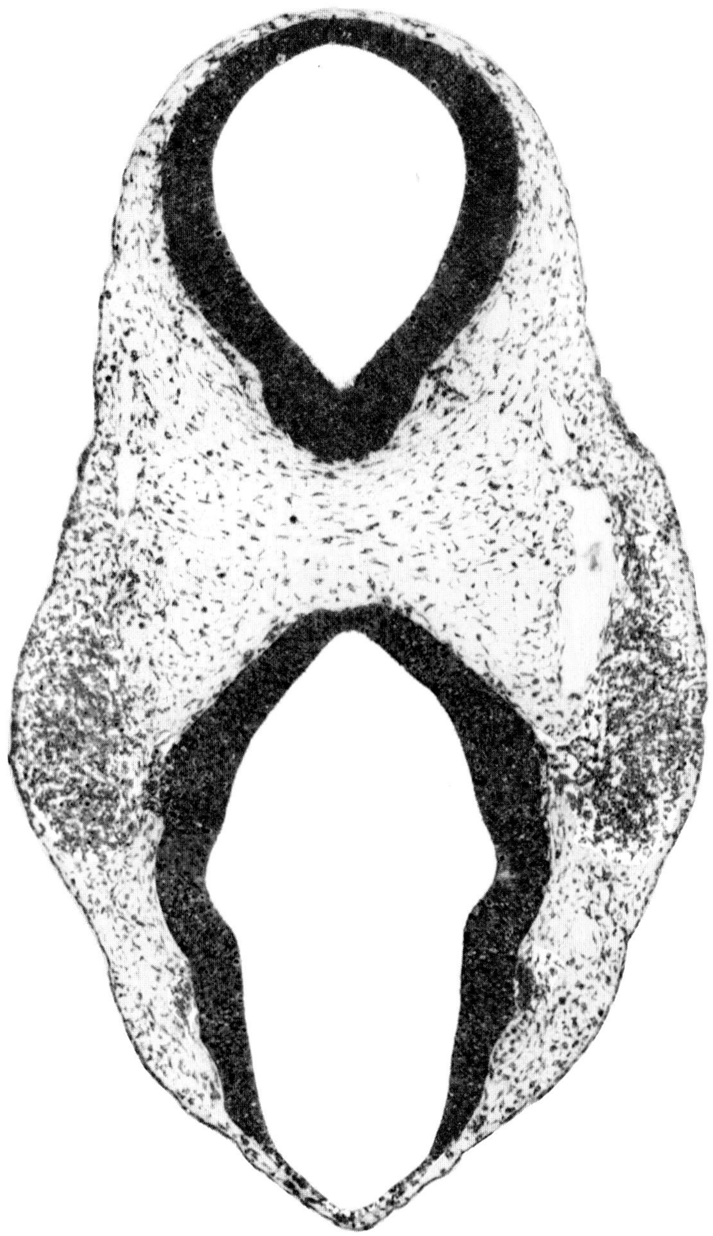

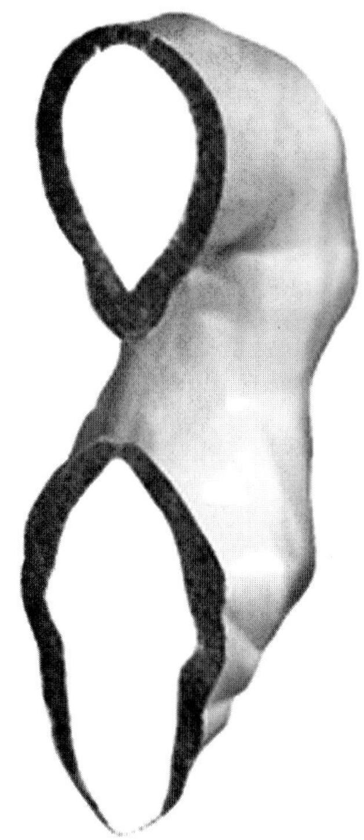

CORONAL FIGURE 11

GLANDS AND PARANEURAL STRUCTURES (g)

g10 velum medullare

MIDBRAIN (M)

M11 tegmental neuroepithelium
M21 pretectal neuroepithelium

NERVES AND NUCLEI, CRANIAL (n)

n5g trigeminal ganglion

VENTRICLES (v)

vapr aqueduct, pretectal
vat aqueduct, tegmental
v4a fourth ventricle, anterior
v4i fourth ventricle, isthmal
v4p fourth ventricle, posterior

PONTINE AREA (X)

X1 pontine neuroepithelium
X31 isthmal neuroepithelium

MEDULLA (Y)

Y1 medullary neuroepithelium

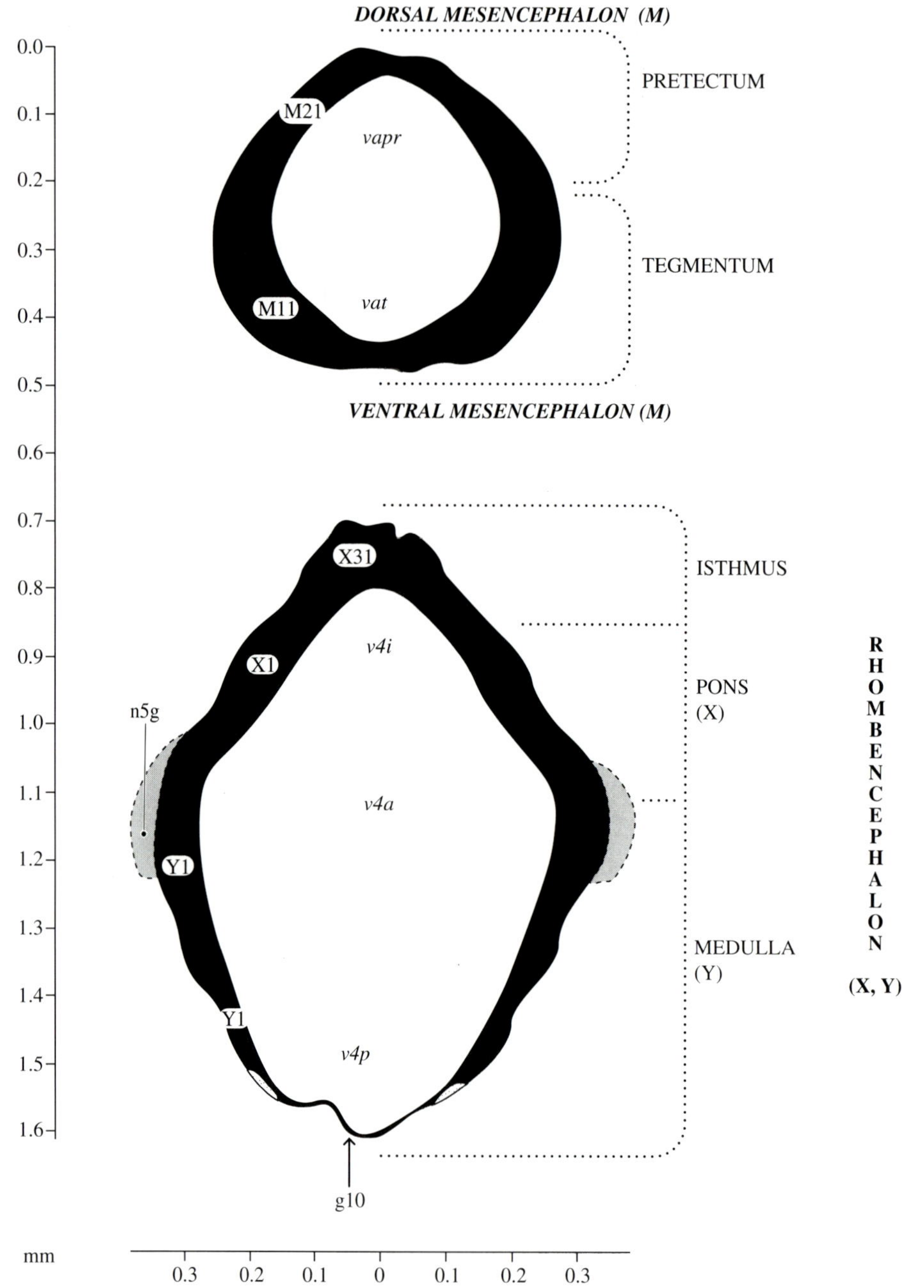

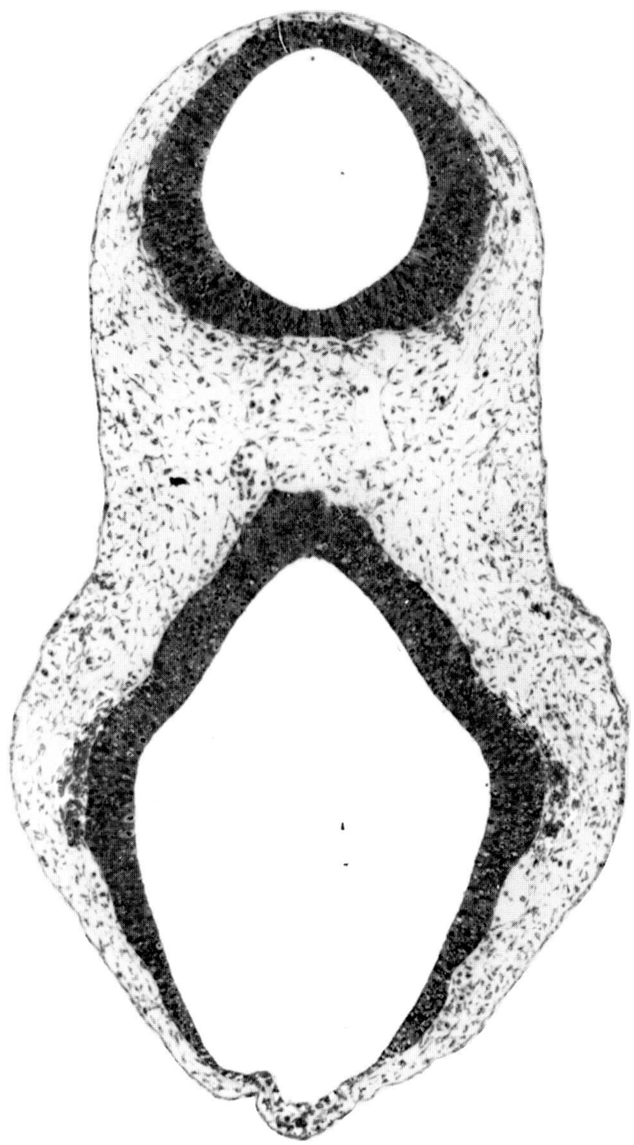

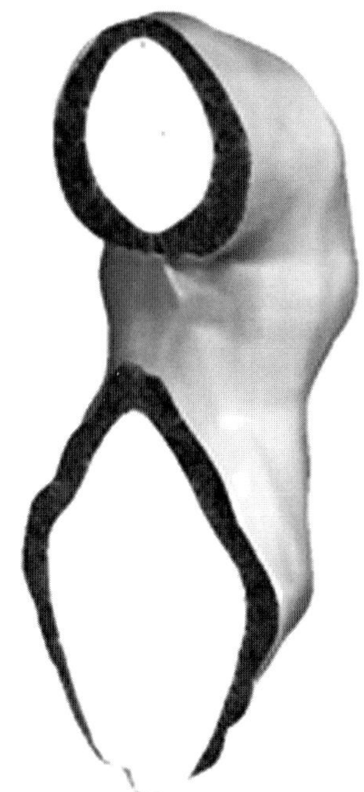

GLANDS AND PARANEURAL STRUCTURES (g)

g10 velum medullare

CEREBELLUM (L)

L1 cerebellar neuroepithelium

MIDBRAIN (M)

M11 tegmental neuroepithelium
M21 pretectal neuroepithelium

VENTRICLES (v)

vapr aqueduct, pretectal
vat aqueduct, tegmental
v4a fourth ventricle, anterior
v4i fourth ventricle, isthmal
v4l fourth ventricle, lateral recess
v4p fourth ventricle, posterior

PONTINE AREA (X)

X1 pontine neuroepithelium
X31 isthmal neuroepithelium

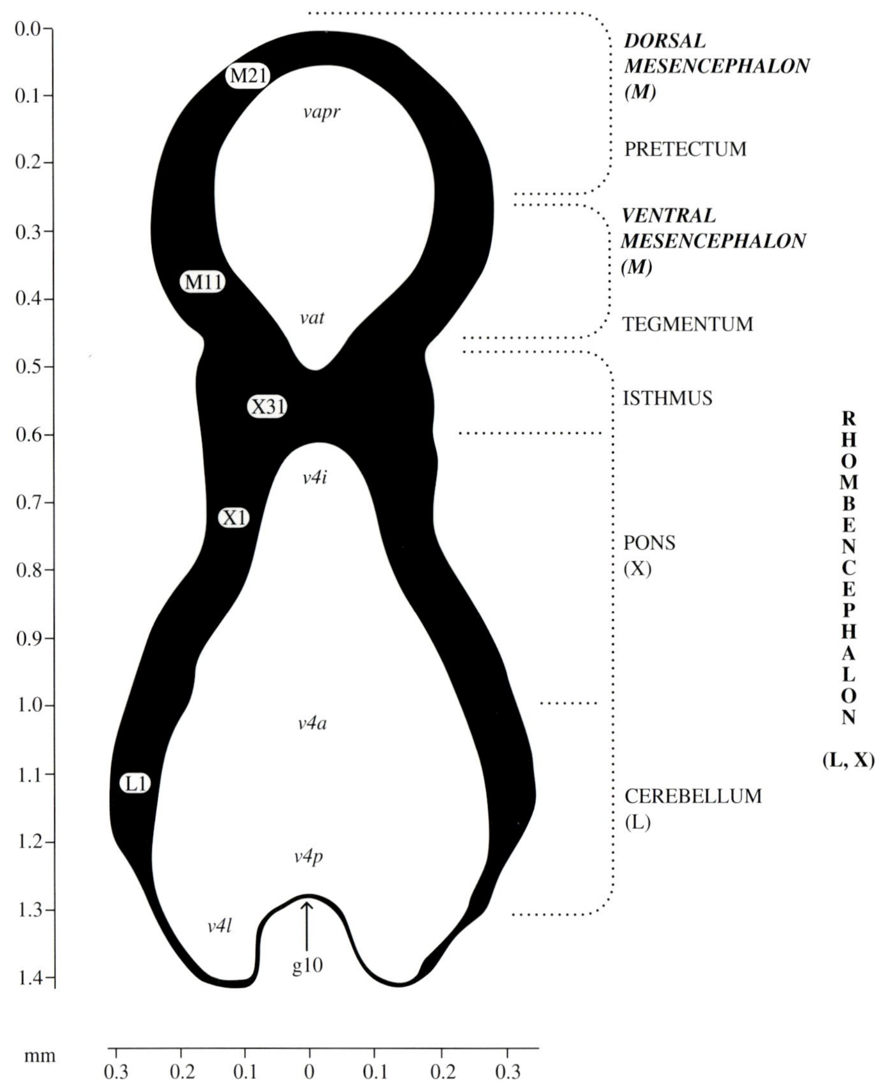

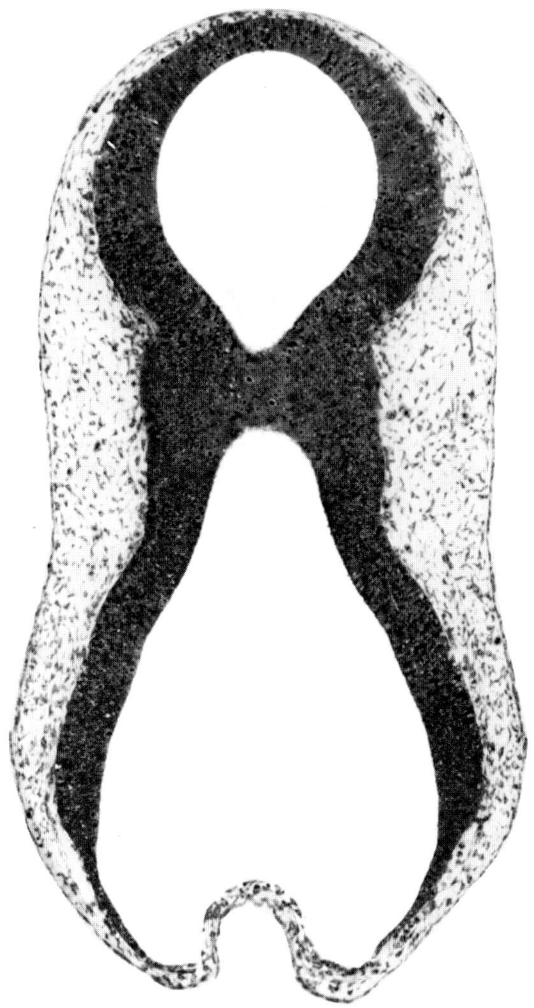

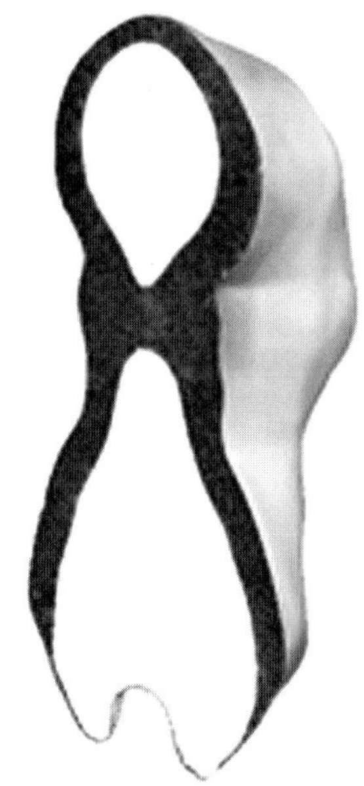

CORONAL FIGURE 13

GLANDS AND PARANEURAL STRUCTURES (g)

g10 velum medullare

CEREBELLUM (L)

L1 cerebellar neuroepithelium

MIDBRAIN (M)

M11 tegmental neuroepithelium
M21 pretectal neuroepithelium

VENTRICLES (v)

vapr aqueduct, pretectal
vat aqueduct, tegmental
v4a fourth ventricle, anterior
v4i fourth ventricle, isthmal
v4l fourth ventricle, lateral recess
v4p fourth ventricle, posterior

PONTINE AREA (X)

X1 pontine neuroepithelium
X31 isthmal neuroepithelium

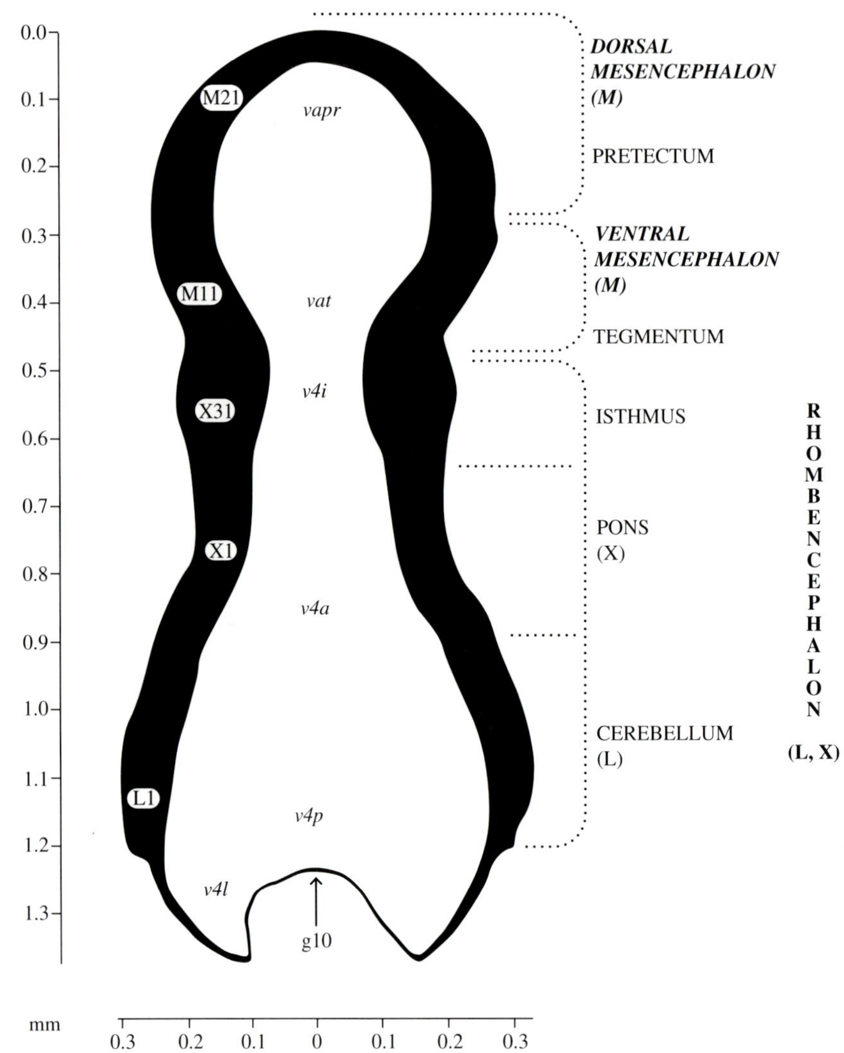

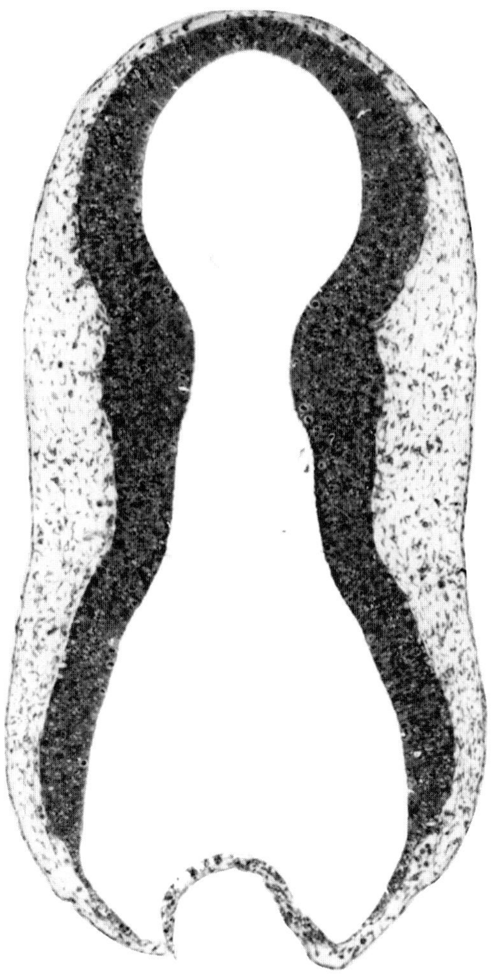

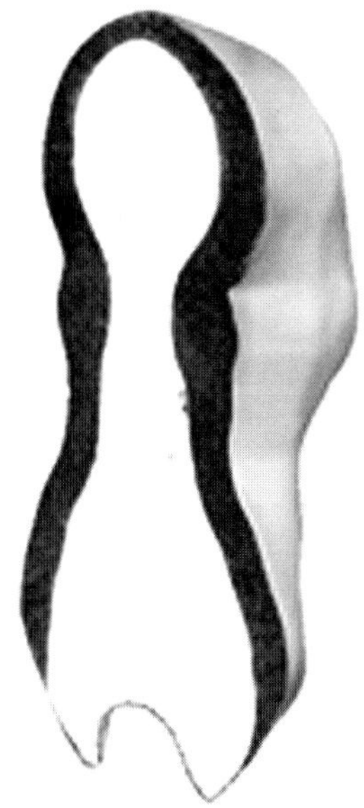

CORONAL FIGURE 14

GLANDS AND PARANEURAL STRUCTURES (g)

g10 velum medullare

CEREBELLUM (L)

L1 cerebellar neuroepithelium

MIDBRAIN (M)

M1 superior tectal neuroepithelium
M11 tegmental neuroepithelium

VENTRICLES (v)

vas aqueduct, superior tectal
vat aqueduct, tegmental
v4a fourth ventricle, anterior
v4i fourth ventricle, isthmal

PONTINE AREA (X)

X31 isthmal neuroepithelium

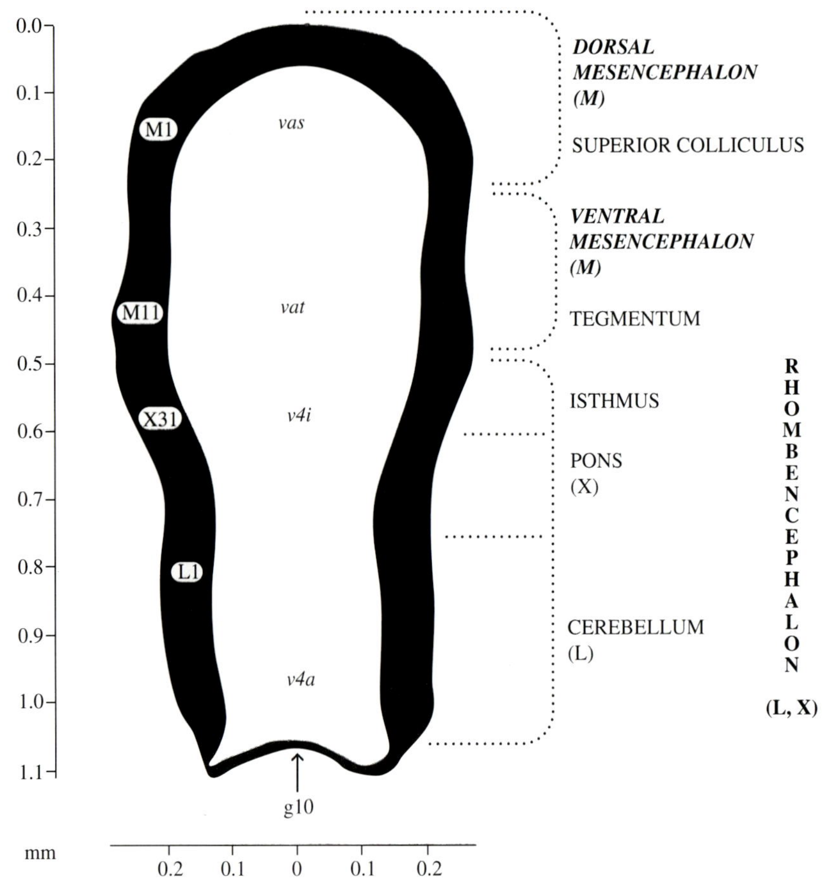

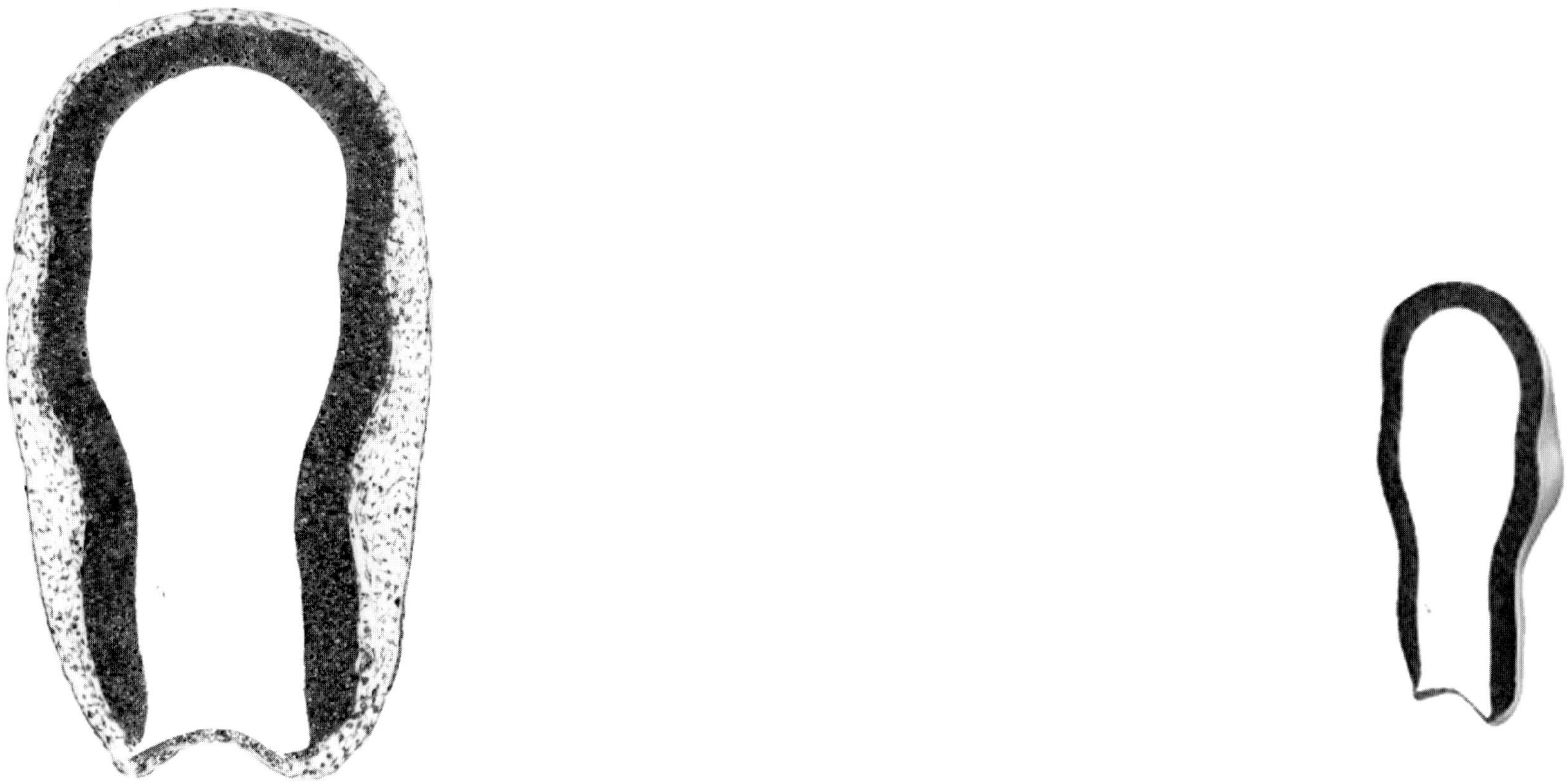

CORONAL FIGURE 15

MIDBRAIN (M)

M1	superior tectal neuroepithelium
M11	tegmental neuroepithelium
M31	inferior tectal neuroepithelium

VENTRICLES (v)

vai	aqueduct, inferior tectal
vas	aqueduct, superior tectal
vat	aqueduct, tegmental

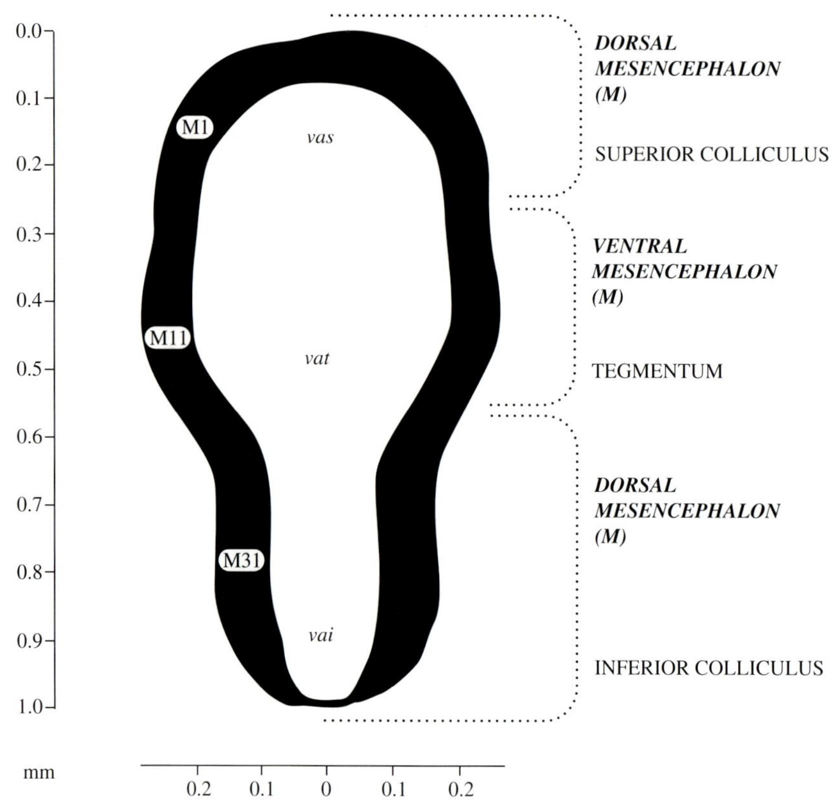

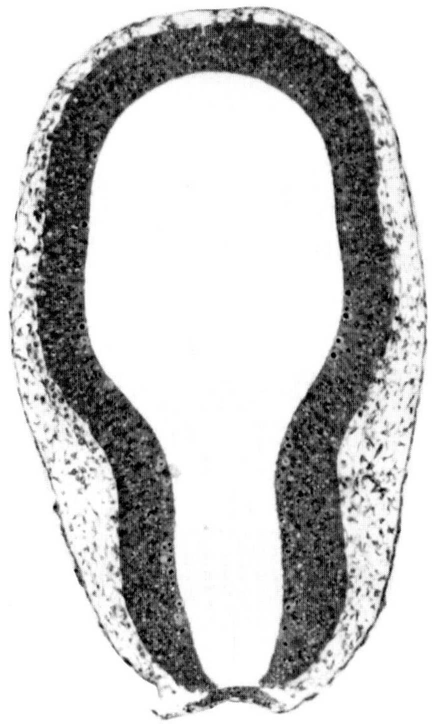

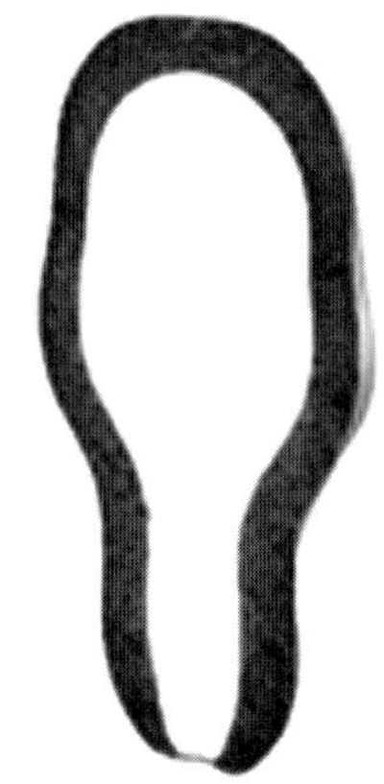

CORONAL FIGURE 16

MIDBRAIN (M)

M1	superior tectal neuroepithelium
M11	tegmental neuroepithelium
M31	inferior tectal neuroepithelium

VENTRICLES (v)

vai	aqueduct, inferior tectal
vas	aqueduct, superior tectal
vat	aqueduct, tegmental

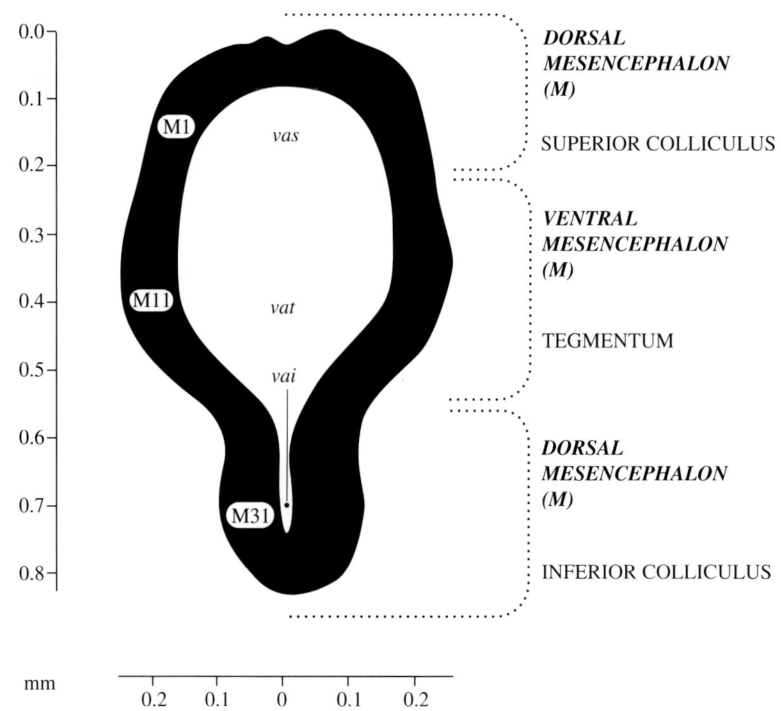

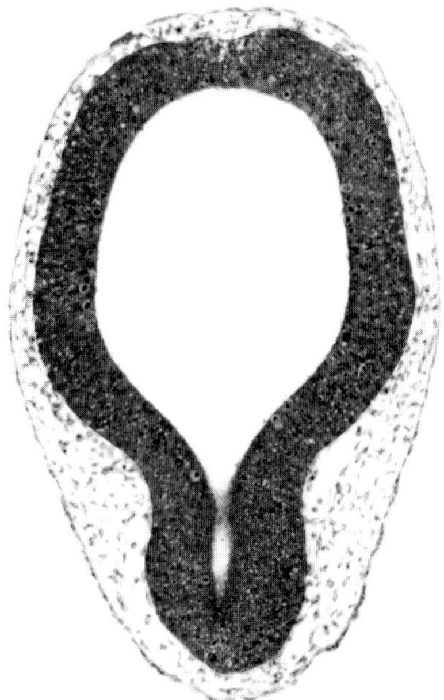

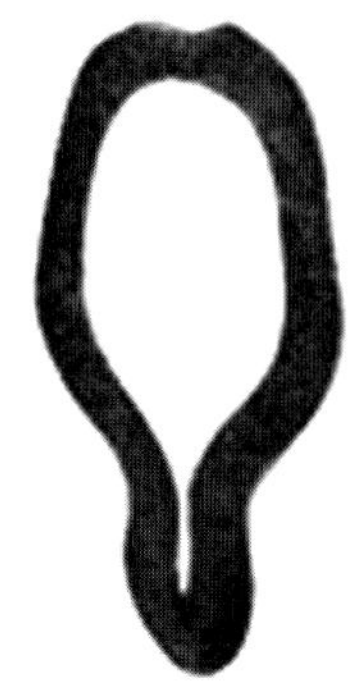

THE E13 BRAIN

COMPUTERIZED 3-D RECONSTRUCTION: 3 VIEWS

RIGHT, POSTERIOR VIEW

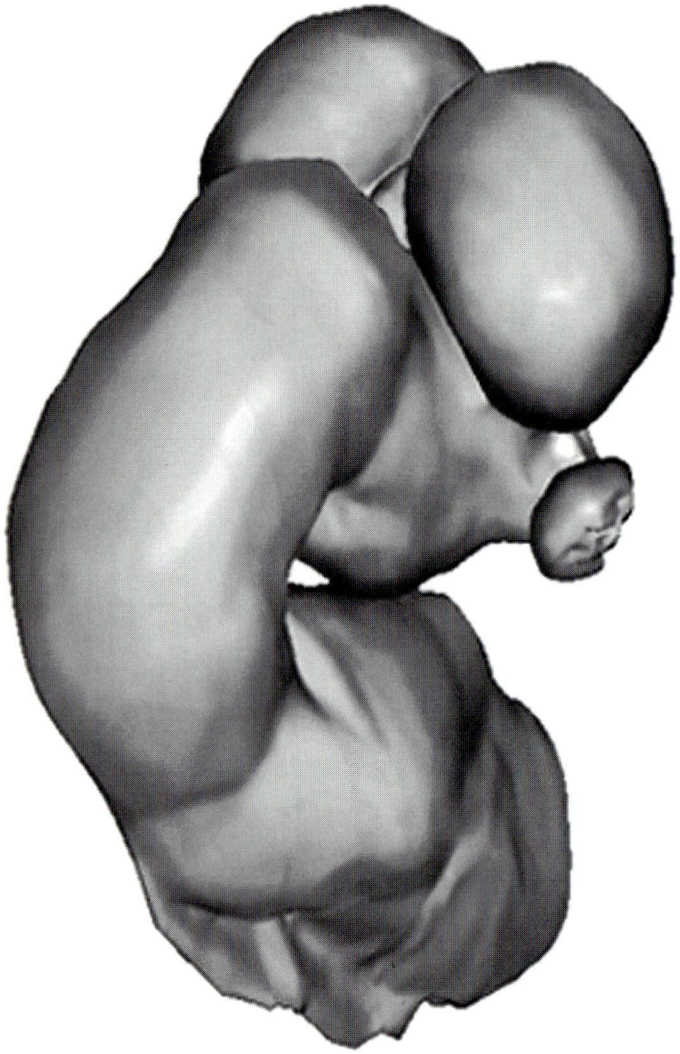

LEFT, ANTERIOR VIEW

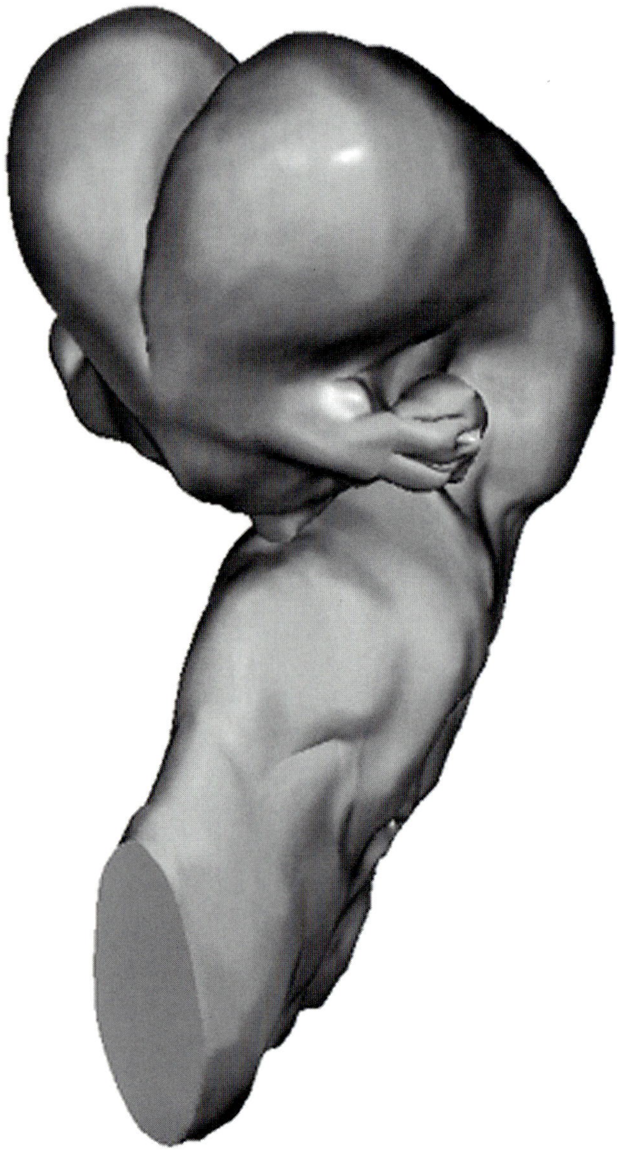

FRONTAL VIEW

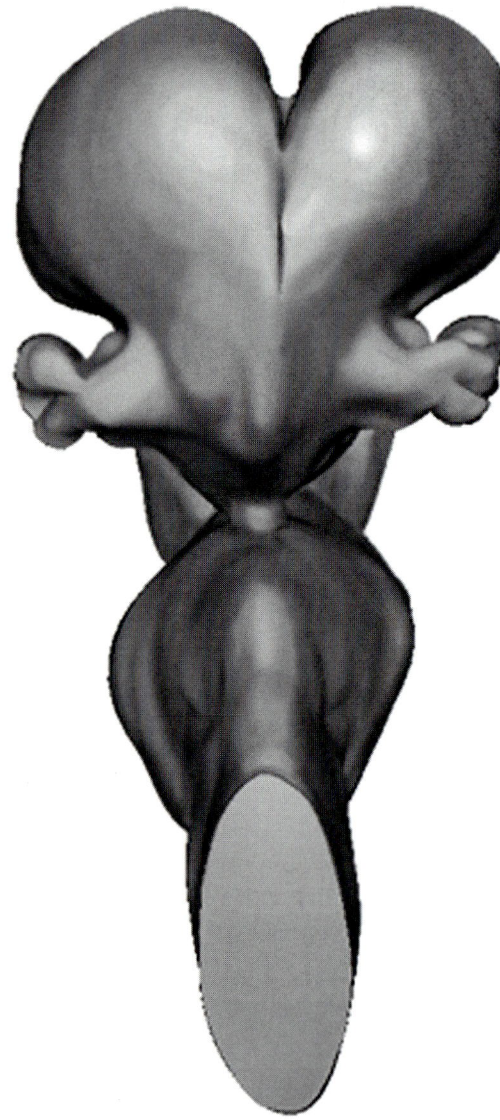

SAGITTAL

FIGURES and PLATES 1-6

SAGITTAL FIGURE 1

CEREBRAL CORTEX (C)

C11 cingulate cortex neuroepithelium

GLANDS AND PARANEURAL STRUCTURES (g)

g6 pituitary gland (primordium)
g10 velum medullare

HYPOTHALAMUS (H)

H1a anterior hypothalamic neuroepithelium
H1m mammillary neuroepithelium
H1v ventral hypothalamic neuroepithelium

HIPPOCAMPAL FORMATION (K)

K11 subicular neuroepithelium

CEREBELLUM (L)

L1 cerebellar neuroepithelium

MIDBRAIN (M)

M1 superior tectal neuroepithelium
M11a anterior tegmental neuroepithelium
M11p posterior tegmental neuroepithelium
M21 pretectal neuroepithelium
M31 inferior tectal neuroepithelium

PREOPTIC AREA AND OPTIC EPITHELIA (P)

P1 preoptic neuroepithelium

RHINENCEPHALON (R)

R1 rhinencephalic neuroepithelium

SEPTAL AREA (S)

S1 septal neuroepithelium

THALAMUS (T)

T1 thalamic neuroepithelium

VENTRICLES (v)

vai aqueduct, inferior tectal
vapr aqueduct, pretectal
vas aqueduct, superior tectal
vat aqueduct, tegmental
vla lateral ventricle, anterior horn
vld lateral ventricle, dorsal
vlm foramen of Monro
vs spinal canal
v3h third ventricle, hypothalamic
v3i third ventricle, infundibular recess
v3pr third ventricle, preoptic recess
v3t third ventricle, thalamic

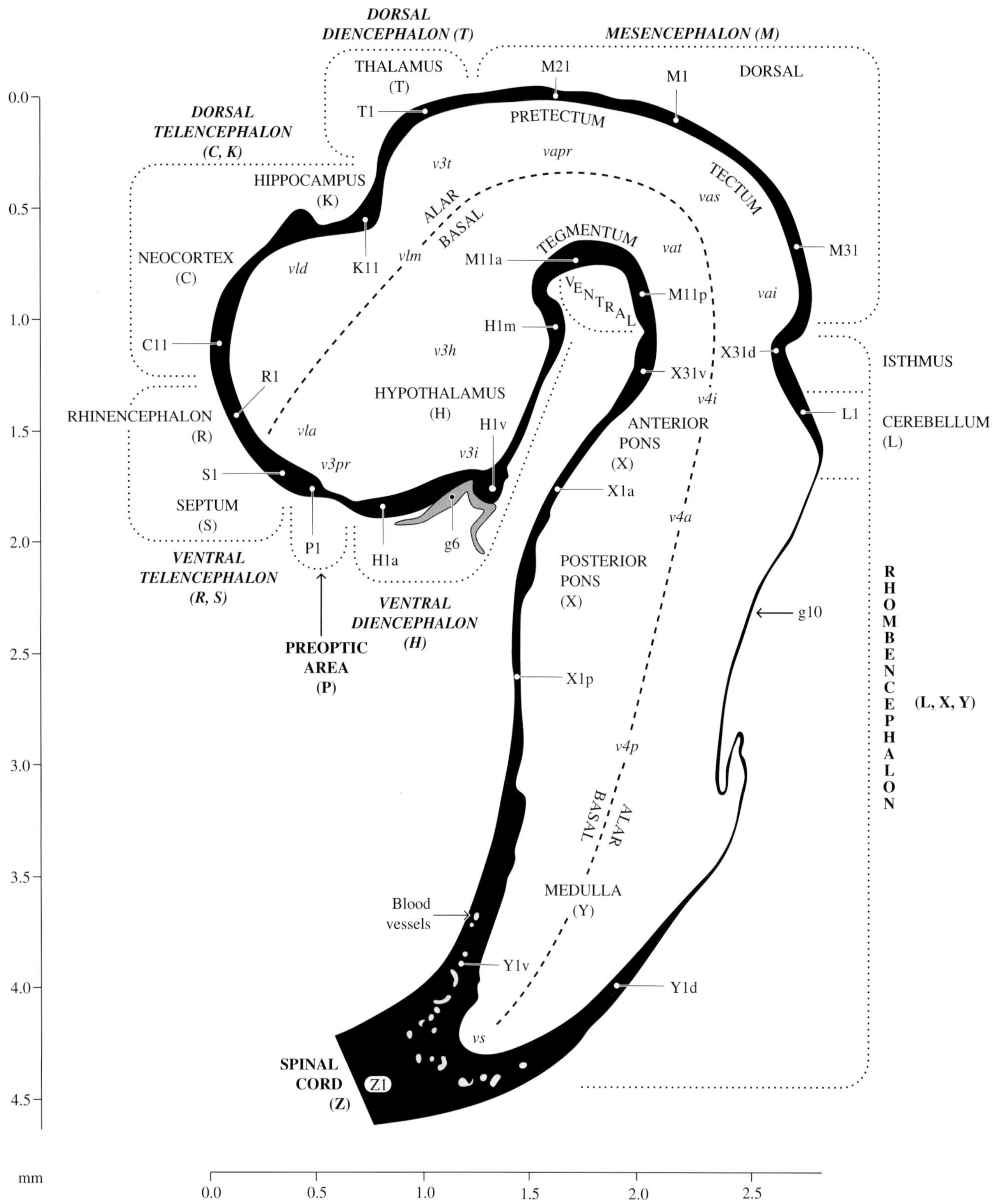

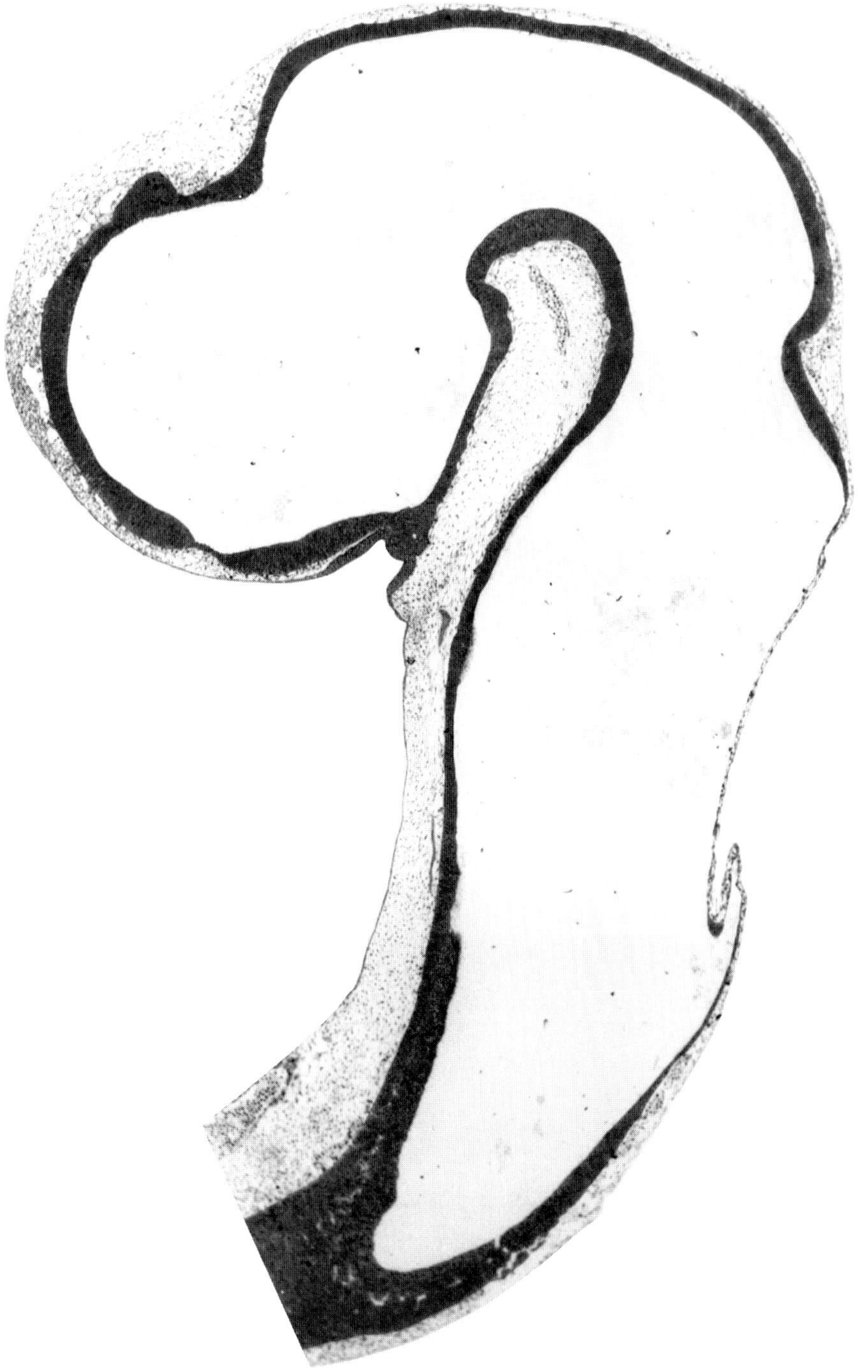

PONTINE AREA (X)

X1a anterior pontine neuroepithelium
X1p posterior pontine neuroepithelium
X31d isthmal neuroepithelium, dorsal
X31v isthmal neuroepithelium, ventral

MEDULLA (Y)

Y1d medullary neuroepithelium, dorsal
Y1v medullary neuroepithelium (rhombomeres), ventral

SPINAL CORD (Z)

Z1 spinal cord neuroepithelium

SAGITTAL FIGURE 2

CEREBRAL CORTEX (C)

C11 cingulate cortex neuroepithelium

GLANDS AND PARANEURAL STRUCTURES (g)

g6 pituitary gland (primordium)
g10 velum medullare

HYPOTHALAMUS (H)

H1a anterior hypothalamic neuroepithelium
H1i intermediate hypothalamic neuroepithelium
H1m mammillary neuroepithelium

HIPPOCAMPAL FORMATION (K)

K11 subicular neuroepithelium

CEREBELLUM (L)

L1 cerebellar neuroepithelium

MIDBRAIN (M)

M1 superior tectal neuroepithelium
M11a anterior tegmental neuroepithelium
M11p posterior tegmental neuroepithelium
M21 pretectal neuroepithelium
M31 inferior tectal neuroepithelium

PREOPTIC AREA AND OPTIC EPITHELIA (P)

P1 preoptic neuroepithelium

RHINENCEPHALON (R)

R1 rhinencephalic neuroepithelium

SEPTAL AREA (S)

S1 septal neuroepithelium

THALAMUS (T)

T1a anterior thalamic neuroepithelium
T1p posterior thalamic neuroepithelium

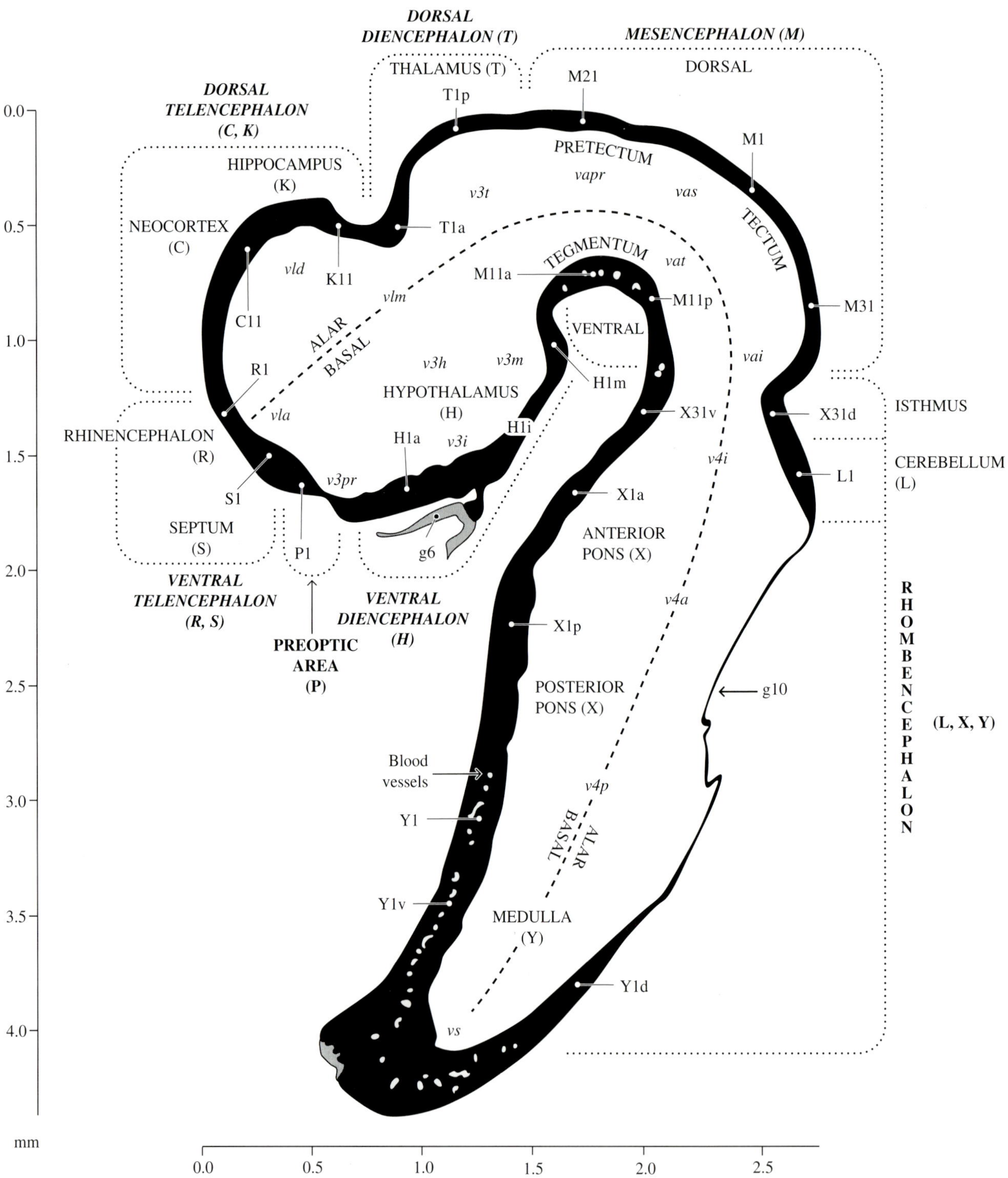

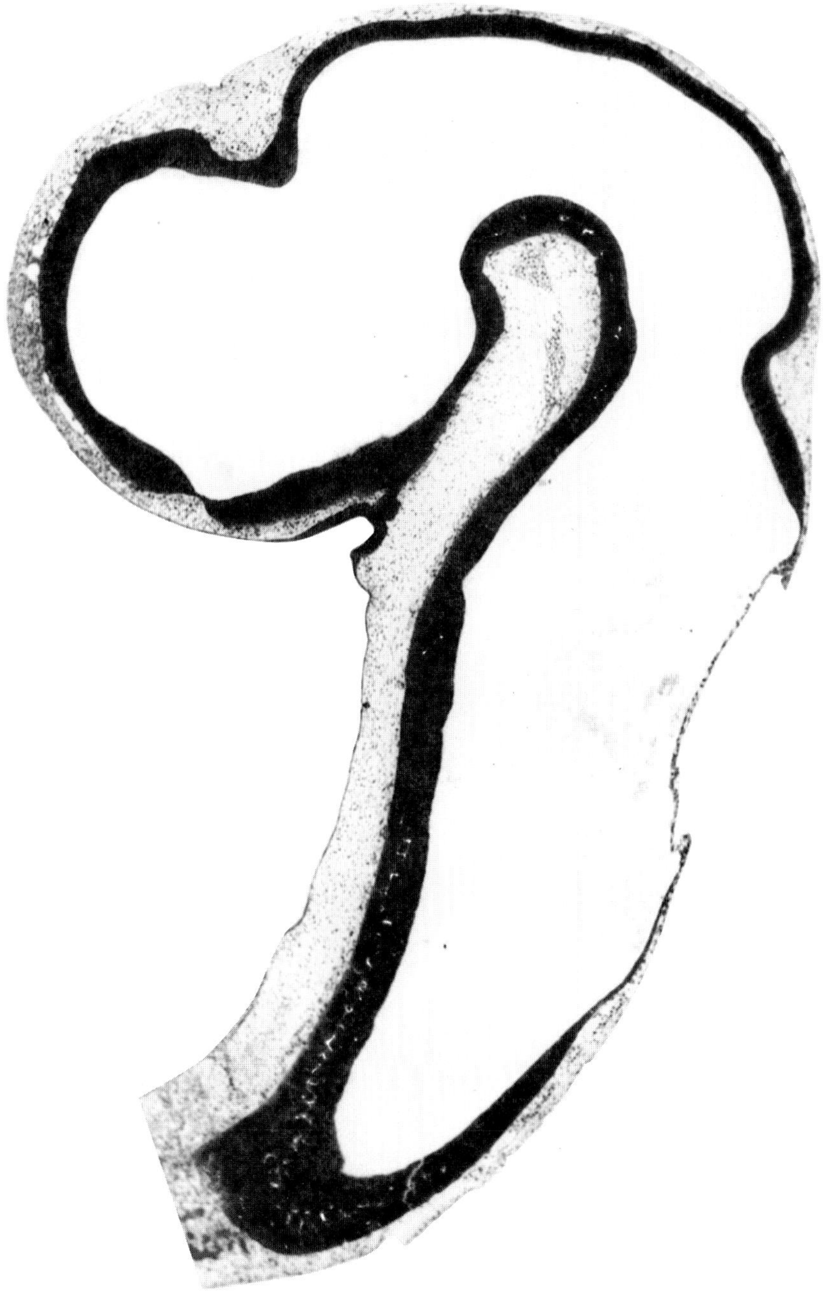

VENTRICLES (v)

vai aqueduct, inferior tectal
vapr aqueduct, pretectal
vas aqueduct, superior tectal
vat aqueduct, tegmental
vla lateral ventricle, anterior horn
vld lateral ventricle, dorsal
vlm foramen of Monro
vs spinal canal
v3h third ventricle, hypothalamic
v3i third ventricle, infundibular recess
v3m third ventricle, mammillary recess
v3pr third ventricle, preoptic recess
v3t third ventricle, thalamic
v4a fourth ventricle, anterior
v4i fourth ventricle, isthmal
v4p fourth ventricle, posterior

PONTINE AREA (X)

X1a anterior pontine neuroepithelium
X1p posterior pontine neuroepithelium
X31d isthmal neuroepithelium, dorsal
X31v isthmal neuroepithelium, ventral

MEDULLA (Y)

Y1 medullary neuroepithelium
Y1d medullary neuroepithelium, dorsal
Y1v medullary neuroepithelium, ventral

SAGITTAL FIGURE 3

CEREBRAL CORTEX (C)

C1 neocortical neuroepithelium

GLANDS AND PARANEURAL STRUCTURES (g)

g10 velum medullare

HYPOTHALAMUS (H)

H1a anterior hypothalamic neuroepithelium
H1i intermediate hypothalamic neuroepithelium
H1m mammillary neuroepithelium
H1p posterior hypothalamic neuroepithelium

HIPPOCAMPAL FORMATION (K)

K1 hippocampal (Ammon's horn) neuroepithelium
K11 subicular neuroepithelium

CEREBELLUM (L)

L1 cerebellar neuroepithelium

MIDBRAIN (M)

M1 superior tectal neuroepithelium
M11a anterior tegmental neuroepithelium
M11p posterior tegmental neuroepithelium
M21 pretectal neuroepithelium
M31 inferior tectal neuroepithelium

PREOPTIC AREA AND OPTIC EPITHELIA (P)

P1 preoptic neuroepithelium

RHINENCEPHALON (R)

R1 rhinencephalic neuroepithelium

SEPTAL AREA (S)

S1 septal neuroepithelium

THALAMUS (T)

T1a anterior thalamic neuroepithelium
T1p posterior thalamic neuroepithelium

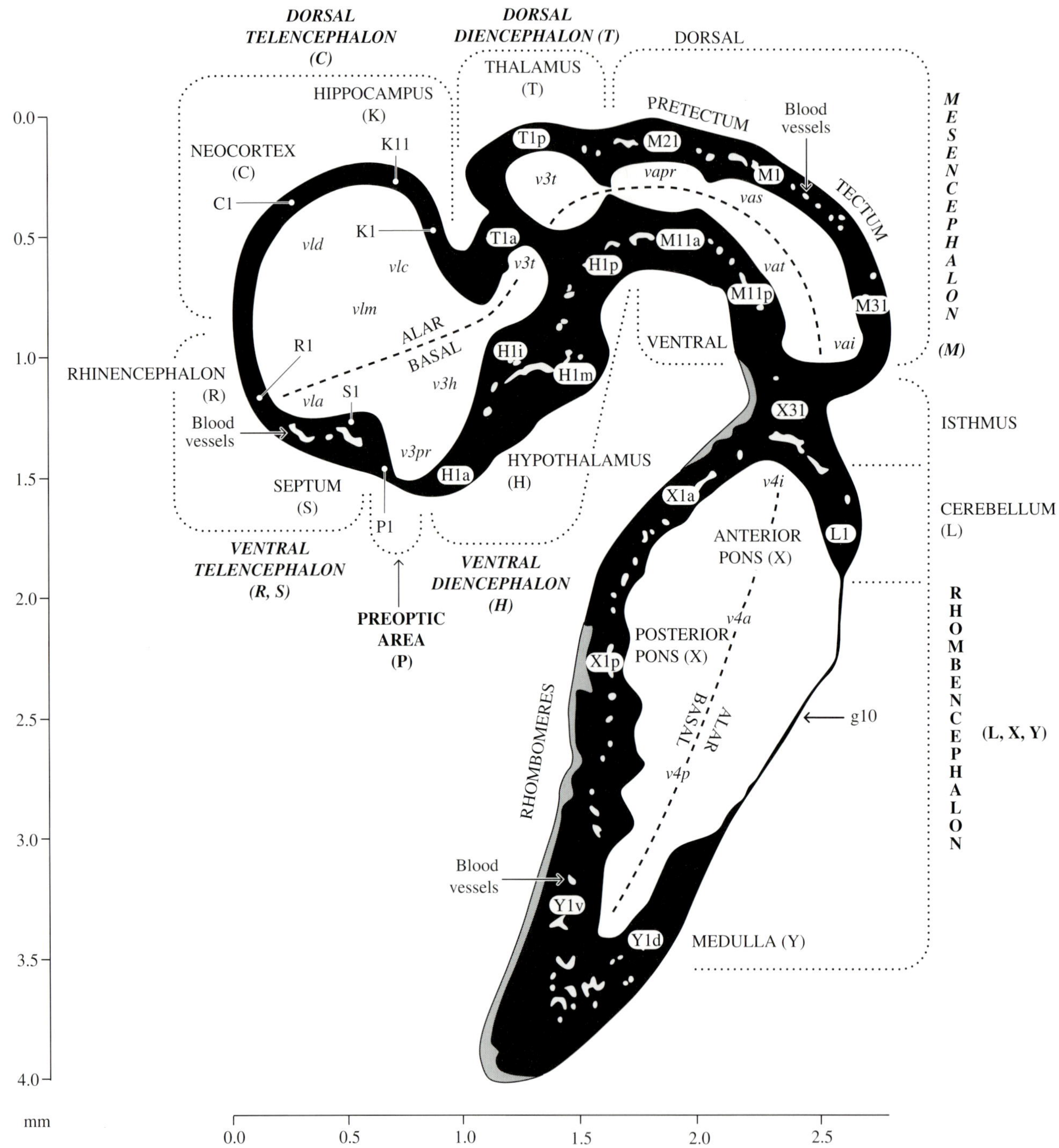

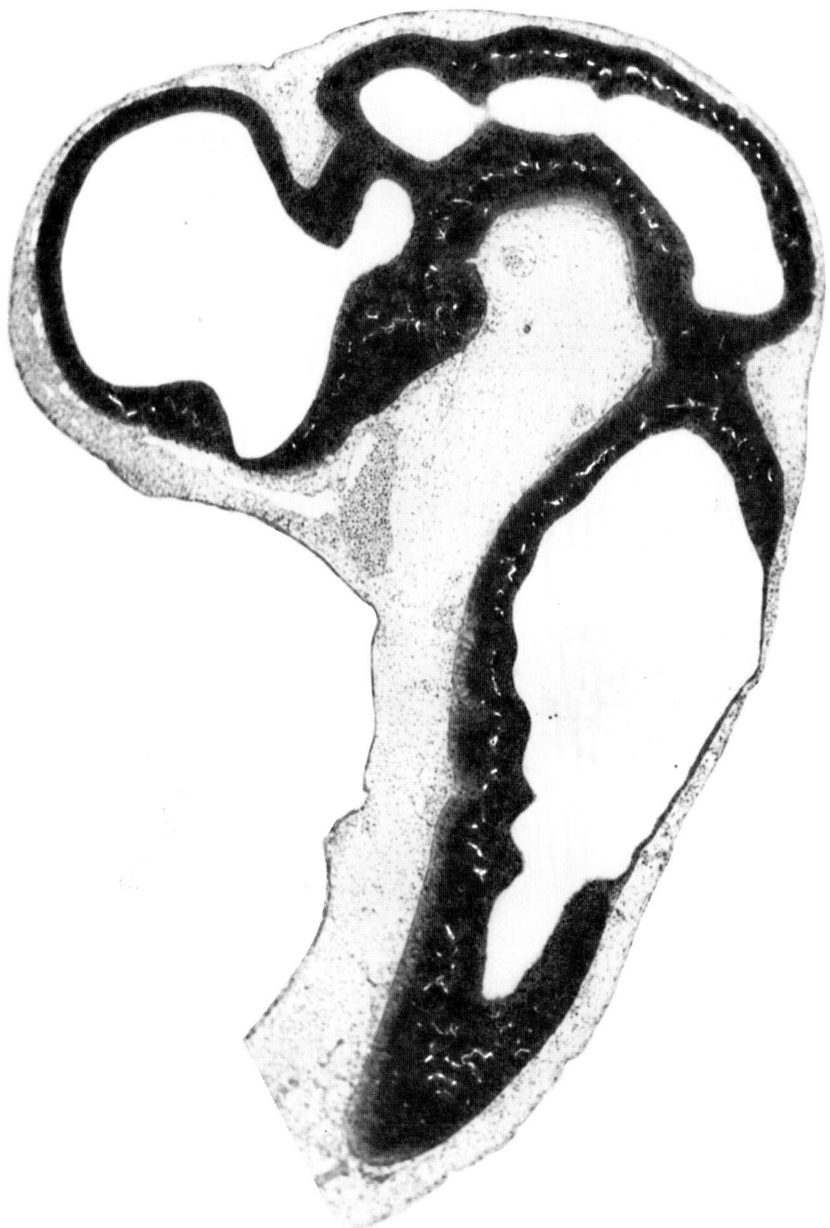

VENTRICLES (v)

vai aqueduct, inferior tectal
vapr aqueduct, pretectal
vas aqueduct, superior tectal
vat aqueduct, tegmental
vla lateral ventricle, anterior horn
vlc lateral ventricle, central
vld lateral ventricle, dorsal
vlm foramen of Monro
v3h third ventricle, hypothalamic
v3pr third ventricle, preoptic recess
v3t third ventricle, thalamic
v4a fourth ventricle, anterior
v4i fourth ventricle, isthmal
v4p fourth ventricle, posterior

PONTINE AREA (X)

X1a anterior pontine neuroepithelium
X1p posterior pontine neuroepithelium
X31 isthmal neuroepithelium

MEDULLA (Y)

Y1d medullary neuroepithelium, dorsal
Y1v medullary neuroepithelium, ventral

SAGITTAL FIGURE 4

AMYGDALA (A)

A1 amygdaloid neuroepithelium

BASAL GANGLIA (B)

B11 pallidal neuroepithelium

CEREBRAL CORTEX (C)

C1 neocortical neuroepithelium

GLANDS AND PARANEURAL STRUCTURES (g)

g10 velum medullare

HYPOTHALAMUS (H)

H1i intermediate hypothalamic neuroepithelium
H1v ventral hypothalamic neuroepithelium

HIPPOCAMPAL FORMATION (K)

K1 hippocampal (Ammon's horn) neuroepithelium
K11 subicular neuroepithelium

CEREBELLUM (L)

L1 cerebellar neuroepithelium

MIDBRAIN (M)

M1 superior tectal neuroepithelium
M11a anterior tegmental neuroepithelium
M11p posterior tegmental neuroepithelium
M21 pretectal neuroepithelium
M31 inferior tectal neuroepithelium

PREOPTIC AREA AND OPTIC EPITHELIA (P)

P1 preoptic neuroepithelium

RHINENCEPHALON (R)

R1 rhinencephalic neuroepithelium

THALAMUS (T)

T1a anterior thalamic neuroepithelium
T1p posterior thalamic neuroepithelium

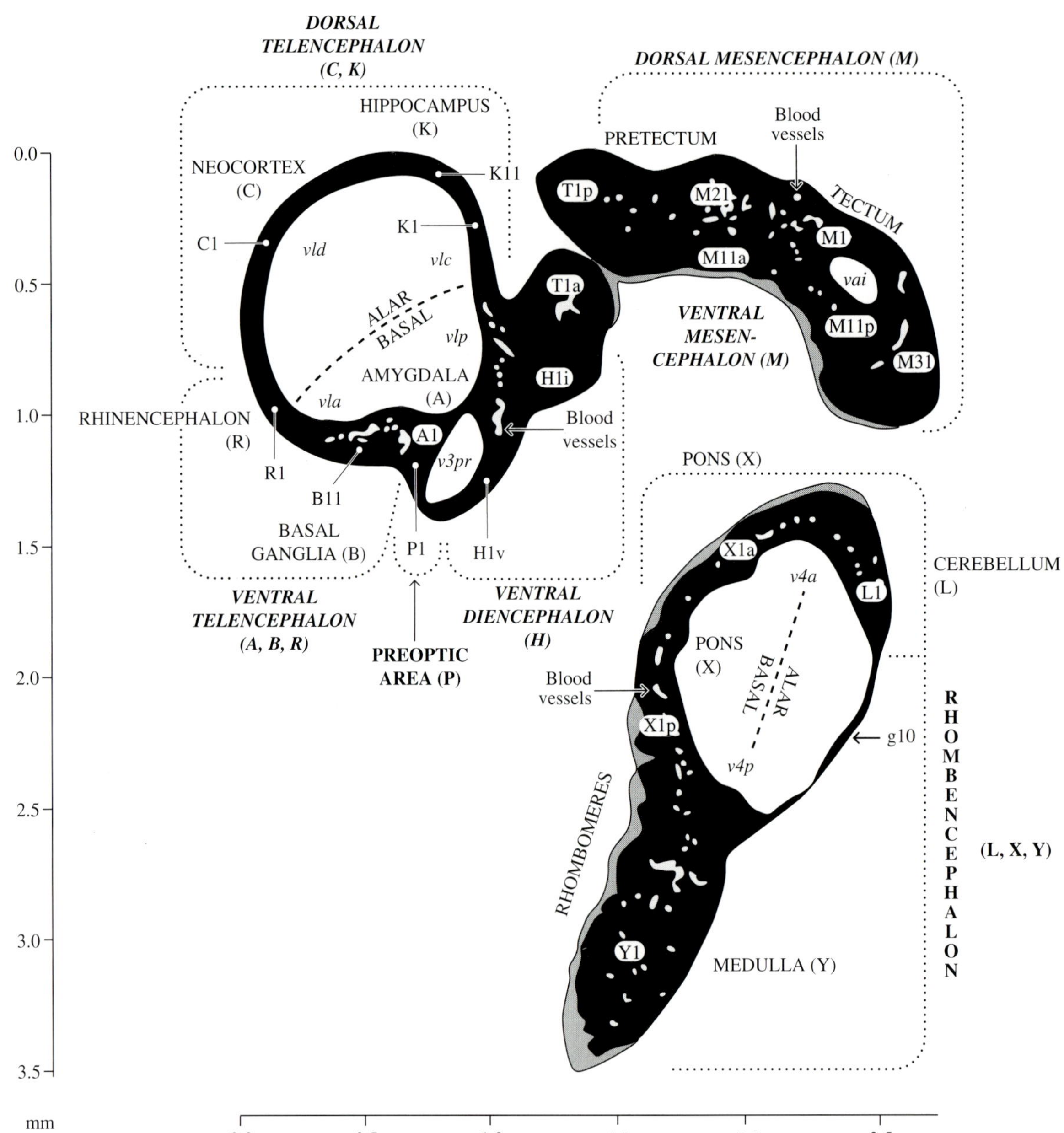

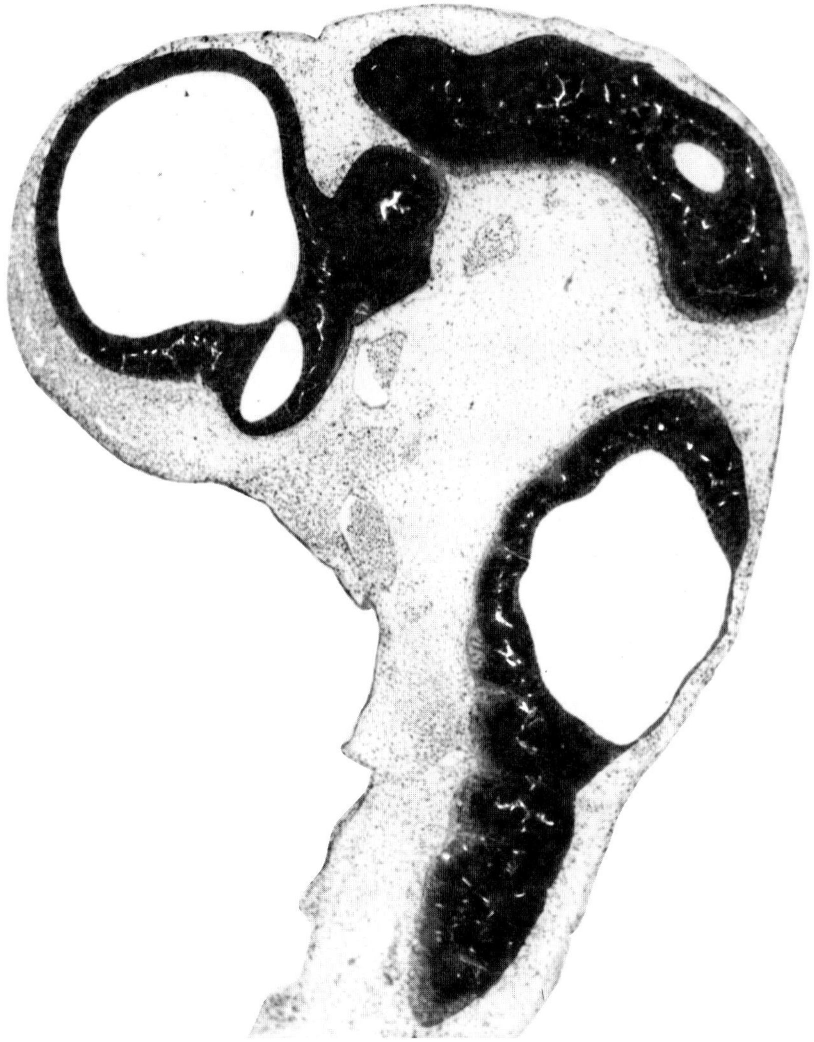

VENTRICLES (v)

vai	aqueduct, inferior tectal
vla	lateral ventricle, anterior horn
vlc	lateral ventricle, central
vld	lateral ventricle, dorsal
vlp	lateral ventricle, posterior horn
v3pr	third ventricle, preoptic recess
v4a	fourth ventricle, anterior
v4p	fourth ventricle, posterior

PONTINE AREA (X)

X1a	anterior pontine neuroepithelium
X1p	posterior pontine neuroepithelium

MEDULLA (Y)

Y1	medullary neuroepithelium

SAGITTAL FIGURE 5

AMYGDALA (A)

A1 amygdaloid neuroepithelium

BASAL GANGLIA (B)

B11 pallidal neuroepithelium

CEREBRAL CORTEX (C)

C1 neocortical neuroepithelium

HYPOTHALAMUS (H)

H1v ventral hypothalamic neuroepithelium

HIPPOCAMPAL FORMATION (K)

K1 hippocampal (Ammon's horn) neuroepithelium
K11 subicular neuroepithelium

CEREBELLUM (L)

L1 cerebellar neuroepithelium

MIDBRAIN (M)

M31 inferior tectal neuroepithelium

NERVES AND NUCLEI, CRANIAL (n)

n8b vestibulocochlear boundary cap
n8g vestibular ganglion

PREOPTIC AREA AND OPTIC EPITHELIA (P)

P1 preoptic neuroepithelium

RHINENCEPHALON (R)

R1 rhinencephalic neuroepithelium

VENTRICLES (v)

vla lateral ventricle, anterior horn
vlc lateral ventricle, central
vld lateral ventricle, dorsal
vlp lateral ventricle, posterior horn
v3pr third ventricle, preoptic recess
v4l fourth ventricle, lateral recess

PONTINE AREA (X)

X1a anterior pontine neuroepithelium
X21 cochlear nuclei neuroepithelium

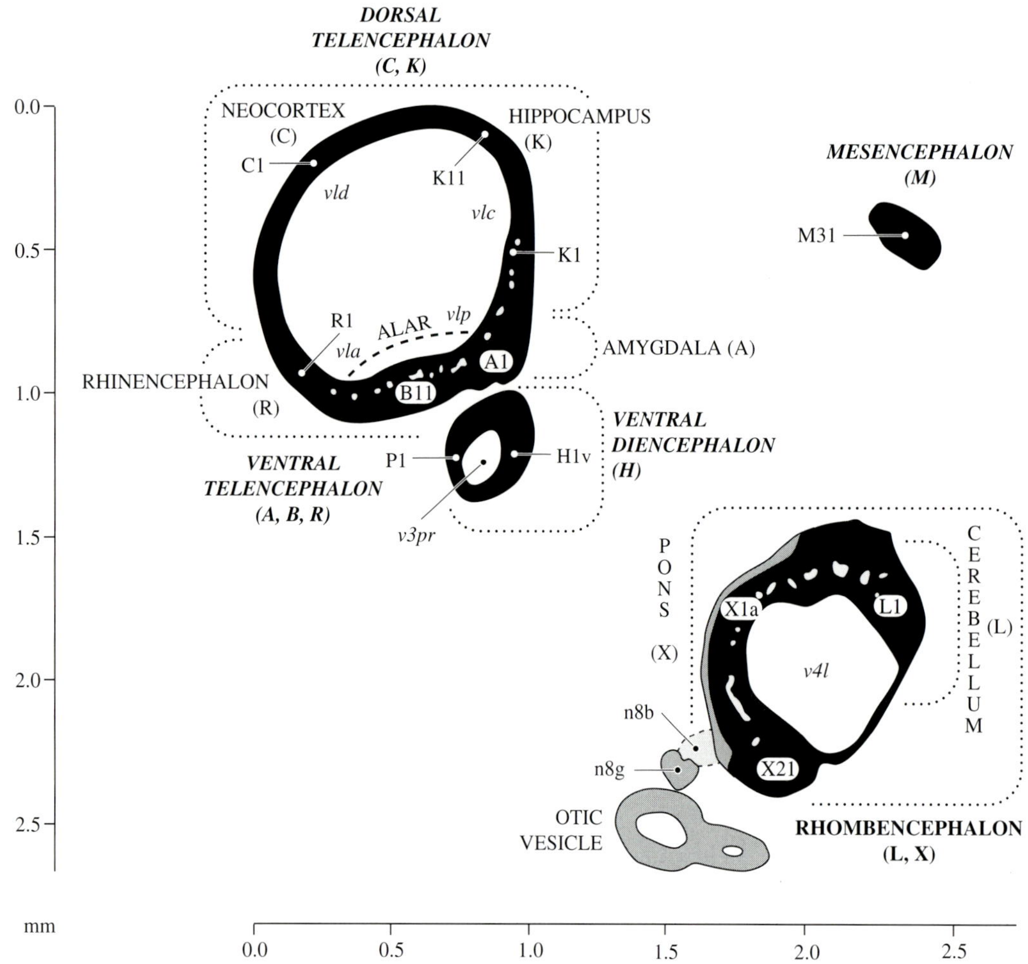

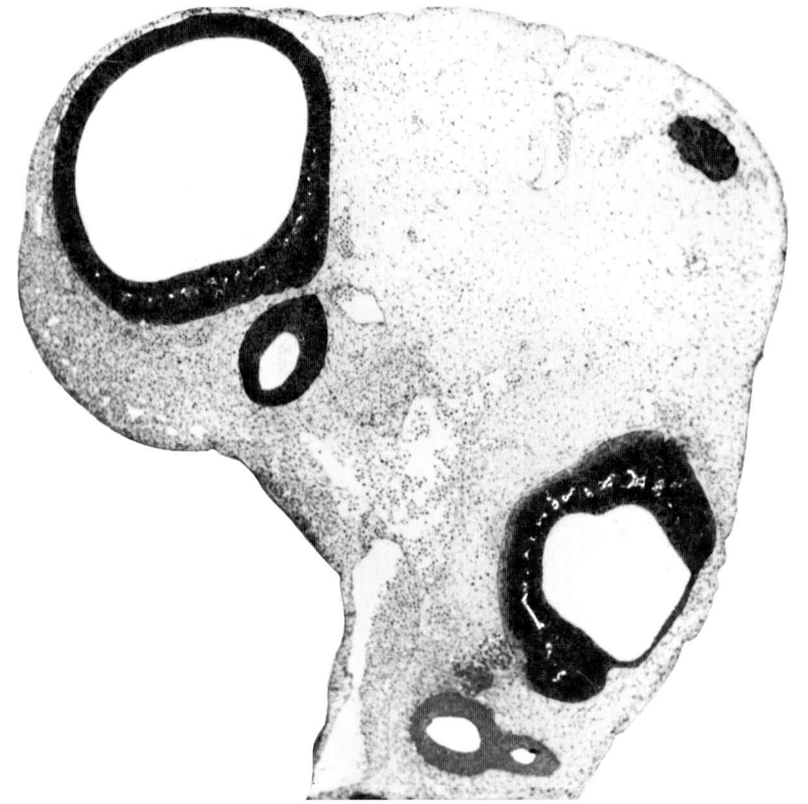

SAGITTAL FIGURE 6

AMYGDALA (A)

A1 amygdaloid neuroepithelium

CEREBRAL CORTEX (C)

C1 neocortical neuroepithelium

HYPOTHALAMUS (H)

H1v ventral hypothalamic neuroepithelium

HIPPOCAMPAL FORMATION (K)

K1 hippocampal (Ammon's horn) neuroepithelium
K11 subicular neuroepithelium

NERVES AND NUCLEI, CRANIAL (n)

n5b trigeminal boundary cap
n5g trigeminal ganglion
n8g vestibular ganglion

PREOPTIC AREA AND OPTIC EPITHELIA (P)

P1 preoptic neuroepithelium

RHINENCEPHALON (R)

R1 rhinencephalic neuroepithelium

VENTRICLES (v)

vla lateral ventricle, anterior horn
vlc lateral ventricle, central
vld lateral ventricle, dorsal
vlp lateral ventricle, posterior horn

PONTINE AREA (X)

X1 pontine neuroepithelium

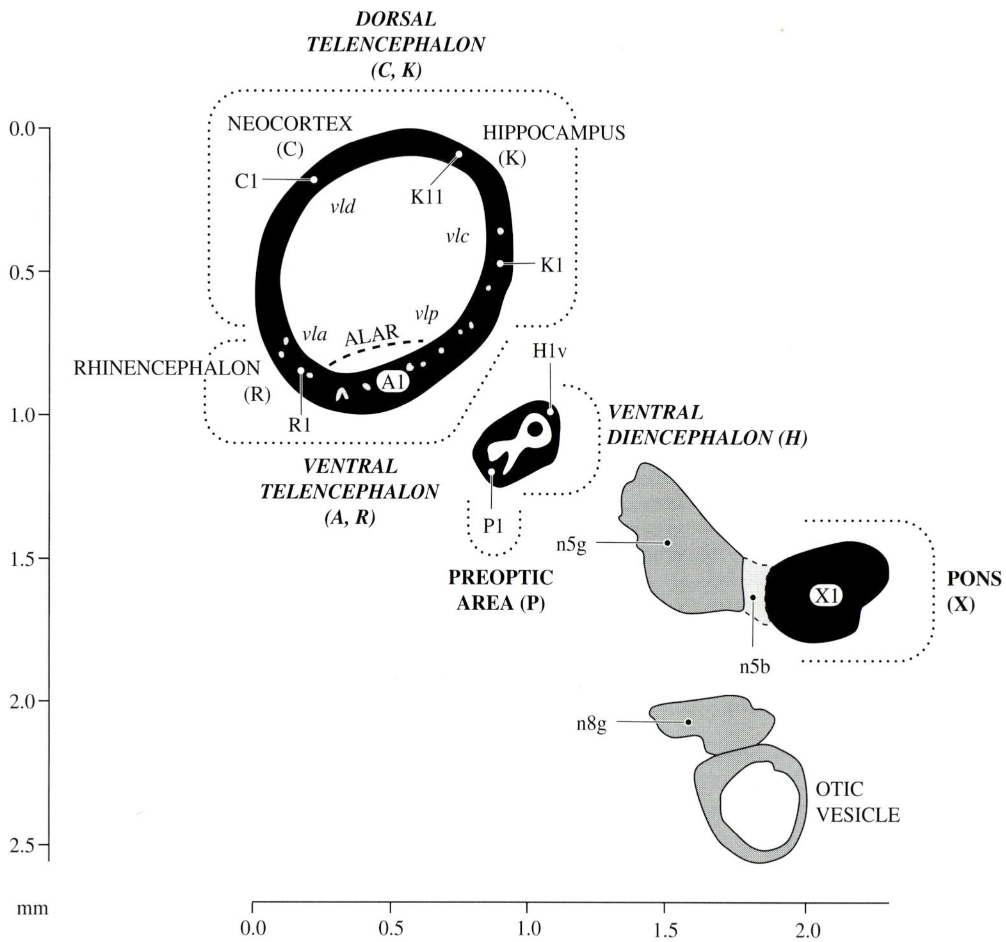

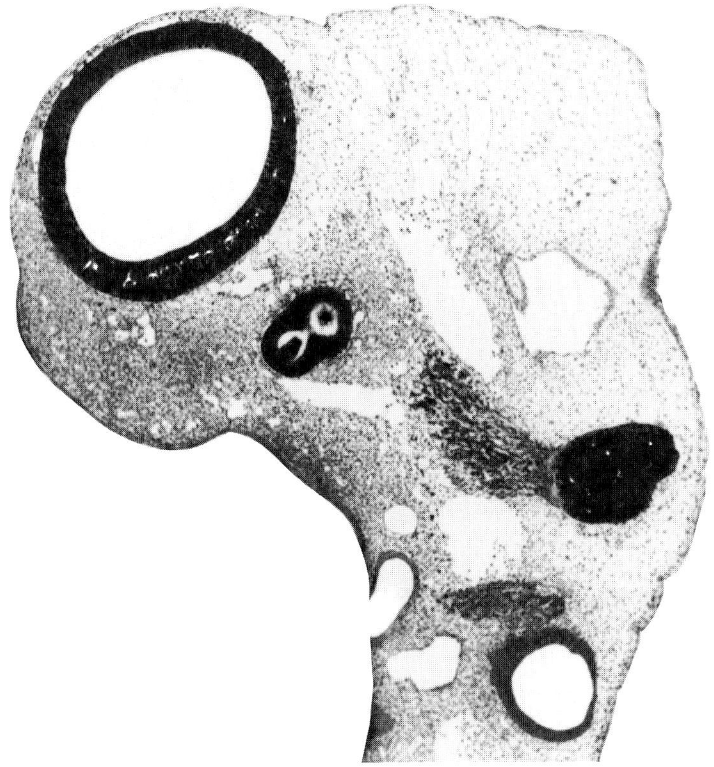

CORONAL

FIGURES and PLATES 1-18

CORONAL FIGURE 1

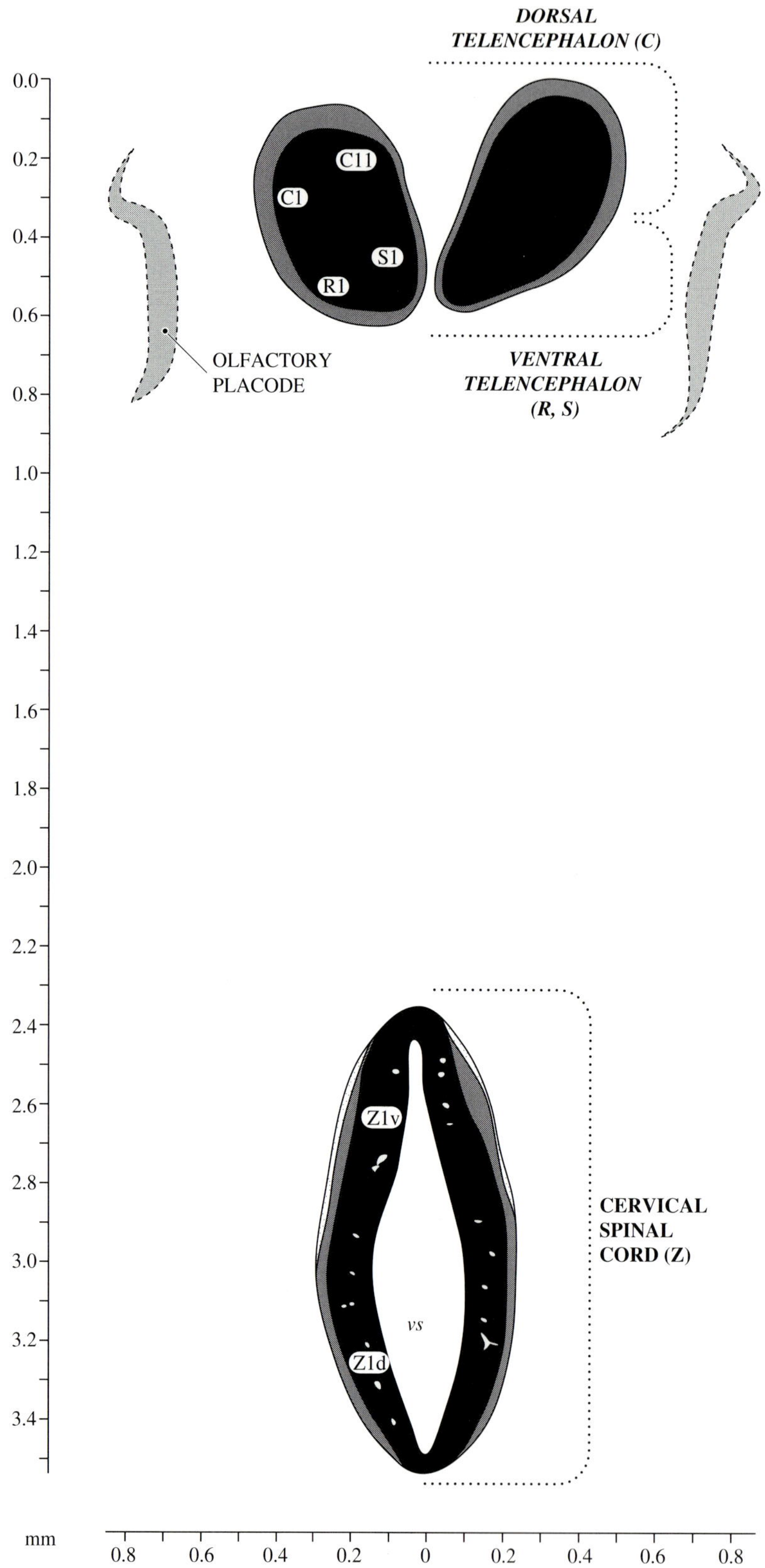

CEREBRAL CORTEX (C)

C1 neocortical neuroepithelium
C11 cingulate cortex neuroepithelium

RHINENCEPHALON (R)

R1 rhinencephalic neuroepithelium

SEPTAL AREA (S)

S1 septal neuroepithelium

VENTRICLES (v)

vs spinal canal

SPINAL CORD (Z)

Z1d spinal cord neuroepithelium, dorsal
Z1v spinal cord neuroepithelium, ventral

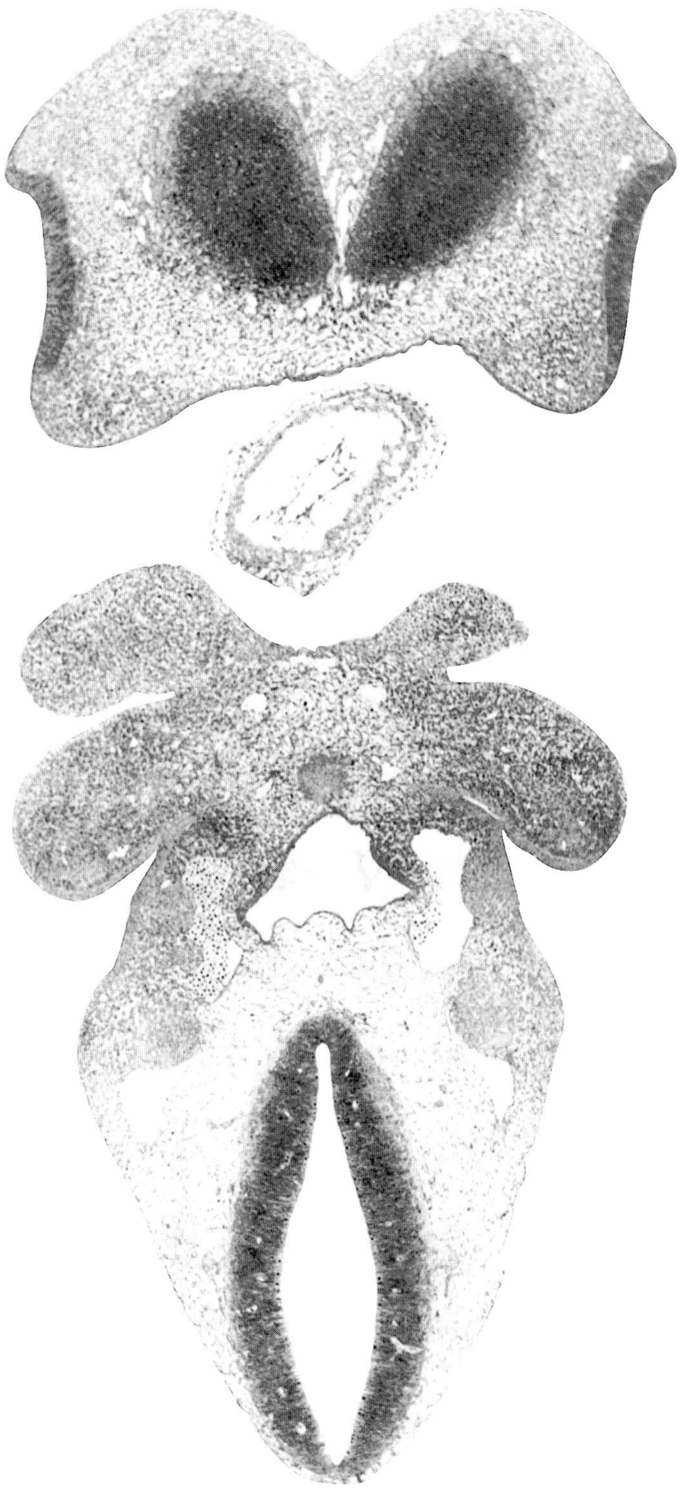

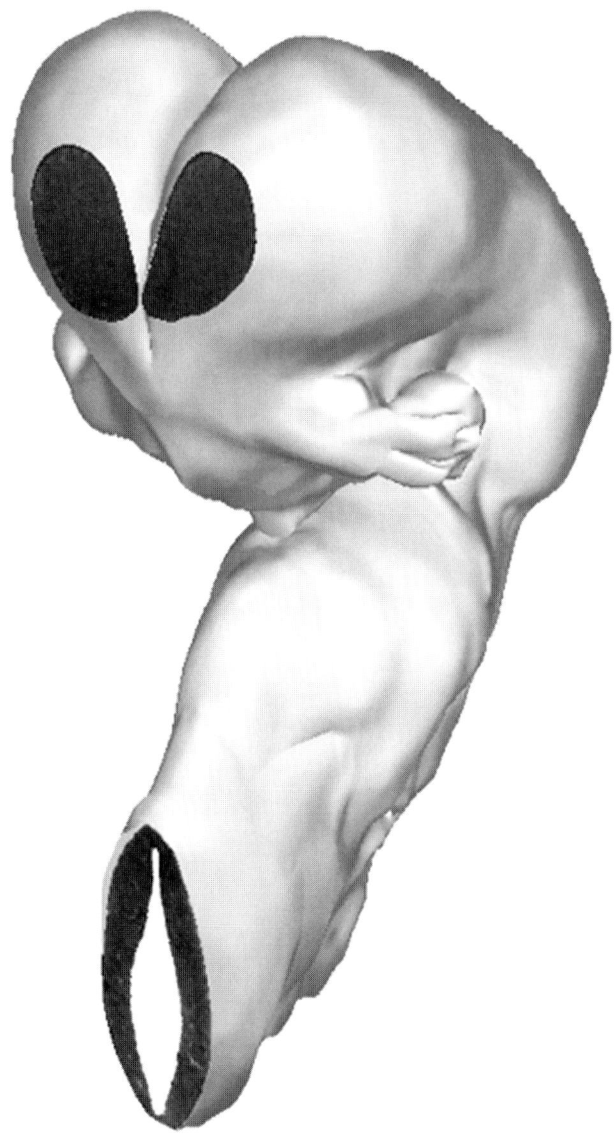

CORONAL FIGURE 2

BASAL GANGLIA (B)

B1 striatal neuroepithelium
B11 pallidal neuroepithelium

CEREBRAL CORTEX (C)

C1 neocortical neuroepithelium
C4 molecular layer (or primordial plexiform layer)
C11 cingulate cortex neuroepithelium

HIPPOCAMPAL FORMATION (K)

K11 subicular neuroepithelium

RHINENCEPHALON (R)

R1 rhinencephalic neuroepithelium

SEPTAL AREA (S)

S1 septal neuroepithelium

VENTRICLES (v)

vla lateral ventricle, anterior horn
vld lateral ventricle, dorsal
vs spinal canal

SPINAL CORD (Z)

Z1d spinal cord neuroepithelium, dorsal
Z1v spinal cord neuroepithelium, ventral
Z3v spinal cord (or differentiating field), ventral

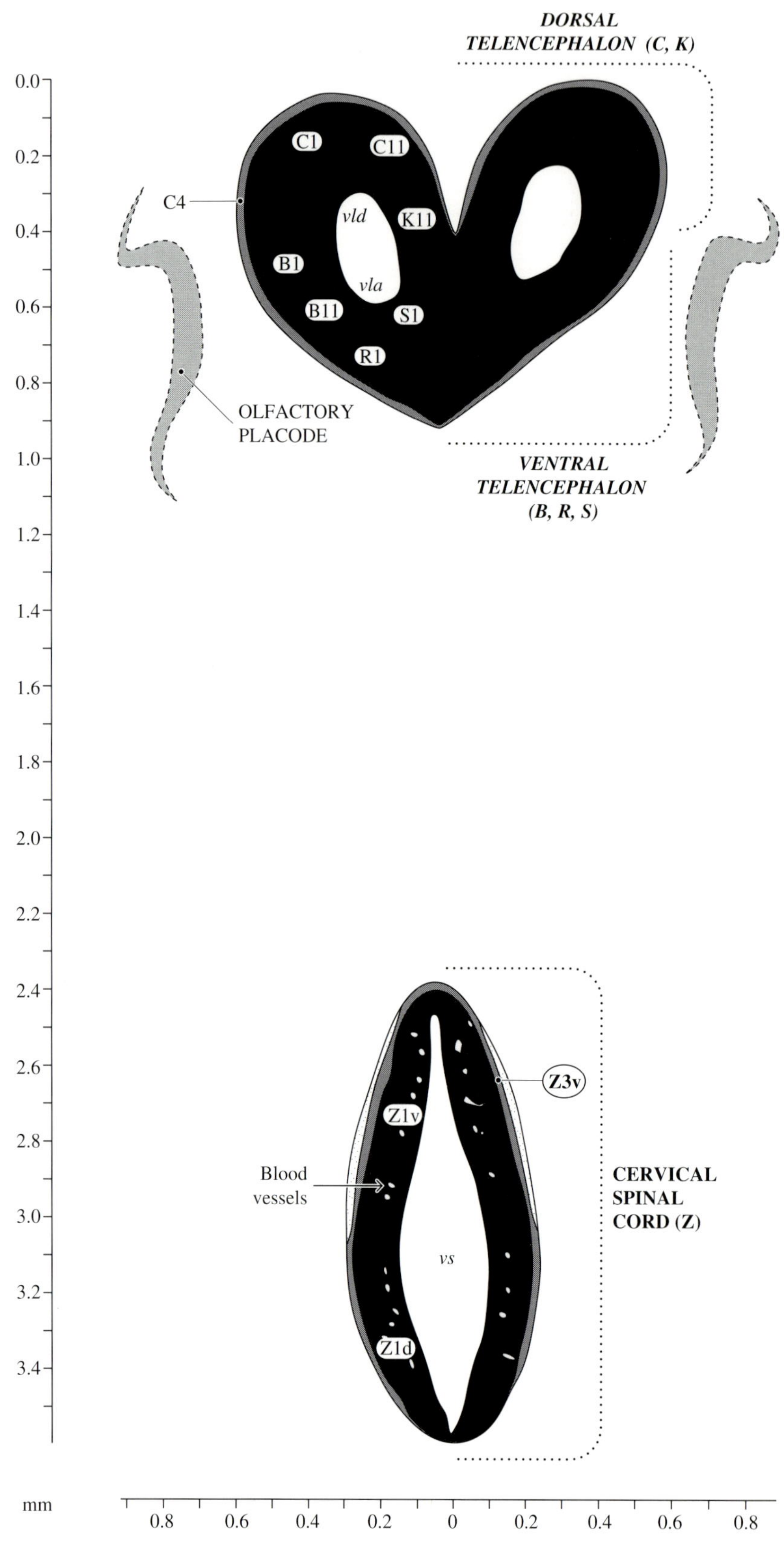

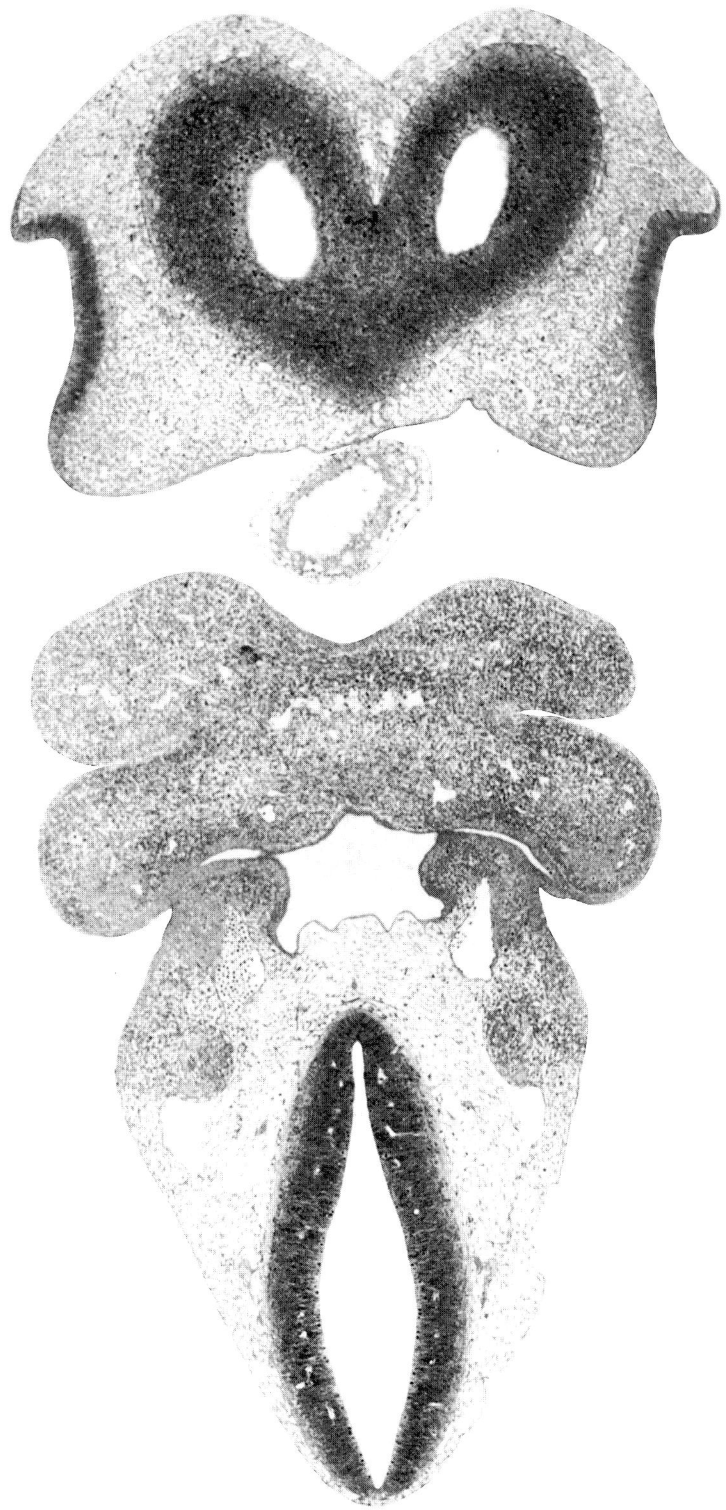

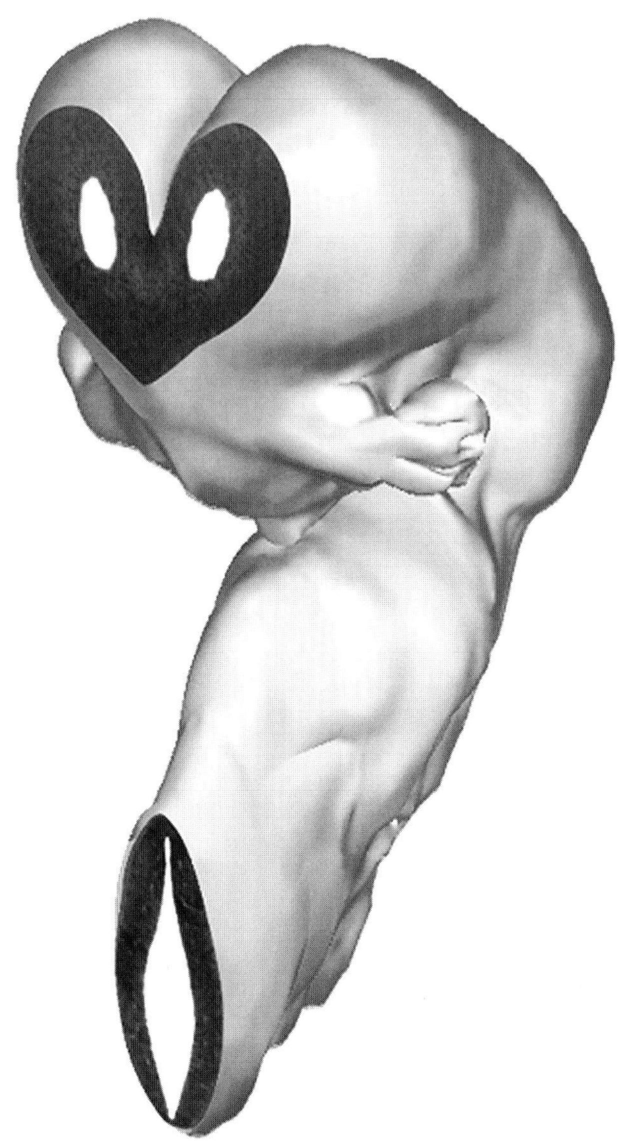

CORONAL FIGURE 3

BASAL GANGLIA (B)

B1 striatal neuroepithelium
B11 pallidal neuroepithelium

CEREBRAL CORTEX (C)

C1 neocortical neuroepithelium
C4 molecular layer (or primordial plexiform layer)
C11 cingulate cortex neuroepithelium

HIPPOCAMPAL FORMATION (K)

K1 hippocampal (Ammon's horn) neuroepithelium
K11 subicular neuroepithelium

RHINENCEPHALON (R)

R1 rhinencephalic neuroepithelium

VENTRICLES (v)

vla lateral ventricle, anterior horn
vlc lateral ventricle, central
vld lateral ventricle, dorsal
vs spinal canal

SPINAL CORD (Z)

Z1d spinal cord neuroepithelium, dorsal
Z1v spinal cord neuroepithelium, ventral
Z3v spinal cord (or differentiating field), ventral

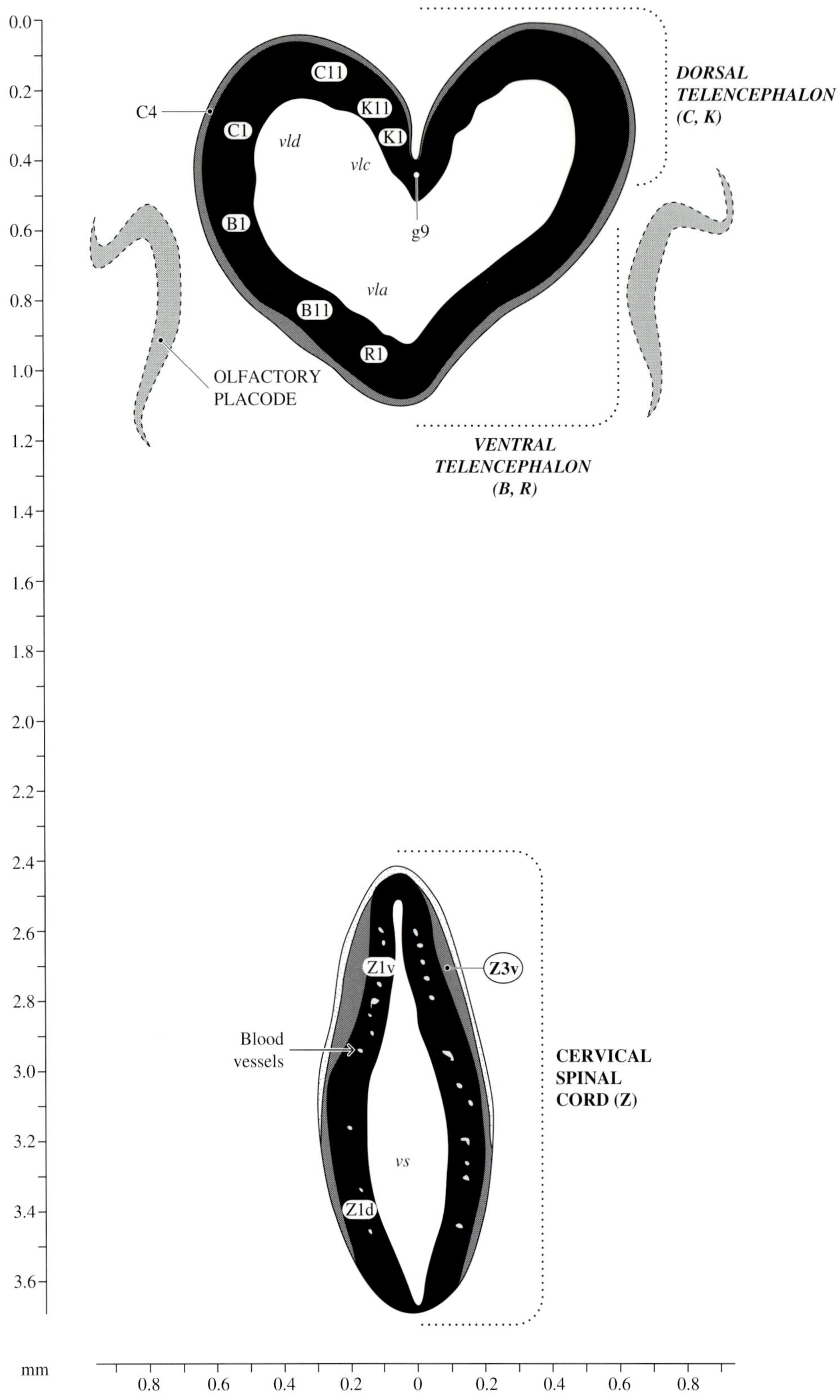

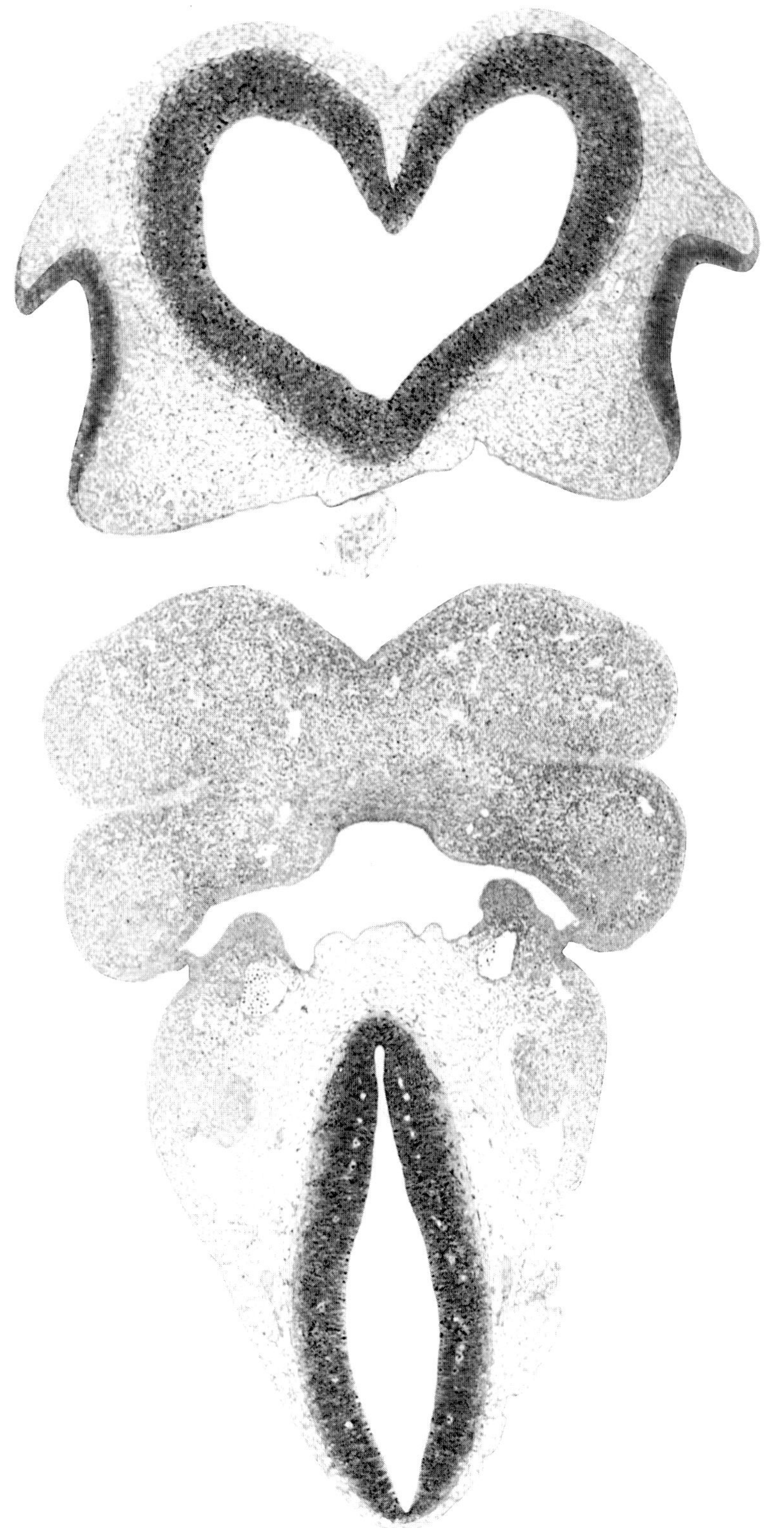

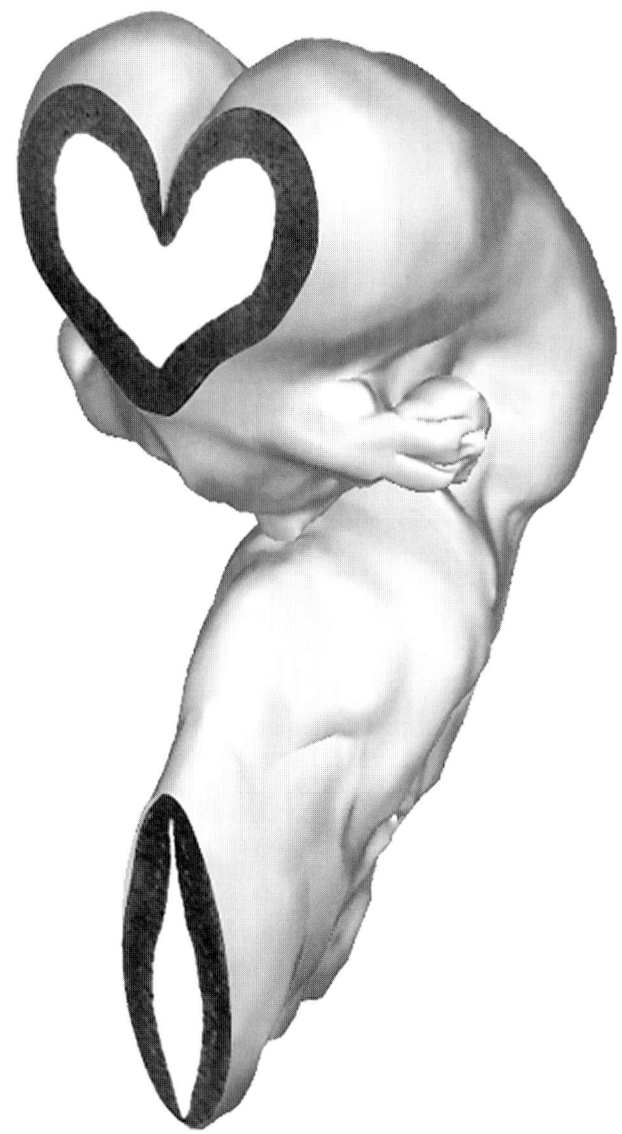

CORONAL FIGURE 4

BASAL GANGLIA (B)

B1	striatal neuroepithelium
B11	pallidal neuroepithelium
B13	pallidum (or differentiating field)

CEREBRAL CORTEX (C)

C1	neocortical neuroepithelium
C11	cingulate cortex neuroepithelium

GLANDS AND PARANEURAL STRUCTURES (g)

g9	lamina terminalis

HIPPOCAMPAL FORMATION (K)

K1	hippocampal (Ammon's horn) neuroepithelium
K11	subicular neuroepithelium

PREOPTIC AREA AND OPTIC EPITHELIA (P)

P1	preoptic neuroepithelium

VENTRICLES (v)

vla	lateral ventricle, anterior horn
vlc	lateral ventricle, central
vld	lateral ventricle, dorsal
vlm	foramen of Monro
vs	spinal canal
v3pr	third ventricle, preoptic recess

SPINAL CORD (Z)

Z1d	spinal cord neuroepithelium, dorsal
Z1v	spinal cord neuroepithelium, ventral
Z3v	spinal cord (or differentiating field), ventral

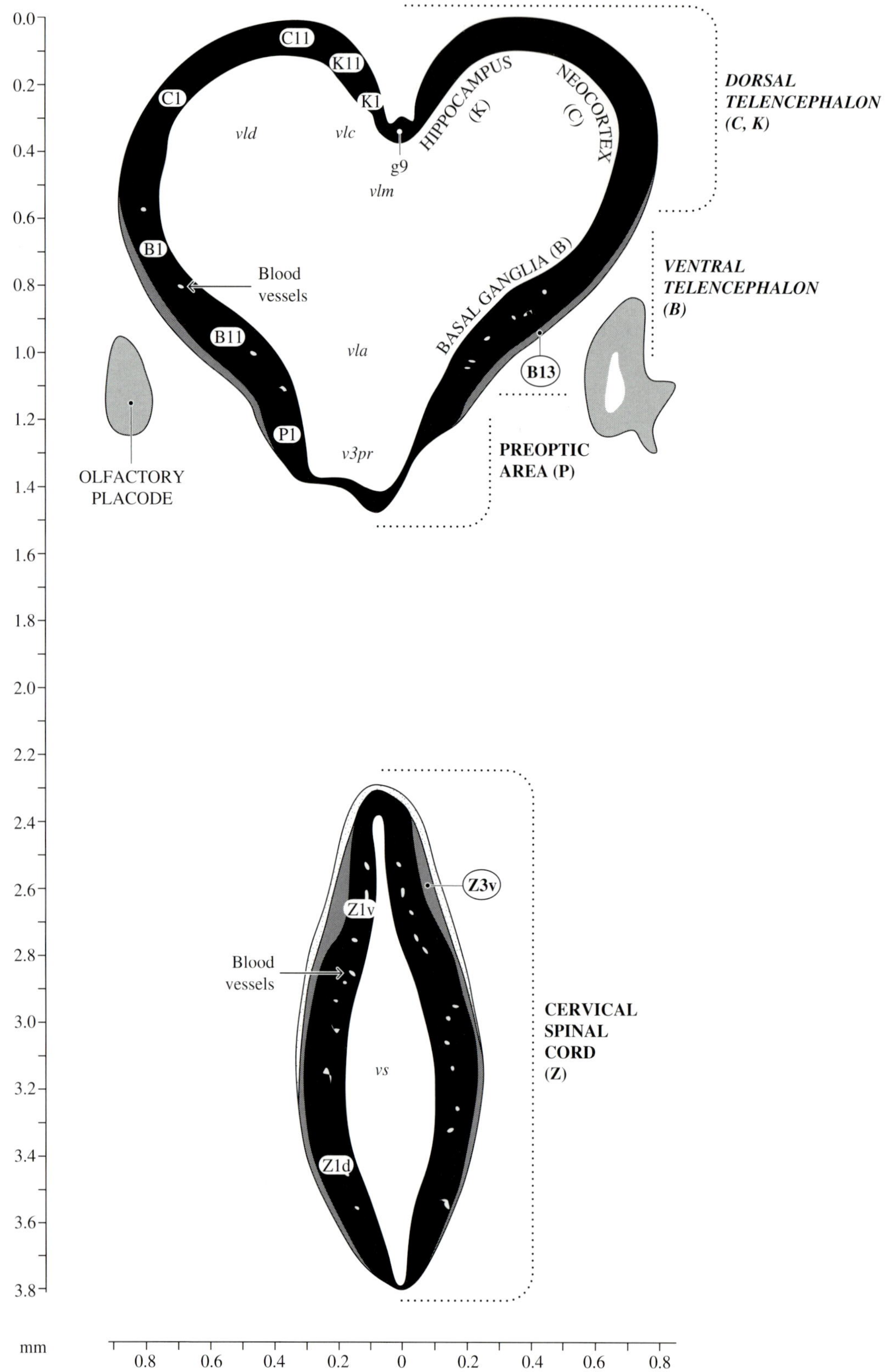

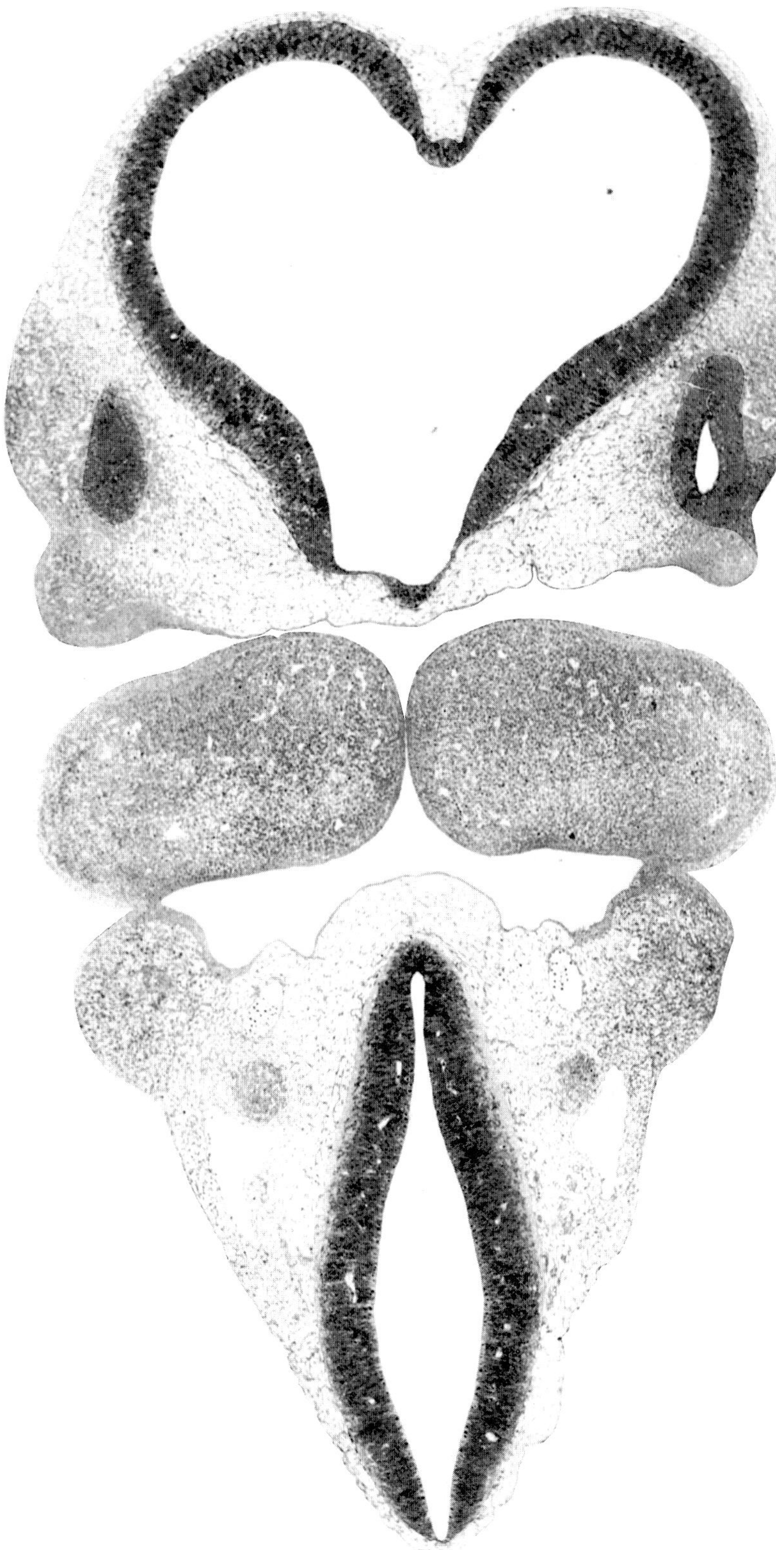

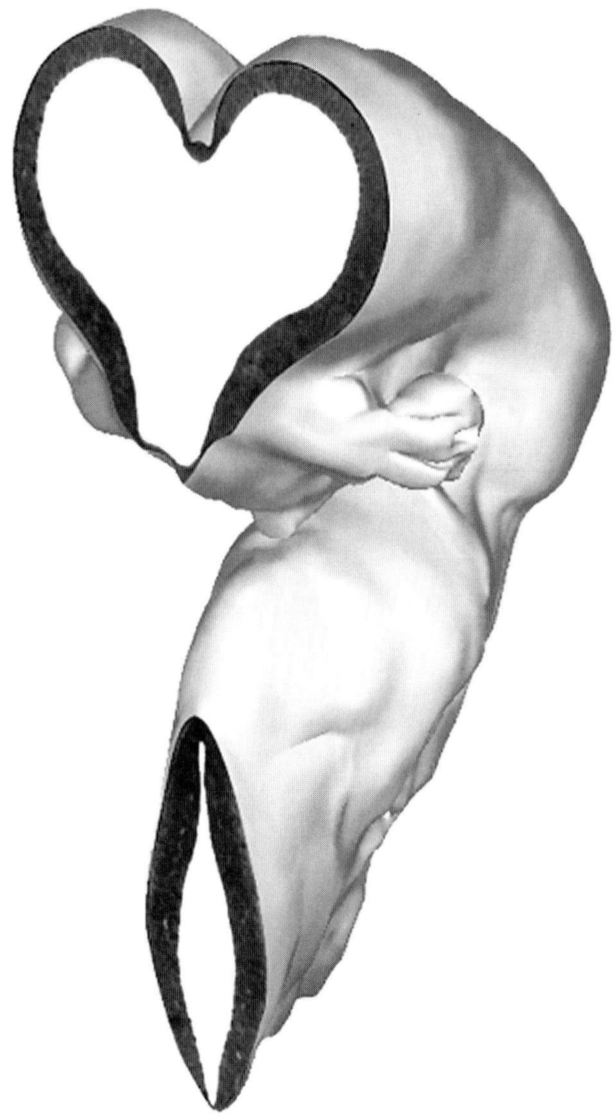

CORONAL FIGURE 5

BASAL GANGLIA (B)

B1 striatal neuroepithelium
B11 pallidal neuroepithelium
B13 pallidum (or differentiating field)

CEREBRAL CORTEX (C)

C1 cortical neuroepithelium
C11 cingulate cortex neuroepithelium

GLANDS AND PARANEURAL STRUCTURES (g)

g9 lamina terminalis

HYPOTHALAMUS (H)

H1a anterior hypothalamic neuroepithelium
H3 hypothalamus (or differentiating field)
H3a anterior hypothalamus (or differentiating field)

HIPPOCAMPAL FORMATION (K)

K11 subicular neuroepithelium

PREOPTIC AREA AND OPTIC EPITHELIA (P)

P1 preoptic neuroepithelium
P12 eye stalk epithelium

VENTRICLES (v)

vlc lateral ventricle, central
vld lateral ventricle, dorsal
vlm foramen of Monro
v3h third ventricle, hypothalamic
v3o third ventricle, optic recess
v3pr third ventricle, preoptic recess
v4a fourth ventricle, anterior
v4p fourth ventricle, posterior

PONTINE AREA (X)

X1 pontine neuroepithelium
X3 pons (or differentiating field)

MEDULLA (Y)

Y1 medullary neuroepithelium
Y3 medulla (or differentiating field)

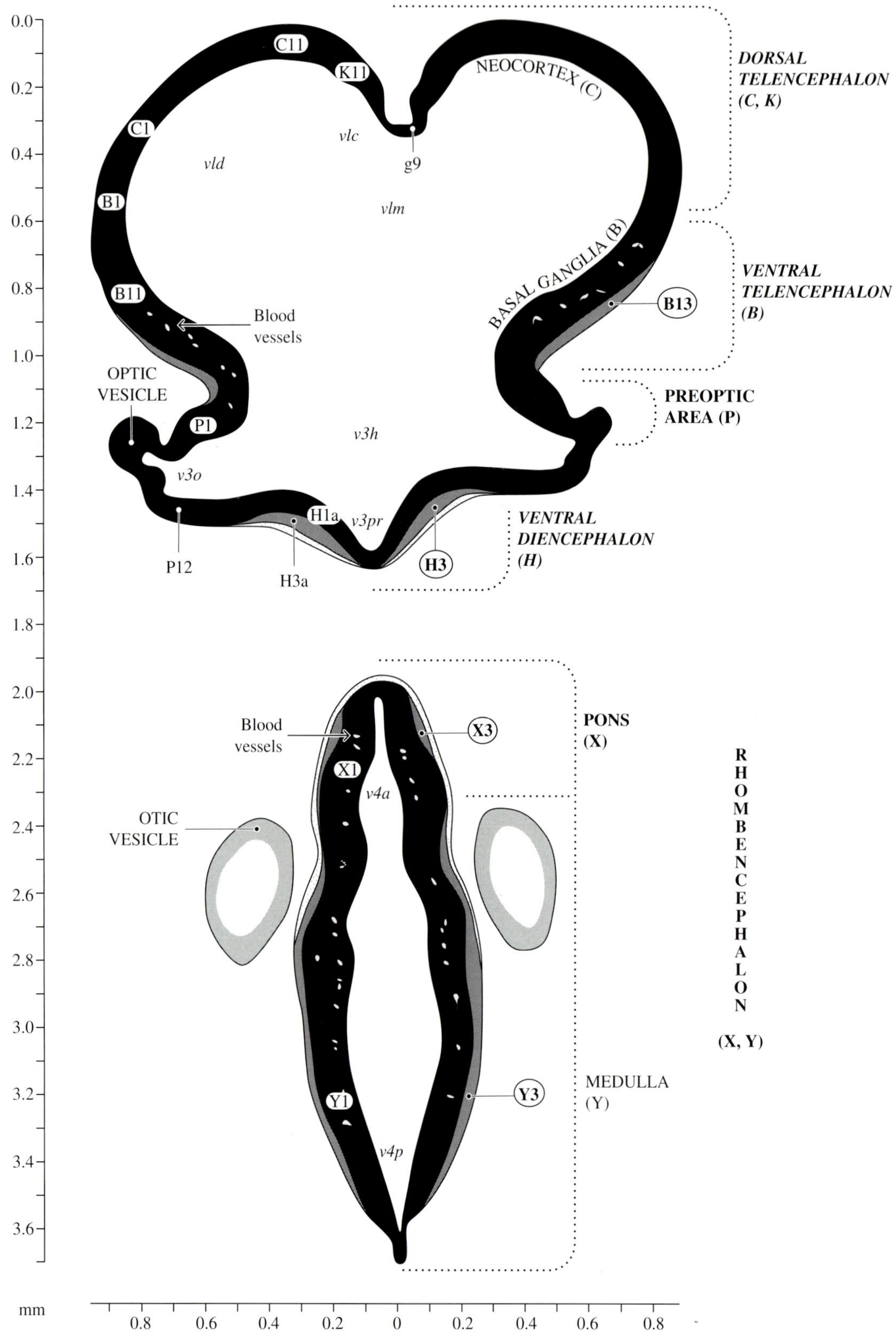

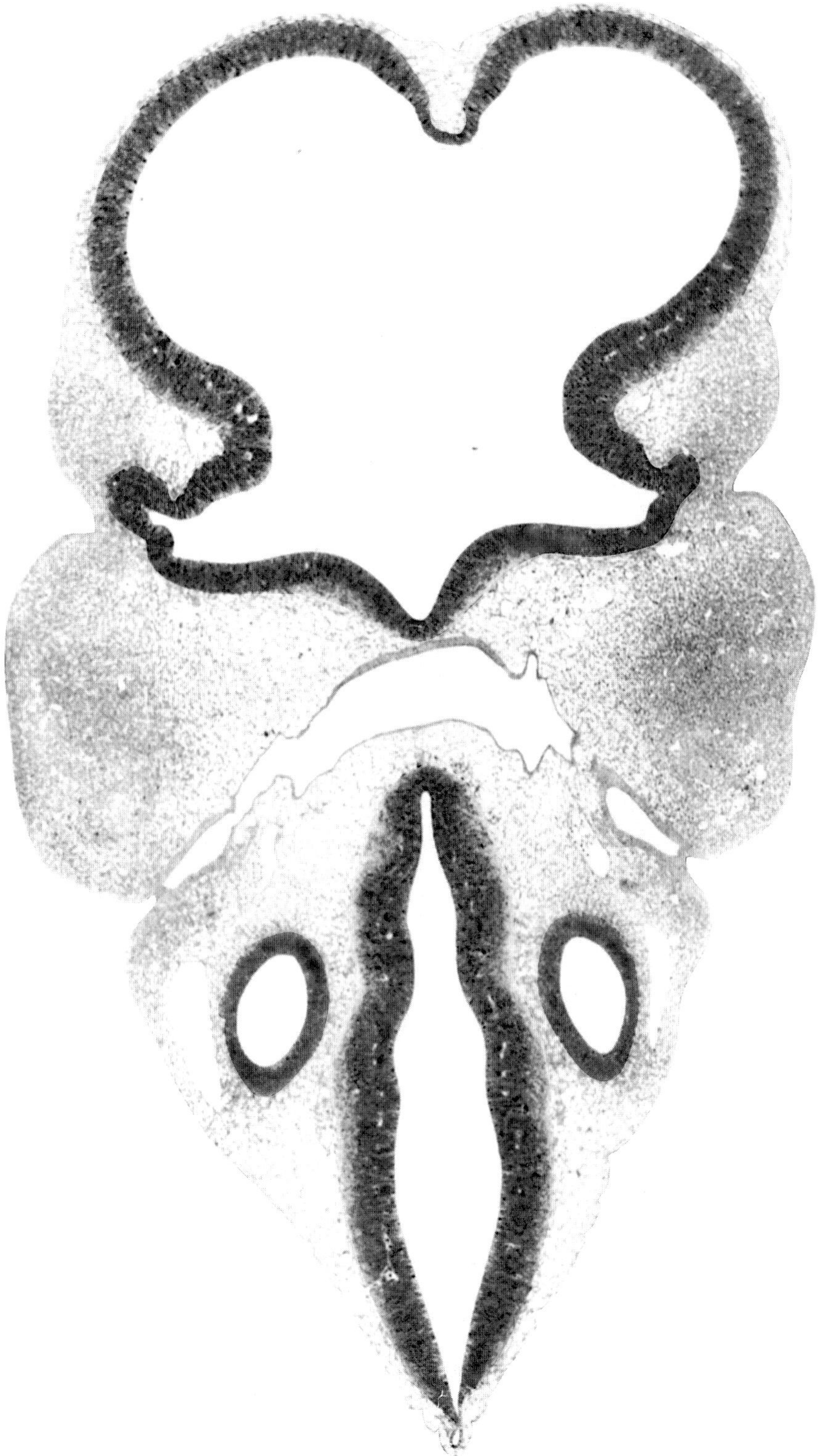

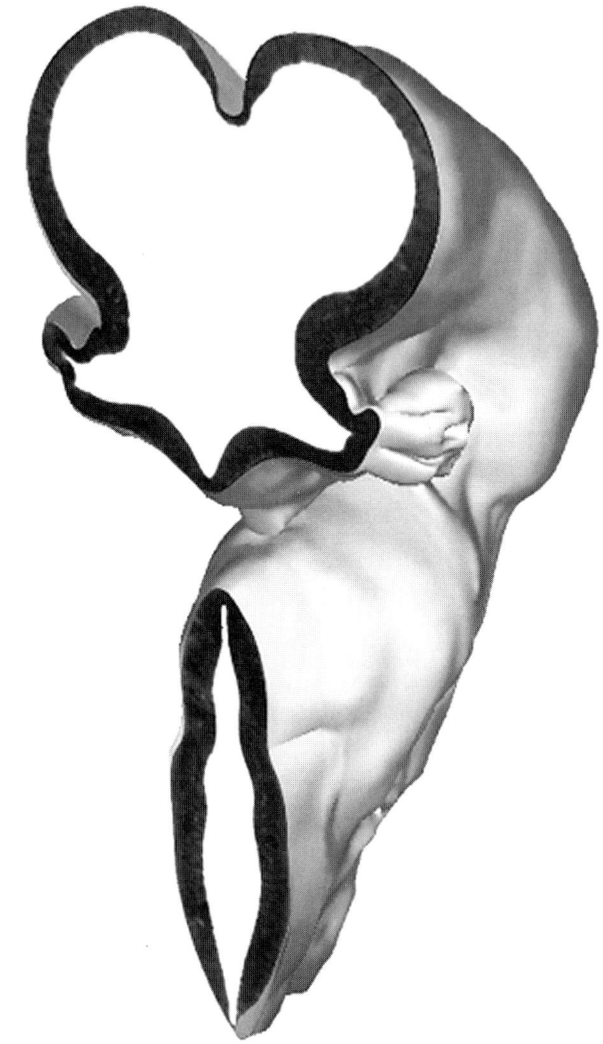

CORONAL FIGURE 6

AMYGDALA (A)

A1 amygdaloid neuroepithelium
A3 amygdala (or differentiating field)

CEREBRAL CORTEX (C)

C1 neocortical neuroepithelium
C11 cingulate cortex neuroepithelium

GLANDS AND PARANEURAL STRUCTURES (g)

g6 pituitary gland (primordium)
g9 lamina terminalis
g10 velum medullare

HYPOTHALAMUS (H)

H1a anterior hypothalamic neuroepithelium
H1v ventral hypothalamic neuroepithelium
H3 hypothalamus (or differentiating field)

NERVES AND NUCLEI, CRANIAL (n)

n8g vestibular ganglion

VENTRICLES (v)

vlc lateral ventricle, central
vld lateral ventricle, dorsal
vlm foramen of Monro
vlp lateral ventricle, posterior horn
v3h third ventricle, hypothalamic
v3i third ventricle, infundibular recess
v4a fourth ventricle, anterior
v4p fourth ventricle, posterior

PONTINE AREA (X)

X1 pontine neuroepithelium
X3 pons (or differentiating field)

MEDULLA (Y)

Y1 medullary neuroepithelium
Y3 medulla (or differentiating field)

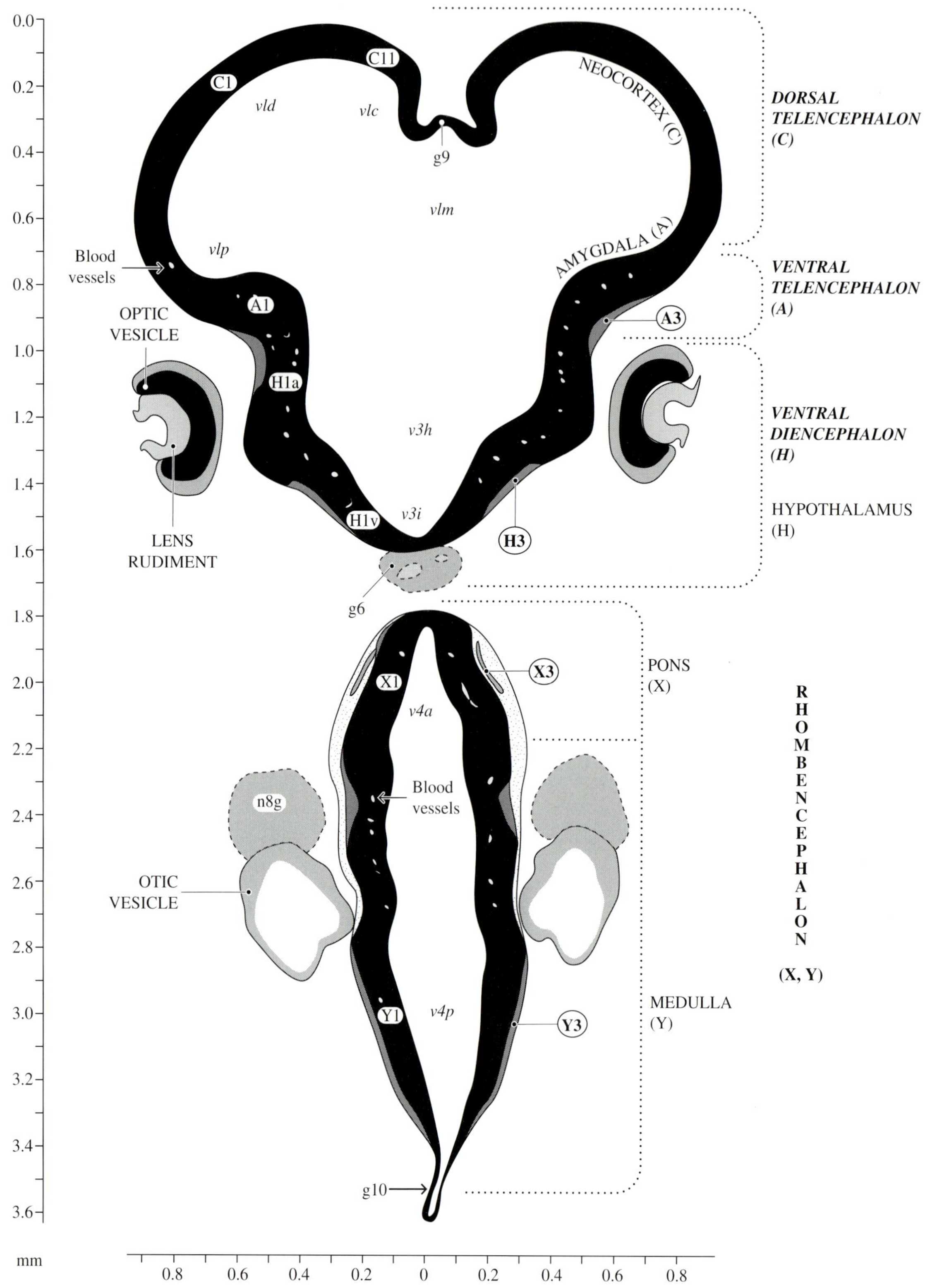

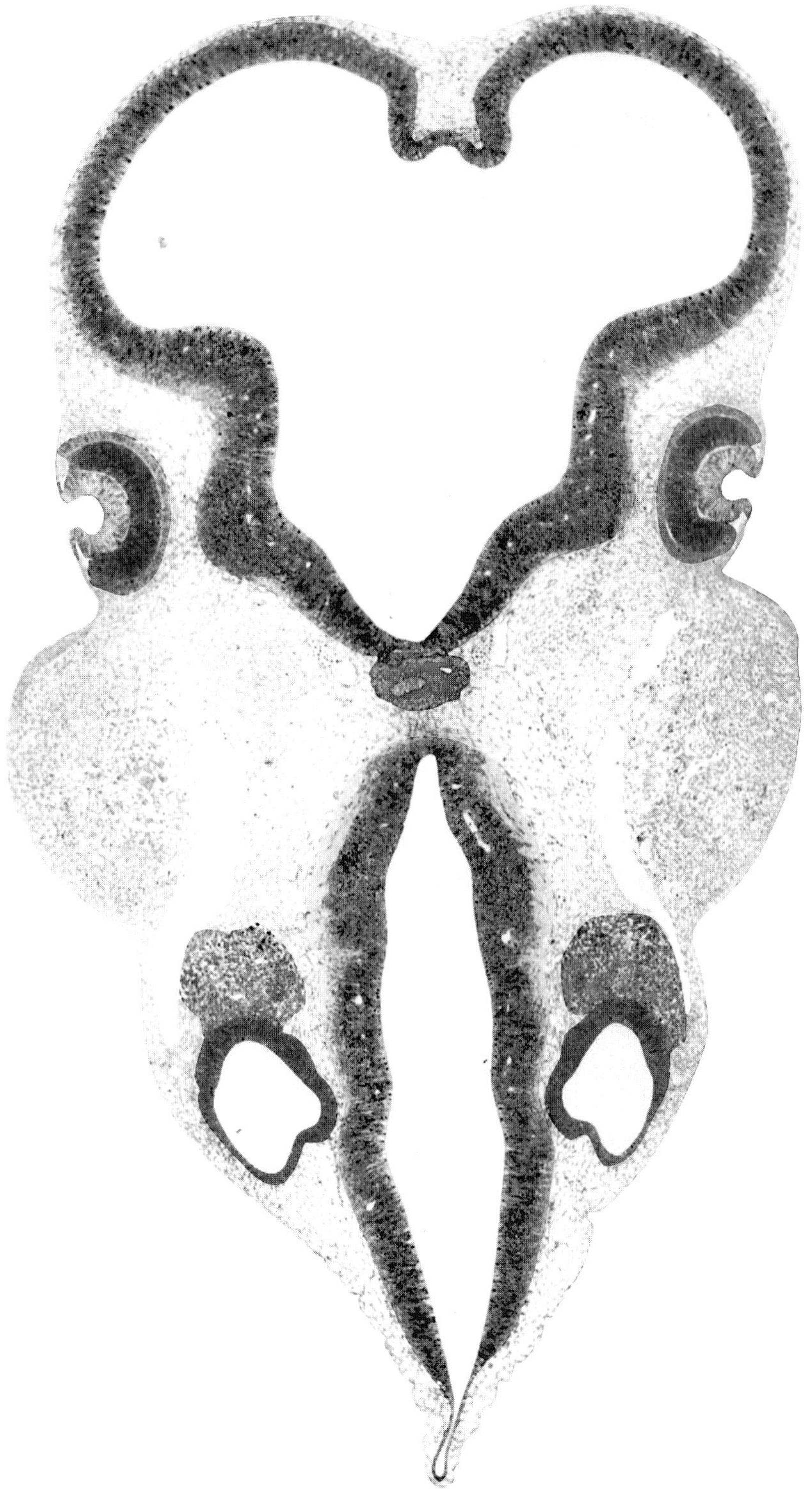

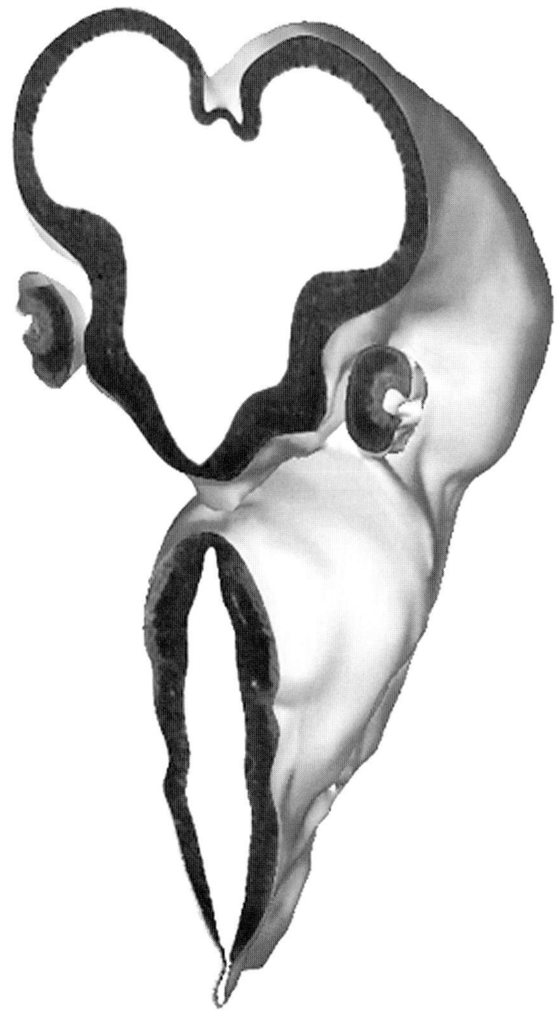

CORONAL FIGURE 7

AMYGDALA (A)

A1 amygdaloid neuroepithelium
A3 amygdala (or differentiating field)

BASAL GANGLIA (B)

B21 strionuclear neuroepithelium

CEREBRAL CORTEX (C)

C1 cortical neuroepithelium
C11 cingulate cortex neuroepithelium

GLANDS AND PARANEURAL STRUCTURES (g)

g9 lamina terminalis
g10 velum medullare

HYPOTHALAMUS (H)

H1a anterior hypothalamic neuroepithelium
H1f Forelian neuroepithelium
H1v ventral hypothalamic neuroepithelium
H3 hypothalamus (or differentiating field)

NERVES AND NUCLEI, CRANIAL (n)

n8b vestibulocochlear boundary cap
n8g vestibular ganglion

VENTRICLES (v)

vld lateral ventricle, dorsal
vlm foramen of Monro
vlp lateral ventricle, posterior horn
v3e third ventricle, epithalamic
v3h third ventricle, hypothalamic
v3i third ventricle, infundibular recess
v3t third ventricle, thalamic
v4a fourth ventricle, anterior
v4p fourth ventricle, posterior

PONTINE AREA (X)

X1 pontine neuroepithelium
X3 pons (or differentiating field)

MEDULLA (Y)

Y1 medullary neuroepithelium
Y3 medulla (or differentiating field)

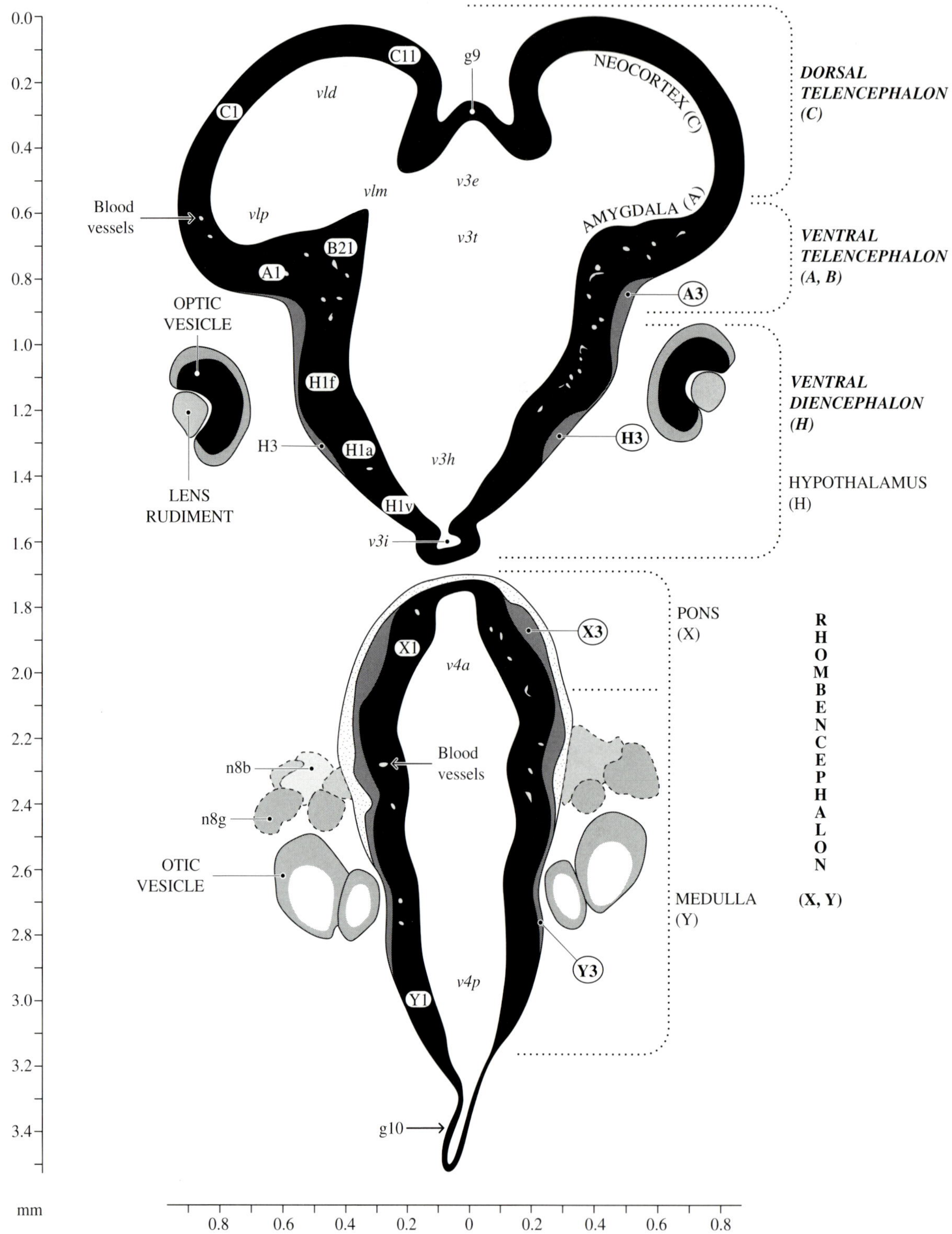

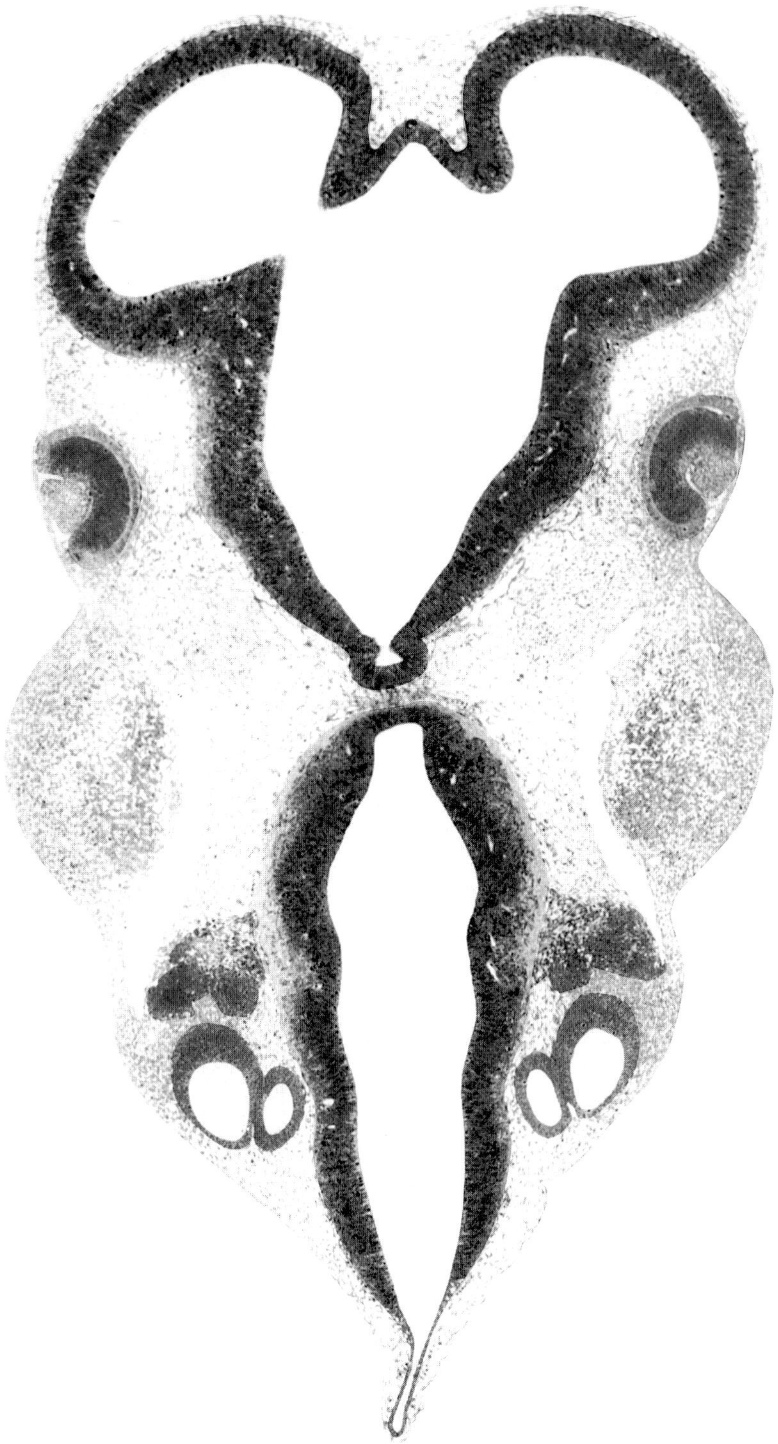

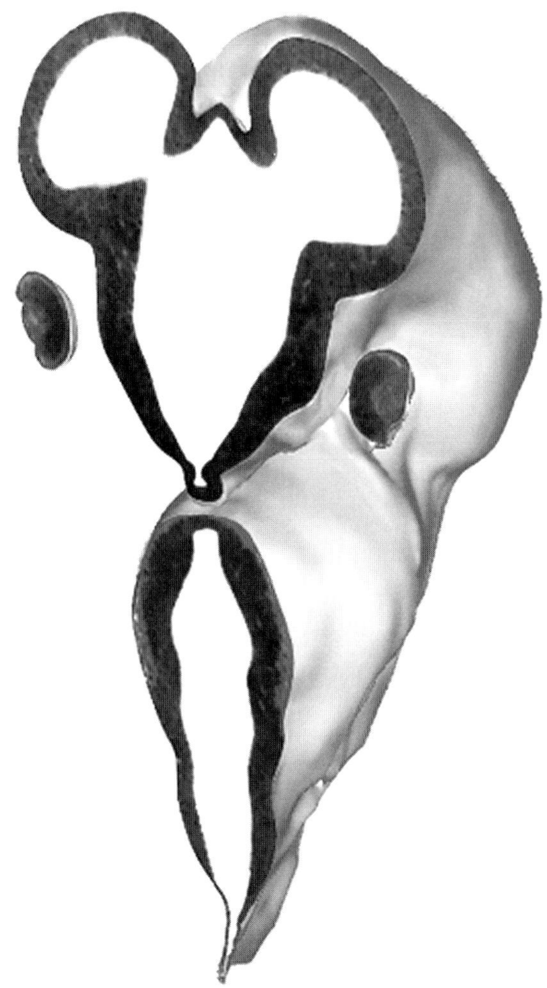

CORONAL FIGURE 8

AMYGDALA (A)

A1 amygdaloid neuroepithelium
A3 amygdala (or differentiating field)

BASAL GANGLIA (B)

B21 strionuclear neuroepithelium

CEREBRAL CORTEX (C)

C1 neocortical neuroepithelium
C11 cingulate cortex neuroepithelium

GLANDS AND PARANEURAL STRUCTURES (g)

g5 pineal gland (primordium)
g10 velum medullare

HYPOTHALAMUS (H)

H1f Forelian neuroepithelium
H1i intermediate hypothalamic neuroepithelium
H1v ventral hypothalamic neuroepithelium
H3 hypothalamus (or differentiating field)
H3i intermediate hypothalamus (or differentiating field)

NERVES AND NUCLEI, CRANIAL (n)

n5g trigeminal ganglion
n8b vestibulocochlear boundary cap
n8g vestibular ganglion

THALAMUS (T)

T1a anterior thalamic neuroepithelium
T1e epithalamic neuroepithelium

VENTRICLES (v)

vld lateral ventricle, dorsal
vlp lateral ventricle, posterior horn
v3e third ventricle, epithalamic
v3h third ventricle, hypothalamic
v3t third ventricle, thalamic
v4a fourth ventricle, anterior
v4p fourth ventricle, posterior

PONTINE AREA (X)

X1 pontine neuroepithelium
X3 pons (or differentiating field)

MEDULLA (Y)

Y1 medullary neuroepithelium
Y3 medulla (or differentiating field)

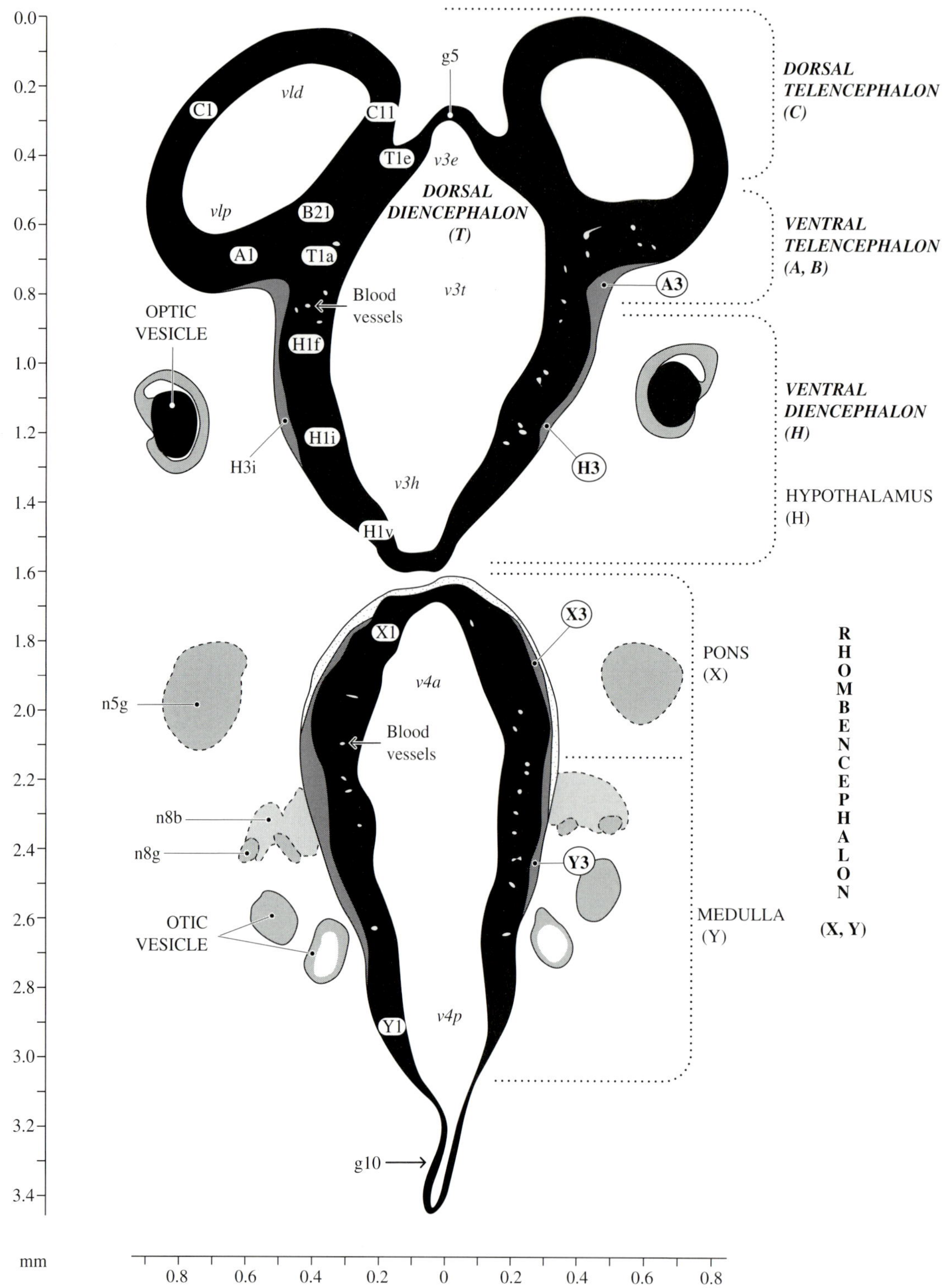

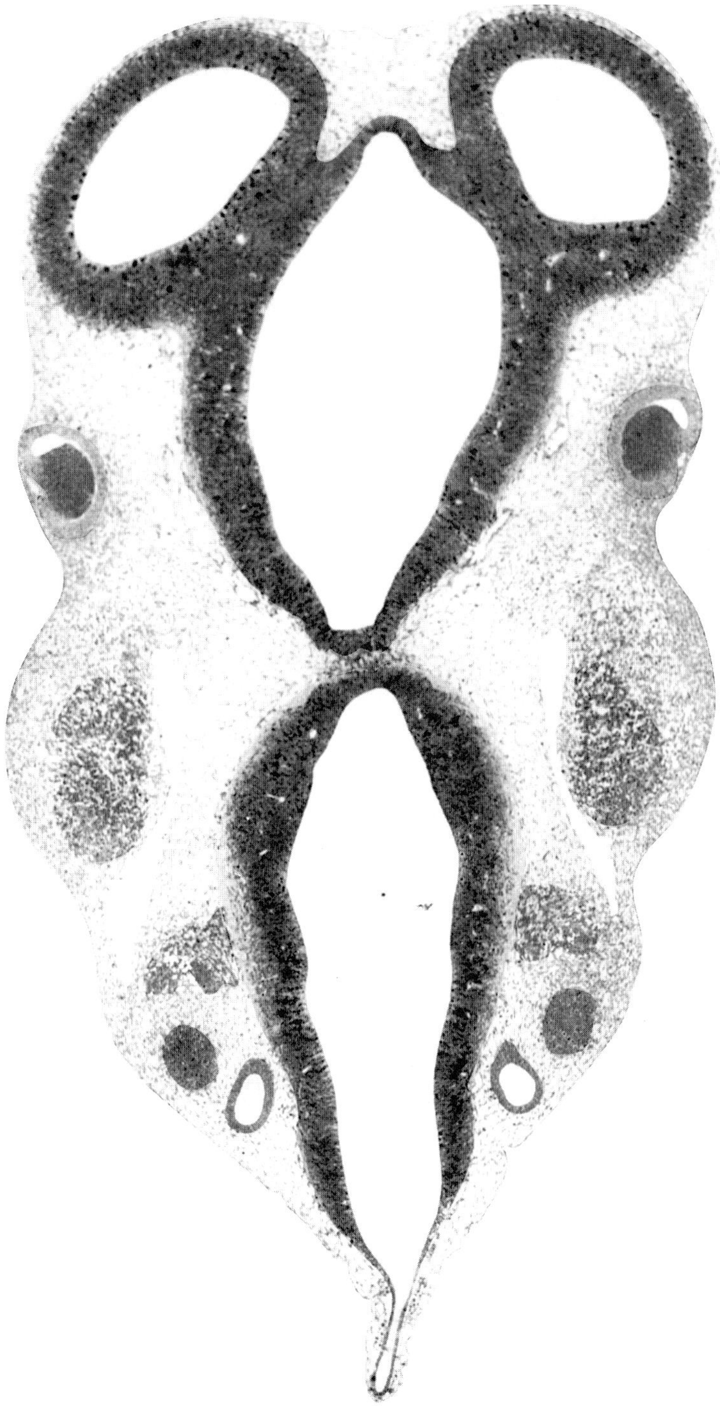

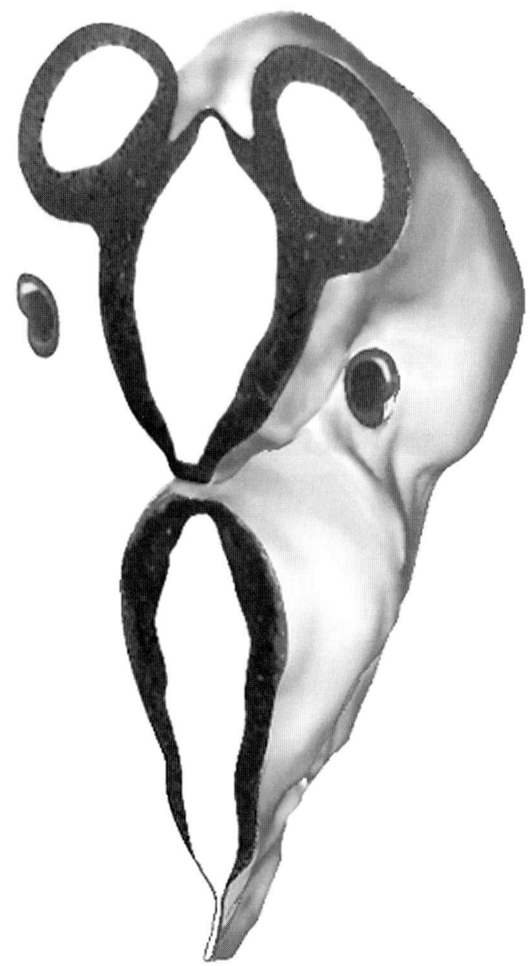

CORONAL FIGURE 9

CEREBRAL CORTEX (C)

C1 neocortical neuroepithelium

GLANDS AND PARANEURAL STRUCTURES (g)

g5 pineal gland (primordium)
g10 velum medullare

HYPOTHALAMUS (H)

H1f Forelian neuroepithelium
H1i intermediate hypothalamic neuroepithelium
H1v ventral hypothalamic neuroepithelium

NERVES AND NUCLEI, CRANIAL (n)

n5g trigeminal ganglion

THALAMUS (T)

T1e epithalamic neuroepithelium
T1i intermediate thalamic neuroepithelium

VENTRICLES (v)

v3e third ventricle, epithalamic
v3h third ventricle, hypothalamic
v3t third ventricle, thalamic
v4a fourth ventricle, anterior
v4p fourth ventricle, posterior

PONTINE AREA (X)

X1 pontine neuroepithelium
X3 pons (or differentiating field)

MEDULLA (Y)

Y1 medullary neuroepithelium

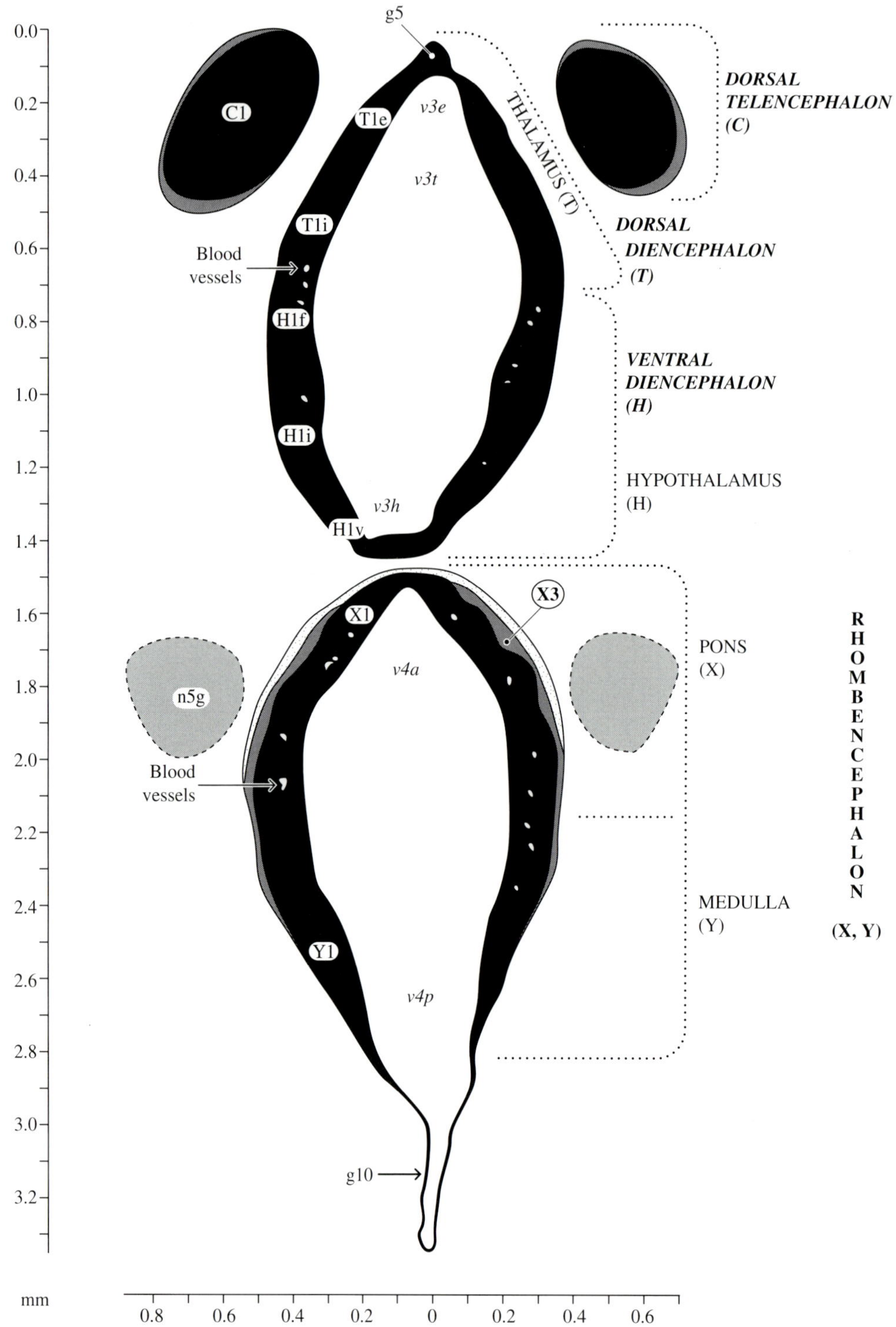

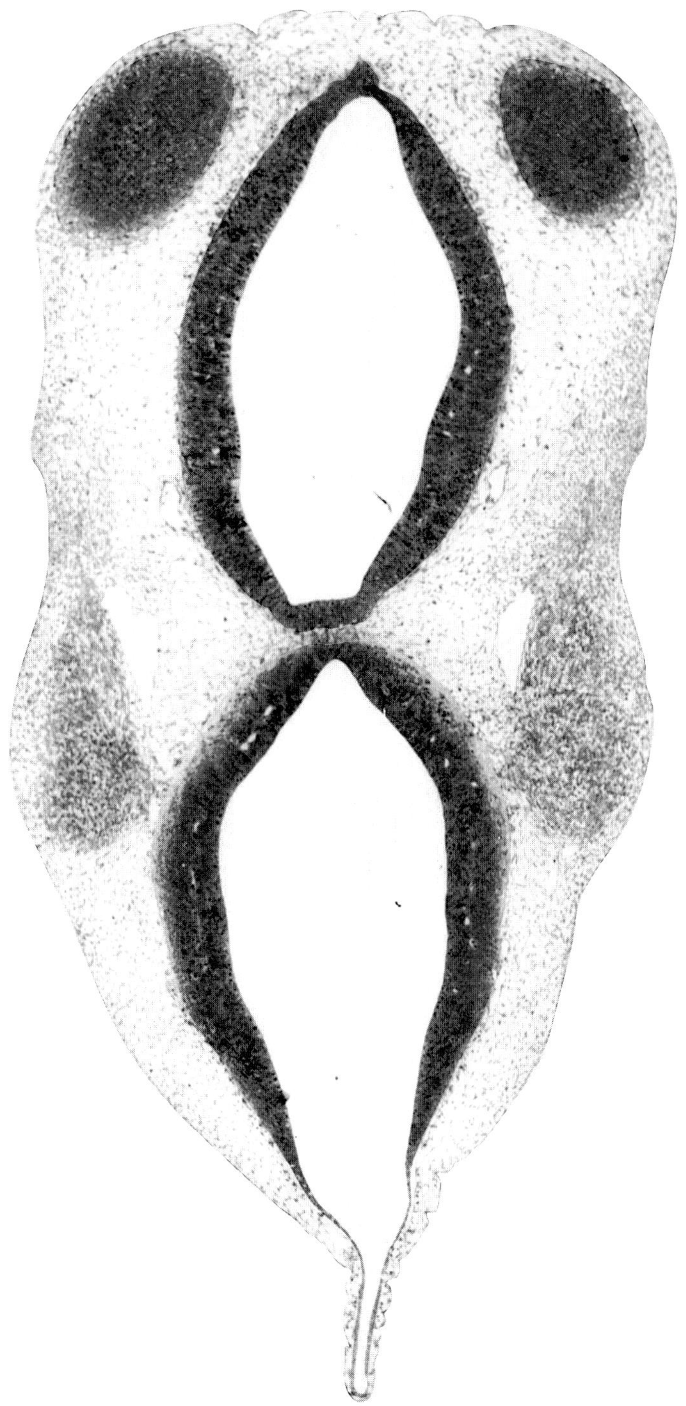

CORONAL FIGURE 10

GLANDS AND PARANEURAL STRUCTURES (g)

g10 velum medullare

HYPOTHALAMUS (H)

H1f Forelian neuroepithelium
H1*l* lateral hypothalamic neuroepithelium
H1v ventral hypothalamic neuroepithelium
H3 hypothalamus (or differentiating field)
H3f fields of Forel (or differentiating field)

NERVES AND NUCLEI, CRANIAL (n)

n5b trigeminal boundary cap

THALAMUS (T)

T1e epithalamic neuroepithelium
T1i intermediate thalamic neuroepithelium

VENTRICLES (v)

v3e third ventricle, epithalamic
v3h third ventricle, hypothalamic
v3t third ventricle, thalamic
v4a fourth ventricle, anterior
v4p fourth ventricle, posterior

PONTINE AREA (X)

X1 pontine neuroepithelium
X3 pons (or differentiating field)

MEDULLA (Y)

Y1 medullary neuroepithelium

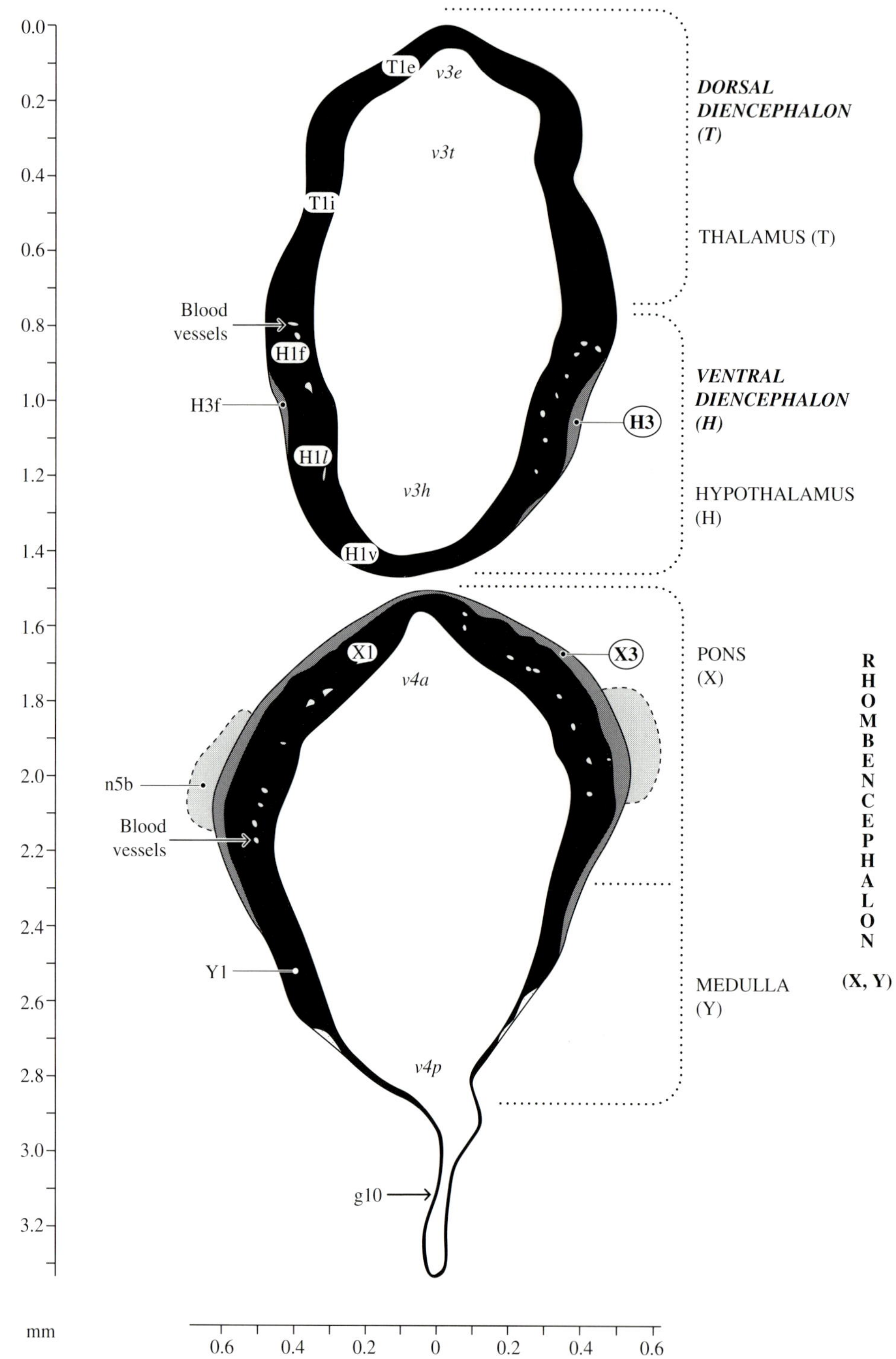

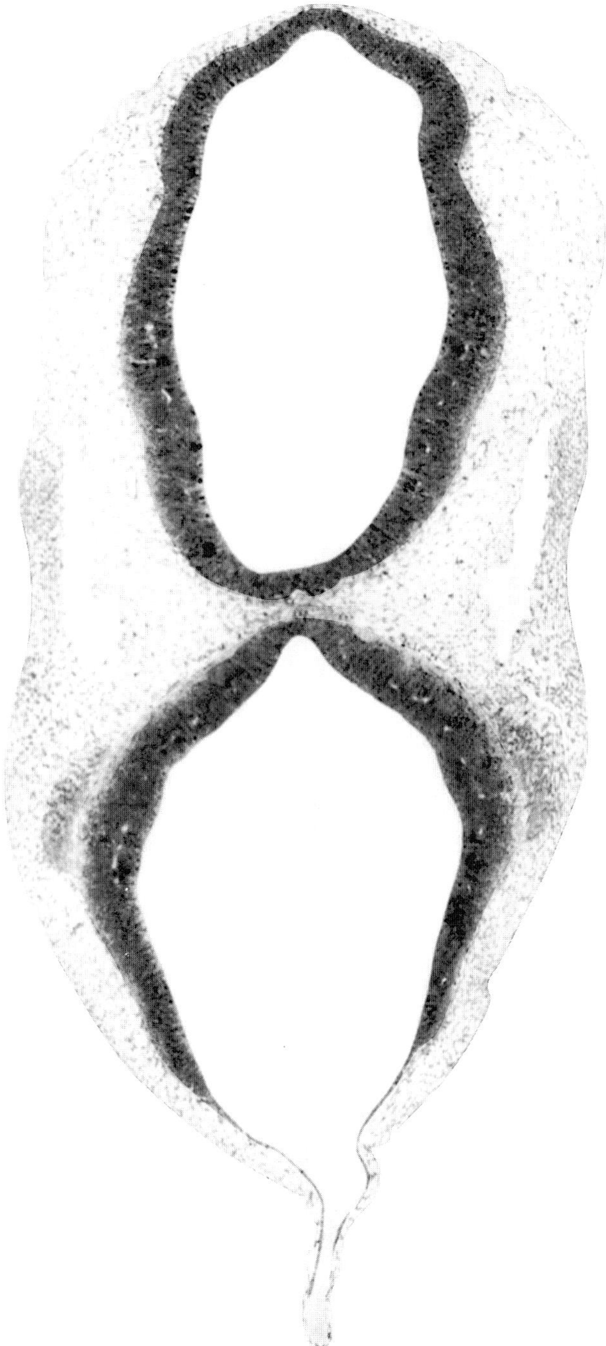

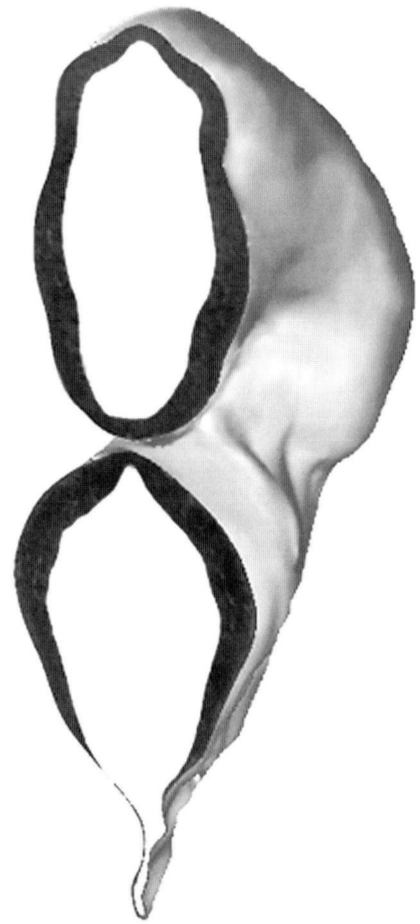

CORONAL FIGURE 11

GLANDS AND PARANEURAL STRUCTURES (g)

g10 velum medullare

HYPOTHALAMUS (H)

H1f Forelian neuroepithelium
H1*l* lateral hypothalamic neuroepithelium
H1v ventral hypothalamic neuroepithelium
H3 hypothalamus (or differentiating field)
H3f fields of Forel (or differentiating field)
H3*l* lateral hypothalamus

CEREBELLUM (L)

L1 cerebellar neuroepithelium
L3 cerebellum (or differentiating field)

THALAMUS (T)

T1i intermediate thalamic neuroepithelium
T1p posterior thalamic neuroepithelium

VENTRICLES (v)

v3h third ventricle, hypothalamic
v3t third ventricle, thalamic
v4a fourth ventricle, anterior
v4p fourth ventricle, posterior

PONTINE AREA (X)

X1 pontine neuroepithelium
X3 pons (or differentiating field)

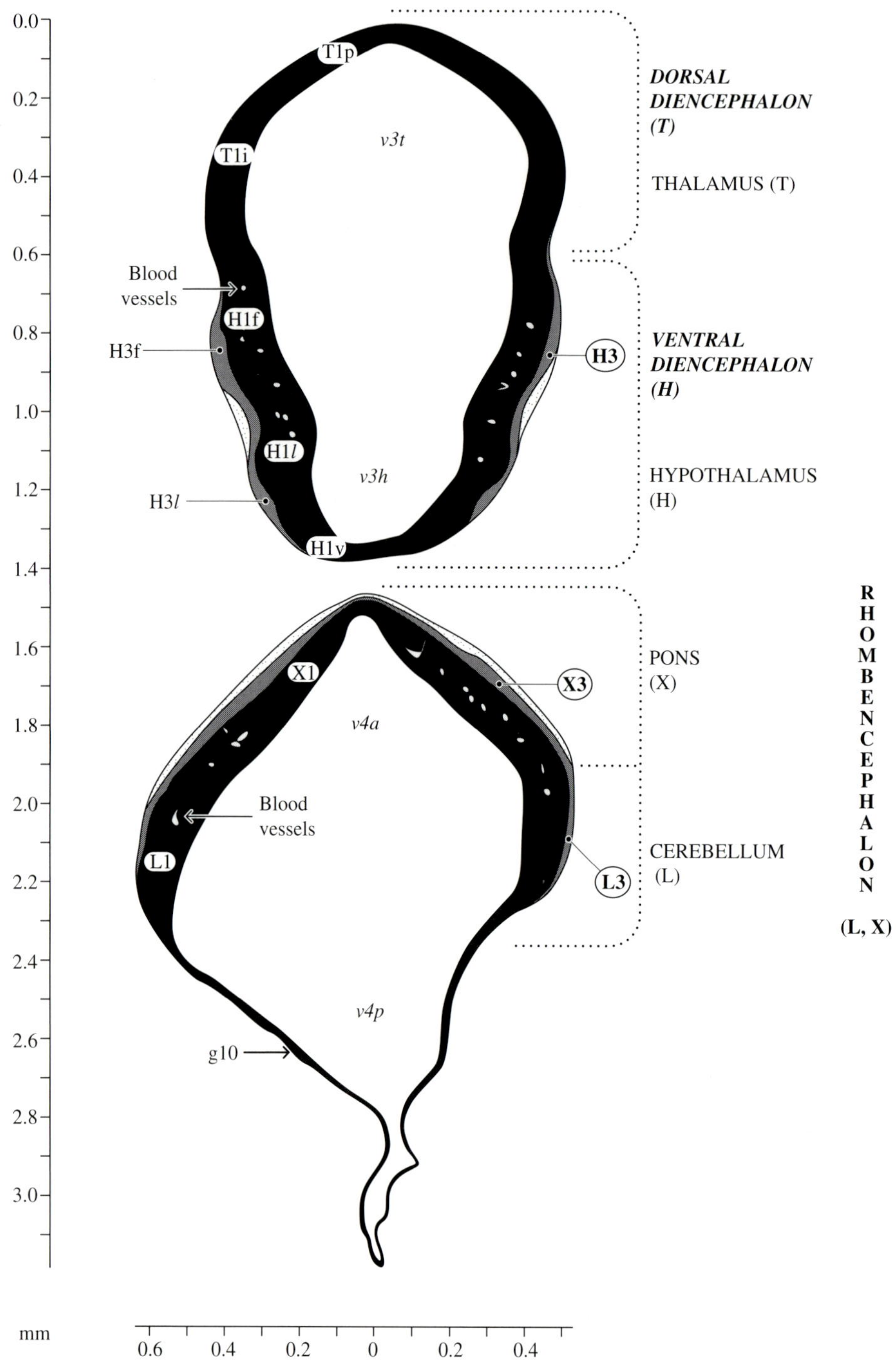

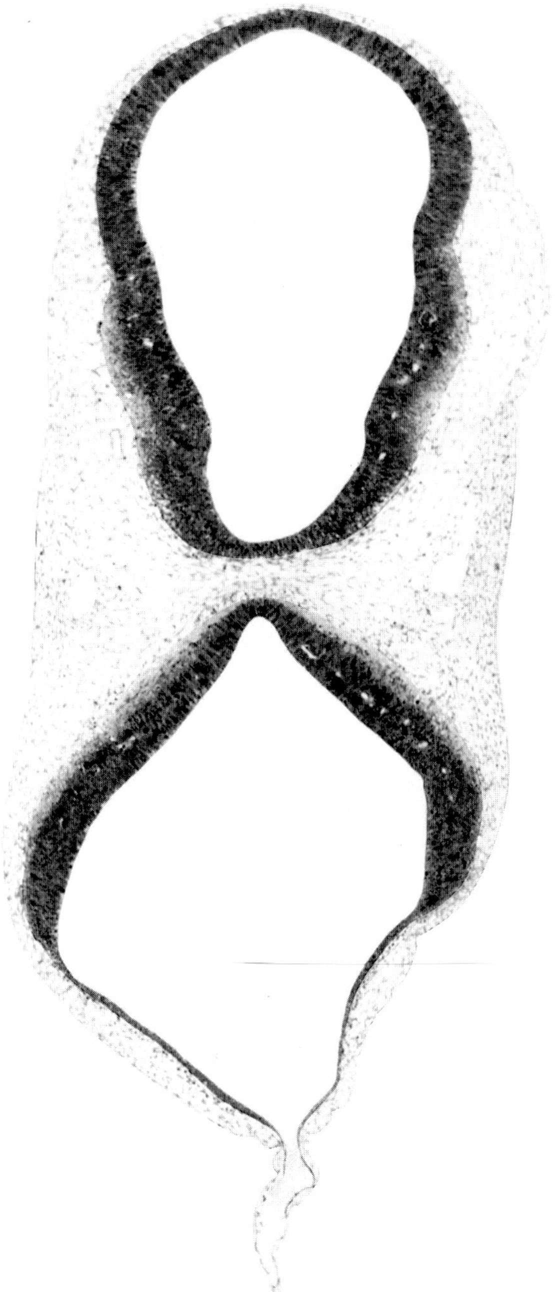

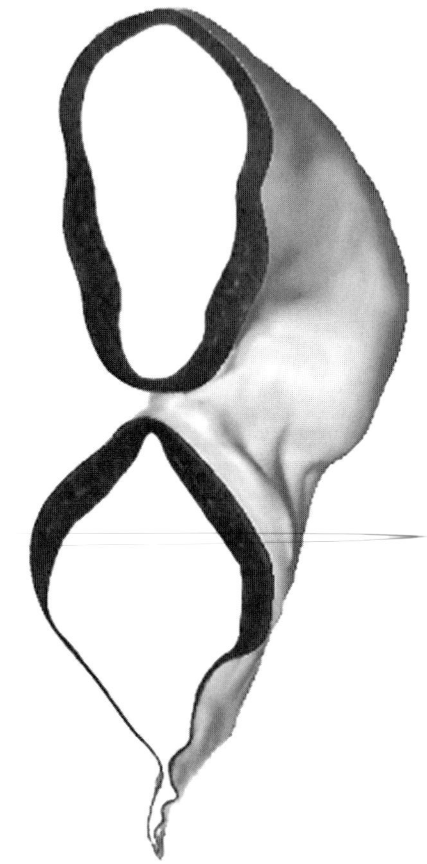

CORONAL FIGURE 12

GLANDS AND PARANEURAL STRUCTURES (g)

g10 velum medullare

HYPOTHALAMUS (H)

H1*l* lateral hypothalamic neuroepithelium
H1m mammillary neuroepithelium
H3 hypothalamus (or differentiating field)
H3*l* lateral hypothalamus (or differentiating field)

CEREBELLUM (L)

L1 cerebellar neuroepithelium
L3 cerebellum (or differentiating field)

MIDBRAIN (M)

M21 pretectal neuroepithelium

THALAMUS (T)

T1i intermediate thalamic neuroepithelium
T1p posterior thalamic neuroepithelium

VENTRICLES (v)

vapr aqueduct, pretectal
v3h third ventricle, hypothalamic
v3m third ventricle, mammillary recess
v3t third ventricle, thalamic
v4a fourth ventricle, anterior
v4p fourth ventricle, posterior

PONTINE AREA (X)

X1 pontine neuroepithelium
X3 pons (or differentiating field)

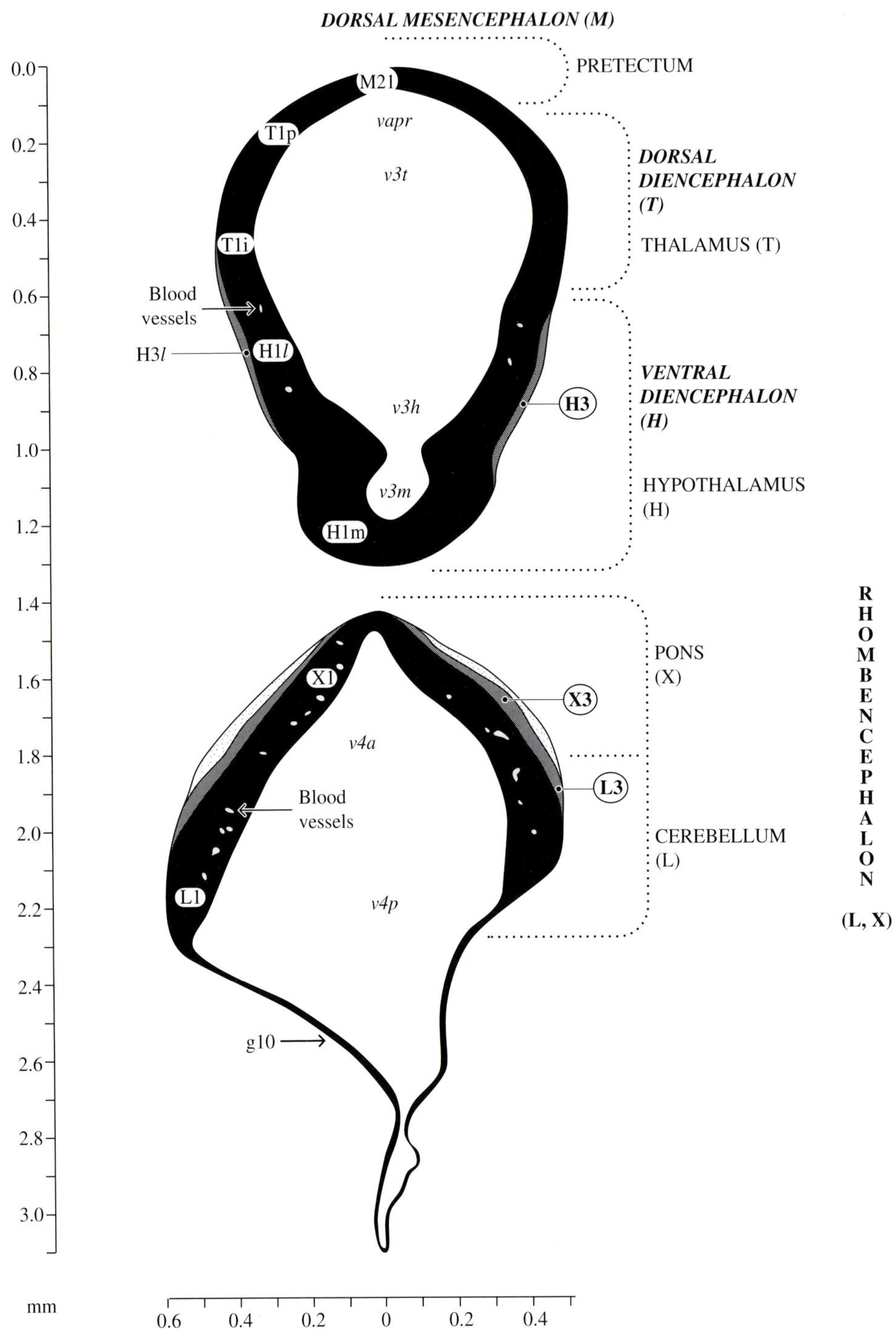

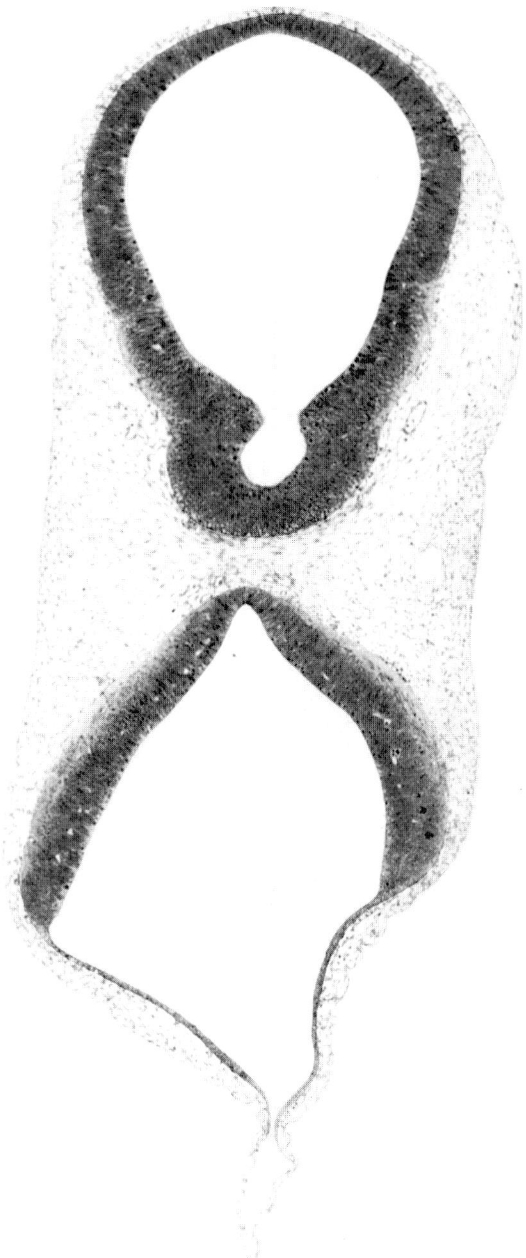

CORONAL FIGURE 13

GLANDS AND PARANEURAL STRUCTURES (g)

g10 velum medullare

HYPOTHALAMUS (H)

H1m mammillary neuroepithelium
H1*l* lateral hypothalamic neuroepithelium
H1pm premammillary neuroepithelium
H1v ventral hypothalamic neuroepithelium
H3 hypothalamus (or differentiating field)

CEREBELLUM (L)

L1 cerebellar neuroepithelium
L3 cerebellum (or differentiating field)

MIDBRAIN (M)

M21 pretectal neuroepithelium

THALAMUS (T)

T1p posterior thalamic neuroepithelium

VENTRICLES (v)

vapr aqueduct, pretectal
v3h third ventricle, hypothalamic
v3m third ventricle, mammillary recess
v3t third ventricle, thalamic
v4a fourth ventricle, anterior
v4p fourth ventricle, posterior

PONTINE AREA (X)

X1 pontine neuroepithelium
X3 pons (or differentiating field)

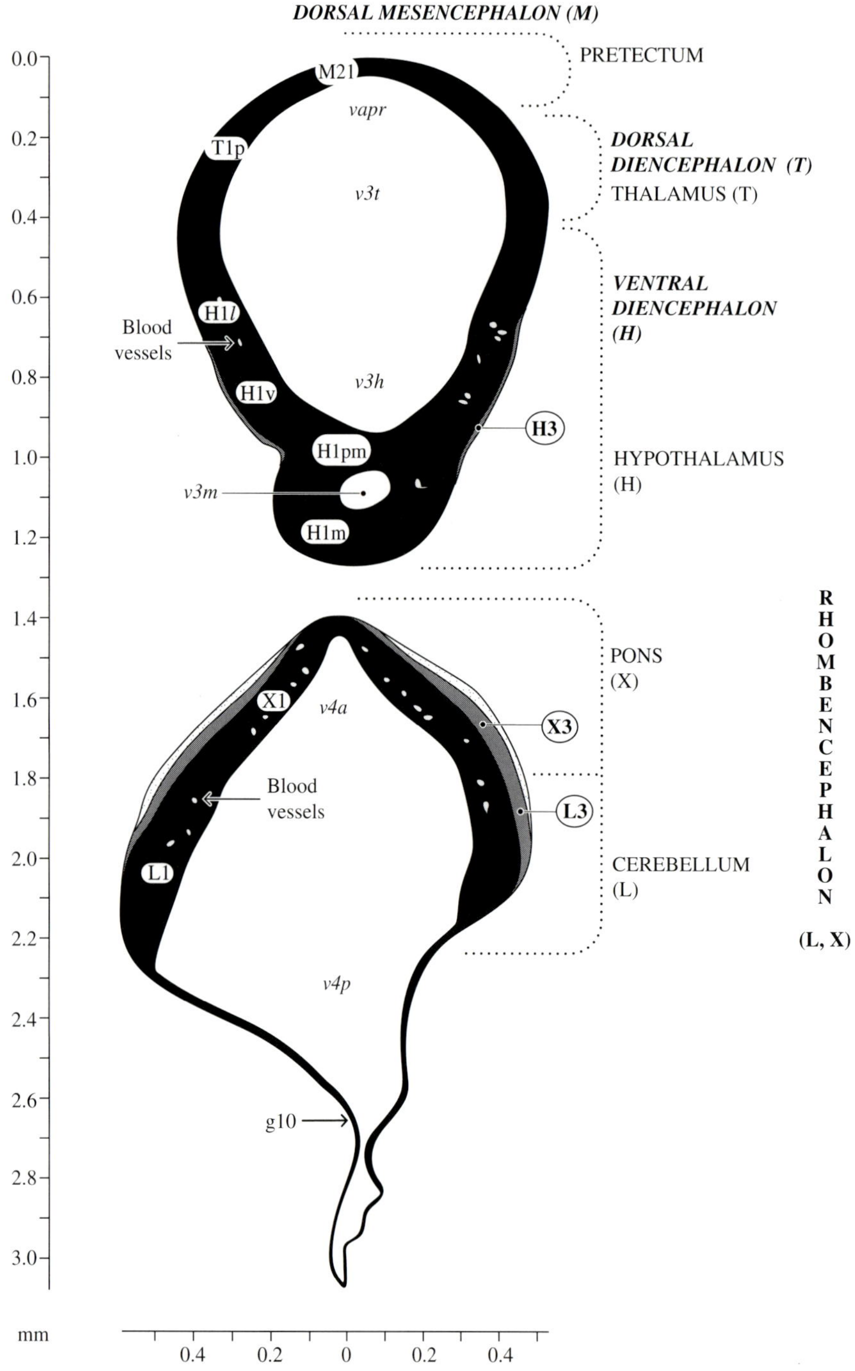

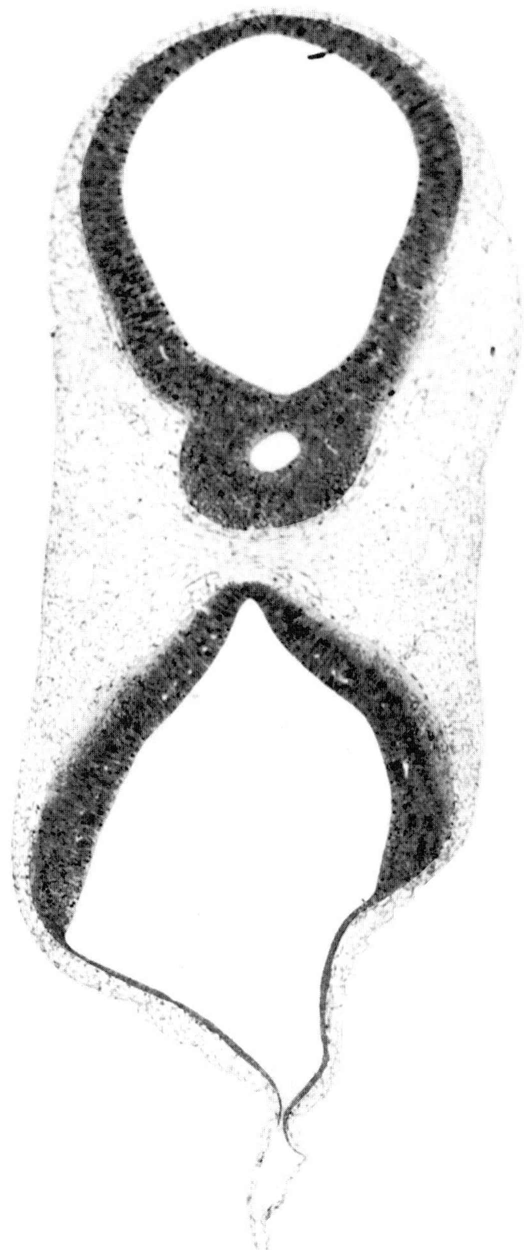

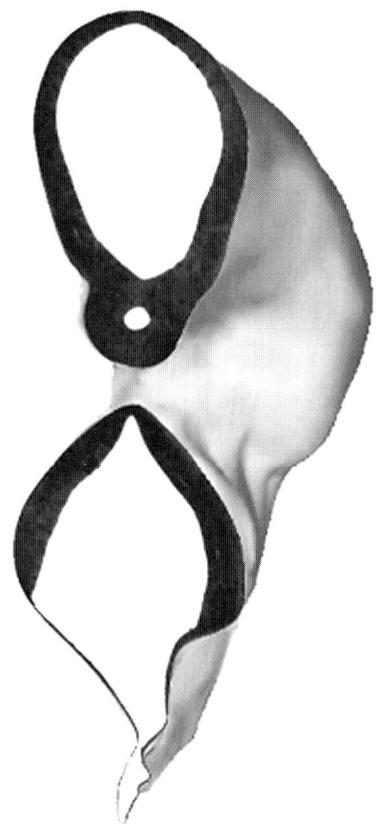

GLANDS AND PARANEURAL STRUCTURES (g)

g10 velum medullare

HYPOTHALAMUS (H)

H1m mammillary neuroepithelium
H1p posterior hypothalamic neuroepithelium
H1pm premammillary neuroepithelium
H1v ventral hypothalamic neuroepithelium
H3 hypothalamus (or differentiating field)

CEREBELLUM (L)

L1 cerebellar neuroepithelium
L3 cerebellum (or differentiating field)

MIDBRAIN (M)

M21 pretectal neuroepithelium

THALAMUS (T)

T1p posterior thalamic neuroepithelium

VENTRICLES (v)

vapr aqueduct, pretectal
v3h third ventricle, hypothalamic
v3m third ventricle, mammillary recess
v3t third ventricle, thalamic
v4a fourth ventricle, anterior
v4p fourth ventricle, posterior

PONTINE AREA (X)

X1 pontine neuroepithelium
X3 pons (or differentiating field)

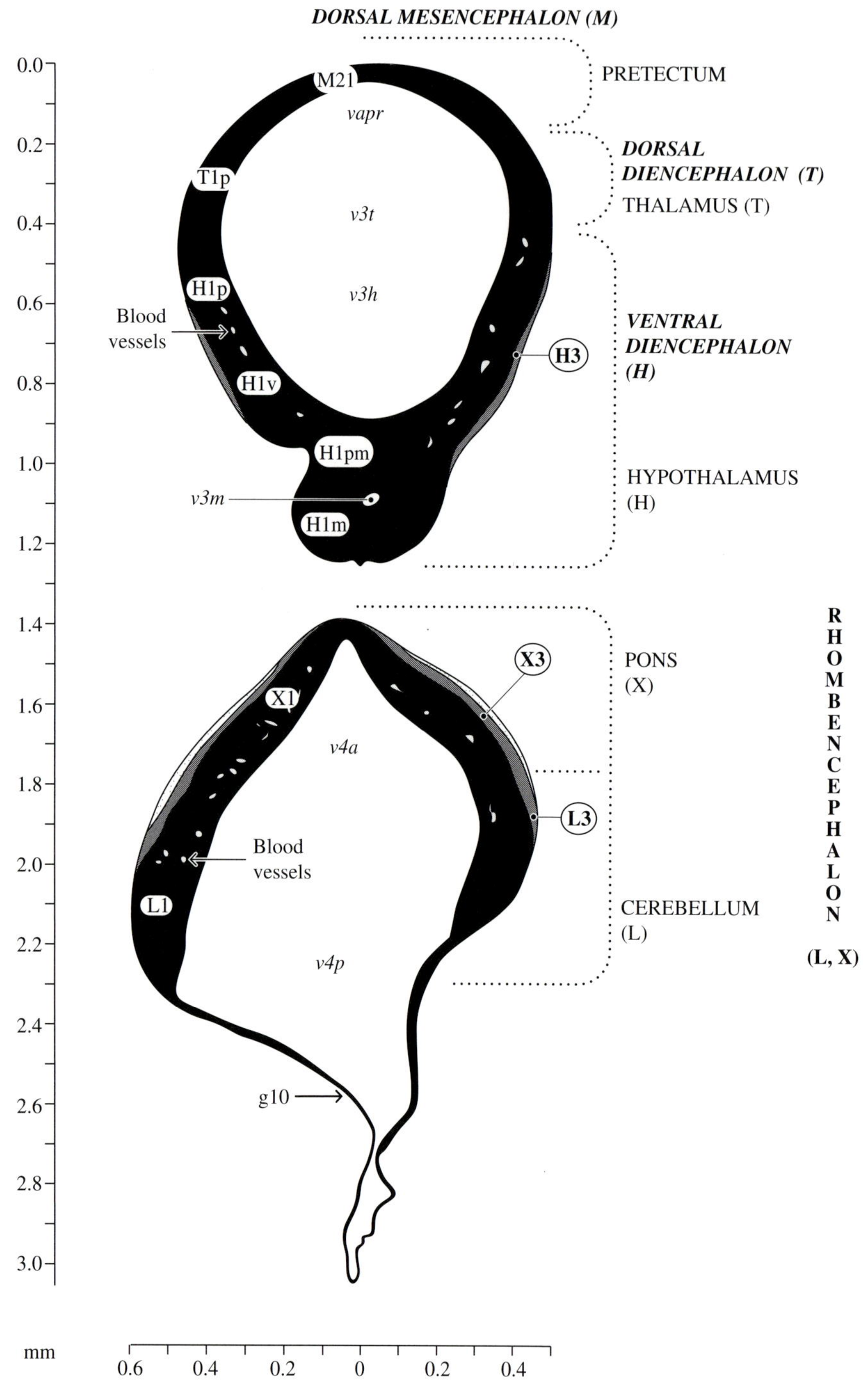

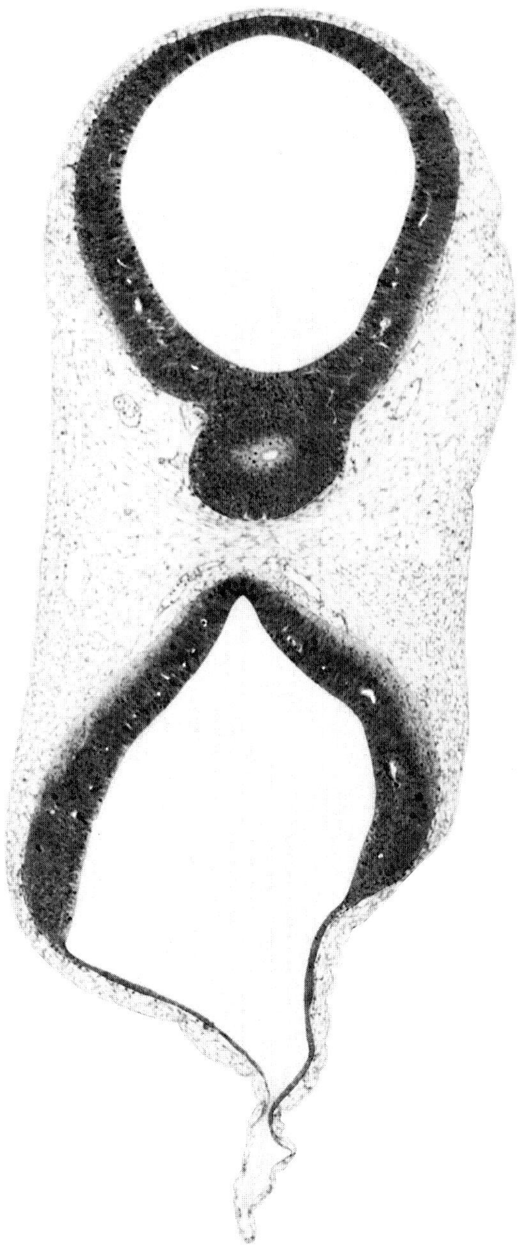

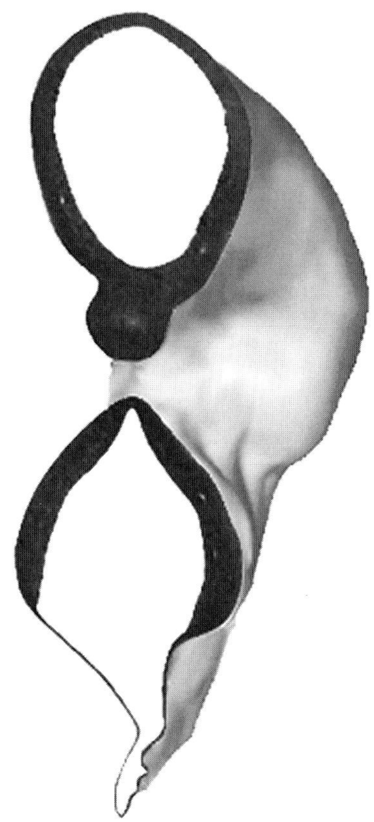

GLANDS AND PARANEURAL STRUCTURES (g)

g10 velum medullare

CEREBELLUM (L)

L1 cerebellar neuroepithelium
L3 cerebellum (or differentiating field)

MIDBRAIN (M)

M11a anterior tegmental neuroepithelium
M13 tegmentum (or differentiating field)
M21 pretectal neuroepithelium

THALAMUS (T)

T1p posterior thalamic neuroepithelium

VENTRICLES (v)

vapr aqueduct, pretectal
vat aqueduct, tegmental
v4a fourth ventricle, anterior
v4p fourth ventricle, posterior

PONTINE AREA (X)

X1 pontine neuroepithelium
X3 pons (or differentiating field)
X31 isthmal neuroepithelium
X33 isthmus (or differentiating field)

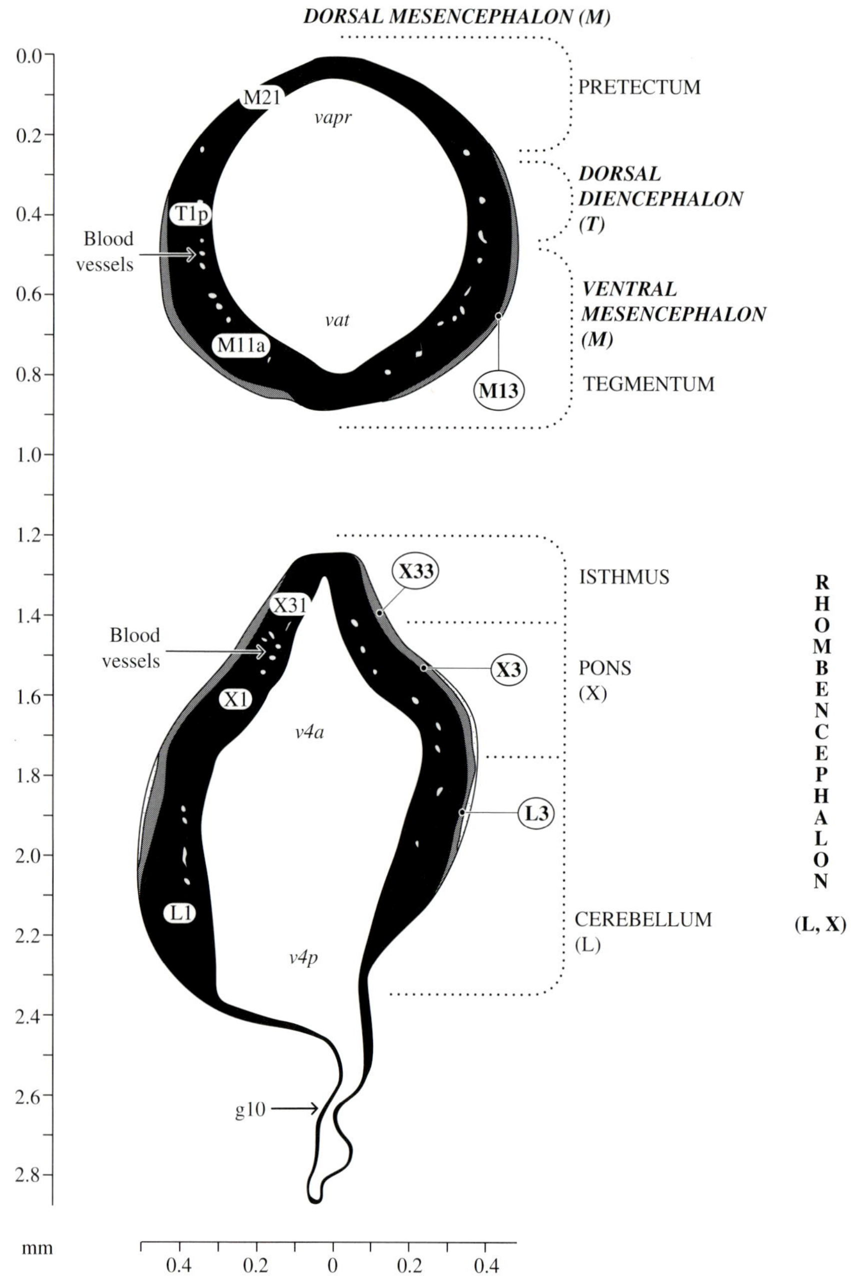

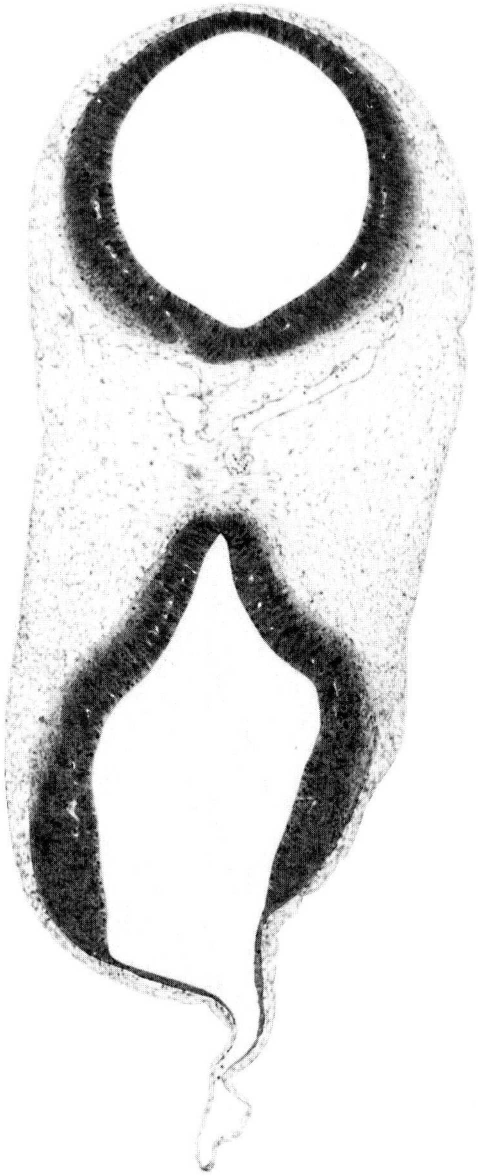

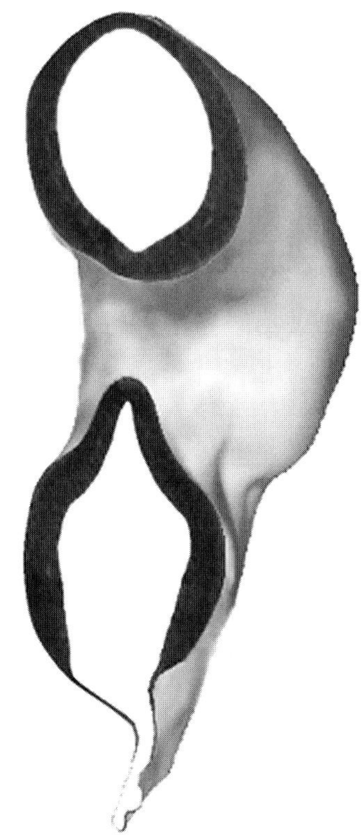

CORONAL FIGURE 16

GLANDS AND PARANEURAL STRUCTURES (g)

g10 velum medullare

CEREBELLUM (L)

L1 cerebellar neuroepithelium
L3 cerebellum (or differentiating field)

MIDBRAIN (M)

M1 superior tectal neuroepithelium
M11p posterior tegmental neuroepithelium
M13 tegmentum (or differentiating field)

VENTRICLES (v)

vas aqueduct, superior tectal
vat aqueduct, tegmental
v4a fourth ventricle, anterior
v4p fourth ventricle, posterior

PONTINE AREA (X)

X1a anterior pontine neuroepithelium
X1p posterior pontine neuroepithelium
X3 pons (or differentiating field)
X31 isthmal neuroepithelium
X33 isthmus (or differentiating field)

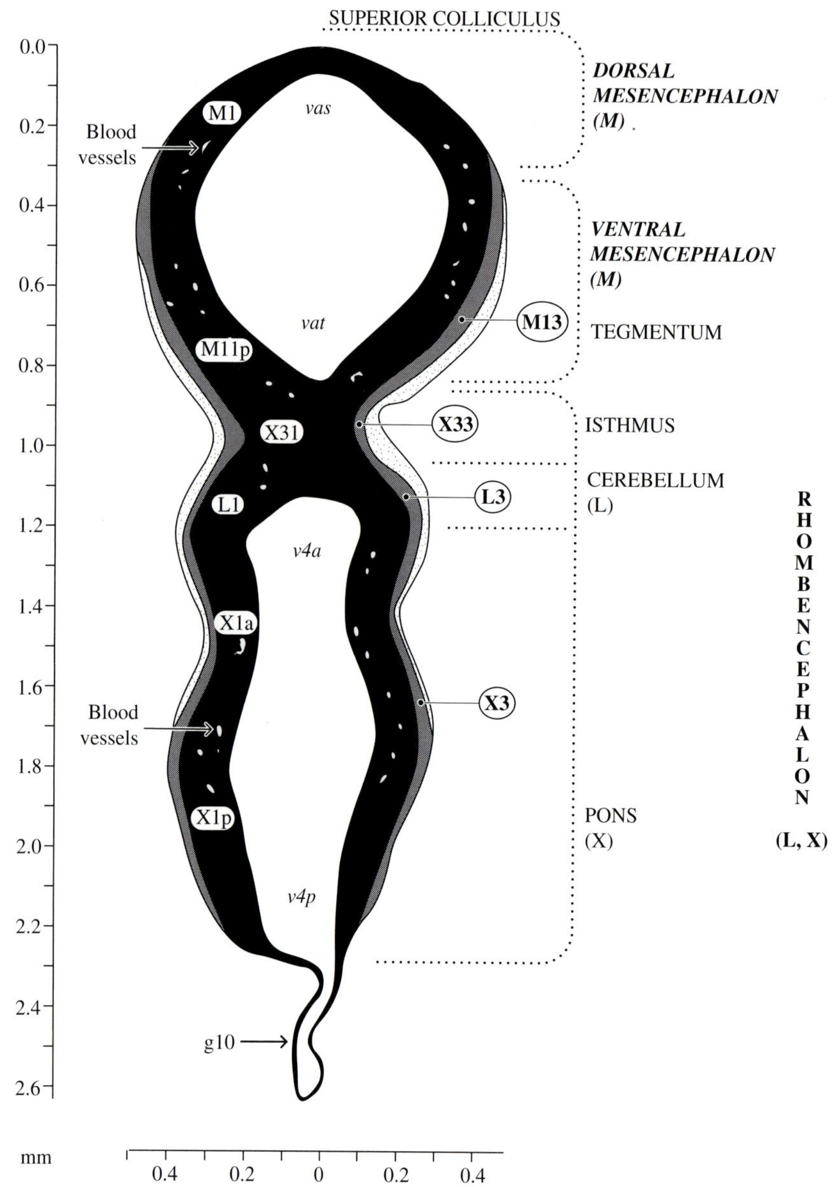

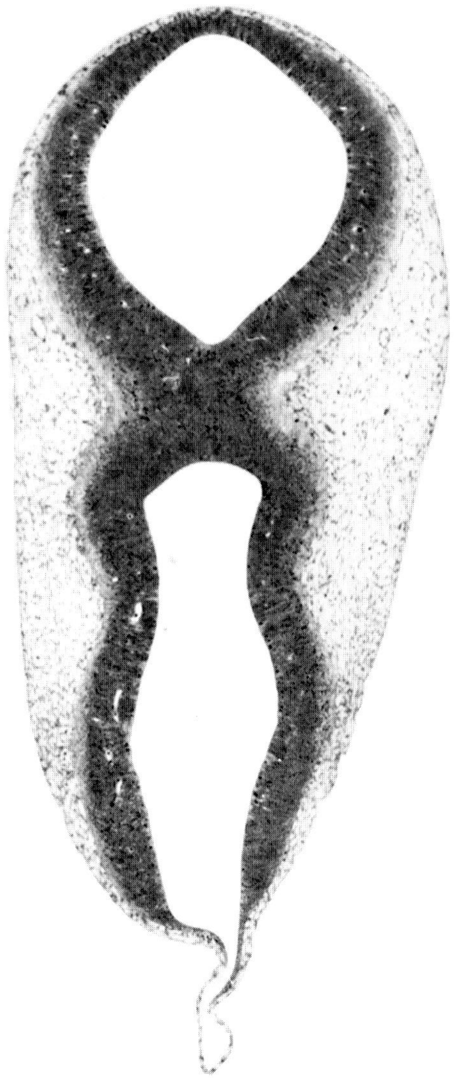

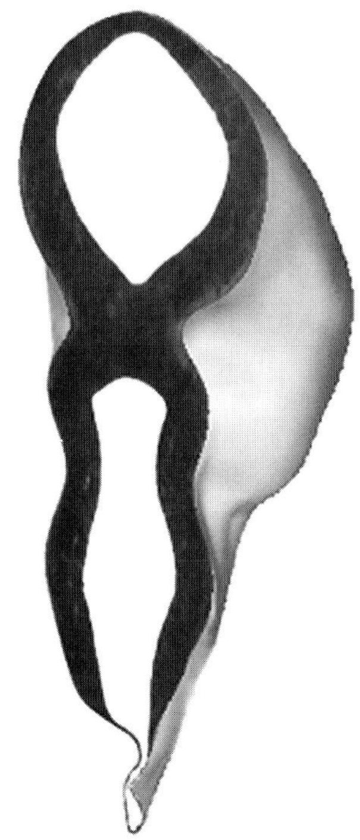

CORONAL FIGURE 17

GLANDS AND PARANEURAL STRUCTURES (g)

g10 velum medullare

MIDBRAIN (M)

M1 superior tectal neuroepithelium
M11p posterior tegmental neuroepithelium
M13 tegmentum (or differentiating field)
M31 inferior tectal neuroepithelium

VENTRICLES (v)

vai aqueduct, inferior tectal
vas aqueduct, superior tectal
vat aqueduct, tegmental
v4a fourth ventricle, anterior
v4i fourth ventricle, isthmal

PONTINE AREA (X)

X1a anterior pontine neuroepithelium
X3 pons (or differentiating field)
X31 isthmal neuroepithelium
X33 isthmus (or differentiating field)

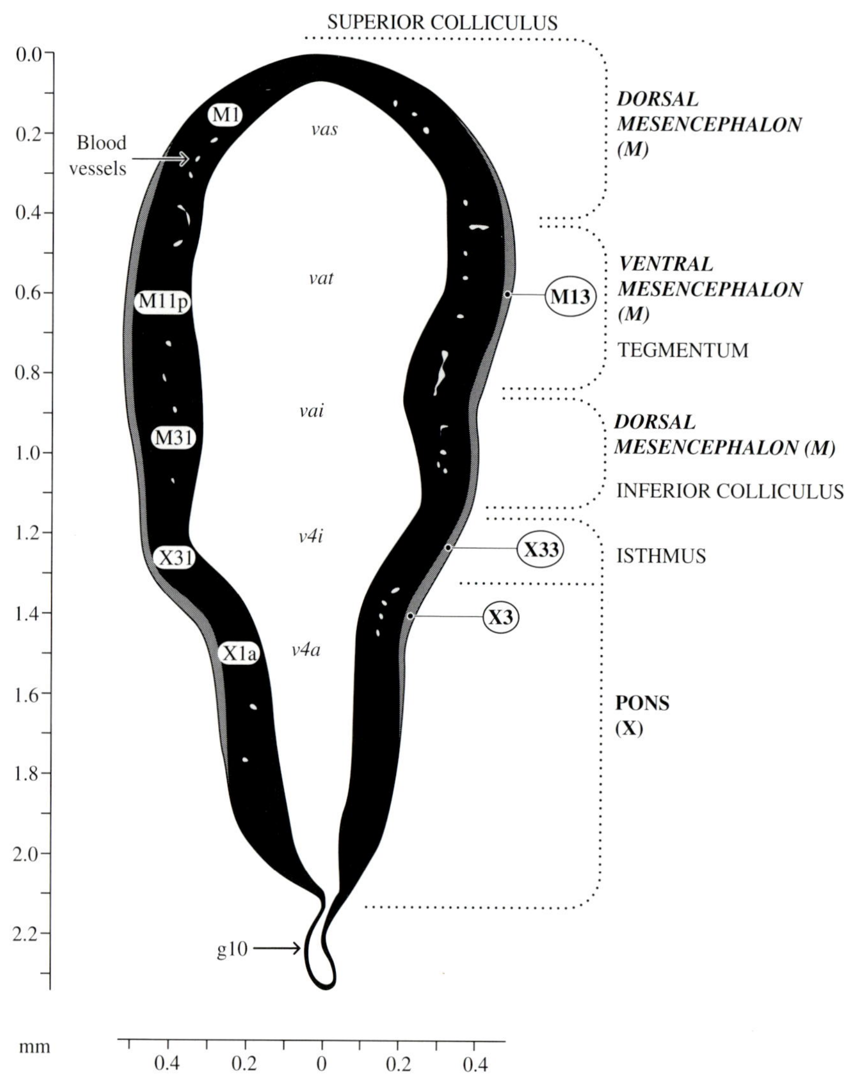

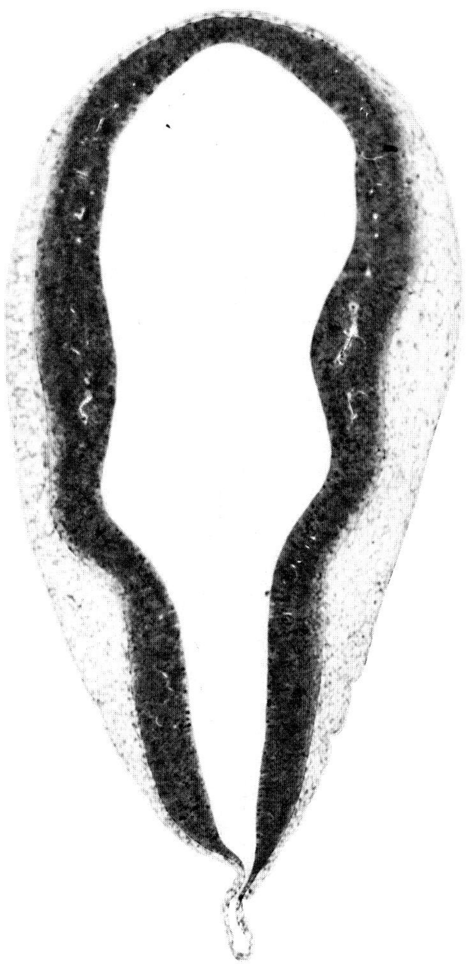

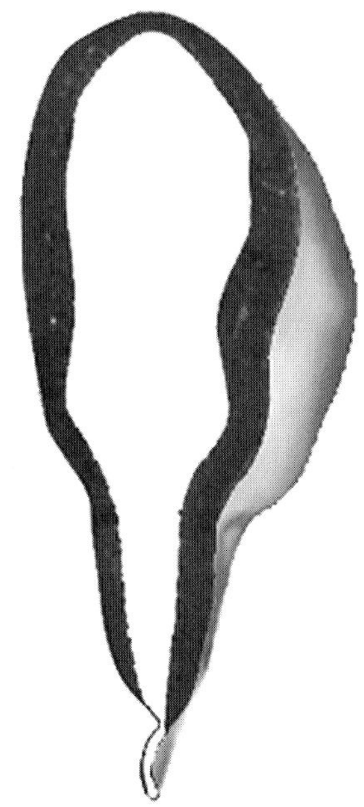

CORONAL FIGURE 18

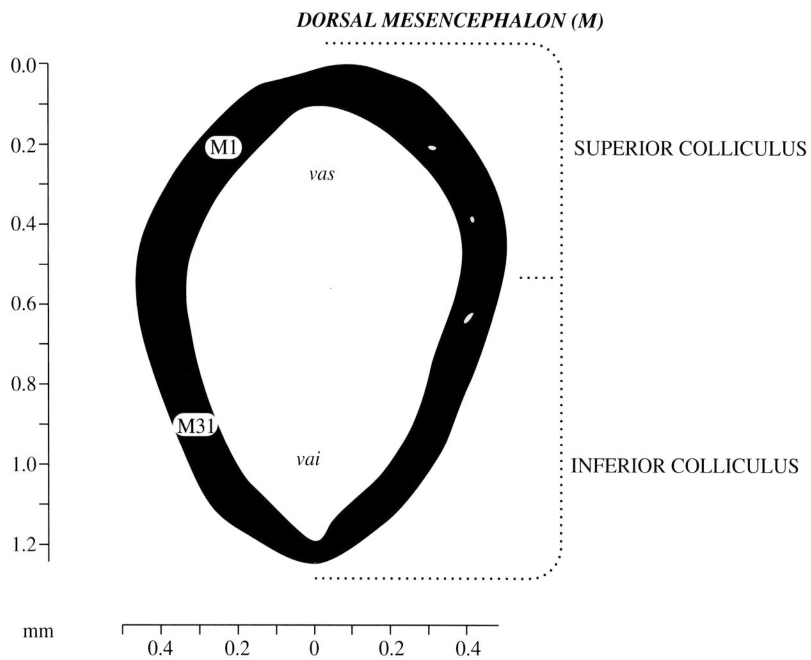

MIDBRAIN (M)

M1	superior tectal neuroepithelium
M31	inferior tectal neuroepithelium

VENTRICLES (v)

vai	aqueduct, inferior tectal
vas	aqueduct, superior tectal

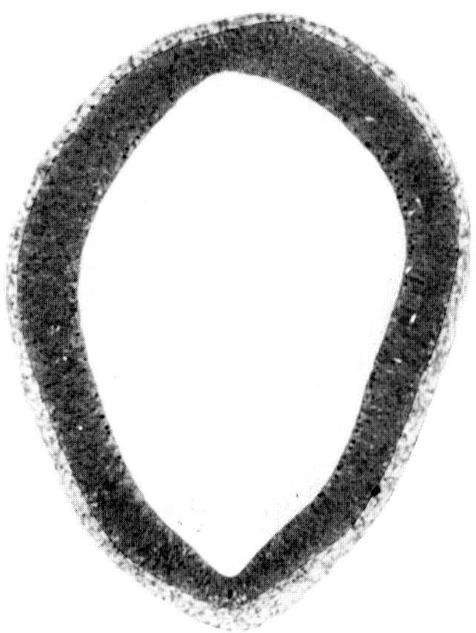

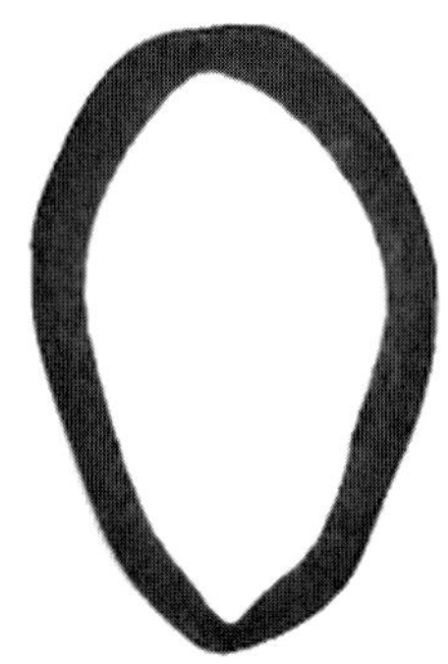

THE E14 BRAIN

LEFT, ANTERIOR VIEW

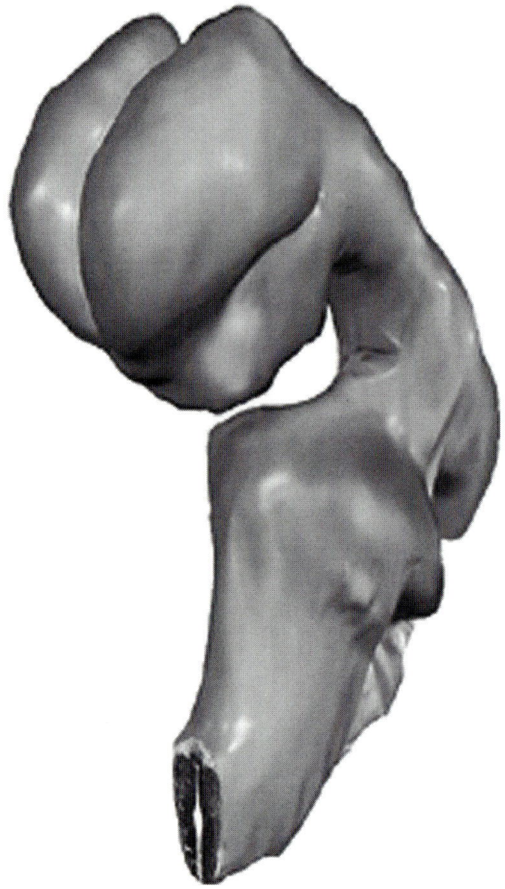

LEFT, POSTERIOR VIEW

DORSAL VIEW

SAGITTAL

FIGURES and PLATES 1-7

SAGITTAL FIGURE 1

CEREBRAL CORTEX (C)

C11 cingulate cortex neuroepithelium
C13 cingulate cortex (or differentiating field)

FIBER TRACTS (f)

f19 posterior commissure

GLANDS AND PARANEURAL STRUCTURES (g)

g6 pituitary gland (primordium)
g9 lamina terminalis
g10 velum medullare

HYPOTHALAMUS (H)

H1a anterior hypothalamic neuroepithelium
H1i intermediate hypothalamic neuroepithelium
H1m mammillary neuroepithelium

CEREBELLUM (L)

L1 cerebellar neuroepithelium

MIDBRAIN (M)

M1 superior tectal neuroepithelium
M11a anterior tegmental neuroepithelium
M11p posterior tegmental neuroepithelium
M21 pretectal neuroepithelium
M31 inferior tectal neuroepithelium

PREOPTIC AREA AND OPTIC EPITHELIA (P)

P1 preoptic neuroepithelium

SEPTAL AREA (S)

S1 septal neuroepithelium
S3 septum (or differentiating field)

THALAMUS (T)

T1 thalamic neuroepithelium

VENTRICLES (v)

vai aqueduct, inferior tectal
vapr aqueduct, pretectal
vas aqueduct, superior tectal
vs spinal canal
v3h third ventricle, hypothalamic
v3m third ventricle, mammillary recess
v3pr third ventricle, preoptic recess
v3t third ventricle, thalamic
v4a fourth ventricle, anterior
v4i fourth ventricle, isthmal
v4p fourth ventricle, posterior

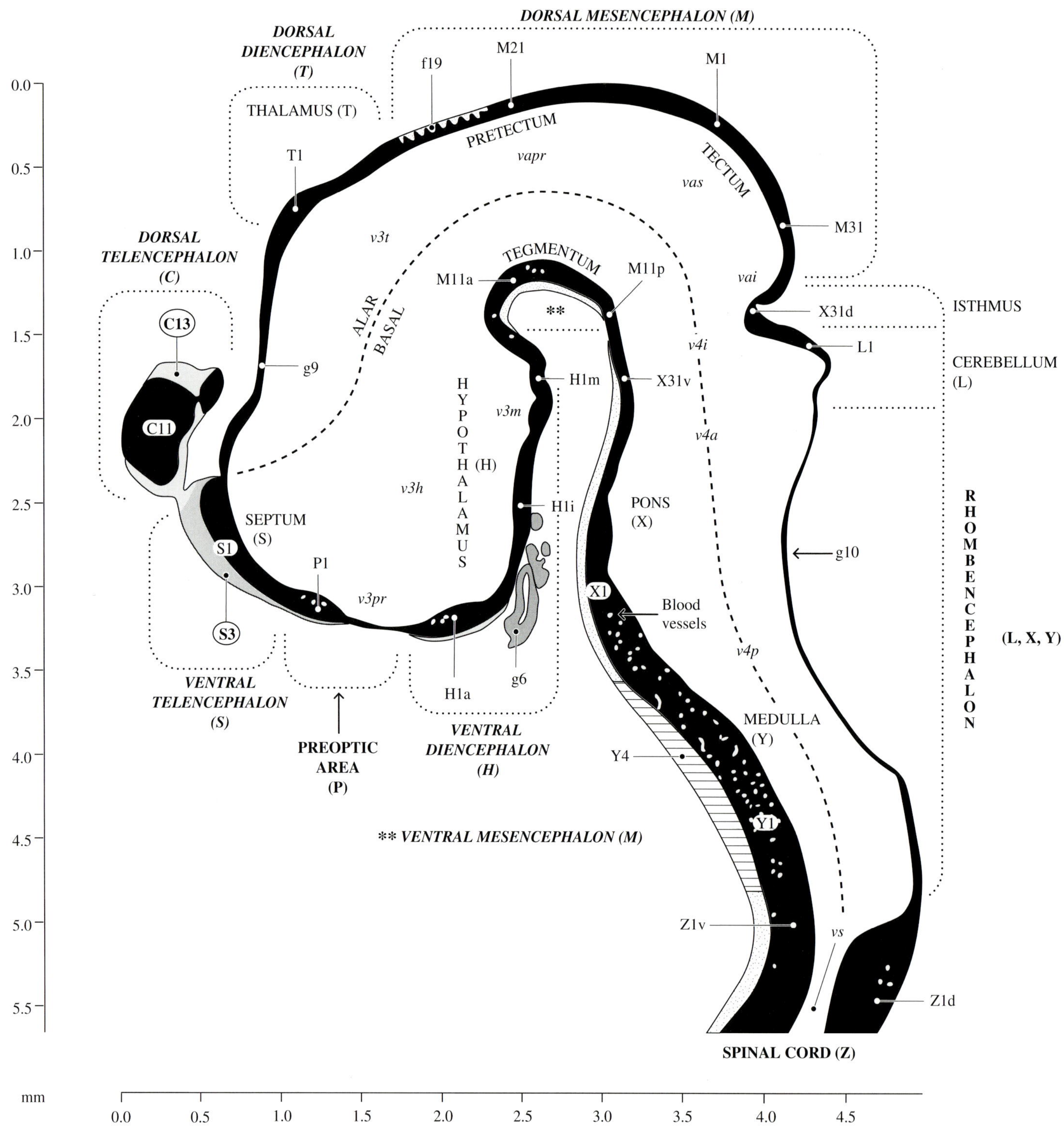

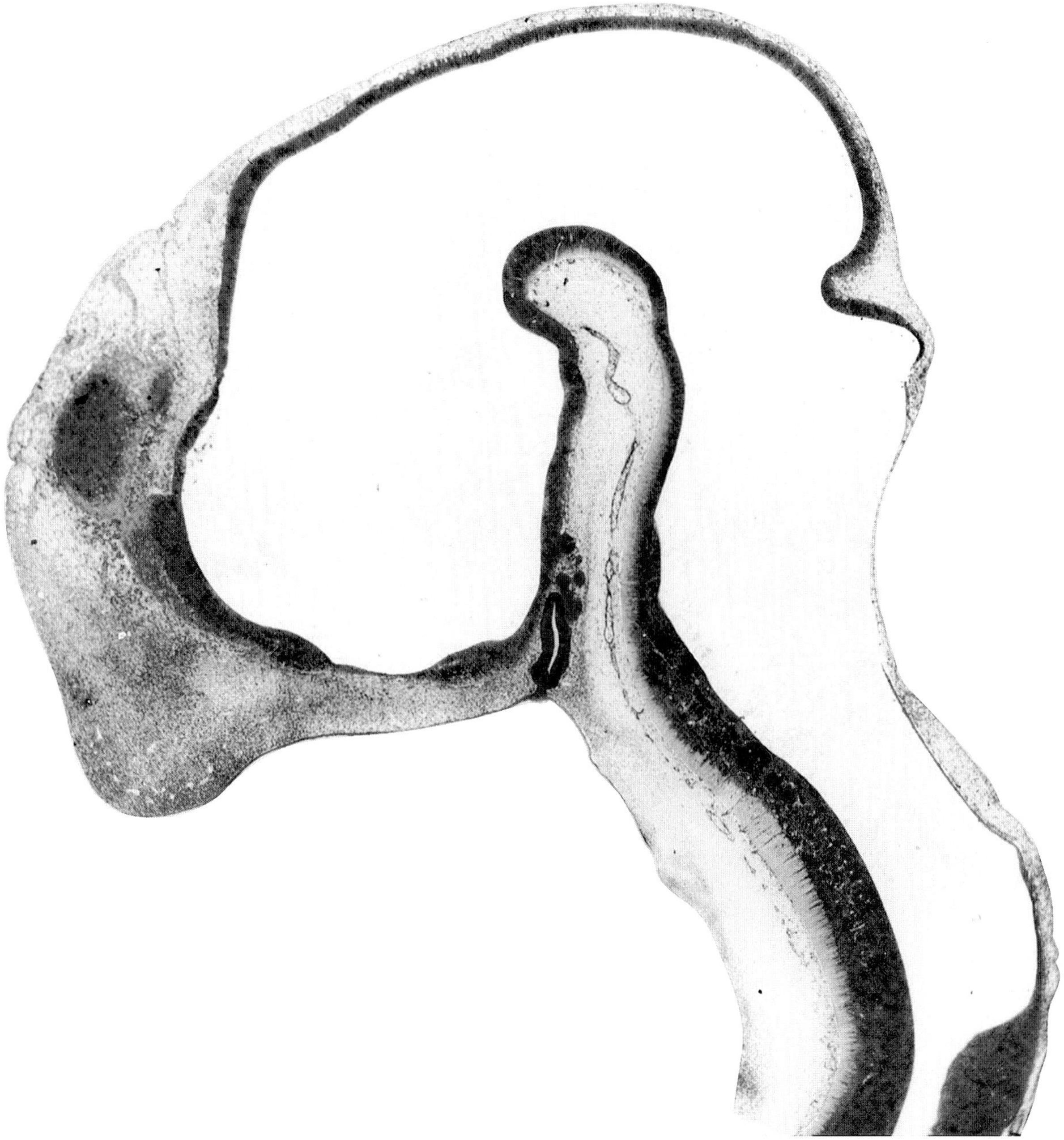

PONTINE AREA (X)

X1 pontine neuroepithelium
X31d isthmal neuroepithelium, dorsal
X31v isthmal neuroepithelium, ventral

MEDULLA (Y)

Y1 medullary neuroepithelium
Y4 medullary glial palisade

SPINAL CORD (Z)

Z1d spinal cord neuroepithelium, dorsal
Z1v spinal cord neuroepithelium, ventral

SAGITTAL FIGURE 2

BASAL GANGLIA (B)

B11 pallidal neuroepithelium

CEREBRAL CORTEX (C)

C11 cingulate cortex neuroepithelium
C13 cingulate cortex (or differentiating field)

FIBER TRACTS (f)

f19 posterior commissure

GLANDS AND PARANEURAL STRUCTURES (g)

g6 pituitary gland (primordium)
g9 lamina terminalis
g10 velum medullare

HYPOTHALAMUS (H)

H1v ventral hypothalamic neuroepithelium

CEREBELLUM (L)

L1 cerebellar neuroepithelium

MIDBRAIN (M)

M1 superior tectal neuroepithelium
M11a anterior tegmental neuroepithelium
M11p posterior tegmental neuroepithelium
M21 pretectal neuroepithelium
M31 inferior tectal neuroepithelium

PREOPTIC AREA AND OPTIC EPITHELIA (P)

P1 preoptic neuroepithelium

SEPTAL AREA (S)

S1 septal neuroepithelium
S3 septum (or differentiating field)

THALAMUS (T)

T1e epithalamic neuroepithelium

VENTRICLES (v)

vai aqueduct, inferior tectal
vapr aqueduct, pretectal
vas aqueduct, superior tectal
vat aqueduct, tegmental
v3h third ventricle, hypothalamic
v3i third ventricle, infundibular recess
v3m third ventricle, mammillary recess
v3pr third ventricle, preoptic recess
v3t third ventricle, thalamic
v4a fourth ventricle, anterior

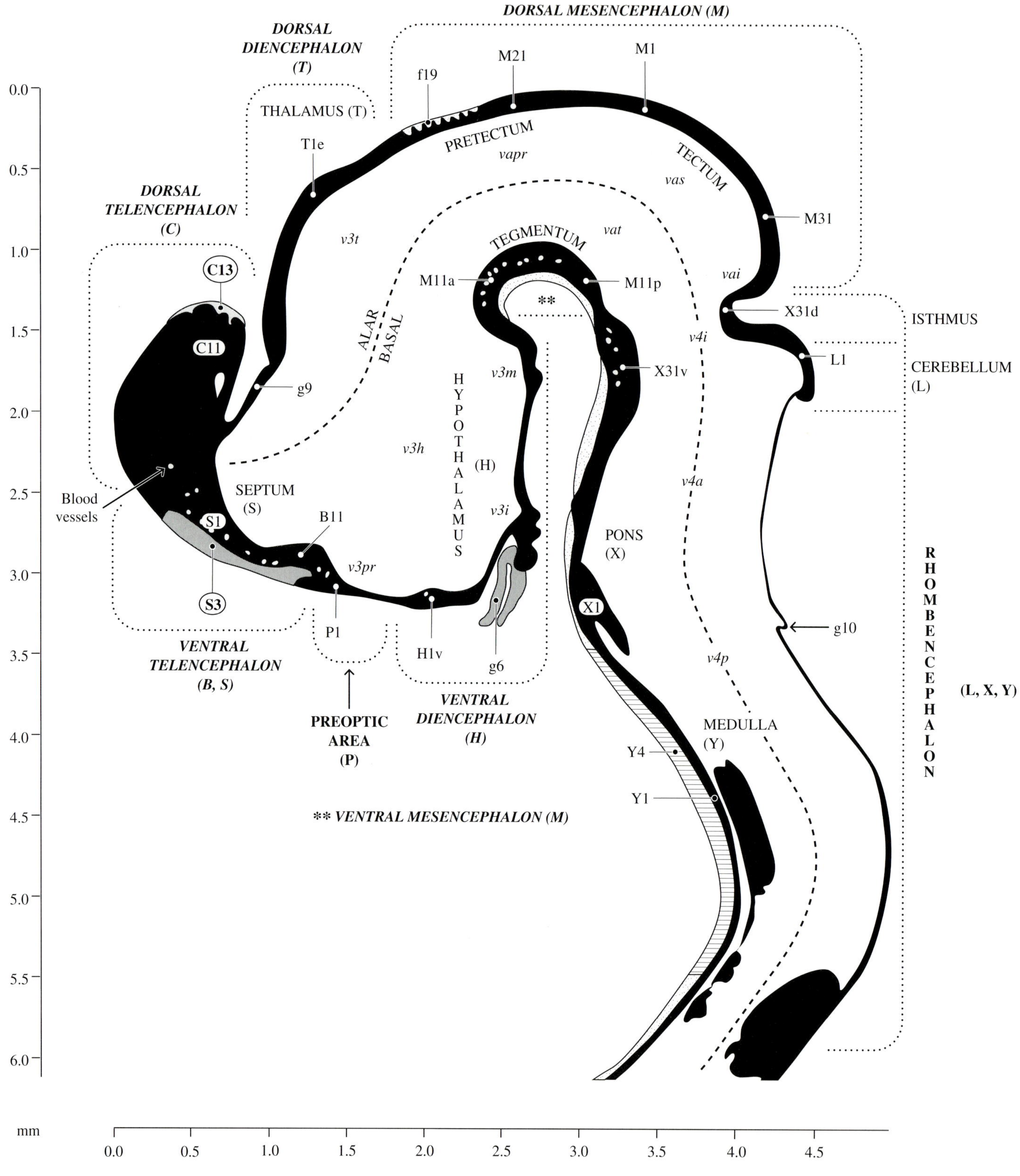

v4i fourth ventricle, isthmal
v4p fourth ventricle, posterior

PONTINE AREA (X)

X1 pontine neuroepithelium
X31d isthmal neuroepithelium, dorsal
X31v isthmal neuroepithelium, ventral

MEDULLA (Y)

Y1 medullary neuroepithelium
Y4 medullary glial palisade

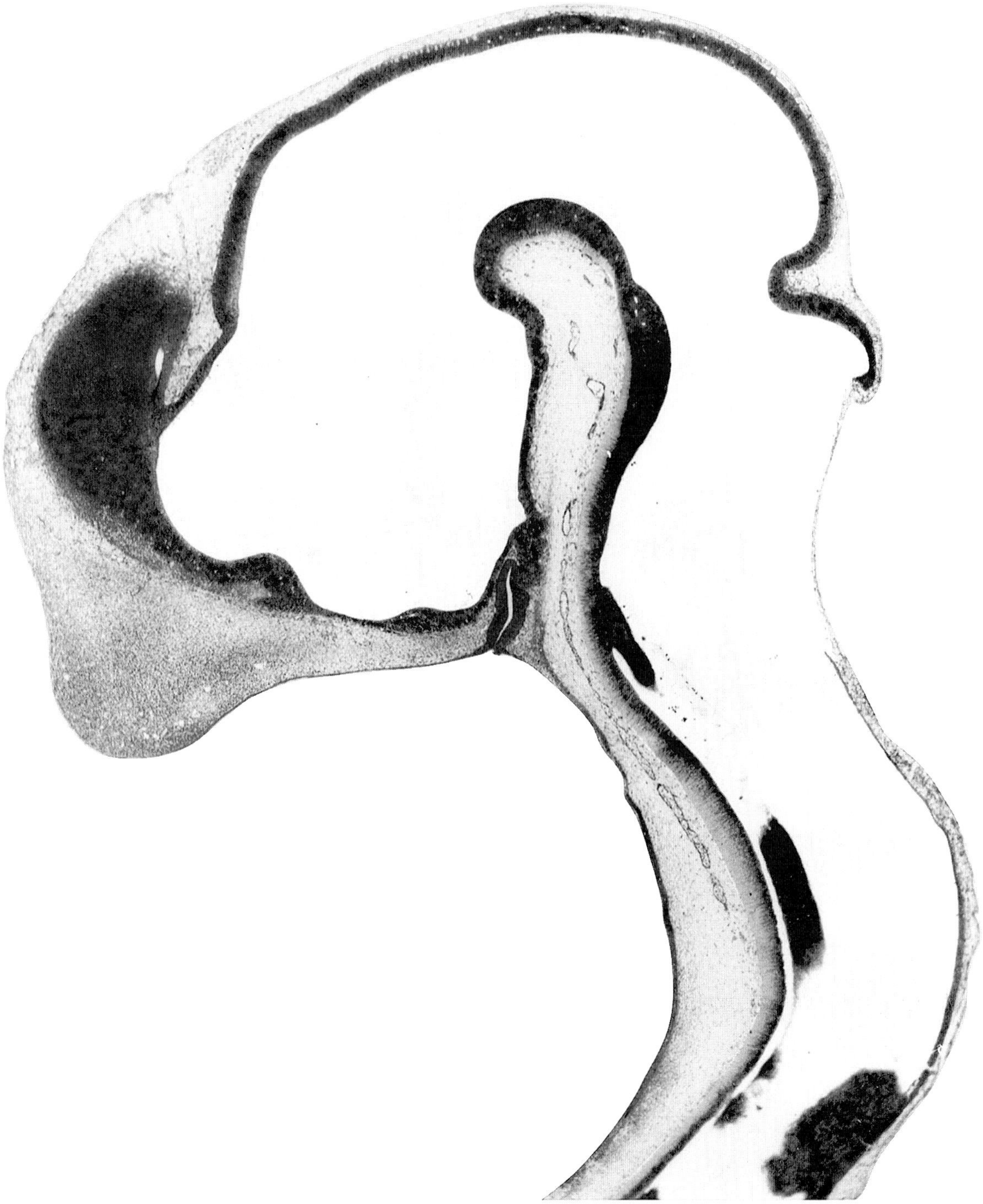

BASAL GANGLIA (B)

B11 pallidal neuroepithelium
B12 pallidal subventricular zone
B13 pallidum (or differentiating field)

CEREBRAL CORTEX (C)

C11 cingulate cortex neuroepithelium

GLANDS AND PARANEURAL STRUCTURES (g)

g10 velum medullare

HYPOTHALAMUS (H)

H1a anterior hypothalamic neuroepithelium
H1i intermediate hypothalamic neuroepithelium
H3 hypothalamus (or differentiating field)

HIPPOCAMPAL FORMATION (K)

K1 hippocampal (Ammon's horn) neuroepithelium
K11 subicular neuroepithelium

CEREBELLUM (L)

L1 cerebellar neuroepithelium
L3 cerebellum (or differentiating field)

MIDBRAIN (M)

M1 superior tectal neuroepithelium
M11a anterior tegmental neuroepithelium
M11p posterior tegmental neuroepithelium
M21 pretectal neuroepithelium
M31 inferior tectal neuroepithelium

NERVES AND NUCLEI, CRANIAL (n)

n1 olfactory nerve

PREOPTIC AREA AND OPTIC EPITHELIA (P)

P1 preoptic neuroepithelium

RHINENCEPHALON (R)

R1 rhinencephalic neuroepithelium

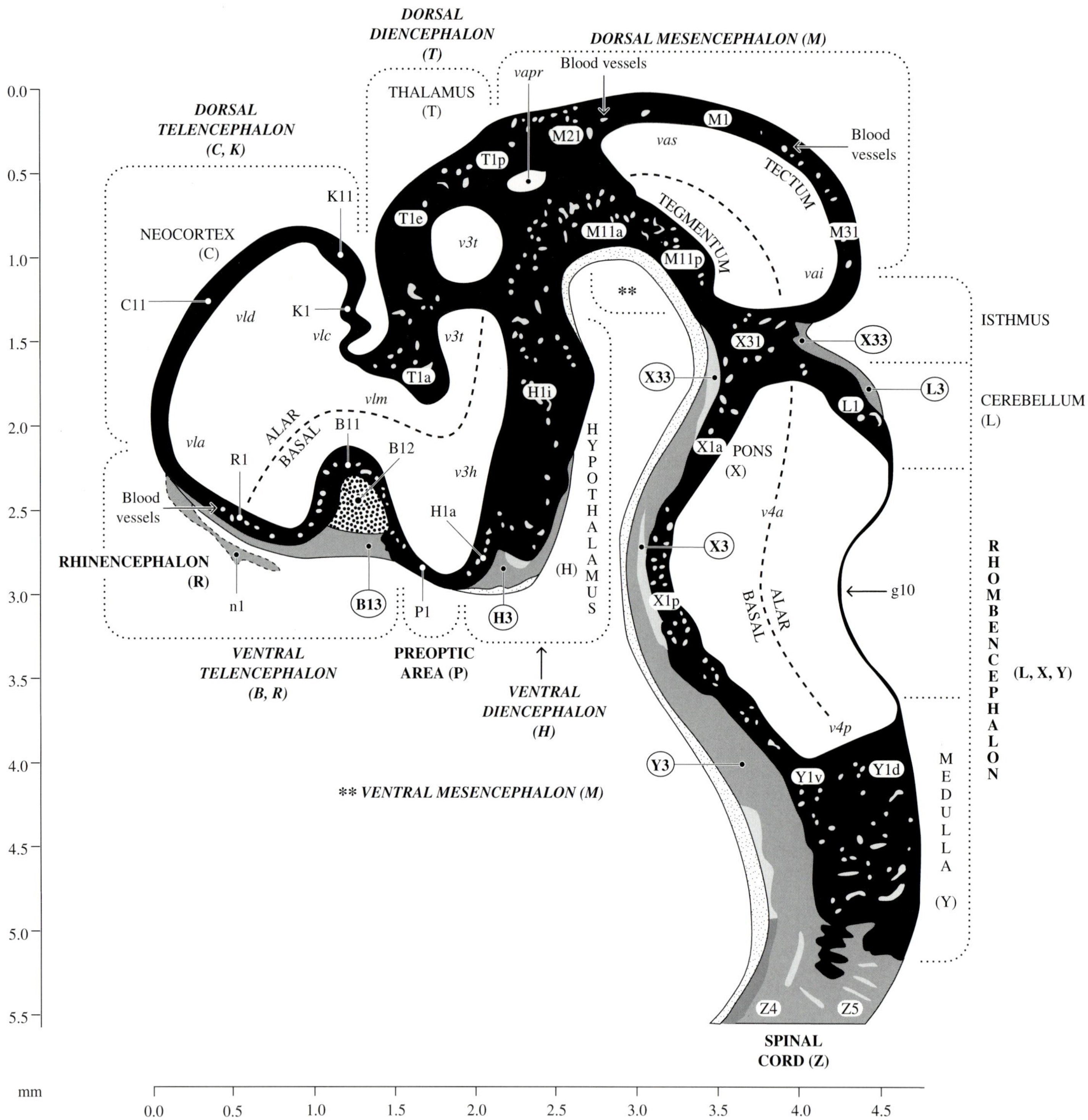

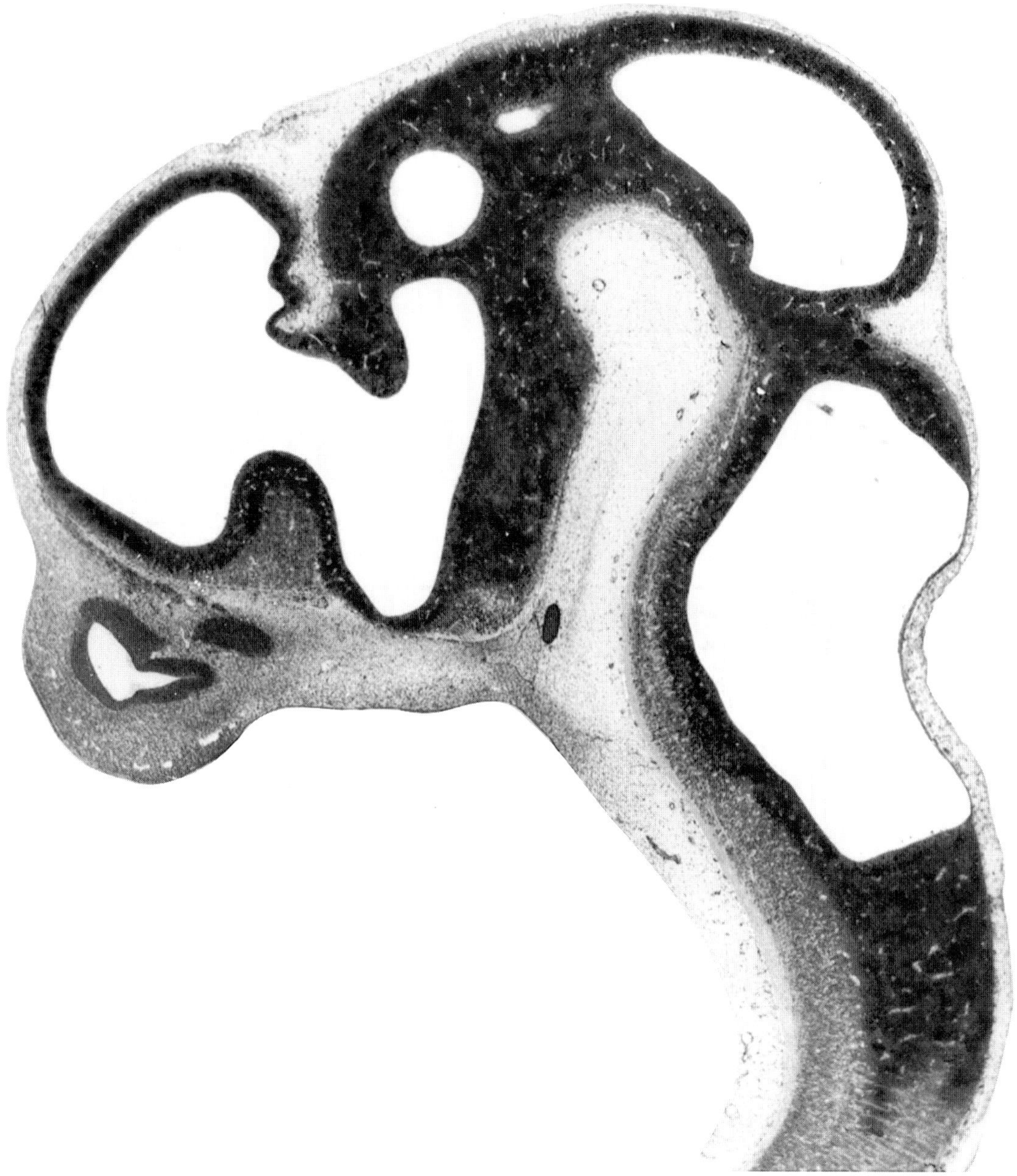

THALAMUS (T)

T1a	anterior thalamic neuroepithelium
T1e	epithalamic neuroepithelium
T1p	posterior thalamic neuroepithelium

VENTRICLES (v)

vai	aqueduct, inferior tectal
vapr	aqueduct, pretectal
vas	aqueduct, superior tectal
vla	lateral ventricle, anterior horn
vlc	lateral ventricle, central
vld	lateral ventricle, dorsal
vlm	foramen of Monro
v3h	third ventricle, hypothalamic
v3t	third ventricle, thalamic
v4a	fourth ventricle, anterior
v4p	fourth ventricle, posterior

PONTINE AREA (X)

X1a	anterior pontine neuroepithelium
X1p	posterior pontine neuroepithelium
X3	pons (or differentiating field)
X31	isthmal neuroepithelium
X33	isthmus (or differentiating field)

MEDULLA (Y)

Y1d	medullary neuroepithelium, dorsal
Y1v	medullary neuroepithelium, ventral
Y3	medulla (or differentiating field)

SPINAL CORD (Z)

Z4	ventral horn
Z5	dorsal horn

BASAL GANGLIA (B)

B11 pallidal neuroepithelium
B12 pallidal subventricular zone
B13 pallidum (or differentiating field)
B21 strionuclear neuroepithelium

CEREBRAL CORTEX (C)

C1 neocortical neuroepithelium

GLANDS AND PARANEURAL STRUCTURES (g)

g10 velum medullare

HYPOTHALAMUS (H)

H1a anterior hypothalamic neuroepithelium
H1f Forelian neuroepithelium
H1m mammillary neuroepithelium
H3 hypothalamus (or differentiating field)

HIPPOCAMPAL FORMATION (K)

K1 hippocampal (Ammon's horn) neuroepithelium
K11 subicular neuroepithelium

CEREBELLUM (L)

L1 cerebellar neuroepithelium
L3 cerebellum (or differentiating field)

MIDBRAIN (M)

M1 superior tectal neuroepithelium
M11a anterior tegmental neuroepithelium
M11p posterior tegmental neuroepithelium
M13 tegmentum (or differentiating field)
M21 pretectal neuroepithelium
M31 inferior tectal neuroepithelium

NERVES AND NUCLEI, CRANIAL (n)

n1 olfactory nerve

PREOPTIC AREA AND OPTIC EPITHELIA (P)

P1 preoptic neuroepithelium

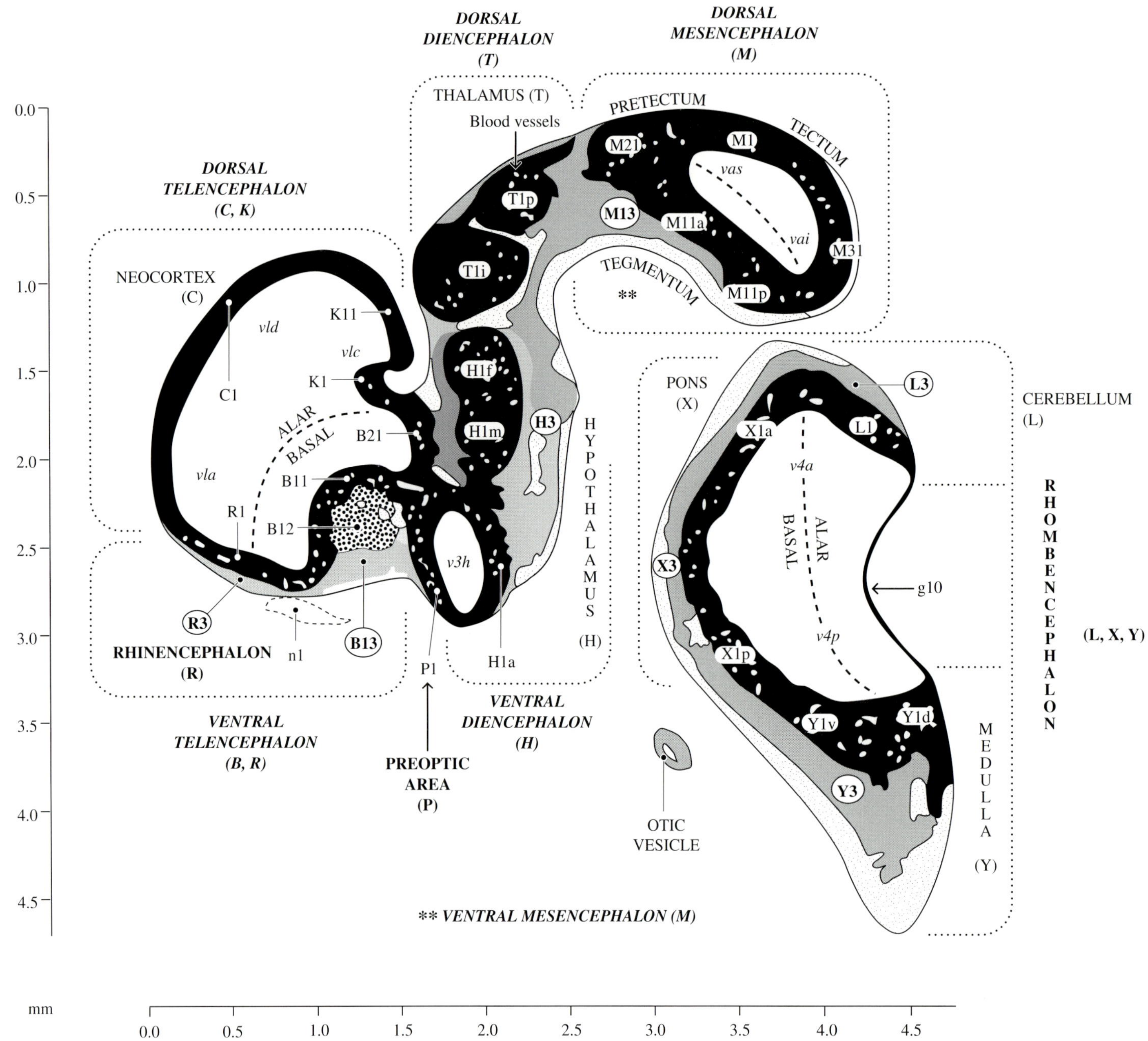

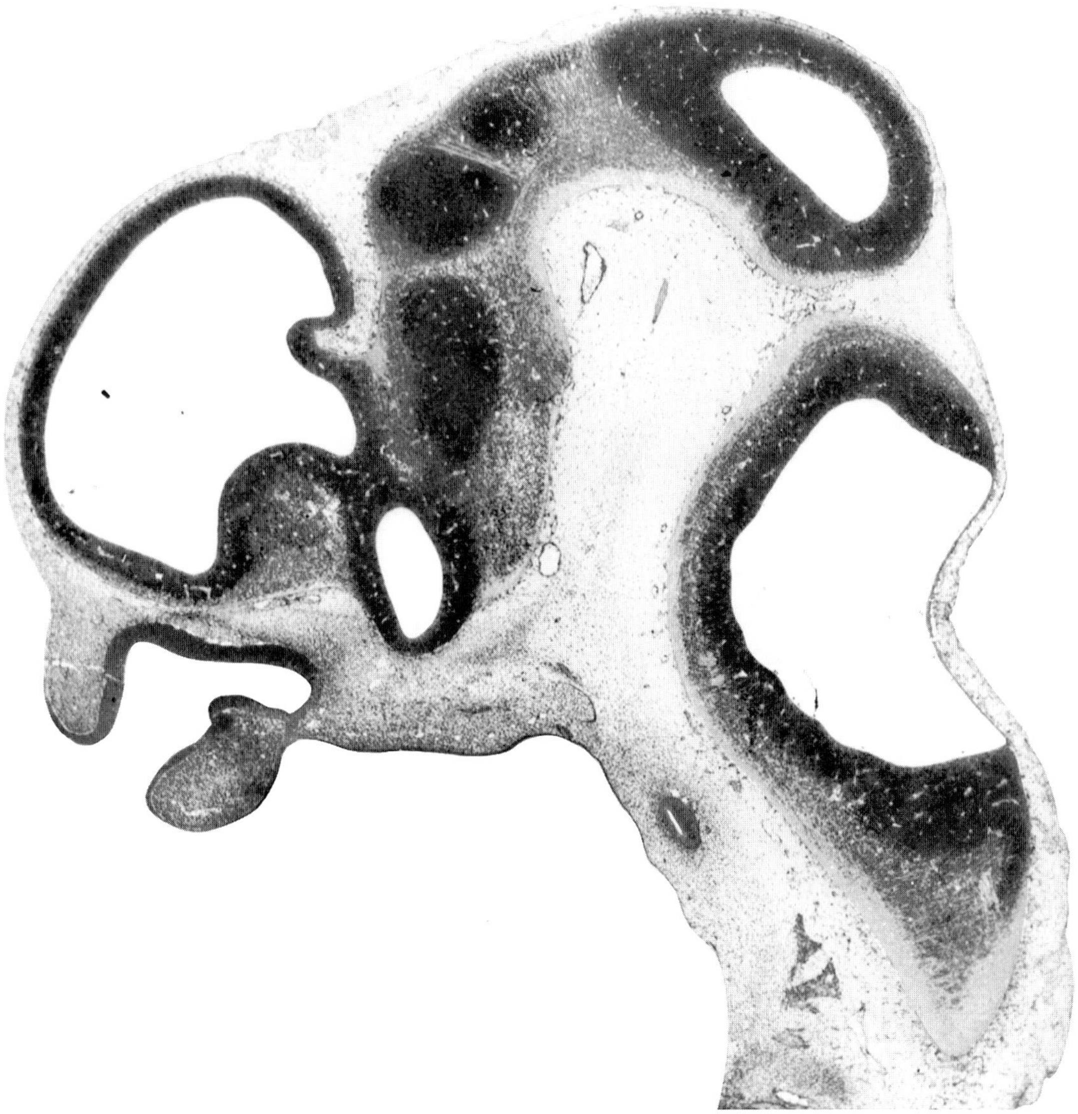

RHINENCEPHALON (R)

R1 rhinencephalic neuroepithelium
R3 rhinencephalon (or differentiating field)

THALAMUS (T)

T1i intermediate thalamic neuroepithelium
T1p posterior thalamic neuroepithelium

VENTRICLES (v)

vai aqueduct, inferior tectal
vas aqueduct, superior tectal
vla lateral ventricle, anterior horn
vlc lateral ventricle, central
vld lateral ventricle, dorsal
v3h third ventricle, hypothalamic
v4a fourth ventricle, anterior
v4p fourth ventricle, posterior

PONTINE AREA (X)

X1a anterior pontine neuroepithelium
X1p posterior pontine neuroepithelium
X3 pons (or differentiating field)

MEDULLA (Y)

Y1d medullary neuroepithelium, dorsal
Y1v medullary neuroepithelium, ventral
Y3 medulla (or differentiating field)

SAGITTAL FIGURE 5

BASAL GANGLIA (B)

B1 striatal neuroepithelium
B3 striatum (or differentiating field)
B11 pallidal neuroepithelium
B13 pallidum (or differentiating field)

CEREBRAL CORTEX (C)

C1 neocortical neuroepithelium

GLANDS AND PARANEURAL STRUCTURES (g)

g10 velum medullare

HYPOTHALAMUS (H)

H1a anterior hypothalamic neuroepithelium
H1i intermediate hypothalamic neuroepithelium
H3 hypothalamus (or differentiating field)

HIPPOCAMPAL FORMATION (K)

K1 hippocampal (Ammon's horn) neuroepithelium
K11 subicular neuroepithelium

CEREBELLUM (L)

L1 cerebellar neuroepithelium
L3 cerebellum (or differentiating field)

MIDBRAIN (M)

M13 tegmentum (or differentiating field)
M31 inferior tectal neuroepithelium

PREOPTIC AREA AND OPTIC EPITHELIA (P)

P1 preoptic neuroepithelium

RHINENCEPHALON (R)

R1 rhinencephalic neuroepithelium
R3 rhinencephalon (or differentiating field)

VENTRICLES (v)

vla lateral ventricle, anterior horn
vlc lateral ventricle, central
vld lateral ventricle, dorsal
vlp lateral ventricle, posterior horn
v3pr third ventricle, preoptic recess
v4a fourth ventricle, anterior
v4p fourth ventricle, posterior

PONTINE AREA (X)

X1a anterior pontine neuroepithelium
X1p posterior pontine neuroepithelium
X3 pons (or differentiating field)

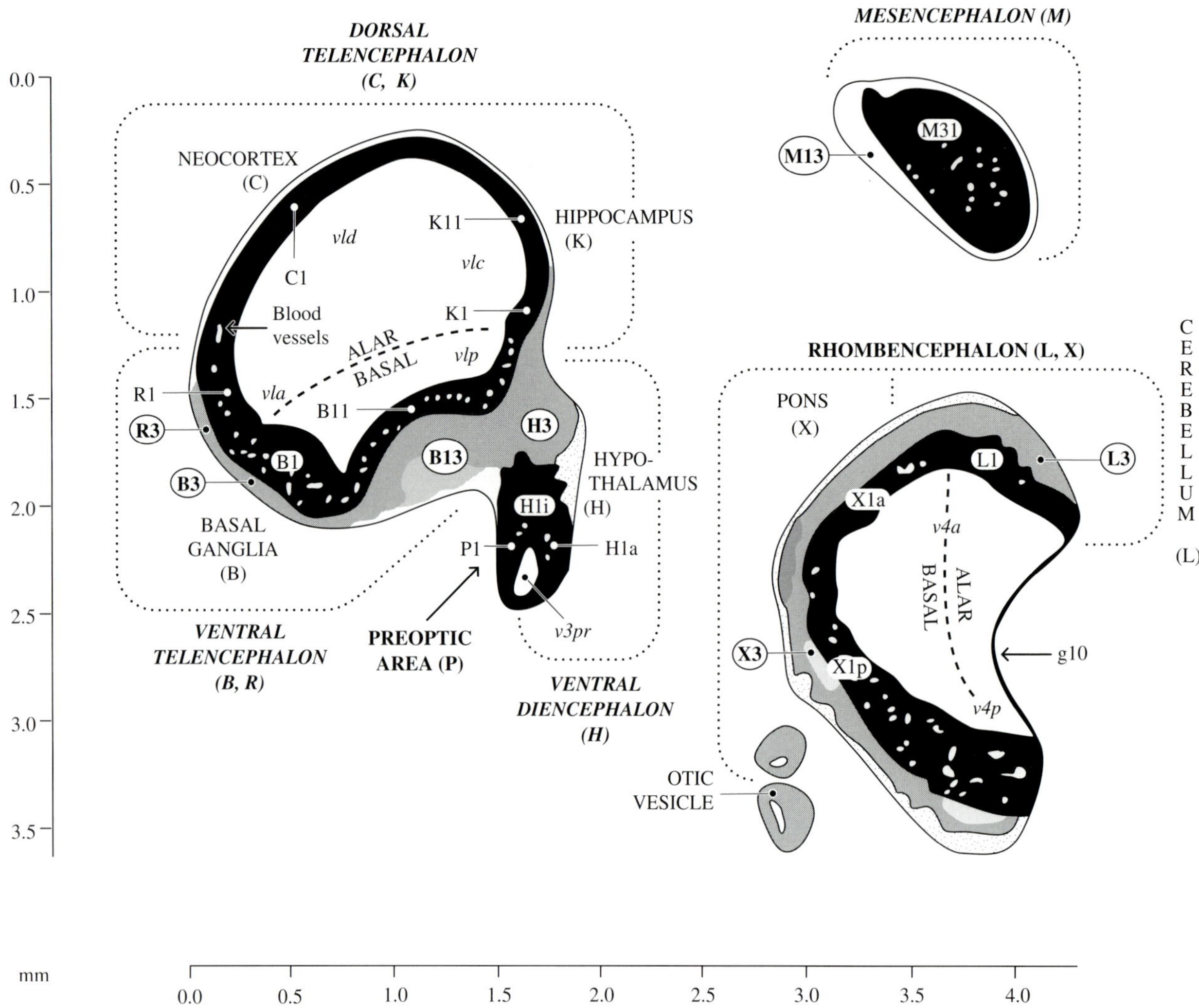

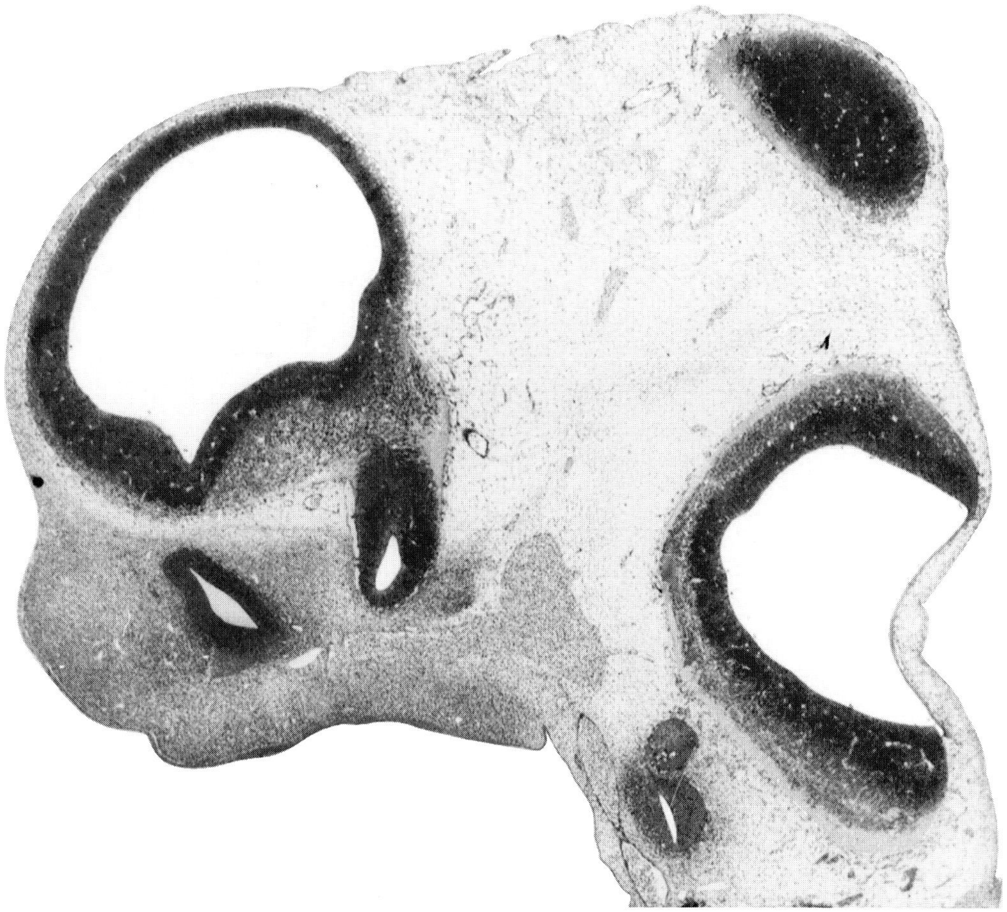

SAGITTAL FIGURE 6

BASAL GANGLIA (B)

B1 striatal neuroepithelium
B11 pallidal neuroepithelium
B13 pallidum (or differentiating field)

CEREBRAL CORTEX (C)

C1 neocortical neuroepithelium

GLANDS AND PARANEURAL STRUCTURES (g)

g10 velum medullare

HYPOTHALAMUS (H)

H1i intermediate hypothalamic neuroepithelium

HIPPOCAMPAL FORMATION (K)

K1 hippocampal (Ammon's horn) neuroepithelium
K11 subicular neuroepithelium

CEREBELLUM (L)

L1 cerebellar neuroepithelium
L3 cerebellum (or differentiating field)

MIDBRAIN (M)

M13 tegmentum (or differentiating field)

NERVES AND NUCLEI, CRANIAL (n)

n8b vestibulocochlear boundary cap
n8g vestibular ganglion

PREOPTIC AREA AND OPTIC EPITHELIA (P)

P1 preoptic neuroepithelium

RHINENCEPHALON (R)

R1 rhinencephalic neuroepithelium
R3 rhinencephalon (or differentiating field)

VENTRICLES (v)

vla lateral ventricle, anterior horn
vlc lateral ventricle, central
vld lateral ventricle, dorsal
vlp lateral ventricle, posterior horn
v3pr third ventricle, preoptic recess
v4a fourth ventricle, anterior
v4p fourth ventricle, posterior

PONTINE AREA (X)

X1a anterior pontine neuroepithelium
X1p posterior pontine neuroepithelium
X3 pons (or differentiating field)

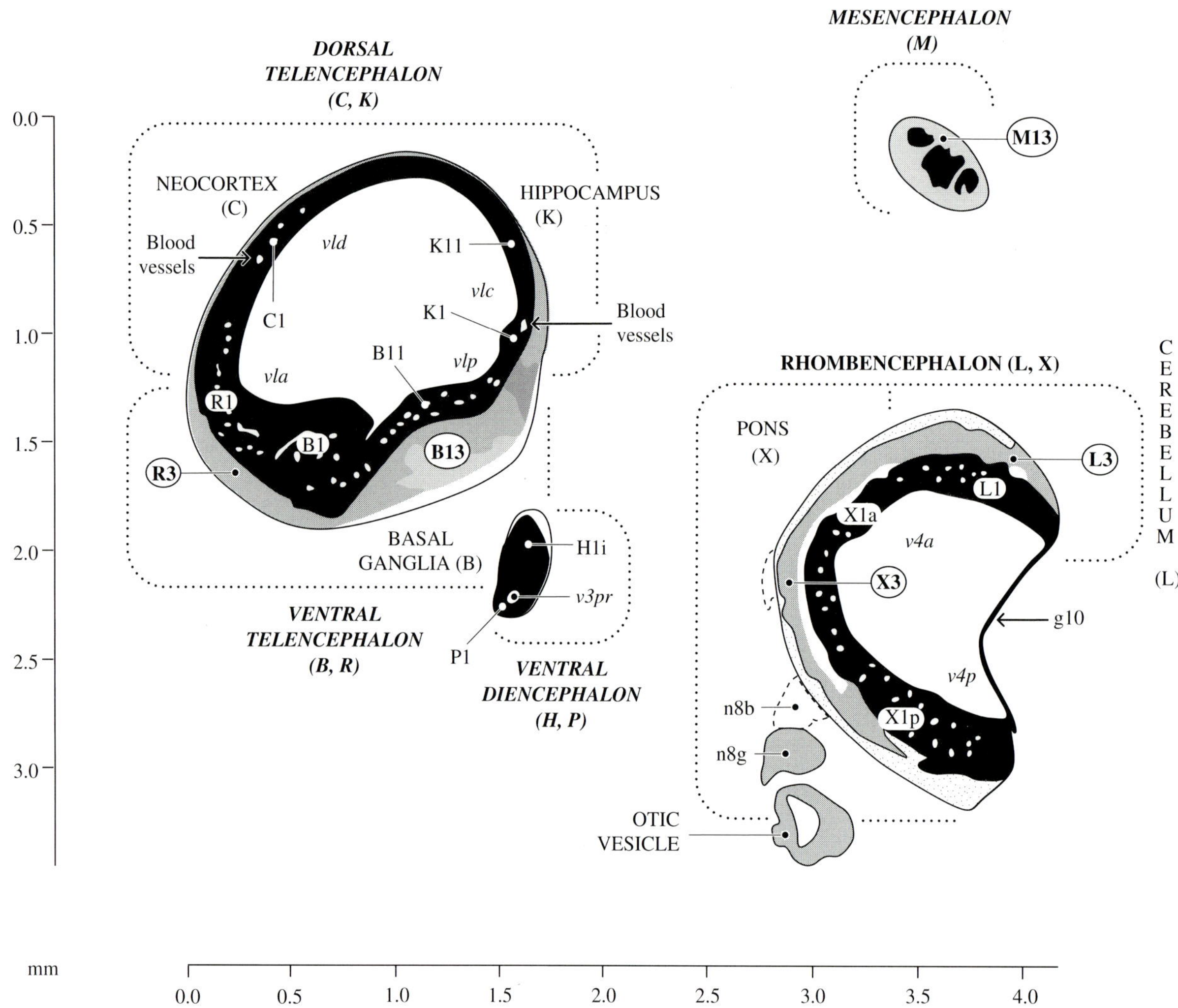

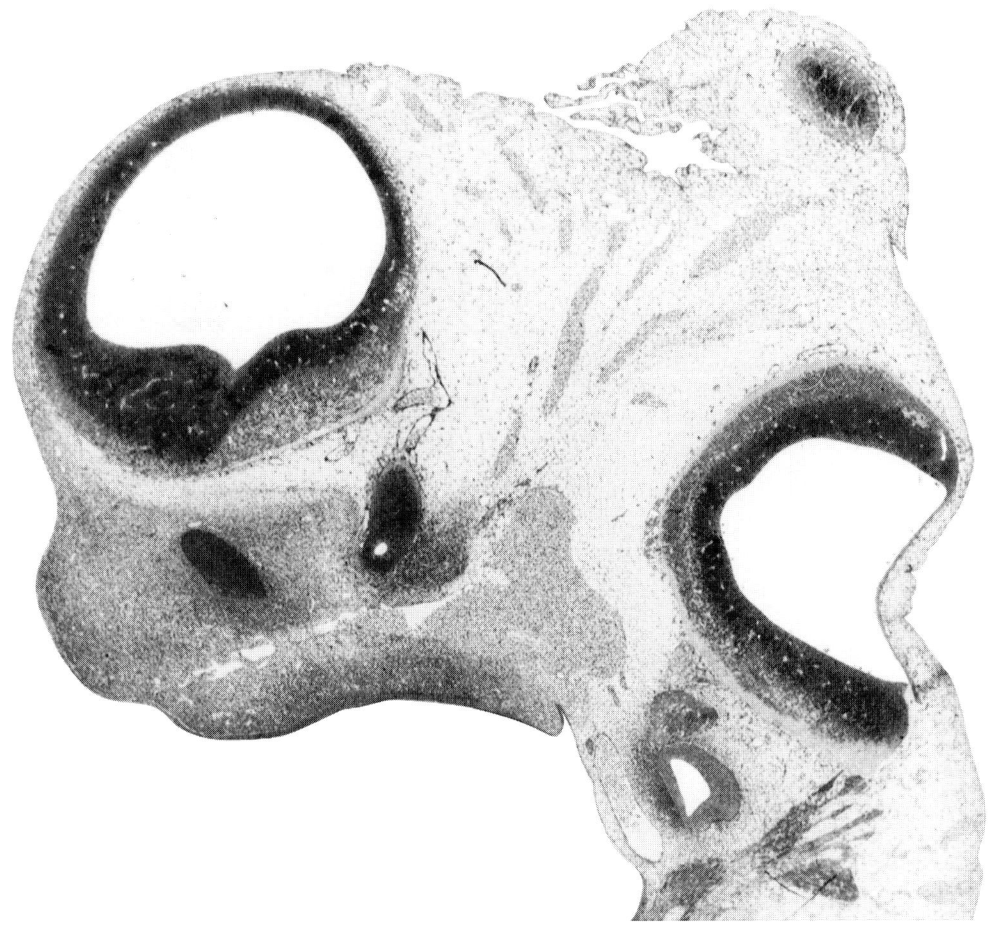

SAGITTAL FIGURE 7

AMYGDALA (A)

A1 amygdaloid neuroepithelium
A3 amygdala (or differentiating field)

BASAL GANGLIA (B)

B11 pallidal neuroepithelium
B13 pallidum (or differentiating field)

CEREBRAL CORTEX (C)

C1 neocortical neuroepithelium

GLANDS AND PARANEURAL STRUCTURES (g)

g10 velum medullare

HIPPOCAMPAL FORMATION (K)

K1 hippocampal (Ammon's horn) neuroepithelium
K11 subicular neuroepithelium

CEREBELLUM (L)

L1 cerebellar neuroepithelium
L3 cerebellum (or differentiating field)

NERVES AND NUCLEI, CRANIAL (n)

n5b trigeminal boundary cap
n5g trigeminal ganglion
n8b vestibulocochlear boundary cap
n8g vestibular ganglion

PREOPTIC AREA AND OPTIC EPITHELIA (P)

P1 preoptic neuroepithelium

RHINENCEPHALON (R)

R1 rhinencephalic neuroepithelium
R3 rhinencephalon (or differentiating field)

VENTRICLES (v)

vlc lateral ventricle, central
vld lateral ventricle, dorsal
vlp lateral ventricle, posterior horn
v3pr third ventricle, preoptic recess
v4l fourth ventricle, lateral recess

PONTINE AREA (X)

X1 pontine neuroepithelium
X21 cochlear nuclei neuroepithelium

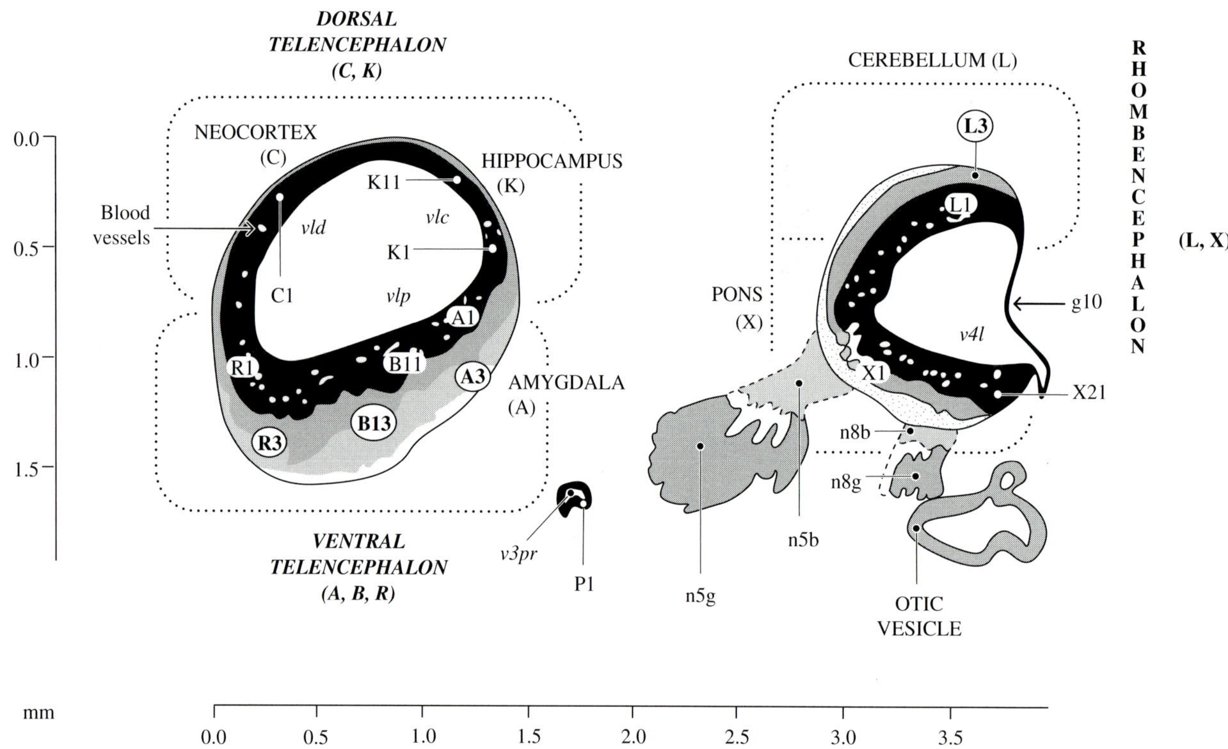

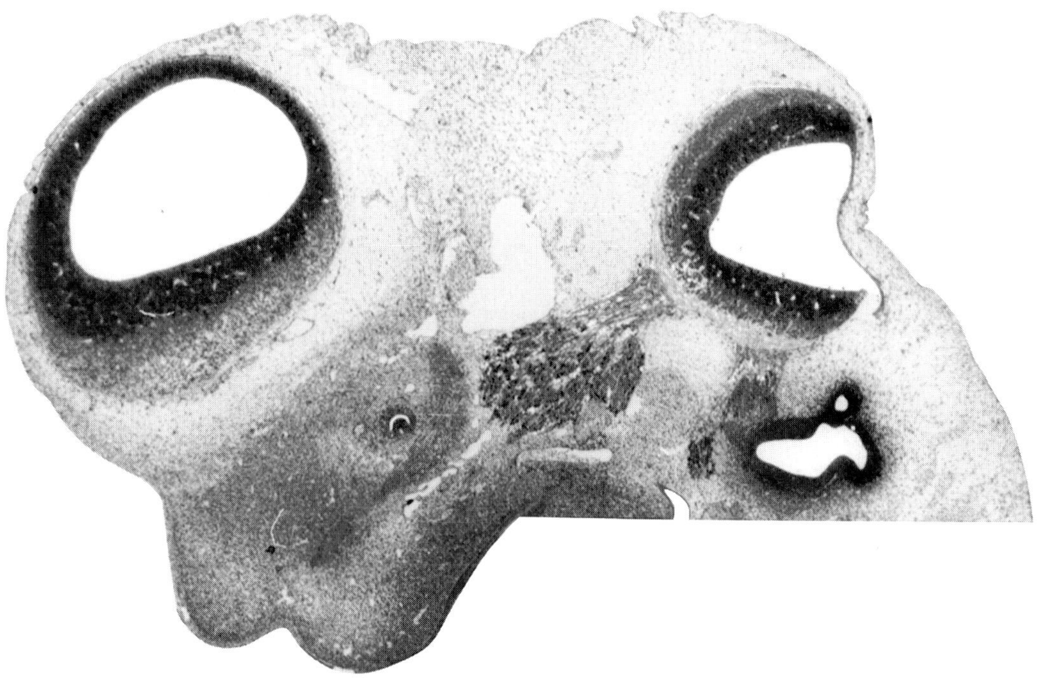

CORONAL

FIGURES and PLATES 1-17

CORONAL FIGURE 1

CEREBRAL CORTEX (C)

C1 neocortical neuroepithelium
C3 neocortex (or differentiating field)
C4 molecular layer (or primordial plexiform layer)
C11 cingulate cortex neuroepithelium
C13 cingulate cortex (or differentiating field)

NERVES AND NUCLEI, CRANIAL (n)

n1 olfactory nerve

OLFACTORY BULB (O)

O1 olfactory bulb neuroepithelium

RHINENCEPHALON (R)

R1 rhinencephalic neuroepithelium
R3 rhinencephalon (or differentiating field)

SEPTAL AREA (S)

S1 septal neuroepithelium
S3 septum (or differentiating field)

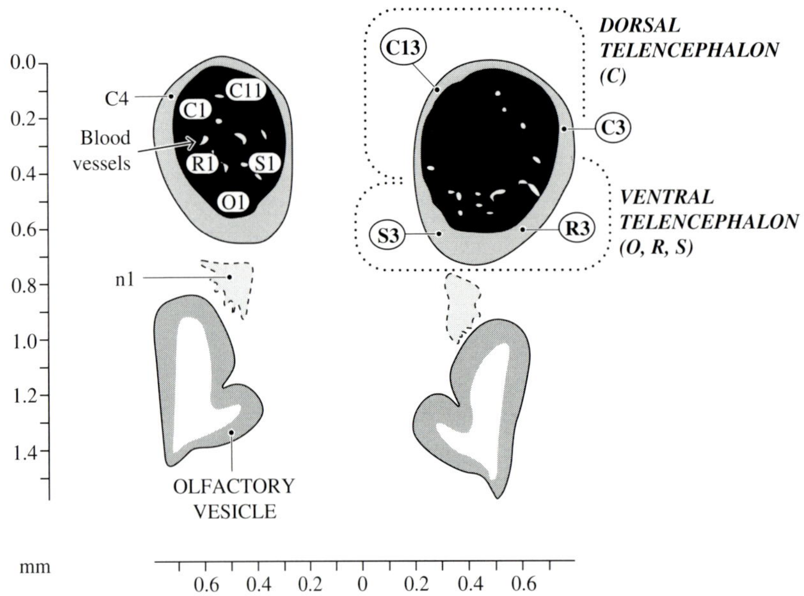

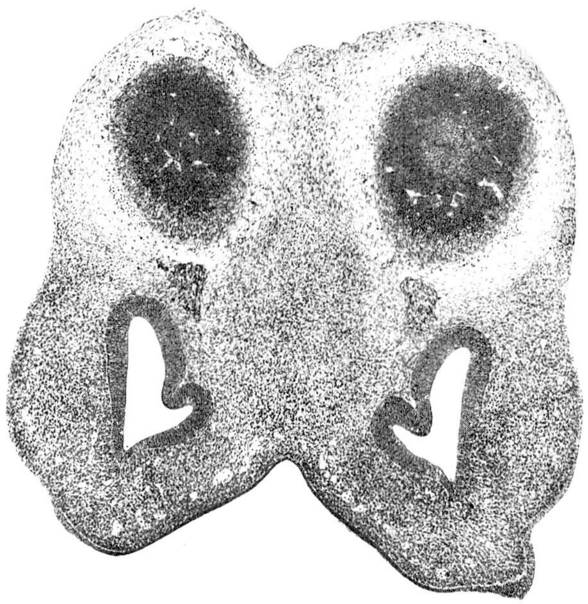

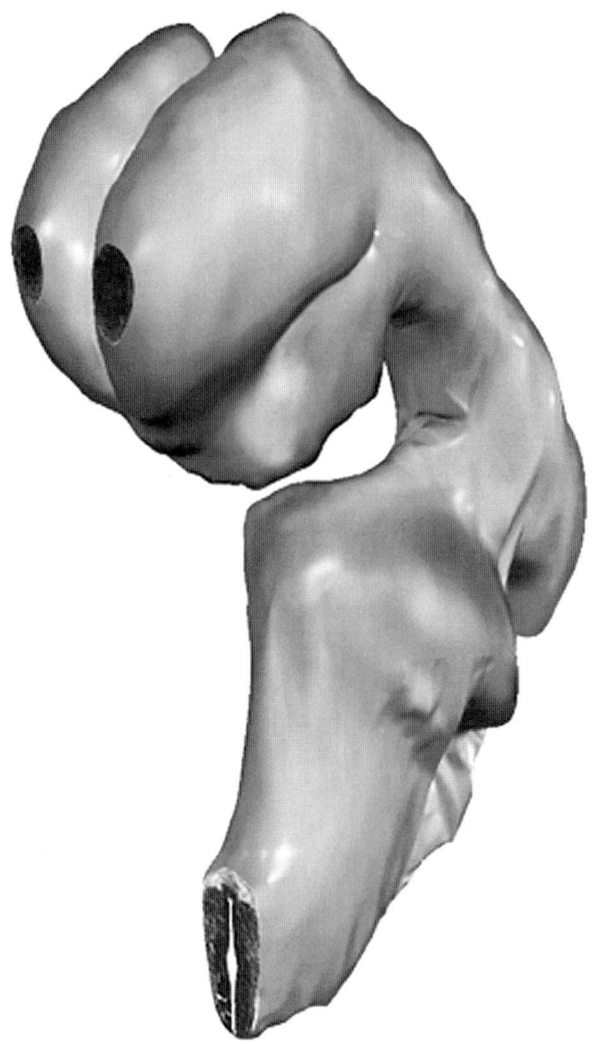

CORONAL FIGURE 2

CEREBRAL CORTEX (C)

C1 neocortical neuroepithelium
C3 neocortex (or differentiating field)
C4 molecular layer (or primordial plexiform layer)
C11 cingulate cortex neuroepithelium
C13 cingulate cortex (or differentiating field)

HIPPOCAMPAL FORMATION (K)

K11 subicular neuroepithelium

NERVES AND NUCLEI, CRANIAL (n)

n1 olfactory nerve

OLFACTORY BULB (O)

O1 olfactory bulb neuroepithelium

RHINENCEPHALON (R)

R1 rhinencephalic neuroepithelium
R3 rhinencephalon (or differentiating field)

SEPTAL AREA (S)

S1 septal neuroepithelium
S3 septum (or differentiating field)

VENTRICLES (v)

vla lateral ventricle, anterior horn
vld lateral ventricle, dorsal

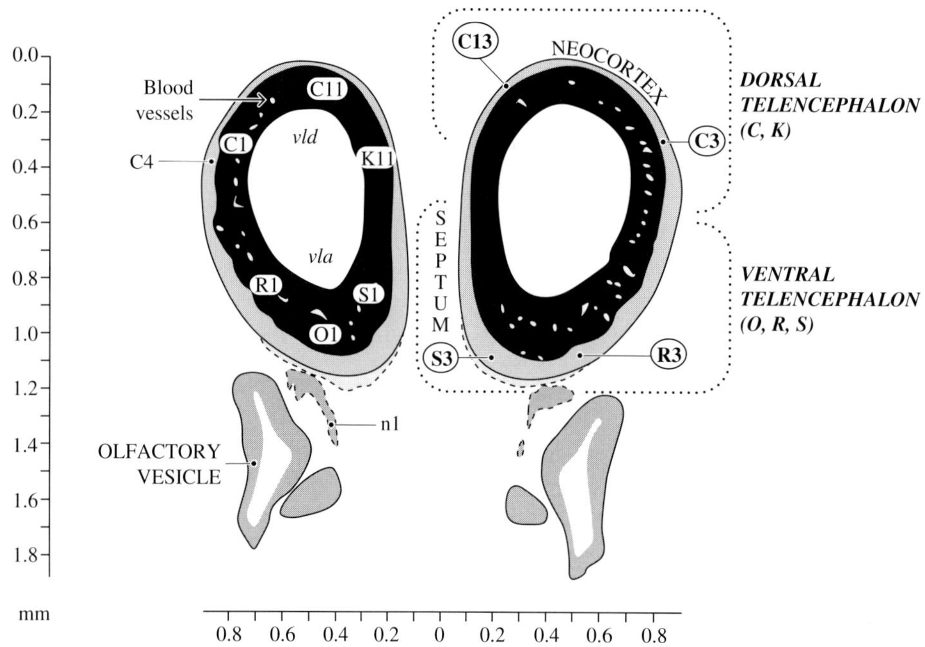

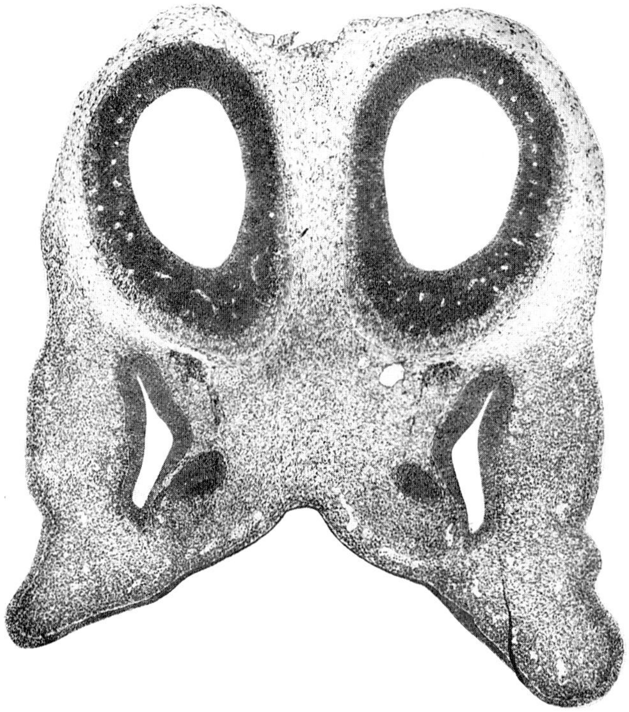

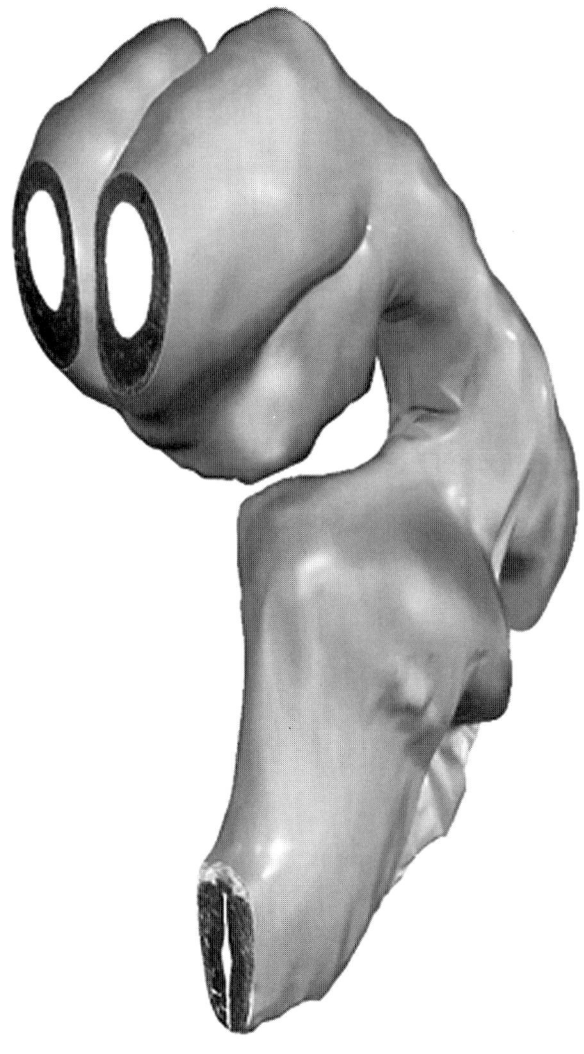

CORONAL FIGURE 3

BASAL GANGLIA (B)

- B1 striatal neuroepithelium
- **B3** striatum (or differentiating field)
- B11 pallidal neuroepithelium
- **B13** pallidum (or differentiating field)

CEREBRAL CORTEX (C)

- C1 neocortical neuroepithelium
- **C3** neocortex (or differentiating field)
- C4 molecular layer (or primordial plexiform layer)
- C11 cingulate cortex neuroepithelium
- **C13** cingulate cortex (or differentiating field)

HIPPOCAMPAL FORMATION (K)

- K1 hippocampal (Ammon's horn) neuroepithelium
- K11 subicular neuroepithelium

RHINENCEPHALON (R)

- R1 rhinencephalic neuroepithelium
- **R3** rhinencephalon (or differentiating field)

VENTRICLES (v)

- *vla* lateral ventricle, anterior horn
- *vlc* lateral ventricle, central
- *vld* lateral ventricle, dorsal

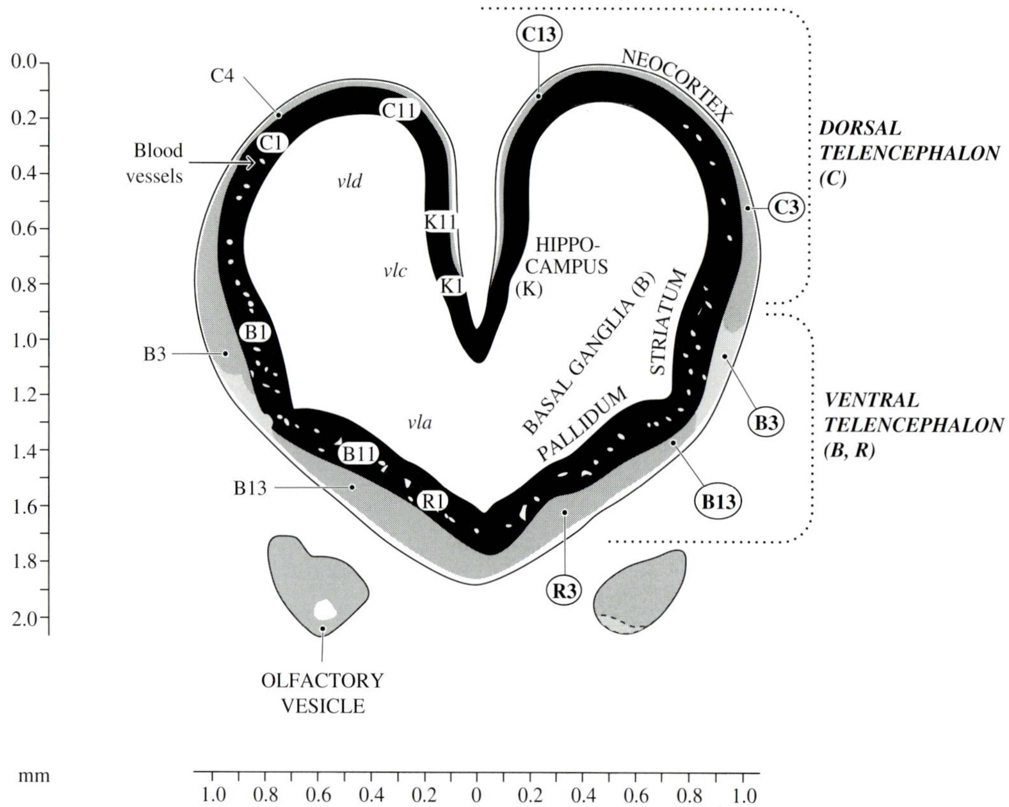

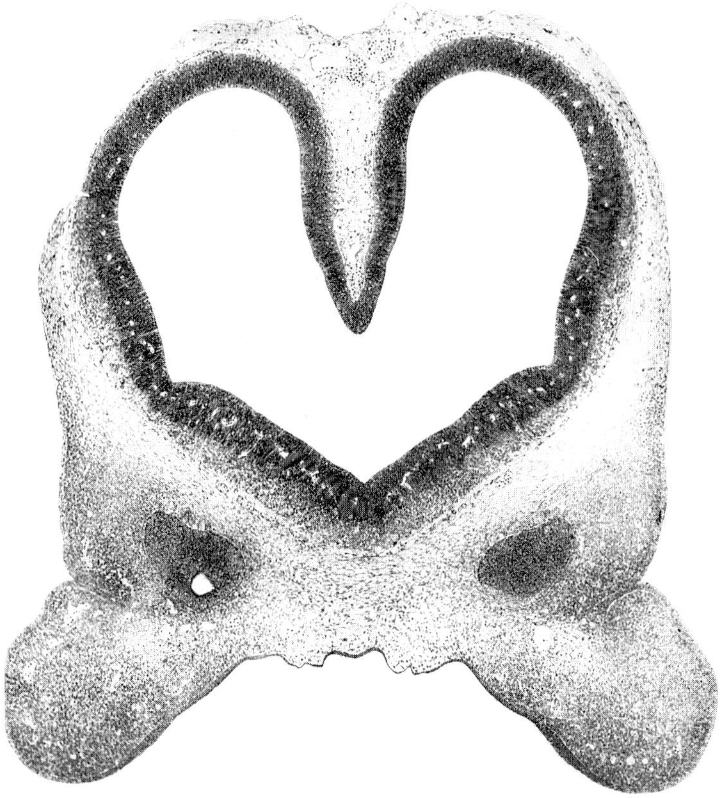

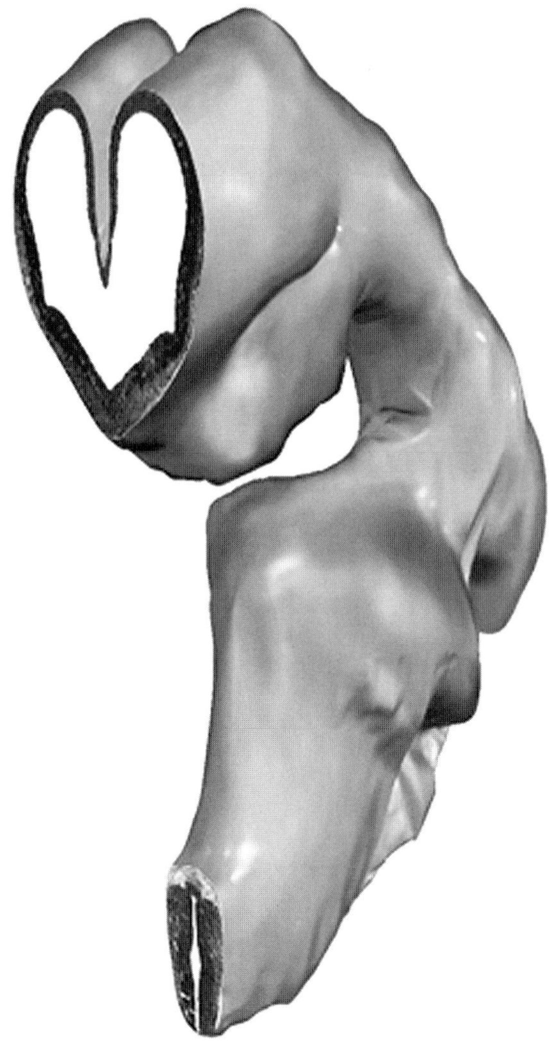

CORONAL FIGURE 4

BASAL GANGLIA (B)

B1 striatal neuroepithelium
B3 striatum (or differentiating field)
B11 pallidal neuroepithelium
B12 pallidal subventricular zone
B13 pallidum (or differentiating field)

CEREBRAL CORTEX (C)

C1 neocortical neuroepithelium
C3 neocortex (or differentiating field)
C4 molecular layer (or primordial plexiform layer)
C11 cingulate cortex neuroepithelium
C13 cingulate cortex (or differentiating field)

FIBER TRACTS (f)

f5 medial forebrain bundle

GLANDS AND PARANEURAL STRUCTURES (g)

g9 lamina terminalis

HIPPOCAMPAL FORMATION (K)

K1 hippocampal (Ammon's horn) neuroepithelium
K11 subicular neuroepithelium
K13 subicular area (or differentiating field)

PREOPTIC AREA AND OPTIC EPITHELIA (P)

P1 preoptic neuroepithelium
P3 preoptic area (or differentiating field)
P12 eye stalk epithelium

VENTRICLES (v)

vla lateral ventricle, anterior horn
vlc lateral ventricle, central
vld lateral ventricle, dorsal
vlfp lateral ventricle, pallidal fork
vlm foramen of Monro
v3pr third ventricle, preoptic recess

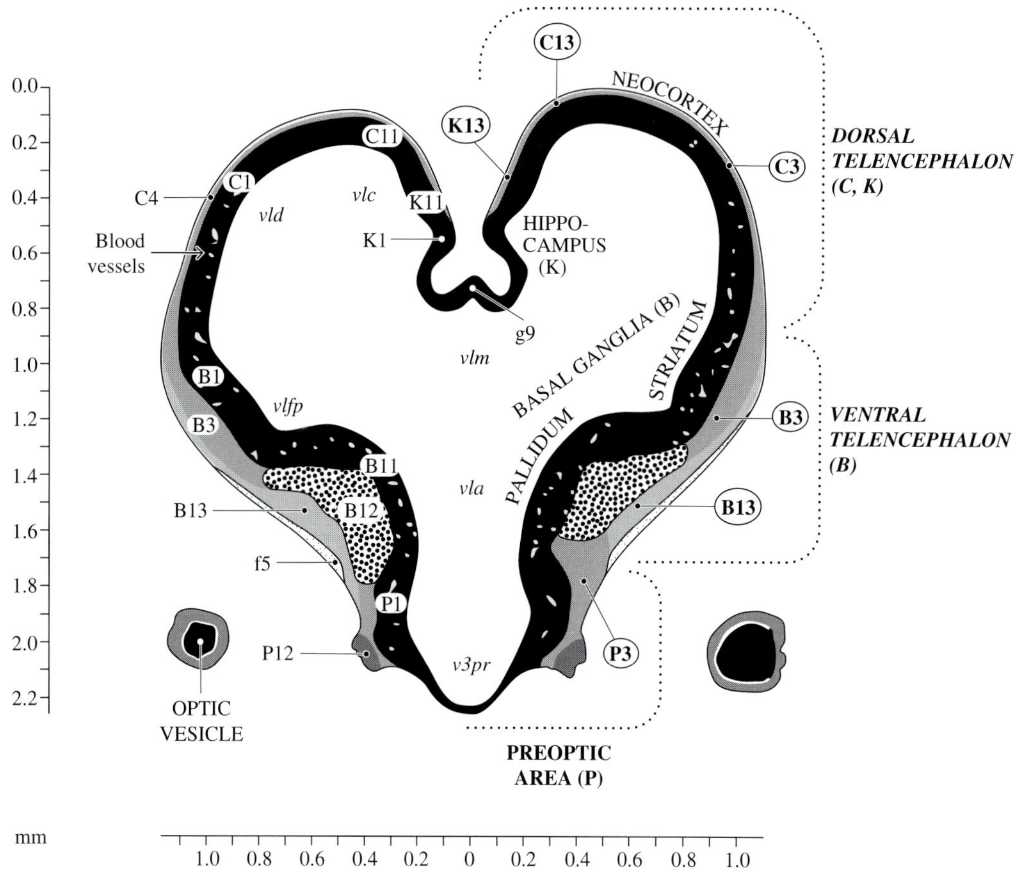

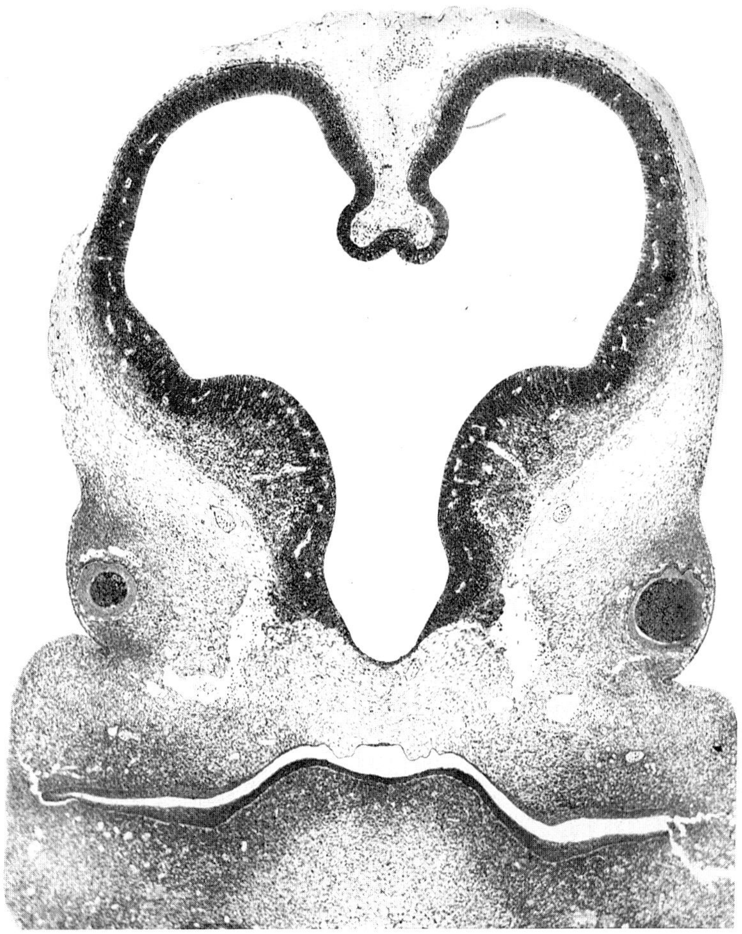

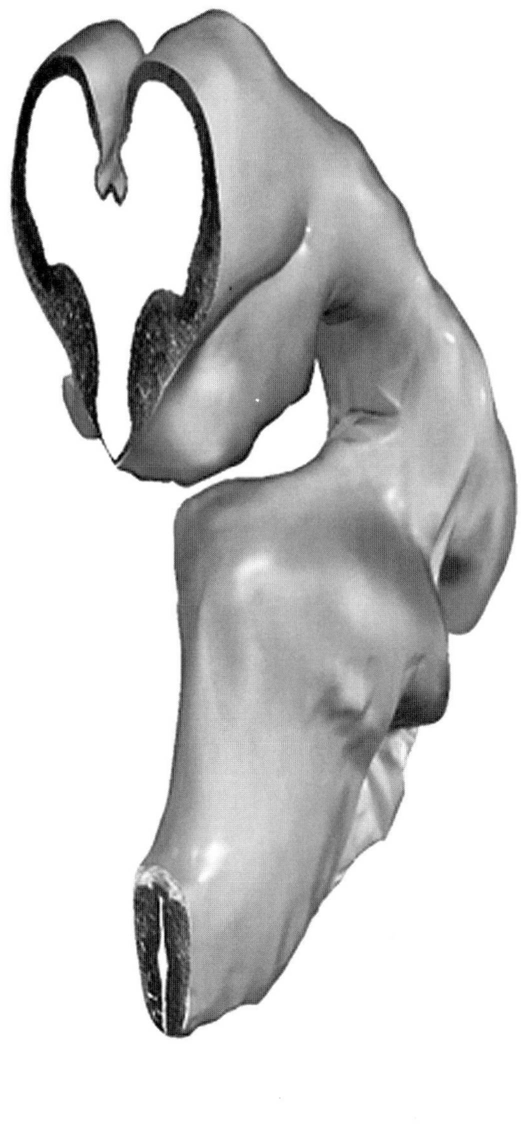

E14

BASAL GANGLIA (B)

B1 striatal neuroepithelium
B3 striatum (or differentiating field)
B11 pallidal neuroepithelium
B12 pallidal subventricular zone
B13 pallidum (or differentiating field)

CEREBRAL CORTEX (C)

C1 neocortical neuroepithelium
C3 neocortex (or differentiating field)
C4 molecular layer (or primordial plexiform layer)
C11 cingulate cortex neuroepithelium
C13 cingulate cortex (or differentiating field)

FIBER TRACTS (f)

f5 medial forebrain bundle

GLANDS AND PARANEURAL STRUCTURES (g)

g9 lamina terminalis

HYPOTHALAMUS (H)

H1a anterior hypothalamic neuroepithelium

HIPPOCAMPAL FORMATION (K)

K1 hippocampal (Ammon's horn) neuroepithelium
K11 subicular neuroepithelium
K13 subicular area (or differentiating field)

PREOPTIC AREA AND OPTIC EPITHELIA (P)

P1 preoptic neuroepithelium
P12 eye stalk epithelium

VENTRICLES (v)

vla lateral ventricle, anterior horn
vlc lateral ventricle, central
vld lateral ventricle, dorsal
vlfp lateral ventricle, pallidal fork
vlm foramen of Monro
v3h third ventricle, hypothalamic
v3o third ventricle, optic recess
v3pr third ventricle, preoptic recess

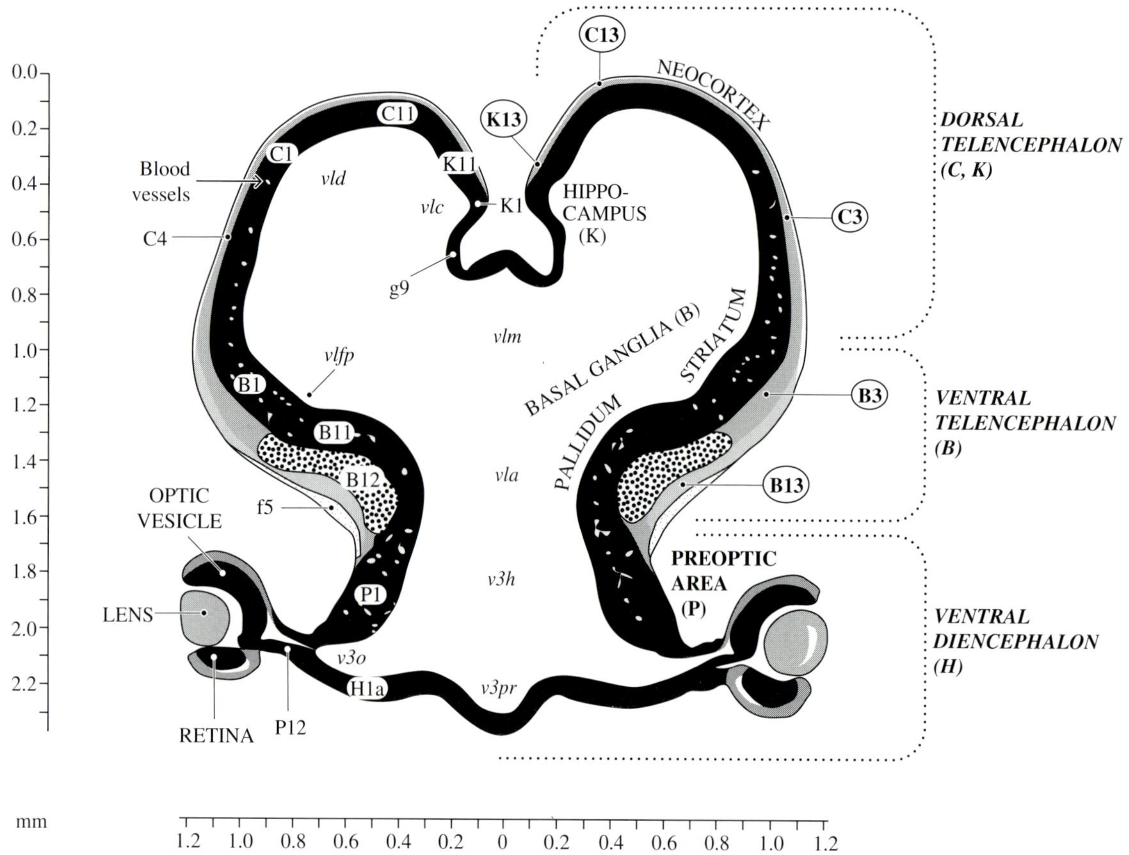

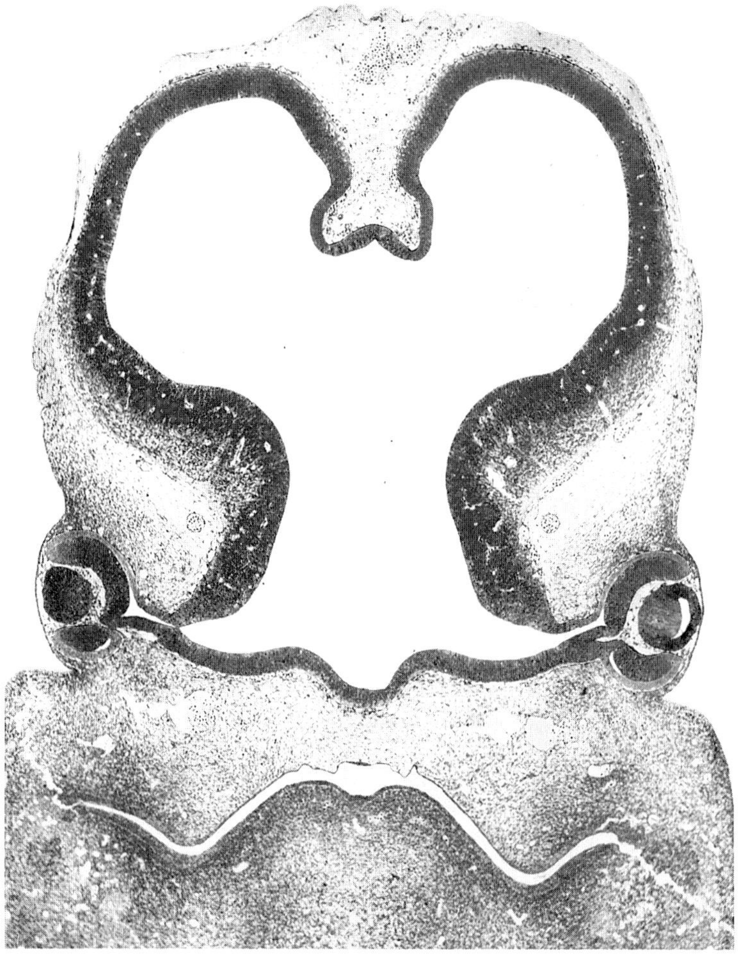

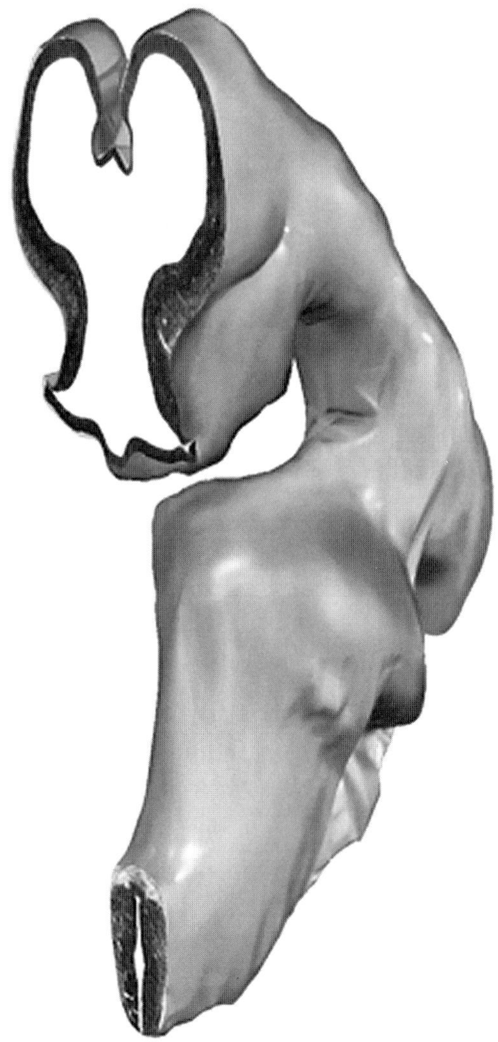

CORONAL FIGURE 6

AMYGDALA (A)

A1 amygdaloid neuroepithelium
A3 amygdala (or differentiating field)

BASAL GANGLIA (B)

B11 pallidal neuroepithelium
B13 pallidum (or differentiating field)
B21 strionuclear neuroepithelium

CEREBRAL CORTEX (C)

C1 neocortical neuroepithelium
C3 neocortex (or differentiating field)
C4 molecular layer (or primordial plexiform layer)
C11 cingulate cortex neuroepithelium
C13 cingulate cortex (or differentiating field)

FIBER TRACTS (f)

f5 medial forebrain bundle

HYPOTHALAMUS (H)

H1i intermediate hypothalamic neuroepithelium
H1v ventral hypothalamic neuroepithelium
H3a anterior hypothalamus (or differentiating field)

HIPPOCAMPAL FORMATION (K)

K1 hippocampal (Ammon's horn) neuroepithelium

NERVES AND NUCLEI, CRANIAL (n)

n5g trigeminal ganglion

THALAMUS (T)

T1a anterior thalamic neuroepithelium
T1e epithalamic neuroepithelium
T1i intermediate thalamic neuroepithelium

VENTRICLES (v)

vld lateral ventricle, dorsal
vlp lateral ventricle, posterior horn
v3e third ventricle, epithalamic
v3h third ventricle, hypothalamic
v3t third ventricle, thalamic

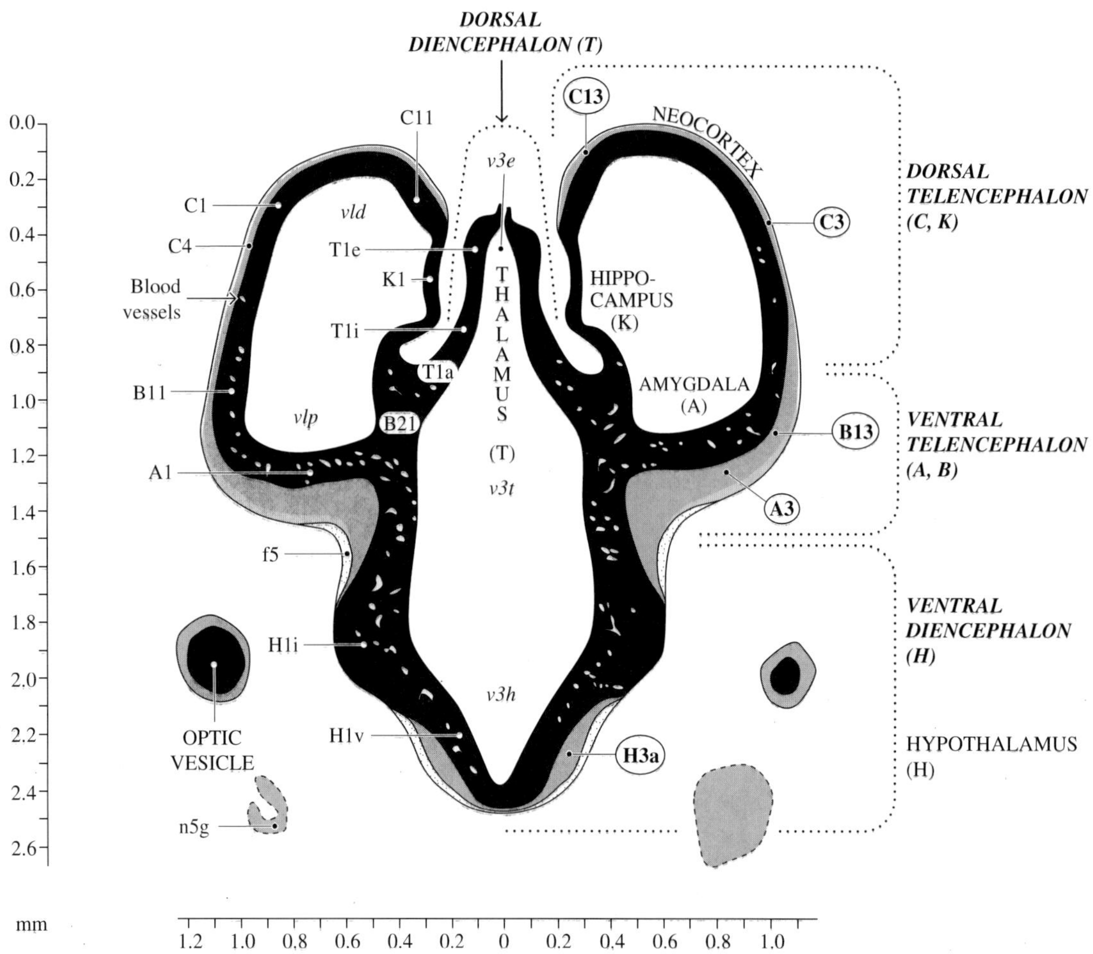

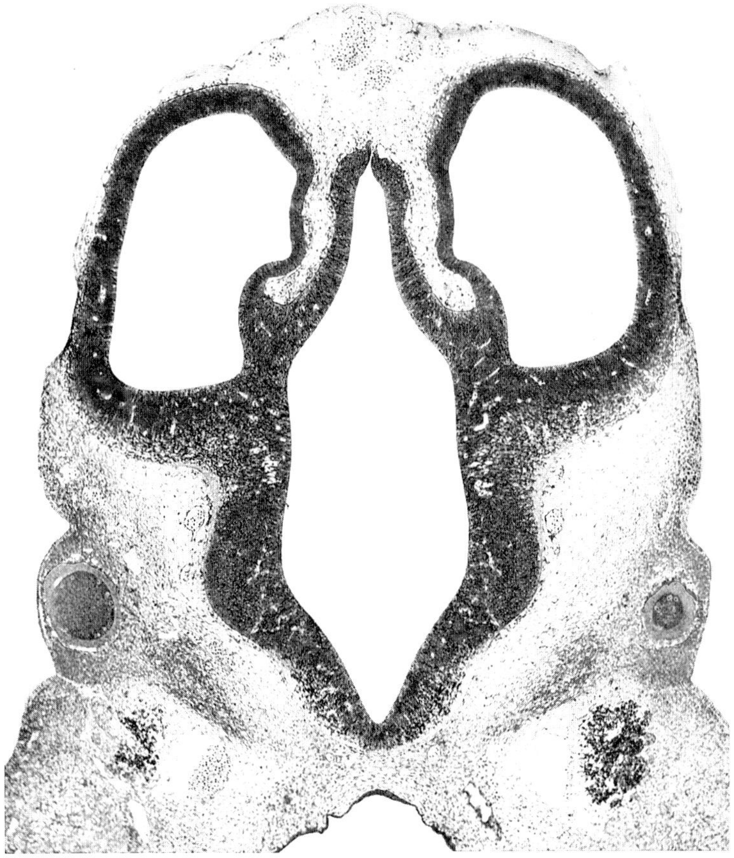

CORONAL FIGURE 7

CEREBRAL CORTEX (C)

C1 neocortical neuroepithelium
C3 neocortex (or differentiating field)
C4 molecular layer (or primordial plexiform layer)
C11 cingulate cortex neuroepithelium
C13 cingulate cortex (or differentiating field)

GLANDS AND PARANEURAL STRUCTURES (g)

g6 pituitary gland (primordium)

HYPOTHALAMUS (H)

H1f Forelian neuroepithelium
H1i intermediate hypothalamic neuroepithelium
H1v ventral hypothalamic neuroepithelium
H3 hypothalamus (or differentiating field)
H3f fields of Forel (or differentiating field)
H3*l* lateral hypothalamus (or differentiating field)

HIPPOCAMPAL FORMATION (K)

K1 hippocampal (Ammon's horn) neuroepithelium
K3 hippocampus (or differentiating field)

NERVES AND NUCLEI, CRANIAL (n)

n5 trigeminal nerve
n5g trigeminal ganglion

THALAMUS (T)

T1a anterior thalamic neuroepithelium
T1e epithalamic neuroepithelium
T1i intermediate thalamic neuroepithelium

VENTRICLES (v)

vld lateral ventricle, dorsal
vlp lateral ventricle, posterior horn
v3e third ventricle, epithalamic
v3h third ventricle, hypothalamic
v3pi third ventricle, pineal recess
v3t third ventricle, thalamic

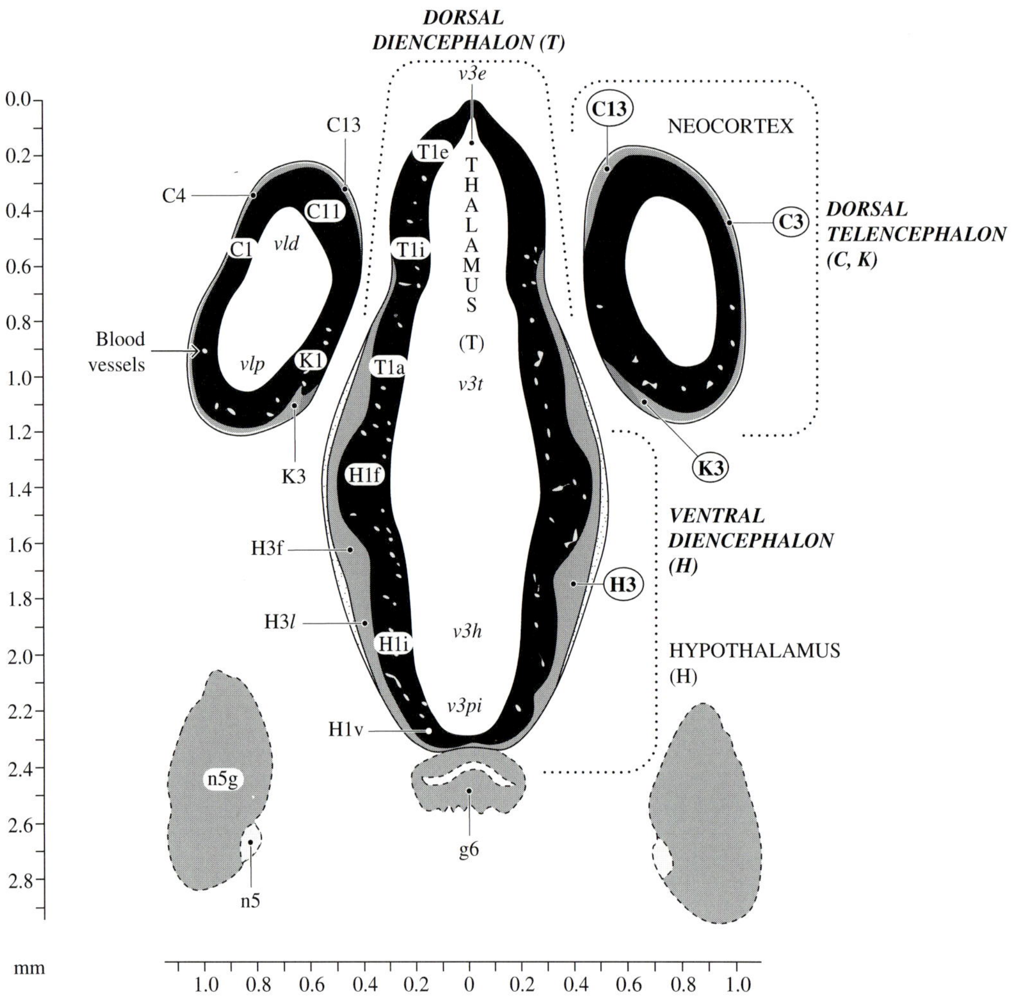

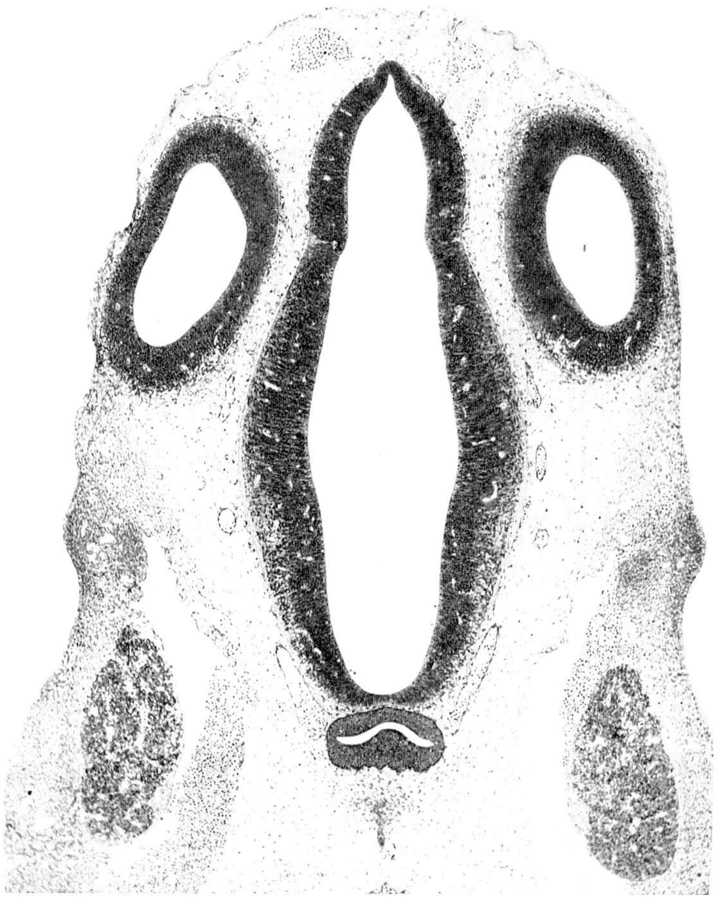

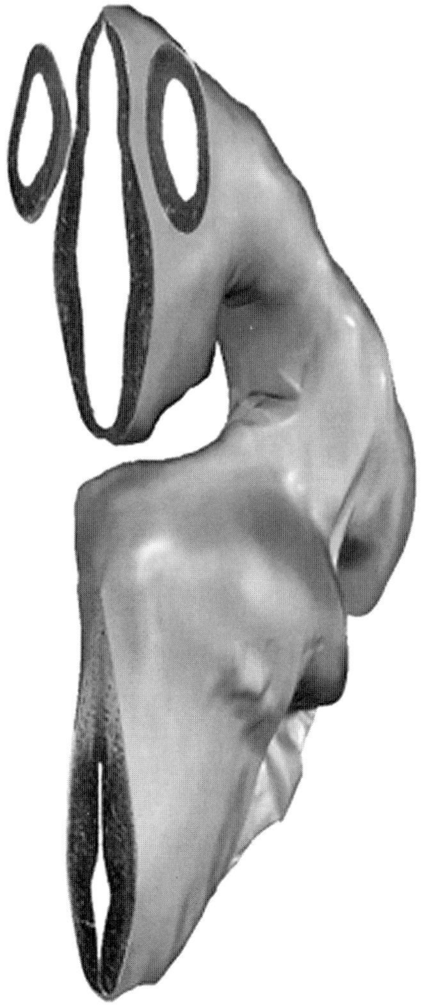

CORONAL FIGURE 8

HYPOTHALAMUS (H)

H1f	Forelian neuroepithelium
H1i	intermediate hypothalamic neuroepithelium
H1v	ventral hypothalamic neuroepithelium
H3	hypothalamus (or differentiating field)
H3f	fields of Forel (or differentiating field)
H3*l*	lateral hypothalamus (or differentiating field)

NERVES AND NUCLEI, CRANIAL (n)

n5g	trigeminal ganglion
n5n	trigeminal motor nucleus
n5p	principal (sensory)trigeminal nucleus
n7n	facial motor nucleus
n8b	vestibulocochlear boundary cap
n8g	vestibular ganglion

THALAMUS (T)

T1e	epithalamic neuroepithelium
T1i	intermediate thalamic neuroepithelium
T1r	reticular (or ventral) thalamic neuroepithelium
T3	thalamus (or differentiating field)
T25	reticular nucleus

VENTRICLES (v)

vs	spinal canal
v3e	third ventricle, epithalamic
v3h	third ventricle, hypothalamic
v3i	third ventricle, infundibular recess
v3t	third ventricle, thalamic
v4a	fourth ventricle, anterior
v4p	fourth ventricle, posterior

PONTINE AREA (X)

X1p	posterior pontine neuroepithelium
X3	pons (or differentiating field)

MEDULLA (Y)

Y1	medullary neuroepithelium
Y3	medulla (or differentiating field)

SPINAL CORD (Z)

Z1	spinal cord neuroepithelium
Z3	spinal cord (or differentiating field)

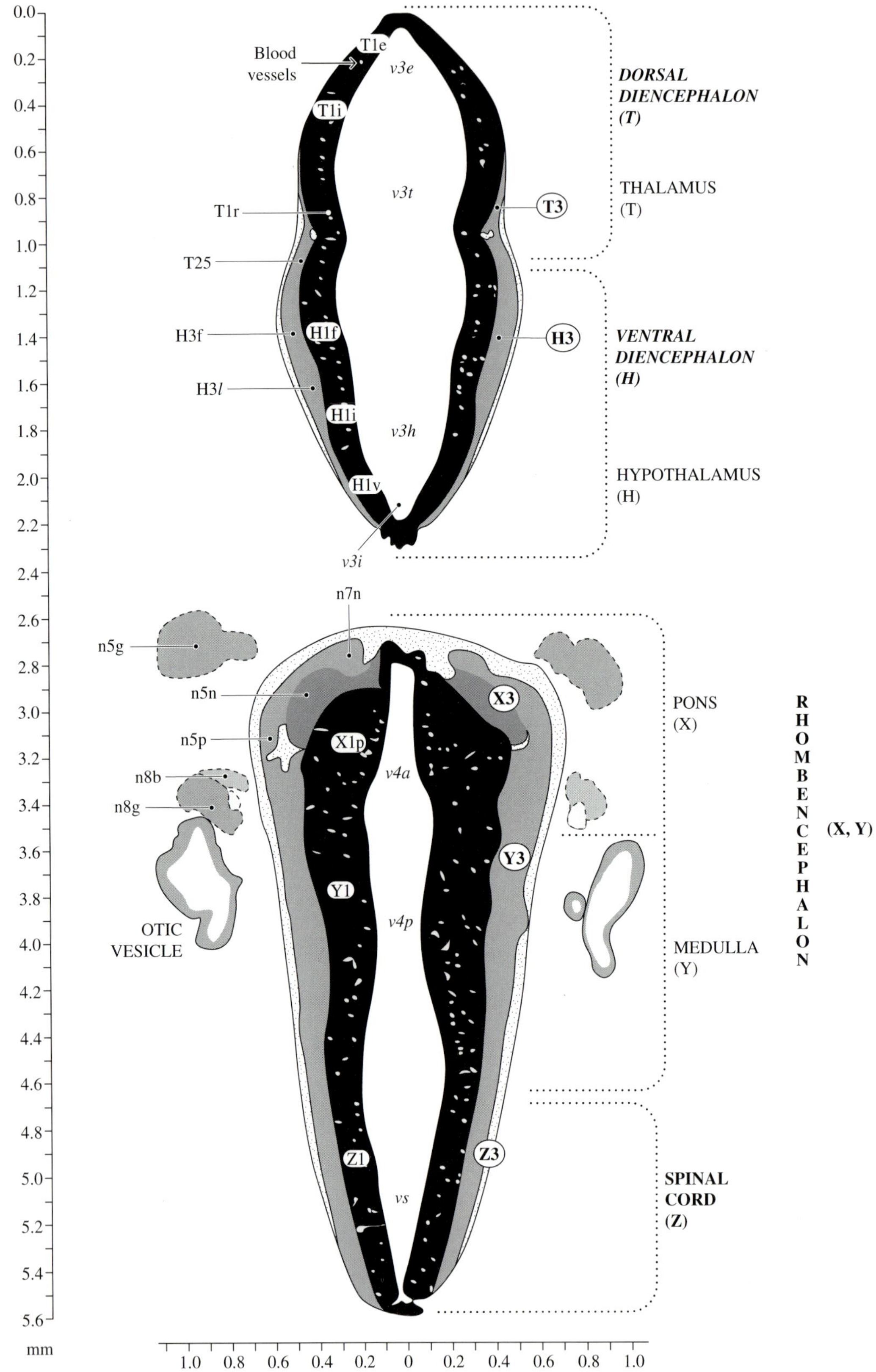

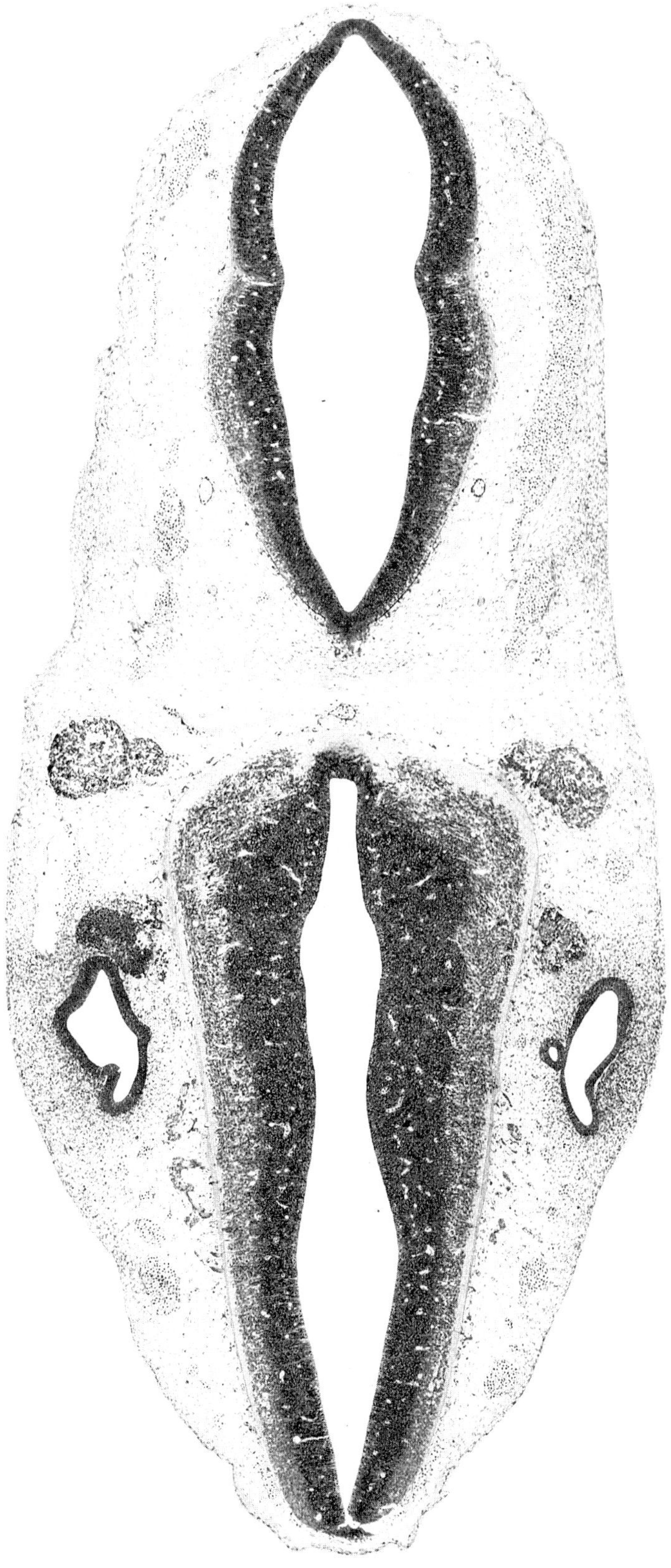

CORONAL FIGURE 9

HYPOTHALAMUS (H)

- H1i intermediate hypothalamic neuroepithelium
- H1v ventral hypothalamic neuroepithelium
- **H3** hypothalamus (or differentiating field)
- H3a anterior hypothalamus (or differentiating field)
- H3f fields of Forel (or differentiating field)

NERVES AND NUCLEI, CRANIAL (n)

- n5 trigeminal nerve
- n5b trigeminal boundary cap
- n5n trigeminal motor nucleus
- n5p principal (sensory) trigeminal nucleus
- n5s spinal trigeminal nucleus
- n7n facial motor nucleus
- n8b vestibulocochlear boundary cap

THALAMUS (T)

- T1e epithalamic neuroepithelium
- T1i intermediate thalamic neuroepithelium
- **T3** thalamus (or differentiating field)
- T25 reticular nucleus

VENTRICLES (v)

- *vs* spinal canal
- *v3h* third ventricle, hypothalamic
- *v3t* third ventricle, thalamic
- *v4a* fourth ventricle, anterior
- *v4p* fourth ventricle, posterior

PONTINE AREA (X)

- X1a anterior pontine neuroepithelium
- X1p posterior pontine neuroepithelium
- **X3** pons (or differentiating field)

MEDULLA (Y)

- Y1 medullary neuroepithelium
- **Y3** medulla (or differentiating field)

SPINAL CORD (Z)

- Z1 spinal cord neuroepithelium
- **Z3** spinal cord (or differentiating field)

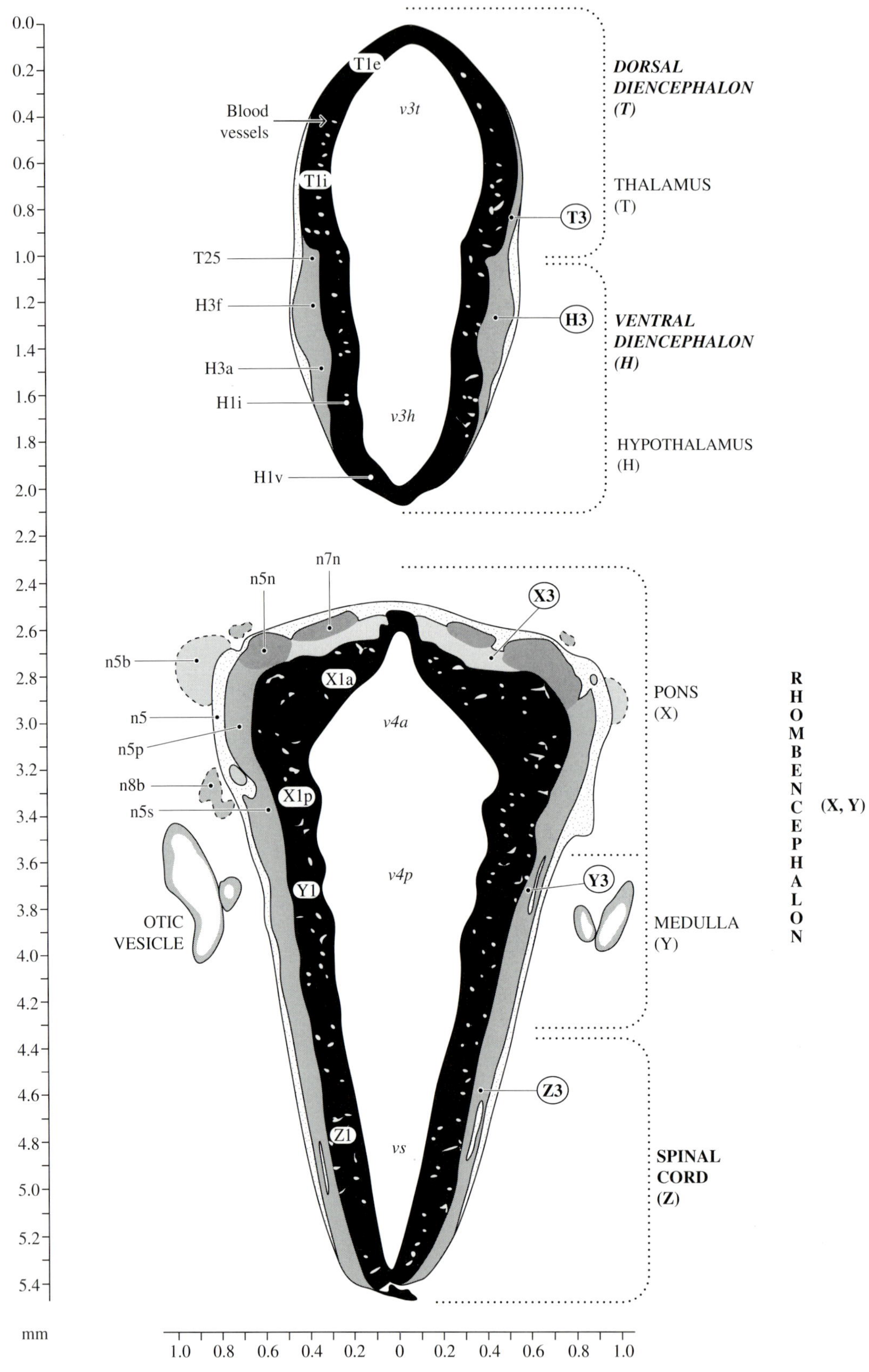

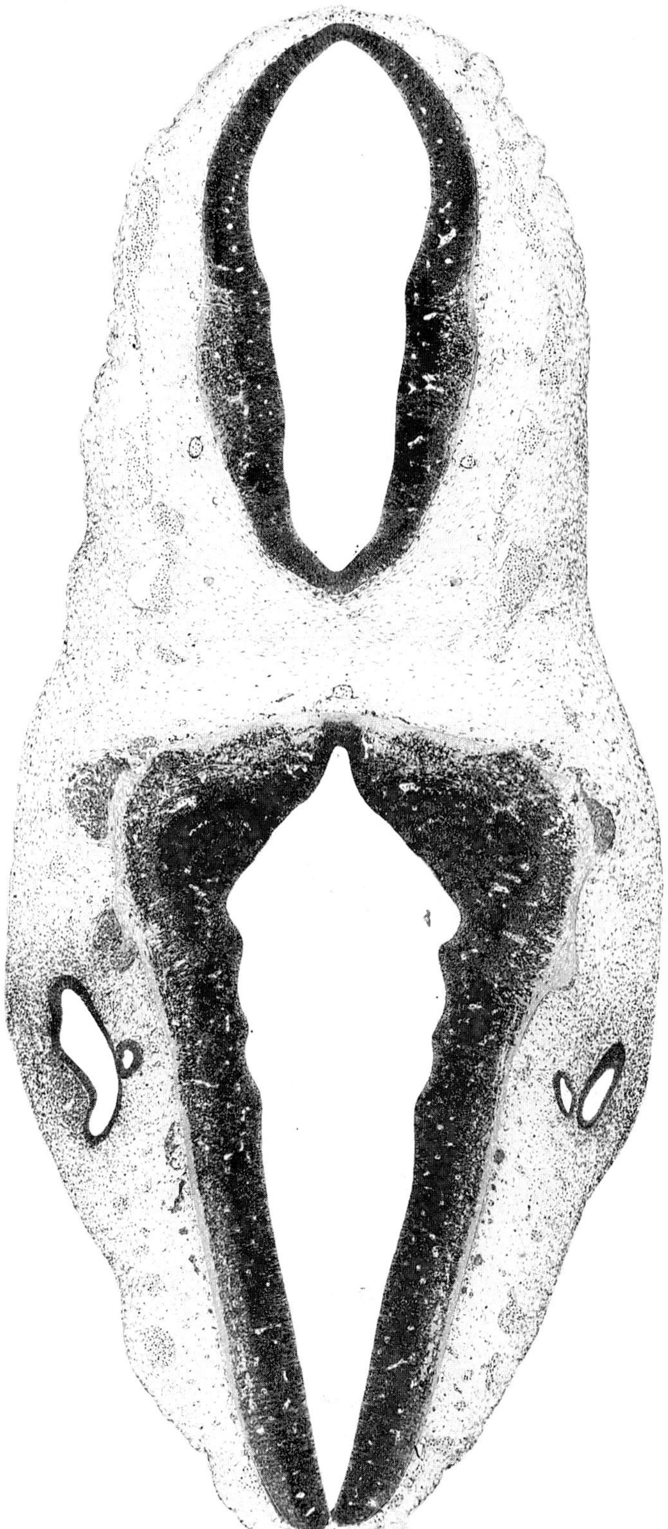

CORONAL FIGURE 10

HYPOTHALAMUS (H)

- H1i intermediate hypothalamic neuroepithelium
- H1m mammillary neuroepithelium
- **H3** hypothalamus (or differentiating field)
- H3*l* lateral hypothalamus (or differentiating field)
- H14 supramammillary nucleus
- H17 lateral mammillary nucleus

NERVES AND NUCLEI, CRANIAL (n)

- n7n facial motor nucleus

THALAMUS (T)

- T1i intermediate thalamic neuroepithelium
- T1p posterior thalamic neuroepithelium
- **T3** thalamus (or differentiating field)
- T25 reticular nucleus

VENTRICLES (v)

- *vs* spinal canal
- *v3h* third ventricle, hypothalamic
- *v3m* third ventricle, mammillary recess
- *v3t* third ventricle, thalamic
- *v4a* fourth ventricle, anterior
- *v4l* fourth ventricle, lateral recess
- *v4p* fourth ventricle, posterior

PONTINE AREA (X)

- X1a anterior pontine neuroepithelium
- X1p posterior pontine neuroepithelium
- **X3** pons (or differentiating field)

MEDULLA (Y)

- Y1 medullary neuroepithelium
- **Y3** medulla (or differentiating field)

SPINAL CORD (Z)

- Z1 spinal cord neuroepithelium
- **Z3** spinal cord (or differentiating field)

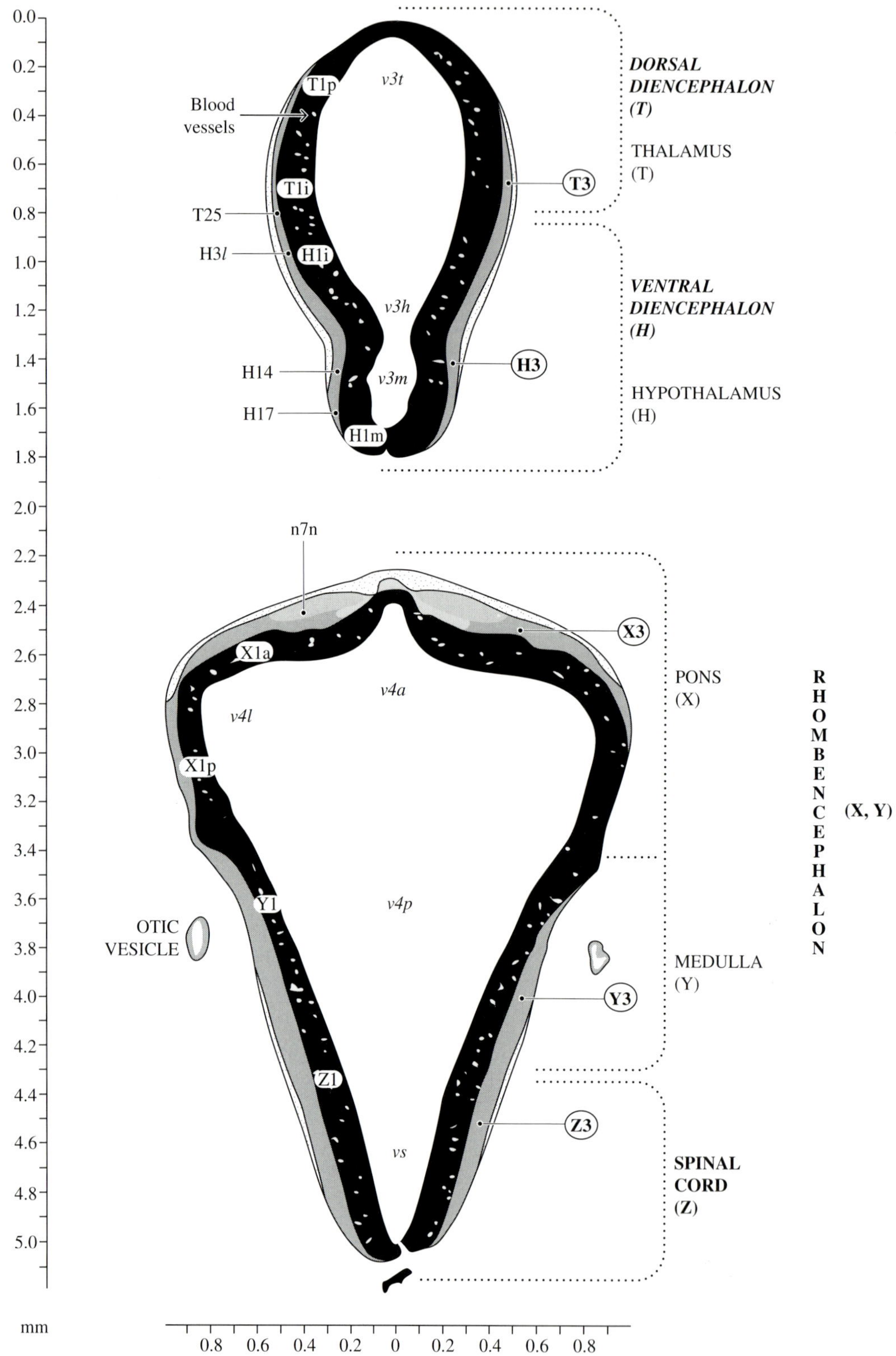

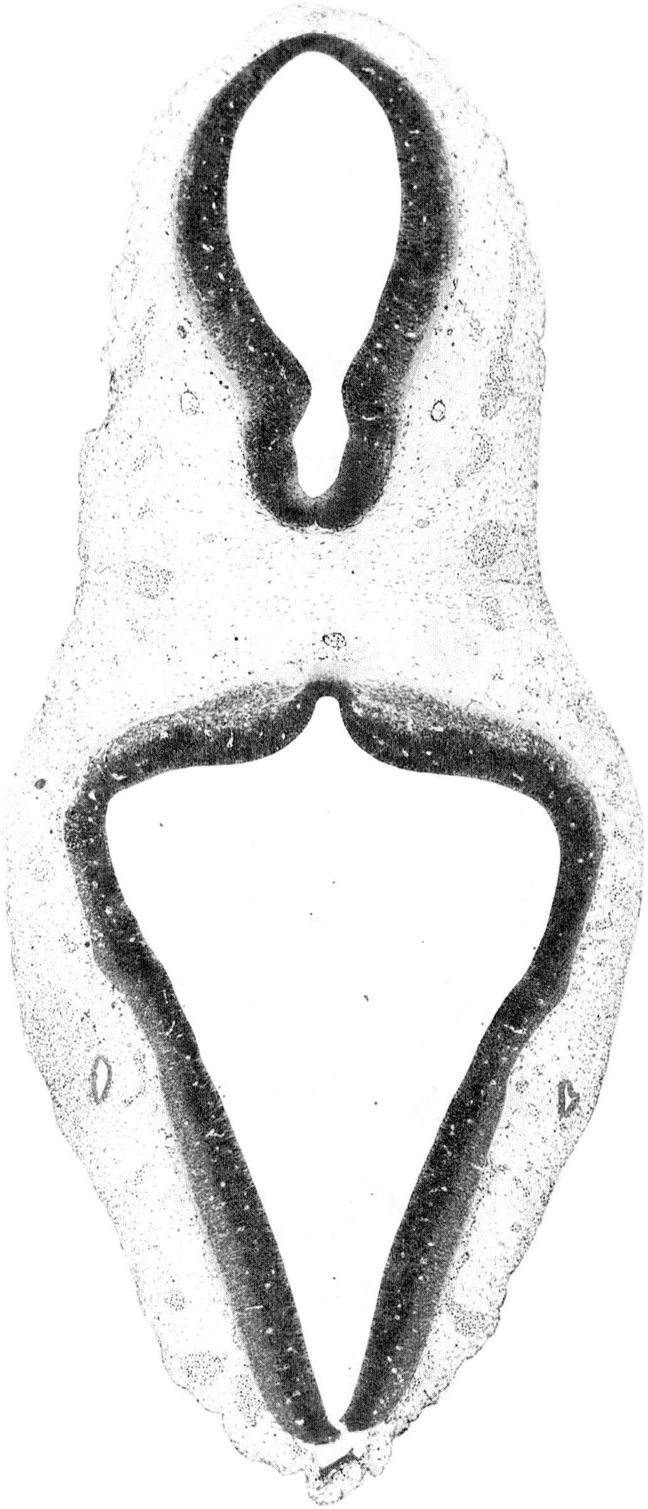

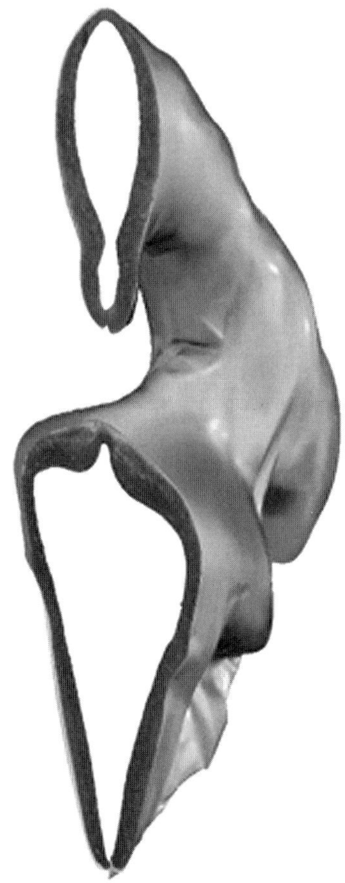

CORONAL FIGURE 11

GLANDS AND PARANEURAL STRUCTURES (g)

g10	velum medullare

HYPOTHALAMUS (H)

H1i	intermediate hypothalamic neuroepithelium
H1m	mammillary neuroepithelium
H1pm	premammillary neuroepithelium
H1v	ventral hypothalamic neuroepithelium
H3	hypothalamus (or differentiating field)
H3*l*	lateral hypothalamus (or differentiating field)
H14	supramammillary nucleus

THALAMUS (T)

T1p	posterior thalamic neuroepithelium
T3	thalamus (or differentiating field)
T34	medial geniculate nucleus

VENTRICLES (v)

vs	spinal canal
v3h	third ventricle, hypothalamic
v3m	third ventricle, mammillary recess
v3t	third ventricle, thalamic
v4a	fourth ventricle, anterior
v4l	fourth ventricle, lateral recess
v4p	fourth ventricle, posterior

PONTINE AREA (X)

X1a	anterior pontine neuroepithelium
X3	pons (or differentiating field)
X31	isthmal neuroepithelium
X33	isthmus (or differentiating field)

MEDULLA (Y)

Y1	medullary neuroepithelium

SPINAL CORD (Z)

Z1	spinal cord neuroepithelium
Z3	spinal cord (or differentiating field)

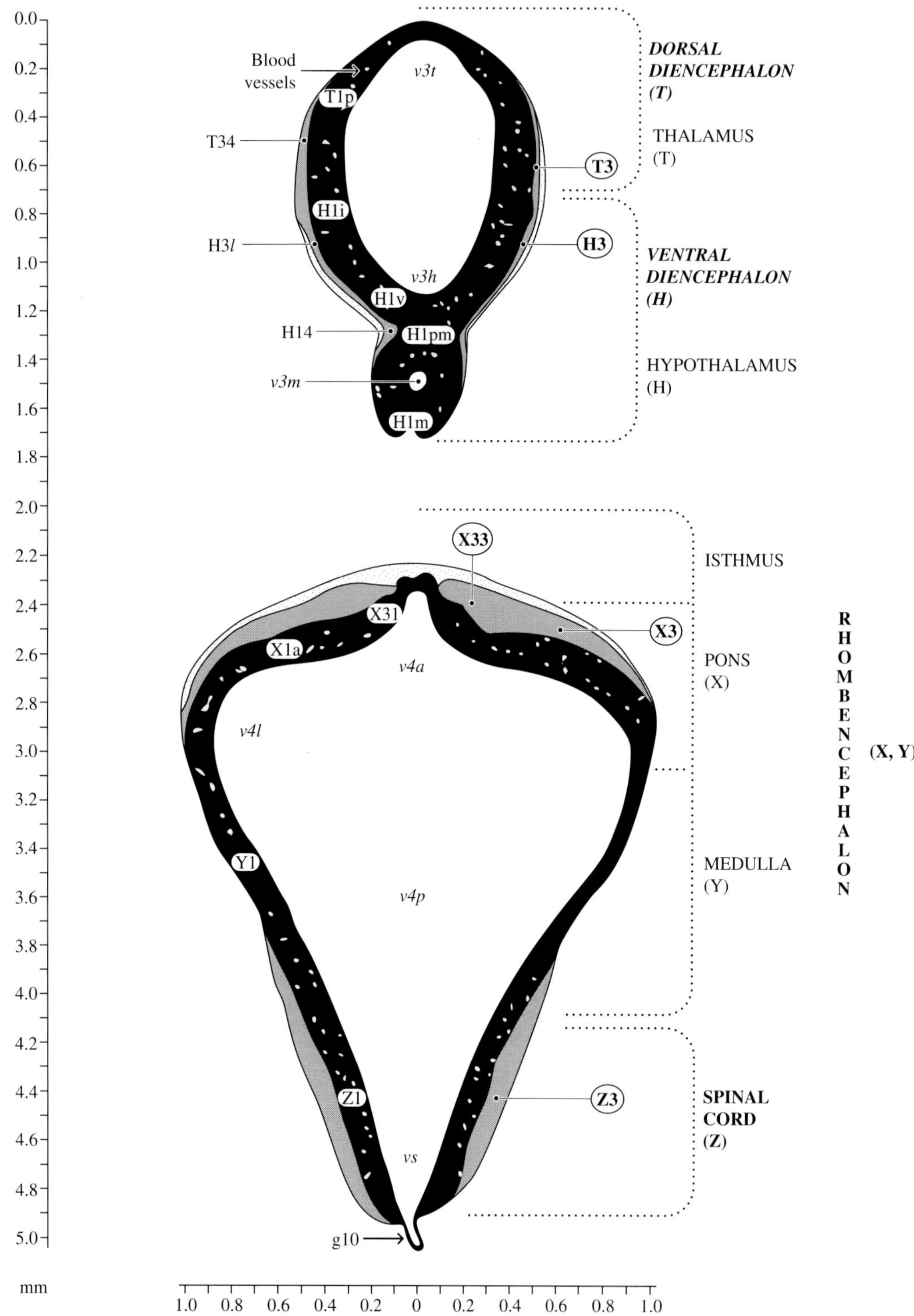

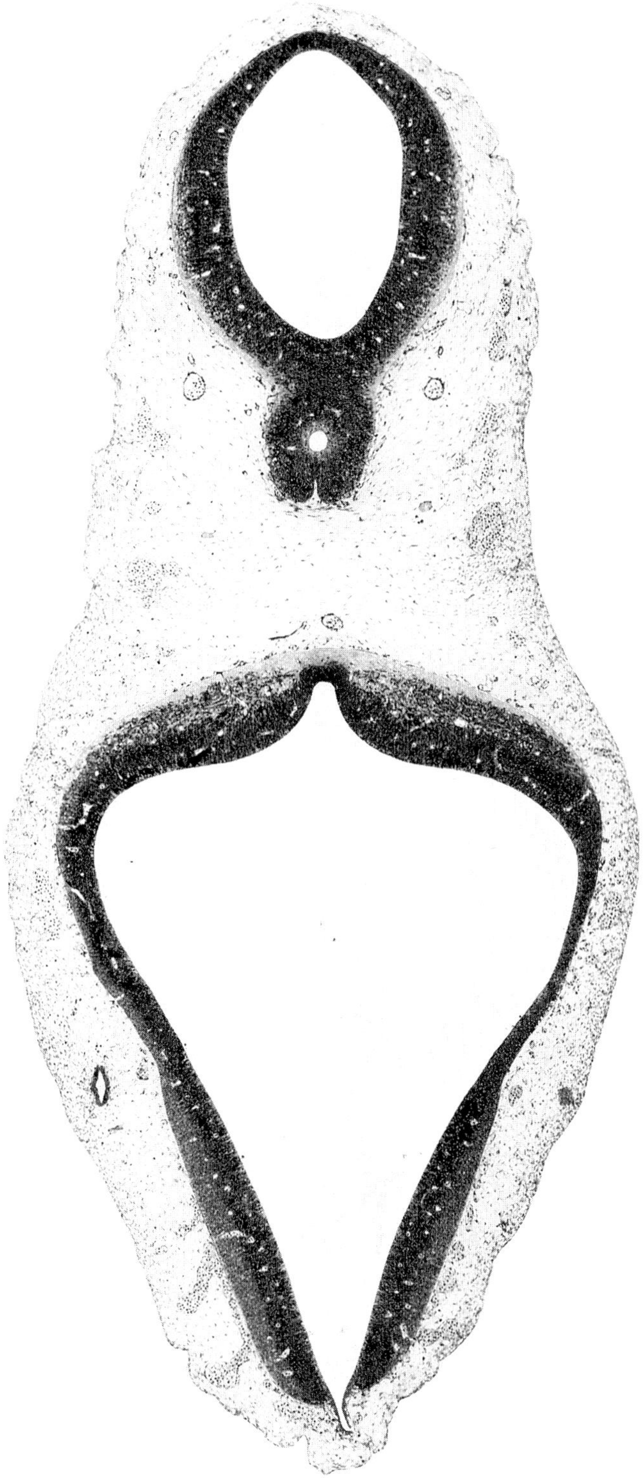

CORONAL FIGURE 12

FIBER TRACTS (f)

f19 posterior commissure

GLANDS AND PARANEURAL STRUCTURES (g)

g10 velum medullare

CEREBELLUM (L)

L1 cerebellar neuroepithelium
L3 cerebellum (or differentiating field)

MIDBRAIN (M)

M11a anterior tegmental neuroepithelium
M13 tegmentum (or differentiating field)
M21 pretectal neuroepithelium
M23 pretectum (or differentiating field)

THALAMUS (T)

T1p posterior thalamic neuroepithelium
T3 thalamus (or differentiating field)
T34 medial geniculate nucleus

VENTRICLES (v)

vapr aqueduct, pretectal
vat aqueduct, tegmental
v4a fourth ventricle, anterior
v4i fourth ventricle, isthmal
v4l fourth ventricle, lateral recess
v4p fourth ventricle, posterior

PONTINE AREA (X)

X1a anterior pontine neuroepithelium
X3 pons (or differentiating field)
X31 isthmal neuroepithelium
X33 isthmus (or differentiating field)

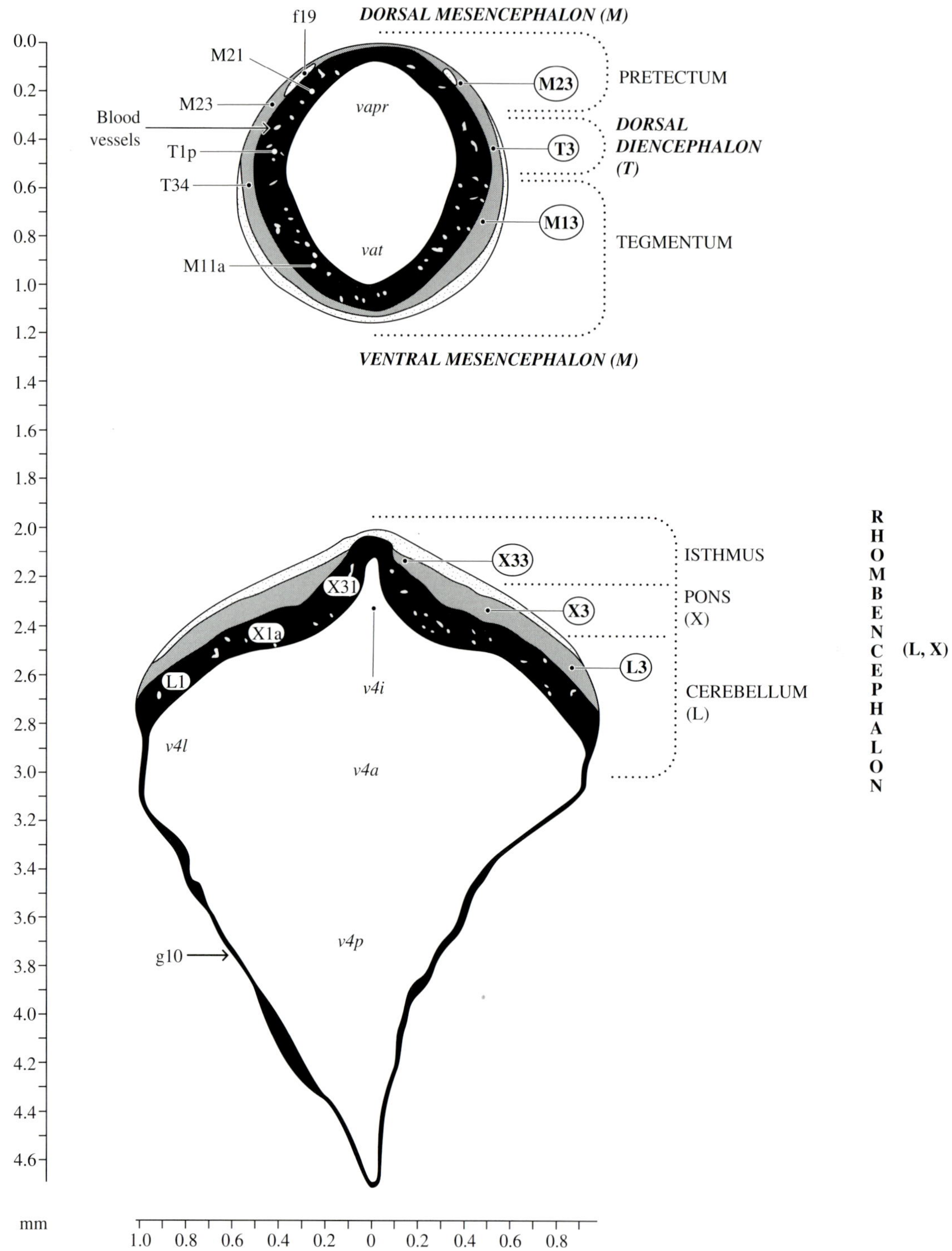

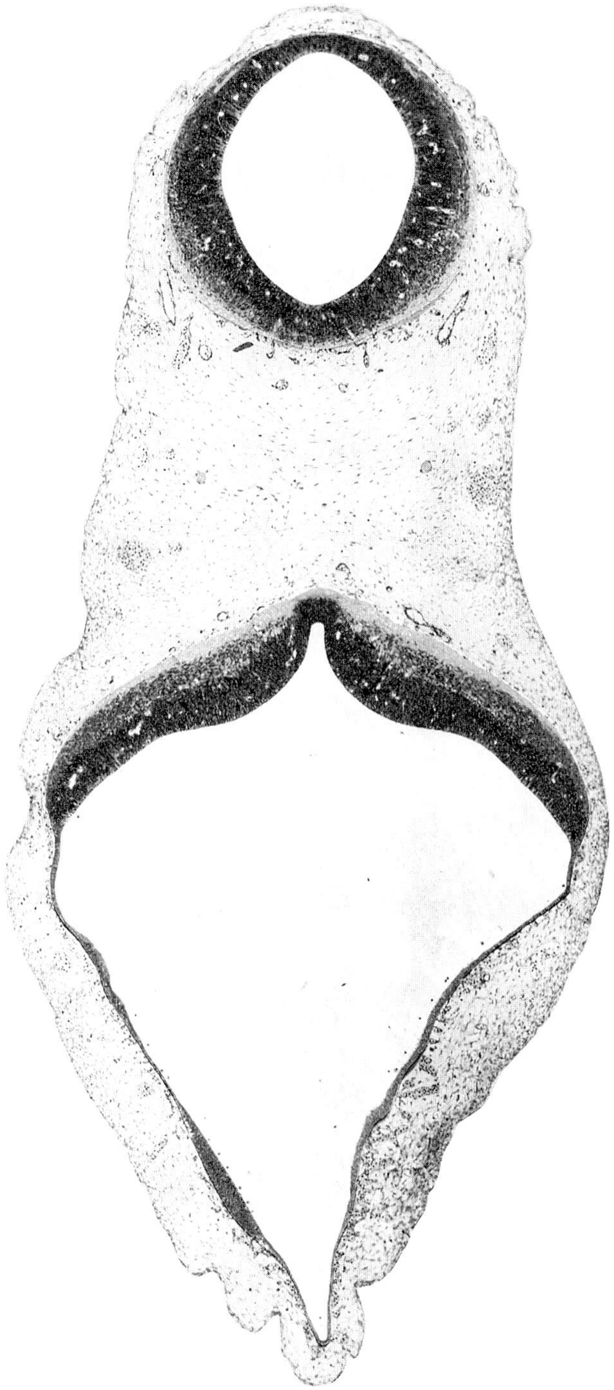

CORONAL FIGURE 13

FIBER TRACTS (f)

f19 posterior commissure

GLANDS AND PARANEURAL STRUCTURES (g)

g10 velum medullare

CEREBELLUM (L)

L1 cerebellar neuroepithelium
L3 cerebellum (or differentiating field)

MIDBRAIN (M)

M11a anterior tegmental neuroepithelium
M13 tegmentum (or differentiating field)
M21 pretectal neuroepithelium
M23 pretectum (or differentiating field)

VENTRICLES (v)

vapr aqueduct, pretectal
vat aqueduct, tegmental
v4a fourth ventricle, anterior
v4i fourth ventricle, isthmal
v4l fourth ventricle, lateral recess
v4p fourth ventricle, posterior

PONTINE AREA (X)

X1a anterior pontine neuroepithelium
X3 pons (or differentiating field)
X31 isthmal neuroepithelium
X33 isthmus (or differentiating field)

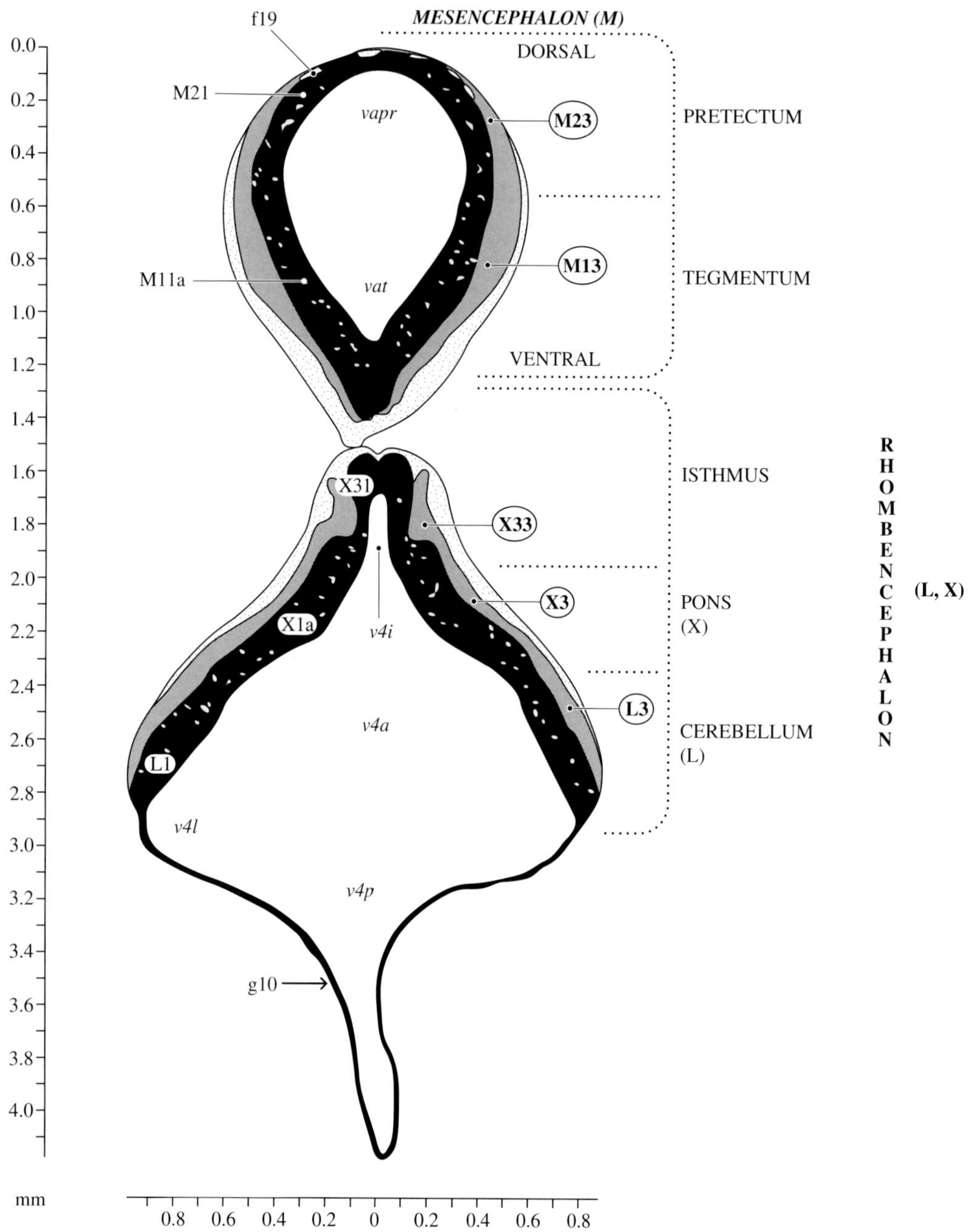

E
1
4

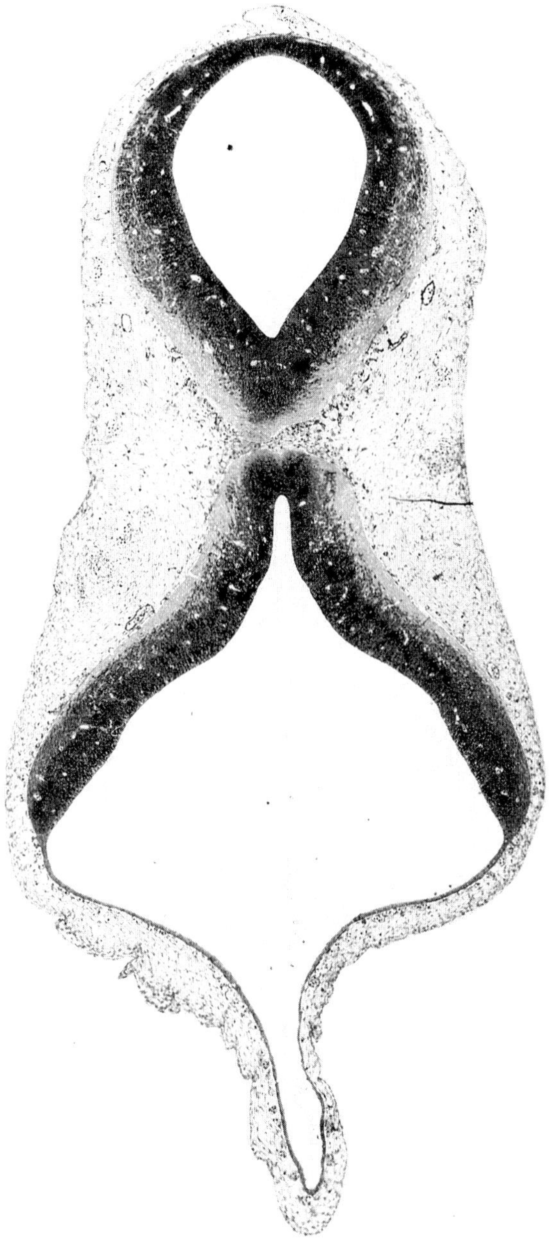

CORONAL FIGURE 14

FIBER TRACTS (f)

f19 posterior commissure

GLANDS AND PARANEURAL STRUCTURES (g)

g10 velum medullare

CEREBELLUM (L)

L1 cerebellar neuroepithelium
L3 cerebellum (or differentiating field)

MIDBRAIN (M)

M11a anterior tegmental neuroepithelium
M13 tegmentum (or differentiating field)
M21 pretectal neuroepithelium
M23 pretectum (or differentiating field)

VENTRICLES (v)

vapr aqueduct, pretectal
vat aqueduct, tegmental
v4a fourth ventricle, anterior
v4i fourth ventricle, isthmal
v4l fourth ventricle, lateral recess
v4p fourth ventricle, posterior

PONTINE AREA (X)

X1a anterior pontine neuroepithelium
X3 pons (or differentiating field)
X31 isthmal neuroepithelium
X33 isthmus (or differentiating field)

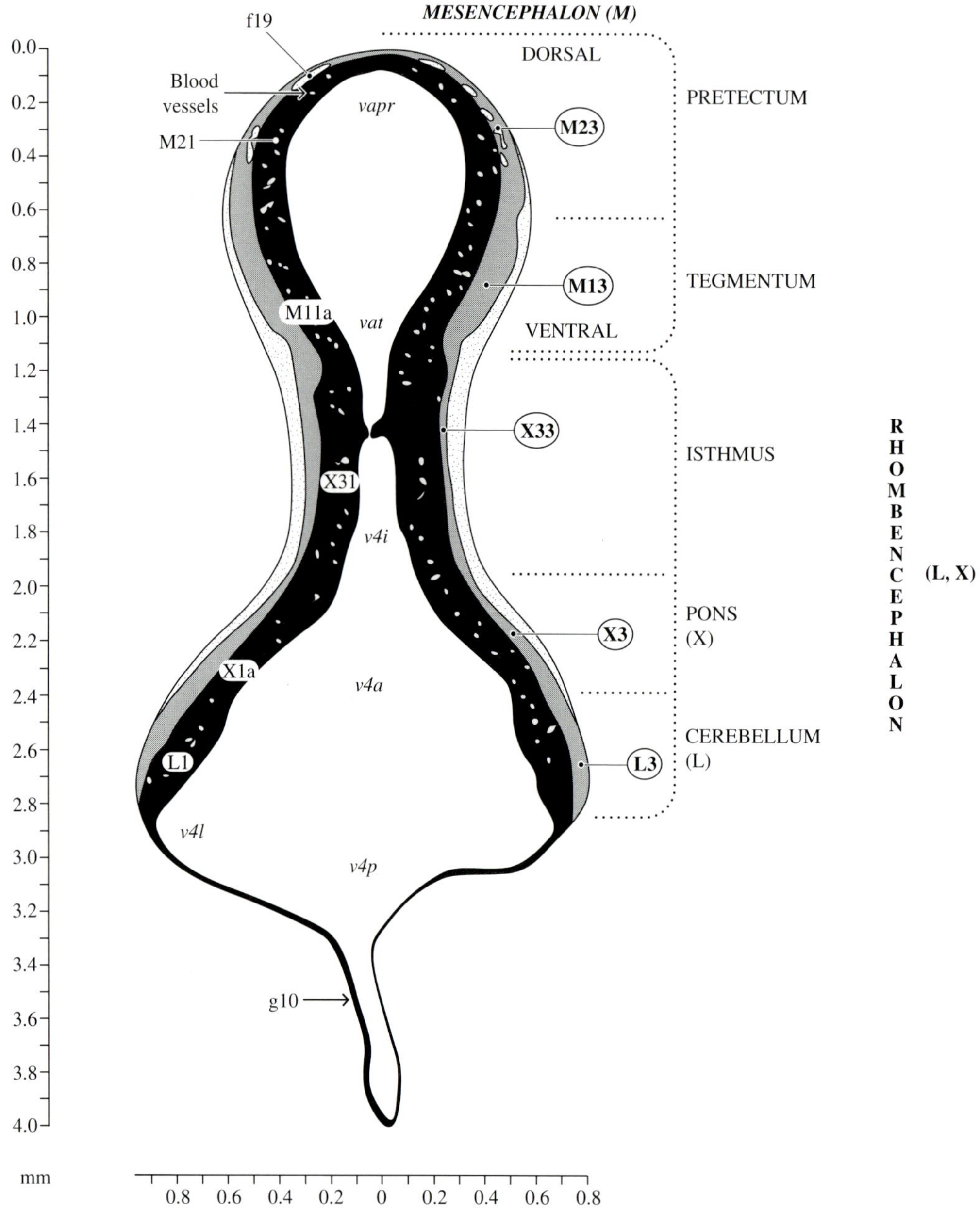

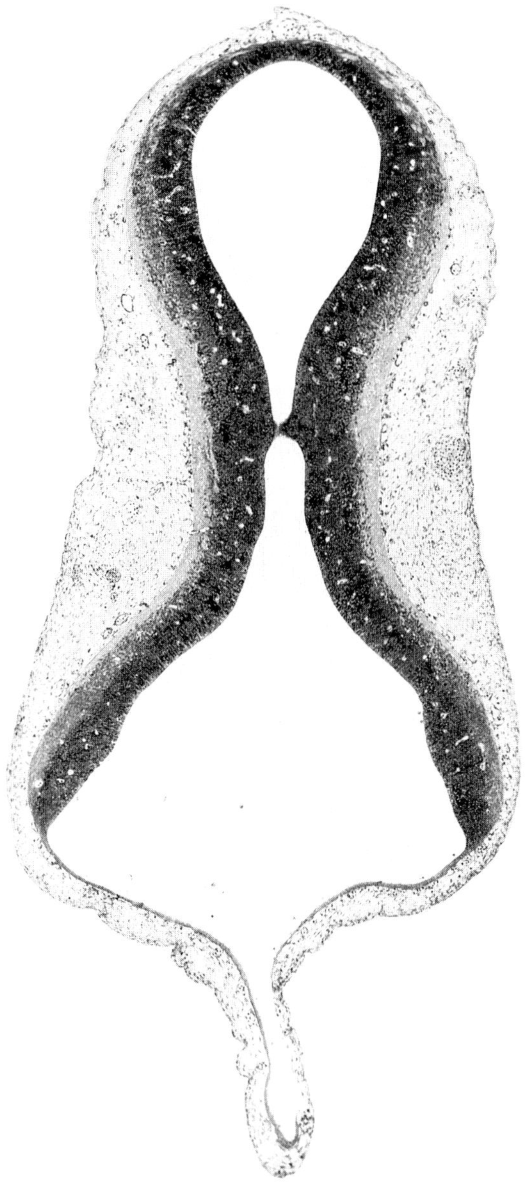

CORONAL FIGURE 15

GLANDS AND PARANEURAL STRUCTURES (g)

g10 velum medullare

CEREBELLUM (L)

L1 cerebellar neuroepithelium
L3 cerebellum (or differentiating field)

MIDBRAIN (M)

M1 superior tectal neuroepithelium
M3 superior colliculus (or differentiating field)
M11p posterior tegmental neuroepithelium
M13 tegmentum (or differentiating field)
M31 inferior tectal neuroepithelium

VENTRICLES (v)

vai aqueduct, inferior tectal
vas aqueduct, superior tectal
vat aqueduct, tegmental
v4 fourth ventricle
v4i fourth ventricle, isthmal

PONTINE AREA (X)

X31 isthmal neuroepithelium
X33 isthmus (or differentiating field)

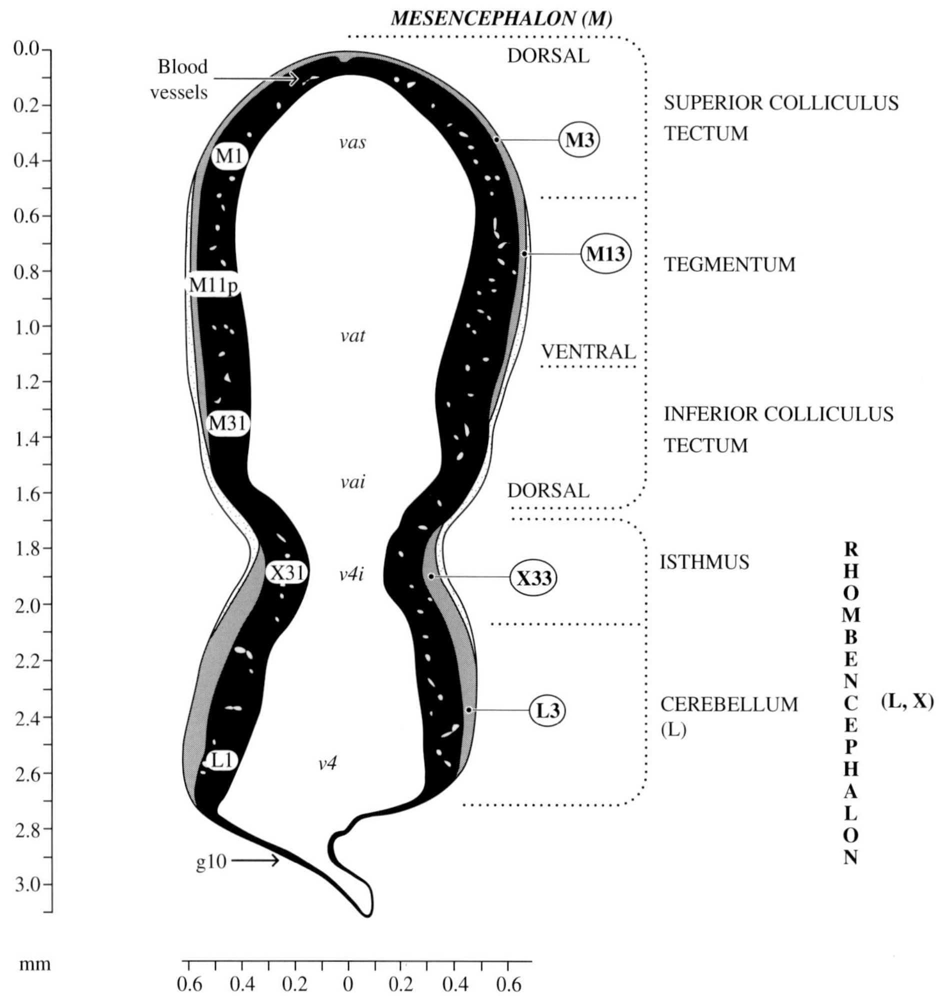

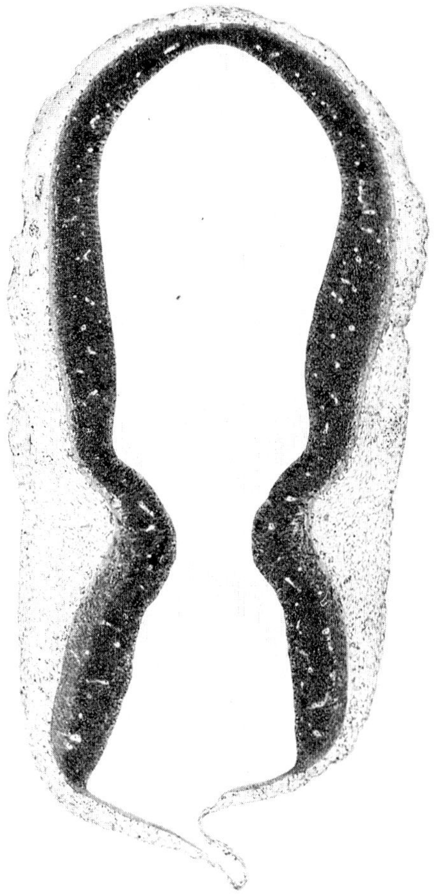

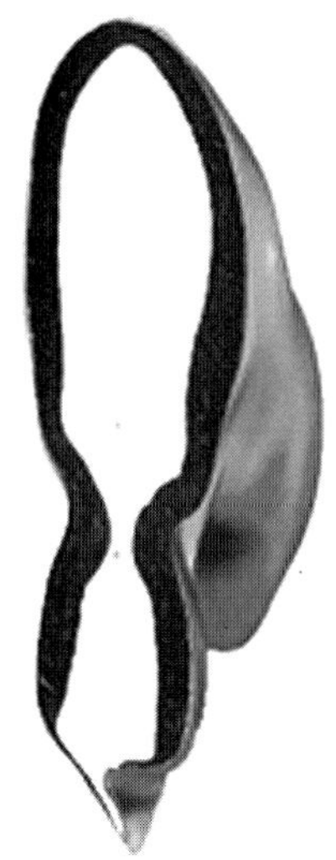

CORONAL FIGURE 16

GLANDS AND PARANEURAL STRUCTURES (g)

g10 velum medullare

CEREBELLUM (L)

L1 cerebellar neuroepithelium
L3 cerebellum (or differentiating field)

MIDBRAIN (M)

M1 superior tectal neuroepithelium
M11p posterior tegmental neuroepithelium
M13 tegmentum (or differentiating field)
M31 inferior tectal neuroepithelium

VENTRICLES (v)

vai aqueduct, inferior tectal
vas aqueduct, superior tectal
vat aqueduct, tegmental
v4 fourth ventricle

PONTINE AREA (X)

X31 isthmal neuroepithelium

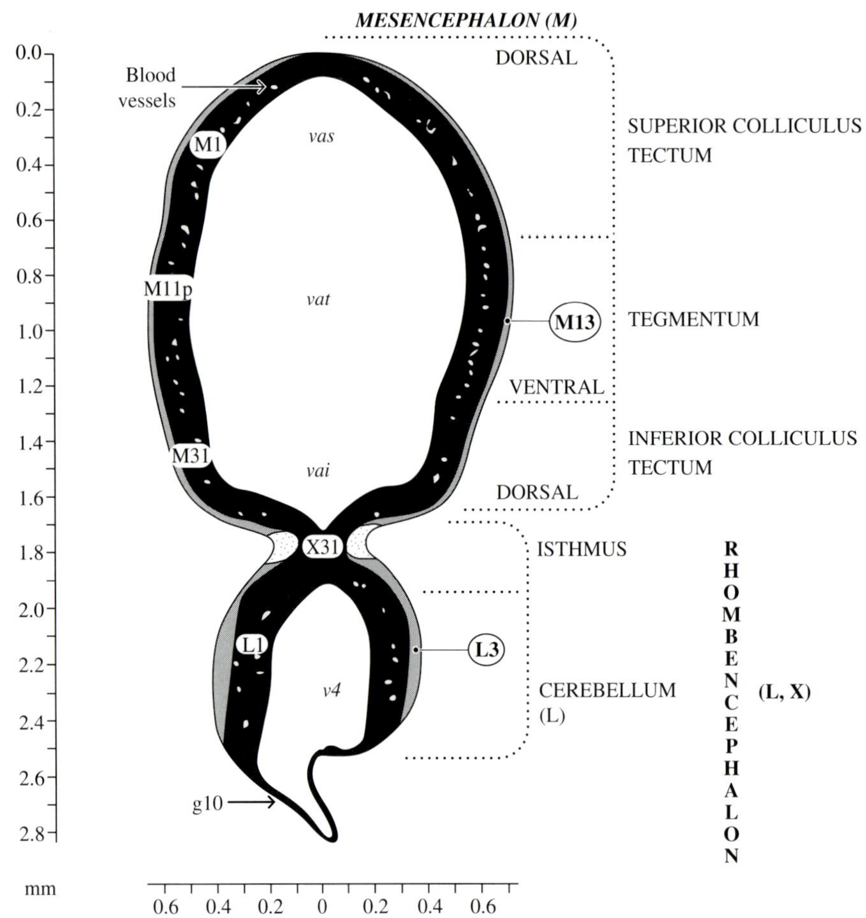

E14

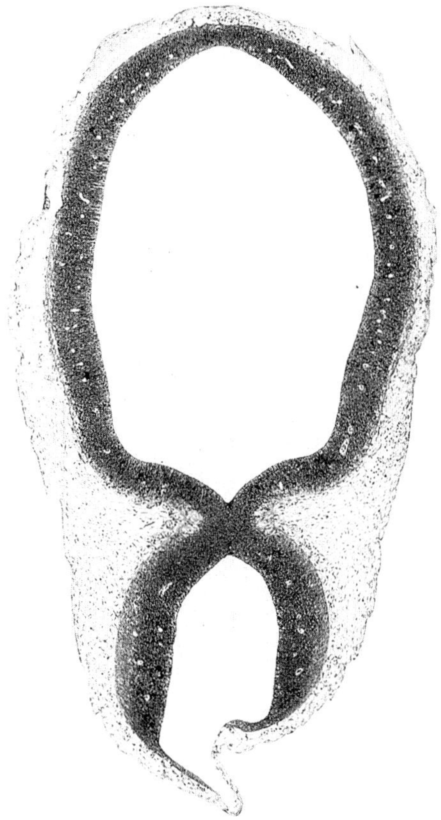

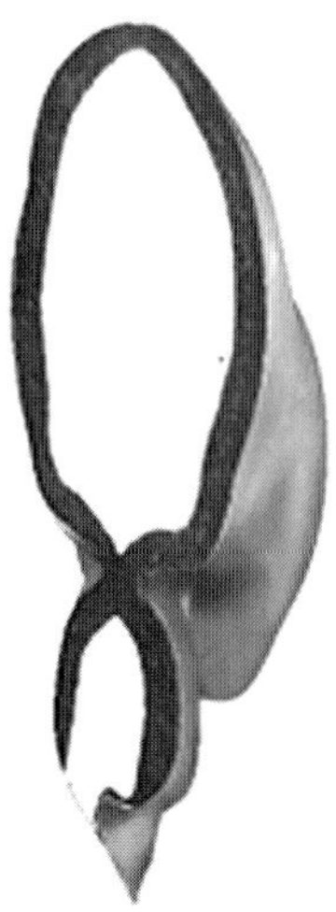

CORONAL FIGURE 17

CEREBELLUM (L)

L1 cerebellar neuroepithelium
L3 cerebellum (or differentiating field)

MIDBRAIN (M)

M1 superior tectal neuroepithelium
M31 inferior tectal neuroepithelium

VENTRICLES (v)

vai aqueduct, inferior tectal
vas aqueduct, superior tectal
v4 fourth ventricle

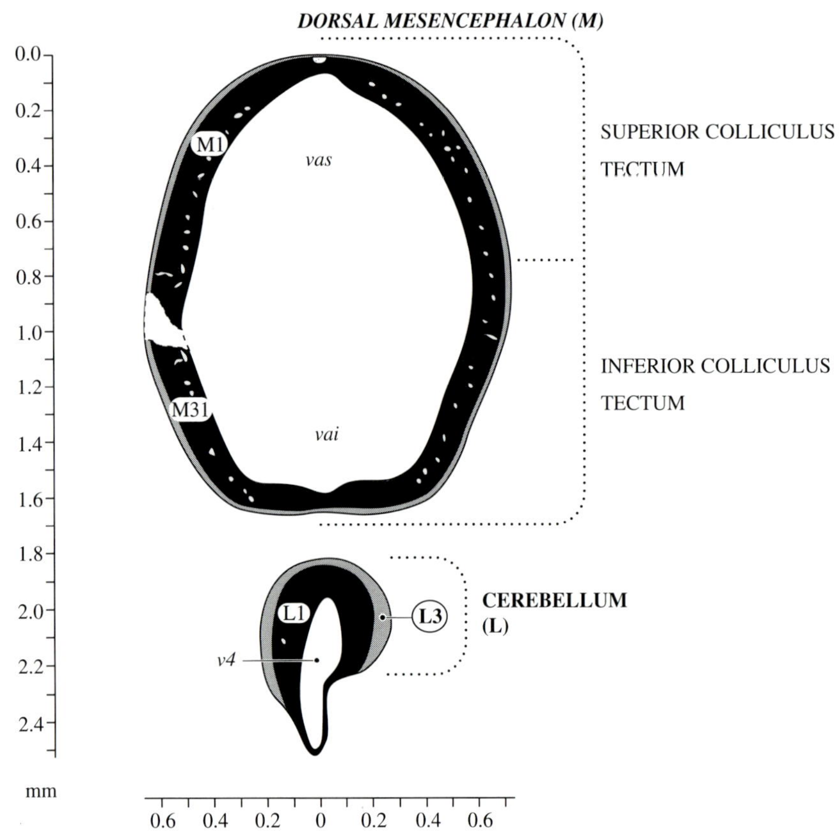

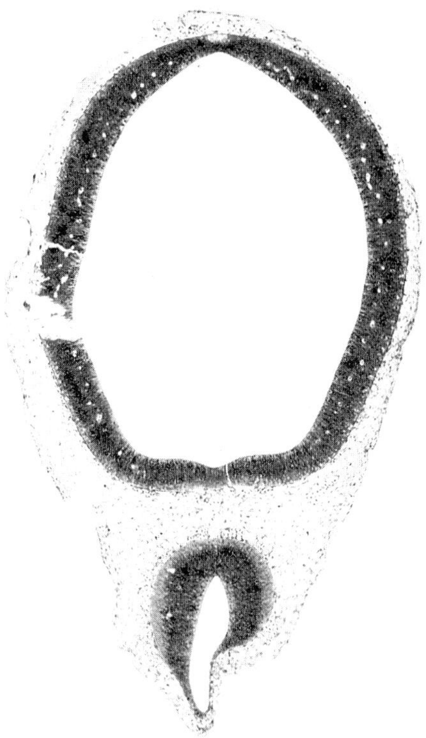

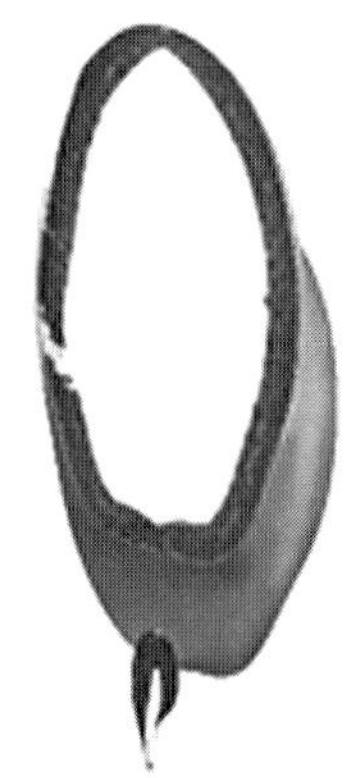

THE E15 BRAIN

COMPUTERIZED 3-D RECONSTRUCTION: 4 VIEWS

LEFT, ANTERIOR VIEW

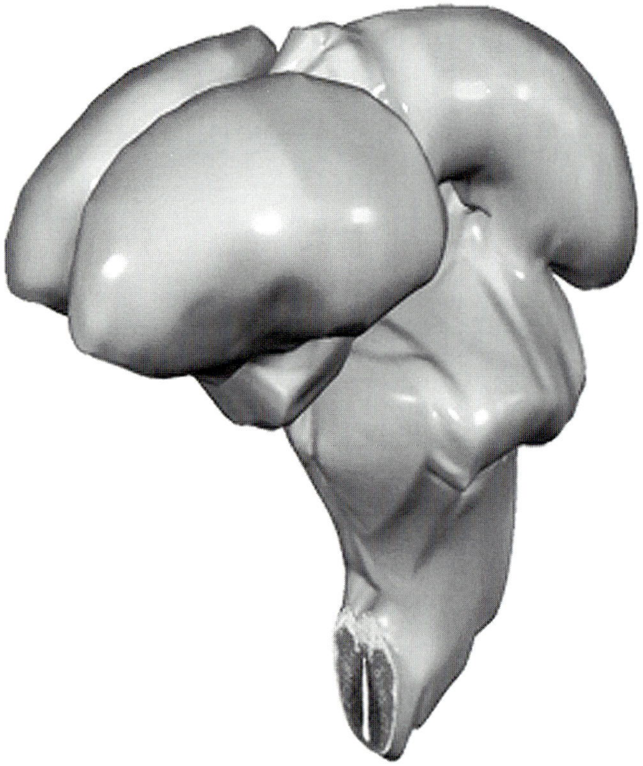

RIGHT, POSTERIOR VIEW

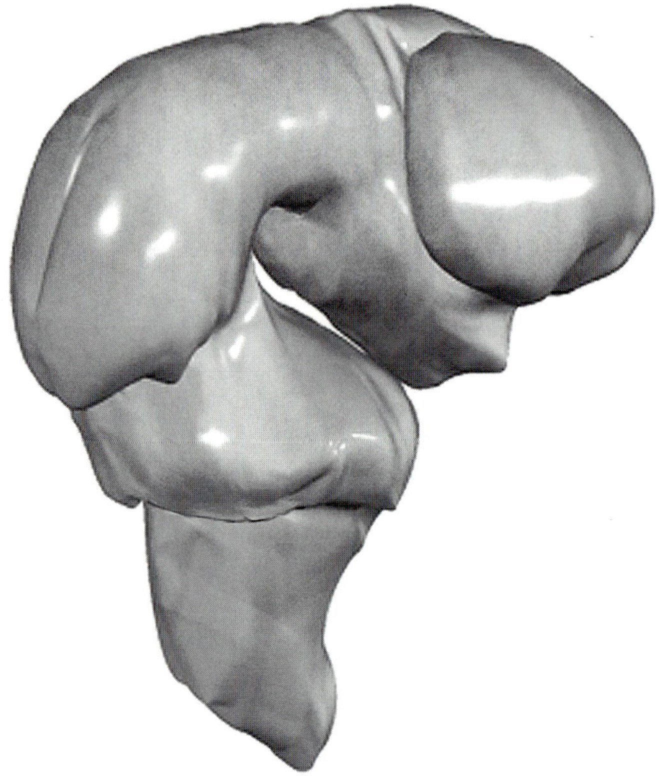

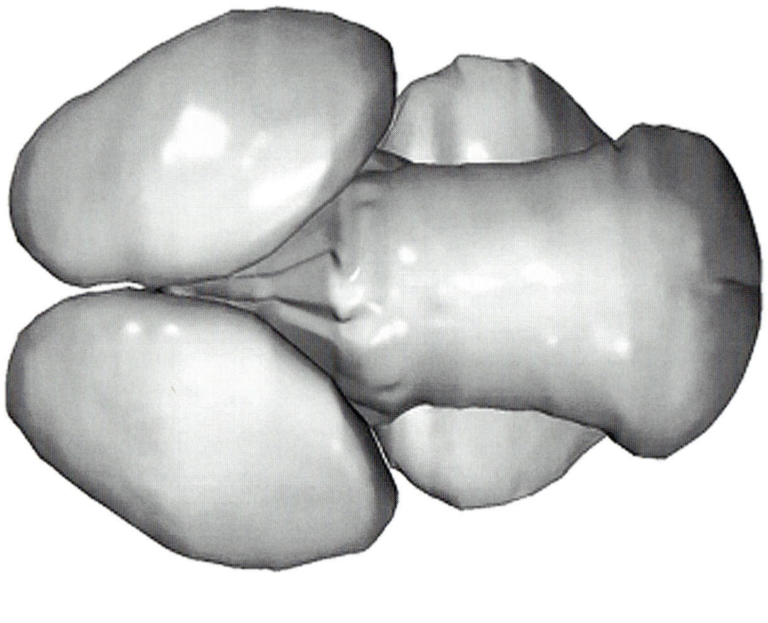

DORSAL VIEW

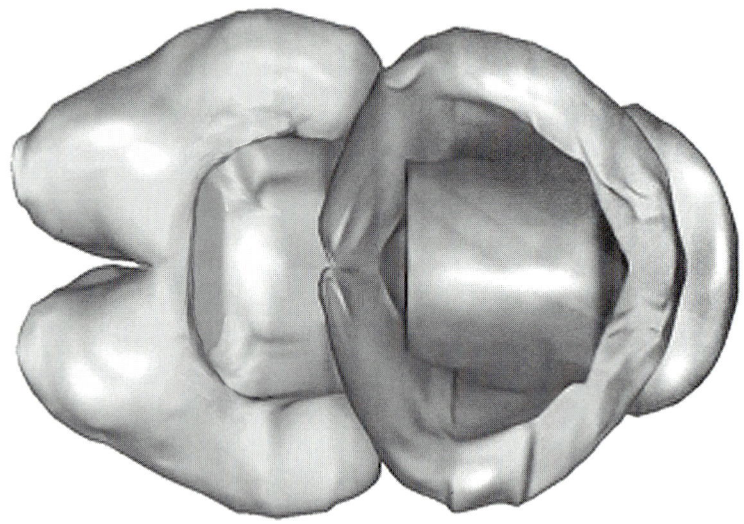

VENTRAL VIEW

SAGITTAL

FIGURES and PLATES 1-7

SAGITTAL FIGURE 1

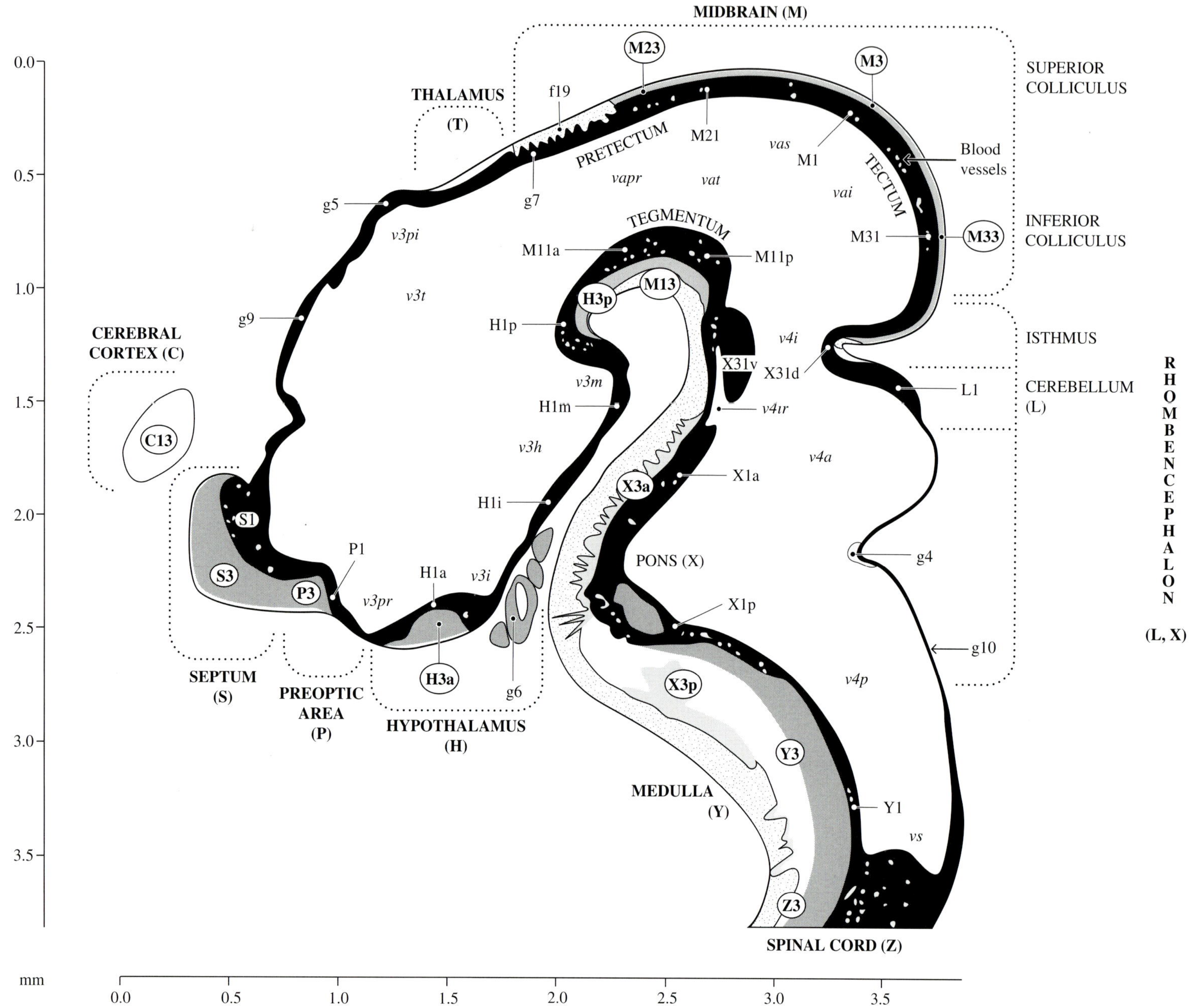

CEREBRAL CORTEX (C)

C13 cingulate cortex (or differentiating field)

FIBER TRACTS (f)

f19 posterior commissure

GLANDS AND PARANEURAL STRUCTURES (g)

g4 choroid plexus of fourth ventricle
g5 pineal gland (primordium)
g6 pituitary gland (primordium)
g7 subcommissural organ
g9 lamina terminalis
g10 velum medullare

HYPOTHALAMUS (H)

H1a anterior hypothalamic neuroepithelium
H1i intermediate hypothalamic neuroepithelium
H1m mammillary neuroepithelium
H1p posterior hypothalamic neuroepithelium
H3a anterior hypothalamus (or differentiating field)
H3p posterior hypothalamus (or differentiating field)

CEREBELLUM (L)

L1 cerebellar neuroepithelium

MIDBRAIN (M)

M1 superior tectal neuroepithelium
M3 superior colliculus (or differentiating field)
M11a anterior tegmental neuroepithelium
M11p posterior tegmental neuroepithelium
M13 tegmentum (or differentiating field)
M21 pretectal neuroepithelium
M23 pretectum (or differentiating field)
M31 inferior tectal neuroepithelium
M33 inferior colliculus (or differentiating field)

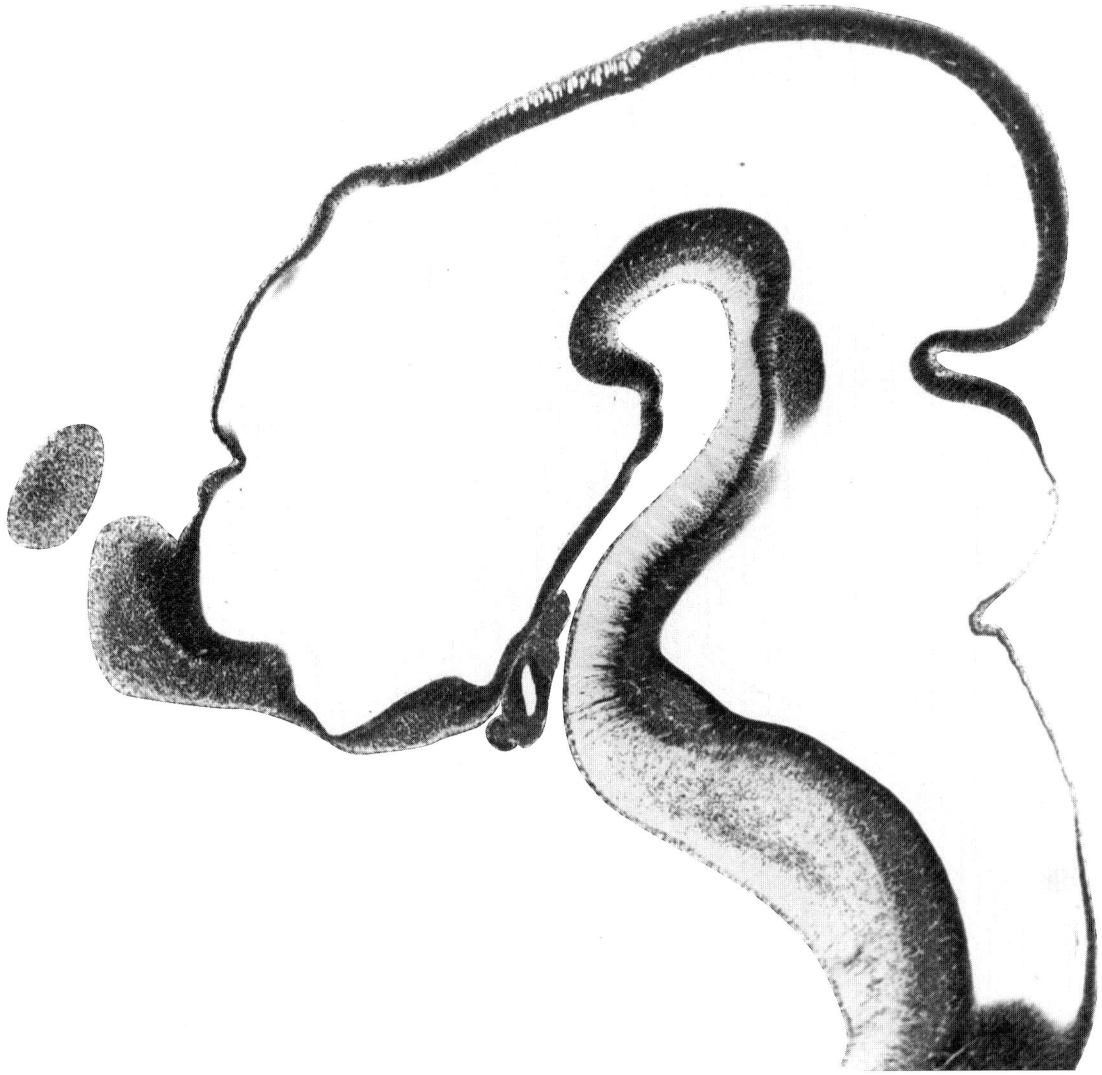

PREOPTIC AREA AND OPTIC EPITHELIA (P)

P1 preoptic neuroepithelium
P3 preoptic area (or differentiating field)

SEPTAL AREA (S)

S1 septal neuroepithelium
S3 septum (or differentiating field)

VENTRICLES (v)

vai aqueduct, inferior tectal
vapr aqueduct, pretectal
vas aqueduct, superior tectal
vat aqueduct, tegmental
vs spinal canal
v3h third ventricle, hypothalamic
v3i third ventricle, infundibular recess
v3m third ventricle, mammillary recess
v3pi third ventricle, pineal recess
v3pr third ventricle, preoptic recess
v3t third ventricle, thalamic
v4a fourth ventricle, anterior
v4i fourth ventricle, isthmal
v4ir fourth ventricle, isthmal recess
v4p fourth ventricle, posterior

PONTINE AREA (X)

X1a anterior pontine neuroepithelium
X1p posterior pontine neuroepithelium
X3a anterior pons (or differentiating field)
X3p posterior pons (or differentiating field)
X31d isthmal neuroepithelium, dorsal
X31v isthmal neuroepithelium, venral

MEDULLA (Y)

Y1 medullary neuroepithelium
Y3 medulla (or differentiating field)

SPINAL CORD (Z)

Z3 spinal cord (or differentiating field)

SAGITTAL FIGURE 2

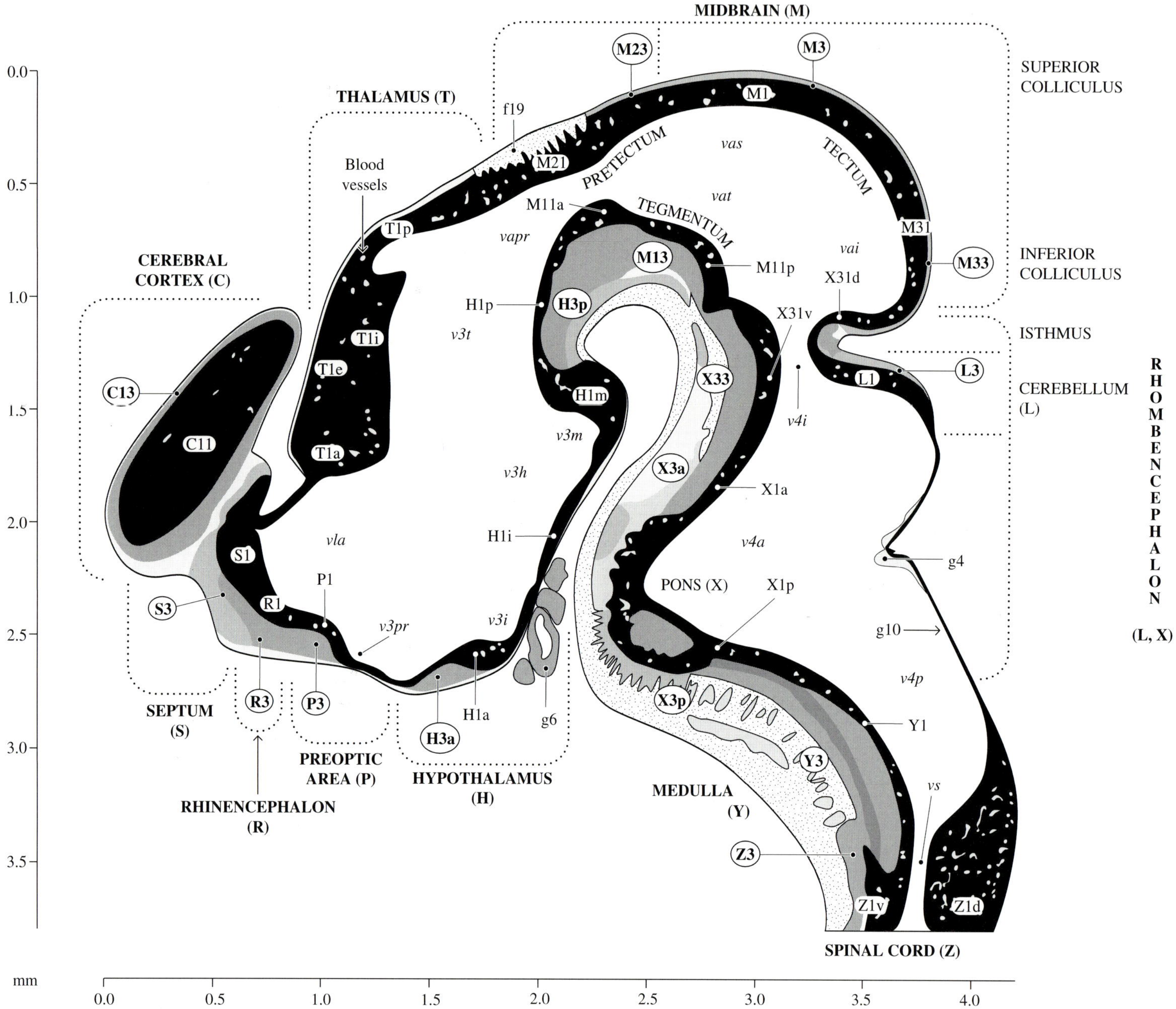

CEREBRAL CORTEX (C)

C11	cingulate cortex neuroepithelium
C13	cingulate cortex (or differentiating field)

FIBER TRACTS (f)

f19	posterior commissure

GLANDS AND PARANEURAL STRUCTURES (g)

g4	choroid plexus of fourth ventricle
g6	pituitary gland (primordium)
g10	velum medullare

HYPOTHALAMUS (H)

H1a	anterior hypothalamic neuroepithelium
H1i	intermediate hypothalamic neuroepithelium
H1m	mammillary neuroepithelium
H1p	posterior hypothalamic neuroepithelium
H3a	anterior hypothalamus (or differentiating field)
H3p	posterior hypothalamus (or differentiating field)

CEREBELLUM (L)

L1	cerebellar neuroepithelium
L3	cerebellum (or differentiating field)

MIDBRAIN (M)

M1	superior tectal neuroepithelium
M3	superior colliculus (or differentiating field)
M11a	anterior tegmental neuroepithelium
M11p	posterior tegmental neuroepithelium
M13	tegmentum (or differentiating field)
M21	pretectal neuroepithelium
M23	pretectum (or differentiating field)
M31	inferior tectal neuroepithelium
M33	inferior colliculus (or differentiating field)

PREOPTIC AREA AND OPTIC EPITHELIA (P)

P1	preoptic neuroepithelium
P3	preoptic area (or differentiating field)

RHINENCEPHALON (R)

R1	rhinencephalic neuroepithelium
R3	rhinencephalon (or differentiating field)

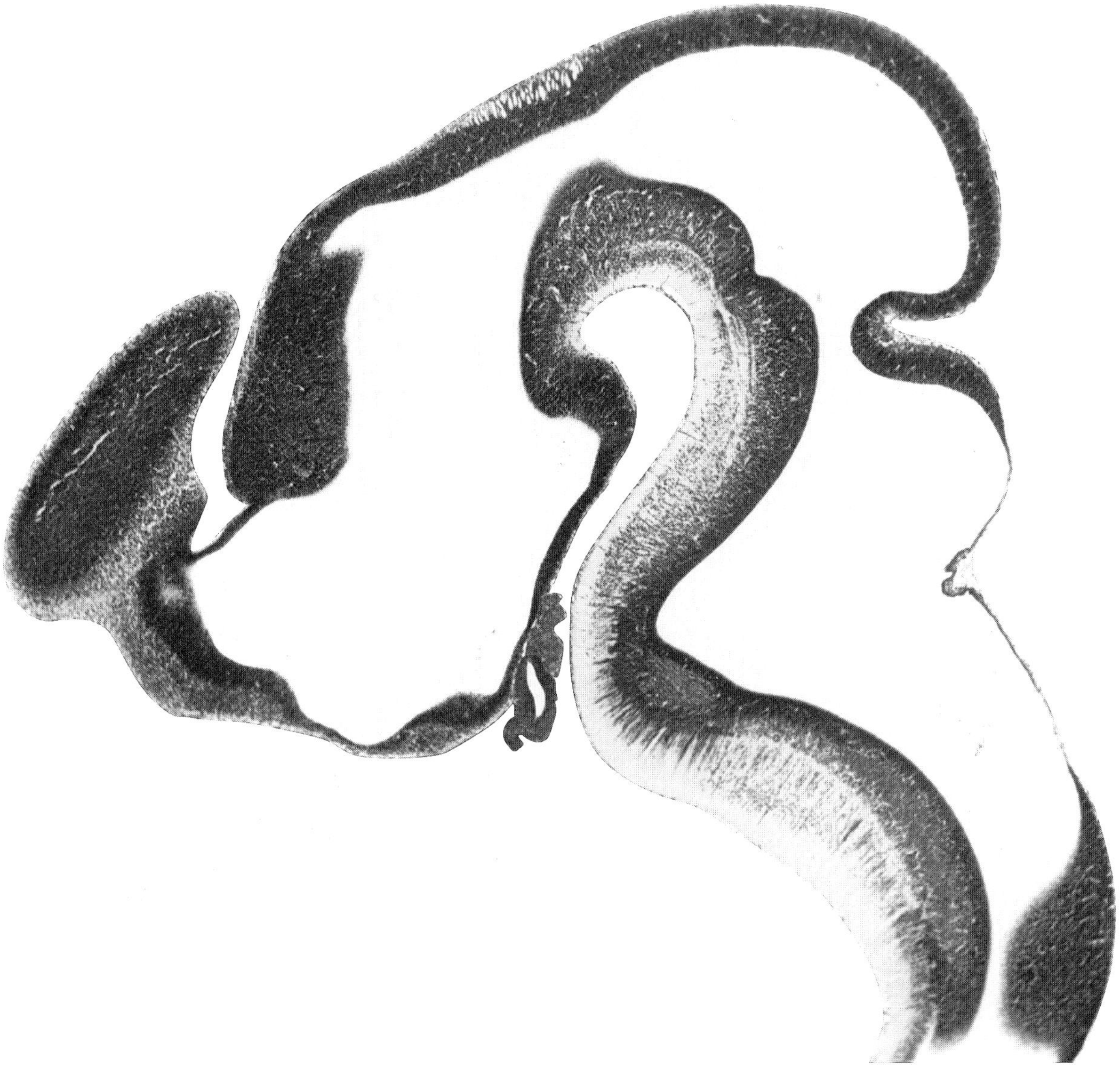

SEPTAL AREA (S)

S1 septal neuroepithelium
S3 septum (or differentiating field)

THALAMUS (T)

T1a anterior thalamic neuroepithelium
T1e epithalamic neuroepithelium
T1i intermediate thalamic neuroepithelium
T1p posterior thalamic neuroepithelium

VENTRICLES (v)

vai aqueduct, inferior tectal
vapr aqueduct, pretectal
vas aqueduct, superior tectal
vat aqueduct, tegmental
vla lateral ventricle, anterior horn
vs spinal canal
v3h third ventricle, hypothalamic
v3i third ventricle, infundibular recess
v3m third ventricle, mammillary recess
v3pr third ventricle, preoptic recess
v3t third ventricle, thalamic
v4a fourth ventricle, anterior
v4i fourth ventricle, isthmal
v4p fourth ventricle, posterior

PONTINE AREA (X)

X1a anterior pontine neuroepithelium
X1p posterior pontine neuroepithelium
X3a anterior pons (or differentiating field)
X3p posterior pons (or differentiating field)
X31d isthmal neuroepithelium, dorsal
X31v isthmal neuroepithelium, ventral
X33 isthmus (or differentiating field)

MEDULLA (Y)

Y1 medullary neuroepithelium
Y3 medulla (or differentiating field)

SPINAL CORD (Z)

Z1d spinal cord neuroepithelium, dorsal
Z1v spinal cord neuroepithelium, ventral
Z3 spinal cord (or differentiating field)

SAGITTAL FIGURE 3

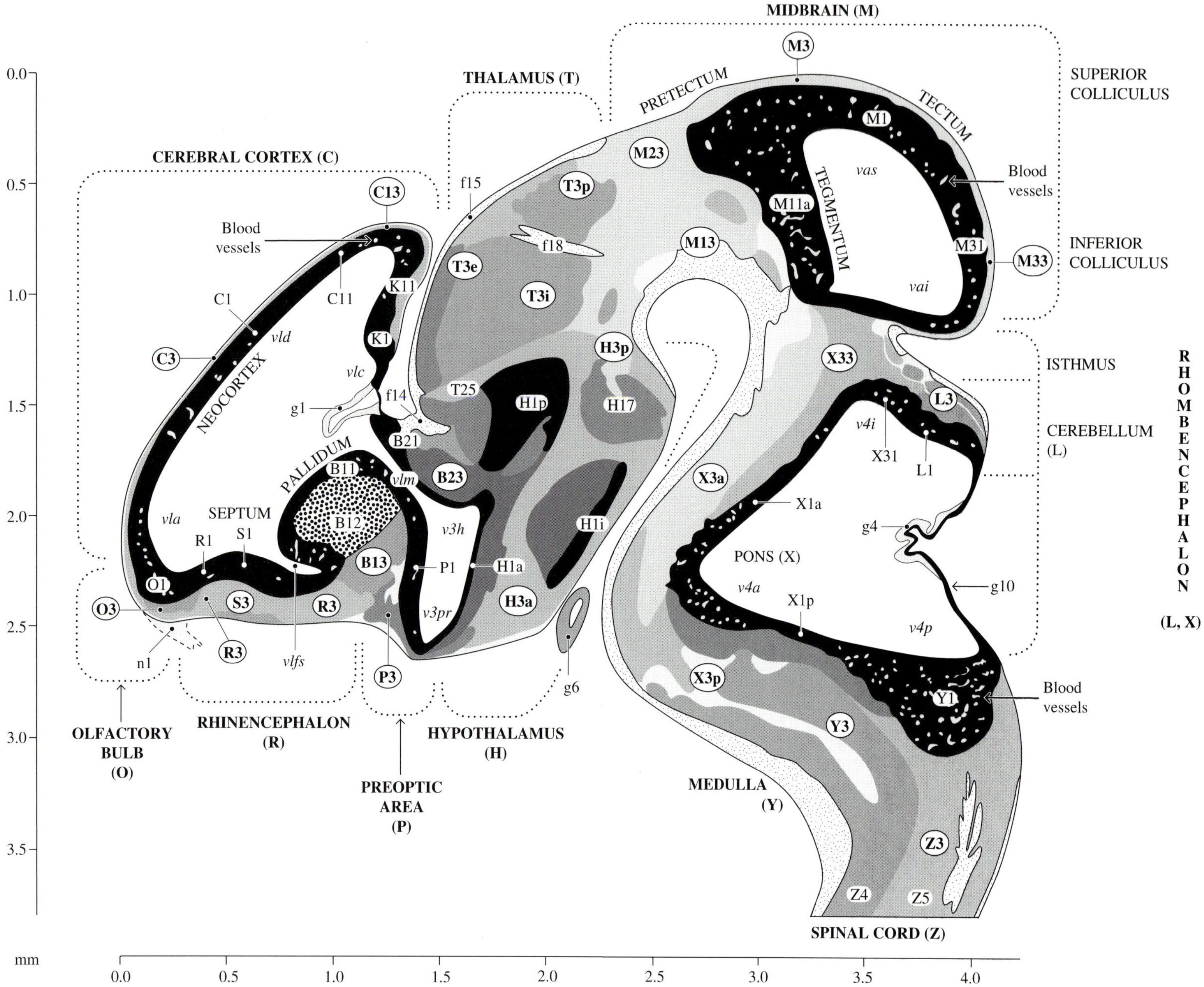

BASAL GANGLIA (B)

- B11 pallidal neuroepithelium
- B12 pallidal subventricular zone
- **B13** pallidum (or differentiating field)
- B21 strionuclear neuroepithelium
- **B23** bed nucleus of stria terminalis (or differentiating field)

CEREBRAL CORTEX (C)

- C1 neocortical neuroepithelium
- **C3** neocortex (or differentiating field)
- C11 cingulate cortex neuroepithelium
- **C13** cingulate cortex (or differentiating field)

FIBER TRACTS (f)

- f14 stria terminalis
- f15 stria medullaris
- f18 habenulopeduncular tract

GLANDS AND PARANEURAL STRUCTURES (g)

- g1 choroid plexus of lateral ventricle
- g4 choroid plexus of fourth ventricle
- g6 pituitary gland (primordium)
- g10 velum medullare

HYPOTHALAMUS (H)

- H1a anterior hypothalamic neuroepithelium
- H1i intermediate hypothalamic neuroepithelium
- H1p posterior hypothalamic neuroepithelium
- **H3a** anterior hypothalamus (or differentiating field)
- **H3p** posterior hypothalamus (or differentiating field)
- H17 lateral mammillary nucleus

HIPPOCAMPAL FORMATION (K)

- K1 hippocampal (Ammon's horn) neuroepithelium
- K11 subicular neuroepithelium

CEREBELLUM (L)

- L1 cerebellar neuroepithelium
- **L3** cerebellum (or differentiating field)

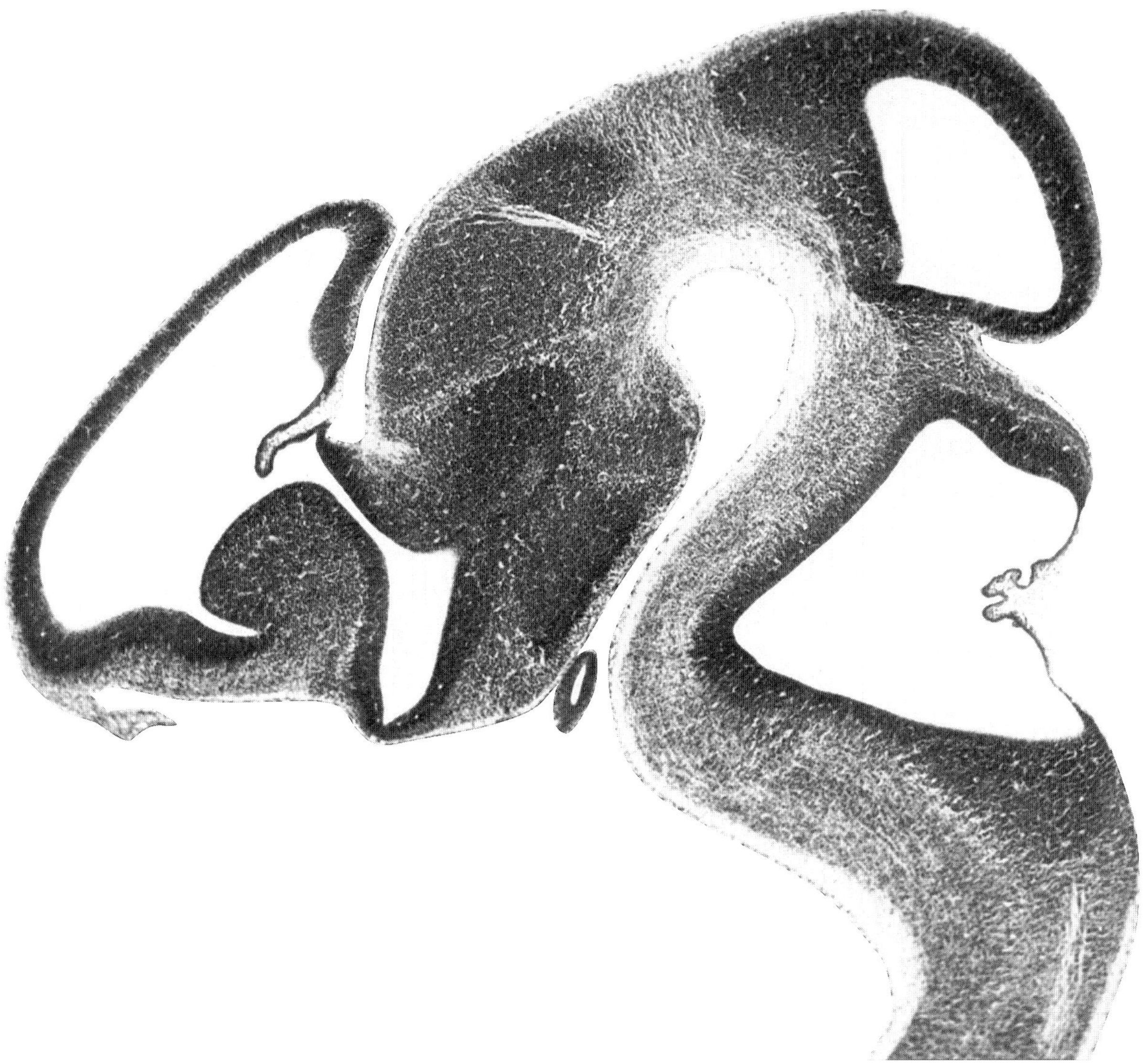

MIDBRAIN (M)

M1 superior tectal neuroepithelium
M3 superior colliculus (or differentiating field)
M11a anterior tegmental neuroepithelium
M13 tegmentum (or differentiating field)
M23 pretectum (or differentiating field)
M31 inferior tectal neuroepithelium
M33 inferior colliculus (or differentiating field)

NERVES AND NUCLEI, CRANIAL (n)

n1 olfactory nerve

OLFACTORY BULB (O)

O1 olfactory bulb neuroepithelium
O3 olfactory bulb (or differentiating field)

PREOPTIC AREA AND OPTIC EPITHELIA (P)

P1 preoptic neuroepithelium
P3 preoptic area (or differentiating field)

RHINENCEPHALON (R)

R1 rhinencephalic neuroepithelium
R3 rhinencephalon (or differentiating field)

SEPTAL AREA (S)

S1 septal neuroepithelium
S3 septum (or differentiating field)

THALAMUS (T)

T3e epithalamus (or differentiating field)
T3i intermediate thalamus (or differentiating field)
T3p posterior thalamus (or differentiating field)
T25 reticular nucleus

VENTRICLES (v)

vai aqueduct, inferior tectal
vas aqueduct, superior tectal
vla lateral ventricle, anterior
vlc lateral ventricle, central
vld lateral ventricle, dorsal
vlfs lateral ventricle, septal fork
v3h third ventricle, hypothalamic
vlm foramen of Monro
v3pr third ventricle, preoptic recess
v4a fourth ventricle, anterior
v4i fourth ventricle, isthmal
v4p fourth ventricle, posterior

PONTINE AREA (X)

X1a anterior pontine neuroepithelium
X1p posterior pontine neuroepithelium
X3a anterior pons (or differentiating field)
X3p posterior pons (or differentiating field)
X31 isthmal neuroepithelium
X33 isthmus (or differentiating field)

MEDULLA (Y)

Y1 medullary neuroepithelium
Y3 medulla (or differentiating field)

SPINAL CORD (Z)

Z3 spinal cord (or differentiating field)
Z4 ventral horn
Z5 dorsal horn

SAGITTAL FIGURE 4

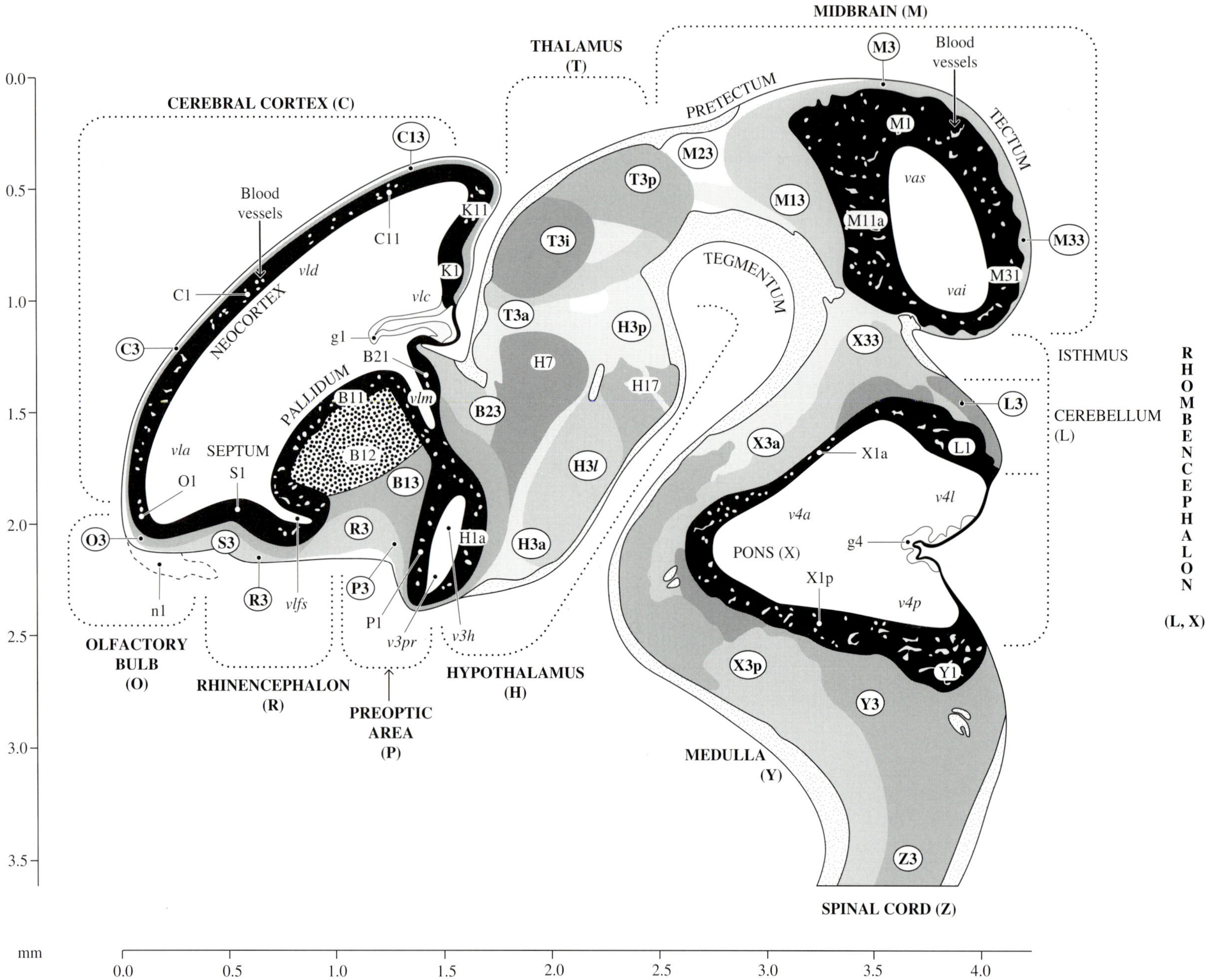

BASAL GANGLIA (B)

B11 pallidal neuroepithelium
B12 pallidal subventricular zone
B13 pallidum (or differentiating field)
B21 strionuclear neuroepithelium
B23 bed nucleus of stria terminalis (or differentiating field)

CEREBRAL CORTEX (C)

C1 neocortical neuroepithelium
C3 neocortex (or differentiating field)
C11 cingulate cortex neuroepithelium
C13 cingulate cortex (or differentiating field)

GLANDS AND PARANEURAL STRUCTURES (g)

g1 choroid plexus of lateral ventricle
g4 choroid plexus of fourth ventricle

HYPOTHALAMUS (H)

H1a anterior hypothalamic neuroepithelium
H3a anterior hypothalamus (or differentiating field)
H3*l* lateral hypothalamus (or differentiating field)
H3p posterior hypothalamus (or differentiating field)
H7 zona incerta
H17 lateral mammillary nucleus

HIPPOCAMPAL FORMATION (K)

K1 hippocampal (Ammon's horn) neuroepithelium
K11 subicular neuroepithelium

CEREBELLUM (L)

L1 cerebellar neuroepithelium
L3 cerebellum (or differentiating field)

MIDBRAIN (M)

M1 superior tectal neuroepithelium
M3 superior colliculus (or differentiating field)
M11a anterior tegmental neuroepithelium
M13 tegmentum (or differentiating field)
M23 pretectum (or differentiating field)
M31 inferior tectal neuroepithelium
M33 inferior colliculus (or differentiating field)

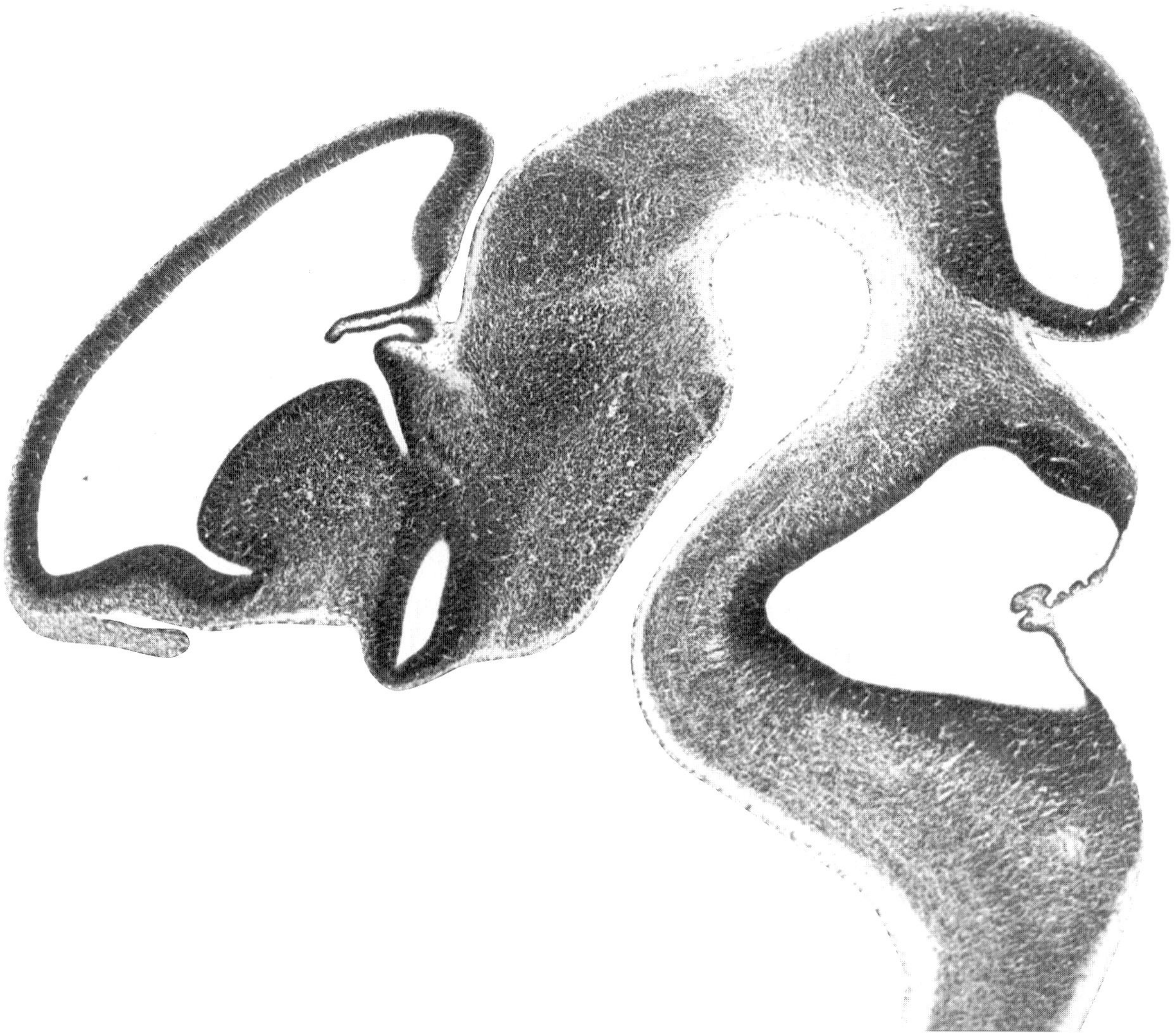

NERVES AND NUCLEI, CRANIAL (n)

n1 olfactory nerve

OLFACTORY BULB (O)

O1 olfactory bulb neuroepithelium
O3 olfactory bulb (or differentiating field)

PREOPTIC AREA AND OPTIC EPITHELIA (P)

P1 preoptic neuroepithelium
P3 preoptic area (or differentiating field)

RHINENCEPHALON (R)

R3 rhinencephalon (or differentiating field)

SEPTAL AREA (S)

S1 septal neuroepithelium
S3 septum (or differentiating field)

THALAMUS (T)

T3a anterior thalamus (or differentiating field)
T3i intermediate thalamus (or differentiating field)
T3p posterior thalamus (or differentiating field)

VENTRICLES (v)

vai aqueduct, inferior tectal
vas aqueduct, superior tectal
vla lateral ventricle, anterior horn
vlc lateral ventricle, central
vld lateral ventricle, dorsal
vlfs lateral ventricle, septal fork
vlm foramen of Monro
v3h third ventricle, hypothalamic
v3pr third ventricle, preoptic recess
v4a fourth ventricle, anterior
v4l fourth ventricle, lateral recess
v4p fourth ventricle, posterior

PONTINE AREA (X)

X1a anterior pontine neuroepithelium
X1p posterior pontine neuroepithelium
X3a anterior pons (or differentiating field)
X3p posterior pons (or differentiating field)
X33 isthmus (or differentiating field)

MEDULLA (Y)

Y1 medullary neuroepithelium
Y3 medulla (or differentiating field)

SPINAL CORD (Z)

Z3 spinal cord (or differentiating field)

SAGITTAL FIGURE 5

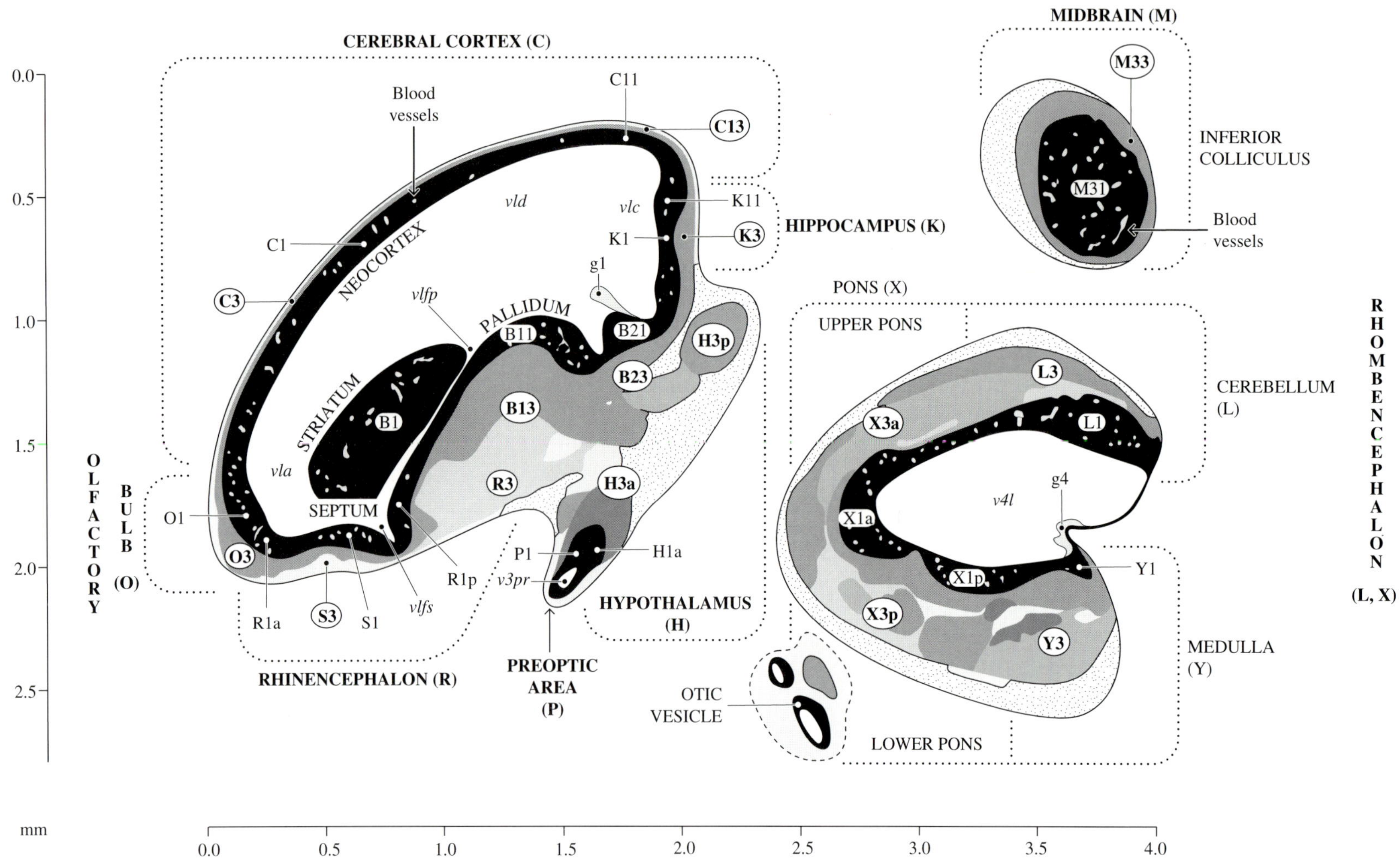

BASAL GANGLIA (B)

B1 striatal neuroepithelium
B11 pallidal neuroepithelium
B13 pallidum (or differentiating field)
B21 strionuclear neuroepithelium
B23 bed nucleus of stria terminalis (or differentiating field)

CEREBRAL CORTEX (C)

C1 neocortical neuroepithelium
C3 neocortex (or differentiating field)
C11 cingulate cortex neuroepithelium
C13 cingulate cortex (or differentiating field)

GLANDS AND PARANEURAL STRUCTURES (g)

g1 choroid plexus of lateral ventricle
g4 choroid plexus of fourth ventricle

HYPOTHALAMUS (H)

H1a anterior hypothalamic neuroepithelium
H3a anterior hypothalamus (or differentiating field)
H3p posterior hypothalamus (or differentiating field)

HIPPOCAMPAL FORMATION (K)

K1 hippocampal (Ammon's horn) neuroepithelium
K3 hippocampus (or differentiating field)
K11 subicular neuroepithelium

CEREBELLUM (L)

L1 cerebellar neuroepithelium
L3 cerebellum (or differentiating field)

MIDBRAIN (M)

M31 inferior tectal neuroepithelium
M33 inferior colliculus (or differentiating field)

OLFACTORY BULB (O)

O1 olfactory bulb neuroepithelium
O3 olfactory bulb (or differentiating field)

PREOPTIC AREA AND OPTIC EPITHELIA (P)

P1 preoptic neuroepithelium

RHINENCEPHALON (R)

R1a anterior rhinencephalic neuroepithelium
R1p posterior rhinencephalic neuroepithelium
R3 rhinencephalon (or differentiating field)

SEPTAL AREA (S)

S1 septal neuroepithelium
S3 septum (or differentiating field)

VENTRICLES (v)

vla lateral ventricle, anterior horn
vlc lateral ventricle, central
vld lateral ventricle, dorsal
vlfp lateral ventricle, pallidal fork
vlfs lateral ventricle, septal fork
v3pr third ventricle, preoptic recess
v4l fourth ventricle, lateral recess

PONTINE AREA (X)

X1a anterior pontine neuroepithelium
X1p posterior pontine neuroepithelium
X3a anterior pons (or differentiating field)
X3p posterior pons (or differentiating field)

MEDULLA (Y)

Y1 medullary neuroepithelium
Y3 medulla (or differentiating field)

SAGITTAL FIGURE 6

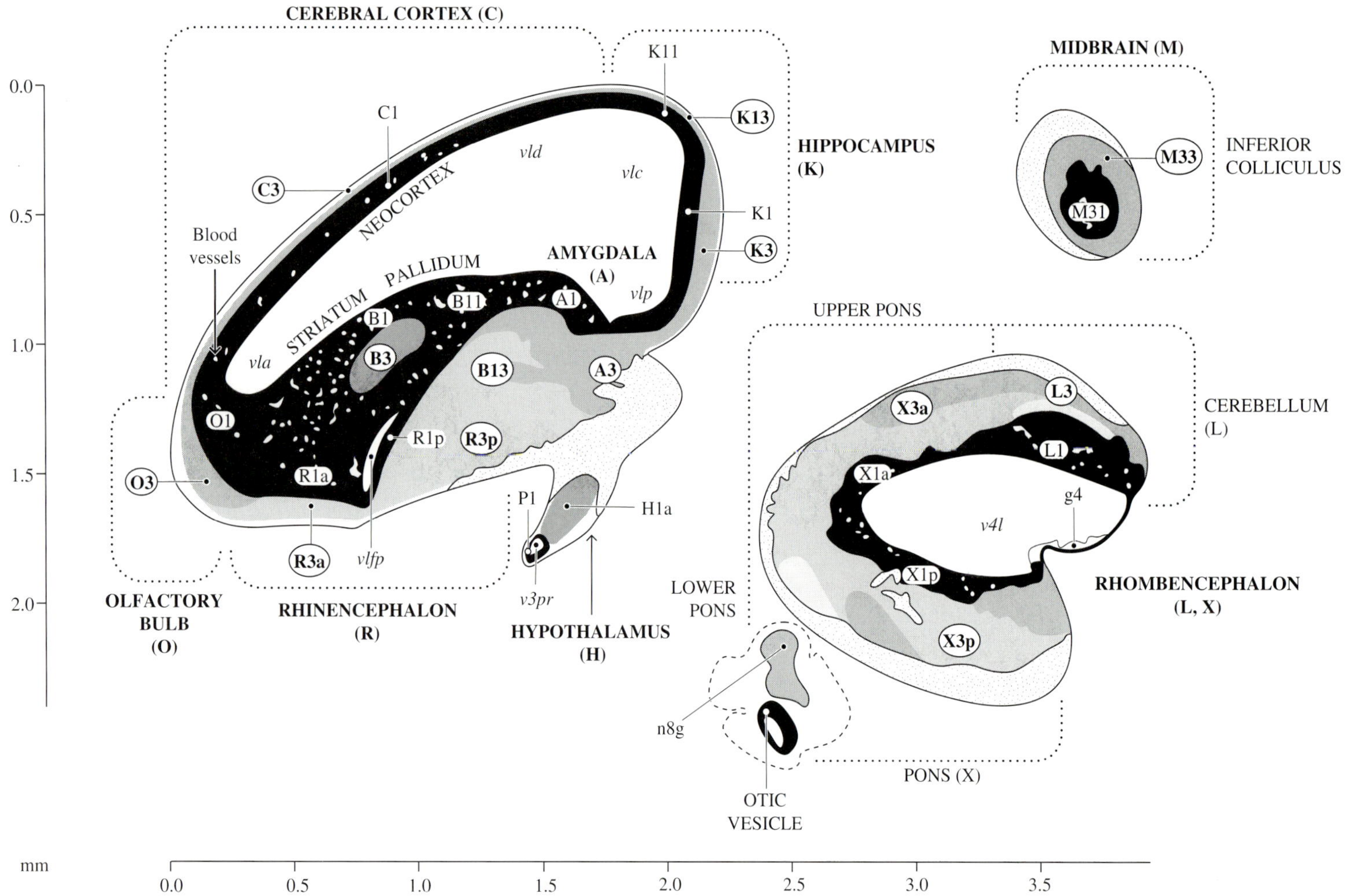

AMYGDALA (A)

A1 amygdaloid neuroepithelium
A3 amygdala (or differentiating field)

BASAL GANGLIA (B)

B1 striatal neuroepithelium
B3 striatum (or differentiating field)
B11 pallidal neuroepithelium
B13 pallidum (or differentiating field)

CEREBRAL CORTEX (C)

C1 neocortical neuroepithelium
C3 neocortex (or differentiating field)

GLANDS AND PARANEURAL STRUCTURES (g)

g4 choroid plexus of fourth ventricle

HYPOTHALAMUS (H)

H1a anterior hypothalamic neuroepithelium

HIPPOCAMPAL FORMATION (K)

K1 hippocampal (Ammon's horn) neuroepithelium
K3 hippocampus (or differentiating field)
K11 subicular neuroepithelium
K13 subicular area (or differentiating field)

CEREBELLUM (L)

L1 cerebellar neuroepithelium
L3 cerebellum (or differentiating field)

MIDBRAIN (M)

M31 inferior tectal neuroepithelium
M33 inferior colliculus (or differentiating field)

NERVES AND NUCLEI, CRANIAL (n)

n8g vestibular ganglion

OLFACTORY BULB (O)

O1 olfactory bulb neuroepithelium
O3 olfactory bulb (or differentiating field)

PREOPTIC AREA AND OPTIC EPITHELIA (P)

P1 preoptic neuroepithelium

RHINENCEPHALON (R)

R1a rhinencephalic neuroepithelium, anterior
R1p rhinencephalic neuroepithelium, posterior
R3a rhinencephalon, anterior
R3p rhinencephalon, posterior

VENTRICLES (v)

vla lateral ventricle, anterior horn
vlc lateral ventricle, central
vld lateral ventricle, dorsal
vlfp lateral ventricle, pallidal fork
vlp lateral ventricle, posterior horn
v3pr third ventricle, preoptic recess
v4l fourth ventricle, lateral recess

PONTINE AREA (X)

X1a anterior pontine neuroepithelium
X1p posterior pontine neuroepithelium
X3a anterior pons (or differentiating field)
X3p posterior pons (or differentiating field)

SAGITTAL FIGURE 7

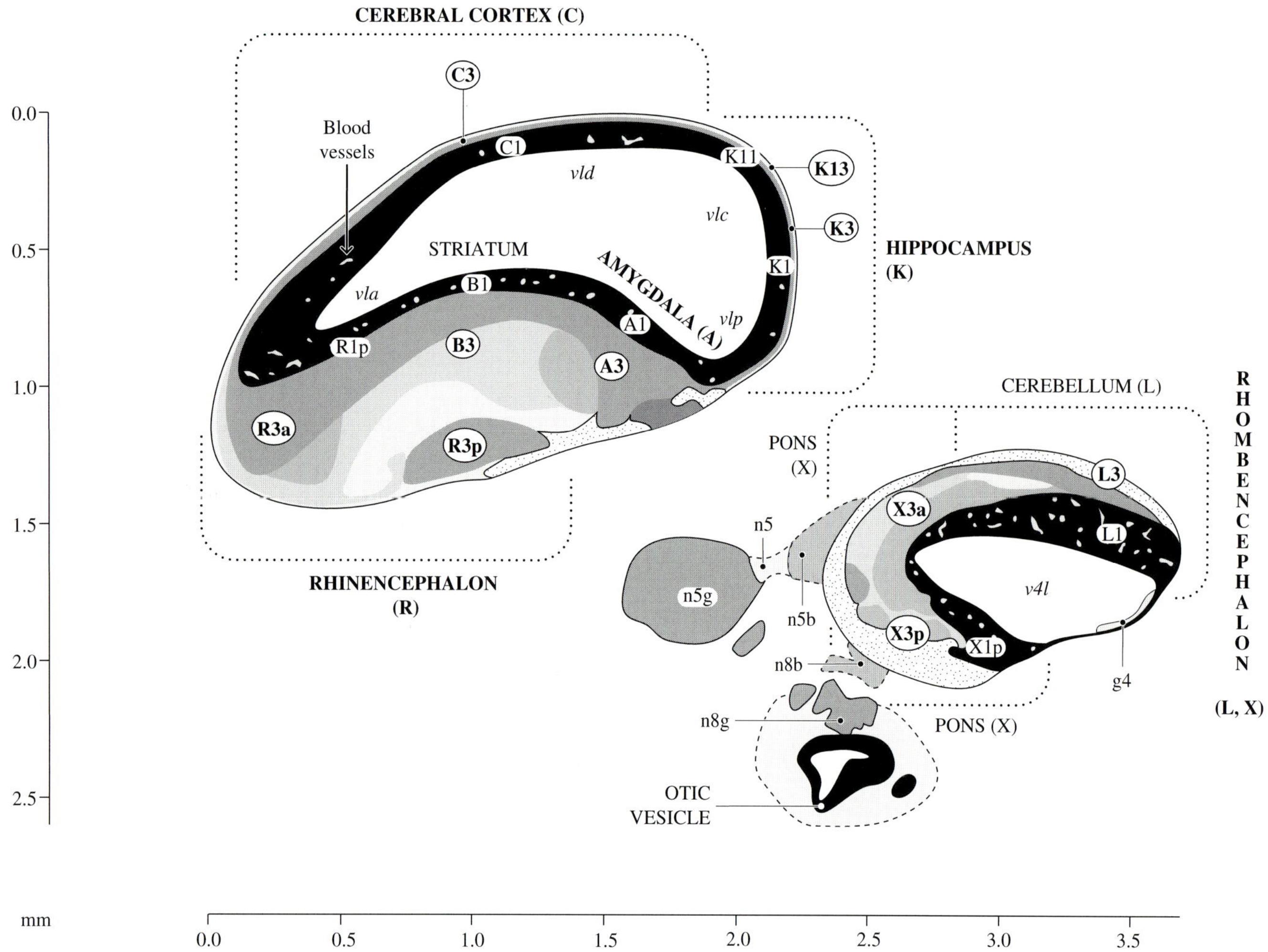

AMYGDALA (A)

A1 amygdaloid neuroepithelium
A3 amygdala (or differentiating field)

BASAL GANGLIA (B)

B1 striatal neuroepithelium
B3 striatum (or differentiating field)

CEREBRAL CORTEX (C)

C1 neocortical neuroepithelium
C3 neocortex (or differentiating field)

GLANDS AND PARANEURAL STRUCTURES (g)

g4 choroid plexus of fourth ventricle

HIPPOCAMPAL FORMATION (K)

K1 hippocampal (Ammon's horn) neuroepithelium
K3 hippocampus (or differentiating field)
K11 subicular neuroepithelium
K13 subicular area (or differentiating field)

CEREBELLUM (L)

L1 cerebellar neuroepithelium
L3 cerebellum (or differentiating field)

NERVES AND NUCLEI, CRANIAL (n)

n5 trigeminal nerve
n5b trigeminal boundary cap
n5g trigeminal ganglion
n8b vestibulocochlear boundary cap
n8g vestibular ganglion

RHINENCEPHALON (R)

R1p rhinencephalic neuroepithelium, posterior
R3a rhinencephalon, anterior (or differentiating field)
R3p rhinencephalon, posterior (or differentiating field)

VENTRICLES (v)

vla lateral ventricle, anterior horn
vlc lateral ventricle, central
vld lateral ventricle, dorsal
vlp lateral ventricle, posterior horn
v4l fourth ventricle, lateral recess

PONTINE AREA (X)

X1p posterior pontine neuroepithelium
X3a anterior pons (or differentiating field)
X3p posterior pons (or differentiating field)

CORONAL

FIGURES and PLATES 1-18

CORONAL FIGURE 1

CEREBRAL CORTEX (C)

C1 neocortical neuroepithelium
C3 neocortex (or differentiating field)

OLFACTORY BULB (O)

O1 olfactory bulb neuroepithelium
O3 olfactory bulb (or differentiating field)
O5 external plexiform layer

RHINENCEPHALON (R)

R1 rhinencephalic neuroepithelium
R3 rhinencephalon (or differentiating field)

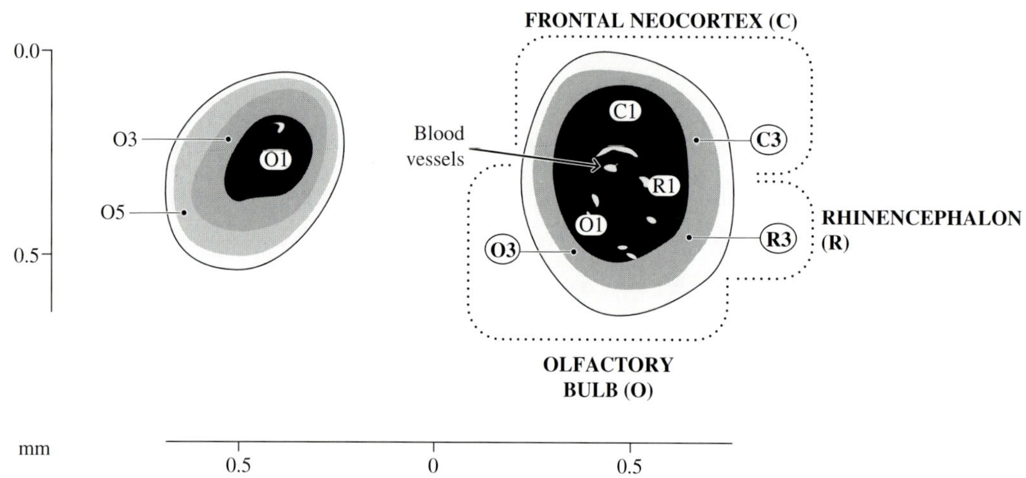

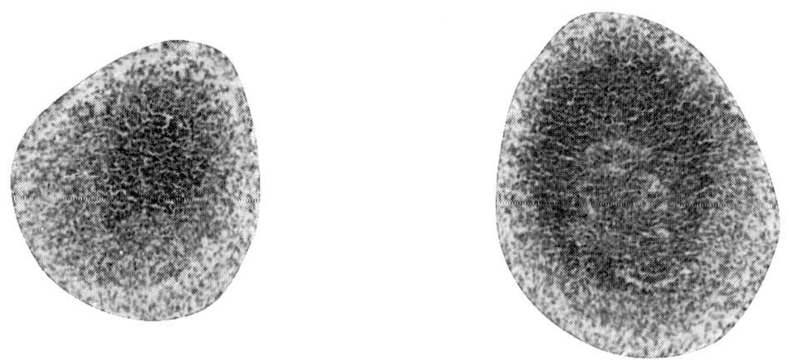

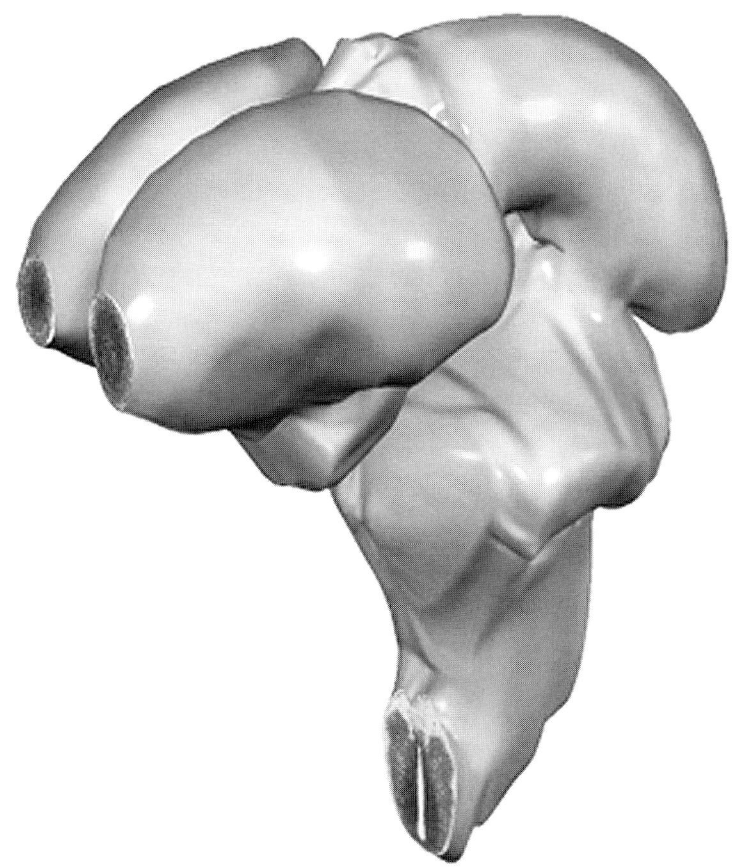

CORONAL FIGURE 2

CEREBRAL CORTEX (C)

C1	neocortical neuroepithelium
C3	neocortex (or differentiating field)
C4	molecular layer (or primordial plexiform layer)
C5	intermediate zone
C11	cingulate cortex neuroepithelium
C13	cingulate cortex (or differentiating field)

NERVES AND NUCLEI, CRANIAL (n)

n1	olfactory nerve

OLFACTORY BULB (O)

O1	olfactory bulb neuroepithelium
O3	olfactory bulb (or differentiating field)

RHINENCEPHALON (R)

R1	rhinencephalic neuroepithelium
R3	rhinencephalon (or differentiating field)

VENTRICLES (v)

vla	lateral ventricle, anterior horn
vld	lateral ventricle, dorsal

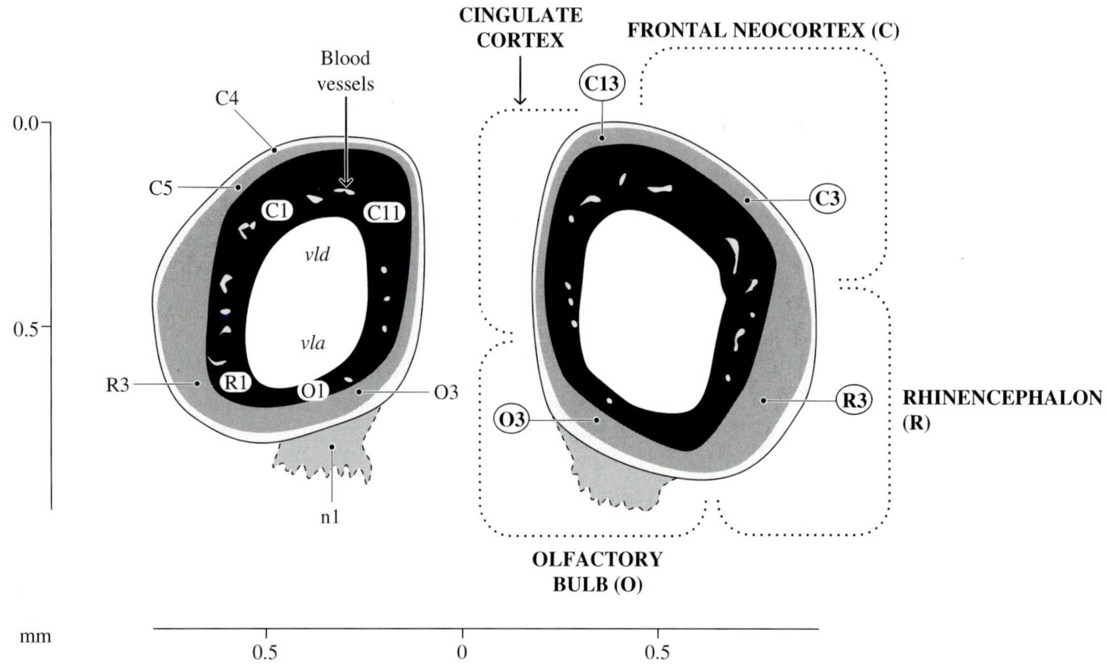

E15

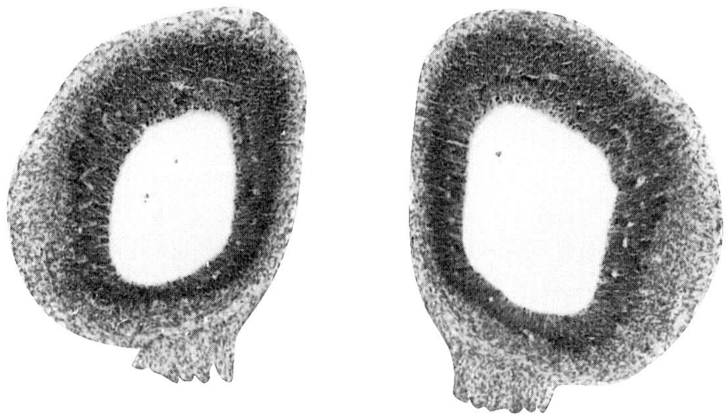

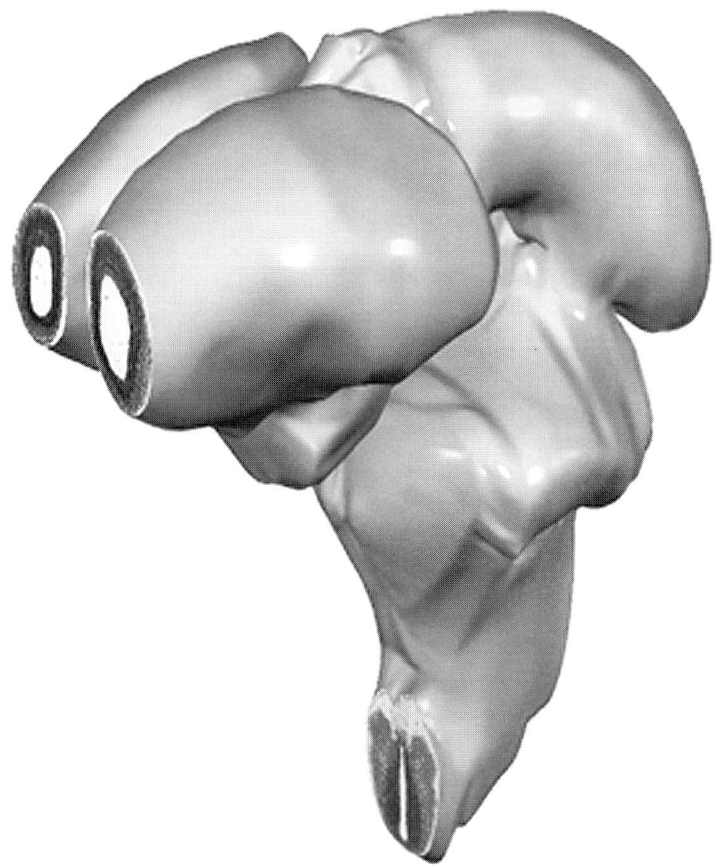

CORONAL FIGURE 3

BASAL GANGLIA (B)

B1 striatal neuroepithelium
B3 striatum (or differentiating field)

CEREBRAL CORTEX (C)

C1 neocortical neuroepithelium
C3 neocortex (or differentiating field)
C4 molecular layer (or primordial plexiform layer)
C5 intermediate zone
C7 white matter
C11 cingulate cortex neuroepithelium
C13 cingulate cortex (or differentiating field)
C23 orbital and insular area (or differentiating field)
C24 insular cortex
C25 orbital cortex

HIPPOCAMPAL FORMATION (K)

K1 hippocampal (Ammon's horn) neuroepithelium
K11 subicular neuroepithelium
K13 subicular area (or differentiating field)

RHINENCEPHALON (R)

R1 rhinencephalic neuroepithelium
R3 rhinencephalon (or differentiating field)
R4 piriform cortex

SEPTAL AREA (S)

S1 septal neuroepithelium
S3 septum (or differentiating field)
S6 medial septum

VENTRICLES (v)

vla lateral ventricle, anterior horn
vlc lateral ventricle, central
vld lateral ventricle, dorsal

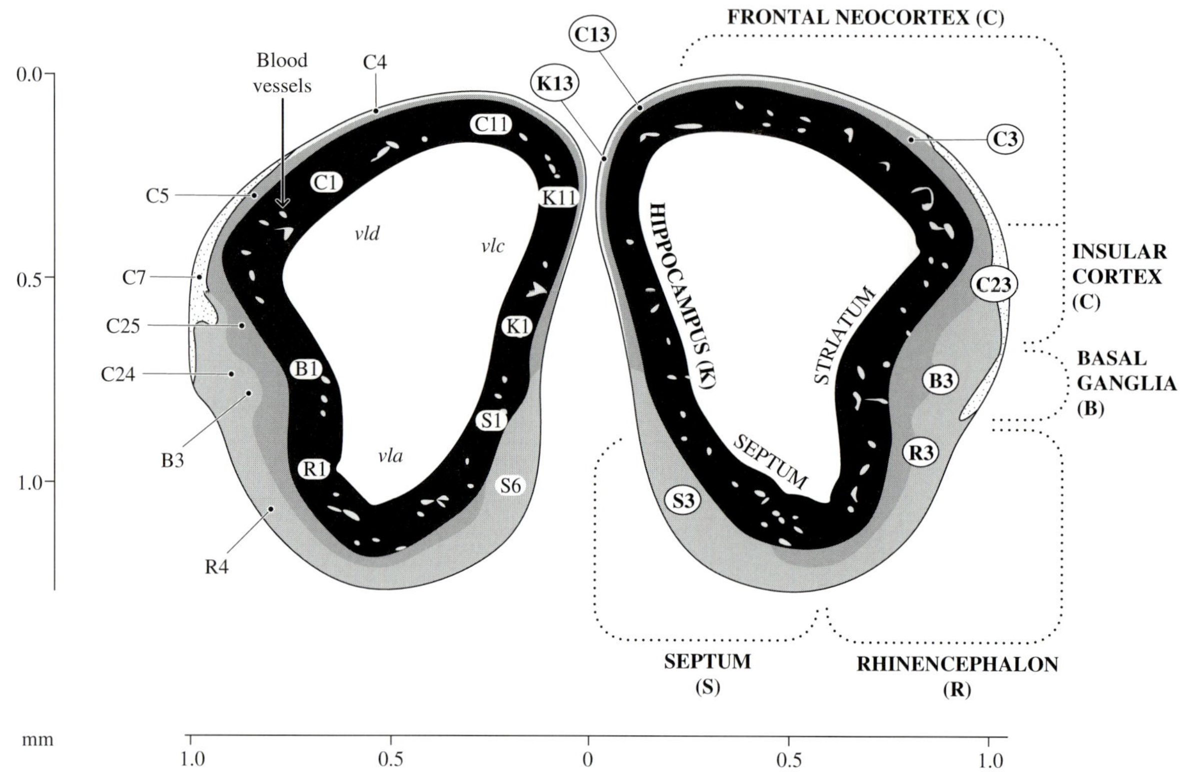

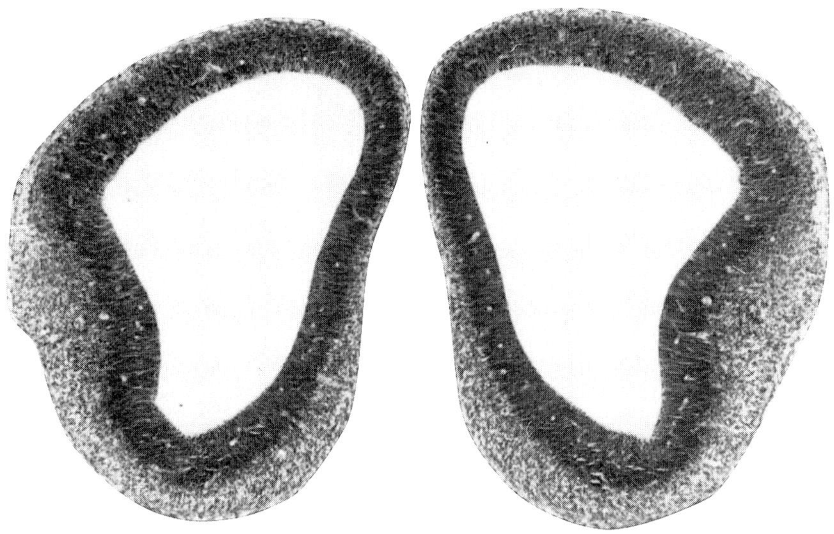

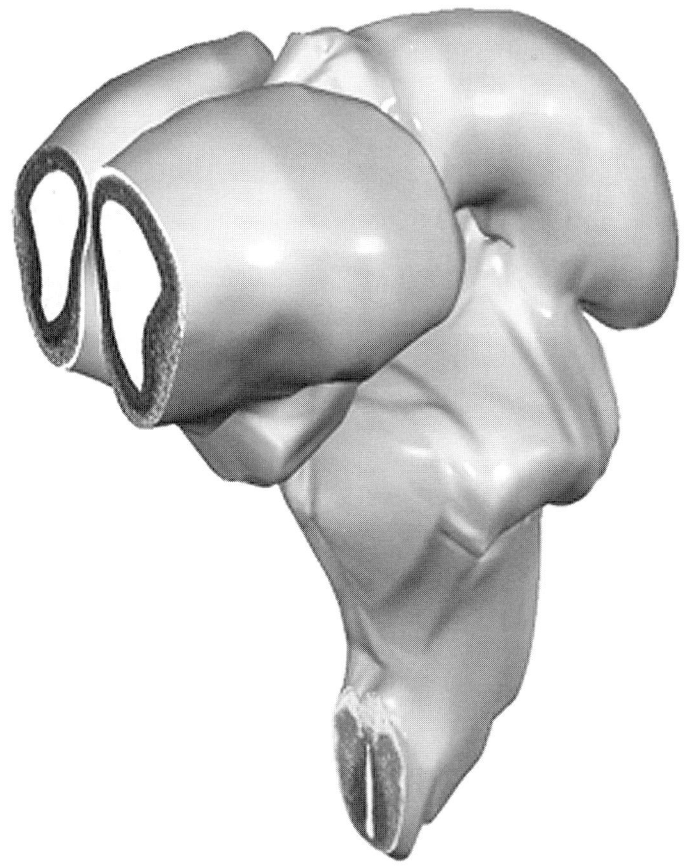

CORONAL FIGURE 4

BASAL GANGLIA (B)

- B1 striatal neuroepithelium
- B2 striatal subventricular zone
- **B3** striatum (or differentiating field)
- B11 pallidal neuroepithelium
- B12 pallidal subventricular zone
- **B13** pallidum (or differentiating field)

CEREBRAL CORTEX (C)

- C1 neocortical neuroepithelium
- **C3** neocortex (or differentiating field)
- C4 molecular layer (or primordial plexiform layer)
- C5 intermediate zone
- C7 white matter
- C11 cingulate cortex neuroepithelium
- **C13** cingulate cortex (or differentiating field)
- **C23** orbital and insular area (or differentiating field)
- C24 insular cortex
- C25 orbital cortex

HIPPOCAMPAL FORMATION (K)

- K1 hippocampal (Ammon's horn) neuroepithelium
- **K3** hippocampus (or differentiating field)
- K11 subicular neuroepithelium
- **K13** subicular area (or differentiating field)

RHINENCEPHALON (R)

- **R3** rhinencephalon (or differentiating field)
- R4 piriform cortex

SEPTAL AREA (S)

- S1 septal neuroepithelium
- **S3** septum (or differentiating field)
- S6 medial septum
- S9 diagonal band (horizontal limb)

VENTRICLES (v)

- *vlc* lateral ventricle, central
- *vld* lateral ventricle, dorsal
- *vlfp* lateral ventricle, pallidal fork
- *vlfs* lateral ventricle, septal fork

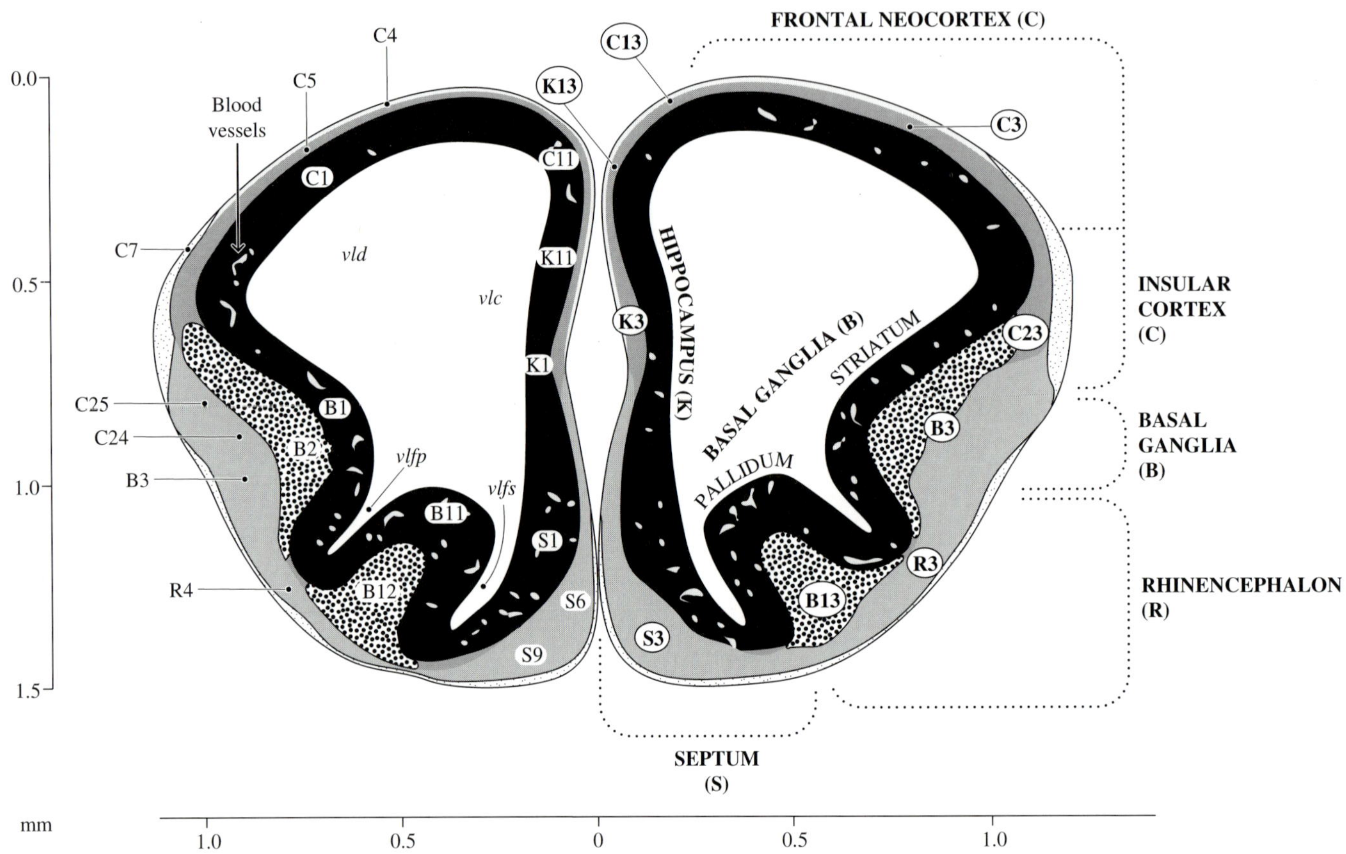

E 1 5

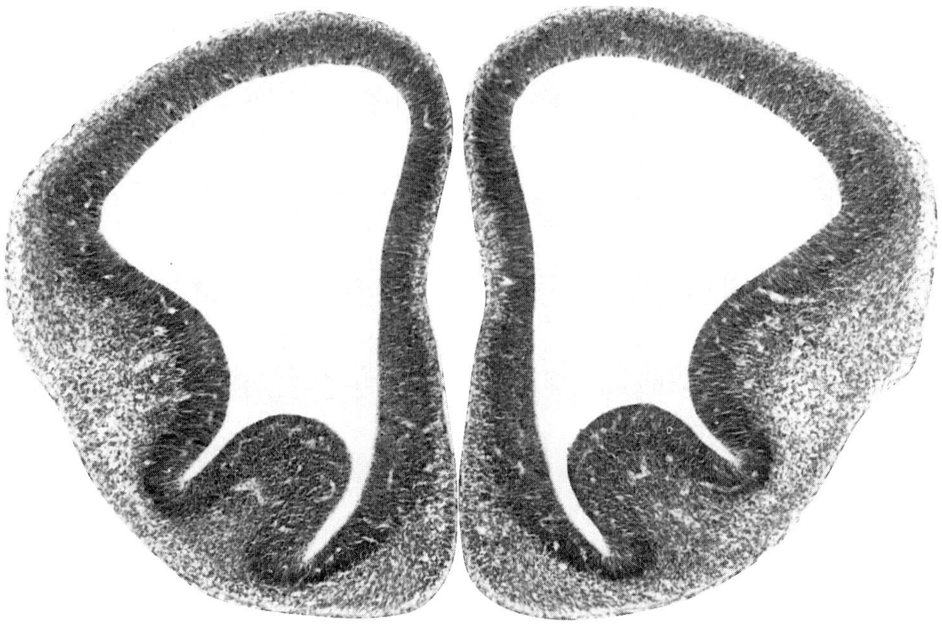

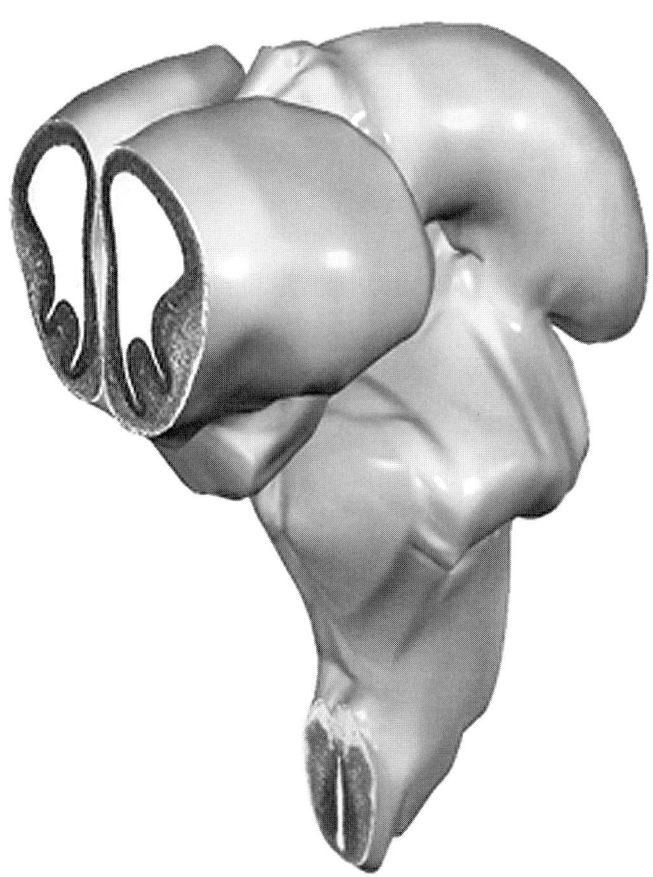

CORONAL FIGURE 5

BASAL GANGLIA (B)

B1 striatal neuroepithelium
B2 striatal subventricular zone
B3 striatum (or differentiating field)
B11 pallidal neuroepithelium
B12 pallidal subventricular zone
B13 pallidum (or differentiating field)

CEREBRAL CORTEX (C)

C1 neocortical neuroepithelium
C3 neocortex (or differentiating field)
C4 molecular layer (or primordial plexiform layer)
C5 intermediate zone
C7 white matter
C11 cingulate cortex neuroepithelium
C13 cingulate cortex (or differentiating field)
C23 orbital and insular area (or differentiating field)
C24 insular cortex

HIPPOCAMPAL FORMATION (K)

K1 hippocampal (Ammon's horn) neuroepithelium
K3 hippocampus (or differentiating field)
K11 subicular neuroepithelium
K13 subicular area (or differentiating field)

PREOPTIC AREA AND OPTIC EPITHELIA (P)

P1 preoptic neuroepithelium
P3 preoptic area (or differentiating field)
P4 medial preoptic region

RHINENCEPHALON (R)

R3 rhinencephalon (or differentiating field)
R4 piriform cortex

VENTRICLES (v)

vlc lateral ventricle, central
vld lateral ventricle, dorsal
vlfp lateral ventricle, pallidal fork
vlfs lateral ventricle, septal fork
v3pr third ventricle, preoptic recess

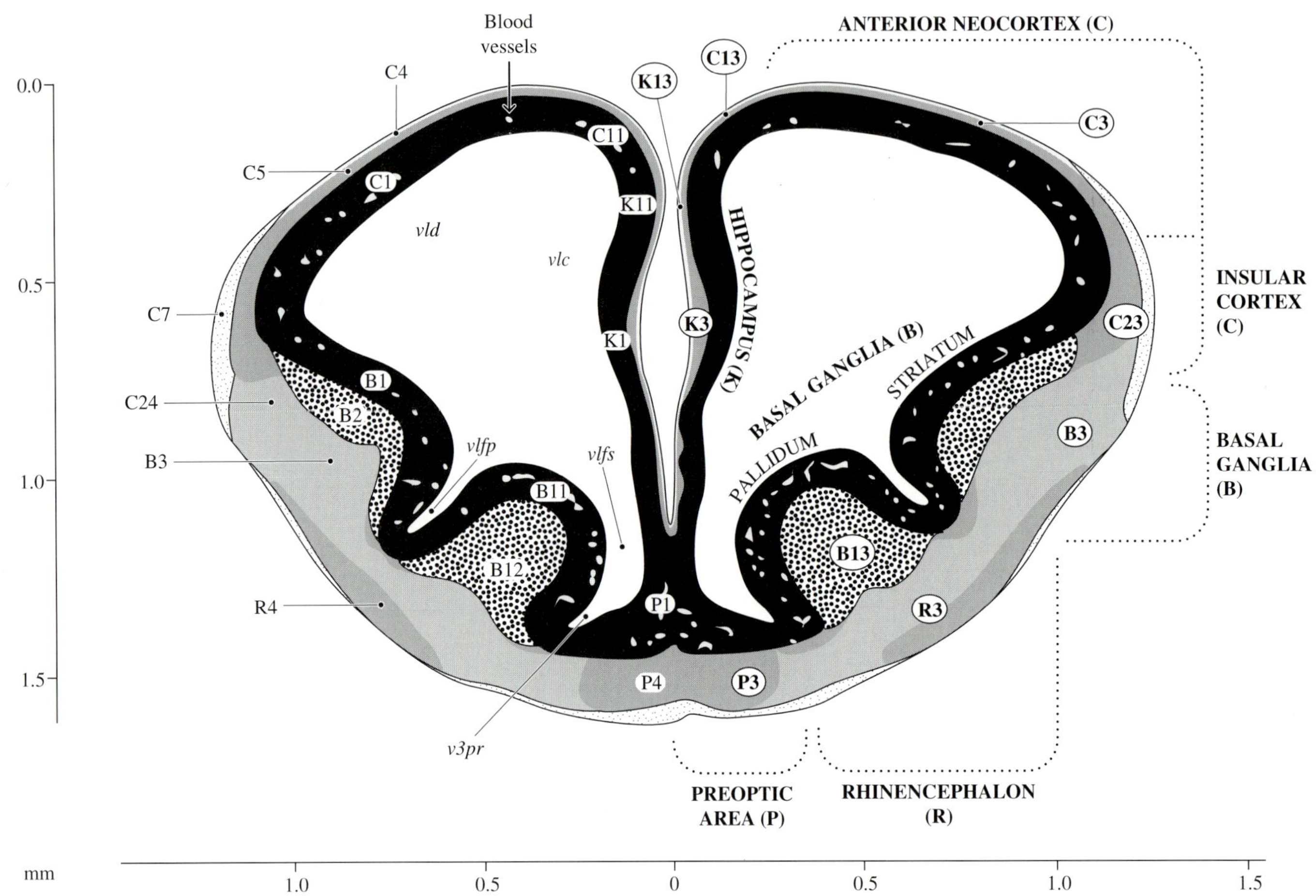

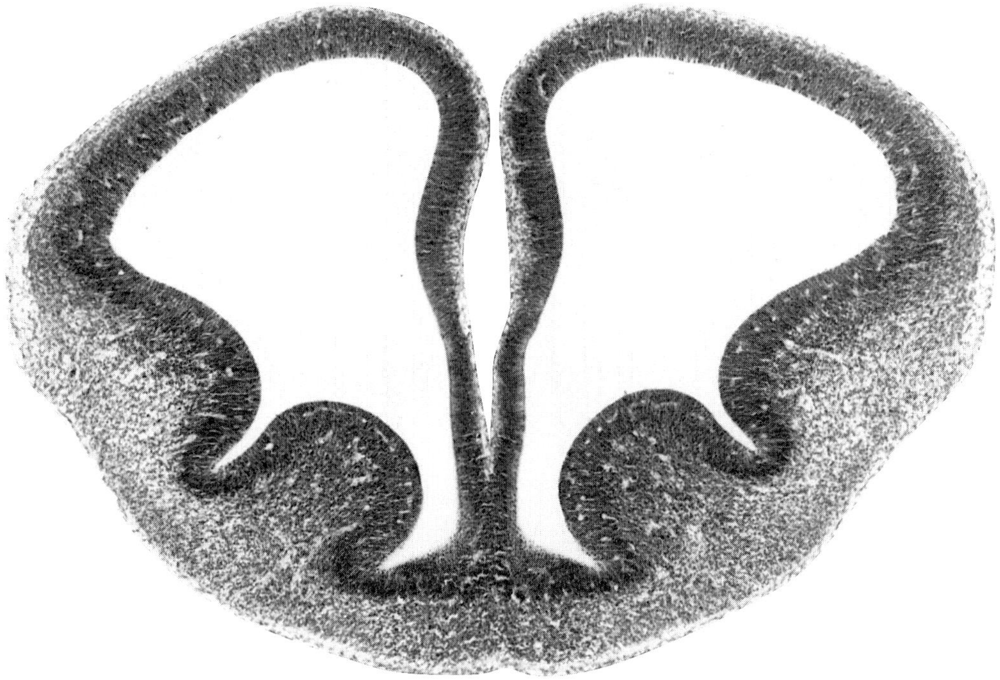

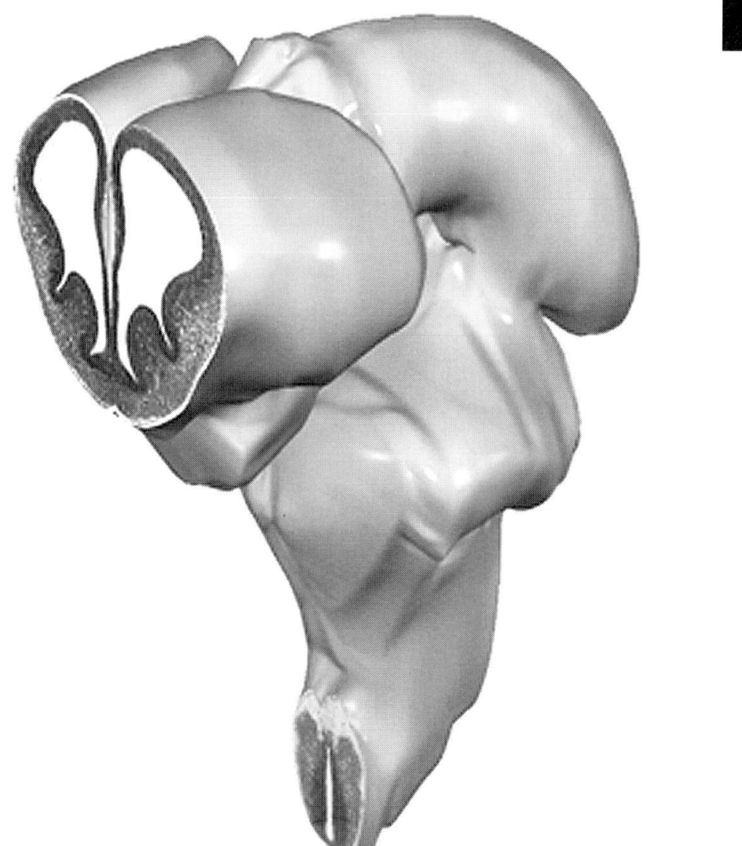

CORONAL FIGURE 6

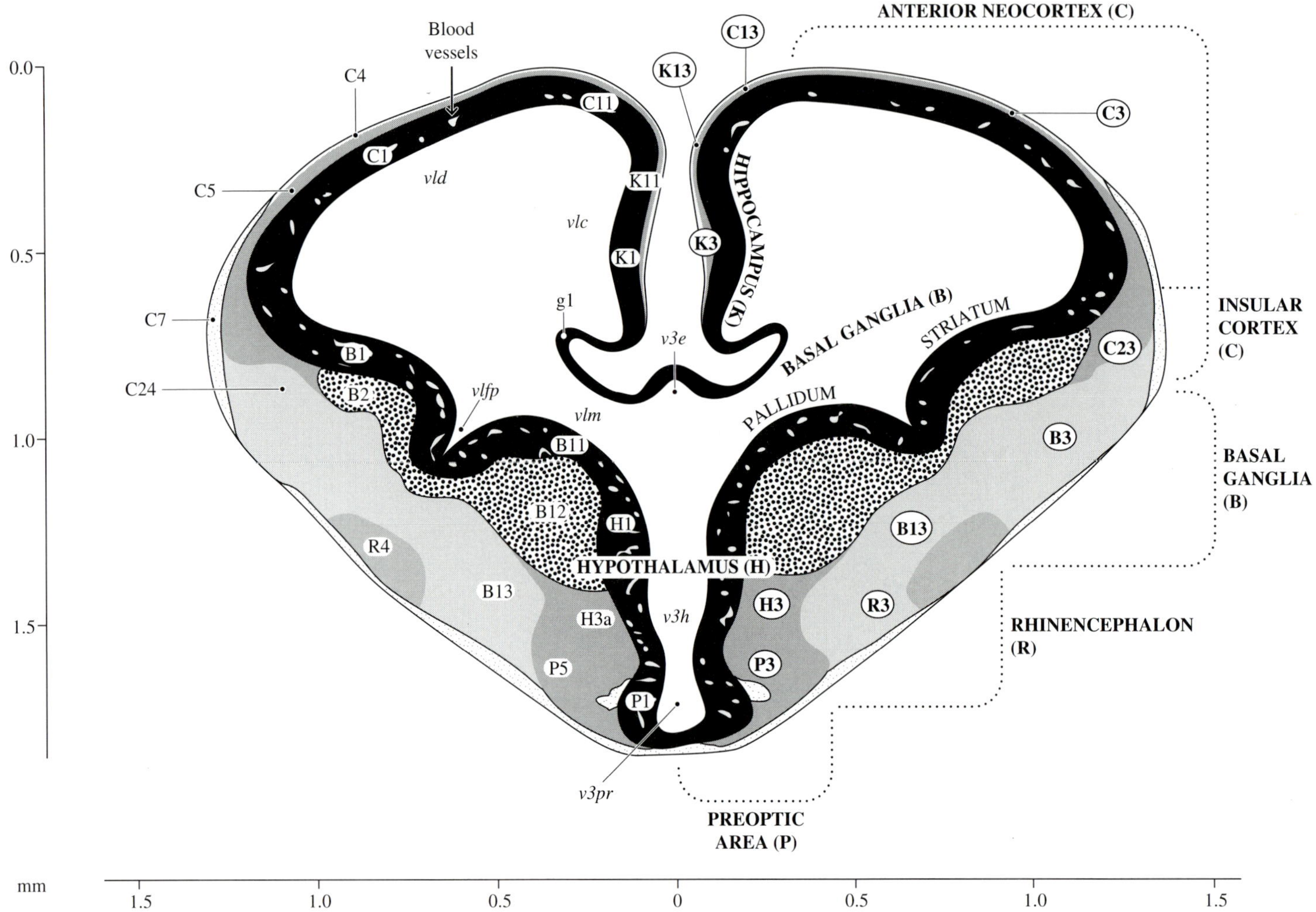

BASAL GANGLIA (B)

B1 striatal neuroepithelium
B2 striatal subventricular zone
B3 striatum (or differentiating field)
B11 pallidal neuroepithelium
B12 pallidal subventricular zone
B13 pallidum (or differentiating field)

CEREBRAL CORTEX (C)

C1 neocortical neuroepithelium
C3 neocortex (or differentiating field)
C4 molecular layer (or primordial plexiform layer)
C5 intermediate zone
C7 white matter
C11 cingulate cortex neuroepithelium
C13 cingulate cortex (or differentiating field)
C23 orbital and insular area (or differentiating field)
C24 insular cortex

GLANDS AND PARANEURAL STRUCTURES (g)

g1 choroid plexus of lateral ventricle

HYPOTHALAMUS (H)

H1 hypothalamic neuroepithelium
H3 hypothalamus (or differentiating field)
H3a anterior hypothalamus (or differentiating field)

HIPPOCAMPAL FORMATION (K)

K1 hippocampal (Ammon's horn) neuroepithelium
K3 hippocampus (or differentiating field)
K11 subicular neuroepithelium
K13 subicular area (or differentiating field)

PREOPTIC AREA AND OPTIC EPITHELIA (P)

P1 preoptic neuroepithelium
P3 preoptic area (or differentiating field)
P5 lateral preoptic region

RHINENCEPHALON (R)

R3 rhinencephalon (or differentiating field)
R4 piriform cortex

VENTRICLES (v)

vlc lateral ventricle, central
vld lateral ventricle, dorsal
vlfp lateral ventricle, pallidal fork
vlm foramen of Monro
v3e third ventricle, epithalamic
v3h third ventricle, hypothalamic
v3pr third ventricle, preoptic recess

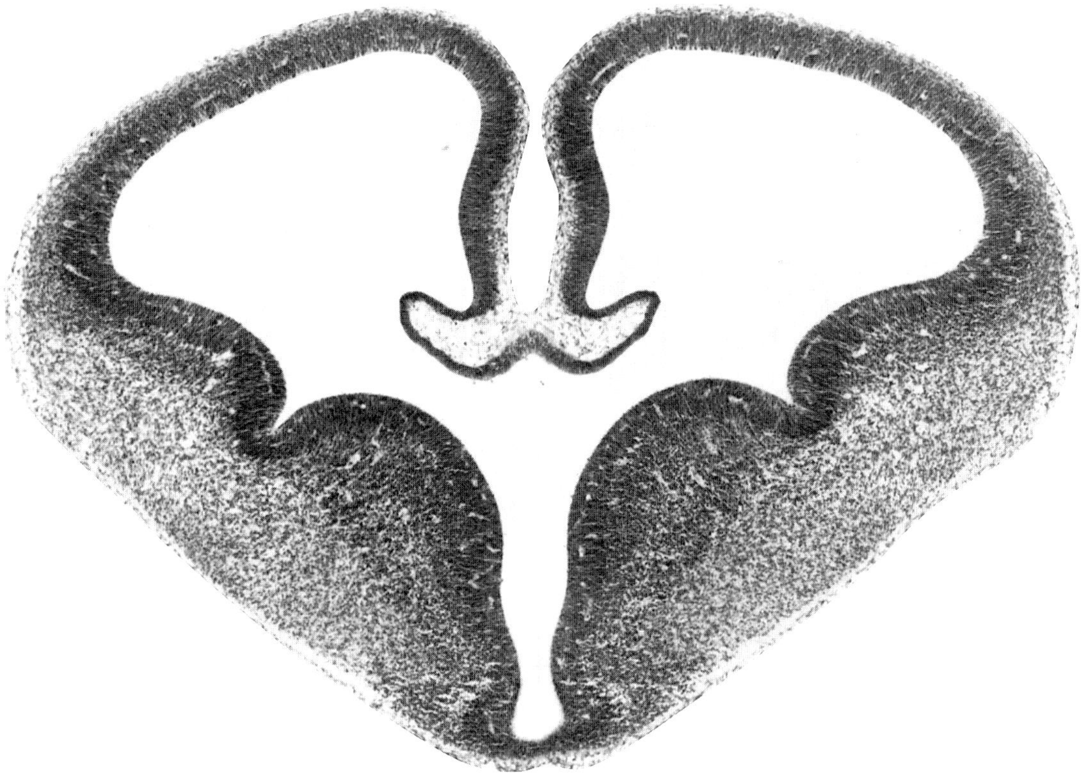

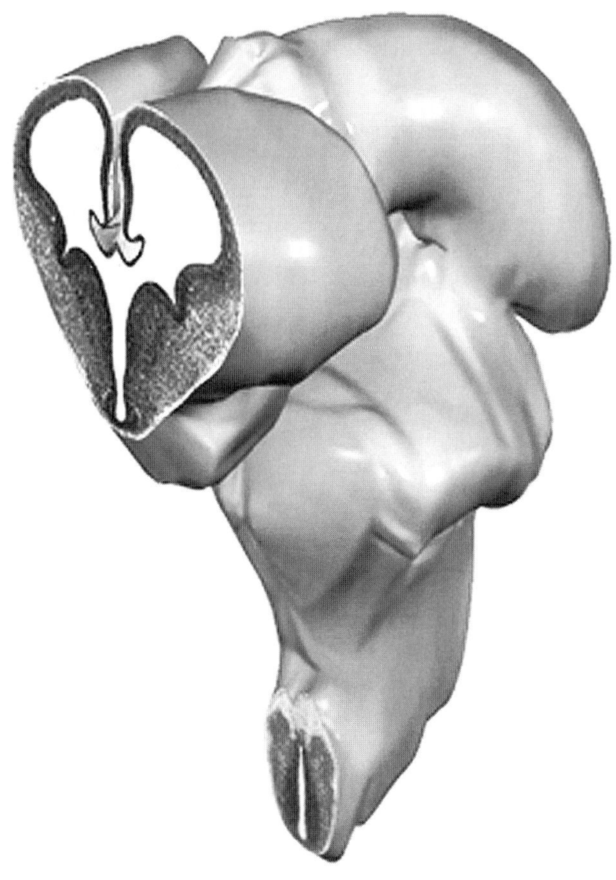

CORONAL FIGURE 7

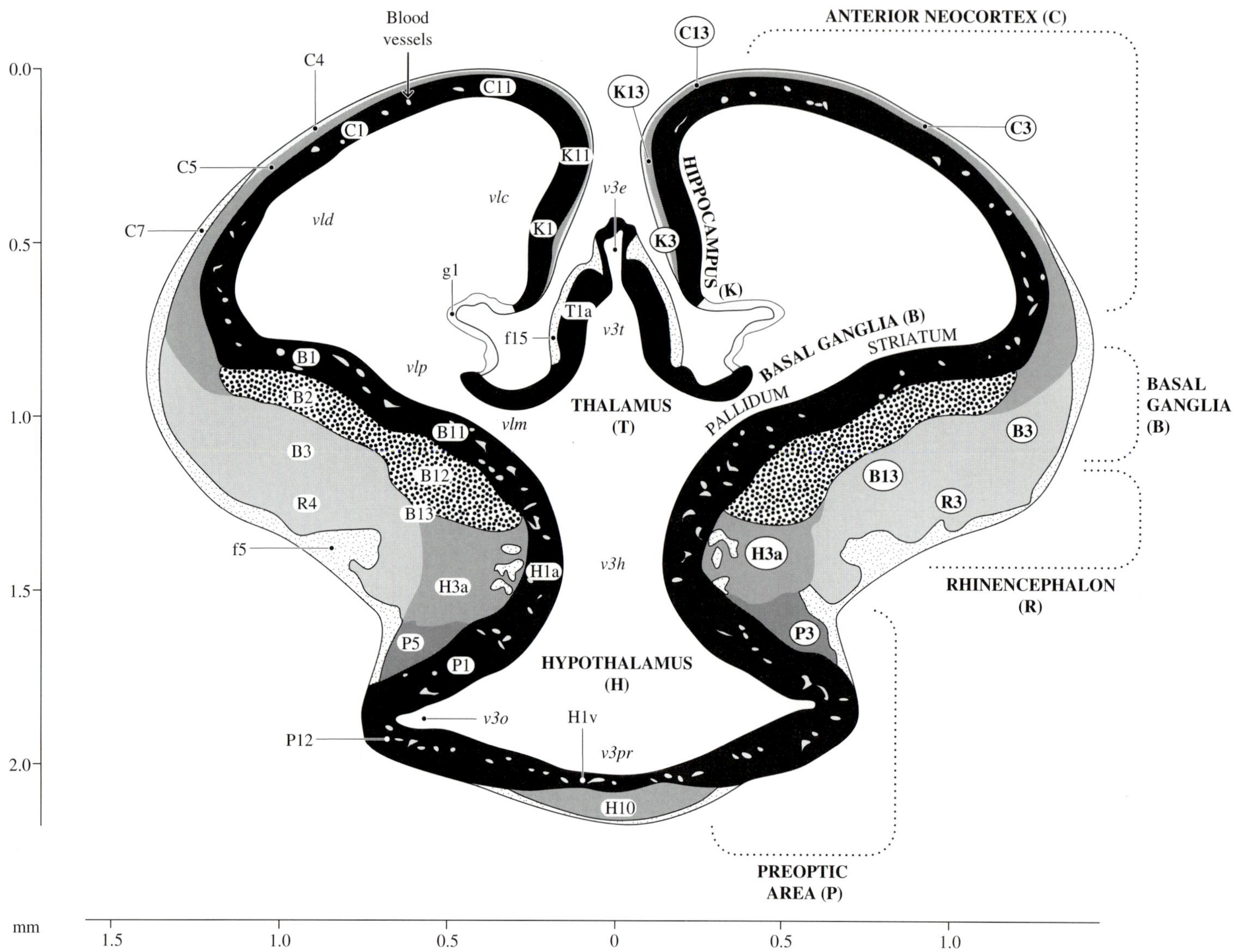

BASAL GANGLIA (B)

B1 striatal neuroepithelium
B2 striatal subventricular zone
B3 striatum (or differentiating field)
B11 pallidal neuroepithelium
B12 pallidal subventricular zone
B13 pallidum (or differentiating field)

CEREBRAL CORTEX (C)

C1 neocortical neuroepithelium
C3 neocortex (or differentiating field)
C4 molecular layer (or primordial plexiform layer)
C5 intermediate zone
C7 white matter
C11 cingulate cortex neuroepithelium
C13 cingulate cortex (or differentiating field)

FIBER TRACTS (f)

f5 medial forebrain bundle
f15 stria medullaris

GLANDS AND PARANEURAL STRUCTURES (g)

g1 choroid plexus of lateral ventricle

HYPOTHALAMUS (H)

H1a anterior hypothalamic neuroepithelium
H1v ventral hypothalamic neuroepithelium
H3a anterior hypothalamus (or differentiating field)
H10 anterobasal nucleus

HIPPOCAMPAL FORMATION (K)

K1 hippocampal (Ammon's horn) neuroepithelium
K3 hippocampus (or differentiating field)
K11 subicular neuroepithelium
K13 subicular area (or differentiating field)

PREOPTIC AREA AND OPTIC EPITHELIA (P)

P1 preoptic neuroepithelium
P3 preoptic area (or differentiating field)
P5 lateral preoptic region
P12 eye stalk epithelium

RHINENCEPHALON (R)

R3 rhinencephalon (or differentiating field)
R4 piriform cortex

THALAMUS (T)

T1a anterior thalamic neuroepithelium

VENTRICLES (v)

vlc lateral ventricle, central
vld lateral ventricle, dorsal
vlm foramen of Monro
vlp lateral ventricle, posterior horn
v3e third ventricle, epithalamic
v3h third ventricle, hypothalamic
v3o third ventricle, optic recess
v3pr third ventricle, preoptic recess
v3t third ventricle, thalamic

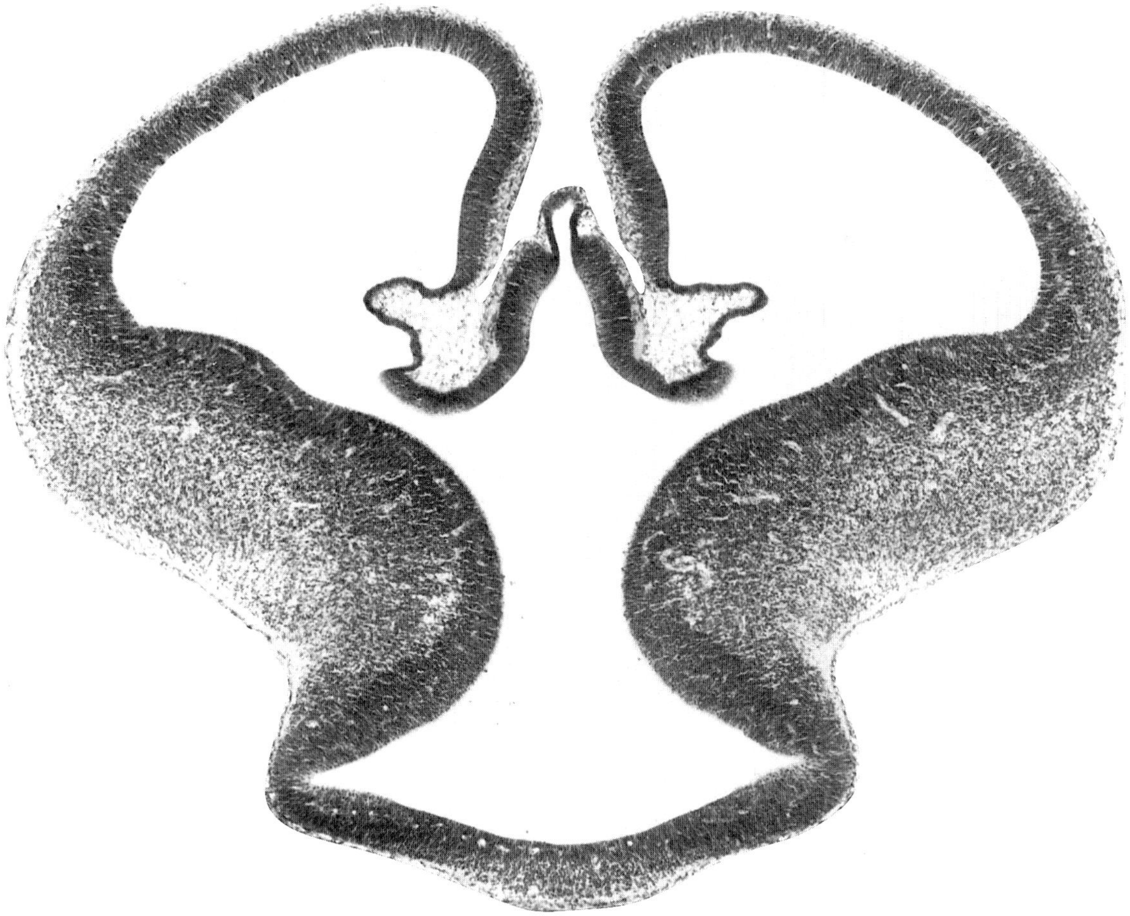

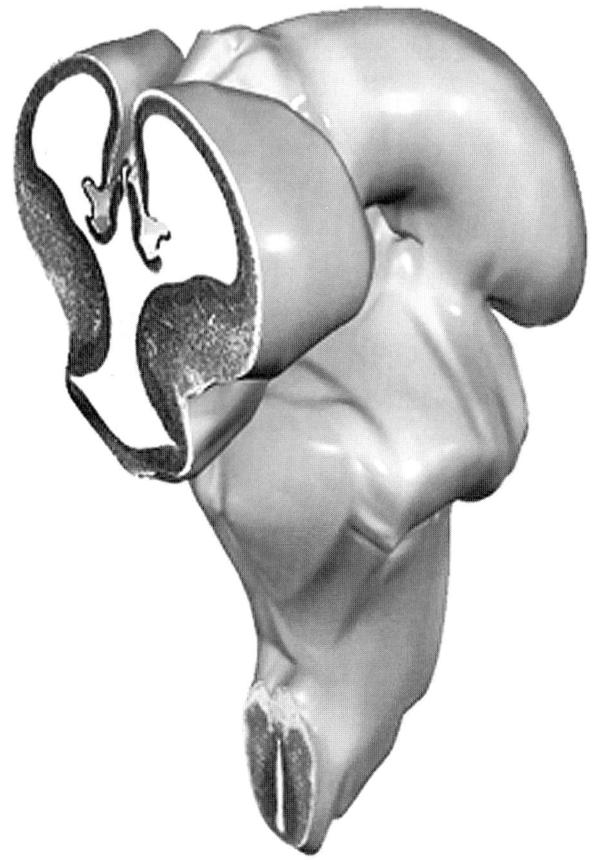

CORONAL FIGURE 8

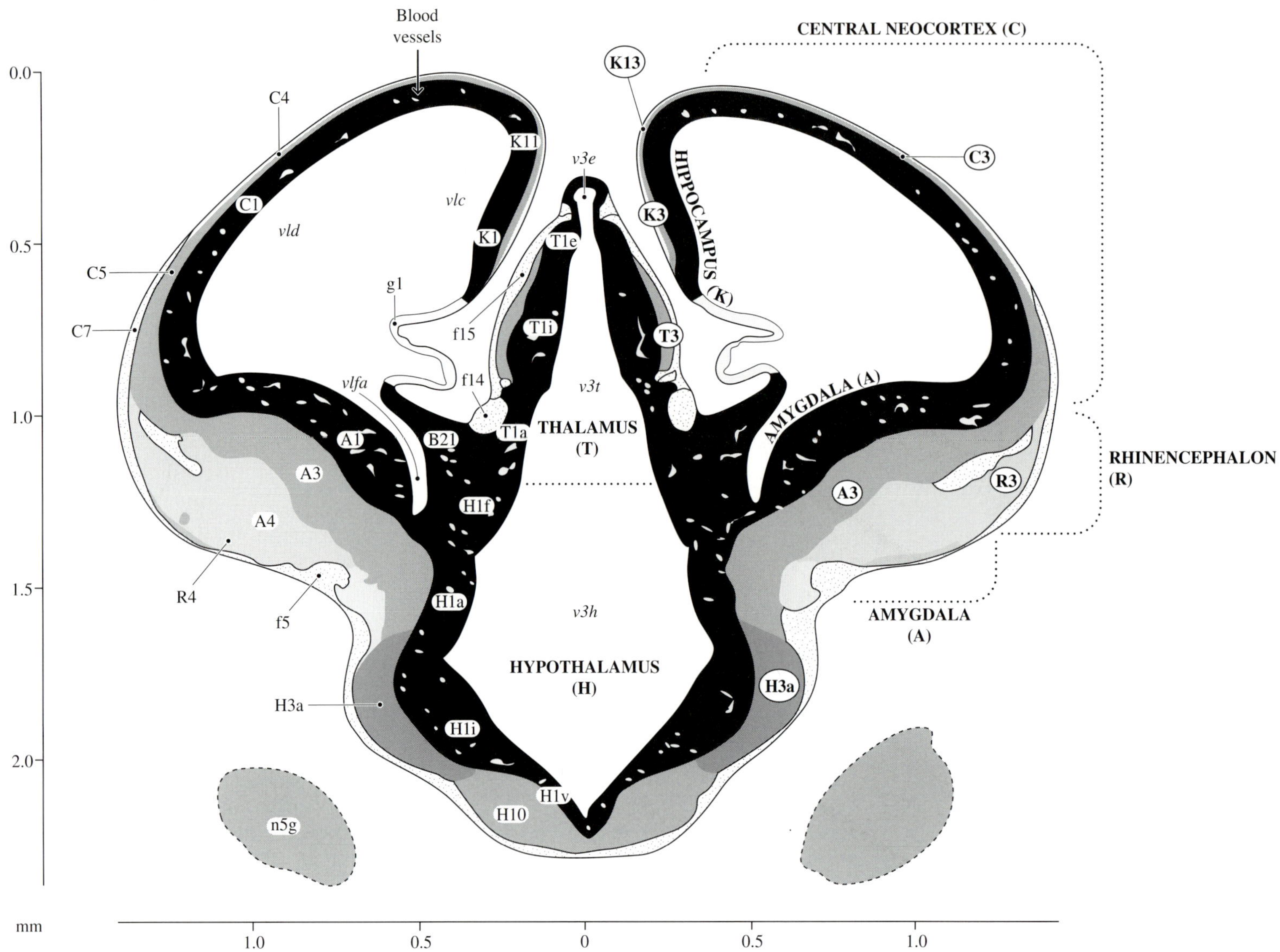

AMYGDALA (A)

A1	amygdaloid neuroepithelium
A3	amygdala (or differentiating field)
A4	anterior amygdaloid area

BASAL GANGLIA (B)

B21	strionuclear neuroepithelium

CEREBRAL CORTEX (C)

C1	neocortical neuroepithelium
C3	neocortex (or differentiating field)
C4	molecular layer (or primordial plexiform layer)
C5	intermediate zone
C7	white matter

FIBER TRACTS (f)

f5	medial forebrain bundle
f14	stria terminalis
f15	stria medullaris

GLANDS AND PARANEURAL STRUCTURES (g)

g1	choroid plexus of lateral ventricle

HYPOTHALAMUS (H)

H1a	anterior hypothalamic neuroepithelium
H1f	Forelian neuroepithelium
H1i	intermediate hypothalamic neuroepithelium
H1v	ventral hypothalamic neuroepithelium
H3a	anterior hypothalamus (or differentiating field)
H10	anterobasal nucleus

HIPPOCAMPAL FORMATION (K)

K1	hippocampal (Ammon's horn) neuroepithelium
K3	hippocampus (or differentiating field)
K11	subicular neuroepithelium
K13	subicular area (or differentiating field)

NERVES AND NUCLEI, CRANIAL (n)

n5g	trigeminal ganglion

RHINENCEPHALON (R)

R3	rhinencephalon (or differentiating field)
R4	piriform cortex

THALAMUS (T)

T1a	anterior thalamic neuroepithelium
T1e	epithalamic neuroepithelium
T1i	intermediate thalamic neuroepithelium
T3	thalamus (or differentiating field)

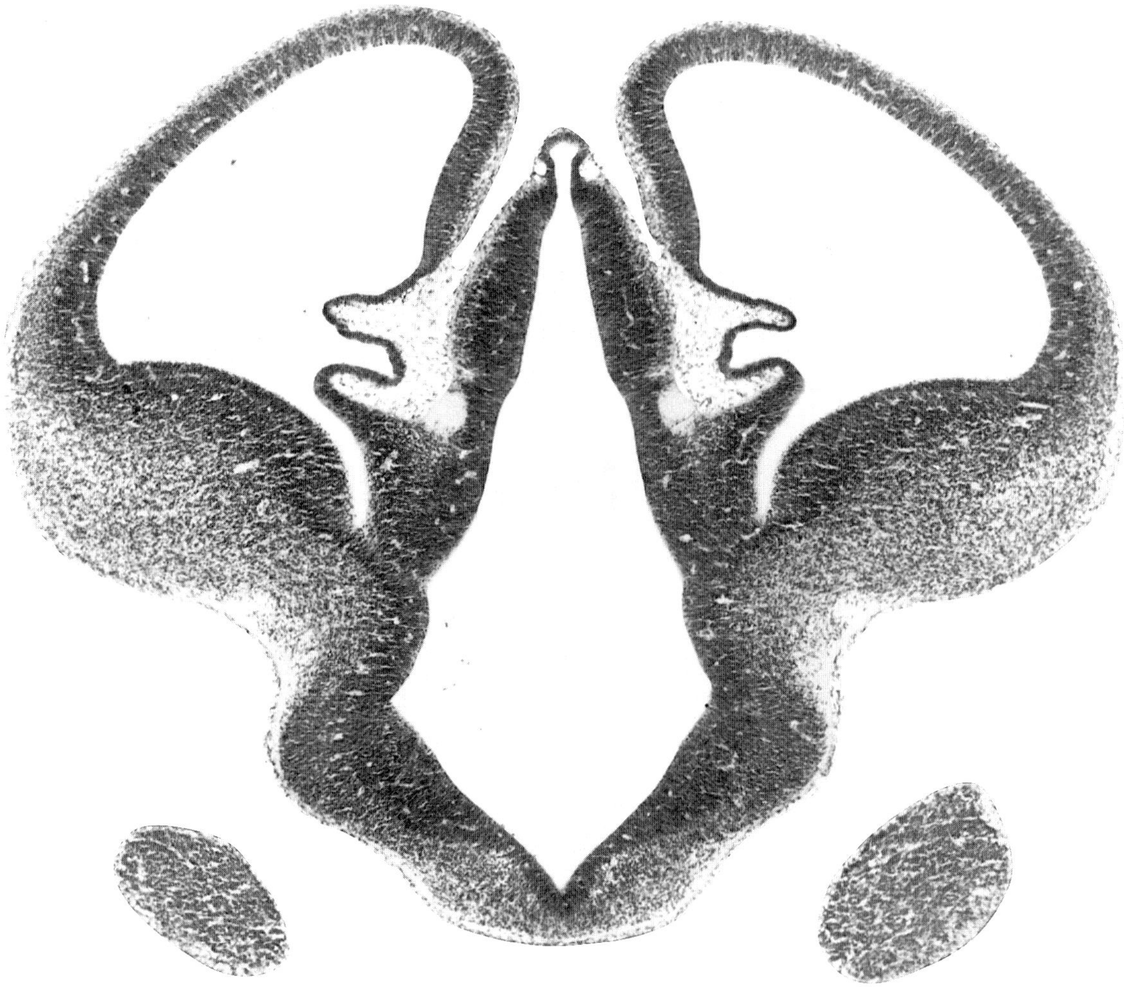

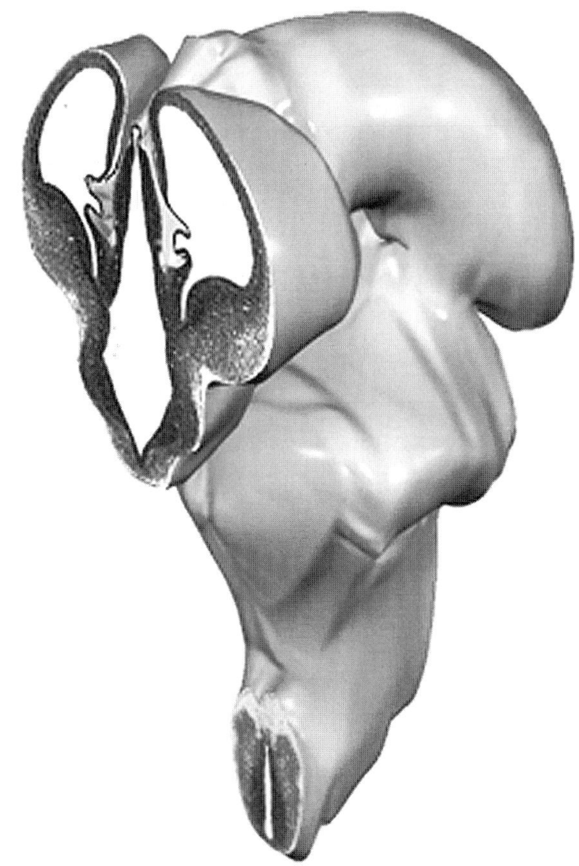

VENTRICLES (v)

vlc	lateral ventricle, central
vld	lateral ventricle, dorsal
vlfa	lateral ventricle, amygdaloid fork
v3e	third ventricle, epithalamic
v3h	third ventricle, hypothalamic
v3t	third ventricle, thalamic

CORONAL FIGURE 9

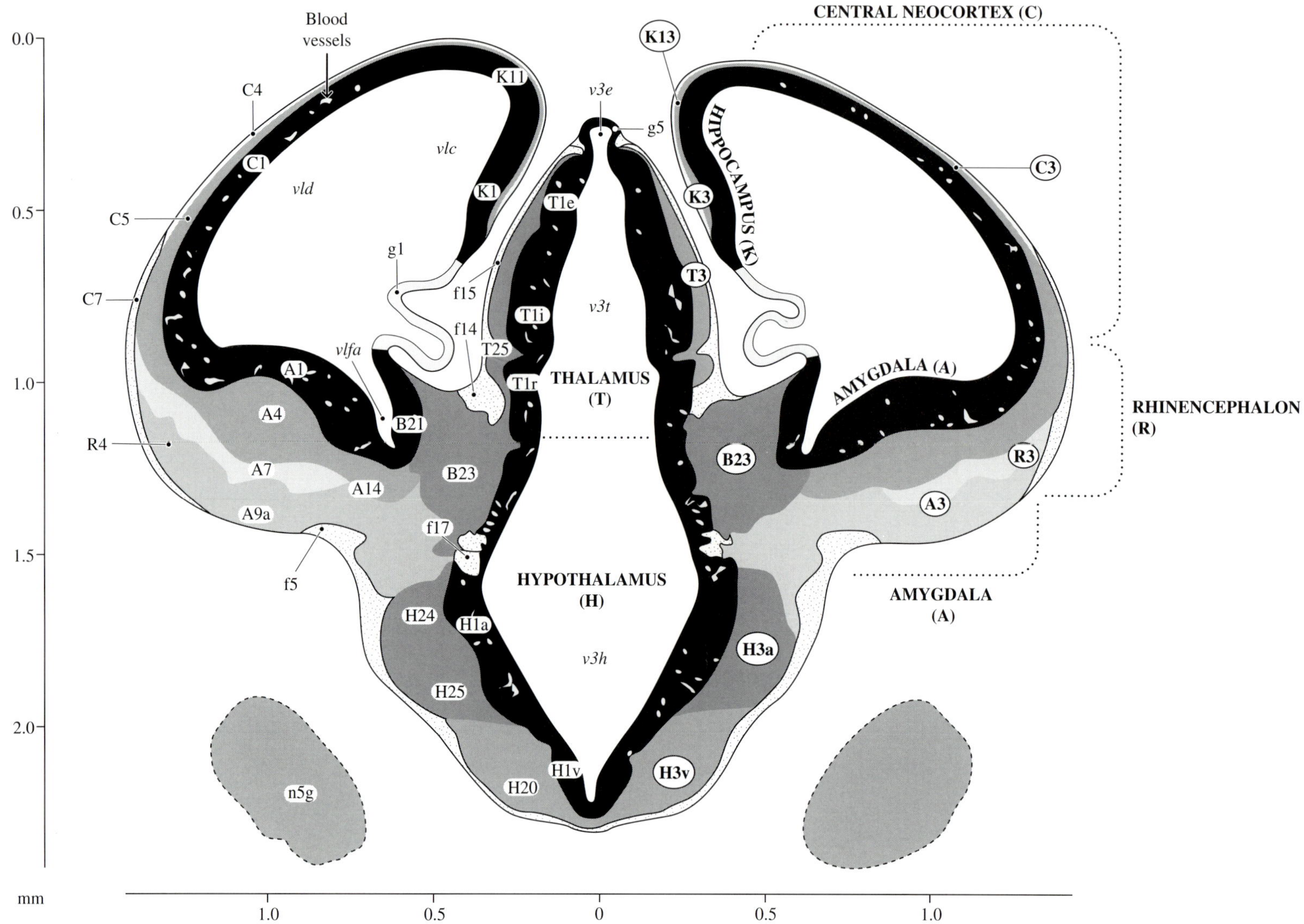

AMYGDALA (A)

A1 amygdaloid neuroepithelium
A3 amygdala (or differentiating field)
A4 anterior amygdaloid area
A7 basomedial amygdaloid nucleus
A9a cortical amygdaloid nucleus, anterior
A14 medial amygdaloid nucleus

BASAL GANGLIA (B)

B21 strionuclear neuroepithelium
B23 bed nucleus of stria terminalis (or differentiating field)

CEREBRAL CORTEX (C)

C1 neocortical neuroepithelium
C3 neocortex (or differentiating field)
C4 molecular layer (or primordial plexiform layer)
C5 intermediate zone
C7 white matter

FIBER TRACTS (f)

f5 medial forebrain bundle
f14 stria terminalis
f15 stria medullaris
f17 mammillothalamic tract

GLANDS AND PARANEURAL STRUCTURES (g)

g1 choroid plexus of lateral ventricle
g5 pineal gland (primordium)

HYPOTHALAMUS (H)

H1a anterior hypothalamic neuroepithelium
H1v ventral hypothalamic neuroepithelium
H3a anterior hypothalamus (or differentiating field)
H3v ventral hypothalamus (or differentiating field)
H20 arcuate nucleus
H24 anterodorsal nucleus
H25 anteroventral nucleus

HIPPOCAMPAL FORMATION (K)

K1 hippocampal (Ammon's horn) neuroepithelium
K3 hippocampus (or differentiating field)
K11 subicular neuroepithelium
K13 subicular area (or differentiating field)

NERVES AND NUCLEI, CRANIAL (n)

n5g trigeminal ganglion

RHINENCEPHALON (R)

R3 rhinencephalon (or differentiating field)
R4 piriform cortex

THALAMUS (T)

T1e epithalamic neuroepithelium
T1i intermediate thalamic neuroepithelium
T1r reticular (or ventral) thalamic neuroepithelium
T3 thalamus (or differentiating field)
T25 reticular nucleus

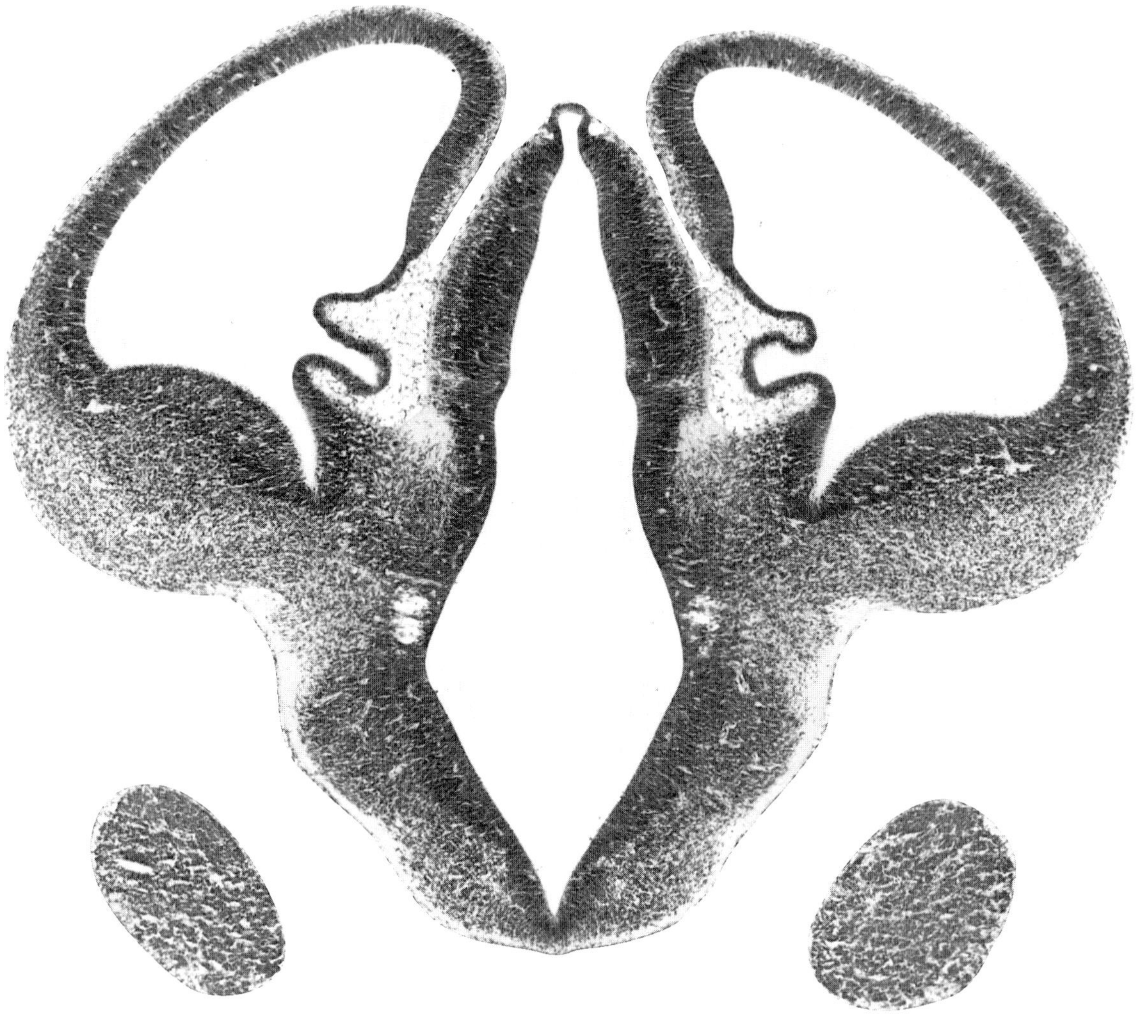

VENTRICLES (v)

vlc lateral ventricle, central
vld lateral ventricle, dorsal
vlfa lateral ventricle, amygdaloid fork
v3e third ventricle, epithalamic
v3h third ventricle, hypothalamic
v3t third ventricle, thalamic

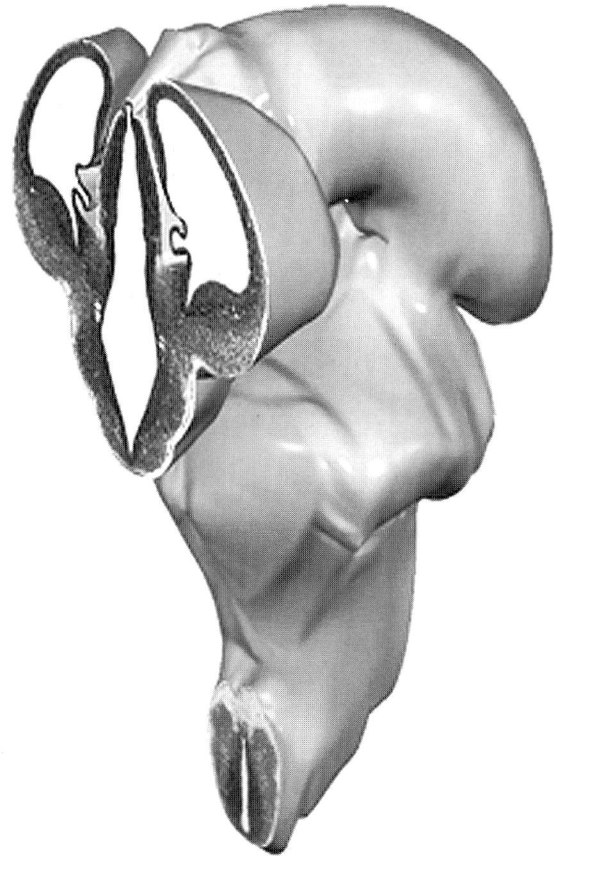

CORONAL FIGURE 10

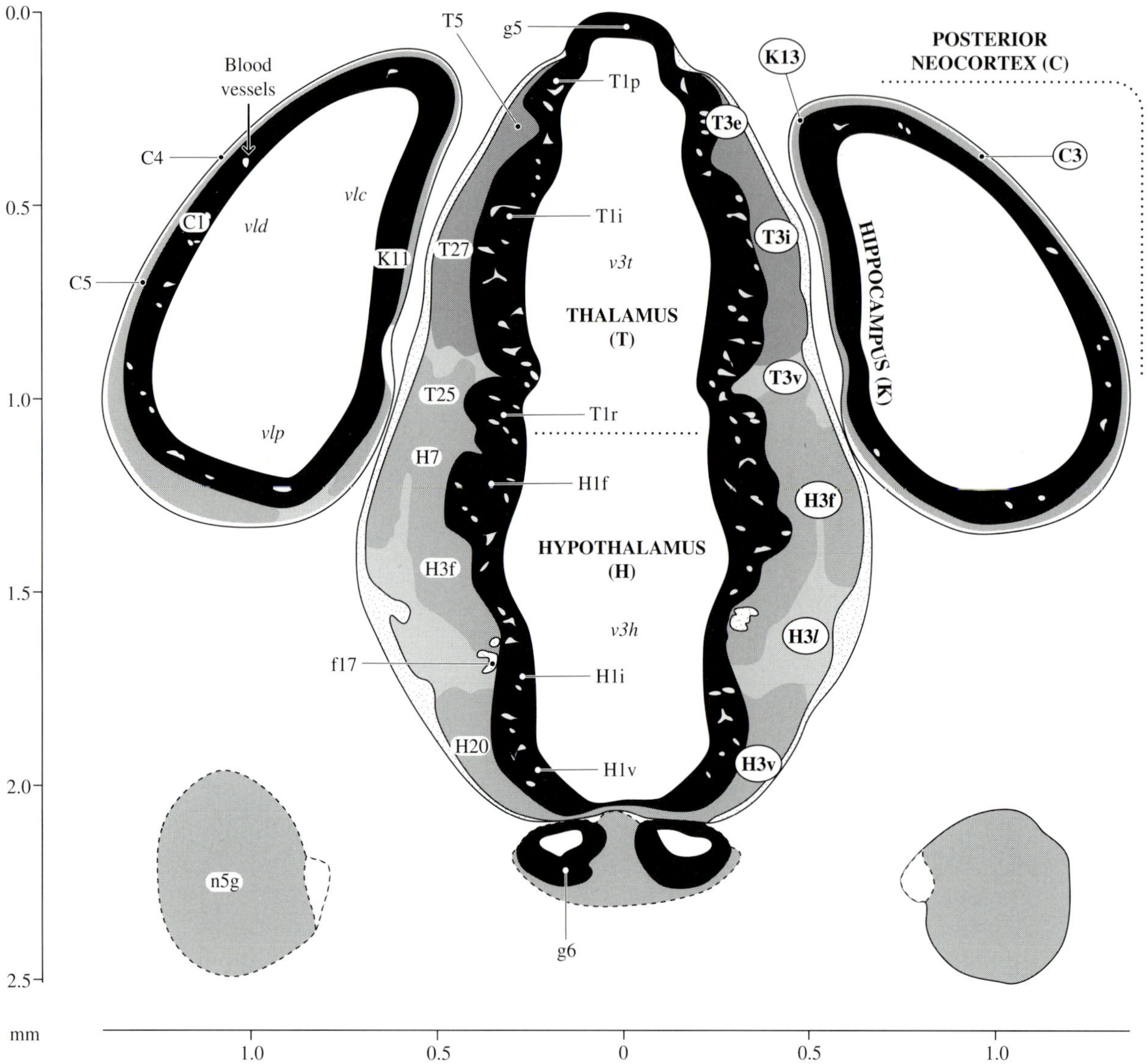

CEREBRAL CORTEX (C)

- C1 neocortical neuroepithelium
- **C3** neocortex (or differentiating field)
- C4 molecular layer (or primordial plexiform layer)
- C5 intermediate zone

FIBER TRACTS (f)

- f17 mammillothalamic tract

GLANDS AND PARANEURAL STRUCTURES (g)

- g5 pineal gland (primordium)
- g6 pituitary gland (primordium)

HYPOTHALAMUS (H)

- H1i intermediate hypothalamic neuroepithelium
- H1f Forelian neuroepithelium
- H1v ventral hypothalamic neuroepithelium
- **H3f** fields of Forel (or differentiating field)
- **H3*l*** lateral hypothalamus (or differentiating field)
- **H3v** ventral hypothalamus (or differentiating field)
- H7 zona incerta
- H20 arcuate nucleus

HIPPOCAMPAL FORMATION (K)

- K11 subicular neuroepithelium
- **K13** subicular area (or differentiating field)

NERVES AND NUCLEI, CRANIAL (n)

- n5g trigeminal ganglion

THALAMUS (T)

- T1i intermediate thalamic neuroepithelium
- T1p posterior thalamic neuroepithelium
- T1r reticular (or ventral) thalamic neuroepithelium
- **T3e** epithalamus (or differentiating field)
- **T3i** intermediate thalamus (or differentiating field)
- **T3v** ventral thalamus (or differentiating field)
- T5 lateral habenular nucleus
- T25 reticular nucleus
- T27 ventrolateral nuclear complex

VENTRICLES (v)

- *vlc* lateral ventricle, central
- *vld* lateral ventricle, dorsal
- *vlp* lateral ventricle, posterior horn
- *v3h* third ventricle, hypothalamic
- *v3t* third ventricle, thalamic

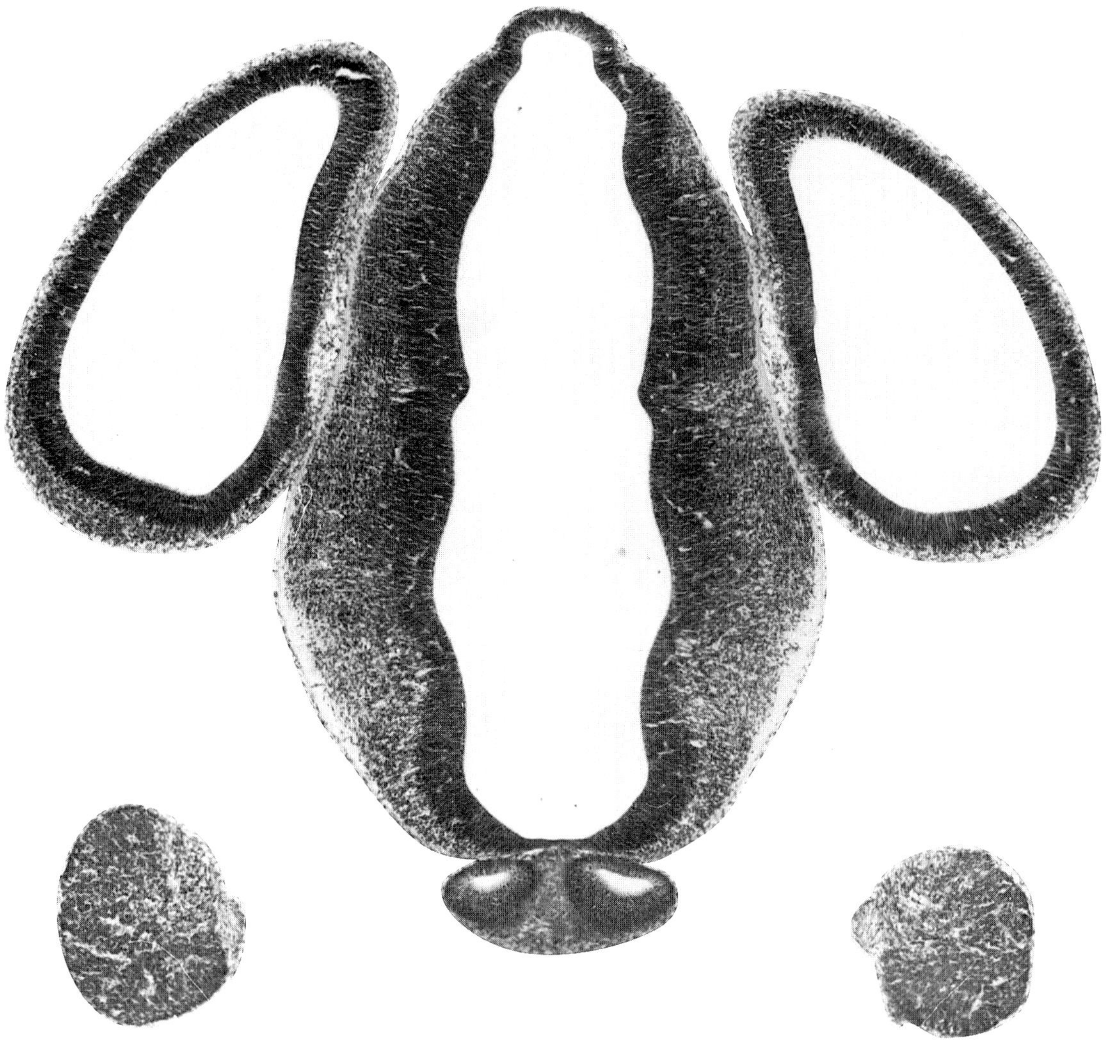

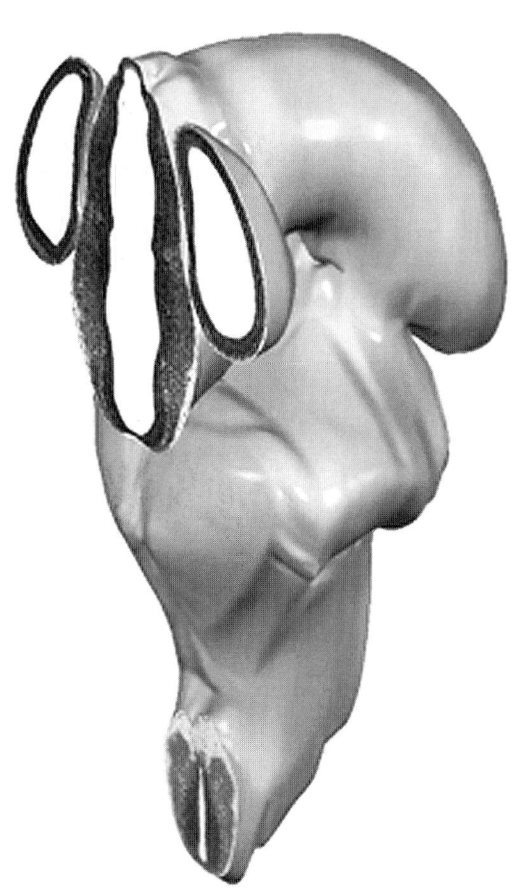

CORONAL FIGURE 11

CEREBRAL CORTEX (C)

C1 neocortical neuroepithelium
C3 neocortex (or differentiating field)
C4 molecular layer (or primordial plexiform layer)
C5 intermediate zone

FIBER TRACTS (f)

f9 cerebral peduncle
f17 mammillothalamic tract
f27 medial longitudinal fasciculus

HYPOTHALAMUS (H)

H1f Forelian neuroepithelium
H1i intermediate hypothalamic neuroepithelium
H1v ventral hypothalamic neuroepithelium
H3f fields of Forel (or differentiating field)
H3*l* lateral hypothalamus (or differentiating field)
H3v ventral hypothalamus (or differentiating field)
H20 arcuate nucleus

MIDBRAIN (M)

M8 red nucleus

NERVES AND NUCLEI, CRANIAL (n)

n5b trigeminal boundary cap
n5n trigeminal motor nucleus
n5p principal (sensory) trigeminal nucleus
n7 facial nerve
n7n facial motor nucleus
n8b vestibulocochlear boundary cap
n8g vestibulocochlear ganglion
n12n hypoglossal nucleus

THALAMUS (T)

T1i intermediate thalamic neuroepithelium
T1p posterior thalamic neuroepithelium
T3i intermediate thalamus (or differentiating field)
T3p posterior thalamus (or differentiating field)
T25 reticular nucleus
T28 ventrobasal nuclear complex
T29 dorsal lateral geniculate nucleus
T34 medial geniculate nucleus

VENTRICLES (v)

v3h third ventricle, hypothalamic
v3t third ventricle, thalamic
v4a fourth ventricle, anterior
v4p fourth ventricle, posterior

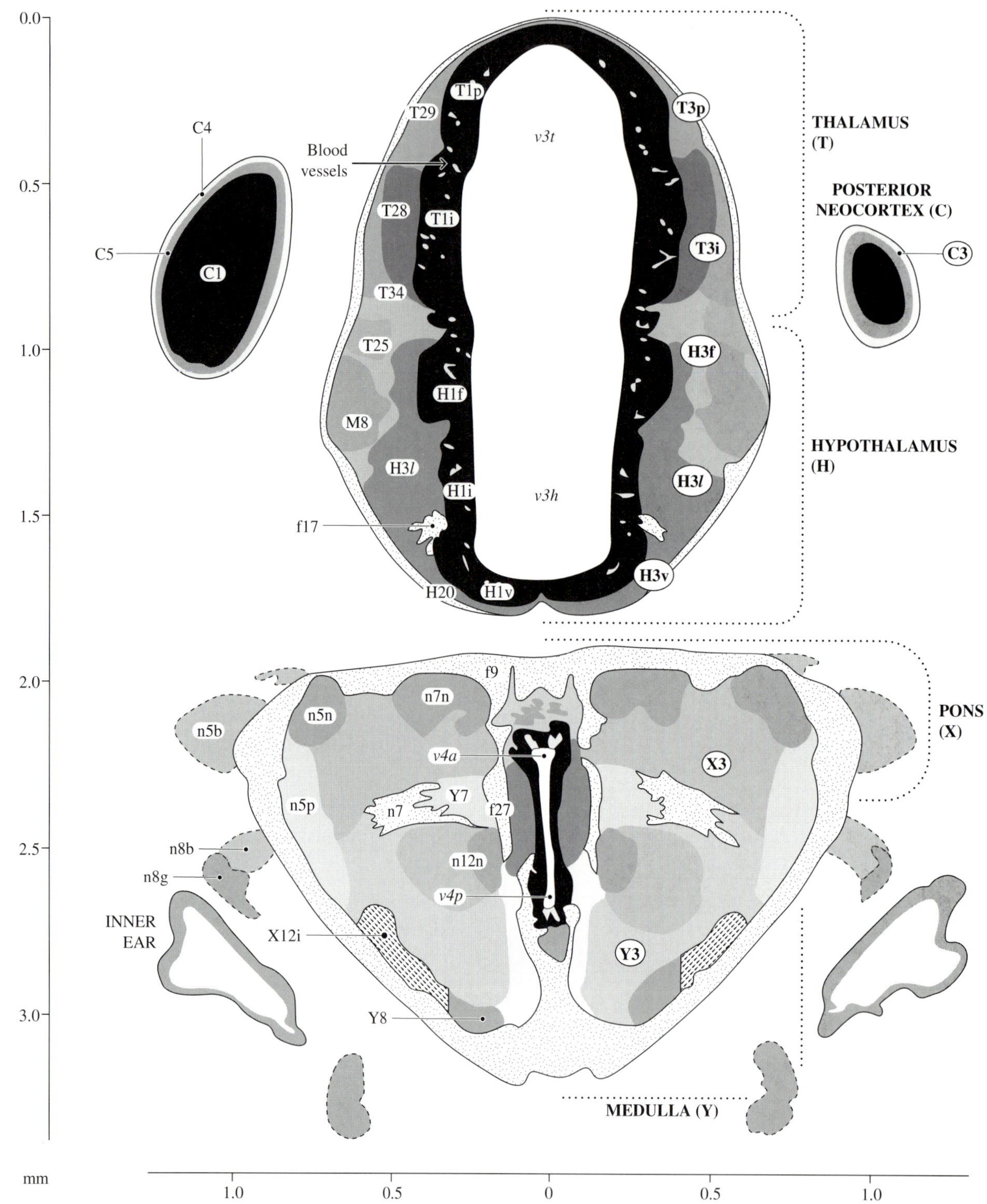

MEDULLA (Y)

Y3 medulla (or differentiating field)
Y7 prepositus nucleus
Y8 inferior olive

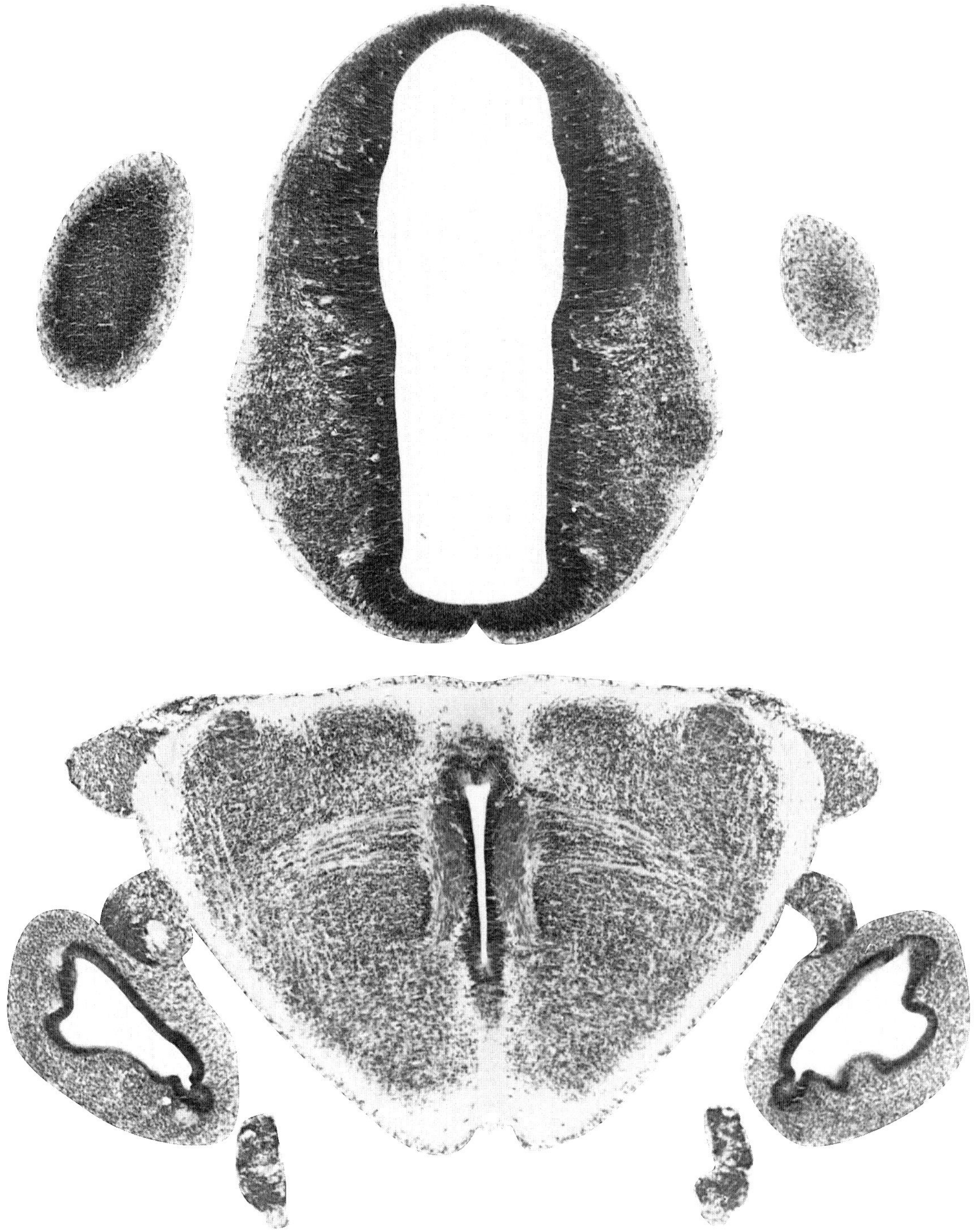

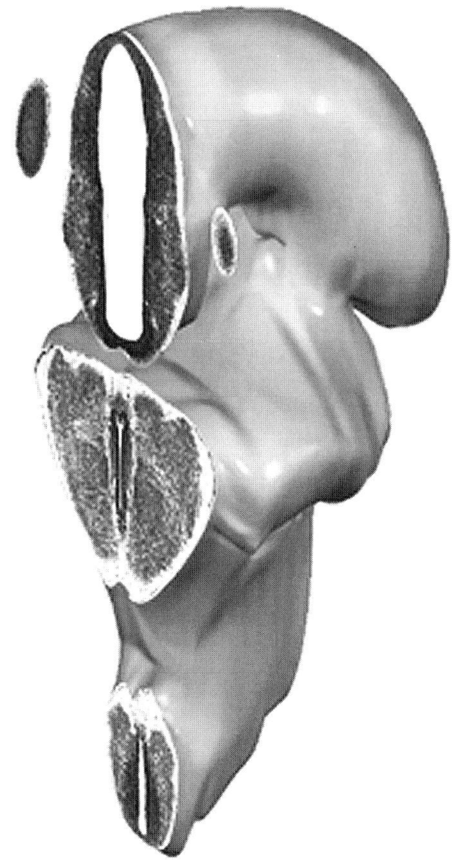

CORONAL FIGURE 12

FIBER TRACTS (f)

f9 cerebral peduncle
f18 habenulopeduncular tract

HYPOTHALAMUS (H)

H1f Forelian neuroepithelium
H1i intermediate hypothalamic neuroepithelium
H1v ventral hypothalamic neuroepithelium
H3i intermediate hypothalamus (or differentiating field)
H3*l* lateral hypothalamus (or differentiating field)
H3v ventral hypothalamus (or differentiating field)
H19 subthalamic nucleus

MIDBRAIN (M)

M8 red nucleus

NERVES AND NUCLEI, CRANIAL (n)

n5n trigeminal motor nucleus
n5p principal (sensory) trigeminal nucleus
n5s spinal trigeminal nucleus
n7n facial motor nucleus
n12n hypoglossal nucleus

THALAMUS (T)

T1p posterior thalamic neuroepithelium
T3p posterior thalamus (or differentiating field)
T29 dorsal lateral geniculate nucleus
T34 medial geniculate nucleus

VENTRICLES (v)

v3h third ventricle, hypothalamic
v3m third ventricle, mammillary recess
v3t third ventricle, thalamic
v4a fourth ventricle, anterior
v4l fourth ventricle, lateral recess
v4p fourth ventricle, posterior

PONTINE AREA (X)

X1a anterior pontine neuroepithelium
X1p posterior pontine neuroepithelium
X3a anterior pons (or differentiating field)
X3p posterior pons (or differentiating field)
X15 lateral vestibular nucleus
X16 superior vestibular nucleus
X17 medial vestibular nucleus
X21 cochlear nuclei neuroepithelium

MEDULLA (Y)

Y1 medullary neuroepithelium
Y3 medulla (or differentiating field)
Y7 prepositus nucleus

SPINAL CORD (Z)

Z3 spinal cord (or differentiating field)

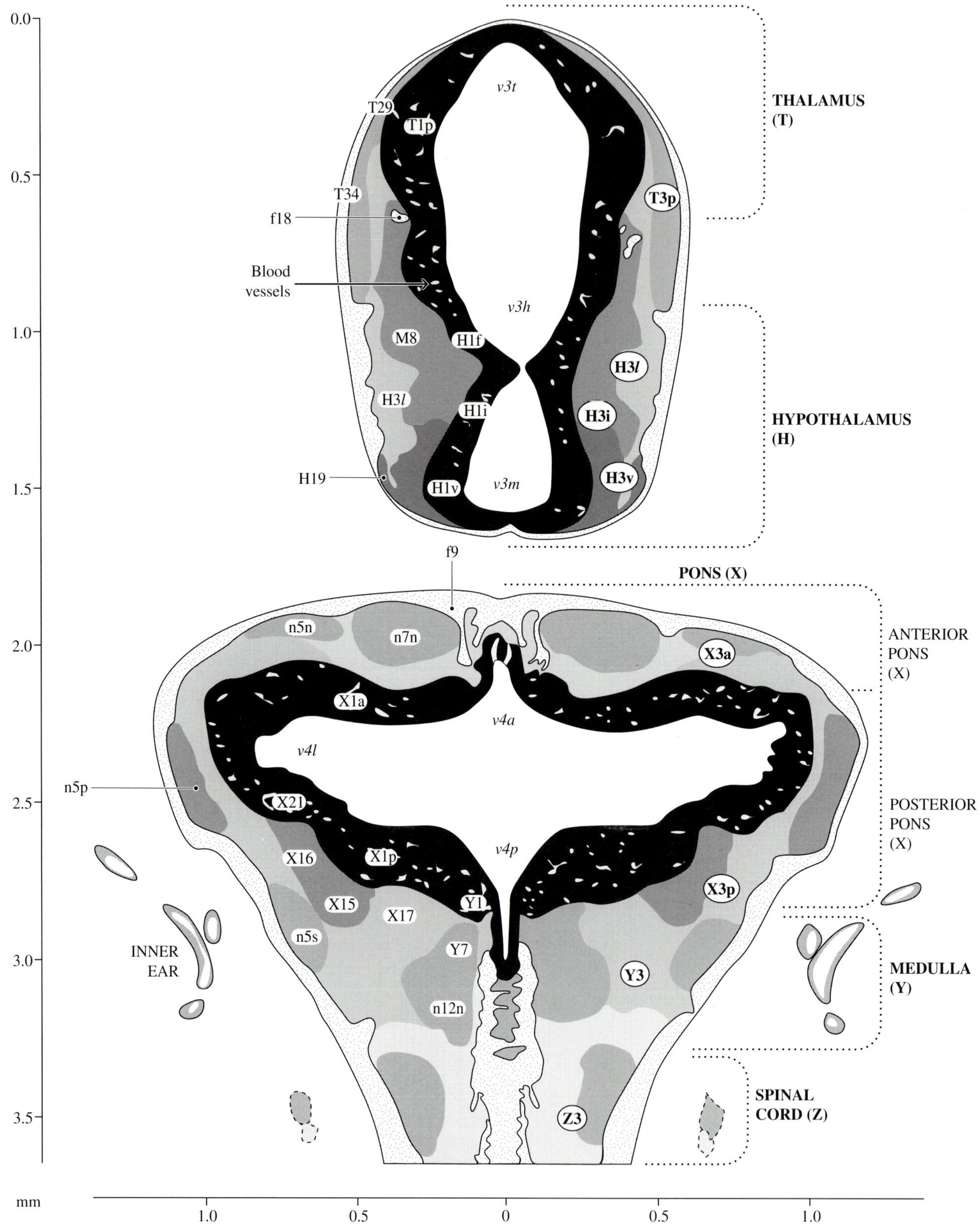

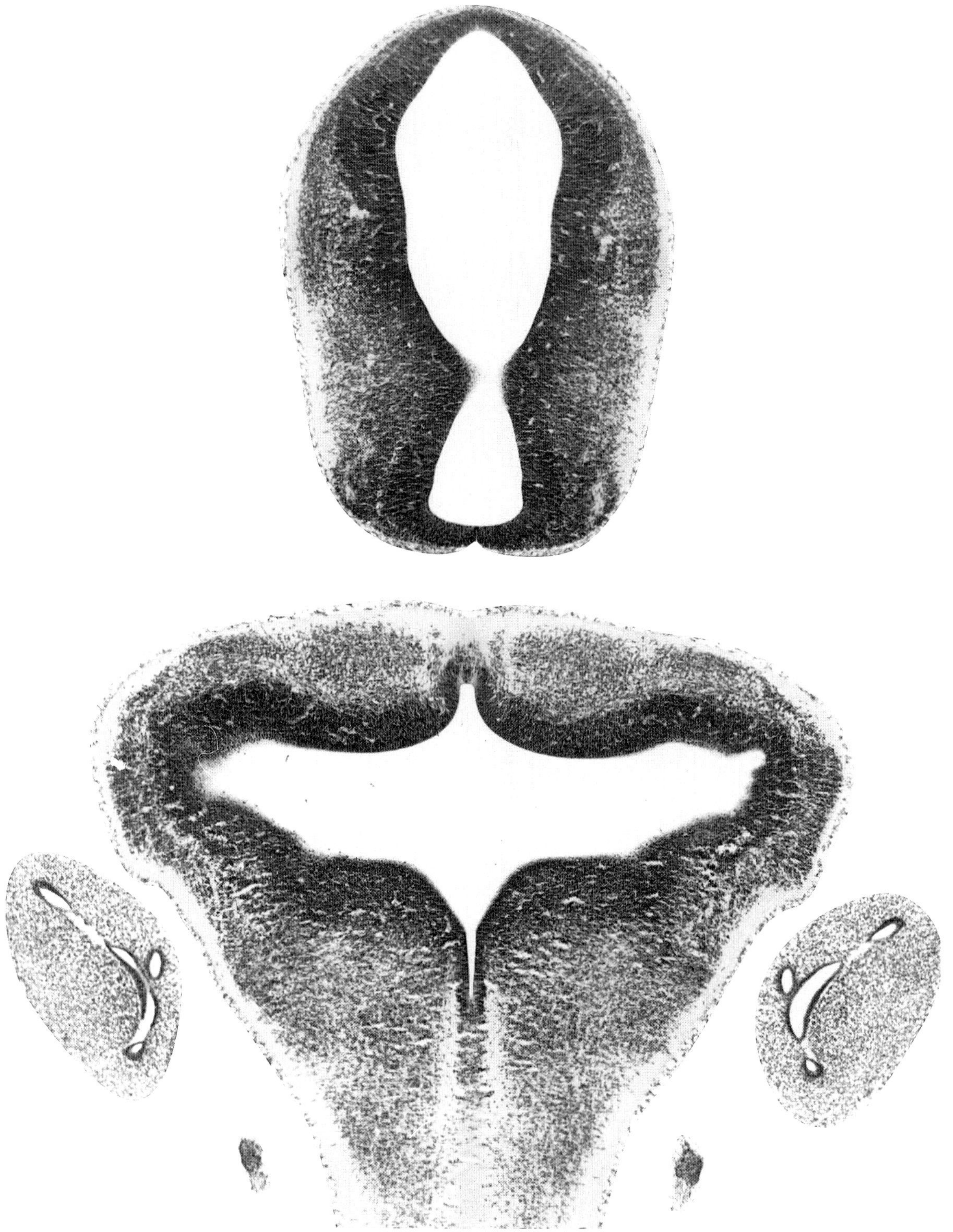

CORONAL FIGURE 13

FIBER TRACTS (f)

f19	posterior commissure

HYPOTHALAMUS (H)

H1i	intermediate hypothalamic neuroepithelium
H1m	mammillary neuroepithelium
H1v	ventral hypothalamic neuroepithelium
H3m	mammillary body (or differentiating field)
H3v	ventral hypothalamus (or differentiating field)
H14	supramammillary nucleus
H19	subthalamic nucleus

MIDBRAIN (M)

M8	red nucleus
M21	pretectal neuroepithelium
M23	pretectum (or differentiating field)

NERVES AND NUCLEI, CRANIAL (n)

n5n	trigeminal motor nucleus
n5s	spinal trigeminal nucleus
n7n	facial motor nucleus

THALAMUS (T)

T1p	posterior thalamic neuroepithelium
T3p	posterior thalamus (or differentiating field)
T29	dorsal lateral geniculate nucleus
T34	medial geniculate nucleus

VENTRICLES (v)

vapr	aqueduct, pretectal
vs	spinal canal
v3h	third ventricle, hypothalamic
v3m	third ventricle, mammillary recess
v3t	third ventricle, thalamic
v4a	fourth ventricle, anterior
v4l	fourth ventricle, lateral recess
v4p	fourth ventricle, posterior

PONTINE AREA (X)

X1a	anterior pontine neuroepithelium
X1p	posterior pontine neuroepithelium
X3a	anterior pons (or differentiating field)
X21	cochlear nuclei neuroepithelium
X23	cochlear nuclei (or differentiating field)

MEDULLA (Y)

Y1	medullary neuroepithelium
Y3	medulla (or differentiating field)
Y16	solitary nucleus

SPINAL CORD (Z)

Z3	spinal cord (or differentiating field)

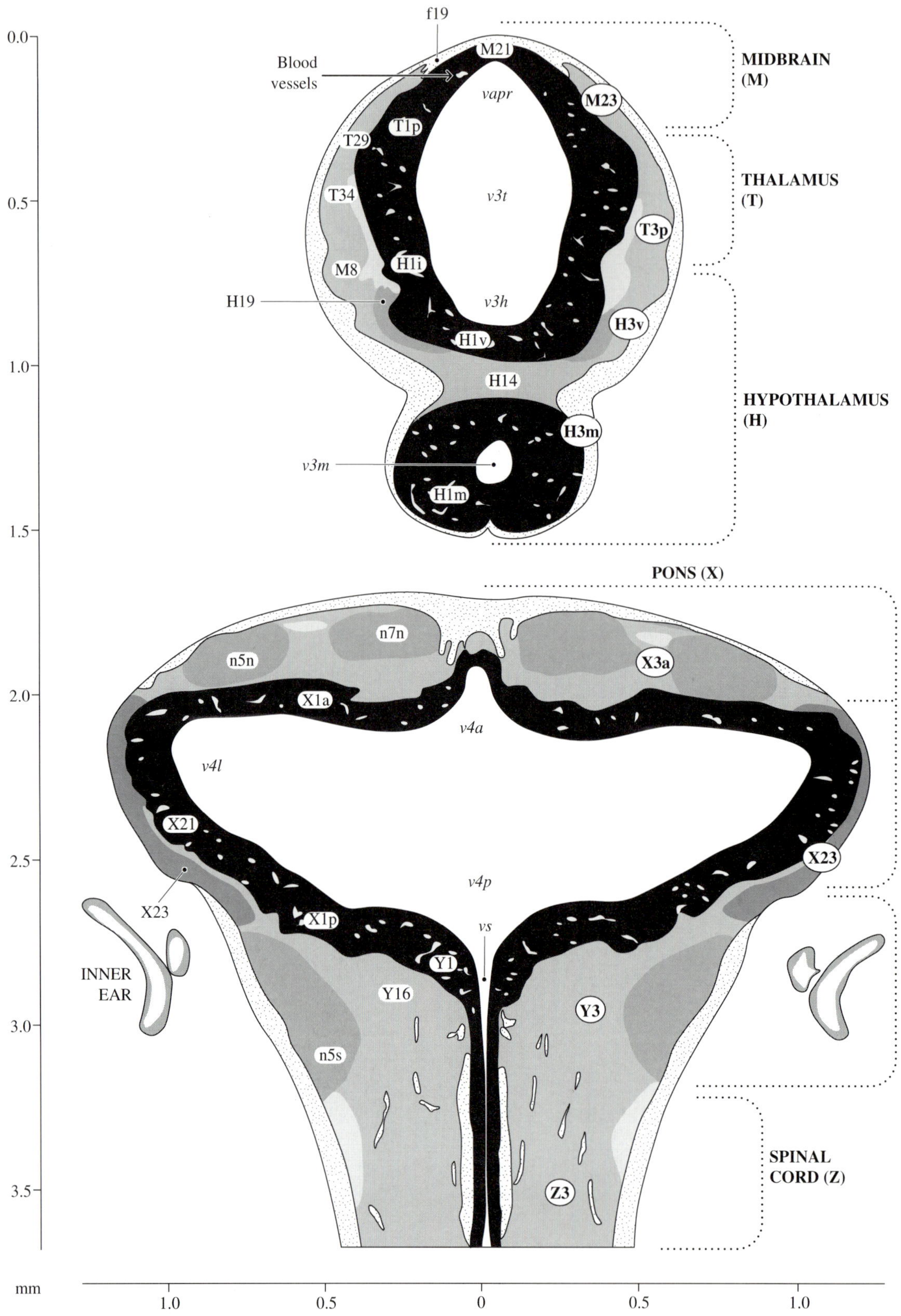

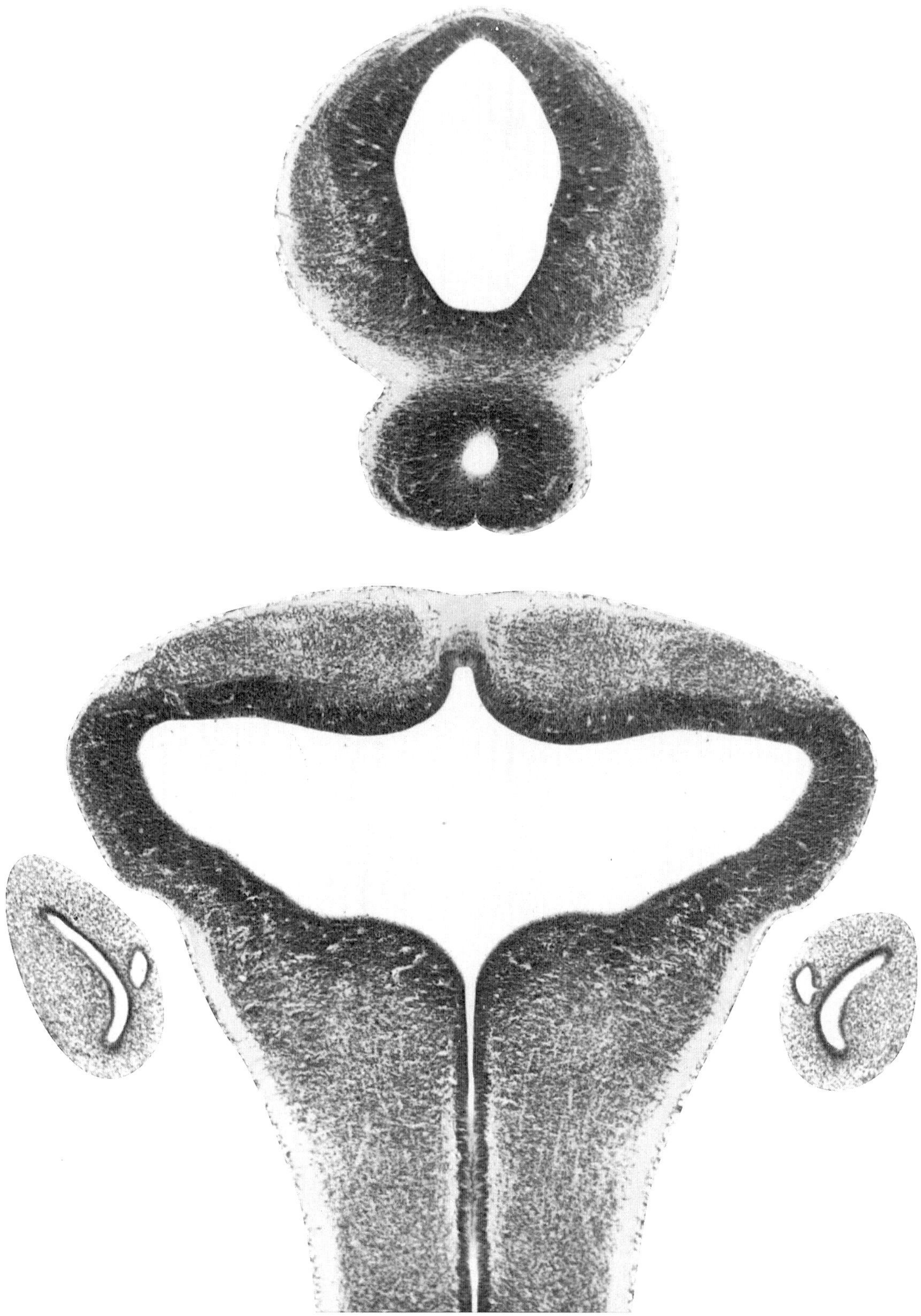

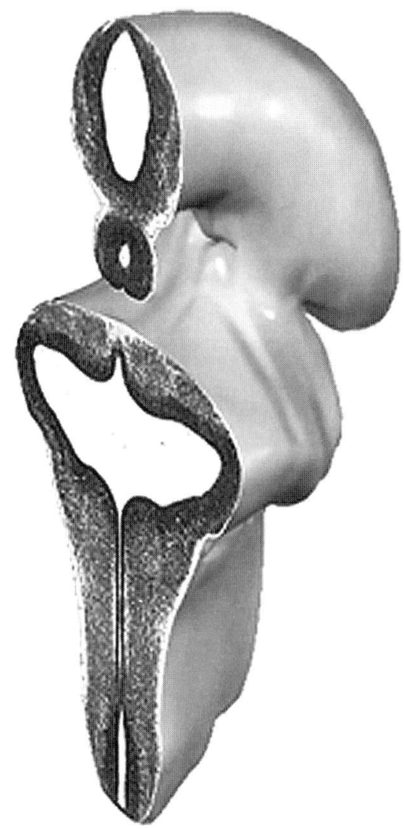

CORONAL FIGURE 14

FIBER TRACTS (f)

f19 posterior commissure

GLANDS AND PARANEURAL STRUCTURES (g)

g4 choroid plexus of fourth ventricle
g10 velum medullare

CEREBELLUM (L)

L1 cerebellar neuroepithelium
L3 cerebellum (or differentiating field)
L7mi lateral deep nucleus, migration

MIDBRAIN (M)

M9 midbrain reticular formation
M11 tegmental neuroepithelium
M13 tegmentum (or differentiating field)
M14 substantia nigra
M15 interpeduncular nucleus
M16 dorsal raphe
M21 pretectal neuroepithelium
M23 pretectum (or differentiating field)

VENTRICLES (v)

va aqueduct
vapr aqueduct, pretectal
vat aqueduct, tegmental
vs spinal canal
v4a fourth ventricle, anterior
v4i fourth ventricle, isthmal
v4l fourth ventricle, lateral recess
v4p fourth ventricle, posterior

PONTINE AREA (X)

X1a anterior pontine neuroepithelium
X3a anterior pons (or differentiating field)
X7 pontine reticular formation
X31 isthmal neuroepithelium
X33 isthmus (or differentiating field)

MEDULLA (Y)

Y1 medullary neuroepithelium
Y3 medulla (or differentiating field)

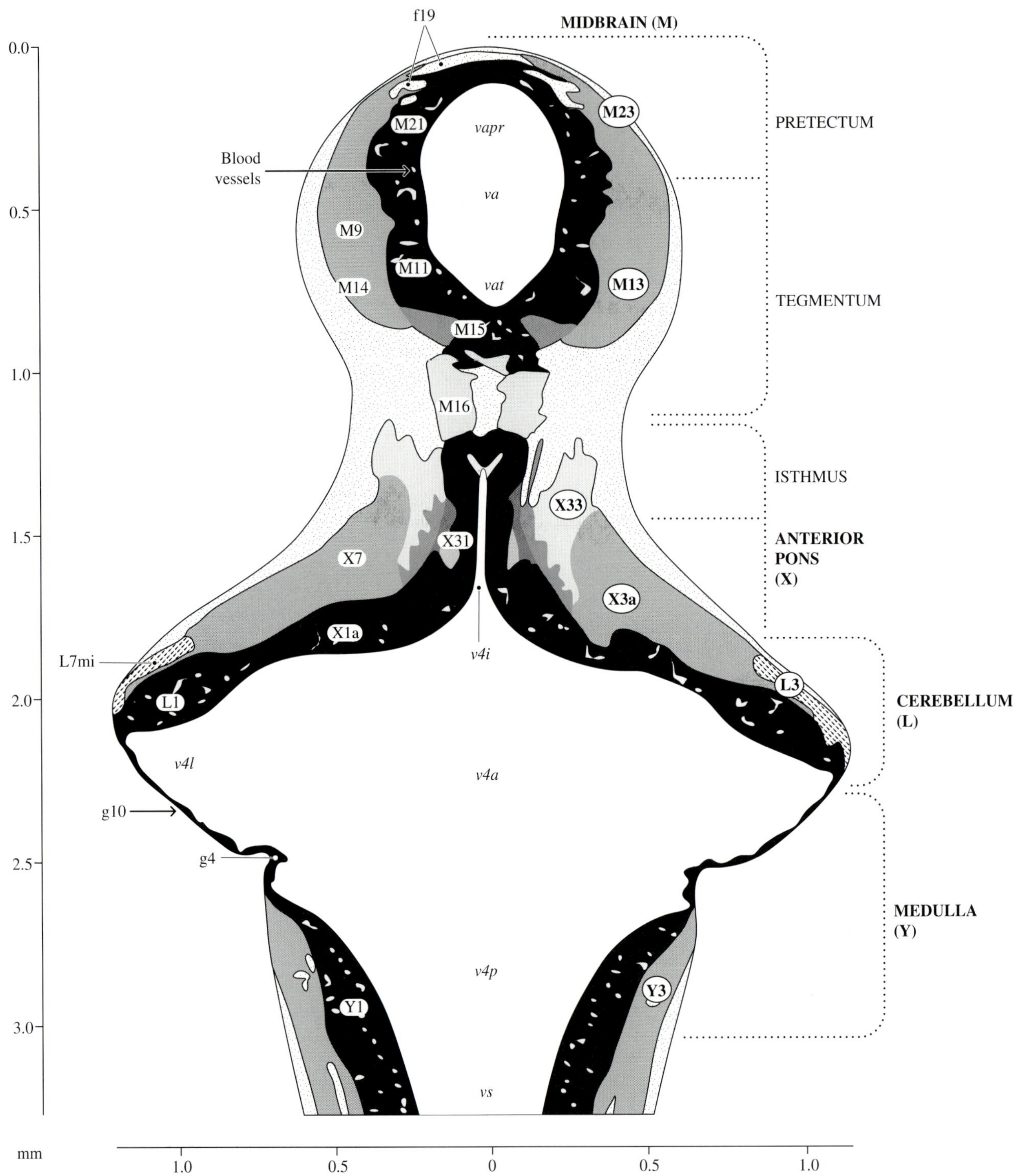

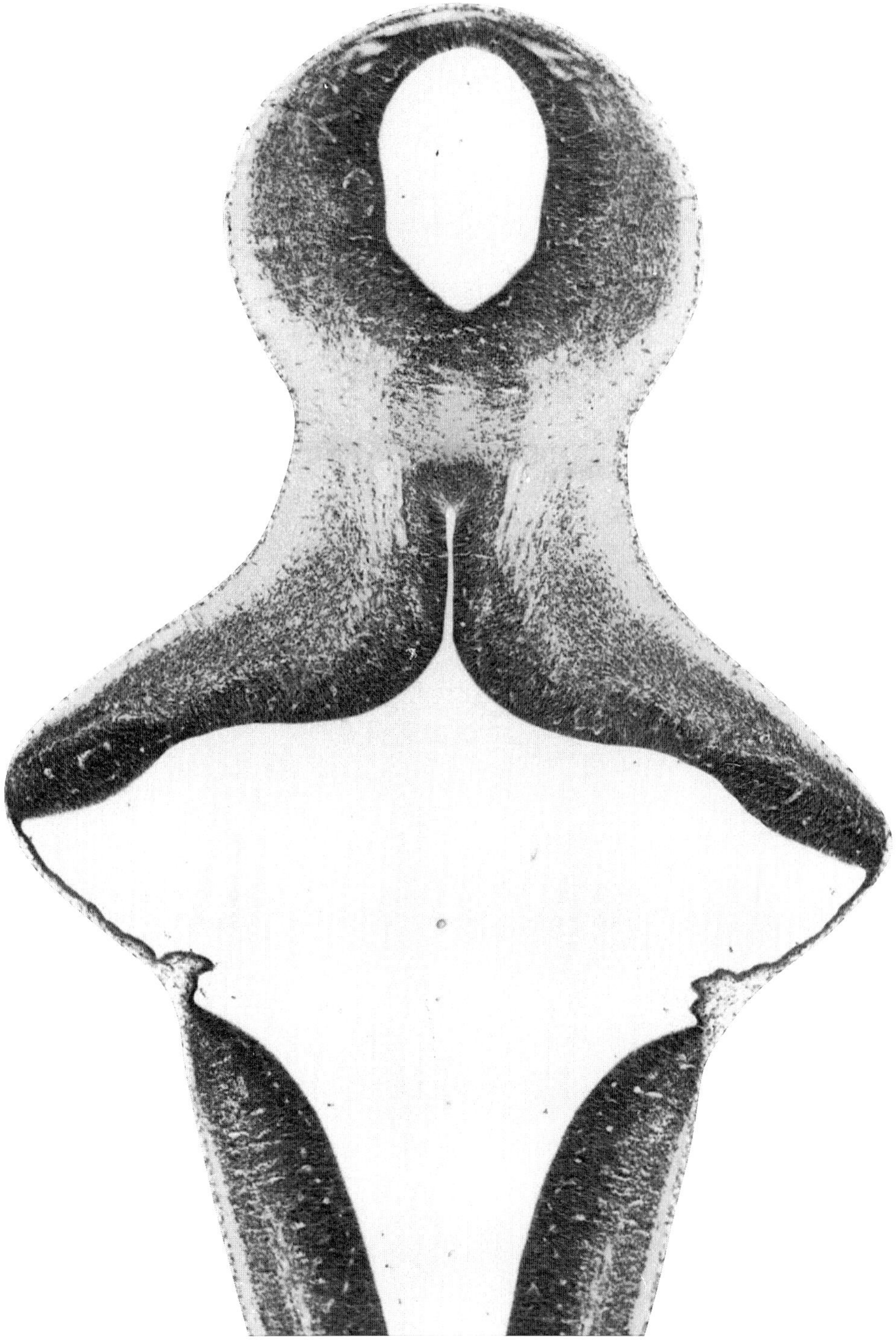

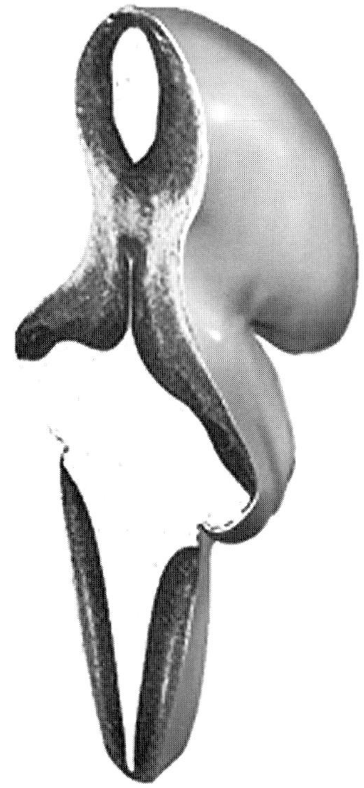

CORONAL FIGURE 15

GLANDS AND PARANEURAL STRUCTURES (g)

g4	choroid plexus of fourth ventricle

CEREBELLUM (L)

L1h	cerebellar hemisphere neuroepithelium
L1v	cerebellar vermis neuroepithelium
L3	cerebellum (or differentiating field)
L3h	cerebellar hemisphere (or differentiating field)

MIDBRAIN (M)

M1	superior tectal neuroepithelium
M3	superior colliculus (or differentiating field)
M9	midbrain reticular formation
M11	tegmental neuroepithelium
M13	tegmentum (or differentiating field)

VENTRICLES (v)

vas	aqueduct, superior tectal
vat	aqueduct, tegmental
vs	spinal canal
v4a	fourth ventricle, anterior
v4i	fourth ventricle, isthmal
v4ir	fourth ventricle, isthmal recess
v4l	fourth ventricle, lateral recess
v4p	fourth ventricle, posterior

PONTINE AREA (X)

X1a	anterior pontine neuroepithelium
X3a	anterior pons (or differentiating field)
X31	isthmal neuroepithelium
X33	isthmus (or differentiating field)
X34	locus coeruleus
X36d	nucleus of lateral lemniscus, dorsal
X36v	nucleus of lateral lemniscus, ventral

MEDULLA (Y)

Y1	medullary neuroepithelium
Y3	medulla (or differentiating field)

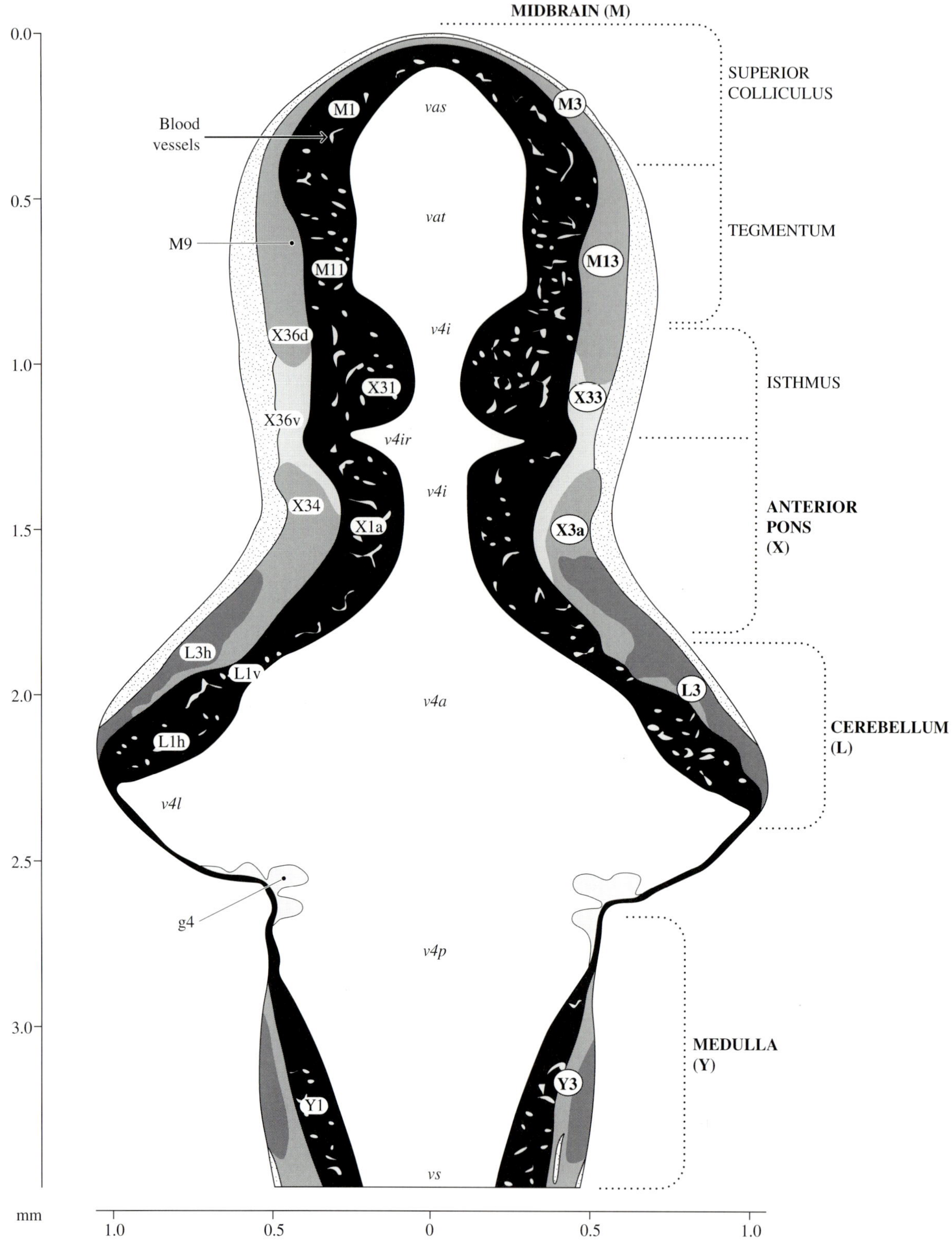

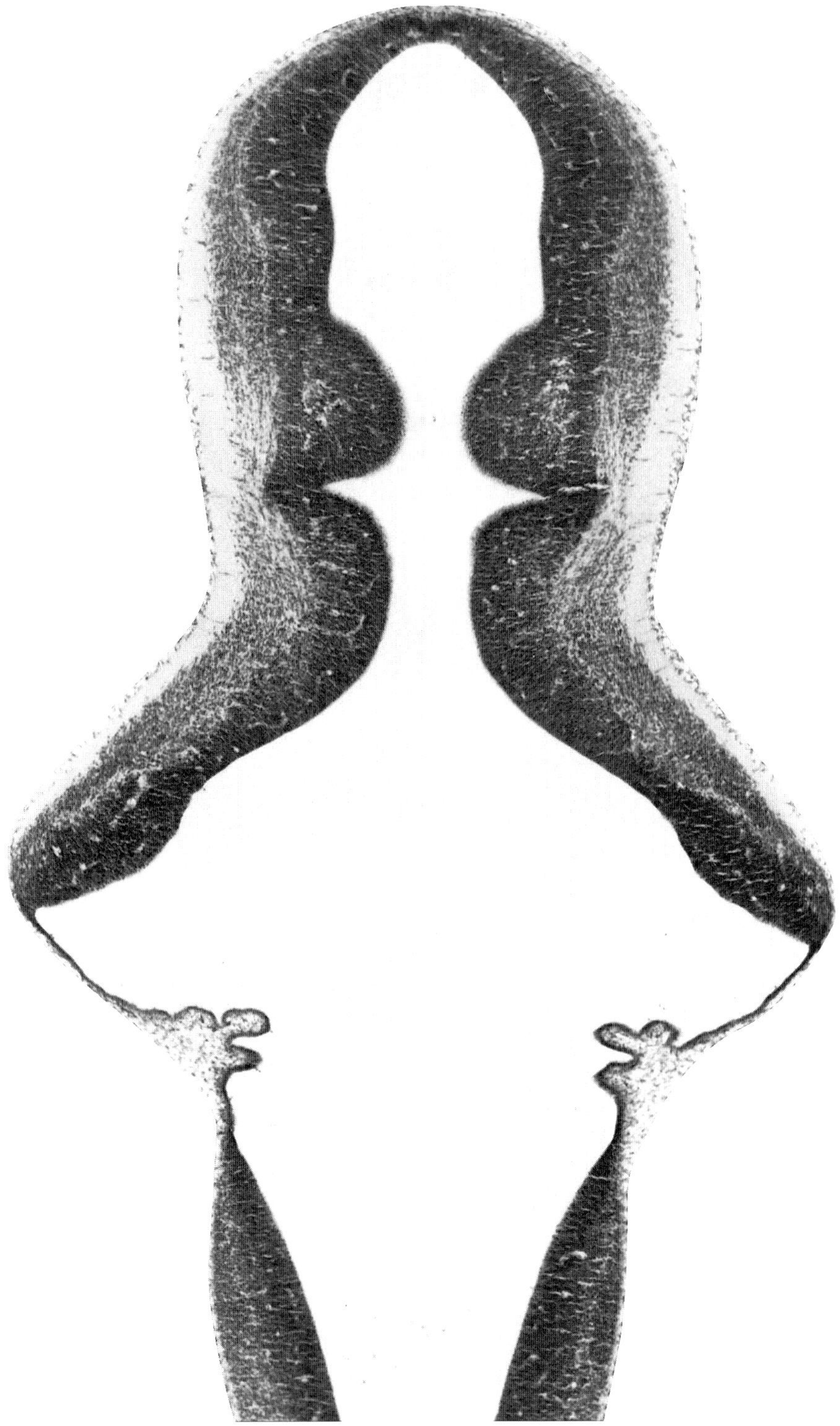

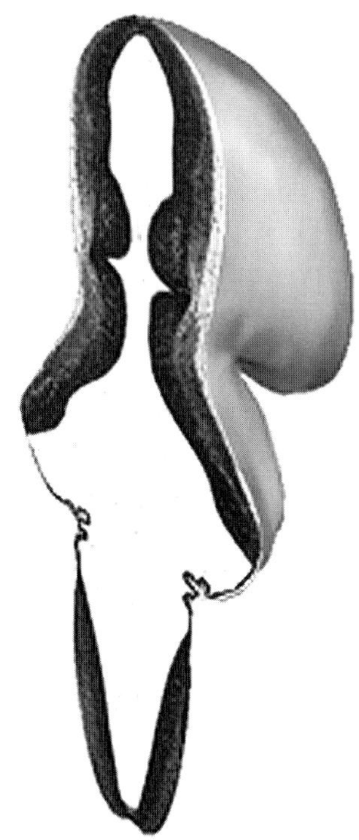

CORONAL FIGURE 16

GLANDS AND PARANEURAL STRUCTURES (g)

g4 choroid plexus of fourth ventricle
g10 velum medullare

CEREBELLUM (L)

L1h cerebellar hemisphere neuroepithelium
L1v cerebellar vermis neuroepithelium
L3 cerebellum (or differentiating field)
L3h cerebellar hemisphere (or differentiating field)

MIDBRAIN (M)

M1 superior tectal neuroepithelium,
M3 superior colliculus (or differentiating field)
M9 midbrain reticular formation
M11 tegmental neuroepithelium
M13 tegmentum (or differentiating field)
M31 inferior tectal neuroepithelium
M33 inferior colliculus (or differentiating field)

VENTRICLES (v)

vai aqueduct, inferior tectal
vas aqueduct, superior tectal
vat aqueduct, tegmental
v4a fourth ventricle, anterior
v4i fourth ventricle, isthmal
v4l fourth ventricle, lateral recess
v4p fourth ventricle, posterior

PONTINE AREA (X)

X31 isthmal neuroepithelium
X33 isthmus (or differentiating field)
X34 locus coeruleus

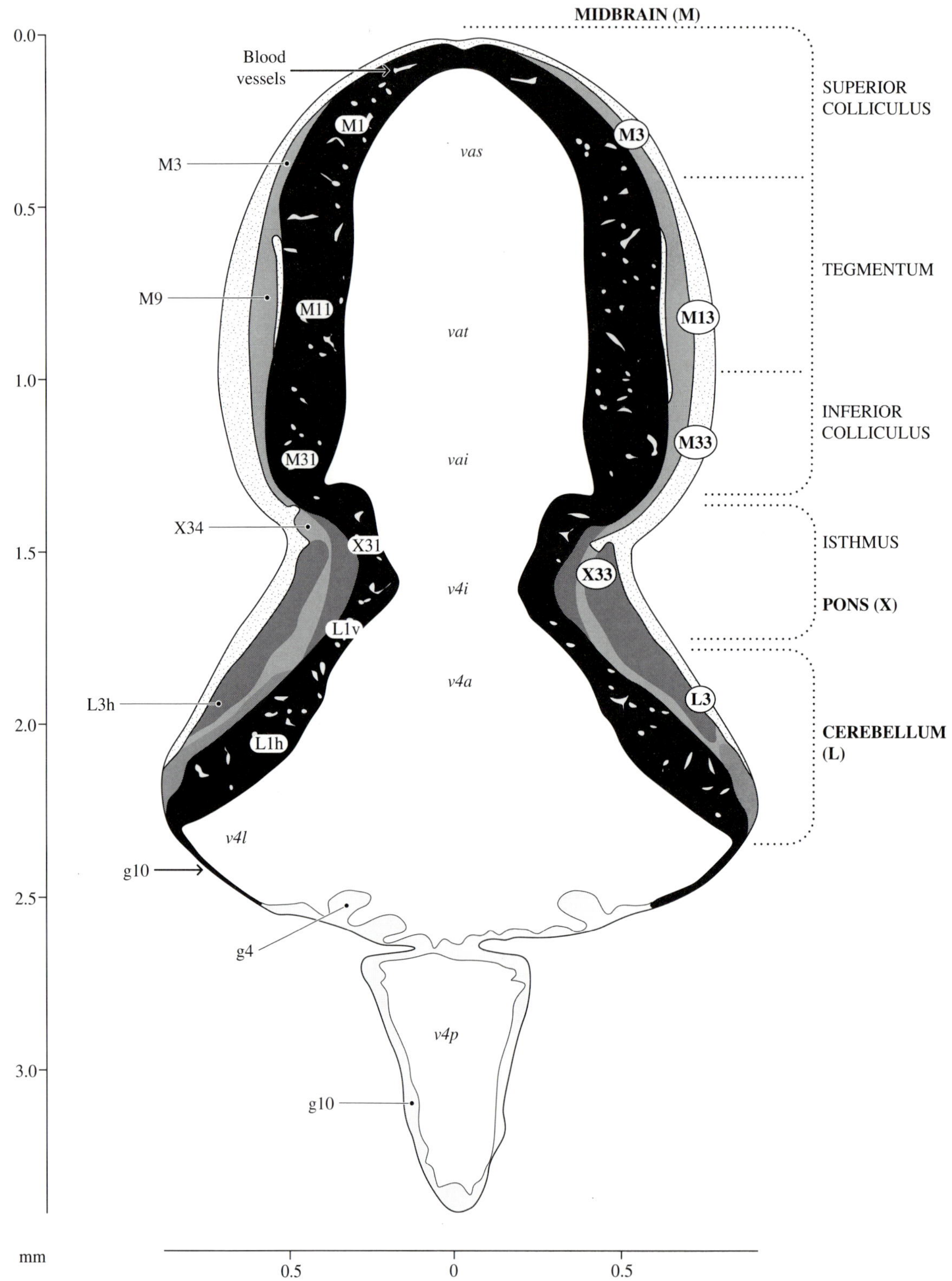

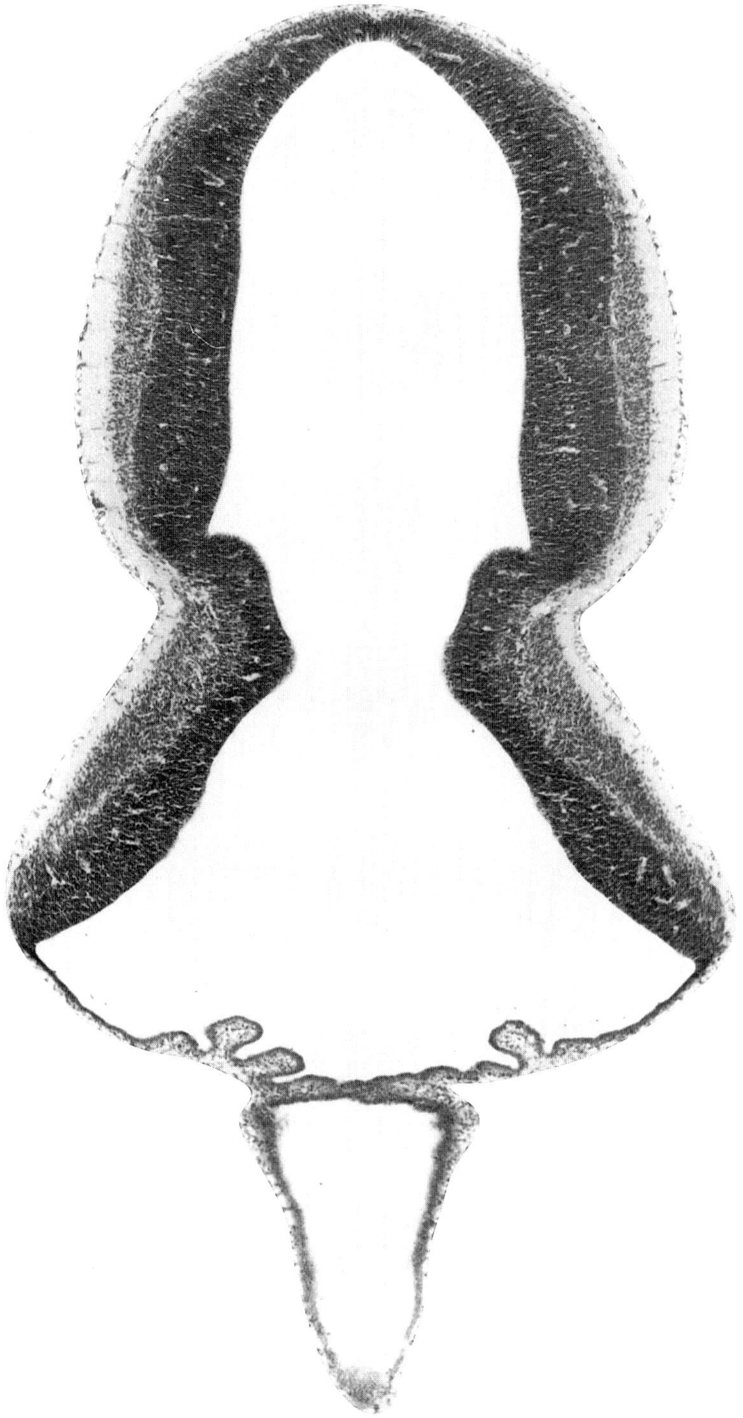

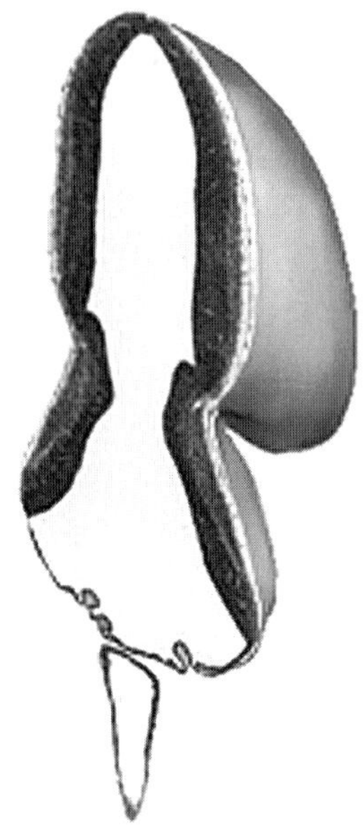

CORONAL FIGURE 17

GLANDS AND PARANEURAL STRUCTURES (g)

g4 choroid plexus of fourth ventricle

CEREBELLUM (L)

L1h cerebellar hemisphere neuroepithelium
L1v cerebellar vermis neuroepithelium
L3 cerebellum (or differentiating field)
L3h cerebellar hemisphere (or differentiating field)
L3v cerebellar vermis (or differentiating field)

MIDBRAIN (M)

M1 superior tectal neuroepithelium
M3 superior colliculus (or differentiating field)
M11 tegmental neuroepithelium
M13 tegmentum (or differentiating field)
M31 inferior tectal neuroepithelium
M33 inferior colliculus (or differentiating field)

VENTRICLES (v)

vai aqueduct, inferior tectal
vas aqueduct, superior tectal
vat aqueduct, tegmental
v4a fourth ventricle, anterior
v4i fourth ventricle, isthmal

PONTINE AREA (X)

X31 isthmal neuroepithelium
X33 isthmus (or differentiating field)

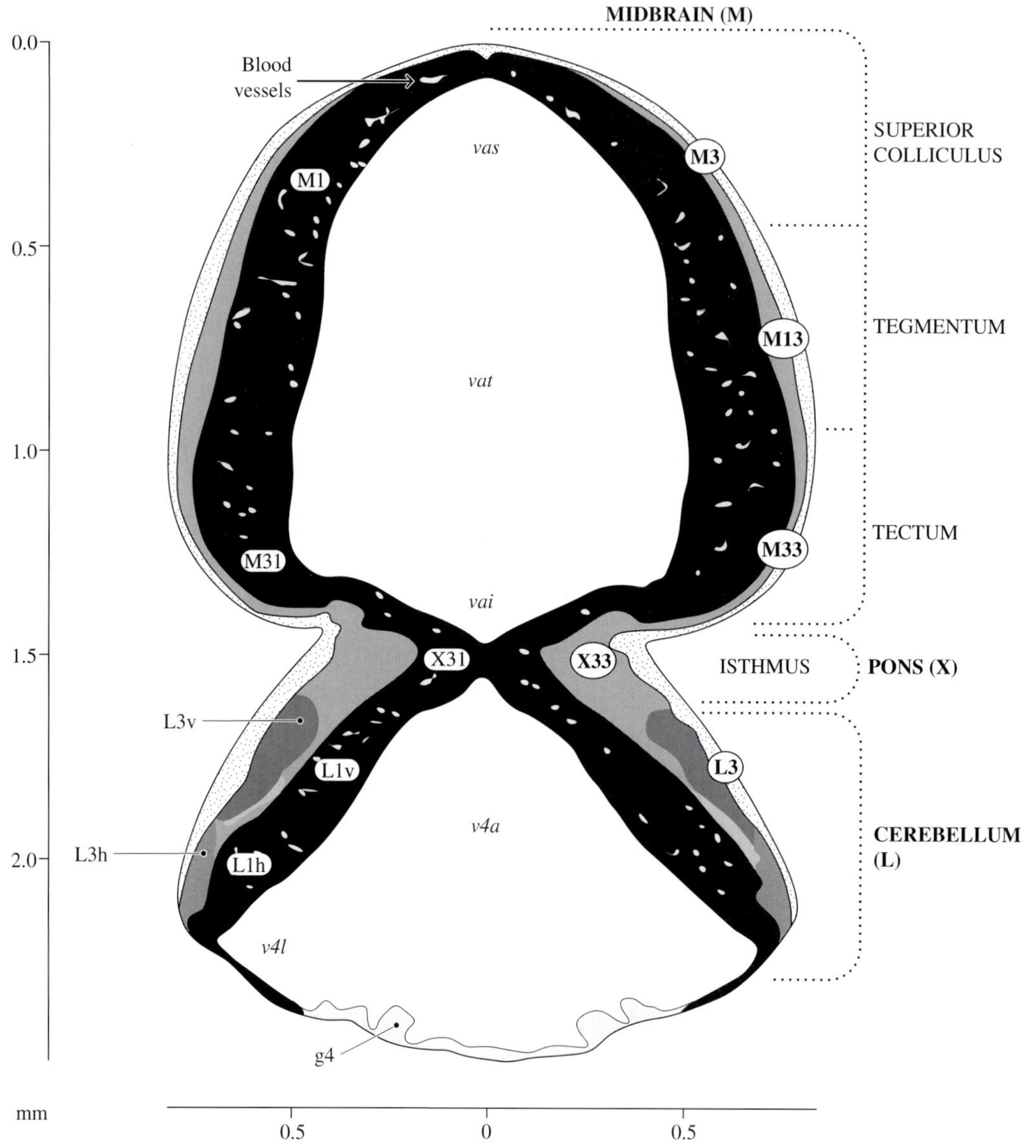

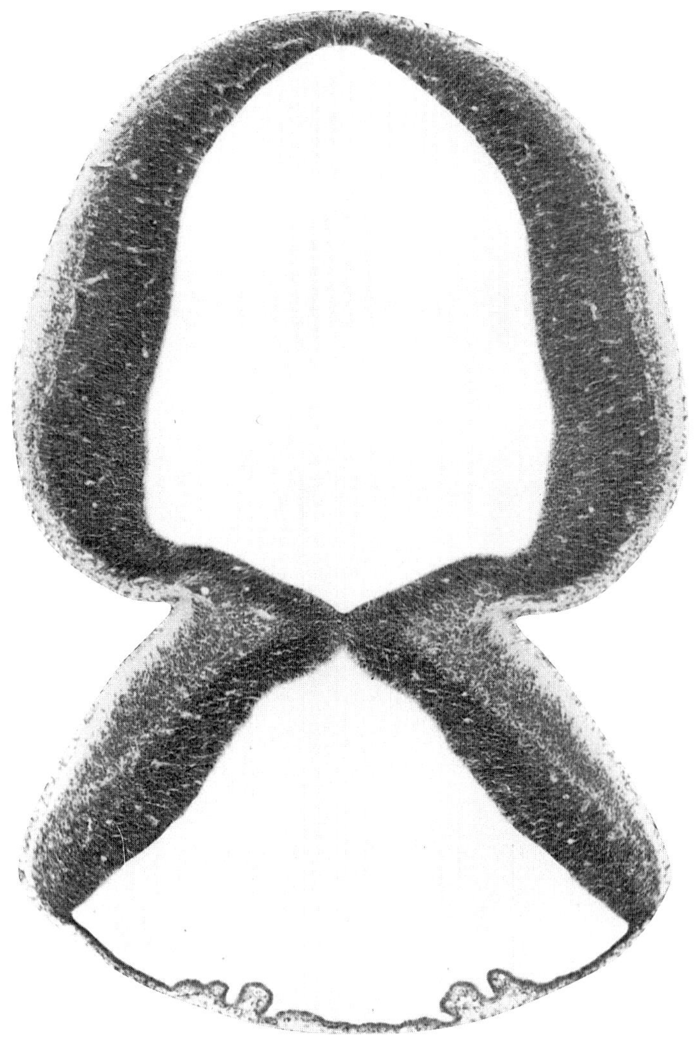

CORONAL FIGURE 18

CEREBELLUM (L)

L1v cerebellar vermis neuroepithelium
L3 cerebellum (or differentiating field)

MIDBRAIN (M)

M1 superior tectal neuroepithelium
M31 inferior tectal neuroepithelium

VENTRICLES (v)

vai aqueduct, inferior tectal
vas aqueduct, superior tectal
v4a fourth ventricle, anterior

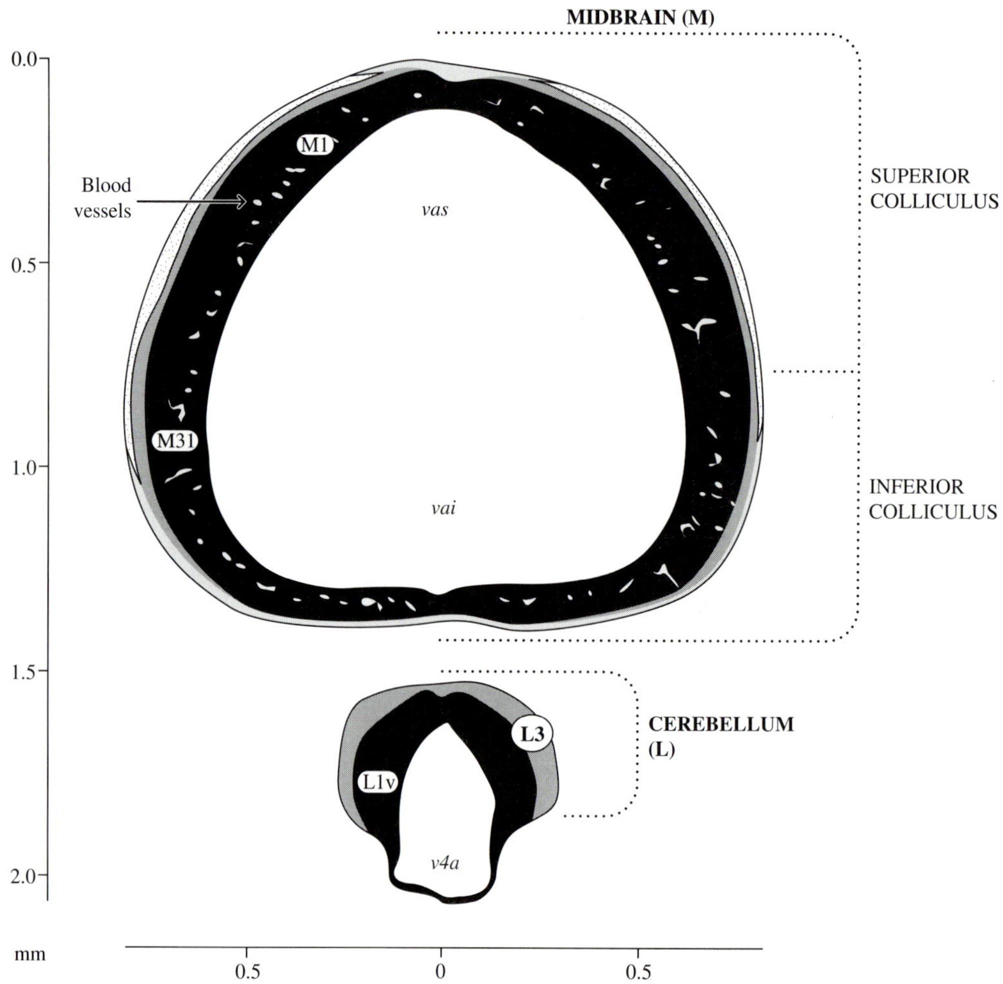

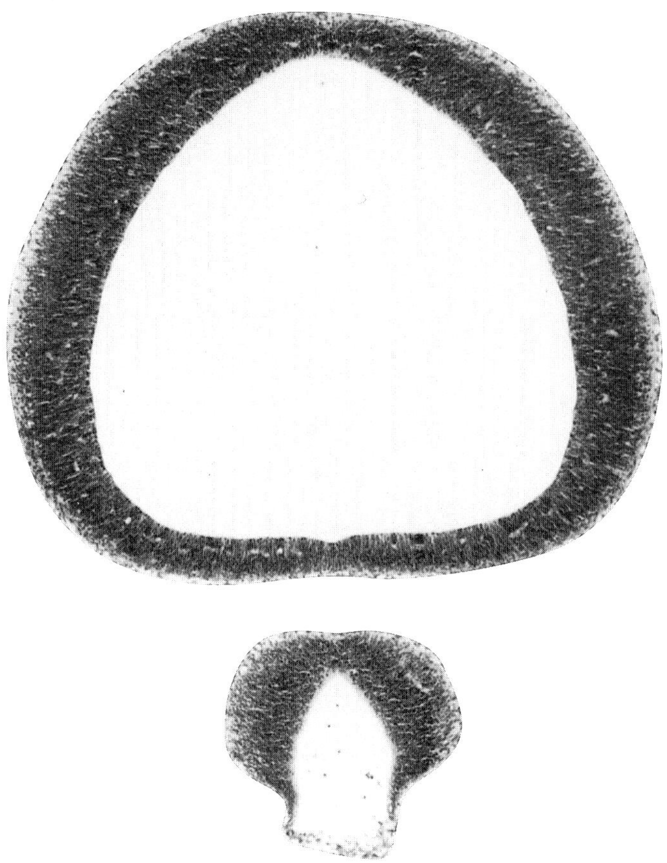

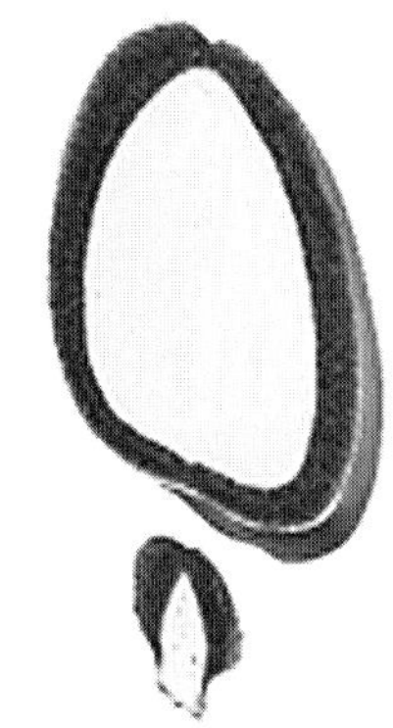

THE E16 BRAIN

LEFT, ANTERIOR VIEW

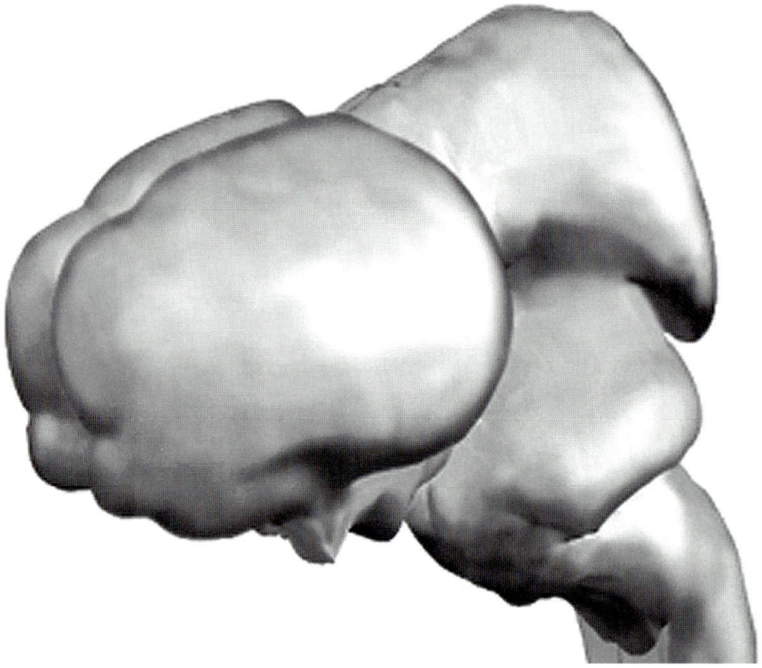

RIGHT, POSTERIOR VIEW

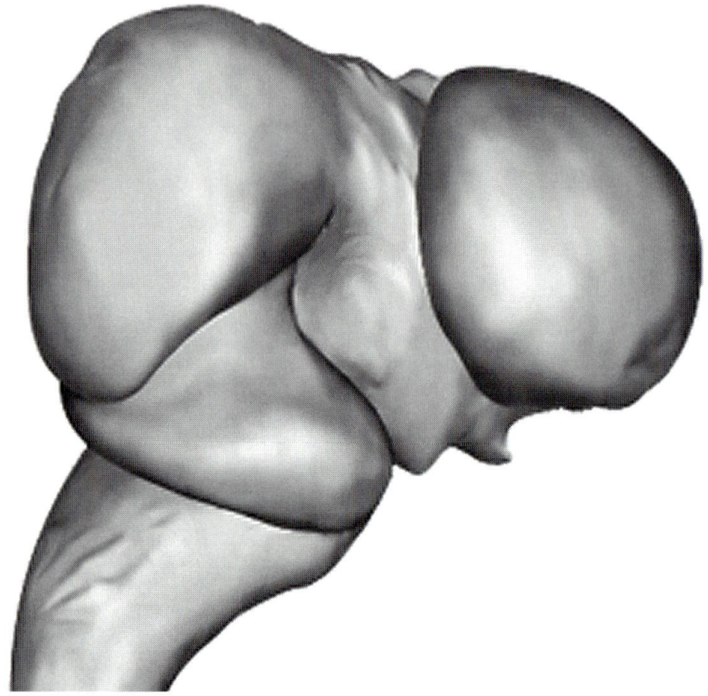

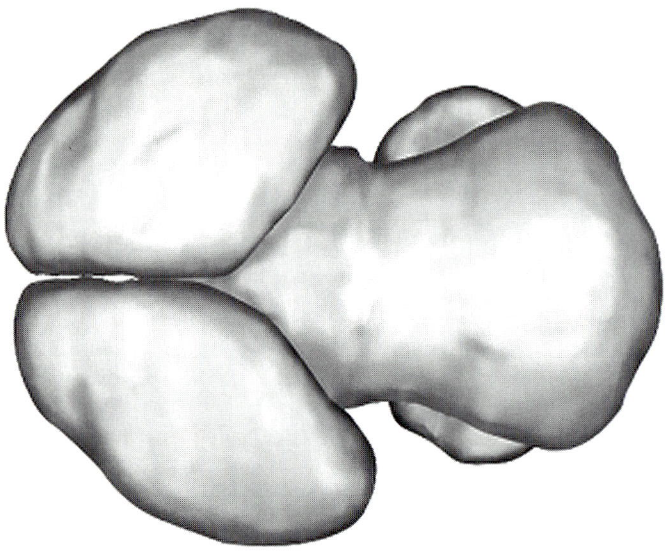

DORSAL VIEW

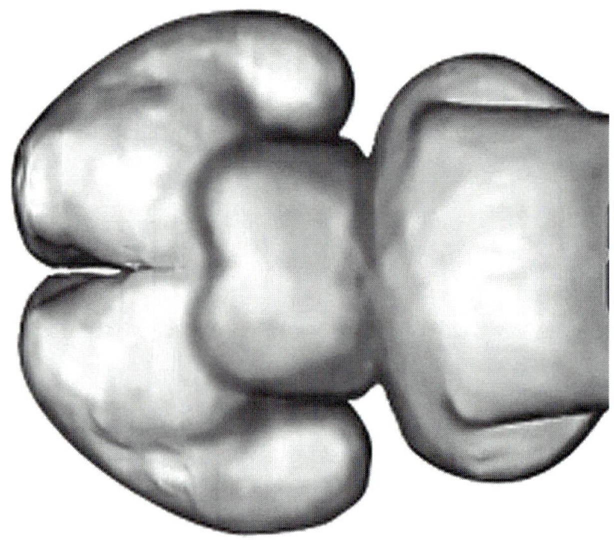

VENTRAL VIEW

SAGITTAL

FIGURES and PLATES 1-8

SAGITTAL FIGURE 1

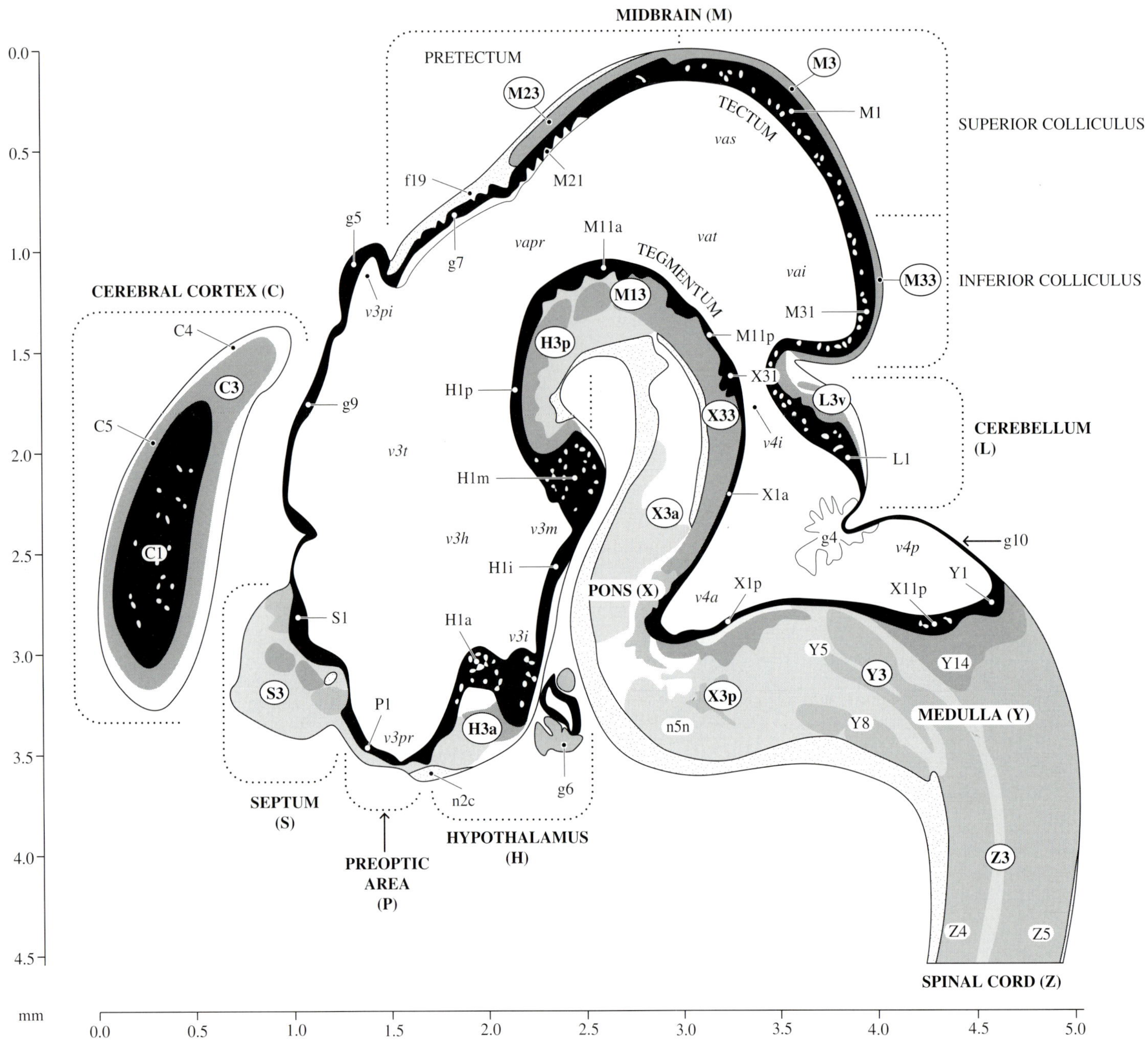

CEREBRAL CORTEX (C)

C1	neocortical neuroepithelium
C3	neocortex (or differentiating field)
C4	molecular layer (or primordial plexiform layer)
C5	intermediate zone

FIBER TRACTS (f)

f19	posterior commissure

GLANDS AND PARANEURAL STRUCTURES (g)

g4	choroid plexus of fourth ventricle
g5	pineal gland (primordium)
g6	pituitary gland (primordium)
g7	subcommissural organ
g9	lamina terminalis
g10	velum medullare

HYPOTHALAMUS (H)

H1a	anterior hypothalamic neuroepithelium
H1i	intermediate hypothalamic neuroepithelium
H1m	mammillary neuroepithelium
H1p	posterior hypothalamic neuroepithelium
H3a	anterior hypothalamus (or differentiating field)
H3p	posterior hypothalamus (or differentiating field)

CEREBELLUM (L)

L1	cerebellar neuroepithelium
L3v	cerebellar vermis (or differentiating field)

MIDBRAIN (M)

M1	superior tectal neuroepithelium
M3	superior colliculus (or differentiating field)
M11a	anterior tegmental neuroepithelium

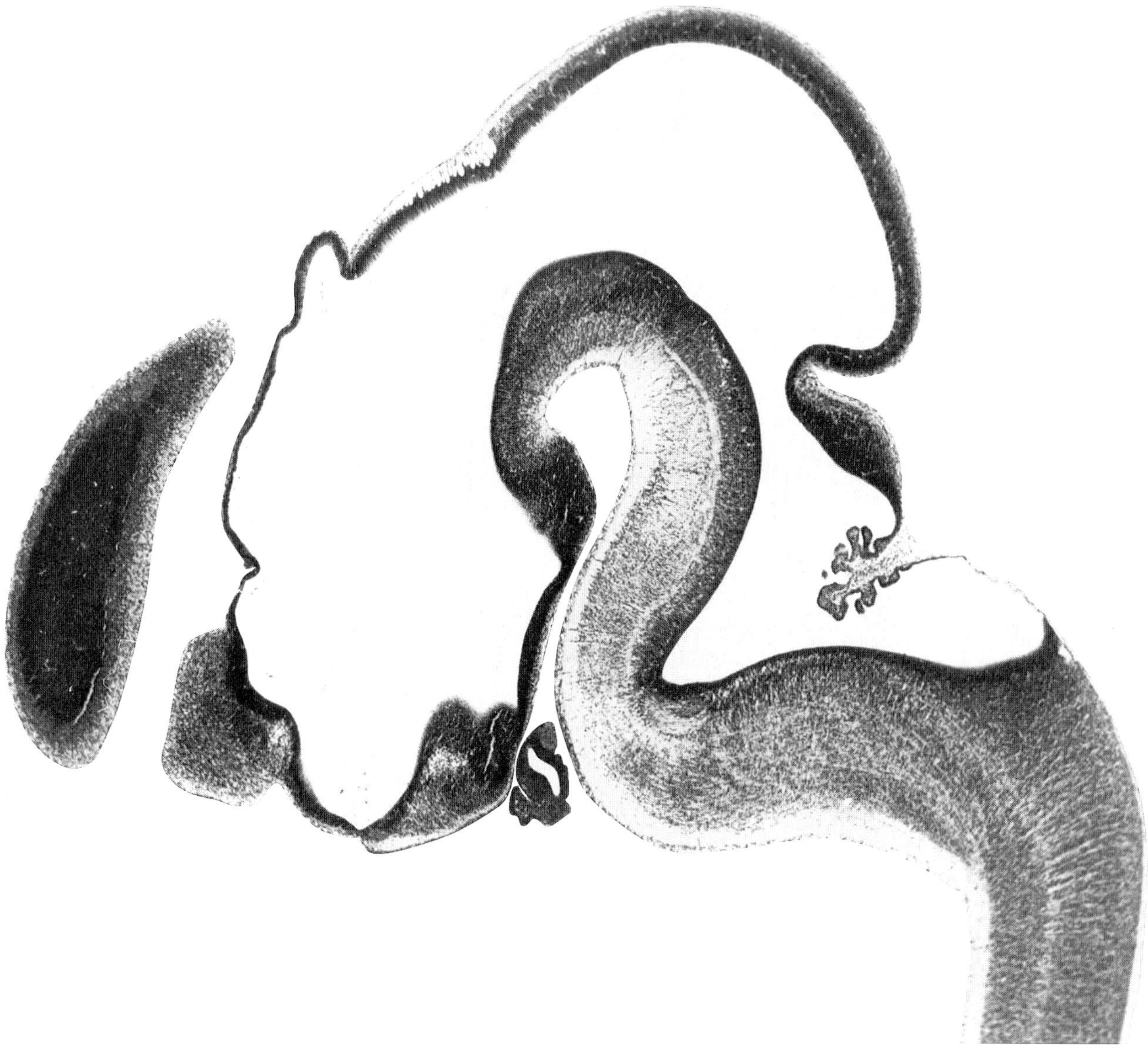

M11p posterior tegmental neuroepithelium
M13 tegmentum (or differentiating field)
M21 pretectal neuroepithelium
M23 pretectum (or differentiating field)
M31 inferior tectal neuroepithelium
M33 inferior colliculus (or differentiating field)

NERVES AND NUCLEI, CRANIAL (n)

n2c optic chiasma
n5n trigeminal motor nucleus

PREOPTIC AREA AND OPTIC EPITHELIA (P)

P1 preoptic neuroepithelium

SEPTAL AREA (S)

S1 septal neuroepithelium
S3 septum (or differentiating field)

VENTRICLES (v)

vai aqueduct, inferior tectal
vapr aqueduct, pretectal
vas aqueduct, superior tectal
vat aqueduct, tegmental
v3h third ventricle, hypothalamic
v3i third ventricle, infundibular recess
v3m third ventricle, mammillary recess
v3pi third ventricle, pineal recess
v3pr third ventricle, preoptic recess
v3t third ventricle, thalamic
v4a fourth ventricle, anterior
v4i fourth ventricle, isthmal
v4p fourth ventricle, posterior

PONTINE AREA (P)

X1a anterior pontine neuroepithelium
X1p posterior pontine neuroepithelium
X3a anterior pons (or differentiating field)
X3p posterior pons (or differentiating field)
X11p posterior precerebellar neuroepithelium
X31 isthmal neuroepithelium
X33 isthmus (or differentiating field)

MEDULLA (Y)

Y1 medullary neuroepithelium
Y3 medulla (or differentiating field)
Y5 medullary raphe
Y8 inferior olive
Y14 gracile nucleus

SPINAL CORD (Z)

Z3 spinal cord (or differentiating field)
Z4 ventral horn
Z5 dorsal horn

SAGITTAL FIGURE 2

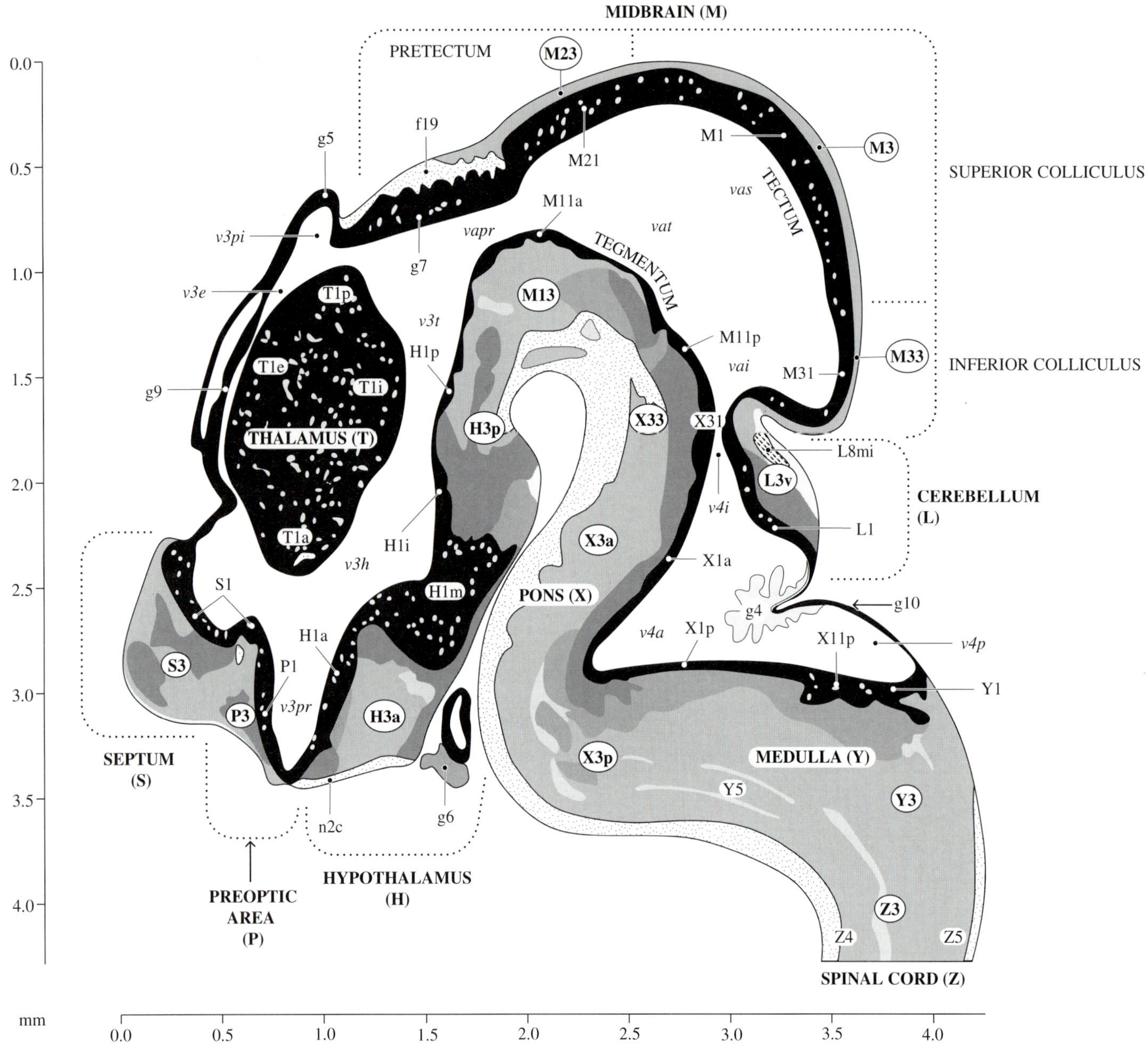

FIBER TRACTS (f)

f19 posterior commissure

GLANDS AND PARANEURAL STRUCTURES (g)

g4 choroid plexus of fourth ventricle
g5 pineal gland (primordium)
g6 pituitary gland (primordium)
g7 subcommissural organ
g9 lamina terminalis
g10 velum medullare

HYPOTHALAMUS (H)

H1a anterior hypothalamic neuroepithelium
H1i intermediate hypothalamic neuroepithelium
H1m mammillary neuroepithelium
H1p posterior hypothalamic neuroepithelium
H3a anterior hypothalamus (or differentiating field)
H3p posterior hypothalamus (or differentiating field)

CEREBELLUM (L)

L1 cerebellar neuroepithelium
L3v cerebellar vermis (or differentiating field)
L8mi medial deep nucleus, migration

MIDBRAIN (M)

M1 superior tectal neuroepithelium
M3 superior colliculus (or differentiating field)
M11a anterior tegmental neuroepithelium
M11p posterior tegmental neuroepithelium
M13 tegmentum (or differentiating field)
M21 pretectal neuroepithelium
M23 pretectum (or differentiating field)
M31 inferior tectal neuroepithelium
M33 inferior colliculus (or differentiating field)

NERVES AND NUCLEI, CRANIAL (n)

n2c optic chiasma

PREOPTIC AREA AND OPTIC EPITHELIA (P)

P1 preoptic neuroepithelium
P3 preoptic area (or differentiating field)

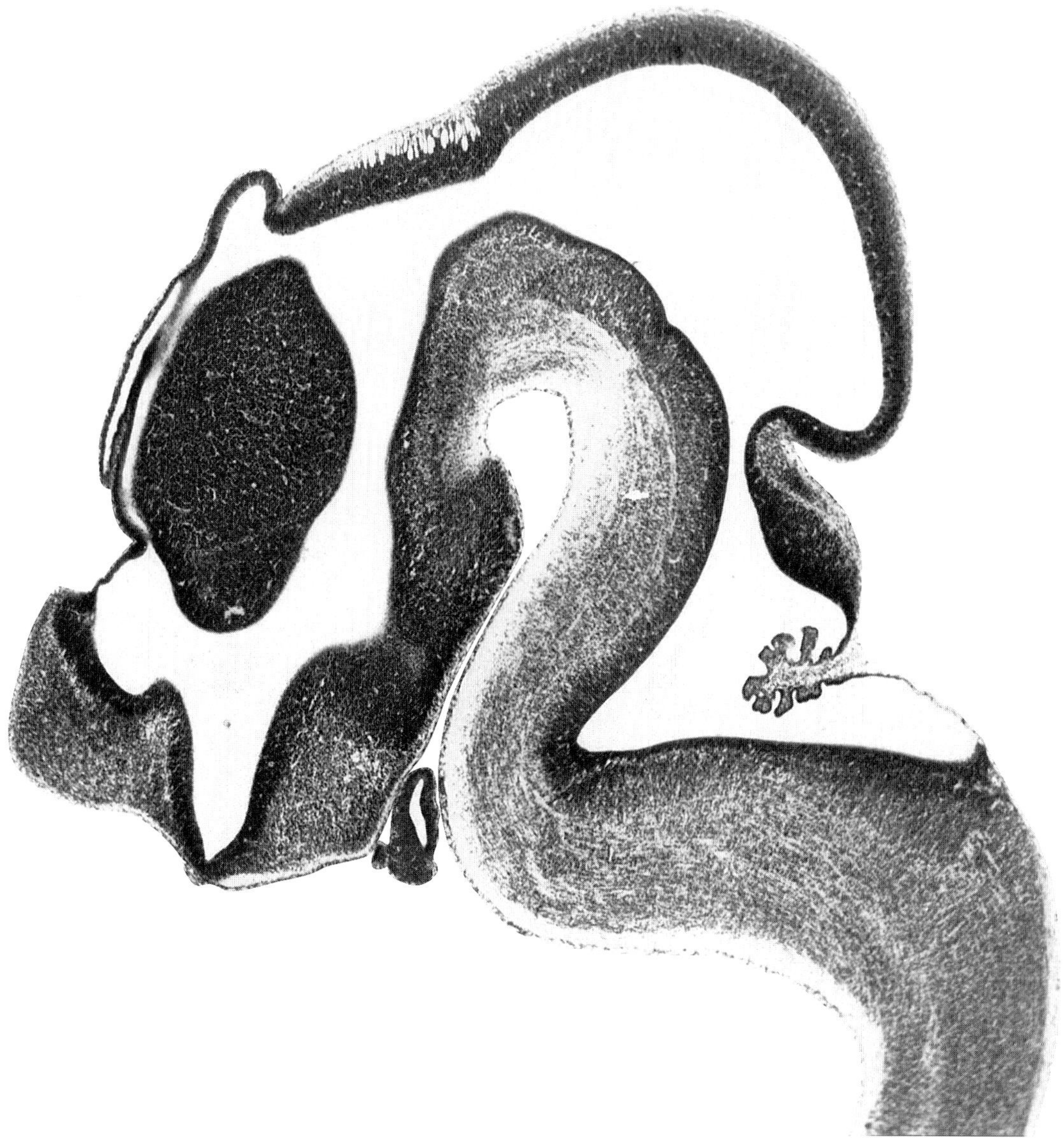

SEPTAL AREA (S)

S1 septal neuroepithelium
S3 septum (or differentiating field)

THALAMUS (T)

T1a anterior thalamic neuroepithelium
T1e epithalamic neuroepithelium
T1i intermediate thalamic neuroepithelium
T1p posterior thalamic neuroepithelium

VENTRICLES (v)

vai aqueduct, inferior tectal
vapr aqueduct, pretectal
vas aqueduct, superior tectal
vat aqueduct, tegmental
v3e third ventricle, epithalamic
v3h third ventricle, hypothalamic
v3pi third ventricle, pineal recess
v3pr third ventricle, preoptic recess
v3t third ventricle, thalamic
v4a fourth ventricle, anterior
v4i fourth ventricle, isthmal
v4p fourth ventricle, posterior

PONTINE AREA (X)

X1a anterior pontine neuroepithelium
X1p posterior pontine neuroepithelium
X3a anterior pons (or differentiating field)
X3p posterior pons (or differentiating field)
X11p posterior precerebellar neuroepithelium
X31 isthmal neuroepithelium
X33 isthmus (or differentiating field)

MEDULLA (Y)

Y1 medullary neuroepithelium
Y3 medulla (or differentiating field)
Y5 medullary raphe

SPINAL CORD (Z)

Z3 spinal cord (or differentiating field)
Z4 ventral horn
Z5 dorsal horn

SAGITTAL FIGURE 3

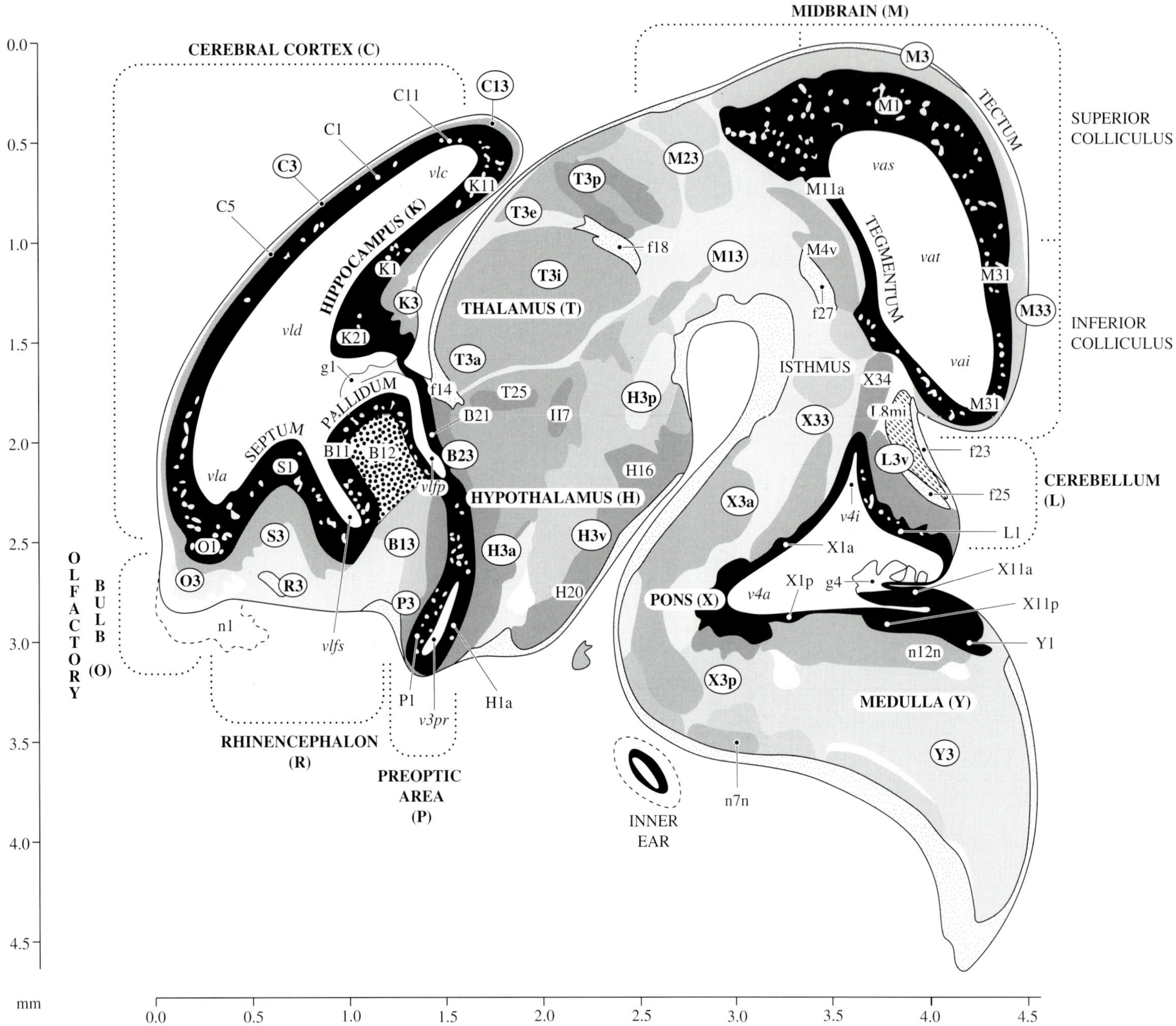

BASAL GANGLIA (B)

B11 pallidal neuroepithelium
B12 pallidal subventricular zone
B13 pallidum (or differentiating field)
B21 strionuclear neuroepithelium
B23 bed nucleus of stria terminalis (or differentiating field)

CEREBRAL CORTEX (C)

C1 neocortical neuroepithelium
C3 neocortex (or differentiating field)
C5 intermediate zone
C11 cingulate cortex neuroepithelium
C13 cingulate cortex (or differentiating field)

FIBER TRACTS (f)

f14 stria terminalis
f18 habenulopeduncular tract
f23 superior cerebellar peduncle
f25 inferior cerebellar peduncle
f27 medial longitudinal fasciculus

GLANDS AND PARANEURAL STRUCTURES (g)

g1 choroid plexus of lateral ventricle
g4 choroid plexus of fourth ventricle

HYPOTHALAMUS (H)

H1a anterior hypothalamic neuroepithelium
H3a anterior hypothalamus (or differentiating field)
H3p posterior hypothalamus (or differentiating field)
H3v ventral hypothalamus (or differentiating field)
H7 zona incerta
H16 medial mammillary nucleus
H20 arcuate nucleus

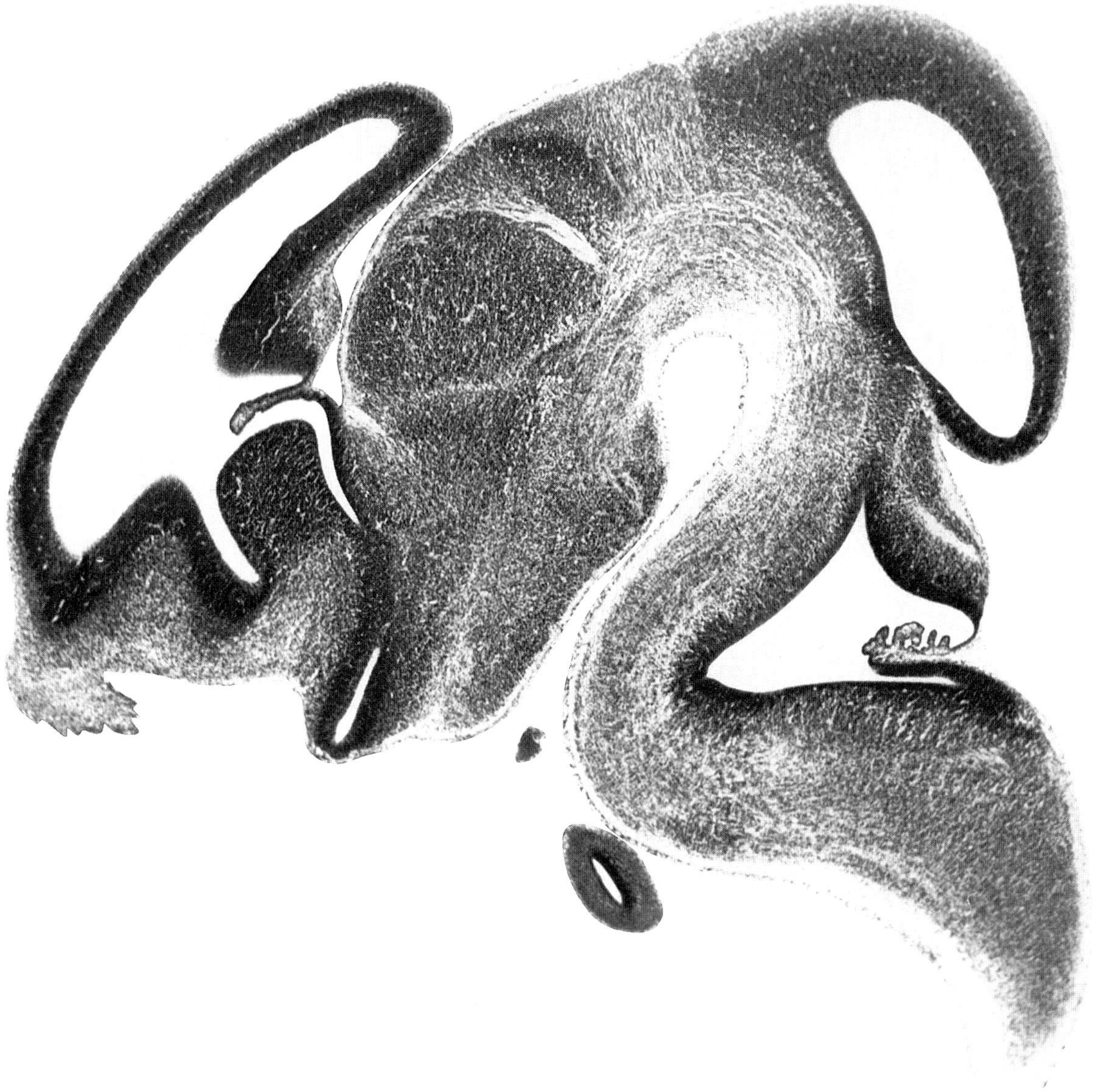

RHINENCEPHALON (R)

R3 rhinencephalon (or differentiating field)

SEPTAL AREA (S)

S1 septal neuroepithelium
S3 septum (or differentiating field)

THALAMUS (T)

T3a anterior thalamus (or differentiating field)
T3e epithalamus (or differentiating field)
T3i intermediate thalamus (or differentiating field)
T3p posterior thalamus (or differentiating field)
T25 reticular nucleus

VENTRICLES (v)

vai aqueduct, inferior tectal
vas aqueduct, superior tectal
vat aqueduct, tegmental
vla lateral ventricle, anterior horn
vlc lateral ventricle, central
vld lateral ventricle, dorsal
vlfp lateral ventricle, pallidal fork
vlfs lateral ventricle, septal fork
v3pr third ventricle, preoptic recess
v4a fourth ventricle, anterior
v4i fourth ventricle, isthmal

HIPPOCAMPAL FORMATION (K)

K1 hippocampal (Ammon's horn) neuroepithelium
K3 hippocampus (or differentiating field)
K11 subicular neuroepithelium
K21 dentate neuroepithelium

CEREBELLUM (L)

L1 cerebellar neuroepithelium
L3v cerebellar vermis (or differentiating field)
L8mi medial deep nucleus, migration

MIDBRAIN (M)

M1 superior tectal neuroepithelium
M3 superior colliculus (or differentiating field)
M4v ventral periaqueductal gray
M11a anterior tegmental neuroepithelium
M13 tegmentum (or differentiating field)
M23 pretectum (or differentiating field)
M31 inferior tectal neuroepithelium
M33 inferior colliculus (or differentiating field)

NERVES AND NUCLEI, CRANIAL (n)

n1 olfactory nerve
n7n facial motor nucleus
n12n hypoglossal nucleus

OLFACTORY BULB (O)

O1 olfactory bulb neuroepithelium
O3 olfactory bulb (or differentiating field)

PREOPTIC AREA AND OPTIC EPITHELIA (P)

P1 preoptic neuroepithelium
P3 preoptic area (or differentiating field)

PONTINE AREA (X)

X1a anterior pontine neuroepithelium
X1p posterior pontine neuroepithelium
X3a anterior pons (or differentiating field)
X3p posterior pons (or differentiating field)
X11a anterior precerebellar neuroepithelium
X11p posterior precerebellar neuroepithelium
X33 isthmus (or differentiating field)
X34 locus coeruleus

MEDULLA (Y)

Y1 medullary neuroepithelium
Y3 medulla (or differentiating field)

SAGITTAL FIGURE 4

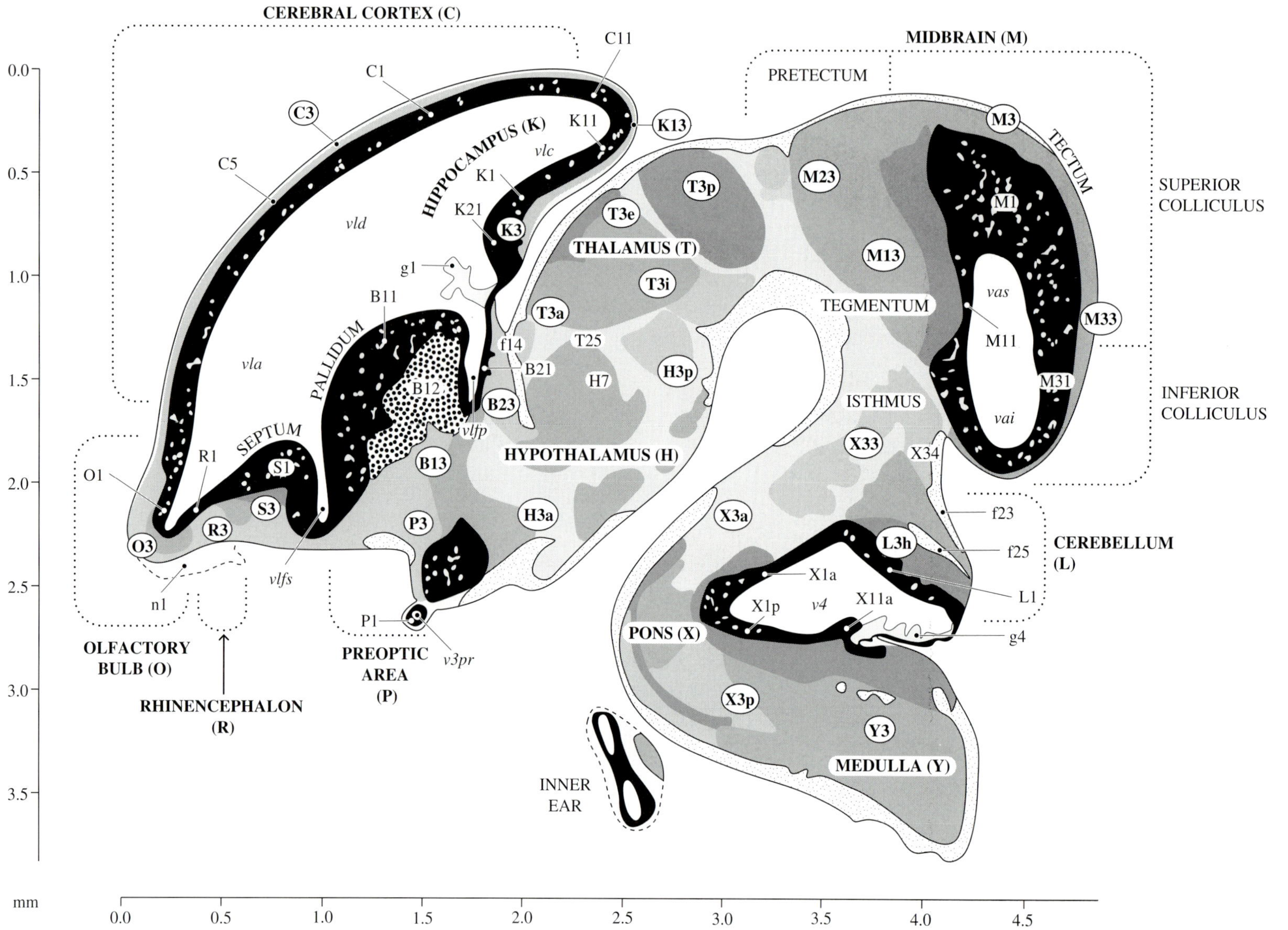

BASAL GANGLIA (B)

- B11 pallidal neuroepithelium
- B12 pallidal subventricular zone
- **B13** pallidum (or differentiating field)
- B21 strionuclear neuroepithelium
- **B23** bed nucleus of stria terminalis (or differentiating field)

CEREBRAL CORTEX (C)

- C1 neocortical neuroepithelium
- **C3** neocortex (or differentiating field)
- C5 intermediate zone
- C11 cingulate cortex neuroepithelium

FIBER TRACTS (f)

- f14 stria terminalis
- f23 superior cerebellar peduncle
- f25 inferior cerebellar peduncle

GLANDS AND PARANEURAL STRUCTURES (g)

- g1 choroid plexus of lateral ventricle
- g4 choroid plexus of fourth ventricle

HYPOTHALAMUS (H)

- **H3a** anterior hypothalamus (or differentiating field)
- **H3p** posterior hypothalamus (or differentiating field)
- H7 zona incerta

HIPPOCAMPAL FORMATION (K)

- K1 hippocampal (Ammon's horn) neuroepithelium
- **K3** hippocampus (or differentiating field)
- K11 subicular neuroepithelium
- **K13** subicular area (or differentiating field)
- K21 dentate neuroepithelium

CEREBELLUM (L)

- L1 cerebellar neuroepithelium
- **L3h** cerebellar hemisphere (or differentiating field)

MIDBRAIN (M)

- M1 superior tectal neuroepithelium
- **M3** superior colliculus (or differentiating field)
- M11 tegmental neuroepithelium
- **M13** tegmentum (or differentiating field)
- **M23** pretectum (or differentiating field)
- M31 inferior tectal neuroepithelium
- **M33** inferior colliculus (or differentiating field)

NERVES AND NUCLEI, CRANIAL (n)

- n1 olfactory nerve

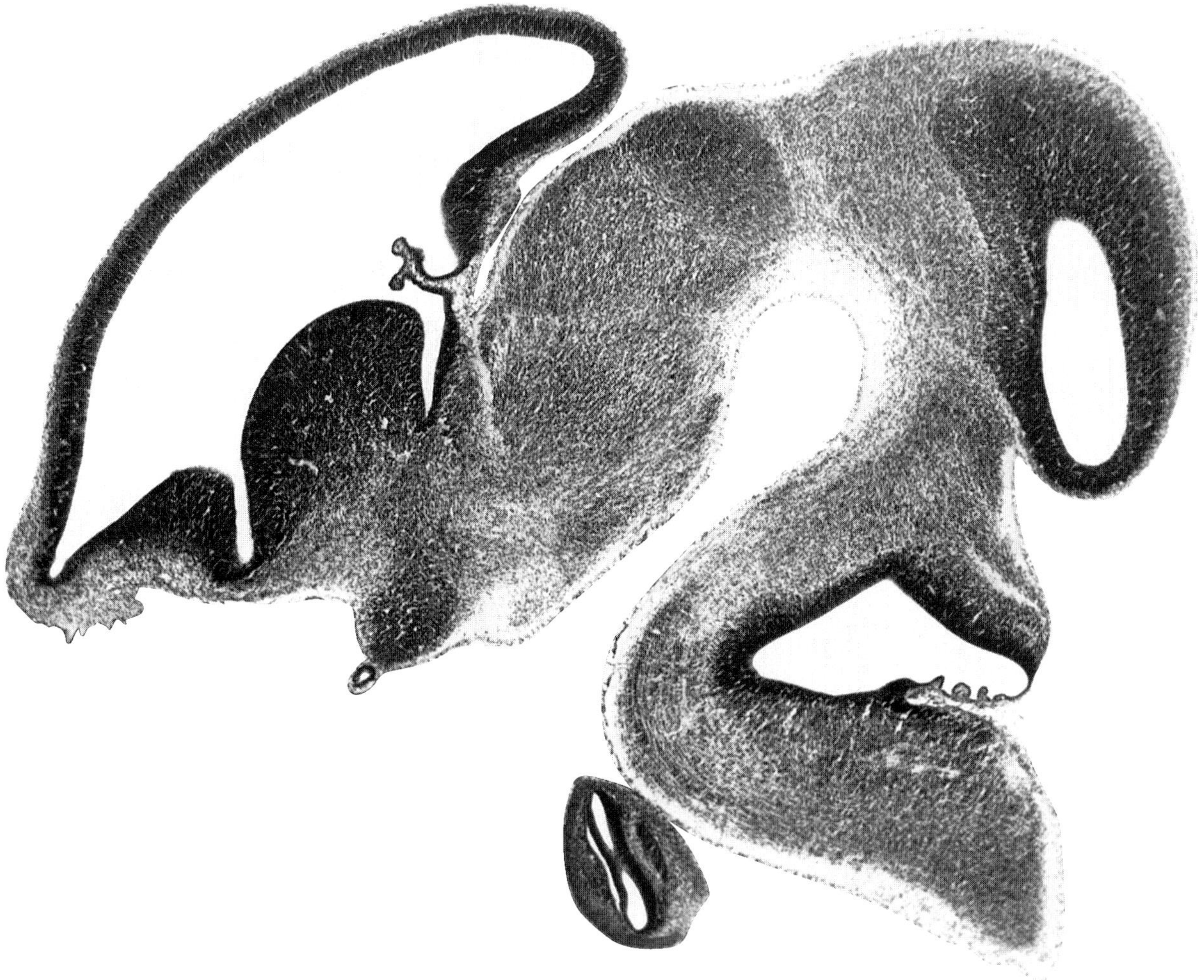

OLFACTORY BULB (O)

O1 olfactory bulb neuroepithelium
O3 olfactory bulb (or differentiating field)

PREOPTIC AREA AND OPTIC EPITHELIA (P)

P1 preoptic neuroepithelium
P3 preoptic area (or differentiating field)

RHINENCEPHALON (R)

R1 rhinencephalic neuroepithelium
R3 rhinencephalon (or differentiating field)

SEPTAL AREA (S)

S1 septal neuroepithelium
S3 septum (or differentiating field)

THALAMUS (T)

T3a anterior thalamus (or differentiating field)
T3e epithalamus (or differentiating field)
T3i intermediate thalamus (or differentiating field)
T3p posterior thalamus (or differentiating field)
T25 reticular nucleus

VENTRICLES (v)

vai aqueduct, inferior tectal
vas aqueduct, superior tectal
vla lateral ventricle, anterior horn
vlc lateral ventricle, central
vld lateral ventricle, dorsal
vlfp lateral ventricle, pallidal fork
vlfs lateral ventricle, septal fork
v3pr third ventricle, preoptic recess
v4 fourth ventricle

PONTINE AREA (X)

X1a anterior pontine neuroepithelium
X1p posterior pontine neuroepithelium
X3a anterior pons (or differentiating field)
X3p posterior pons (or differentiating field)
X11a anterior precerebellar neuroepithelium
X33 isthmus (or differentiating field)
X34 locus coeruleus

MEDULLA (Y)

Y3 medulla (or differentiating field)

SAGITTAL FIGURE 5

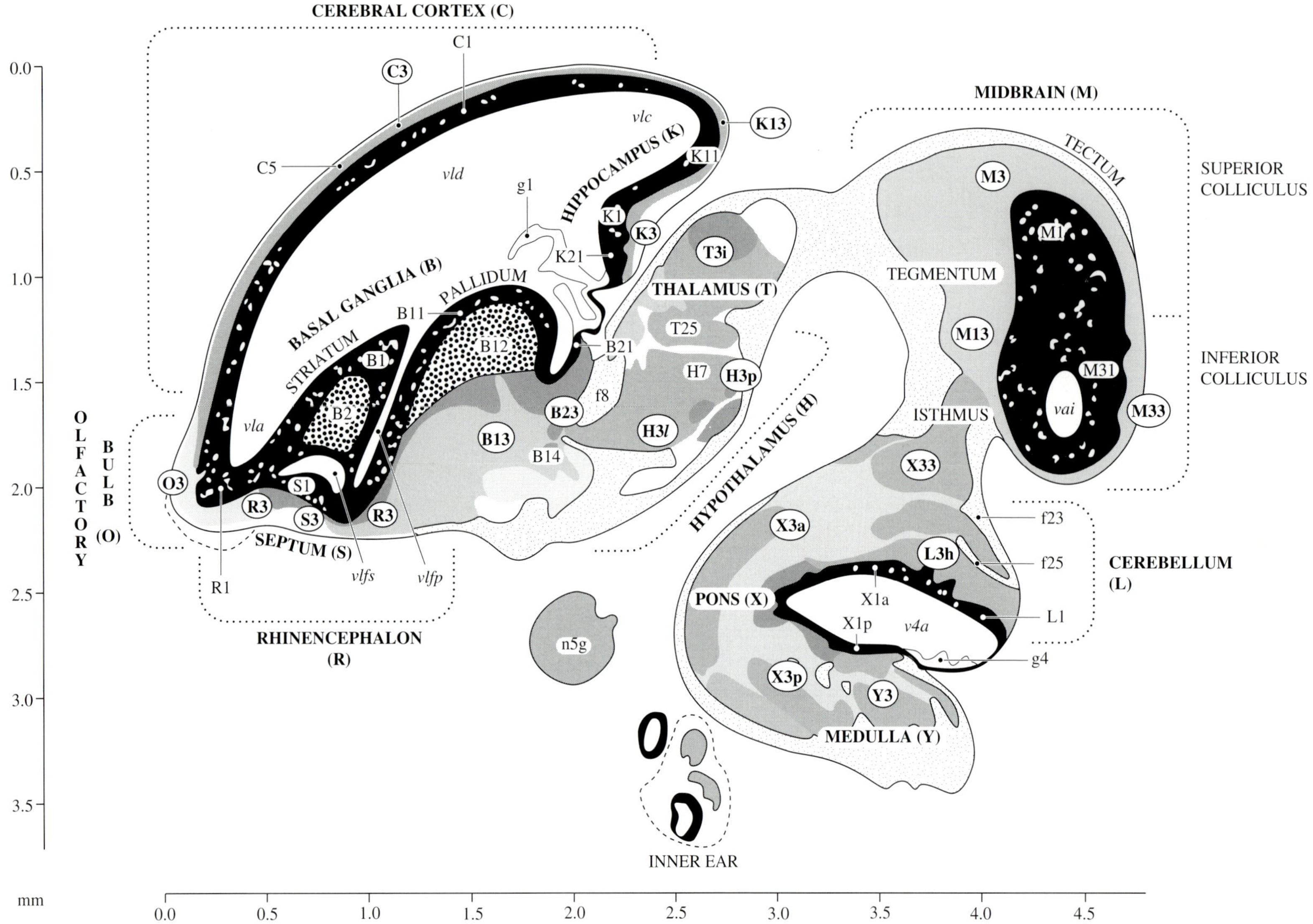

BASAL GANGLIA (B)

B1 striatal neuroepithelium
B2 striatal subventricular zone
B11 pallidal neuroepithelium
B12 pallidal subventricular zone
B13 pallidum (or differentiating field)
B14 entopeduncular nucleus
B21 strionuclear neuroepithelium
B23 bed nucleus of stria terminalis (or differentiating field)

CEREBRAL CORTEX (C)

C1 neocortical neuroepithelium
C3 neocortex (or differentiating field)
C5 intermediate zone

FIBER TRACTS (f)

f8 internal capsule
f23 superior cerebellar peduncle
f25 inferior cerebellar peduncle

GLANDS AND PARANEURAL STRUCTURES (g)

g1 choroid plexus of lateral ventricle
g4 choroid plexus of fourth ventricle

HYPOTHALAMUS (H)

H3*l* lateral hypothalamus
H3p posterior hypothalamus (or differentiating field)
H7 zona incerta

HIPPOCAMPAL FORMATION (K)

K1 hippocampal (Ammon's horn) neuroepithelium
K3 hippocampus (or differentiating field)
K11 subicular neuroepithelium
K13 subicular area (or differentiating field)
K21 dentate neuroepithelium

CEREBELLUM (L)

L1 cerebellar neuroepithelium
L3h cerebellar hemisphere (or differentiating field)

MIDBRAIN (M)

M1 superior tectal neuroepithelium
M3 superior colliculus (or differentiating field)
M13 tegmentum (or differentiating field)
M31 inferior tectal neuroepithelium
M33 inferior colliculus (or differentiating field)

NERVES AND NUCLEI, CRANIAL (n)

n5g trigeminal ganglion

OLFACTORY BULB (O)

O3 olfactory bulb (or differentiating field)

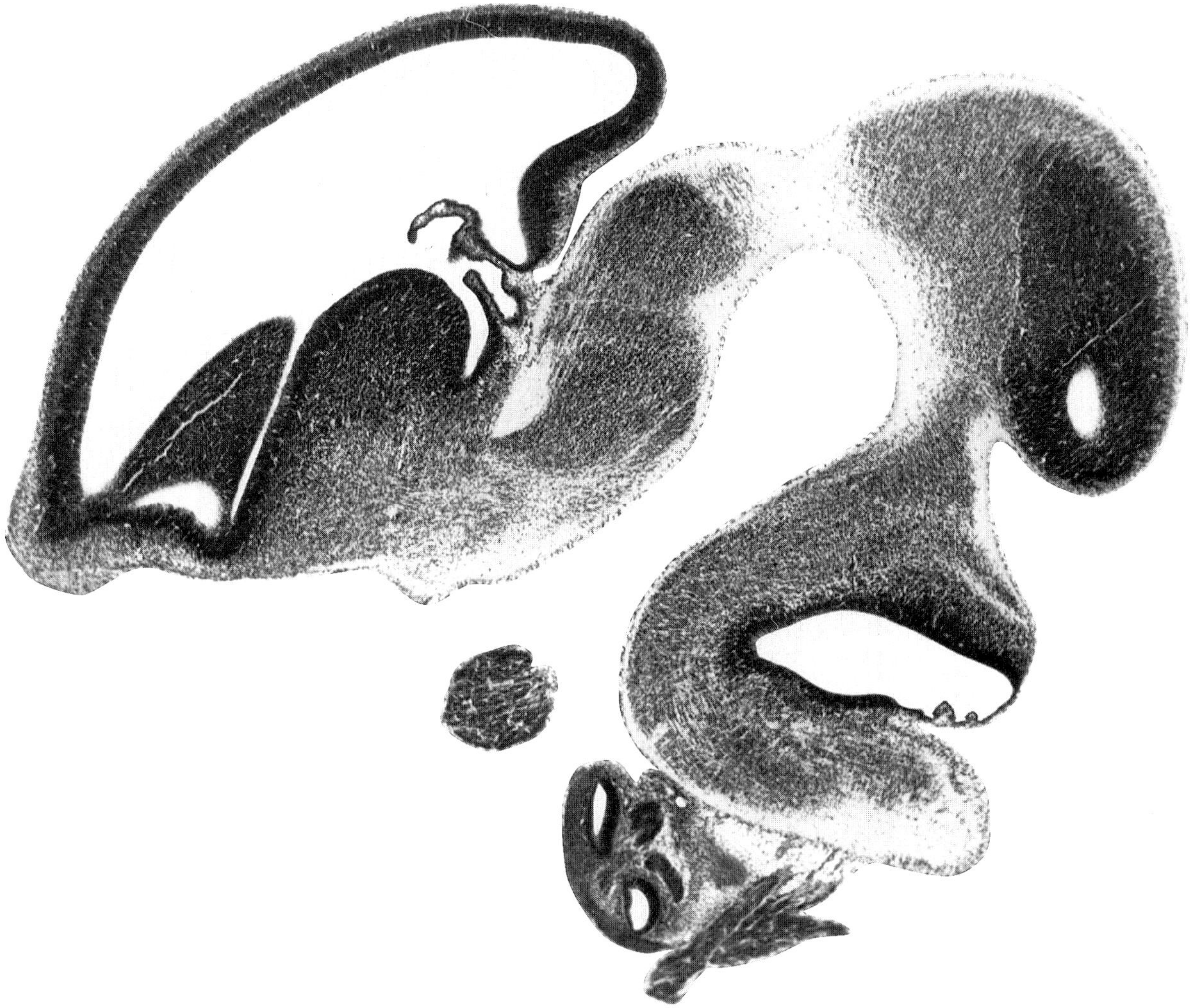

RHINENCEPHALON (R)

R1 rhinencephalic neuroepithelium
R3 rhinencephalon (or differentiating field)

SEPTAL AREA (S)

S1 septal neuroepithelium
S3 septum (or differentiating field)

THALAMUS (T)

T3i intermediate thalamus (or differentiating field)
T25 reticular nucleus

VENTRICLES (v)

vai aqueduct, inferior tectal
vla lateral ventricle, anterior horn
vlc lateral ventricle, central
vld lateral ventricle, dorsal
vlfp lateral ventricle, pallidal fork
vlfs lateral ventricle, septal fork
v4a fourth ventricle, anterior

PONTINE AREA (X)

X1a anterior pontine neuroepithelium
X1p posterior pontine neuroepithelium
X3a anterior pons (or differentiating field)
X3p posterior pons (or differentiating field)
X33 isthmus (or differentiating field)

MEDULLA (Y)

Y3 medulla (or differentiating field)

SAGITTAL FIGURE 6

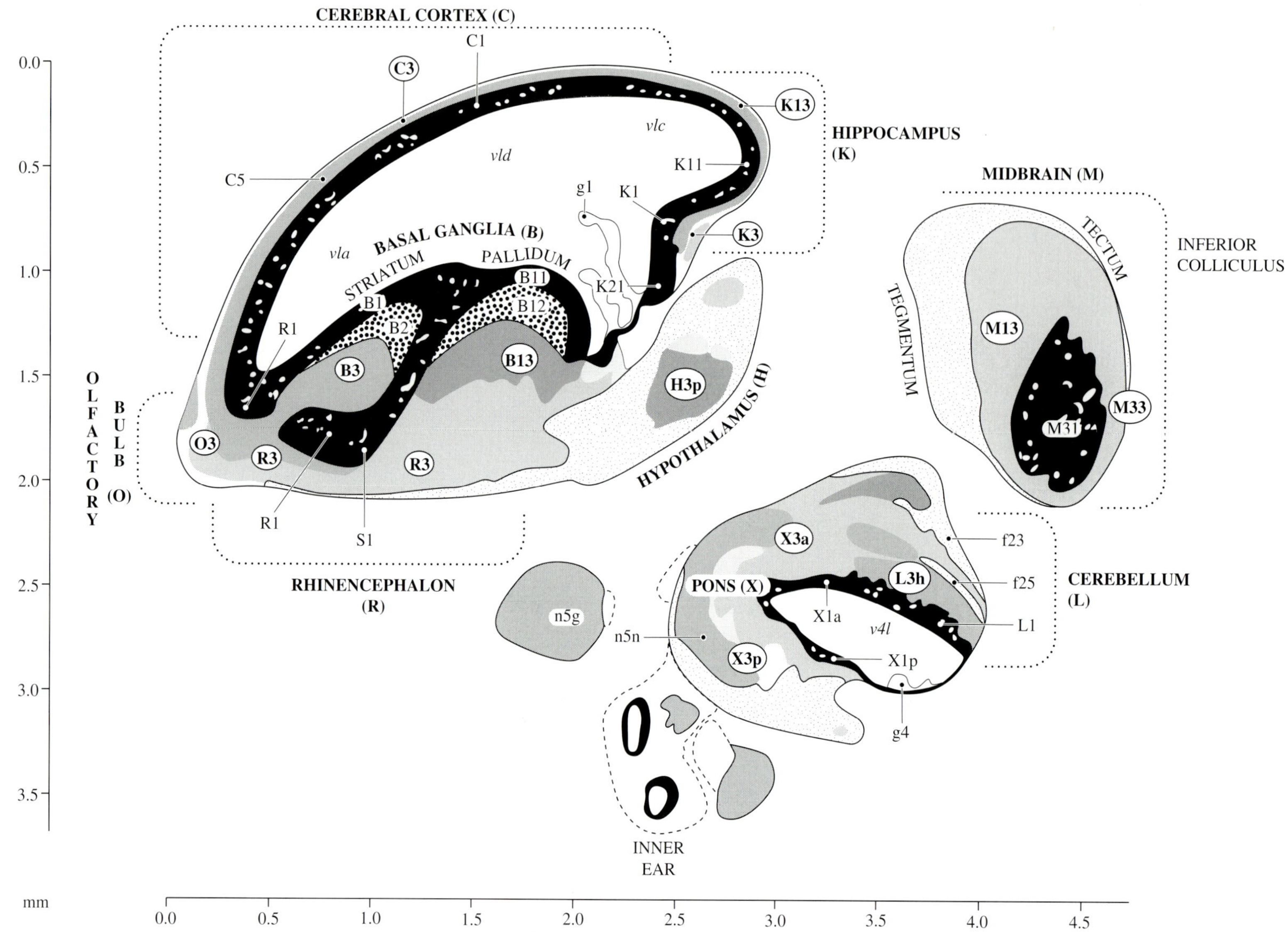

BASAL GANGLIA (B)

B1 striatal neuroepithelium
B2 striatal subventricular zone
B3 striatum (or differentiating field)
B11 pallidal neuroepithelium
B12 pallidal subventricular zone
B13 pallidum (or differentiating field)

CEREBRAL CORTEX (C)

C1 neocortical neuroepithelium
C3 neocortex (or differentiating field)
C5 intermediate zone

FIBER TRACTS (f)

f23 superior cerebellar peduncle
f25 inferior cerebellar peduncle

GLANDS AND PARANEURAL STRUCTURES (g)

g1 choroid plexus of lateral ventricle
g4 choroid plexus of fourth ventricle

HYPOTHALAMUS (H)

H3p posterior hypothalamus (or differentiating field)

HIPPOCAMPAL FORMATION (K)

K1 hippocampal (Ammon's horn) neuroepithelium
K3 hippocampus (or differentiating field)
K11 subicular neuroepithelium
K13 subicular area (or differentiating field)
K21 dentate neuroepithelium

CEREBELLUM (L)

L1 cerebellar neuroepithelium
L3h cerebellar hemisphere (or differentiating field)

MIDBRAIN (M)

M13 tegmentum (or differentiating field)
M31 inferior tectal neuroepithelium
M33 inferior colliculus (or differentiating field)

NERVES AND NUCLEI, CRANIAL (n)

n5g trigeminal ganglion
n5n trigeminal motor nucleus

OLFACTORY BULB (O)

O3 olfactory bulb (or differentiating field)

RHINENCEPHALON (R)

R1 rhinencephalic neuroepithelium
R3 rhinencephalon (or differentiating field)

E16

SEPTAL AREA (S)

S1 septal neuroepithelium

VENTRICLES (v)

vla lateral ventricle, anterior horn
vlc lateral ventricle, central
vld lateral ventricle, dorsal
v4l fourth ventricle, lateral recess

PONTINE AREA (X)

X1a anterior pontine neuroepithelium
X1p posterior pontine neuroepithelium
X3a anterior pons (or differentiating field)
X3p posterior pons (or differentiating field)

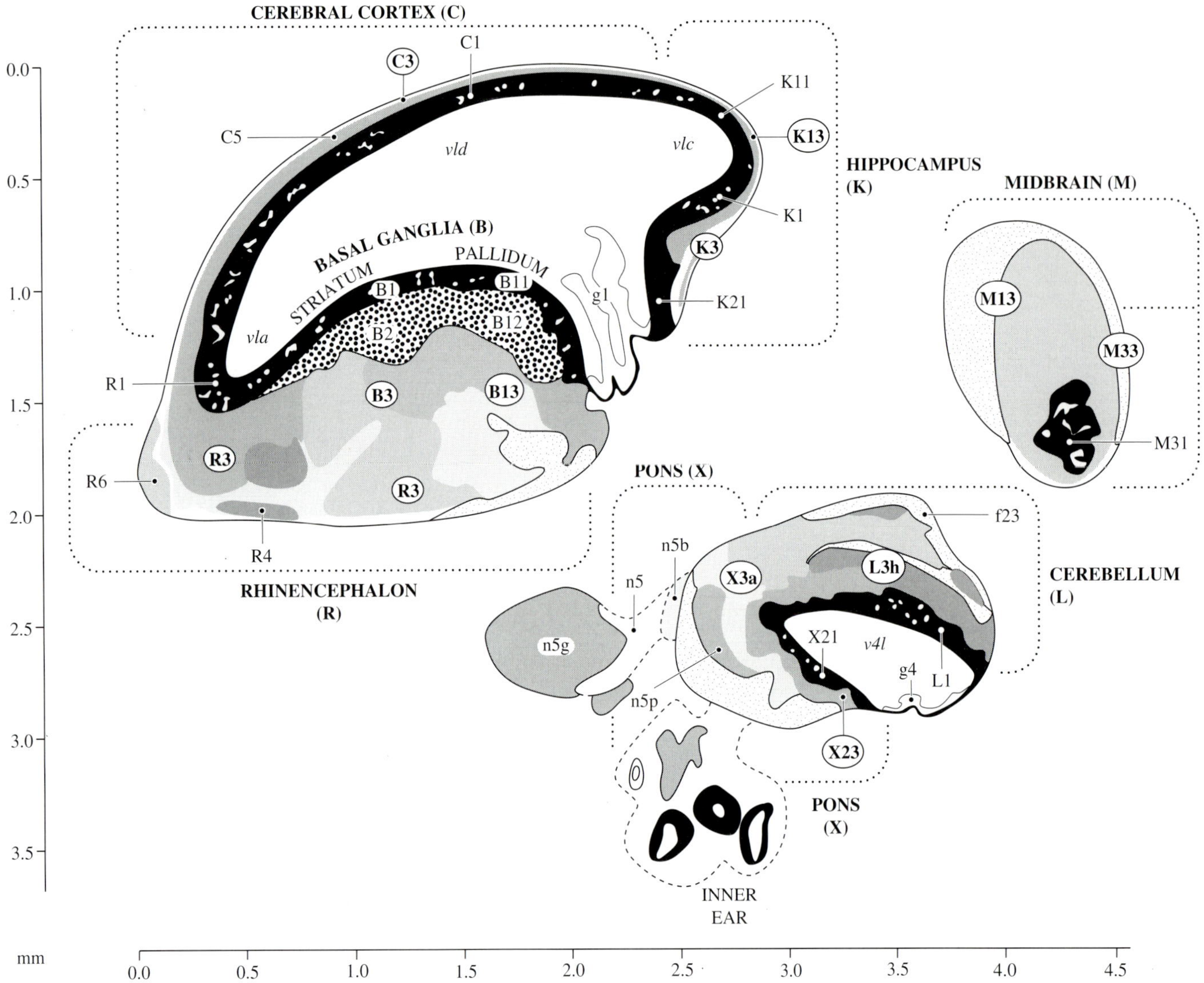

BASAL GANGLIA (B)

- B1 striatal neuroepithelium
- B2 striatal subventricular zone
- **B3** striatum (or differentiating field)
- B11 pallidal neuroepithelium
- B12 pallidal subventricular zone
- **B13** pallidum (or differentiating field)

CEREBRAL CORTEX (C)

- C1 neocortical neuroepithelium
- **C3** neocortex (or differentiating field)
- C5 intermediate zone

FIBER TRACTS (f)

- f23 superior cerebellar peduncle

GLANDS AND PARANEURAL STRUCTURES (g)

- g1 choroid plexus of lateral ventricle
- g4 choroid plexus of fourth ventricle

HIPPOCAMPAL FORMATION (K)

- K1 hippocampal (Ammon's horn) neuroepithelium
- **K3** hippocampus (or differentiating field)
- K11 subicular neuroepithelium
- **K13** subicular area (or differentiating field)
- K21 dentate neuroepithelium

CEREBELLUM (L)

- L1 cerebellar neuroepithelium
- **L3h** cerebellar hemisphere (or differentiating field)

MIDBRAIN (M)

- **M13** tegmentum (or differentiating field)
- M31 inferior tectal neuroepithelium
- **M33** inferior colliculus (or differentiating field)

NERVES AND NUCLEI, CRANIAL (n)

- n5 trigeminal nerve
- n5b trigeminal boundary cap
- n5g trigeminal ganglion
- n5p principal (sensory) trigeminal nucleus

RHINENCEPHALON (R)

- R1 rhinencephalic neuroepithelium
- **R3** rhinencephalon (or differentiating field)
- R4 piriform cortex
- R6 anterior olfactory nucleus

VENTRICLES (v)

- *vla* lateral ventricle, anterior horn
- *vlc* lateral ventricle, central
- *vld* lateral ventricle, dorsal
- *v4l* fourth ventricle, lateral recess

PONTINE AREA (X)

- **X3a** anterior pons (or differentiating field)
- X21 cochlear nuclei neuroepithelium
- **X23** cochlear nuclei (or differentiating field)

SAGITTAL FIGURE 8

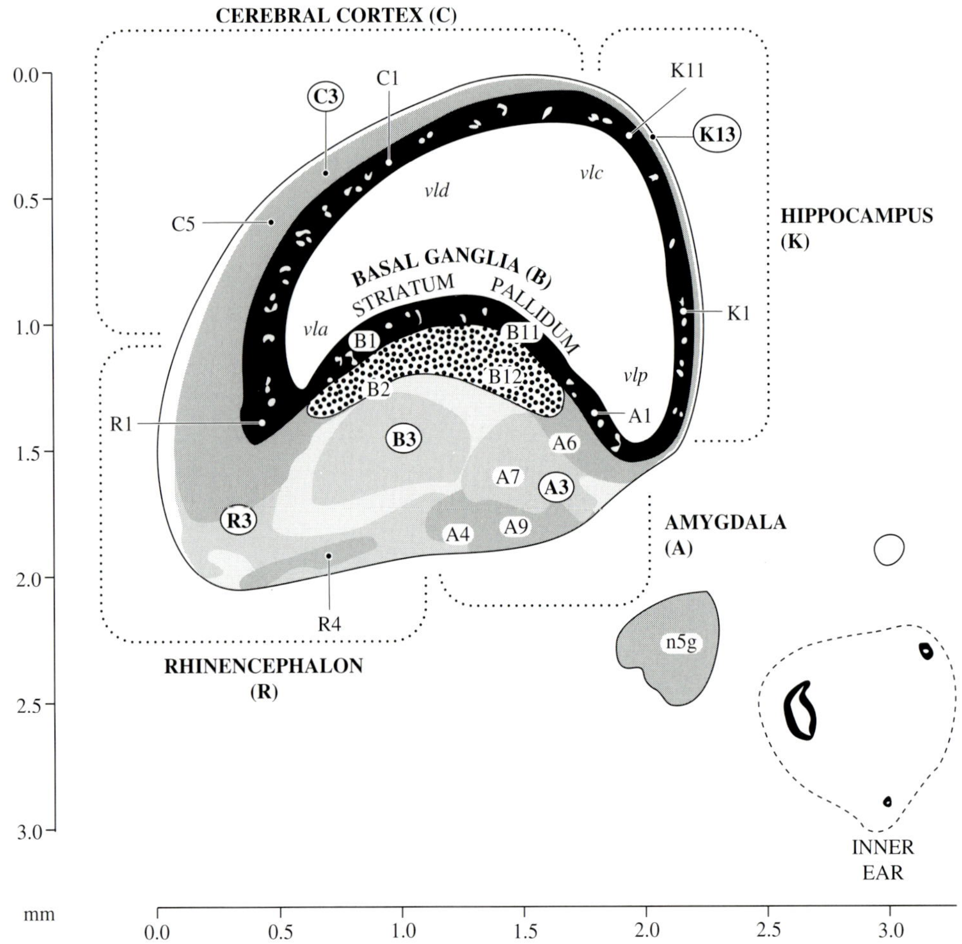

AMYGDALA (A)

A1 amygdaloid neuroepithelium
A3 amygdala (or differentiating field)
A4 anterior amygdaloid area
A6 central amygdaloid nucleus
A7 basomedial amygdaloid nucleus
A9 cortical amygdaloid nucleus

BASAL GANGLIA (B)

B1 striatal neuroepithelium
B2 striatal subventricular zone
B3 striatum (or differentiating field)
B11 pallidal neuroepithelium
B12 pallidal subventricular zone

CEREBRAL CORTEX (C)

C1 neocortical neuroepithelium
C3 neocortex (or differentiating field)
C5 intermediate zone

HIPPOCAMPAL FORMATION (K)

K1 hippocampal (Ammon's horn) neuroepithelium
K11 subicular neuroepithelium
K13 subicular area (or differentiating field)

NERVES AND NUCLEI, CRANIAL (n)

n5g trigeminal ganglion

RHINENCEPHALON (R)

R1 rhinencephalic neuroepithelium
R3 rhinencephalon (or differentiating field)
R4 piriform cortex

VENTRICLES (v)

vla lateral ventricle, anterior horn
vlc lateral ventricle, central
vld lateral ventricle, dorsal
vlp lateral ventricle, posterior horn

CORONAL

FIGURES and PLATES 1-22

CORONAL FIGURE 1

CEREBRAL CORTEX (C)

C1 neocortical neuroepithelium
C3 neocortex (or differentiating field)
C4 molecular layer (or primordial plexiform layer)
C5 intermediate zone
C11 cingulate cortex neuroepithelium
C13 cingulate cortex (or differentiating field)

NERVES AND NUCLEI, CRANIAL (n)

n1 olfactory nerve

OLFACTORY BULB (O)

O3 olfactory bulb (or differentiating field)

RHINENCEPHALON (R)

R3 rhinencephalon (or differentiating field)

SEPTAL AREA (S)

S3 septum (or differentiating field)

VENTRICLES (v)

vla lateral ventricle, anterior horn

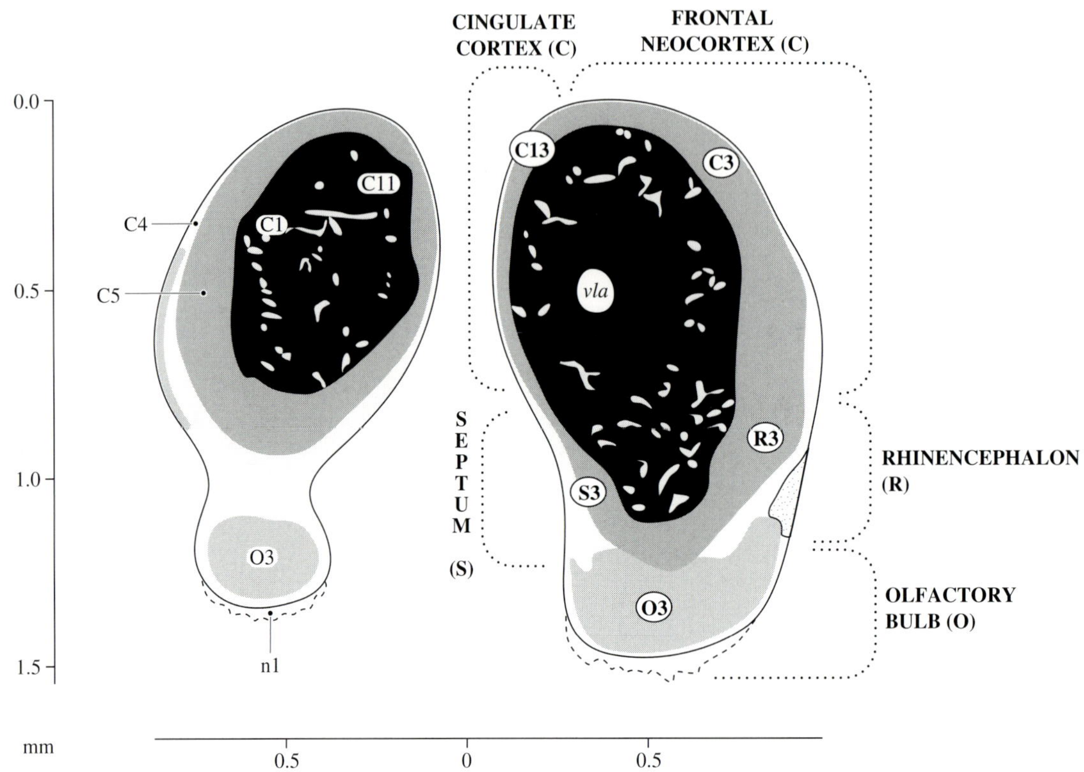

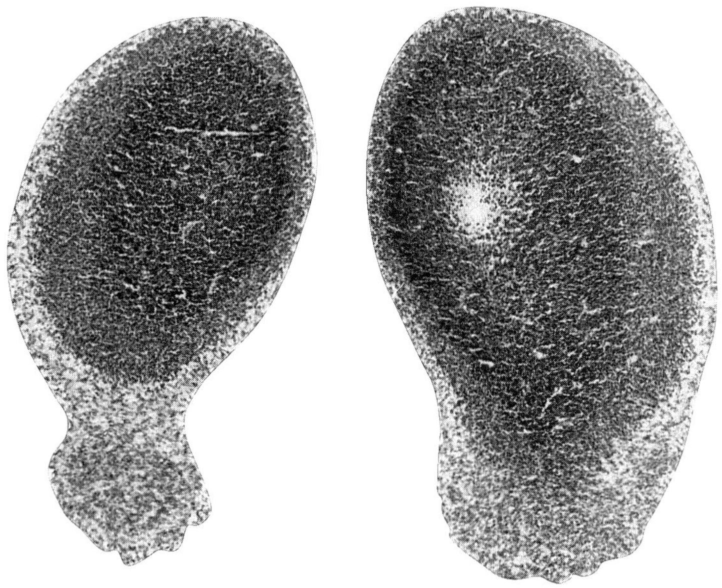

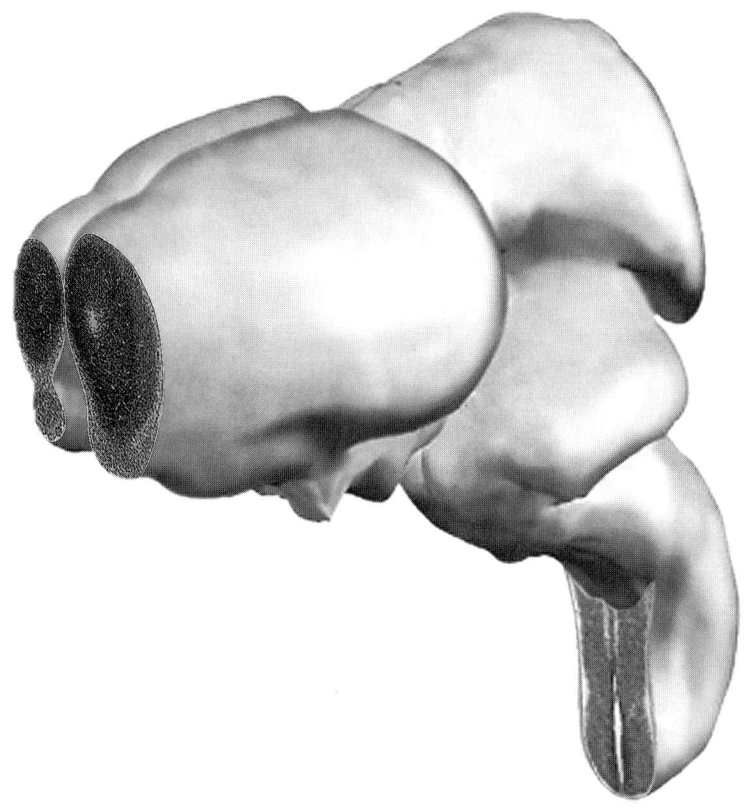

CORONAL FIGURE 2

CEREBRAL CORTEX (C)

C1 neocortical neuroepithelium
C3 neocortex (or differentiating field)
C4 molecular layer (or primordial plexiform layer)
C5 intermediate zone
C11 cingulate cortex neuroepithelium
C13 cingulate cortex (or differentiating field)

NERVES AND NUCLEI, CRANIAL (n)

n1 olfactory nerve

OLFACTORY BULB (O)

O1 olfactory bulb neuroepithelium
O3 olfactory bulb (or differentiating field)

RHINENCEPHALON (R)

R1 rhinencephalic neuroepithelium
R3 rhinencephalon (or differentiating field)
R4 piriform cortex

SEPTAL AREA (S)

S1 septal neuroepithelium
S3 septum (or differentiating field)

VENTRICLES (v)

vla lateral ventricle, anterior horn
vld lateral ventricle, dorsal

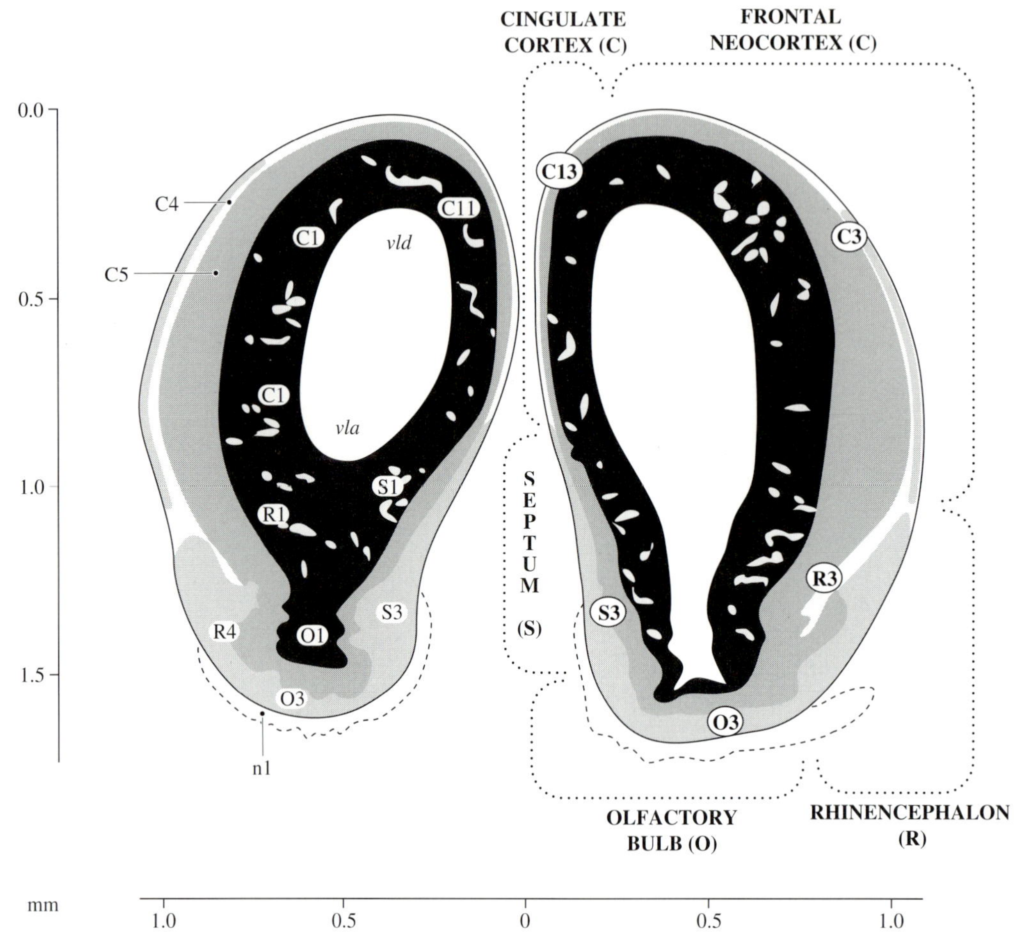

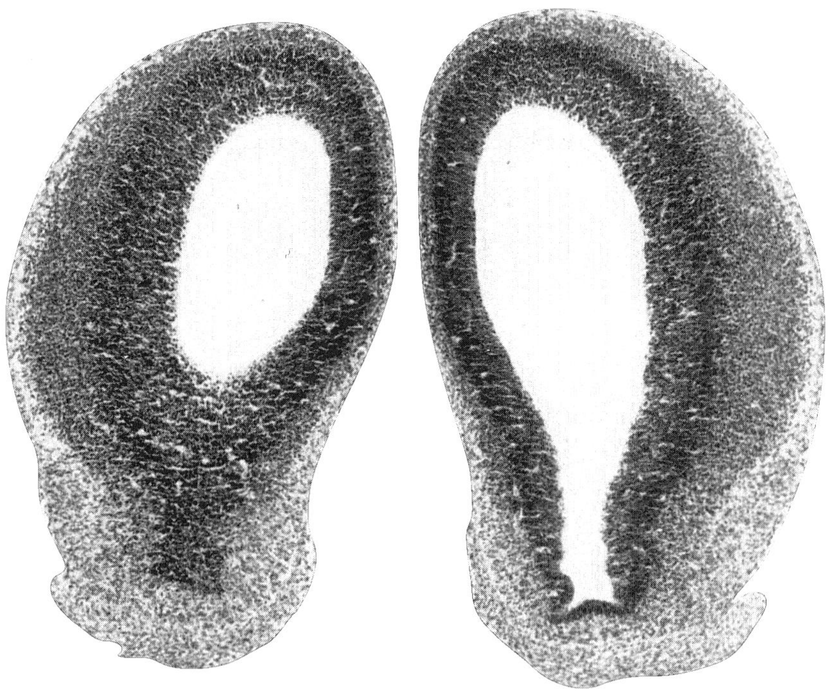

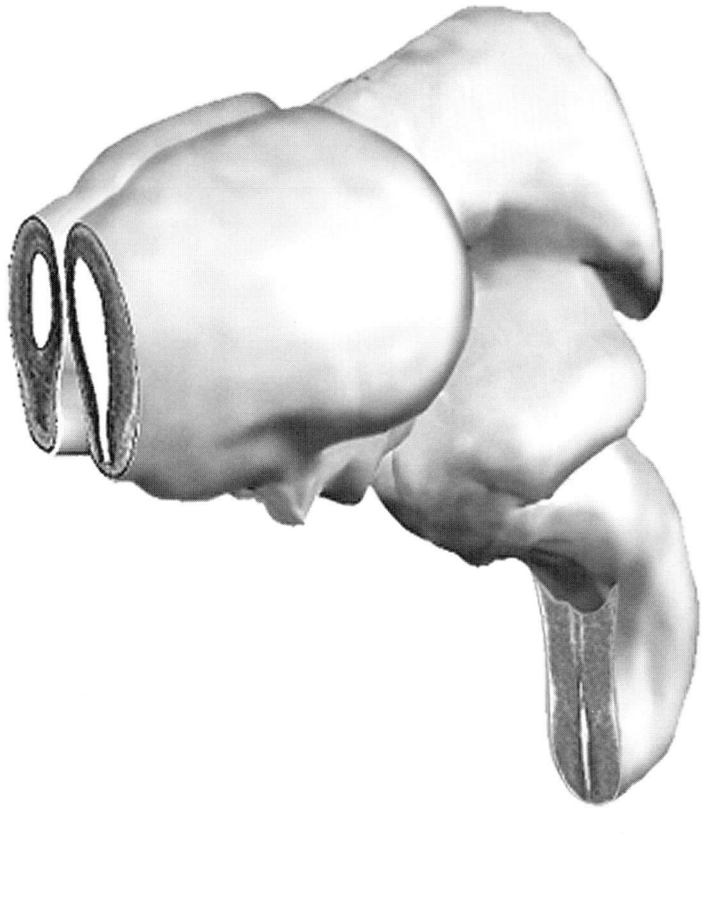

CORONAL FIGURE 3

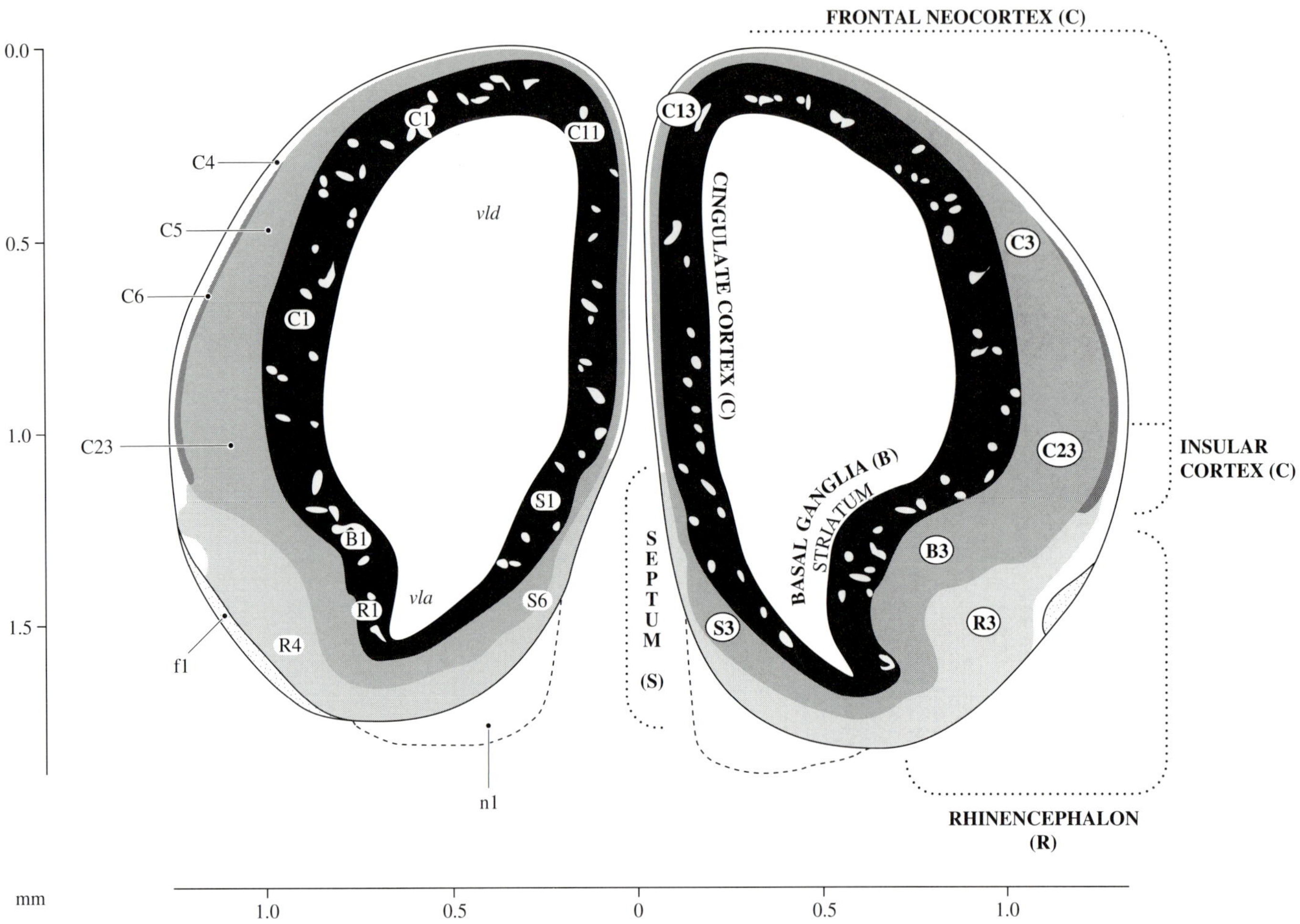

BASAL GANGLIA (B)

B1 striatal neuroepithelium
B3 striatum (or differentiating field)

CEREBRAL CORTEX (C)

C1 neocortical neuroepithelium
C3 neocortex (or differentiating field)
C4 molecular layer (or primordial plexiform layer)
C5 intermediate zone
C6 cortical plate
C11 cingulate cortex neuroepithelium
C13 cingulate cortex (or differentiating field)
C23 orbital and insular area (or differentiating field)

FIBER TRACTS (f)

f1 lateral olfactory tract

NERVES AND NUCLEI, CRANIAL (n)

n1 olfactory nerve

RHINENCEPHALON (R)

R1 rhinencephalic neuroepithelium
R3 rhinencephalon (or differentiating field)
R4 piriform cortex

SEPTAL AREA (S)

S1 septal neuroepithelium
S3 septum (or differentiating field)
S6 medial septum

VENTRICLES (v)

vla lateral ventricle, anterior horn
vld lateral ventricle, dorsal

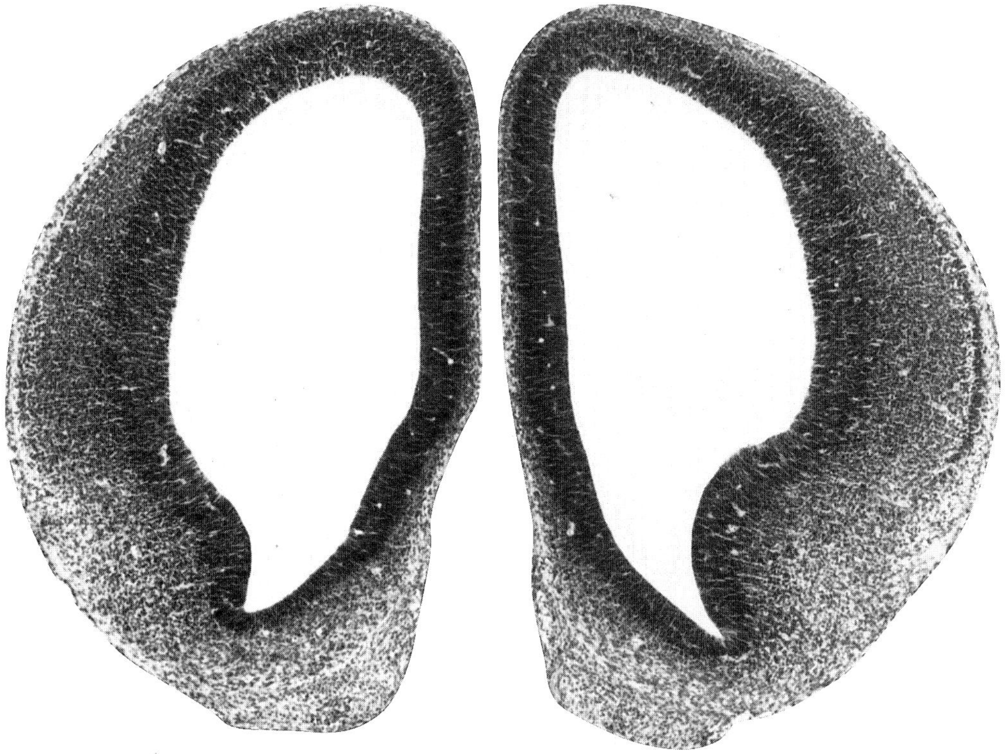

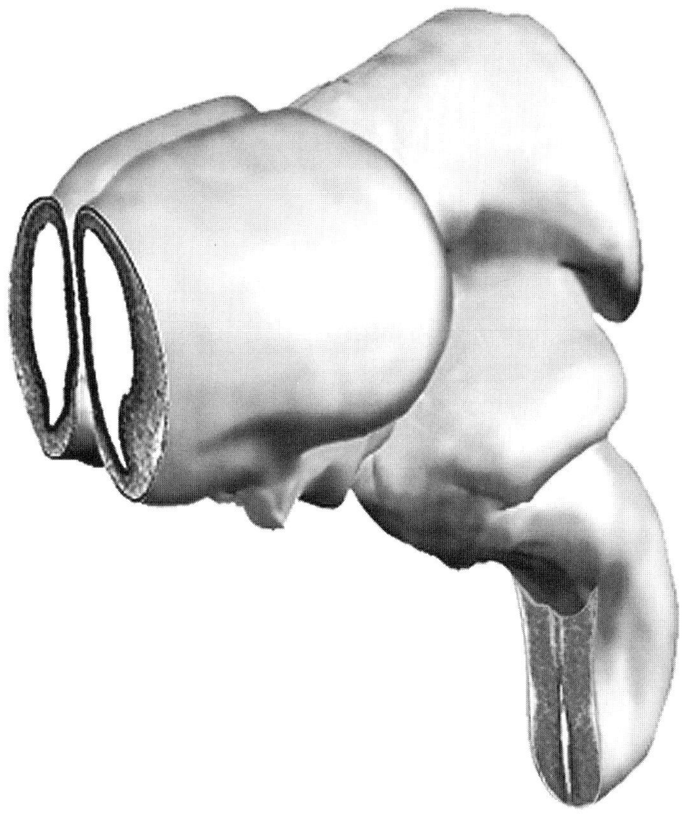

CORONAL FIGURE 4

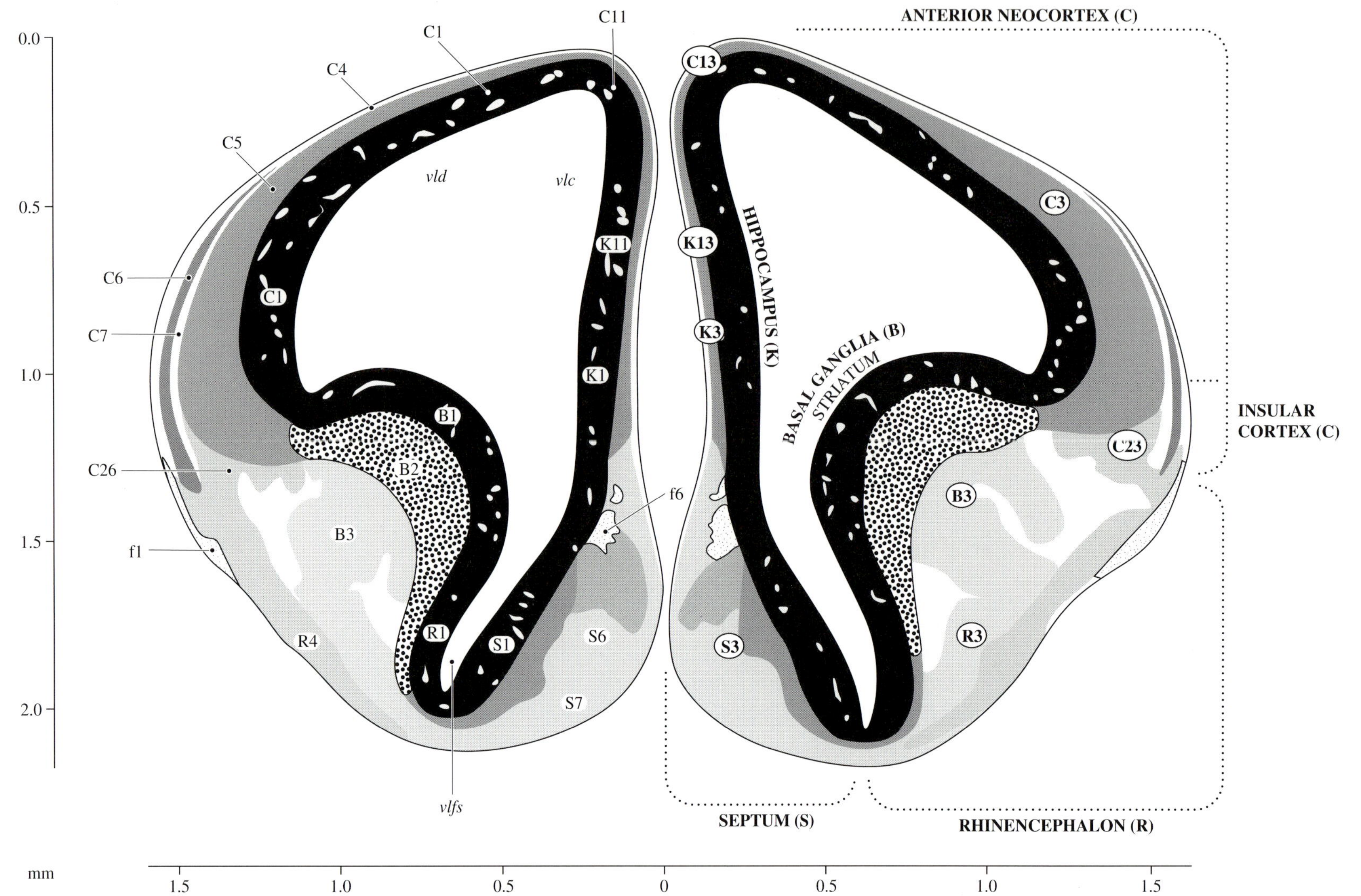

BASAL GANGLIA (B)

B1 striatal neuroepithelium
B2 striatal subventricular zone
B3 striatum (or differentiating field)

CEREBRAL CORTEX (C)

C1 neocortical neuroepithelium
C3 neocortex (or differentiating field)
C4 molecular layer (or primordial plexiform layer)
C5 intermediate zone
C6 cortical plate
C7 white matter
C11 cingulate cortex neuroepithelium
C13 cingulate cortex (or differentiating field)
C23 orbital and insular area (or differentiating field)
C26 claustrum

FIBER TRACTS (f)

f1 lateral olfactory tract
f6 fornix

HIPPOCAMPAL FORMATION (K)

K1 hippocampal (Ammon's horn) neuroepithelium
K3 hippocampus (or differentiating field)
K11 subicular neuroepithelium
K13 subicular area (or differentiating field)

RHINENCEPHALON (R)

R1 rhinencephalic neuroepithelium
R3 rhinencephalon (or differentiating field)
R4 piriform cortex

SEPTAL AREA (S)

S1 septal neuroepithelium
S3 septum (or differentiating field)
S6 medial septum
S7 diagonal band (or differentiating field)

VENTRICLES (v)

vlc lateral ventricle, central
vld lateral ventricle, dorsal
vlfs lateral ventricle, septal fork

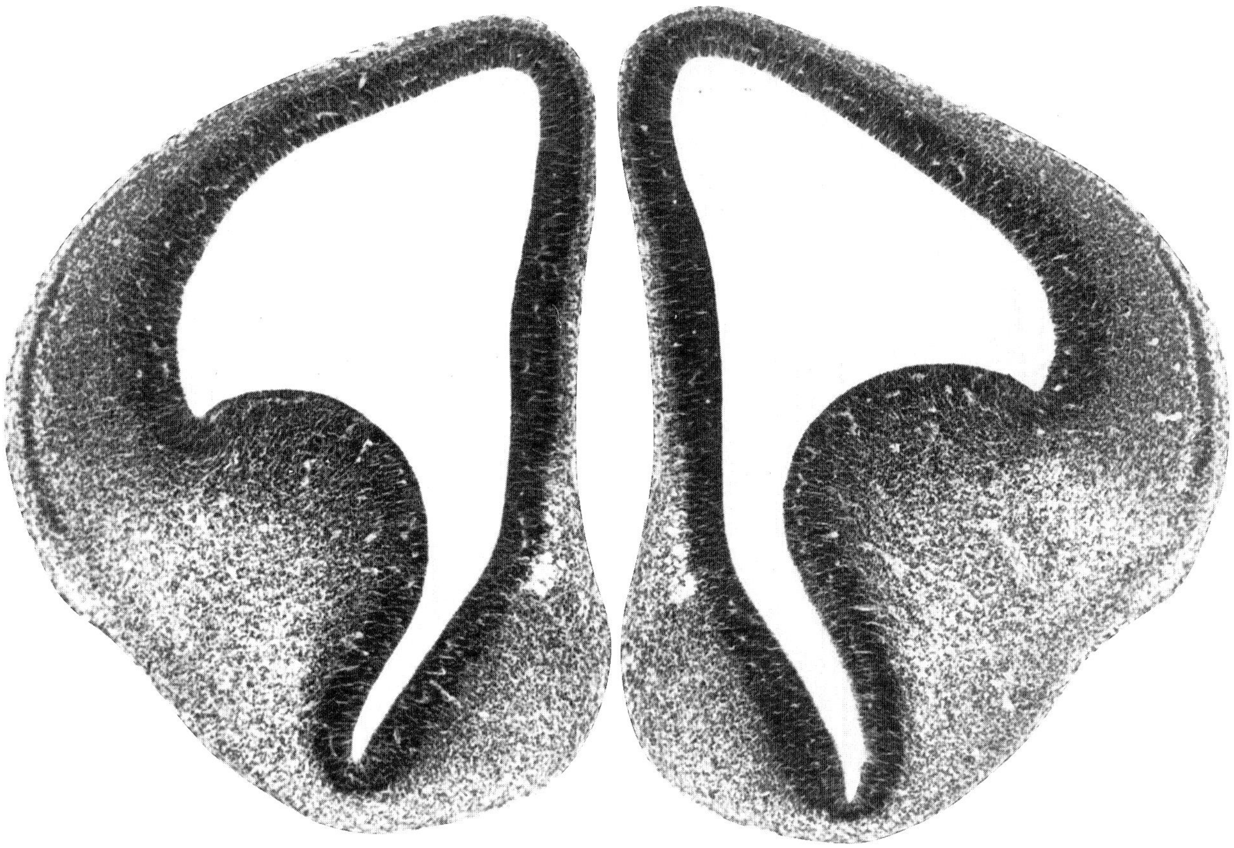

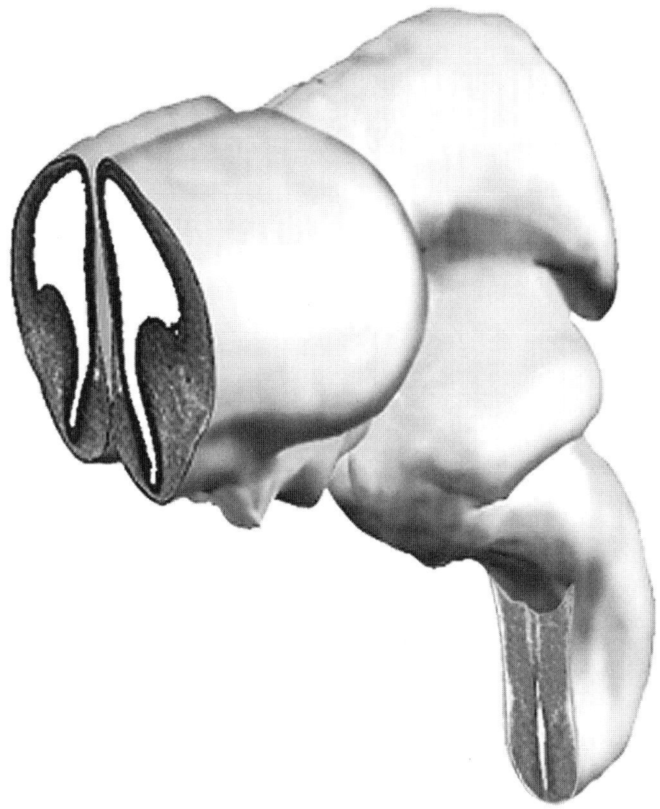

CORONAL FIGURE 5

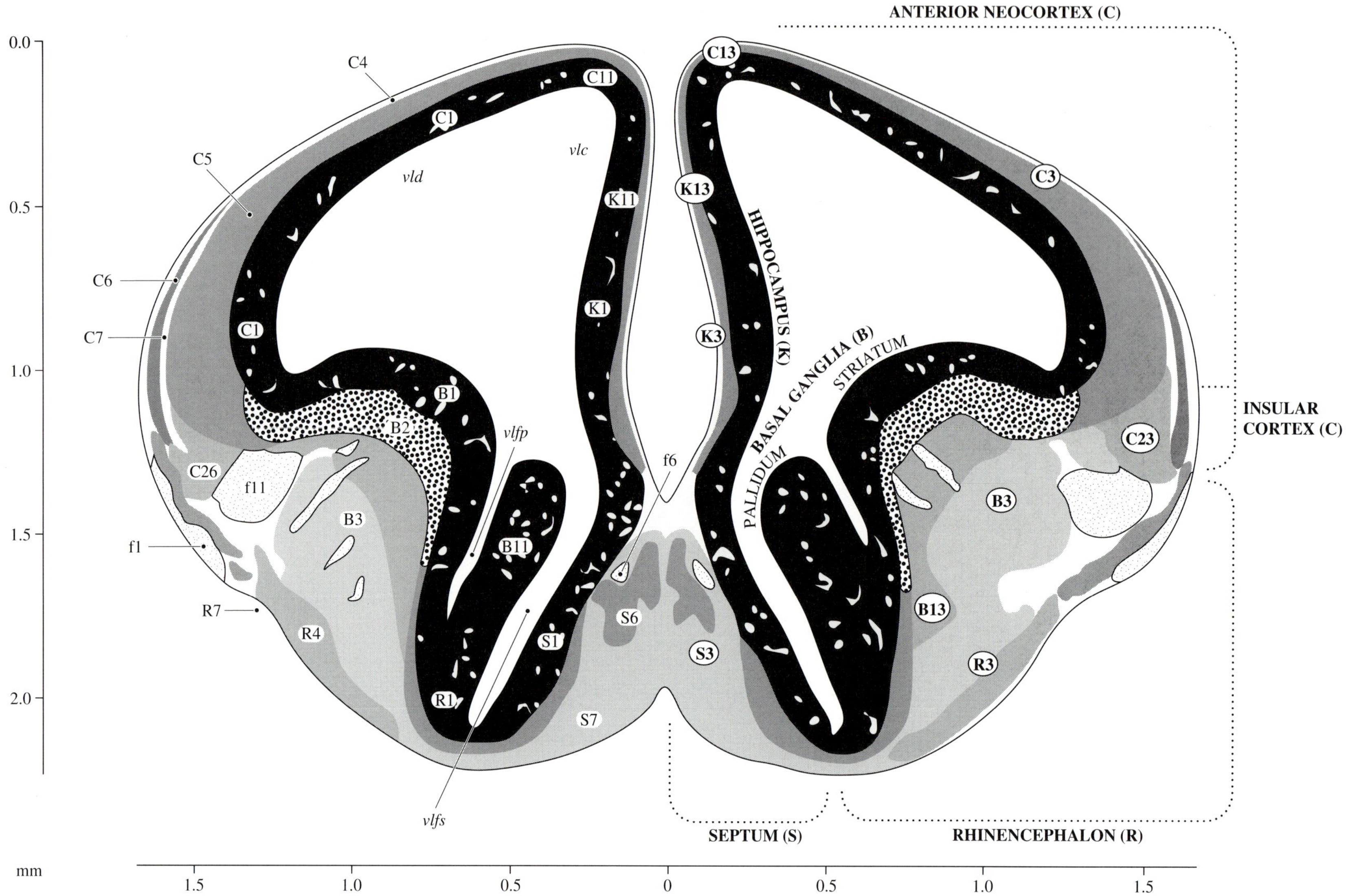

BASAL GANGLIA (B)

- B1 striatal neuroepithelium
- B2 striatal subventricular zone
- **B3** striatum (or differentiating field)
- B11 pallidal neuroepithelium
- **B13** pallidum (or differentiating field)

CEREBRAL CORTEX (C)

- C1 neocortical neuroepithelium
- **C3** neocortex (or differentiating field)
- C4 molecular layer (or primordial plexiform layer)
- C5 intermediate zone
- C6 cortical plate
- C7 white matter
- C11 cingulate cortex neuroepithelium
- **C13** cingulate cortex (or differentiating field)
- **C23** orbital and insular area (or differentiating field)
- C26 claustrum

FIBER TRACTS (f)

- f1 lateral olfactory tract
- f6 fornix
- f11 external capsule

HIPPOCAMPAL FORMATION (K)

- K1 hippocampal (Ammon's horn) neuroepithelium
- **K3** hippocampus (or differentiating field)
- K11 subicular neuroepithelium
- **K13** subicular area (or differentiating field)

RHINENCEPHALON (R)

- R1 rhinencephalic neuroepithelium
- **R3** rhinencephalon (or differentiating field)
- R4 piriform cortex
- R7 rhinal fissure

SEPTAL AREA (S)

- S1 septal neuroepithelium
- **S3** septum (or differentiating field)
- S6 medial septum
- S7 diagonal band (or differentiating field)

VENTRICLES (v)

- *vlc* lateral ventricle, central
- *vld* lateral ventricle, dorsal
- *vlfp* lateral ventricle, pallidal fork
- *vlfs* lateral ventricle, septal fork

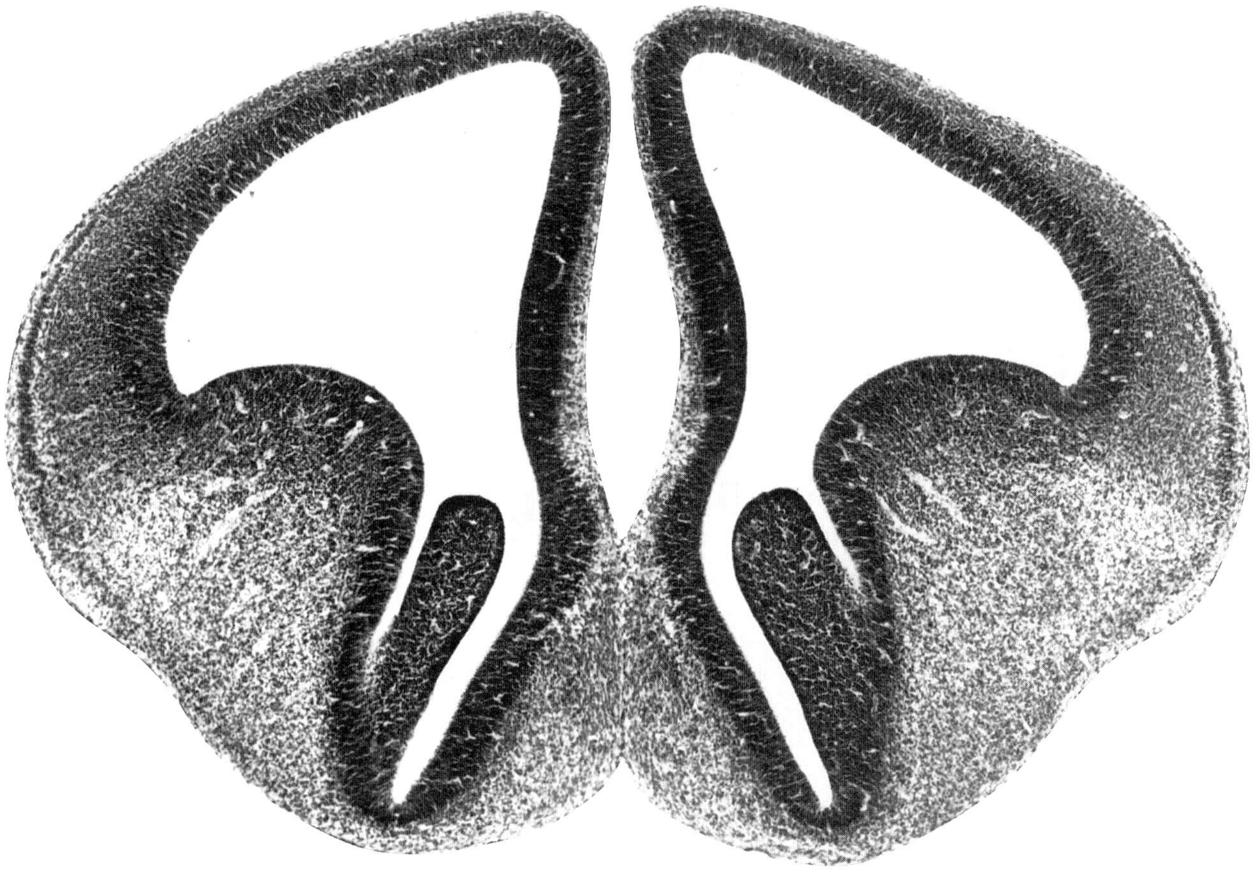

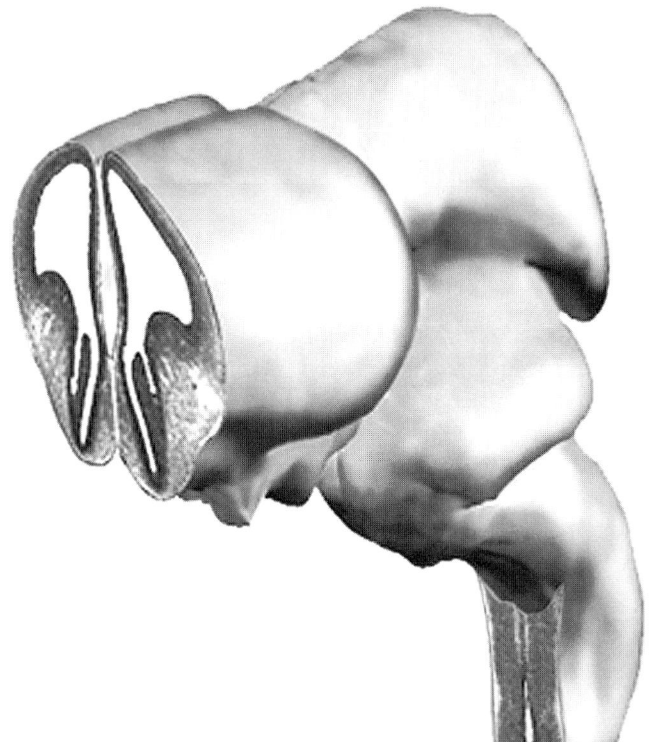

CORONAL FIGURE 6

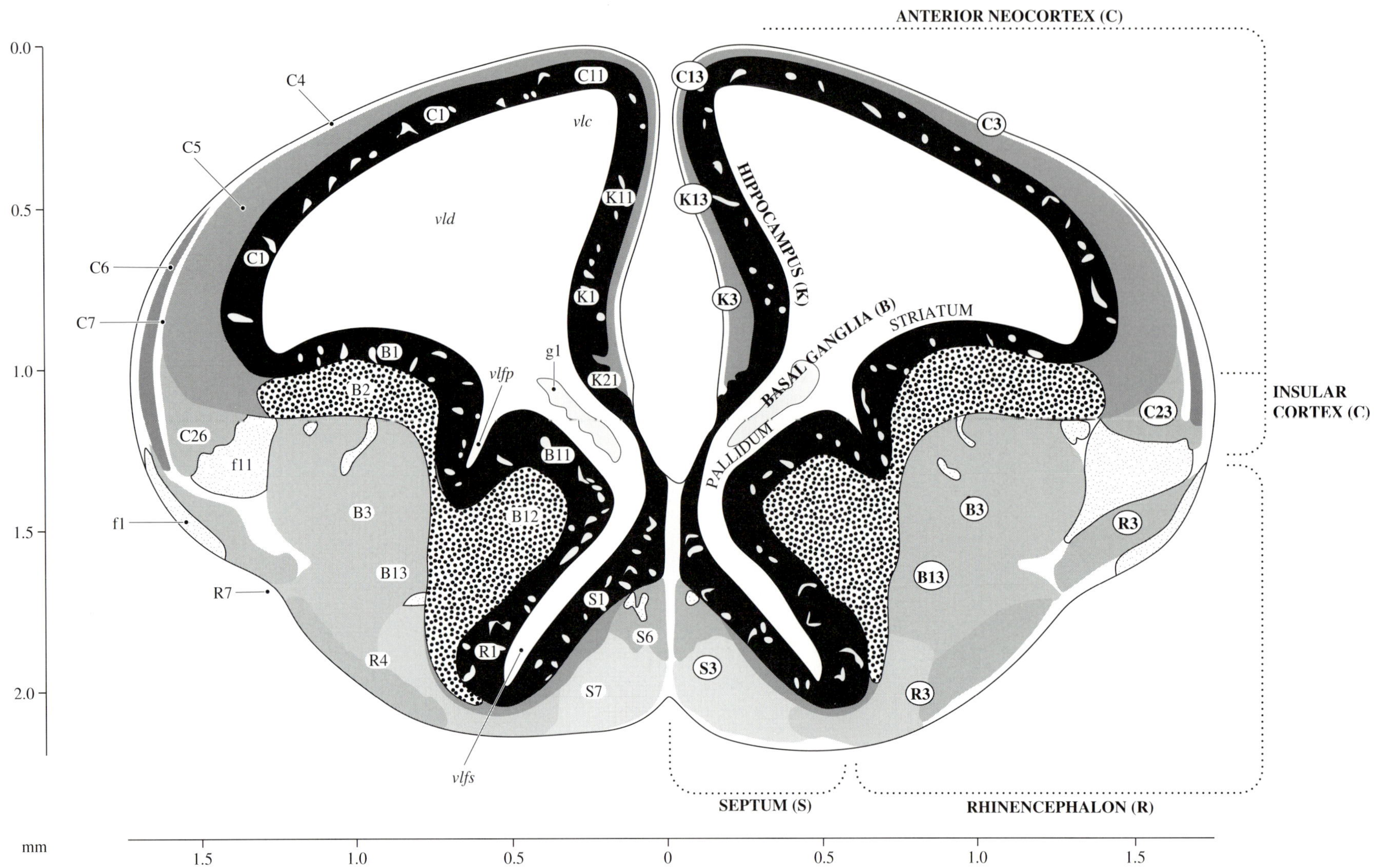

BASAL GANGLIA (B)

B1	striatal neuroepithelium
B2	striatal subventricular zone
B3	striatum (or differentiating field)
B11	pallidal neuroepithelium
B12	pallidal subventricular zone
B13	pallidum (or differentiating field)

CEREBRAL CORTEX (C)

C1	neocortical neuroepithelium
C3	neocortex (or differentiating field)
C4	molecular layer (or primordial plexiform layer)
C5	intermediate zone
C6	cortical plate
C7	white matter
C11	cingulate cortex neuroepithelium
C13	cingulate cortex (or differentiating field)
C23	orbital and insular area (or differentiating field)
C26	claustrum

FIBER TRACTS (f)

f1	lateral olfactory tract
f11	external capsule

GLANDS AND PARANEURAL STRUCTURES (g)

g1	choroid plexus of lateral ventricle

HIPPOCAMPAL FORMATION (K)

K1	hippocampal (Ammon's horn) neuroepithelium
K3	hippocampus (or differentiating field)
K11	subicular neuroepithelium
K13	subicular area (or differentiating field)
K21	dentate gyrus neuroepithelium

RHINENCEPHALON (R)

R1	rhinencephalic neuroepithelium
R3	rhinencephalon (or differentiating field)
R4	piriform cortex
R7	rhinal fissure

SEPTAL AREA (S)

S1	septal neuroepithelium
S3	septum (or differentiating field)
S6	medial septum
S7	diagonal band (or differentiating field)

VENTRICLES (v)

vlc	lateral ventricle, central
vld	lateral ventricle, dorsal
vlfp	lateral ventricle, pallidal fork
vlfs	lateral ventricle, septal fork

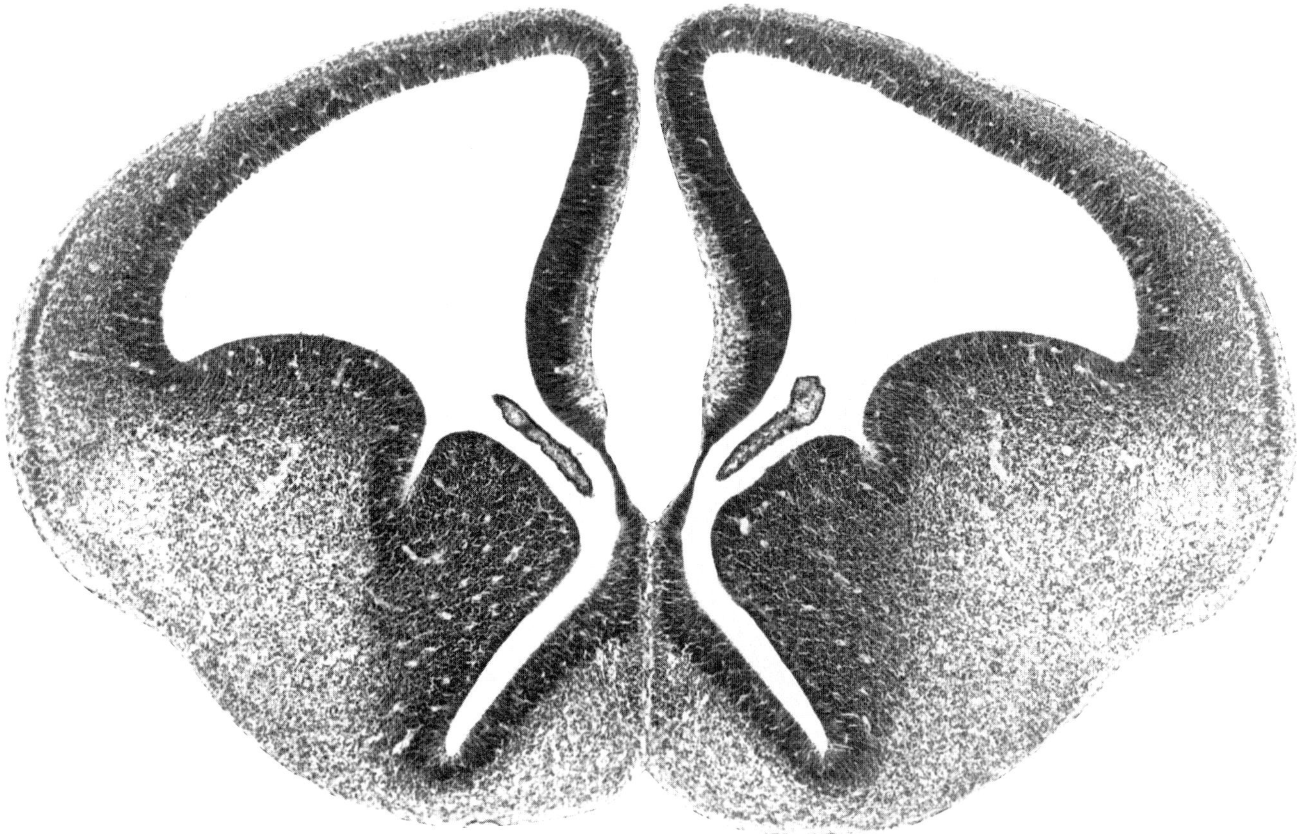

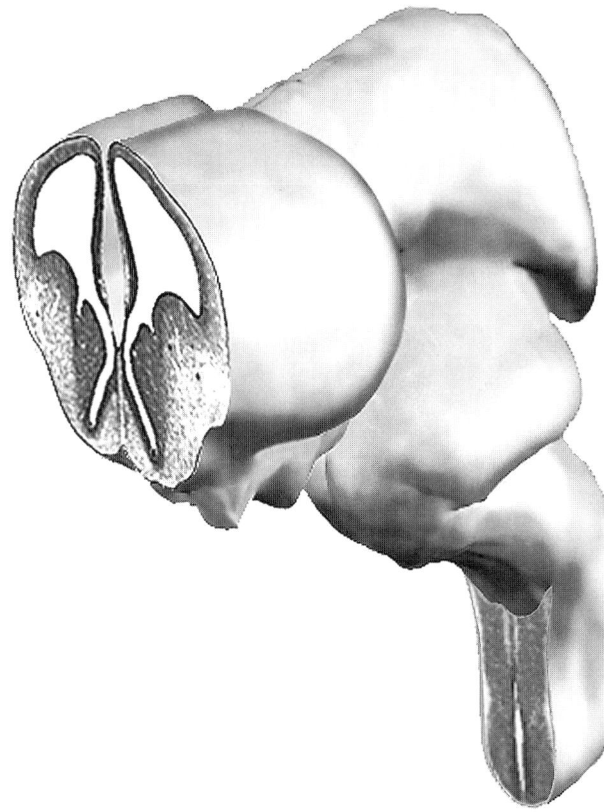

CORONAL FIGURE 7

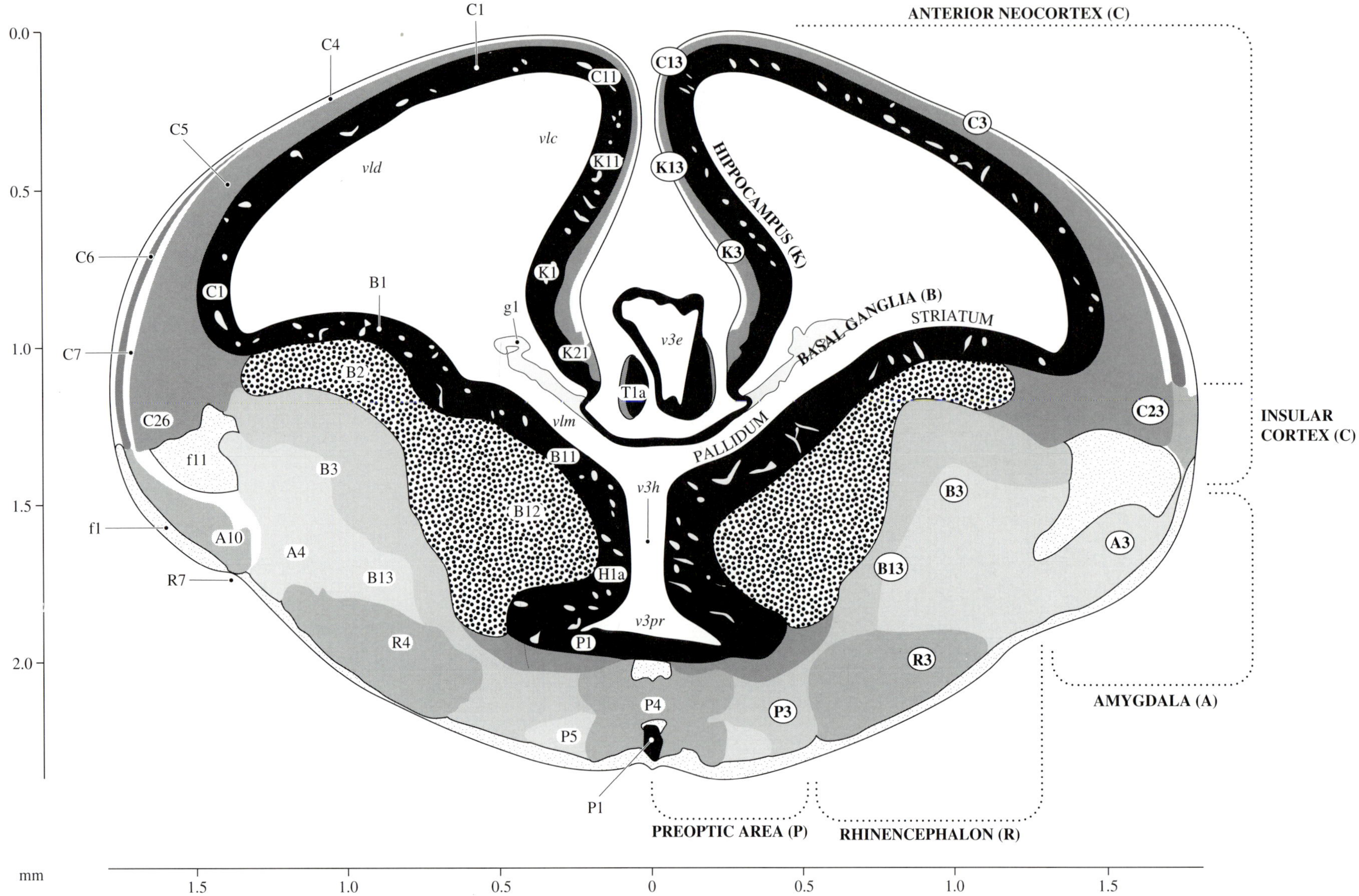

E16

AMYGDALA (A)

A3 amygdala (or differentiating field)
A4 anterior amygdaloid area
A10 lateral amygdaloid nucleus

BASAL GANGLIA (B)

B1 striatal neuroepithelium
B2 striatal subventricular zone
B3 striatum (or differentiating field)
B11 pallidal neuroepithelium
B12 pallidal subventricular zone
B13 pallidum (or differentiating field)

CEREBRAL CORTEX (C)

C1 neocortical neuroepithelium
C3 neocortex (or differentiating field)
C4 molecular layer (or primordial plexiform layer)
C5 intermediate zone
C6 cortical plate
C7 white matter
C11 cingulate cortex neuroepithelium
C13 cingulate cortex (or differentiating field)
C23 orbital and insular area (or differentiating field)
C26 claustrum

FIBER TRACTS (f)

f1 lateral olfactory tract
f11 external capsule

GLANDS AND PARANEURAL STRUCTURES (g)

g1 choroid plexus of lateral ventricle

HYPOTHALAMUS (H)

H1a anterior hypothalamic neuroepithelium

HIPPOCAMPAL FORMATION (K)

K1 hippocampal (Ammon's horn) neuroepithelium
K3 hippocampus (or differentiating field)
K11 subicular neuroepithelium
K13 subicular area (or differentiating field)
K21 dentate gyrus neuroepithelium

PREOPTIC AREA AND OPTIC EPITHELIA (P)

P1 preoptic neuroepithelium
P3 preoptic area (or differentiating field)
P4 medial preoptic region
P5 lateral preoptic region

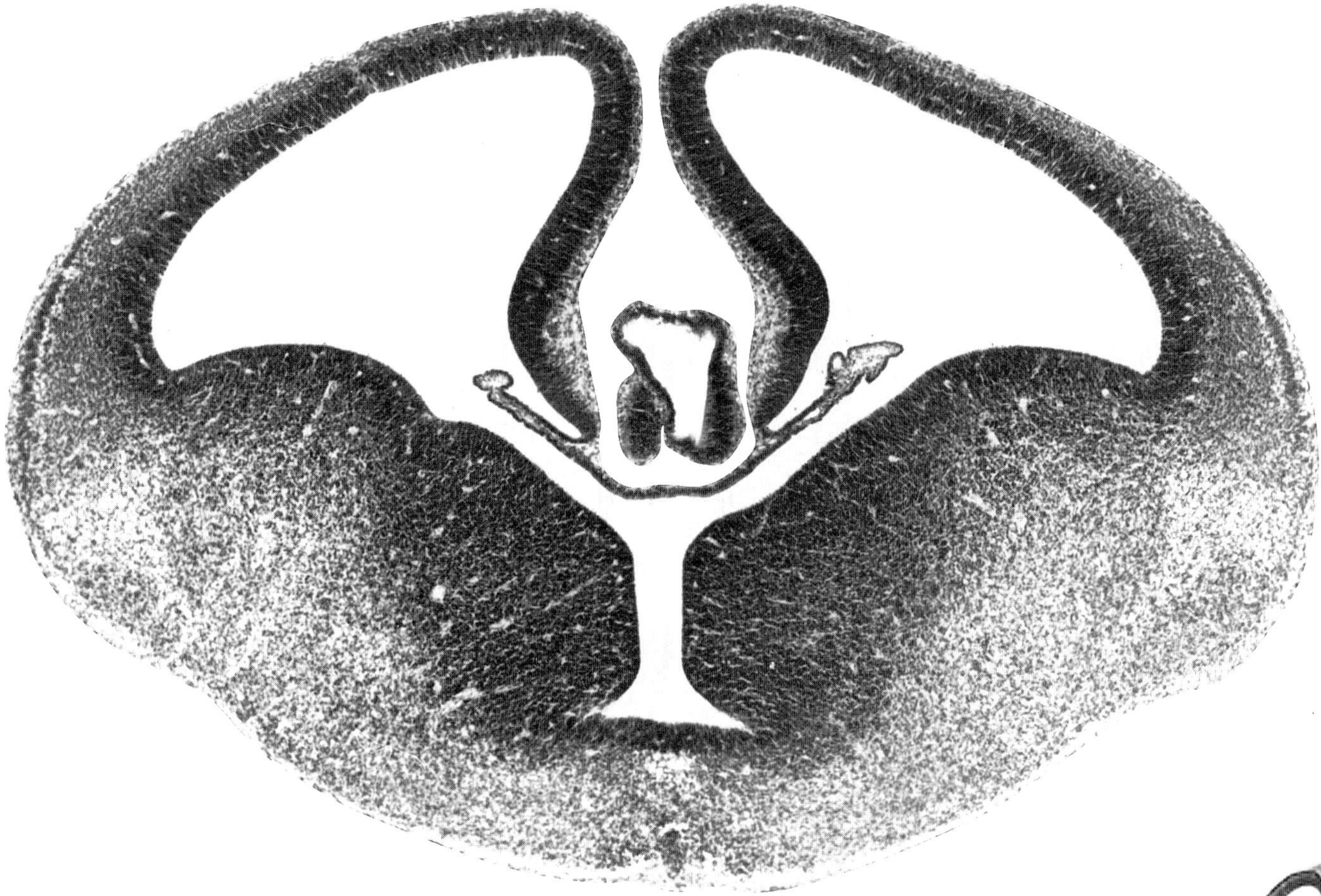

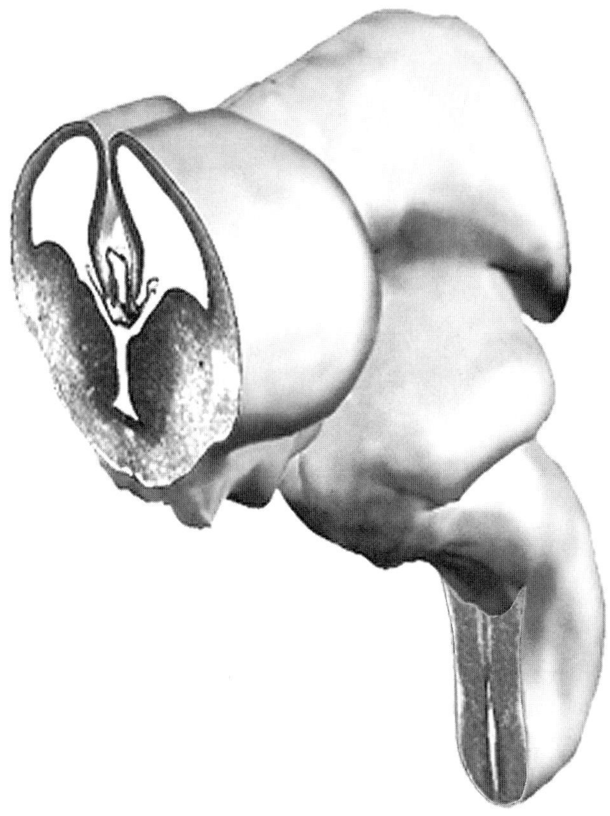

RHINENCEPHALON (R)

R3 rhinencephalon (or differentiating field)
R4 piriform cortex
R7 rhinal fissure

THALAMUS (T)

T1a anterior thalamic neuroepithelium

VENTRICLES (v)

vlc lateral ventricle, central
vld lateral ventricle, dorsal
vlm foramen of Monro
v3e third ventricle, epithalamic
v3h third ventricle, hypothalamic
v3pr third ventricle, preoptic recess

CORONAL FIGURE 8

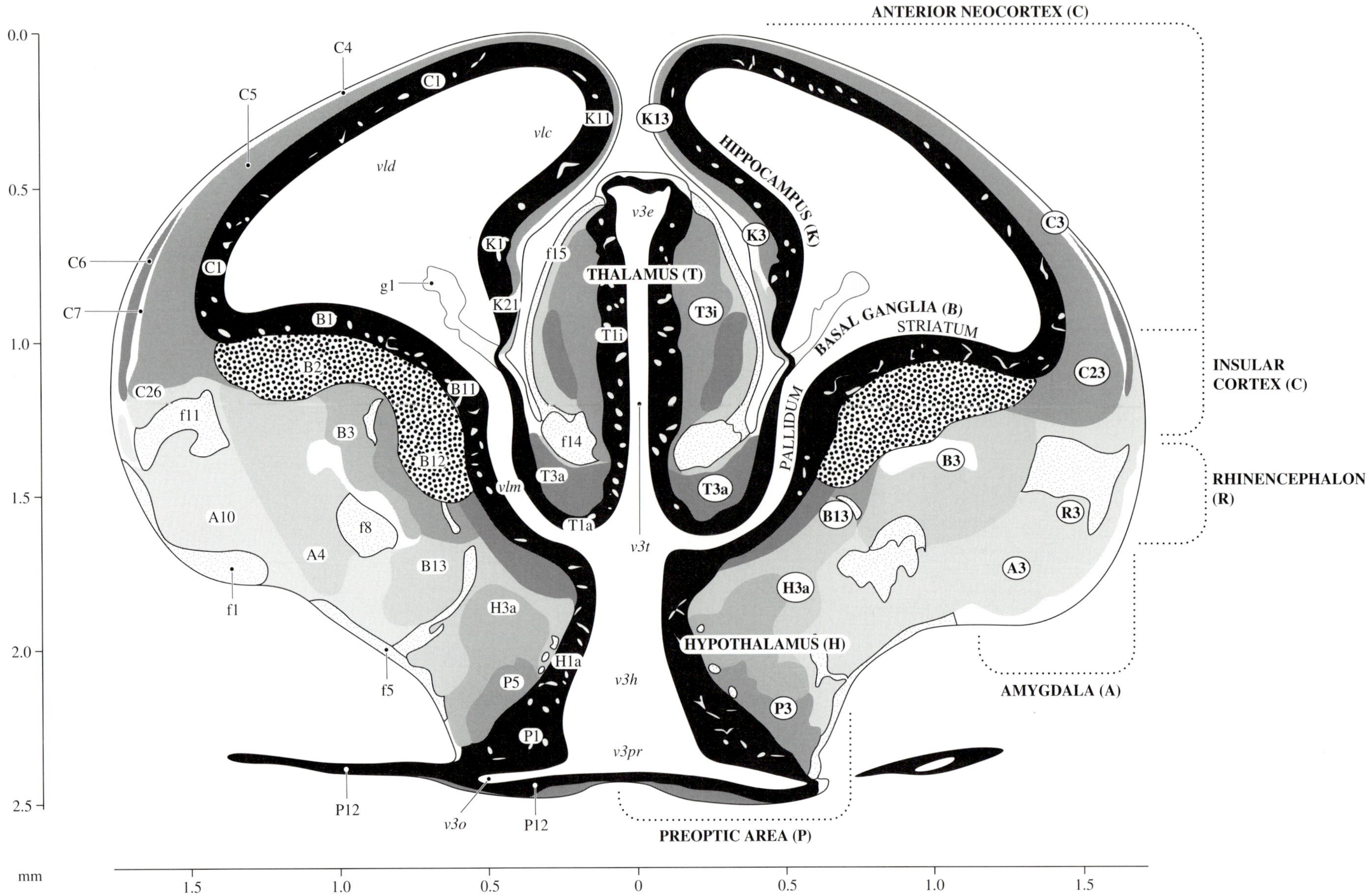

AMYGDALA (A)

A3 amygdala (or differentiating field)
A4 anterior amygdaloid area
A10 lateral amygdaloid nucleus

BASAL GANGLIA (B)

B1 striatal neuroepithelium
B2 striatal subventricular zone
B3 striatum (or differentiating field)
B11 pallidal neuroepithelium
B12 pallidal subventricular zone
B13 pallidum (or differentiating field)

CEREBRAL CORTEX (C)

C1 neocortical neuroepithelium
C3 neocortex (or differentiating field)
C4 molecular layer (or primordial plexiform layer)
C5 intermediate zone
C6 cortical plate
C7 white matter
C23 orbital and insular area (or differentiating field)
C26 claustrum

FIBER TRACTS (f)

f1 lateral olfactory tract
f5 medial forebrain bundle
f8 internal capsule
f11 external capsule
f14 stria terminalis
f15 stria medullaris

GLANDS AND PARANEURAL STRUCTURES (g)

g1 choroid plexus of lateral ventricle

HYPOTHALAMUS (H)

H1a anterior hypothalamic neuroepithelium
H3a anterior hypothalamic area (or differentiating field)

HIPPOCAMPAL FORMATION (K)

K1 hippocampal (Ammon's horn) neuroepithelium
K3 hippocampus (or differentiating field)
K11 subicular neuroepithelium
K13 subicular area (or differentiating field)
K21 dentate gyrus neuroepithelium

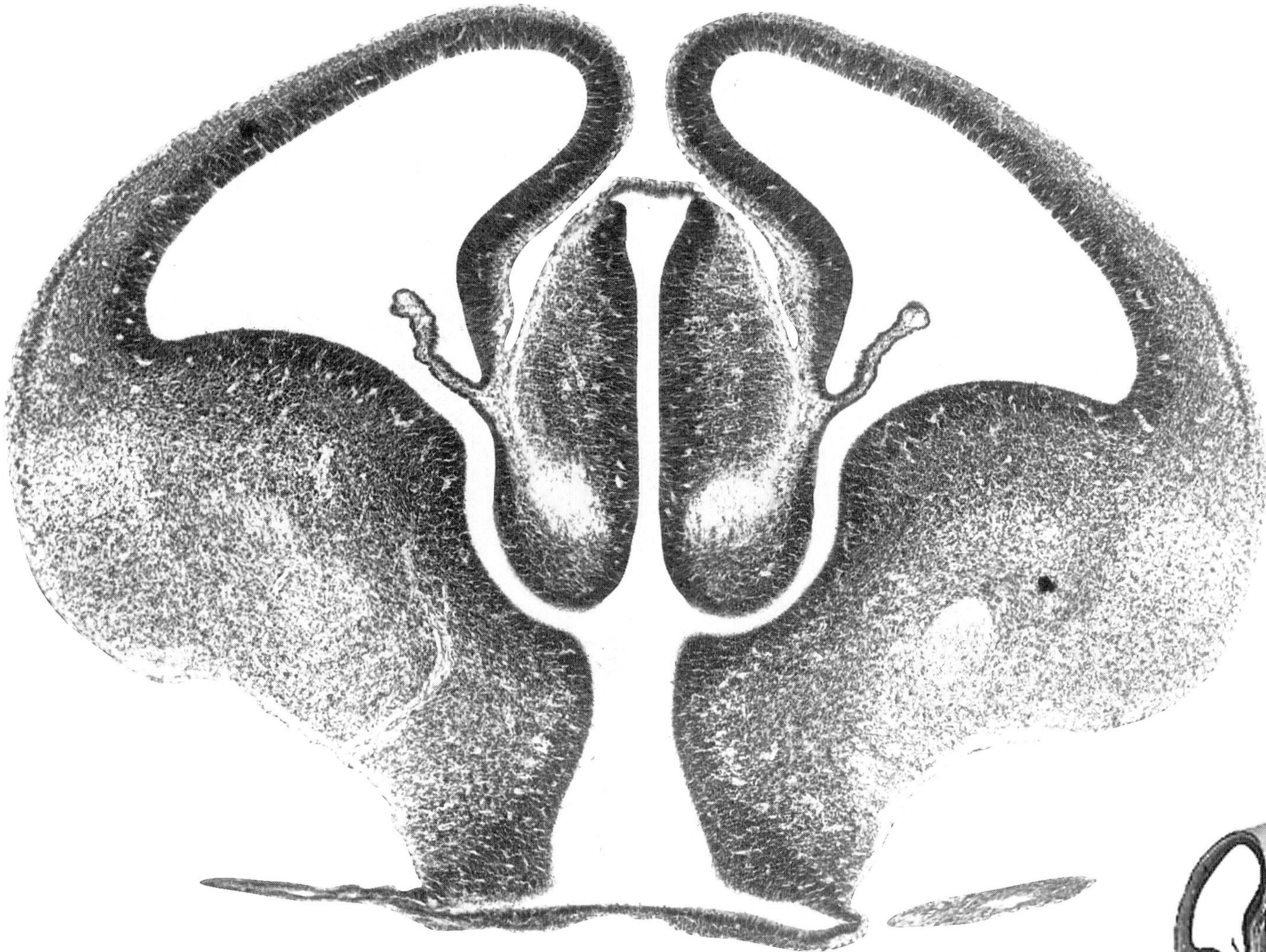

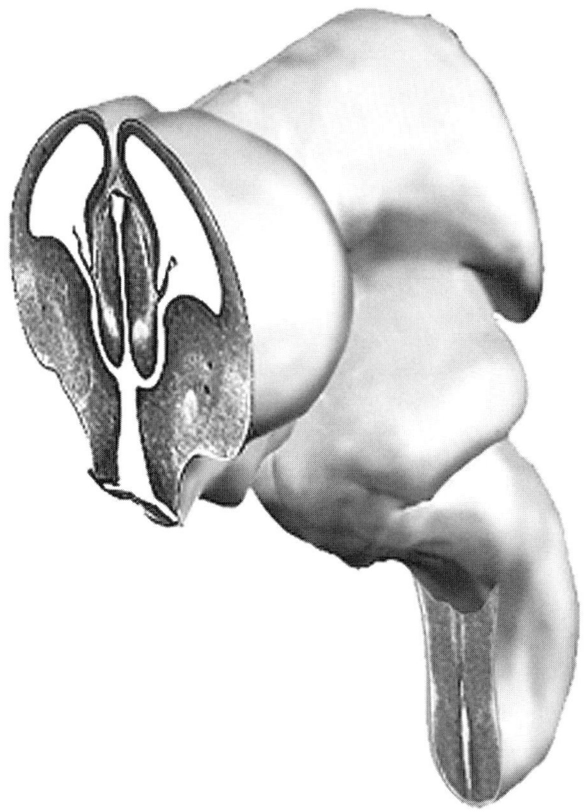

PREOPTIC AREA AND OPTIC EPITHELIA (P)

P1 preoptic neuroepithelium
P3 preoptic area (or differentiating field)
P5 lateral preoptic region
P12 eye stalk epithelium

RHINENCEPHALON (R)

R3 rhinencephalon (or differentiating field)
R4 piriform cortex

THALAMUS (T)

T1a anterior thalamic neuroepithelium
T1i intermediate thalamic neuroepithelium
T3a anterior thalamus (or differentiating field)
T3i intermediate thalamus (or differentiating field)

VENTRICLES (v)

vlc lateral ventricle, central
vld lateral ventricle, dorsal
vlm foramen of Monro
v3e third ventricle, epithalamic
v3h third ventricle, hypothalamic
v3o third ventricle, optic recess
v3pr third ventricle, preoptic recess
v3t third ventricle, thalamic

CORONAL FIGURE 9

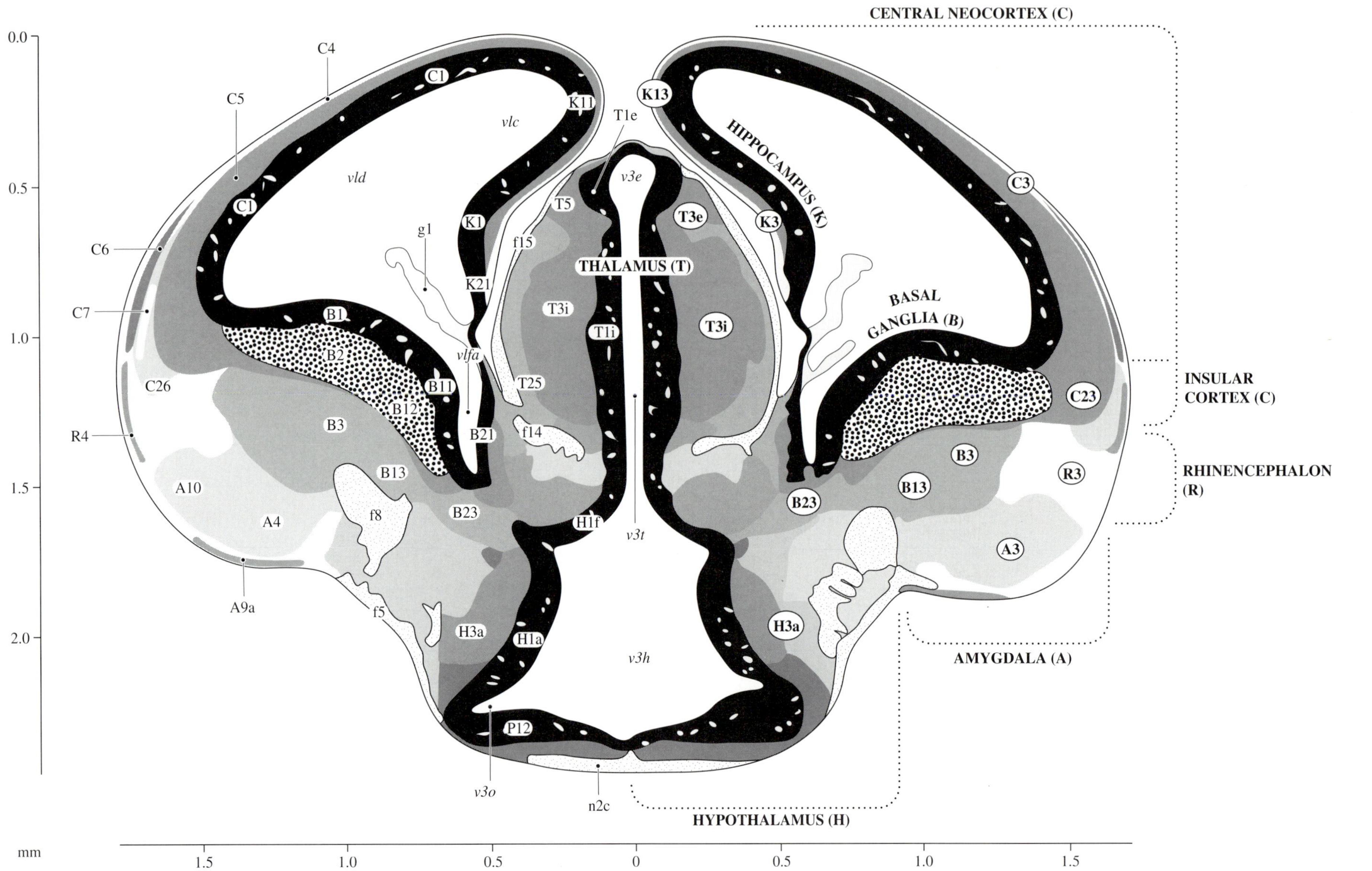

AMYGDALA (A)

- **A3** amygdala (or differentiating field)
- A4 anterior amygdaloid area
- A9a cortical amygdaloid nucleus, anterior
- A10 lateral amygdaloid nucleus

BASAL GANGLIA (B)

- B1 striatal neuroepithelium
- B2 striatal subventricular zone
- **B3** striatum (or differentiating field)
- B11 pallidal neuroepithelium
- B12 pallidal subventricular zone
- **B13** pallidum (or differentiating field)
- B21 strionuclear neuroepithelium
- **B23** bed nucleus of stria terminalis (or differentiating field)

CEREBRAL CORTEX (C)

- C1 neocortical neuroepithelium
- **C3** neocortex (or differentiating field)
- C4 molecular layer (or primordial plexiform layer)
- C5 intermediate zone
- C6 cortical plate
- C7 white matter
- **C23** orbital and insular area (or differentiating field)
- C26 claustrum

FIBER TRACTS (f)

- f5 medial forebrain bundle
- f8 internal capsule
- f14 stria terminalis
- f15 stria medullaris

GLANDS AND PARANEURAL STRUCTURES (g)

- g1 choroid plexus of lateral ventricle

HYPOTHALAMUS (H)

- H1a anterior hypothalamic neuroepithelium
- H1f Forelian neuroepithelium
- **H3a** anterior hypothalamus (or differentiating field)

HIPPOCAMPAL FORMATION (K)

- K1 hippocampal (Ammon's horn) neuroepithelium
- **K3** hippocampus (or differentiating field)
- K11 subicular neuroepithelium
- **K13** subicular area (or differentiating field)
- K21 dentate gyrus neuroepithelium

NERVES AND NUCLEI, CRANIAL (n)

- n2c optic chiasma

PREOPTIC AREA AND OPTIC EPITHELIA (P)

- P12 eye stalk epithelium

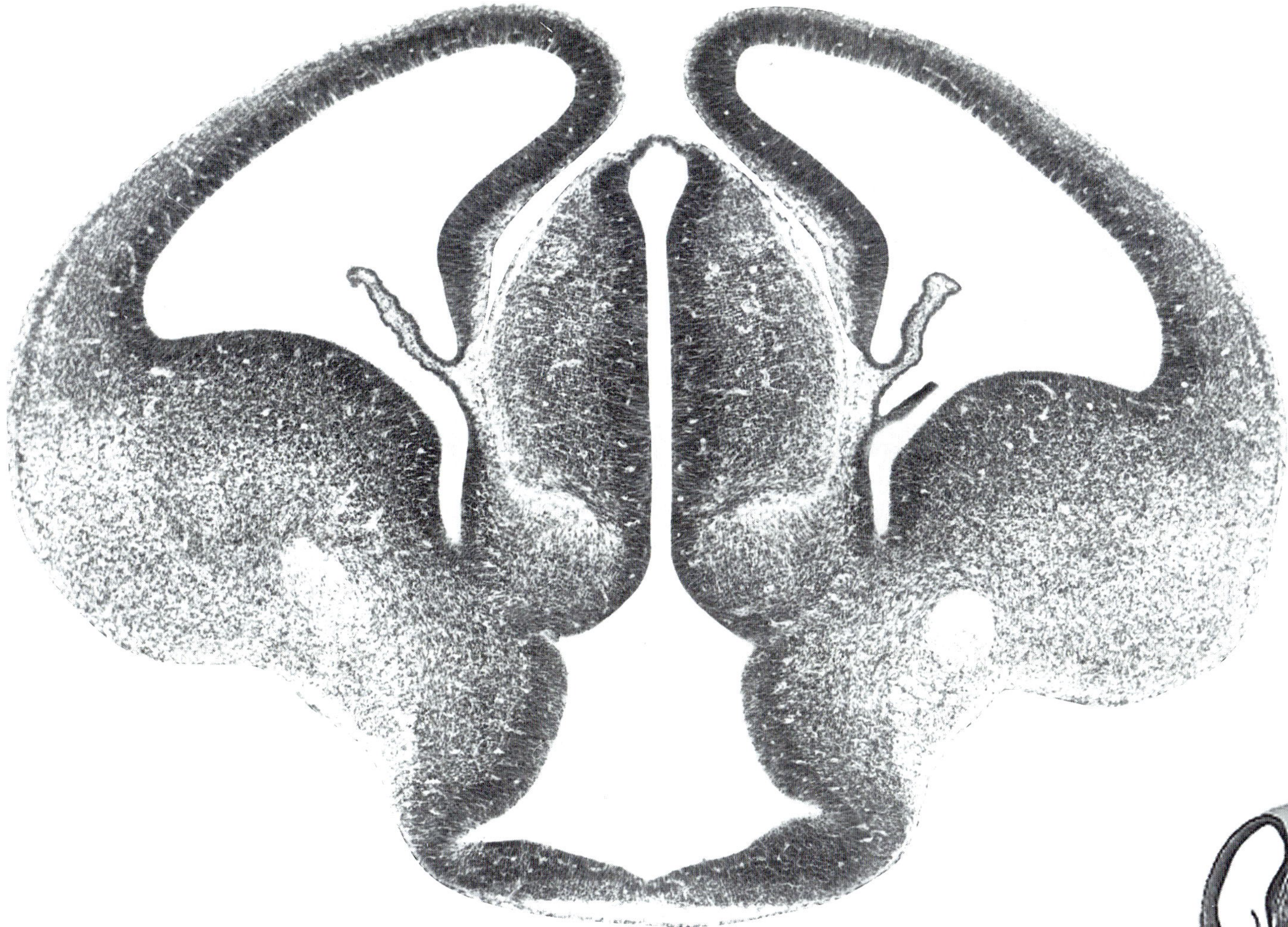

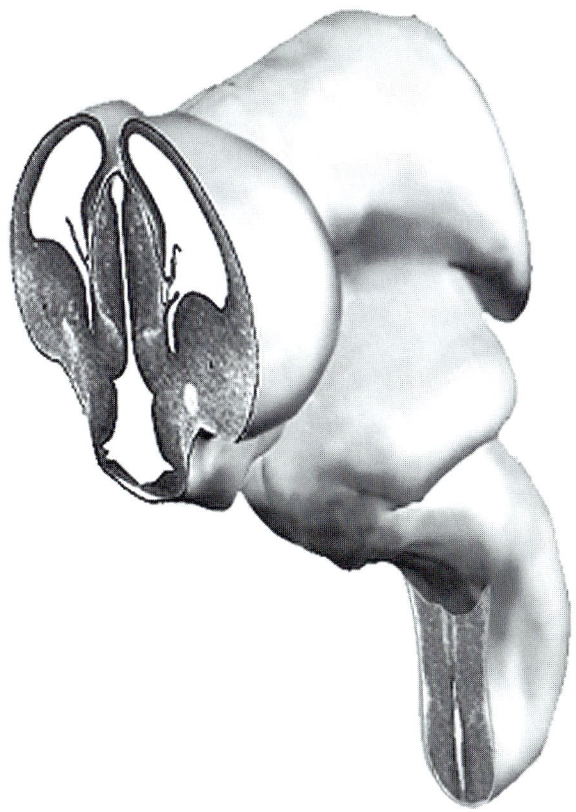

RHINENCEPHALON (R)

R3 rhinencephalon (or differentiating field)
R4 piriform cortex

THALAMUS (T)

T1e epithalamic neuroepithelium
T1i intermediate thalamic neuroepithelium
T3e epithalamus (or differentiating field)
T3i intermediate thalamus (or differentiating field)
T5 lateral habenular nucleus
T25 reticular nucleus

VENTRICLES (v)

vlc lateral ventricle, central
vld lateral ventricle, dorsal
vlfa lateral ventricle, amygdaloid fork
v3e third ventricle, epithalamic
v3h third ventricle, hypothalamic
v3o third ventricle, optic recess
v3t third ventricle, thalamic

CORONAL FIGURE 10

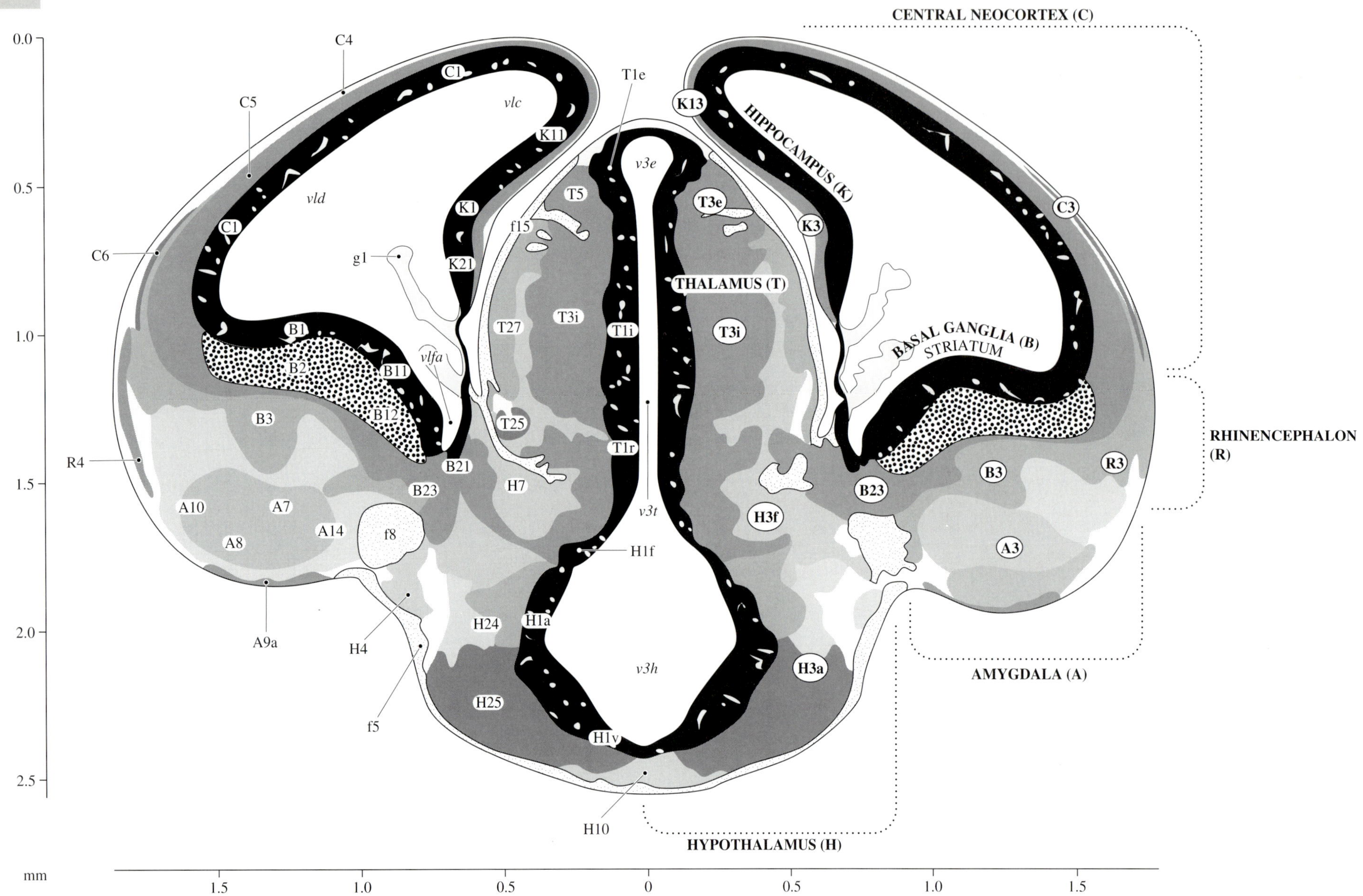

AMYGDALA (A)

A3 amygdala (or differentiating field)
A7 basomedial amygdaloid nucleus
A8 basolateral amygdaloid nucleus
A9a cortical amygdaloid nucleus, anterior
A10 lateral amygdaloid nucleus
A14 medial amygdaloid nucleus

BASAL GANGLIA (B)

B1 striatal neuroepithelium
B2 striatal subventricular zone
B3 striatum (or differentiating field)
B11 pallidal neuroepithelium
B12 pallidal subventricular zone
B21 strionuclear neuroepithelium
B23 bed nucleus of stria terminalis (or differentiating field)

CEREBRAL CORTEX (C)

C1 neocortical neuroepithelium
C3 neocortex (or differentiating field)
C4 molecular layer (or primordial plexiform layer)
C5 intermediate zone
C6 cortical plate

FIBER TRACTS (f)

f5 medial forebrain bundle
f8 internal capsule
f15 stria medullaris

GLANDS AND PARANEURAL STRUCTURES (g)

g1 choroid plexus of lateral ventricle

HYPOTHALAMUS (H)

H1a anterior hypothalamic neuroepithelium
H1f Forelian neuroepithelium
H1v ventral hypothalamic neuroepithelium
H3a anterior hypothalamus (or differentiating field)
H3f fields of Forel (or differentiating field)
H4 supraoptic nucleus
H7 zona incerta
H10 anterobasal nucleus
H24 anterodorsal nucleus
H25 anteroventral nucleus

HIPPOCAMPAL FORMATION (K)

K1 hippocampal (Ammon's horn) neuroepithelium
K3 hippocampus (or differentiating field)
K11 subicular neuroepithelium
K13 subicular area (or differentiating field)
K21 dentate gyrus neuroepithelium

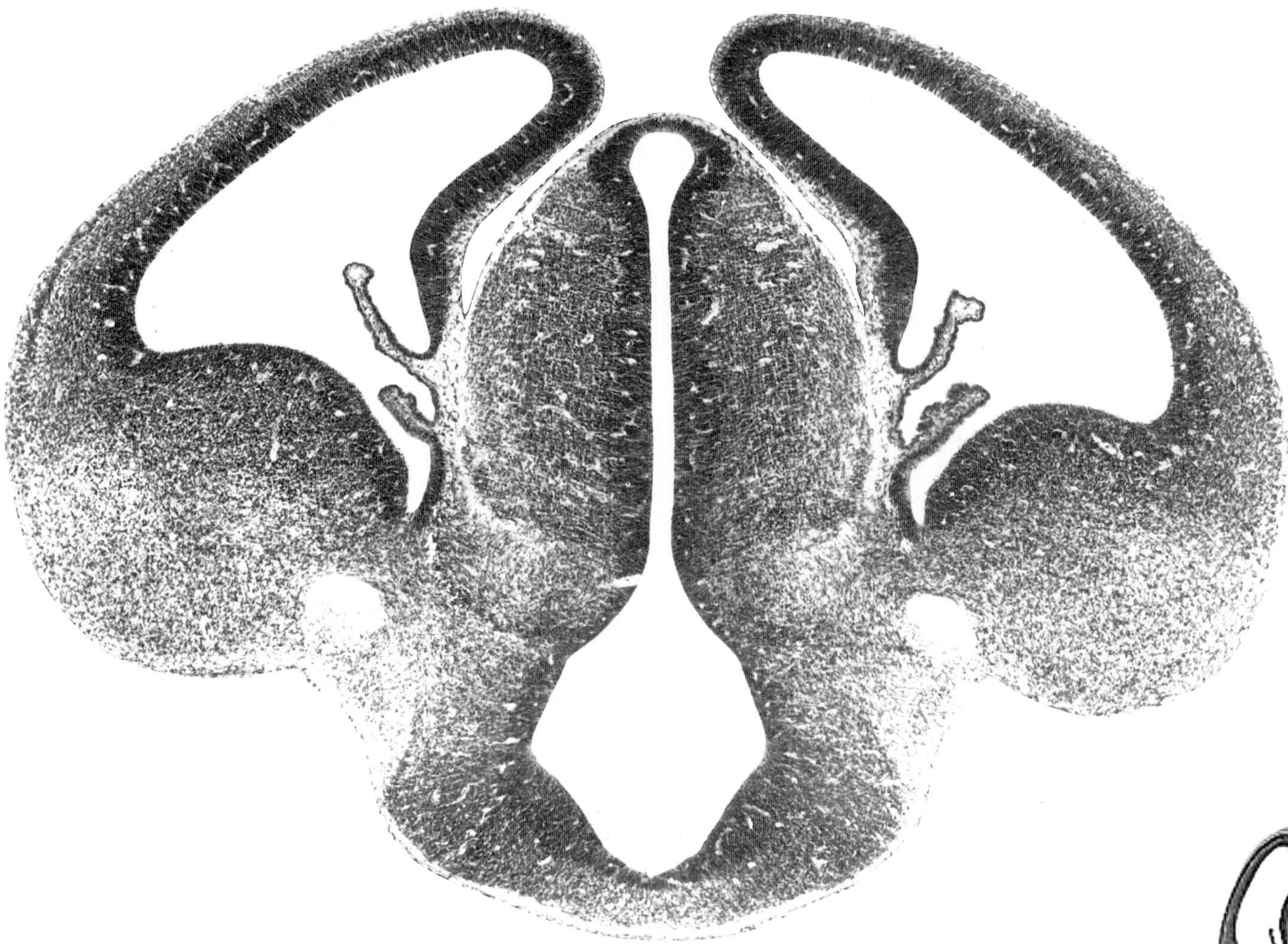

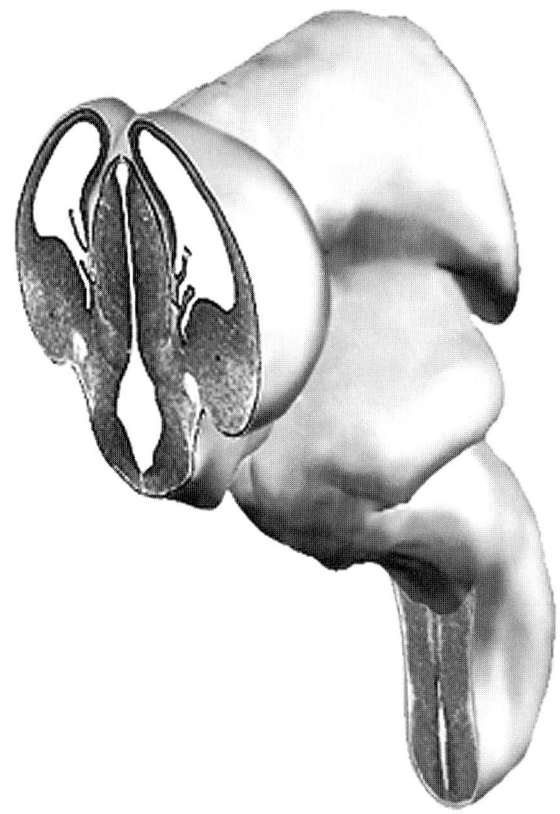

RHINENCEPHALON (R)

R3 rhinencephalon (or differentiating field)
R4 piriform cortex

THALAMUS (T)

T1e epithalamic neuroepithelium
T1i intermediate thalamic neuroepithelium
T1r reticular (or ventral) thalamic neuroepithelium
T3e epithalamus (or differentiating field)
T3i intermediate thalamus (or differentiating field)
T5 lateral habenular nucleus
T25 reticular nucleus
T27 ventrolateral nuclear complex

VENTRICLES (v)

vlc lateral ventricle, central
vld lateral ventricle, dorsal
vlfa lateral ventricle, amygdaloid fork
v3e third ventricle, epithalamic
v3h third ventricle, hypothalamic
v3t third ventricle, thalamic

CORONAL FIGURE 11

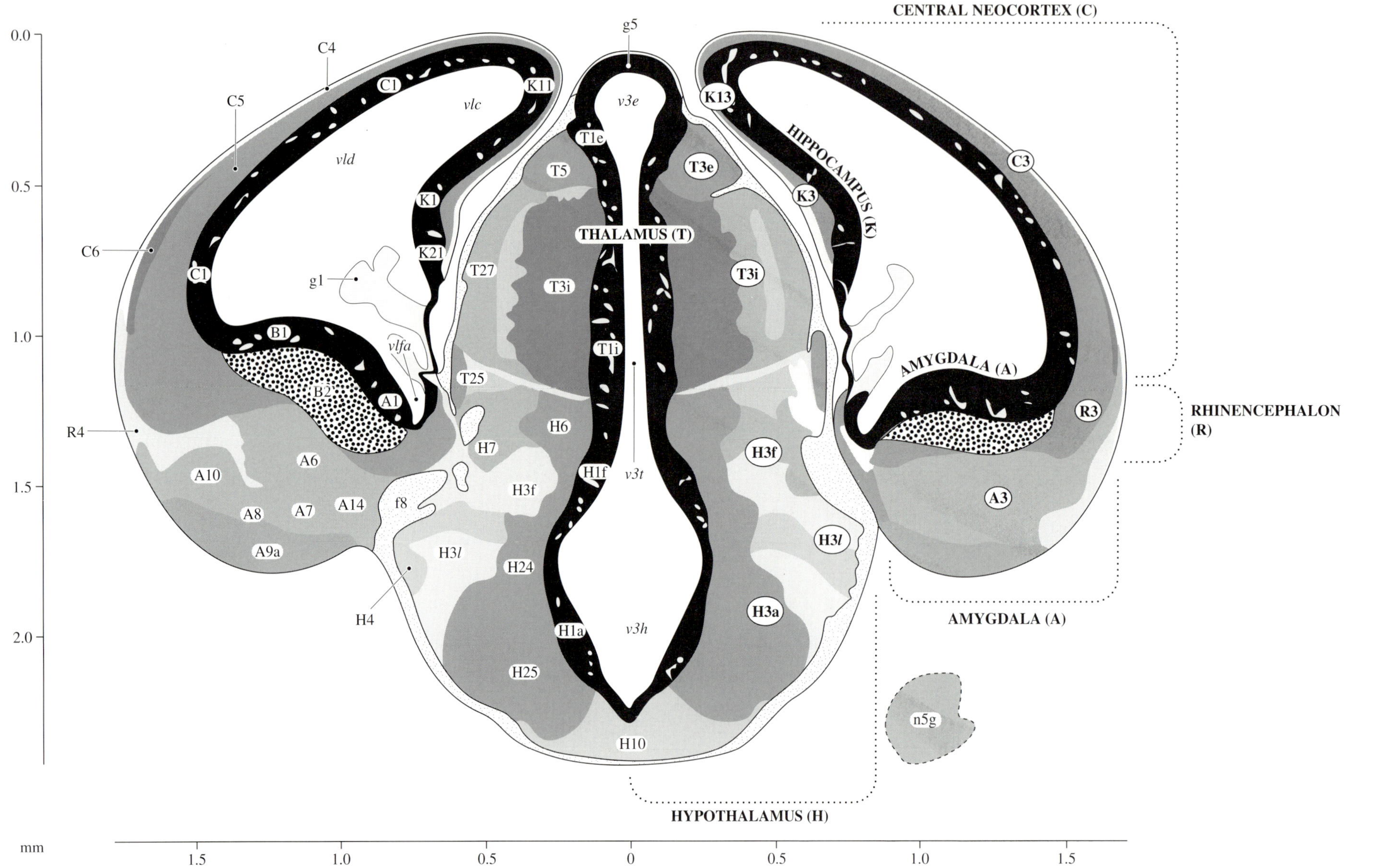

AMYGDALA (A)

- A1 amygdaloid neuroepithelium
- **A3** amygdala (or differentiating field)
- A6 central amygdaloid nucleus
- A7 basomedial amygdaloid nucleus
- A8 basolateral amygdaloid nucleus
- A9a cortical amygdaloid nucleus, anterior
- A10 lateral amygdaloid nucleus
- A14 medial amygdaloid nucleus

BASAL GANGLIA (B)

- B1 striatal neuroepithelium
- B2 striatal subventricular zone

CEREBRAL CORTEX (C)

- C1 neocortical neuroepithelium
- **C3** neocortex (or differentiating field)
- C4 molecular layer (or primordial plexiform layer)
- C5 intermediate zone
- C6 cortical plate

FIBER TRACTS (f)

- f8 internal capsule

GLANDS AND PARANEURAL STRUCTURES (g)

- g1 choroid plexus of lateral ventricle
- g5 pineal gland (primordium)

HYPOTHALAMUS (H)

- H1a anterior hypothalamic neuroepithelium
- H1f Forelian neuroepithelium
- **H3a** anterior hypothalamus (or differentiating field)
- **H3f** fields of Forel (or differentiating field)
- **H3*l*** lateral hypothalamic area (or differentiating field)
- H4 supraoptic nucleus
- H6 paraventricular nucleus
- H7 zona incerta
- H10 anterobasal nucleus
- H24 anterodorsal nucleus
- H25 anteroventral nucleus

HIPPOCAMPAL FORMATION (K)

- K1 hippocampal (Ammon's horn) neuroepithelium
- **K3** hippocampus (or differentiating field)
- K11 subicular neuroepithelium
- **K13** subicular area (or differentiating field)
- K21 dentate gyrus neuroepithelium

NERVES AND NUCLEI, CRANIAL (n)

- n5g trigeminal ganglion

RHINENCEPHALON (R)

- **R3** rhinencephalon (or differentiating field)
- R4 piriform cortex

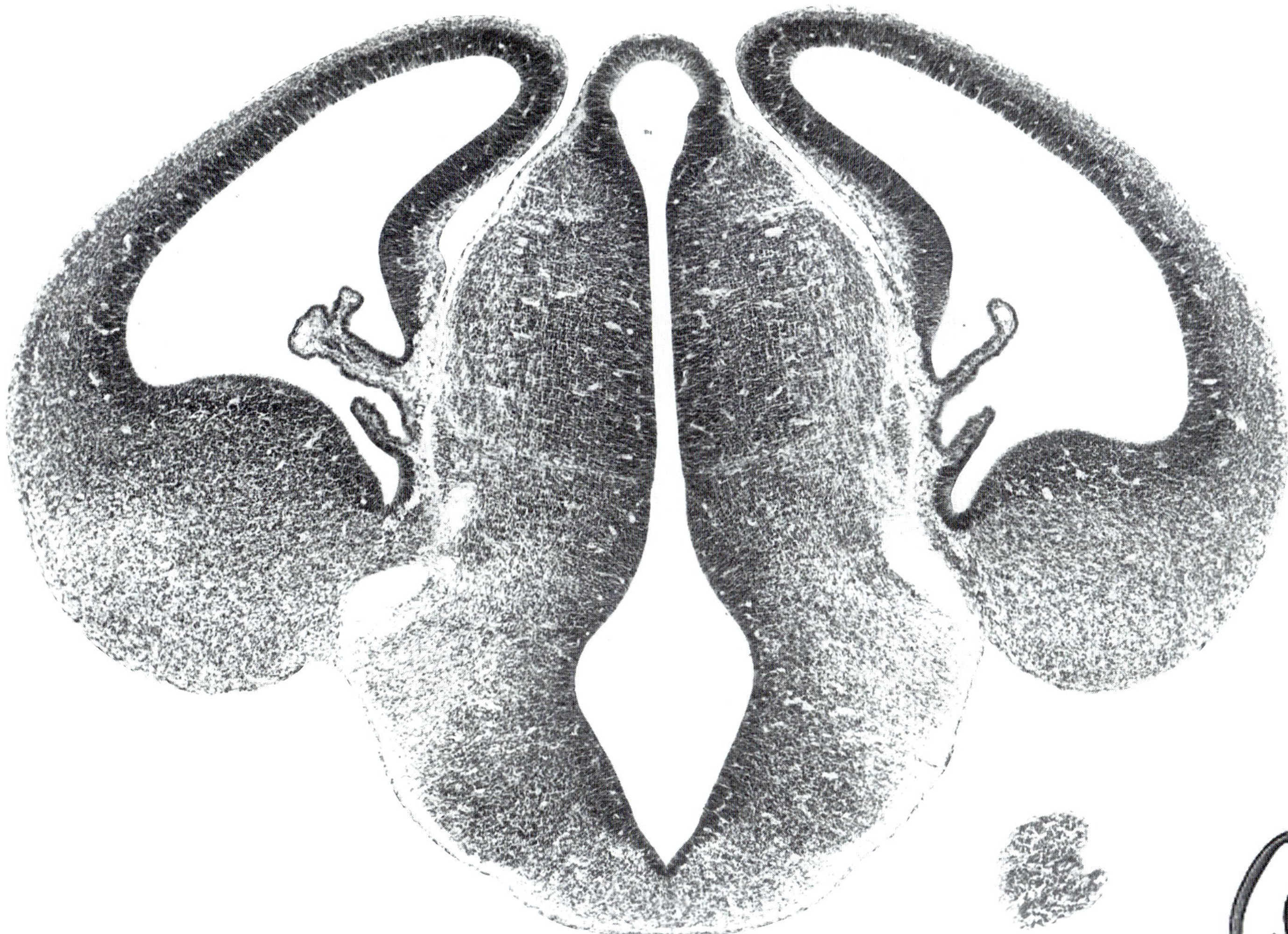

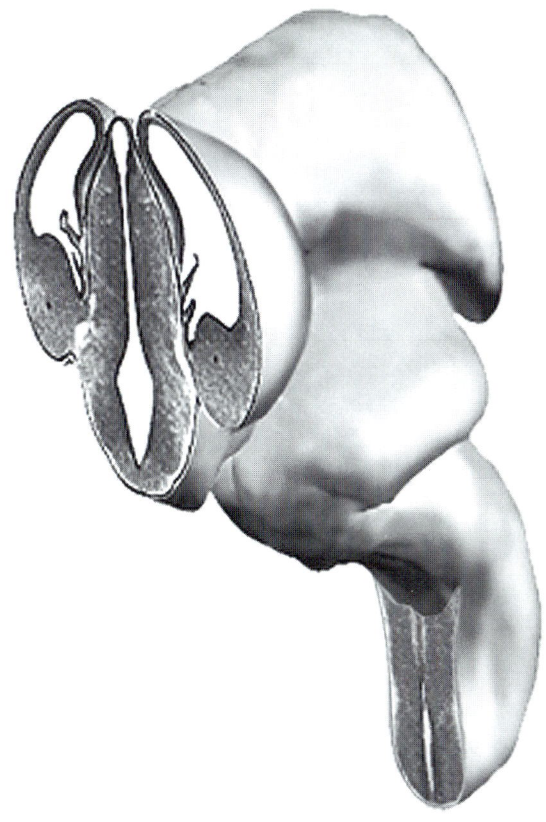

THALAMUS (T)

T1e epithalamic neuroepithelium
T1i intermediate thalamic neuroepithelium
T3e epithalamus (or differentiating field)
T3i intermediate thalamus (or differentiating field)
T5 lateral habenular nucleus
T25 reticular nucleus
T27 ventrolateral nuclear complex

VENTRICLES (v)

vlc lateral ventricle, central
vld lateral ventricle, dorsal
vlfa lateral ventricle, amygdaloid fork
v3e third ventricle, epithalamic
v3h third ventricle, hypothalamic
v3t third ventricle, thalamic

CORONAL FIGURE 12

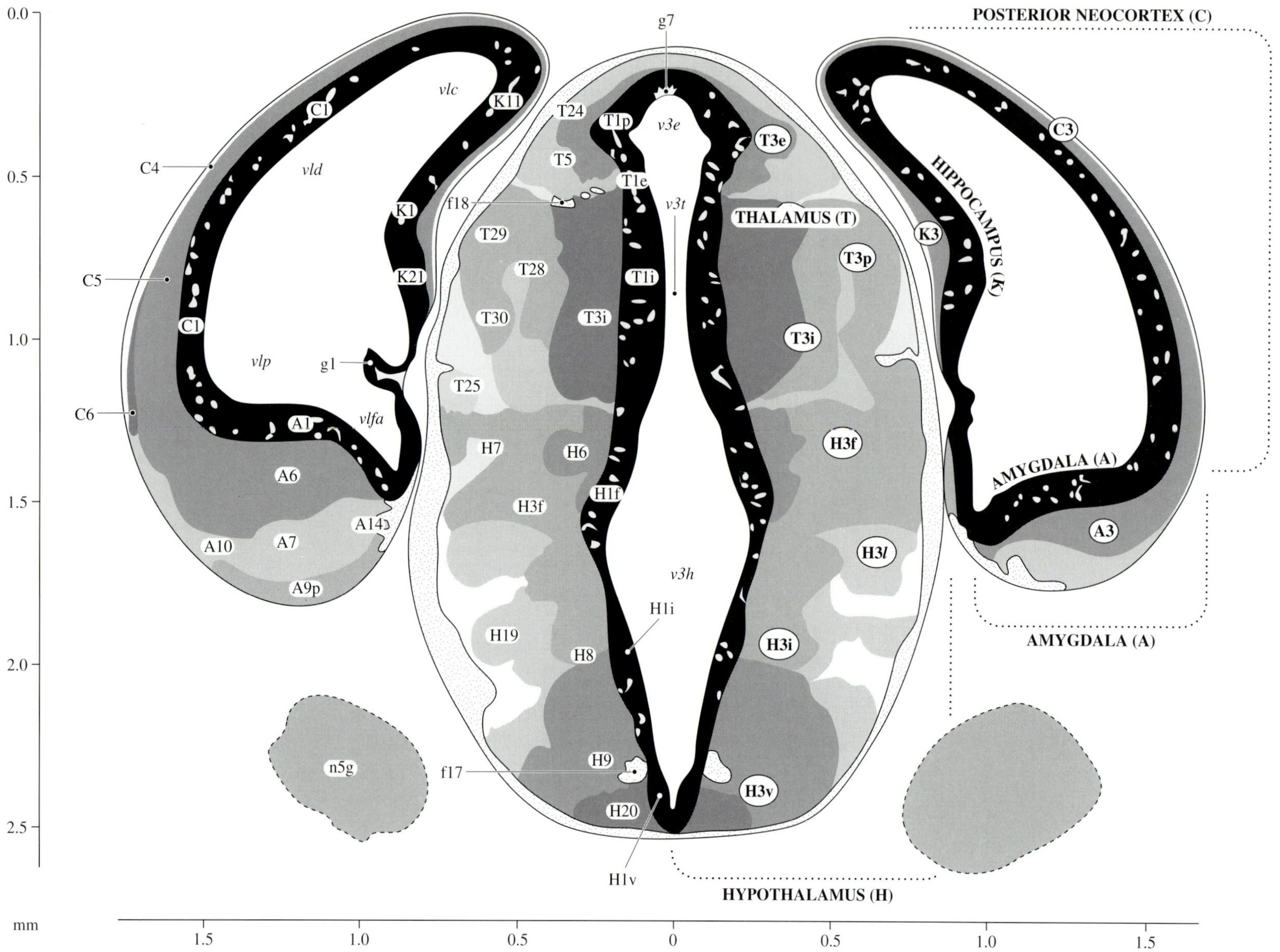

AMYGDALA (A)

A1 amygdaloid neuroepithelium
A3 amygdala (or differentiating field)
A6 central amygdaloid nucleus
A7 basomedial amygdaloid nucleus
A9p cortical amygdaloid nucleus, posterior
A10 lateral amygdaloid nucleus
A14 medial amygdaloid nucleus

CEREBRAL CORTEX (C)

C1 neocortical neuroepithelium
C3 neocortex (or differentiating field)
C4 molecular layer (or primordial plexiform layer)
C5 intermediate zone
C6 cortical plate

FIBER TRACTS (f)

f17 mammillothalamic tract
f18 habenulopeduncular tract

GLANDS AND PARANEURAL STRUCTURES (g)

g1 choroid plexus of lateral ventricle
g7 subcommissural organ

HYPOTHALAMUS (H)

H1f Forelian neuroepithelium
H1i intermediate hypothalamic neuroepithelium
H1v ventral hypothalamic neuroepithelium
H3f fields of Forel (or differentiating field)
H3i intermediate hypothalamus (or differentiating field)
H3*l* lateral hypothalamus (or differentiating field)
H3v ventral hypothalamus (or differentiating field)
H6 paraventricular nucleus
H7 zona incerta
H8 dorsomedial nucleus
H9 ventromedial nucleus
H19 subthalamic nucleus
H20 arcuate nucleus

HIPPOCAMPAL FORMATION (K)

K1 hippocampal (Ammon's horn) neuroepithelium
K3 hippocampus (or differentiating field)
K11 subicular neuroepithelium
K21 dentate gyrus neuroepithelium

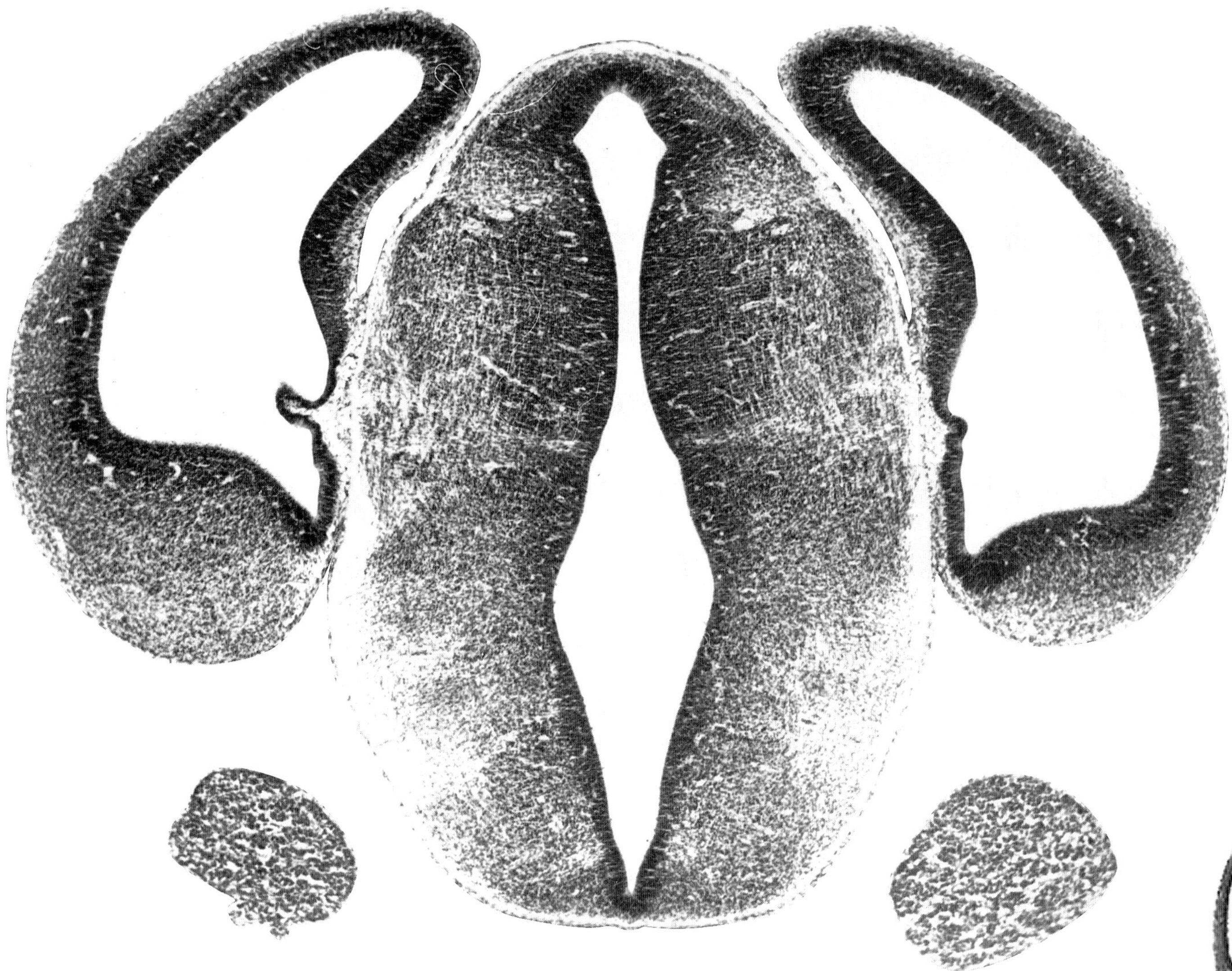

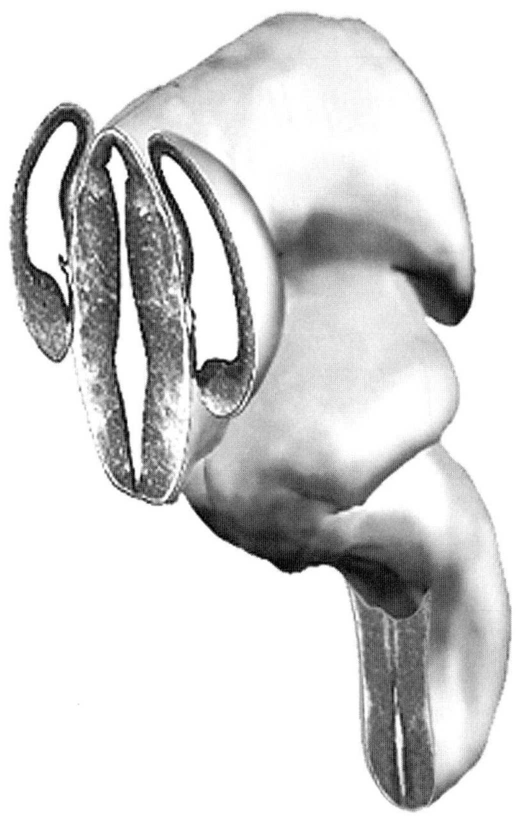

NERVES AND NUCLEI, CRANIAL (n)

n5g trigeminal ganglion

THALAMUS (T)

T1e epithalamic neuroepithelium
T1i intermediate thalamic neuroepithelium
T1p posterior thalamic neuroepithelium
T3e epithalamus (or differentiating field)
T3i intermediate thalamus (or differentiating field)
T3p posterior thalamus (or differentiating field)
T5 lateral habenular nucleus
T24 lateral posterior nucleus
T25 reticular nucleus
T28 ventrobasal nuclear complex
T29 dorsal lateral geniculate nucleus
T30 ventral lateral geniculate nucleus

VENTRICLES (v)

vlc lateral ventricle, central
vld lateral ventricle, dorsal
vlfa lateral ventricle, amygdaloid fork
vlp lateral ventricle, posterior horn
v3e third ventricle, epithalamic
v3h third ventricle, hypothalamic
v3t third ventricle, thalamic

CORONAL FIGURE 13

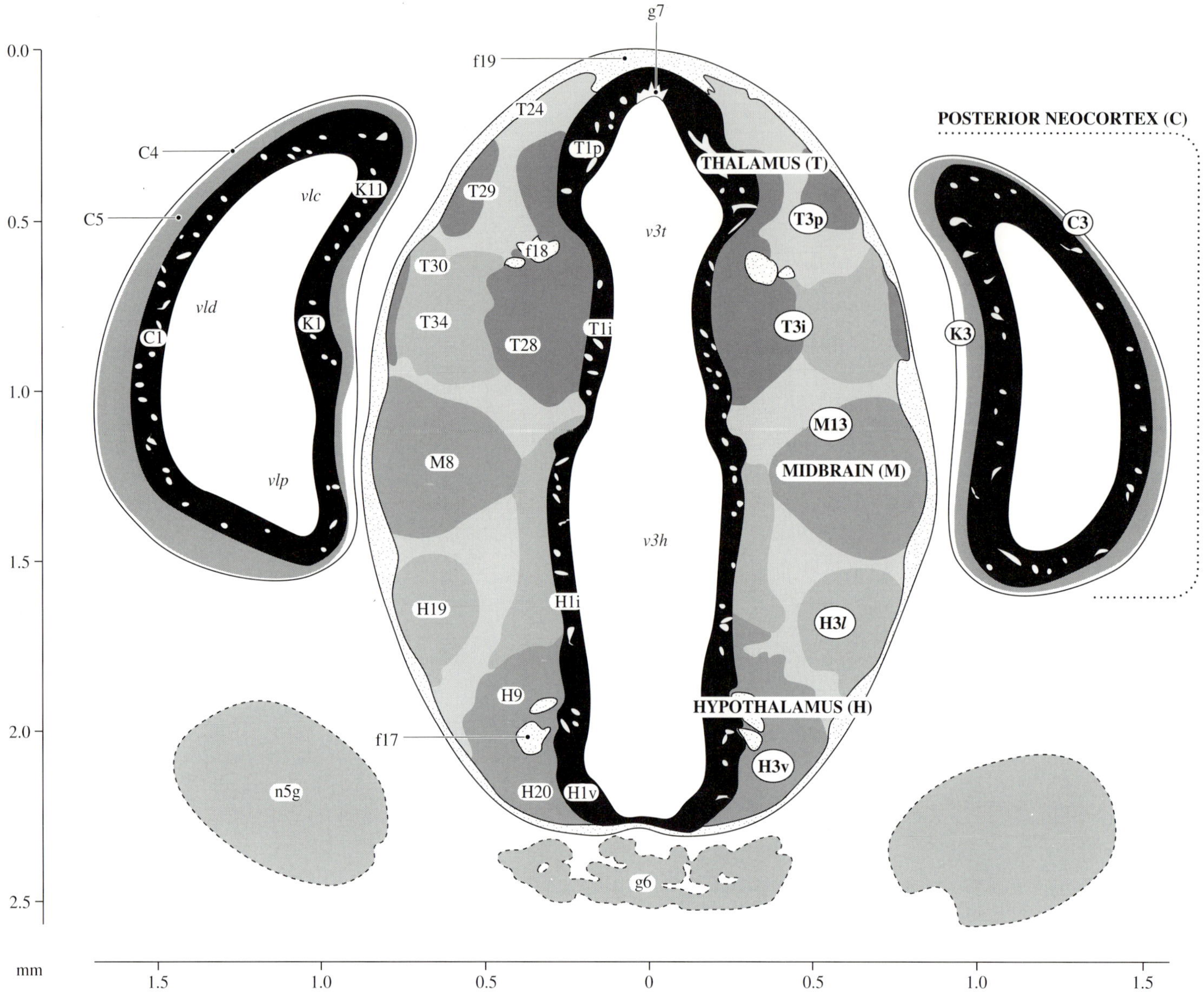

CEREBRAL CORTEX (C)

C1 neocortical neuroepithelium
C3 neocortex (or differentiating field)
C4 molecular layer (or primordial plexiform layer)
C5 intermediate zone

FIBER TRACTS (f)

f17 mammillothalamic tract
f18 habenulopeduncular tract
f19 posterior commissure

GLANDS AND PARANEURAL STRUCTURES (g)

g6 pituitary gland (primordium)
g7 subcommissural organ

HYPOTHALAMUS (H)

H1i intermediate hypothalamic neuroepithelium
H1v ventral hypothalamic neuroepithelium
H3*l* lateral hypothalamus (or differentiating field)
H3v ventral hypothalamus (or differentiating field)
H9 ventromedial nucleus
H19 subthalamic nucleus
H20 arcuate nucleus

HIPPOCAMPAL FORMATION (K)

K1 hippocampal (Ammon's horn) neuroepithelium
K3 hippocampus (or differentiating field)
K11 subicular neuroepithelium

MIDBRAIN (M)

M8 red nucleus
M13 tegmentum (or differentiating field)

NERVES AND NUCLEI, CRANIAL (n)

n5g trigeminal ganglion

THALAMUS (T)

T1i intermediate thalamic neuroepithelium
T1p posterior thalamic neuroepithelium
T3i intermediate thalamus (or differentiating field)
T3p posterior thalamus (or differentiating field)
T24 lateral posterior nucleus
T28 ventrobasal nuclear complex
T29 dorsal lateral geniculate nucleus
T30 ventral lateral geniculate nucleus
T34 medial geniculate nucleus

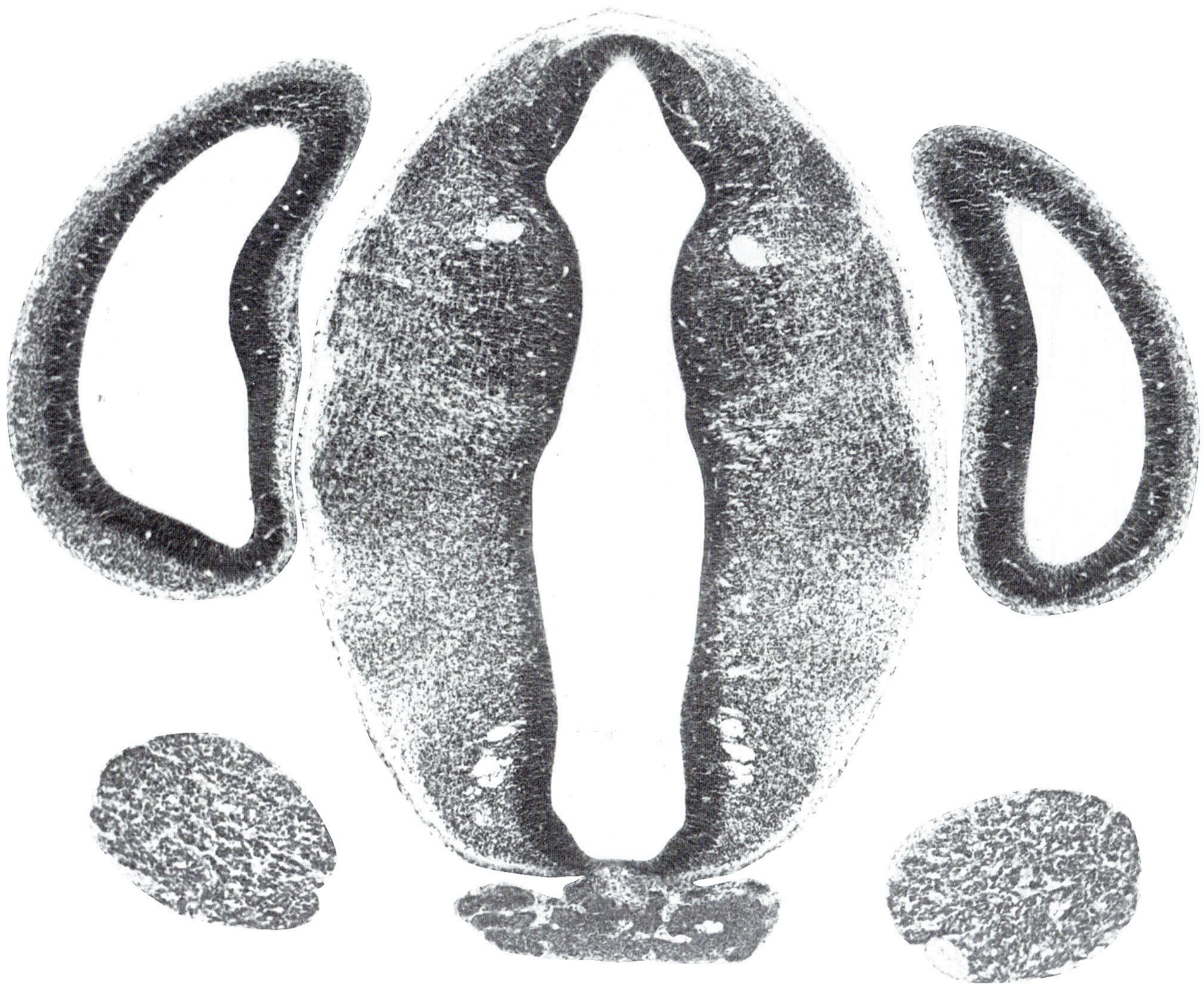

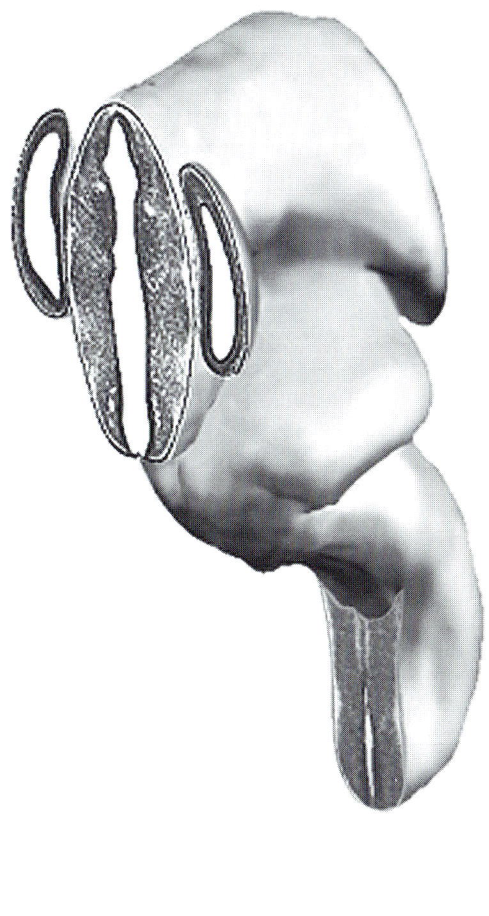

VENTRICLES (v)

vlc	lateral ventricle, central
vld	lateral ventricle, dorsal
vlp	lateral ventricle, posterior horn
v3h	third ventricle, hypothalamic
v3t	third ventricle, thalamic

CORONAL FIGURE 14

CEREBRAL CORTEX (C)

C3 neocortex (or differentiating field)

FIBER TRACTS (f)

f17 mammillothalamic tract
f18 habenulopeduncular tract
f19 posterior commissure

GLANDS AND PARANEURAL STRUCTURES (g)

g7 subcommissural organ

HYPOTHALAMUS (H)

H1i intermediate hypothalamic neuroepithelium
H1v ventral hypothalamic neuroepithelium
H3*l* lateral hypothalamus (or differentiating field)
H3v ventral hypothalamus (or differentiating field)
H15 premammillary nucleus
H19 subthalamic nucleus
H20 arcuate nucleus

MIDBRAIN (M)

M4v ventral periaqueductal gray
M8 red nucleus
M11 tegmental neuroepithelium
M13 tegmentum (or differentiating field)
M21 pretectal neuroepithelium
M23 pretectum (or differentiating field)
M24 anterior pretectal nucleus

NERVES AND NUCLEI, CRANIAL (n)

n5 trigeminal nerve
n5g trigeminal ganglion
n5n trigeminal motor nucleus
n5p principal (sensory) trigeminal nucleus

THALAMUS (T)

T3p posterior thalamus (or differentiating field)
T34 medial geniculate nucleus

VENTRICLES (v)

vapr aqueduct, pretectal
v3h third ventricle, hypothalamic
v3t third ventricle, thalamic

PONTINE AREA (X)

X3a anterior pons (or differentiating field)
X12a anterior precerebellar extramural migration

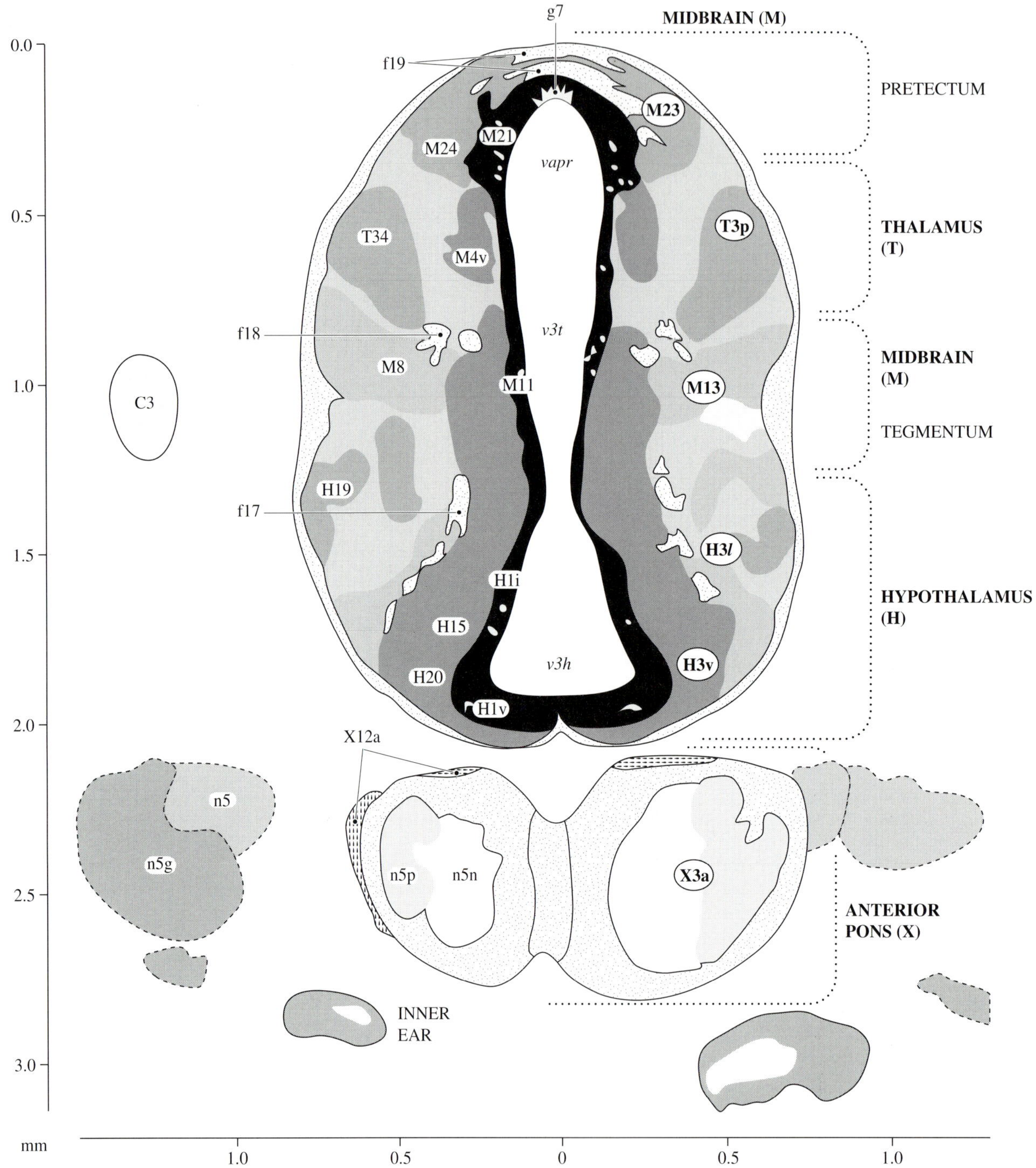

E16

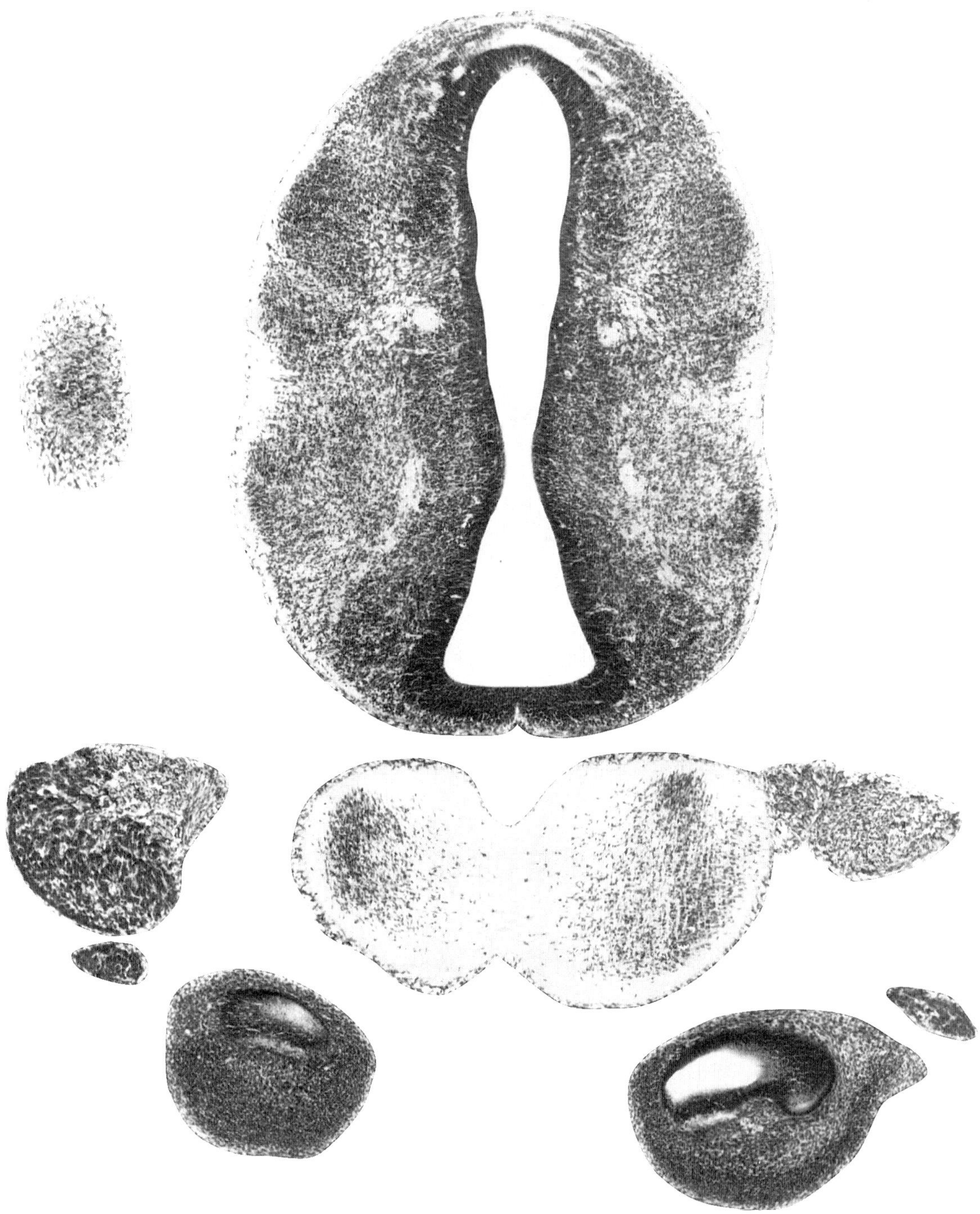

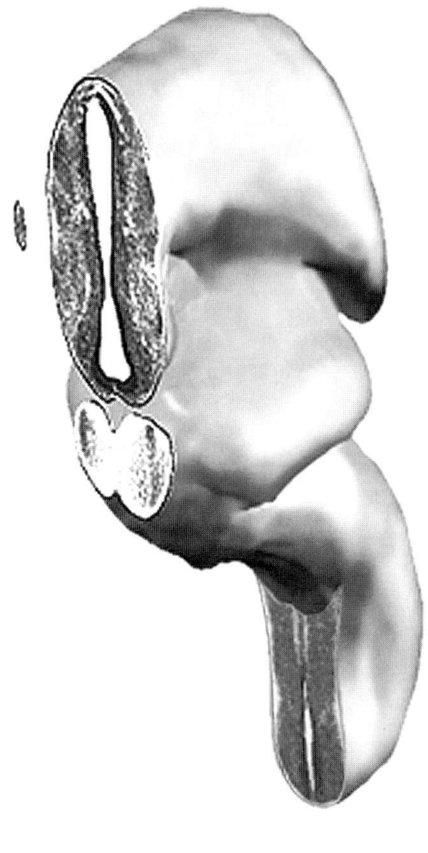

CORONAL FIGURE 15

FIBER TRACTS (f)

f17 mammillothalamic tract
f18 habenulopeduncular tract
f19 posterior commissure

GLANDS AND PARANEURAL STRUCTURES (g)

g7 subcommissural organ

HYPOTHALAMUS (H)

H1m mammillary neuroepithelium
H3m mammillary body (or differentiating field)
H14 supramammillary nucleus
H16 medial mammillary nucleus
H17 lateral mammillary nucleus
H19 subthalamic nucleus

MIDBRAIN (M)

M4v ventral periaqueductal gray
M8 red nucleus
M11 tegmental neuroepithelium
M13 tegmentum (or differentiating field)
M14 substantia nigra
M16 dorsal raphe
M21 pretectal neuroepithelium
M23 pretectum (or differentiating field)
M26 olivary pretectal nucleus

NERVES AND NUCLEI, CRANIAL (n)

n5g trigeminal ganglion
n5n trigeminal motor nucleusn
n5p principal (sensory) trigeminal nucleus

THALAMUS (T)

T3p posterior thalamus (or differentiating field)
T34 medial geniculate nucleus

VENTRICLES (v)

vapr aqueduct, pretectal
v3m third ventricle, mammillary recess

PONTINE AREA (X)

X3a anterior pons (or differentiating field)
X7 pontine reticular formation
X8 superior central raphe nucleus
X35 parabrachial nucleus

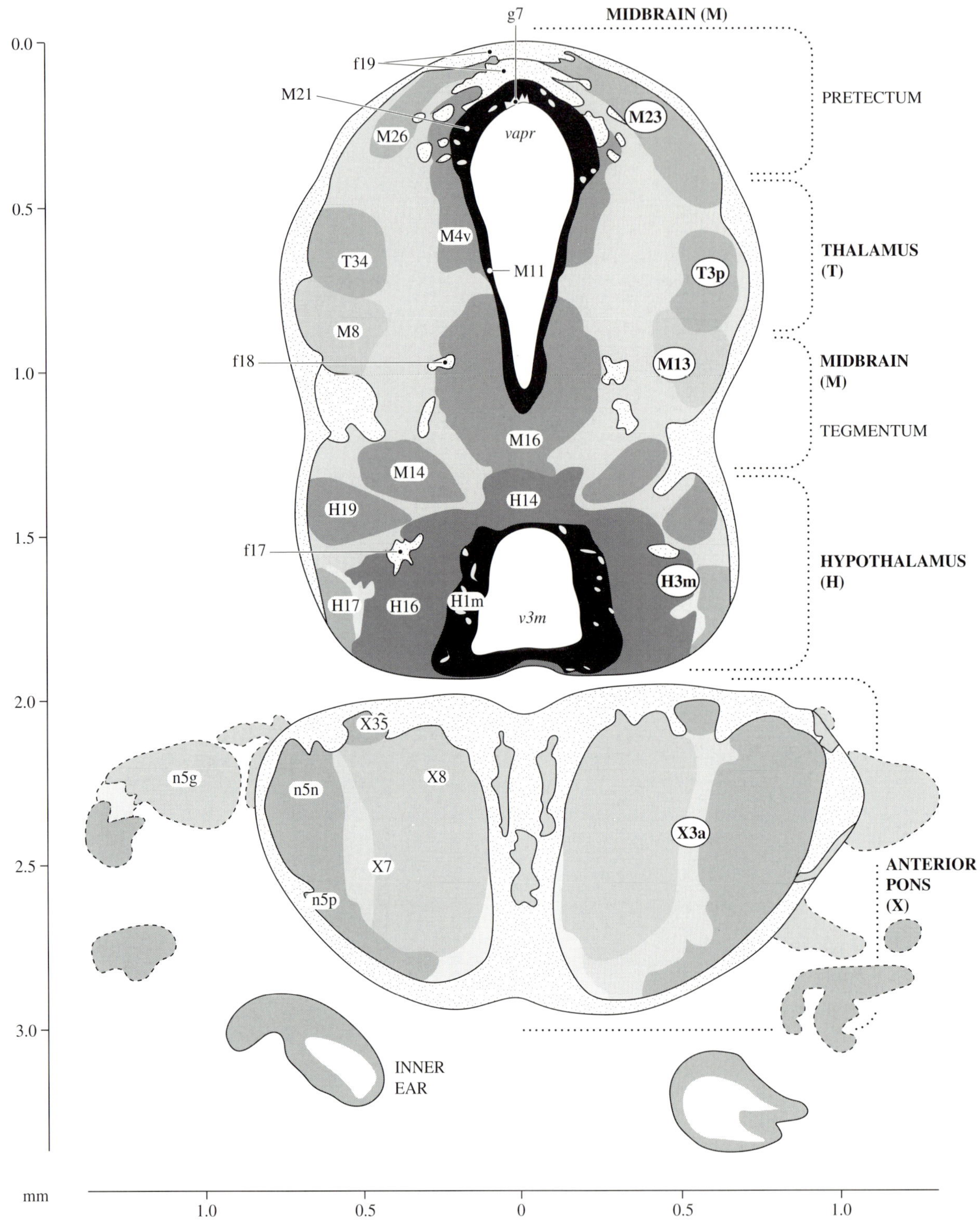

E16

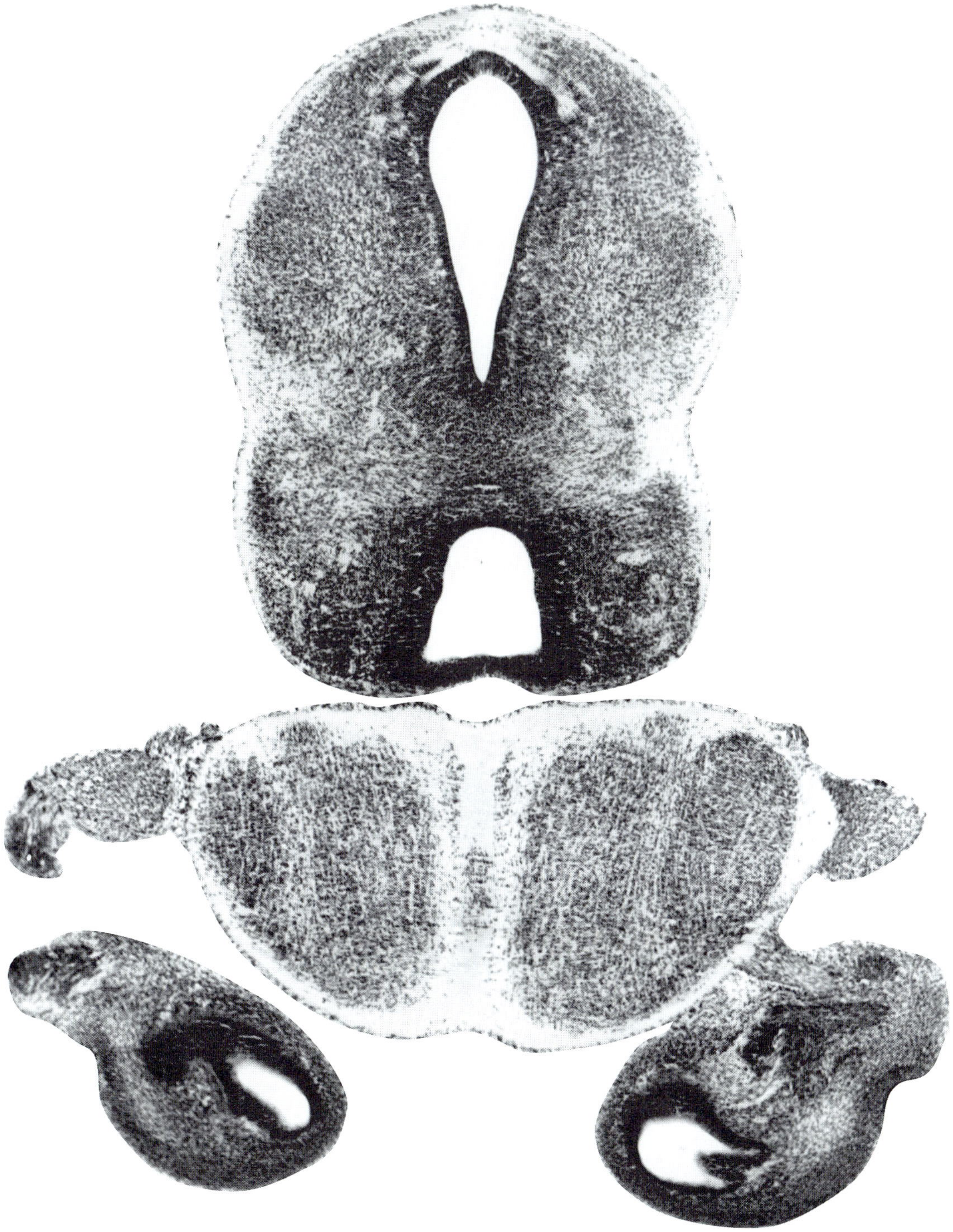

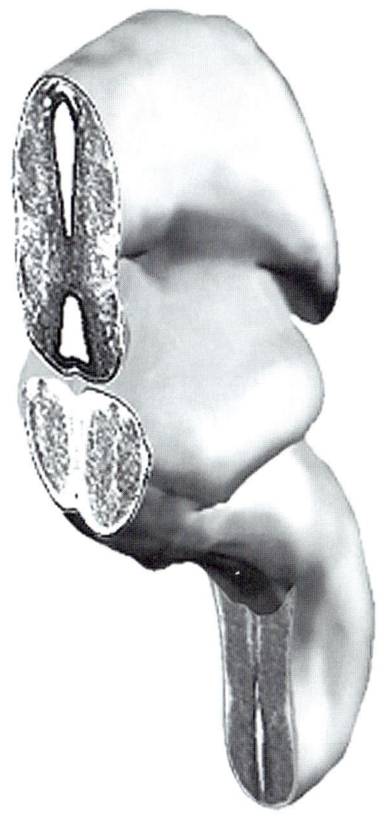

CORONAL FIGURE 16

FIBER TRACTS (f)

f19 posterior commissure

MIDBRAIN (M)

M1 superior tectal neuroepithelium
M3 superior colliculus (or differentiating field)
M4v ventral periaqueductal gray
M6 superior colliculus, deep layers
M9 midbrain reticular formation
M11 tegmental neuroepithelium
M13 tegmentum (or differentiating field)
M15 interpeduncular nucleus
M17 ventral tegmental nucleus (mesencephalic)

NERVES AND NUCLEI, CRANIAL (n)

n3n oculomotor nucleus
n5n trigeminal motor nucleus
n5p principal (sensory) trigeminal nucleus
n5s spinal trigeminal nucleus
n7n facial motor nucleus

VENTRICLES (v)

vas aqueduct, superior (tectal)
vat aqueduct, tegmental
v4a fourth ventricle, anterior

PONTINE AREA (X)

X2a anterior pons (subventricular zone)
X2p posterior pons (subventricular zone)
X3a anterior pons (or differentiating field)
X3p posterior pons (or differentiating field)
X7 pontine reticular formation
X8 superior central raphe nucleus
X24 superior olivary nucleus
X35 parabrachial nucleus

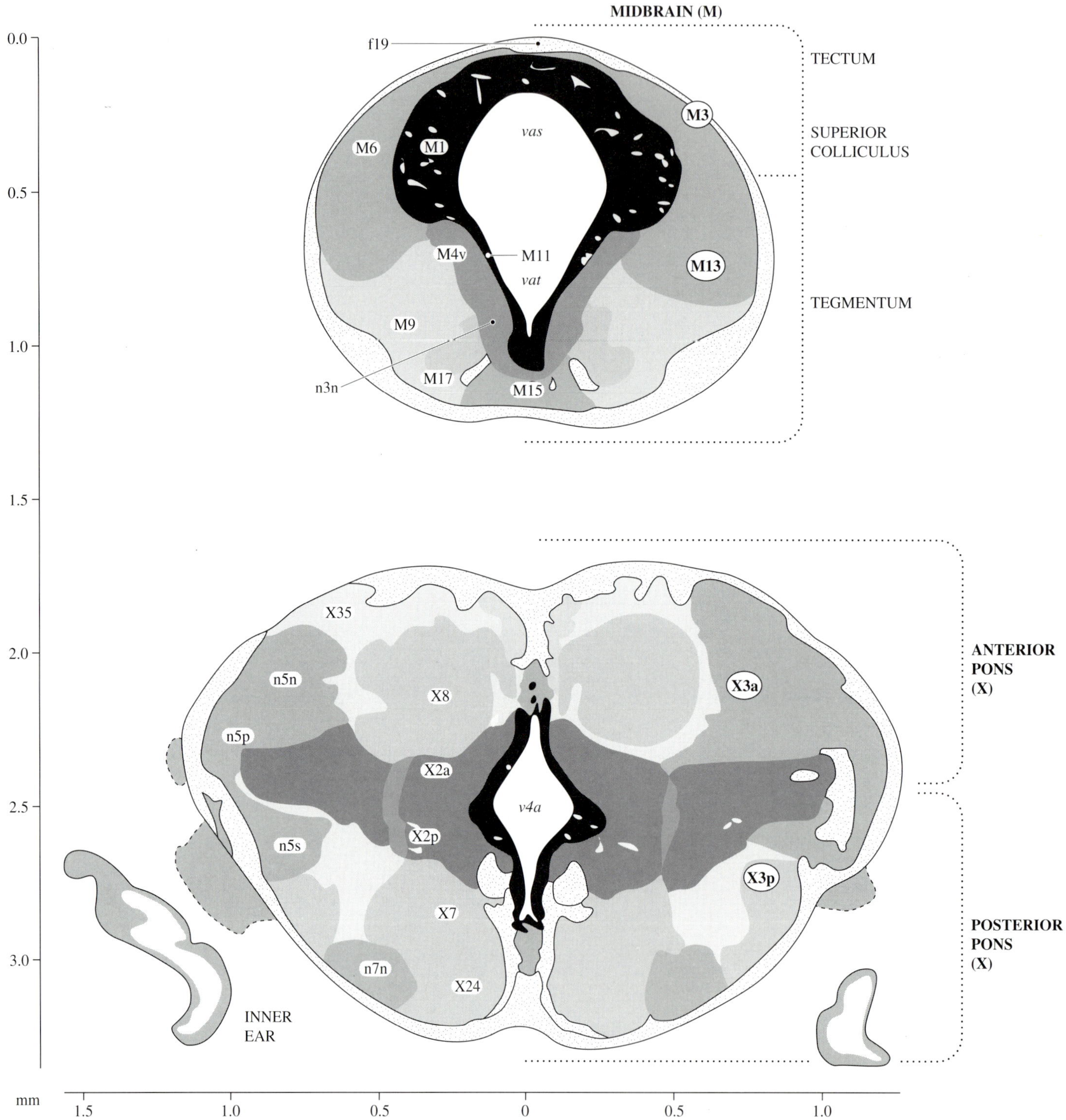

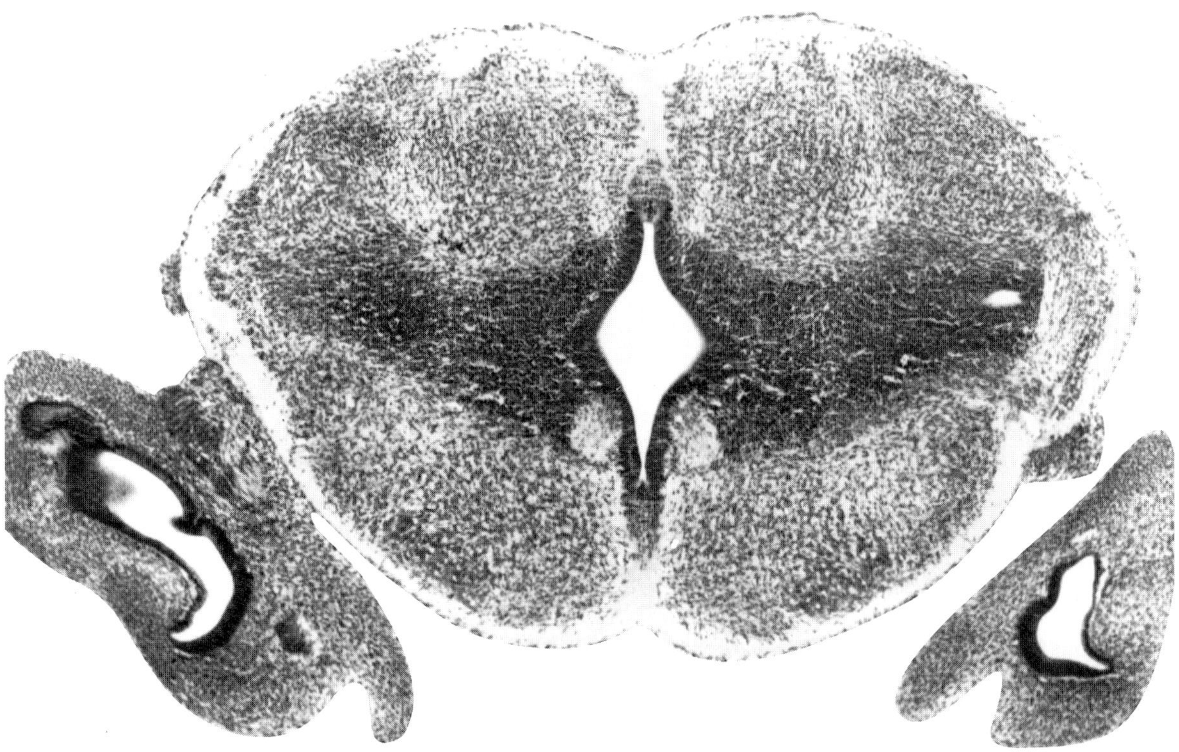

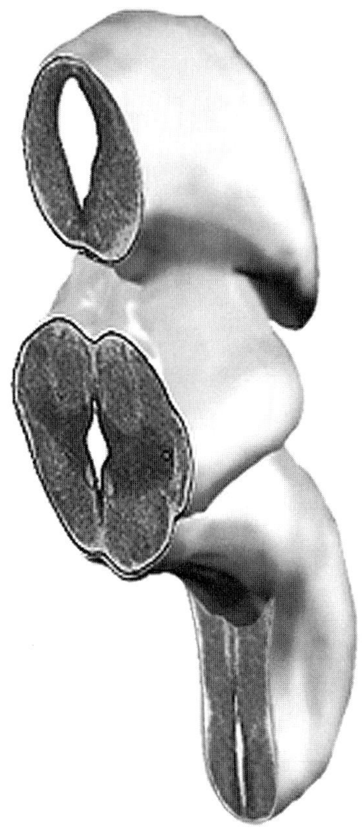

CORONAL FIGURE 17

CEREBELLUM (L)

L1 cerebellar neuroepithelium
L3h cerebellar hemisphere (or differentiating field)
L7mi lateral deep nucleus, migration

MIDBRAIN (M)

M1 superior tectal neuroepithelium
M3 superior colliculus (or differentiating field)
M4v ventral periaqueductal gray
M6 superior colliculus, deep layers
M9 midbrain reticular formation
M11 tegmental neuroepithelium
M13 tegmentum (or differentiating field)
M15 interpeduncular nucleus

NERVES AND NUCLEI, CRANIAL (n)

n5s spinal trigeminal nucleus
n7n facial motor nucleus
n8g vestibular ganglion

VENTRICLES (v)

vas aqueduct, superior tectal
vat aqueduct, tegmental
v4a fourth ventricle, anterior
v4i fourth ventricle, isthmal
v4l fourth ventricle, lateral recess
v4p fourth ventricle, posterior

PONTINE AREA (X)

X1a anterior pontine neuroepithelium
X1p posterior pontine neuroepithelium
X3a anterior pons (or differentiating field)
X3p posterior pons (or differentiating field)
X7 pontine reticular formation
X8 superior central raphe nucleus
X10 ventral tegmental nucleus (pontine)
X14 spinal vestibular nucleus
X15 lateral vestibular nucleus
X16 superior vestibular nucleus
X17 medial vestibular nucleus
X21 cochlear nuclei neuroepithelium
X23 cochlear nuclei (or differentiating field)
X33 isthmus (or differentiating field)
X35 parabrachial nucleus

MEDULLA (Y)

Y3 medulla (or differentiating field)
Y6 medullary reticular formation
Y18 magnocellular reticular nucleus

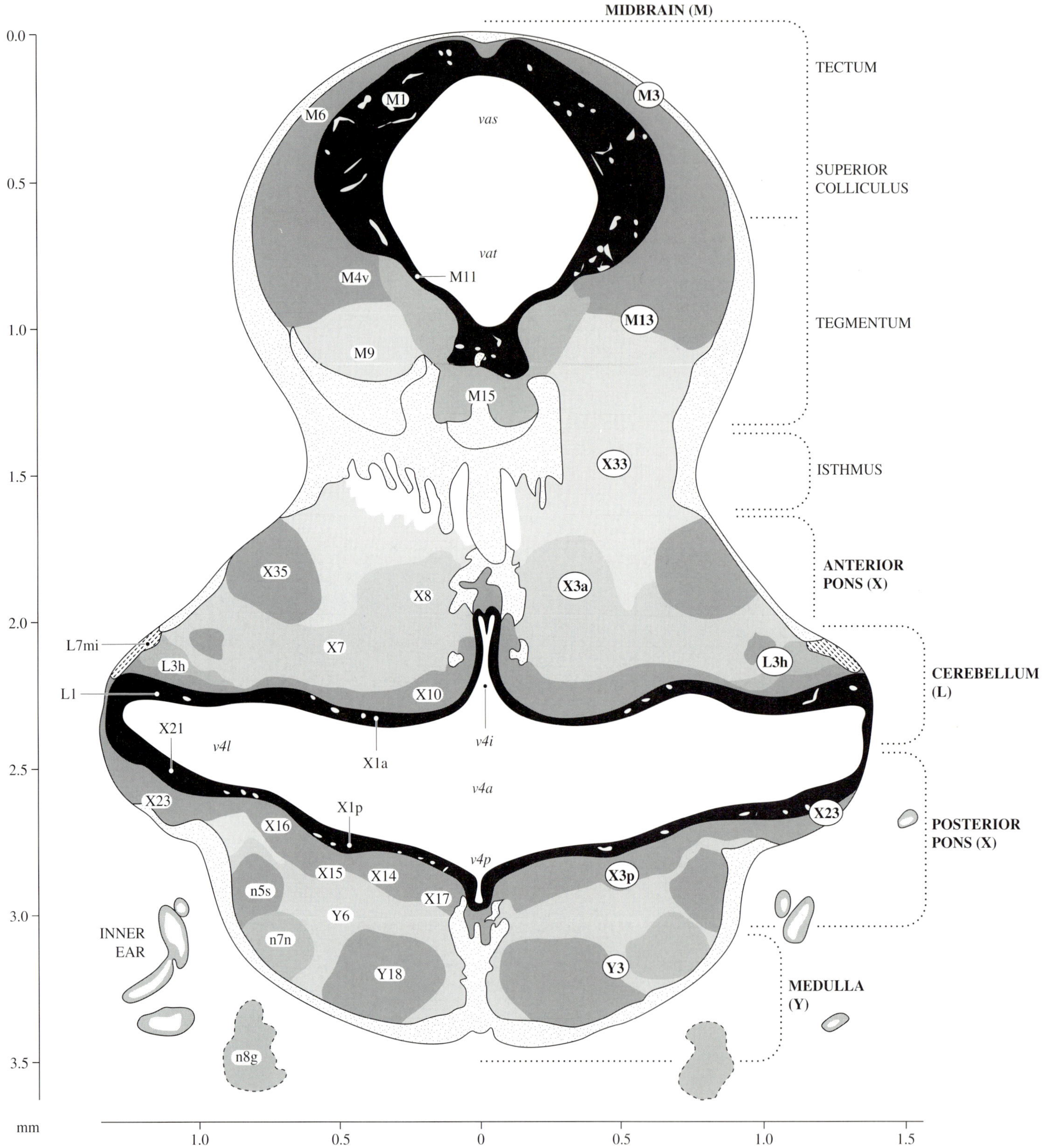

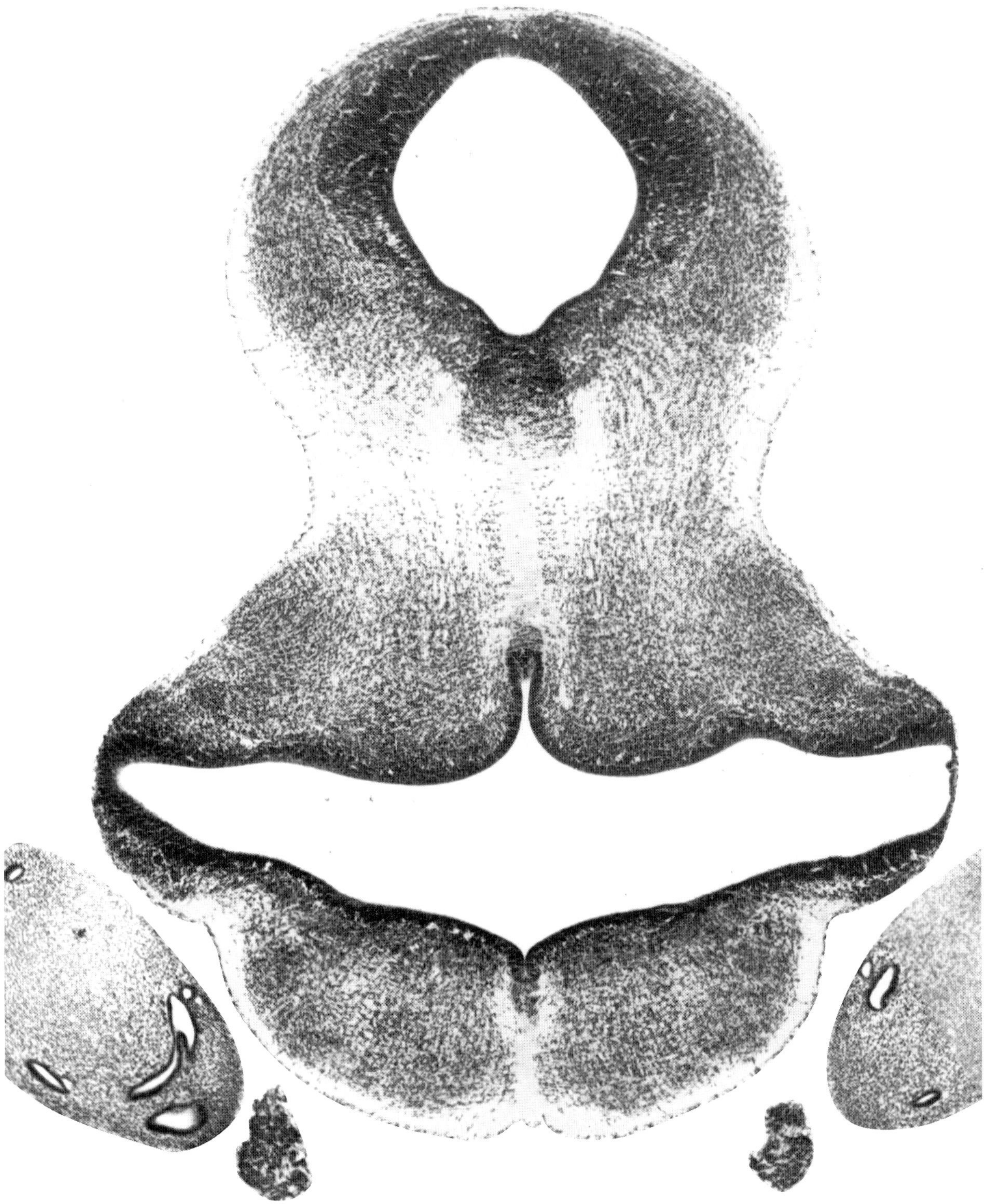

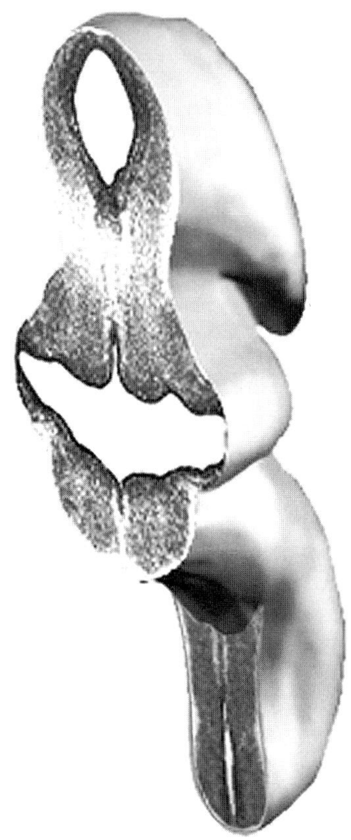

CORONAL FIGURE 18

CEREBELLUM (L)

L1h cerebellar hemisphere neuroepithelium
L3h cerebellar hemisphere (or differentiating field)
L7mi lateral deep nucleus, migration

MIDBRAIN (M)

M1 superior tectal neuroepithelium
M3 superior colliculus (or differentiating field)
M4v ventral periaqueductal gray
M6 superior colliculus, deep layers
M9 midbrain reticular formation
M11 tegmental neuroepithelium
M13 tegmentum

NERVES AND NUCLEI, CRANIAL (n)

n5s spinal trigeminal nucleus

VENTRICLES (v)

vas aqueduct, superior (tectal)
vat aqueduct, tegmental
v4a fourth ventricle, anterior
v4i fourth ventricle, isthmal
v4ir fourth ventricle, isthmal recess
v4l fourth ventricle, lateral recess
v4p fourth ventricle, posterior

PONTINE AREA (X)

X1a anterior pontine neuroepithelium
X1p posterior pontine neuroepithelium
X3a anterior pons (or differentiating field)
X3p posterior pons (or differentiating field)
X7 pontine reticular formation
X8 superior central raphe nucleus
X9 dorsal tegmental nucleus (pontine)
X10 ventral tegmental nucleus (pontine)
X12i precerebellar intramural migration
X12p posterior precerebellar extramural migration
X14 spinal vestibular nucleus
X15 lateral vestibular nucleus
X17 medial vestibular nucleus
X21 cochlear nuclei neuroepithelium
X23 cochlear nuclei (or differentiating field)
X31 isthmal neuroepithelium
X33 isthmus (or differentiating field)
X35 parabrachial nucleus
X36 nucleus of lateral lemniscus

MEDULLA (Y)

Y3 medulla (or differentiating field)
Y6 medullary reticular formation
Y8 inferior olive

E16

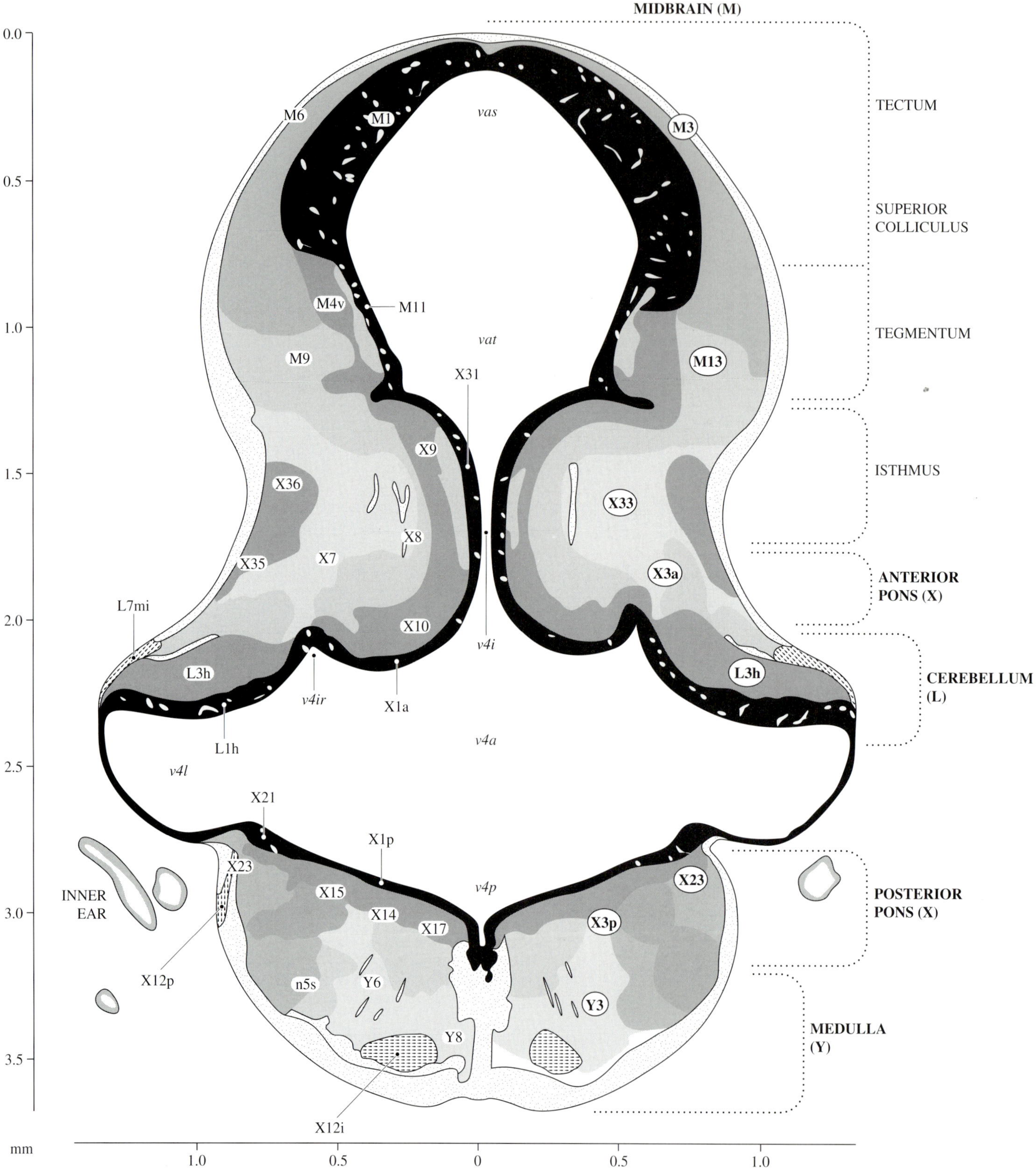

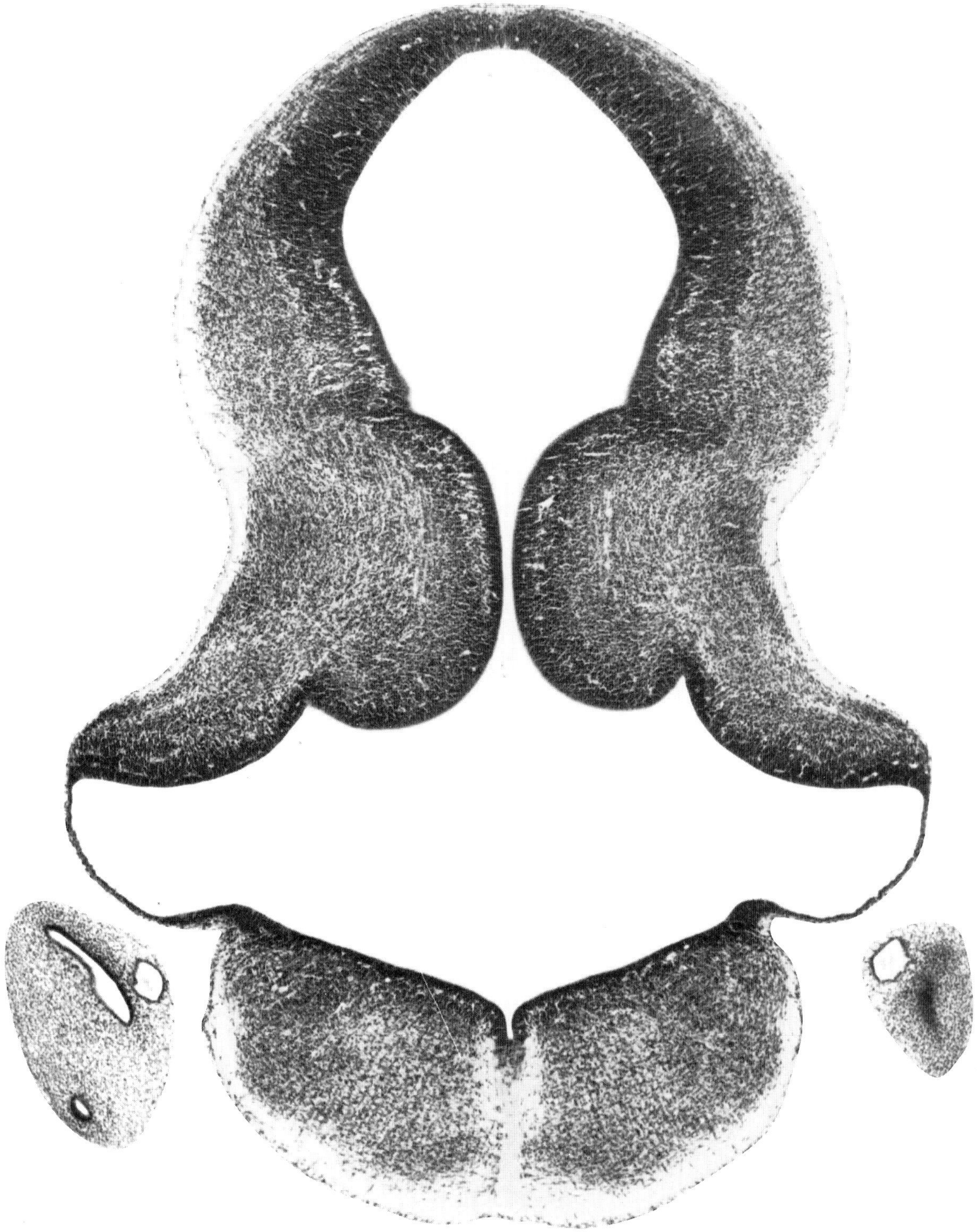

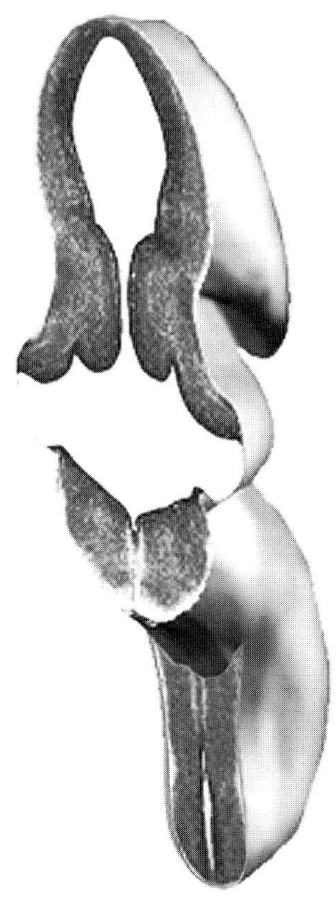

CORONAL FIGURE 19

CEREBELLUM (L)

L1h cerebellar hemisphere neuroepithelium
L1v cerebellar vermis neuroepithelium
L3h cerebellar hemisphere (or differentiating field)
L3v cerebellar vermis (or differentiating field)
L7mi lateral deep nucleus, migration

MIDBRAIN (M)

M1 superior tectal neuroepithelium
M3 superior colliculus (or differentiating field)
M6 superior colliculus, deep layers
M9 midbrain reticular formation
M11 tegmental neuroepithelium
M13 tegmentum (or differentiating field)
M31 inferior tectal neuroepithelium
M33 inferior colliculus (or differentiating field)

NERVES AND NUCLEI, CRANIAL (n)

n5s spinal trigeminal nucleus

VENTRICLES (v)

vai aqueduct, inferior tectal
vas aqueduct, superior tectal
vat aqueduct, tegmental
v4a fourth ventricle, anterior
v4i fourth ventricle, isthmal
v4ir fourth ventricle, isthmal recess
v4l fourth ventricle, lateral recess
v4p fourth ventricle, posterior

PONTINE AREA (X)

X1p posterior pontine neuroepithelium
X3p posterior pons (or differentiating field)
X11p posterior precerebellar neuroepithelium
X12p posterior precerebellar extramural migration
X12i precerebellar intramural migration
X14 spinal vestibular nucleus
X15 lateral vestibular nucleus
X17 medial vestibular nucleus
X31 isthmal neuroepithelium
X33 isthmus (or differentiating field)
X34 locus coeruleus

MEDULLA (Y)

Y3 medulla (or differentiating field)
Y6 medullary reticular formation
Y8 inferior olive

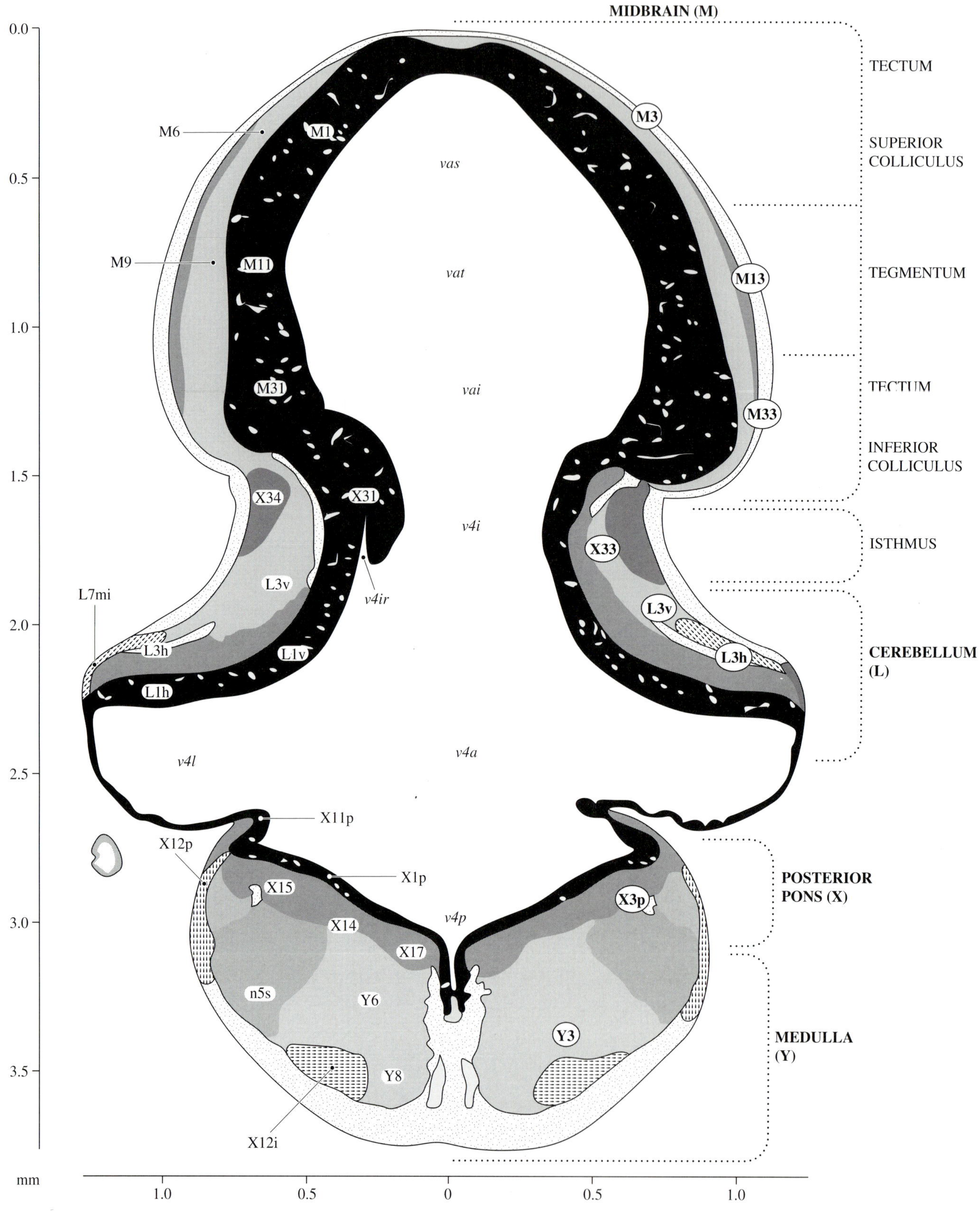

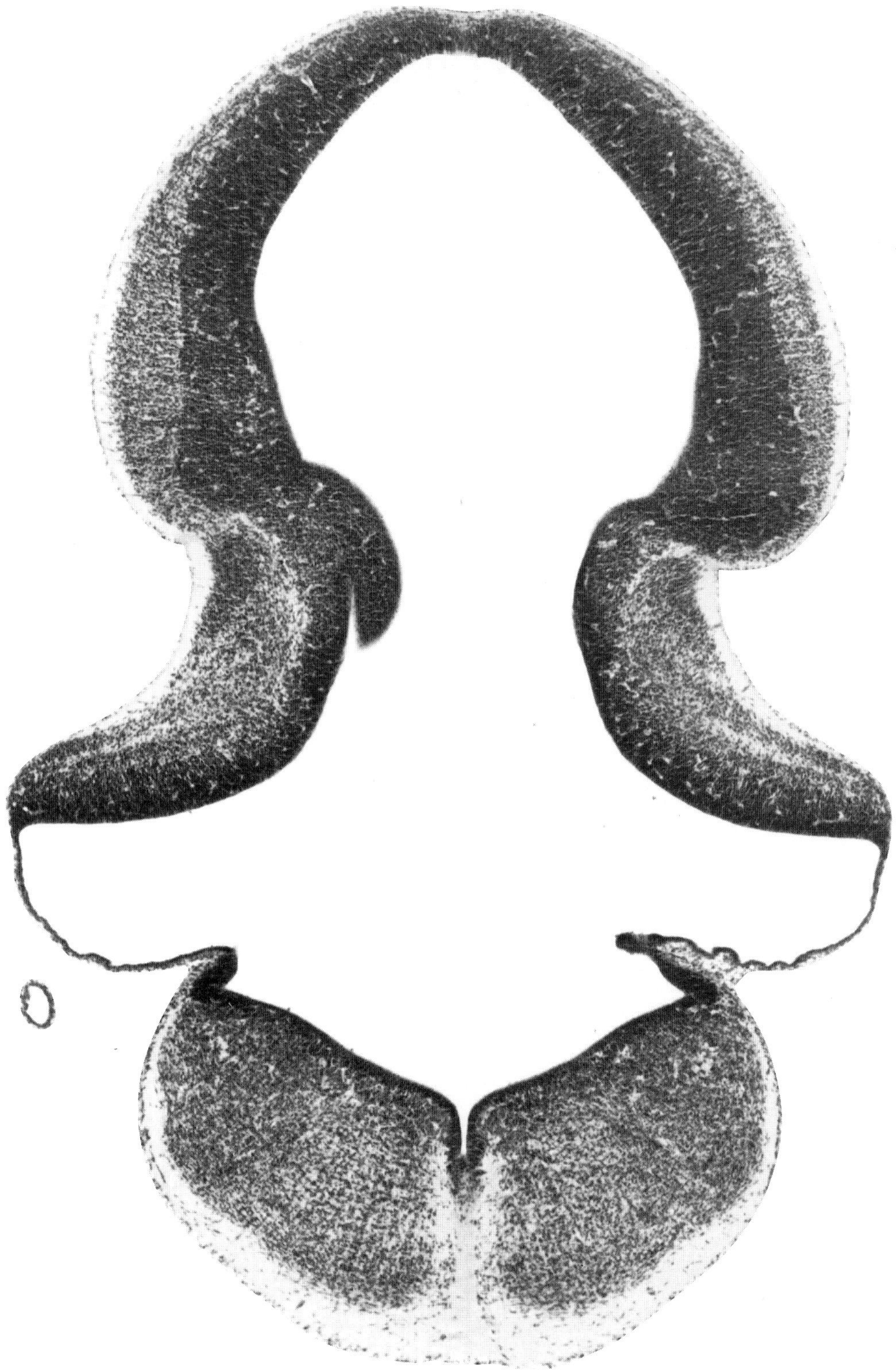

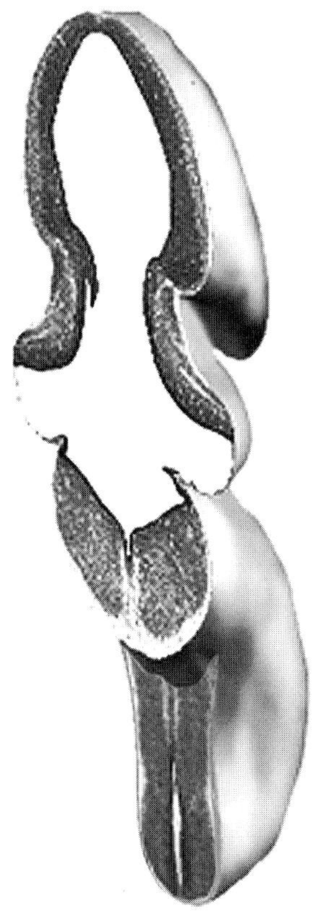

CORONAL FIGURE 20

FIBER TRACTS (f)

f23 superior cerebellar peduncle
f25 inferior cerebellar peduncle

GLANDS AND PARANEURAL STRUCTURES (g)

g4 choroid plexus of fourth ventricle

CEREBELLUM (L)

L1h cerebellar hemisphere neuroepithelium
L1v cerebellar vermis neuroepithelium
L3h cerebellar hemisphere (or differentiating field)
L3v cerebellar vermis (or differentiating field)
L8mi medial deep nucleus, migration

MIDBRAIN(M)

M1 superior tectal neuroepithelium
M3 superior colliculus (or differentiating field)
M6 superior colliculus, deep layers
M31 inferior tectal neuroepithelium
M33 inferior colliculus (or differentiating field)

NERVES AND NUCLEI, CRANIAL (n)

n5s spinal trigeminal nucleus
n12n hypoglossal nucleus

VENTRICLES (v)

vai aqueduct, inferior tectal
vas aqueduct, superior tectal
vat aqueduct, tegmental
v4a fourth ventricle, anterior
v4i fourth ventricle, isthmal
v4l fourth ventricle, lateral recess
v4p fourth ventricle, posterior

PONTINE AREA (X)

X11p posterior precerebellar neuroepithelium
X12i precerebellar intramural migration
X12p posterior precerebellar extramural migration
X34 locus coeruleus

MEDULLA (Y)

Y1 medullary neuroepithelium
Y3 medulla (or differentiating field)
Y6 medullary reticular formation
Y7 prepositus nucleus
Y8 inferior olive
Y16 solitary nucleus

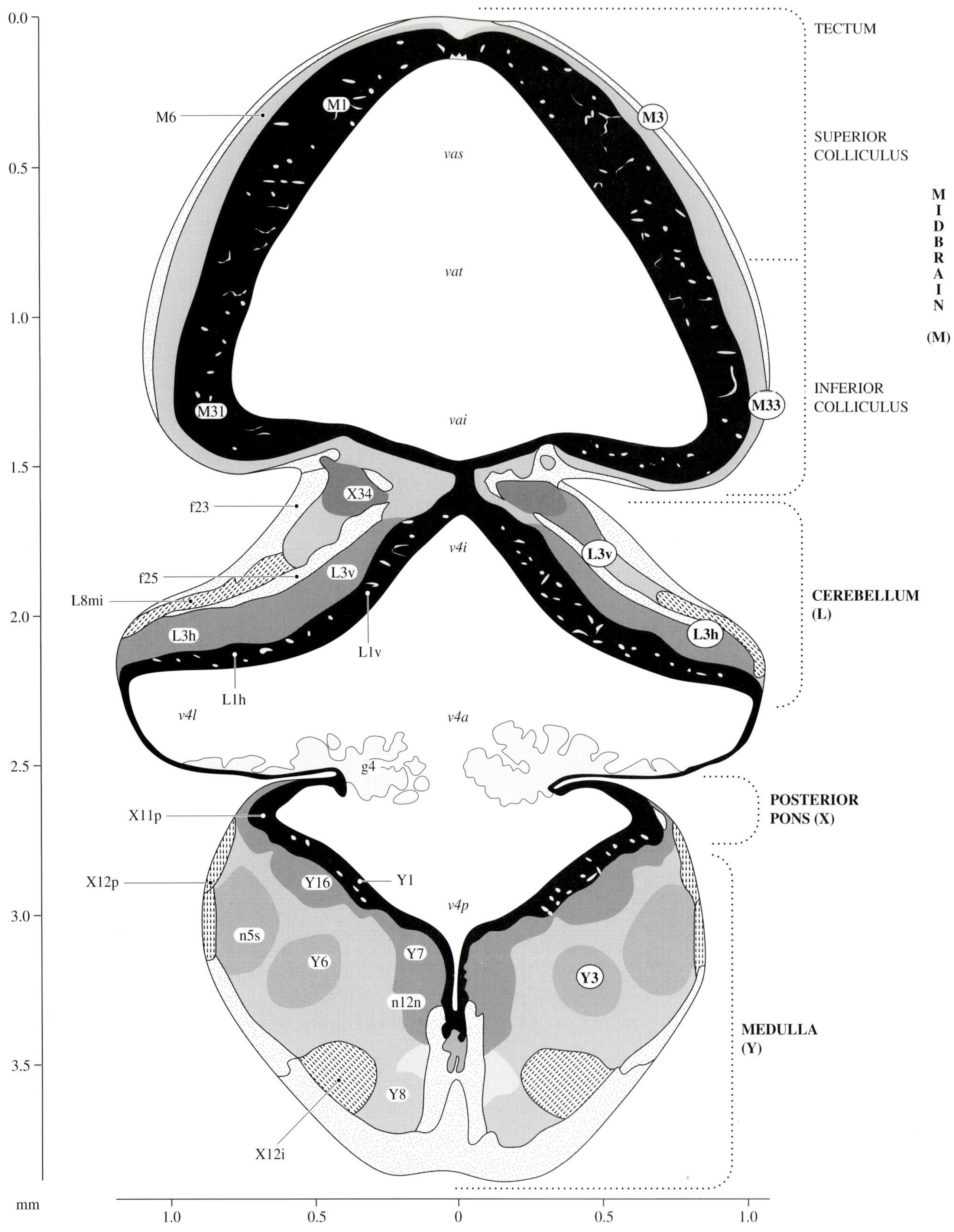

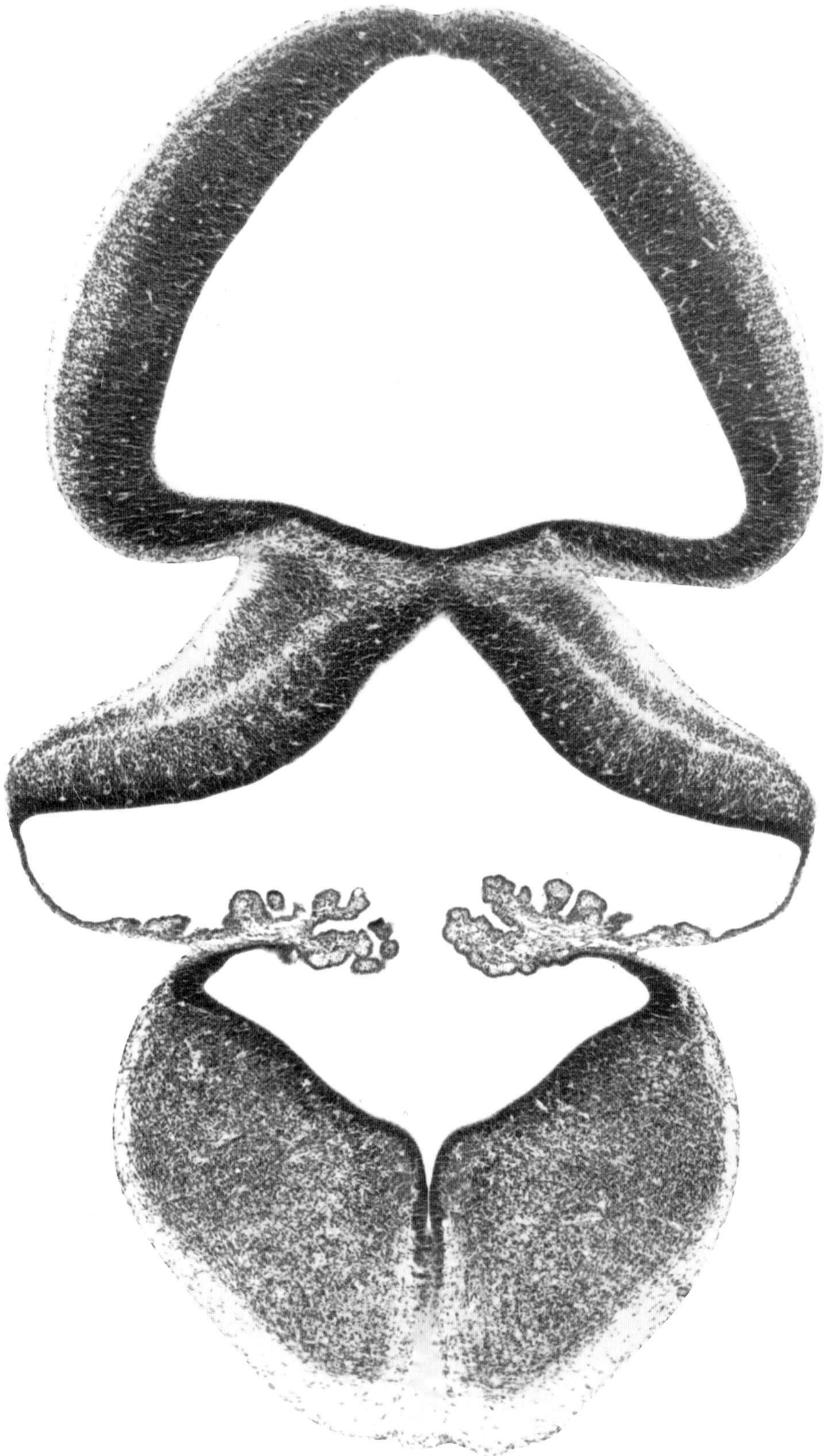

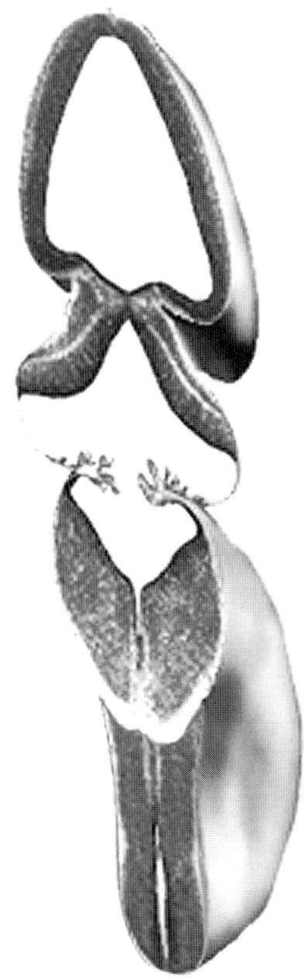

CORONAL FIGURE 21

FIBER TRACTS (f)

f23 superior cerebellar peduncle
f25 inferior cerebellar peduncle

GLANDS AND PARANEURAL STRUCTURES (g)

g4 choroid plexus of fourth ventricle

CEREBELLUM (L)

L1h cerebellar hemisphere neuroepithelium
L1v cerebellar vermis neuroepithelium
L3h cerebellar hemisphere (or differentiating field)
L3v cerebellar vermis (or differentiating field)
L8mi medial deep nucleus, migration

MIDBRAIN (M)

M1 superior tectal neuroepithelium
M3 superior colliculus (or differentiating field)
M6 superior colliculus, deep layers
M31 inferior tectal neuroepithelium
M33 inferior colliculus (or differentiating field)

NERVES AND NUCLEI, CRANIAL (n)

n5s spinal trigeminal nucleus
n12n hypoglossal nucleus

VENTRICLES (v)

vai aqueduct, inferior tectal
vas aqueduct, superior tectal
v4a fourth ventricle, anterior
v4l fourth ventricle, lateral recess
v4p fourth ventricle, posterior

MEDULLA (Y)

Y1 medullary neuroepithelium
Y3 medulla (or differentiating field)
Y6 medullary reticular formation
Y7 prepositus nucleus
Y14 gracile nucleus
Y15 cuneate nucleus
Y16 solitary nucleus

SPINAL CORD (Z)

Z3 spinal cord (or differentiating field)

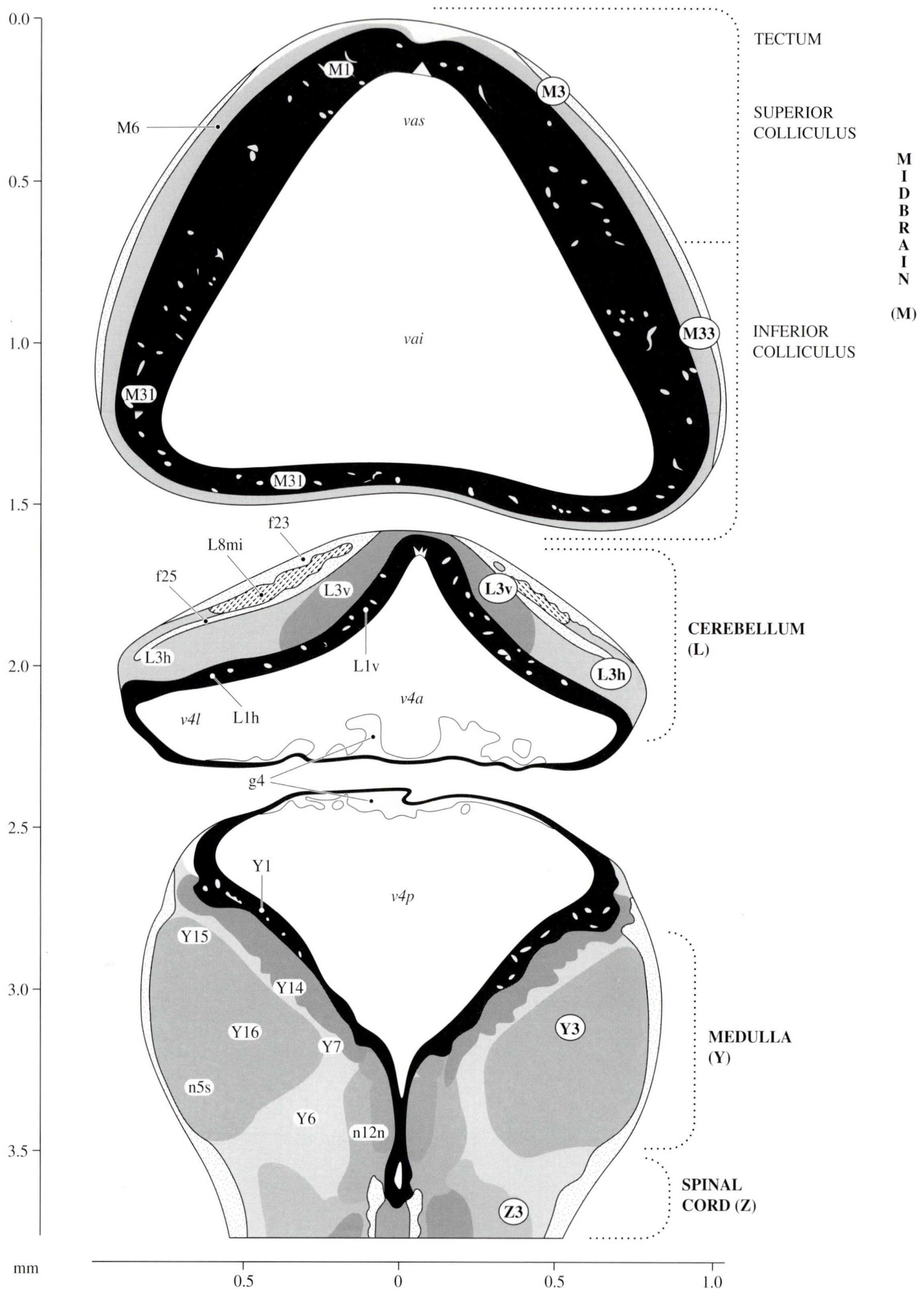

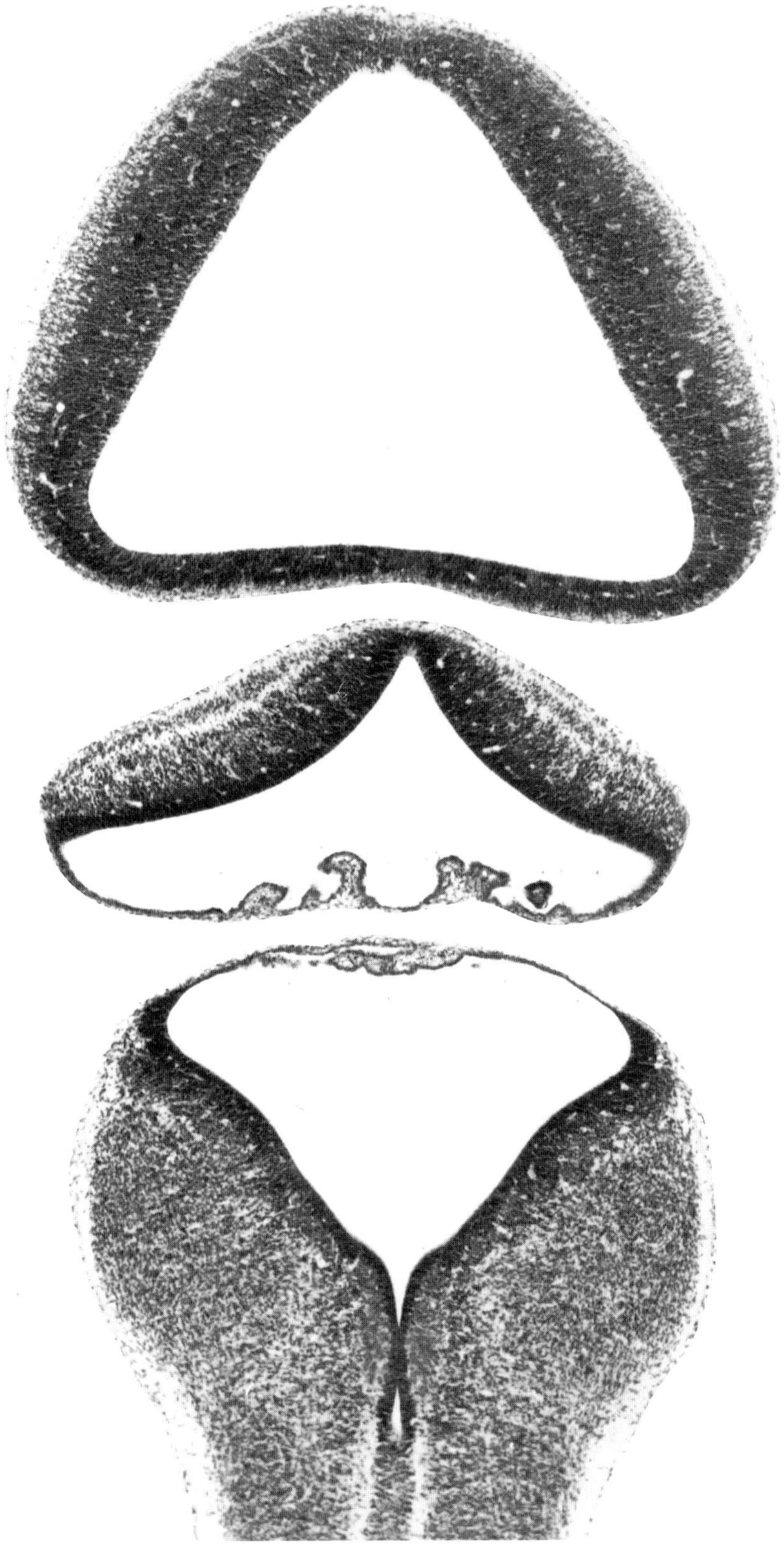

CORONAL FIGURE 22

CEREBELLUM (L)

L1v cerebellar vermis neuroepithelium
L3v cerebellar vermis (or differentiating field)

MIDBRAIN (M)

M1 superior tectal neuroepithelium
M3 superior colliculus (or differentiating field)
M6 superior colliculus, deep layers
M31 inferior tectal neuroepithelium
M33 inferior colliculus (or differentiating field)

NERVES AND NUCLEI, CRANIAL (n)

n5s spinal trigeminal nucleus
n12n hypoglossal nucleus

VENTRICLES (v)

vai aqueduct, inferior tectal
vas aqueduct, superior tectal
vs spinal canal
v4a fourth ventricle, anterior
v4p fourth ventricle, posterior

MEDULLA (Y)

Y1 medullary neuroepithelium
Y3 medulla (or differentiating field)
Y7 prepositus nucleus
Y14 gracile nucleus
Y15 cuneate nucleus
Y16 solitary nucleus

SPINAL CORD (Z)

Z3 spinal cord (or differentiating field)

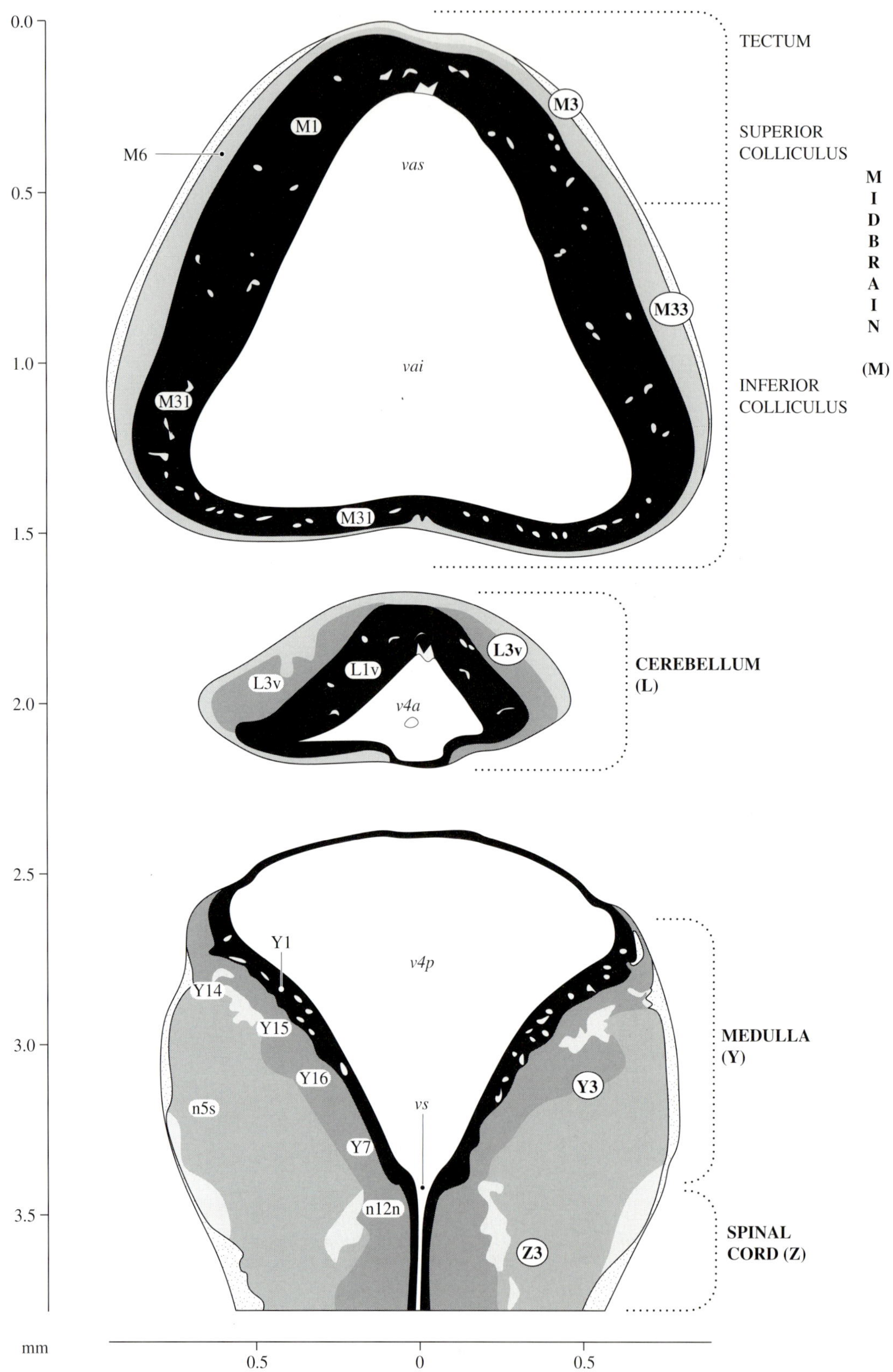

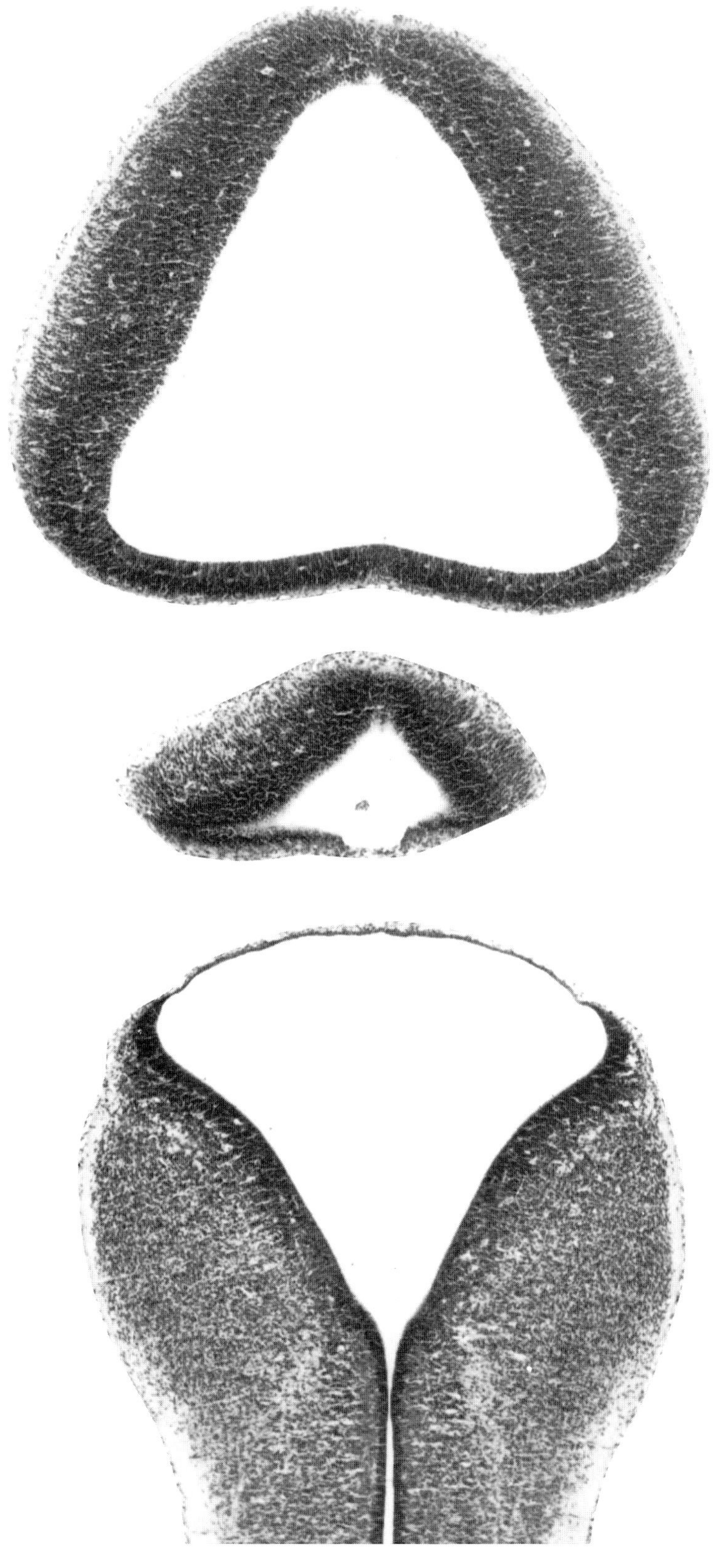

THE E17 BRAIN

COMPUTERIZED 3-D RECONSTRUCTION: 4 VIEWS

LEFT, ANTERIOR VIEW

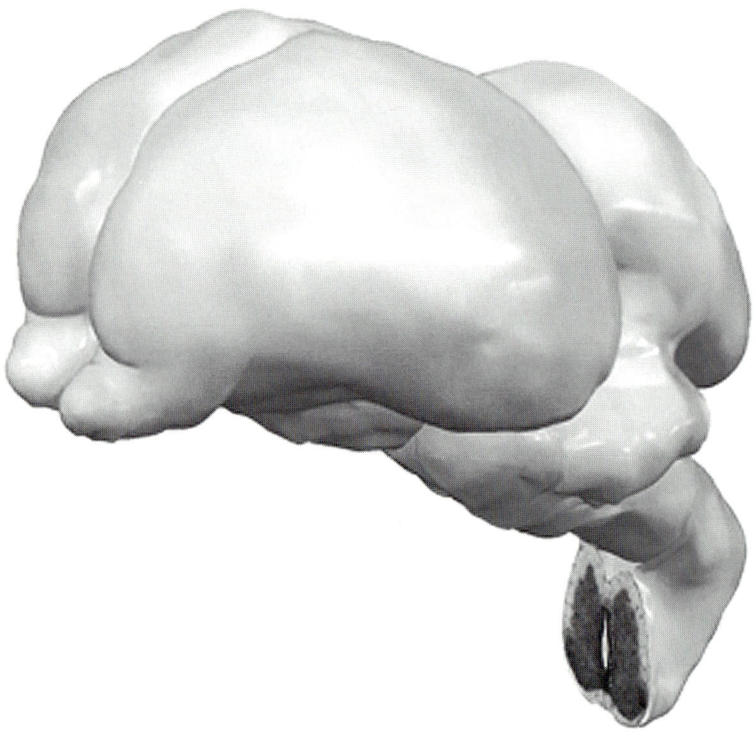

RIGHT, POSTERIOR VIEW

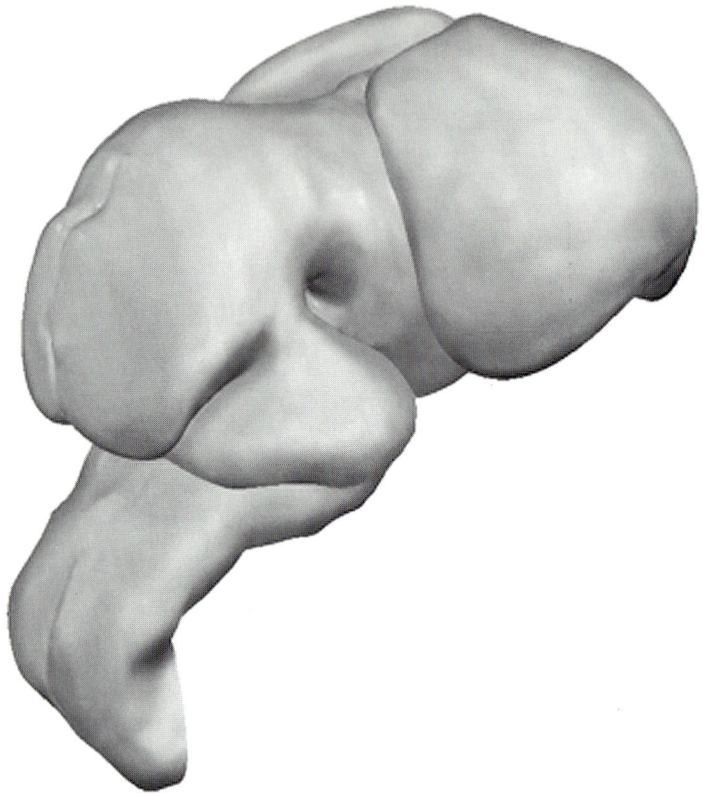

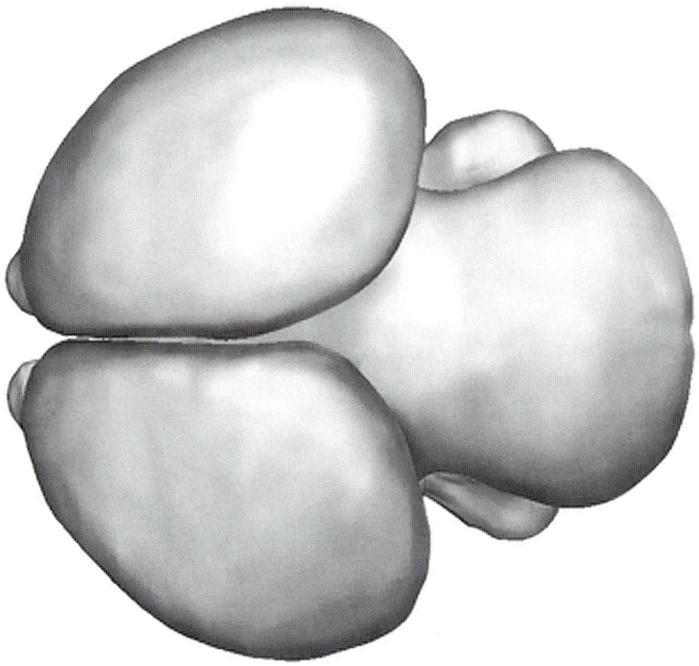

DORSAL VIEW

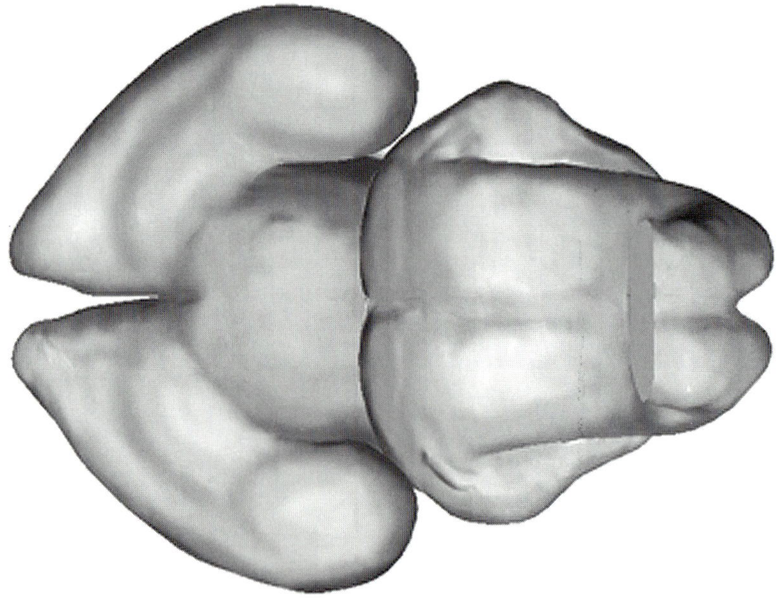

VENTRAL VIEW

SAGITTAL

FIGURES and PLATES 1-8

SAGITTAL FIGURE 1

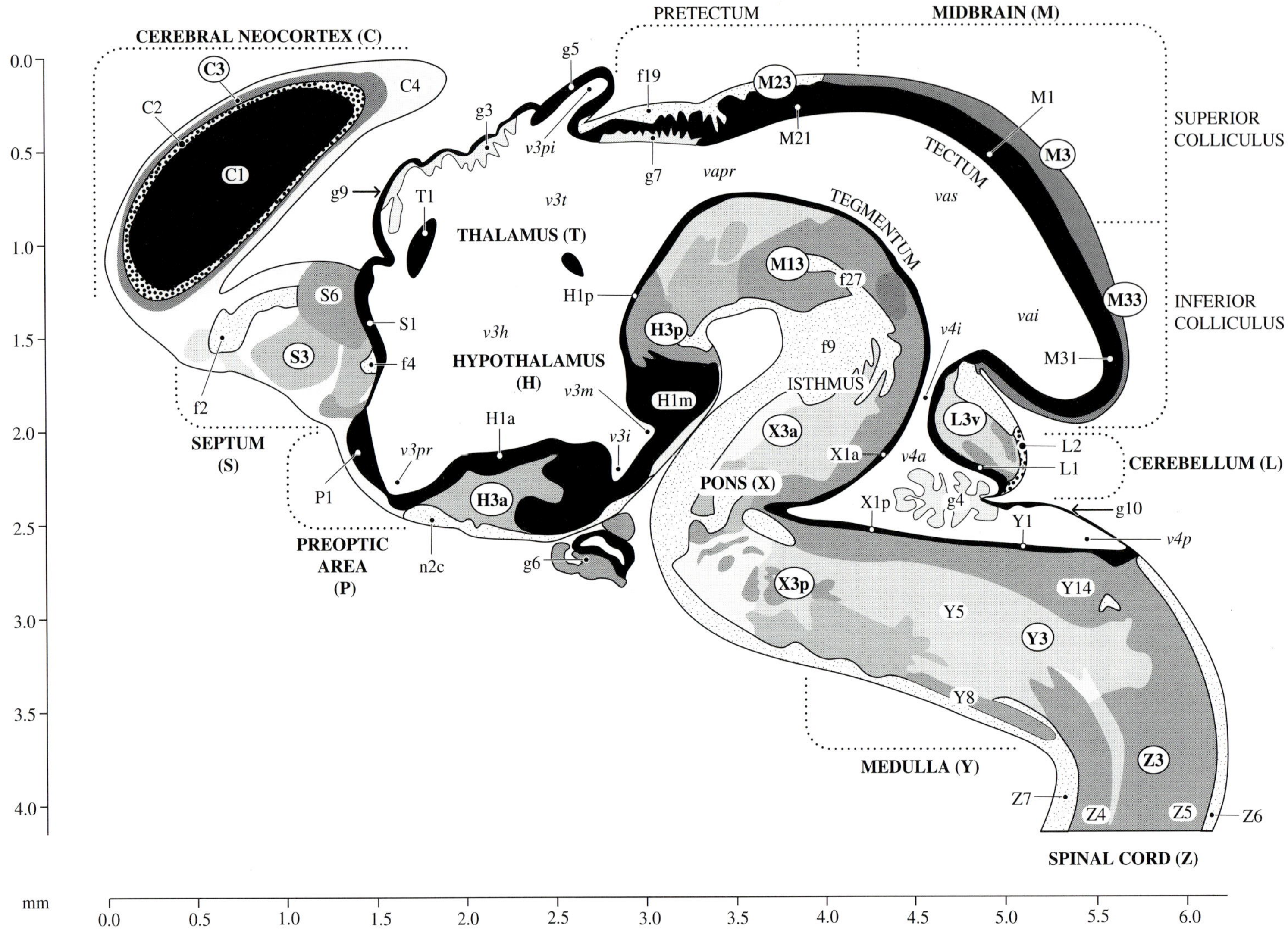

CEREBRAL CORTEX (C)

C1 neocortical neuroepithelium
C2 neocortical subventricular zone
C3 neocortex (or differentiating field)
C4 molecular layer (or primordial plexiform layer)

FIBER TRACTS (f)

f2 medial olfactory tract
f4 anterior commissure
f9 cerebral peduncle
f19 posterior commissure
f27 medial longitudinal fasciculus

GLANDS AND PARANEURAL STRUCTURES (g)

g3 choroid plexus of lateral ventricle
g4 choroid plexus of fourth ventricle
g5 pineal gland (primordium)
g6 pituitary gland (primordium)
g7 subcommissural organ
g9 lamina terminalis
g10 velum medullare

HYPOTHALAMUS (H)

H1a anterior hypothalamic neuroepithelium
H1m mammillary neuroepithelium
H1p posterior hypothalamic neuroepithelium
H3a anterior hypothalamus (or differentiating field)
H3p posterior hypothalamus (or differentiating field)

CEREBELLUM (L)

L1 cerebellar neuroepithelium
L2 external germinal layer
L3v cerebellar vermis (or differentiating field)

MIDBRAIN (M)

M1 superior tectal neuroepithelium
M3 superior colliculus (or differentiating field)
M13 tegmentum (or differentiating field)
M21 pretectal neuroepithelium
M23 pretectum (or differentiating field)
M31 inferior tectal neuroepithelium
M33 inferior colliculus (or differentiating field)

NERVES AND NUCLEI, CRANIAL (n)

n2c optic chiasma

PREOPTIC AREA AND OPTIC EPITHELIA (P)

P1 preoptic neuroepithelium

SEPTAL AREA (S)

S1 septal neuroepithelium
S3 septum (or differentiating field)
S6 medial septum

THALAMUS (T)

T1 thalamic neuroepithelium

VENTRICLES (v)

vai aqueduct, inferior tectal
vapr aqueduct, pretectal
vas aqueduct, superior tectal
v3h third ventricle, hypothalamic
v3i third ventricle, infundibular recess
v3m third ventricle, mammillary recess
v3pi third ventricle, pineal recess
v3pr third ventricle, preoptic recess
v3t third ventricle, thalamic
v4a fourth ventricle, anterior
v4i fourth ventricle, isthmal
v4p fourth ventricle, posterior

PONTINE AREA (X)

X1a anterior pontine neuroepithelium
X1p posterior pontine neuroepithelium
X3a anterior pons (or differentiating field)
X3p posterior pons (or differentiating field)

MEDULLA (Y)

Y1 medullary neuroepithelium
Y3 medulla (or differentiating field)
Y5 medullary raphe
Y8 inferior olive
Y14 gracile nucleus

SPINAL CORD (Z)

Z3 spinal cord (or differentiating field)
Z4 ventral horn
Z5 dorsal horn
Z6 dorsal funiculus
Z7 ventral funiculus

SAGITTAL FIGURE 2

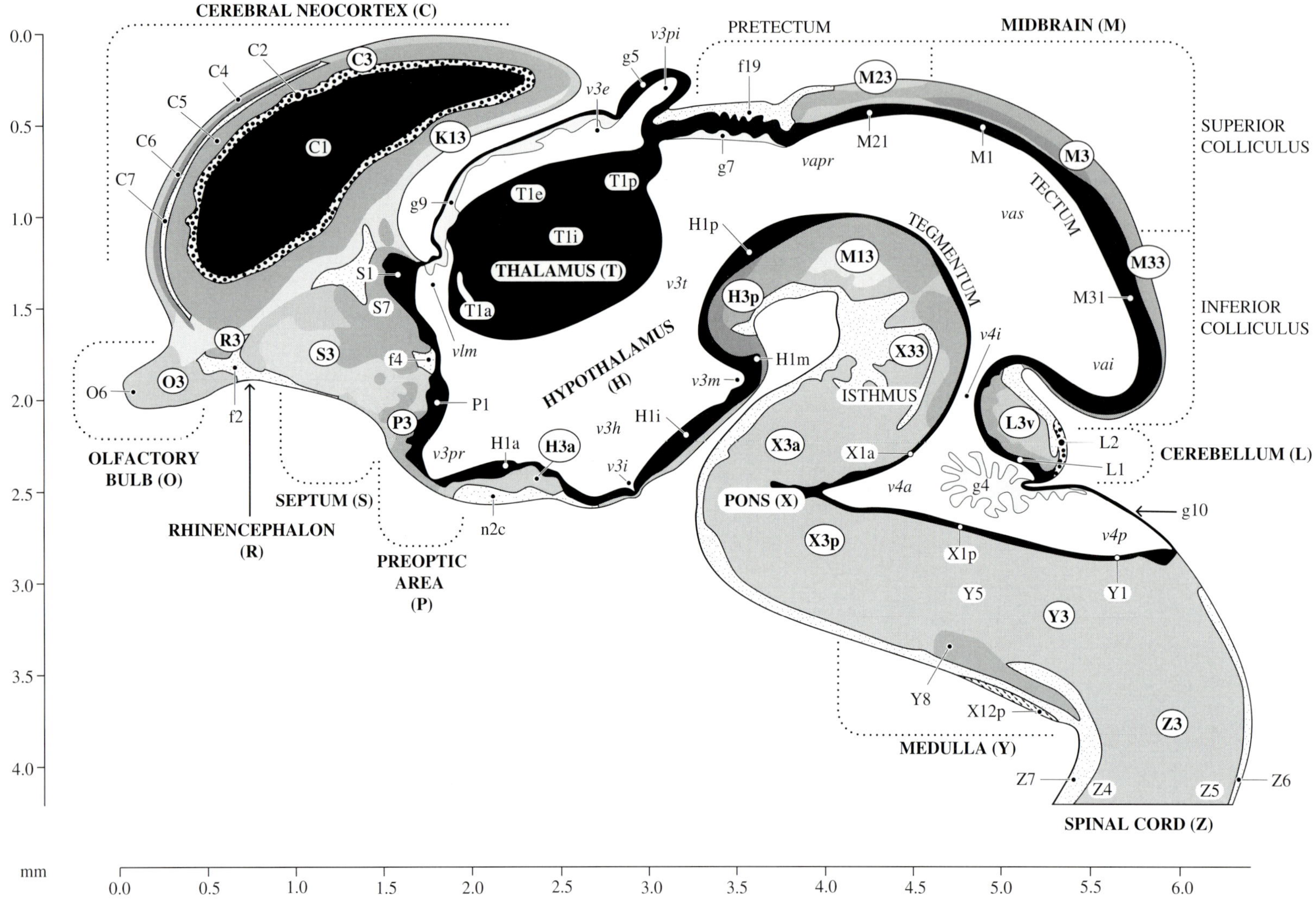

CEREBRAL CORTEX (C)

C1 neocortical neuroepithelium
C2 neocortical subventricular zone
C3 neocortex (or differentiating field)
C4 molecular layer (or primordial plexiform layer)
C5 intermediate zone
C6 cortical plate
C7 white matter

FIBER TRACTS (f)

f2 medial olfactory tract
f4 anterior commissure
f19 posterior commissure

GLANDS AND PARANEURAL STRUCTURES (g)

g4 choroid plexus of fourth ventricle
g5 pineal gland (primordium)
g7 subcommissural organ
g9 lamina terminalis
g10 velum medullare

HYPOTHALAMUS (H)

H1a anterior hypothalamic neuroepithelium
H1i intermediate hypothalamic neuroepithelium
H1m mammillary neuroepithelium
H1p posterior hypothalamic neuroepithelium
H3a anterior hypothalamus (or differentiating field)
H3p posterior hypothalamus (or differentiating field)

HIPPOCAMPAL FORMATION (K)

K13 subicular area (or differentiating field)

CEREBELLUM (L)

L1 cerebellar neuroepithelium
L2 external germinal layer
L3v cerebellar vermis (or differentiating field)

MIDBRAIN (M)

M1 superior tectal neuroepithelium
M3 superior colliculus (or differentiating field)
M13 tegmentum (or differentiating field)
M21 pretectal neuroepithelium
M23 pretectum (or differentiating field)
M31 inferior tectal neuroepithelium
M33 inferior colliculus (or differentiating field)

NERVES AND NUCLEI, CRANIAL (n)

n2c optic chiasma

OLFACTORY BULB (O)

O3 olfactory bulb (or differentiating field)
O6 mitral cell layer

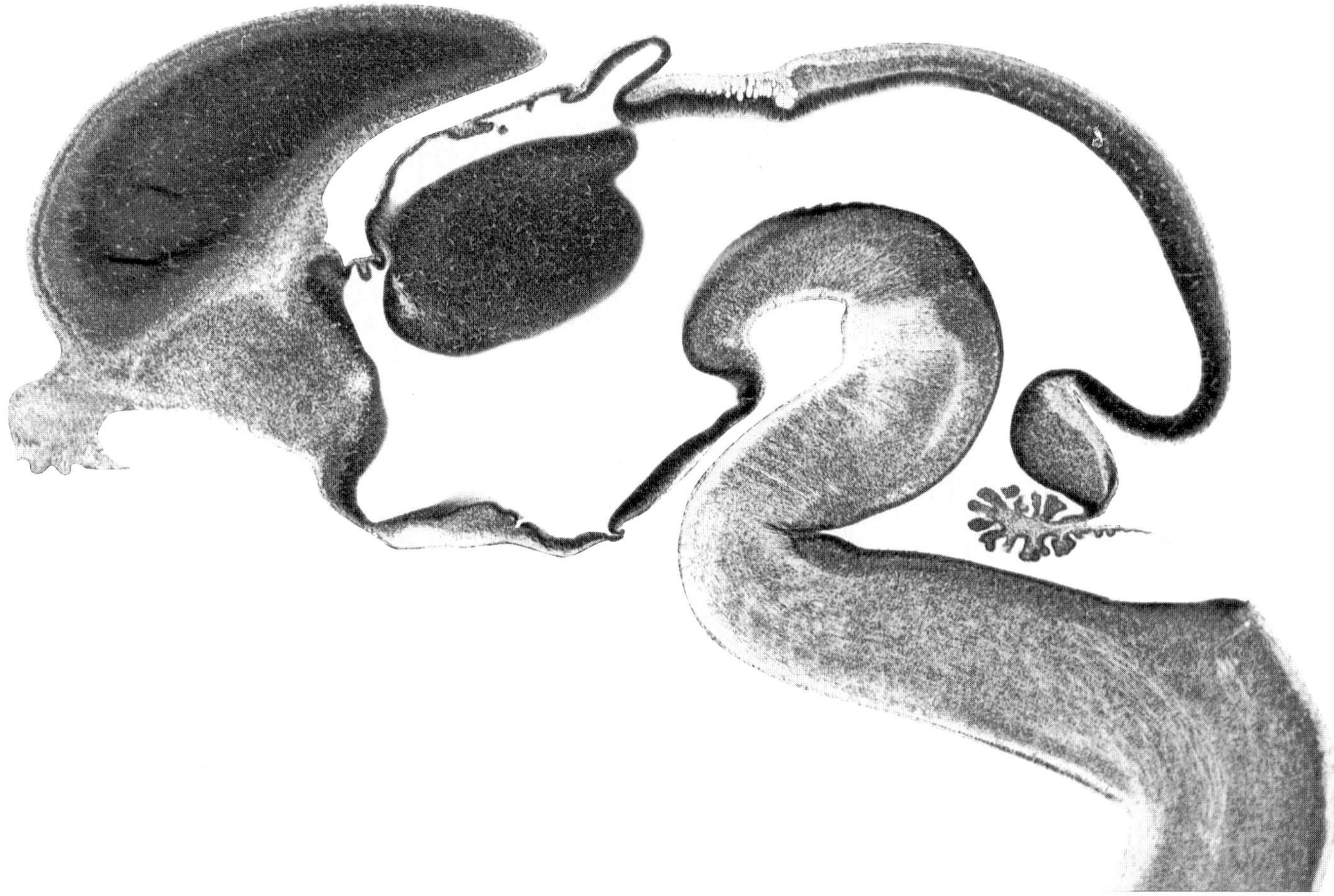

PREOPTIC AREA AND OPTIC EPITHELIA (P)

P1 preoptic neuroepithelium
P3 preoptic area (or differentiating field)

RHINENCEPHALON (R)

R3 rhinencephalon (or differentiating field)

SEPTAL AREA (S)

S1 septal neuroepithelium
S3 septum (or differentiating field)
S7 diagonal band (or differentiating field)

THALAMUS (T)

T1a anterior thalamic neuroepithelium
T1e epithalamic neuroepithelium
T1i intermediate thalamic neuroepithelium
T1p posterior thalamic neuroepithelium

VENTRICLES (v)

vai aqueduct, inferior tectal
vapr aqueduct, pretectal
vas aqueduct, superior tectal
vlm foramen of Monro
v3e third ventricle, epithalamic
v3h third ventricle, hypothalamic
v3i third ventricle, infundibular recess
v3m third ventricle, mammillary recess
v3pi third ventricle, pineal recess
v3pr third ventricle, preoptic recess
v3t third ventricle, thalamic
v4a fourth ventricle, anterior
v4i fourth ventricle, isthmal
v4p fourth ventricle, posterior

PONTINE AREA (X)

X1a anterior pontine neuroepithelium
X1p posterior pontine neuroepithelium
X3a anterior pons (or differentiating field)
X3p posterior pons (or differentiating field)
X12p posterior precerebellar extramural migration
X33 isthmus (or differentiating field)

MEDULLA (Y)

Y1 medullary neuroepithelium
Y3 medulla (or differentiating field)
Y5 medullary raphe
Y8 inferior olive

SPINAL CORD (Z)

Z3 spinal cord (or differentiating field)
Z4 ventral horn
Z5 dorsal horn
Z6 dorsal funiculus
Z7 ventral funiculus

SAGITTAL FIGURE 3

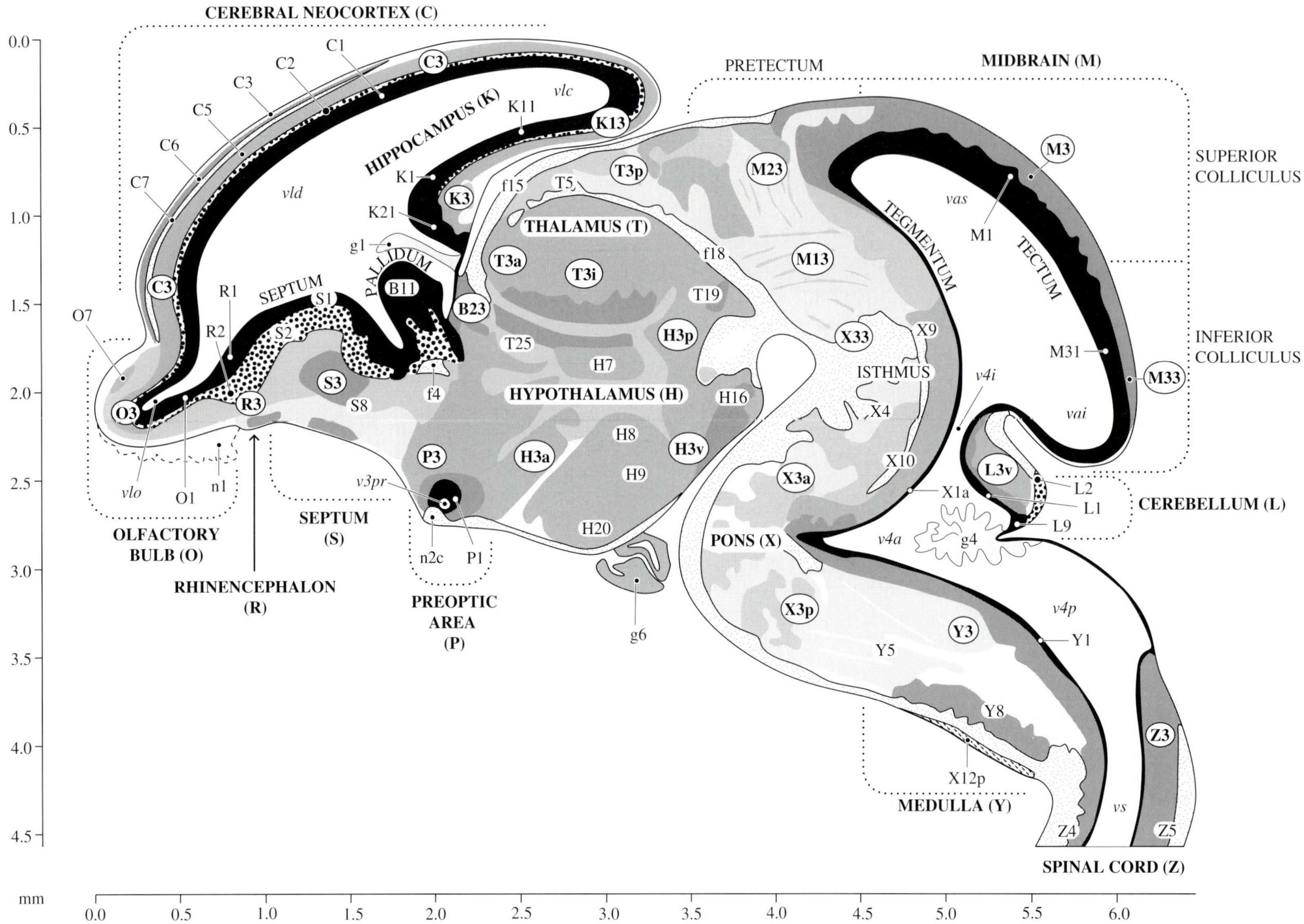

BASAL GANGLIA (B)

- B11 pallidal neuroepithelium
- **B23** bed nucleus of stria terminalis (or differentiating field)

CEREBRAL CORTEX (C)

- C1 neocortical neuroepithelium
- C2 neocortical subventricular zone
- **C3** neocortex (or differentiating field)
- C4 molecular layer (or primordial plexiform layer)
- C5 intermediate zone
- C6 cortical plate
- C7 white matter

FIBER TRACTS (f)

- f4 anterior commissure
- f15 stria medullaris
- f18 habenulopeduncular tract

GLANDS AND PARANEURAL STRUCTURES (g)

- g1 choroid plexus of lateral ventricle
- g4 choroid plexus of fourth ventricle
- g6 pituitary gland (primordium)

HYPOTHALAMUS (H)

- **H3a** anterior hypothalamus (or differentiating field)
- **H3p** posterior hypothalamus (or differentiating field)
- **H3v** ventral hypothalamus (or differentiating field)
- H7 zona incerta
- H8 dorsomedial nucleus
- H9 ventromedial nucleus
- H16 medial mammillary nucleus
- H20 arcuate nucleus

HIPPOCAMPAL FORMATION (K)

- K1 hippocampal (Ammon's horn) neuroepithelium
- **K3** hippocampus (or differentiating field)
- K11 subicular neuroepithelium
- **K13** subicular area (or differentiating field)
- K21 dentate gyrus neuroepithelium

CEREBELLUM (L)

- L1 cerebellar neuroepithelium
- L2 external germinal layer
- **L3v** cerebellar vermis (or differentiating field)
- L9 cerebellar germinal trigone

MIDBRAIN (M)

- M1 superior tectal neuroepithelium
- **M3** superior colliculus (or differentiating field)
- **M13** tegmentum (or differentiating field)
- **M23** pretectum (or differentiating field)
- M31 inferior tectal neuroepithelium
- **M33** inferior colliculus (or differentiating field)

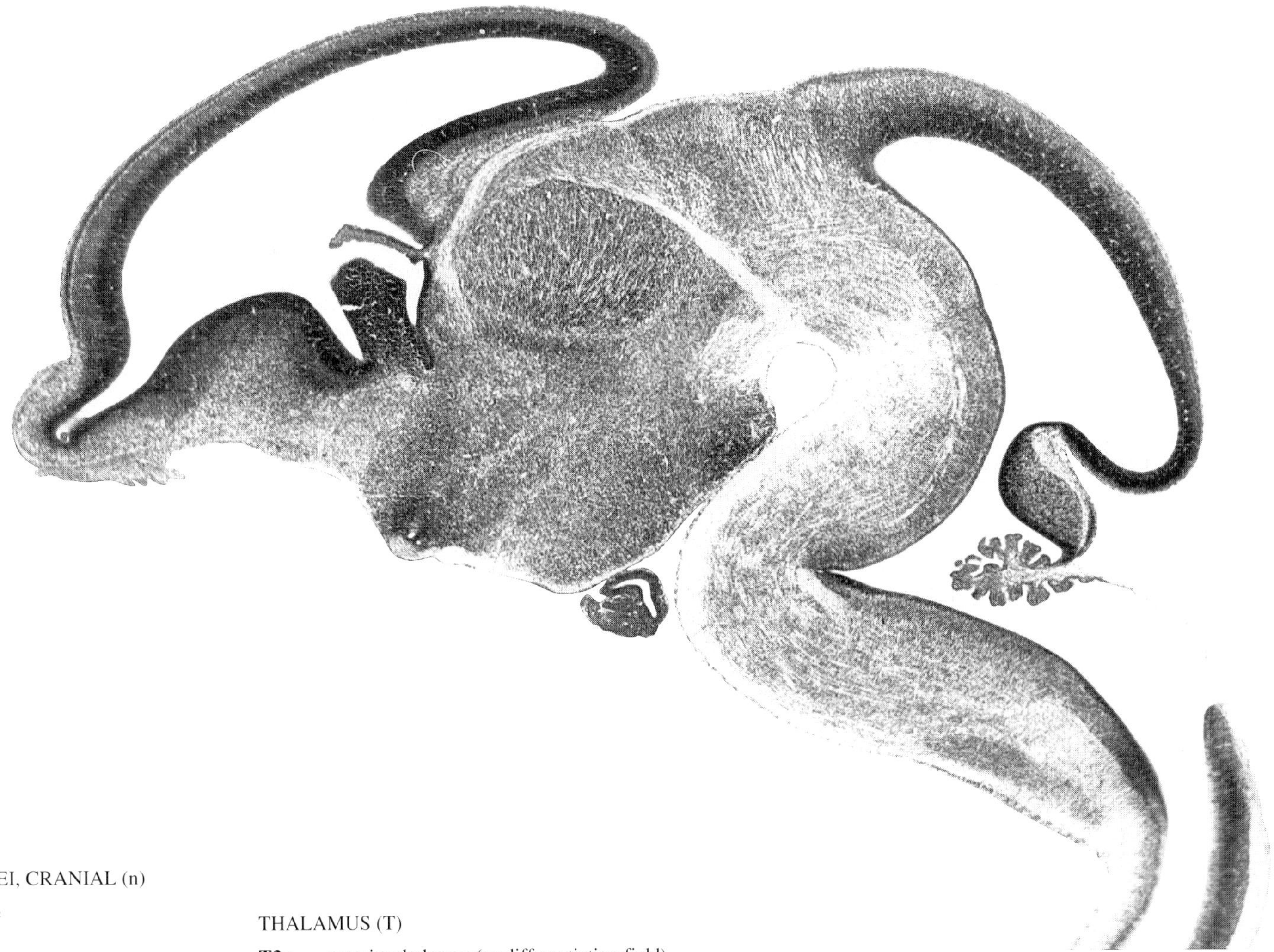

NERVES AND NUCLEI, CRANIAL (n)

n1 olfactory nerve
n2c optic chiasma

OLFACTORY BULB (O)

O1 olfactory bulb neuroepithelium
O3 olfactory bulb (or differentiating field)
O7 internal plexiform layer

PREOPTIC AREA AND OPTIC EPITHELIA (P)

P1 preoptic neuroepithelium
P3 preoptic area (or differentiating field)

RHINENCEPHALON (R)

R1 rhinencephalic neuroepithelium
R2 rhinencephalic subventricular zone
R3 rhinencephalon (or differentiating field)

SEPTAL AREA (S)

S1 septal neuroepithelium
S2 septal subventricular zone
S3 septum (or differentiating field)
S8 diagonal band (vertical limb)

THALAMUS (T)

T3a anterior thalamus (or differentiating field)
T3i intermediate thalamus (or differentiating field)
T3p posterior thalamus (or differentiating field)
T5 lateral habenular nucleus
T19 parafascicular nucleus
T25 reticular nucleus

VENTRICLES (v)

vai aqueduct, inferior tectal
vas aqueduct, superior tectal
vlc lateral ventricle, central
vld lateral ventricle, dorsal
vlo olfactory ventricle
vs spinal canal
v3pr third ventricle, preoptic recess
v4a fourth ventricle, anterior
v4i fourth ventricle, isthmal
v4p fourth ventricle, posterior

PONTINE AREA (X)

X1a anterior pontine neuroepithelium
X3a anterior pons (or differentiating field)
X3p posterior pons (or differentiating field)
X4 dorsal raphe nucleus
X9 dorsal tegmental nucleus
X10 ventral tegmental nucleus
X12p posterior precerebellar extramural migration
X33 isthmus (or differentiating field)

MEDULLA (Y)

Y1 medullary neuroepithelium
Y3 medulla (or differentiating field)
Y5 medullary raphe
Y8 inferior olive

SPINAL CORD (Z)

Z3 spinal cord (or differentiating field)
Z4 ventral horn
Z5 dorsal horn

SAGITTAL FIGURE 4

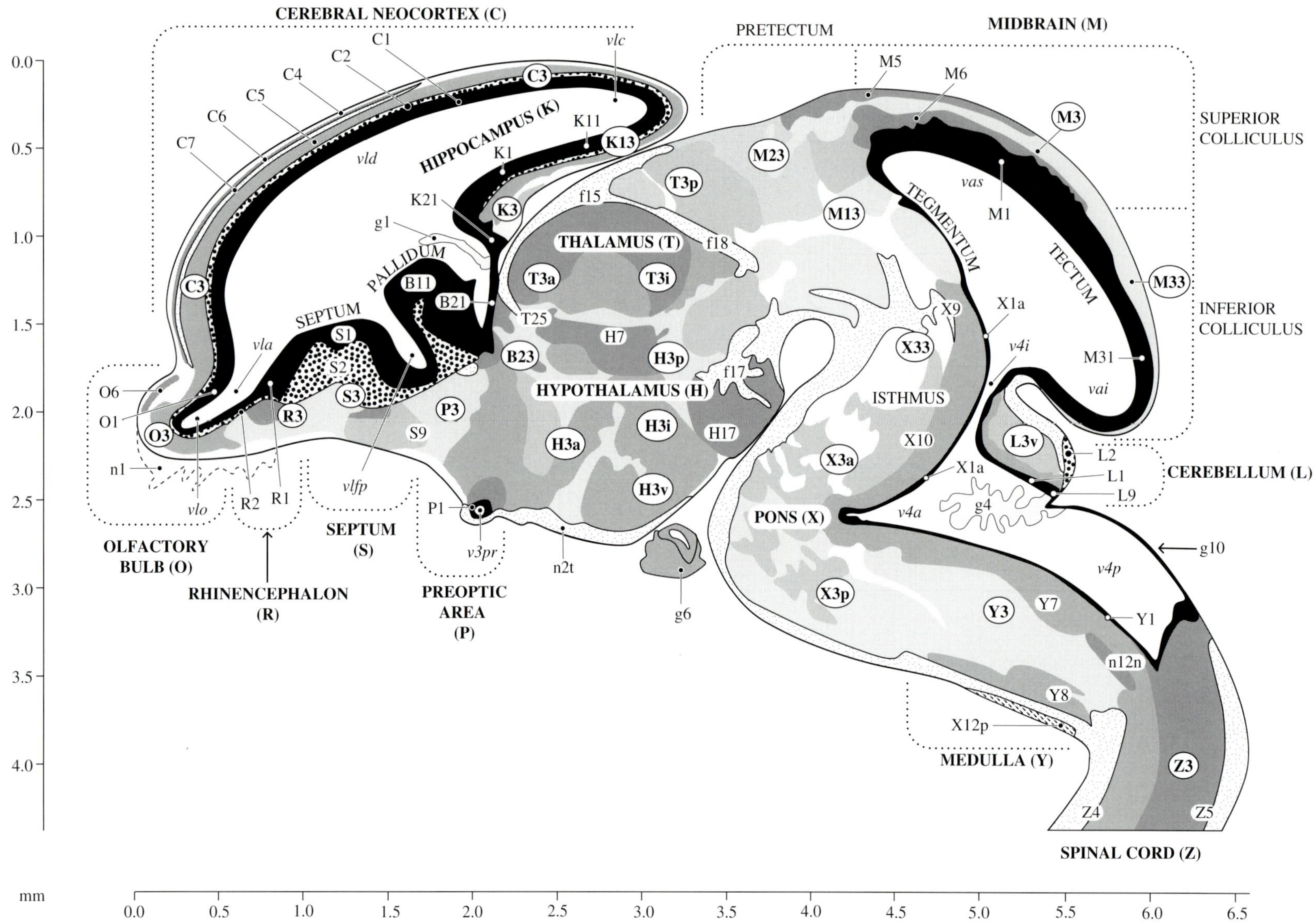

BASAL GANGLIA (B)

B11	pallidal neuroepithelium
B21	strionuclear neuroepithelium
B23	bed nucleus of stria terminalis (or differentiating field)

CEREBRAL CORTEX (C)

C1	neocortical neuroepithelium
C2	neocortical subventricular zone
C3	neocortex (or differentiating field)
C4	molecular layer (or primordial plexiform layer)
C5	intermediate zone
C6	cortical plate
C7	white matter

FIBER TRACTS (f)

f15	stria medullaris
f17	mammillothalamic tract
f18	habenulopeduncular tract

GLANDS AND PARANEURAL STRUCTURES (g)

g1	choroid plexus of lateral ventricle
g4	choroid plexus of fourth ventricle
g6	pituitary gland (primordium)
g10	velum medullare

HYPOTHALAMUS (H)

H3a	anterior hypothalamus (or differentiating field)
H3i	intermediate hypothalamus (or differentiating field)
H3v	ventral hypothalamus (or differentiating field)
H7	zona incerta
H17	lateral mammillary nucleus

HIPPOCAMPAL FORMATION (K)

K1	hippocampal (Ammon's horn) neuroepithelium
K3	hippocampus (or differentiating field)
K11	subicular neuroepithelium
K13	subicular area (or differentiating field)
K21	dentate gyrus neuroepithelium

CEREBELLUM (L)

L1	cerebellar neuroepithelium
L2	external germinal layer
L3v	cerebellar vermis (or differentiating field)
L9	cerebellar germinal trigone

MIDBRAIN (M)

M1	superior tectal neuroepithelium
M3	superior colliculus (or differentiating field)
M5	superior colliculus, superficial layers
M6	superior colliculus, deep layers
M13	tegmentum (or differentiating field)
M23	pretectum (or differentiating field)
M31	inferior tectal neuroepithelium
M33	inferior colliculus (or differentiating field)

NERVES AND NUCLEI, CRANIAL (n)

n1 olfactory nerve
n2t optic tract
n12n hypoglossal nucleus

OLFACTORY BULB (O)

O1 olfactory bulb neuroepithelium
O3 olfactory bulb (or differentiating field)
O6 mitral cell layer

PREOPTIC AREA AND OPTIC EPITHELIA (P)

P1 preoptic neuroepithelium
P3 preoptic area (or differentiating field)

RHINENCEPHALON (R)

R1 rhinencephalic neuroepithelium
R2 rhinencephalic subventricular zone
R3 rhinencephalon (or differentiating field)

SEPTAL AREA (S)

S1 septal neuroepithelium
S2 septal subventricular zone
S3 septum (or differentiating field)
S9 diagonal band (horizontal limb)

THALAMUS (T)

T3a anterior thalamus (or differentiating field)
T3i intermediate thalamus (or differentiating field)
T3p posterior thalamus (or differentiating field)
T25 reticular nucleus

VENTRICLES (v)

vai aqueduct, inferior tectal
vas aqueduct, superior tectal
vla lateral ventricle, anterior horn
vlc lateral ventricle, central
vld lateral ventricle, dorsal
vlfp lateral ventricle, pallidal fork
vlo olfactory ventricle
vlp lateral ventricle, posterior horn
v3pr third ventricle, preoptic recess
v4a fourth ventricle, anterior
v4i fourth ventricle, isthmal
v4p fourth ventricle, posterior

PONTINE AREA (X)

X1a anterior pontine neuroepithelium
X3a anterior pons (or differentiating field)
X3p posterior pons (or differentiating field)
X9 dorsal tegmental nucleus
X10 ventral tegmental nucleus
X12p posterior precerebellar extramural migration
X33 isthmus (or differentiating field)

MEDULLA (Y)

Y1 medullary neuroepithelium
Y3 medulla (or differentiating field)
Y7 prepositus nucleus
Y8 inferior olive

SPINAL CORD (Z)

Z3 spinal cord (or differentiating field)
Z4 ventral horn
Z5 dorsal horn

SAGITTAL FIGURE 5

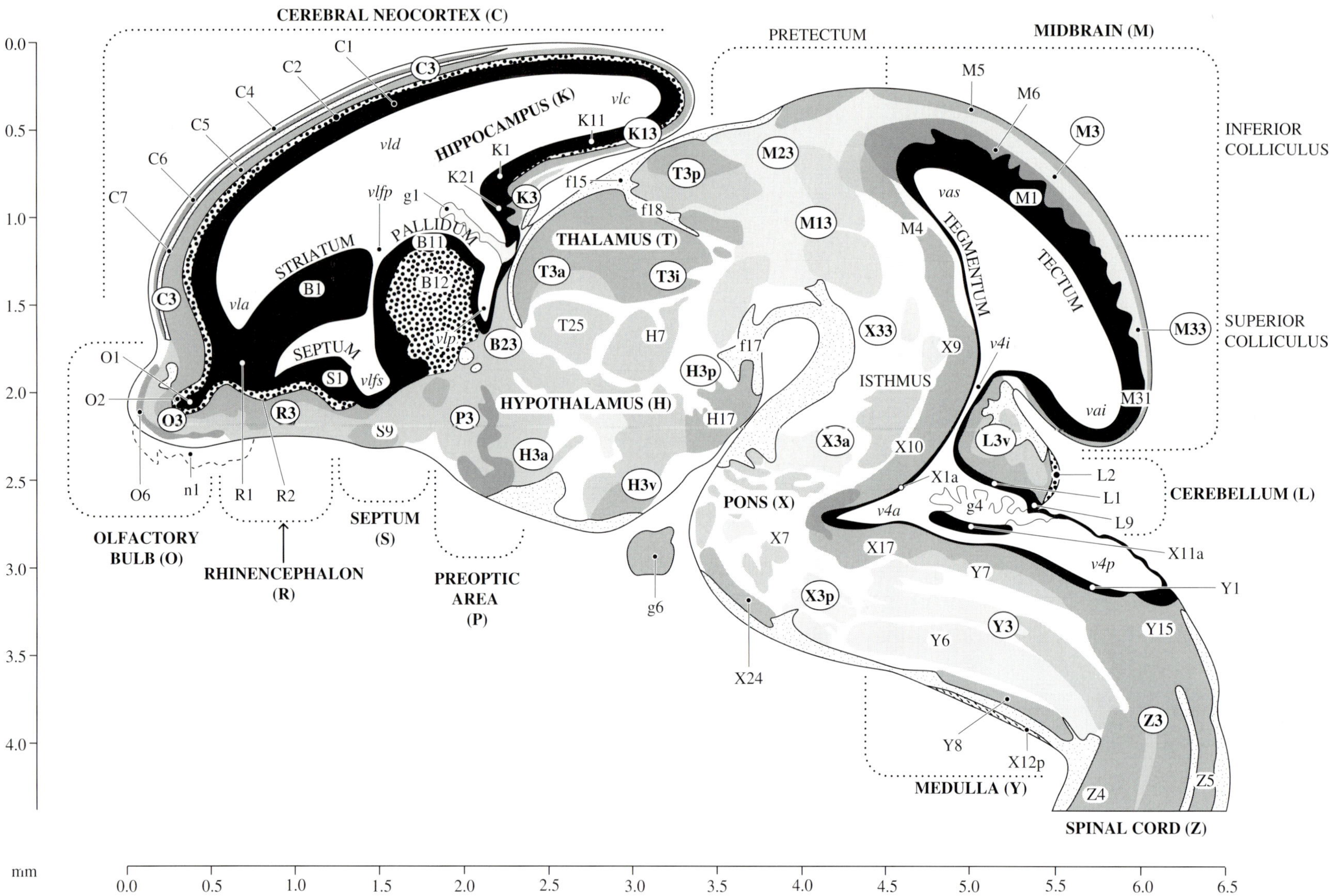

E17

BASAL GANGLIA (B)

- B1 striatal neuroepithelium
- B11 pallidal neuroepithelium
- B12 pallidal subventricular zone
- **B23** bed nucleus of stria terminalis (or differentiating field)

CEREBRAL CORTEX (C)

- C1 neocortical neuroepithelium
- C2 neocortical subventricular zone
- **C3** neocortex (or differentiating field)
- C4 molecular layer (or primordial plexiform layer)
- C5 intermediate zone
- C6 cortical plate
- C7 white matter

FIBER TRACTS (f)

- f15 stria medullaris
- f17 mammillothalamic tract
- f18 habenulopeduncular tract

GLANDS AND PARANEURAL STRUCTURES (g)

- g1 choroid plexus of lateral ventricle
- g4 choroid plexus of fourth ventricle
- g6 pituitary gland (primordium)

HYPOTHALAMUS (H)

- **H3a** anterior hypothalamus (or differentiating field)
- **H3p** posterior hypothalamus (or differentiating field)
- **H3v** ventral hypothalamus (or differentiating field)
- H7 zona incerta
- H17 lateral mammillary nucleus

HIPPOCAMPAL FORMATION (K)

- K1 hippocampal (Ammon's horn) neuroepithelium
- **K3** hippocampus (or differentiating field)
- K11 subicular neuroepithelium
- **K13** subicular area (or differentiating field)
- K21 dentate gyrus neuroepithelium

CEREBELLUM (L)

- L1 cerebellar neuroepithelium
- L2 external germinal layer
- **L3v** cerebellar vermis (or differentiating field)
- L9 cerebellar germinal trigone

MIDBRAIN (M)

- M1 superior tectal neuroepithelium
- **M3** superior colliculus (or differentiating field)
- M4 periaqueductal gray
- M5 superior colliculus, superficial layers
- M6 superior colliculus, deep layers
- **M13** tegmentum (or differentiating field)
- **M23** pretectum (or differentiating field)
- M31 inferior tectal neuroepithelium
- **M33** inferior colliculus (or differentiating field)

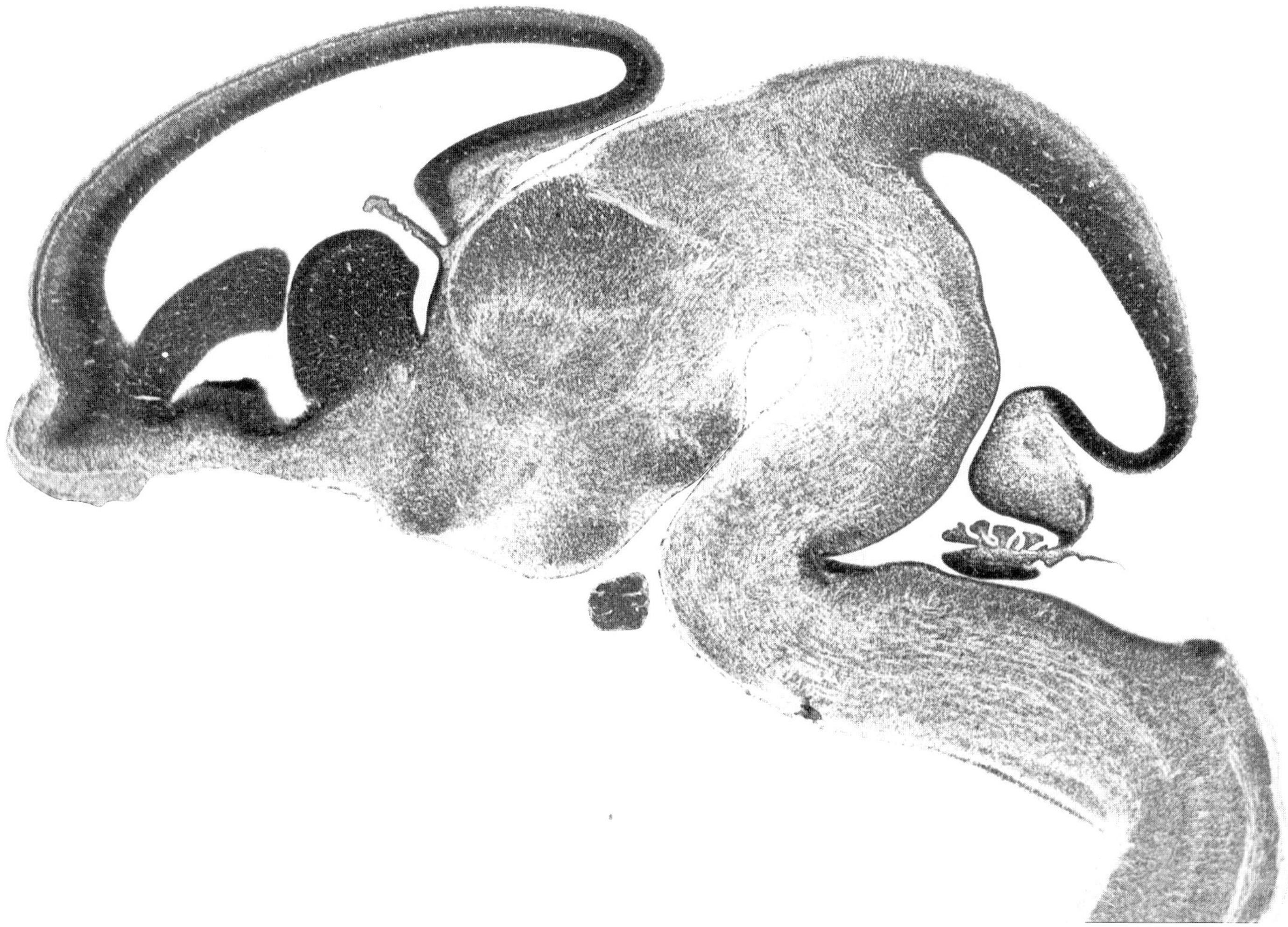

NERVES AND NUCLEI, CRANIAL (n)

n1 olfactory nerve

OLFACTORY BULB (O)

O1 olfactory bulb neuroepithelium
O2 olfactory bulb subventricular zone
O3 olfactory bulb (or differentiating field)
O6 mitral cell layer

PREOPTIC AREA AND OPTIC EPITHELIA (P)

P3 preoptic area (or differentiating field)

RHINENCEPHALON (R)

R1 rhinencephalic neuroepithelium
R2 rhinencephalic subventricular zone
R3 rhinencephalon (or differentiating field)

SEPTAL AREA (S)

S1 septal neuroepithelium
S9 diagonal band (horizontal limb)

THALAMUS (T)

T3a anterior thalamus (or differentiating field)
T3i intermediate thalamus (or differentiating field)
T3p posterior thalamus (or differentiating field)
T25 reticular nucleus

VENTRICLES (v)

vai aqueduct, inferior tectal
vas aqueduct, superior tectal
vla lateral ventricle, anterior horn
vlc lateral ventricle, central
vld lateral ventricle, dorsal
vlfp lateral ventricle, pallidal fork
vlfs lateral ventricle, septal fork
vlp lateral ventricle, posterior horn
v4a fourth ventricle, anterior
v4i fourth ventricle, isthmal
v4p fourth ventricle, posterior

PONTINE AREA (X)

X1a anterior pontine neuroepithelium
X3a anterior pons (or differentiating field)
X3p posterior pons (or differentiating field)
X7 pontine reticular formation
X9 dorsal tegmental nucleus
X10 ventral tegmental nucleus
X11a anterior precerebellar neuroepithelium
X12p posterior precerebellar extramural migration
X17 medial vestibular nucleus
X24 superior olivary nucleus
X33 isthmus (or differentiating field)

MEDULLA (Y)

Y1 medullary neuroepithelium
Y3 medulla (or differentiating field)
Y6 medullary reticular formation
Y7 prepositus nucleus
Y8 inferior olive
Y15 cuneate nucleus

SPINAL CORD (Z)

Z3 spinal cord (or differentiating field)
Z4 ventral horn
Z5 dorsal horn

SAGITTAL FIGURE 6

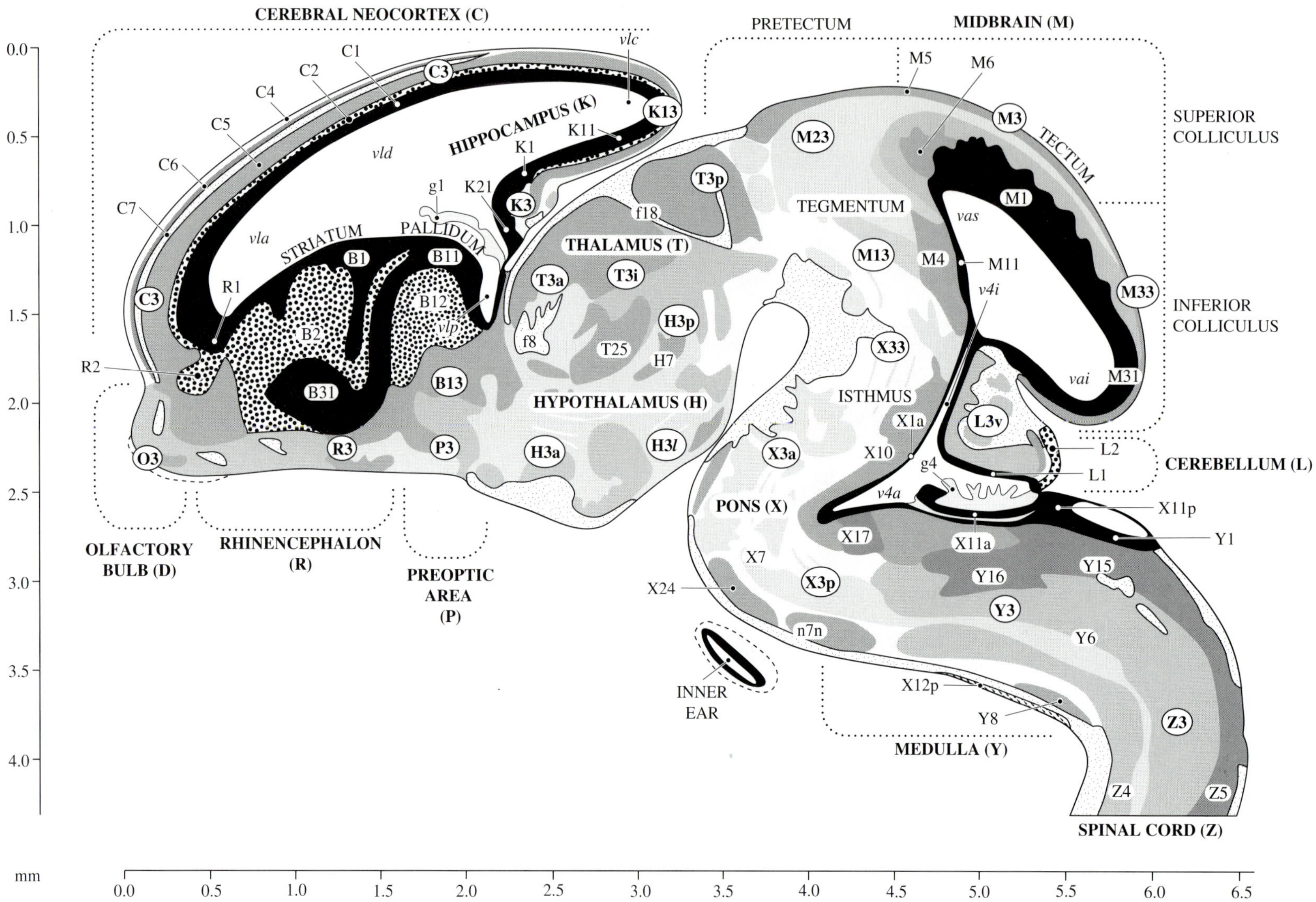

BASAL GANGLIA (B)

B1	striatal neuroepithelium
B2	striatal subventricular zone
B11	pallidal neuroepithelium
B12	pallidal subventricular zone
B13	pallidum (or differentiating field)
B31	accumbens neuroepithelium

CEREBRAL CORTEX (C)

C1	neocortical neuroepithelium
C2	neocortical subventricular zone
C3	neocortex (or differentiating field)
C4	molecular layer (or primordial plexiform layer)
C5	intermediate zone
C6	cortical plate
C7	white matter

FIBER TRACTS (f)

f8	internal capsule
f18	habenulopeduncular tract

GLANDS AND PARANEURAL STRUCTURES (g)

g1	choroid plexus of lateral ventricle
g4	choroid plexus of fourth ventricle

HYPOTHALAMUS (H)

H3a	anterior hypothalamus (or differentiating field)
H3*l*	lateral hypothalamus (or differentiating field)
H3p	posterior hypothalamus (or differentiating field)
H7	zona incerta

HIPPOCAMPAL FORMATION (K)

K1	hippocampal (Ammon's horn) neuroepithelium
K3	hippocampus (or differentiating field)
K11	subicular neuroepithelium
K13	subicular area (or differentiating field)
K21	dentate gyrus neuroepithelium

CEREBELLUM (L)

L1	cerebellar neuroepithelium
L2	external germinal layer
L3v	cerebellar vermis (or differentiating field)

MIDBRAIN (M)

M1	superior tectal neuroepithelium
M3	superior colliculus (or differentiating field)
M4	periaqueductal gray
M5	superior colliculus, superficial layers
M6	superior colliculus, deep layers
M11	tegmental neuroepithelium
M13	tegmentum (or differentiating field)
M23	pretectum (or differentiating field)
M31	inferior tectal neuroepithelium
M33	inferior colliculus (or differentiating field)

NERVES AND NUCLEI, CRANIAL (n)

n7n	facial motor nucleus

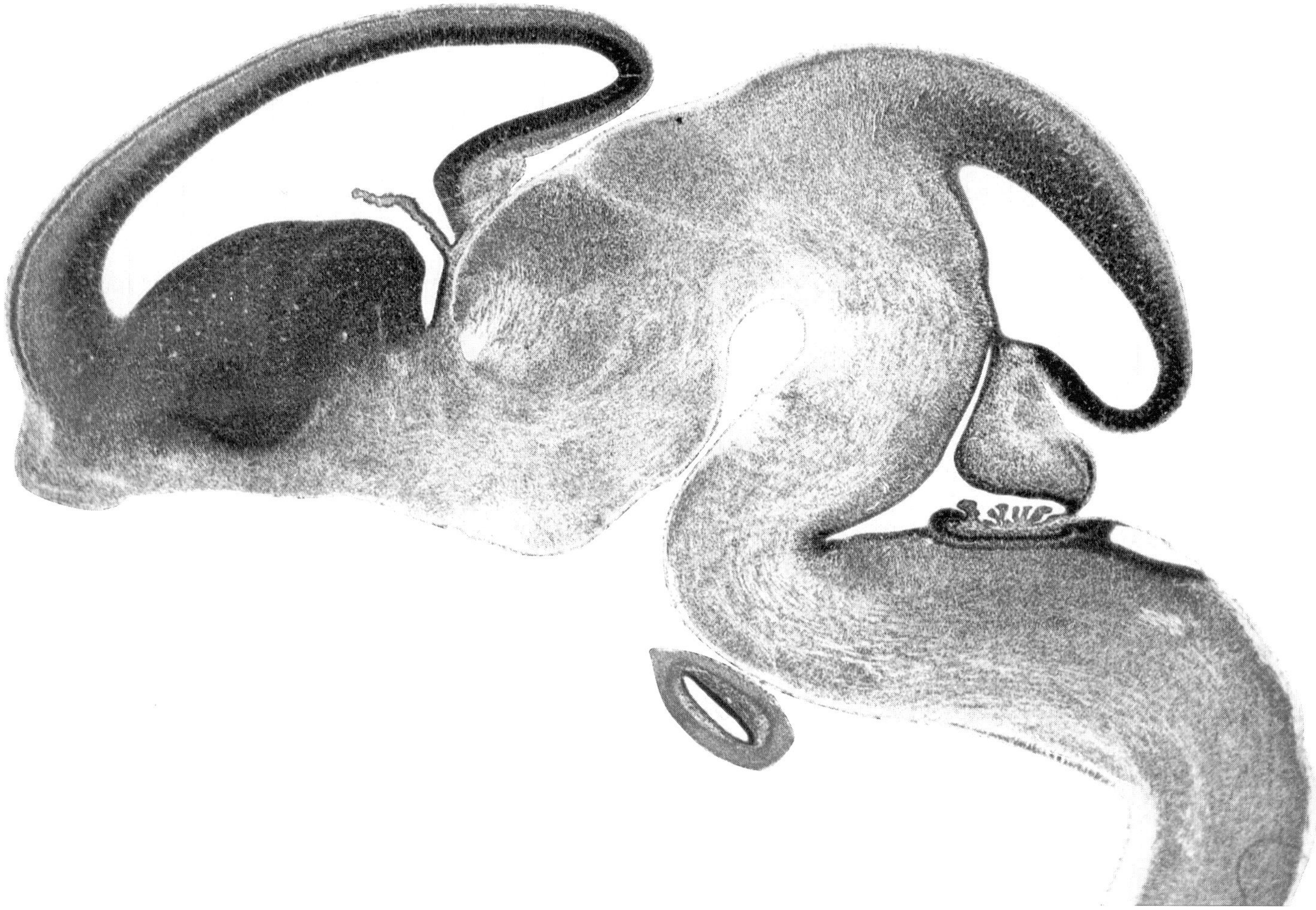

OLFACTORY BULB (O)

O3 olfactory bulb (or differentiating field)

PREOPTIC AREA AND OPTIC EPITHELIA (P)

P3 preoptic area (or differentiating field)

RHINENCEPHALON (R)

R1 rhinencephalic neuroepithelium
R2 rhinencephalic subventricular zone
R3 rhinencephalon (or differentiating field)

THALAMUS (T)

T3a anterior thalamus (or differentiating field)
T3i intermediate thalamus (or differentiating field)
T3p posterior thalamus (or differentiating field)
T25 reticular nucleus

VENTRICLES (v)

vai aqueduct, inferior tectal
vas aqueduct, superior tectal
vla lateral ventricle, anterior horn
vlc lateral ventricle, central
vld lateral ventricle, dorsal
vlp lateral ventricle, posterior horn
v4a fourth ventricle, anterior
v4i fourth ventricle, isthmal

PONTINE AREA (X)

X1a anterior pontine neuroepithelium
X3a anterior pons (or differentiating field)
X3p posterior pons (or differentiating field)
X7 pontine reticular formation
X10 ventral tegmental nucleus (pontine)
X11a anterior precerebellar neuroepithelium
X11p posterior precerebellar neuroepithelium
X12p posterior precerebellar extramural migration
X17 medial vestibular nucleus
X24 superior olivary nucleus
X33 isthmus (or differentiating field)

MEDULLA (Y)

Y1 medullary neuroepithelium
Y3 medulla (or differentiating field)
Y6 medullary reticular formation
Y8 inferior olive
Y15 cuneate nucleus
Y16 solitary nucleus

SPINAL CORD (Z)

Z3 spinal cord (or differentiating field)
Z4 ventral horn
Z5 dorsal horn

SAGITTAL FIGURE 7

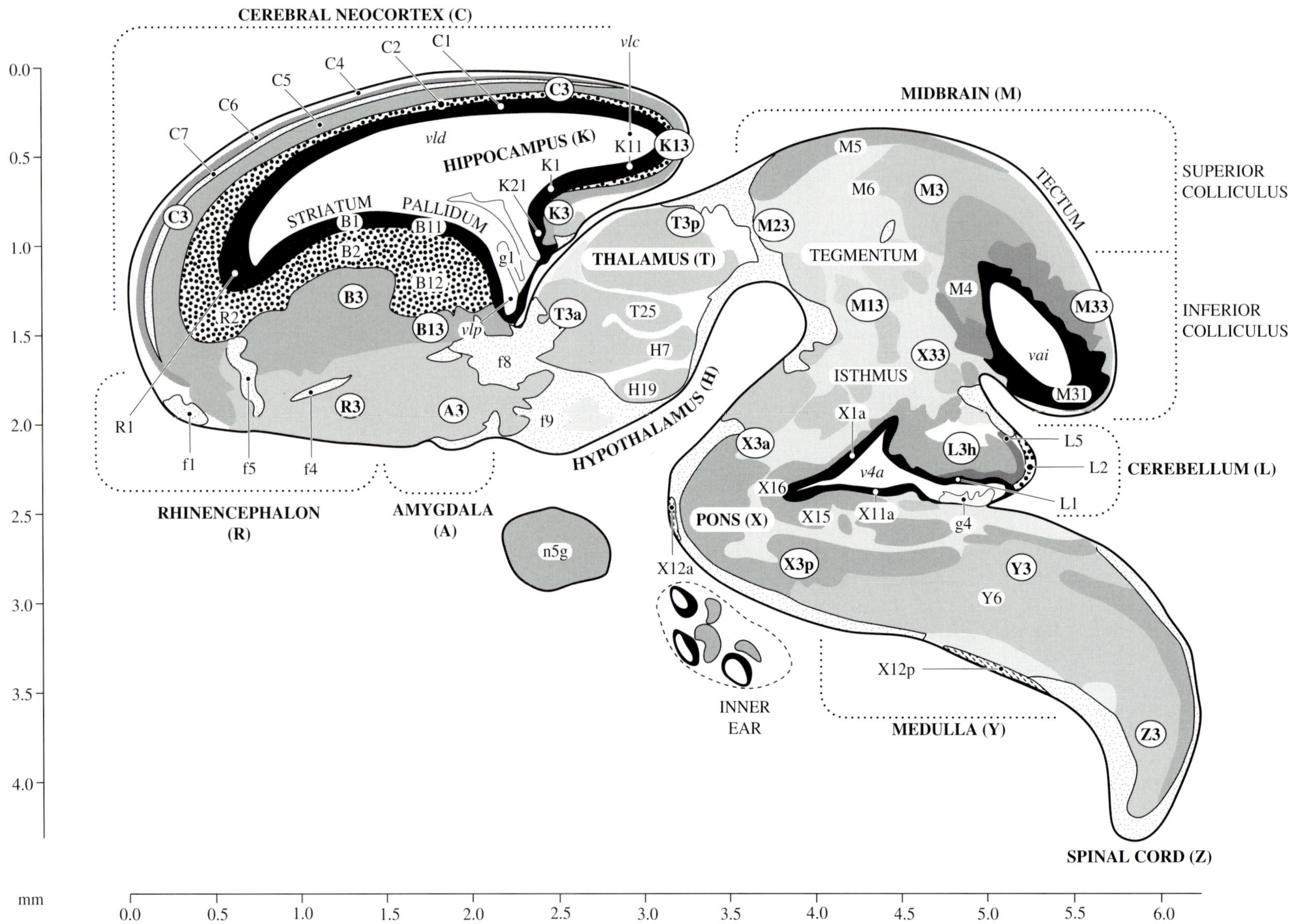

AMYGDALA (A)

A3 amygdala (or differentiating field)

BASAL GANGLIA (B)

B1 striatal neuroepithelium
B2 striatal subventricular zone
B3 striatum (or differentiating field)
B11 pallidal neuroepithelium
B12 pallidal subventricular zone
B13 pallidum (or differentiating field)

CEREBRAL CORTEX (C)

C1 neocortical neuroepithelium
C2 neocortical subventricular zone
C3 neocortex (or differentiating field)
C4 molecular layer (or primordial plexiform layer)
C5 intermediate zone
C6 cortical plate
C7 white matter

FIBER TRACTS (f)

f1 lateral olfactory tract
f4 anterior commissure
f5 medial forebrain bundle
f8 internal capsule
f9 cerebral peduncle

GLANDS AND PARANEURAL STRUCTURES (g)

g1 choroid plexus of lateral ventricle
g4 choroid plexus of fourth ventricle

HYPOTHALAMUS (H)

H7 zona incerta
H19 subthalamic nucleus

HIPPOCAMPAL FORMATION (K)

K1 hippocampal (Ammon's horn) neuroepithelium
K3 hippocampus (or differentiating field)
K11 subicular neuroepithelium
K13 subicular area (or differentiating field)
K21 dentate gyrus neuroepithelium

CEREBELLUM (L)

L1 cerebellar neuroepithelium
L2 external germinal layer
L3h cerebellar hemisphere (or differentiating field)
L5 Purkinje cell layer

MIDBRAIN (M)

M3 superior colliculus (or differentiating field)
M4 periaqueductal gray
M5 superior colliculus, superficial layer
M6 superior colliculus, deep layers
M13 tegmentum (or differentiating field)
M23 pretectum (or differentiating field)

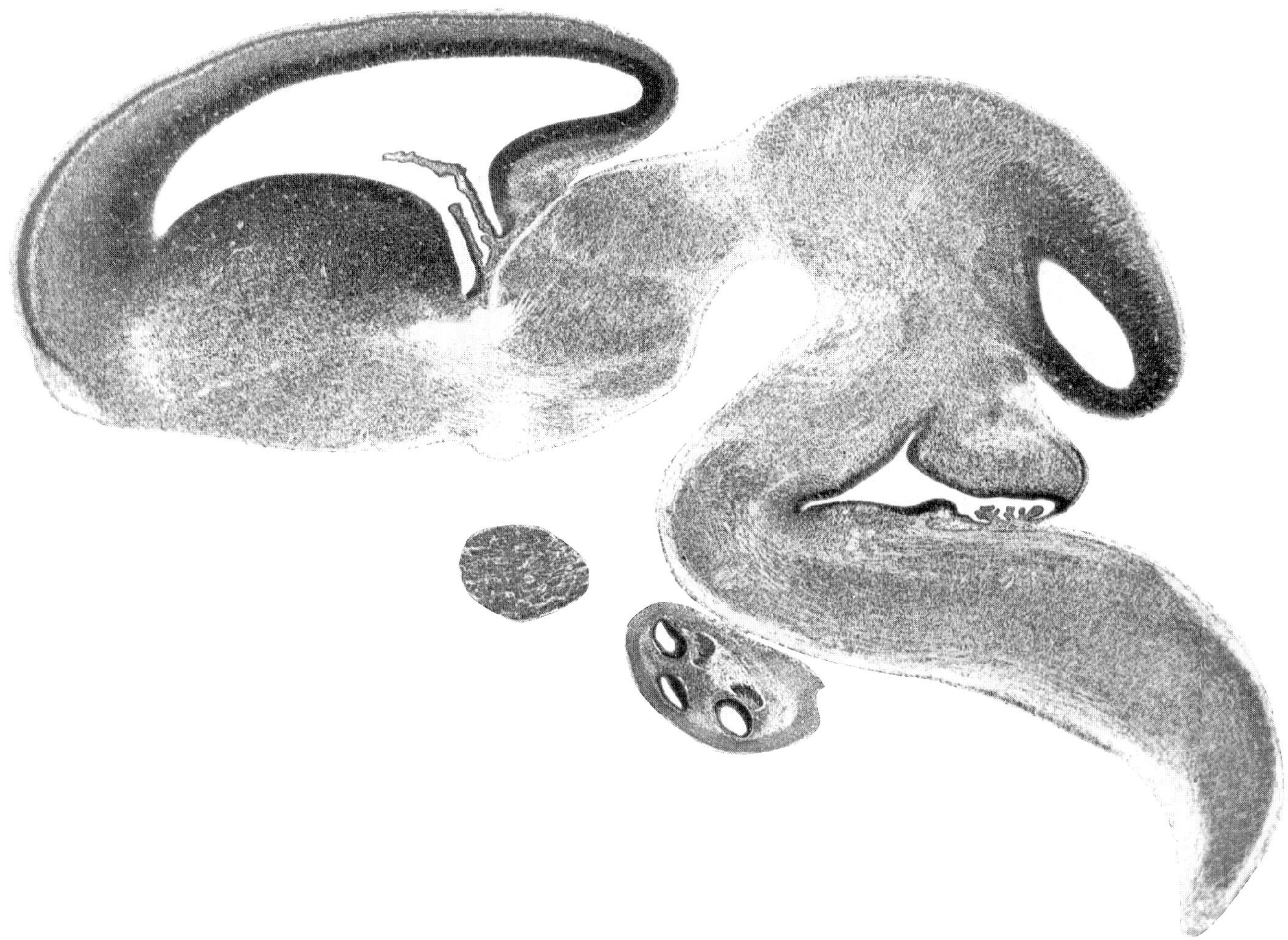

M31 tectal neuroepithelium
M33 inferior colliculus (or differentiating field)

NERVES AND NUCLEI, CRANIAL (n)

n5g trigeminal ganglion

RHINENCEPHALON (R)

R1 rhinencephalic neuroepithelium
R2 rhinencephalic subventricular zone
R3 rhinencephalon (or differentiating field)

THALAMUS (T)

T3a anterior thalamus (or differentiating field)
T3p posterior thalamus (or differentiating field)
T25 reticular nucleus

VENTRICLES (v)

vai aqueduct, inferior tectal
vlc lateral ventricle, central
vld lateral ventricle, dorsal
vlp lateral ventricle, posterior horn
v4a fourth ventricle, anterior

PONTINE AREA (X)

X1a pontine neuroepithelium, anterior
X3a anterior pons (or differentiating field)
X3p posterior pons (or differentiating field)
X11a anterior precerebellar neuroepithelium
X12a anterior precerebellar extramural migration
X12p posterior precerebellar extramural migration
X15 spinal vestibular nucleus
X16 superior vestibular nucleus
X33 isthmus (or differentiating field)

MEDULLA (Y)

Y3 medulla (or differentiating field)
Y6 medullary reticular formation

SPINAL CORD (Z)

Z3 spinal cord (or differentiating field)

SAGITTAL FIGURE 8

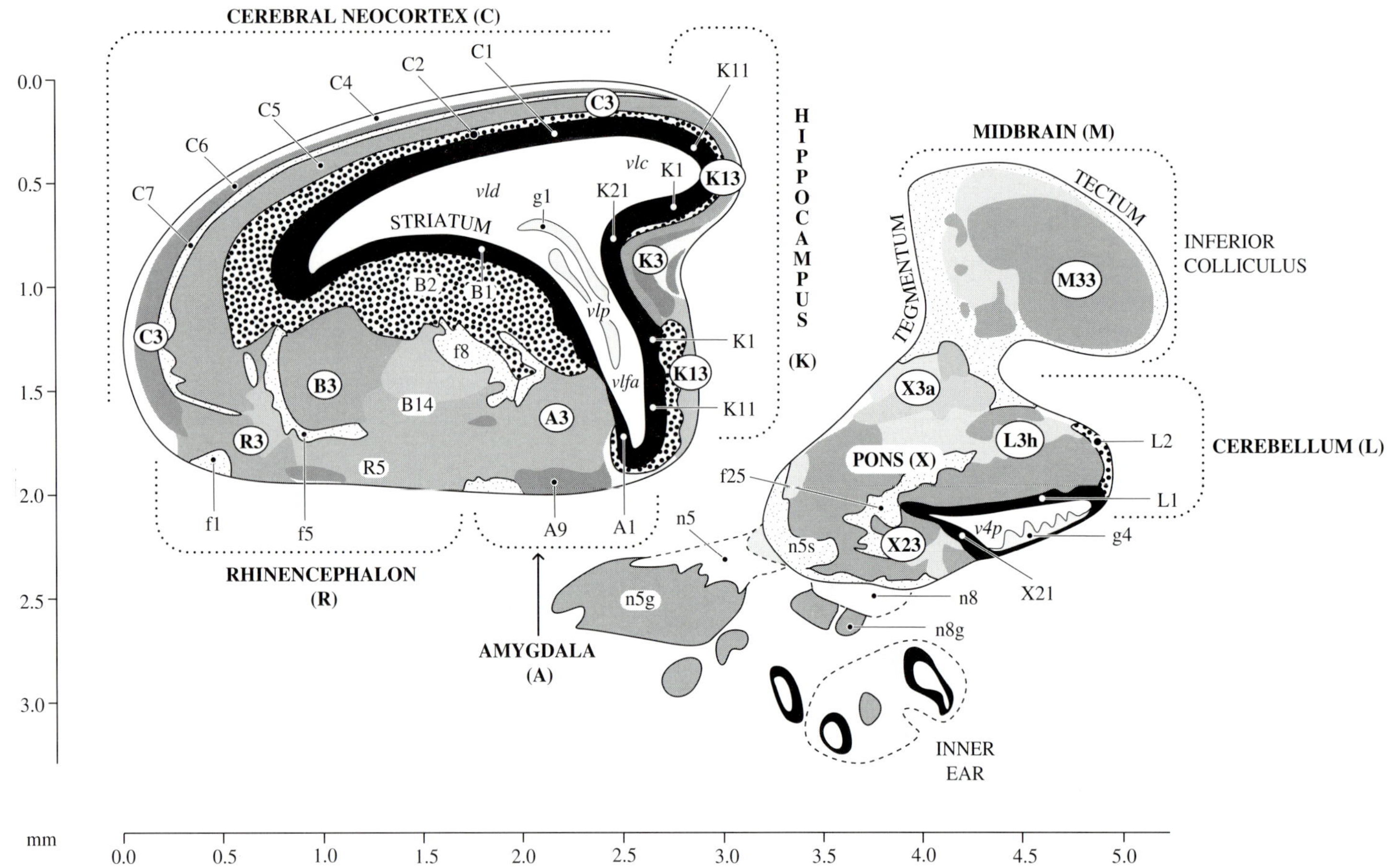

AMYGDALA (A)

A1	amygdaloid neuroepithelium
A3	amygdala (or differentiating field)
A9	cortical amygdaloid nucleus

BASAL GANGLIA (B)

B1	striatal neuroepithelium
B2	striatal subventricular zone
B3	striatum (or differentiating field)
B14	entopeduncular nucleus

CEREBRAL CORTEX (C)

C1	neocortical neuroepithelium
C2	neocortical subventricular zone
C3	neocortex (or differentiating field)
C4	molecular layer (or primordial plexiform layer)
C5	intermediate zone
C6	cortical plate
C7	white matter

FIBER TRACTS (f)

f1	lateral olfactory tract
f5	medial forebrain bundle
f8	internal capsule
f25	inferior cerebellar peduncle

GLANDS AND PARANEURAL STRUCTURES (g)

g1	choroid plexus of lateral ventricle
g4	choroid plexus of fourth ventricle

HIPPOCAMPAL FORMATION (K)

K1	hippocampal (Ammon's horn) neuroepithelium
K3	hippocampus (or differentiating field)
K11	subicular neuroepithelium
K13	subicular area (or differentiating field)
K21	dentate gyrus neuroepithelium

CEREBELLUM (L)

L1	cerebellar neuroepithelium
L2	external germinal layer
L3h	cerebellar hemisphere (or differentiating field)

MIDBRAIN (M)

M33	inferior colliculus (or differentiating field)

NERVES AND NUCLEI, CRANIAL (n)

n5	trigeminal nerve
n5g	trigeminal ganglion
n5s	spinal tract of trigeminal nucleus
n8	vestibulocochlear nerve
n8g	vestibular ganglion

RHINENCEPHALON (R)

R3	rhinencephalon (or differentiating field)
R5	endopiriform nucleus

VENTRICLES (v)

vlc	lateral ventricle, central
vld	lateral ventricular, dorsal
vlfa	lateral ventricle, amygdaloid fork
vlp	lateral ventricle, posterior horn
v4p	fourth ventricle, posterior

PONTINE AREA (X)

X3a	anterior pons (or differentiating field)
X21	cochlear nuclei neuroepithelium
X23	cochlear nuclei (or differentiating field)

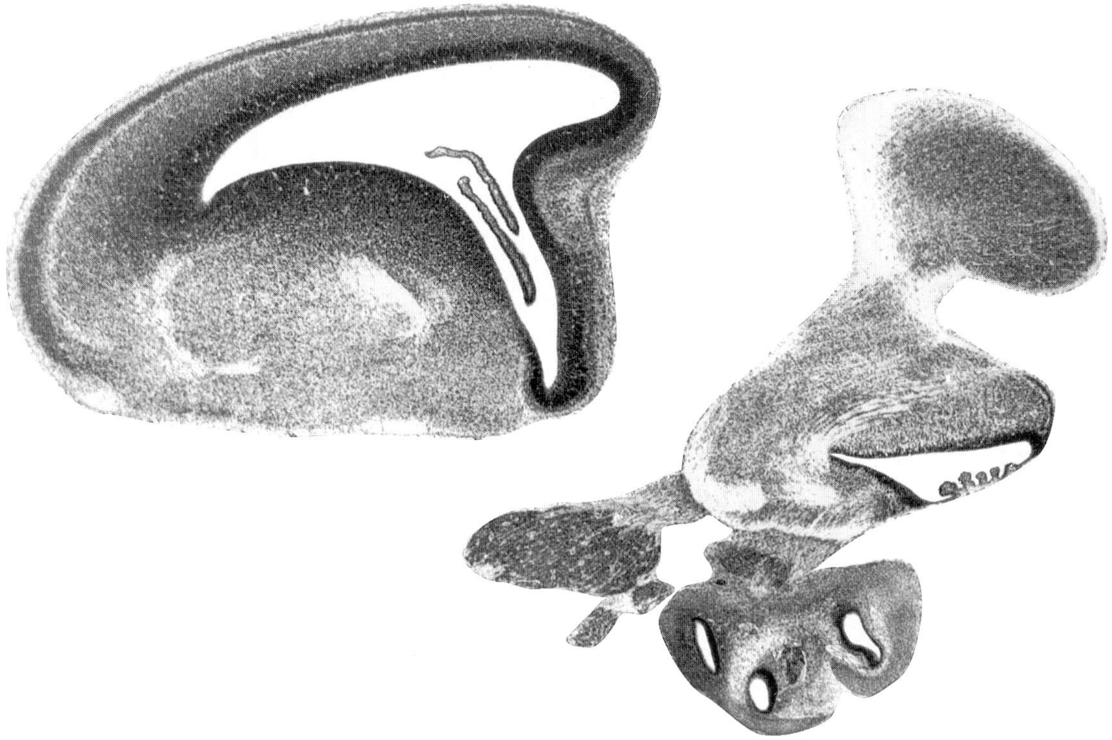

CORONAL

FIGURES and PLATES 1-21

CORONAL FIGURE 1

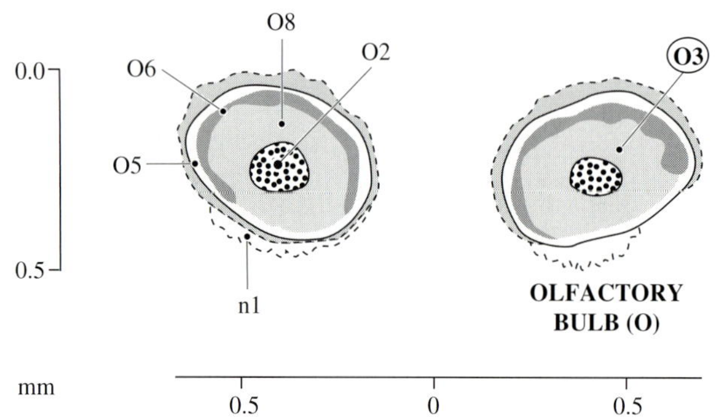

NERVES AND NUCLEI, CRANIAL (n)

n1 olfactory nerve

OLFACTORY BULB (O)

O2 olfactory bulb subventricular zone
O3 olfactory bulb (or differentiating field)
O5 external plexiform layer
O6 mitral cell layer
O8 granular layer

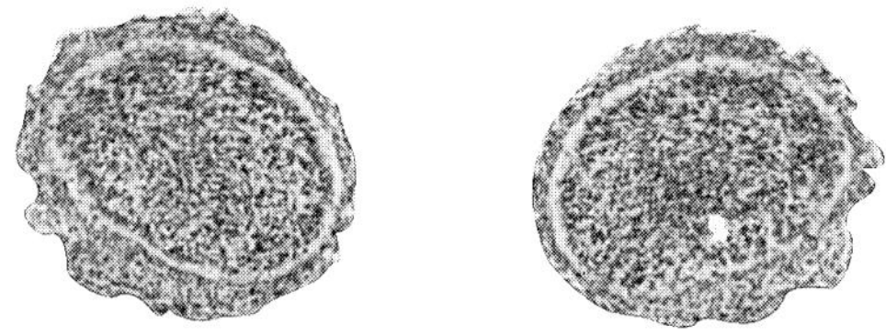

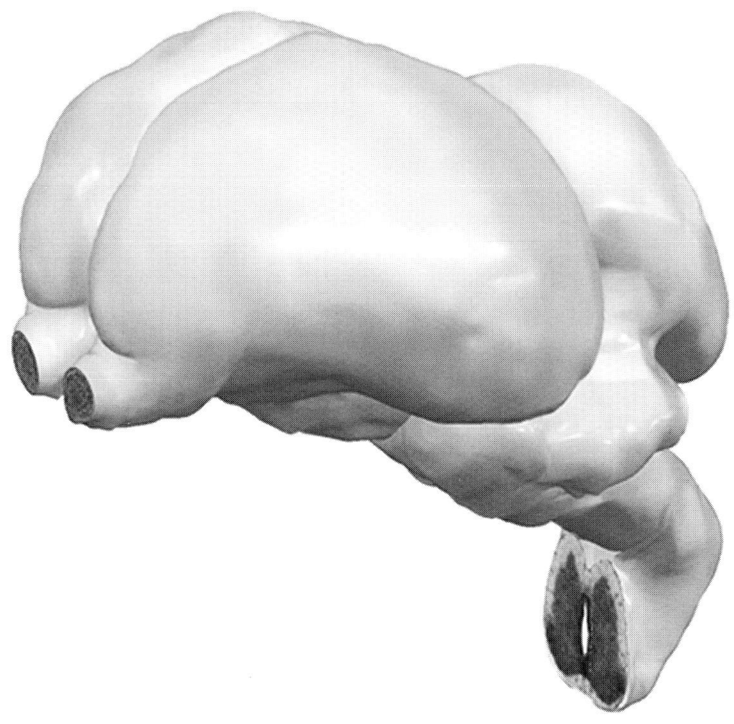

CORONAL FIGURE 2

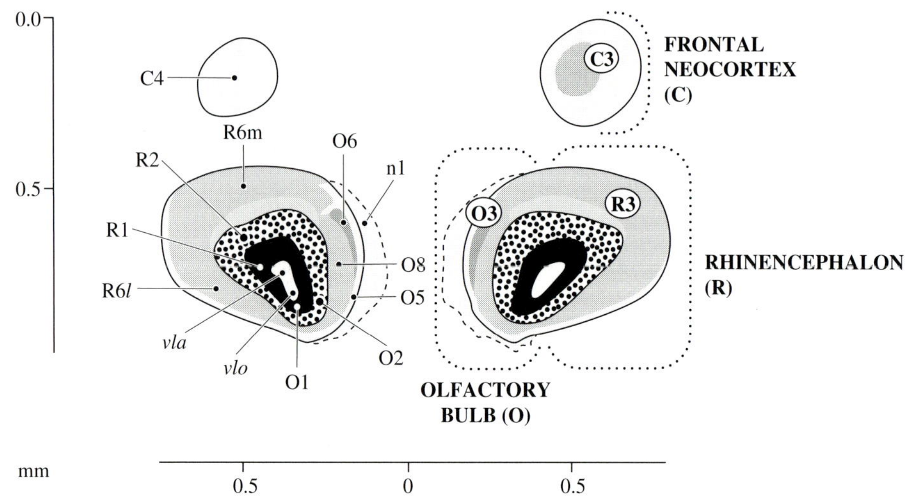

CEREBRAL CORTEX (C)

C3 neocortex (or differentiating field)
C4 molecular layer (or primordial plexiform layer)

NERVES AND NUCLEI, CRANIAL (n)

n1 olfactory nerve

OLFACTORY BULB (O)

O1 olfactory bulb neuroepithelium
O2 olfactory bulb subventricular zone
O3 olfactory bulb (or differentiating field)
O5 external plexiform layer
O6 mitral cell layer
O8 granular layer

RHINENCEPHALON (R)

R1 rhinencephalic neuroepithelium
R2 rhinencephalic subventricular zone
R3 rhinencephalon (or differentiating field)
R6*l* anterior olfactory nucleus, lateral part
R6m anterior olfactory nucleus, medial part

VENTRICLES (v)

vla lateral ventricle, anterior horn
vlo olfactory ventricle

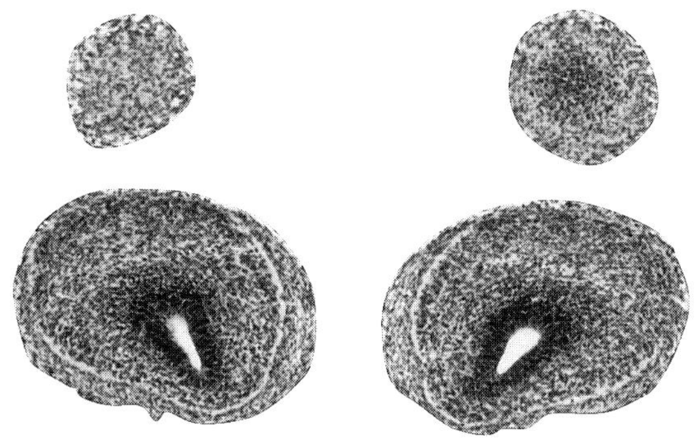

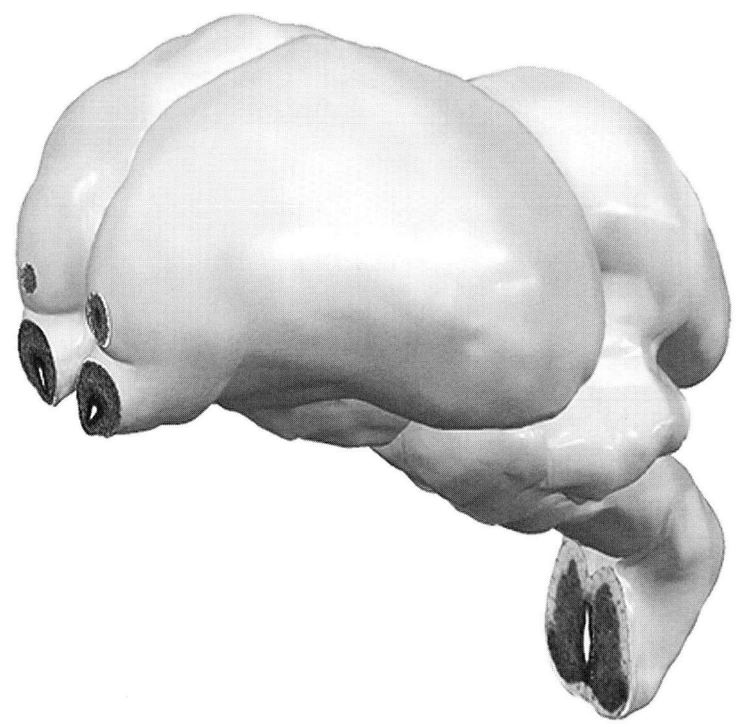

CORONAL FIGURE 3

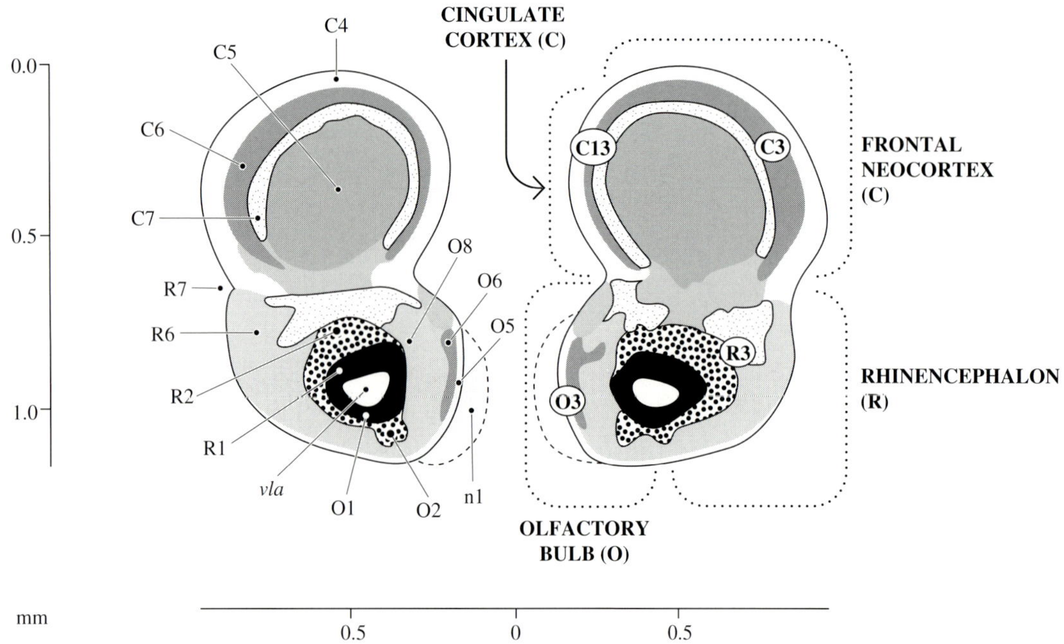

CEREBRAL CORTEX (C)

C3 neocortex (or differentiating field)
C4 molecular layer (or primordial plexiform layer)
C5 intermediate zone
C6 cortical plate
C7 white matter
C13 cingulate cortex (or differentiating field)

NERVES AND NUCLEI, CRANIAL (n)

n1 olfactory nerve

OLFACTORY BULB (O)

O1 olfactory bulb neuroepithelium
O2 olfactory bulb subventricular zone
O3 olfactory bulb (or differentiating field)
O5 external plexiform layer
O6 mitral cell layer
O8 granular layer

RHINENCEPHALON (R)

R1 rhinencephalic neuroepithelium
R2 rhinencephalic subventricular zone
R3 rhinencephalon (or differentiating field)
R6 anterior olfactory nucleus
R7 rhinal fissure

VENTRICLES (v)

vla lateral ventricle, anterior horn

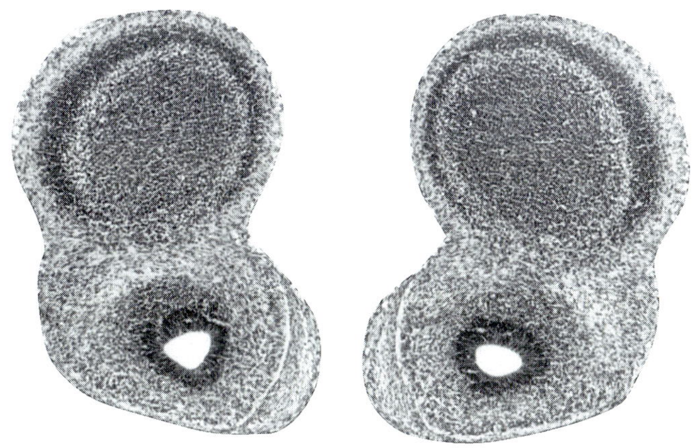

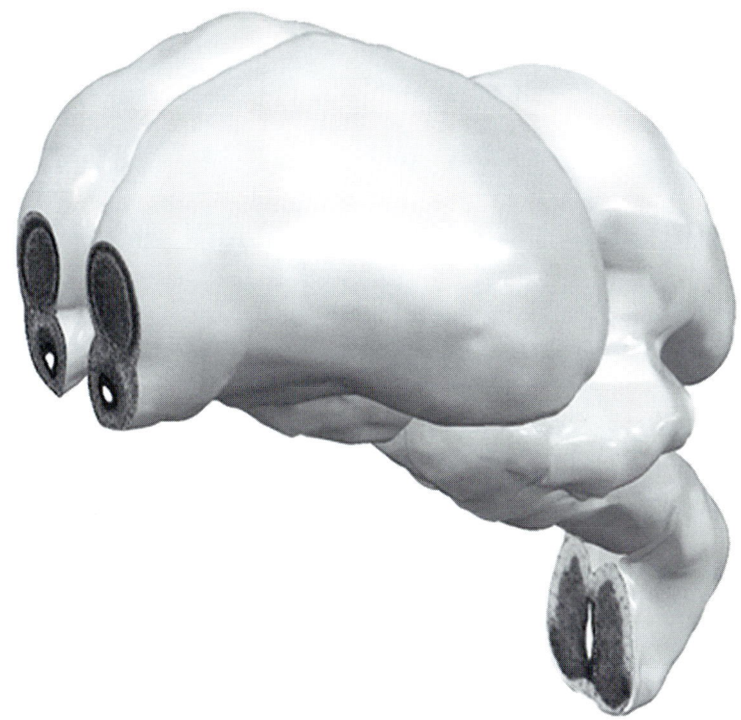

CORONAL FIGURE 4

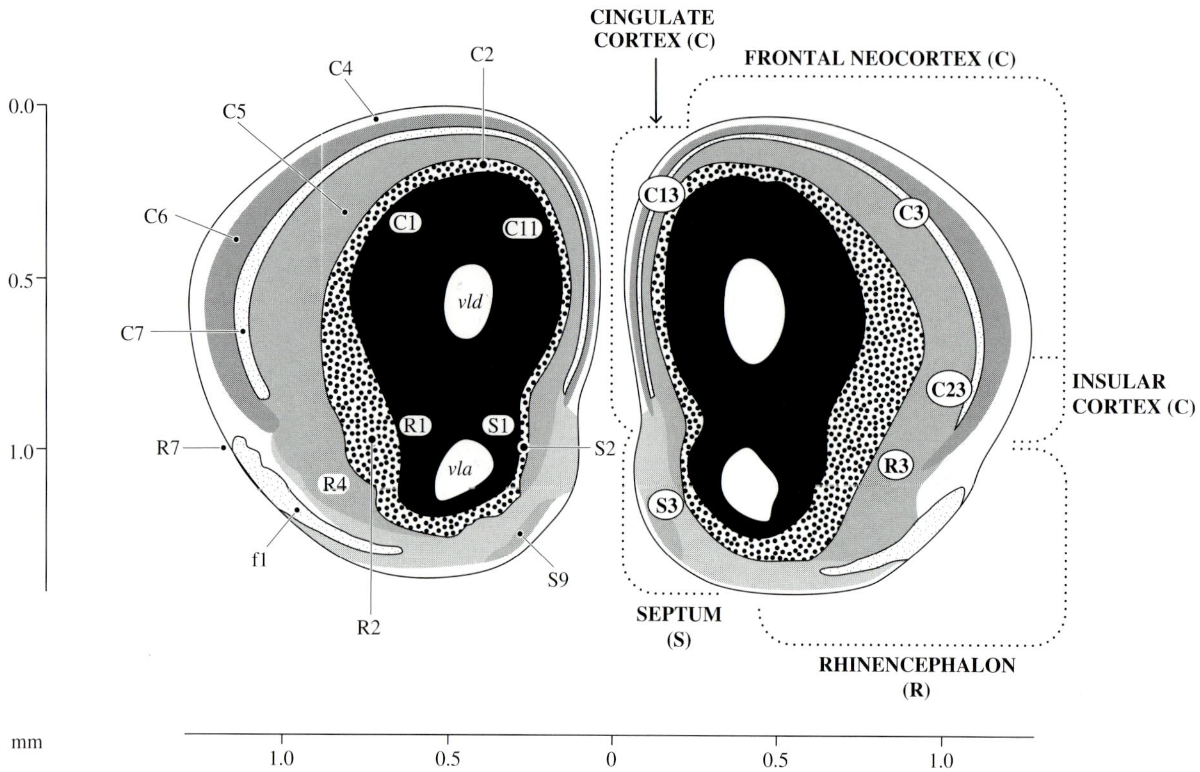

CEREBRAL CORTEX (C)

C1 neocortical neuroepithelium
C2 neocortical subventricular zone
C3 neocortex (or differentiating field)
C4 molecular layer (or primordial plexiform layer)
C5 intermediate zone
C6 cortical plate
C7 white matter
C11 cingulate cortex neuroepithelium
C13 cingulate cortex (or differentiating field)
C23 orbital and insular area (or differentiating field)

FIBER TRACTS (f)

f1 lateral olfactory tract

RHINENCEPHALON (R)

R1 rhinencephalic neuroepithelium
R2 rhinencephalic subventricular zone
R3 rhinencephalon (or differentiating field)
R4 piriform cortex
R7 rhinal fissure

SEPTAL AREA (S)

S1 septal neuroepithelium
S2 septal subventricular zone
S3 septum (or differentiating field)
S9 diagonal band (horizontal limb)

VENTRICLES (v)

vla lateral ventricle, anterior horn
vld lateral ventricle, dorsal

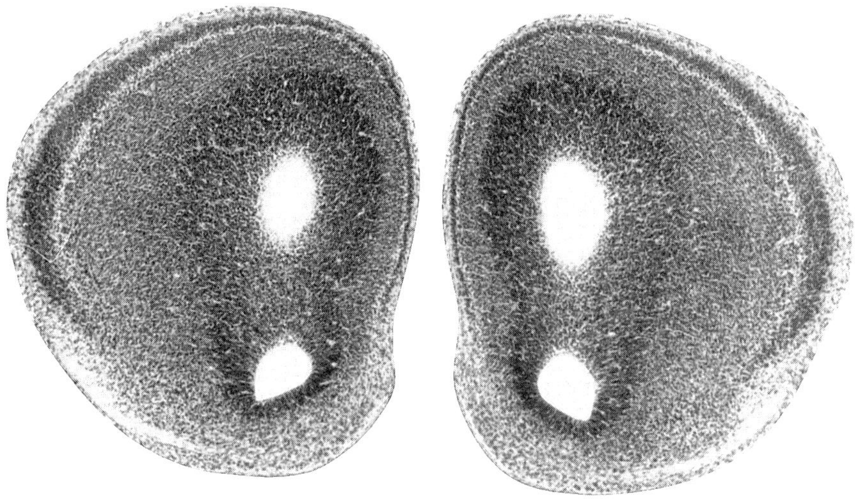

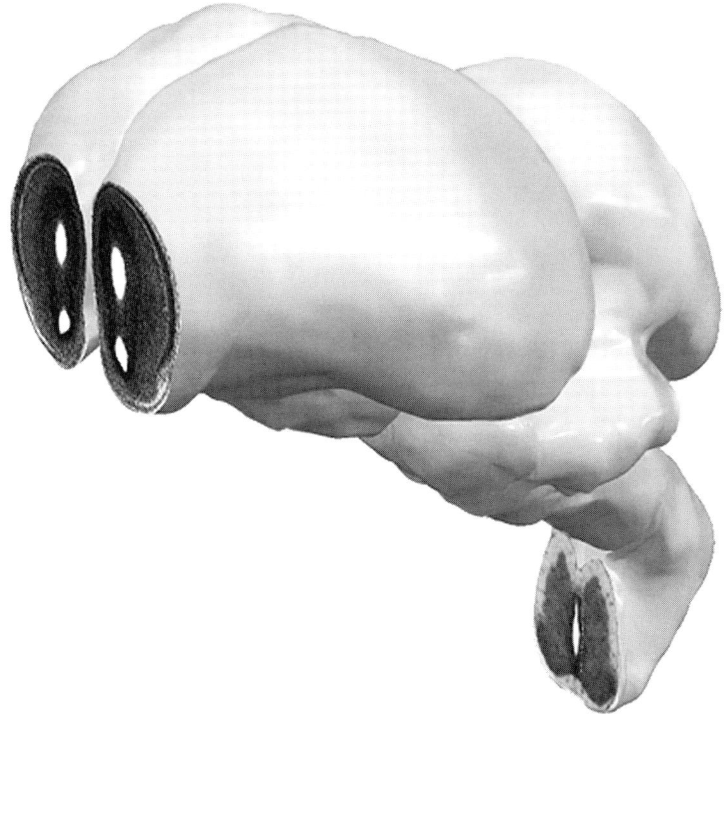

CORONAL FIGURE 5

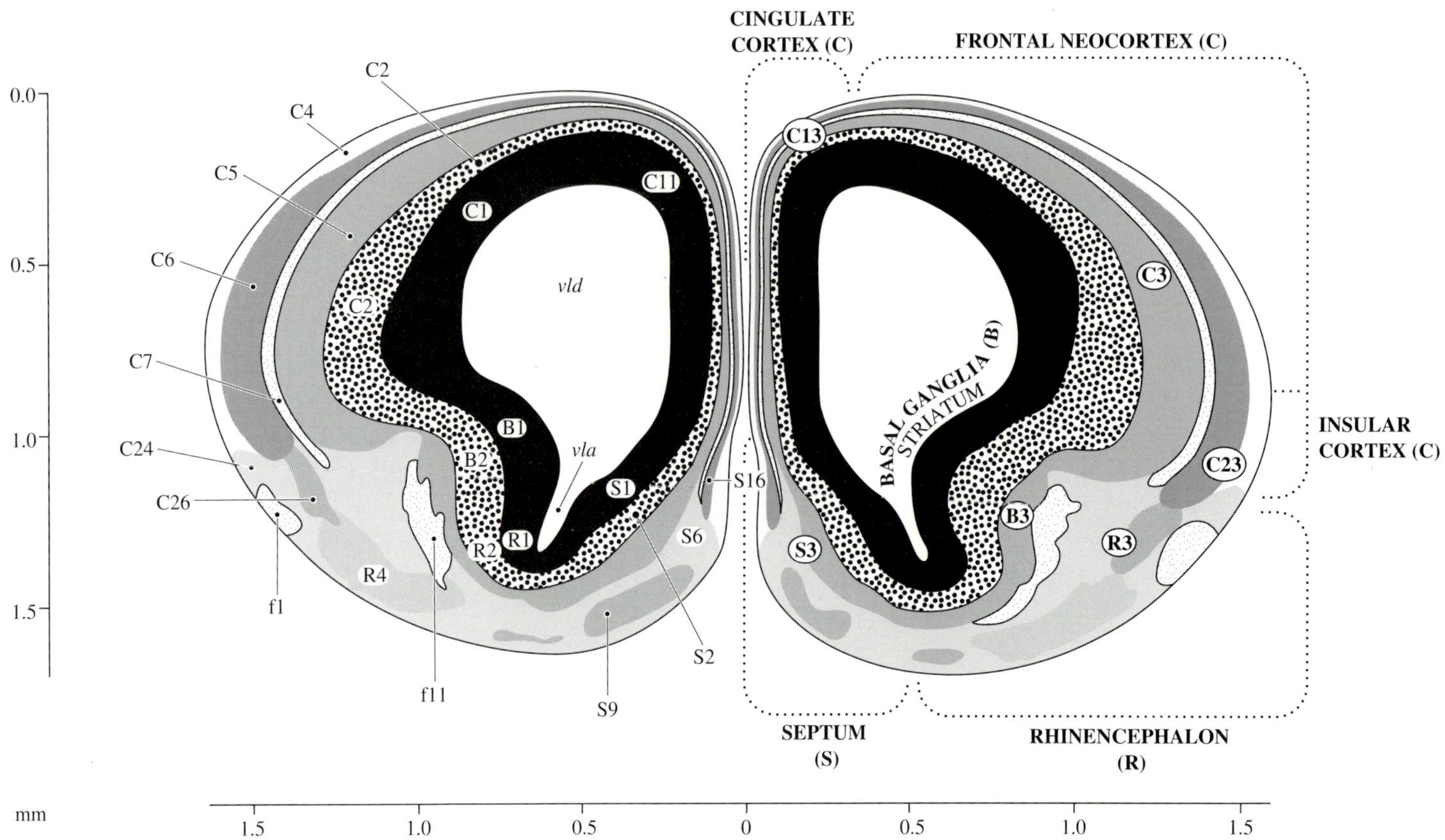

BASAL GANGLIA (B)

B1	striatal neuroepithelium
B2	striatal subventricular zone
B3	striatum (or differentiating field)

CEREBRAL CORTEX (C)

C1	neocortical neuroepithelium
C2	neocortical subventricular zone
C3	neocortex (or differentiating field)
C4	molecular layer (or primordial plexiform layer)
C5	intermediate zone
C6	cortical plate
C7	white matter
C11	cingulate cortex neuroepithelium
C13	cingulate cortex (or differentiating field)
C23	orbital and insular area (or differentiating field)
C24	insular cortex
C26	claustrum

FIBER TRACTS (f)

f1	lateral olfactory tract
f11	external capsule

RHINENCEPHALON (R)

R1	rhinencephalic neuroepithelium
R2	rhinencephalic subventricular zone
R3	rhinencephalon (or differentiating field)
R4	piriform cortex

SEPTAL AREA (S)

S1	septal neuroepithelium
S2	septal subventricular zone
S3	septum (or differentiating field)
S6	medial septum
S9	diagonal band (horizontal limb)
S16	taenia tecta

VENTRICLES (v)

vla	lateral ventricle, anterior horn
vld	lateral ventricle, dorsal

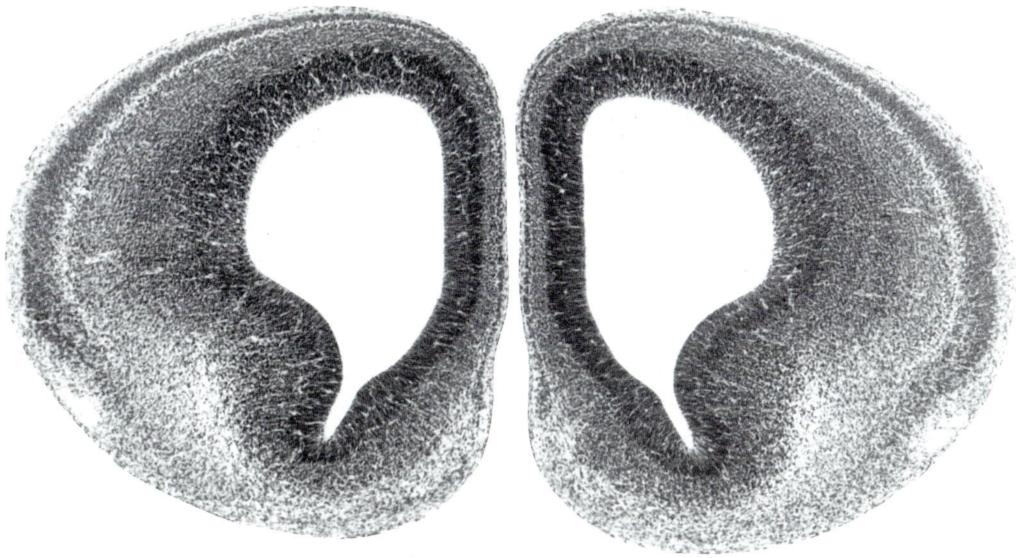

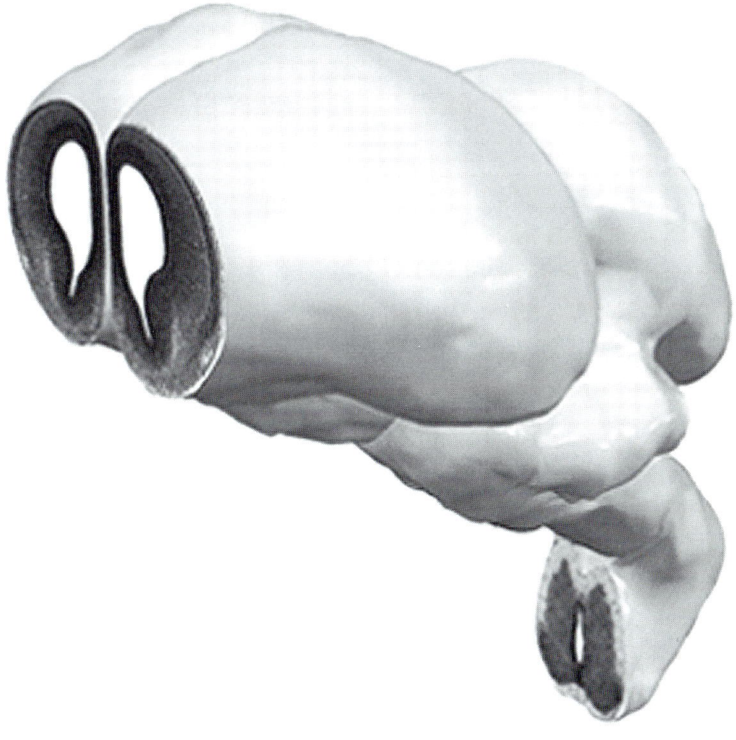

CORONAL FIGURE 6

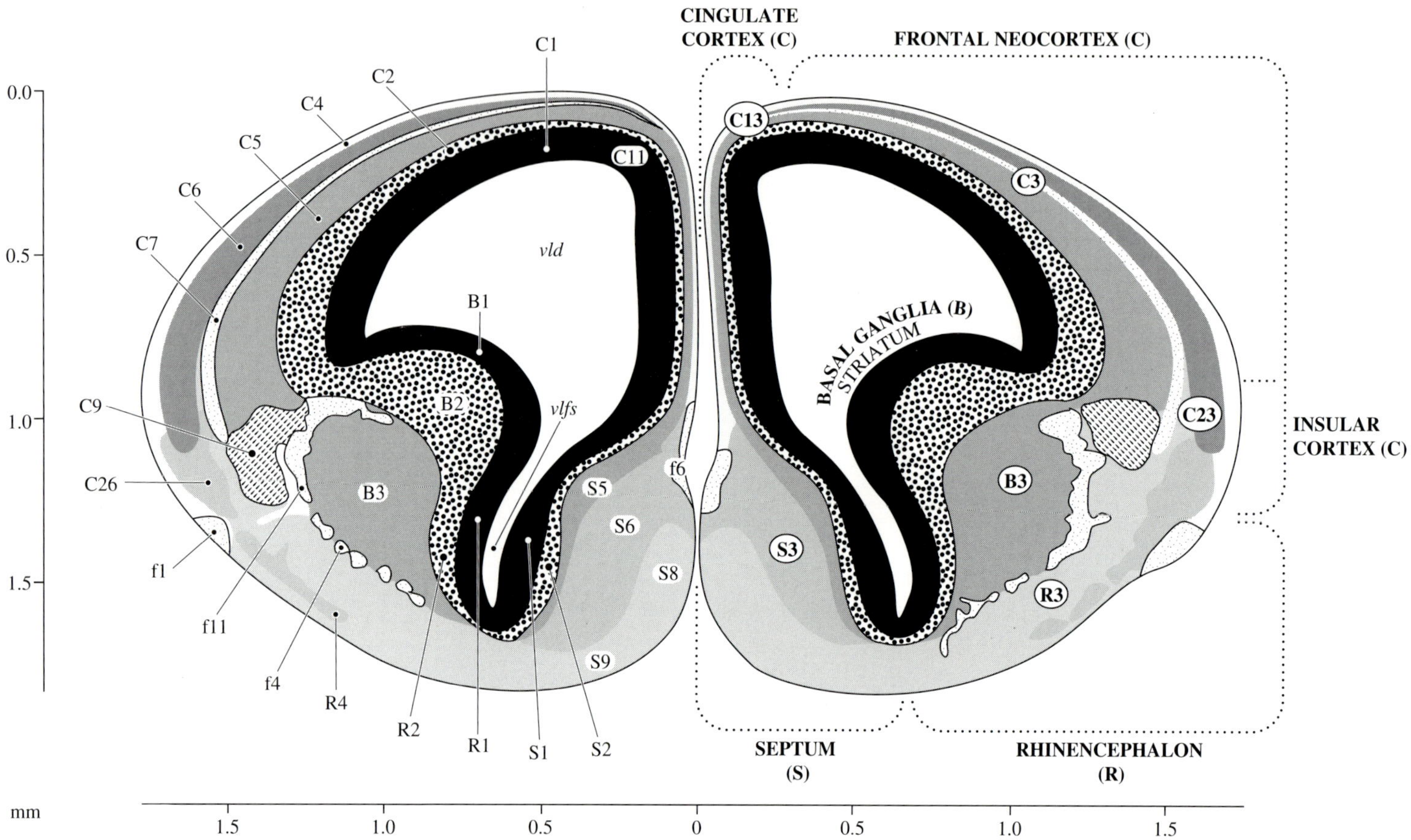

BASAL GANGLIA (B)

B1	striatal neuroepithelium
B2	striatal subventricular zone
B3	striatum (or differentiating field)

CEREBRAL CORTEX (C)

C1	neocortical neuroepithelium
C2	neocortical subventricular zone
C3	neocortex (or differentiating field)
C4	molecular layer (or primordial plexiform layer)
C5	intermediate zone
C6	cortical plate
C7	white matter
C9	lateral migratory stream
C11	cingulate cortex neuroepithelium
C13	cingulate cortex (or differentiating field)
C23	orbital and insular area (or differentiating field)
C26	claustrum

FIBER TRACTS (f)

f1	lateral olfactory tract
f4	anterior commissure
f6	fornix
f11	external capsule

RHINENCEPHALON (R)

R1	rhinencephalic neuroepithelium
R2	rhinencephalic subventricular zone
R3	rhinencephalon (or differentiating field)
R4	piriform cortex

SEPTAL AREA (S)

S1	septal neuroepithelium
S2	septal subventricular zone
S3	septum (or differentiating field)
S5	lateral septum
S6	medial septum
S8	diagonal band (vertical limb)
S9	diagonal band (horizontal limb)

VENTRICLES (v)

vld	lateral ventricle, dorsal
vlfs	lateral ventricle, septal fork

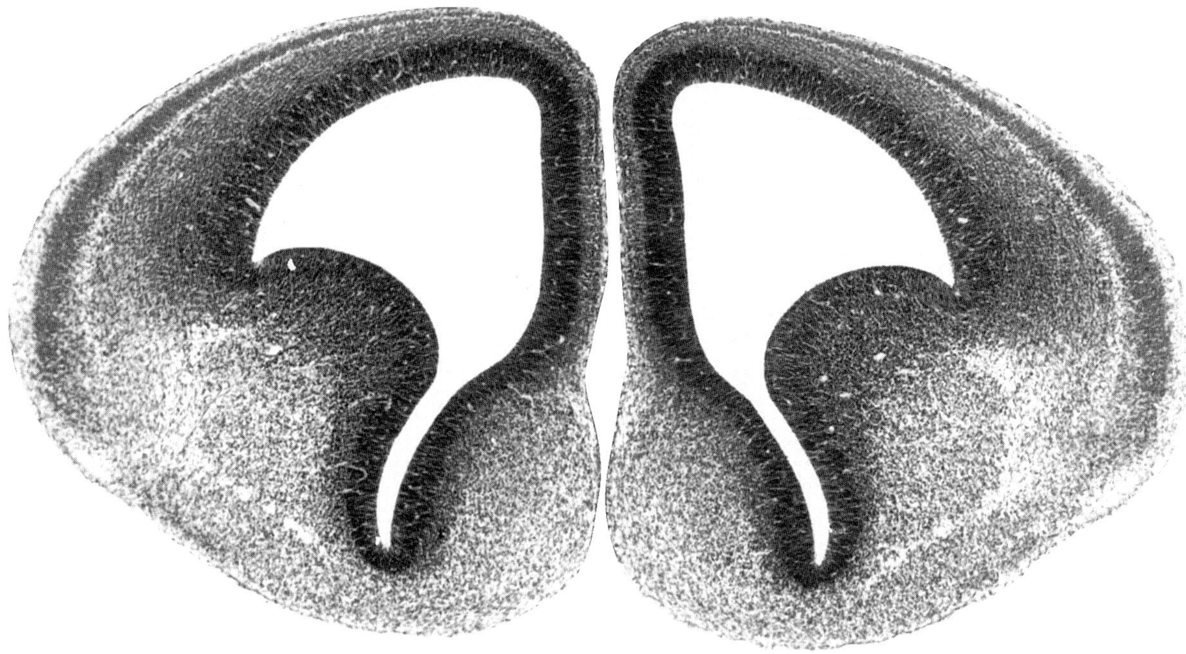

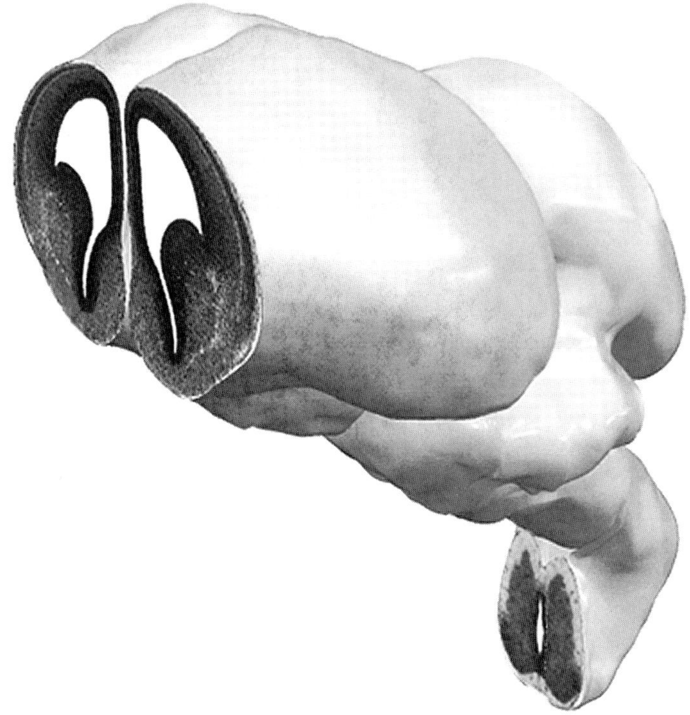

CORONAL FIGURE 7

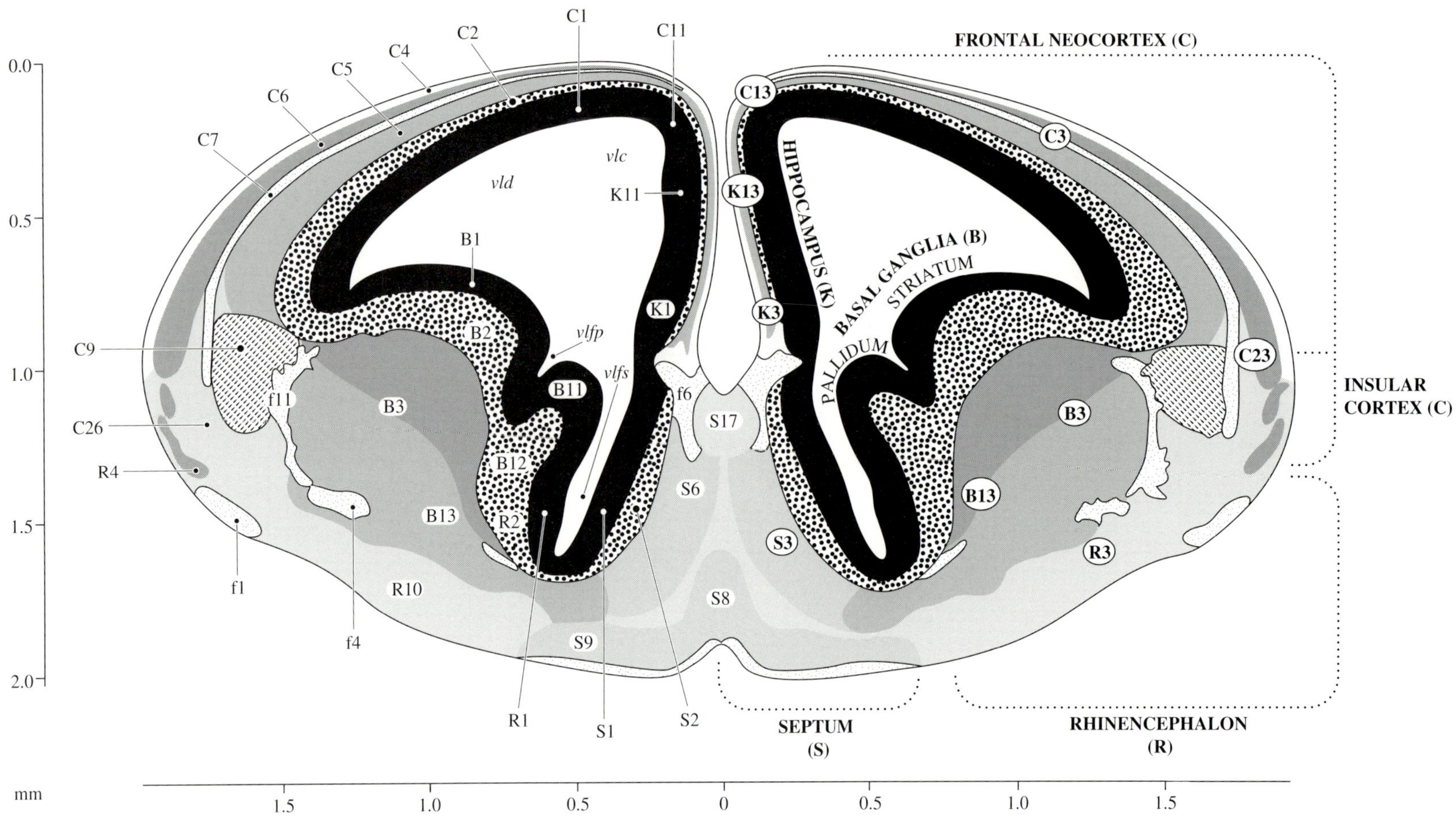

BASAL GANGLIA (B)

B1 striatal neuroepithelium
B2 striatal subventricular zone
B3 striatum (or differentiating field)
B11 pallidal neuroepithelium
B12 pallidal subventricular zone
B13 pallidum (or differentiating field)

CEREBRAL CORTEX (C)

C1 neocortical neuroepithelium
C2 neocortical subventricular zone
C3 neocortex (or differentiating field)
C4 molecular layer (or primordial plexiform layer)
C5 intermediate zone
C6 cortical plate
C7 white matter
C9 lateral migratory stream
C11 cingulate cortex neuroepithelium
C13 cingulate cortex (or differentiating field)
C23 orbital and insular area (or differentiating field)
C26 claustrum

FIBER TRACTS (f)

f1 lateral olfactory tract
f4 anterior commissure
f6 fornix
f11 external capsule

HIPPOCAMPAL FORMATION (K)

K1 hippocampal (Ammon's horn) neuroepithelium
K3 hippocampus (or differentiating field)
K11 subicular neuroepithelium
K13 subicular area (or differentiating field)

RHINENCEPHALON (R)

R1 rhinencephalic neuroepithelium
R2 rhinencephalic subventricular zone
R3 rhinencephalon (or differentiating field)
R4 piriform cortex
R10 olfactory tubercle

SEPTAL AREA (S)

S1 septal neuroepithelium
S2 septal subventricular zone
S3 septum (or differentiating field)
S6 medial septum
S8 diagonal band (vertical limb)
S9 diagonal band (horizontal limb)
S17 septofimbrial nucleus

VENTRICLES (v)

vlc lateral ventricle, central
vld lateral ventricle, dorsal
vlfp lateral ventricle, pallidal fork
vlfs lateral ventricle, septal fork

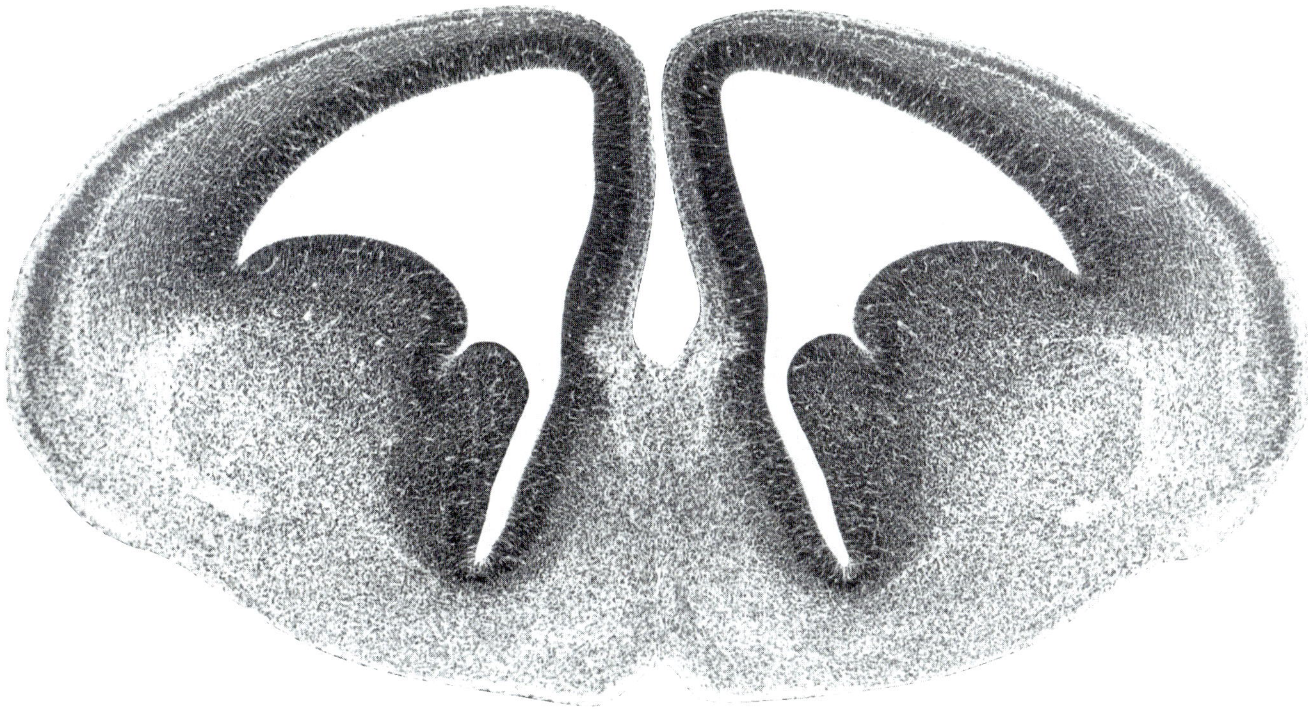

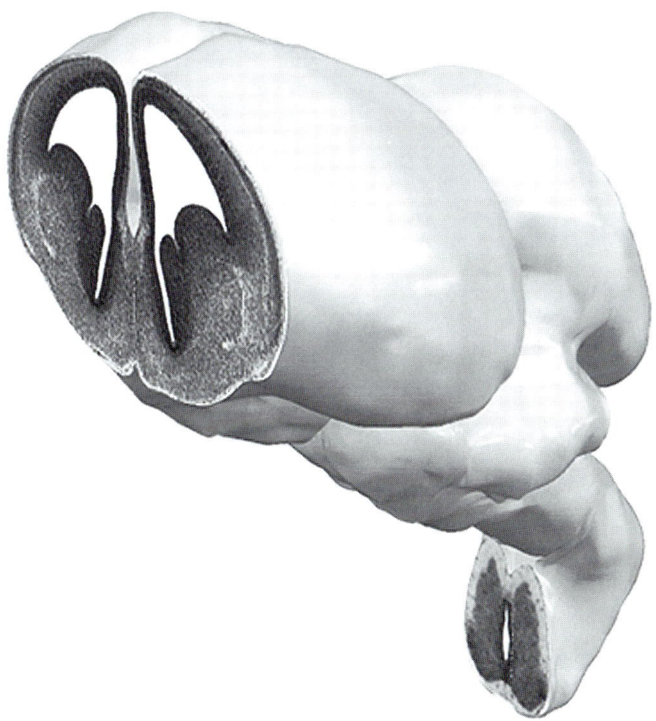

CORONAL FIGURE 8

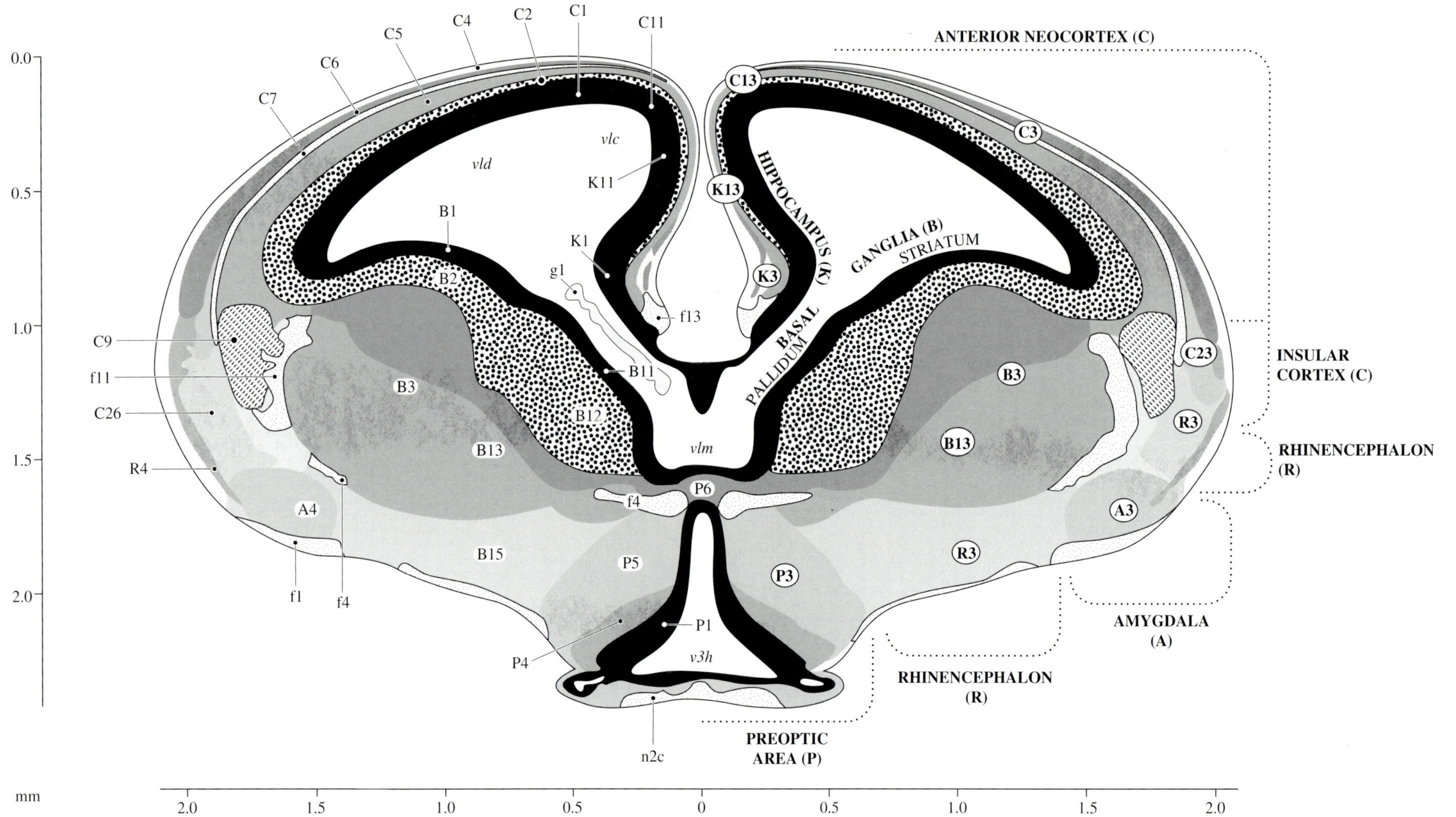

AMYGDALA (A)

- **A3** amygdala (or differentiating field)
- A4 anterior amygdaloid area

BASAL GANGLIA (B)

- B1 striatal neuroepithelium
- B2 striatal subventricular zone
- **B3** striatum (or differentiating field)
- B11 pallidal neuroepithelium
- B12 pallidal subventricular zone
- **B13** pallidum (or differentiating field)
- B15 substantia innominata

CEREBRAL CORTEX (C)

- C1 neocortical neuroepithelium
- C2 neocortical subventricular zone
- **C3** neocortex (or differentiating field)
- C4 molecular layer (or primordial plexiform layer)
- C5 intermediate zone
- C6 cortical plate
- C7 white matter
- C9 lateral migratory stream
- C11 cingulate cortex neuroepithelium
- **C13** cingulate cortex (or differentiating field)
- **C23** orbital and insular area (or differentiating field)
- C26 claustrum

FIBER TRACTS (f)

- f1 lateral olfactory tract
- f4 anterior commissure
- f11 external capsule
- f13 fimbria

GLANDS AND PARANEURAL STRUCTURES (g)

- g1 choroid plexus of lateral ventricle

HIPPOCAMPAL FORMATION (K)

- K1 hippocampal (Ammon's horn) neuroepithelium
- **K3** hippocampus (or differentiating field)
- K11 subicular neuroepithelium
- **K13** subicular area (or differentiating field)

NERVES AND NUCLEI, CRANIAL (n)

- n2c optic chiasma

PREOPTIC AREA AND OPTIC EPITHELIA (P)

- P1 preoptic neuroepithelium
- **P3** preoptic area (or differentiating field)
- P4 medial preoptic region
- P5 lateral preoptic region
- P6 superior preoptic region

RHINENCEPHALON (R)

- **R3** rhinencephalon (or differentiating field)
- R4 piriform cortex

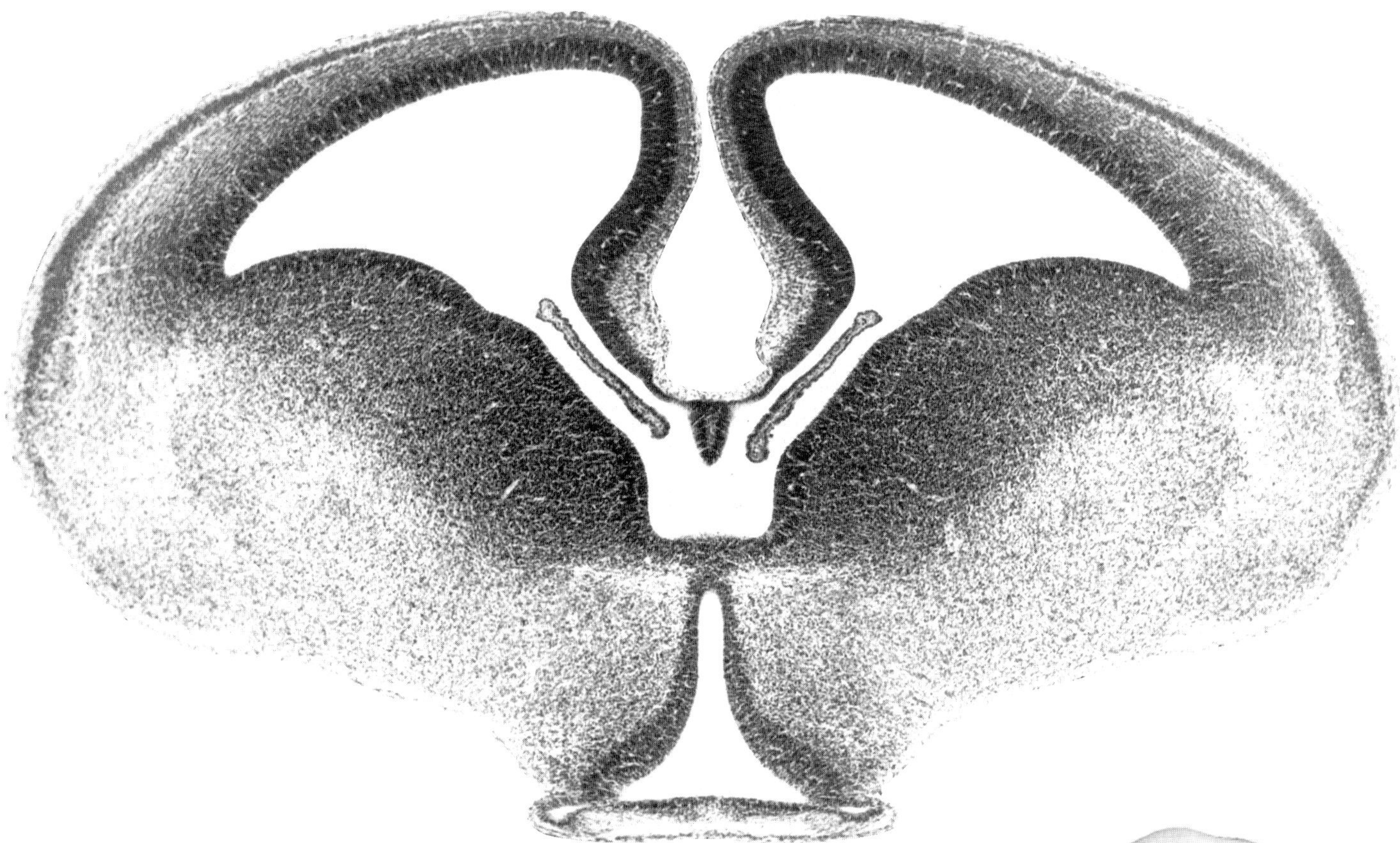

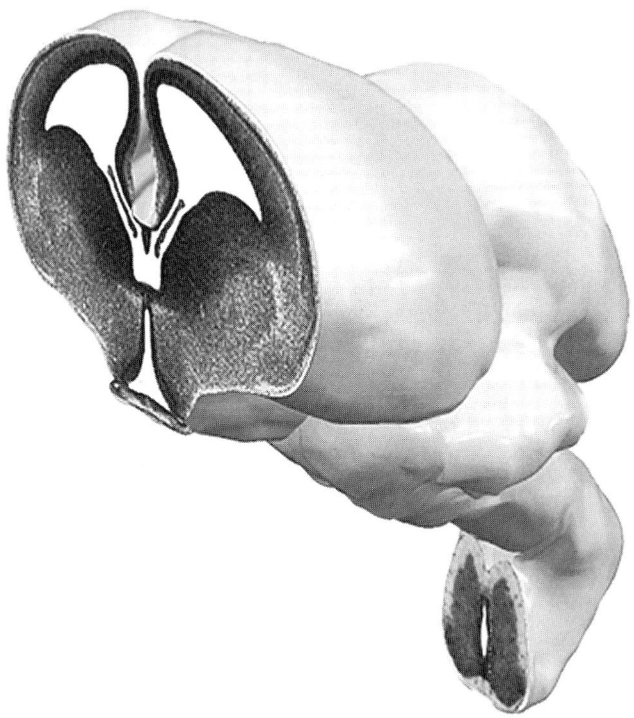

VENTRICLES (v)

vlc lateral ventricle, central
vld lateral ventricle, dorsal
vlm foramen of Monro
v3h third ventricle, hypothalamic

CORONAL FIGURE 9

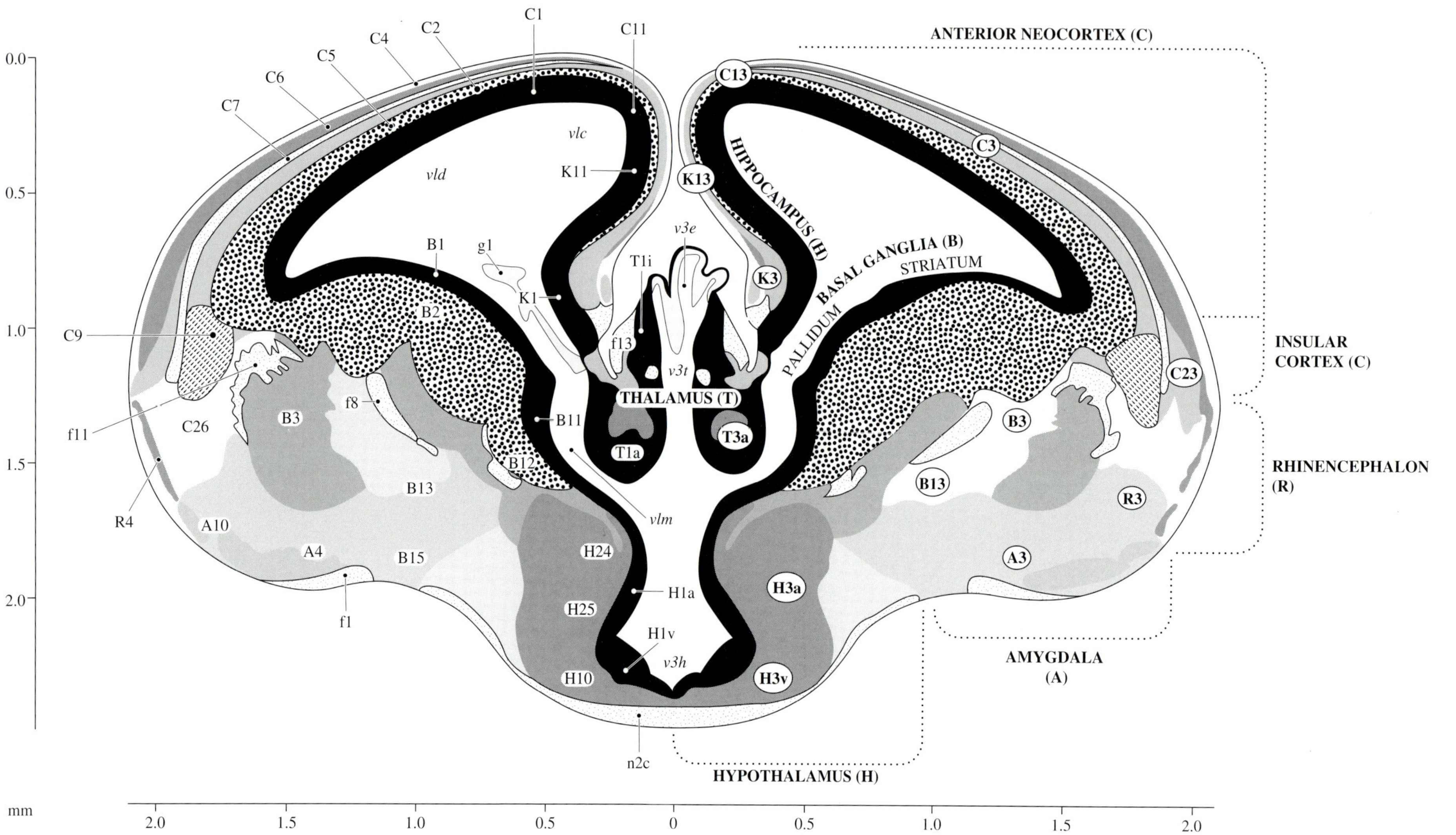

AMYGDALA (A)

A3 amygdala (or differentiating field)
A4 anterior amygdaloid area
A10 lateral amygdaloid nucleus

BASAL GANGLIA (B)

B1 striatal neuroepithelium
B2 striatal subventricular zone
B3 striatum (or differentiating field)
B11 pallidal neuroepithelium
B12 pallidal subventricular zone
B13 pallidum (or differentiating field)
B15 substantia innominata

CEREBRAL CORTEX (C)

C1 neocortical neuroepithelium
C2 neocortical subventricular zone
C3 neocortex (or differentiating field)
C4 molecular layer (or primordial plexiform layer)
C5 intermediate zone
C6 cortical plate
C7 white matter
C9 lateral migratory stream
C11 cingulate cortex neuroepithelium
C13 cingulate cortex (or differentiating field)
C23 orbital and insular area (or differentiating field)
C26 claustrum

FIBER TRACTS (f)

f1 lateral olfactory tract
f8 internal capsule
f11 external capsule
f13 fimbria

GLANDS AND PARANEURAL STRUCTURES (g)

g1 choroid plexus of lateral ventricle

HYPOTHALAMUS (H)

H1a anterior hypothalamic neuroepithelium
H1v ventral hypothalamic neuroepithelium
H3a anterior hypothalamus (or differentiating field)
H3v ventral hypothalamus (or differentiating field)
H10 anterobasal nucleus
H24 anterodorsal nucleus
H25 anteroventral nucleus

HIPPOCAMPAL FORMATION (K)

K1 hippocampal (Ammon's horn) neuroepithelium
K3 hippocampus (or differentiating field)
K11 subicular neuroepithelium
K13 subicular area (or differentiating field)

NERVES AND NUCLEI, CRANIAL (n)

n2c optic chiasma

RHINENCEPHALON (R)

R3 rhinencephalon (or differentiating field)
R4 piriform cortex

THALAMUS (T)

T1a anterior thalamic neuroepithelium
T1i intermediate thalamic neuroepithelium
T3a anterior thalamus (or differentiating field)

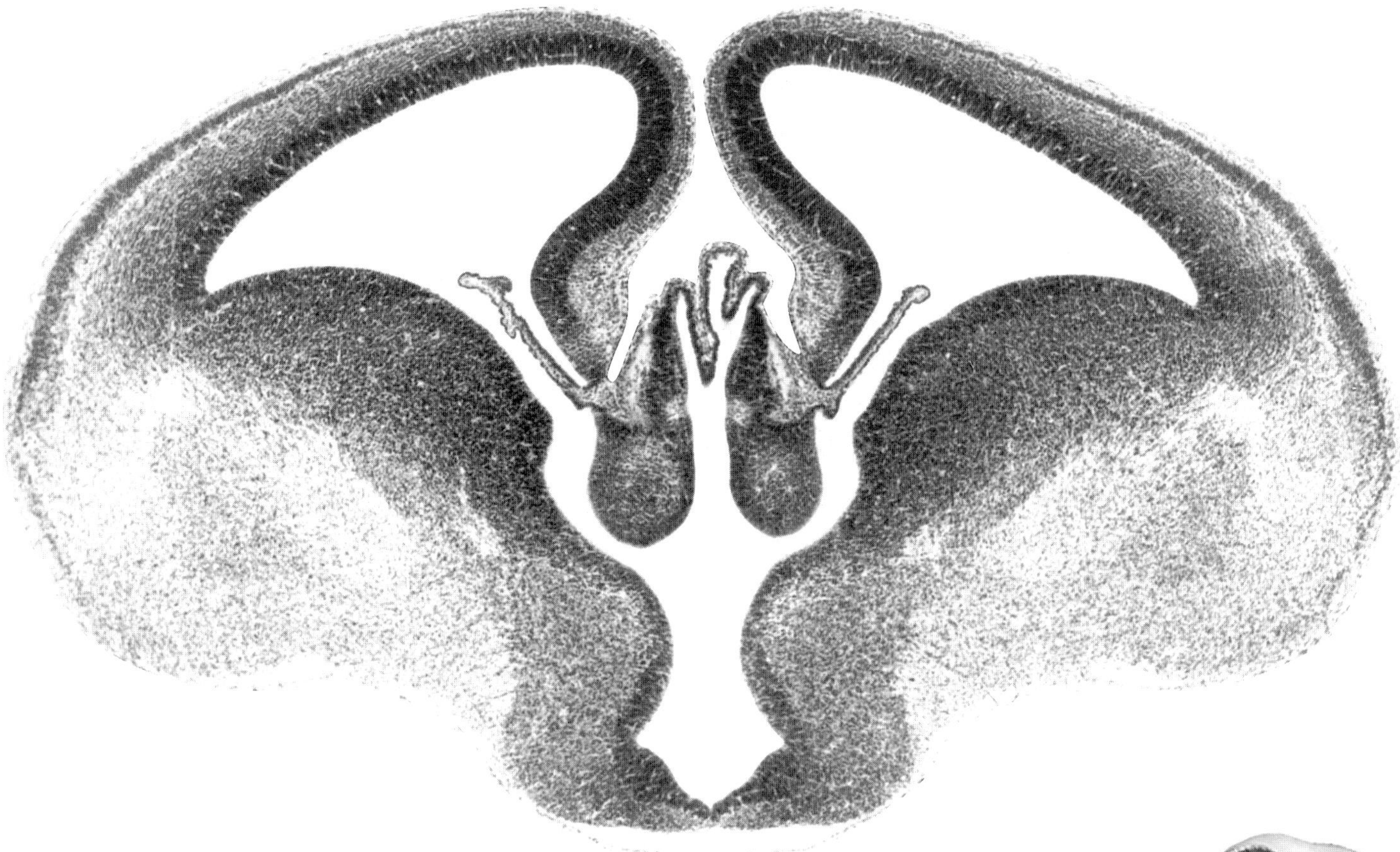

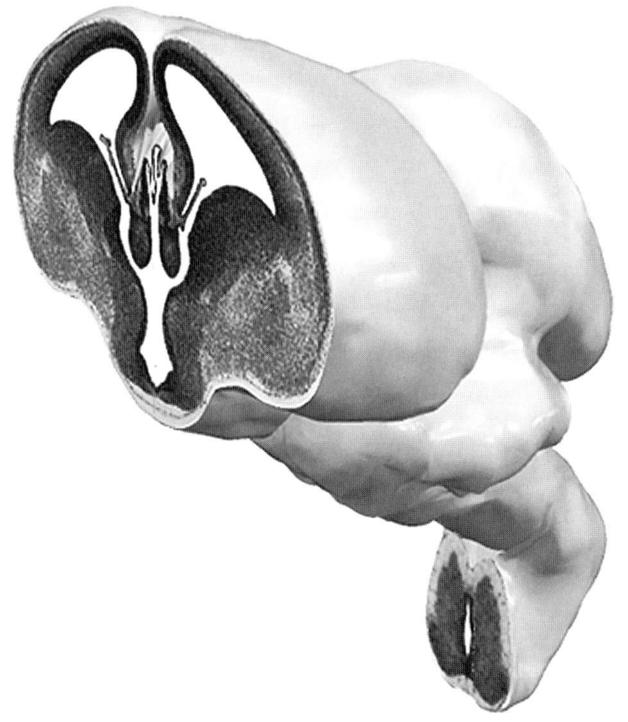

VENTRICLES (v)

vlc	lateral ventricle, central
vld	lateral ventricle, dorsal
vlm	foramen of Monro
v3e	third ventricle, epithalamic
v3h	third ventricle, hypothalamic
v3t	third ventricle, thalamic

CORONAL FIGURE 10

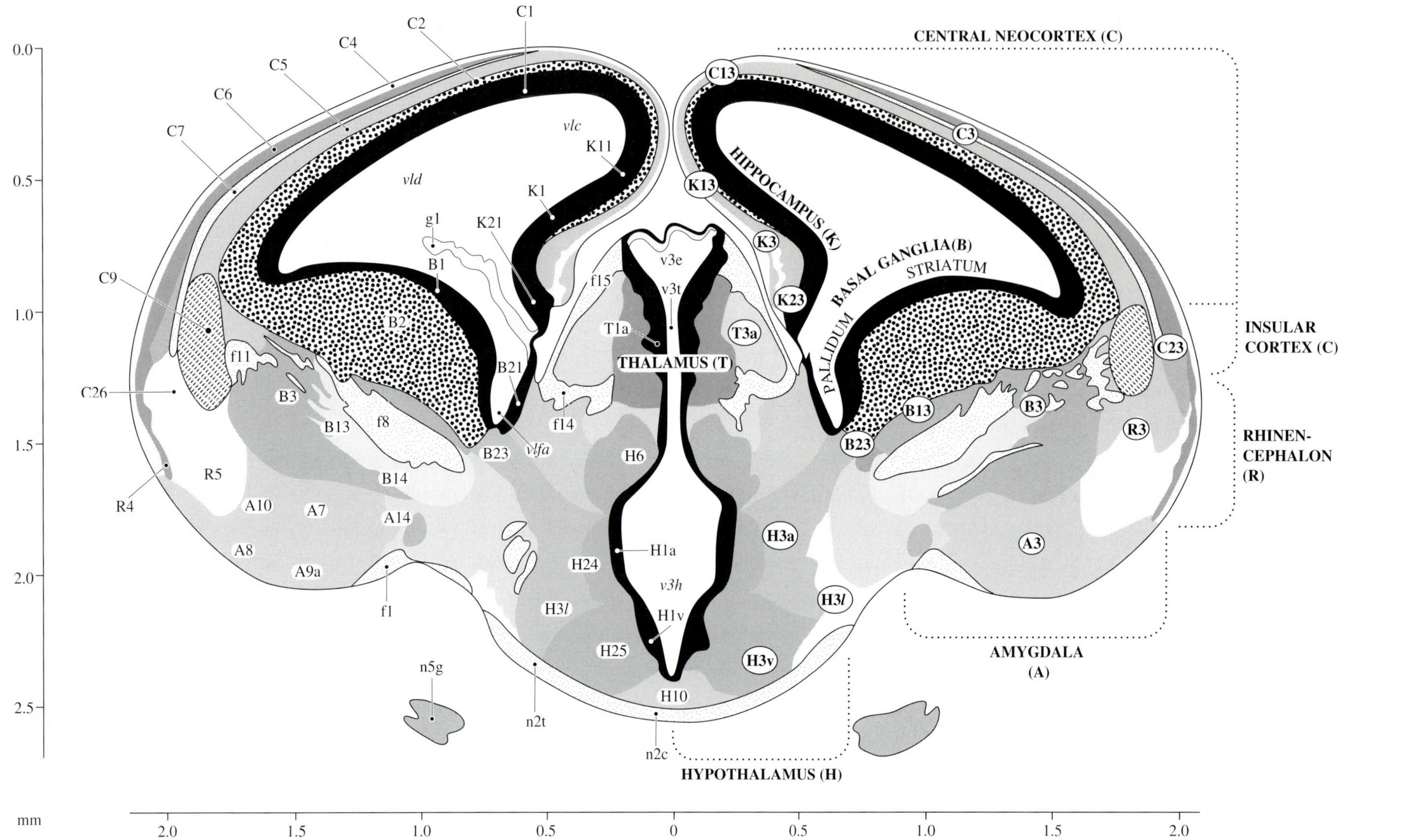

AMYGDALA (A)

A3 amygdala (or differentiating field)
A7 basomedial amygdaloid nucleus
A8 basolateral amygdaloid nucleus
A9a cortical amygdaloid nucleus, anterior
A10 lateral amygdaloid nucleus
A14 medial amygdaloid nucleus

BASAL GANGLIA (B)

B1 striatal neuroepithelium
B2 striatal subventricular zone
B3 striatum (or differentiating field)
B13 pallidum (or differentiating field)
B14 entopeduncular nucleus
B21 strionuclear neuroepithelium
B23 bed nucleus of stria terminalis (or differentiating field)

CEREBRAL CORTEX (C)

C1 neocortical neuroepithelium
C2 neocortical subventricular zone
C3 neocortex (or differentiating field)
C4 molecular layer (primordial plexiform layer)
C5 intermediate zone
C6 cortical plate
C7 white matter
C9 lateral migratory stream
C23 orbital and insular area (or differentiating field)
C26 claustrum

FIBER TRACTS (f)

f1 lateral olfactory tract
f8 internal capsule
f11 external capsule
f14 stria terminalis
f15 stria medullaris

GLANDS AND PARANEURAL STRUCTURES (g)

g1 choroid plexus of lateral ventricle

HYPOTHALAMUS (H)

H1a anterior hypothalamic neuroepithelium
H1v ventral hypothalamic neuroepithelium
H3a anterior hypothalamus (or differentiating field)
H3*l* lateral hypothalamus (or differentiating field)
H3v ventral hypothalamus (or differentiating field)
H6 paraventricular nucleus
H10 anterobasal nucleus
H24 anterodorsal nucleus
H25 anteroventral nucleus

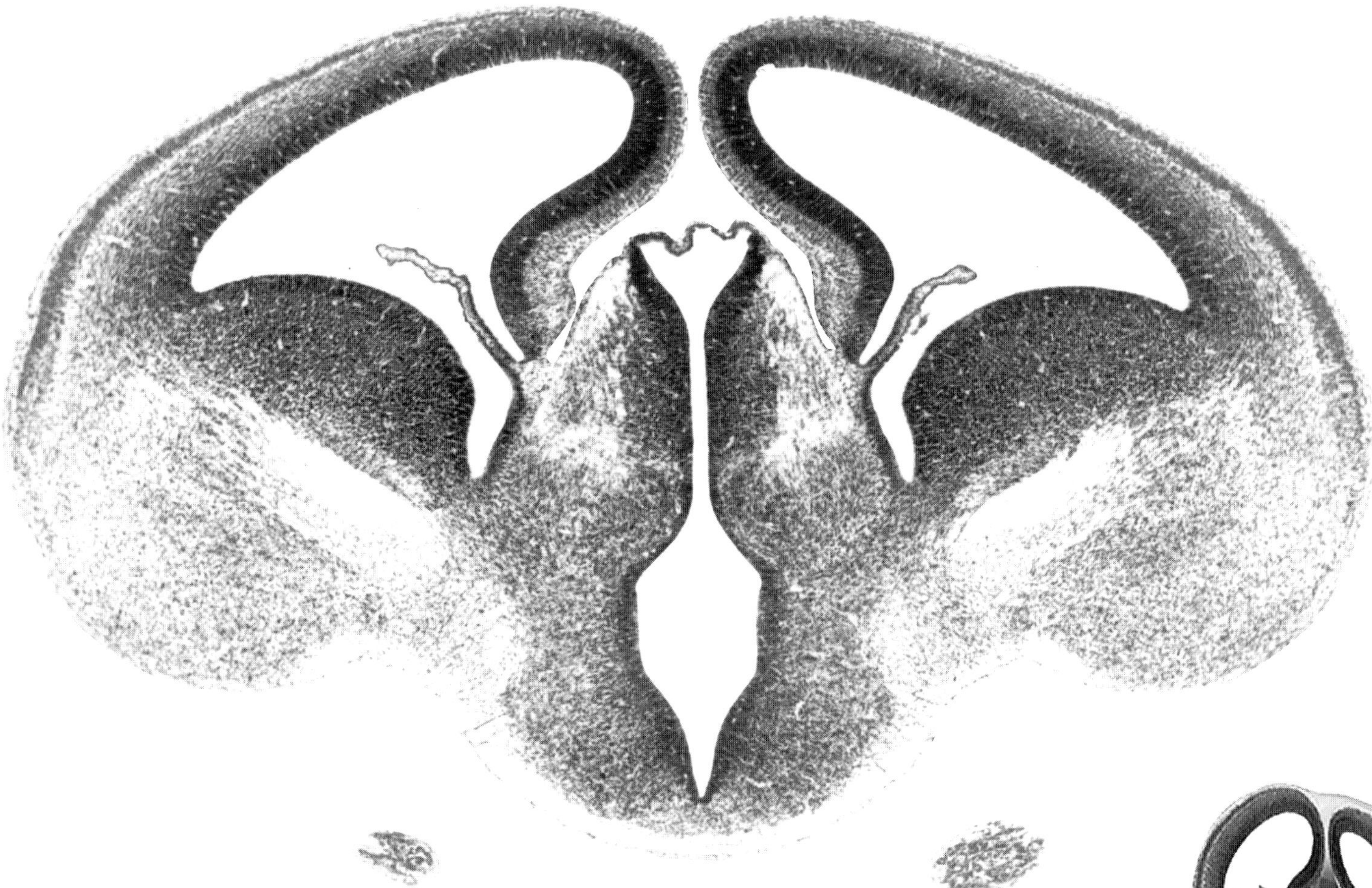

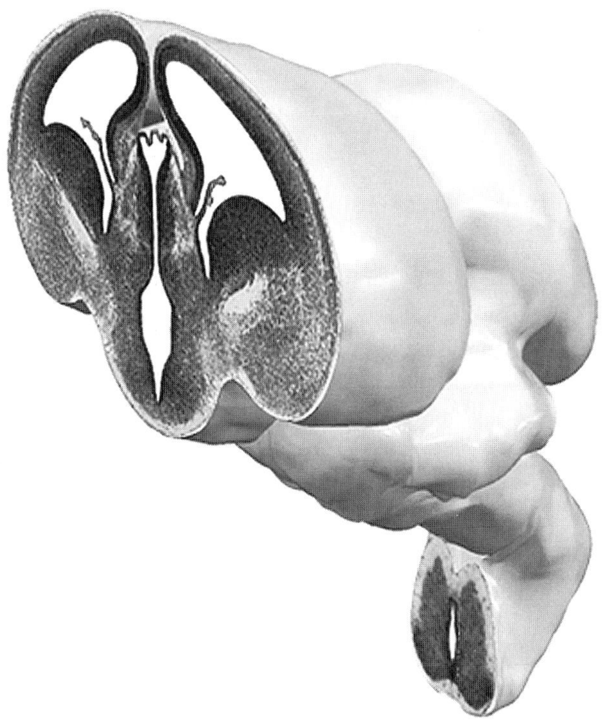

HIPPOCAMPAL FORMATION (K)

K1 hippocampal (Ammon's horn) neuroepithelium
K3 hippocampus (or differentiating field)
K11 subicular neuroepithelium
K13 subicular area (or differentiating field)
K21 dentate gyrus neuroepithelium
K23 dentate gyrus (or differentiating field)

NERVES AND NUCLEI, CRANIAL (n)

n2c optic chiasma
n2t optic tract
n5g trigeminal ganglion

RHINENCEPHALON (R)

R3 rhinencephalon (or differentiating field)
R4 piriform cortex
R5 endopiriform nucleus

THALAMUS (T)

T1a anterior thalamic neuroepithelium
T3a anterior thalamus (or differentiating field)

VENTRICLES (v)

vlc lateral ventricle, central
vld lateral ventricle, dorsal
vlfa lateral ventricle, amygdaloid fork
v3e third ventricle, epithalamic
v3h third ventricle, hypothalamic
v3t third ventricle, thalamic

CORONAL FIGURE 11

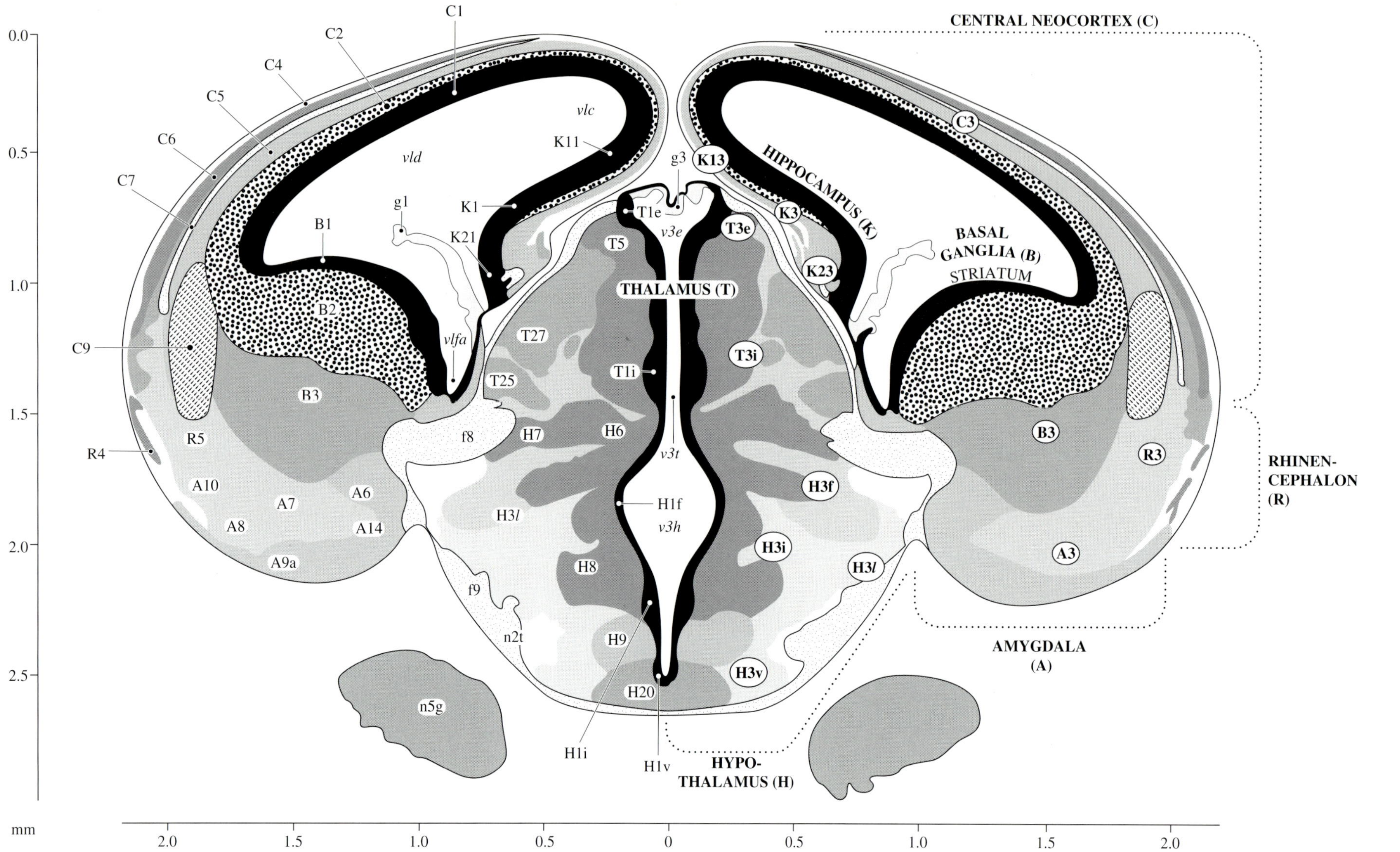

AMYGDALA (A)

A3 amygdala (or differentiating field)
A6 central amygdaloid nucleus
A7 basomedial amygdaloid nucleus
A8 basolateral amygdaloid nucleus
A9a cortical amygdaloid nucleus, anterior
A10 lateral amygdaloid nucleus
A14 medial amygdaloid nucleus

BASAL GANGLIA (B)

B1 striatal neuroepithelium
B2 striatal subventricular zone
B3 striatum (or differentiating field)

CEREBRAL CORTEX (C)

C1 neocortical neuroepithelium
C2 neocortical subventricular zone
C3 neocortex (or differentiating field)
C4 molecular layer (or primordial plexiform layer)
C5 intermediate zone
C6 cortical plate
C7 white matter
C9 lateral migratory stream

FIBER TRACTS (f)

f8 internal capsule
f9 cerebral peduncle

GLANDS AND PARANEURAL STRUCTURES (g)

g1 choroid plexus of lateral ventricle
g3 choroid plexus of third ventricle

HYPOTHALAMUS (H)

H1f Forelian neuroepithelium
H1i intermediate hypothalamic neuroepithelium
H1v ventral hypothalamic neuroepithelium
H3i intermediate hypothalamus (or differentiating field)
H3f fields of Forel (or differentiating field)
H3*l* lateral hypothalamus (or differentiating field)
H3v ventral hypothalamus (or differentiating field)
H6 paraventricular nucleus
H7 zona incerta
H8 dorsomedial nucleus
H9 ventromedial nucleus
H20 arcuate nucleus

HIPPOCAMPAL FORMATION (K)

K1 hippocampal (Ammon's horn) neuroepithelium
K3 hippocampus (or differentiating field)
K11 subicular neuroepithelium
K13 subicular area (or differentiating field)
K21 dentate gyrus neuroepithelium
K23 dentate gyrus (or differentiating field)

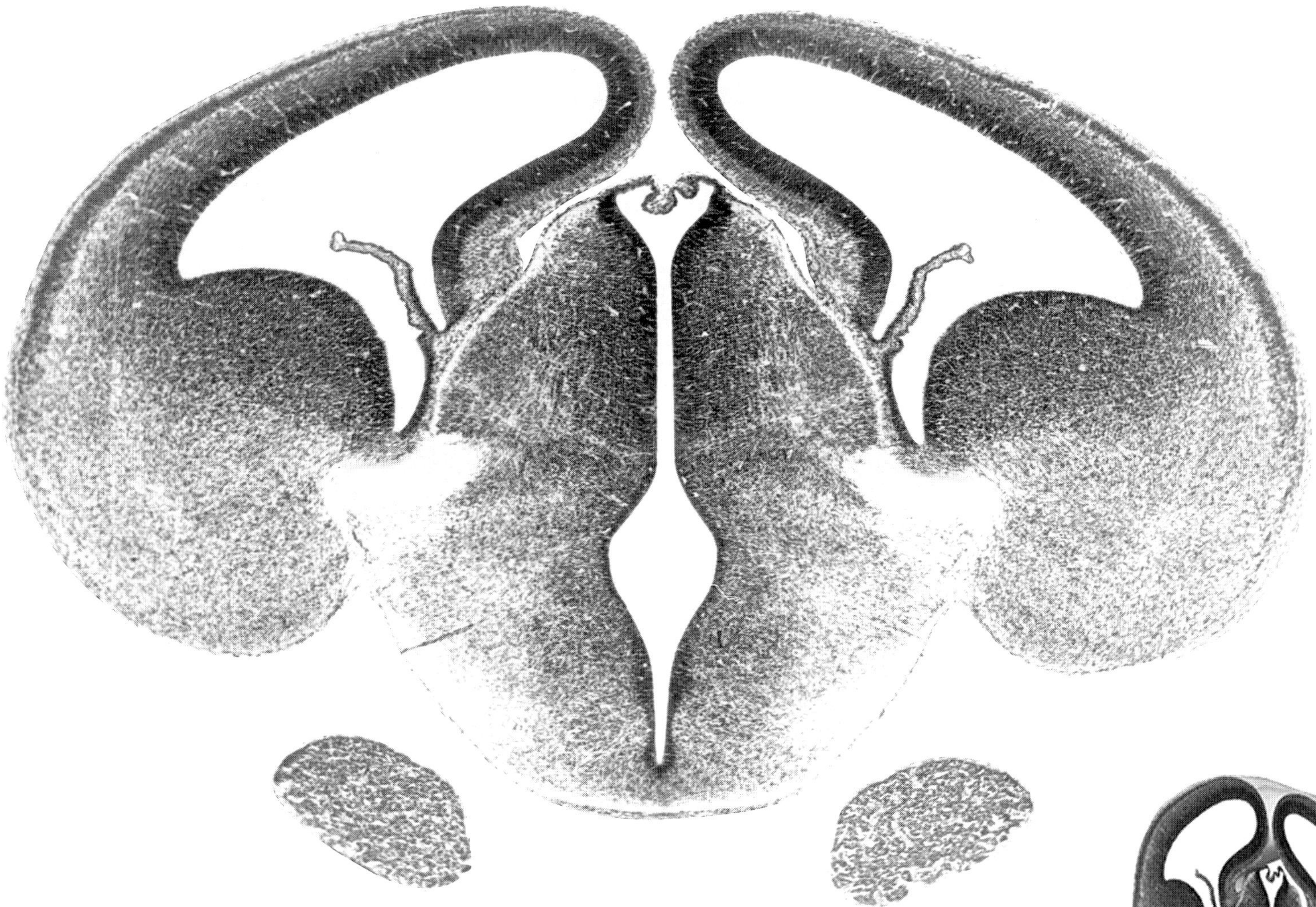

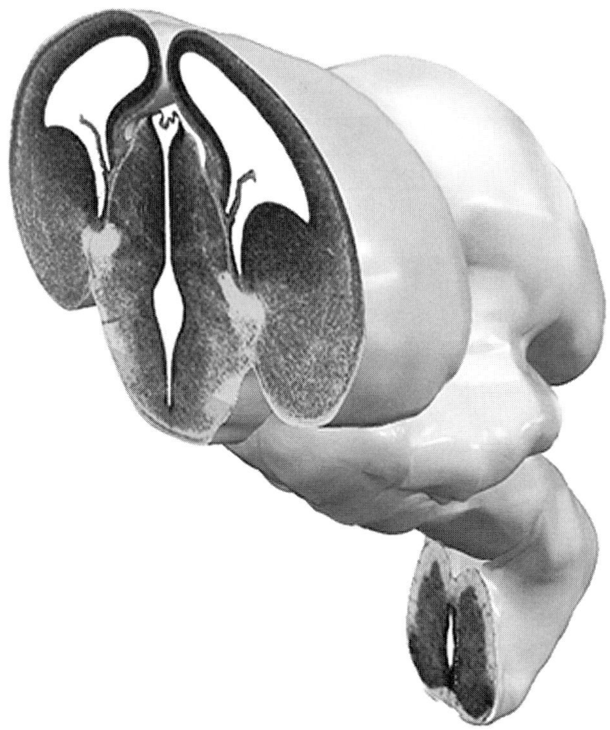

NERVES AND NUCLEI, CRANIAL (n)

n2t optic tract
n5g trigeminal ganglion

RHINENCEPHALON (R)

R3 rhinencephalon (or differentiating field)
R4 piriform cortex
R5 endopiriform nucleus

THALAMUS (T)

T1e epithalamic neuroepithelium
T1i intermediate thalamic neuroepithelium
T3e epithalamus (or differentiating field)
T3i intermediate thalamus (or differentiating field)
T5 lateral habenular nucleus
T25 reticular nucleus
T27 ventrolateral nuclear complex

VENTRICLES (v)

vlc lateral ventricle, central
vld lateral ventricle, dorsal
vlfa lateral ventricle, amygdaloid fork
v3e third ventricle, epithalamic
v3h third ventricle, hypothalamic
v3t third ventricle, thalamic

CORONAL FIGURE 12

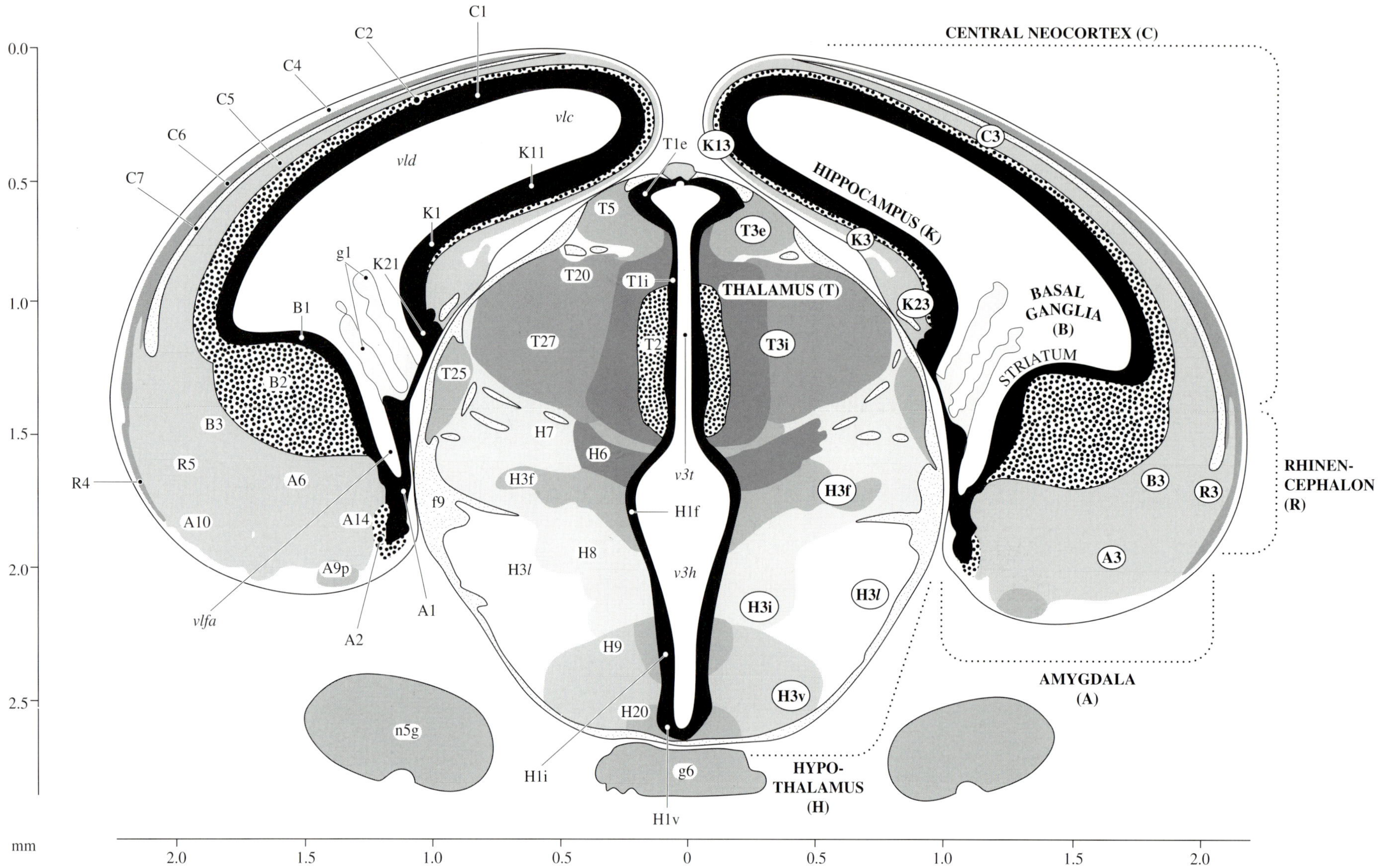

AMYGDALA (A)

A1 amygdaloid neuroepithelium
A2 amygdaloid subventricular zone
A3 amygdala (or differentiating field)
A6 medial amygdaloid nucleus
A9p cortical amygdaloid nucleus, posterior
A10 lateral amygdaloid nucleus
A14 medial amygdaloid nucleus

BASAL GANGLIA (B)

B1 striatal neuroepithelium
B2 striatal subventricular zone
B3 striatum (or differentiating field)

CEREBRAL CORTEX (C)

C1 neocortical neuroepithelium
C2 neocortical subventricular zone
C3 neocortex (or differentiating field)
C4 molecular layer (or primordial plexiform layer)
C5 intermediate zone
C6 cortical plate
C7 white matter

FIBER TRACTS (f)

f9 cerebral peduncle

GLANDS AND PARANEURAL STRUCTURES (g)

g1 choroid plexus of lateral ventricle
g6 pituitary gland (primordium)

HYPOTHALAMUS (H)

H1f Forelian neuroepithelium
H1i intermediate hypothalamic neuroepithelium
H1v ventral hypothalamic neuroepithelium
H3f fields of Forel (or differentiating field)
H3i intermediate hypothalamus (or differentiating field)
H3*l* lateral hypothalamus (or differentiating field)
H3v ventral hypothalamus (or differentiating field)
H6 paraventricular nucleus
H7 zona incerta
H8 dorsomedial nucleus
H9 ventromedial nucleus
H20 arcuate nucleus

HIPPOCAMPAL FORMATION (K)

K1 hippocampal (Ammon's horn) neuroepithelium
K3 hippocampus (or differentiating field)
K11 subicular neuroepithelium
K13 subicular area (or differentiating field)
K21 dentate gyrus neuroepithelium
K23 dentate gyrus (or differentiating field)

NERVES AND NUCLEI, CRANIAL (n)

n5g trigeminal ganglion

RHINENCEPHALON (R)

R3 rhinencephalon (or differentiating field)
R4 piriform cortex
R5 endopiriform nucleus

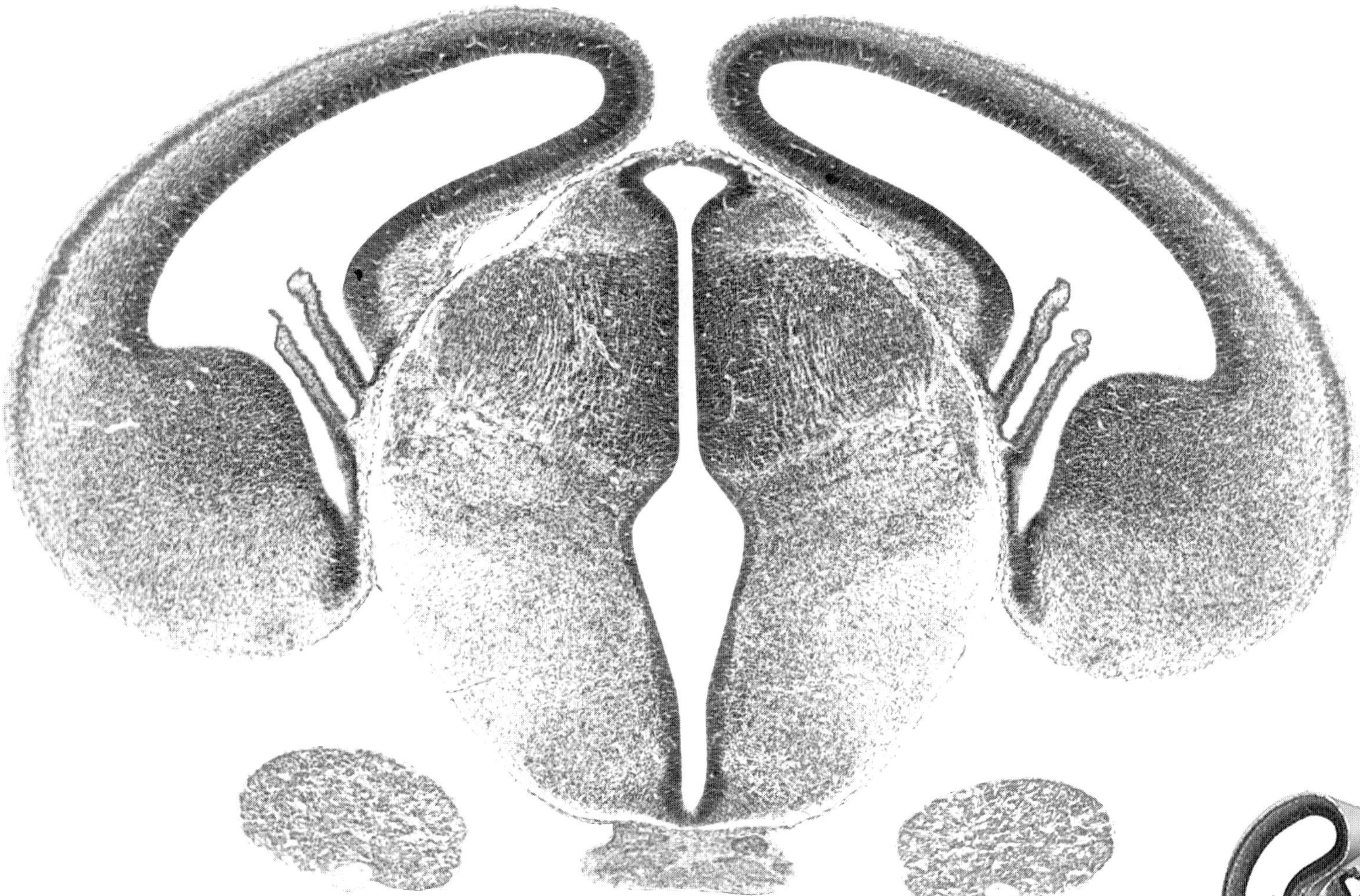

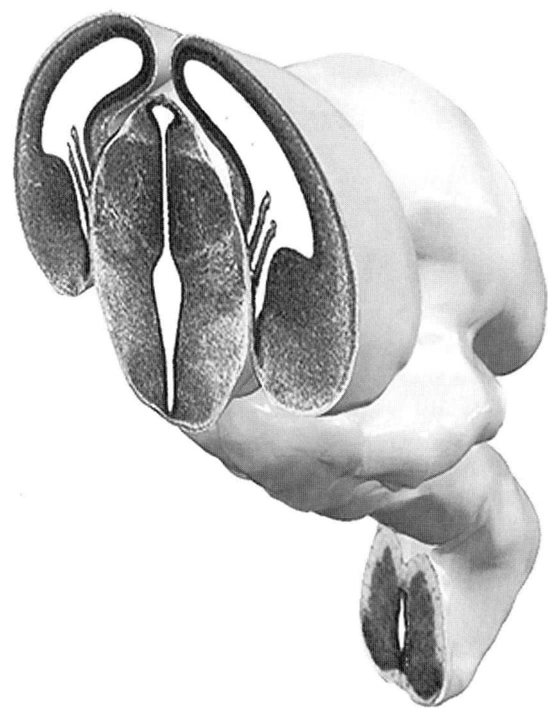

THALAMUS (T)

T1e epithalamic neuroepithelium
T1i intermediate thalamic neuroepithelium
T2 thalamic subventricular zone
T3e epithalamus (or differentiating field)
T3i intermediate thalamus (or differentiating field)
T5 lateral habenular nucleus
T20 laterodorsal nucleus
T25 reticular nucleus
T27 ventrolateral nuclear complex

VENTRICLES (v)

vlc lateral ventricle, central
vld lateral ventricle, dorsal
vlfa lateral ventricle, amygdaloid fork
v3h third ventricle, hypothalamic
v3t third ventricle, thalamic

CORONAL FIGURE 13

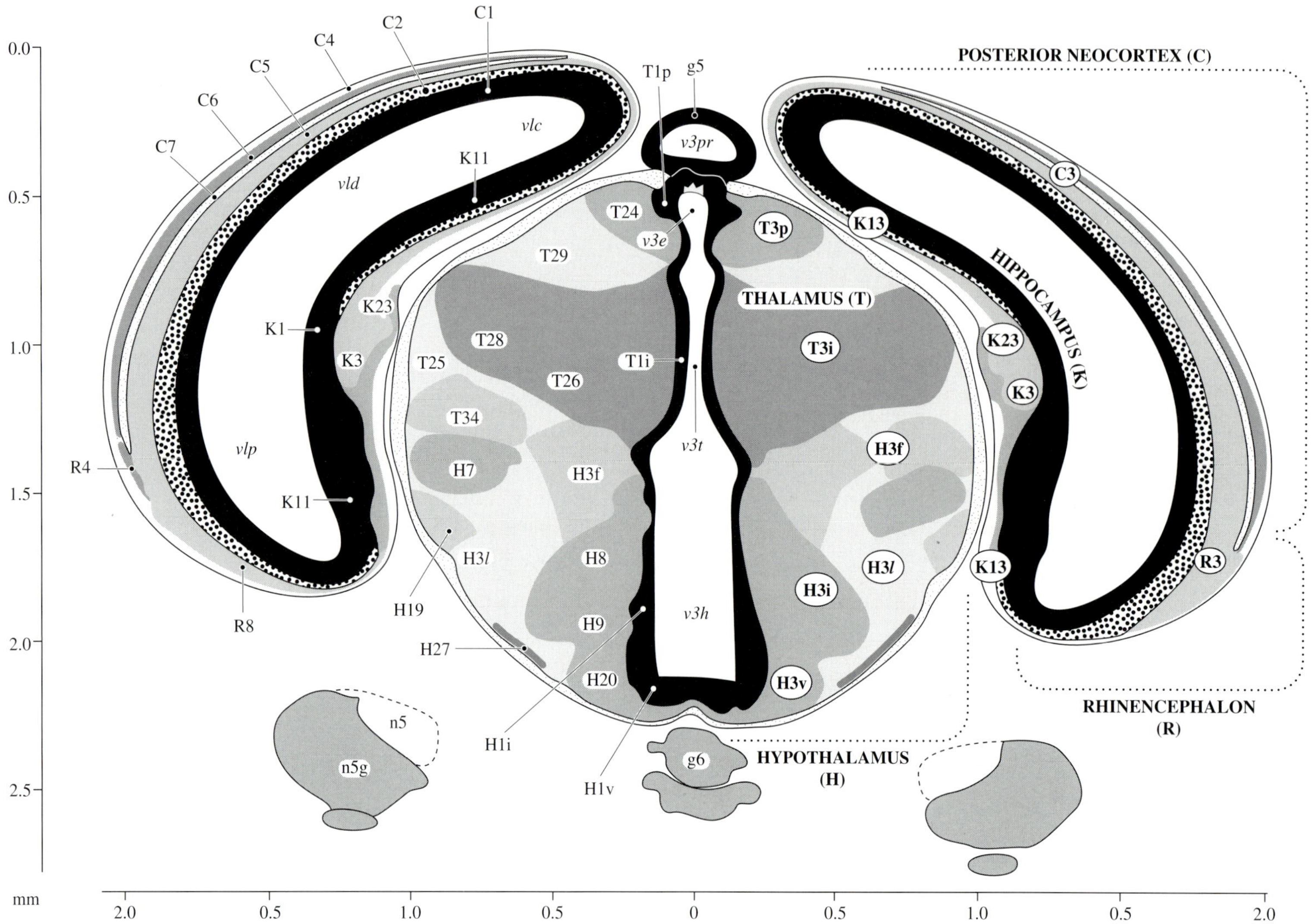

CEREBRAL CORTEX (C)

C1 neocortical neuroepithelium
C2 neocortical subventricular zone
C3 neocortex (or differentiating field)
C4 molecular layer (or primordial plexiform layer)
C5 intermediate zone
C6 cortical plate
C7 white matter

FIBER TRACTS (f)

f9 cerebral peduncle

GLANDS AND PARANEURAL STRUCTURES (g)

g5 pineal gland (primordium)
g6 pituitary gland (primordium)

HYPOTHALAMUS (H)

H1i intermediate hypothalamic neuroepithelium
H1v ventral hypothalamic neuroepithelium
H3f fields of Forel (or differentiating field)
H3i intermediate hypothalamus (or differentiating field)
H3*l* lateral hypothalamus (or differentiating field)
H3v ventral hypothalamus (or differentiating field)
H7 zona incerta
H8 dorsomedial nucleus
H9 ventromedial nucleus
H19 subthalamic nucleus
H20 arcuate nucleus
H27 tuberomammillary nucleus

HIPPOCAMPAL FORMATION (K)

K1 hippocampal (Ammon's horn) neuroepithelium
K3 hippocampus (or differentiating field)
K11 subicular neuroepithelium
K13 subicular area (or differentiating field)
K23 dentate gyrus (or differentiating field)

NERVES AND NUCLEI, CRANIAL (n)

n5 trigeminal nerve
n5g trigeminal ganglion

RHINENCEPHALON (R)

R3 rhinencephalon (or differentiating field)
R4 piriform cortex
R8 entorhinal area

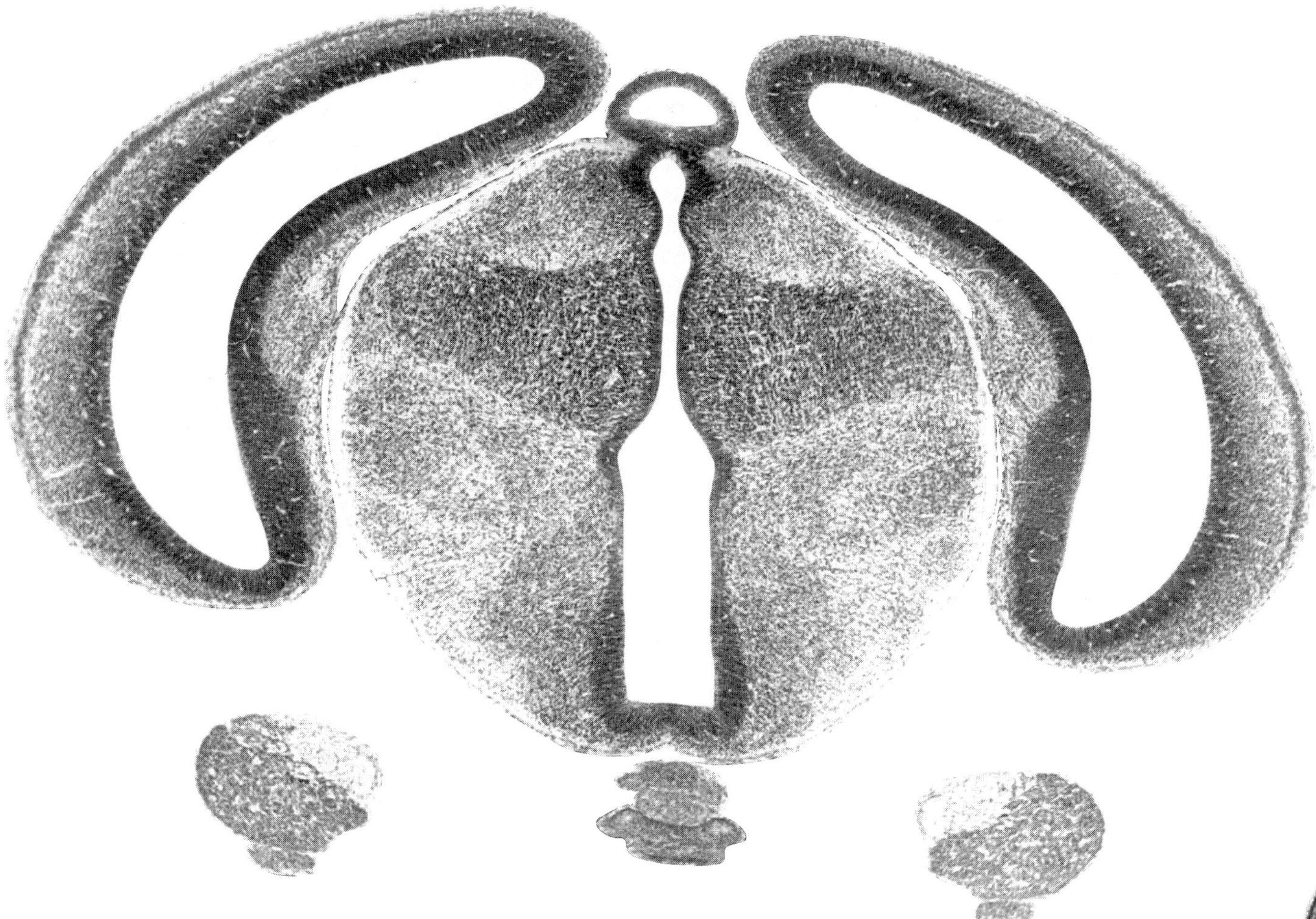

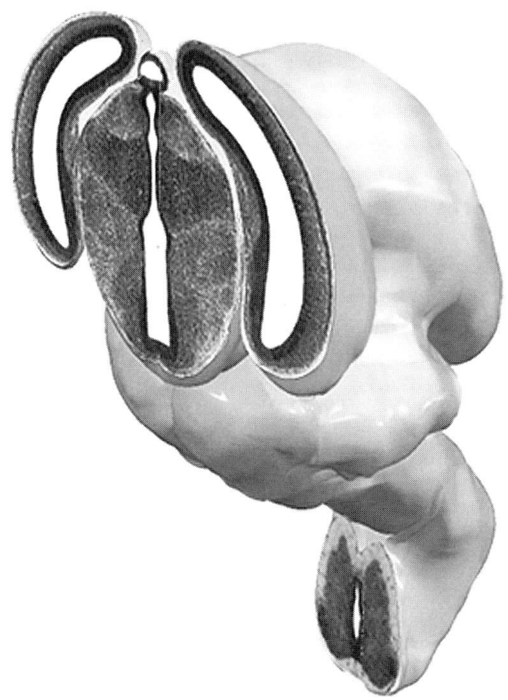

THALAMUS (T)

T1i intermediate thalamic neuroepithelium
T1p posterior thalamic neuroepithelium
T2 thalamic subventricular zone
T3i intermediate thalamus (or differentiating field)
T3p posterior thalamus (or differentiating field)
T24 lateral posterior nucleus
T25 reticular nucleus
T26 ventral medial nucleus
T28 ventrobasal nuclear complex
T29 dorsal lateral geniculate nucleus
T34 medial geniculate nucleus

VENTRICLES (v)

vlc lateral ventricle, central
vld lateral ventricle, dorsal
vlp lateral ventricle, posterior horn
v3e third ventricle, epithalamic
v3h third ventricle, hypothalamic
v3pr third ventricle, preoptic recess
v3t third ventricle, thalamic

CORONAL FIGURE 14

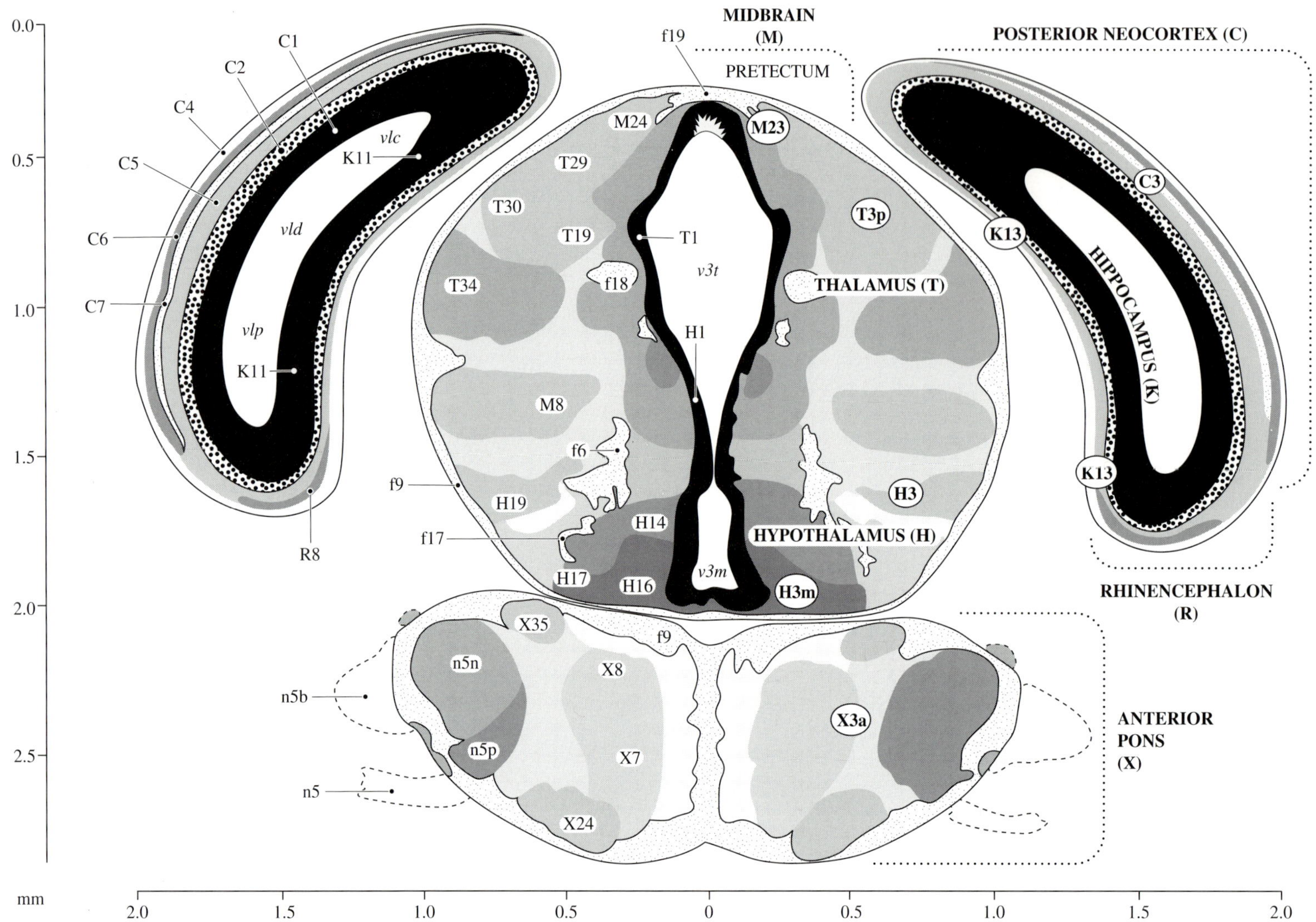

CEREBRAL CORTEX (C)

- C1 neocortical neuroepithelium
- C2 neocortical subventricular zone
- **C3** neocortex (or differentiating field)
- C4 molecular layer (or primordial plexiform layer)
- C5 intermediate zone
- C6 cortical plate
- C7 white matter

FIBER TRACTS (f)

- f6 fornix
- f9 cerebral peduncle
- f17 mammillothalamic tract
- f18 habenulopeduncular tract
- f19 posterior commissure

HYPOTHALAMUS (H)

- H1 hypothalamic neuroepithelium
- **H3** hypothalamus (or differentiating field)
- **H3m** mammillary body
- H14 supramammillary nucleus
- H16 medial mammillary nucleus
- H17 lateral mammillary nucleus
- H19 subthalamic nucleus

HIPPOCAMPAL FORMATION (K)

- K11 subicular neuroepithelium
- **K13** subicular area (or differentiating field)

MIDBRAIN (M)

- M8 red nucleus
- **M23** pretectum (or differentiating field)
- M24 anterior pretectal nucleus

NERVES AND NUCLEI, CRANIAL (n)

- n5 trigeminal nerve
- n5b trigeminal boundary cap
- n5n trigeminal motor nucleus
- n5p principal trigeminal nucleus

RHINENCEPHALON (R)

- R8 entorhinal area

THALAMUS (T)

- T1 thalamic neuroepithelium
- **T3p** posterior thalamus (or differentiating field)
- T19 parafascicular nucleus
- T29 dorsal lateral geniculate nucleus
- T30 ventral lateral geniculate nucleus
- T34 medial geniculate nucleus

VENTRICLES (v)

- *vlc* lateral ventricle, central
- *vld* lateral ventricle, dorsal
- *vlp* lateral ventricle, posterior horn
- *v3m* third ventricle, mammillary recess
- *v3t* third ventricle, thalamic

PONTINE AREA (X)

- **X3a** anterior pons (or differentiating field)
- X7 pontine reticular formation
- X8 superior central raphe nucleus
- X24 superior olivary nucleus
- X35 parabrachial nucleus

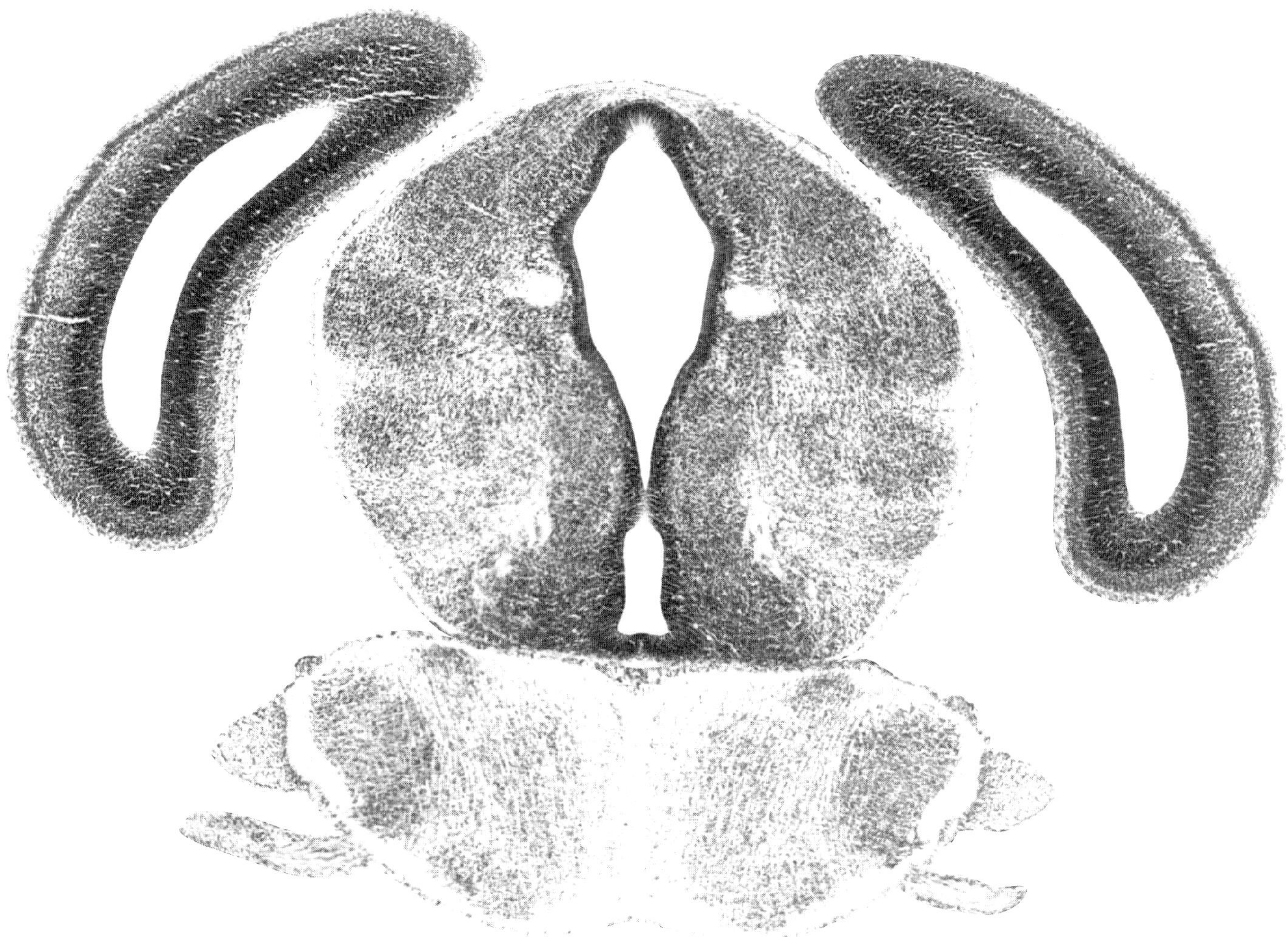

CORONAL FIGURE 15

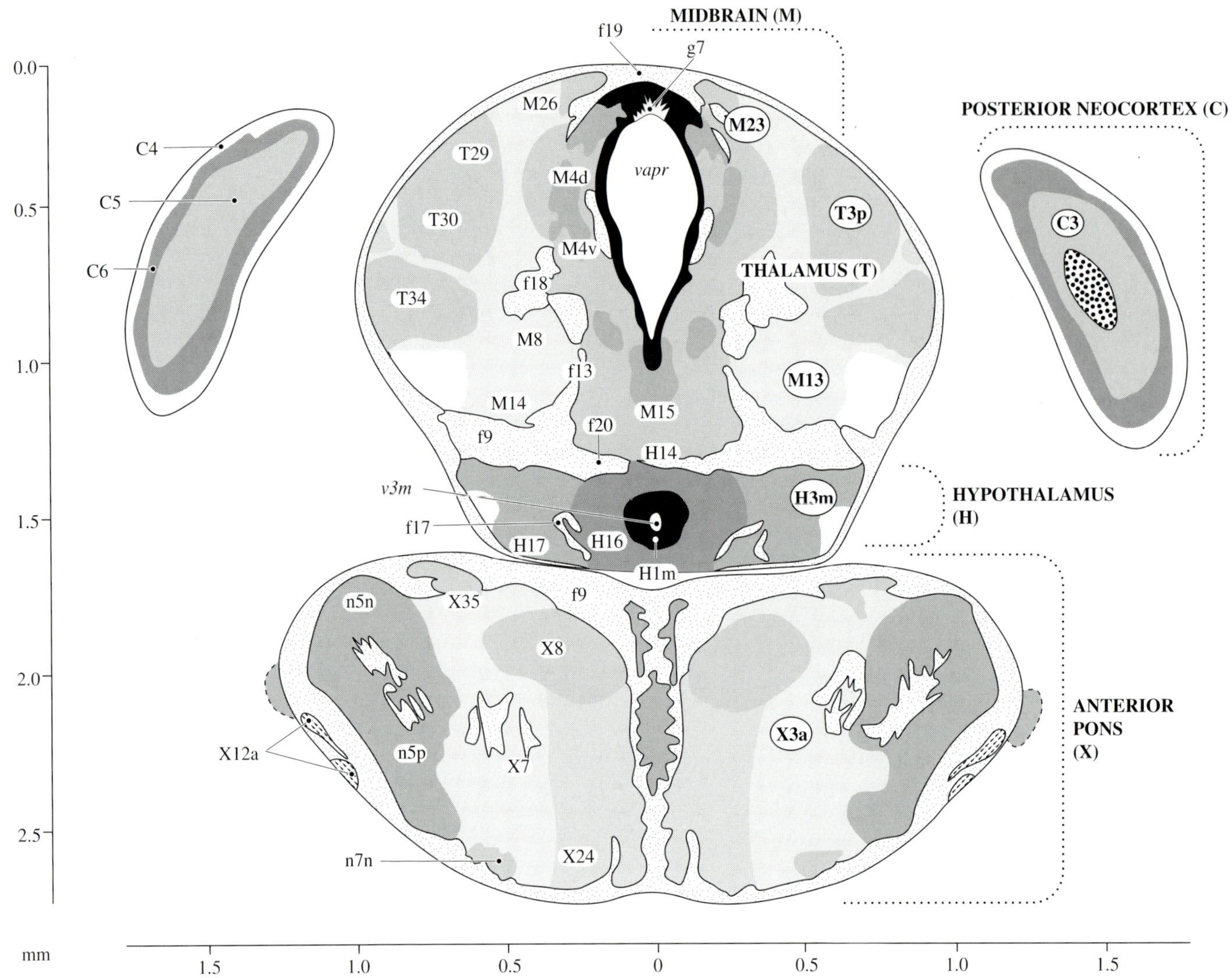

CEREBRAL CORTEX (C)

C3 neocortex (or differentiating field)
C4 molecular layer (or primordial plexiform layer)
C5 intermediate zone
C6 cortical plate

FIBER TRACTS (f)

f9 cerebral peduncle
f13 fimbria
f17 mammillothalamic tract
f18 habenulopeduncular tract
f19 posterior commissure
f20 supramammillary decussation

GLANDS AND PARANEURAL STRUCTURES (g)

g7 subcommissural organ

HYPOTHALAMUS (H)

H1m mammillary neuroepithelium
H3m mammillary body (or differentiating field)
H14 supramammillary nucleus
H16 medial mammillary nucleus
H17 lateral mammillary nucleus

MIDBRAIN (M)

M4d dorsal periaqueductal gray
M4v ventral periaqueductal gray
M8 red nucleus
M13 tegmentum (or differentiating field)
M14 substantia nigra
M15 interpeduncular nucleus
M23 pretectum (or differentiating field)
M26 olivary pretectal nucleus

NERVES AND NUCLEI, CRANIAL (n)

n5n trigeminal motor nucleus
n5p principal trigeminal nucleus
n7n facial motor nucleus

THALAMUS (T)

T3p posterior thalamus (or differentiating field)
T29 dorsal lateral geniculate nucleus
T30 ventral lateral geniculate nucleus
T34 medial geniculate nucleus

VENTRICLES (v)

vapr aqueduct, pretectal
v3m third ventricle, mammillary recess

PONTINE AREA (X)

X3a anterior pons (or differentiating field)
X7 pontine reticular formation
X8 superior central raphe nucleus
X24 superior olivary nucleus
X35 parabrachial nucleus

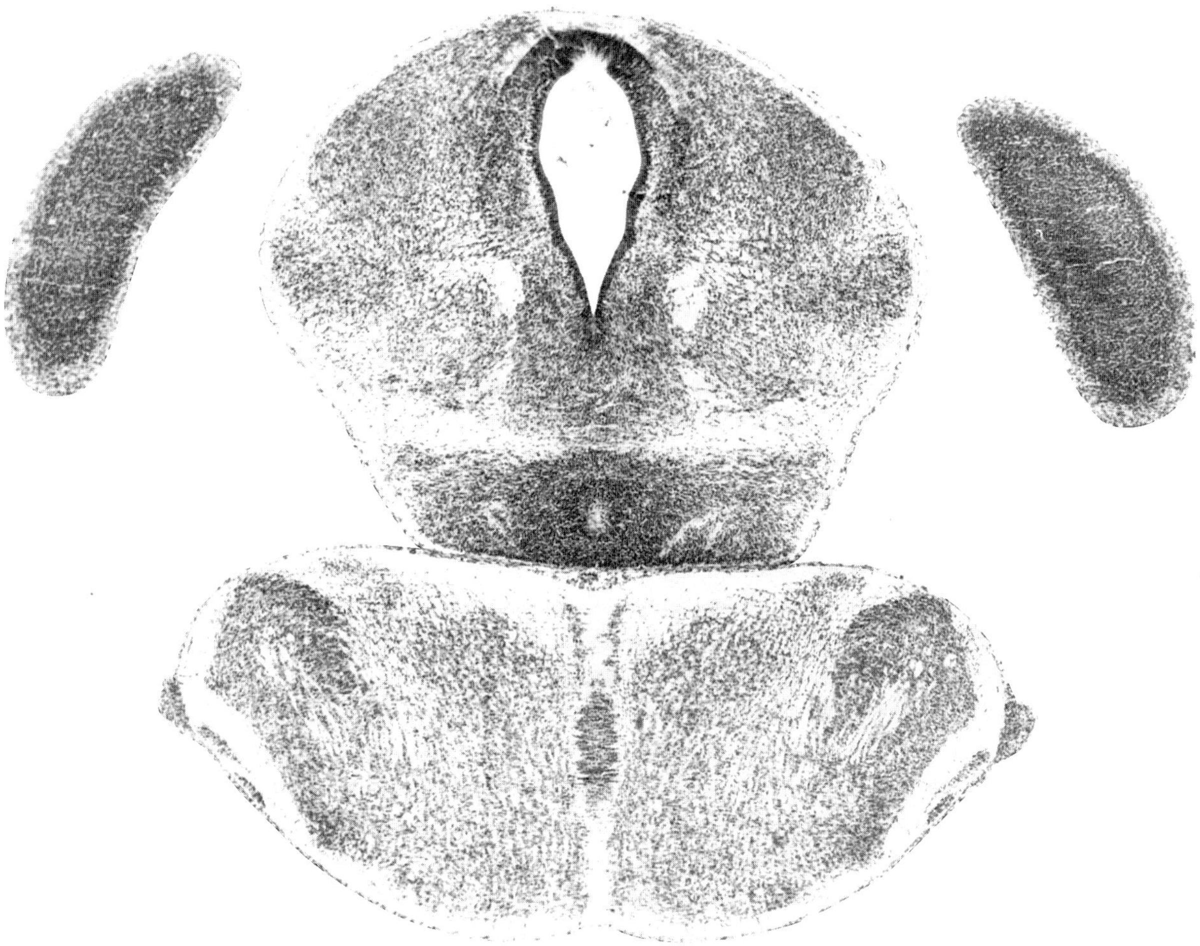

CORONAL FIGURE 16

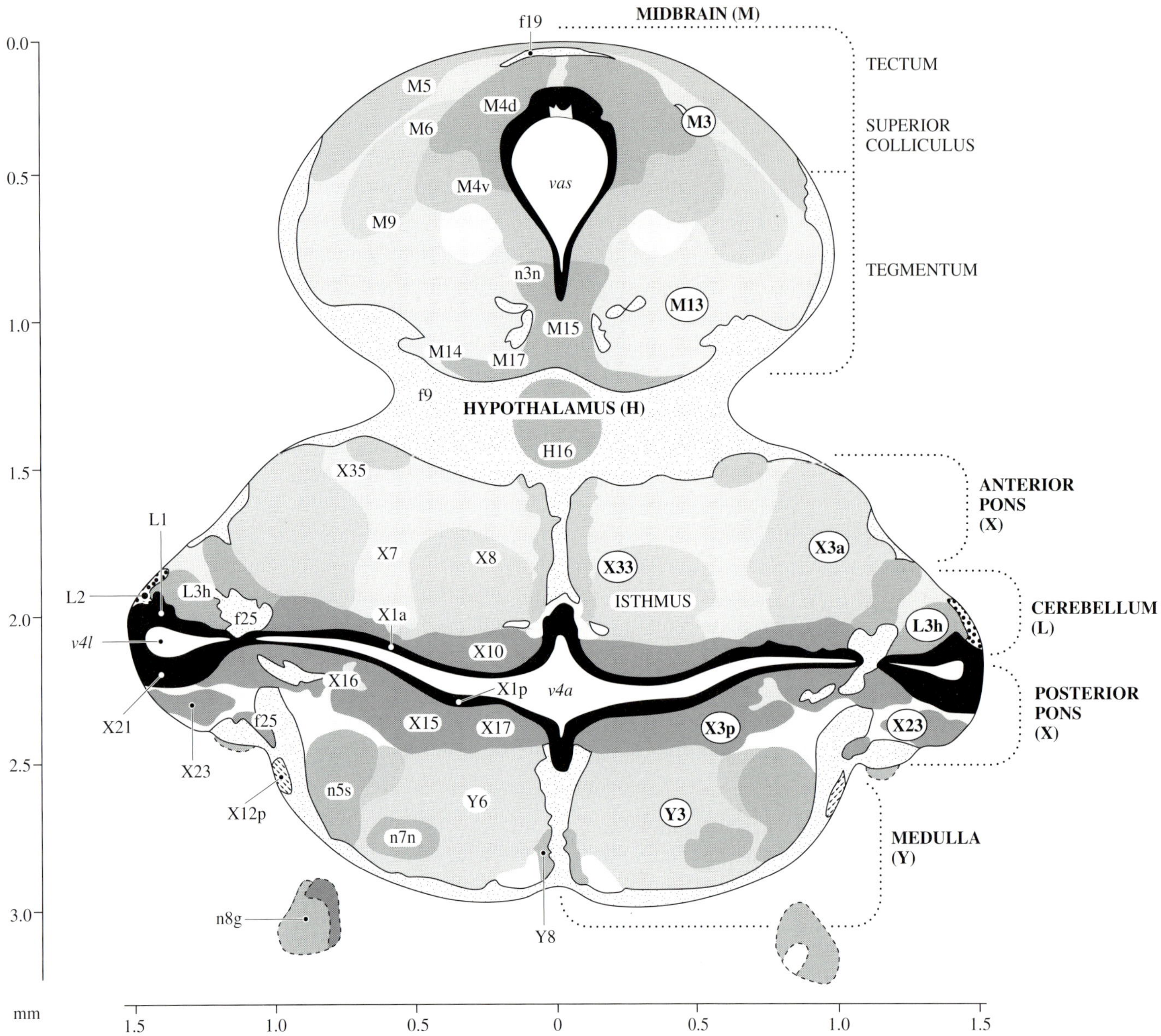

FIBER TRACTS (f)

f9 cerebral peduncle
f19 posterior commissure
f25 inferior cerebellar peduncle

HYPOTHALAMUS (H)

H16 medial mammillary nucleus

CEREBELLUM (L)

L1 cerebellar neuroepithelium
L2 external germinal layer
L3h cerebellar hemisphere (or differentiating field)

MIDBRAIN (M)

M3 superior colliculus (or differentiating field)
M4d dorsal periaqueductal gray
M4v ventral periaqueductal gray
M5 superior colliculus, superficial layers
M6 superior colliculus, deep layers
M9 midbrain reticular formation
M13 tegmentum (or differentiating field)
M14 substantia nigra
M15 interpeduncular nucleus
M17 ventral tegmental area

NERVES AND NUCLEI, CRANIAL (n)

n3n oculomotor nucleus
n5g principal trigeminal nucleus
n5s spinal trigeminal nucleus
n7n facial motor nucleus
n8g vestibular ganglion

VENTRICLES (v)

vas aqueduct, superior
vat aqueduct, tegmental
v4a fourth ventricle, anterior
v4l fourth ventricle, lateral recess

PONTINE AREA (X)

X1a anterior pontine neuroepithelium
X1p posterior pontine neuroepithelium
X3a anterior pons (or differentiating field)
X3p posterior pons (or differentiating field)
X7 pontine reticular formation
X8 superior central raphe nucleus
X10 ventral tegmental nucleus
X12p posterior precerebellar extramural migration
X15 lateral vestibular nucleus
X16 superior vestibular nucleus
X17 medial vestibular nucleus

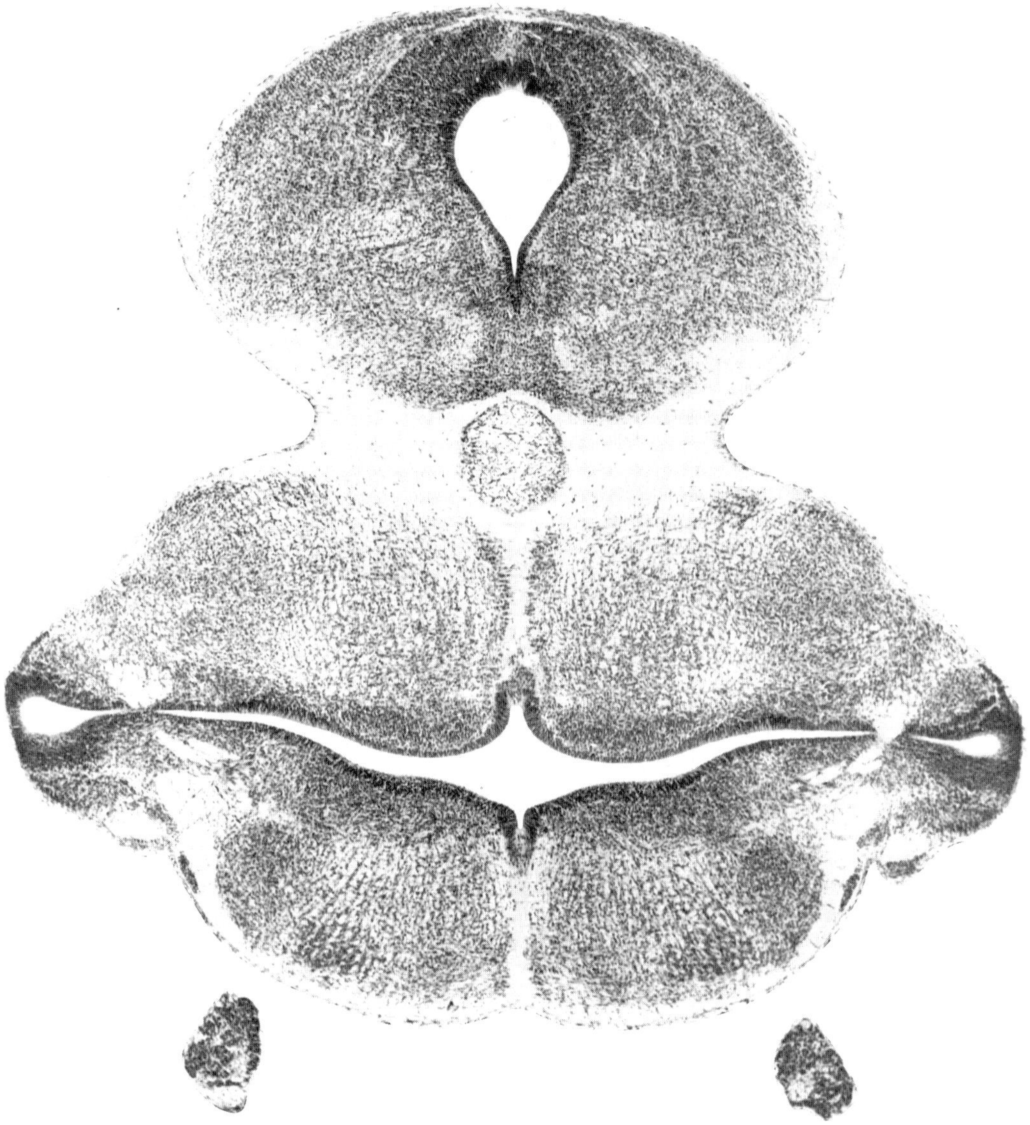

X33 isthmus (or differentiating field)
X35 parabrachial nucleus

MEDULLA (Y)

Y3 medulla (or differentiating field)
Y6 medullary reticular formation
Y8 inferior olive

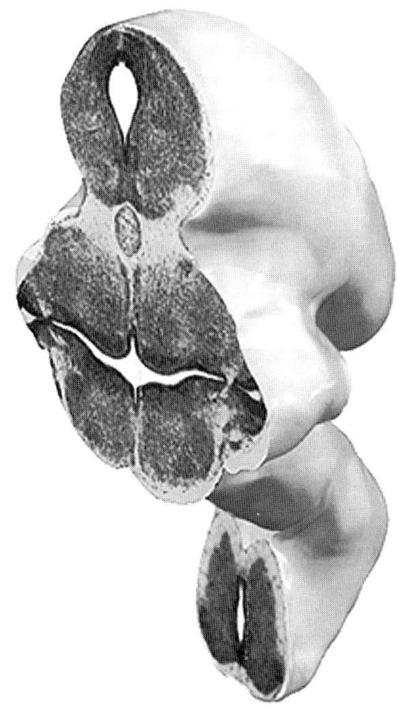

CORONAL FIGURE 17

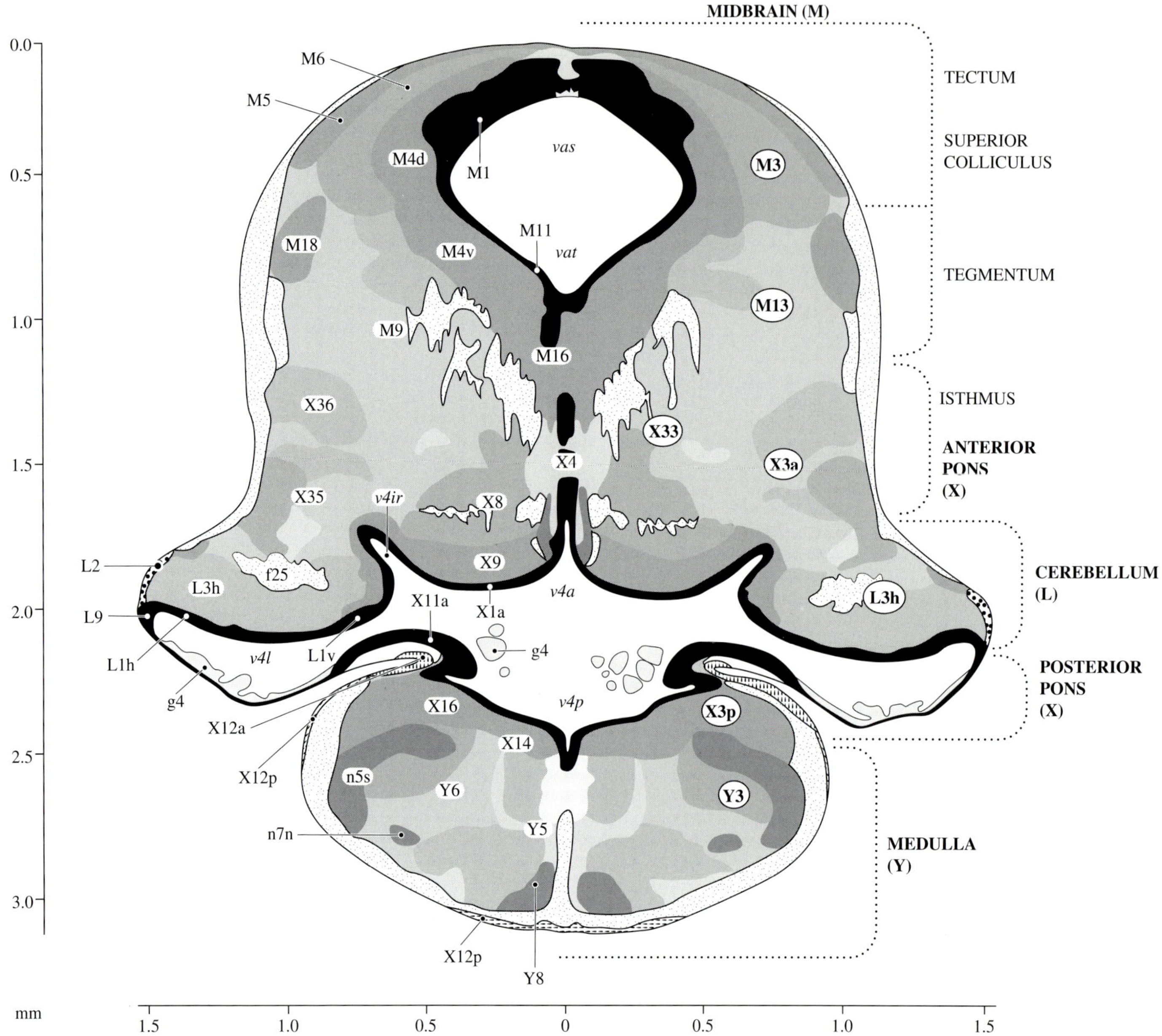

FIBER TRACTS (f)

f25 inferior cerebellar peduncle

GLANDS AND PARANEURAL STRUCTURES (g)

g4 choroid plexus of fourth ventricle

CEREBELLUM (L)

L1h cerebellar hemisphere neuroepithelium
L1v cerebellar vermis neuroepithelium
L2 external germinal layer
L3h cerebellar hemisphere (or differentiating field)
L9 cerebellar germinal trigone

MIDBRAIN (M)

M1 superior tectal neuroepithelium
M3 superior colliculus (or differentiating field)
M4d dorsal periaqueductal gray
M4v ventral periaqueductal gray
M5 superior colliculus, superficial layers
M6 superior colliculus, deep layers
M9 midbrain reticular formation
M13 tegmentum (or differentiating field)
M16 dorsal raphe (mesencephalic)
M18 parabigeminal nucleus

NERVES AND NUCLEI, CRANIAL (n)

n5s spinal trigeminal nucleus
n7n facial motor nucleus

VENTRICLES (v)

vas aqueduct, superior tectal
vat aqueduct, tegmental
v4a fourth ventricle, anterior
v4ir fourth ventricle, isthmal recess
v4l fourth ventricle, lateral recess
v4p fourth ventricle, posterior

E17

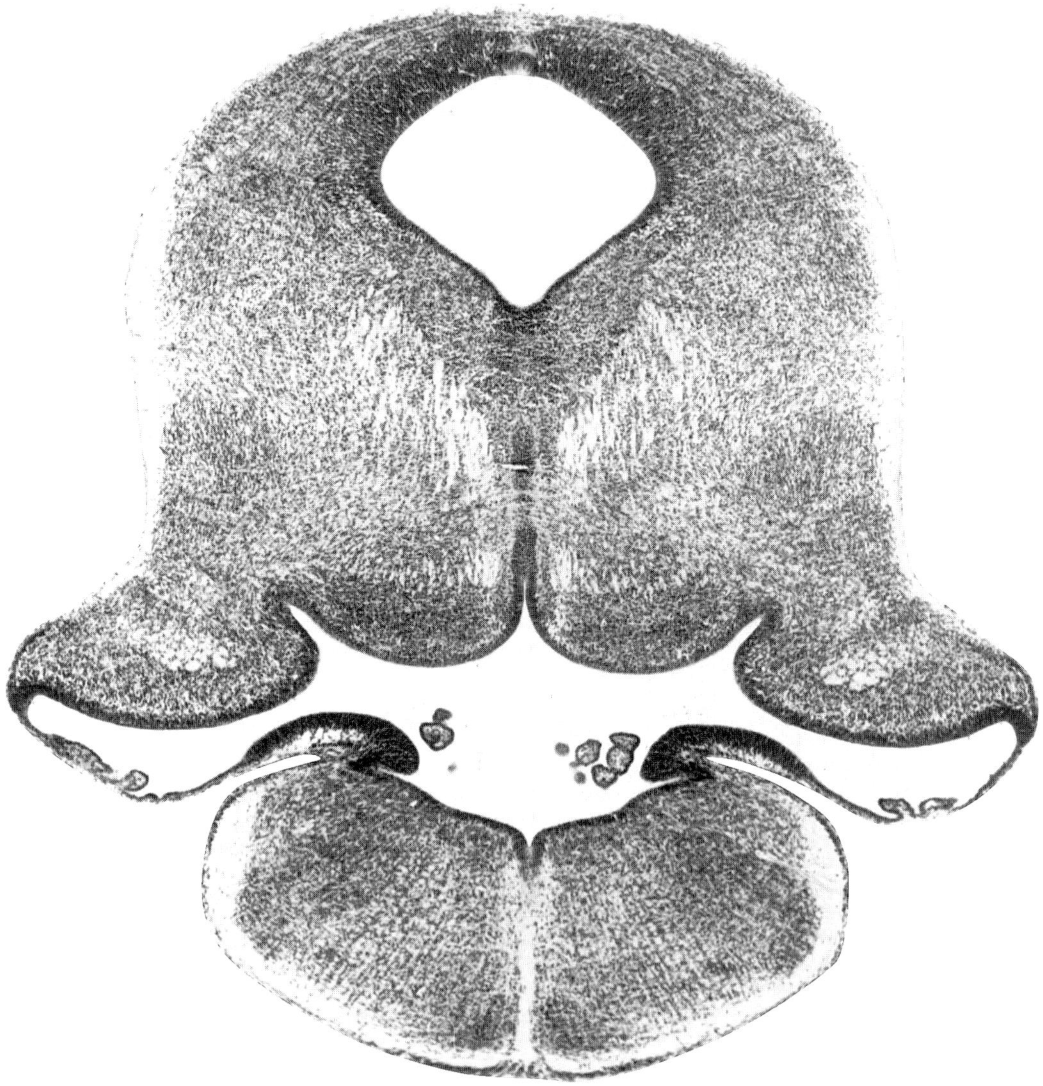

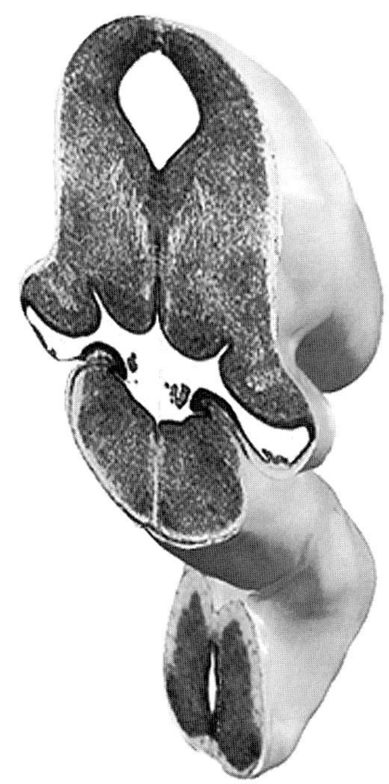

PONTINE AREA (X)

X1a anterior pontine neuroepithelium
X3a anterior pons (or differentiating field)
X3p posterior pons (or differentiating field)
X4 dorsal raphe nucleus (pontine)
X8 superior central raphe nucleus
X9 dorsal tegmental nucleus
X11a anterior precerebellar neuroepithelium
X12a anterior precerebellar extramural migration
X12p posterior precerebellar extramural migration
X14 spinal vestibular nucleus
X16 superior vestibular nucleus
X33 isthmus (or differentiating field)
X35 parabrachial nucleus
X36 nucleus of lateral lemniscus

MEDULLA (Y)

Y3 medulla (or differentiating field)
Y5 medullary raphe
Y6 medullary reticular formation
Y8 inferior olive

CORONAL FIGURE 18

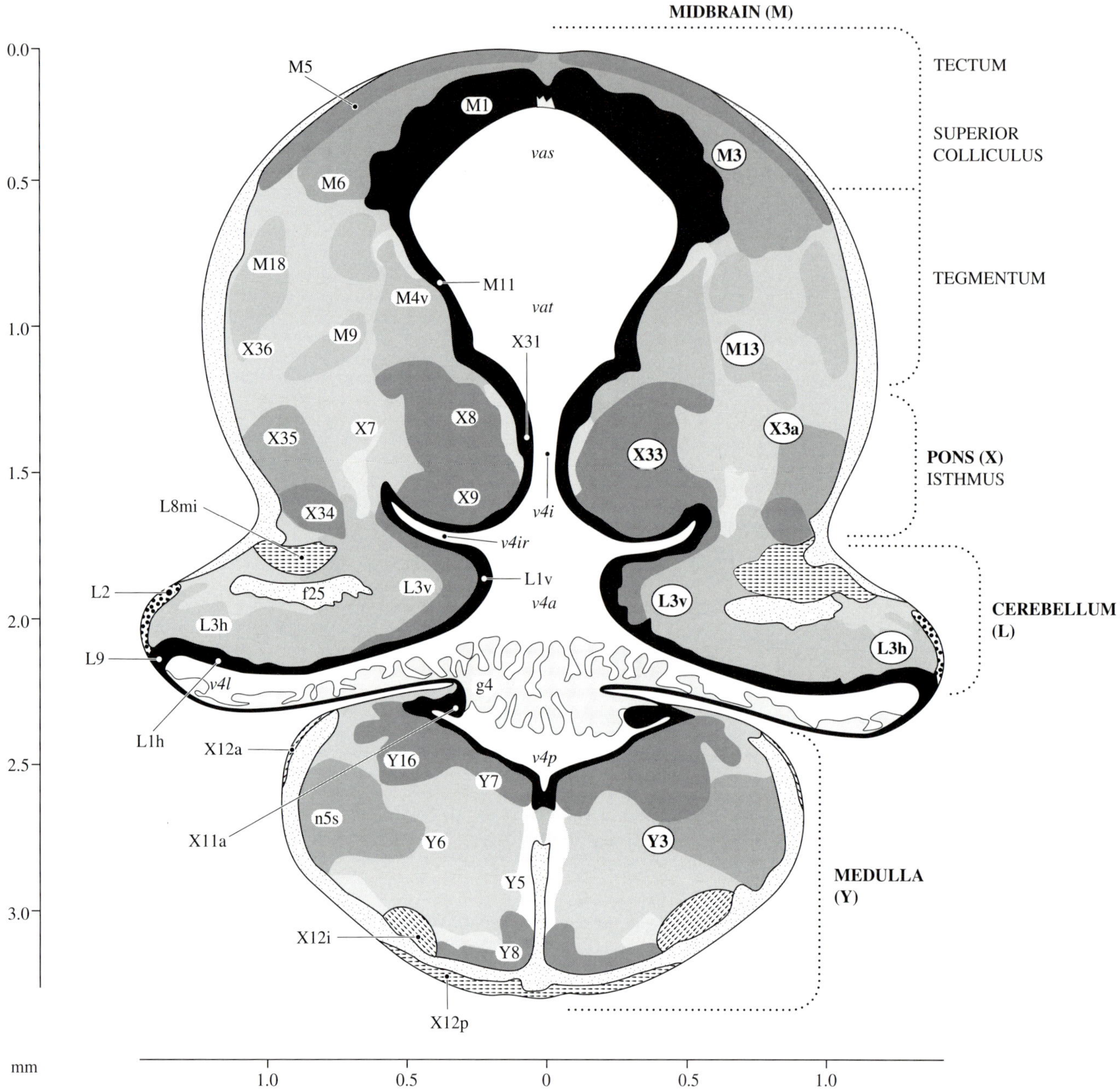

FIBER TRACTS (f)

f25 inferior cerebellar peduncle

GLANDS AND PARANEURAL STRUCTURES (g)

g4 choroid plexus of fourth ventricle

CEREBELLUM (L)

L1h cerebellar hemisphere neuroepithelium
L1v cerebellar vermis neuroepithelium
L2 external germinal layer
L3h cerebellar hemisphere (or differentiating field)
L3v cerebellar vermis (or differentiating field)
L8mi medial deep nucleus, migration
L9 cerebellar germinal trigone

MIDBRAIN (M)

M1 superior tectal neuroepithelium
M3 superior colliculus (or differentiating field)
M4v ventral periaqueductal gray
M5 superior colliculus, superficial layers
M6 superior colliculus, deep layers
M9 midbrain reticular formation
M13 tegmentum (or differentiating field)
M18 parabigeminal nucleus

NERVES AND NUCLEI, CRANIAL (n)

n5s spinal trigeminal nucleus

VENTRICLES (v)

vas aqueduct, superior tectal
vat aqueduct, tegmental
v4a fourth ventricle, anterior
v4i fourth ventricle, isthmal
v4ir fourth ventricle, isthmal recess
v4l fourth ventricle, lateral recess
v4p fourth ventricle, posterior

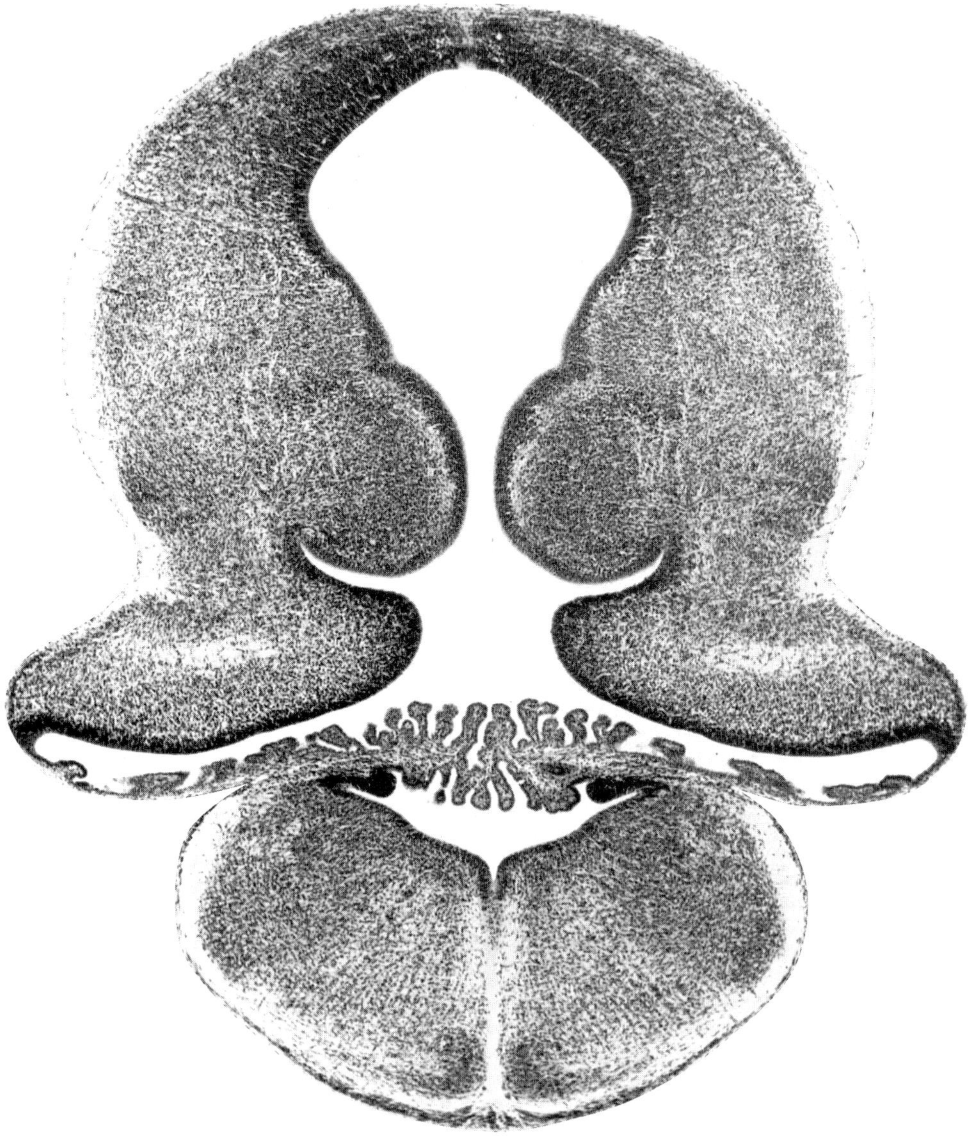

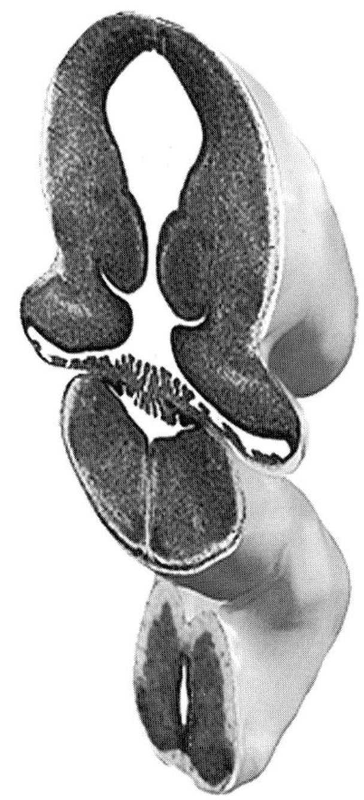

PONTINE AREA (X)

X3a anterior pons (or differentiating field)
X7 pontine reticular formation
X8 superior central raphe nucleus
X9 dorsal tegmental nucleus
X11a anterior precerebellar neuroepithelium
X12a anterior precerebellar extramural migration
X12i precerebellar intramural migration
X12p posterior precerebellar extramural migration
X31 isthmal neuroepithelium
X33 isthmus (or differentiating field)
X34 locus coeruleus
X35 parabrachial nucleus
X36 nucleus of lateral lemniscus

MEDULLA (Y)

Y3 medulla (or differentiating field)
Y5 medullary raphe
Y6 medullary reticular formation
Y7 prepositus nucleus
Y8 inferior olive
Y16 solitary nucleus

CORONAL FIGURE 19

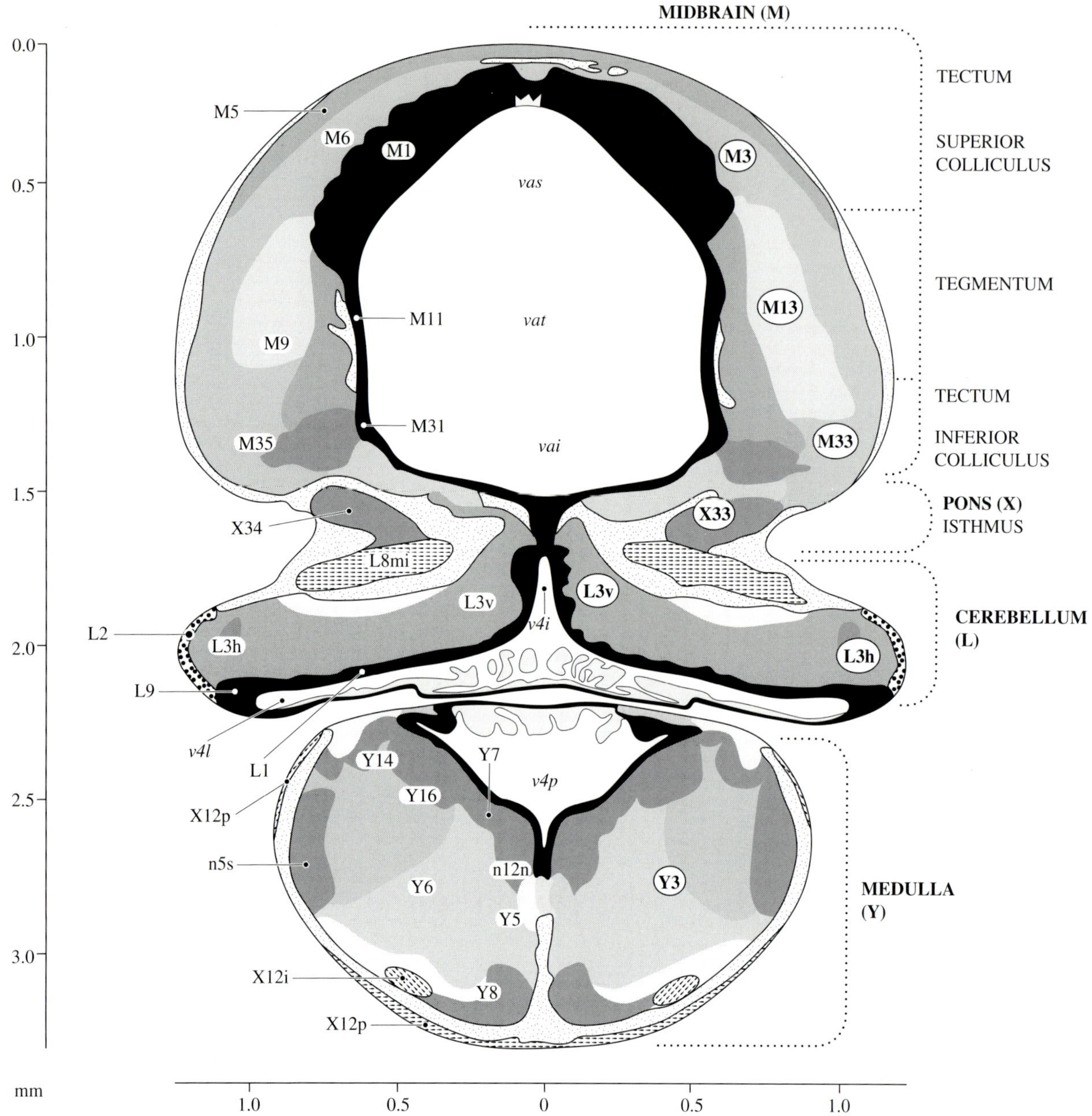

CEREBELLUM (L)

- L1 cerebellar neuroepithelium
- L2 external germinal layer
- **L3h** cerebellar hemisphere (or differentiating field)
- **L3v** cerebellar vermis (or differentiating field)
- L8mi medial deep nucleus, migration
- L9 cerebellar germinal trigone

MIDBRAIN (M)

- M1 superior tectal neuroepithelium
- **M3** superior colliculus (or differentiating field)
- M5 superior colliculus, superficial layers
- M6 superior colliculus, deep layers
- M9 midbrain reticular formation
- M11 tegmental neuroepithelium
- **M13** tegmentum (or differentiating field)
- M31 inferior tectal neuroepithelium
- **M33** inferior colliculus (or differentiating field)
- M35 central nucleus of inferior colliculus

NERVES AND NUCLEI, CRANIAL (n)

- n5s spinal trigeminal nucleus
- n12n hypoglossal nucleus

VENTRICLES (v)

- *vai* aqueduct, inferior tectal
- *vas* aqueduct, superior tectal
- *vat* aqueduct, tegmental
- *v4i* fourth ventricle, isthmal recess
- *v4l* fourth ventricle, lateral recess
- *v4p* fourth ventricle, posterior

PONTINE AREA (X)

- X12i precerebellar intramural migration
- X12p posterior precerebellar extramural migration
- **X33** isthmus (or differentiating field)
- X34 locus coeruleus

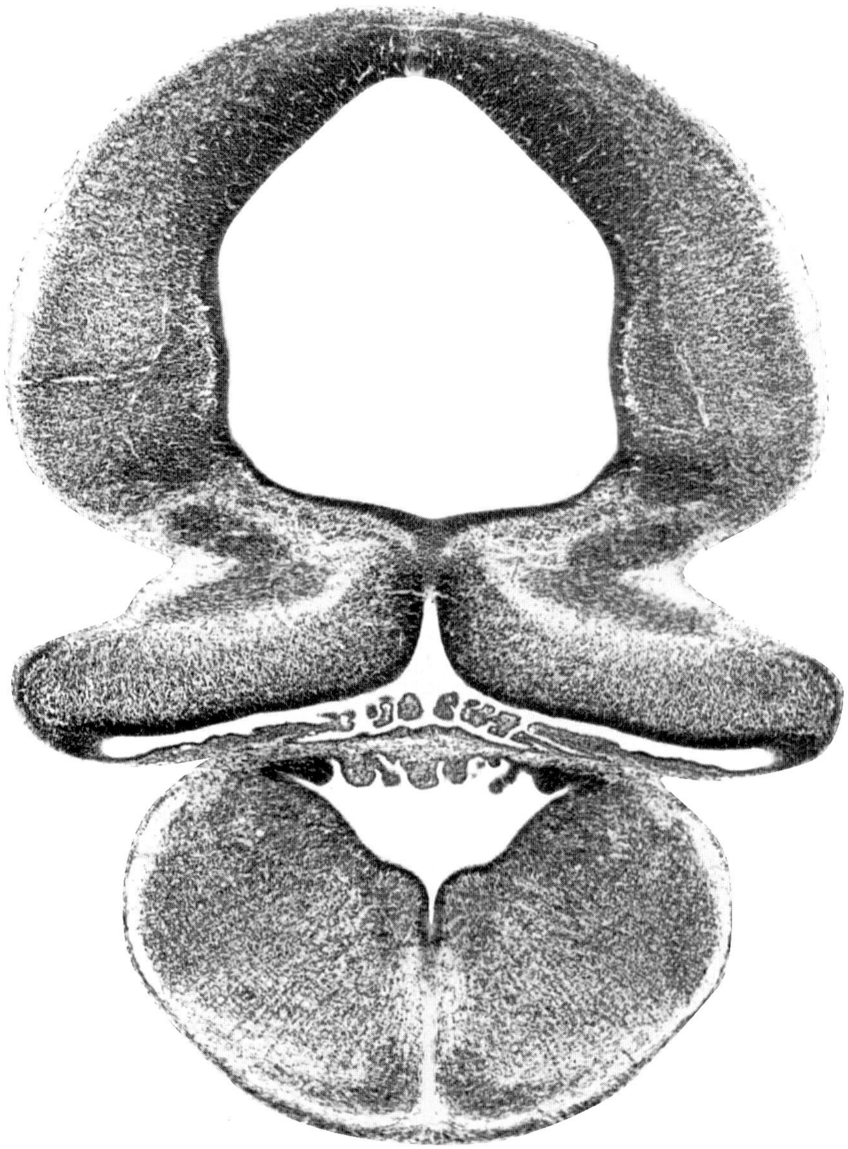

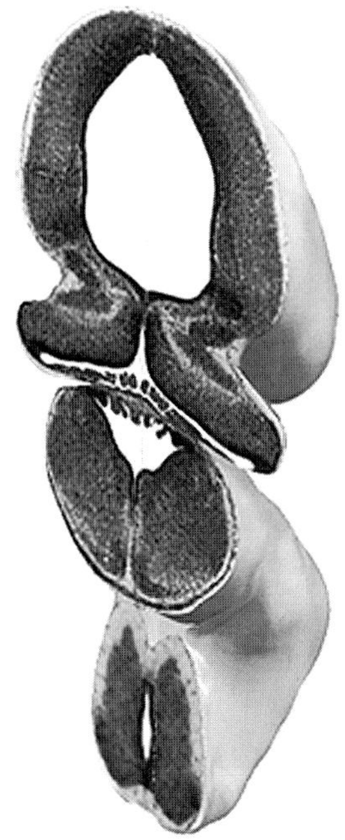

MEDULLA (Y)

Y3 medulla (or differentiating field)
Y5 medullary raphe
Y6 medullary reticular formation
Y7 prepositus nucleus
Y8 inferior olive
Y14 gracile nucleus
Y16 solitary nucleus

CORONAL FIGURE 20

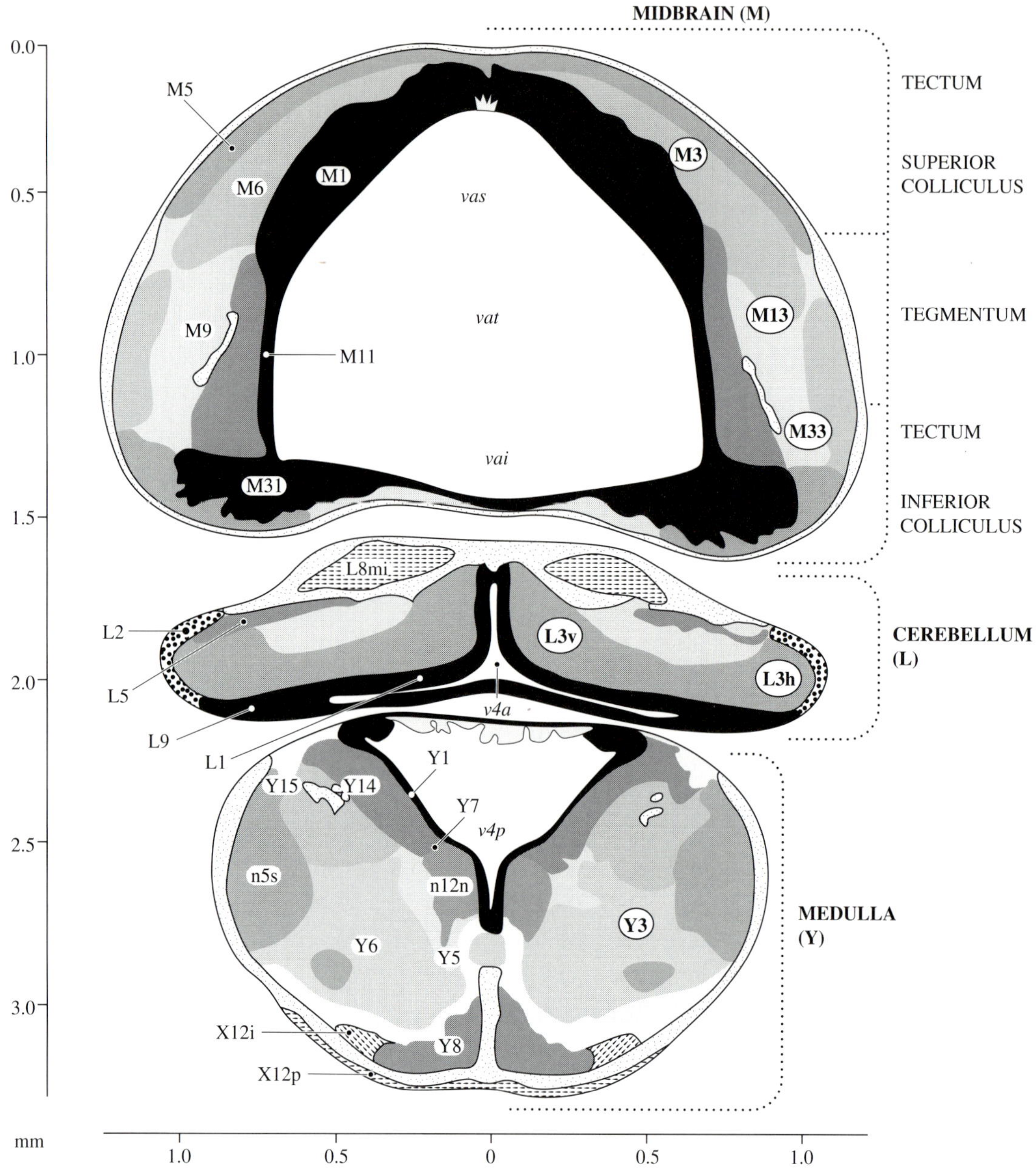

CEREBELLUM (L)

L1 cerebellar neuroepithelium
L2 external germinal layer
L3h cerebellar hemisphere (or differentiating field)
L3v cerebellar vermis (or differentiating field)
L5 Purkinje cell layer
L8mi medial deep nucleus, migration
L9 cerebellar germinal trigone

MIDBRAIN (M)

M1 superior tectal neuroepithelium
M3 superior colliculus (or differentiating field)
M5 superior colliculus, superficial layers
M6 superior colliculus, deep layers
M9 midbrain reticular formation
M11 tegmental neuroepithelium
M13 tegmentum (or differentiating field)
M31 inferior tectal neuroepithelium
M33 inferior colliculus (or differentiating field)

NERVES AND NUCLEI, CRANIAL (n)

n5s spinal trigeminal nucleus
n12n hypoglossal nucleus

VENTRICLES (v)

vai aqueduct, inferior tectal
vas aqueduct, superior tectal
vat aqueduct, tegmental
v4a fourth ventricle, anterior
v4p fourth ventricle, posterior

PONTINE AREA (X)

X12i precerebellar intramural migration
X12p posterior precerebellar extramural migration

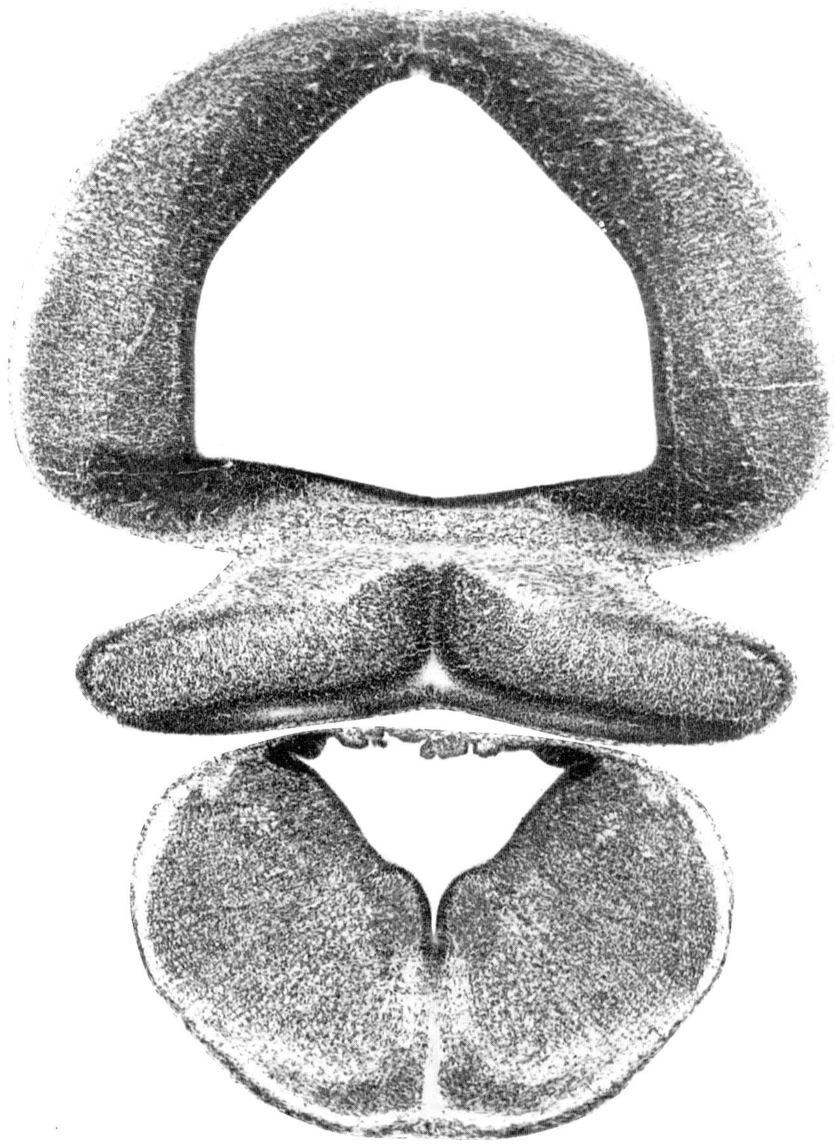

MEDULLA (Y)

Y1 medullary neuroepithelium
Y3 medulla (or differentiating field)
Y5 medullary raphe
Y6 medullary reticular formation
Y7 prepositus nucleus
Y8 inferior olive
Y14 gracile nucleus
Y15 cuneate nucleus

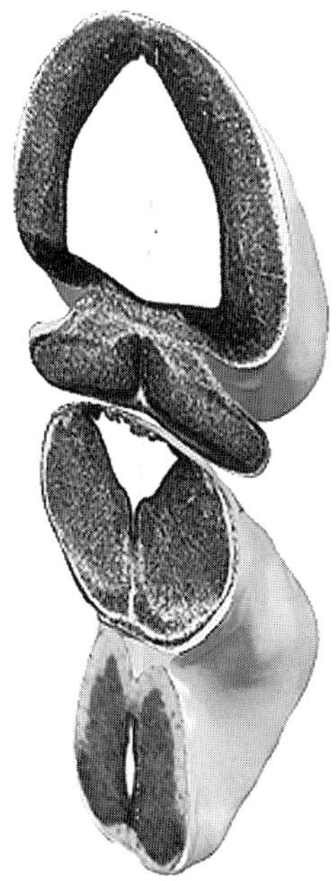

CORONAL FIGURE 21

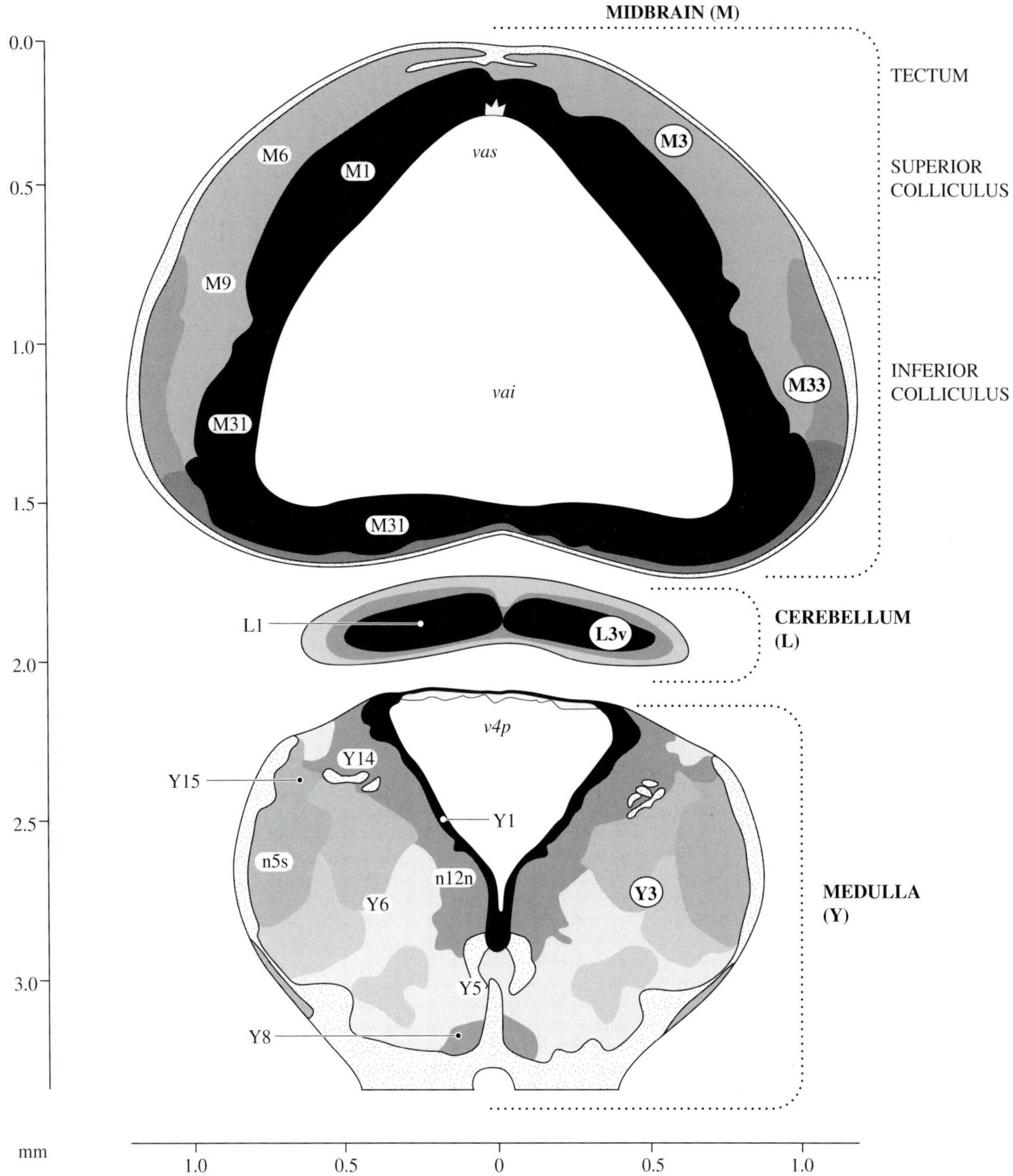

CEREBELLUM (L)

L1	cerebellar neuroepithelium
L3v	cerebellar vermis (or differentiating field)

MIDBRAIN (M)

M1	superior tectal neuroepithelium
M3	superior colliculus (or differentiating field)
M6	superior colliculus, deep layers
M9	midbrain reticular formation
M31	inferior tectal neuroepithelium
M33	inferior colliculus (or differentiating field)

NERVES AND NUCLEI, CRANIAL (n)

n5s	spinal trigeminal nucleus
n12n	hypoglossal nucleus

VENTRICLES (v)

vai	aqueduct, inferior tectal
vas	aqueduct, superior tectal
v4p	fourth ventricle, posterior

MEDULLA (Y)

Y1	medullary neuroepithelium
Y3	medulla (or differentiating field)
Y5	medullary raphe
Y6	medullary reticular formation
Y8	inferior olive
Y14	gracile nucleus
Y15	cuneate nucleus

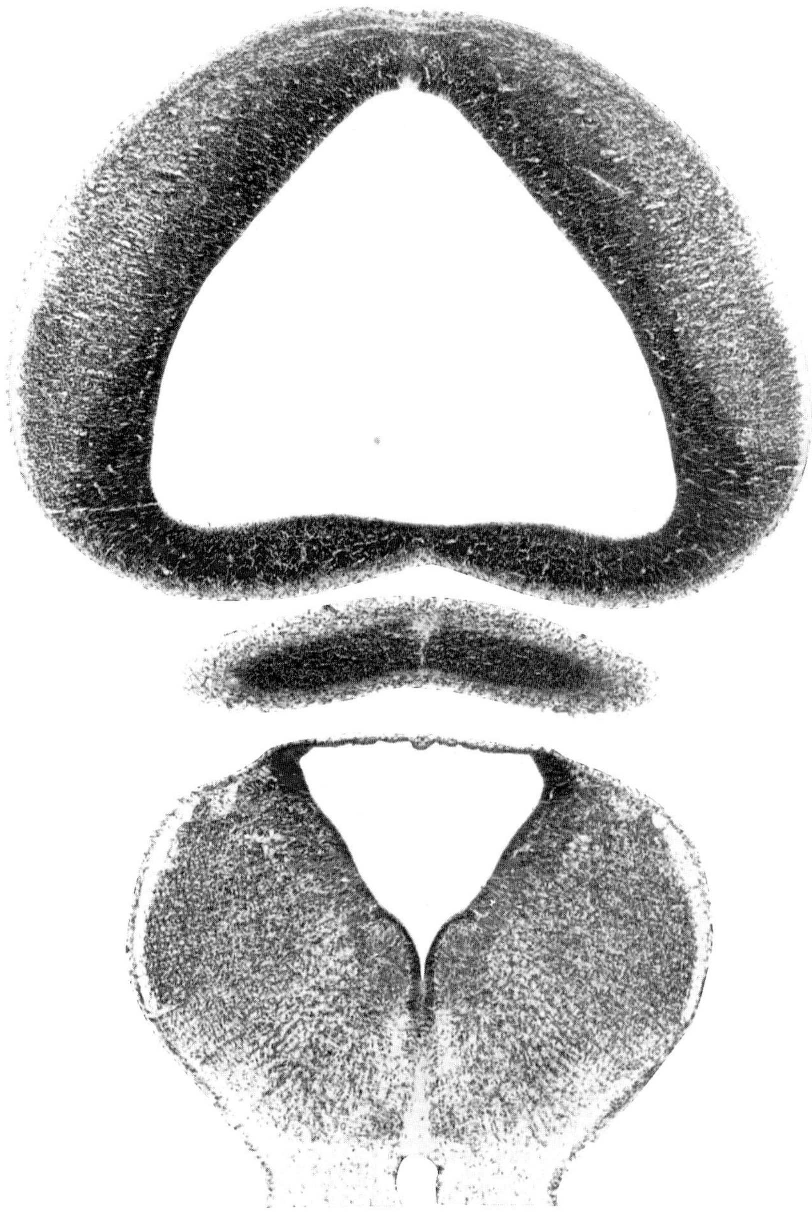

THE E18 BRAIN

COMPUTERIZED 3-D RECONSTRUCTION: 4 VIEWS

LEFT, ANTERIOR VIEW

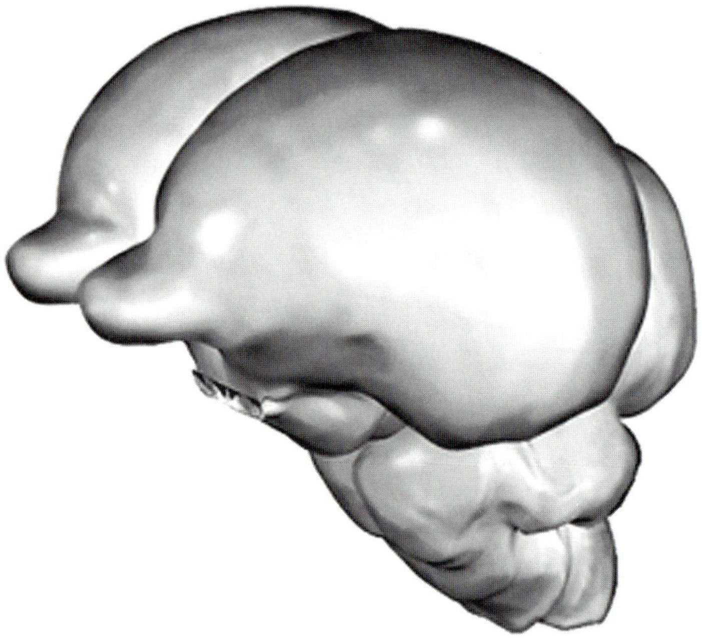

RIGHT, POSTERIOR VIEW

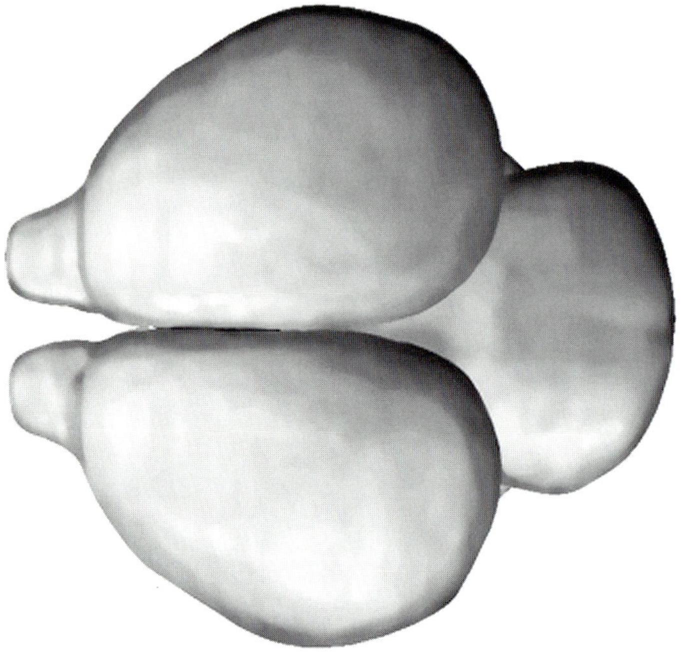

DORSAL VIEW

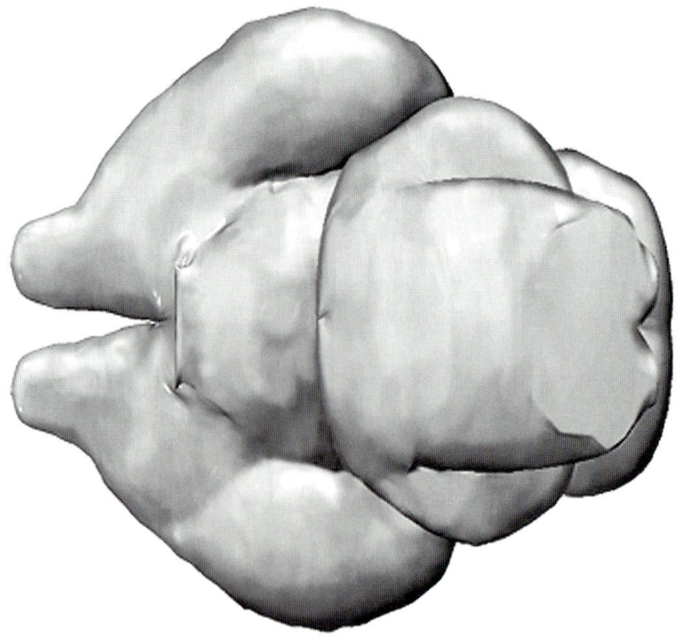

VENTRAL VIEW

SAGITTAL

FIGURES and PLATES 1-8

SAGITTAL FIGURE 1

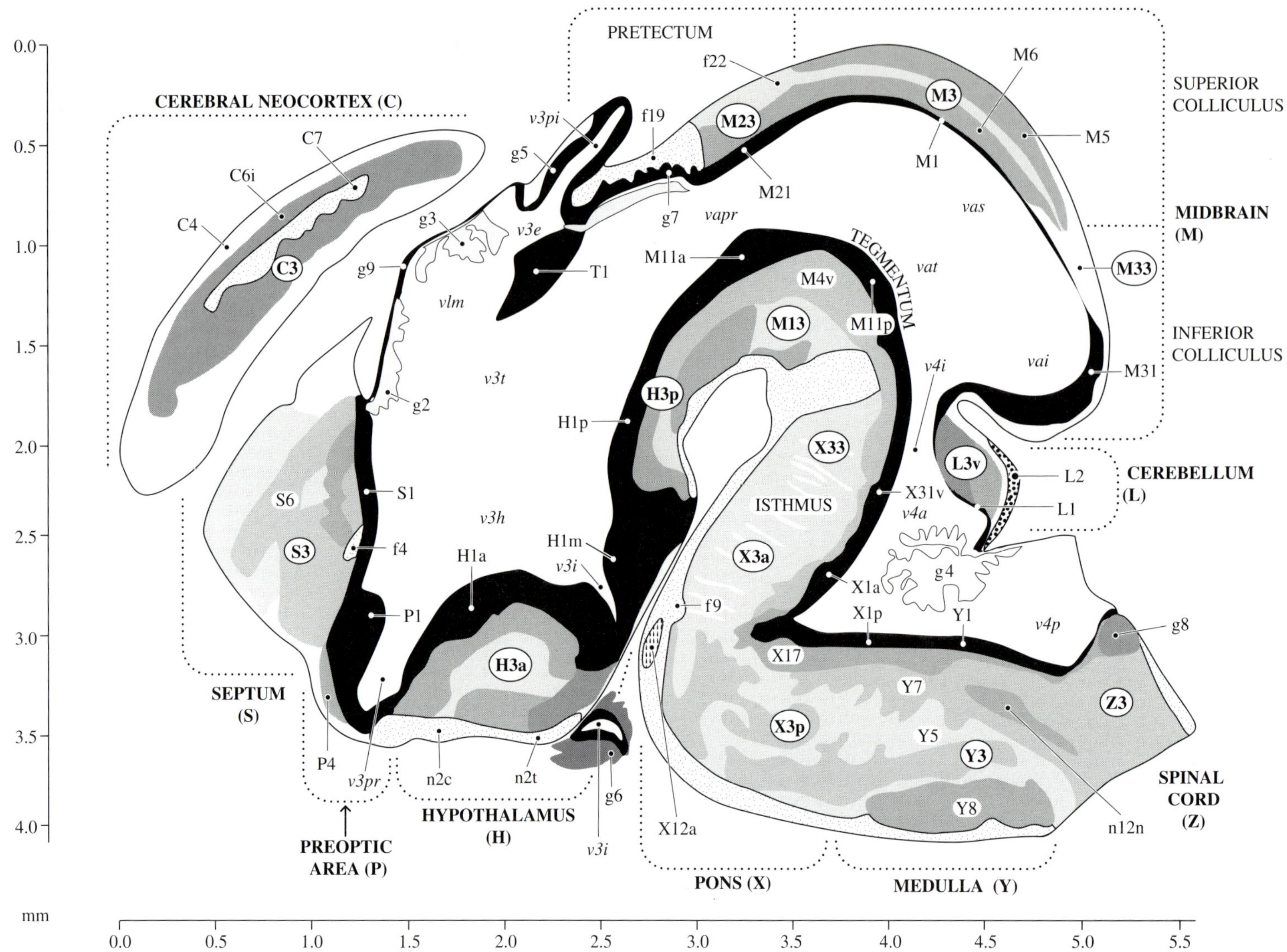

CEREBRAL CORTEX (C)

C3 neocortex (or differentiating field)
C4 molecular layer (or primordial plexiform layer)
C6i cortical plate, infragranular part
C7 white matter

FIBER TRACTS (f)

f4 anterior commissure
f9 cerebral peduncle
f19 posterior commissure
f22 commissure of superior colliculus

GLANDS AND PARANEURAL STRUCTURES (g)

g2 subfornical organ
g3 choroid plexus of third ventricle
g4 choroid plexus of fourth ventricle
g5 pineal gland (primordium)
g6 pituitary gland (primordium)
g7 subcommissural organ
g8 area postrema (primordium)
g9 lamina terminalis

HYPOTHALAMUS (H)

H1a anterior hypothalamic neuroepithelium
H1m mammillary neuroepithelium
H1p posterior hypothalamic neuroepithelium
H3a anterior hypothalamus (or differentiating field)
H3p posterior hypothalamus (or differentiating field)

CEREBELLUM (L)

L1 cerebellar neuroepithelium
L2 external germinal layer
L3v cerebellar vermis (or differentiating field)

MIDBRAIN (M)

M1 superior tectal neuroepithelium
M3 superior colliculus (or differentiating field)
M4v ventral periaqueductal gray
M5 superior colliculus, superficial layers
M6 superior colliculus, deep layers
M11a anterior tegmental neuroepithelium
M11p posterior tegmental neuroepithelium
M13 tegmentum (or differentiating field)
M21 pretectal neuroepithelium
M23 pretectum (or differentiating field)
M31 inferior tectal neuroepithelium
M33 inferior colliculus (or differentiating field)

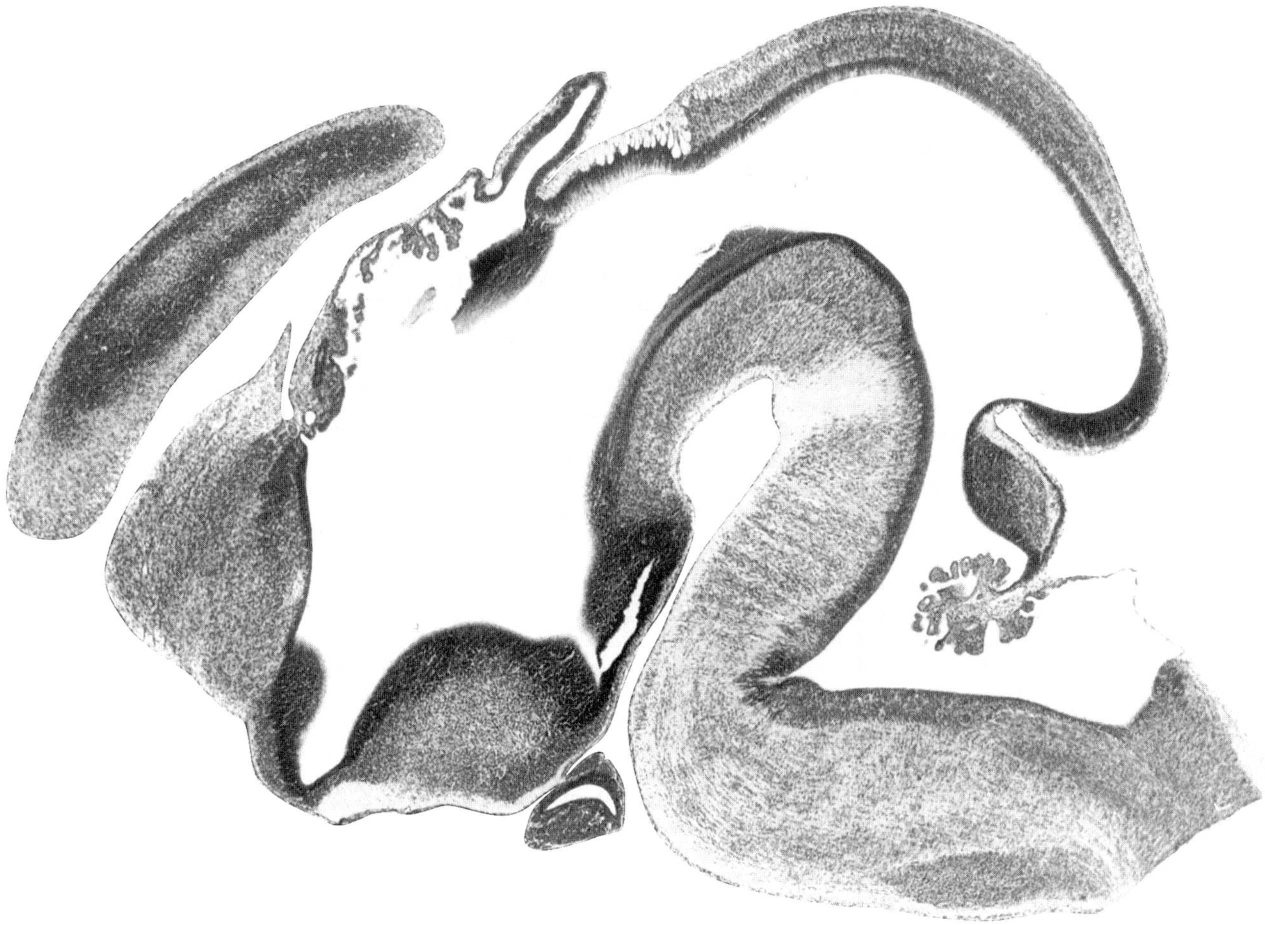

NERVES AND NUCLEI, CRANIAL (n)

n2c optic chiasma
n2t optic tract
n12n hypoglossal nucleus

PREOPTIC AREA AND OPTIC EPITHELIA (P)

P1 preoptic neuroepithelium
P4 medial preoptic region

SEPTAL AREA (S)

S1 septal neuroepithelium
S3 septum (or differentiating field)
S6 medial septum

THALAMUS (T)

T1 thalamic neuroepithelium

VENTRICLES (v)

vai aqueduct, inferior tectal
vapr aqueduct, pretectal
vas aqueduct, superior tectal
vat aqueduct, tegmental
vlm foramen of Monro
v3e third ventricle, epithalamic
v3h third ventricle, hypothalamic
v3i third ventricle, infundibular recess
v3pi third ventricle, pineal recess
v3pr third ventricle, preoptic recess
v3t third ventricle, thalamic
v4a fourth ventricle, anterior
v4i fourth ventricle, isthmal
v4p fourth ventricle, posterior

PONTINE AREA (X)

X1a anterior pontine neuroepithelium
X1p posterior pontine neuroepithelium
X3a anterior pons (differentiating field)
X3p posterior pons (differentiating field)
X12a anterior precerebellar extramural migration
X17 medial vestibular nucleus
X31v isthmal neuroepithelium, ventral
X33 isthmus (or differentiating field)

MEDULLA (Y)

Y1 medullary neuroepithelium
Y3 medulla (or differentiating field)
Y5 medullary raphe
Y7 prepositus nucleus
Y8 inferior olive

SPINAL CORD (Z)

Z3 spinal cord (or differentiating field)

SAGITTAL FIGURE 2

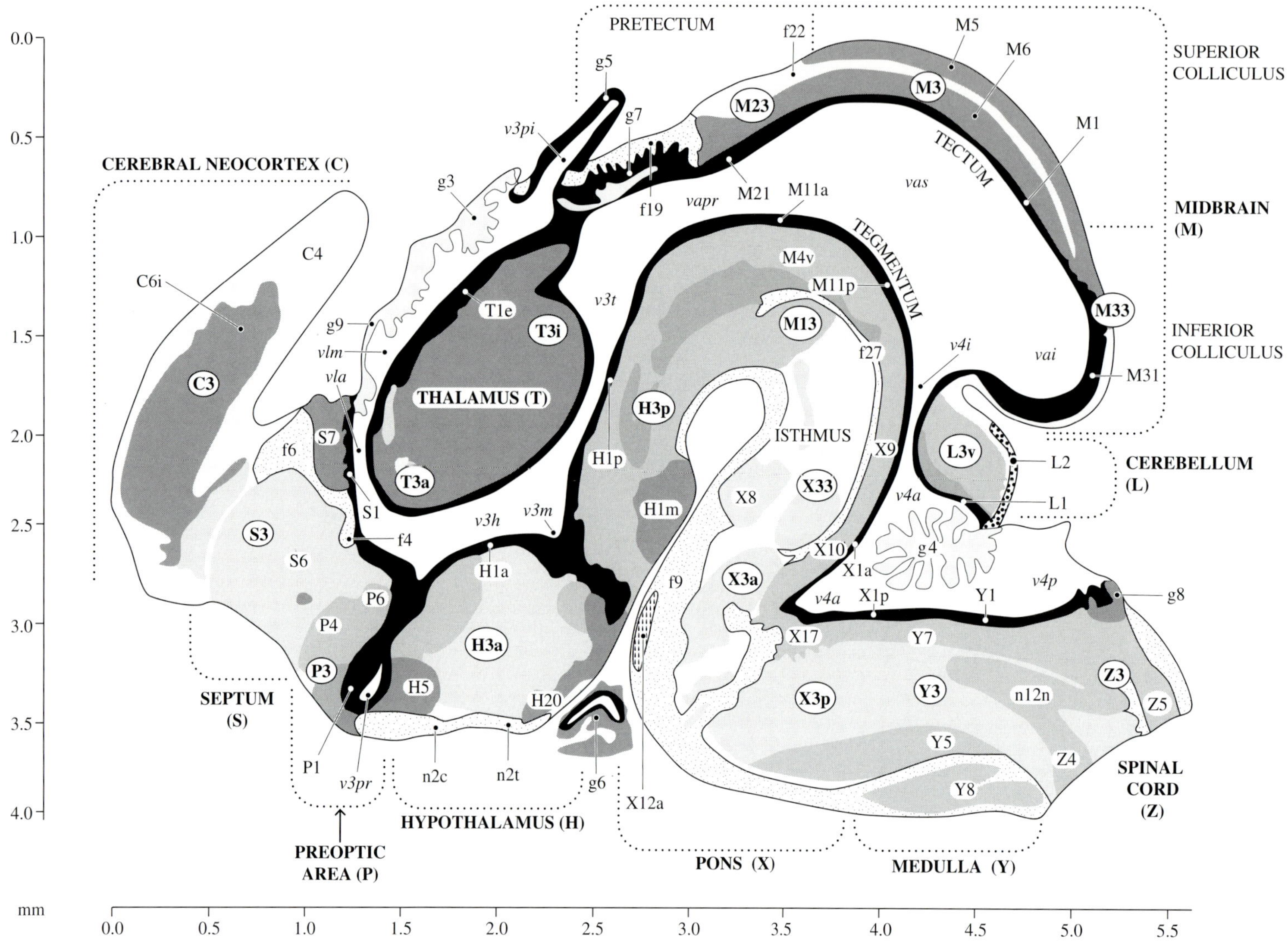

CEREBRAL CORTEX (C)

C3	neocortex (or differentiating field)
C4	molecular layer (or primordial plexiform layer)
C6i	cortical plate, infragranular part

FIBER TRACTS (f)

f4	anterior commissure
f6	fornix
f9	cerebral peduncle
f19	posterior commissure
f22	commissure of superior colliculus
f27	medial longitudinal fasciculus

GLANDS AND PARANEURAL STRUCTURES (g)

g3	choroid plexus of third ventricle
g4	choroid plexus of fourth ventricle
g5	pineal gland (primordium)
g6	pituitary gland (primordium)
g7	subcommissural organ
g8	area postrema
g9	lamina terminalis

HYPOTHALAMUS (H)

H1a	anterior hypothalamic neuroepithelium
H1m	mammillary neuroepithelium
H1p	posterior hypothalamic neuroepithelium
H3a	anterior hypothalamus (or differentiating field)
H3p	posterior hypothalamus (or differentiating field)
H5	suprachiasmatic nucleus
H20	arcuate nucleus

CEREBELLUM (L)

L1	cerebellar neuroepithelium
L2	external germinal layer
L3v	cerebellar vermis (or differentiating field)

MIDBRAIN (M)

M1	superior tectal neuroepithelium
M3	superior colliculus (or differentiating field)
M4v	ventral periaqueductal gray
M5	superior colliculus, superficial layers
M6	superior colliculus, deep layers
M11a	anterior tegmental neuroepithelium
M11p	posterior tegmental neuroepithelium
M13	tegmentum (or differentiating field)
M21	pretectal neuroepithelium
M23	pretectum (or differentiating field)
M31	inferior tectal neuroepithelium
M33	inferior colliculus (or differentiating field)

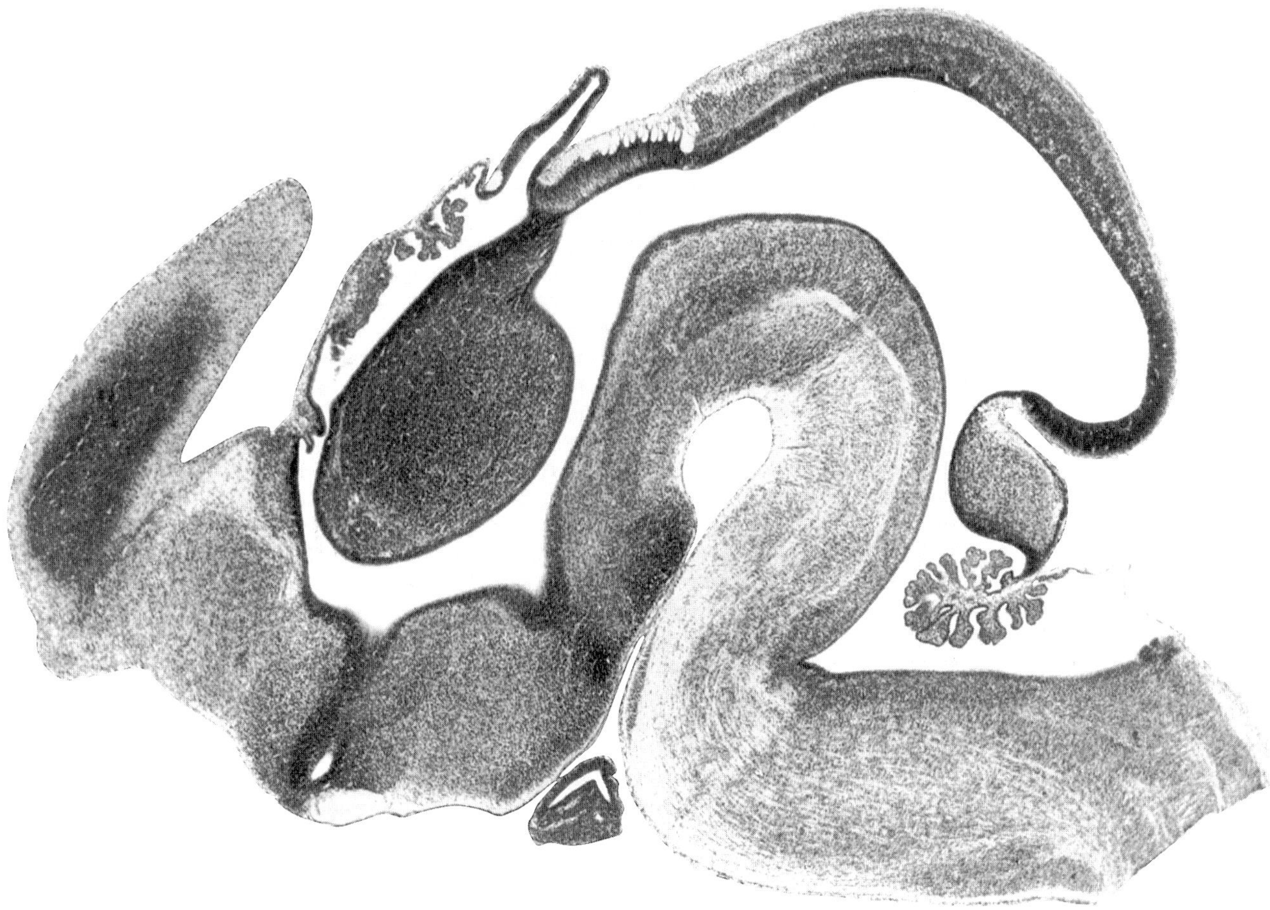

NERVES AND NUCLEI, CRANIAL (n)

n2c optic chiasma
n2t optic tract
n12n hypoglossal nucleus

PREOPTIC AREA AND OPTIC EPITHELIA (P)

P1 preoptic neuroepithelium
P3 preoptic area (or differentiating field)
P4 medial preoptic region
P6 superior preoptic region

SEPTAL AREA (S)

S1 septal neuroepithelium
S3 septum (or differentiating field)
S6 medial septum
S7 diagonal band (or differentiating field)

THALAMUS (T)

T1e epithalamic neuroepithelium
T3a anterior thalamus (or differentiating field)
T3i intermediate thalamus (or differentiating field)

VENTRICLES (v)

vai aqueduct, inferior tectal
vapr aqueduct, pretectal
vas aqueduct, superior tectal
v1a lateral ventricle, anterior horn
vlm foramen of Monro
v3h third ventricle, hypothalamic
v3m third ventricle, mammillary recess
v3pi third ventricle, pineal recess
v3pr third ventricle, preoptic recess
v3t third ventricle, thalamic
v4a fourth ventricle, anterior
v4i fourth ventricle, isthmal
v4p fourth ventricle, posterior

PONTINE AREA (X)

X1a anterior pontine neuroepithelium
X1p posterior pontine neuroepithelium
X3a anterior pons (or differentiating field)
X3p posterior pons (or differentiating field)
X8 superior central raphe nucleus
X9 dorsal tegmental nucleus
X10 ventral tegmental nucleus
X12a anterior precerebellar extramural migration
X17 medial vestibular nucleus
X33 isthmus (or differentiating field)

MEDULLA (Y)

Y1 medullary neuroepithelium
Y3 medulla (or differentiating field)
Y5 medullary raphe
Y7 prepositus nucleus
Y8 inferior olive

SPINAL CORD (Z)

Z3 spinal cord (or differentiating field)
Z4 ventral horn
Z5 dorsal horn

SAGITTAL FIGURE 3

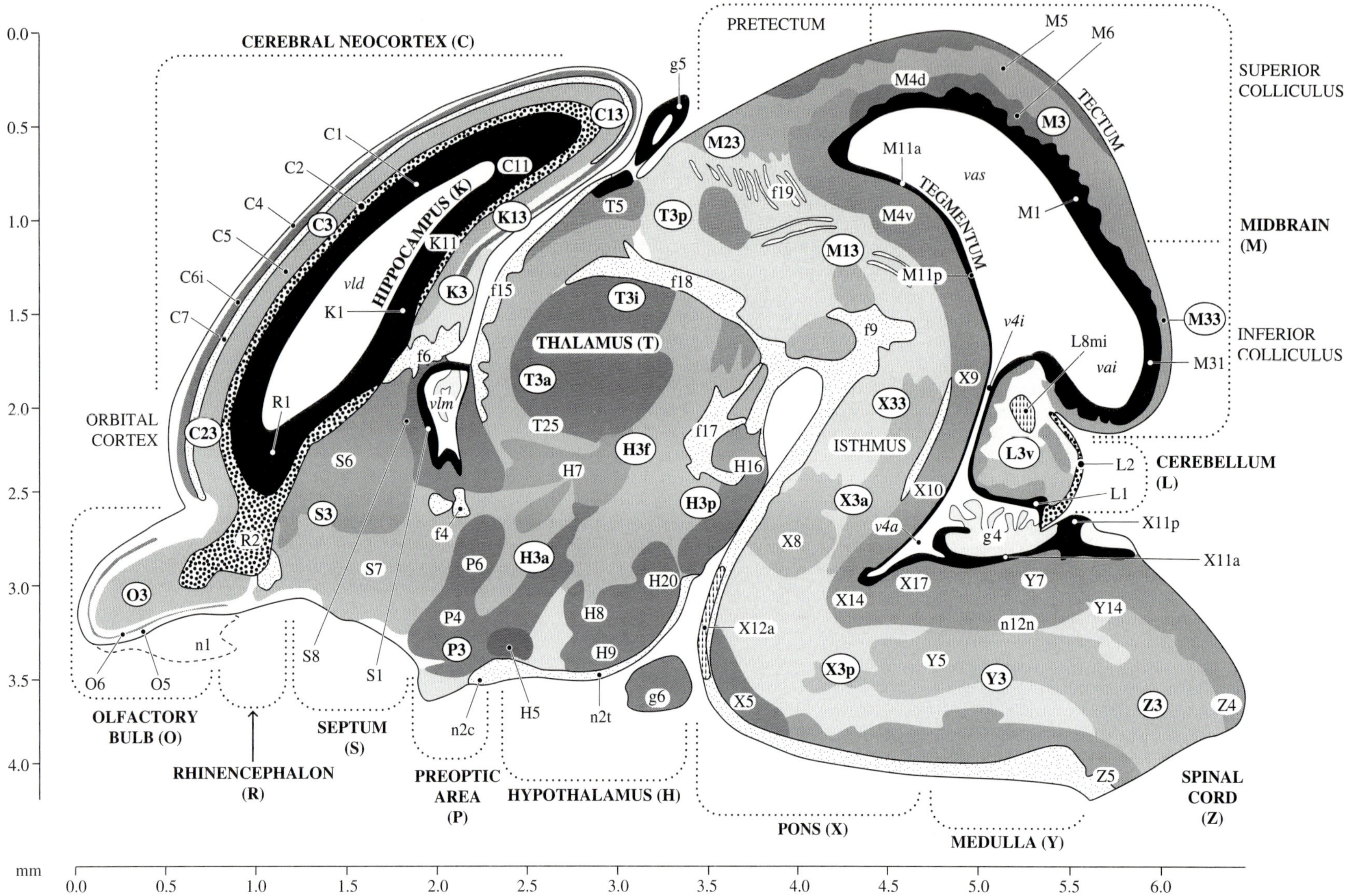

CEREBRAL CORTEX (C)

- C1 neocortical neuroepithelium
- C2 neocortical subventricular zone
- **C3** neocortex (or differentiating field)
- C4 molecular layer (or primordial plexiform layer)
- C5 intermediate zone
- C6i cortical plate, infragranular part
- C7 white matter
- C11 cingulate cortex neuroepithelium
- **C13** cingulate cortex (or differentiating field)
- **C23** orbital and insular area (or differentiating field)

FIBER TRACTS (f)

- f4 anterior commissure
- f6 fornix
- f9 cerebral peduncle
- f15 stria medullaris
- f17 mammillothalamic tract
- f18 habenulopeduncular tract
- f19 posterior commissure

GLANDS AND PARANEURAL STRUCTURES (g)

- g4 choroid plexus of fourth ventricle
- g5 pineal gland (primordium)
- g6 pituitary gland (primordium)

HYPOTHALAMUS (H)

- **H3a** anterior hypothalamus (or differentiating field)
- **H3f** fields of Forel (or differentiating field)
- **H3p** posterior hypothalamus (or differentiating field)
- H5 suprachiasmatic nucleus
- H7 zona incerta
- H8 dorsomedial nucleus
- H9 ventromedial nucleus
- H16 medial mammillary nucleus
- H20 arcuate nucleus

HIPPOCAMPAL FORMATION (K)

- K1 hippocampal (Ammon's horn) neuroepithelium
- **K3** hippocampus (or differentiating field)
- K11 subicular neuroepithelium
- **K13** subicular area (or differentiating field)

CEREBELLUM (L)

- L1 cerebellar neuroepithelium
- L2 external germinal layer
- **L3v** cerebellar vermis (or differentiating field)
- L8mi medial deep nucleus, migration

MIDBRAIN (M)

M1 superior tectal neuroepithelium
M3 superior colliculus (or differentiating field)
M4d dorsal periaqueductal gray
M4v ventral periaqueductal gray
M5 superior colliculus, superficial layers
M6 superior colliculus, deep layers
M11a anterior tegmental neuroepithelium
M11p posterior tegmental neuroepithelium
M13 tegmentum (or differentiating field)
M23 pretectum (or differentiating field)
M31 inferior tectal neuroepithelium
M33 inferior colliculus (or differentiating field)

NERVES AND NUCLEI, CRANIAL (n)

n1 olfactory nerve
n2c optic chiasma
n2t optic tract
n12n hypoglossal nucleus

OLFACTORY BULB (O)

O3 olfactory bulb (or differentiating field)
O5 external plexiform layer
O6 mitral cell layer

PREOPTIC AREA AND OPTIC EPITHELIA (P)

P3 preoptic area (or differentiating field)
P4 medial preoptic region
P6 superior preoptic region

RHINENCEPHALON (R)

R1 rhinencephalic neuroepithelium
R2 rhinencephalic subventricular zone

SEPTAL AREA (S)

S1 septal neuroepithelium
S3 septum (or differentiating field)
S6 medial septum
S7 diagonal band (or differentiating field)
S8 diagonal band (vertical limb)

THALAMUS (T)

T3a anterior thalamus (or differentiating field)
T3i intermediate thalamus (or differentiating field)
T3p posterior thalamus (or differentiating field)
T5 lateral habenular nucleus
T25 reticular nucleus

VENTRICLES (v)

vai aqueduct, inferior tectal
vas aqueduct, superior tectal
vld lateral ventricle, dorsal
vlm foramen of Monro
v4a fourth ventricle, anterior
v4i fourth ventricle, esthmal

PONTINE AREA (X)

X3a anterior pons (or differentiating field)
X3p posterior pons (or differentiating field)
X5 pontine gray nucleus
X8 superior central raphe nucleus
X9 dorsal tegmental nucleus
X10 ventral tegmental nucleus
X11a anterior precerebellar neuroepithelium
X11p posterior precerebellar neuroepithelium
X12a anterior precerebellar extramural migration
X14 spinal vestibular nucleus
X17 medial vestibular nucleus
X33 isthmus (or differentiating field)

MEDULLA (Y)

Y3 medulla (or differentiating field)
Y5 medullary raphe
Y7 prepositus nucleus
Y14 gracile nucleus

SPINAL CORD (Z)

Z3 spinal cord (or differentiating field)
Z4 ventral horn
Z5 dorsal horn

SAGITTAL FIGURE 4

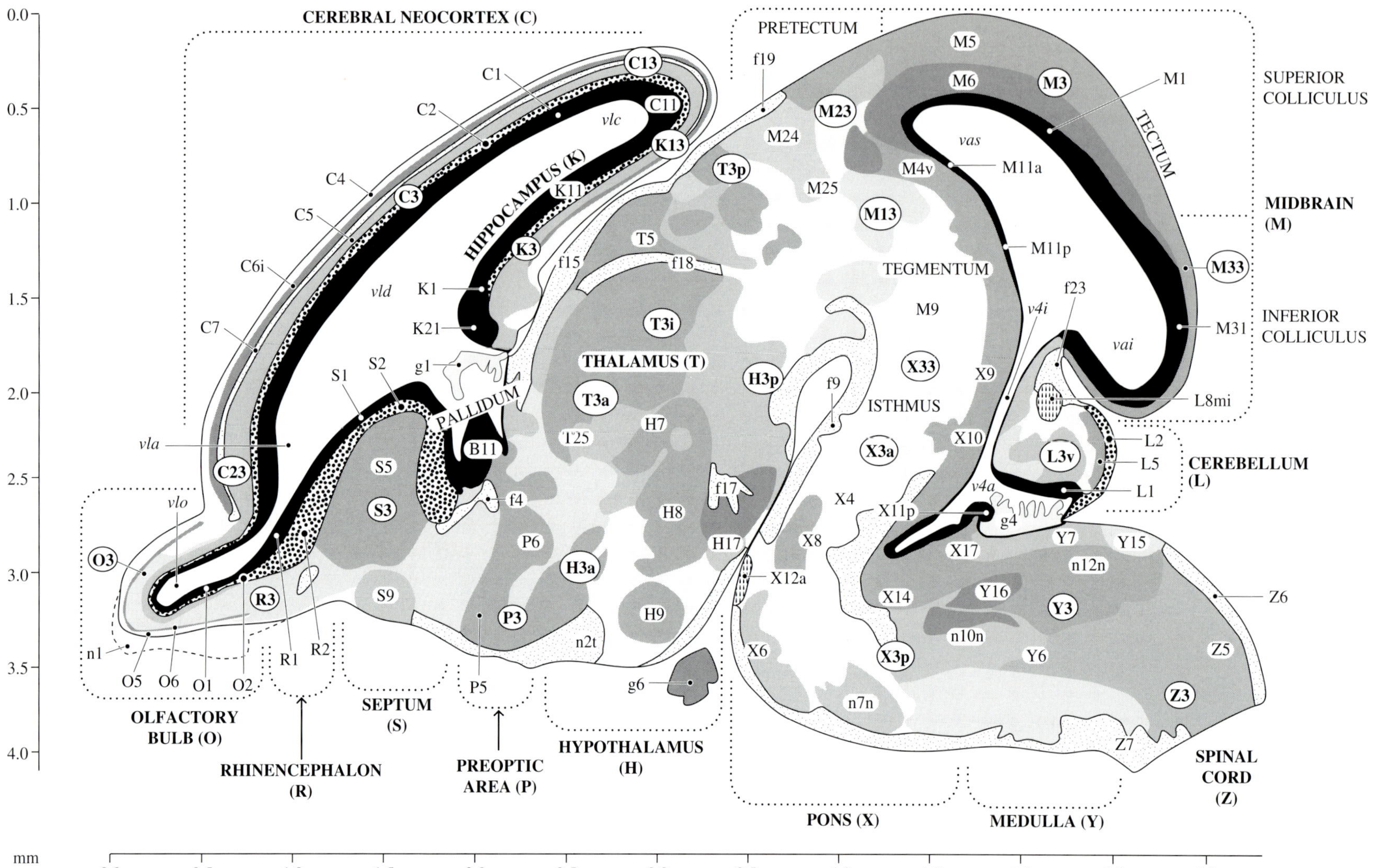

BASAL GANGLIA (B)

B11	pallidal neuroepithelium

CEREBRAL CORTEX (C)

C1	neocortical neuroepithelium
C2	neocortical subventricular zone
C3	neocortex (or differentiating field)
C4	molecular layer (or primordial plexiform layer)
C5	intermediate zone
C6i	cortical plate, infragranular part
C7	white matter
C11	cingulate cortex neuroepithelium
C13	cingulate cortex (or differentiating field)
C23	orbital and insular area (or differentiating field)

FIBER TRACTS (f)

f4	anterior commissure
f9	cerebral peduncle
f15	stria medullaris
f17	mammillothalamic tract
f18	habenulopeduncular tract
f19	posterior commissure
f23	superior cerebellar peduncle

GLANDS AND PARANEURAL STRUCTURES (g)

g1	choroid plexus of lateral ventricle
g4	choroid plexus of fourth ventricle
g6	pituitary gland (primordium)

HYPOTHALAMUS (H)

H3a	anterior hypothalamus (or differentiating field)
H3f	fields of Forel (or differentiating field)
H3p	posterior hypothalamus (or differentiating field)
H7	zona incerta
H8	dorsomedial nucleus
H9	ventromedial nucleus
H17	lateral mammillary nucleus

HIPPOCAMPAL FORMATION (K)

K1	hippocampal (Ammon's horn) neuroepithelium
K3	hippocampus (or differentiating field)
K11	subicular neuroepithelium
K13	subicular area (or differentiating field)
K21	dentate gyrus neuroepithelium

CEREBELLUM (L)

L1	cerebellar neuroepithelium
L2	external germinal layer
L3v	cerebellar vermis (or differentiating field)
L5	Purkinje cell layer
L8mi	medial deep nucleus, migration

MIDBRAIN (M)

M1	superior tectal neuroepithelium
M3	superior colliculus (or differentiating field)
M4v	ventral periaqueductal gray
M5	superior colliculus, superficial layers
M6	superior colliculus, deep layers

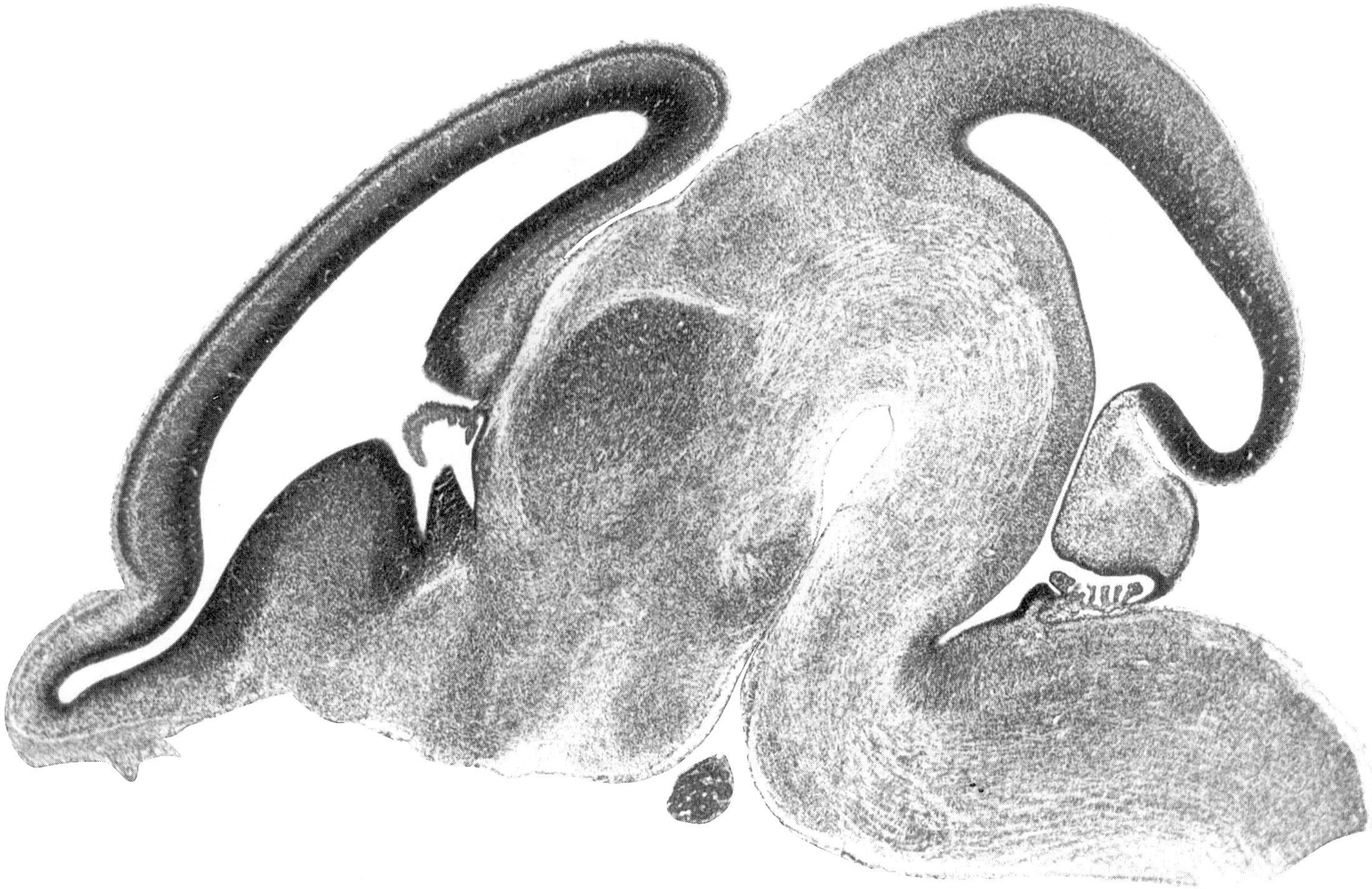

M9 midbrain reticular formation
M11a anterior tegmental neuroepithelium
M11p posterior tegmental neuroepithelium
M13 tegmentum (or differentiating field)
M23 pretectum (or differentiating field)
M24 anterior pretectal nucleus
M25 nucleus of the posterior commissure
M31 inferior tectal neuroepithelium
M33 inferior colliculus (or differentiating field)

NERVES AND NUCLEI, CRANIAL (n)

n1 olfactory nerve
n2t optic tract
n7n facial motor nucleus
n10n dorsal motor nucleus of vagus
n12n hypoglossal nucleus

OLFACTORY BULB (O)

O1 olfactory bulb neuroepithelium
O2 olfactory bulb subventricular zone
O3 olfactory bulb (or differentiating field)
O5 external plexiform layer
O6 mitral cell layer

PREOPTIC AREA AND OPTIC EPITHELIA (P)

P3 preoptic area (or differentiating field)
P5 anterolateral preoptic area
P6 superior preoptic region

RHINENCEPHALON (R)

R1 rhinencephalic neuroepithelium
R2 rhinencephalic subventricular zone
R3 rhinencephalon (or differentiating field)

SEPTAL AREA (S)

S1 septal neuroepithelium
S2 septal subventricular zone
S3 septum (or differentiating field)
S5 lateral septum
S9 diagonal band (horizontal limb)

THALAMUS (T)

T3a anterior thalamus (or differentiating field)
T3i intermediate thalamus (or differentiating field)
T3p posterior thalamus (or differentiating field)
T5 lateral habenular nucleus
T25 reticular nucleus

VENTRICLES (v)

vai aqueduct, inferior tectal
vas aqueduct, superior tectal
vla lateral ventricle, anterior horn
vlc lateral ventricle, central
vld lateral ventricle, dorsal
vlo olfactory ventricle
v4a fourth ventricle, anterior
v4i fourth ventricle, isthmal

PONTINE AREA (X)

X3a anterior pons (or differentiating field)
X3p posterior pons (or differentiating field)
X4 dorsal raphe nucleus
X6 pontine reticulotegmental nucleus
X8 superior central raphe nucleus
X9 dorsal tegmental nucleus
X10 ventral tegmental nucleus
X11p posterior precerebellar neuroepithelium
X12a anterior precerebellar extramural migration
X14 spinal vestibular nucleus
X17 medial vestibular nucleus
X33 isthmus (or differentiating field)

MEDULLA (Y)

Y3 medulla (or differentiating field)
Y6 medullary reticular formation
Y7 prepositus nucleus
Y15 cuneate nucleus
Y16 solitary nucleus

SPINAL CORD (Z)

Z3 spinal cord (or differentiating field)
Z5 dorsal horn
Z6 dorsal funiculus
Z7 ventral funiculus

SAGITTAL FIGURE 5

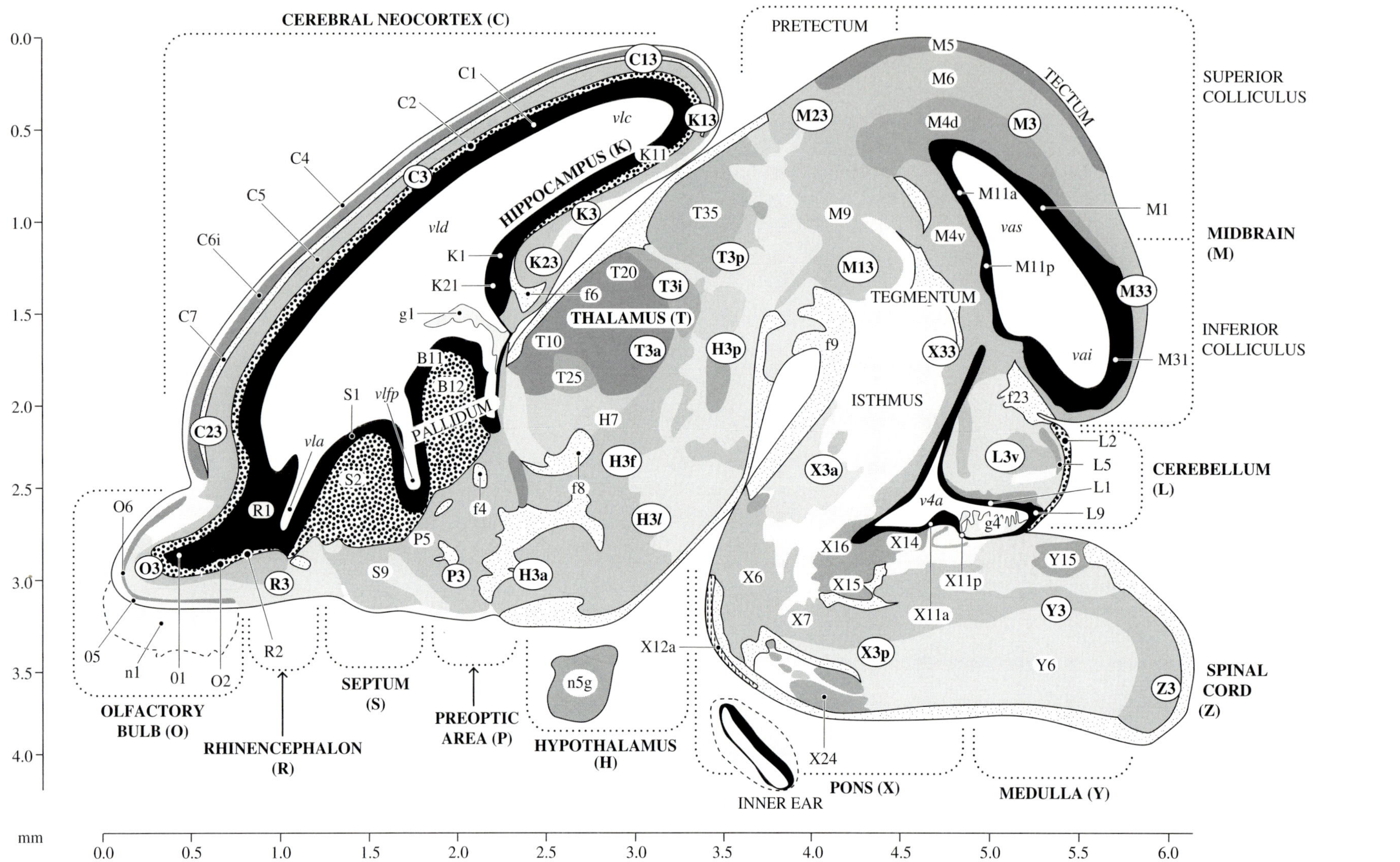

BASAL GANGLIA (B)

- B11 pallidal neuroepithelium
- B12 pallidal subventricular zone

CEREBRAL CORTEX (C)

- C1 neocortical neuroepithelium
- C2 neocortical subventricular zone
- **C3** neocortex (or differentiating field)
- C4 molecular layer (or primordial plexiform layer)
- C5 intermediate zone
- C6i cortical plate, infragranular part
- C7 white matter
- **C13** cingulate cortex (or differentiating field)
- **C23** orbital and insular area (or differentiating field)

FIBER TRACTS (f)

- f4 anterior commissure
- f6 fornix
- f8 internal capsule
- f9 cerebral peduncle
- f23 superior cerebellar peduncle

GLANDS AND PARANEURAL STRUCTURES (g)

- g1 choroid plexus of lateral ventricle
- g4 choroid plexus of fourth ventricle

HYPOTHALAMUS (H)

- **H3a** anterior hypothalamus (or differentiating field)
- **H3f** fields of Forel (or differentiating field)
- **H3*l*** lateral hypothalamus (or differentiating field)
- **H3p** posterior hypothalamus (or differentiating field)
- H7 zona incerta

HIPPOCAMPAL FORMATION (K)

- K1 hippocampal (Ammon's horn) neuroepithelium
- **K3** hippocampus (or differentiating field)
- K11 subicular neuroepithelium
- **K13** subicular area (or differentiating field)
- K21 dentate gyrus neuroepithelium
- **K23** dentate gyrus (or differentiating field)

CEREBELLUM (L)

- L1 cerebellar neuroepithelium
- L2 external germinal layer
- **L3v** cerebellar vermis (or differentiating field)
- L5 Purkinje cell layer
- L9 cerebellar germinal trigone

MIDBRAIN (M)

- M1 superior tectal neuroepithelium
- **M3** superior colliculus (or differentiating field)
- M4d dorsal periaqueductal gray
- M4v ventral periaqueductal gray
- M5 superior colliculus, superficial layers
- M6 superior colliculus, deep layers
- M9 midbrain reticular formation
- M11a anterior tegmental neuroepithelium
- M11p posterior tegmental neuroepithelium
- **M13** tegmentum (or differentiating field)
- **M23** pretectum (or differentiating field)
- M31 inferior tectal neuroepithelium

NERVES AND NUCLEI, CRANIAL (n)

n1 olfactory nerve
n5g trigeminal ganglion

OLFACTORY BULB (O)

O1 olfactory bulb neuroepithelium
O2 olfactory bulb subventricular zone
O3 olfactory bulb (or differentiating field)
O5 external plexiform layer
O6 mitral cell layer

PREOPTIC AREA AND OPTIC EPITHELIA (P)

P3 preoptic area (or differentiating field)
P5 lateral preoptic region

RHINENCEPHALON (R)

R1 rhinencephalic neuroepithelium
R2 rhinencephalic subventricular zone
R3 rhinencephalon (or differentiating field)

SEPTAL AREA (S)

S1 septal neuroepithelium
S2 septal subventricular zone
S9 diagonal band (horizontal limb)

THALAMUS (T)

T3a anterior thalamus (or differentiating field)
T3i intermediate thalamus (or differentiating field)
T3p posterior thalamus (or differentiating field)
T10 anterodorsal nucleus
T20 laterodorsal nucleus
T25 reticular nucleus
T35 posterior nuclear complex

VENTRICLES (v)

vai aqueduct, inferior
vas aqueduct, superior
vla lateral ventricle, anterior horn
vlc lateral ventricle, central
vld lateral ventricle, dorsal
vlfp lateral ventricle, pallidal fork
v4a fourth ventricle, anterior

PONTINE AREA (X)

X3a anterior pons (or differentiating field)
X3p posterior pons (or differentiating field)
X6 pontine reticulotegmental nucleus
X7 pontine reticular formation
X11a anterior precerebellar neuroepithelium
X11p posterior precerebellar neuroepithelium
X12a anterior precerebellar extramural migration
X14 spinal vestibular nucleus
X15 lateral vestibular nucleus
X16 superior vestibular nucleus
X24 superior olivary nucleus
X33 isthmus (or differentiating field)

MEDULLA (Y)

Y3 medulla (or differentiating field)
Y6 medullary reticular formation
Y15 cuneate nucleus

SPINAL CORD (Z)

Z3 spinal cord (or differentiating field)

SAGITTAL FIGURE 6

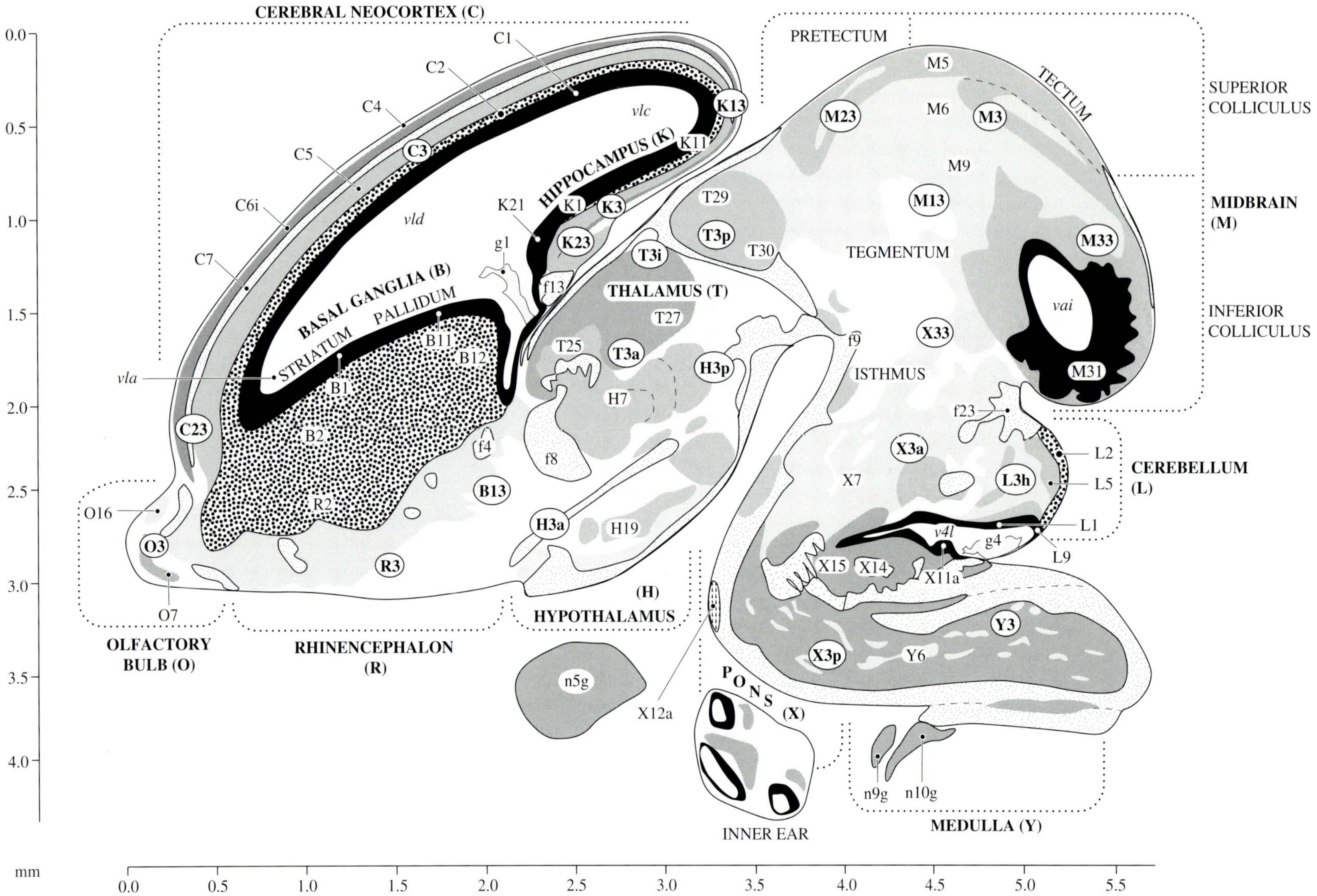

BASAL GANGLIA (B)

- B1 striatal neuroepithelium
- B2 striatal subventricular zone
- B11 pallidal neuroepithelium
- B12 pallidal subventricular zone
- **B13** pallidum (or differentiating field)

CEREBRAL CORTEX (C)

- C1 neocortical neuroepithelium
- C2 neocortical subventricular zone
- **C3** neocortex (or differentiating field)
- C4 molecular layer (or primordial plexiform layer)
- C5 intermediate zone
- C6i cortical plate, infragranular part
- C7 white matter
- **C23** orbital and insular area (or differentiating field)

FIBER TRACTS (f)

- f4 anterior commissure
- f8 internal capsule
- f9 cerebral peduncle
- f13 fimbria
- f23 superior cerebellar peduncle

GLANDS AND PARANEURAL STRUCTURES (g)

- g1 choroid plexus of lateral ventricle
- g4 choroid plexus of fourth ventricle

HYPOTHALAMUS (H)

- **H3a** anterior hypothalamus (or differentiating field)
- **H3p** posterior hypothalamus (or differentiating field)
- H7 zona incerta
- H19 subthalamic nucleus

HIPPOCAMPAL FORMATION (K)

- K1 hippocampal (Ammon's horn) neuroepithelium
- **K3** hippocampus (or differentiating field)
- K11 subicular neuroepithelium
- **K13** subicular area (or differentiating field)
- K21 dentate gyrus neuroepithelium
- **K23** dentate gyrus (or differentiating field)

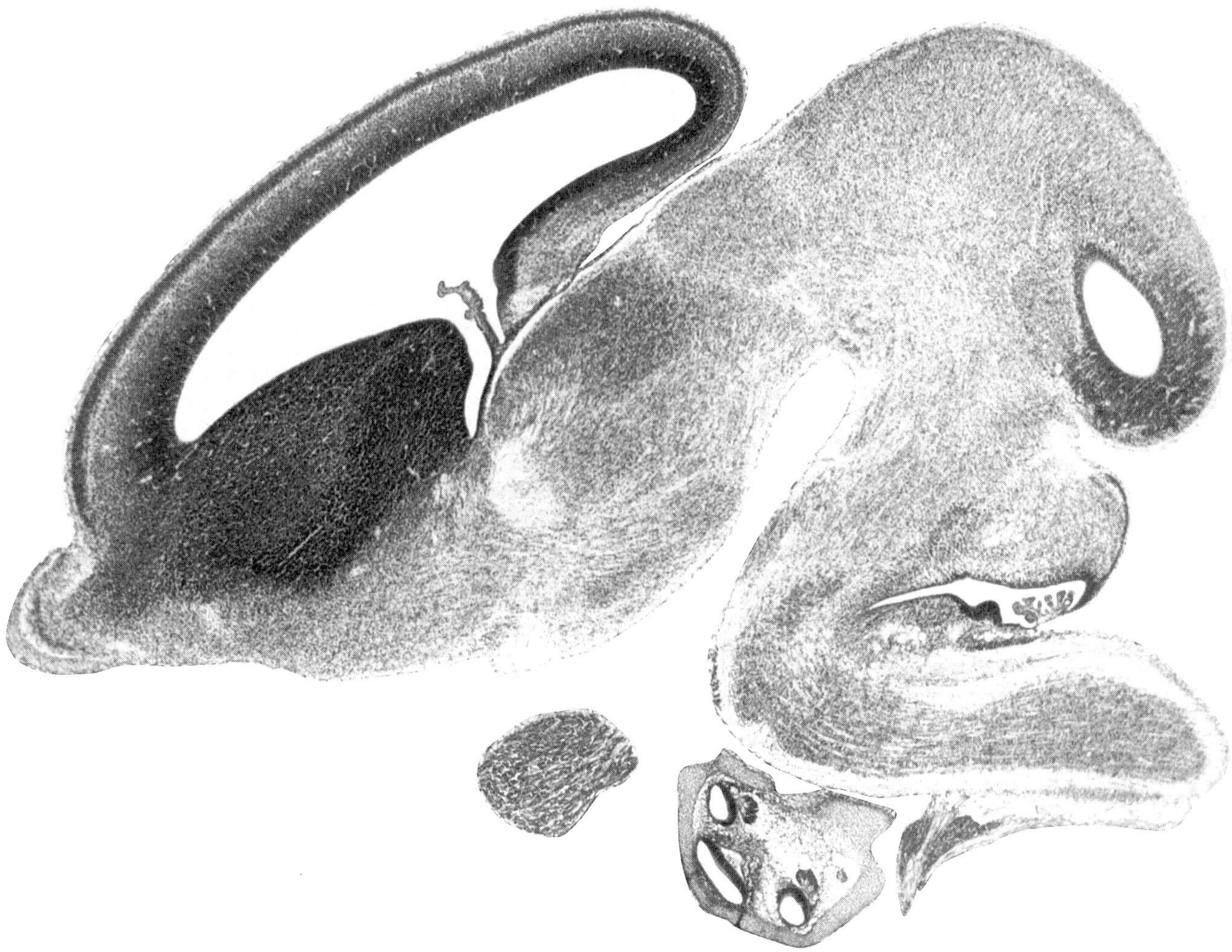

CEREBELLUM (L)

L1 cerebellar neuroepithelium
L2 external germinal layer
L3h cerebellar hemisphere (or differentiating field)
L5 Purkinje cell layer
L9 cerebellar germinal trigone

MIDBRAIN (M)

M3 superior colliculus (or differentiating field)
M5 superior colliculus, superficial layers
M6 superior colliculus, deep layers
M9 midbrain reticular formation
M13 tegmentum (or differentiating field)
M23 pretectum (or differentiating field)
M31 inferior tectal neuroepithelium
M33 inferior colliculus (or differentiating field)

NERVES AND NUCLEI, CRANIAL (n)

n5g trigeminal ganglion
n9g ganglion of glossopharyngeal nerve
n10g ganglion of the vagus

OLFACTORY BULB (O)

O3 olfactory bulb (or differentiating field)
O7 mitral cell layer
O16 mitral cell layer (accessory olfactory bulb)

RHINENCEPHALON (R)

R2 rhinencephalic subventricular zone
R3 rhinencephalon (or differentiating field)

THALAMUS (T)

T3a anterior thalamus (or differentiating field)
T3i intermediate thalamus (or differentiating field)
T3p posterior thalamus (or differentiating field)
T25 reticular nucleus
T27 ventrolateral nuclear complex
T29 dorsal lateral geniculate nucleus
T30 ventral lateral geniculate nucleus

VENTRICLES (v)

vai aqueduct, inferior tectal
vla lateral ventricle, anterior horn
vlc lateral ventricle, central
vld lateral ventricle, dorsal
v4l fourth ventricle, lateral recess

PONTINE AREA (X)

X3a anterior pons (or differentiating field)
X3p posterior pons (or differentiating field)
X7 pontine reticular formation
X11a anterior precerebellar neuroepithelium
X12a anterior precerebellar extramural migration
X14 spinal vestibular nucleus
X15 lateral vestibular nucleus
X33 isthmus (or differentiating field)

MEDULLA (Y)

Y3 medulla (or differentiating field)
Y6 medullary reticular formation

SAGITTAL FIGURE 7

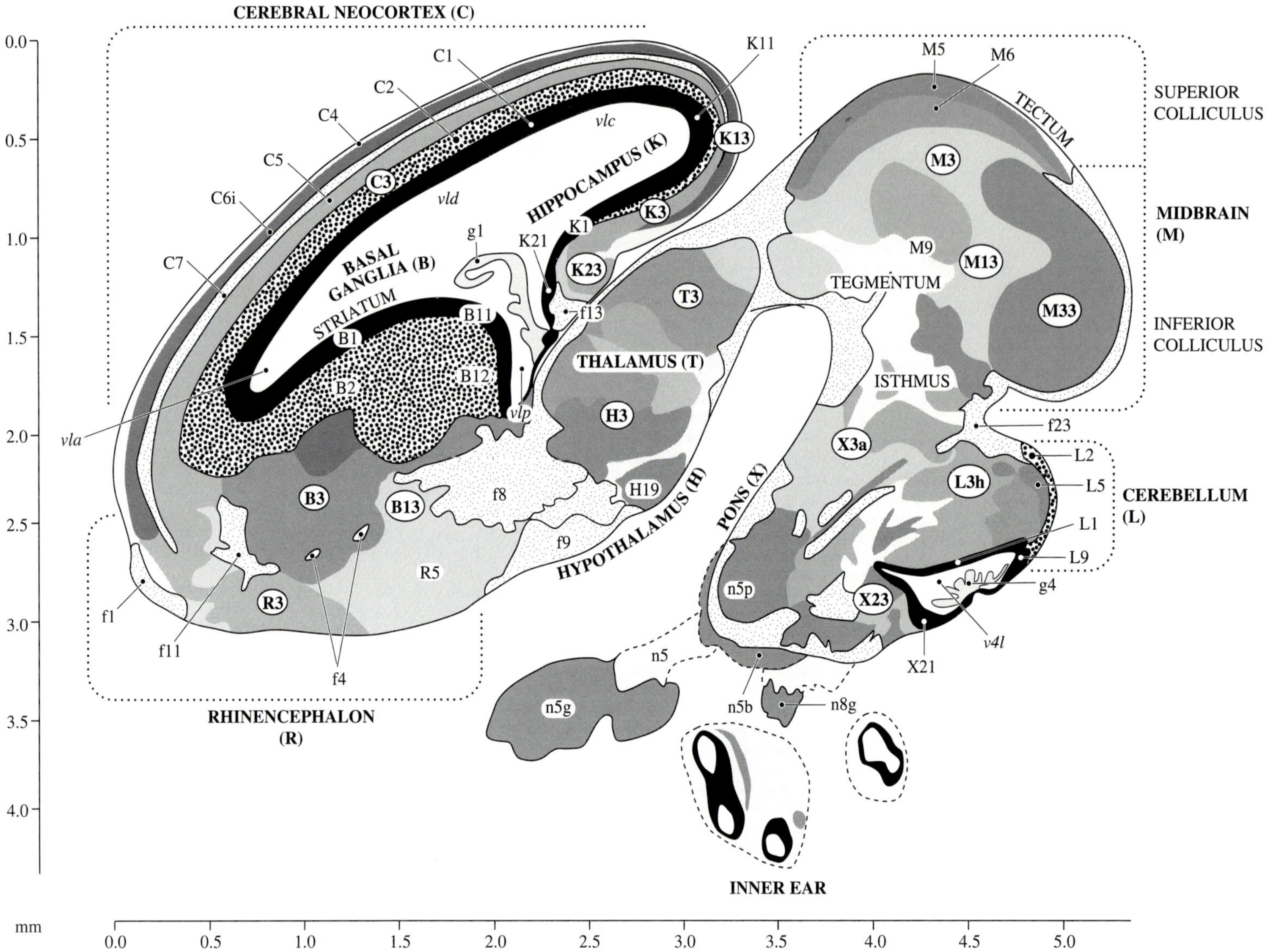

BASAL GANGLIA (B)

- B1 striatal neuroepithelium
- B2 striatal subventricular zone
- **B3** striatum (or differentiating field)
- B11 pallidal neuroepithelium
- B12 pallidal subventricular zone
- **B13** pallidum (or differentiating field)

CEREBRAL CORTEX (C)

- C1 neocortical neuroepithelium
- C2 neocortical subventricular zone
- **C3** neocortex (or differentiating field)
- C4 molecular layer (or primordial plexiform layer)
- C5 intermediate zone
- C6i cortical plate, infragranular part
- C7 white matter

FIBER TRACTS (f)

- f1 lateral olfactory tract
- f4 anterior commissure
- f8 internal capsule
- f9 cerebral peduncle
- f11 external capsule
- f13 fimbria
- f23 superior cerebellar peduncle

GLANDS AND PARANEURAL STRUCTURES (g)

- g1 choroid plexus of lateral ventricle
- g4 choroid plexus of fourth ventricle

HYPOTHALAMUS (H)

- **H3** hypothalamus (or differentiating field)
- H19 subthalamic nucleus

HIPPOCAMPAL FORMATION (K)

- K1 hippocampal (Ammon's horn) neuroepithelium
- **K3** hippocampus (or differentiating field)
- K11 subicular neuroepithelium
- **K13** subicular area (or differentiating field)
- K21 dentate gyrus neuroepithelium
- **K23** dentate gyrus (or differentiating field)

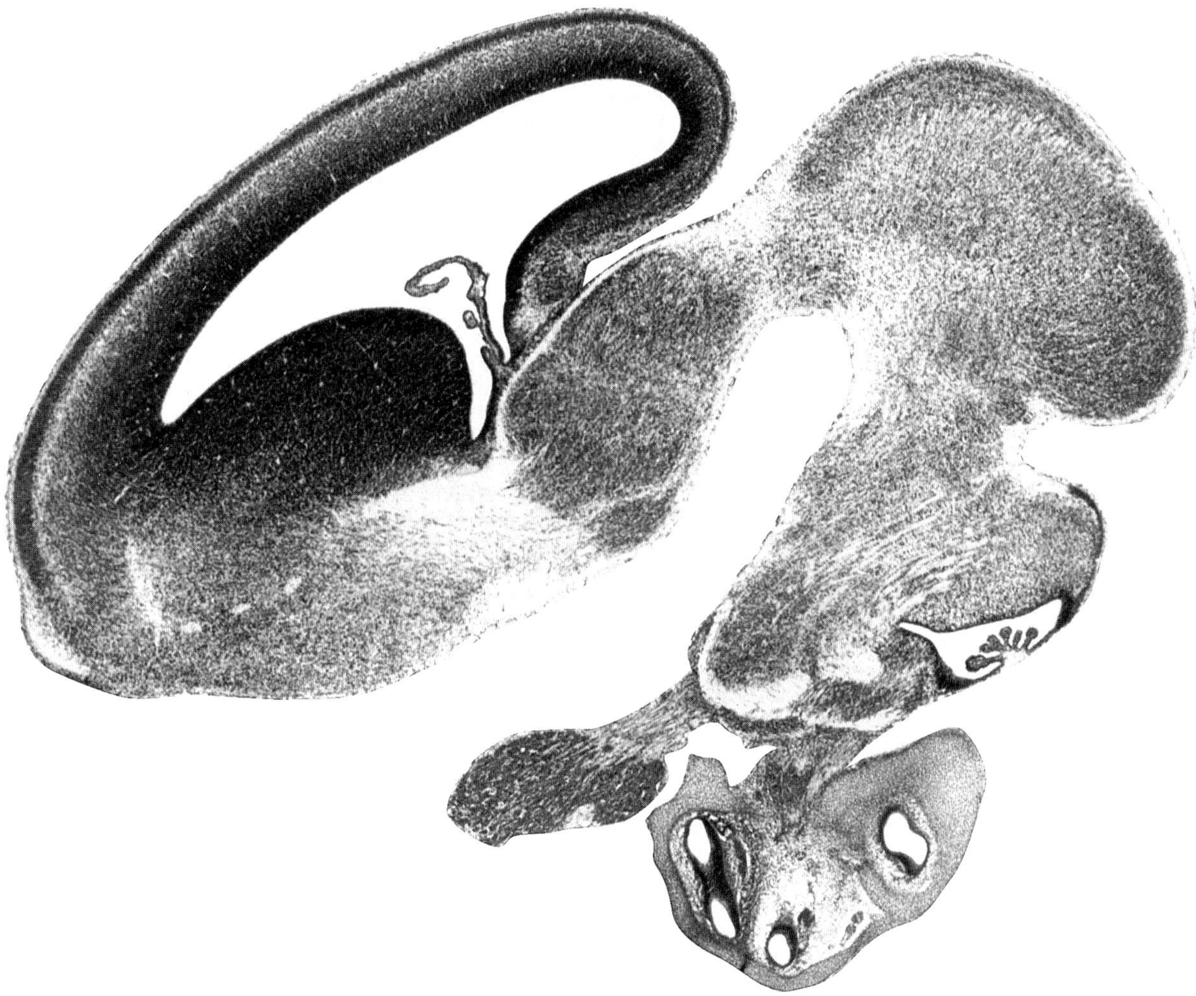

CEREBELLUM (L)

L1 cerebellar neuroepithelium
L2 external germinal layer
L3h cerebellar hemisphere (or differentiating field)
L5 Purkinje cell layer
L9 cerebellar germinal trigone

MIDBRAIN (M)

M3 superior colliculus (or differentiating field)
M5 superior colliculus, superficial layers
M6 superior colliculus, deep layers
M9 midbrain reticular formation
M13 tegmentum (or differentiating field)
M33 inferior colliculus (or differentiating field)

NERVES AND NUCLEI, CRANIAL (n)

n5 trigeminal nerve
n5b trigeminal boundary cap
n5g trigeminal ganglion
n5p principal trigeminal nucleus
n8g vestibular ganglion

RHINENCEPHALON (R)

R3 rhinencephalon (or differentiating field)
R5 endopiriform nucleus

THALAMUS (T)

T3 thalamus (or differentiating field)

VENTRICLES (v)

vla lateral ventricle, anterior horn
vlc lateral ventricle, central
vld lateral ventricle, dorsal
vlp lateral ventricle, posterior horn
v4l fourth ventricle, lateral recess

PONTINE AREA (X)

X3a anterior pons (or differentiating field)
X21 cochlear nuclei neuroepithelium
X23 cochlear nuclei (or differentiating field)

SAGITTAL FIGURE 8

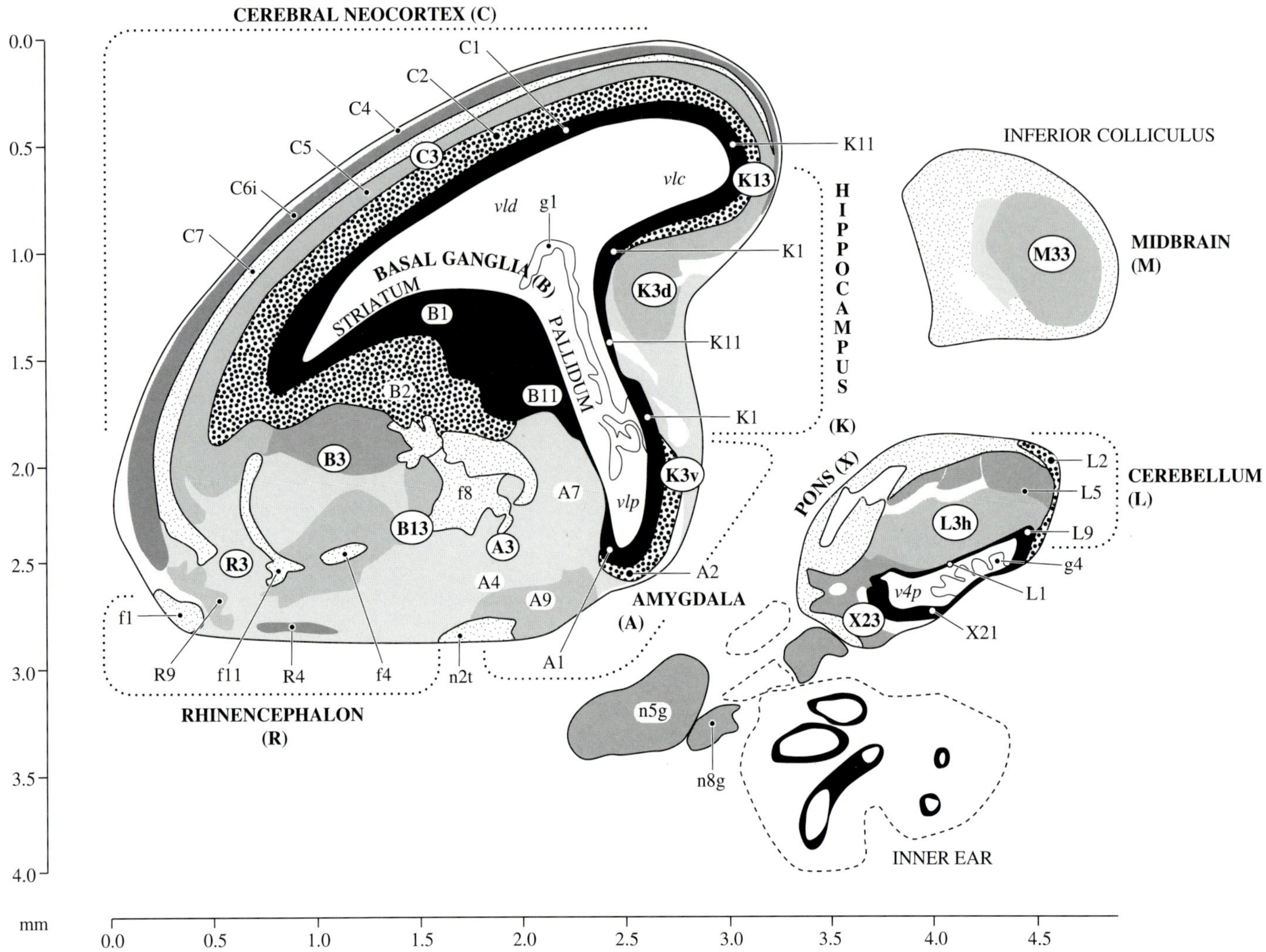

AMYGDALA (A)

- A1 amygdaloid neuroepithelium
- A2 amygdaloid subventricular zone
- **A3** amygdala (or differentiating field)
- A4 anterior amygdaloid area
- A7 basomedial amygdaloid nucleus
- A9 cortical amygdaloid nucleus

BASAL GANGLIA (B)

- B1 striatal neuroepithelium
- B2 striatal subventricular zone
- **B3** striatum (or differentiating field)
- B11 pallidal neuroepithelium
- **B13** pallidum (or differentiating field)

CEREBRAL CORTEX (C)

- C1 neocortical neuroepithelium
- C2 neocortical subventricular zone
- **C3** neocortex (or differentiating field)
- C4 molecular layer (or primordial plexiform layer)
- C5 intermediate zone
- C6i cortical plate, infragranular part
- C7 white matter

FIBER TRACTS (f)

- f1 lateral olfactory tract
- f4 anterior commissure
- f8 internal capsule
- f11 external capsule

GLANDS AND PARANEURAL STRUCTURES (g)

- g1 choroid plexus of lateral ventricle
- g4 choroid plexus of fourth ventricle

HIPPOCAMPAL FORMATION (K)

- K1 hippocampal (Ammon's horn) neuroepithelium
- **K3d** dorsal hippocampus (or differentiating field)
- **K3v** ventral hippocampus (or differentiating field)
- K11 subicular neuroepithelium
- **K13** subicular area (or differentiating field)

CEREBELLUM (L)

- L1 cerebellar neuroepithelium
- L2 external germinal layer
- **L3h** cerebellar hemisphere (or differentiating field)
- L5 Purkinje cell layer
- L9 cerebellar germinal trigone

MIDBRAIN (M)

- **M33** inferior colliculus (or differentiating field)

NERVES AND NUCLEI, CRANIAL (n)

- n2t optic tract
- n5g trigeminal ganglion
- n8g vestibular ganglion

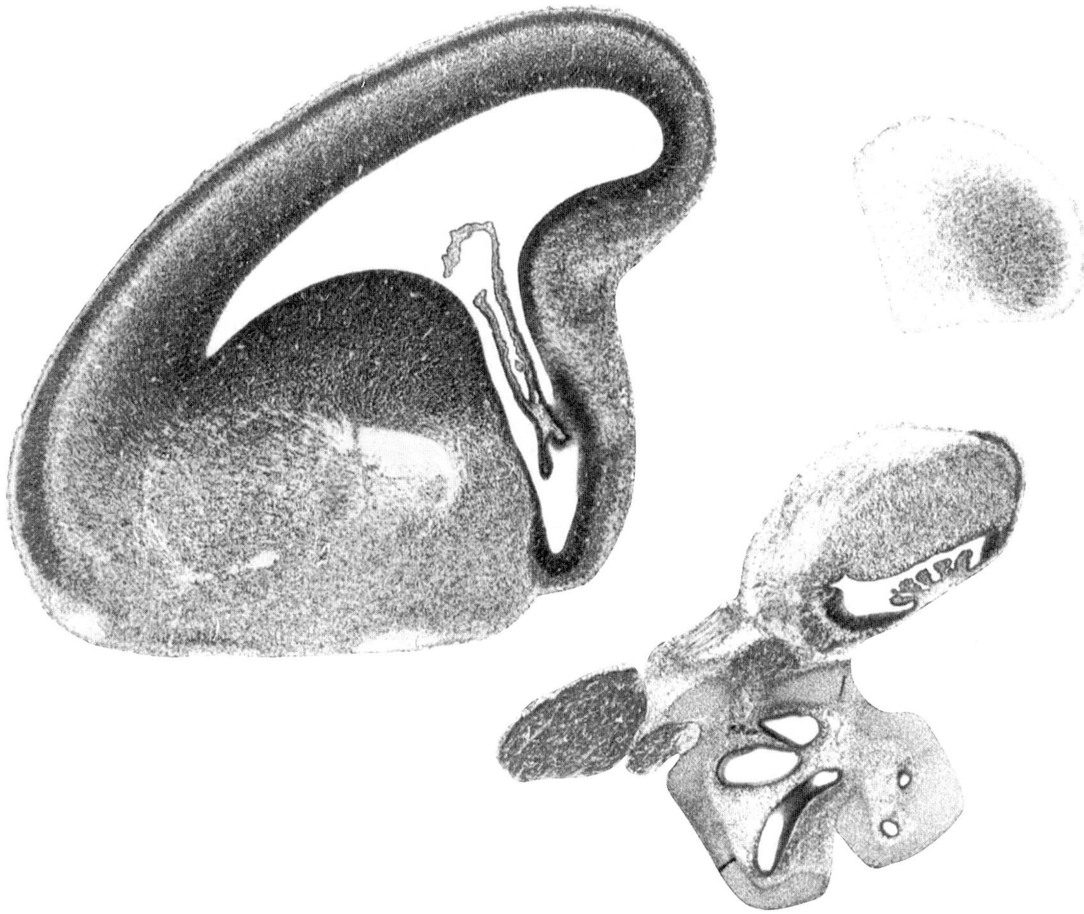

RHINENCEPHALON (R)

R3 rhinencephalon (or differentiating field)
R4 piriform cortex
R9 olfactory tubercle

VENTRICLES (v)

vlc lateral ventricle, central
vld lateral ventricle, dorsal
vlp lateral ventricle, posterior horn
v4p fourth ventricle, posterior

PONTINE AREA (X)

X21 cochlear nuclei neuroepithelium
X23 cochlear nuclei (or differentiating field)

CORONAL

FIGURES and PLATES 1-19

CORONAL FIGURE 1

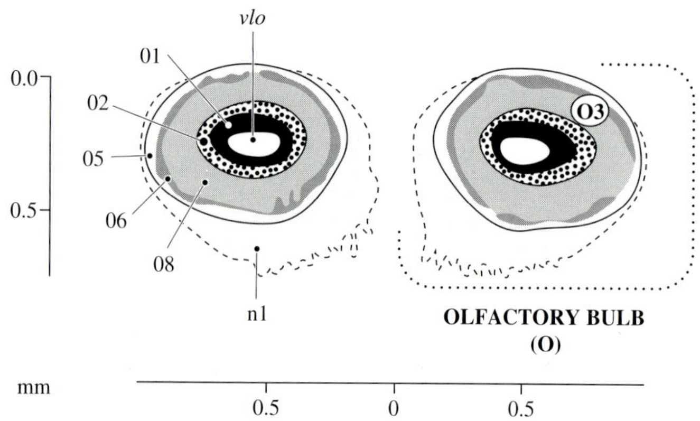

OLFACTORY BULB
(O)

NERVES AND NUCLEI, CRANIAL (n)

n1 olfactory nerve

OLFACTORY BULB (O)

O1 olfactory bulb neuroepithelium
O2 olfactory bulb subventricular zone
O3 olfactory bulb (or differentiating field)
O5 external plexiform layer
O6 mitral cell layer
O8 granular layer

VENTRICLES (v)

vlo olfactory ventricle

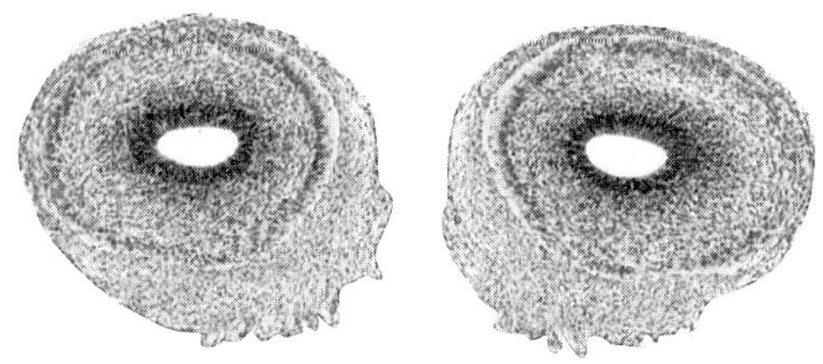

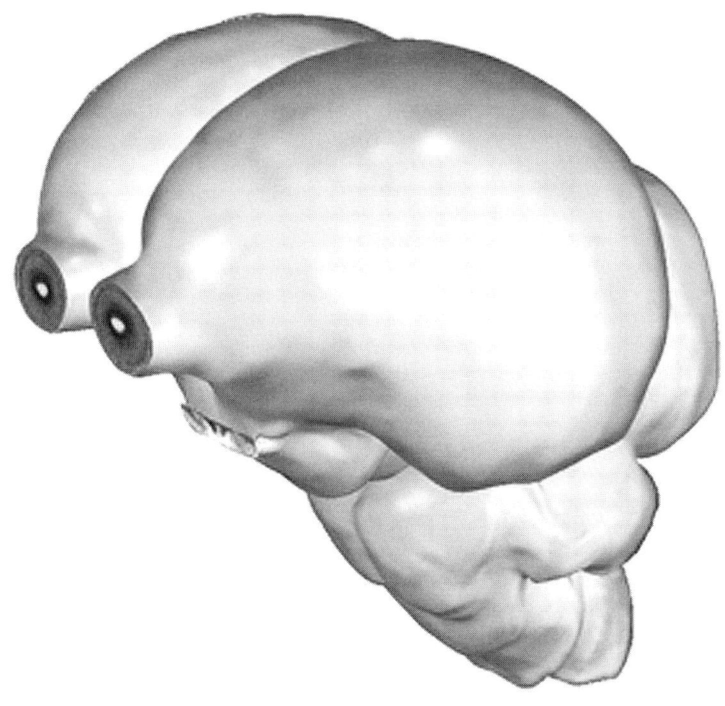

CORONAL FIGURE 2

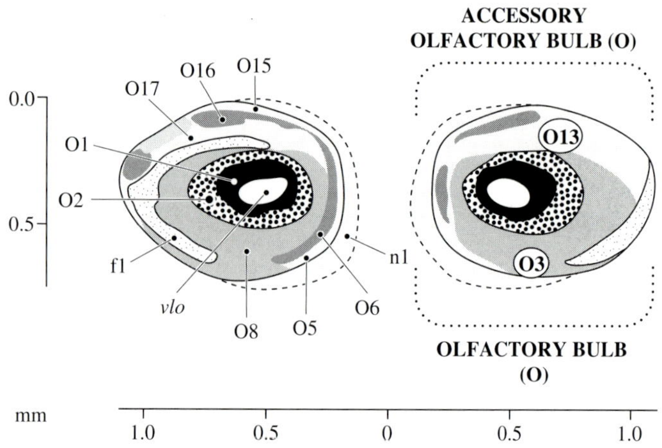

FIBER TRACTS (f)

f1 lateral olfactory tract

NERVES AND NUCLEI, CRANIAL (n)

n1 olfactory nerve

OLFACTORY BULB (O)

O1 olfactory bulb neuroepithelium
O2 olfactory bulb subventricular zone
O3 olfactory bulb (or differentiating field)
O5 external plexiform layer
O6 mitral cell layer
O8 granular layer
O13 accessory olfactory bulb (or differentiating field)
O15 external plexiform layer (accessory olfactory bulb)
O16 mitral cell ayer (accessory olfactory bulb)
O17 internal plexiform layer (accessory olfactory bulb)

VENTRICLES (v)

vlo olfactory ventricle

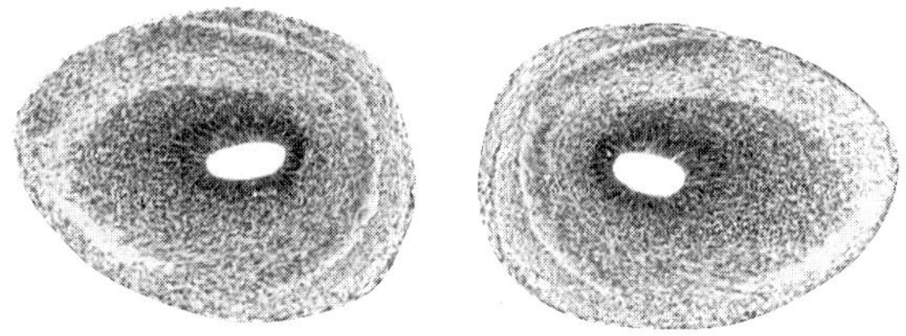

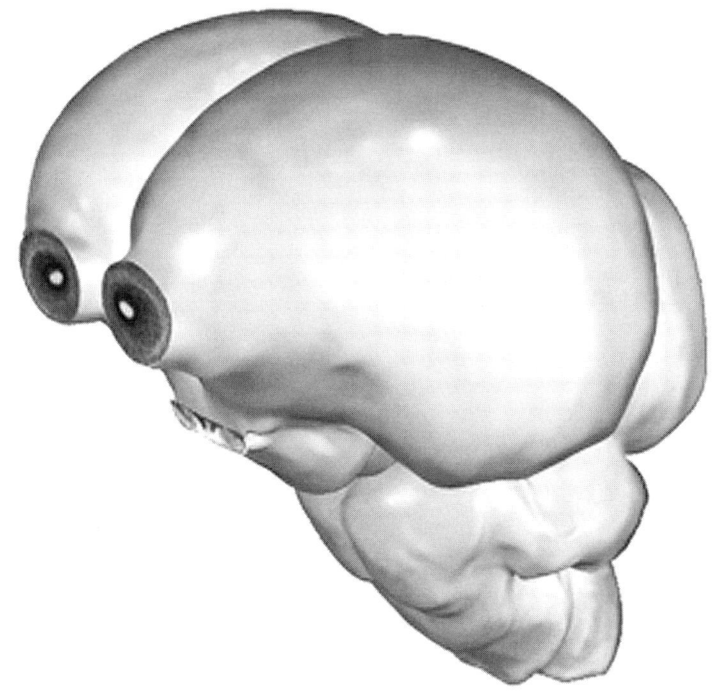

CORONAL FIGURE 3

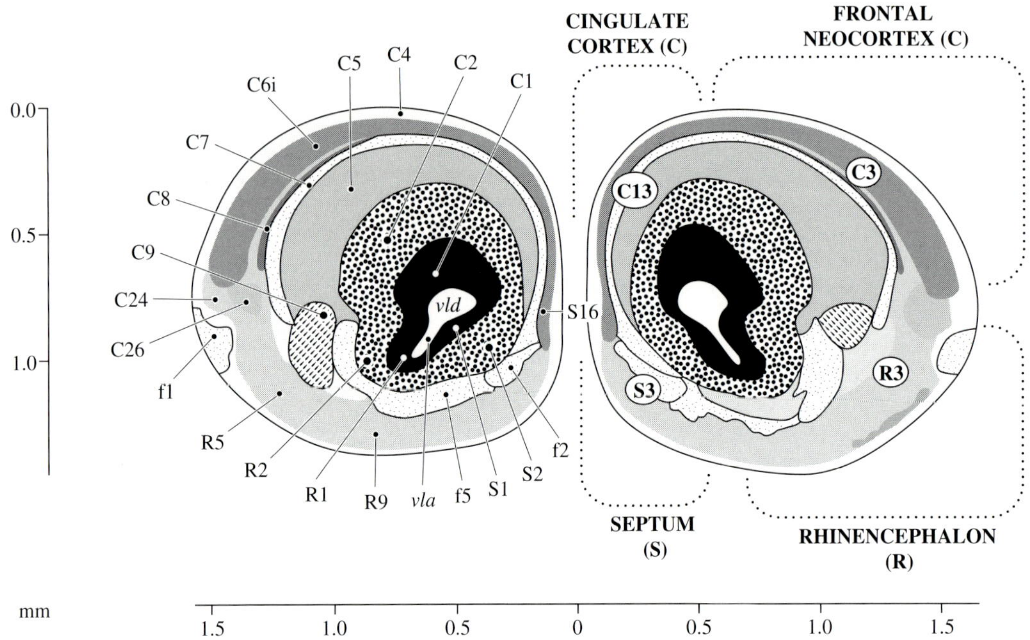

CEREBRAL CORTEX (C)

C1	neocortical neuroepithelium
C2	neocortical subventricular zone
C3	neocortex (or differentiating field)
C4	molecular layer (or primordial plexiform layer)
C5	intermediate zone
C6i	cortical plate, infragranular part
C7	white matter
C8	subplate
C9	lateral migratory stream
C13	cingulate cortex (or differentiating field)
C24	insular cortex
C26	claustrum

FIBER TRACTS (f)

f1	lateral olfactory tract
f2	medial olfactory tract
f5	medial forebrain bundle

RHINENCEPHALON (R)

R1	rhinencephalic neuroepithelium
R2	rhinencephalic subventricular zone
R3	rhinencephalon (or differentiating field)
R5	endopiriform nucleus
R9	olfactory tubercle

SEPTAL AREA (S)

S1	septal neuroepithelium
S2	septal subventricular zone
S3	septum (or differentiating field)
S16	taenia tecta

VENTRICLES (v)

vla	lateral ventricle, anterior horn
vld	lateral ventricle, dorsal

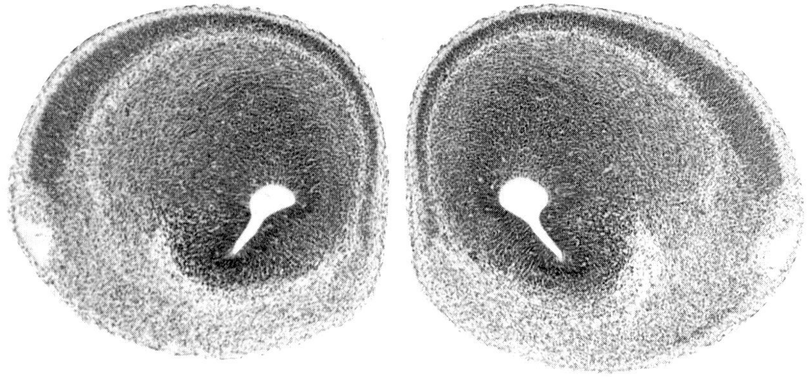

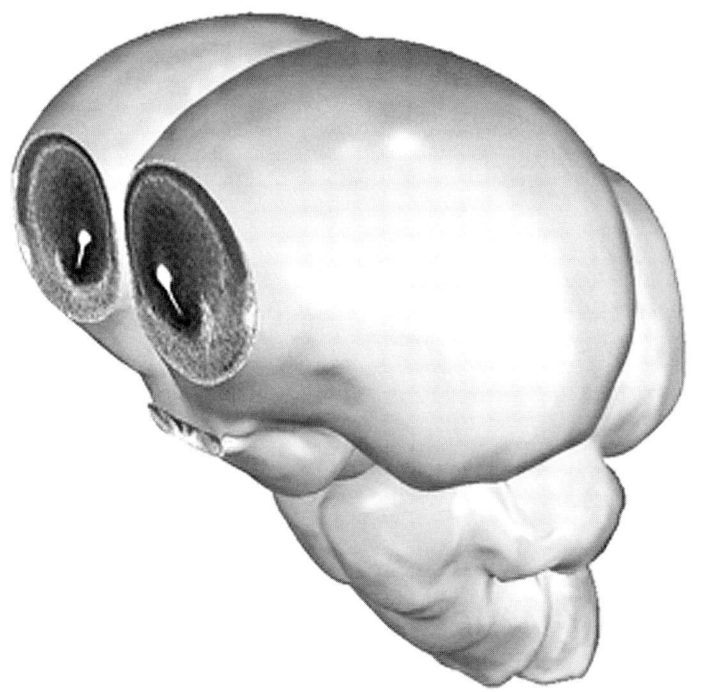

CORONAL FIGURE 4

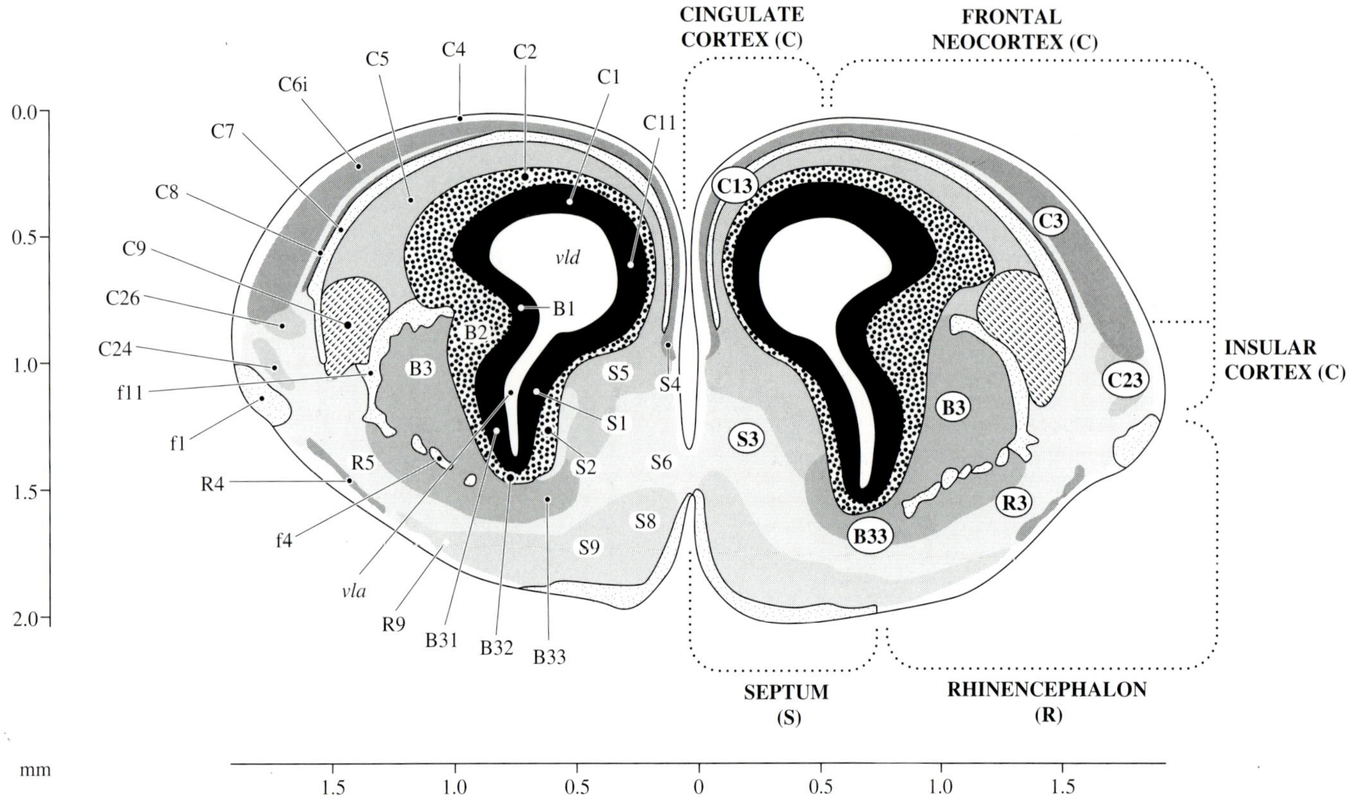

BASAL GANGLIA (B)

- B1 striatal neuroepithelium
- B2 striatal subventricular zone
- **B3** striatum (or differentiating field)
- B31 accumbens neuroepithelium
- B32 accumbens subventricular zone
- **B33** nucleus accumbens (or differentiating field)

CEREBRAL CORTEX (C)

- C1 neocortical neuroepithelium
- C2 neocortical subventricular zone
- **C3** neocortex (or differentiating field)
- C4 molecular layer (or primordial plexiform layer)
- C5 intermediate zone
- C6i cortical plate, infragranular part
- C7 white matter
- C8 subplate
- C9 lateral migratory stream
- C11 cingulate cortex neuroepithelium
- **C13** cingulate cortex (or differentiating field)
- **C23** orbital and insular area (or differentiating field)
- C24 insular cortex
- C26 claustrum

FIBER TRACTS (f)

- f1 lateral olfactory tract
- f4 anterior commissure
- f11 external capsule

RHINENCEPHALON (R)

- **R3** rhinencephalon (or differentiating field)
- R4 piriform cortex
- R5 endopiriform nucleus
- R9 olfactory tubercle

SEPTAL AREA (S)

- S1 septal neuroepithelium
- S2 septal subventricular zone
- **S3** septum (or differentiating field)
- S4 indesium griseum
- S5 lateral septum
- S6 medial septum
- S8 diagonal band (vertical limb)
- S9 diagonal band (horizontal limb)

VENTRICLES (v)

- *vla* lateral ventricle, anterior horn
- *vld* lateral ventricle, dorsal

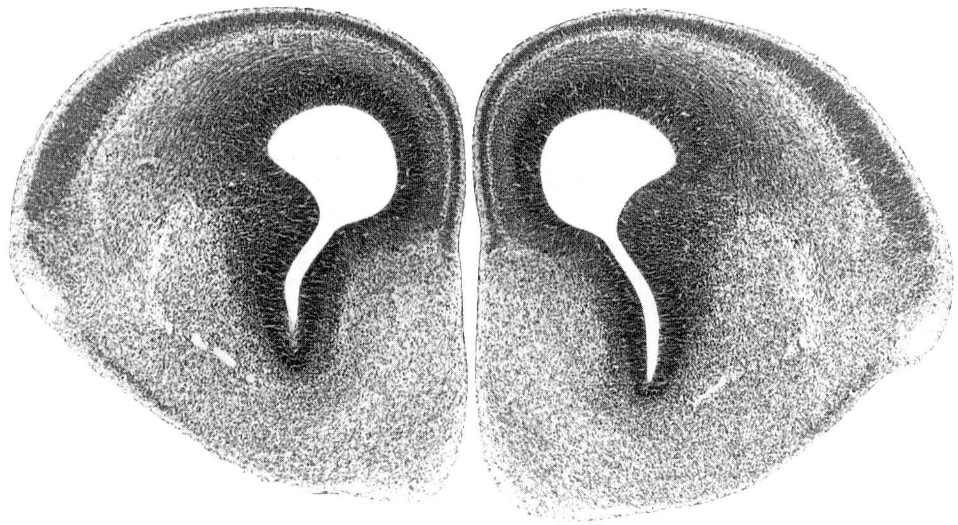

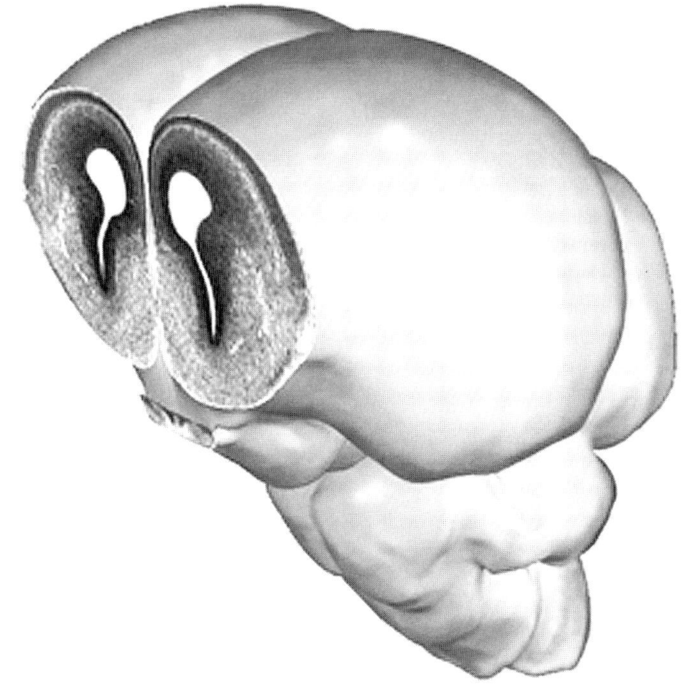

CORONAL FIGURE 6

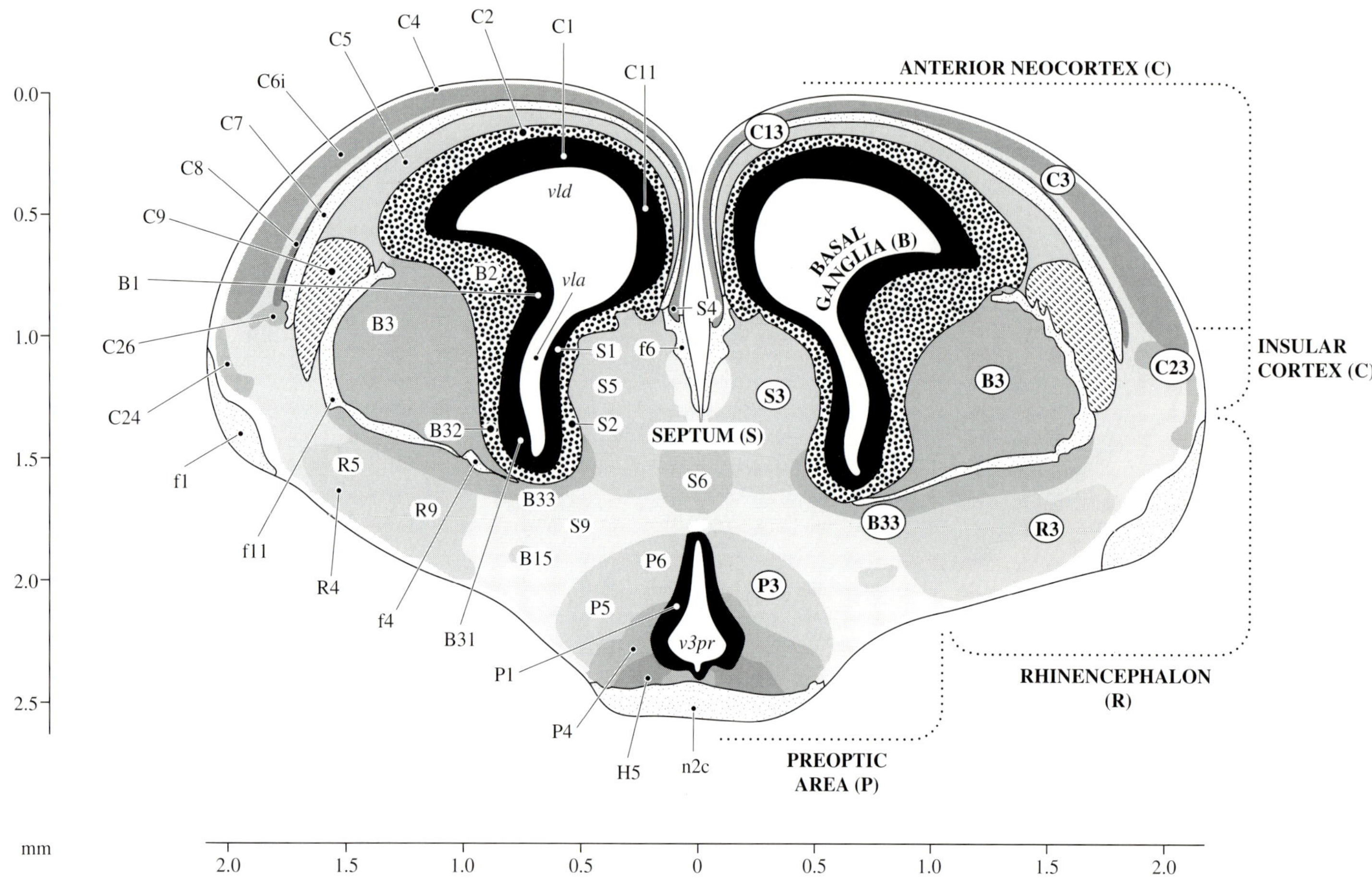

BASAL GANGLIA (B)

B1	striatal neuroepithelium
B2	striatal subventricular zone
B3	striatum (or differentiating field)
B15	substantia innominata
B31	accumbens neuroepithelium
B32	accumbens subventricular zone
B33	nucleus accumbens (or differentiating field)

CEREBRAL CORTEX (C)

C1	neocortical neuroepithelium
C2	neocortical subventricular zone
C3	neocortex (or differentiating field)
C4	molecular layer (or primordial plexiform layer)
C5	intermediate zone
C6i	cortical plate, infragranular part
C7	white matter
C8	subplate
C9	lateral migratory stream
C11	cingulate cortex neuroepithelium
C13	cingulate cortex (or differentiating field)
C23	orbital and insular area (or differentiating field)
C24	insular cortex
C26	claustrum

FIBER TRACTS (f)

f1	lateral olfactory tract
f4	anterior commissure
f6	fornix
f11	external capsule

HYPOTHALAMUS (H)

H5	suprachiasmatic nucleus

NERVES AND NUCLEI, CRANIAL (n)

n2c	optic chiasma

PREOPTIC AREA AND OPTIC EPITHELIA (P)

P1	preoptic neuroepithelium
P3	preoptic area (or differentiating field)
P4	medial preoptic region
P5	lateral preoptic region
P6	superior preoptic region

RHINENCEPHALON (R)

R3	rhinencephalon (or differentiating field)
R4	piriform cortex
R5	endopiriform nucleus
R9	olfactory tubercle

SEPTAL AREA (S)

S1	septal neuroepithelium
S2	septal subventricular zone
S3	septum (or differentiating field)
S4	indesium griseum
S5	lateral septum
S6	medial septum
S9	diagonal band (horizontal limb)

VENTRICLES (v)

vla	lateral ventricle, anterior horn
vld	lateral ventricle, dorsal
v3pr	third ventricle, preoptic recess

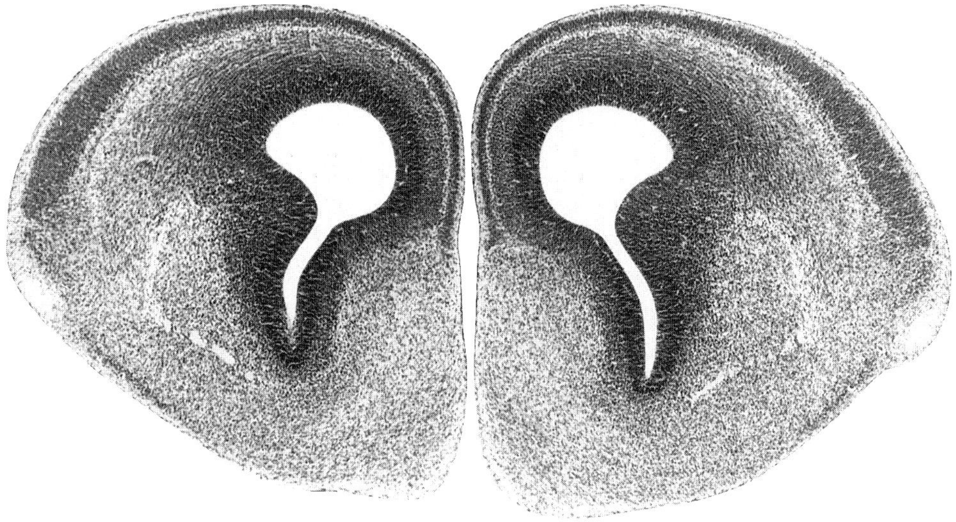

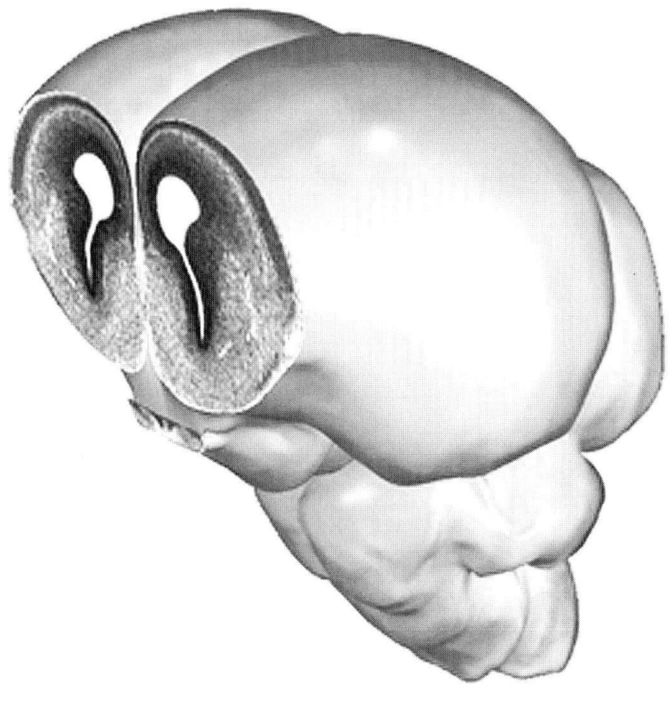

CORONAL FIGURE 5

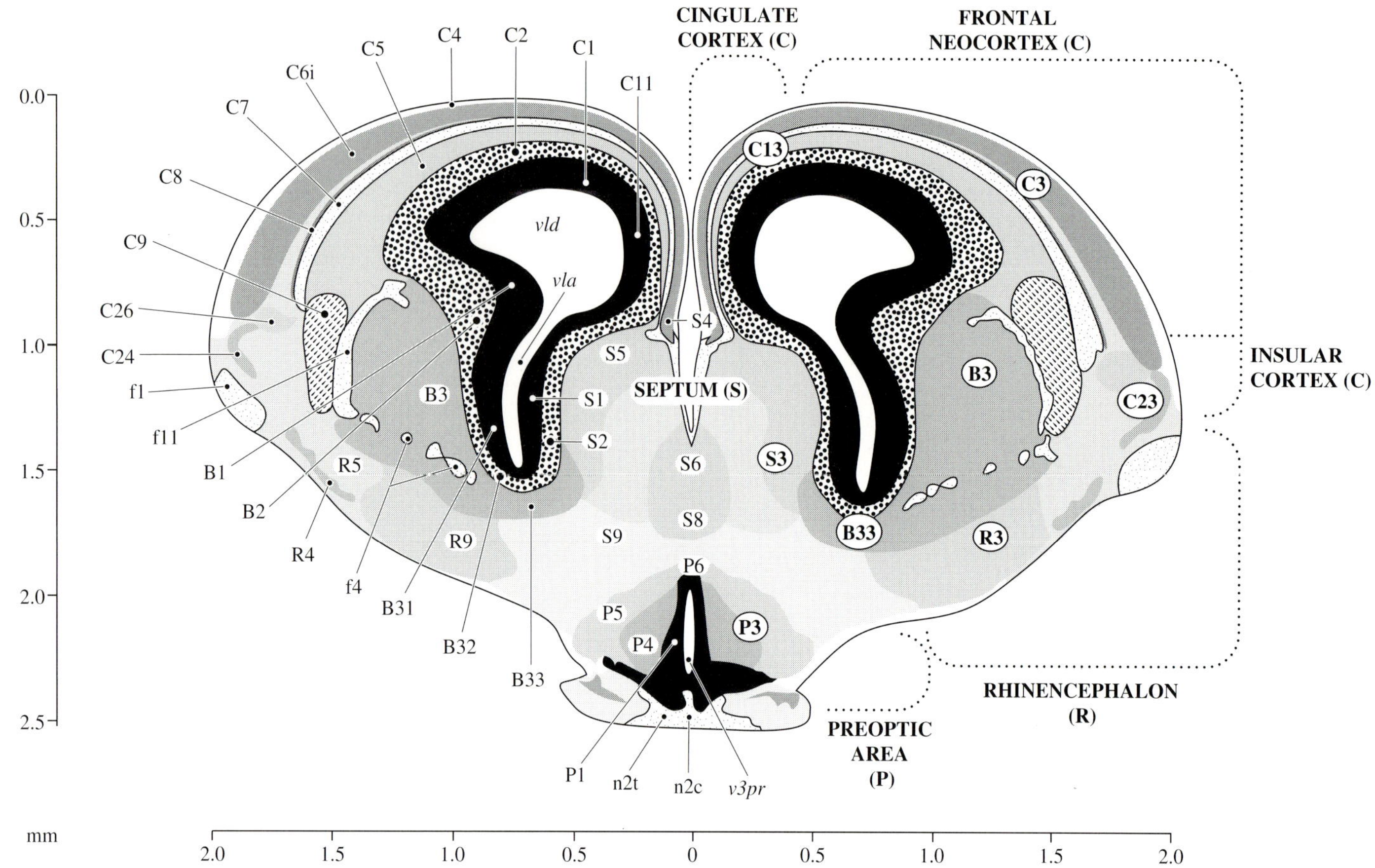

BASAL GANGLIA (B)

- B1 striatal neuroepithelium
- B2 striatal subventricular zone
- **B3** striatum (or differentiating field)
- B31 accumbens neuroepithelium
- B32 accumbens subventricular zone
- **B33** nucleus accumbens (or differentiating field)

CEREBRAL CORTEX (C)

- C1 neocortical neuroepithelium
- C2 neocortical subventricular zone
- **C3** neocortex (or differentiating field)
- C4 molecular layer (or primordial plexiform layer)
- C5 intermediate zone
- C6i cortical plate, infragranular part
- C7 white matter
- C8 subplate
- C9 lateral migratory stream
- C11 cingulate cortex neuroepithelium
- **C13** cingulate cortex (or differentiating field)
- **C23** orbital and insular area (or differentiating field)
- C24 insular cortex
- C26 claustrum

FIBER TRACTS (f)

- f1 lateral olfactory tract
- f4 anterior commissure
- f11 external capsule

NERVES AND NUCLEI, CRANIAL (n)

- n2c optic chiasma
- n2t optic tract

PREOPTIC AREA AND OPTIC EPITHELIA (P)

- P1 preoptic neuroepithelium
- **P3** preoptic area (or differentiating field)
- P4 medial preoptic region
- P5 lateral preoptic region
- P6 superior preoptic region

RHINENCEPHALON (R)

- **R3** rhinencephalon (or differentiating field)
- R4 piriform cortex
- R5 endopiriform nucleus
- R9 olfactory tubercle

SEPTAL AREA (S)

- S1 septal neuroepithelium
- S2 septal subventricular zone
- **S3** septum (or differentiating field)
- S4 indesium griseum
- S5 lateral septum
- S6 medial septum
- S8 diagonal band (vertical limb)
- S9 diagonal band (horizontal limb)

VENTRICLES (v)

- *vla* lateral ventricle, anterior horn
- *vld* lateral ventricle, dorsal
- *v3pr* third ventricle, preoptic recess

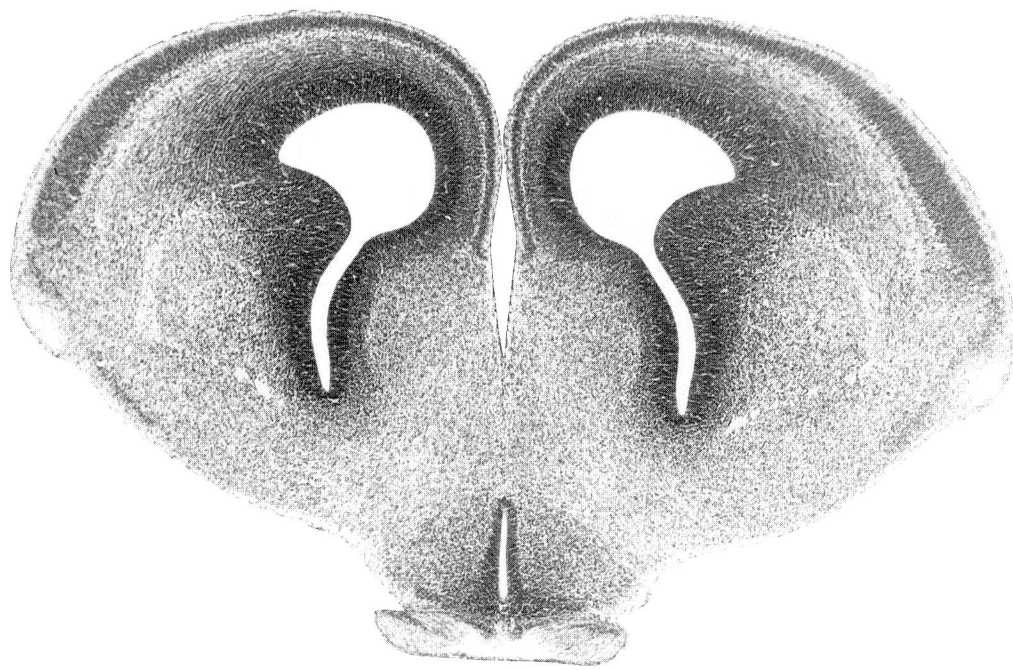

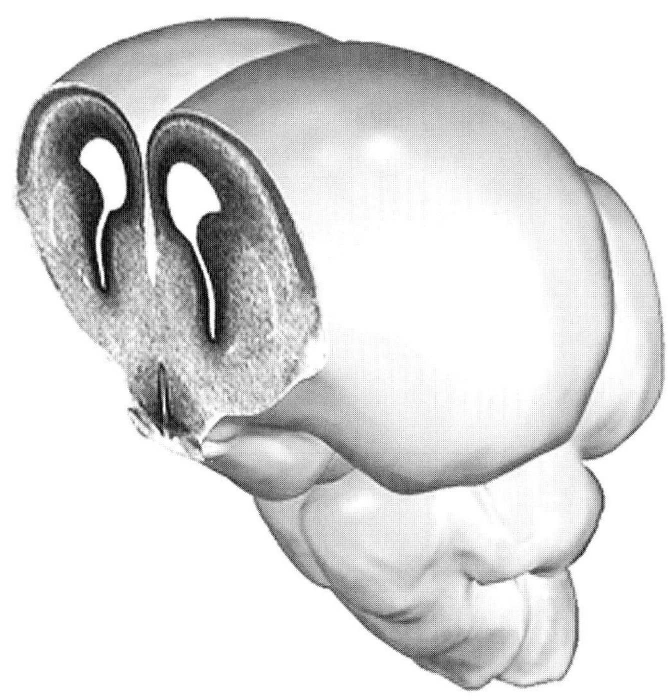

CORONAL FIGURE 6

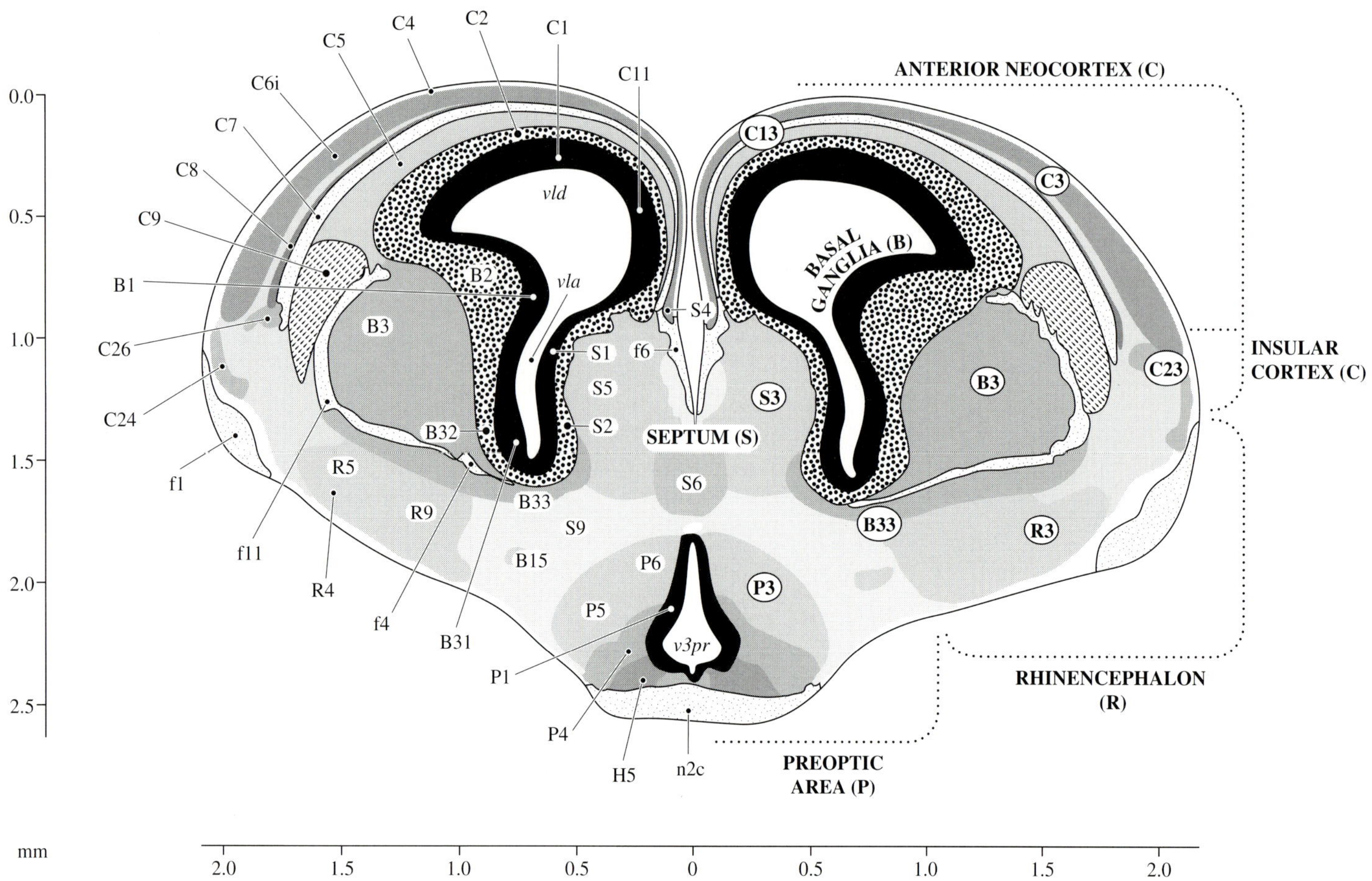

BASAL GANGLIA (B)

B1 striatal neuroepithelium
B2 striatal subventricular zone
B3 striatum (or differentiating field)
B15 substantia innominata
B31 accumbens neuroepithelium
B32 accumbens subventricular zone
B33 nucleus accumbens (or differentiating field)

CEREBRAL CORTEX (C)

C1 neocortical neuroepithelium
C2 neocortical subventricular zone
C3 neocortex (or differentiating field)
C4 molecular layer (or primordial plexiform layer)
C5 intermediate zone
C6i cortical plate, infragranular part
C7 white matter
C8 subplate
C9 lateral migratory stream
C11 cingulate cortex neuroepithelium
C13 cingulate cortex (or differentiating field)
C23 orbital and insular area (or differentiating field)
C24 insular cortex
C26 claustrum

FIBER TRACTS (f)

f1 lateral olfactory tract
f4 anterior commissure
f6 fornix
f11 external capsule

HYPOTHALAMUS (H)

H5 suprachiasmatic nucleus

NERVES AND NUCLEI, CRANIAL (n)

n2c optic chiasma

PREOPTIC AREA AND OPTIC EPITHELIA (P)

P1 preoptic neuroepithelium
P3 preoptic area (or differentiating field)
P4 medial preoptic region
P5 lateral preoptic region
P6 superior preoptic region

RHINENCEPHALON (R)

R3 rhinencephalon (or differentiating field)
R4 piriform cortex
R5 endopiriform nucleus
R9 olfactory tubercle

SEPTAL AREA (S)

S1 septal neuroepithelium
S2 septal subventricular zone
S3 septum (or differentiating field)
S4 indesium griseum
S5 lateral septum
S6 medial septum
S9 diagonal band (horizontal limb)

VENTRICLES (v)

vla lateral ventricle, anterior horn
vld lateral ventricle, dorsal
v3pr third ventricle, preoptic recess

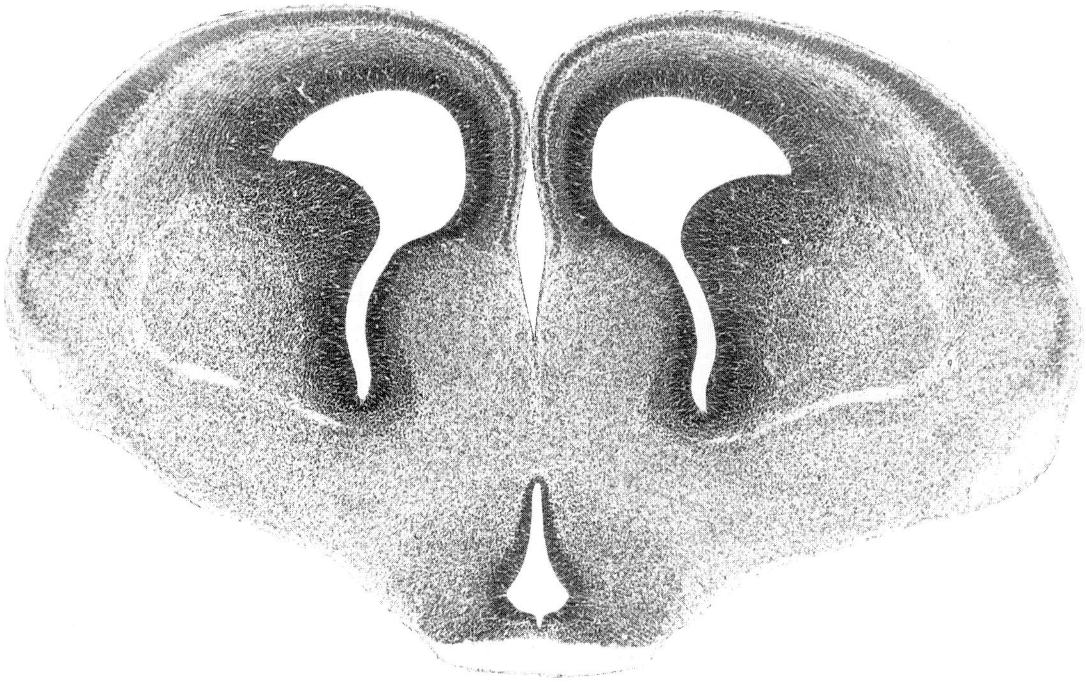

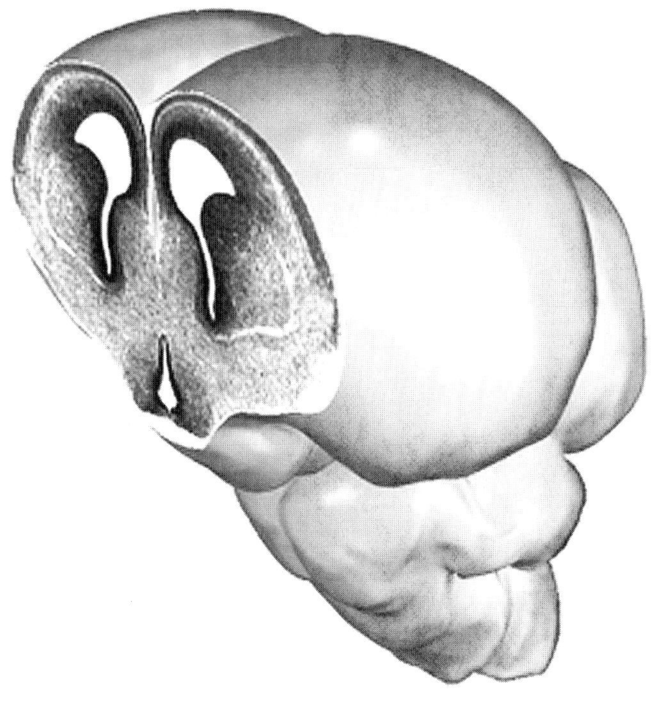

CORONAL FIGURE 7

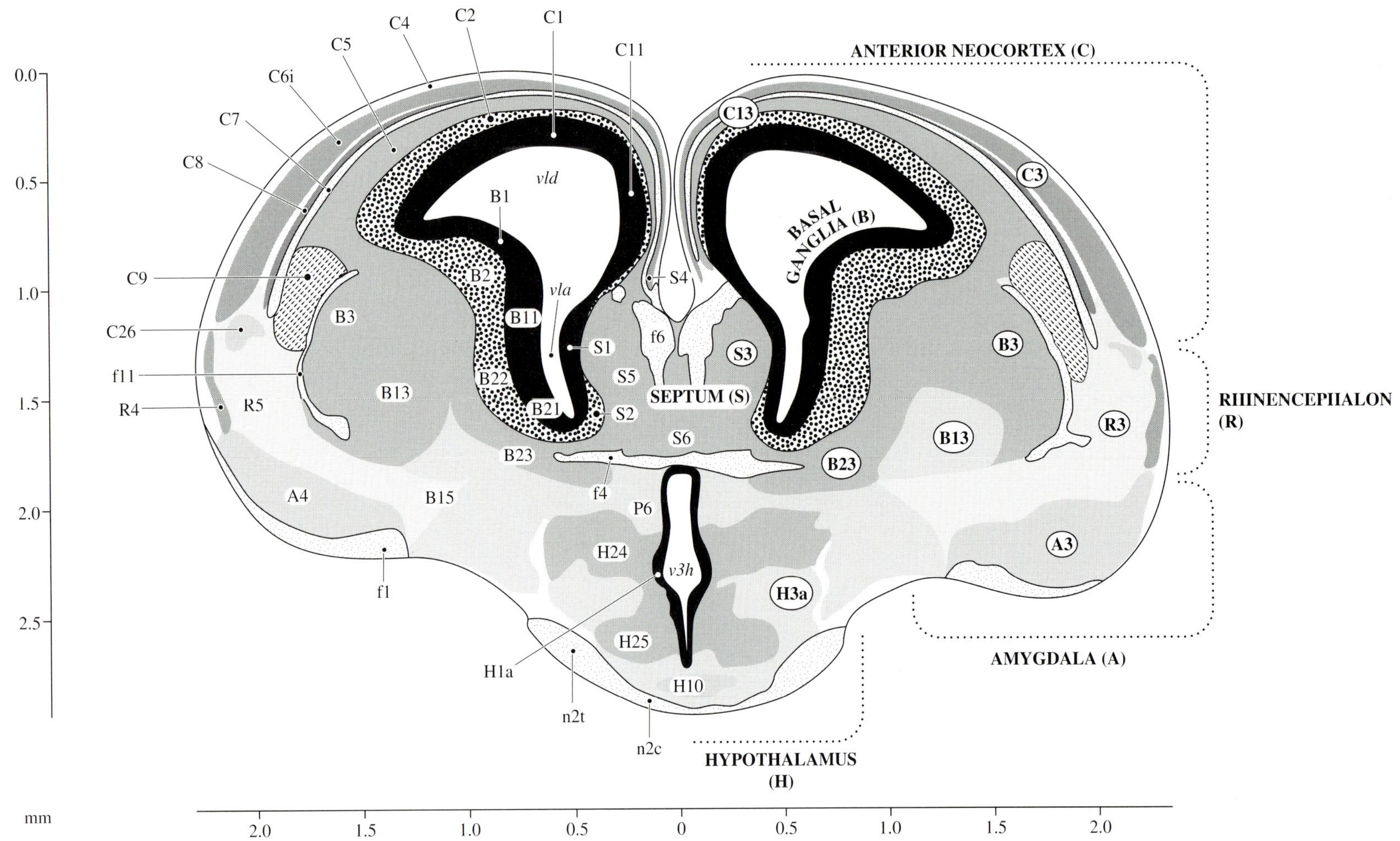

AMYGDALA (A)

A3 amygdala (or differentiating field)
A4 anterior amygdaloid area

BASAL GANGLIA (B)

B1 striatal neuroepithelium
B2 striatal subventricular zone
B3 striatum (or differentiating field)
B11 pallidal neuroepithelium
B13 pallidum (or differentiating field)
B15 substantia innominata
B21 strionuclear neuroepithelium
B22 strionuclear subventricular zone
B23 bed nucleus of stria terminalis (or differentiating field)

CEREBRAL CORTEX (C)

C1 neocortical neuroepithelium
C2 neocortical subventricular zone
C3 neocortex (or differentiating field)
C4 molecular layer (or primordial plexiform layer)
C5 intermediate zone
C6i cortical plate, infragranular part
C7 white matter
C8 subplate
C9 lateral migratory stream
C11 cingulate cortex neuroepithelium
C13 cingulate cortex (or differentiating field)
C26 claustrum

FIBER TRACTS (f)

f1 lateral olfactory tract
f4 anterior commissure
f6 fornix
f11 external capsule

HYPOTHALAMUS (H)

H1a hypothalamic neuroepithelium, anterior
H3a anterior hypothalamus (or differentiating field)
H10 anterobasal nucleus
H24 anterodorsal nucleus
H25 anteroventral nucleus

NERVES AND NUCLEI, CRANIAL (n)

n2c optic chiasma
n2t optic tract

PREOPTIC AREA AND OPTIC EPITHELIA (P)

P6 superior preoptic region

RHINENCEPHALON (R)

R3 rhinencephalon (or differentiating field)
R4 piriform cortex
R5 endopiriform nucleus

SEPTAL AREA (S)

S1 septal neuroepithelium
S2 septal subventricular zone
S3 septum (or differentiating field)
S4 indesium griseum
S5 lateral septum
S6 medial septum

VENTRICLES (v)

vla lateral ventricle, anterior horn
vld lateral ventricle, dorsal
v3h third ventricle, hypothalamic

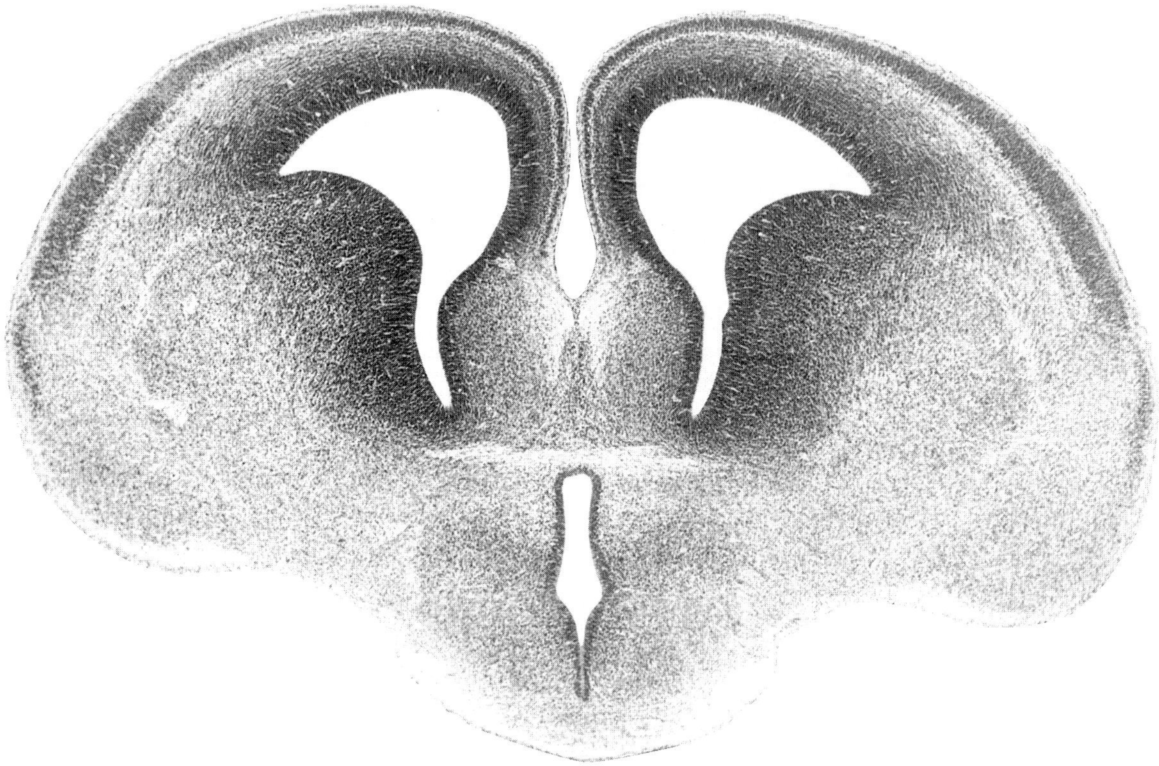

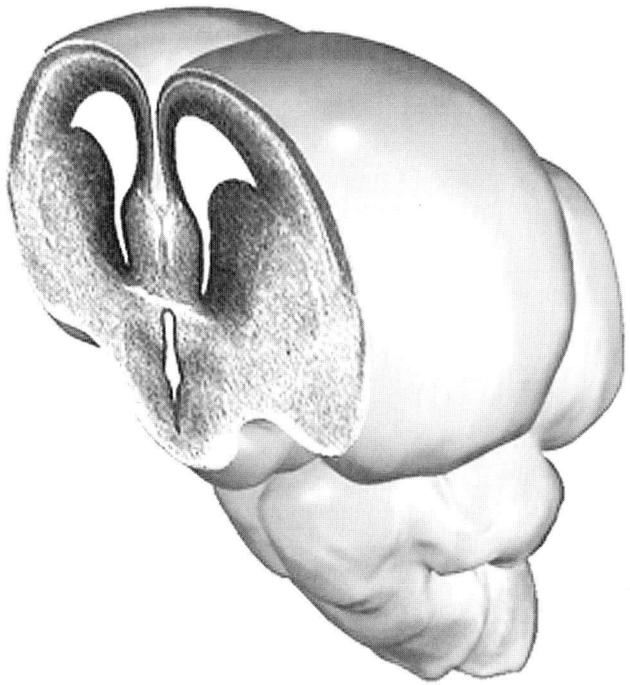

CORONAL FIGURE 8

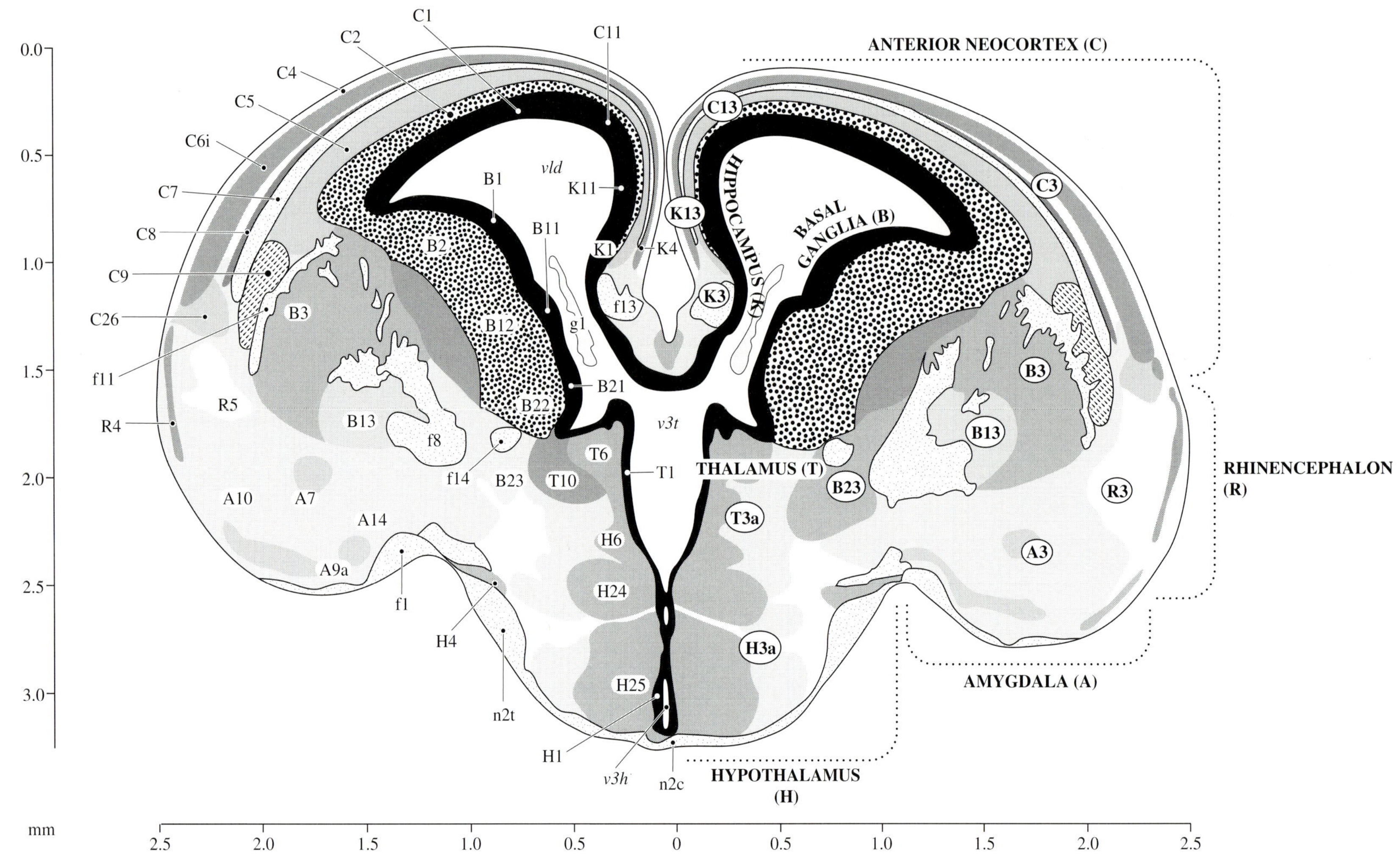

AMYGDALA (A)

A3 amygdala (or differentiating field)
A7 basomedial amygdaloid nucleus
A9a cortical amygdaloid nucleus, anterior
A10 lateral amygdaloid nucleus
A14 medial amygdaloid nucleus

BASAL GANGLIA (B)

B1 striatal neuroepithelium
B2 striatal subventricular zone
B3 striatum (or differentiating field)
B11 pallidal neuroepithelium
B12 pallidal subventricular zone
B13 pallidum (or differentiating field)
B21 strionuclear neuroepithelium
B22 strionuclear subventricular zone
B23 bed nucleus of stria terminalis (or differentiating field)

CEREBRAL CORTEX (C)

C1 neocortical neuroepithelium
C2 neocortical subventricular zone
C3 neocortex (or differentiating field)
C4 molecular layer (or primordial plexiform layer)
C5 intermediate zone
C6i cortical plate, infragranular part
C7 white matter
C8 subplate
C9 lateral migratory stream
C11 cingulate cortex neuroepithelium
C13 cingulate cortex (or differentiating field)
C26 claustrum

FIBER TRACTS (f)

f1 lateral olfactory tract
f8 internal capsule
f11 external capsule
f13 fimbria
f14 stria terminalis

GLANDS AND PARANEURAL STRUCTURES (g)

g1 choroid plexus of lateral ventricle

HYPOTHALAMUS (H)

H1 hypothalamic neuroepithelium
H3a anterior hypothalamus (or differentiating field)
H4 supraoptic nucleus
H6 paraventricular nucleus
H24 anterodorsal nucleus
H25 anteroventral nucleus

HIPPOCAMPAL FORMATION (K)

K1 hippocampal (Ammon's horn) neuroepithelium
K3 hippocampus (or differentiating field)
K4 CA1 field
K11 subicular neuroepithelium
K13 subicular area (or differentiating field)

NERVES AND NUCLEI, CRANIAL (n)

n2c optic chiasma
n2t optic tract

RHINENCEPHALON (R)

R3 rhinencephalon (or differentiating field)
R4 piriform cortex
R5 endopiriform nucleus

THALAMUS (T)

T1 thalamic neuroepithelium
T3a anterior thalamus (or differentiating field)
T6 paraventricular nucleus
T10 anterodorsal nucleus

VENTRICLES (v)

vld lateral ventricle, dorsal
v3h third ventricle, hypothalamic
v3t third ventricle, thalamic

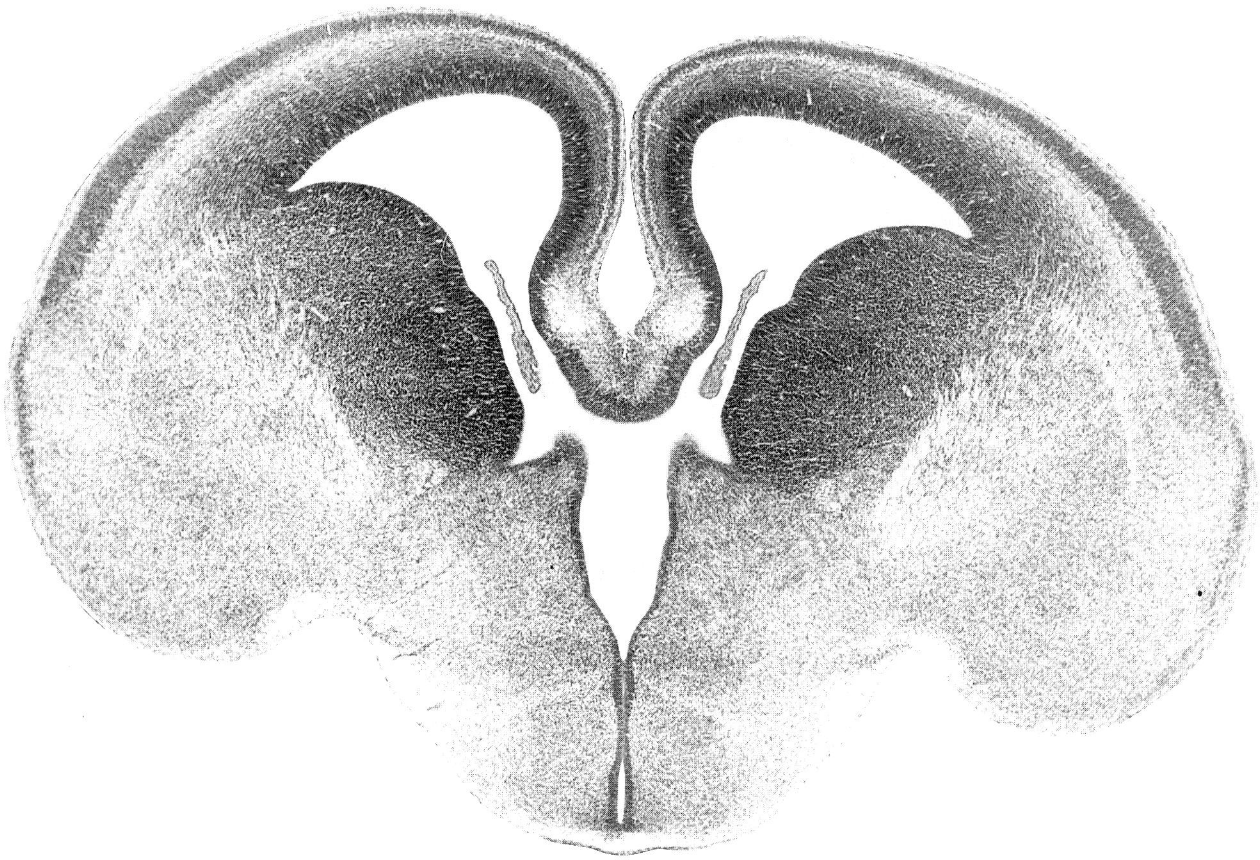

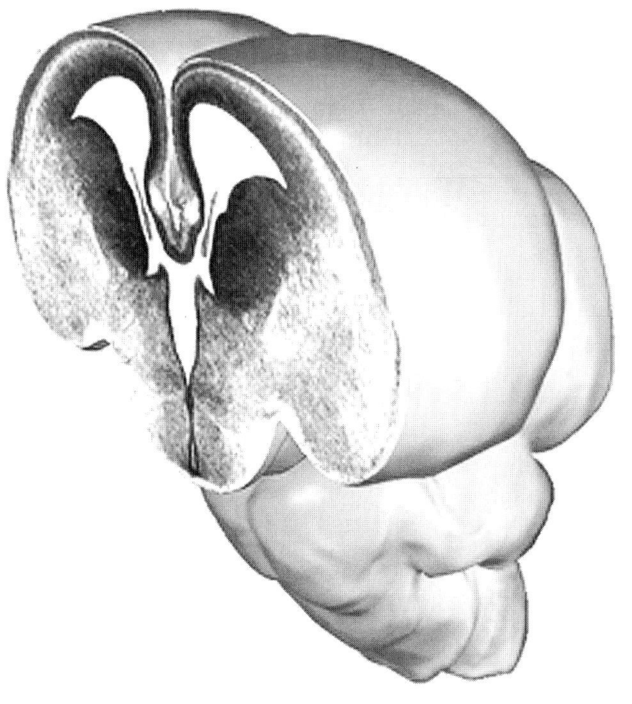

CORONAL FIGURE 9

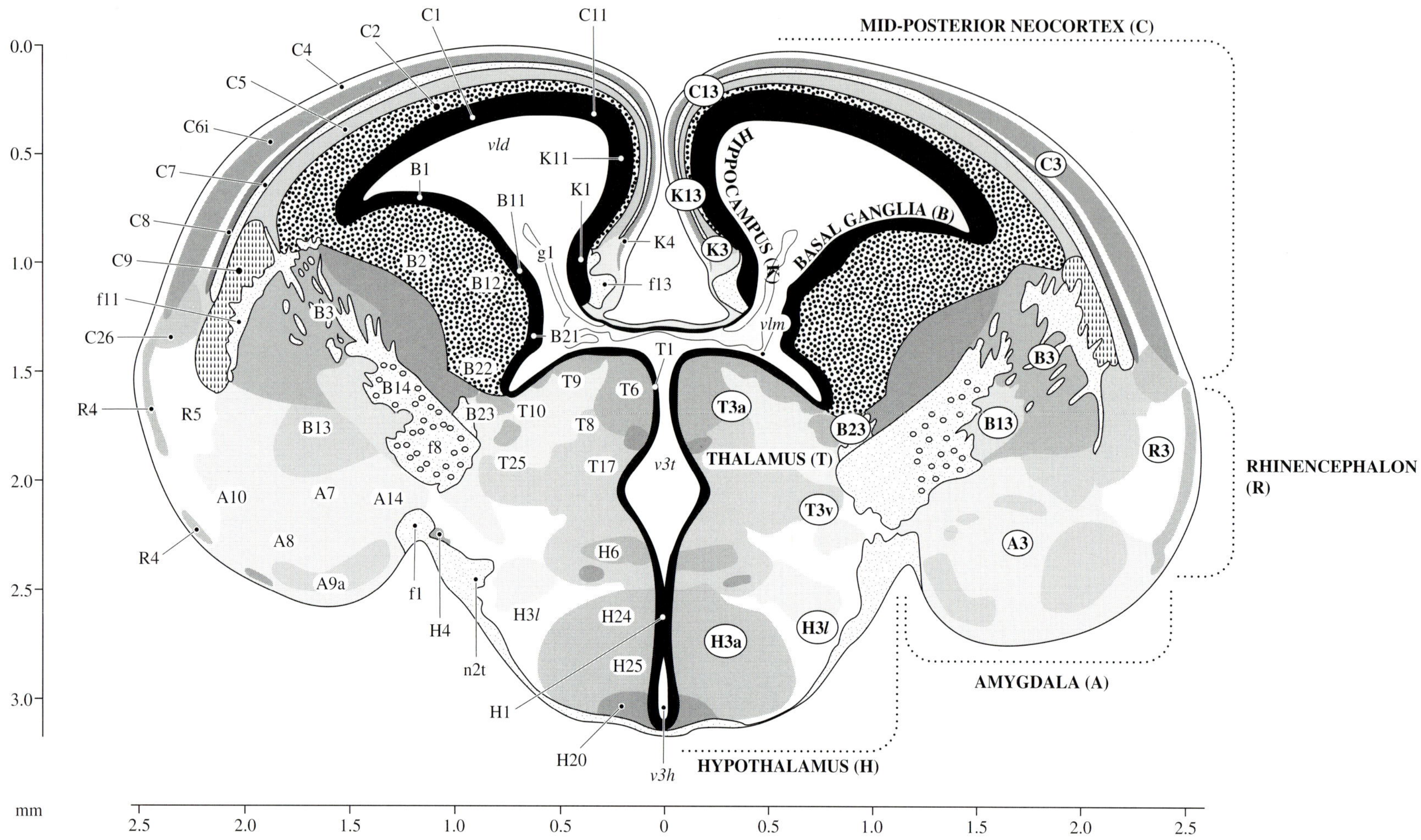

AMYGDALA (A)

- **A3** amygdala (or differentiating field)
- A7 basomedial amygdaloid nucleus
- A8 basolateral amygdaloid nucleus
- A9a cortical amygdaloid nucleus, anterior
- A10 lateral amygdaloid nucleus
- A14 medial amygdaloid nucleus

BASAL GANGLIA (B)

- B1 striatal neuroepithelium
- B2 striatal subventricular zone
- **B3** striatum (or differentiating field)
- B11 pallidal neuroepithelium
- B12 pallidal subventricular zone
- **B13** pallidum (or differentiating field)
- B14 entopeduncular nucleus
- B21 strionuclear neuroepithelium
- B22 strionuclear subventricular zone
- **B23** bed nucleus of stria terminalis (or differentiating field)

CEREBRAL CORTEX (C)

- C1 neocortical neuroepithelium
- C2 neocortical subventricular zone
- **C3** neocortex (or differentiating field)
- C4 molecular layer (or primordial plexiform layer)
- C5 intermediate zone
- C6i cortical plate, infragranular part
- C7 white matter
- C8 subplate
- C9 lateral migratory stream
- C11 cingulate cortex neuroepithelium
- **C13** cingulate cortex (or differentiating field)
- C26 claustrum

FIBER TRACTS (f)

- f1 lateral olfactory tract
- f8 internal capsule
- f11 external capsule
- f13 fimbria

GLANDS AND PARANEURAL STRUCTURES (g)

- g1 choroid plexus of lateral ventricle

HYPOTHALAMUS (H)

- H1 hypothalamic neuroepithelium
- **H3a** anterior hypothalamus (or differentiating field)
- **H3*l*** lateral hypothalamus (or differentiating field)
- H4 supraoptic nucleus
- H6 paraventricular nucleus
- H20 arcuate nucleus
- H24 anterodorsal nucleus
- H25 anteroventral nucleus

HIPPOCAMPAL FORMATION (K)

- K1 hippocampal (Ammon's horn) neuroepithelium
- **K3** hippocampus (or differentiating field)
- K4 CA1 field
- K11 subicular neuroepithelium
- **K13** subicular area (or differentiating field)

NERVES AND NUCLEI, CRANIAL (n)

- n2t optic tract

RHINENCEPHALON (R)

- **R3** rhinencephalon (or differentiating field)
- R4 piriform cortex
- R5 endopiriform nucleus

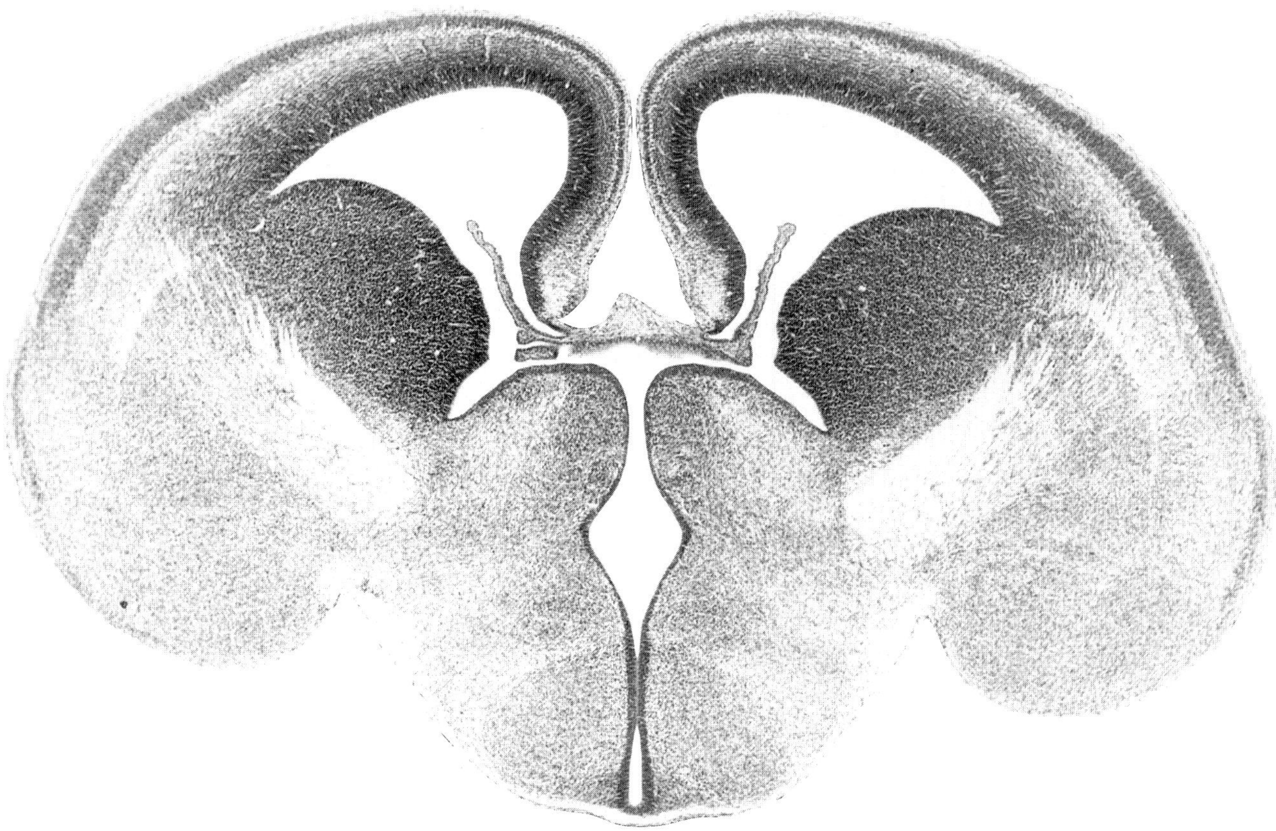

THALAMUS (T)

T1 thalamic neuroepithelium
T3a anterior thalamus (or differentiating field)
T3v ventral thalamus (or differentiating field)
T6 paraventricular nucleus
T8 anteromedial nucleus
T9 anteroventral nucleus
T10 anterodorsal nucleus
T17 rhomboid nucleus
T25 reticular nucleus

VENTRICLES (v)

vld lateral ventricle, dorsal
vlm foramen of Monro
v3h third ventricle, hypothalamic
v3t third ventricle, thalamic

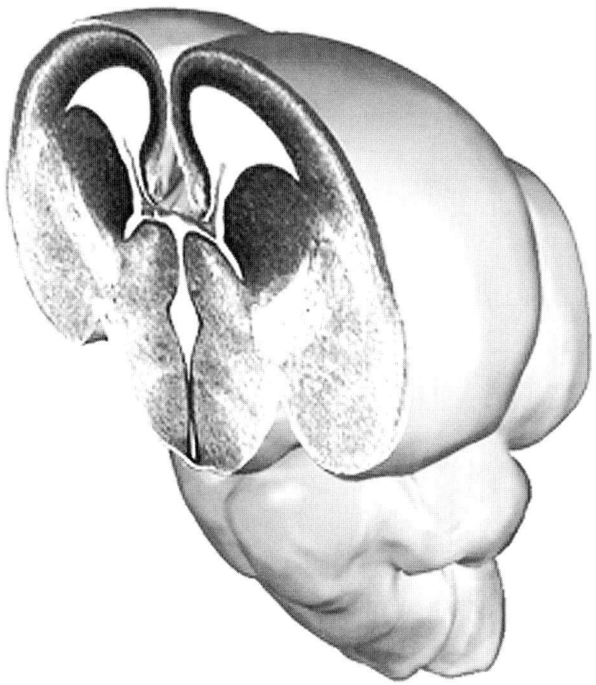

CORONAL FIGURE 10

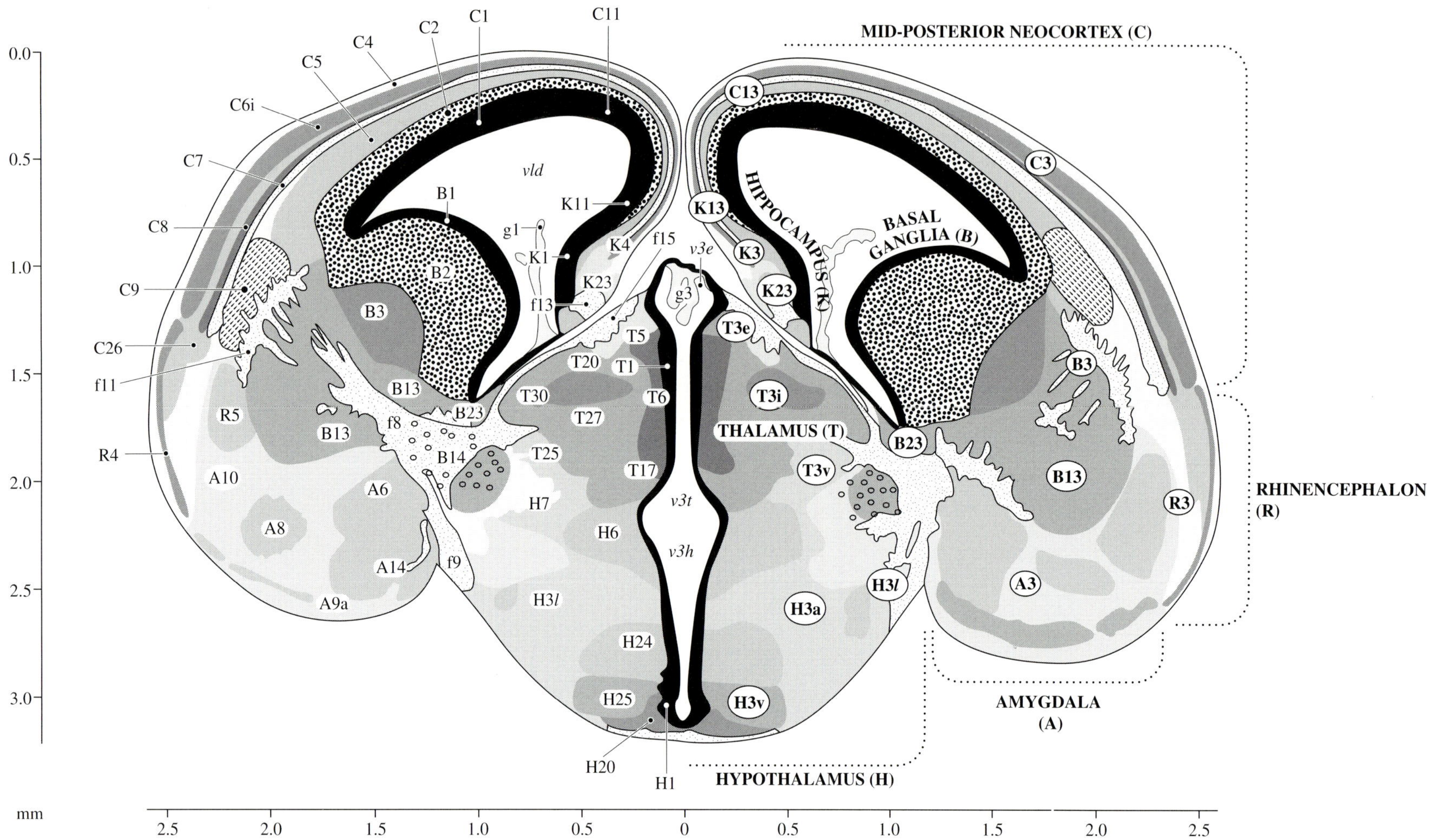

AMYGDALA (A)

A3 amygdala (or differentiating field)
A6 central amygdaloid nucleus
A8 basolateral amygdaloid nucleus
A9a cortical amygdaloid nucleus, anterior
A10 lateral amygdaloid nucleus
A14 medial amygdaloid nucleus

BASAL GANGLIA (B)

B1 striatal neuroepithelium
B2 striatal subventricular zone
B3 striatum (or differentiating field)
B13 pallidum (or differentiating field)
B14 entopeduncular nucleus
B23 bed nucleus of stria terminalis (or differentiating field)

CEREBRAL CORTEX (C)

C1 neocortical neuroepithelium
C2 neocortical subventricular zone
C3 neocortex (or differentiating field)
C4 molecular layer (or primordial plexiform layer)
C5 intermediate zone
C6i cortical plate, infragranular part
C7 white matter
C8 subplate
C9 lateral migratory stream
C11 cingulate cortex neuroepithelium
C13 cingulate cortex (or differentiating field)
C26 claustrum

FIBER TRACTS (f)

f8 internal capsule
f9 cerebral peduncle
f11 external capsule
f13 fimbria
f15 stria medullaris

GLANDS AND PARANEURAL STRUCTURES (g)

g1 choroid plexus of lateral ventricle
g3 choroid plexus of third ventricle

HYPOTHALAMUS (H)

H1 hypothalamic neuroepithelium
H3a anterior hypothalamus (or differentiating field)
H3*l* lateral hypothalamus (or differentiating field)
H3v ventral hypothalamus (or differentiating field)
H6 paraventricular nucleus
H7 zona incerta
H20 arcuate nucleus
H24 anterodorsal nucleus
H25 anteroventral nucleus

HIPPOCAMPAL FORMATION (K)

K1 hippocampal (Ammon's horn) neuroepithelium
K3 hippocampus (or differentiating field)
K4 CA1 field
K11 subicular neuroepithelium
K13 subicular area (or differentiating field)
K23 dentate gyrus (or differentiating field)

RHINENCEPHALON (R)

R3 rhinencephalon (or differentiating field)
R4 piriform cortex
R5 endopiriform nucleus

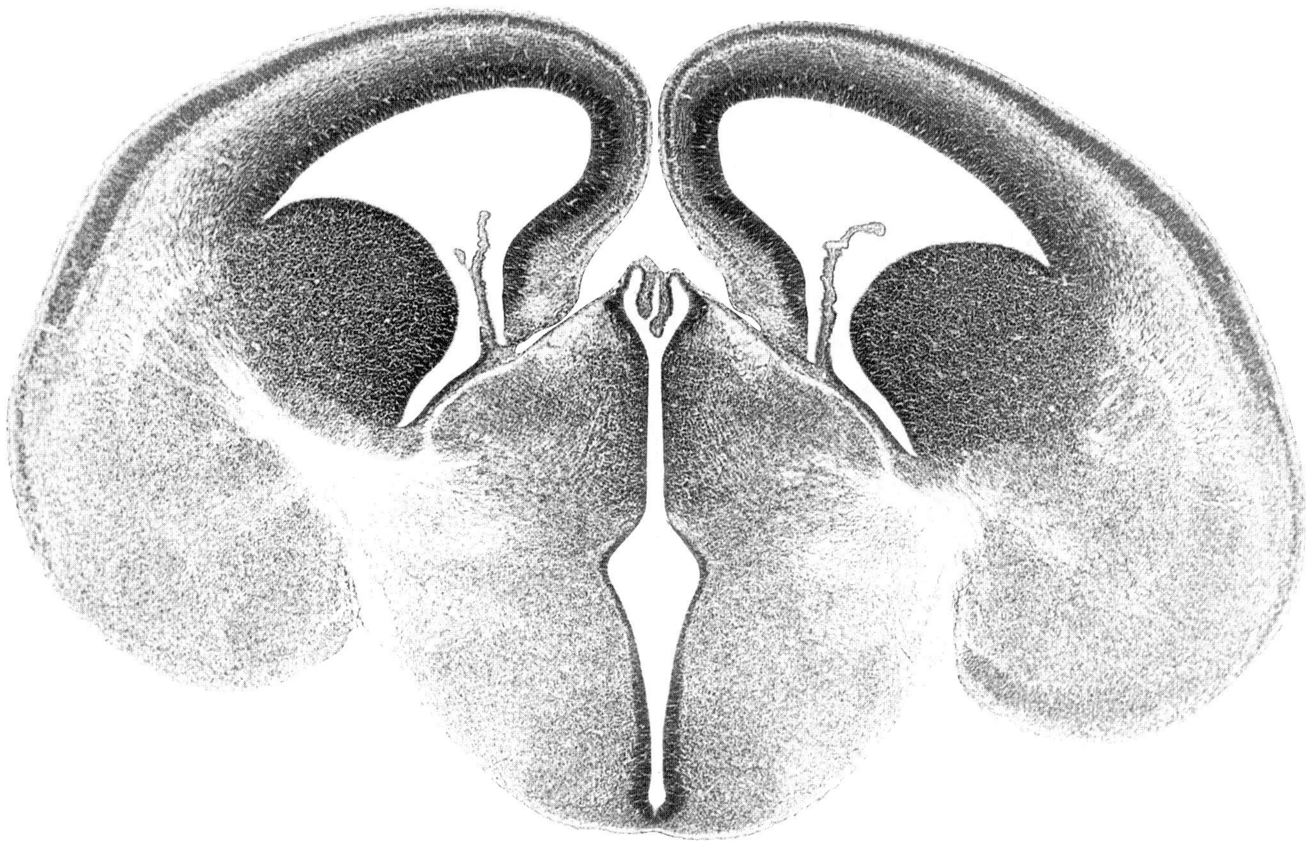

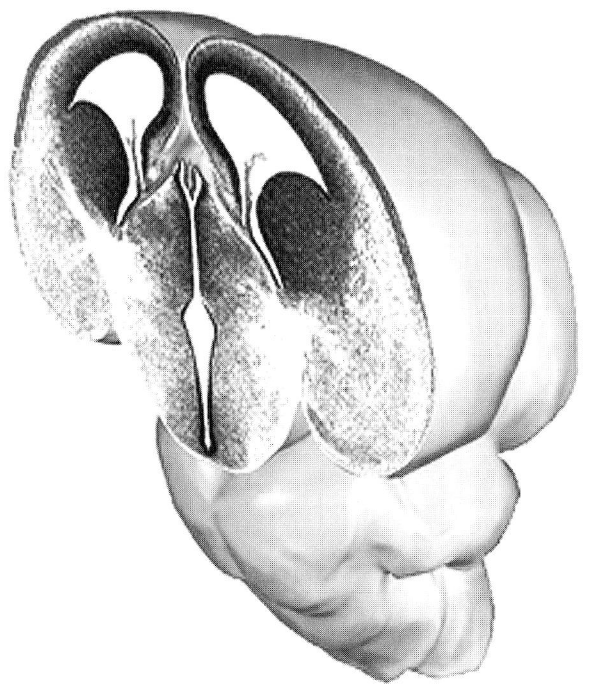

THALAMUS (T)

T1 thalamic neuroepithelium
T3e epithalamus (or differentiating field)
T3i intermediate thalamus (or differentiating field)
T3v ventral thalamus (or differentiating field)
T5 lateral habenular nucleus
T6 paraventricular nucleus
T17 rhomboid nucleus
T20 laterodorsal nucleus
T25 reticular nucleus
T27 ventrolateral nuclear complex
T30 ventral lateral geniculate nucleus

VENTRICLES (v)

vld lateral ventricle, dorsal
v3e third ventricle, epithalamic
v3h third ventricle, hypothalamic
v3t third ventricle, thalamic

CORONAL FIGURE 11

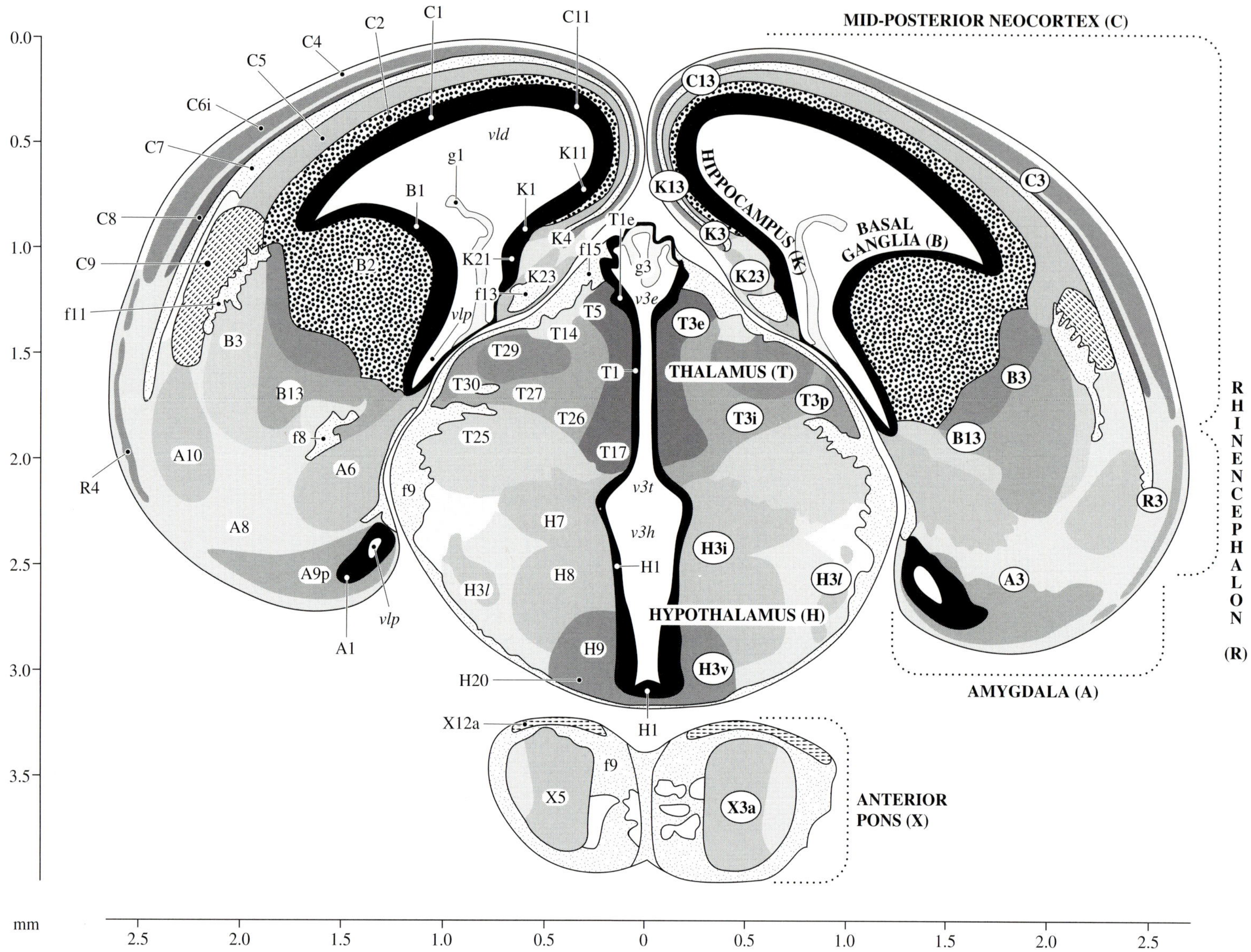

AMYGDALA (A)

- A1 amygdaloid neuroepithelium
- **A3** amygdala (or differentiating field)
- A6 central amygdaloid nucleus
- A8 basolateral amygdaloid nucleus
- A9p cortical amygdaloid nucleus, posterior
- A10 lateral amygdaloid nucleus

BASAL GANGLIA (B)

- B1 striatal neuroepithelium
- B2 striatal subventricular zone
- **B3** striatum (or differentiating field)
- **B13** pallidum (or differentiating field)

CEREBRAL CORTEX (C)

- C1 neocortical neuroepithelium
- C2 neocortical subventricular zone
- **C3** neocortex (or differentiating field)
- C4 molecular layer (or primordial plexiform layer)
- C5 intermediate zone
- C6i cortical plate, infragranular part
- C7 white matter
- C8 subplate
- C9 lateral migratory stream
- C11 cingulate cortex neuroepithelium
- **C13** cingulate cortex (or differentiating field)

FIBER TRACTS (f)

- f8 internal capsule
- f9 cerebral peduncle
- f11 external capsule
- f13 fimbria
- f15 stria medullaris

GLANDS AND PARANEURAL STRUCTURES (g)

- g1 choroid plexus of lateral ventricle
- g3 choroid plexus of third ventricle

HYPOTHALAMUS (H)

- H1 hypothalamic neuroepithelium
- **H3i** intermediate hypothalamus (or differentiating field)
- **H3*l*** lateral hypothalamus (or differentiating field)
- **H3v** ventral hypothalamus (or differentiating field)
- H7 zona incerta
- H8 dorsomedial nucleus
- H9 ventromedial nucleus
- H20 arcuate nucleus

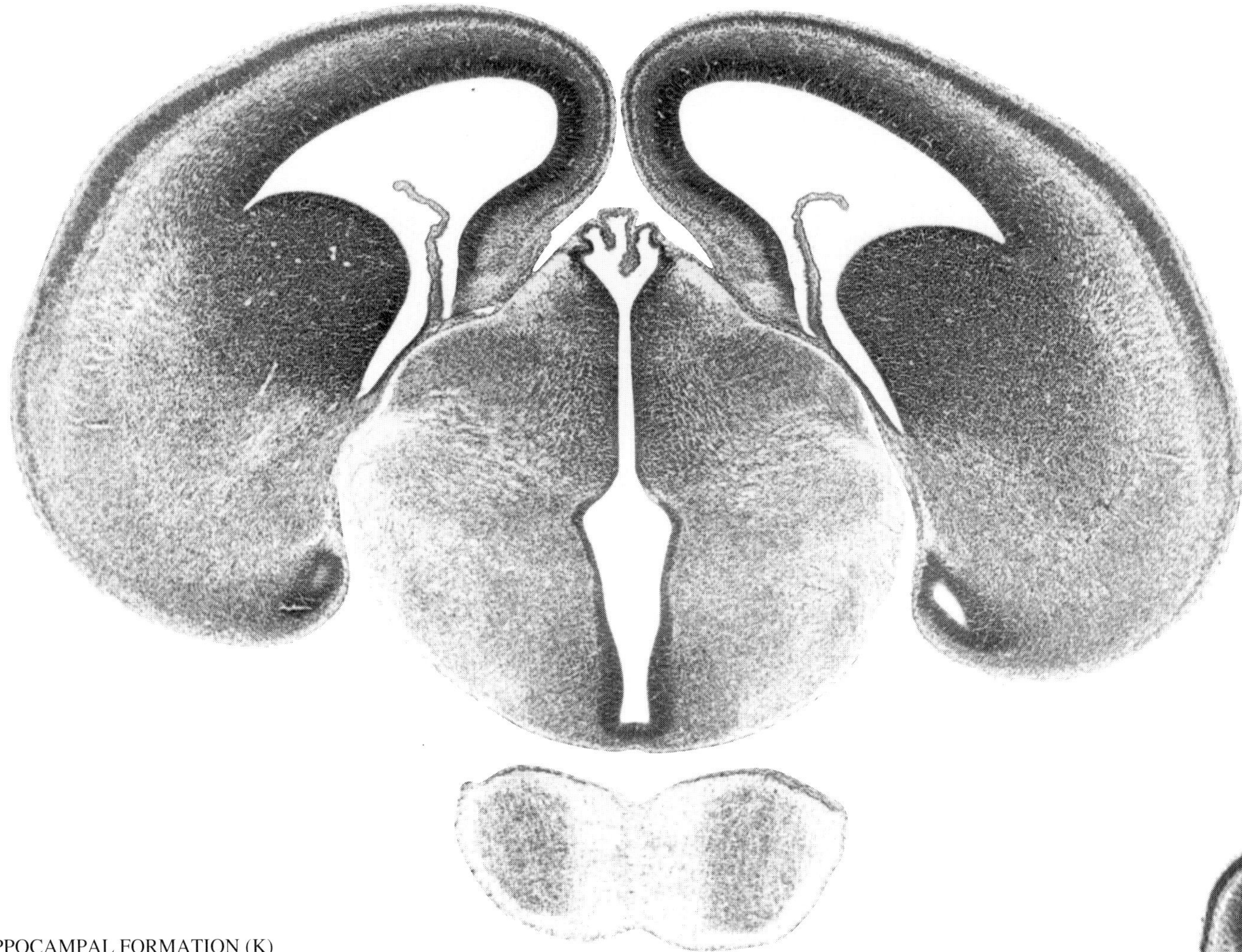

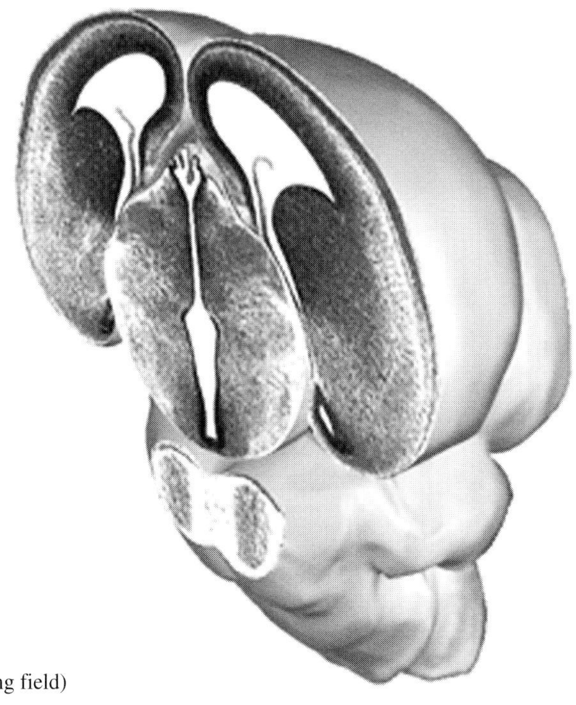

HIPPOCAMPAL FORMATION (K)

K1 hippocampal (Ammon's horn) neuroepithelium
K3 hippocampus (or differentiating field)
K4 CA1 field
K11 subicular neuroepithelium
K13 subicular area (or differentiating field)
K21 dentate gyrus neuroepithelium
K23 dentate gyrus (or differentiating field)

RHINENCEPHALON (R)

R3 rhinencephalon (or differentiating field)
R4 piriform cortex

THALAMUS (T)

T1 thalamic neuroepithelium
T1e epithalamic neuroepithelium
T3e epithalamus (or differentiating field)
T3i intermediate thalamus (or differentiating field)
T3p posterior thalamus (or differentiating field)
T5 lateral habenular nucleus
T14 centromedian nucleus
T17 rhomboid nucleus
T25 reticular nucleus
T26 ventral medial nucleus
T27 ventrolateral nuclear complex
T29 dorsal lateral geniculate nucleus
T30 ventral lateral geniculate nucleus

VENTRICLES (v)

vld lateral ventricle, dorsal
vlp lateral ventricle, posterior horn
v3e third ventricle, epithalamic
v3h third ventricle, hypothalamic
v3t third ventricle, thalamic

PONTINE AREA (X)

X3a anterior pons (or differentiating field)
X5 pontine gray nucleus
X12a anterior precerebellar extramural migration

CORONAL FIGURE 12

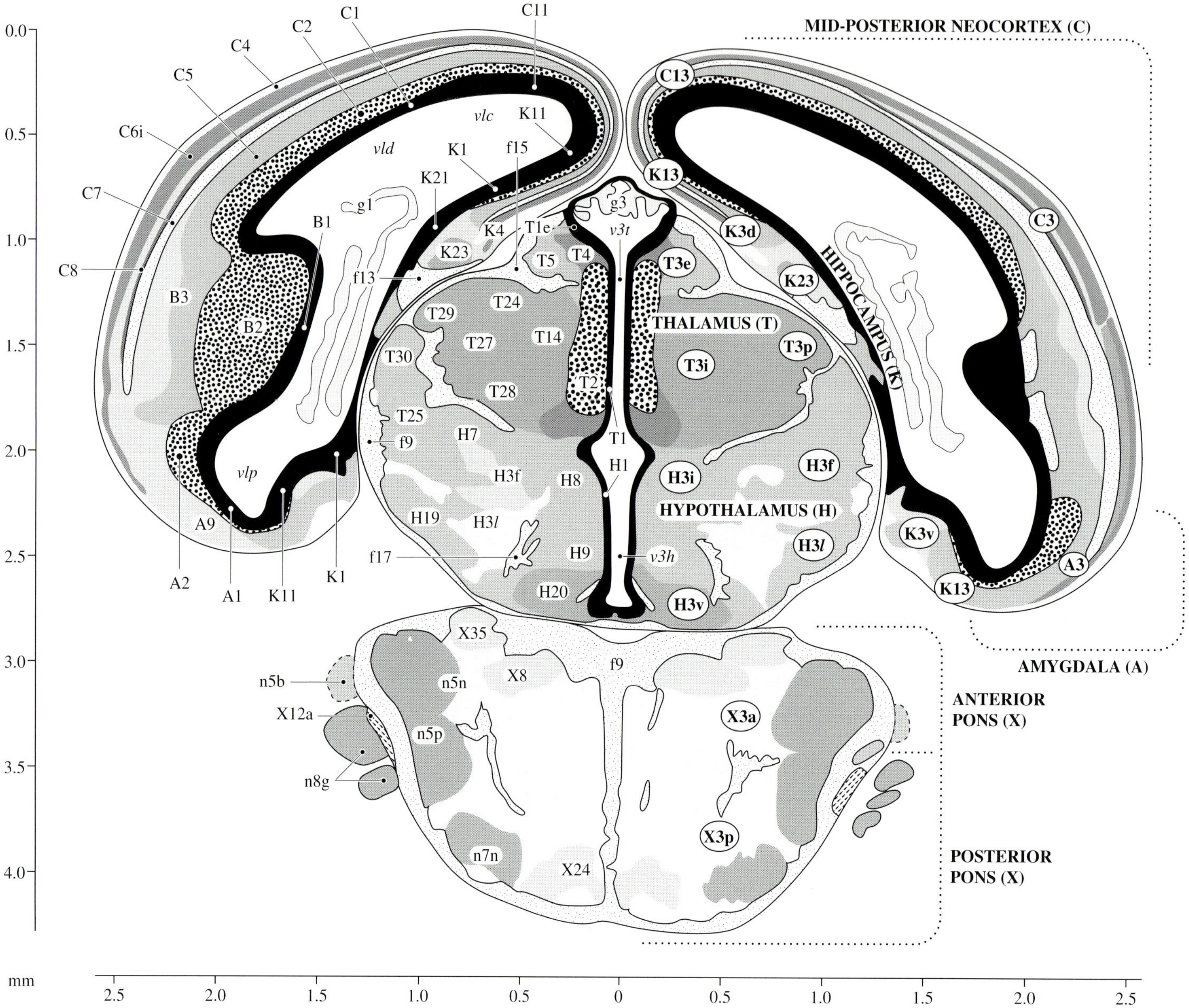

AMYGDALA (A)

A1	amygdaloid neuroepithelium
A2	amygdaloid subventricular zone
A3	amygdala (or differentiating field)
A9	cortical amygdaloid nucleus

BASAL GANGLIA (B)

B1	striatal neuroepithelium
B2	striatal subventricular zone
B3	striatum (or differentiating field)

CEREBRAL CORTEX (C)

C1	neocortical neuroepithelium
C2	neocortical subventricular zone
C3	neocortex (or differentiating field)
C4	molecular layer (or primordial plexiform layer)
C5	intermediate zone
C6i	cortical plate, infragranular part
C7	white matter
C8	subplate
C11	cingulate cortex neuroepithelium
C13	cingulate cortex (or differentiating field)

FIBER TRACTS (f)

f9	cerebral peduncle
f13	fimbria
f15	stria medullaris
f17	mammillothalamic tract

GLANDS AND PARANEURAL STRUCTURES (g)

g1	choroid plexus of lateral ventricle
g3	choroid plexus of third ventricle

HYPOTHALAMUS (H)

H1	hypothalamic neuroepithelium
H3f	fields of Forel (or differentiating field)
H3i	intermediate hypothalamus (or differentiating field)
H3l	lateral hypothalamus (or differentiating field)
H3v	ventral hypothalamus (or differentiating field)
H7	zona incerta
H8	dorsomedial nucleus
H9	ventromedial nucleus
H19	subthalamic nucleus
H20	arcuate nucleus

HIPPOCAMPAL FORMATION (K)

K1	hippocampal (Ammon's horn) neuroepithelium
K3d	dorsal hippocampus (or differentiating field)
K3v	ventral hippocampus (or differentiating field)
K4	CA1 field
K11	subicular neuroepithelium
K13	subicular area (or differentiating field)
K21	dentate gyrus neuroepithelium
K23	dentate gyrus (or differentiating field)

NERVES AND NUCLEI, CRANIAL (n)

n5b	trigeminal boundary cap
n5n	trigeminal motor nucleus
n5p	principal (sensory) trigeminal nucleus
n7n	facial motor nucleus
n8g	vestibular ganglion

THALAMUS (T)

T1	thalamic neuroepithelium
T1e	epithalamic neuroepithelium
T2	thalamic subventricular zone
T3e	epithalamus (or differentiating field)
T3i	intermediate thalamus (or differentiating field)
T3p	posterior thalamus (or differentiating field)
T4	medial habenular nucleus
T5	lateral habenular nucleus
T14	centromedian nucleus
T24	lateral posterior nucleus
T25	reticular nucleus
T27	ventrolateral nuclear complex
T28	ventrobasal nuclear complex
T29	dorsal lateral geniculate nucleus
T30	ventral lateral geniculate nucleus

VENTRICLES (v)

vlc	lateral ventricle, central
vld	lateral ventricle, dorsal
vlp	lateral ventricle, posterior horn
v3h	third ventricle, hypothalamic
v3t	third ventricle, thalamic

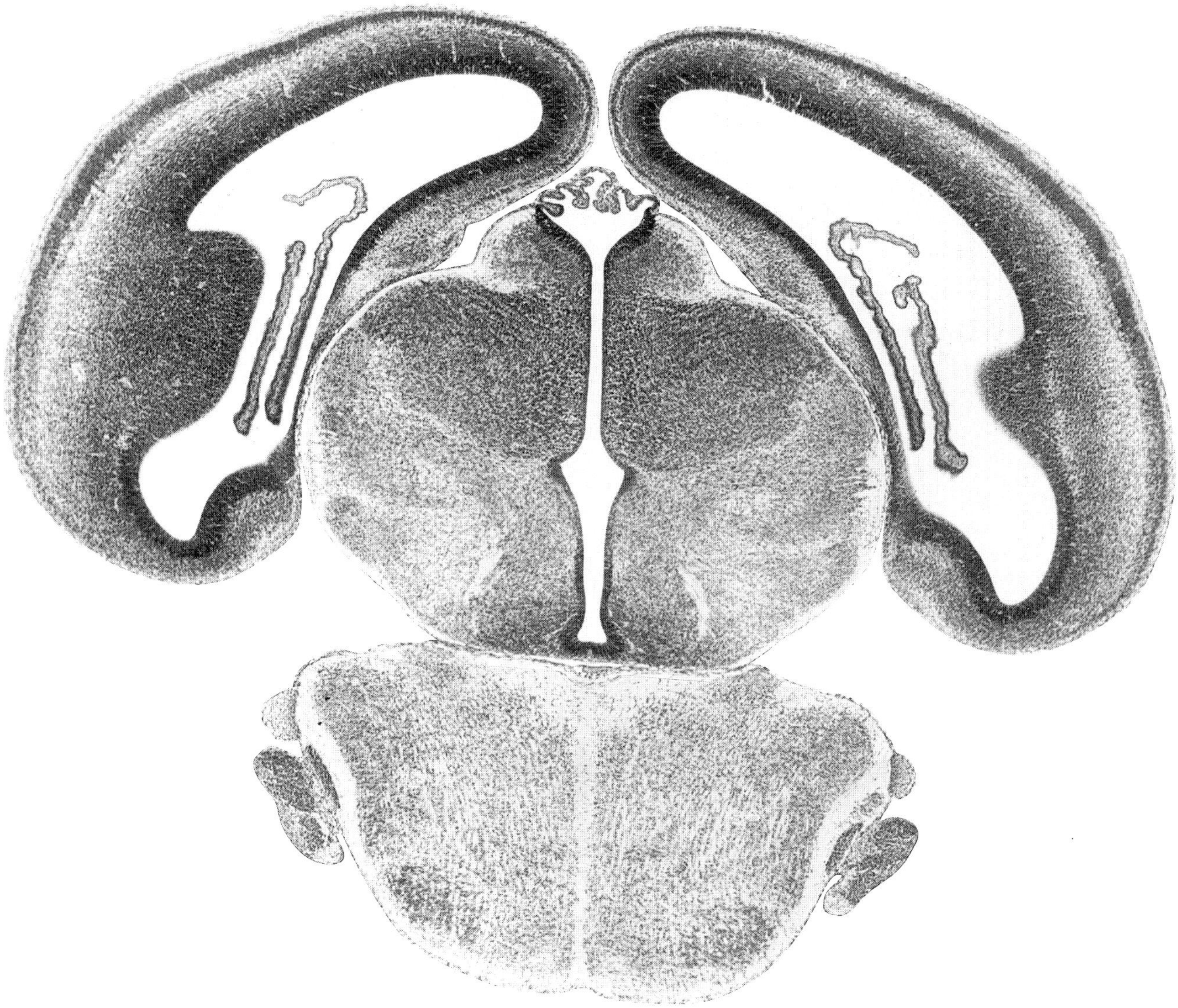

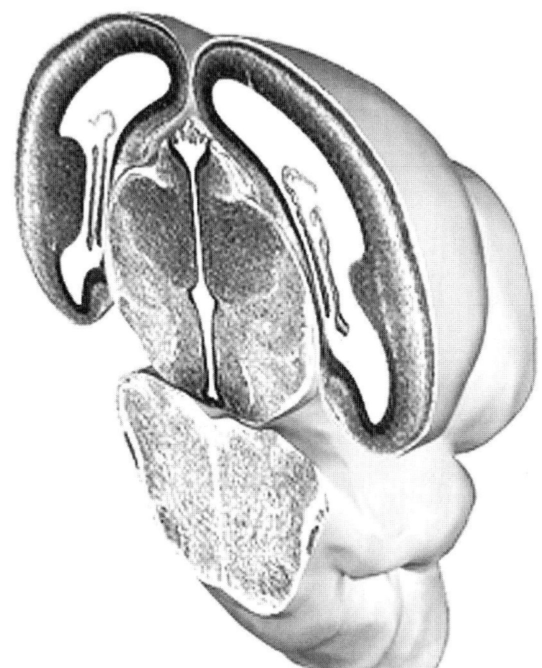

PONTINE AREA (X)

X3a anterior pons (or differentiating field)
X3p posterior pons (or differentiating field)
X8 superior central raphe nucleus
X12a anterior precerebellar extramural migration
X24 superior olivary nucleus
X35 parabrachial nucleus

CORONAL FIGURE 13

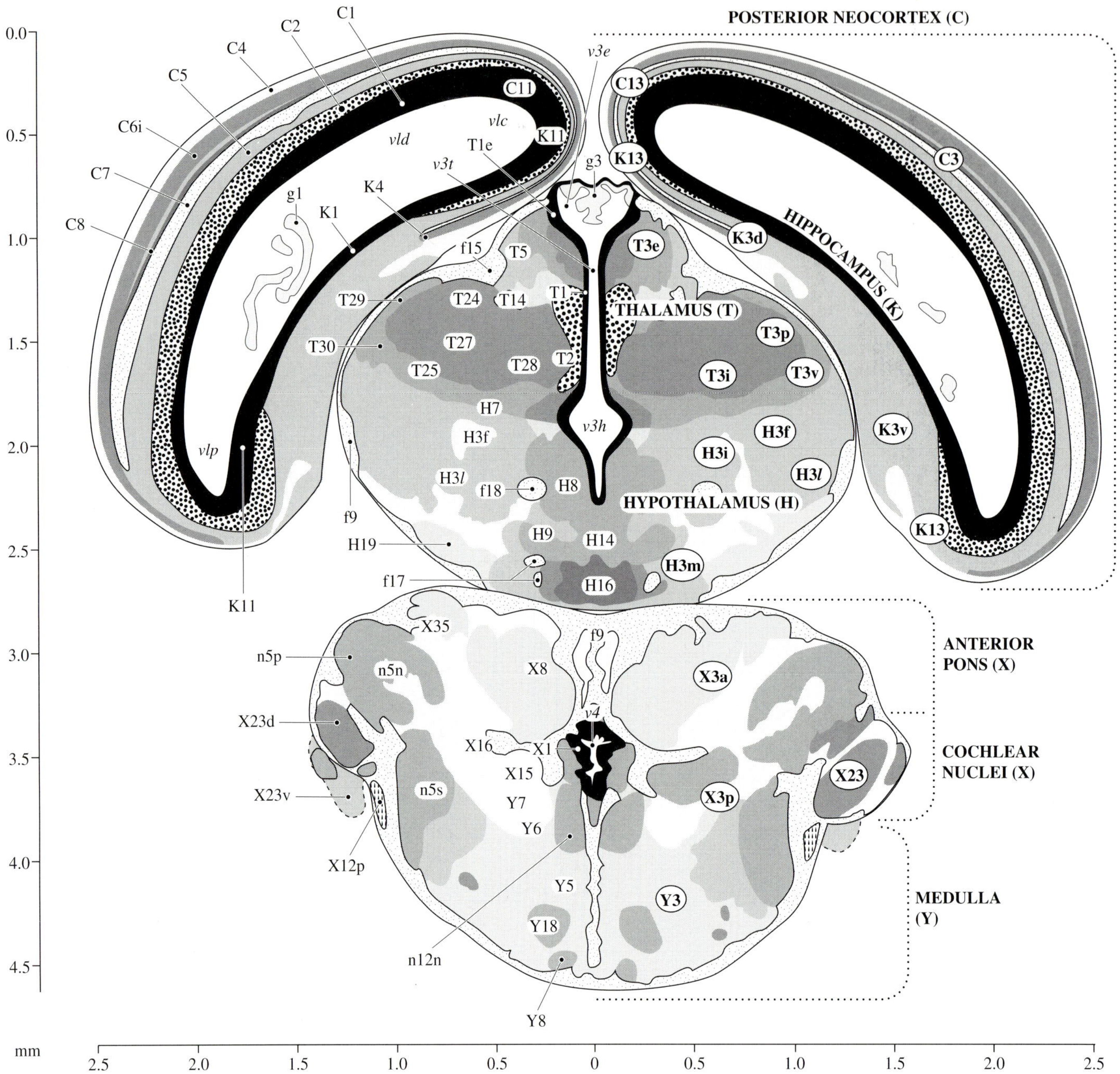

CEREBRAL CORTEX (C)

C1 neocortical neuroepithelium
C2 neocortical subventricular zone
C3 neocortex (or differentiating field)
C4 molecular layer (or primordial plexiform layer)
C5 intermediate zone
C6i cortical plate, infragranular part
C7 white matter
C8 subplate
C11 cingulate cortex neuroepithelium
C13 cingulate cortex (or differentiating field)

FIBER TRACTS (f)

f9 cerebral peduncle
f15 stria medullaris
f17 mammillothalamic tract
f18 habenulopeduncular tract

GLANDS AND PARANEURAL STRUCTURES (g)

g1 choroid plexus of lateral ventricle
g3 choroid plexus of third ventricle

HYPOTHALAMUS (H)

H3f fields of Forel (or differentiating field)
H3i intermediate hypothalamus (or differentiating field)
H3*l* lateral hypothalamus (or differentiating field)
H3m mammillary body (or differentiating field)
H7 zona incerta
H8 dorsomedial nucleus
H9 ventromedial nucleus
H14 supramammillary nucleus
H16 medial mammillary nucleus
H19 subthalamic nucleus

HIPPOCAMPAL FORMATION (K)

K1 hippocampal (Ammon's horn) neuroepithelium
K3d dorsal hippocampus (or differentiating field)
K3v ventral hippocampus (or differentiating field)
K4 CA1 field
K11 subicular neuroepithelium
K13 subicular area (or differentiating field)

NERVES AND NUCLEI, CRANIAL (n)

n5n trigeminal motor nucleus
n5p principal (sensory)trigeminal nucleus
n5s spinal trigeminal nucleus
n12n hypoglossal nucleus

THALAMUS (T)

T1 thalamic neuroepithelium
T1e epithalamic neuroepithelium
T2 thalamic subventricular zone
T3e epithalamus (or differentiating field)
T3i intermediate thalamus (or differentiating field)
T3p posterior thalamus (or differentiating field)
T3v ventral thalamus (or differentiating field)
T5 lateral habenular nucleus
T14 centromedian nucleus
T24 lateral posterior nucleus
T25 reticular nucleus
T27 ventrolateral nuclear complex
T28 ventrobasal nuclear complex
T29 dorsal lateral geniculate nucleus
T30 ventral lateral geniculate nucleus

VENTRICLES (v)

vlc lateral ventricle, central
vld lateral ventricle, dorsal
vlp lateral ventricle, posterior horn
v3e third ventricle, epithalamic
v3h third ventricle, hypothalamic
v3t third ventricle, thalamic
v4 fourth ventricle

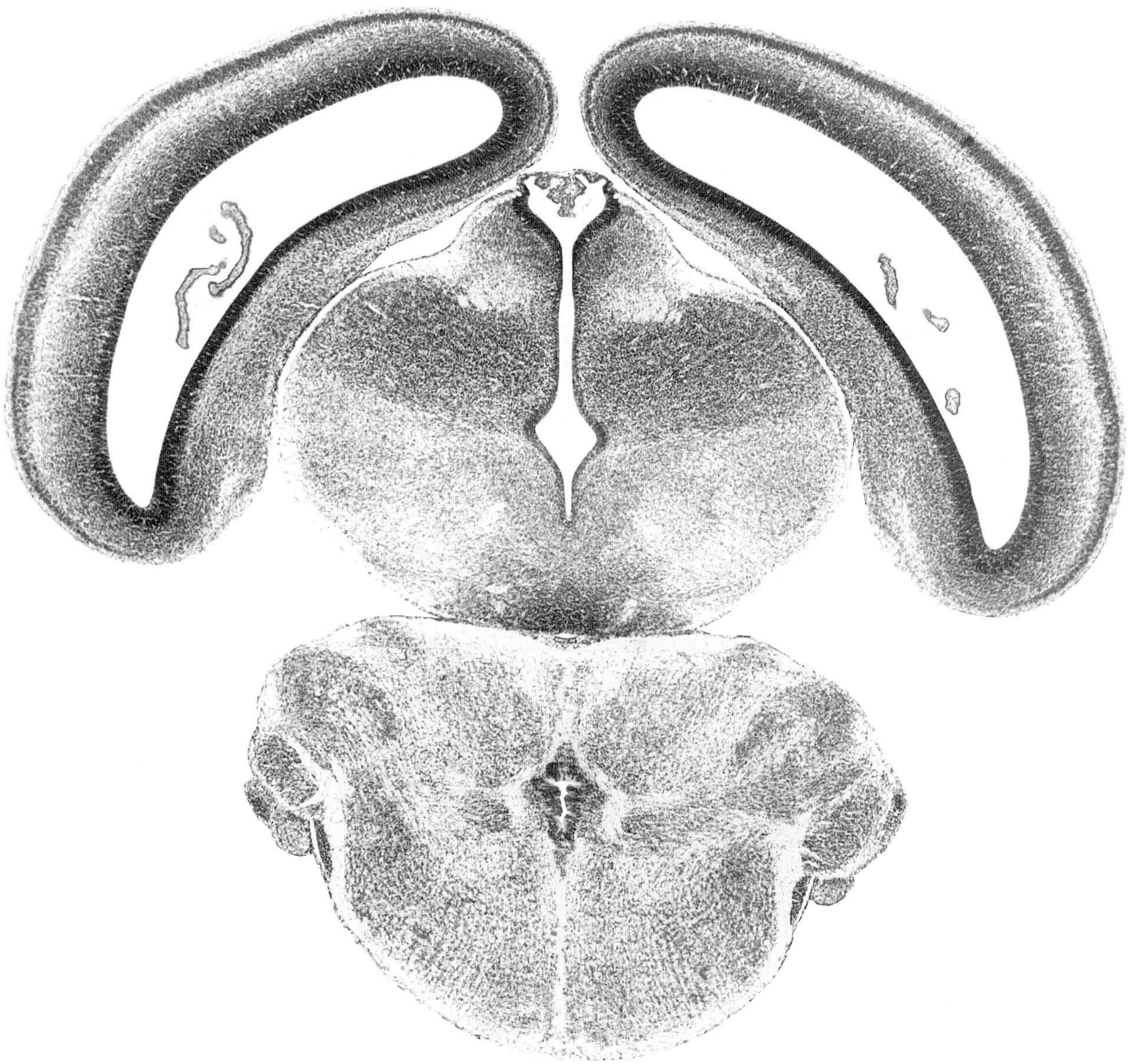

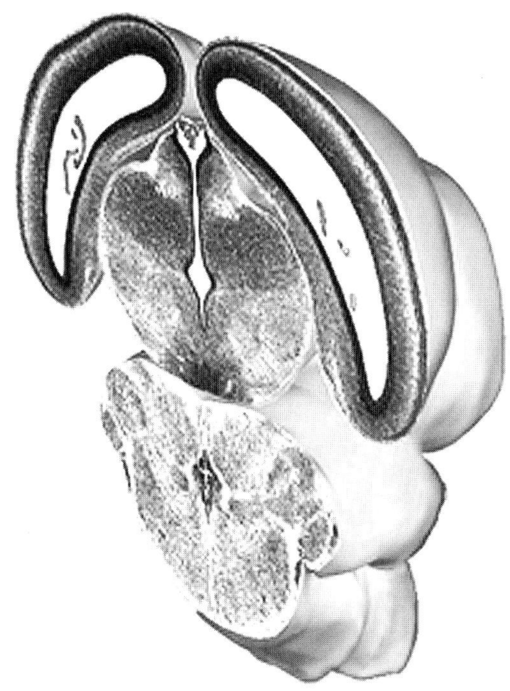

PONTINE AREA (X)

X1 pontine neuroepithelium
X3a anterior pons (or differentiating field)
X3p posterior pons (or differentiating field)
X8 superior central raphe nucleus
X12p posterior precerebellar extramural migration
X15 lateral vestibular nucleus
X16 superior vestibular nucleus
X23 cochlear nuclei (or differentiating field)
X23d dorsal cochlear nucleus (or differentiating field)
X23v ventral cochlear nucleus (or differentiating field)
X35 parabrachial nucleus

MEDULLA (Y)

Y3 medulla (or differentiating field)
Y5 medullary raphe
Y6 medullary reticular formation
Y7 prepositus nucleus
Y8 inferior olive
Y18 magnocellular reticular nucleus

CORONAL FIGURE 14

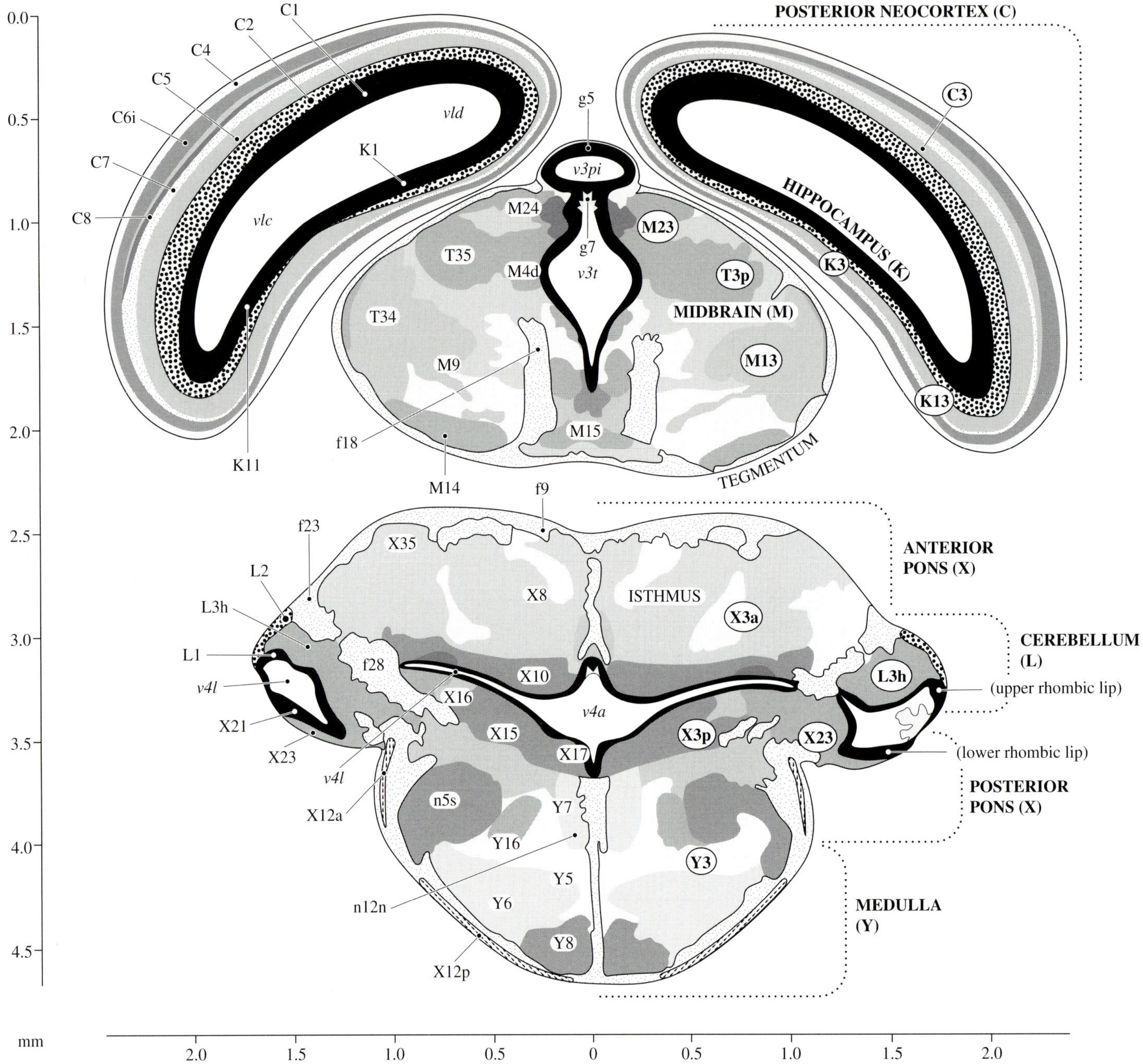

CEREBRAL CORTEX (C)

- C1 neocortical neuroepithelium
- C2 neocortical subventricular zone
- **C3** neocortex (or differentiating field)
- C4 molecular layer (or primordial plexiform layer)
- C5 intermediate zone
- C6i cortical plate, infragranular part
- C7 white matter
- C8 subplate

FIBER TRACTS (f)

- f9 cerebral peduncle
- f18 habenulopeduncular tract
- f23 superior cerebellar peduncle
- f28 lateral lemniscus

GLANDS AND PARANEURAL STRUCTURES (g)

- g5 pineal gland (primordium)
- g7 subcommissural organ

HIPPOCAMPAL FORMATION (K)

- K1 hippocampal (Ammon's horn) neuroepithelium
- **K3** hippocampus (or differentiating field)
- K11 subicular neuroepithelium
- **K13** subicular area (or differentiating field)

CEREBELLUM (L)

- L1 cerebellar neuroepithelium
- L2 external germinal layer
- **L3h** cerebellar hemisphere (or differentiating field)

MIDBRAIN (M)

- M4d dorsal periaqueductal gray
- M9 midbrain reticular formation
- **M13** tegmentum (or differentiating field)
- M14 substantia nigra
- M15 interpeduncular nucleus
- **M23** pretectum (or differentiating field)
- M24 anterior pretectal nucleus

NERVES AND NUCLEI, CRANIAL (n)

- n5s spinal trigeminal nucleus
- n12n hypoglossal nucleus

THALAMUS (T)

- **T3p** posterior thalamus (or differentiating field)
- T34 medial geniculate nucleus
- T35 posterior nuclear complex

VENTRICLES (v)

- *vlc* lateral ventricle, central
- *vld* lateral ventricle, dorsal
- *v3pi* third ventricle, pineal recess
- *v3t* third ventricle, thalamic
- *v4a* fourth ventricle, anterior
- *v4l* fourth ventricle, lateral recess

PONTINE AREA (X)

- **X3a** anterior pons (or differentiating field)
- **X3p** posterior pons (or differentiating field)
- X8 superior central raphe nucleus
- X10 ventral tegmental nucleus
- X12a anterior precerebellar extramural migration
- X12p posterior precerebellar extramural migration
- X15 lateral vestibular nucleus
- X16 superior vestibular nucleus
- X17 medial vestibular nucleus
- X21 cochlear nuclei neuroepithelium
- **X23** cochlear nuclei (or differentiating field)
- X35 parabrachial nucleus

MEDULLA (Y)

- **Y3** medulla (or differentiating field)
- Y5 medullary raphe
- Y6 medullary reticular formation
- Y7 prepositus nucleus
- Y8 inferior olive
- Y16 solitary nucleus

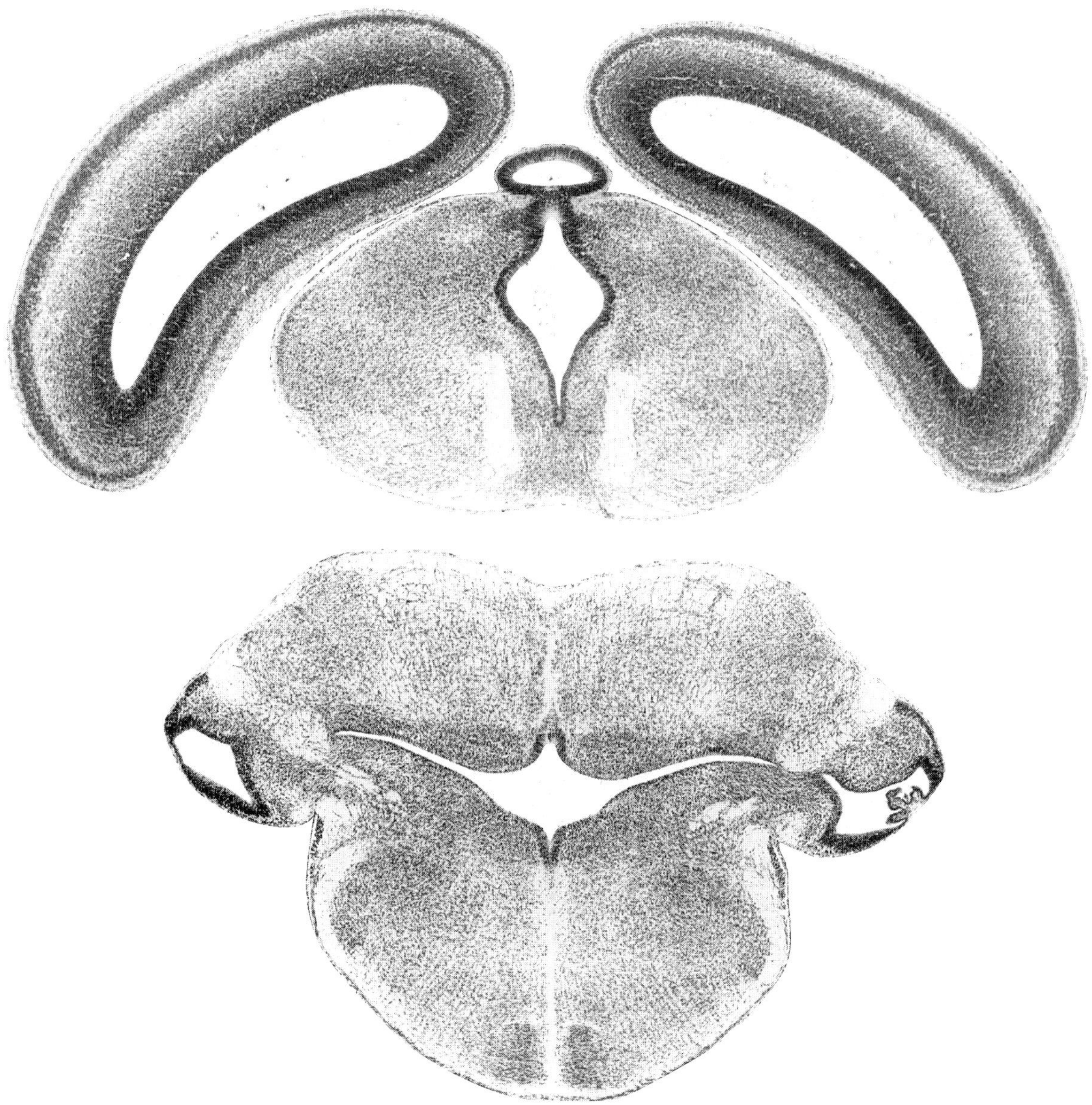

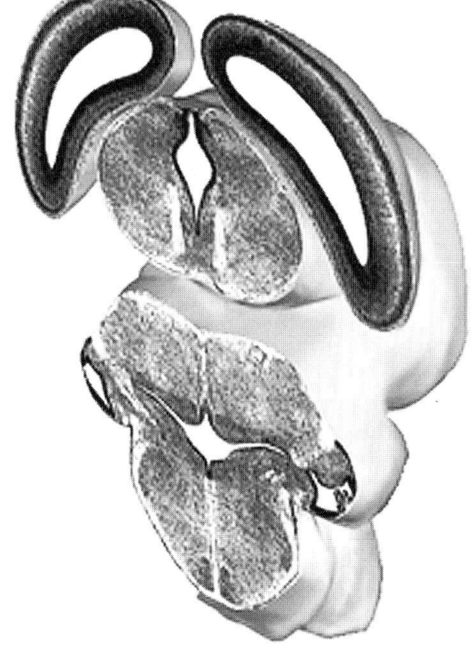

CORONAL FIGURE 15

CEREBRAL CORTEX (C)

- C1 neocortical neuroepithelium
- C2 neocortical subventricular zone
- **C3** neocortex (or differentiating field)
- C4 molecular layer (or primordial plexiform layer)
- C5 intermediate zone
- C6i cortical plate, infragranular part
- C7 white matter
- C8 subplate

FIBER TRACTS (f)

- f9 cerebral peduncle
- f19 posterior commissure
- f23 superior cerebellar peduncle

GLANDS AND PARANEURAL STRUCTURES (g)

- g4 choroid plexus of fourth ventricle
- g5 pineal gland (primordium)
- g7 subcommissural organ

CEREBELLUM (L)

- L1 cerebellar neuroepithelium
- L2 external germinal layer
- **L3h** cerebellar hemisphere (or differentiating field)
- L5 Purkinje cell layer
- L9 cerebellar germinal trigone

MIDBRAIN (M)

- M4 periaqueductal gray
- M8 red nucleus
- M9 midbrain reticular formation
- **M13** tegmentum (or differentiating field)
- M14 substantia nigra
- M15 interpeduncular nucleus
- M16 dorsal raphe
- M17 ventral tegmental nucleus
- M18 parabigeminal nucleus
- M21 pretectal neuroepithelium
- **M23** pretectum (or differentiating field)
- M25 nucleus of the posterior commissure
- M28 posterior pretectal nucleus

NERVES AND NUCLEI, CRANIAL (n)

- n3n oculomotor nucleus
- n5s spinal trigeminal nucleus
- n12n hypoglossal nucleus

THALAMUS (T)

- **T3p** posterior thalamus (or differentiating field)
- T34 medial geniculate nucleus

VENTRICLES (v)

- *vap* aqueduct, pretectal
- *vld* lateral ventricle, dorsal
- *v4a* fourth ventricle, anterior
- *v4l* fourth ventricle, lateral recess
- *v4p* fourth ventricle, posterior

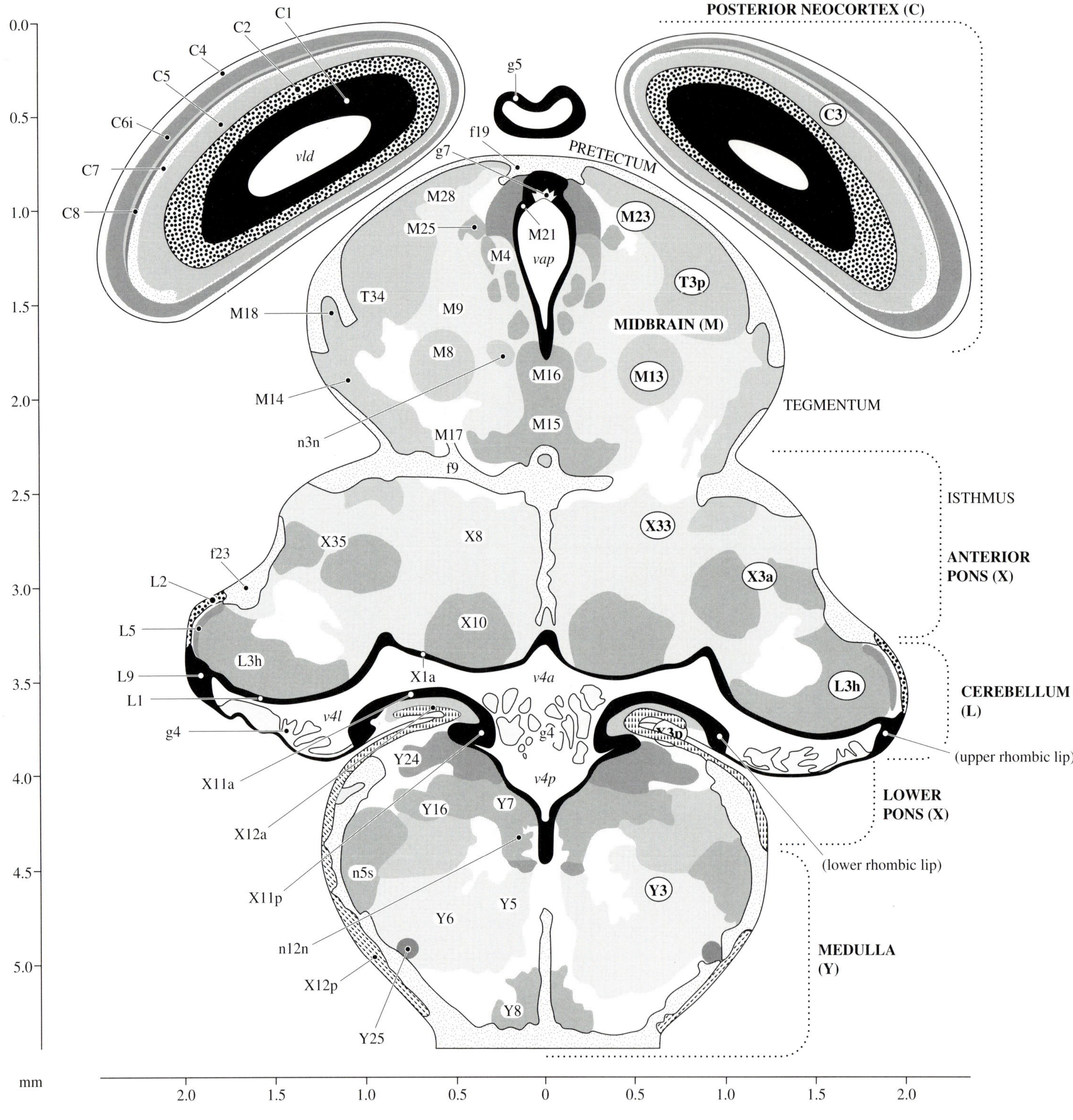

E18

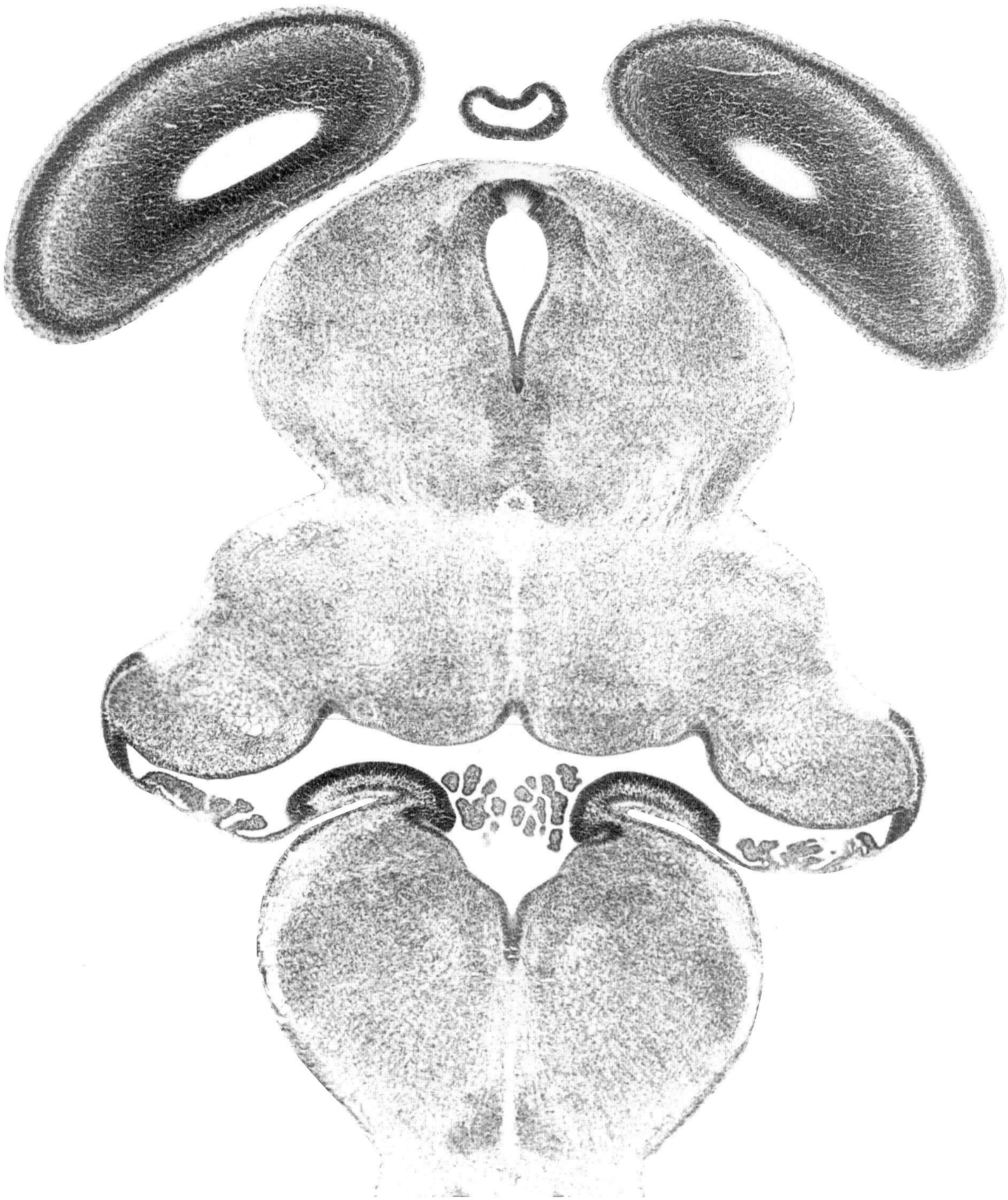

PONTINE AREA (X)

X1a pontine neuroepithelium, anterior
X3a anterior pons (or differentiating field)
X3p posterior pons (or differentiating field)
X8 superior central raphe nucleus
X10 ventral tegmental nucleus
X11a anterior precerebellar neuroepithelium
X11p posterior precerebellar neuroepithelium
X12a anterior precerebellar extramural migration
X12p posterior precerebellar extramural migration
X33 isthmus (or differentiating field)
X35 parabrachial nucleus

MEDULLA (Y)

Y3 medulla (or differentiating field)
Y5 medullary raphe
Y6 medullary reticular formation
Y7 prepositus nucleus
Y8 inferior olive
Y16 solitary nucleus
Y24 external cuneate nucleus
Y25 lateral reticular nucleus

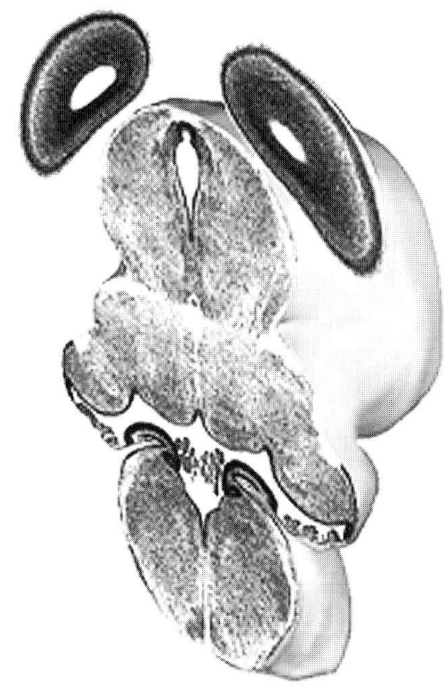

CORONAL FIGURE 16

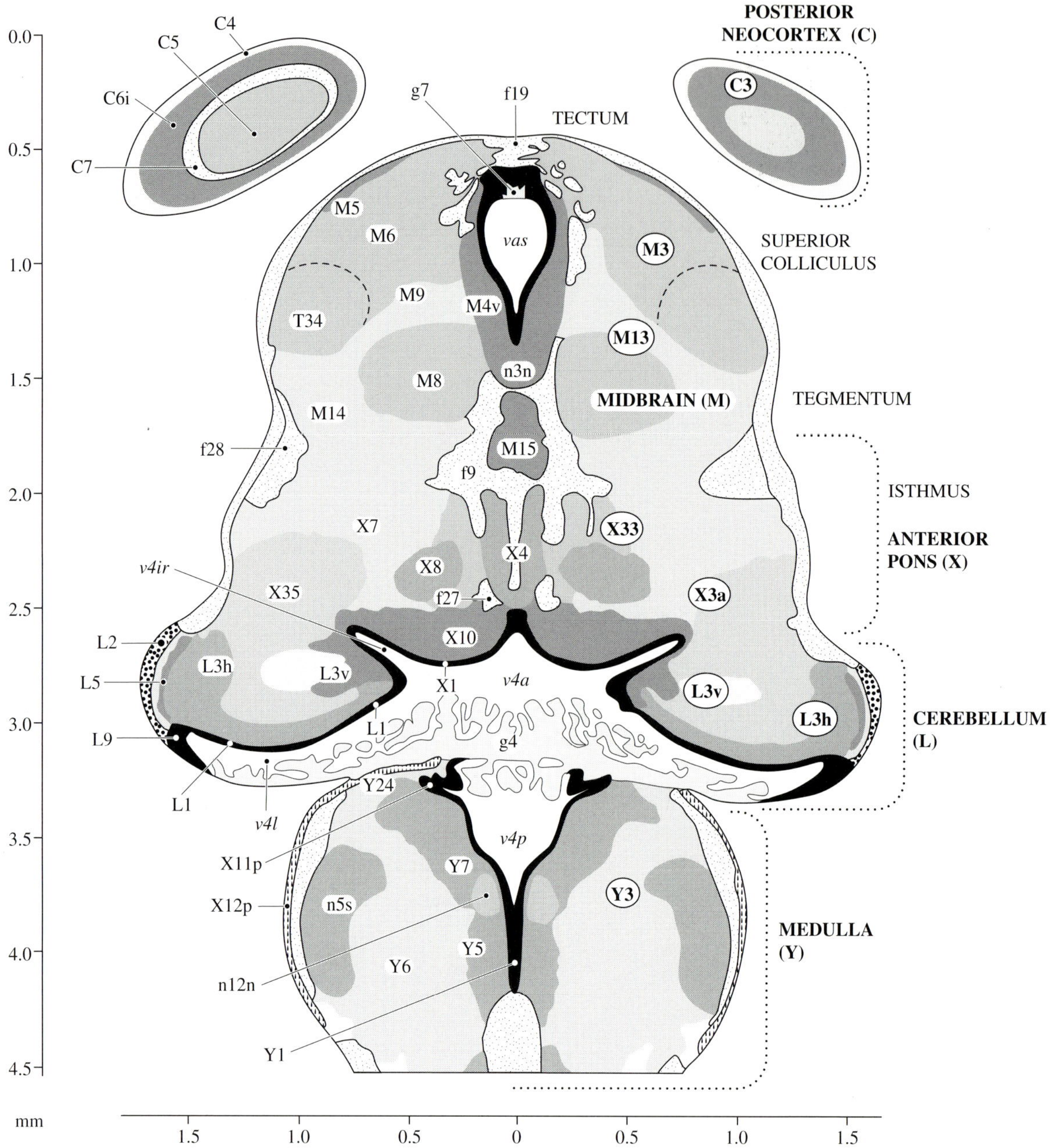

CEREBRAL CORTEX (C)

C3 neocortex (or differentiating field)
C4 molecular layer (or primordial plexiform layer)
C5 intermediate zone
C6i cortical plate, infragranular part
C7 white matter

FIBER TRACTS (f)

f9 cerebral peduncle
f19 posterior commissure
f27 medial longitudinal fasciculus
f28 lateral lemniscus

GLANDS AND PARANEURAL STRUCTURES (g)

g4 choroid plexus of fourth ventricle
g7 subcommissural organ

CEREBELLUM (L)

L1 cerebellar neuroepithelium
L2 external germinal layer
L3h cerebellar hemisphere (or differentiating field)
L3v cerebellar vermis (or differentiating field)
L5 Purkinje cell layer
L9 cerebellar germinal trigone

MIDBRAIN (M)

M3 superior colliculus (or differentiating field)
M4v ventral periaqueductal gray
M5 superior colliculus, superficial layers
M6 superior colliculus, deep layers
M8 red nucleus
M9 midbrain reticular formation
M13 tegmentum (or differentiating field)
M14 substantia nigra
M15 interpeduncular nucleus

NERVES AND NUCLEI, CRANIAL (n)

n3n oculomotor nucleus
n5s spinal trigeminal nucleus
n12n hypoglossal nucleus

THALAMUS (T)

T34 medial geniculate nucleus

VENTRICLES (v)

vas aqueduct, superior tectal
v4a fourth ventricle, anterior
v4ir fourth ventricle, isthmal recess
v4l fourth ventricle, lateral recess
v4p fourth ventricle, posterior

PONTINE AREA (X)

X1 pontine neuroepithelium
X3a anterior pons (or differentiating field)
X4 dorsal raphe nucleus
X7 pontine reticular formation
X8 superior central raphe nucleus
X10 ventral tegmental nucleus
X11p posterior precerebellar neuroepithelium
X12p posterior precerebellar extramural migration
X33 isthmus (or differentiating field)
X35 parabrachial nucleus

MEDULLA (Y)

Y1 medullary neuroepithelium
Y3 medulla (or differentiating field)
Y5 medullary raphe
Y6 medullary reticular formation
Y7 prepositus nucleus
Y24 external cuneate nucleus

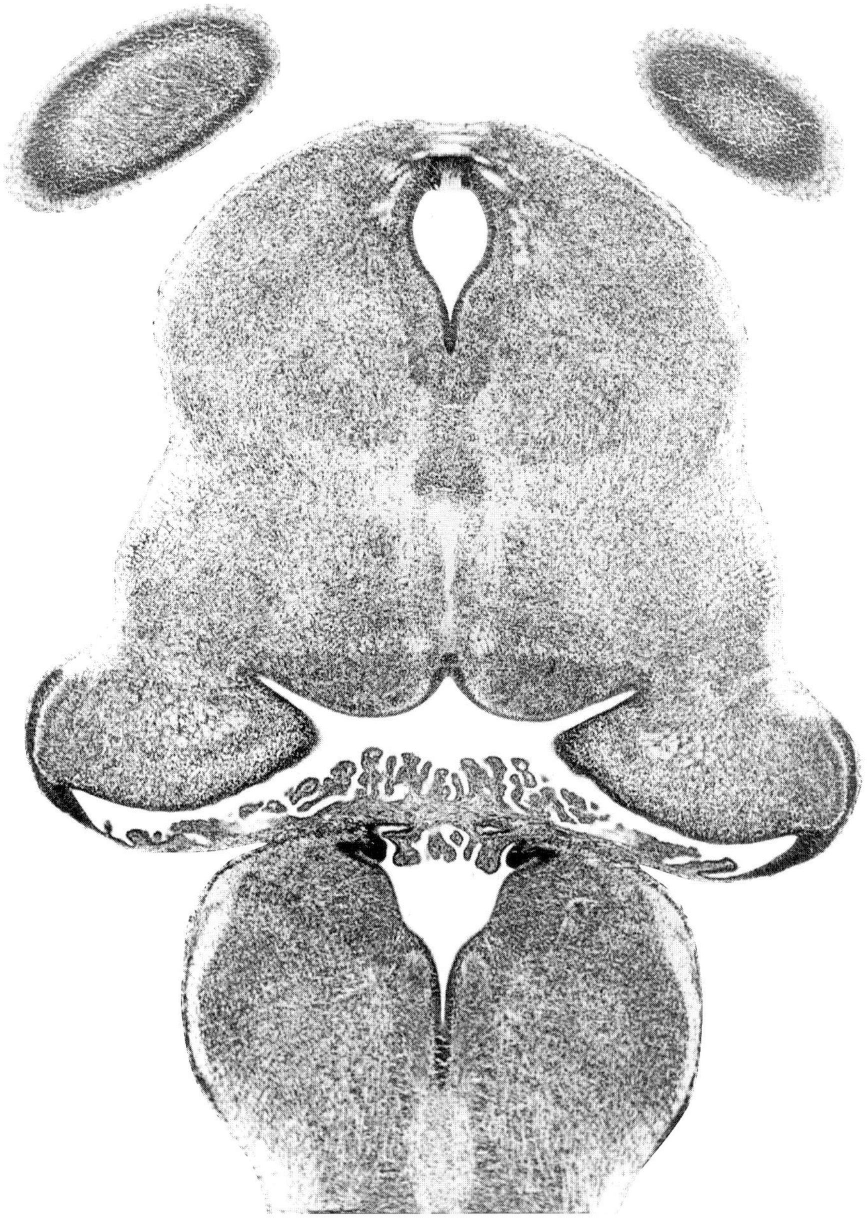

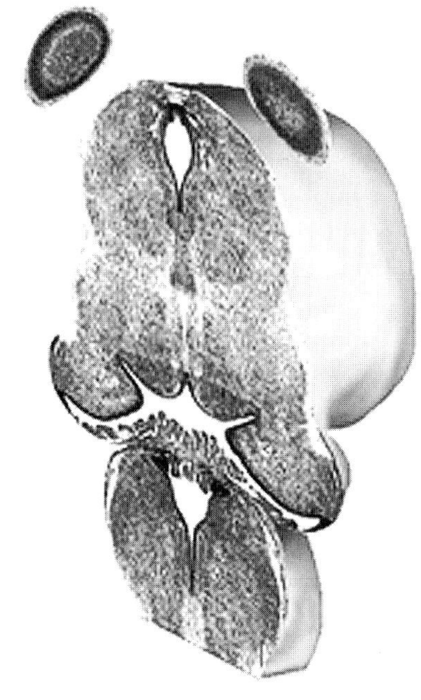

CORONAL FIGURE 17

FIBER TRACTS (f)

f22 commissure of superior colliculus

GLANDS AND PARANEURAL STRUCTURES (g)

g4 choroid plexus of fourth ventricle

CEREBELLUM (L)

L1 cerebellar neuroepithelium
L2 external germinal layer
L3h cerebellar hemisphere (or differentiating field)
L3v cerebellar vermis (or differentiating field)
L5 Purkinje cell layer
L8mi medial deep nucleus, migration

MIDBRAIN (M)

M1 superior tectal neuroepithelium
M3 superior colliculus (or differentiating field)
M4d dorsal periaqueductal gray
M4v ventral periaqueductal gray
M5 superior colliculus, superficial layers
M6 superior colliculus, deep layers
M9 midbrain reticular formation
M13 tegmentum (or differentiating field)

NERVES AND NUCLEI, CRANIAL (n)

n3n oculomotor nucleus
n5s spinal trigeminal nucleus

THALAMUS (T)

T34 medial geniculate nucleus

VENTRICLES (v)

vas aqueduct, superior tectal
v4a fourth ventricle, anterior
v4ir fourth ventricle, isthmal recess
v4p fourth ventricle, posterior

PONTINE AREA (X)

X1 pontine neuroepithelium
X3a anterior pons (or differentiating field)
X4 dorsal raphe nucleus
X7 pontine reticular formation
X8 superior central raphe nucleus
X9 dorsal tegmental nucleus
X33 isthmus (or differentiating field)
X34 locus coeruleus
X36 nucleus of lateral lemniscus

MEDULLA (Y)

Y1 medullary neuroepithelium
Y3 medulla (or differentiating field)
Y5 medullary raphe
Y6 medullary reticular formation
Y14 gracile nucleus
Y15 cuneate nucleus

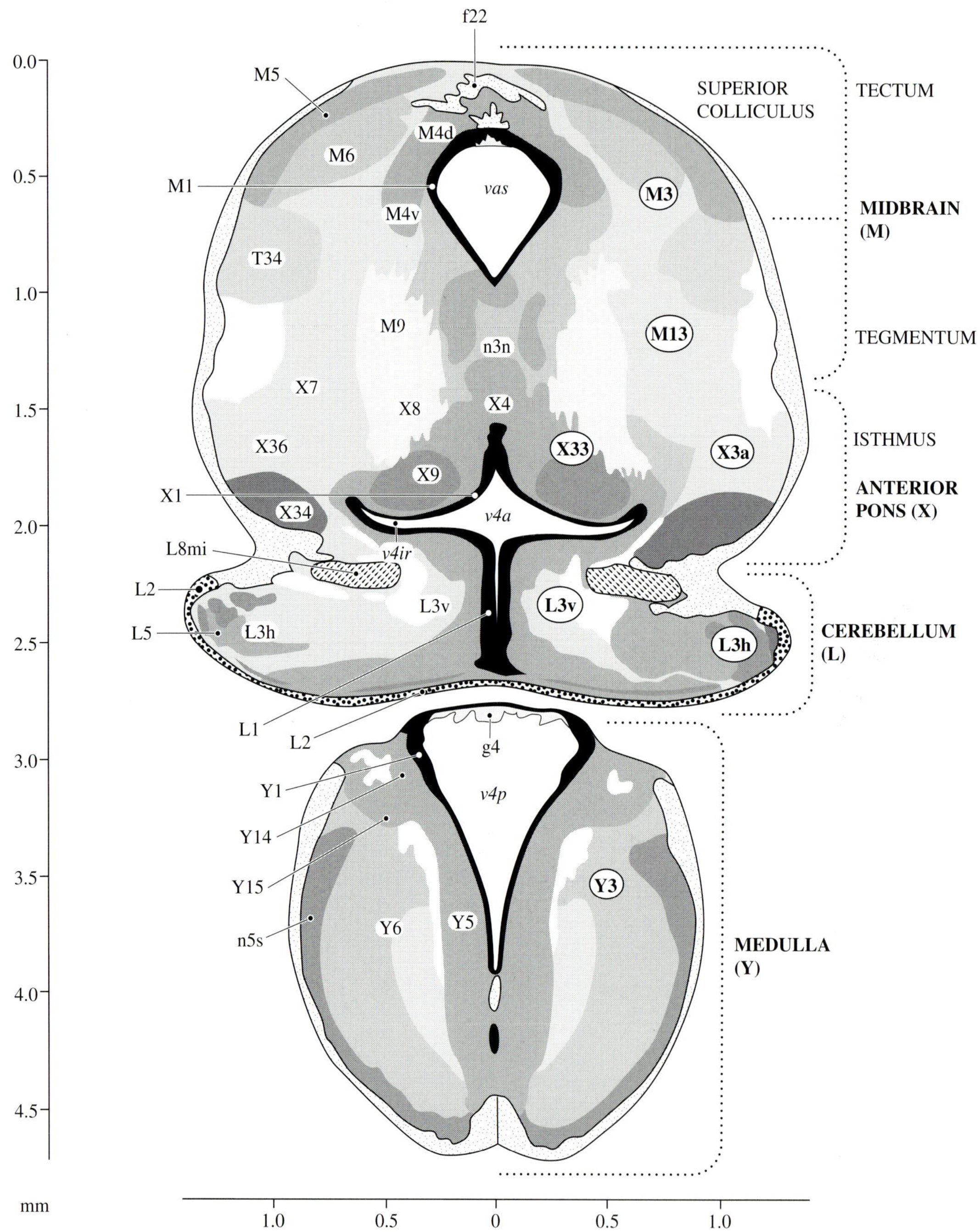

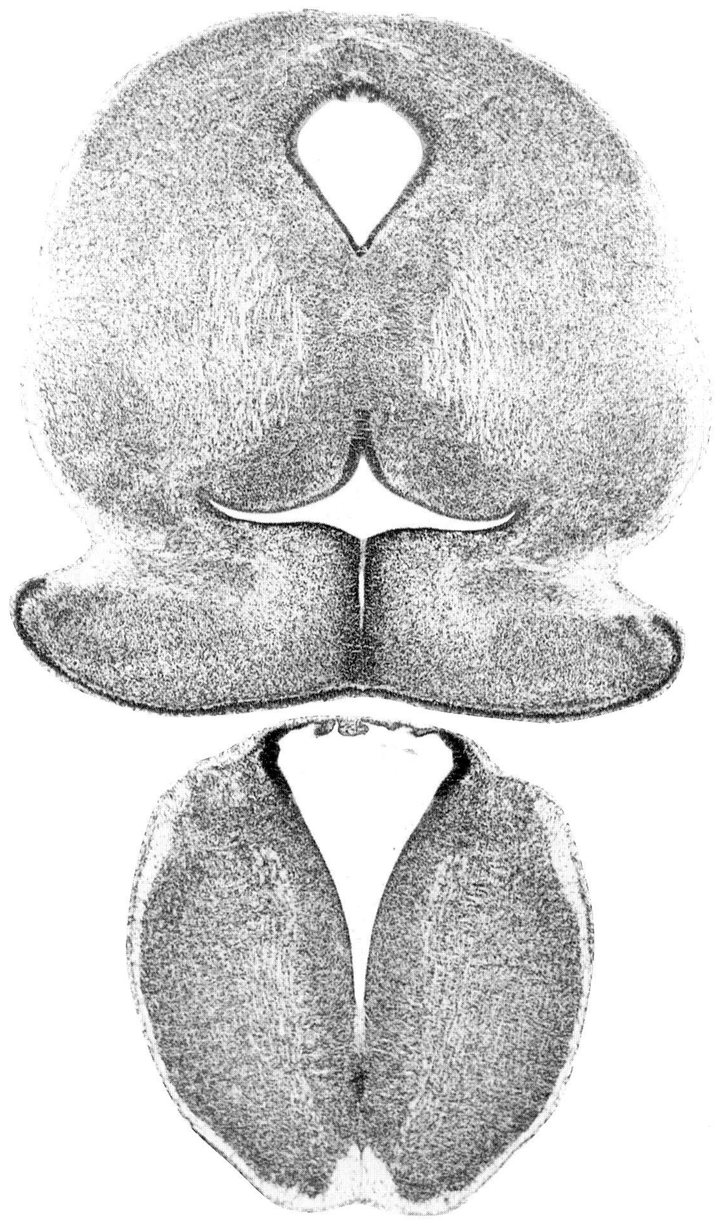

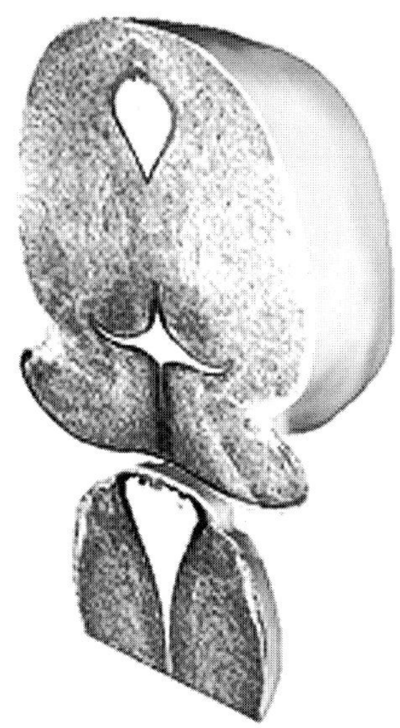

CORONAL FIGURE 18

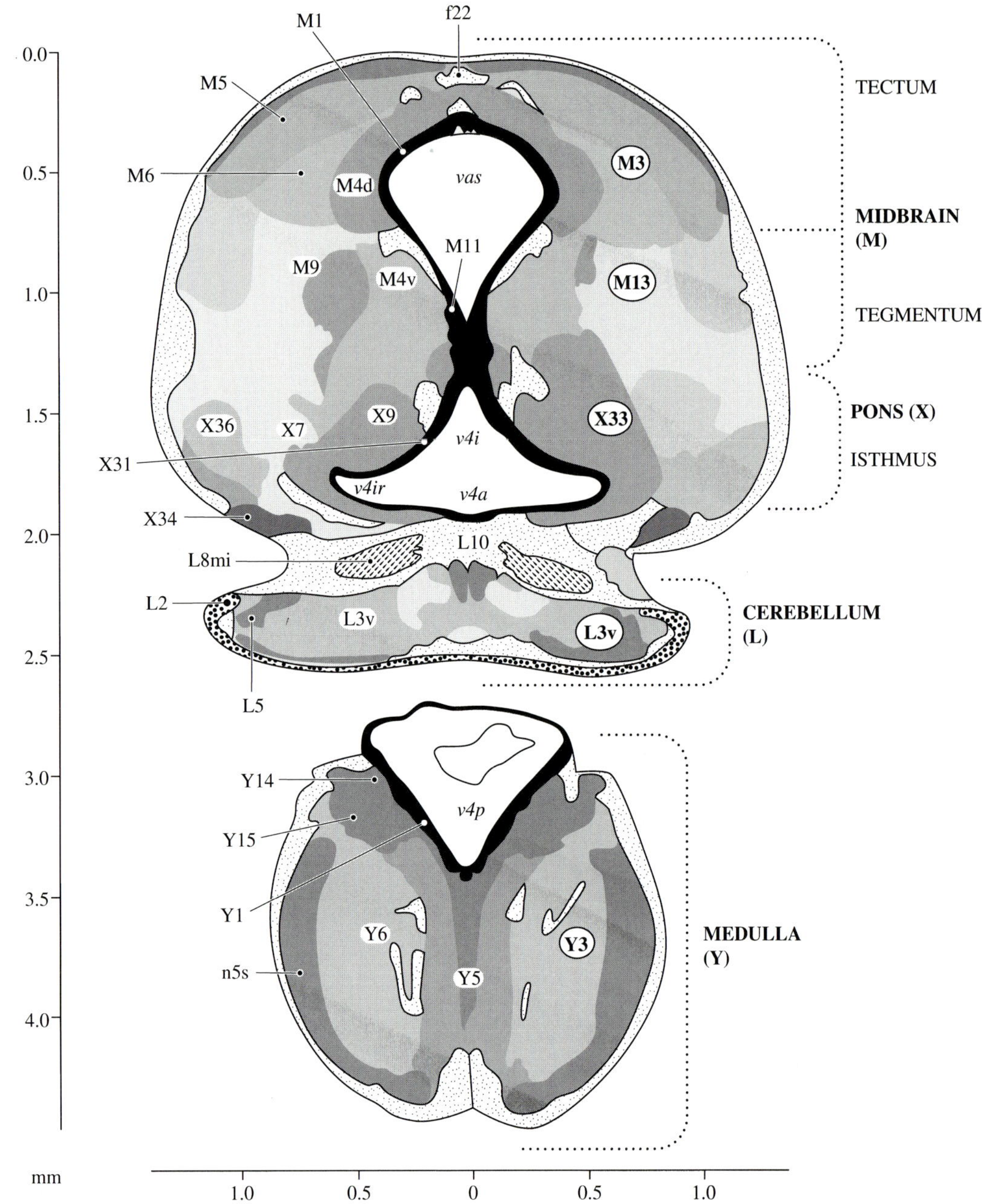

FIBER TRACTS (f)

f22 commissure of superior colliculus

CEREBELLUM (L)

L2 external germinal layer
L3v cerebellar vermis (or differentiating field)
L5 Purkinje cell layer
L8mi medial deep nucleus, migration
L10 cerebellar commissure

MIDBRAIN (M)

M1 superior tectal neuroepithelium
M3 superior colliculus (or differentiating field)
M4d dorsal periaqueductal gray
M4v ventral periaqueductal gray
M5 superior colliculus, superficial layers
M6 superior colliculus, deep layers
M9 midbrain reticular formation
M11 tegmental neuroepithelium
M13 tegmentum (or differentiating field)

NERVES AND NUCLEI, CRANIAL (n)

n5s spinal trigeminal nucleus

VENTRICLES (v)

vas aqueduct, superior tectal
v4a fourth ventricle, anterior
v4i fourth ventricle, isthmal
v4ir fourth ventricle, isthmal recess
v4p fourth ventricle, posterior

PONTINE AREA (X)

X7 pontine reticular formation
X9 dorsal tegmental nucleus
X31 isthmal neuroepithelium
X33 isthmus (or differentiating field)
X34 locus coeruleus
X36 nucleus of lateral lemniscus

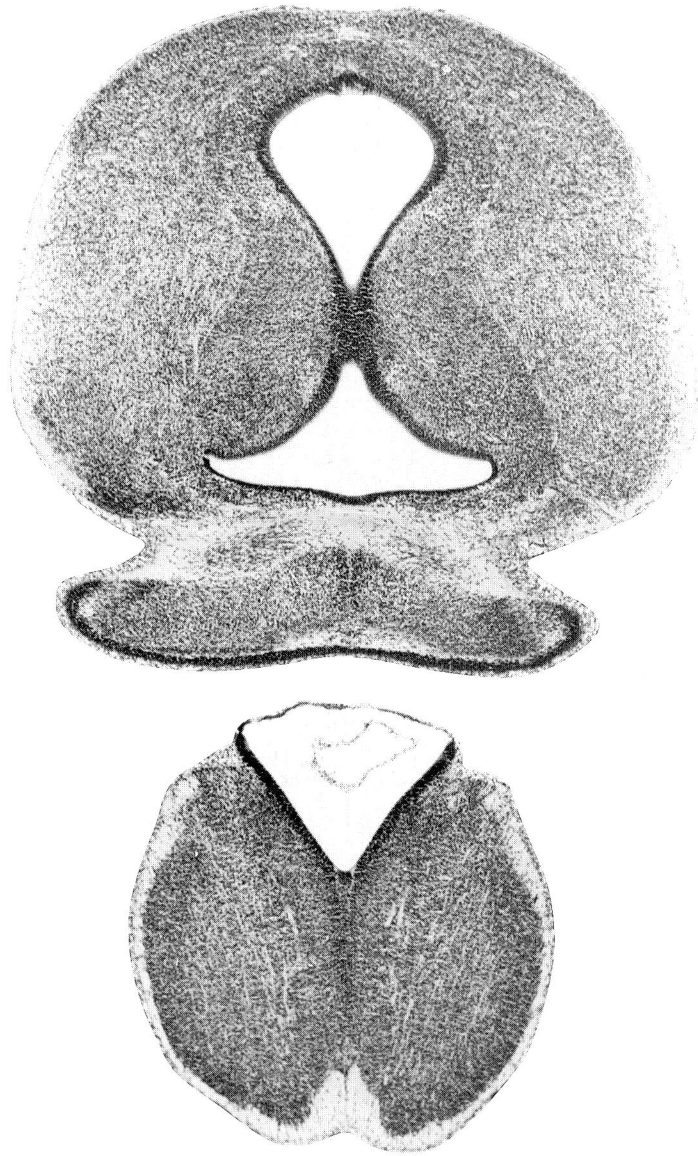

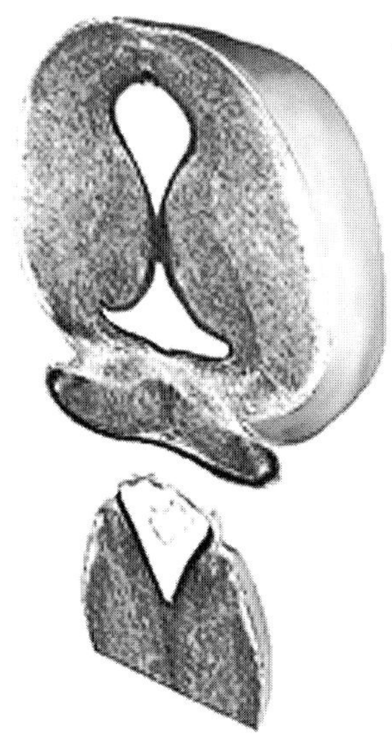

MEDULLA (Y)

Y1	medullary neuroepithelium
Y3	medulla (or differentiating field)
Y5	medullary raphe
Y6	medullary reticular formation
Y14	gracile nucleus
Y15	cuneate nucleus

CORONAL FIGURE 19

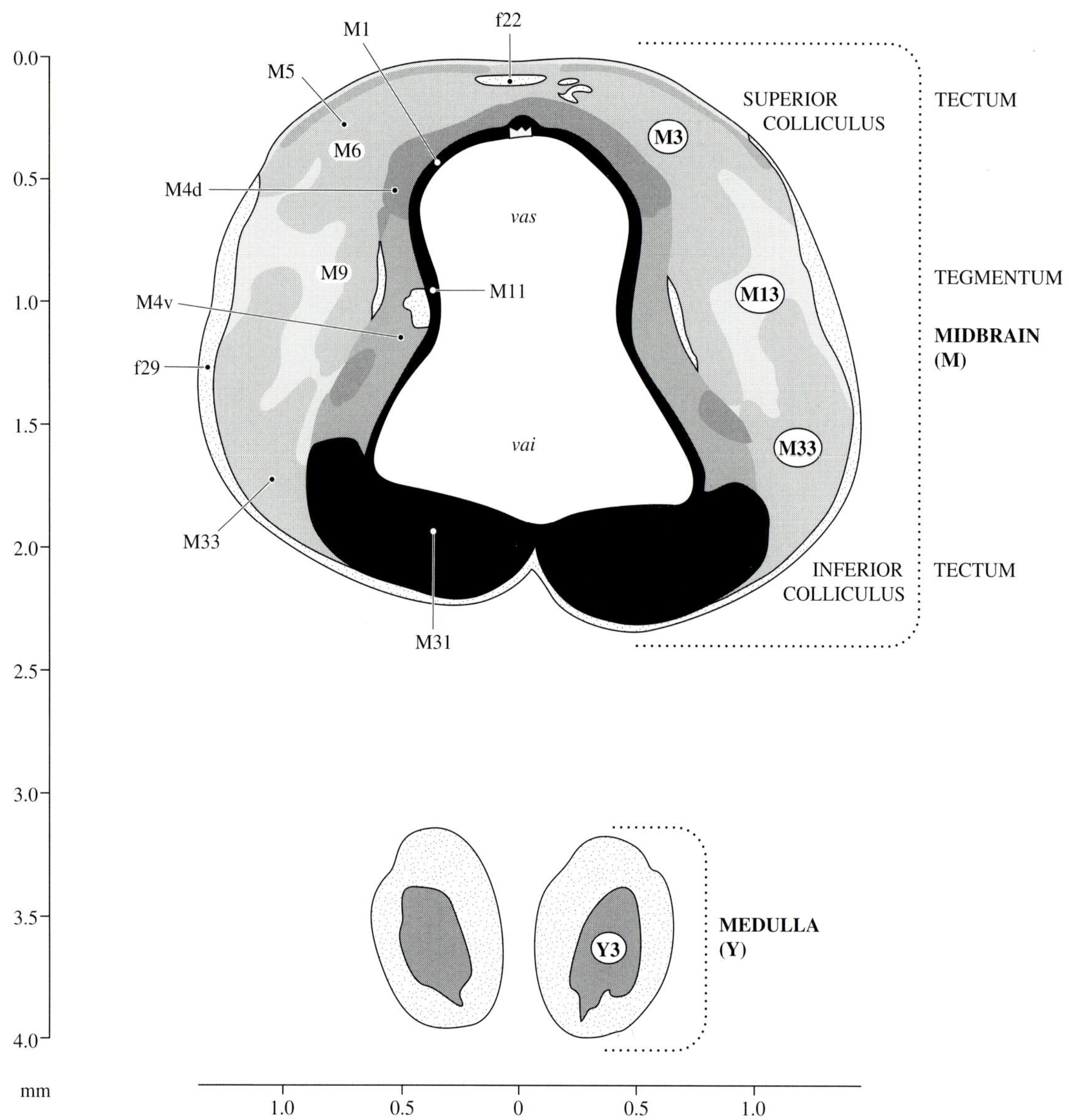

FIBER TRACTS (f)

f22	commissure of superior colliculus
f29	brachium of inferior colliculus

MIDBRAIN (M)

M1	superior tectal neuroepithelium
M3	superior colliculus (or differentiating field)
M4d	dorsal periaqueductal gray
M4v	ventral periaqueductal gray
M5	superior colliculus, superficial layers
M6	superior colliculus, deep layers
M9	midbrain reticular formation
M11	tegmental neuroepithelium
M13	tegmentum (or differentiating field)
M31	inferior tectal neuroepithelium
M33	inferior colliculus (or differentiating field)

VENTRICLES (v)

vai	aqueduct, inferior tectal
vas	aqueduct, superior tectal

MEDULLA (Y)

Y3	medulla (or differentiating field)

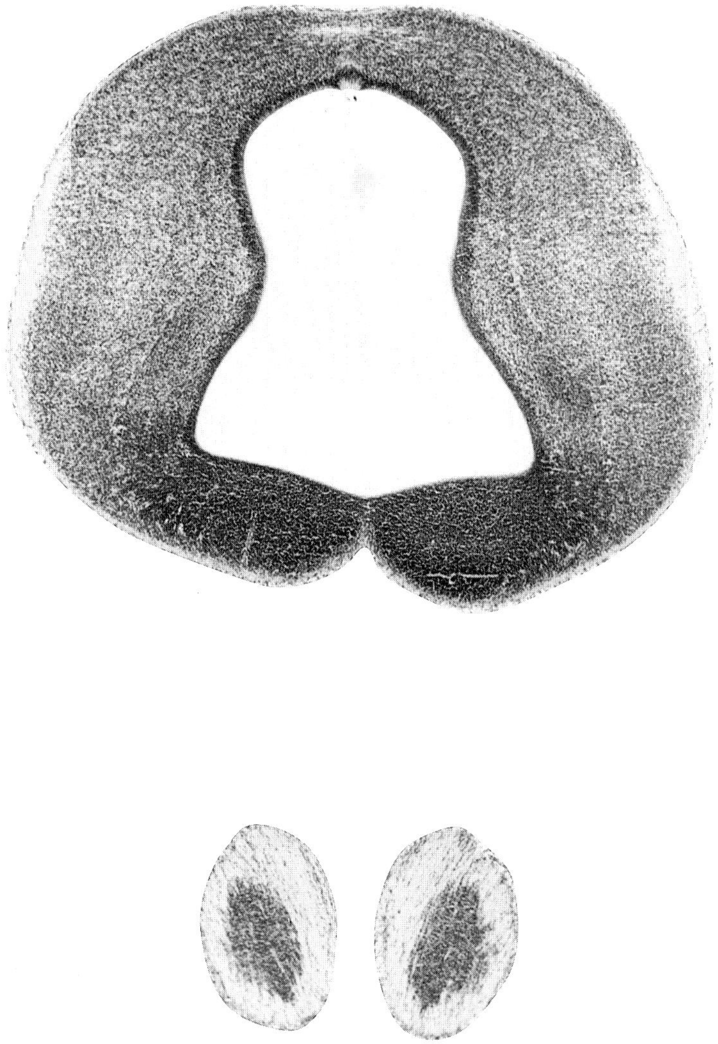

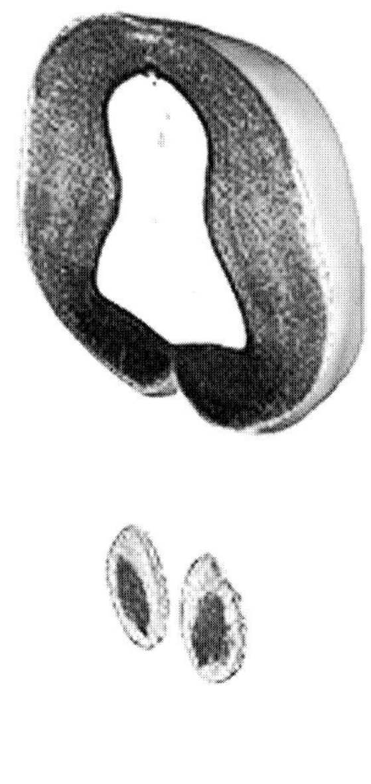

THE E20 BRAIN

LEFT, ANTERIOR VIEW

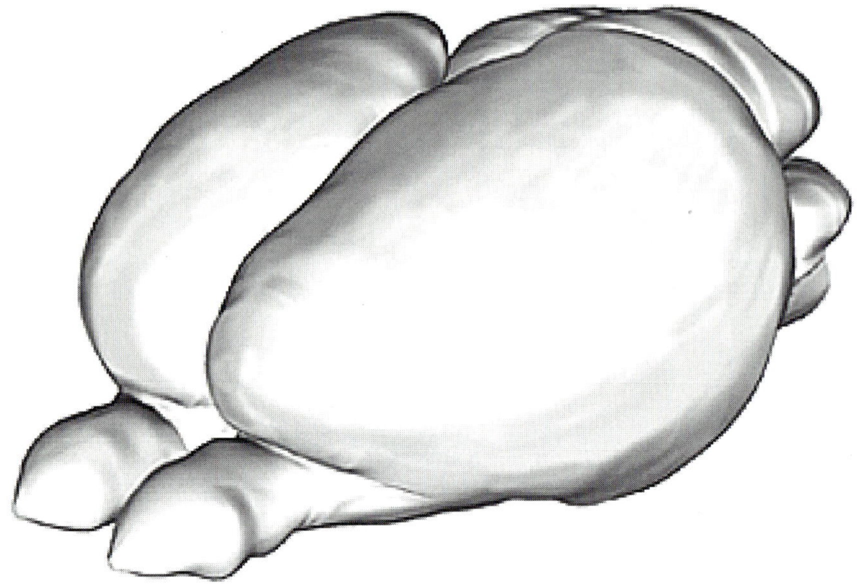

RIGHT, POSTERIOR VIEW

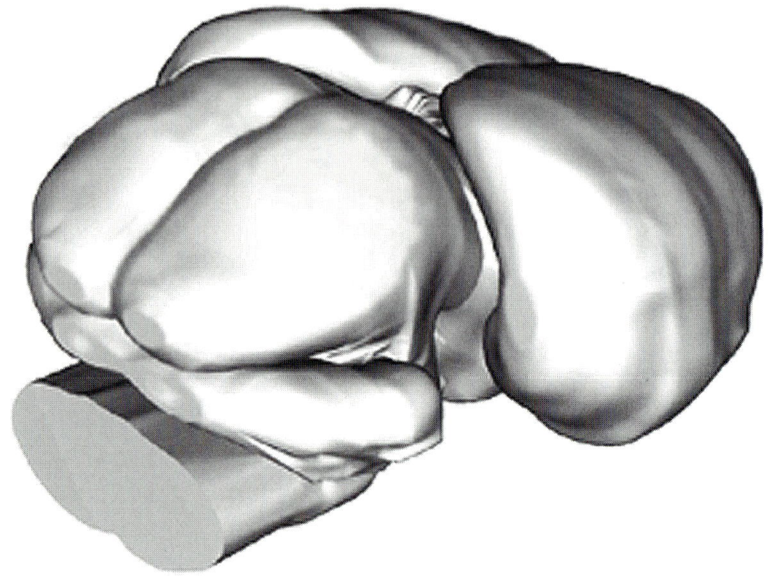

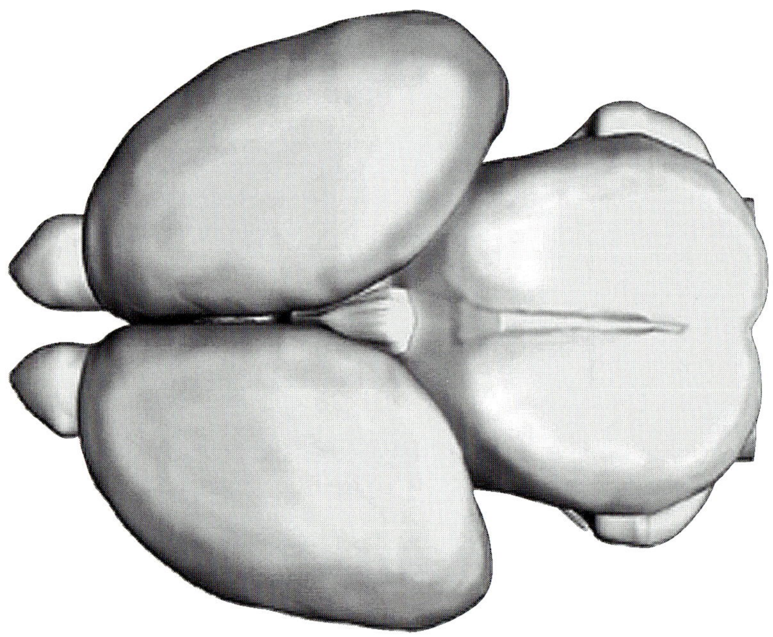

DORSAL VIEW

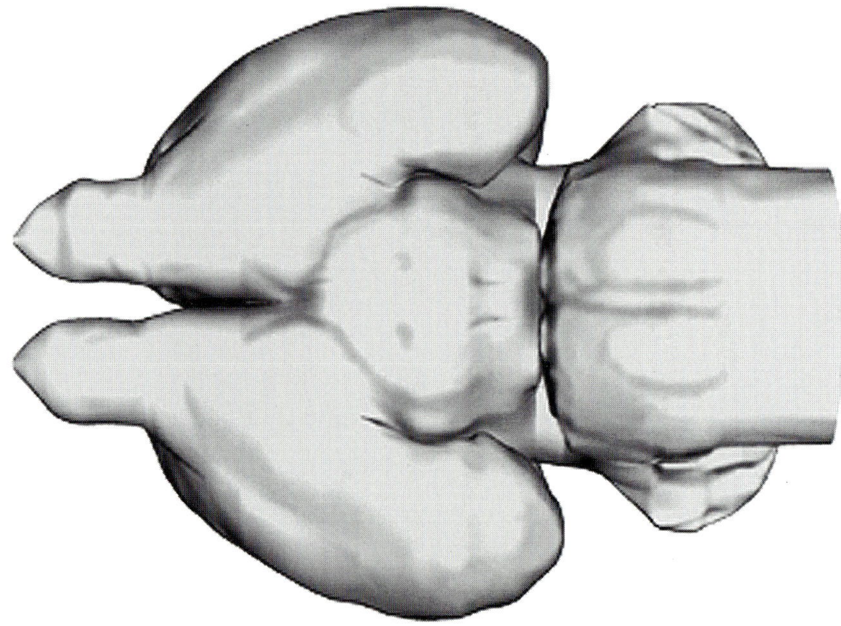

VENTRAL VIEW

SAGITTAL

FIGURES and PLATES 1-8

SAGITTAL FIGURE 1

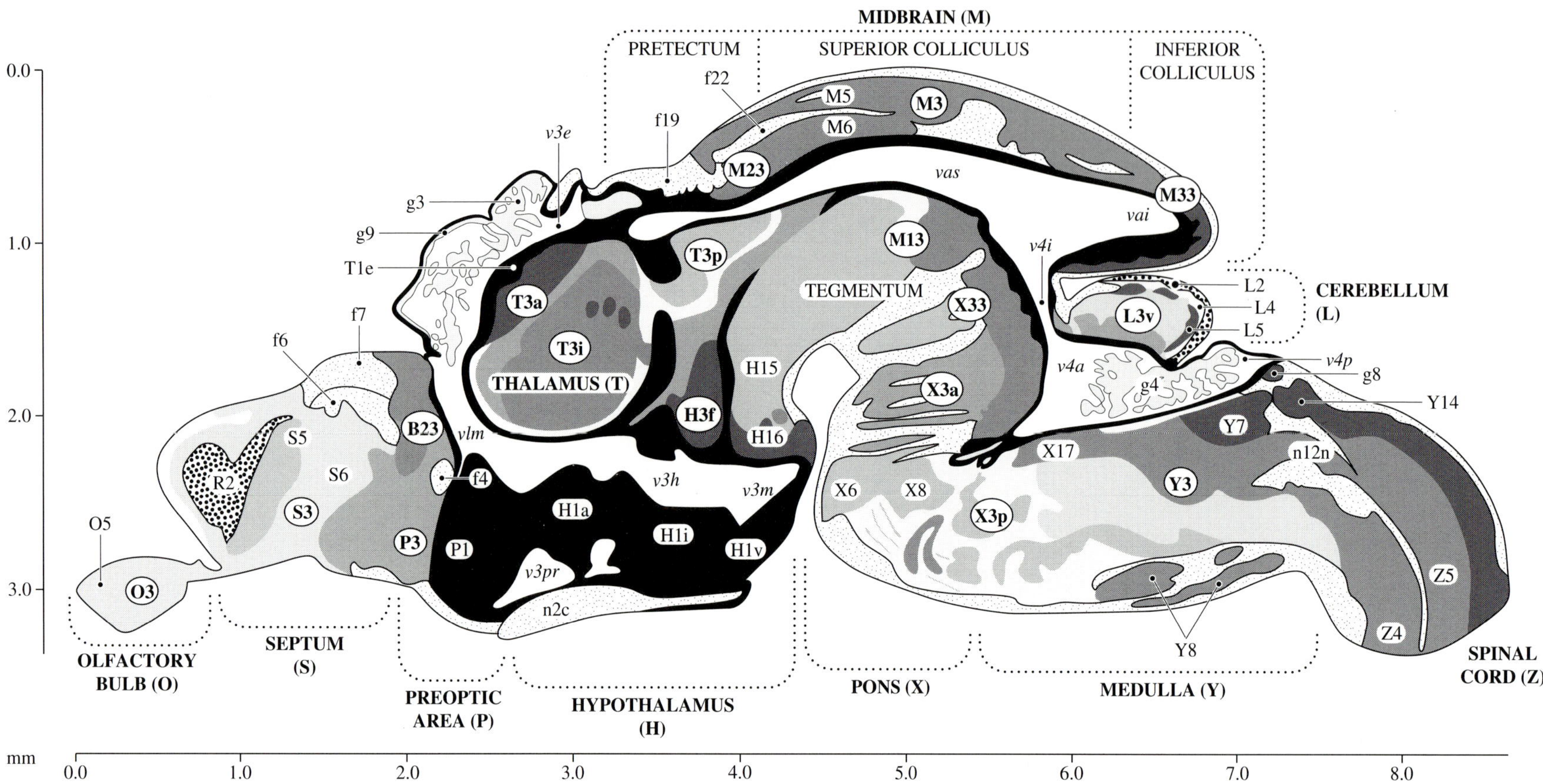

BASAL GANGLIA (B)

B23 bed nucleus of stria terminalis (or differentiating field)

FIBER TRACTS (f)

f4 anterior commissure
f6 fornix
f7 hippocampal commissure
f19 posterior commissure
f22 commissure of superior colliculus

GLANDS AND PARANEURAL STRUCTURES (g)

g3 choroid plexus of third ventricle
g4 choroid plexus of fourth ventricle,
g8 area postrema
g9 lamina terminalis

HYPOTHALAMUS (H)

H1a anterior hypothalamic neuroepithelium
H1i intermediate hypothalamic neuroepithelium
H1v ventral hypothalamic neuroepithelium
H3f fields of Forel (or differentiating field)
H15 premammillary nucleus
H16 medial mammillary nucleus

CEREBELLUM (L)

L2 external germinal layer
L3v cerebellar vermis (or differentiating field)
L4 molecular layer
L5 Purkinje cell layer

MIDBRAIN (M)

M3 superior colliculus (or differentiating field)
M5 superior colliculus, superficial layers
M6 superior colliculus, deep layers
M13 tegmentum (or differentiating field)
M23 pretectum (or differentiating field)
M33 inferior colliculus (or differentiating field)

NERVES AND NUCLEI, CRANIAL (n)

n2c optic chiasma
n12n hypoglossal nucleus

OLFACTORY BULB (O)

O3 olfactory bulb (or differentiating field)
O5 external plexiform layer

PREOPTIC AREA AND OPTIC EPITHELIA (P)

P1 preoptic neuroepithelium
P3 preoptic area (or differentiating field)

RHINENCEPHALON (R)

R2 rhinencephalic subventricular zone

SEPTAL AREA (S)

S3 septum (or differentiating field)
S5 lateral septum
S6 medial septum

THALAMUS (T)

T1e epithalamic neuroepithelium
T3a anterior thalamus (or differentiating field)
T3i intermediate thalamus (or differentiating field)
T3p posterior thalamus (or differentiating field)

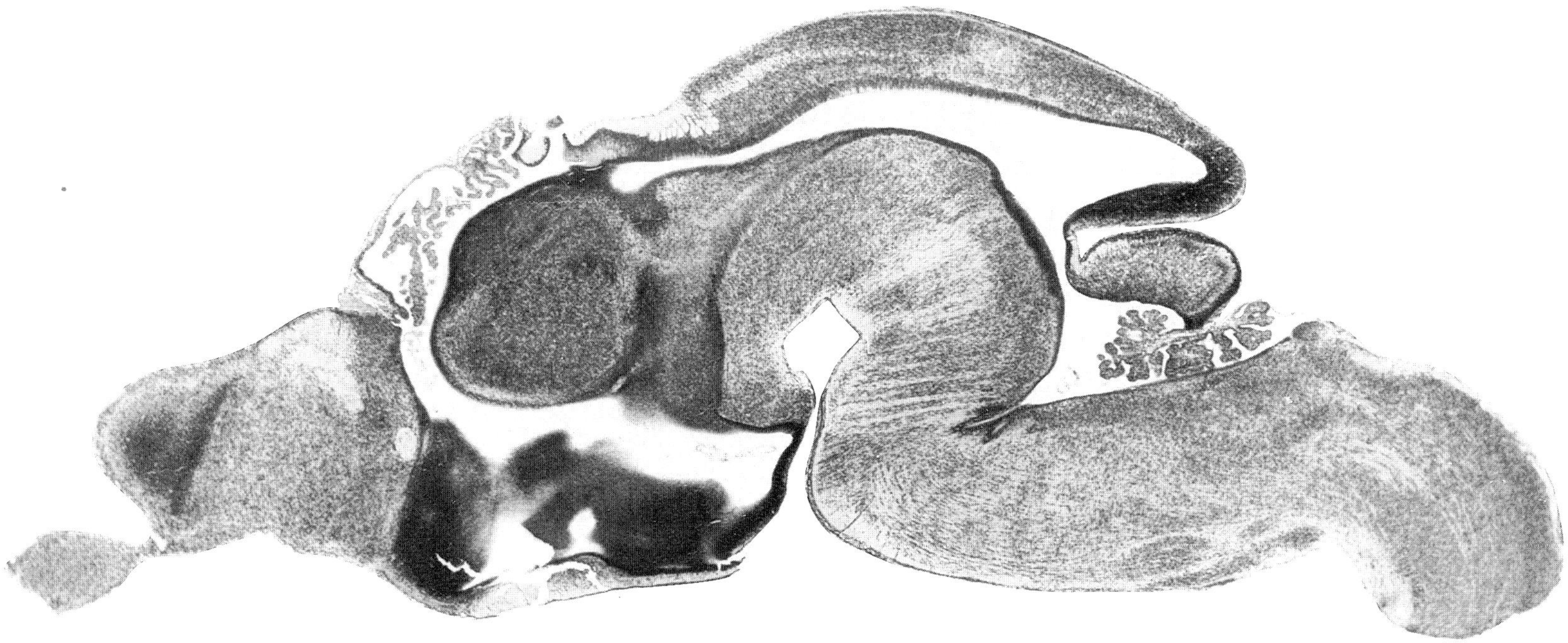

VENTRICLES (v)

vai	aqueduct, inferior tectal
vas	aqueduct, superior tectal
vlm	foramen of Monro
v3e	third ventricle, epithalamic
v3h	third ventricle, hypothalamic
v3m	third ventricle, mammillary recess
v3pr	third ventricle, preoptic recess
v4a	fourth ventricle, anterior
v4i	fourth ventricle, isthmal
v4p	fourth ventricle, posterior

PONTINE AREA (X)

X3a	anterior pons (or differentiating field)
X3p	posterior pons (or differentiating field)
X6	pontine reticulotegmental nucleus
X8	superior central raphe nucleus
X17	medial vestibular nucleus
X33	isthmus (or differentiating field)

MEDULLA (Y)

Y3	medulla (or differentiating field)
Y7	prepositus nucleus
Y8	inferior olive
Y14	gracile nucleus

SPINAL CORD (Z)

Z4	ventral horn
Z5	dorsal horn

SAGITTAL FIGURE 2

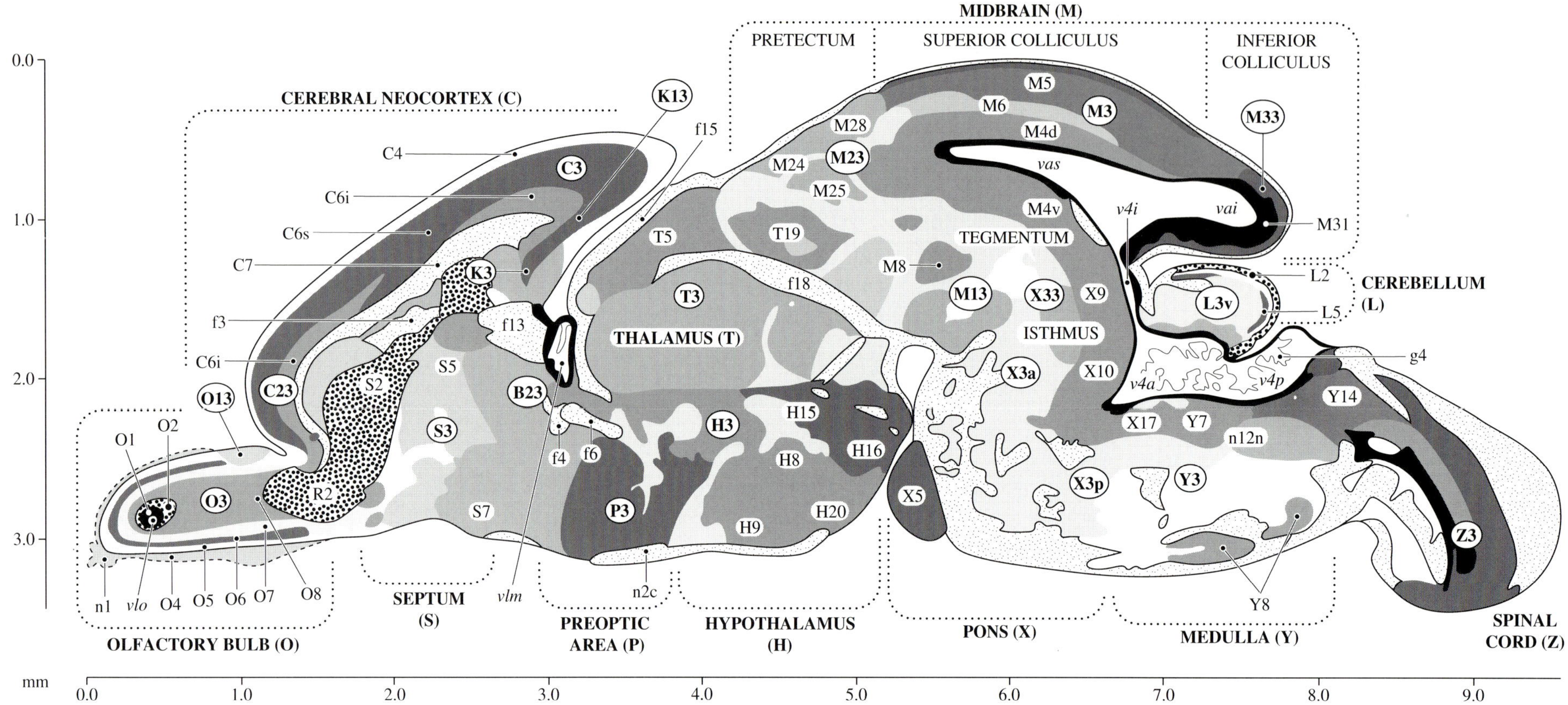

BASAL GANGLIA (B)

B23 bed nucleus of stria terminalis (or differentiating field)

CEREBRAL CORTEX (C)

C3 neocortex (or differentiating field)
C4 molecular layer (or primordial plexiform layer)
C6i cortical plate, infragranular part
C6s cortical plate, supragranular part
C7 white matter
C23 orbital and insular area (or differentiating field)

FIBER TRACTS (f)

f3 corpus callosum
f4 anterior commissure
f6 fornix
f13 fimbria
f15 stria medullaris
f18 habenulopeduncular tract

GLANDS AND PARANEURAL STRUCTURES (g)

g4 choroid plexus of fourth ventricle,

HYPOTHALAMUS (H)

H3 hypothalamus (or differentiating field)
H8 dorsomedial nucleus
H9 ventromedial nucleus
H15 premammillary nucleus
H16 medial mammillary nucleus
H20 arcuate nucleus

HIPPOCAMPAL FORMATION (K)

K3 hippocampus (or differentiating field)
K13 subicular area (or differentiating field)

CEREBELLUM (L)

L2 external germinal layer
L3v cerebellar vermis (or differentiating field)
L5 Purkinje cell layer

MIDBRAIN (M)

M3 superior colliculus (or differentiating field)
M4d dorsal periaqueductal gray
M4v ventral periaqueductal gray
M5 superior colliculus, superficial layers
M6 superior colliculus, deep layers
M8 red nucleus
M13 tegmentum (or differentiating field)
M23 pretectum (or differentiating field)
M24 anterior pretectal nucleus
M25 nucleus of the posterior commissure
M28 posterior pretectal nucleus
M31 inferior tectal neuroepithelium
M33 inferior colliculus (or differentiating field)

NERVES AND NUCLEI, CRANIAL (n)

n1 olfactory nerve
n2c optic chiasma
n12n hypoglossal nucleus

OLFACTORY BULB (O)

O1 olfactory bulb neuroepithelium
O2 olfactory bulb subventricular zone
O3 olfactory bulb (or differentiating field)
O4 glomerular layer
O5 external plexiform layer
O6 mitral cell layer
O7 internal plexiform layer
O8 granular layer
O13 accessory olfactory bulb (or differentiating field)

PREOPTIC AREA AND OPTIC EPITHELIA (P)

P3 preoptic area (or differentiating field)

RHINENCEPHALON (R)

R2 rhinencephalic subventricular zone

SEPTAL AREA (S)

S2 septal subventricular zone
S3 septum (or differentiating field)
S5 lateral septum
S7 diagonal band (or differentiating field)

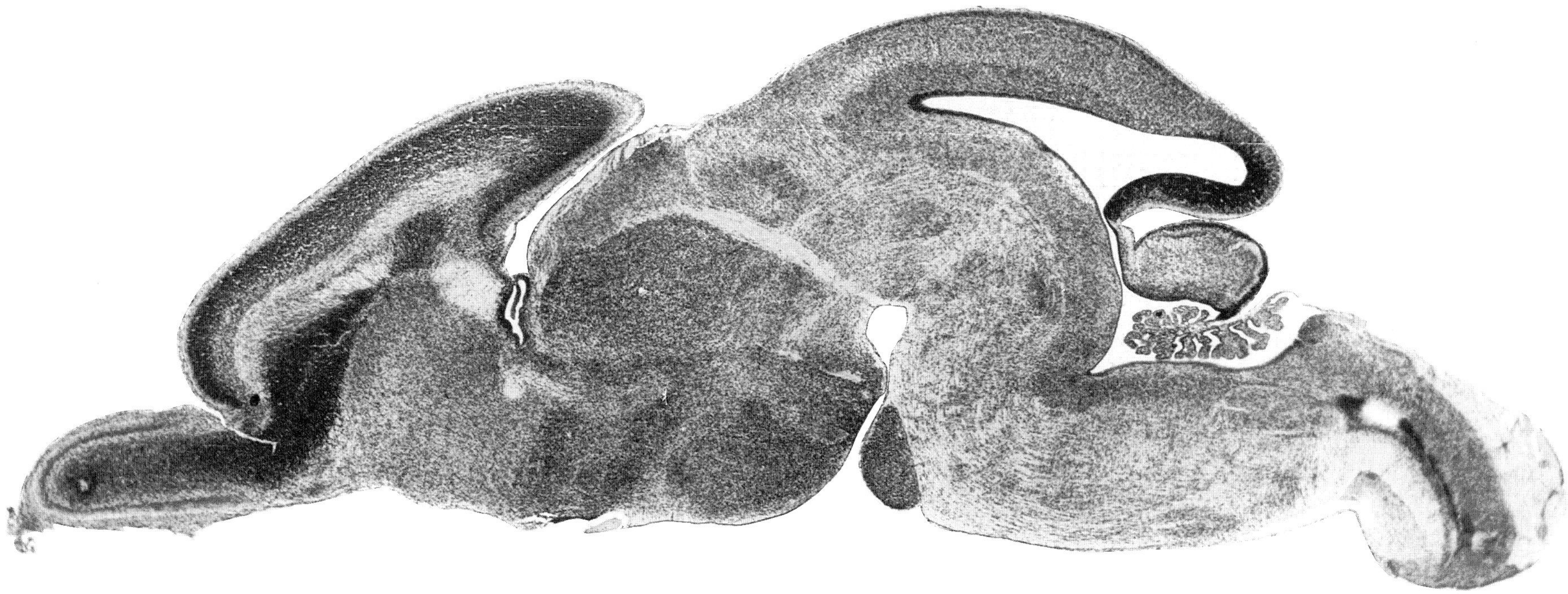

THALAMUS (T)

T3	thalamus (or differentiating field)
T5	lateral habenular nucleus
T19	parafascicular nucleus

VENTRICLES (v)

vai	aqueduct, inferior tectal
vas	aqueduct, superior tectal
vlm	foramen of Monro
vlo	olfactory ventricle
v4a	fourth ventricle, anterior
v4i	fourth ventricle, isthmal
v4p	fourth ventricle, posterior

PONTINE AREA (X)

X3a	anterior pons (or differentiating field)
X3p	posterior pons (or differentiating field)
X5	pontine gray nucleus
X9	dorsal tegmental nucleus
X10	ventral tegmental nucleus
X17	medial vestibular nucleus
X33	isthmus (or differentiating field)

MEDULLA (Y)

Y3	medulla (or differentiating field)
Y7	prepositus nucleus
Y8	inferior olive
Y14	gracile nucleus

SPINAL CORD (Z)

Z3	spinal cord (or differentiating field)

SAGITTAL FIGURE 3

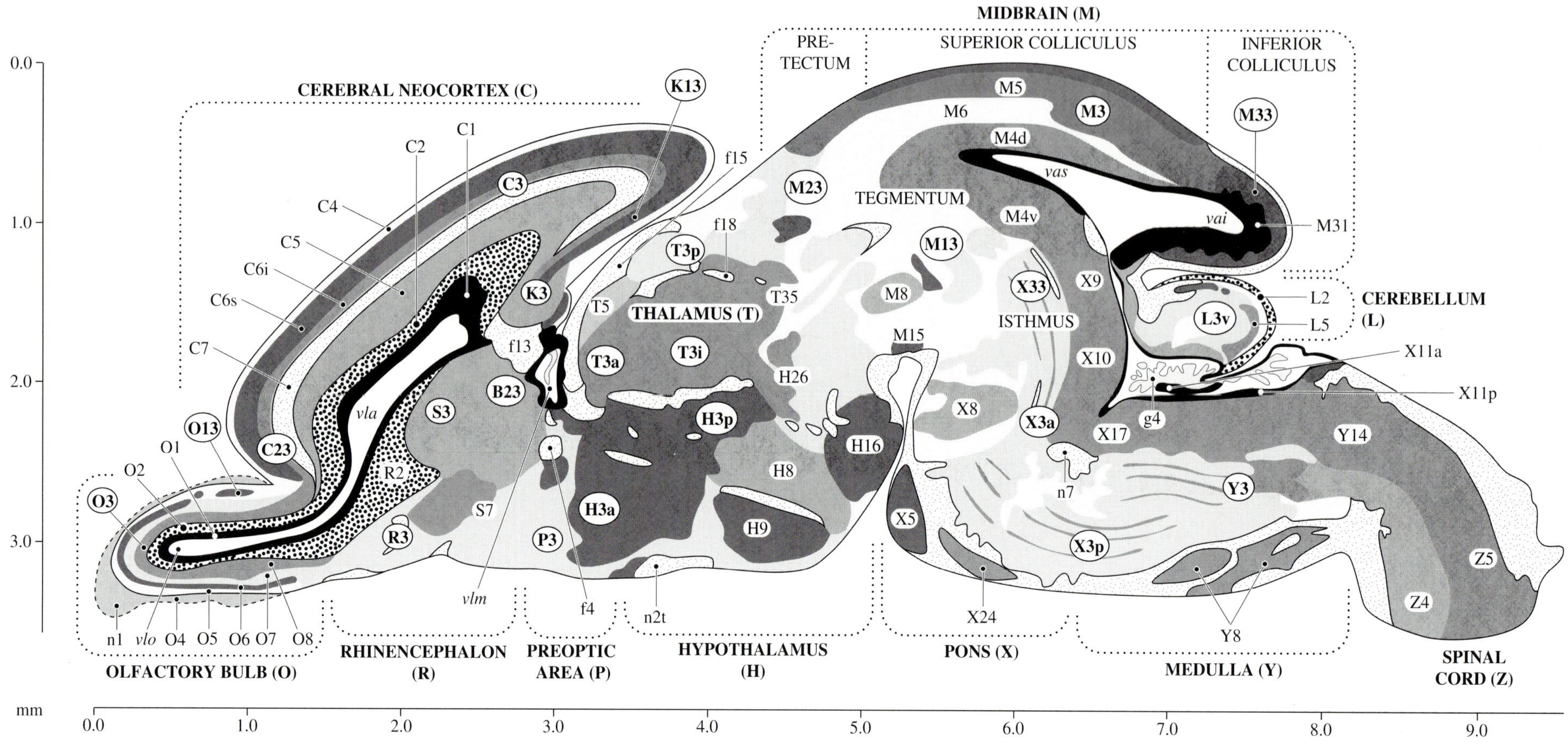

BASAL GANGLIA (B)

B23	bed nucleus of stria terminalis (or differentiating field)

CEREBRAL CORTEX (C)

C1	neocortical neuroepithelium
C2	neocortical subventricular zone
C3	neocortex (or differentiating field)
C4	molecular layer (or primordial plexiform layer)
C5	intermediate zone
C6i	cortical plate, infragranular part
C6s	cortical plate, supragranular part
C7	white matter
C23	orbital and insular area (or differentiating field)

FIBER TRACTS (f)

f4	anterior commissure
f13	fimbria
f15	stria medullaris
f18	habenulopeduncular tract

GLANDS AND PARANEURAL STRUCTURES (g)

g4	choroid plexus of fourth ventricle

HYPOTHALAMUS (H)

H3a	anterior hypothalamus (or differentiating field)
H3p	posterior hypothalamus (or differentiating field)
H8	dorsomedial nucleus
H9	ventromedial nucleus
H16	medial mammillary nucleus
H26	posterior hypothalamic area

HIPPOCAMPAL FORMATION (K)

K3	hippocampus (or differentiating field)
K13	subicular area (or differentiating field)

CEREBELLUM (L)

L2	external germinal layer
L3v	cerebellar vermis (or differentiating field)
L5	Purkinje cell layer

MIDBRAIN (M)

M3	superior colliculus (or differentiating field)
M4d	dorsal periaqueductal gray
M4v	ventral periaqueductal gray
M5	superior colliculus, superficial layers
M6	superior colliculus, deep layers
M8	red nucleus
M13	tegmentum (or differentiating field)
M15	interpeduncular nucleus
M23	pretectum (or differentiating field)
M31	inferior tectal neuroepithelium
M33	inferior colliculus (or differentiating field)

NERVES AND NUCLEI, CRANIAL (n)

n1	olfactory nerve
n2t	optic tract
n7	facial nerve

OLFACTORY BULB (O)

O1	olfactory bulb neuroepithelium
O2	olfactory bulb subventricular zone
O3	olfactory bulb (or differentiating field)
O4	glomerular layer
O5	external plexiform layer
O6	mitral cell layer
O7	internal plexiform layer
O8	granular layer
O13	accessory olfactory bulb (or differentiating field)

PREOPTIC AREA AND OPTIC EPITHELIA (P)

P3	preoptic area (or differentiating field)

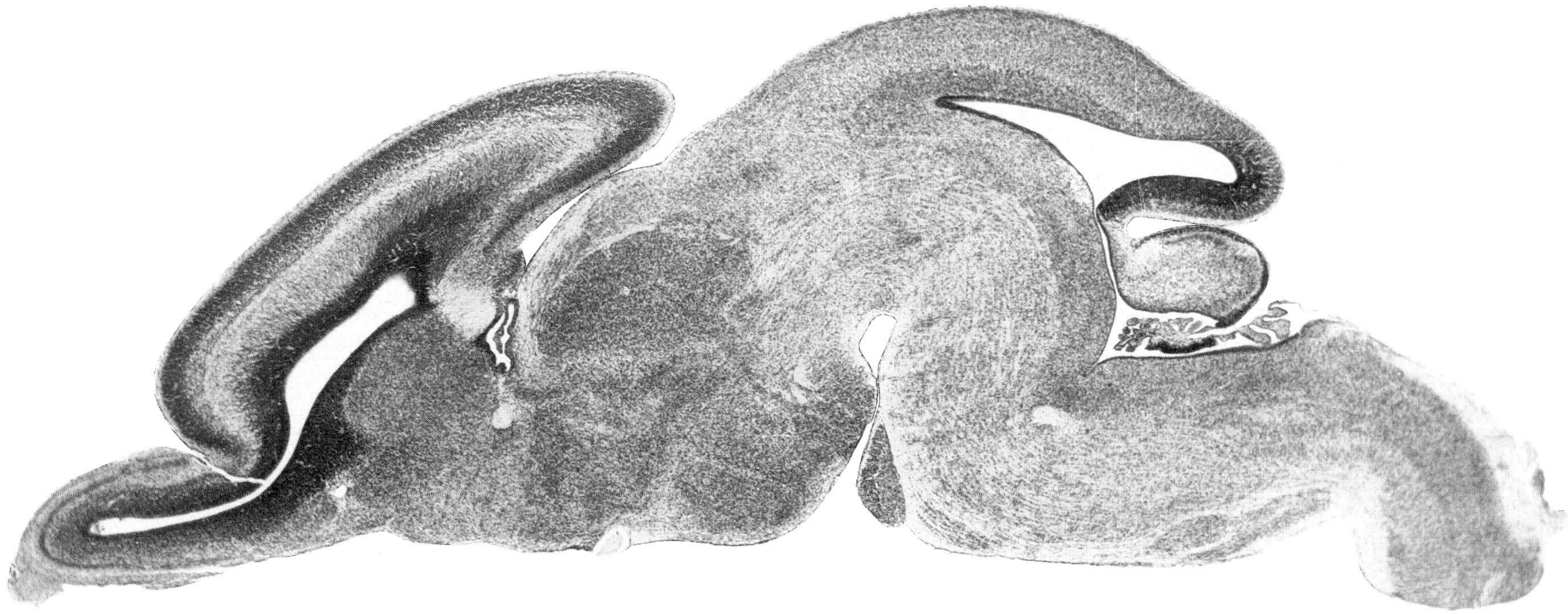

RHINENCEPHALON (R)

R2 rhinencephalic subventricular zone
R3 rhinencephalon (or differentiating field)

SEPTAL AREA (S)

S3 septum (or differentiating field)
S7 diagonal band (or differentiating field)

THALAMUS (T)

T3a anterior thalamus (or differentiating field)
T3i intermediate thalamus (or differentiating field)
T3p posterior thalamus (or differentiating field)
T5 lateral habenular nucleus
T35 posterior nuclear complex

VENTRICLES (v)

vai aqueduct, inferior tectal
vas aqueduct, superior tectal
vla lateral ventricle, anterior horn
vlm foramen of Monro
vlo olfactory ventricle

PONTINE AREA (X)

X3a anterior pons (or differentiating field)
X3p posterior pons (or differentiating field)
X5 pontine gray nucleus
X8 superior central raphe nucleus
X9 dorsal tegmental nucleus
X10 ventral tegmental nucleus
X11a anterior precerebellar neuroepithelium
X11p posterior precerebellar neuroepithelium
X17 medial vestibular nucleus
X24 superior olivary nucleus
X33 isthmus (or differentiating field)

MEDULLA (Y)

Y3 medulla (or differentiating field)
Y8 inferior olive
Y14 gracile nucleus

SPINAL CORD (Z)

Z4 ventral horn
Z5 dorsal horn

SAGITTAL FIGURE 4

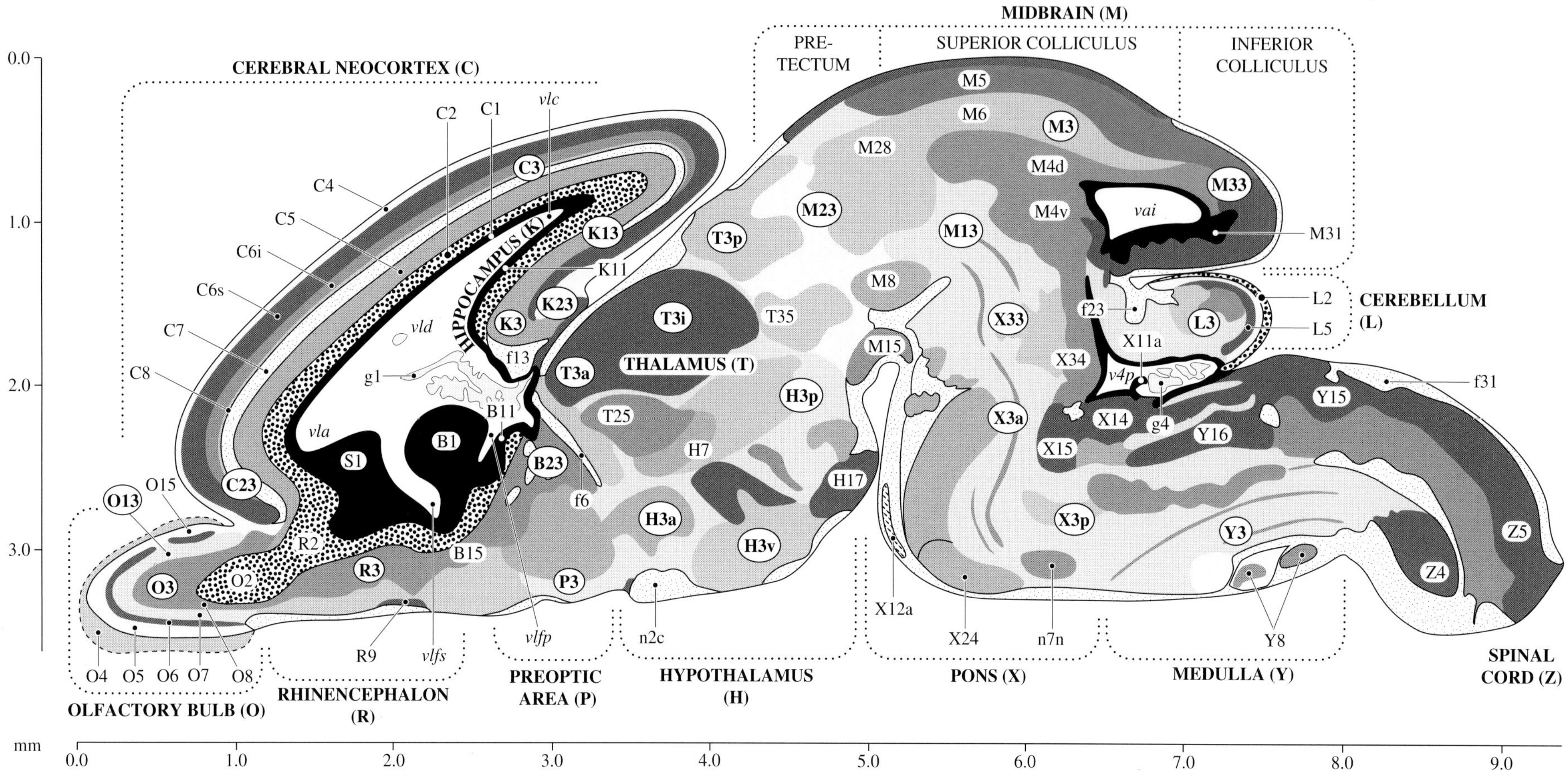

BASAL GANGLIA (B)

B1 striatal neuroepithelium
B11 pallidal neuroepithelium
B15 substantia innominata
B23 bed nucleus of stria terminalis (or differentiating field)

CEREBRAL CORTEX (C)

C1 neocortical neuroepithelium
C2 neocortical subventricular zone
C3 neocortex (or differentiating field)
C4 molecular layer (or primordial plexiform layer)
C5 intermediate zone
C6i cortical plate, infragranular part
C6s cortical plate, supragranular part
C7 white matter
C8 subplate
C23 orbital and insular area (or differentiating field)

FIBER TRACTS (f)

f6 fornix
f13 fimbria
f23 superior cerebellar peduncle
f31 cuneate fasciculus

GLANDS AND PARANEURAL STRUCTURES (g)

g1 choroid plexus of lateral ventricle
g4 choroid plexus of fourth ventricle

HYPOTHALAMUS (H)

H3a anterior hypothalamus (or differentiating field)
H3p posterior hypothalamus (or differentiating field)
H3v ventral hypothalamus (or differentiating field)
H7 zona incerta
H17 lateral mammillary nucleus

HIPPOCAMPAL FORMATION (K)

K3 hippocampus (or differentiating field)
K11 subicular neuroepithelium
K13 subicular area (or differentiating field)
K23 dentate gyrus (or differentiating field)

CEREBELLUM (L)

L2 external germinal layer
L3 cerebellum (or differentiating field)
L5 Purkinje cell layer

MIDBRAIN (M)

M3 superior colliculus (or differentiating field)
M4d dorsal periaqueductal gray
M4v ventral periaqueductal gray
M5 superior colliculus, superficial layers
M6 superior colliculus, deep layers
M8 red nucleus
M13 tegmentum (or differentiating field)
M15 interpeduncular nucleus
M23 pretectum (or differentiating field)
M28 posterior pretectal nucleus
M31 inferior tectal neuroepithelium
M33 inferior colliculus (or differentiating field)

NERVES AND NUCLEI, CRANIAL (n)

n2c optic chiasma
n7n facial motor nucleus

OLFACTORY BULB (O)

O2 olfactory bulb subventricular zone
O3 olfactory bulb (or differentiating field)
O4 glomerular layer
O5 external plexiform layer
O6 mitral cell layer
O7 internal plexiform layer
O8 granular layer
O13 accessory olfactory bulb (or differentiating field)
O15 external plexiform layer (accessory olfactory bulb)

PREOPTIC AREA AND OPTIC EPITHELIA (P)

P3 preoptic area (or differentiating field)

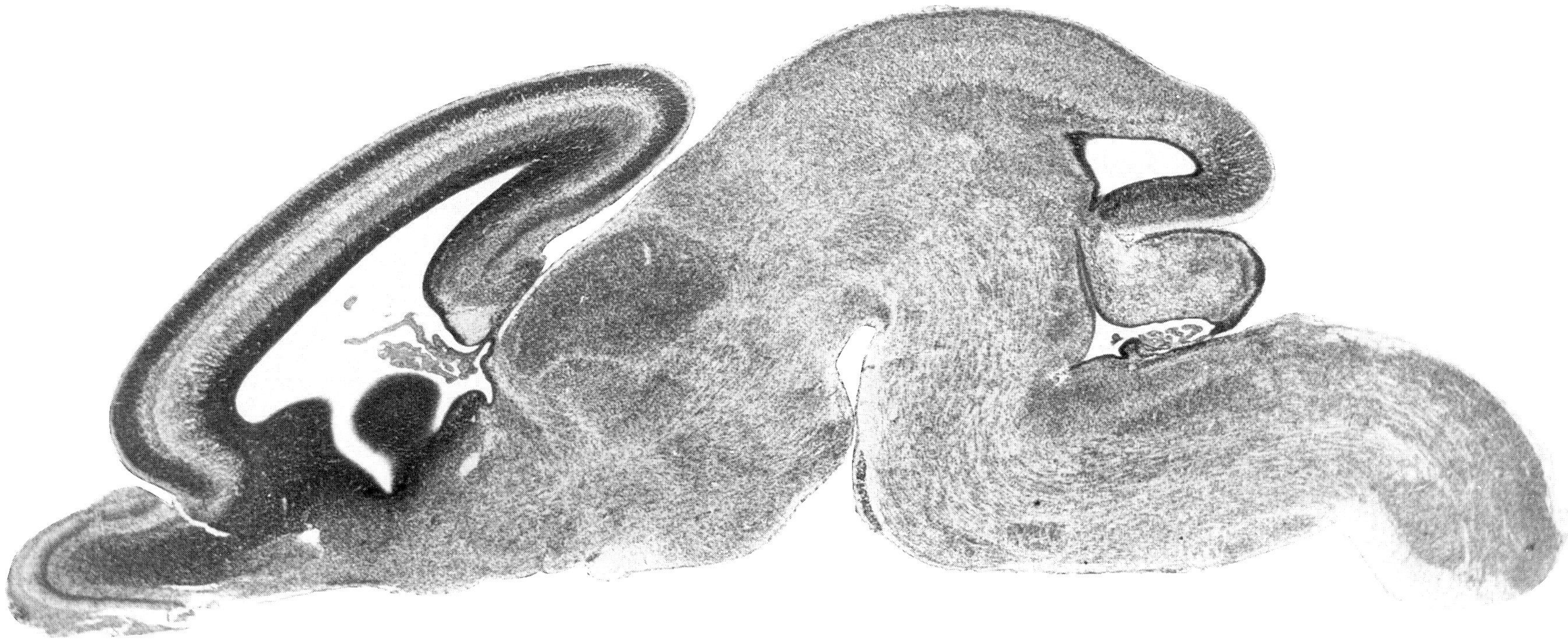

RHINENCEPHALON (R)

R2 rhinencephalic subventricular zone
R3 rhinencephalon (or differentiating field)
R9 olfactory tubercle

SEPTAL AREA (S)

S1 septal neuroèpithelium

THALAMUS (T)

T3a anterior thalamus (or differentiating field)
T3i intermediate thalamus (or differentiating field)
T3p posterior thalamus (or differentiating field)
T25 reticular nucleus
T35 posterior nuclear complex

VENTRICLES (v)

vai aqueduct, inferior tectal
vla lateral ventricle, anterior horn
vlc lateral ventricle, central
vld lateral ventricle, dorsal
vlfp lateral ventricle, pallidal fork
vlfs lateral ventricle, septal fork
v4p fourth ventricle, posterior

PONTINE AREA (X)

X3a anterior pons (or differentiating field)
X3p posterior pons (or differentiating field)
X11a anterior precerebellar neuroepithelium
X12a anterior precerebellar extramural migration
X14 spinal vestibular nucleus
X15 lateral vestibular nucleus
X24 superior olivary nucleus
X33 isthmus (or differentiating field)
X34 locus coeruleus

MEDULLA (Y)

Y3 medulla (or differentiating field)
Y8 inferior olive
Y15 cuneate nucleus
Y16 solitary nucleus

SPINAL CORD (Z)

Z4 ventral horn
Z5 dorsal horn

SAGITTAL FIGURE 5

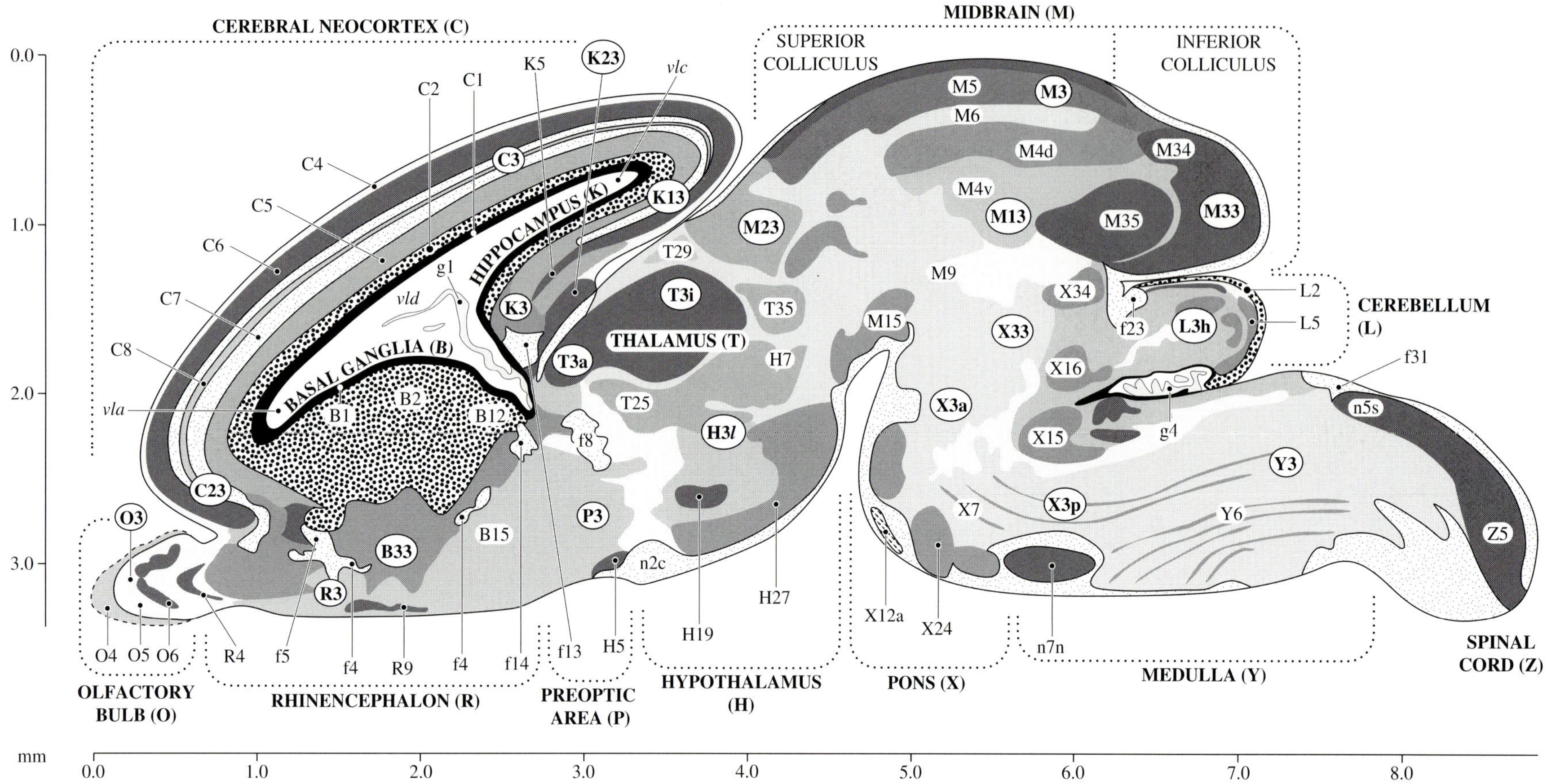

BASAL GANGLIA (B)

B1 striatal neuroepithelium
B2 striatal subventricular zone
B12 pallidal subventricular zone
B15 substantia innominata
B33 nucleus accumbens (or differentiating field)

CEREBRAL CORTEX (C)

C1 neocortical neuroepithelium
C2 neocortical subventricular zone
C3 neocortex (or differentiating field)
C4 molecular layer (or primordial plexiform layer)
C5 intermediate zone
C6 cortical plate
C7 white matter
C8 subplate
C23 orbital and insular area (or differentiating field)

FIBER TRACTS (f)

f4 anterior commissure
f5 medial forebrain bundle
f8 internal capsule
f13 fimbria
f14 stria terminalis
f23 superior cerebellar peduncle
f31 cuneate fasciculus

GLANDS AND PARANEURAL STRUCTURES (g)

g1 choroid plexus of lateral ventricle
g4 choroid plexus of fourth ventricle

HYPOTHALAMUS (H)

H3*l* lateral hypothalamus (or differentiating field)
H5 suprachiasmatic nucleus
H7 zona incerta
H19 subthalamic nucleus
H27 tuberomammillary nucleus

HIPPOCAMPAL FORMATION (K)

K3 hippocampus (or differentiating field)
K5 CA3 field
K13 subicular area (or differentiating field)
K23 dentate gyrus (or differentiating field)

CEREBELLUM (L)

L2 external germinal layer
L3h cerebellar hemisphere (or differentiating field)
L5 Purkinje cell layer

MIDBRAIN (M)

M3 superior colliculus (or differentiating field)
M4d dorsal periaqueductal gray
M4v ventral periaqueductal gray
M5 superior colliculus, superficial layers
M6 superior colliculus, deep layers
M9 midbrain reticular formation
M13 tegmentum (or differentiating field)
M15 interpeduncular nucleus
M23 pretectum (or differentiating field)
M33 inferior colliculus (or differentiating field)
M34 external nucleus of inferior colliculus
M35 central nucleus of inferior colliculus

NERVES AND NUCLEI, CRANIAL (n)

n2c optic chiasma
n5s spinal trigeminal nucleus
n7n facial motor nucleus

OLFACTORY BULB (O)

O3 olfactory bulb (or differentiating field)
O4 glomerular layer
O5 external plexiform layer
O6 mitral cell layer

PREOPTIC AREA AND OPTIC EPITHELIA (P)

P3 preoptic area (or differentiating field)

RHINENCEPHALON (R)

R3 rhinencephalon (or differentiating field)
R4 piriform cortex
R9 olfactory tubercle

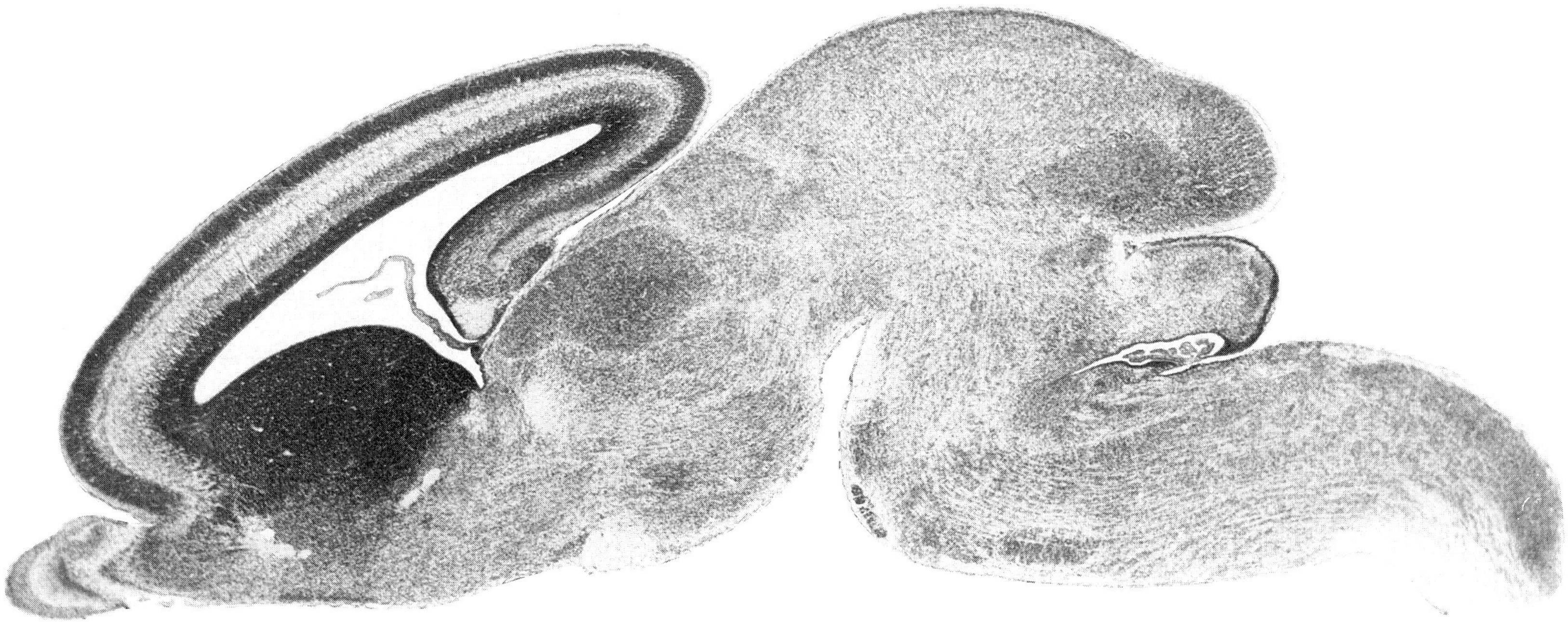

THALAMUS (T)

T3a anterior thalamus (or differentiating field)
T3i intermediate thalamus (or differentiating field)
T25 reticular nucleus
T29 dorsal lateral geniculate nucleus
T35 posterior nuclear complex

VENTRICLES (v)

vla lateral ventricle, anterior horn
vlc lateral ventricle, central
vld lateral ventricle, dorsal

PONTINE AREA (X)

X3a anterior pons (or differentiating field)
X3p posterior pons (or differentiating field)
X7 pontine reticular formation
X12a anterior precerebellar extramural migration
X15 lateral vestibular nucleus
X16 superior vestibular nucleus
X24 superior olivary nucleus
X33 isthmus (or differentiating field)
X34 locus coeruleus

MEDULLA (Y)

Y3 medulla (or differentiating field)
Y6 medullary reticular formation

SPINAL CORD (Z)

Z5 dorsal horn

SAGITTAL FIGURE 6

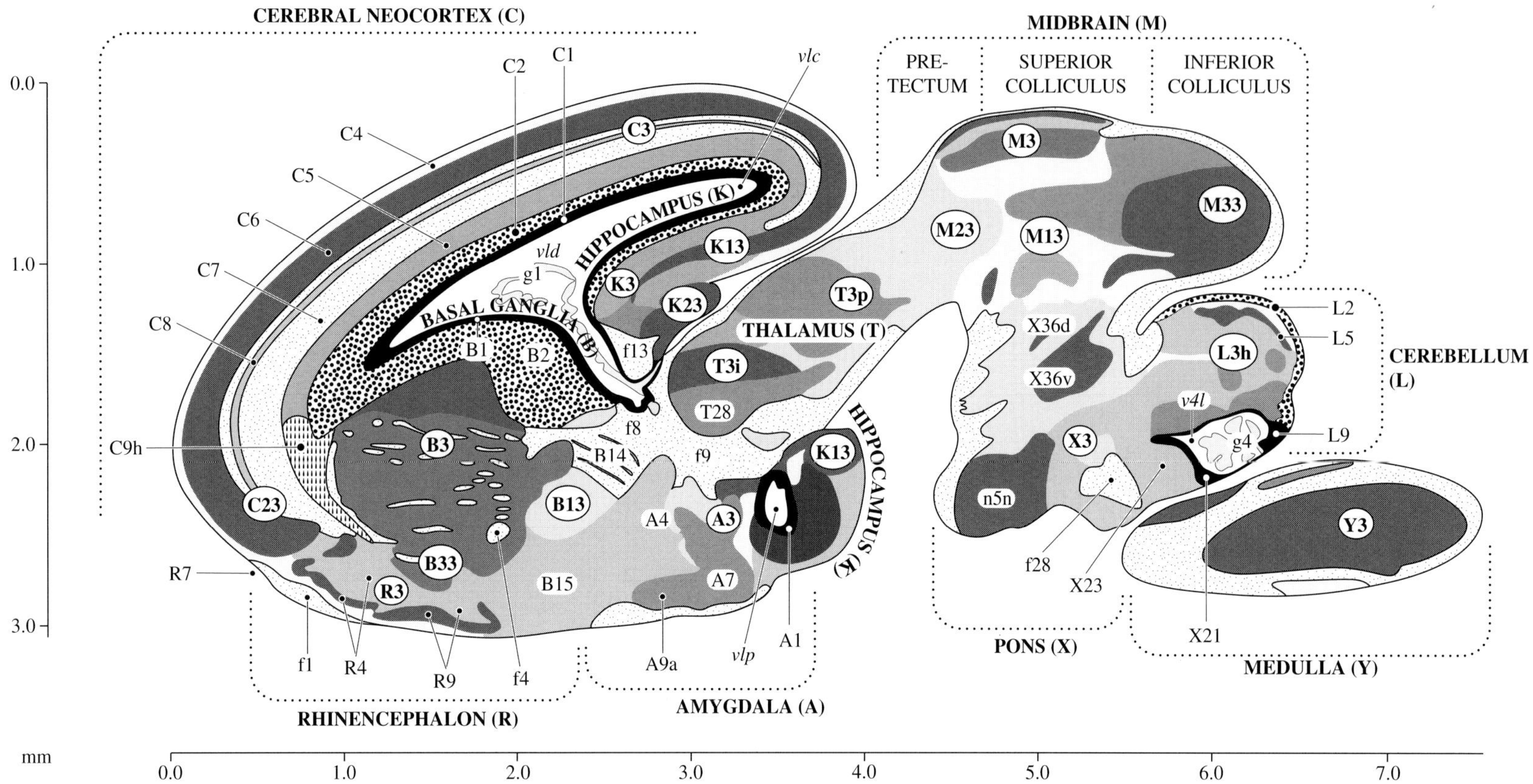

AMYGDALA (A)

A1 amygdaloid neuroepithelium
A3 amygdala (or differentiating field)
A4 anterior amygdaloid area
A7 basomedial amygdaloid nucleus
A9a cortical amygdaloid nucleus, anterior

BASAL GANGLIA (B)

B1 striatal neuroepithelium
B2 striatal subventricular zone
B3 striatum (or differentiating field)
B13 pallidum (or differentiating field)
B14 entopeduncular nucleus
B15 substantia innominata
B33 nucleus accumbens (or differentiating field)

CEREBRAL CORTEX (C)

C1 neocortical neuroepithelium
C2 neocortical subventricular zone
C3 neocortex (or differentiating field)
C4 molecular layer (or primordial plexiform layer)
C5 intermediate zone
C6 cortical plate
C7 white matter
C8 subplate
C9h lateral migratory stream, head
C23 orbital and insular area (or differentiating field)

FIBER TRACTS (f)

f1 lateral olfactory tract
f4 anterior commissure
f8 internal capsule
f9 cerebral peduncle
f13 fimbria
f28 lateral lemniscus

GLANDS AND PARANEURAL STRUCTURES (g)

g1 choroid plexus of lateral ventricle
g4 choroid plexus of fourth ventricle

HIPPOCAMPAL FORMATION (K)

K3 hippocampus (or differentiating field)
K13 subicular area (or differentiating field)
K23 dentate gyrus (or differentiating field)

CEREBELLUM (L)

L2 external germinal layer
L3h cerebellar hemisphere (or differentiating field)
L5 Purkinje cell layer
L9 cerebellar germinal trigone

MIDBRAIN (M)

M3 superior colliculus (or differentiating field)
M13 tegmentum (or differentiatin field)
M23 pretectum (or differentiating field)
M33 inferior colliculus (or differentiating field)

NERVES AND NUCLEI, CRANIAL (n)

n5n trigeminal motor nucleus

RHINENCEPHALON (R)

R3 rhinencephalon (or differentiating field)
R4 piriform cortex
R7 rhinal fissure
R9 olfactory tubercle

THALAMUS (T)

T3i intermediate thalamus (or differentiating field)
T3p posterior thalamus (or differentiating field)
T28 ventrobasal nuclear complex

VENTRICLES (v)

vlc lateral ventricle, central
vld lateral ventricle, dorsal
vlp lateral ventricle, posterior horn
v4l fourth ventricle, lateral recess

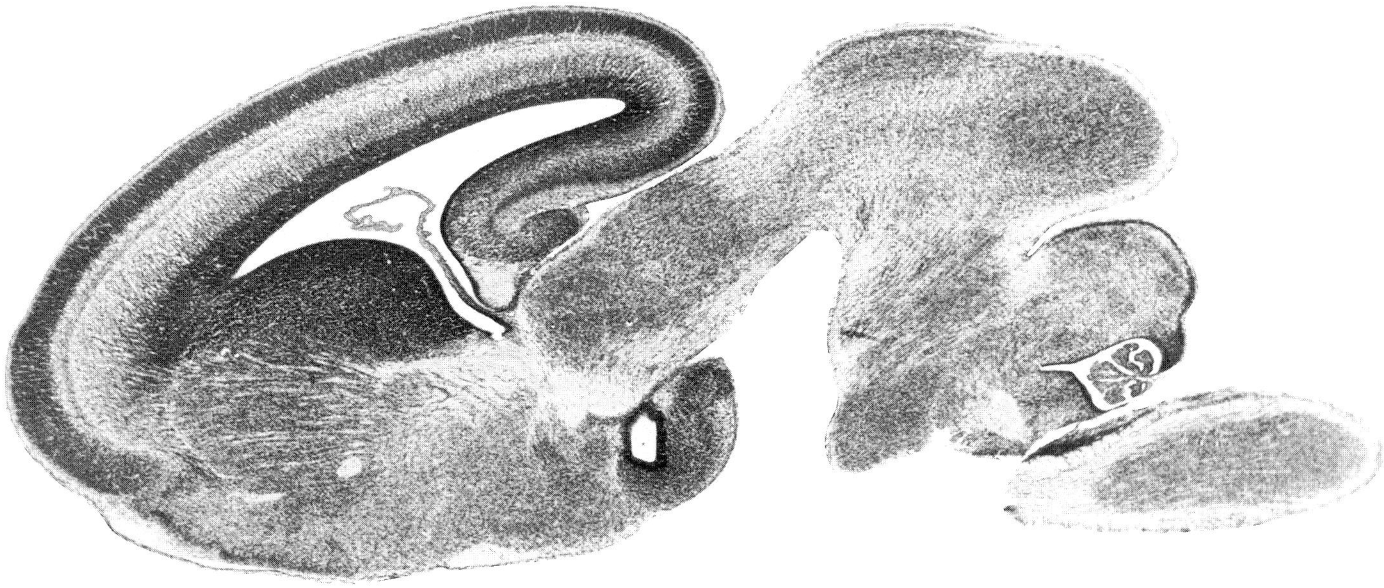

PONTINE AREA (X)

X3 pons (or differentiating field)
X21 cochlear nuclei neuroepithelium
X23 cochlear nuclei (or differentiating field)
X36d nucleus of lateral lemniscus, dorsal
X36v nucleus of lateral lemniscus, ventral

MEDULLA (Y)

Y3 medulla (or differentiating field)

SAGITTAL FIGURE 7

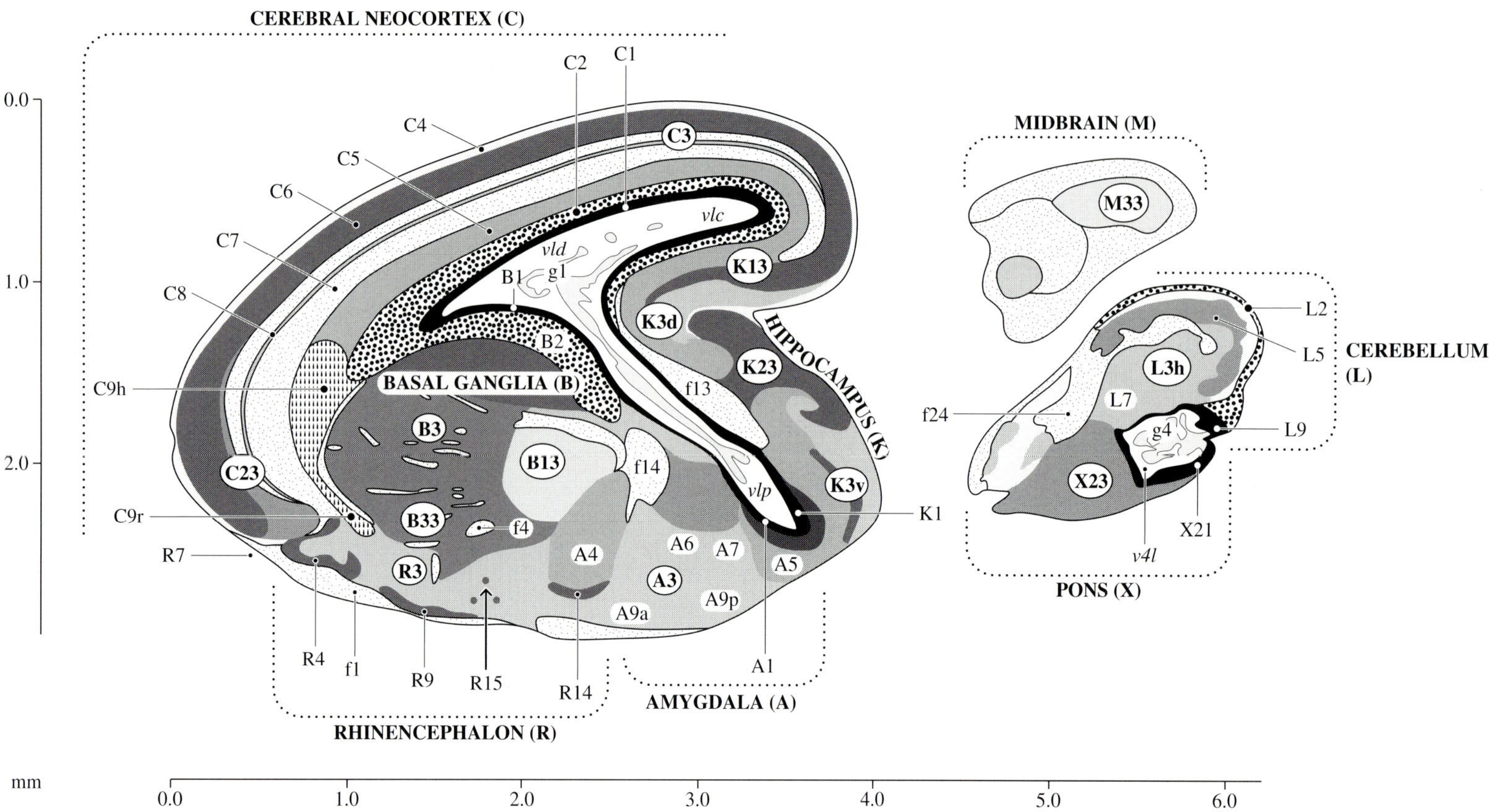

AMYGDALA (A)

A1 amygdaloid neuroepithelium
A3 amygdala (or differentiating field)
A4 anterior amygdaloid area
A5 amygdala-hippocampal area
A6 central amygdaloid nucleus
A7 basomedial amygdaloid nucleus
A9a cortical amygdaloid nucleus, anterior
A9p cortical amygdaloid nucleus, posterior

BASAL GANGLIA (B)

B1 striatal neuroepithelium
B2 striatal subventricular zone
B3 striatum (or differentiating field)
B13 pallidum (or differentiating field)
B33 nucleus accumbens (or differentiating field)

CEREBRAL CORTEX (C)

C1 neocortical neuroepithelium
C2 neocortical subventricular zone
C3 neocortex (or differentiating field)
C4 molecular layer (or primordial plexiform layer)
C5 intermediate zone
C6 cortical plate
C7 white matter
C8 subplate
C9h lateral migratory stream, head
C9r lateral migratory stream, reservoir
C23 orbital and insular area (or differentiating field)

FIBER TRACTS (f)

f1 lateral olfactory tract
f4 anterior commissure
f13 fimbria
f14 stria terminalis
f24 middle cerebellar peduncle

GLANDS AND PARANEURAL STRUCTURES (g)

g1 choroid plexus of lateral ventricle
g4 choroid plexus of fourth ventricle

HIPPOCAMPAL FORMATION (K)

K1 hippocampal (Ammon's horn) neuroepithelium
K3d dorsal hippocampus (or differentiating field)
K3v ventral hippocampus (or differentiating field)
K13 subicular area (or differentiating field)
K23 dentate gyrus (or differentiating field)

CEREBELLUM (L)

L2 external germinal layer
L3h cerebellar hemisphere (or differentiating field)
L5 Purkinje cell layer
L7 lateral deep nucleus (dentate)
L9 cerebellar germinal trigone

MIDBRAIN (M)

M33 inferior colliculus (or differentiating field)

RHINENCEPHALON (R)

R3 rhinencephalon (or differentiating field)
R4 piriform cortex
R7 rhinal fissure
R9 olfactory tubercle
R14 nucleus of the lateral olfactory tract
R15 islands of Calleja

VENTRICLES (v)

vlc lateral ventricle, central
vld lateral ventricle, dorsal
vlp lateral ventricle, posterior horn
v4l fourth ventricle, lateral recess

PONTINE AREA (X)

X21 cochlear nuclei neuroepithelium
X23 cochlear nuclei (or differentiating field)

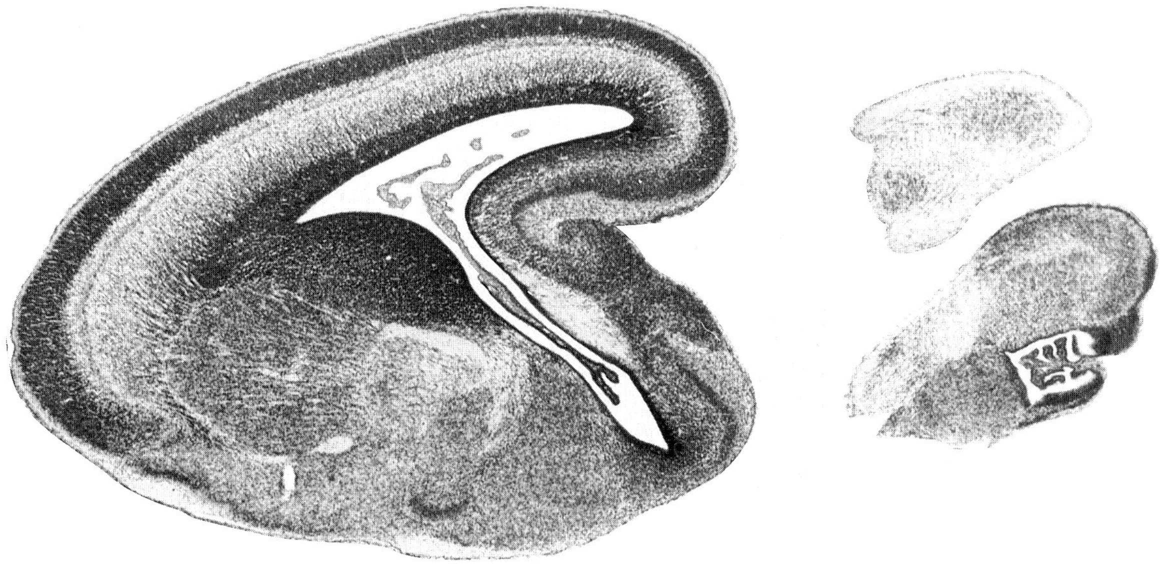

SAGITTAL FIGURE 8

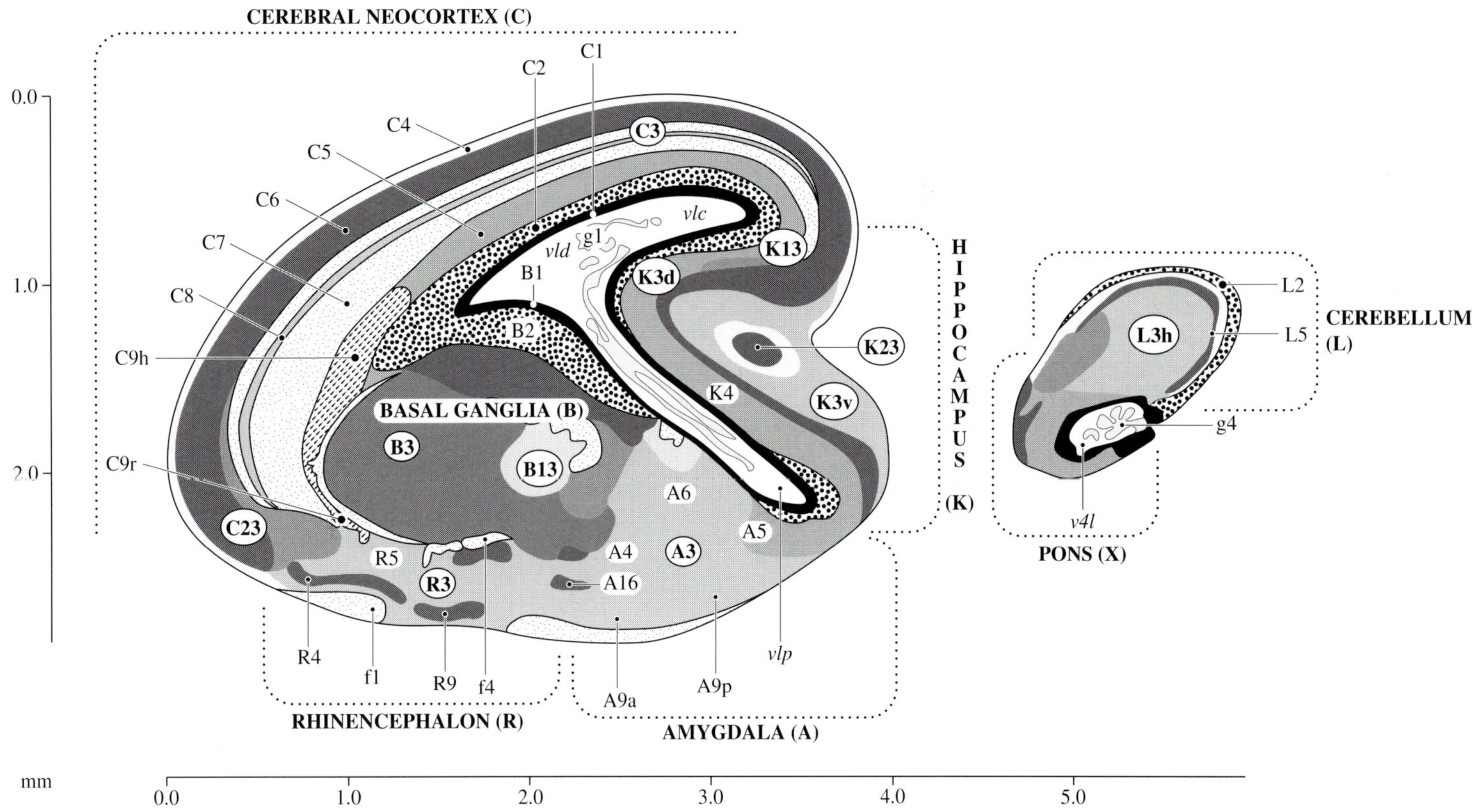

AMYGDALA (A)

A3 amygdala (or differentiating field)
A4 anterior amygdaloid area
A5 amygdala-hippocampal area
A6 central amygdaloid nucleus
A9a cortical amygdaloid nucleus, anterior
A9p cortical amygdaloid nucleus, posterior
A16 intercalated masses

BASAL GANGLIA (B)

B1 striatal neuroepithelium
B2 striatal subventricular zone
B3 striatum (or differentiating field)
B13 pallidum (or differentiating field)

CEREBRAL CORTEX (C)

C1 neocortical neuroepithelium
C2 neocortical subventricular zone
C3 neocortex (or differentiating field)
C4 molecular layer (or primordial plexiform layer)
C5 intermediate zone
C6 cortical plate
C7 white matter
C8 subplate
C9h lateral migratory stream, head
C9r lateral migratory stream, reservoir
C23 orbital and insular area (or differentiating field)

FIBER TRACTS (f)

f1 lateral olfactory tract
f4 anterior commissure

GLANDS AND PARANEURAL STRUCTURES (g)

g1 choroid plexus of lateral ventricle
g4 choroid plexus of fourth ventricle

HIPPOCAMPAL FORMATION (K)

K3d dorsal hippocampus (or differentiating field)
K3v ventral hippocampus (or differentiating field)
K4 CA1 field
K13 subicular area (or differentiating field)
K23 dentate gyrus (or differentiating field)

CEREBELLUM (L)

L2 external germinal layer
L3h cerebellar hemisphere (or differentiating field)
L5 Purkinje cell layer

RHINENCEPHALON (R)

R3 rhinencephalon (or differentiating field)
R4 piriform cortex
R5 endopiriform nucleus
R9 olfactory tubercle

VENTRICLES (v)

vlc lateral ventricle, central
vld lateral ventricle, dorsal
vlp lateral ventricle, posterior horn
v4l fourth ventricle, lateral recess

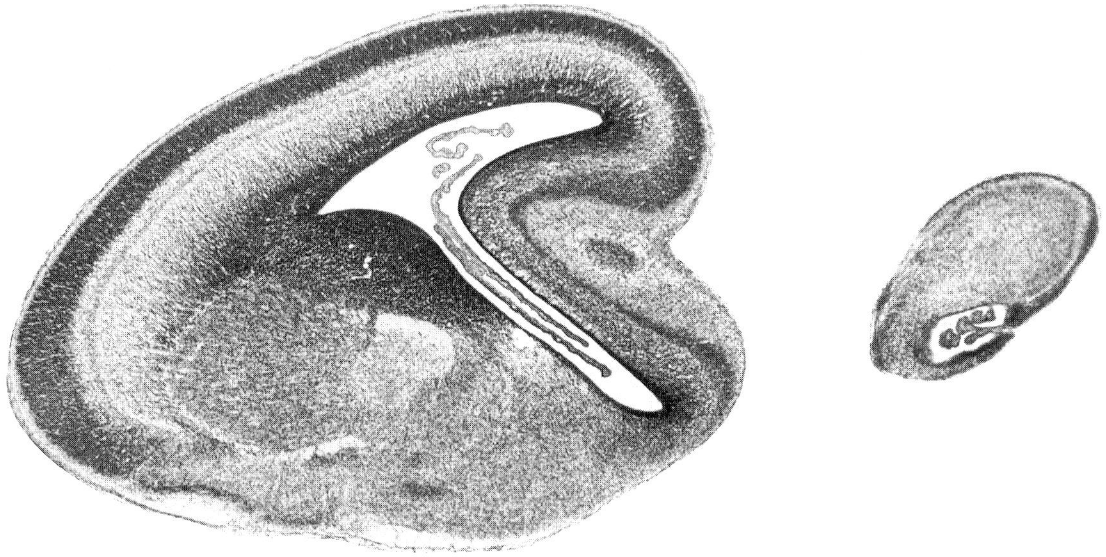

CORONAL

FIGURES and PLATES 1-22

CORONAL FIGURE 1

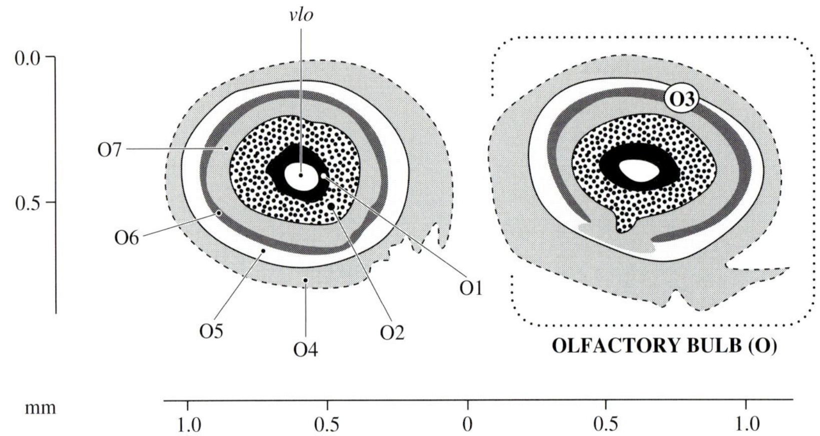

OLFACTORY BULB (O)

O1 olfactory bulb neuroepithelium
O2 olfactory bulb subventricular zone
O3 olfactory bulb (or differentiating field)
O4 glomerular layer
O5 external plexiform layer
O6 mitral cell layer
O7 internal plexiform layer

VENTRICLES (v)

vlo olfactory ventricle

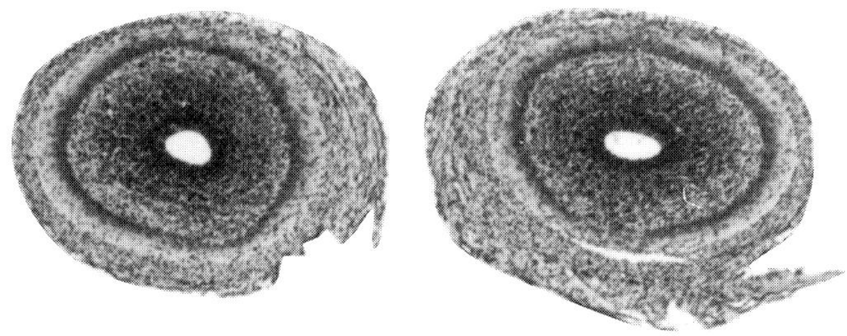

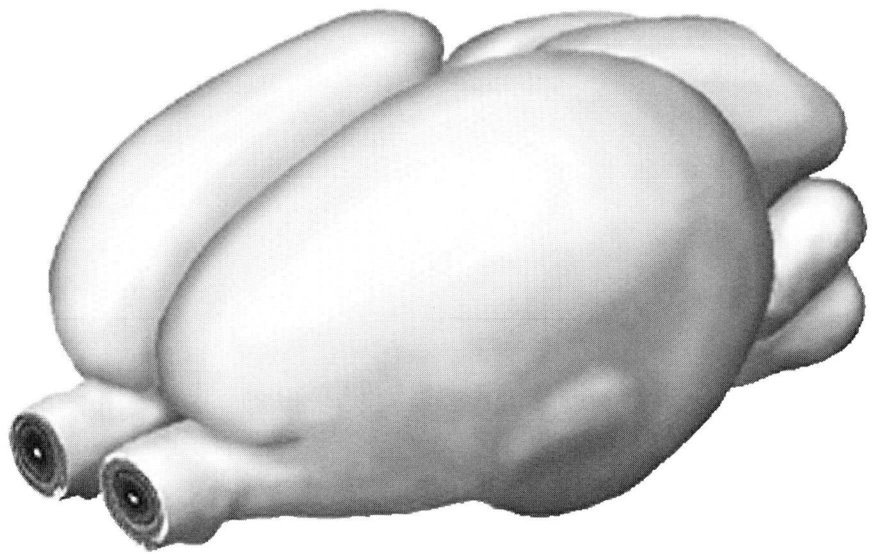

CORONAL FIGURE 2

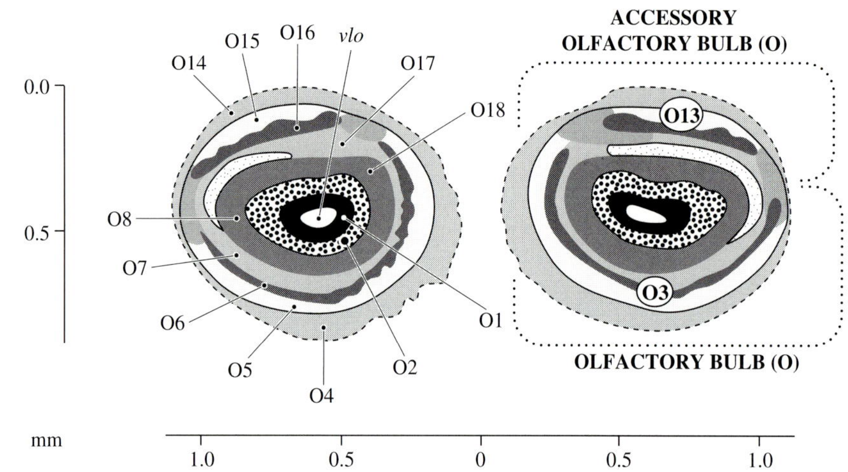

OLFACTORY BULB (O)

O1	olfactory bulb neuroepithelium
O2	olfactory bulb subventricular zone
O3	olfactory bulb (or differentiating field)
O4	glomerular layer
O5	external plexiform layer
O6	mitral cell layer
O7	internal plexiform layer
O8	granular layer
O13	accessory olfactory bulb (or differentiating field)
O14	glomerular layer (accessory olfactory bulb)
O15	external plexiform layer (accessory olfactory bulb)
O16	mitral cell layer (accessory olfactory bulb)
O17	internal plexiform layer (accessory olfactory bulb)
O18	granular layer (accessory olfactory bulb)

VENTRICLES (v)

vlo	olfactory ventricle

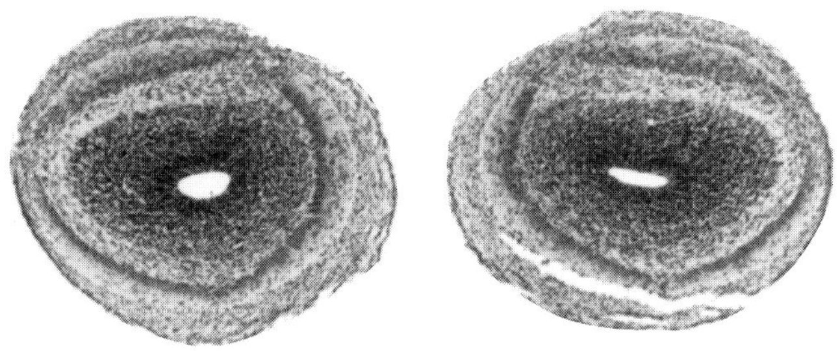

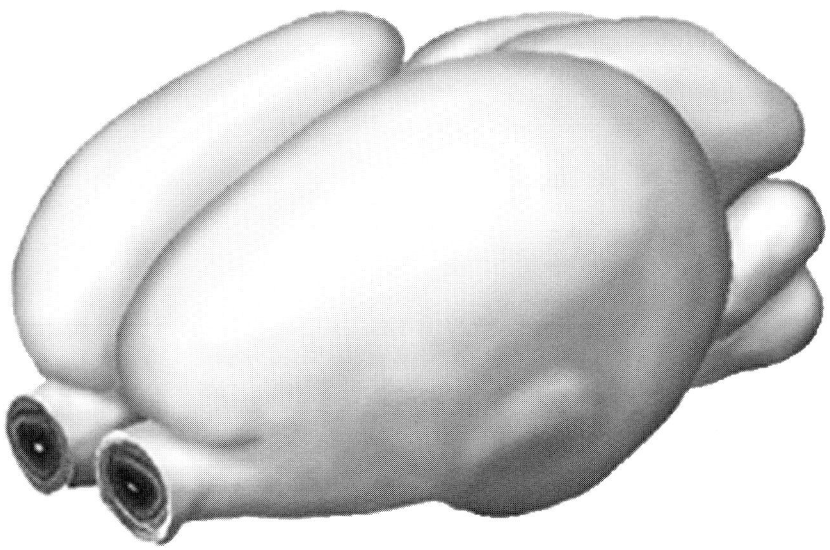

CORONAL FIGURE 3

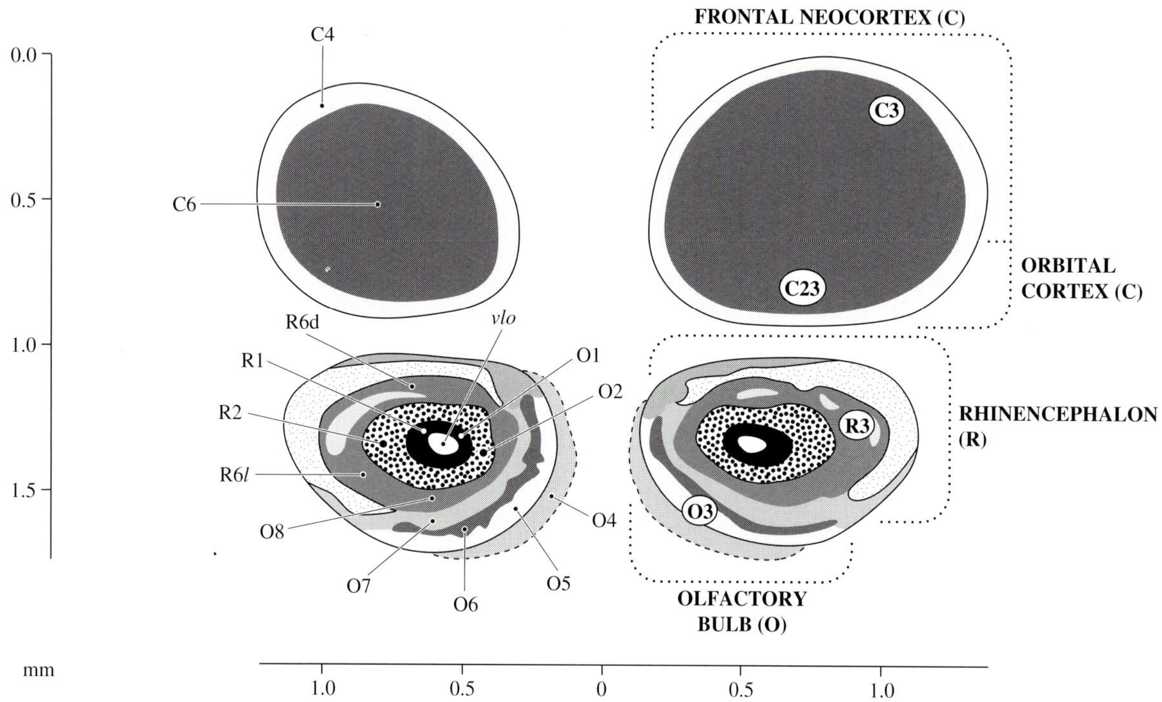

CEREBRAL CORTEX (C)

C3 neocortex (or differentiating field)
C4 molecular layer (or primordial plexiform layer)
C6 cortical plate
C23 orbital and insular area (or differentiating field)

OLFACTORY BULB (O)

O1 olfactory bulb neuroepithelium
O2 olfactory bulb subventricular zone
O3 olfactory bulb (or differentiating field)
O4 glomerular layer
O5 external plexiform layer
O6 mitral cell layer
O7 internal plexiform layer
O8 granular layer

RHINENCEPHALON (R)

R1 rhinencephalic neuroepithelium
R2 rhinencephalic subventricular zone
R3 rhinencephalon (or differentiating field)
R6d anterior olfactory nucleus, dorsal
R6*l* anterior olfactory nucleus, lateral

VENTRICLES (v)

vlo olfactory ventricle

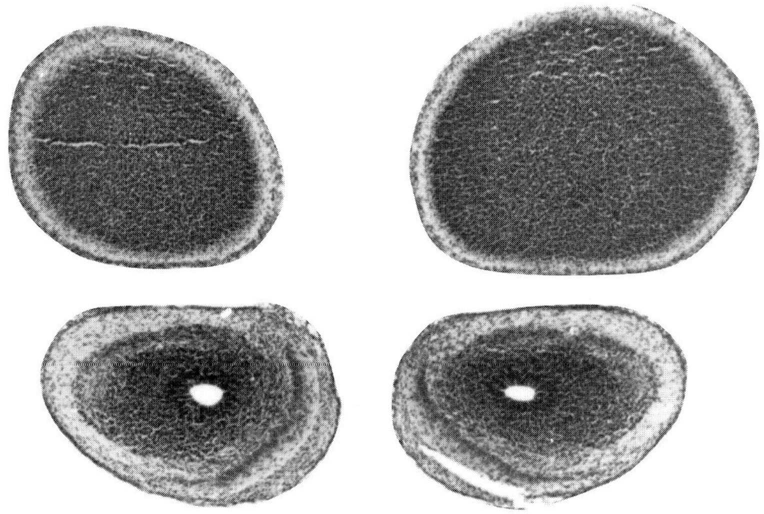

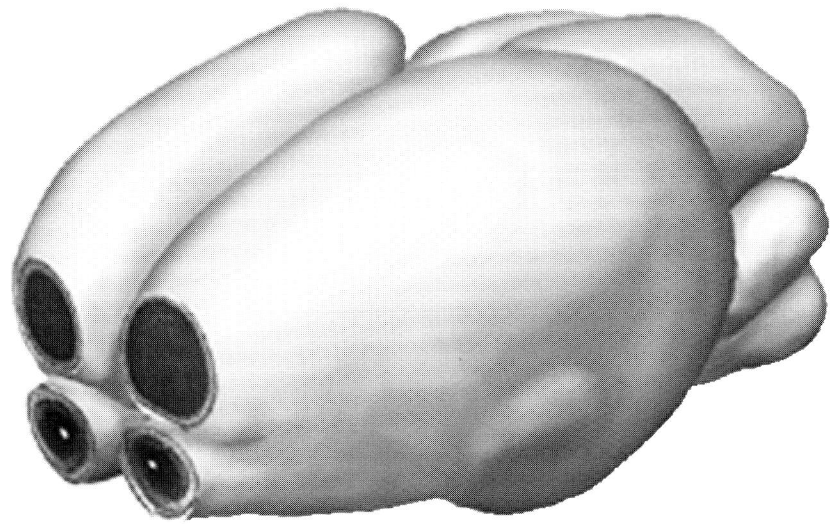

CORONAL FIGURE 4

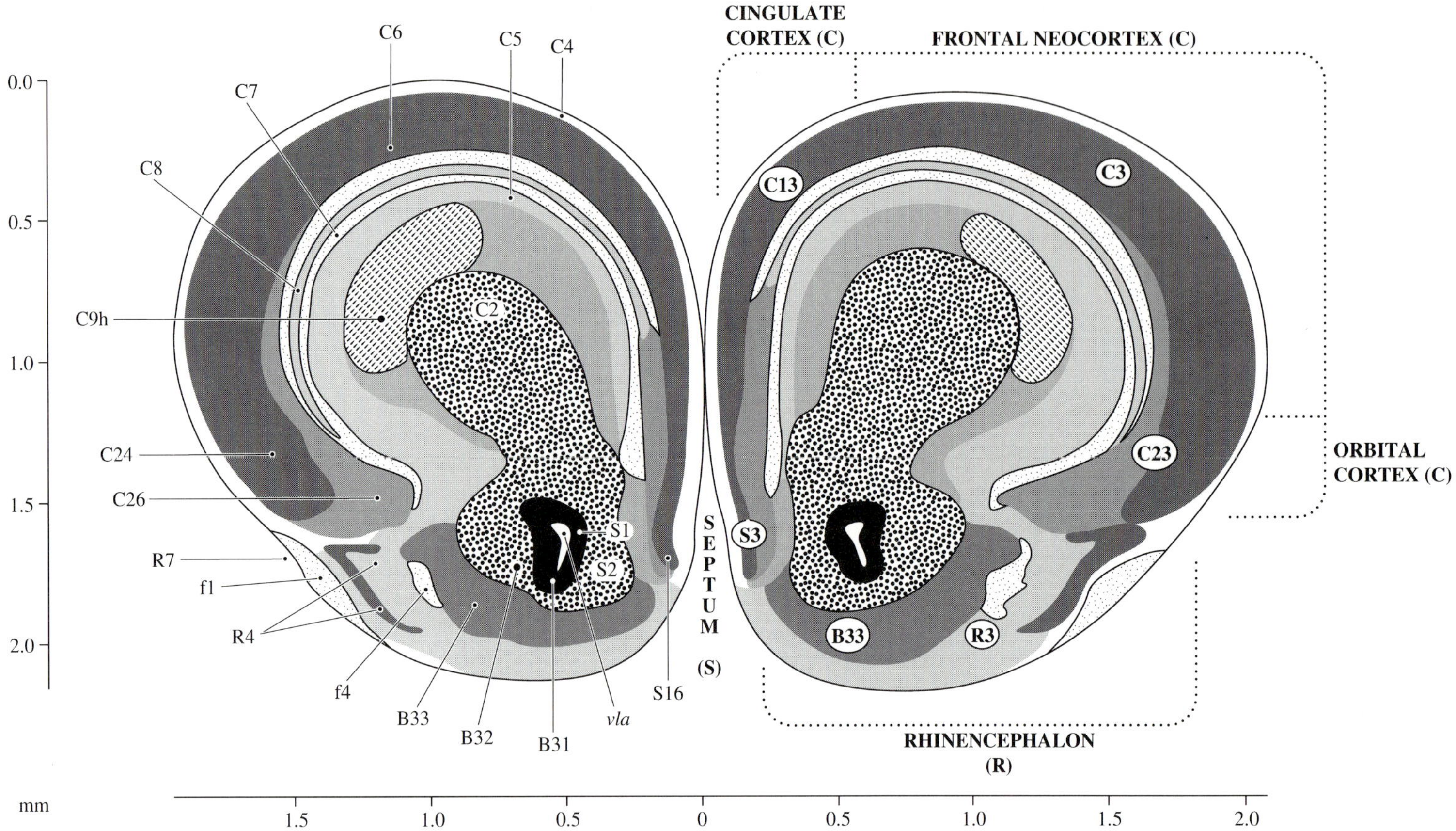

BASAL GANGLIA (B)

B31 accumbens neuroepithelium
B32 accumbens subventricular zone
B33 nucleus accumbens (or differentiating field)

CEREBRAL CORTEX (C)

C2 neocortical subventricular zone
C3 neocortex (or differentiating field)
C4 molecular layer (or primordial plexiform layer)
C5 intermediate zone
C6 cortical plate
C7 white matter
C8 subplate
C9h lateral migratory stream, head
C13 cingulate cortex (or differentiating field)
C23 orbital and insular area (or differentiating field)
C24 insular cortex
C26 orbital cortex

FIBER TRACTS (f)

f1 lateral olfactory tract
f4 anterior commissure

RHINENCEPHALON (R)

R3 rhinencephalon (or differentiating field)
R4 piriform cortex
R7 rhinal fissure

SEPTAL AREA (S)

S1 septal neuroepithelium
S2 septal subventricular zone
S3 septum (or differentiating field)
S16 taenia tecta

VENTRICLES (v)

vla lateral ventricle, anterior horn

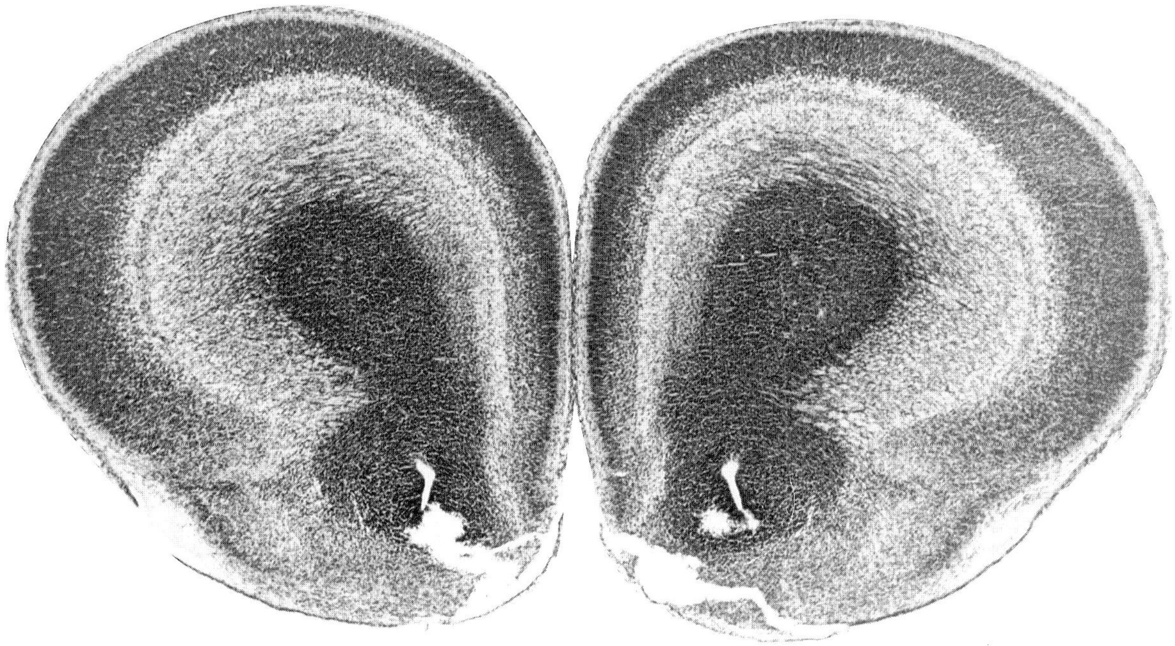

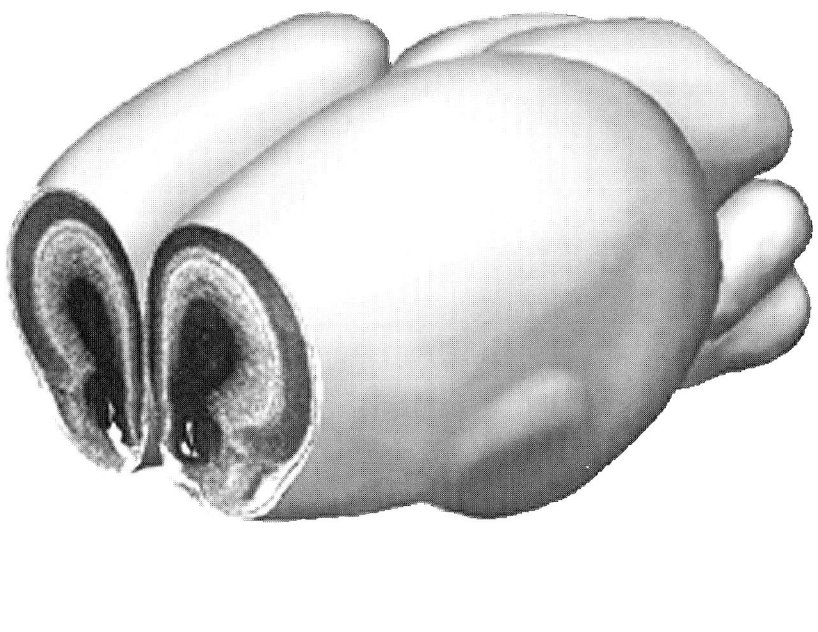

CORONAL FIGURE 5

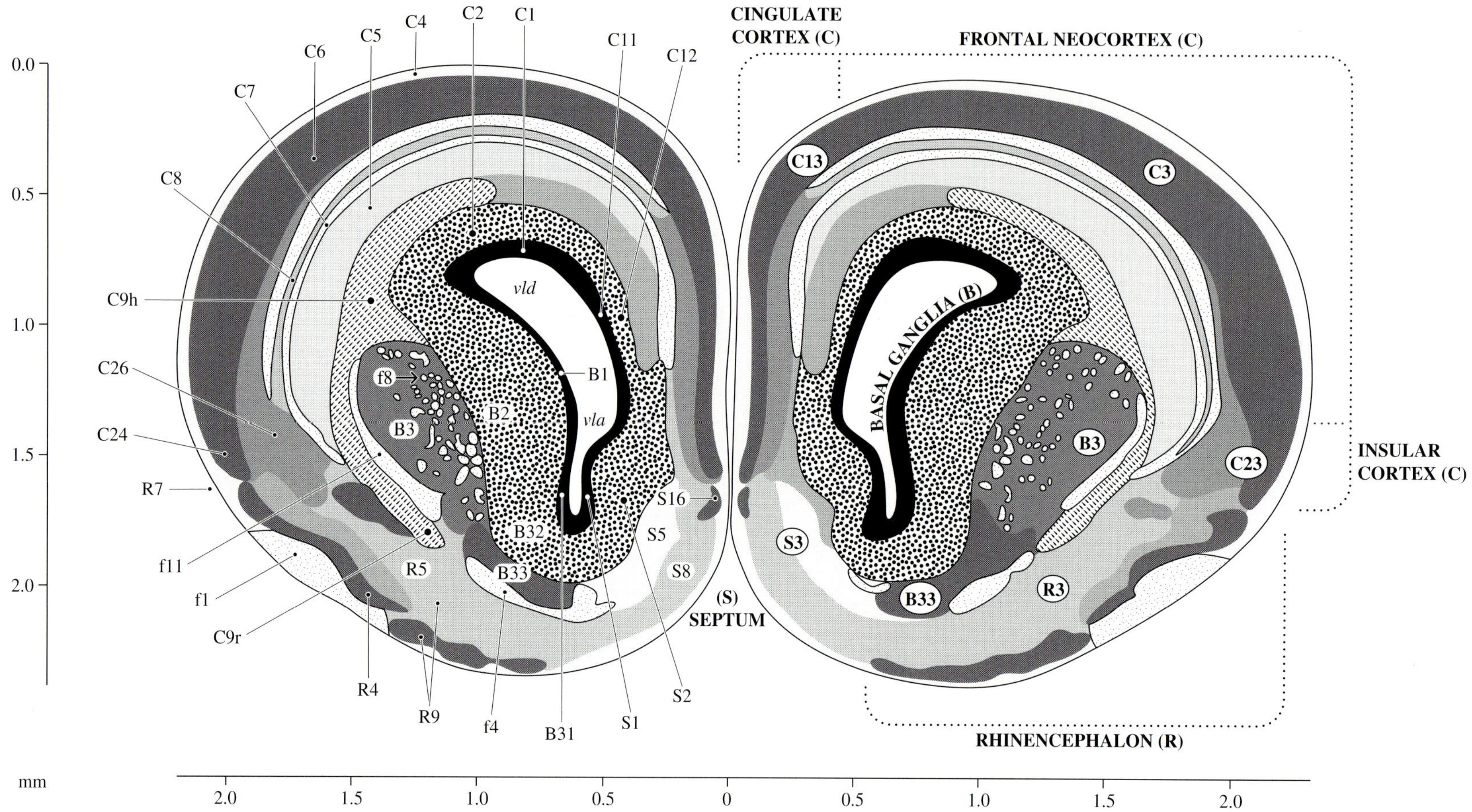

BASAL GANGLIA (B)

B1 striatal neuroepithelium
B2 striatal subventricular zone
B3 striatum (or differentiating field)
B31 accumbens neuroepithelium
B32 accumbens subventricular zone
B33 nucleus accumbens (or differentiating field)

CEREBRAL CORTEX (C)

C1 neocortical neuroepithelium
C2 neocortical subventricular zone
C3 neocortex (or differentiating field)
C4 molecular layer (or primordial plexiform layer)
C5 intermediate zone
C6 cortical plate
C7 white matter
C8 subplate
C9h lateral migratory stream, head
C9r lateral migratory stream, reservoir
C11 cingulate cortex neuroepithelium
C12 cingulate cortex subventricular zone
C13 cingulate cortex (or differentiating field)
C23 orbital and insular area (or differentiating field)
C24 insular cortex
C26 claustrum

FIBER TRACTS (f)

f1 lateral olfactory tract
f4 anterior commissure
f8 internal capsule
f11 external capsule

RHINENCEPHALON (R)

R3 rhinencephalon (or differentiating field)
R4 piriform cortex
R5 endopiriform nucleus
R7 rhinal fissure
R9 olfactory tubercle

SEPTAL AREA (S)

S1 septal neuroepithelium
S2 septal subventricular zone
S3 septum (or differentiating field)
S5 lateral septum
S8 diagonal band (vertical limb)
S16 taenia tecta

VENTRICLES (v)

vla lateral ventricle, anterior horn
vld lateral ventricle, dorsal

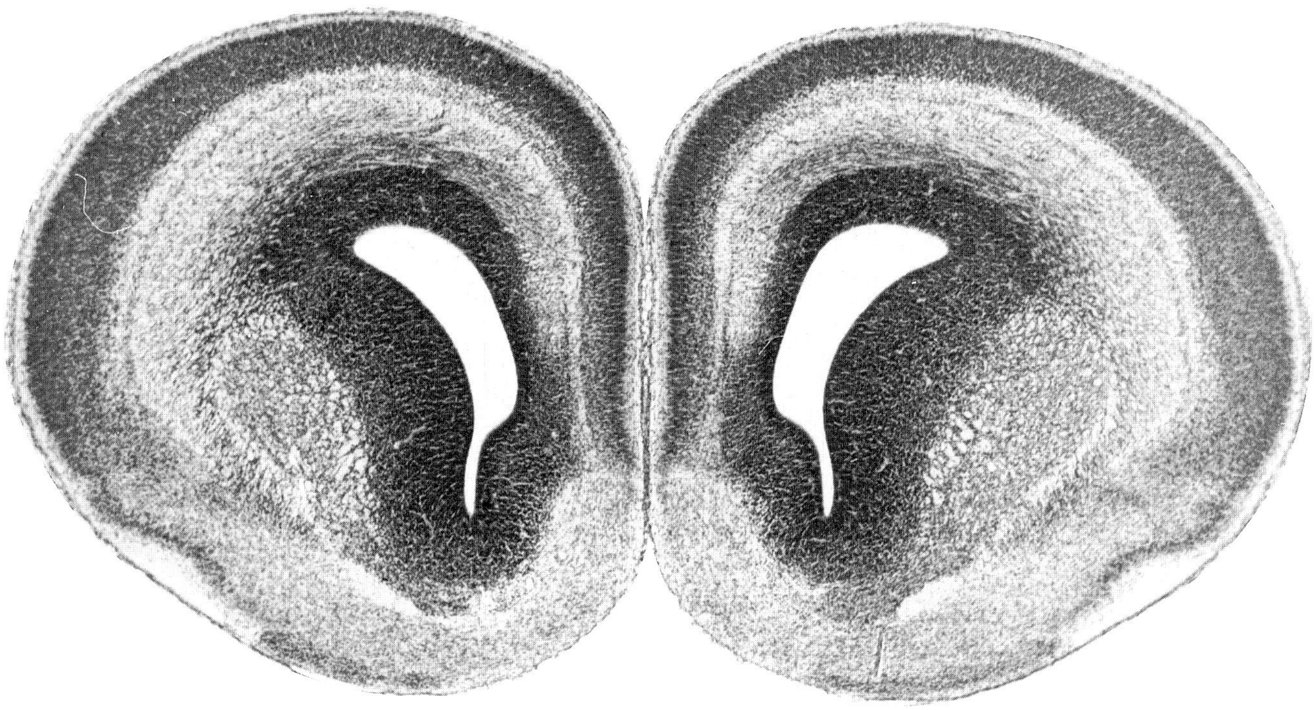

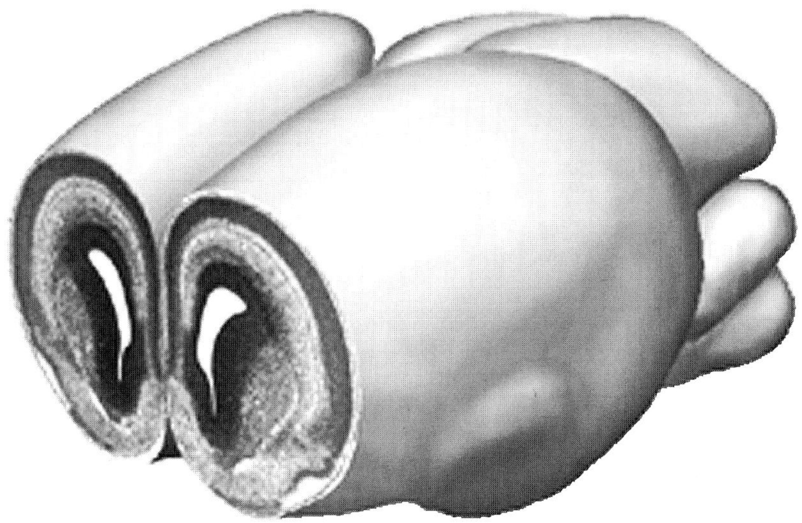

CORONAL FIGURE 6

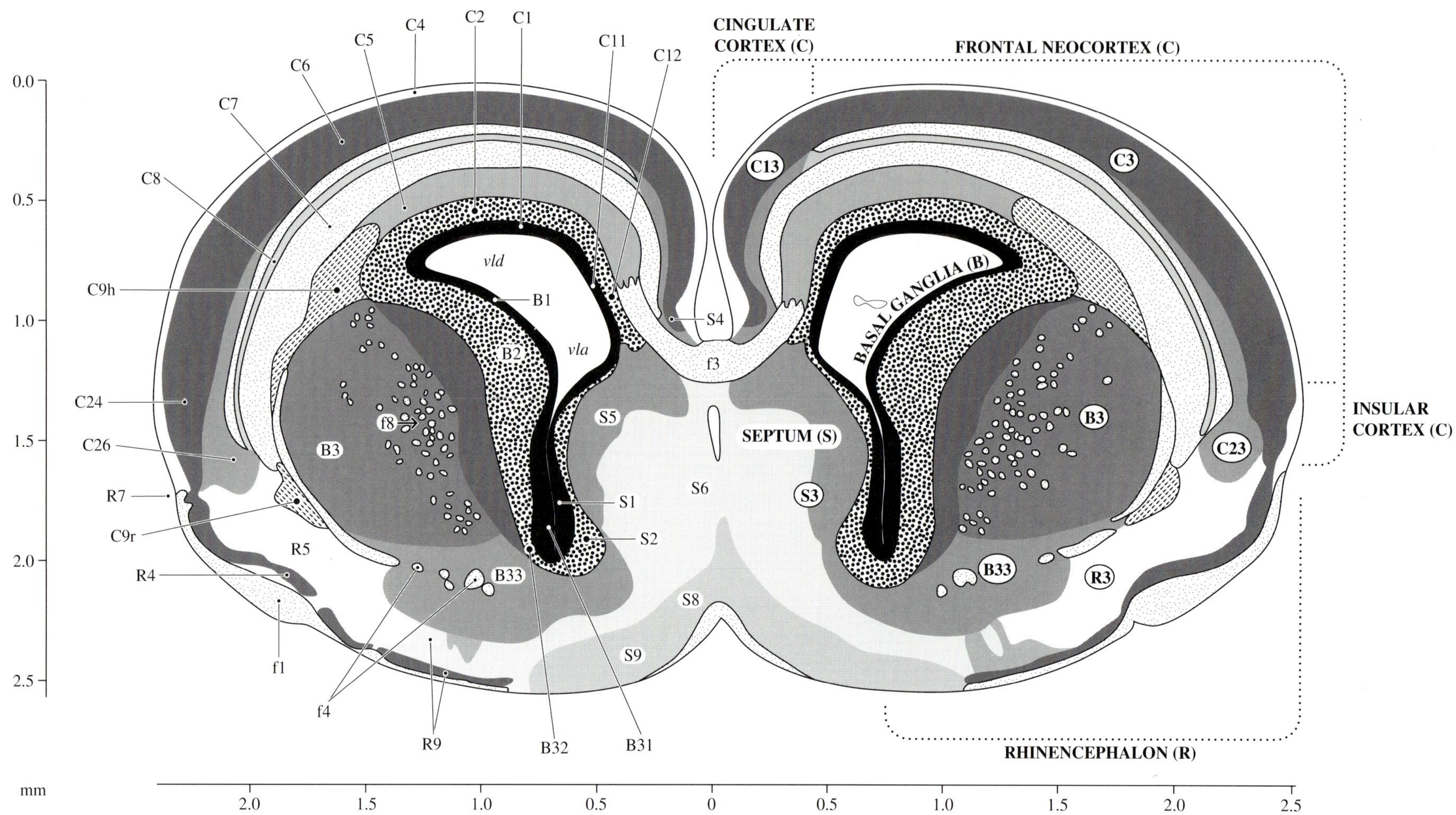

BASAL GANGLIA (B)

B1	striatal neuroepithelium
B2	striatal subventricular zone
B3	striatum (or differentiating field)
B31	accumbens neuroepithelium
B32	accumbens subventricular zone
B33	nucleus accumbens (or differentiating field)

CEREBRAL CORTEX (C)

C1	neocortical neuroepithelium
C2	neocortical subventricular zone
C3	neocortex (or differentiating field)
C4	molecular layer (or primordial plexiform layer)
C5	intermediate zone
C6	cortical plate
C7	white matter
C8	subplate
C9h	lateral migratory stream, head
C9r	lateral migratory stream, reservoir
C11	cingulate cortex neuroepithelium
C12	cingulate cortex subventricular zone
C13	cingulate cortex (or differentiating field)
C23	orbital and insular area (or differentiating field)
C24	insular cortex
C26	claustrum

FIBER TRACTS (f)

f1	lateral olfactory tract
f3	corpus callosum
f4	anterior commissure
f8	internal capsule

RHINENCEPHALON (R)

R3	rhinencephalon (or differentiating field)
R4	piriform cortex
R5	endopiriform nucleus
R7	rhinal fissure
R9	olfactory tubercle

SEPTAL AREA (S)

S1	septal neuroepithelium
S2	septal subventricular zone
S3	septum (or differentiating field)
S4	indesium griseum
S5	lateral septum
S6	medial septum
S8	diagonal band (vertical limb)
S9	diagonal band (horizontal limb)

VENTRICLES (v)

vla	lateral ventricle, anterior horn
vld	lateral ventricle, dorsal

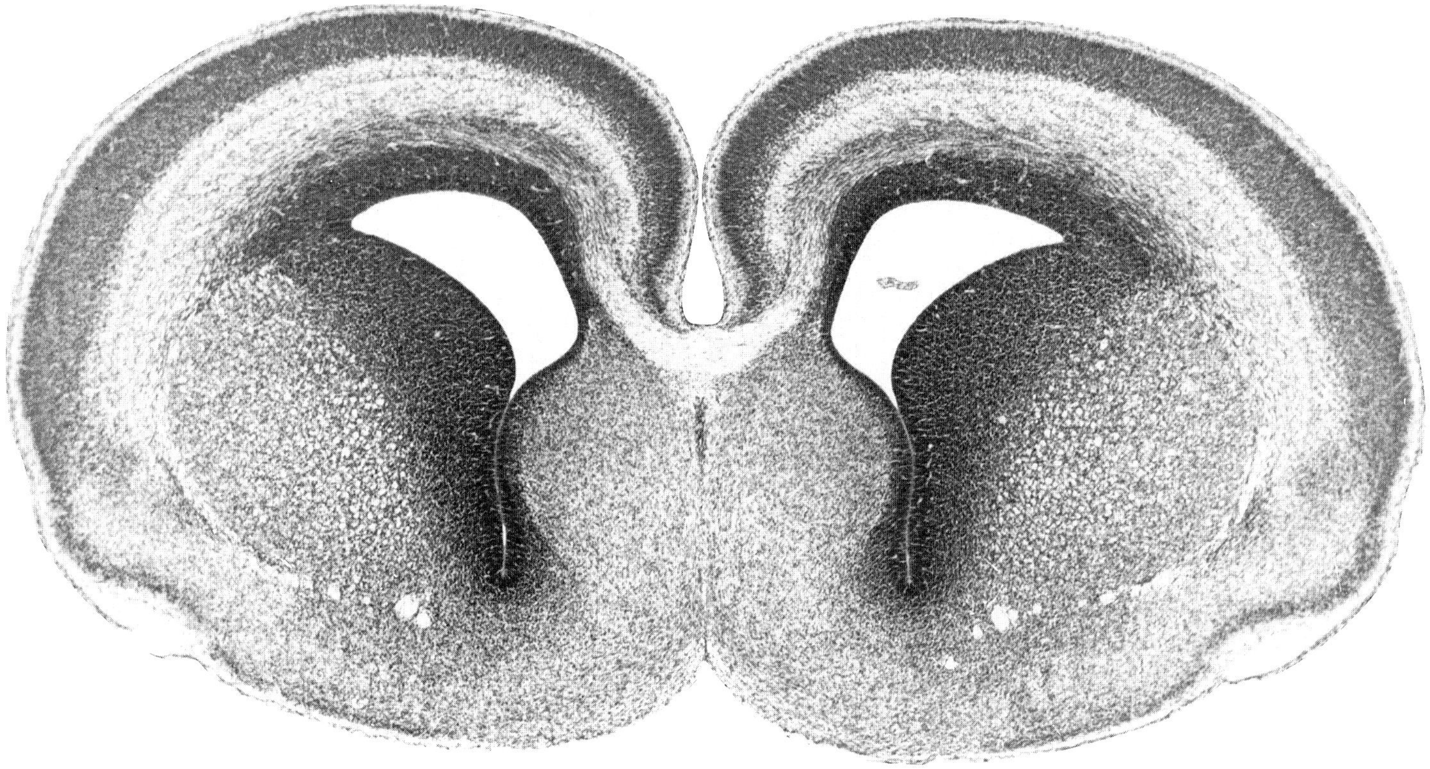

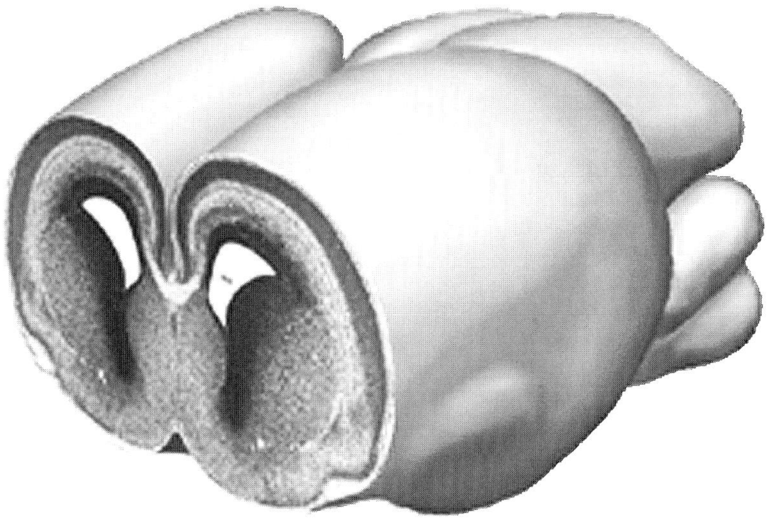

CORONAL FIGURE 7

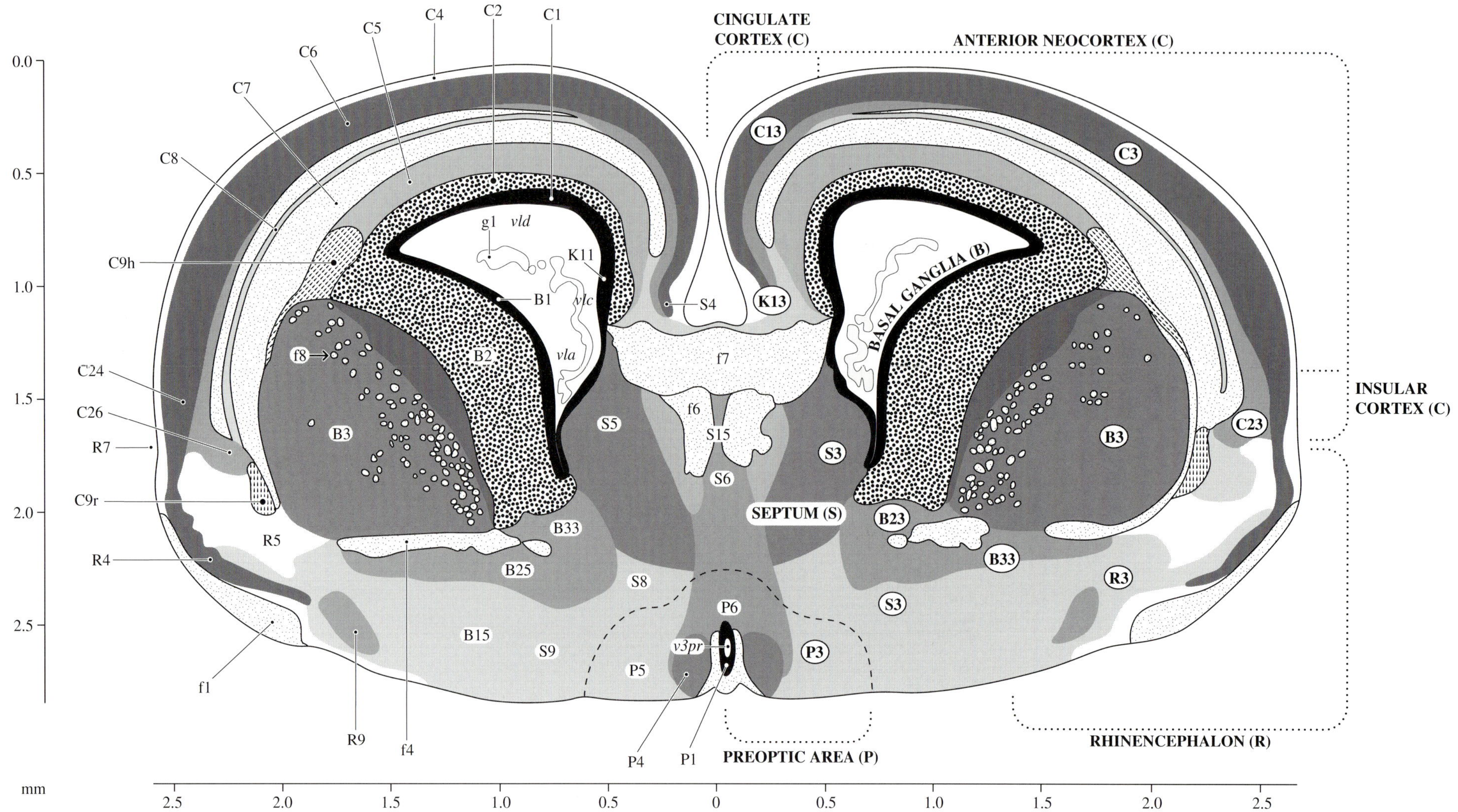

BASAL GANGLIA (B)

- B1 striatal neuroepithelium
- B2 striatal subventricular zone
- **B3** striatum (or differentiating field)
- B15 substantia innominata
- **B23** bed nucleus of stria terminalis (or differentiating field)
- B25 bed nucleus of stria terminalis, ventral division
- **B33** nucleus accumbens (or differentiating field)

CEREBRAL CORTEX (C)

- C1 neocortical neuroepithelium
- C2 neocortical subventricular zone
- **C3** neocortex (or differentiating field)
- C4 molecular layer (or primordial plexiform layer)
- C5 intermediate zone
- C6 cortical plate
- C7 white matter
- C8 subplate
- C9h lateral migratory stream, head
- C9r lateral migratory stream, reservoir
- **C13** cingulate cortex (or differentiating field)
- **C23** orbital and insular area (or differentiating field)
- C24 insular cortex
- C26 claustrum

FIBER TRACTS (f)

- f1 lateral olfactory tract
- f4 anterior commissure
- f6 fornix
- f7 hippocampal commissure
- f8 internal capsule

GLANDS AND PARANEURAL STRUCTURES (g)

- g1 choroid plexus of lateral ventricle

HIPPOCAMPAL FORMATION (K)

- K11 subicular neuroepithelium
- **K13** subicular area (or differentiating field)

PREOPTIC AREA AND OPTIC EPITHELIA (P)

- P1 preoptic neuroepithelium
- **P3** preoptic area (or differentiating field)
- P4 medial preoptic region
- P5 lateral preoptic region
- P6 superior preoptic region

RHINENCEPHALON (R)

- **R3** rhinencephalon (or differentiating field)
- R4 piriform cortex
- R5 endopiriform nucleus
- R7 rhinal fissure
- R9 olfactory tubercle

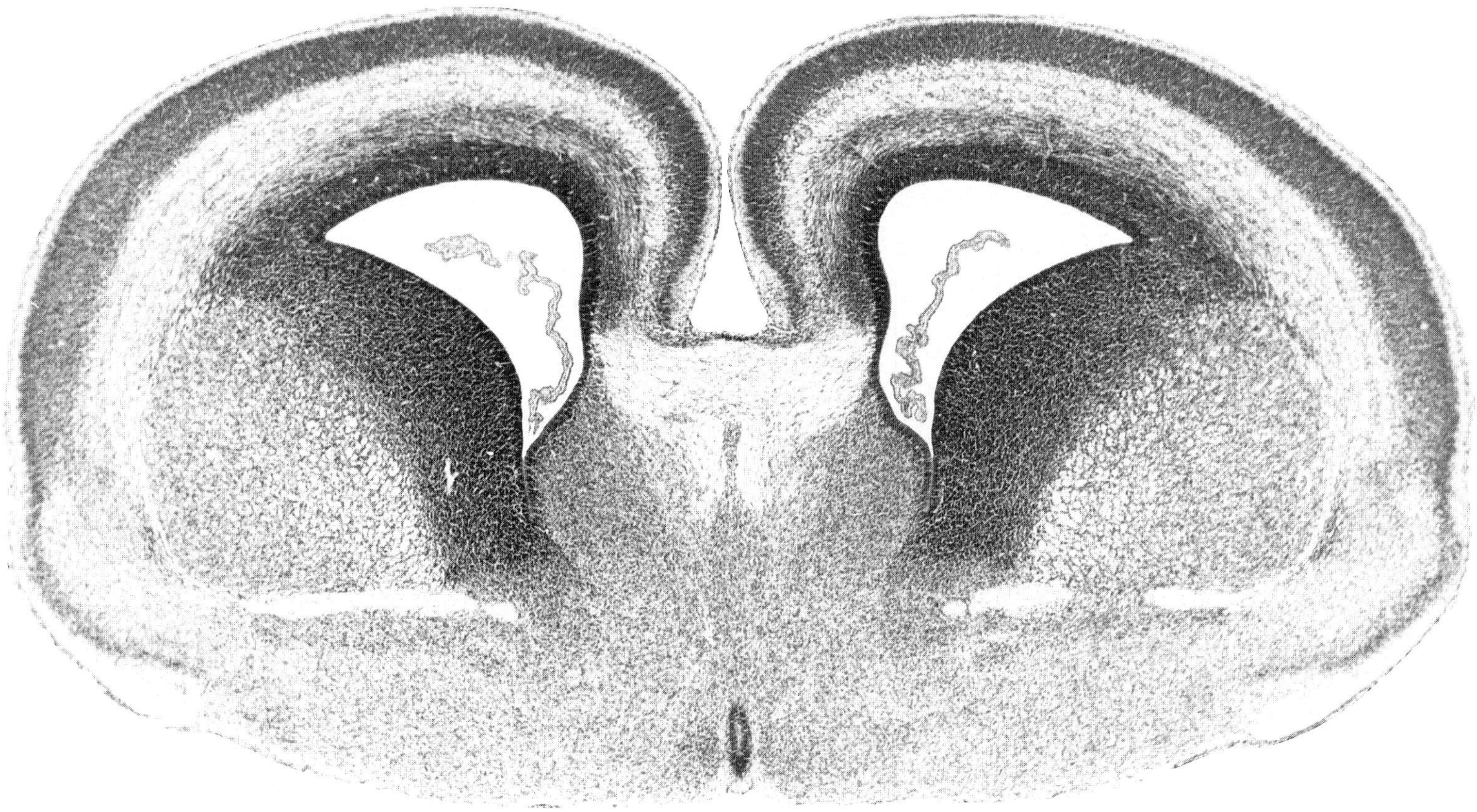

SEPTAL AREA (S)

S3 septum (or differentiating field)
S4 indesium griseum
S5 lateral septum
S6 medial septum
S8 diagonal band (vertical limb)
S9 diagonal band (horizontal limb)
S15 triangular septal nucleus

VENTRICLES (v)

vla lateral ventricle, anterior horn
vlc lateral ventricle, central
vld lateral ventricle, dorsal
v3pr third ventricle, preoptic recess

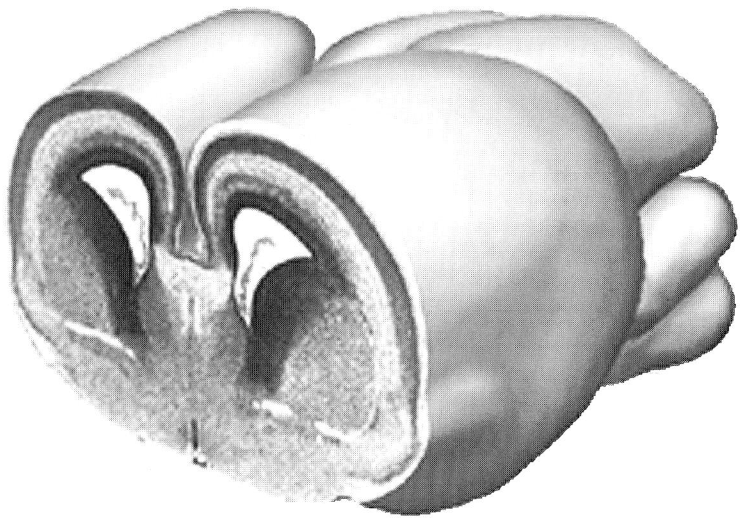

CORONAL FIGURE 8

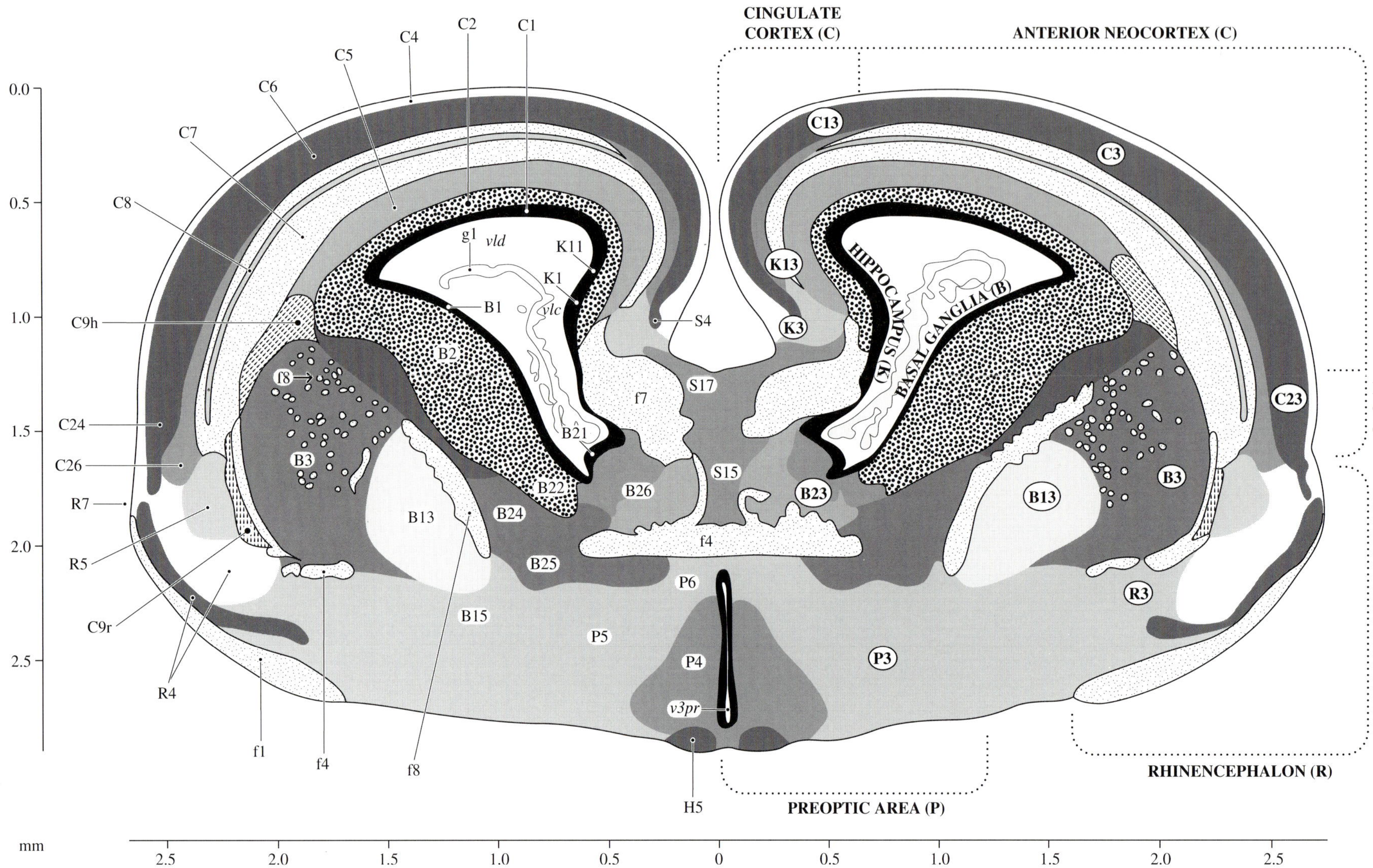

BASAL GANGLIA (B)

- B1 striatal neuroepithelium
- B2 striatal subventricular zone
- **B3** striatum (or differentiating field)
- **B13** pallidum (or differentiating field)
- B15 substantia innominata
- B21 strionuclear neuroepithelium
- B22 strionuclear subventricular zone
- **B23** bed nucleus of stria terminalis (or differentiating field)
- B24 bed nucleus of stria terminalis, dorsal division
- B25 bed nucleus of stria terminalis, ventral division
- B26 bed nucleus of stria terminalis, medial division

CEREBRAL CORTEX (C)

- C1 neocortical neuroepithelium
- C2 neocortical subventricular zone
- **C3** neocortex (or differentiating field)
- C4 molecular layer (or primordial plexiform layer)
- C5 intermediate zone
- C6 cortical plate
- C7 white matter
- C8 subplate
- C9h lateral migratory stream, head
- C9r lateral migratory stream, reservoir
- **C13** cingulate cortex (or differentiating field)
- **C23** orbital and insular area (or differentiating field)
- C24 insular cortex
- C26 claustrum

FIBER TRACTS (f)

- f1 lateral olfactory tract
- f4 anterior commissure
- f7 hippocampal commissure
- f8 internal capsule

GLANDS AND PARANEURAL STRUCTURES (g)

- g1 choroid plexus of lateral ventricle

HYPOTHALAMUS (H)

- H5 suprachiasmatic nucleus

HIPPOCAMPAL FORMATION (K)

- K1 hippocampal (Ammon's horn) neuroepithelium
- **K3** hippocampus (or differentiating field)
- K11 subicular neuroepithelium
- **K13** subicular area (or differentiating field)

PREOPTIC AREA AND OPTIC EPITHELIA (P)

- **P3** preoptic area (or differentiating field)
- P4 medial preoptic region
- P5 lateral preoptic region
- P6 superior preoptic region

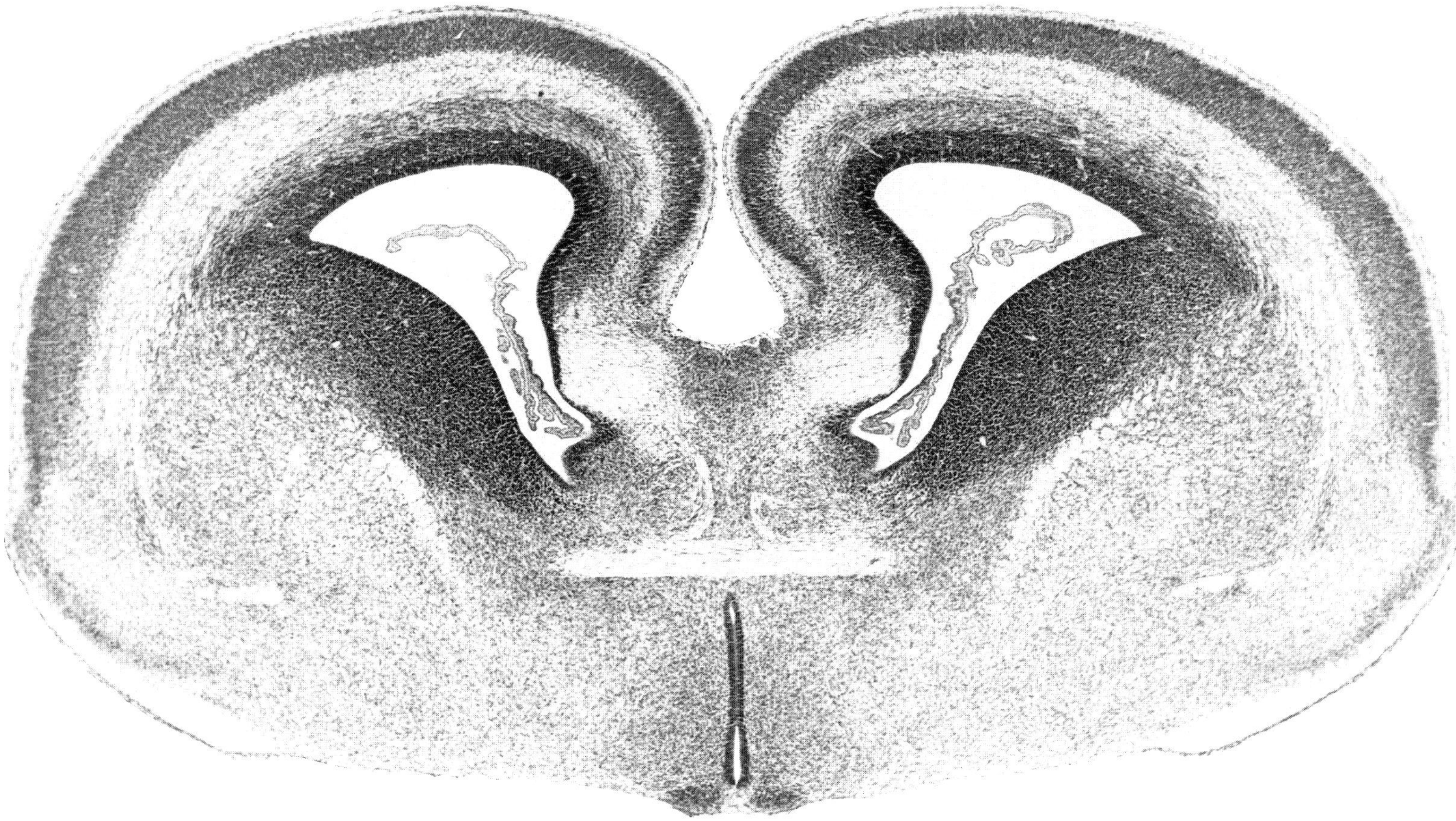

RHINENCEPHALON (R)

R3 rhinencephalon (or differentiating field)
R4 piriform cortex
R5 endopiriform nucleus
R7 rhinal fissure

SEPTAL AREA (S)

S4 indesium griseum
S15 triangular septal nucleus
S17 septofimbrial nucleus

VENTRICLES (v)

vlc lateral ventricle, central
vld lateral ventricle, dorsal
v3pr third ventricle, preoptic recess

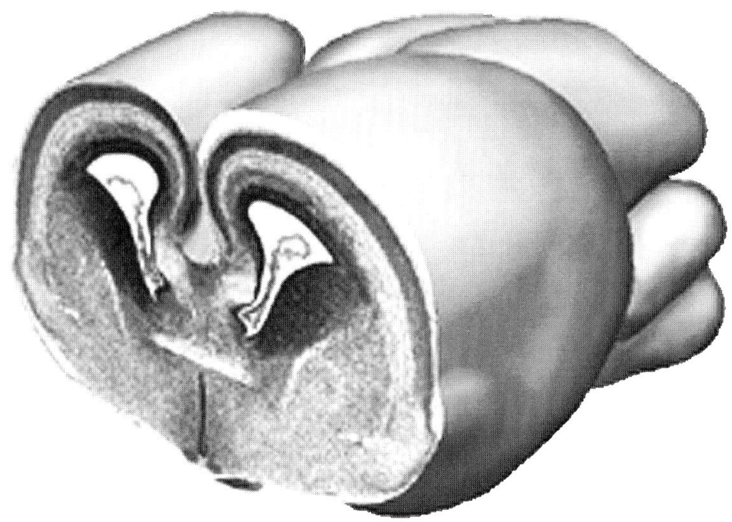

CORONAL FIGURE 9

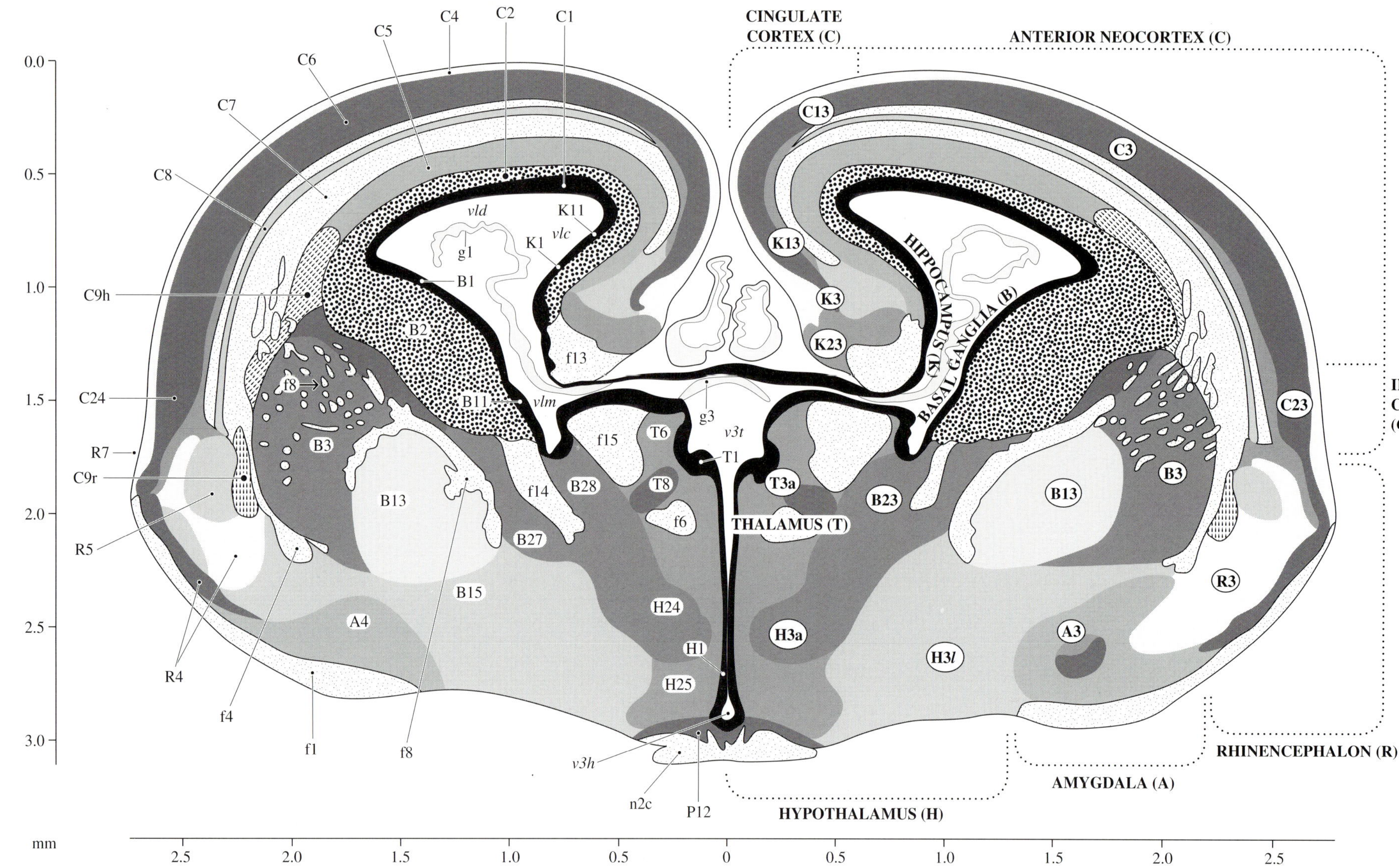

AMYGDALA (A)

A3 amygdala (or differentiating field)
A4 anterior amygdaloid area

BASAL GANGLIA (B)

B1 striatal neuroepithelium
B2 striatal subventricular zone
B3 striatum (or differentiating field)
B11 pallidal neuroepithelium
B13 pallidum (or differentiating field)
B15 substantia innominata
B23 bed nucleus of stria terminalis (or differentiating field)
B27 bed nucleus of stria terminalis, lateroventral division
B28 bed nucleus of stria terminalis, intermediate division

CEREBRAL CORTEX (C)

C1 neocortical neuroepithelium
C2 neocortical subventricular zone
C3 neocortex (or differentiating field)
C4 molecular layer (or primordial plexiform layer)
C5 intermediate zone
C6 cortical plate
C7 white matter
C8 subplate
C9h lateral migratory stream, head
C9r lateral migratory stream, reservoir
C13 cingulate cortex (or differentiating field)
C23 orbital and insular area (or differentiating field)
C24 insular cortex

FIBER TRACTS (f)

f1 lateral olfactory tract
f6 fornix
f8 internal capsule
f13 fimbria
f14 stria terminalis
f15 stria medullaris

GLANDS AND PARANEURAL STRUCTURES (g)

g1 choroid plexus of lateral ventricle
g3 choroid plexus of third ventricle

HYPOTHALAMUS (H)

H1 hypothalamic neuroepithelium
H3a anterior hypothalamus (or differentiating field)
H3*l* lateral hypothalamus (or differentiating field)
H24 anterodorsal nucleus
H25 anteroventral nucleus

HIPPOCAMPAL FORMATION (K)

K1 hippocampal (Ammon's horn) neuroepithelium
K3 hippocampus (or differentiating field)
K11 subicular neuroepithelium
K13 subicular area (or differentiating field)
K23 dentate gyrus (or differentiating field)

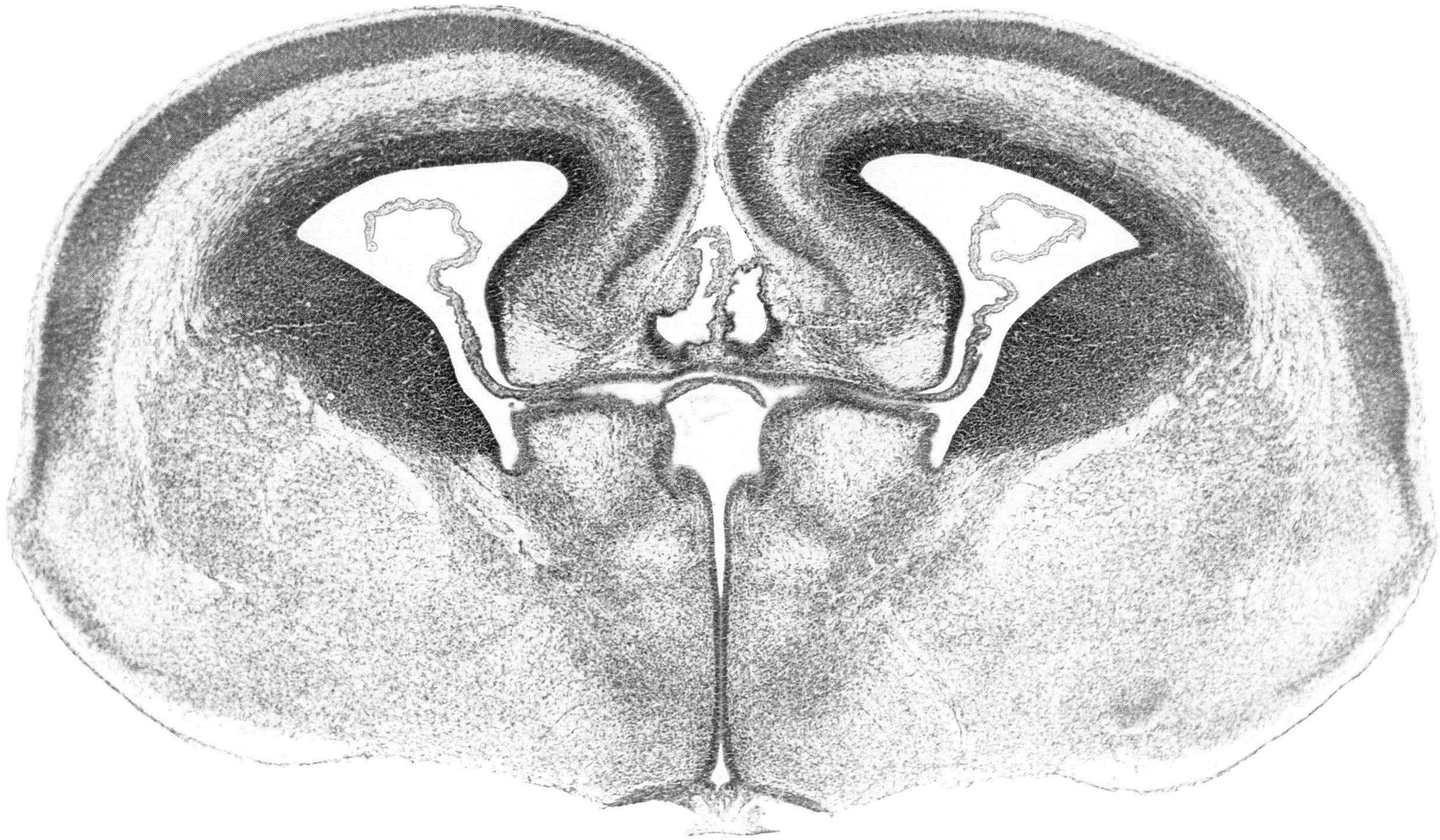

NERVES AND NUCLEI, CRANIAL (n)

n2c optic chiasma

PREOPTIC AREA AND OPTIC EPITHELIA (P)

P12 eye stalk epithelium

RHINENCEPHALON (R)

R3 rhinencephalon (or differentiating field)
R4 piriform cortex
R5 endopiriform nucleus
R7 rhinal fissure

THALAMUS (T)

T1 thalamic neuroepithelium
T3a anterior thalamus (or differentiating field)
T6 paraventricular nucleus
T8 anteromedial nucleus

VENTRICLES (v)

vlc lateral ventricle, central
vld lateral ventricle, dorsal
vlm foramen of Monro
v3h third ventricle, hypothalamic
v3t third ventricle, thalamic

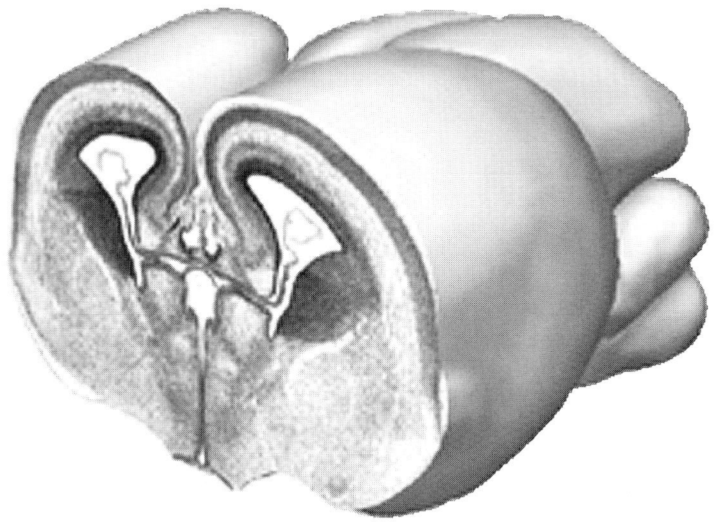

CORONAL FIGURE 10

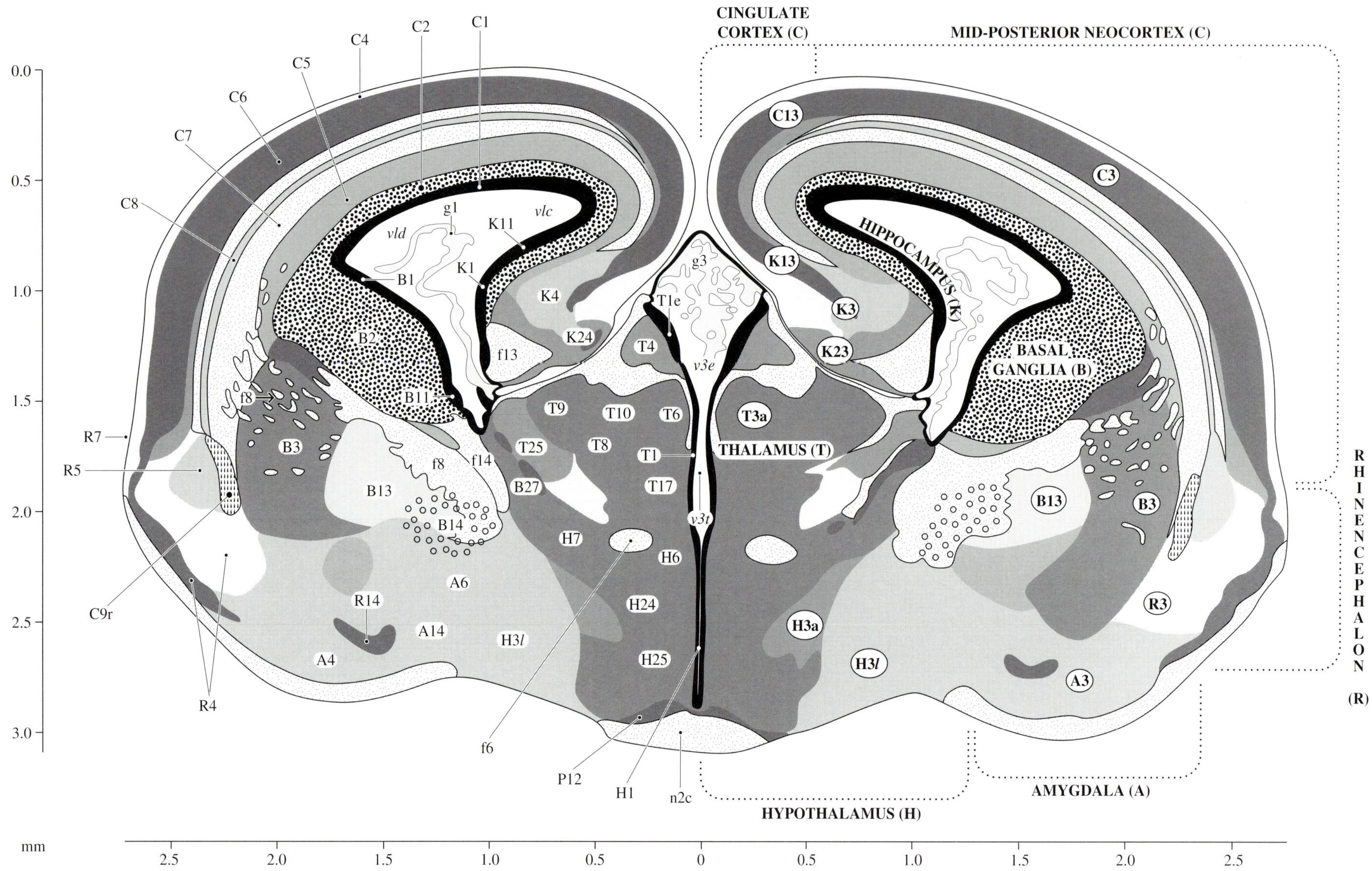

AMYGDALA (A)

A3 amygdala (or differentiating field)
A4 anterior amygdaloid area
A6 central amygdaloid nucleus
A14 medial amygdaloid nucleus

BASAL GANGLIA (B)

B1 striatal neuroepithelium
B2 striatal subventricular zone
B3 striatum (or differentiating field)
B11 pallidal neuroepithelium
B13 pallidum (or differentiating field)
B14 entopeduncular nucleus
B27 bed nucleus of stria terminalis, lateroventral division

CEREBRAL CORTEX (C)

C1 neocortical neuroepithelium
C2 neocortical subventricular zone
C3 neocortex (or differentiating field)
C4 molecular layer (or primordial plexiform layer)
C5 intermediate zone
C6 cortical plate
C7 white matter
C8 subplate
C9r lateral migratory stream, reservoir
C13 cingulate cortex (or differentiating field)

FIBER TRACTS (f)

f6 fornix
f8 internal capsule
f13 fimbria
f14 stria terminalis

GLANDS AND PARANEURAL STRUCTURES (g)

g1 choroid plexus of lateral ventricle
g3 choroid plexus of third ventricle

HYPOTHALAMUS (H)

H1 hypothalamic neuroepithelium
H3a anterior hypothalamus (or differentiating field)
H3*l* lateral hypothalamus (or differentiating field)
H6 paraventricular nucleus
H7 zona incerta
H24 anterodorsal nucleus
H25 anteroventral nucleus

HIPPOCAMPAL FORMATION (K)

K1 hippocampal (Ammon's horn) neuroepithelium
K3 hippocampus (or differentiating field)
K4 CA1 field
K11 subicular neuroepithelium
K13 subicular area (or differentiating field)
K23 dentate gyrus (or differentiating field)
K24 dentate granular layer

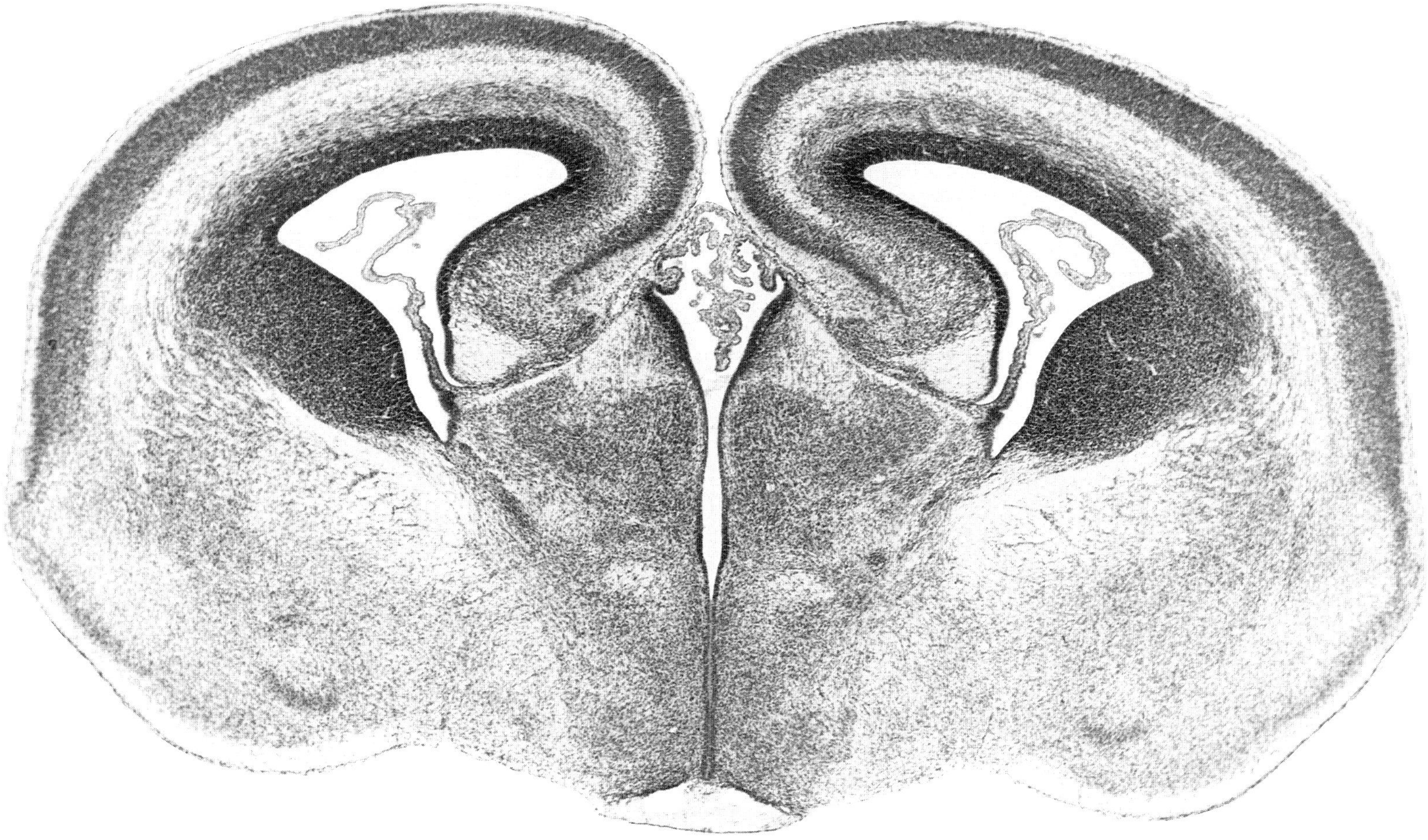

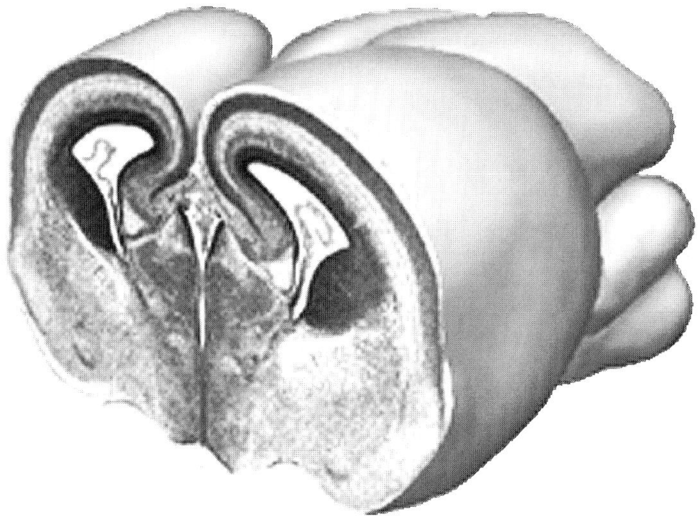

NERVES AND NUCLEI, CRANIAL (n)

n2c optic chiasma

PREOPTIC AREA AND OPTIC EPITHELIA (P)

P12 eye stalk epithelium

RHINENCEPHALON (R)

R3 rhinencephalon (or differentiating field)
R4 piriform cortex
R5 endopiriform nucleus
R7 rhinal fissure
R14 nucleus of the lateral olfactory tract

THALAMUS (T)

T1 thalamic neuroepithelium
T1e epithalamic neuroepithelium
T3a anterior thalamus (or differentiating field)
T4 medial habenular nucleus
T6 paraventricular nucleus
T8 anteromedial nucleus
T9 anteroventral nucleus
T10 anterodorsal nucleus
T17 rhomboid nucleus
T25 reticular nucleus

VENTRICLES (v)

vlc lateral ventricle, central
vld lateral ventricle, dorsal
v3e third ventricle, epithalamic
v3t third ventricle, thalamic

CORONAL FIGURE 11

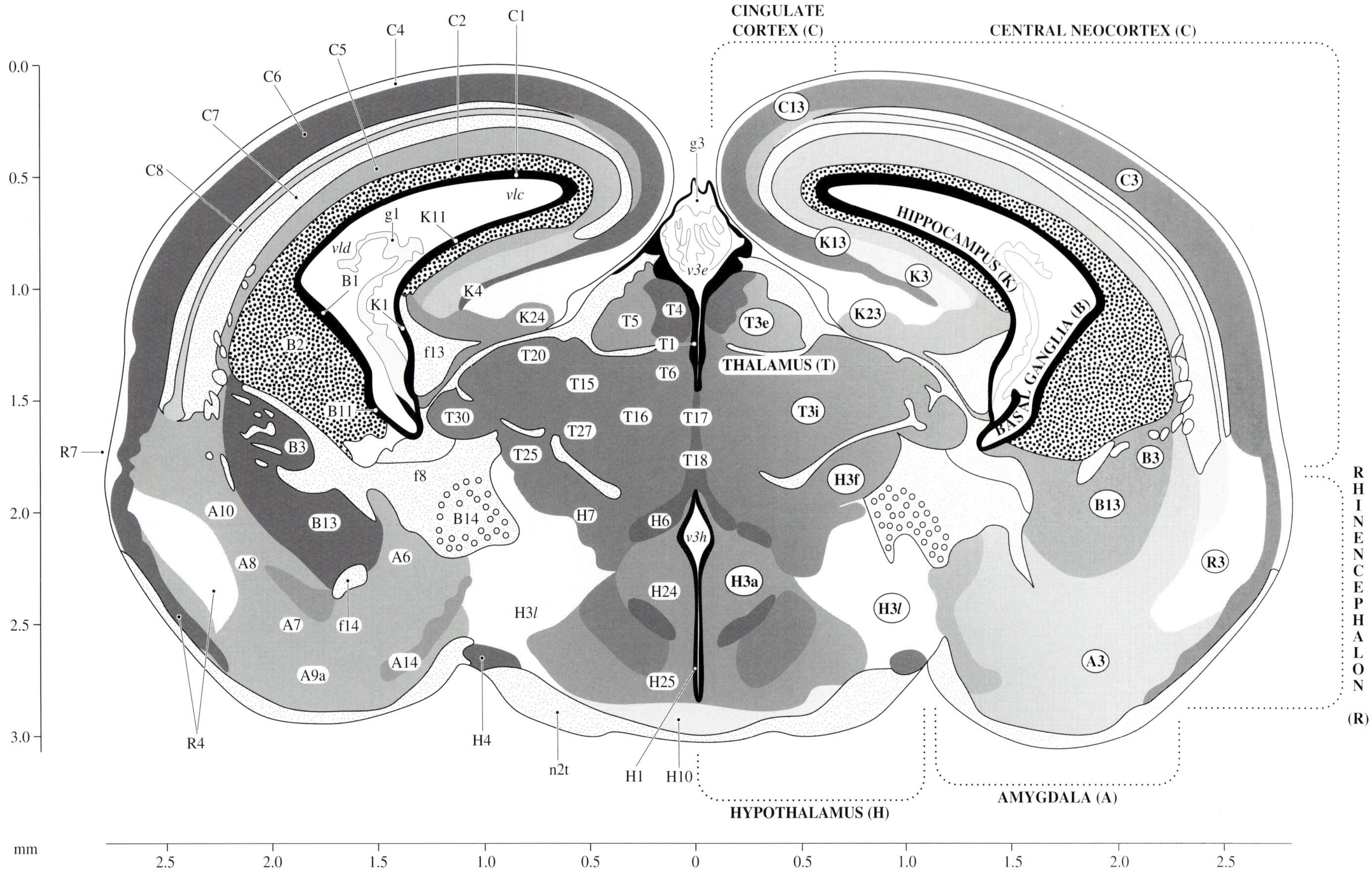

AMYGDALA (A)

A3 amygdala (or differentiating field)
A6 central amygdaloid nucleus
A7 basomedial amygdaloid nucleus
A8 basolateral amygdaloid nucleus
A9a cortical amygdaloid nucleus, anterior
A10 lateral amygdaloid nucleus
A14 medial amygdaloid nucleus

BASAL GANGLIA (B)

B1 striatal neuroepithelium
B2 striatal subventricular zone
B3 striatum (or differentiating field)
B11 pallidal neuroepithelium
B13 pallidum (or differentiating field)
B14 entopeduncular nucleus

CEREBRAL CORTEX (C)

C1 neocortical neuroepithelium
C2 neocortical subventricular zone
C3 neocortex (or differentiating field)
C4 molecular layer (or primordial plexiform layer)
C5 intermediate zone
C6 cortical plate
C7 white matter
C8 subplate
C13 cingulate cortex (or differentiating field)

FIBER TRACTS (f)

f8 internal capsule
f13 fimbria
f14 stria terminalis

GLANDS AND PARANEURAL STRUCTURES (g)

g1 choroid plexus of lateral ventricle
g3 choroid plexus of third ventricle

HYPOTHALAMUS (H)

H1 hypothalamic neuroepithelium
H3a anterior hypothalamus (or differentiating field)
H3f fields of Forel (or differentiating field)
H3*l* lateral hypothalamus (or differentiating field)
H4 supraoptic nucleus
H6 paraventricular nucleus
H7 zona incerta
H10 anterobasal nucleus
H24 anterodorsal nucleus
H25 anteroventral nucleus

HIPPOCAMPAL FORMATION (K)

K1 hippocampal (Ammon's horn) neuroepithelium
K3 hippocampus (or differentiating field)
K4 CA1 field
K11 subicular neuroepithelium
K13 subicular area (or differentiating field)
K23 dentate gyrus (or differentiating field)
K24 dentate granular layer

NERVES AND NUCLEI, CRANIAL (n)

n2t optic tract

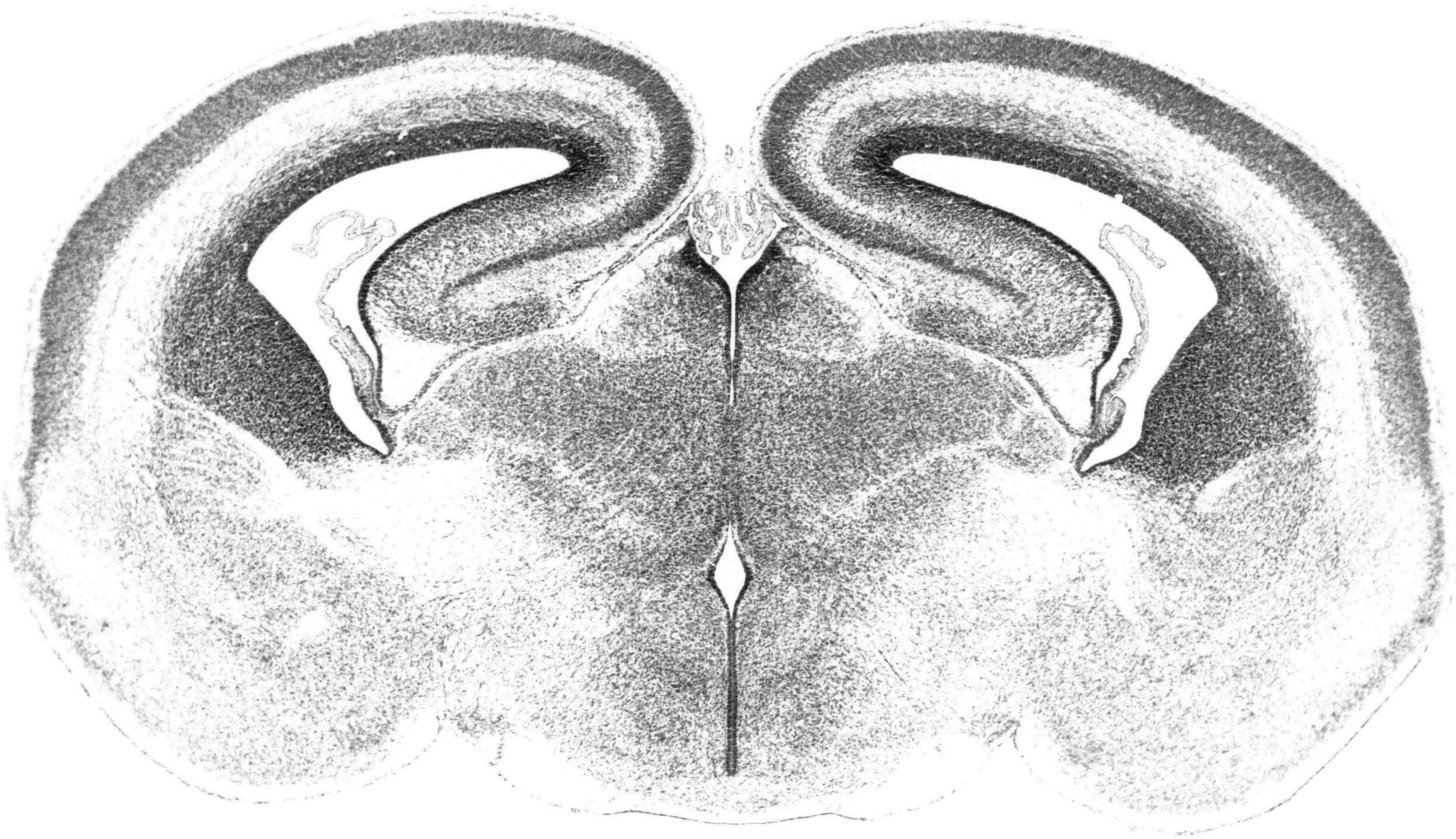

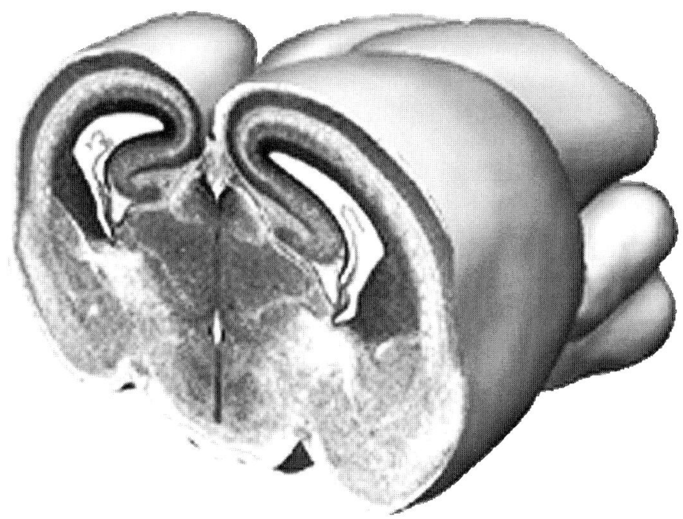

RHINENCEPHALON (R)

R3 rhinencephalon (or differentiating field)
R4 piriform cortex
R7 rhinal fissure

THALAMUS (T)

T1 thalamic neuroepithelium
T3e epithalamus (or differentiating field)
T3i intermediate thalamus (or differentiating field)
T4 medial habenular nucleus
T5 lateral habenular nucleus
T6 paraventricular nucleus
T15 centrolateral nucleus
T16 mediodorsal nucleus
T17 rhomboid nucleus
T18 reuniens nucleus
T20 laterodorsal nucleus
T25 reticular nucleus
T27 ventrolateral nuclear complex
T30 ventral lateral geniculate nucleus

VENTRICLES (v)

vlc lateral ventricle, central
vld lateral ventricle, dorsal
v3e third ventricle, epithalamic
v3h third ventricle, hypothalamic

CORONAL FIGURE 12

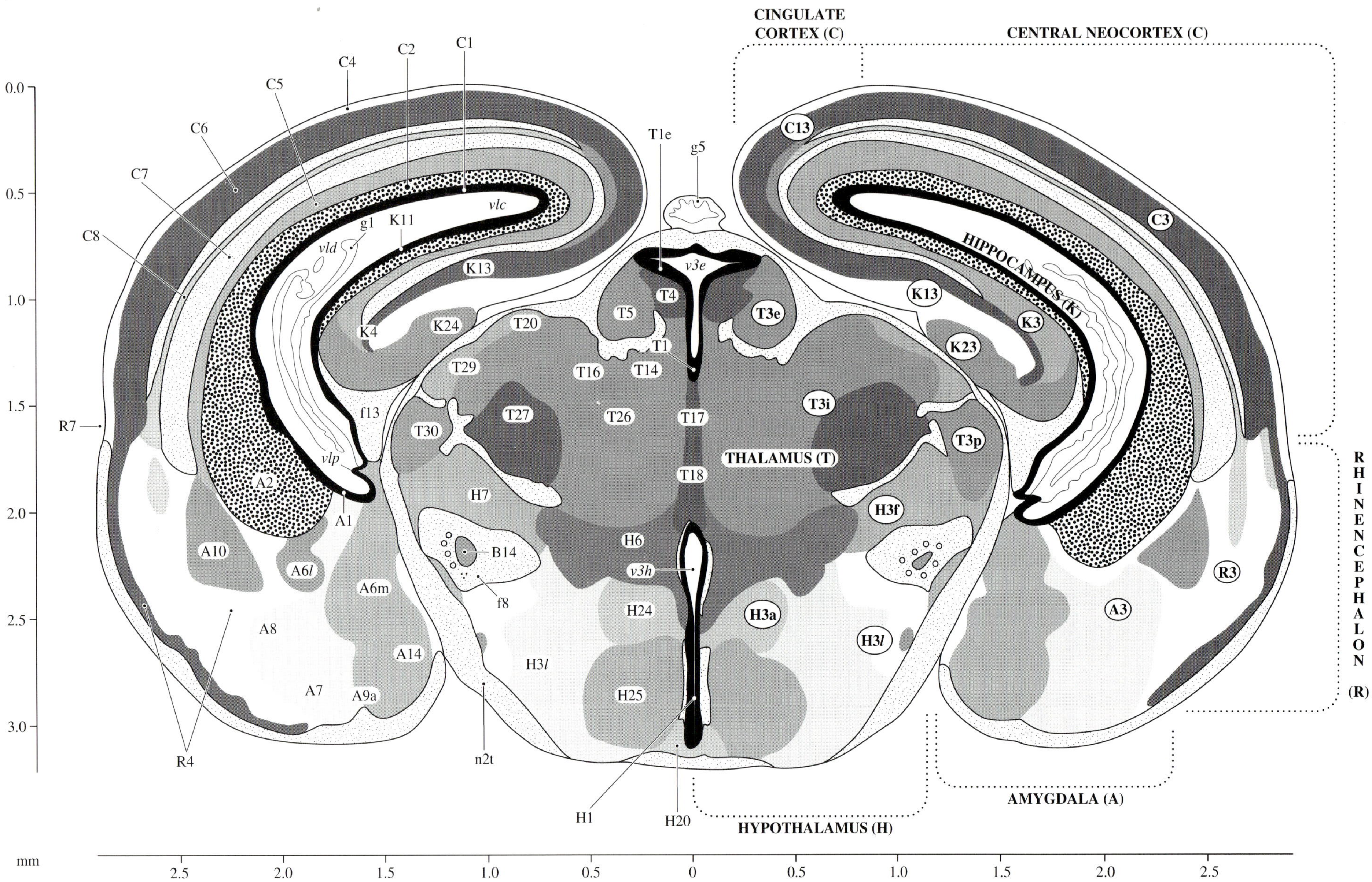

AMYGDALA (A)

- A1 amygdaloid neuroepithelium
- A2 amygdaloid subventricular zone
- **A3** amygdala (or differentiating field)
- A6*l* central amygdaloid nucleus, lateral
- A6m central amygdaloid nucleus, medial
- A7 basomedial amygdaloid nucleus
- A8 basolateral amygdaloid nucleus
- A9a cortical amygdaloid nucleus, anterior
- A10 lateral amygdaloid nucleus
- A14 medial amygdaloid nucleus

BASAL GANGLIA (B)

- B14 entopeduncular nucleus

CEREBRAL CORTEX (C)

- C1 neocortical neuroepithelium
- C2 neocortical subventricular zone
- **C3** neocortex (or differentiating field)
- C4 molecular layer (or primordial plexiform layer)
- C5 intermediate zone
- C6 cortical plate
- C7 white matter
- C8 subplate
- **C13** cingulate cortex (or differentiating field)

FIBER TRACTS (f)

- f8 internal capsule
- f13 fimbria

GLANDS AND PARANEURAL STRUCTURES (g)

- g1 choroid plexus of lateral ventricle
- g5 pineal gland (primordium)

HYPOTHALAMUS (H)

- H1 hypothalamic neuroepithelium
- **H3a** anterior hypothalamus (or differentiating field)
- **H3f** fields of Forel (or differentiating field)
- **H3*l*** lateral hypothalamus (or differentiating field)
- H6 paraventricular nucleus
- H7 zona incerta
- H20 arcuate nucleus
- H24 anterodorsal nucleus
- H25 anteroventral nucleus

HIPPOCAMPAL FORMATION (K)

- **K3** hippocampus (or differentiating field)
- K4 CA1 field
- K11 subicular neuroepithelium
- **K13** subicular area (or differentiating field)
- **K23** dentate gyrus (or differentiating field)
- K24 dentate granular layer

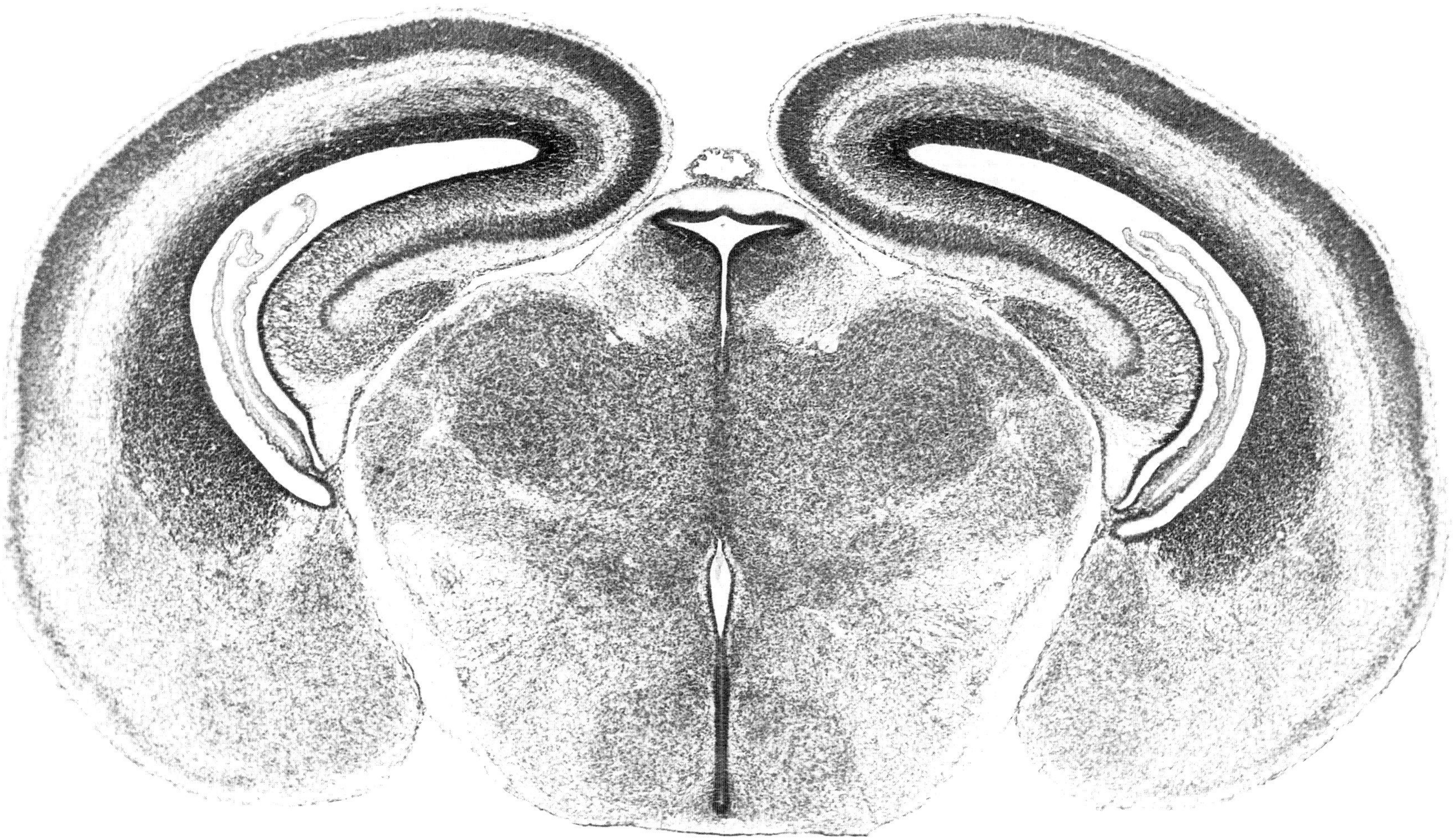

NERVES AND NUCLEI, CRANIAL (n)

n2t optic tract

RHINENCEPHALON (R)

R3 rhinencephalon (or differentiating field)
R4 piriform cortex
R7 rhinal fissure

THALAMUS (T)

T1 thalamic neuroepithelium
T1e epithalamic neuroepithelium
T3e epithalamus (or differentiating field)
T3i intermediate thalamus (or differentiating field)
T3p posterior thalamus (or differentiating field)
T4 medial habenular nucleus
T5 lateral habenular nucleus
T14 centromedial nucleus
T16 mediodorsal nucleus
T17 rhomboid nucleus
T18 reuniens nucleus
T20 laterodorsal nucleus
T26 ventral medial nucleus
T27 ventrolateral nuclear complex
T29 dorsal lateral geniculate nucleus
T30 ventral lateral geniculate nucleus

VENTRICLES (v)

vlc lateral ventricle, central
vld lateral ventricle, dorsal
vlp lateral ventricle, posterior horn
v3e third ventricle, epithalamic
v3h third ventricle, hypothalamic

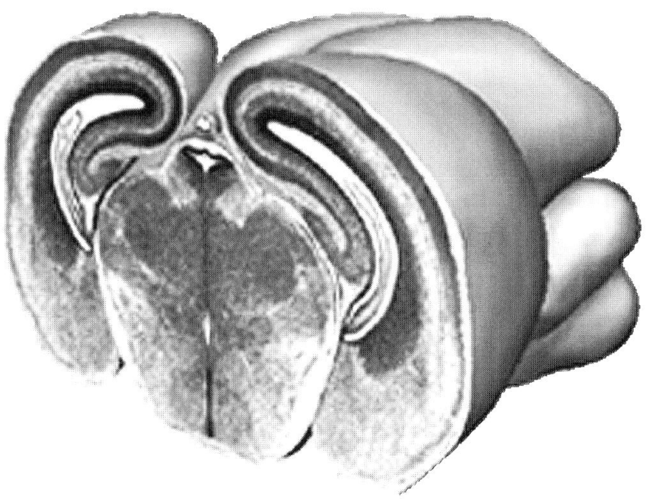

CORONAL FIGURE 13

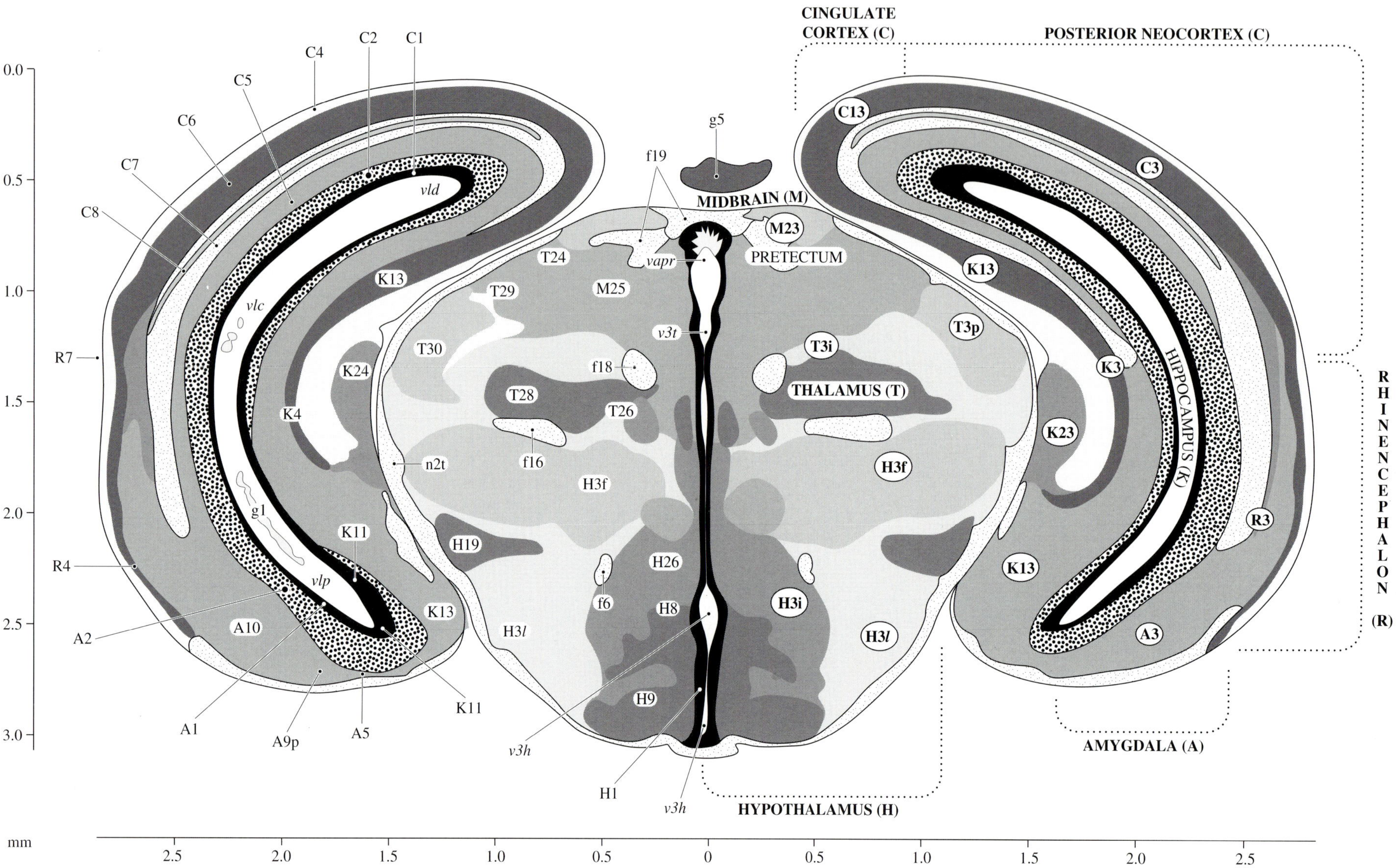

AMYGDALA (A)

- A1 amygdaloid neuroepithelium
- A2 amygdaloid subventricular zone
- **A3** amygdala (or differentiating field)
- A5 amygdala hippocampal area
- A9p cortical amygdaloid nucleus, posterior
- A10 lateral amygdaloid nucleus

CEREBRAL CORTEX (C)

- C1 neocortical neuroepithelium
- C2 neocortical subventricular zone
- **C3** neocortex (or differentiating field)
- C4 molecular layer (or primordial plexiform layer)
- C5 intermediate zone
- C6 cortical plate
- C7 white matter
- C8 subplate
- **C13** cingulate cortex (or differentiating field)

FIBER TRACTS (f)

- f6 fornix
- f16 medial lemniscus
- f18 habenulopeduncular tract
- f19 posterior commissure

GLANDS AND PARANEURAL STRUCTURES (g)

- g1 choroid plexus of lateral ventricle
- g5 pineal gland (primordium)

HYPOTHALAMUS (H)

- H1 hypothalamic neuroepithelium
- **H3f** fields of Forel (or differentiating field)
- **H3i** intermediate hypothalamus (or differentiating field)
- **H3*l*** lateral hypothalamus (or differentiating field)
- H8 dorsomedial nucleus
- H9 ventromedial nucleus
- H19 subthalamic nucleus
- H26 posterior hypothalamic area

HIPPOCAMPAL FORMATION (K)

- **K3** hippocampus (or differentiating field)
- K4 CA1 field
- K11 subicular neuroepithelium
- **K13** subicular area (or differentiating field)
- **K23** dentate gyrus (or differentiating field)
- K24 dentate granular layer

MIDBRAIN (M)

- **M23** pretectum (or differentiating field)
- M25 nucleus of the posterior commissure

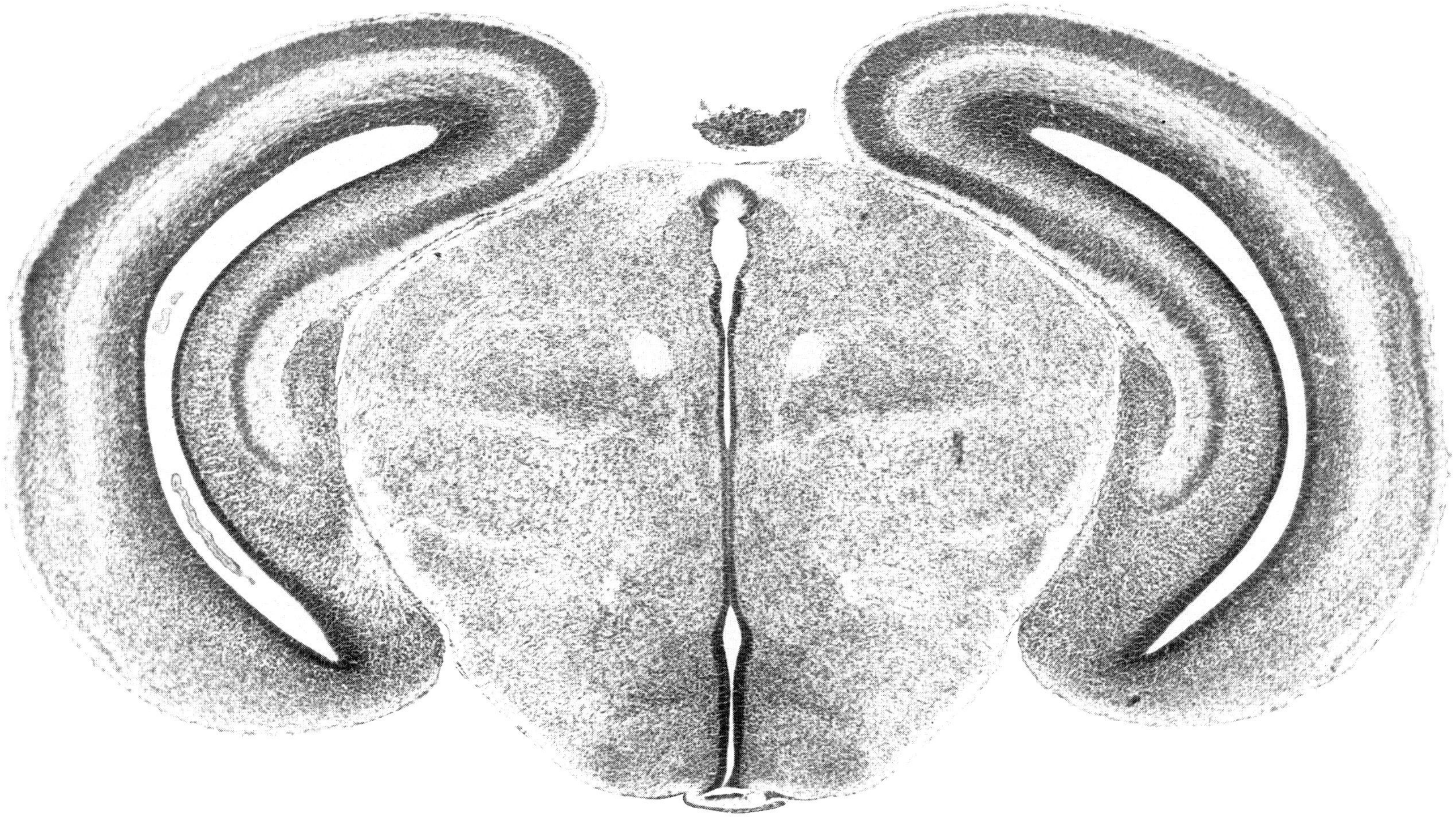

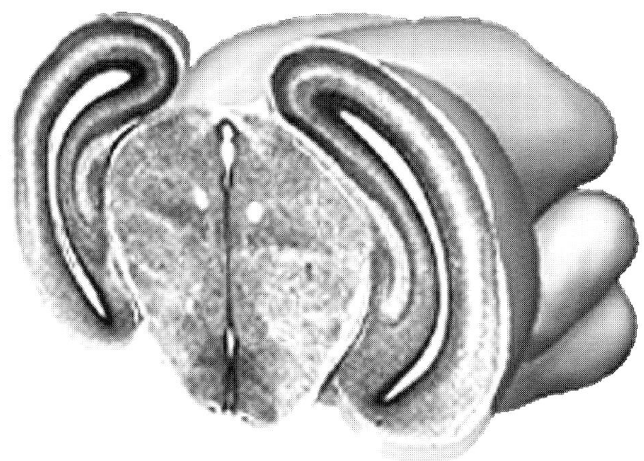

NERVES AND NUCLEI, CRANIAL (n)

n2t optic tract

RHINENCEPHALON (R)

R3 rhinencephalon (or differentiating field)
R4 piriform cortex
R7 rhinal fissure

THALAMUS (T)

T3i intermediate thalamus (or differentiating field)
T3p posterior thalamus (or differentiating field)
T24 lateral posterior nucleus
T26 ventral medial nucleus
T28 ventrobasal nuclear complex
T29 dorsal lateral geniculate nucleus
T30 ventral lateral geniculate nucleus

VENTRICLES (v)

vapr aqueduct, pretectal
vlc lateral ventricle, central
vld lateral ventricle, dorsal
vlp lateral ventricle, posterior horn
v3h third ventricle, hypothalamic
v3t third ventricle, thalamic

CORONAL FIGURE 14

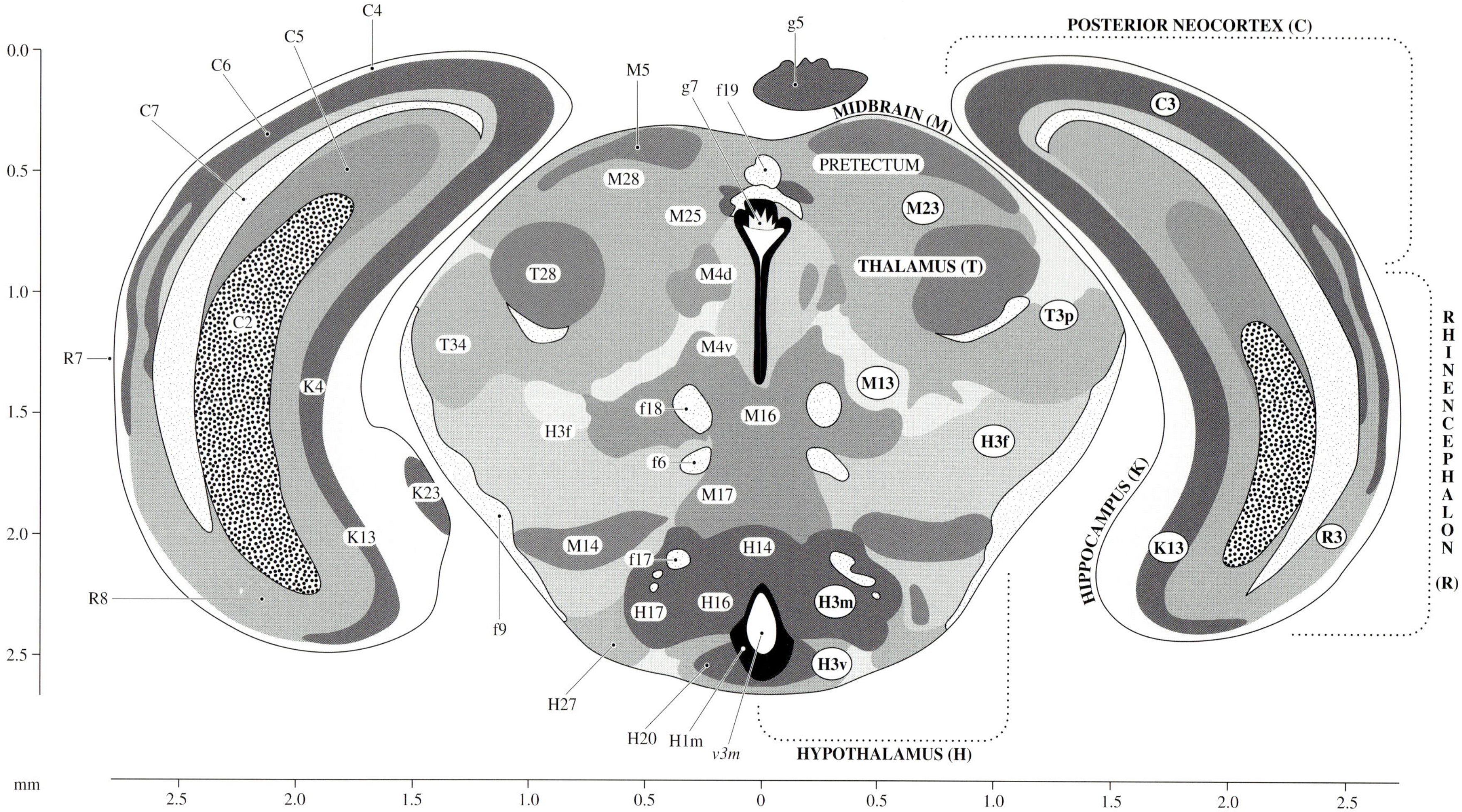

CEREBRAL CORTEX (C)

C2 neocortical subventricular zone
C3 neocortex (or differentiating field)
C4 molecular layer (or primordial plexiform layer)
C5 intermediate zone
C6 cortical plate
C7 white matter

FIBER TRACTS (f)

f6 fornix
f9 cerebral peduncle
f17 mammillothalamic tract
f18 habenulopeduncular tract
f19 posterior commissure

GLANDS AND PARANEURAL STRUCTURES (g)

g5 pineal gland (primordium)
g7 subcommissural organ

HYPOTHALAMUS (H)

H1m mammillary neuroepithelium
H3f fields of Forel (or differentiating field)
H3m mammillary body (or differentiating field)
H3v ventral hypothalamus (or differentiating field)
H14 supramammillary nucleus
H16 medial mammillary nucleus
H17 lateral mammillary nucleus
H20 arcuate nucleus
H27 tuberomammillary nucleus

HIPPOCAMPAL FORMATION (K)

K4 CA1 field
K13 subicular area (or differentiating field)
K23 dentate gyrus (or differentiating field)

MIDBRAIN (M)

M4d dorsal periaqueductal gray
M4v ventral periaqueductal gray
M5 superior colliculus, superficial layers
M13 tegmentum (or differentiating field)
M14 substantia nigra
M16 dorsal raphe
M17 ventral tegmental nucleus
M23 pretectum (or differentiating field)
M25 nucleus of the posterior commissure
M28 posterior pretectal nucleus

RHINENCEPHALON (R)

R3 rhinencephalon (or differentiating field)
R7 rhinal fissure
R8 entorhinal cortex

THALAMUS (T)

T3p posterior thalamus (or differentiating field)
T28 ventrobasal nuclear complex
T34 medial geniculate nucleus

VENTRICLES (v)

v3m third ventricle, mammillary recess

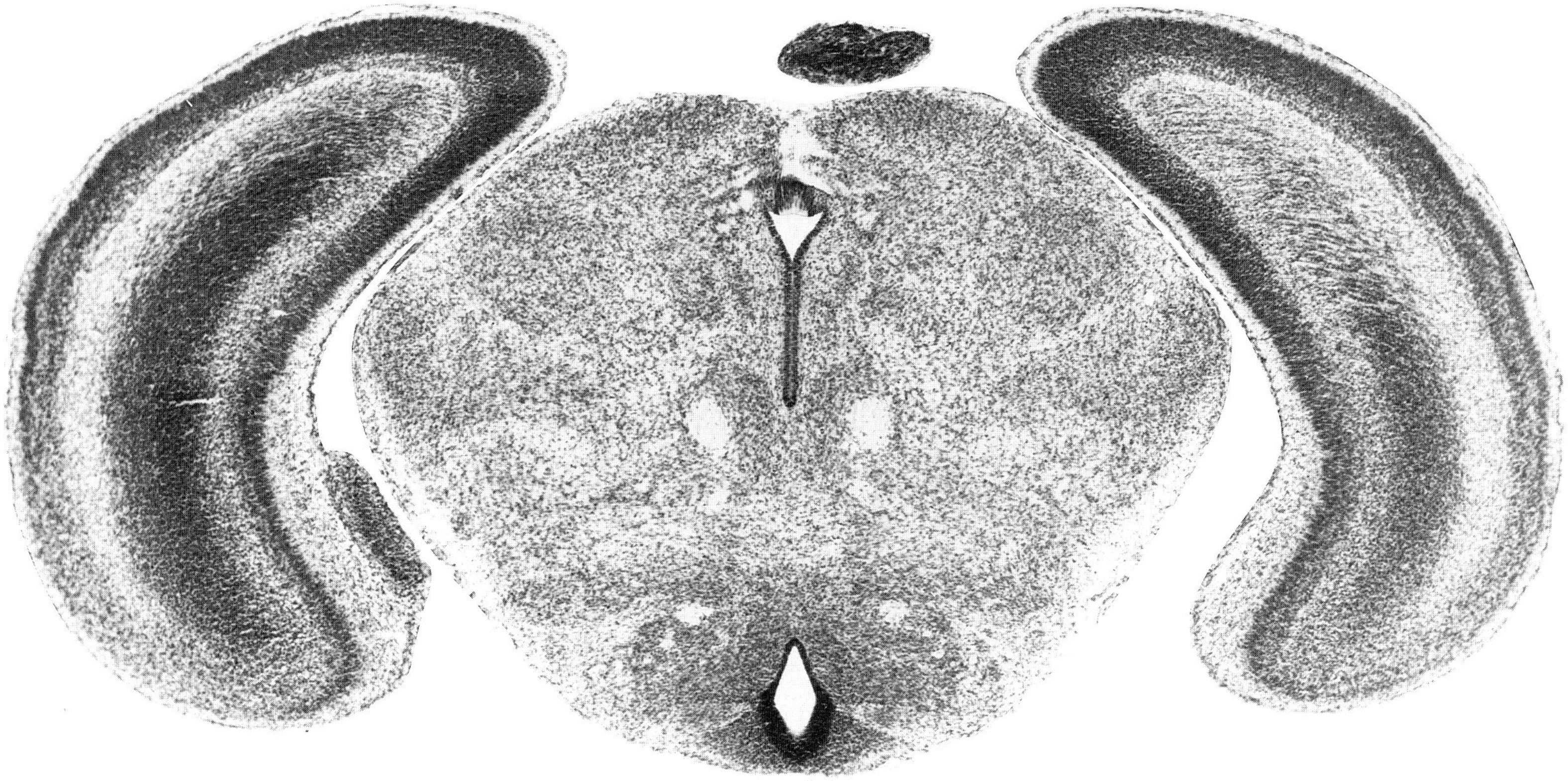

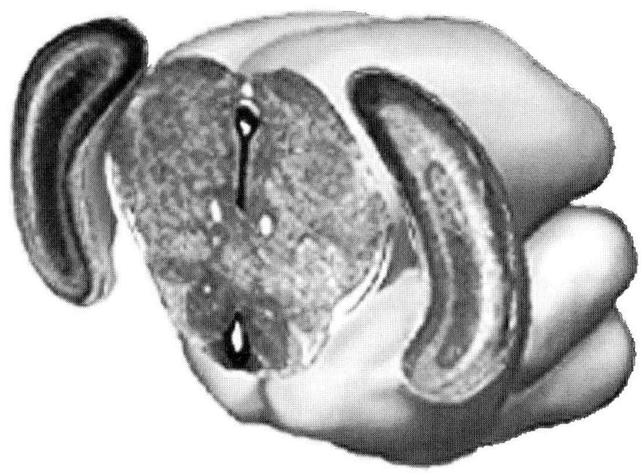

CORONAL FIGURE 15

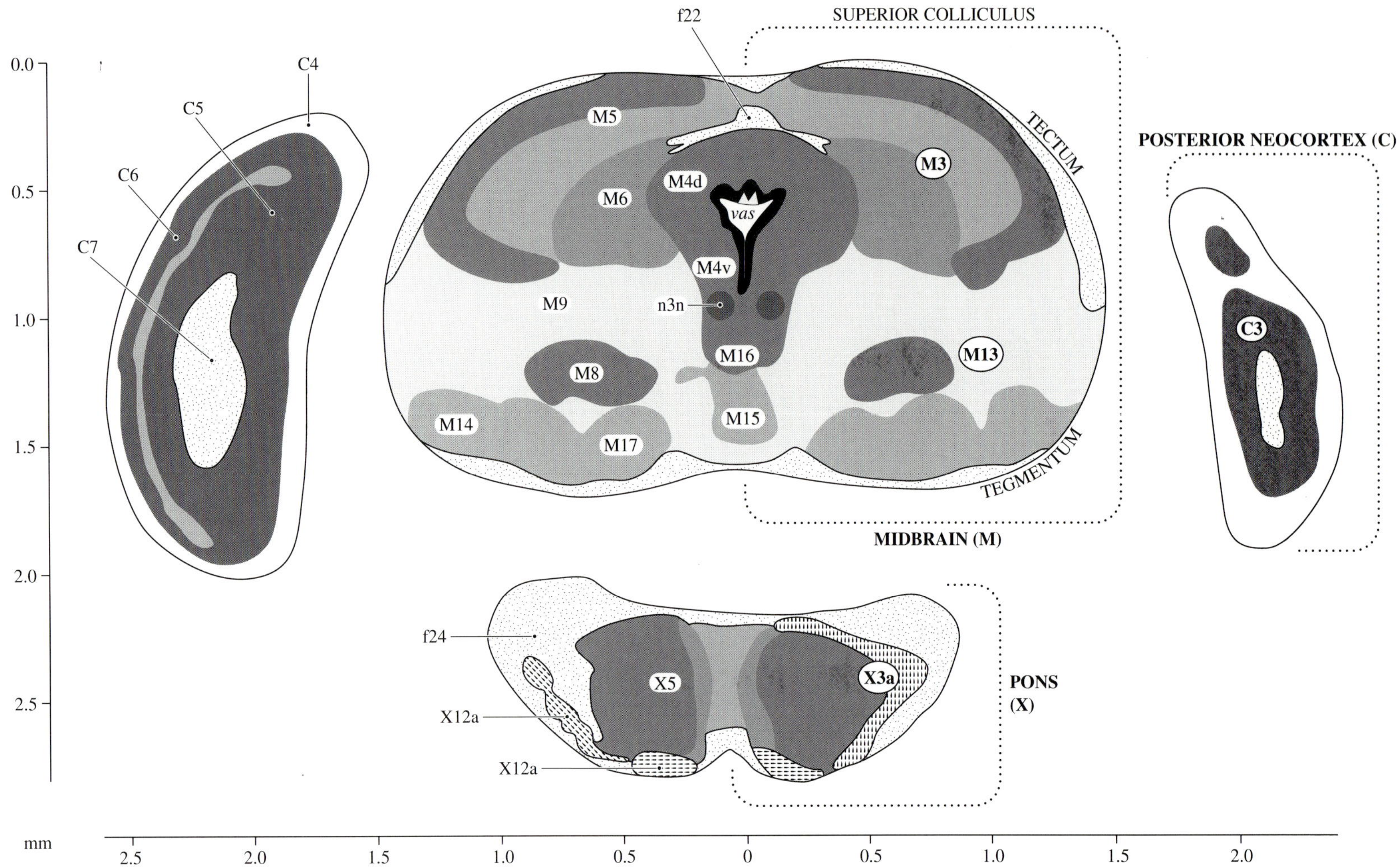

CEREBRAL CORTEX (C)

C3	neocortex (or differentiating field)
C4	molecular layer (or primordial plexiform layer)
C5	intermediate zone
C6	cortical plate
C7	white matter

FIBER TRACTS (f)

f22	commissure of superior colliculus
f24	middle cerebellar peduncle

MIDBRAIN (M)

M3	superior colliculus (or differentiating field)
M4d	dorsal periaqueductal gray
M4v	ventral periaqueductal gray
M5	superior colliculus, superficial layers
M6	superior colliculus, deep layers
M8	red nucleus
M9	midbrain reticular formation
M13	tegmentum (or differentiating field)
M14	substantia nigra
M15	interpeduncular nucleus
M16	dorsal raphe
M17	ventral tegmental nucleus

NERVES AND NUCLEI, CRANIAL (n)

n3n	oculomotor nucleus

VENTRICLES (v)

vas	aqueduct, superior tectal

PONTINE AREA (X)

X3a	anterior pons (or differentiating field)
X5	pontine gray nucleus
X12a	anterior precerebellar extramural migration

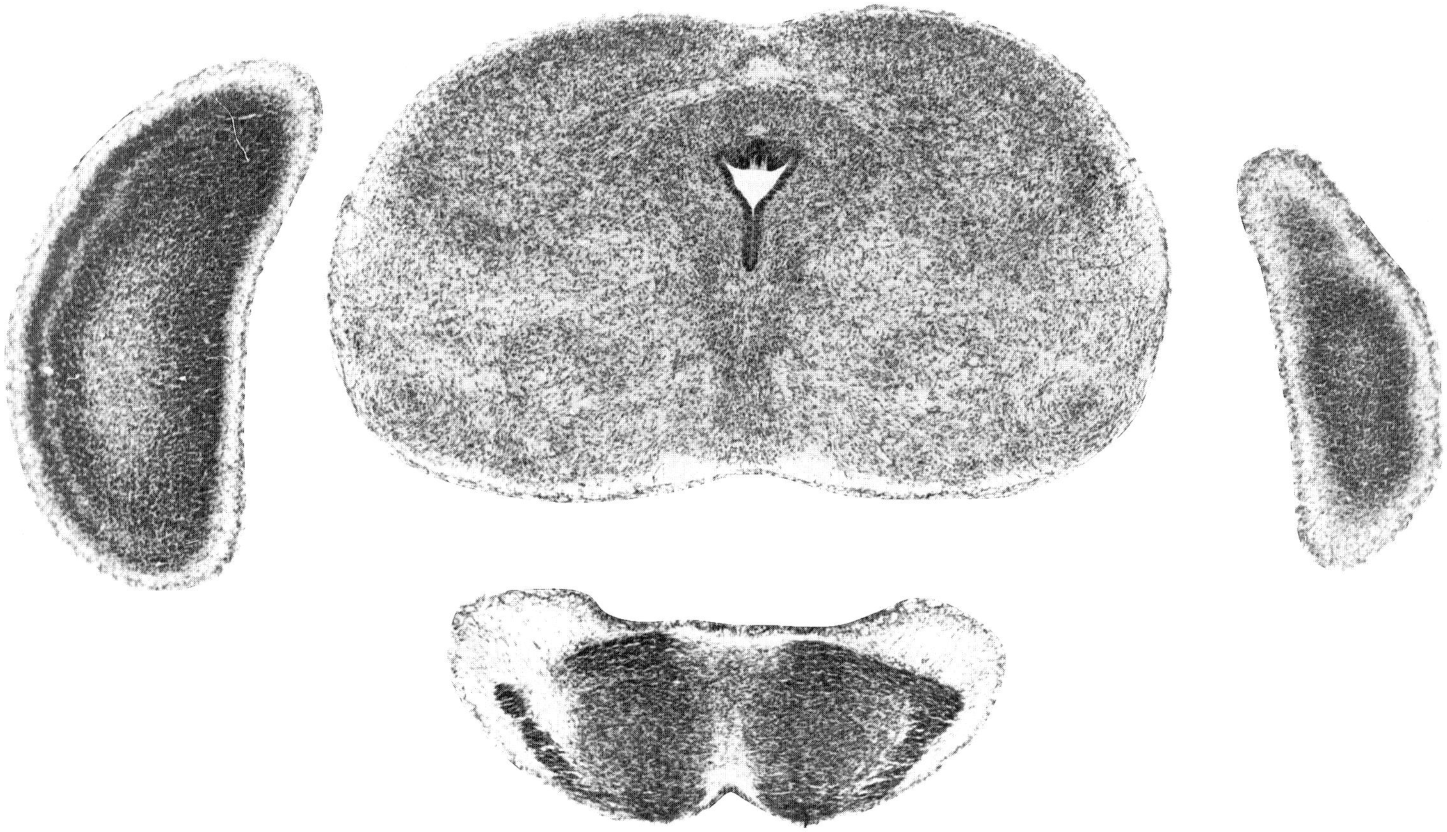

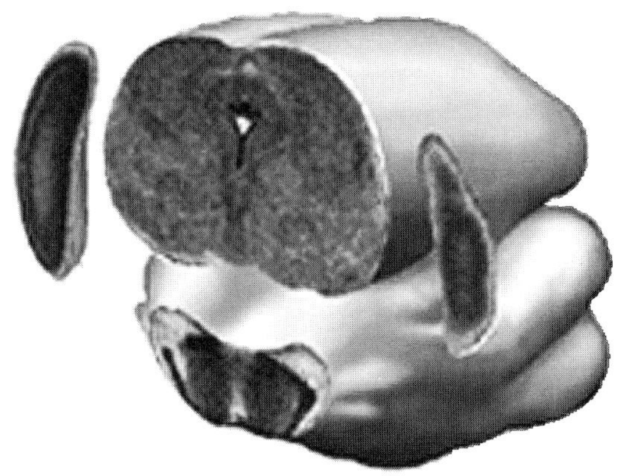

CORONAL FIGURE 16

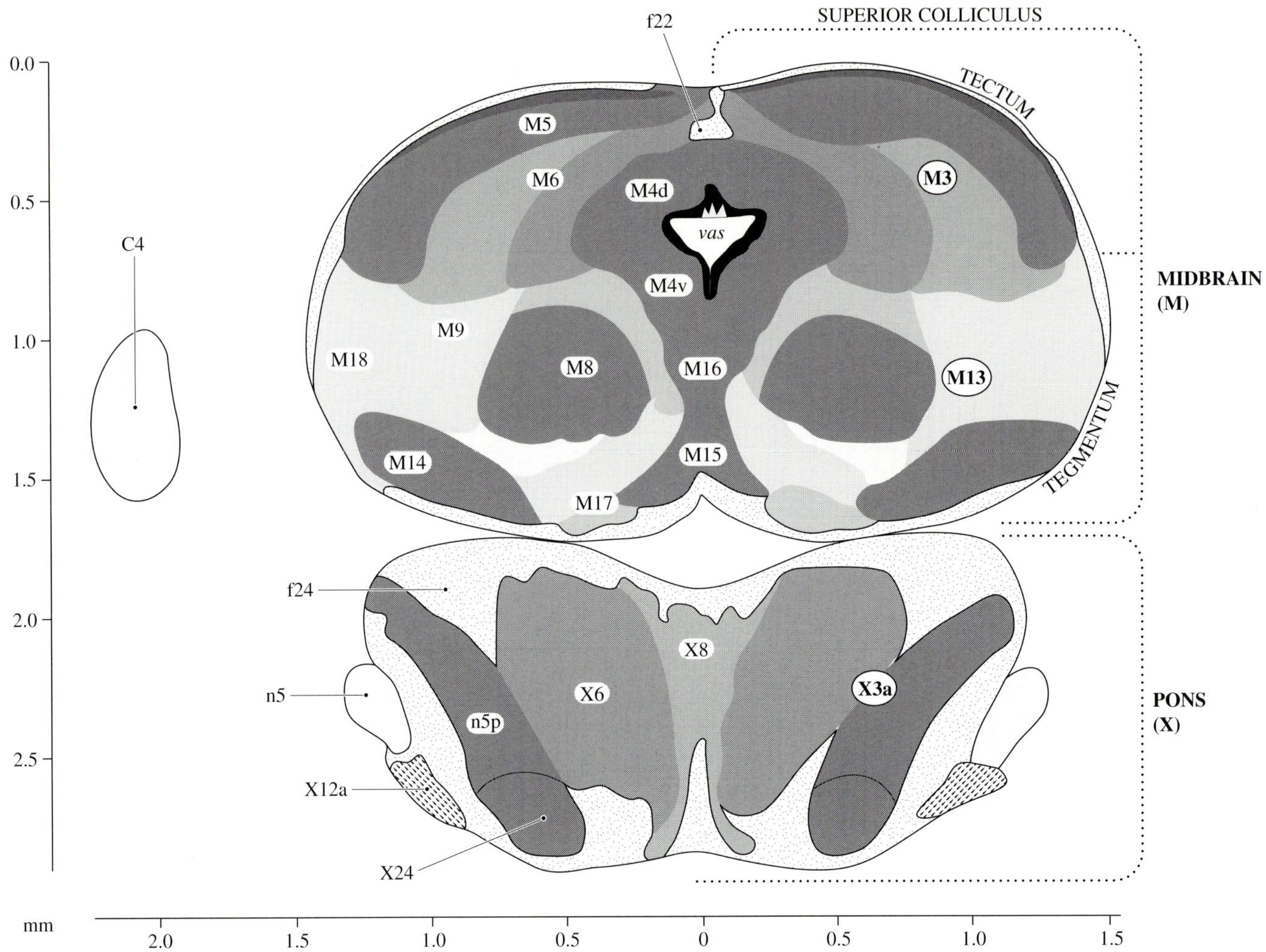

CEREBRAL CORTEX (C)

C4	molecular layer (or primordial plexiform layer)

FIBER TRACTS (f)

f22	commissure of superior colliculus
f24	middle cerebellar peduncle

MIDBRAIN (M)

M3	superior colliculus (or differentiating field)
M4d	dorsal periaqueductal gray
M4v	ventral periaqueductal gray
M5	superior colliculus, superficial layers
M6	superior colliculus, deep layers
M8	red nucleus
M9	midbrain reticular formation
M13	tegmentum (or differentiating field)
M14	substantia nigra
M15	interpeduncular nucleus
M16	dorsal raphe
M17	ventral tegmental nucleus
M18	parabigeminal nucleus

NERVES AND NUCLEI, CRANIAL (n)

n5	trigeminal nerve
n5p	principal (sensory) trigeminal nucleus

VENTRICLES (v)

vas	aqueduct, superior tectal

PONTINE AREA (X)

X3a	anterior pons (or differentiating field)
X6	pontine reticulotegmental nucleus
X8	superior central raphe nucleus
X12a	anterior precerebellar extramural migration
X24	superior olivary nucleus

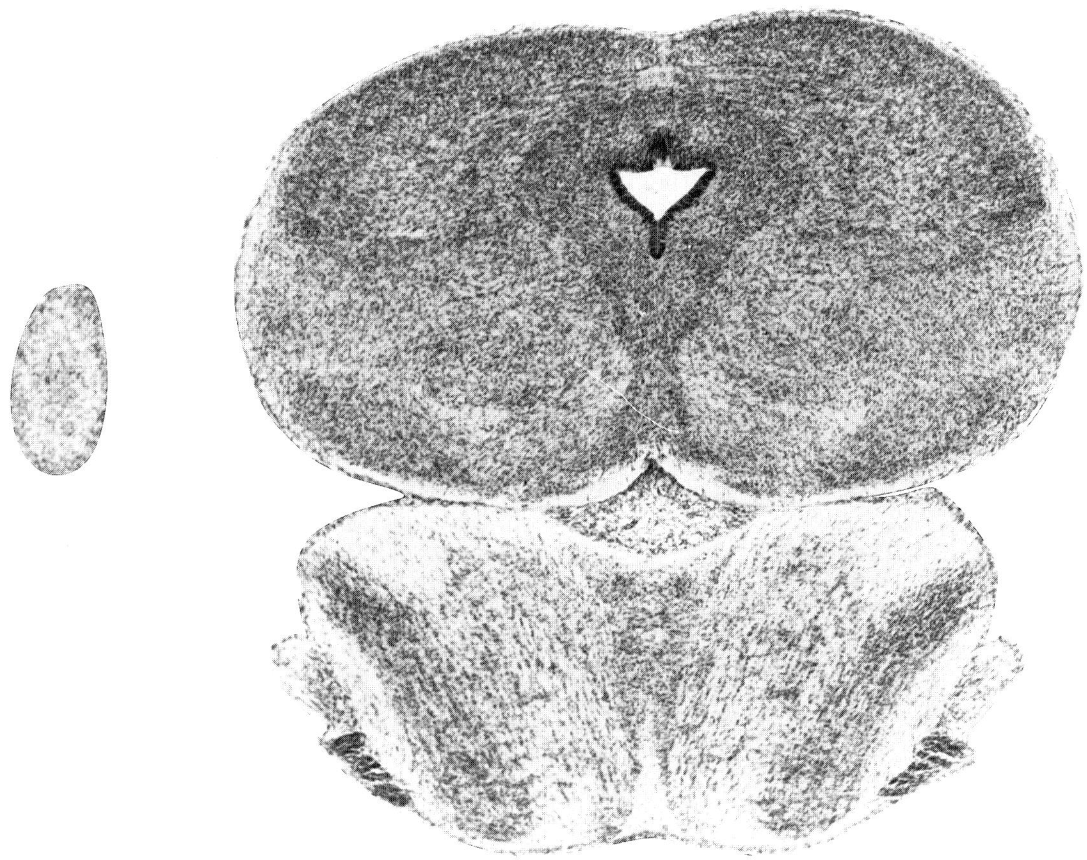

CORONAL FIGURE 17

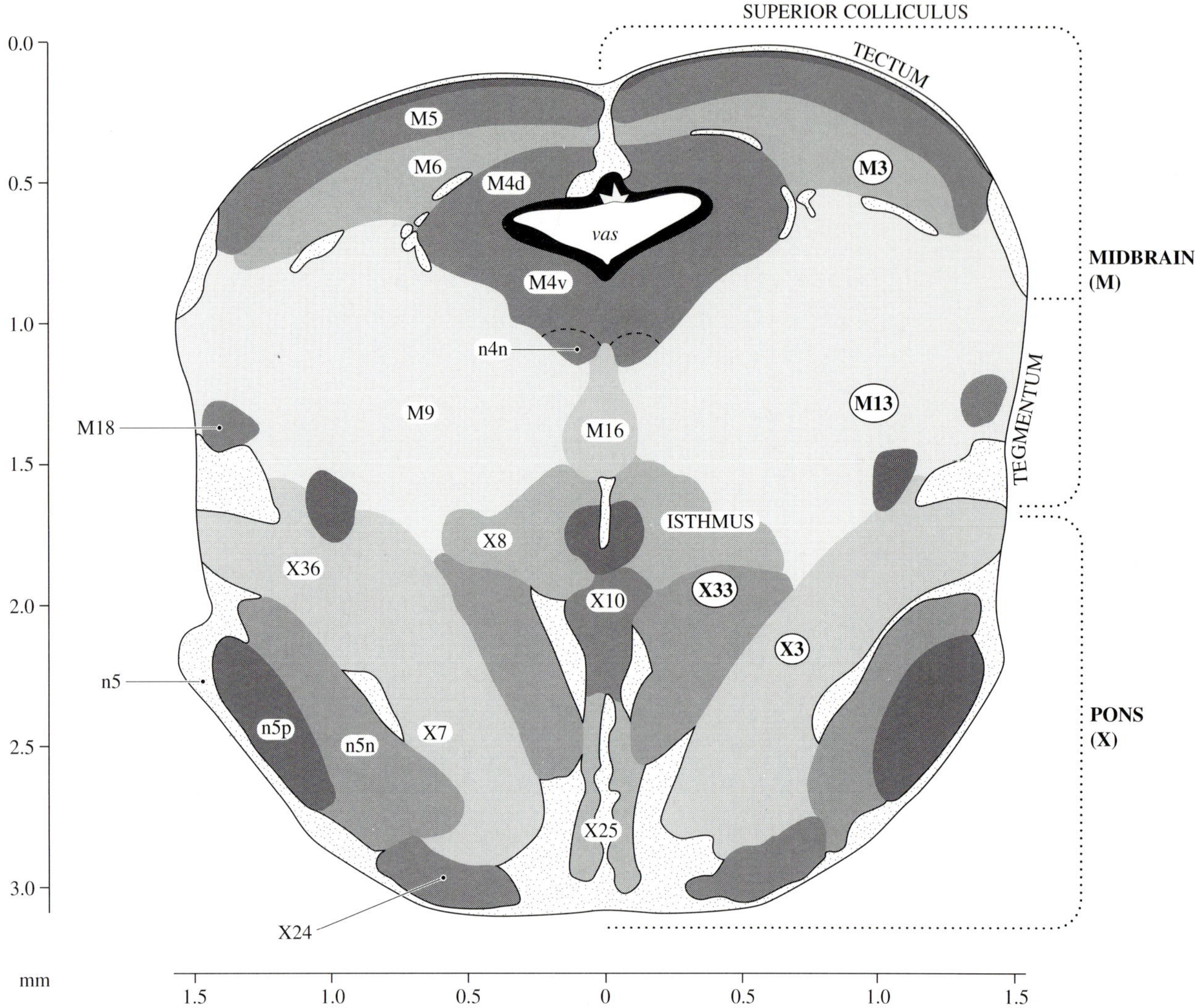

MIDBRAIN (M)

M3 superior colliculus (or differentiating field)
M4d dorsal periaqueductal gray
M4v ventral periaqueductal gray
M5 superior colliculus, superficial layers
M6 superior colliculus, deep layers
M9 midbrain reticular formation
M13 tegmentum (or differentiating field)
M16 dorsal raphe
M18 parabigeminal nucleus

NERVES AND NUCLEI, CRANIAL (n)

n4n trochlear nucleus
n5 trigeminal nerve
n5n trigeminal motor nucleus
n5p principal (sensory) trigeminal nucleus

VENTRICLES (v)

vas aqueduct, superior tectal

PONTINE AREA (X)

X3 pons (or differentiating field)
X7 pontine reticular formation
X8 superior central raphe nucleus
X10 ventral tegmental nucleus
X24 superior olivary nucleus
X25 trapezoid body
X33 isthmus (or differentiating field)
X36 nucleus of lateral lemniscus

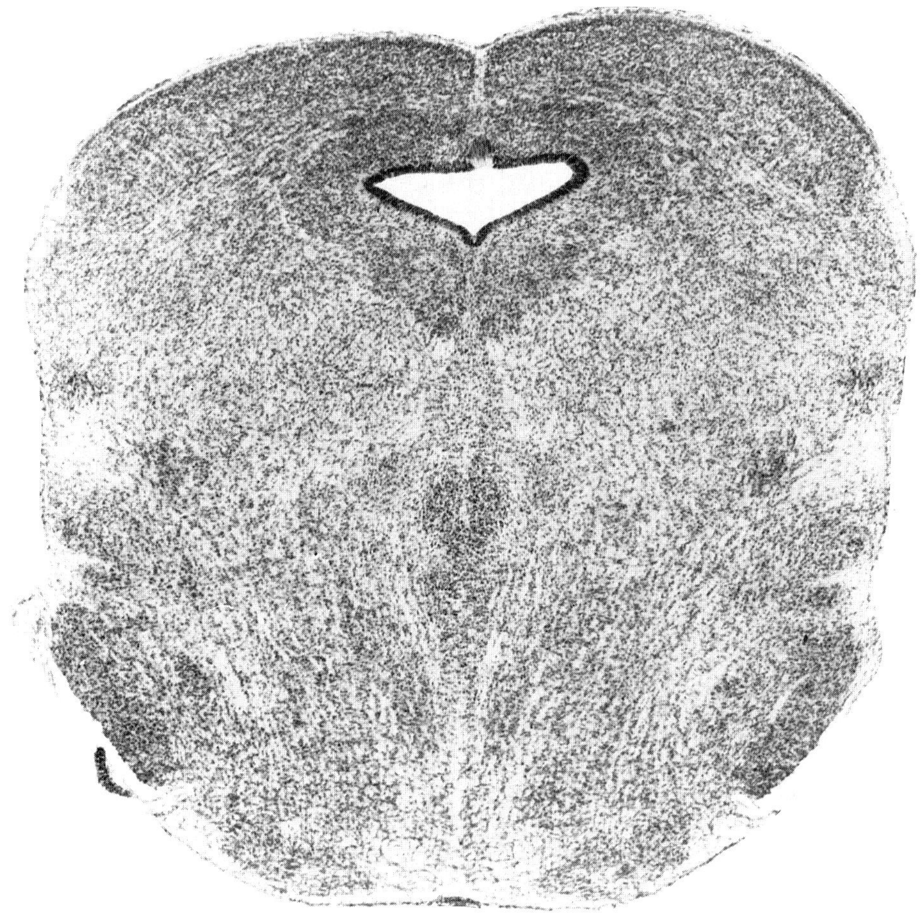

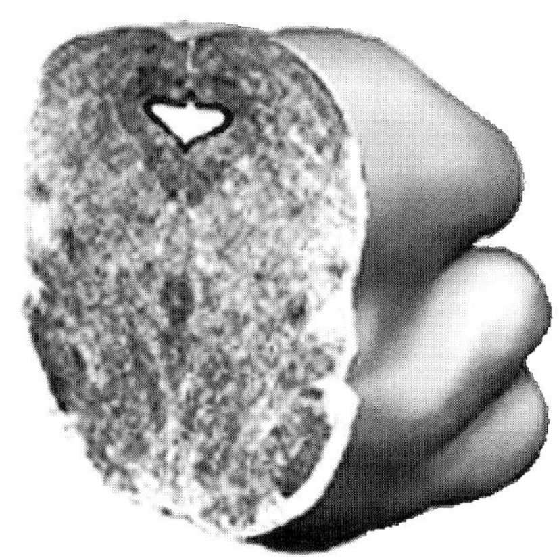

CORONAL FIGURE 18

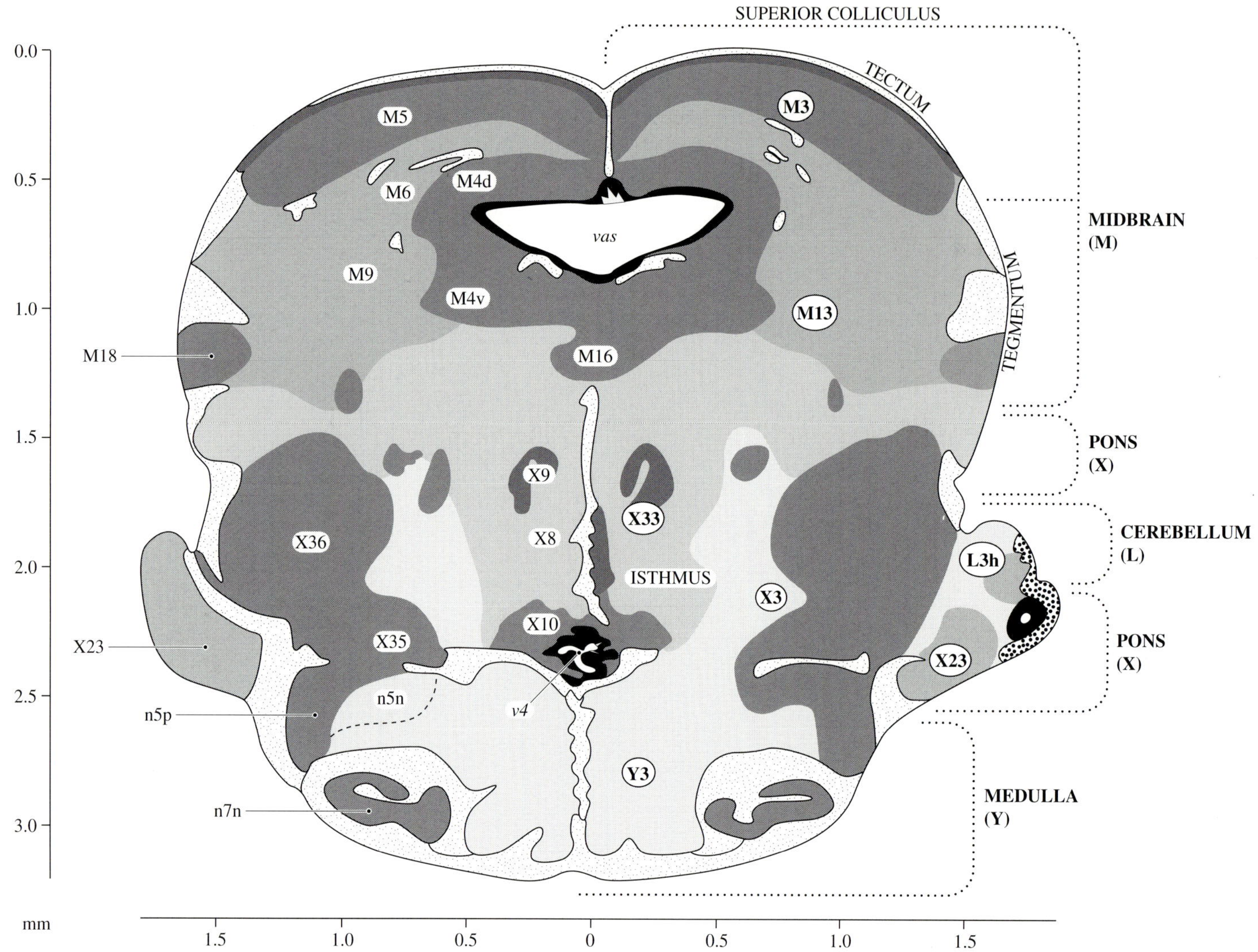

CEREBELLUM (L)

L3h cerebellar hemisphere (or differentiating field)

MIDBRAIN (M)

M3 superior colliculus (or differentiating field)
M4d dorsal periaqueductal gray
M4v ventral periaqueductal gray
M5 superior colliculus, superficial layers
M6 superior colliculus, deep layers
M9 midbrain reticular formation
M13 tegmentum (or differentiating field)
M16 dorsal raphe
M18 parabigeminal nucleus

NERVES AND NUCLEI, CRANIAL (n)

n5n trigeminal motor nucleus
n5p principal (sensory) trigeminal nucleus
n7n facial motor nucleus

VENTRICLES (v)

vas aqueduct, superior tectal
v4 fourth ventricle

PONTINE AREA (X)

X3 pons (or differentiating field)
X8 superior central raphe nucleus
X9 dorsal tegmental nucleus
X10 ventral tegmental nucleus
X23 cochlear nuclei (or differentiating field)
X33 isthmus (or differentiating field)
X35 parabrachial nucleus
X36 nucleus of lateral lemniscus

MEDULLA (Y)

Y3 medulla (or differentiating field)

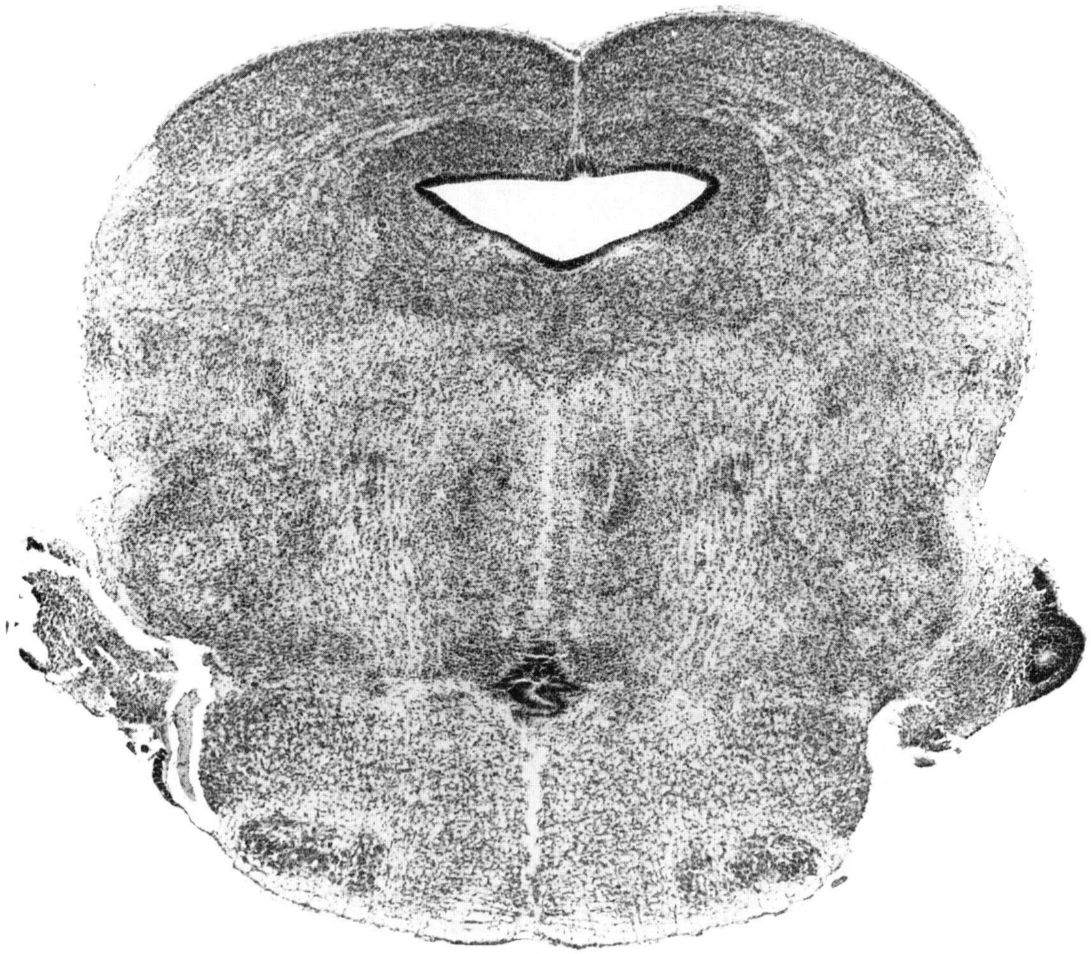

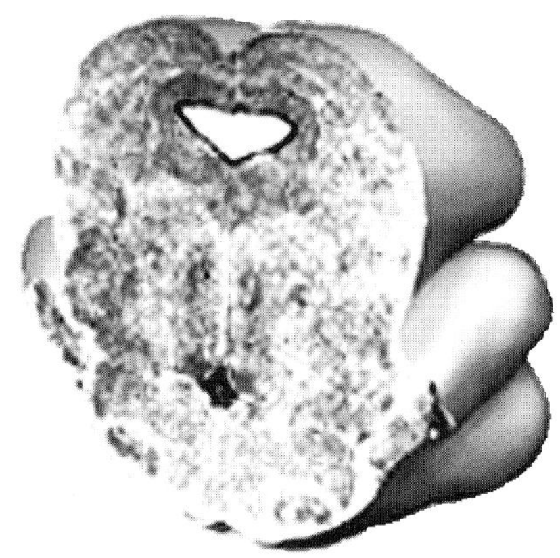

CORONAL FIGURE 19

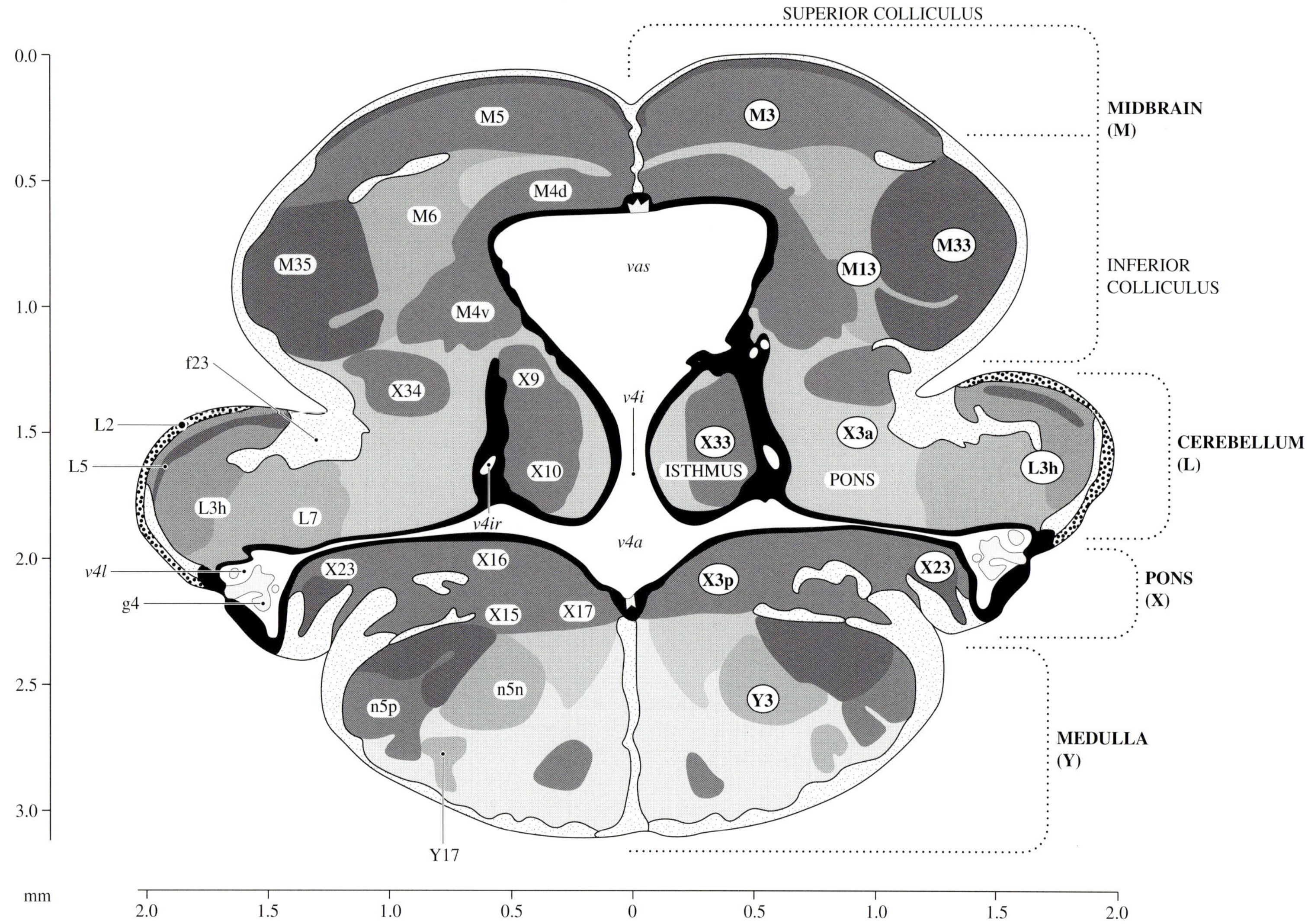

FIBER TRACTS (f)

f23	superior cerebellar peduncle

GLANDS AND PARANEURAL STRUCTURES (g)

g4	choroid plexus of fourth ventricle

CEREBELLUM (L)

L2	external germinal layer
L3h	cerebellar hemisphere (or differentiating field)
L5	Purkinje cell layer
L7	lateral deep nucleus (dentate)

MIDBRAIN (M)

M3	superior colliculus (or differentiating field)
M4d	dorsal periaqueductal gray
M4v	ventral periaqueductal gray
M5	superior colliculus, superficial layers
M6	superior colliculus, deep layers
M13	tegmentum (or differentiating field)
M33	inferior colliculus (or differentiating field)
M35	central nucleus of inferior colliculus

NERVES AND NUCLEI, CRANIAL (n)

n5n	trigeminal motor nucleus
n5p	principal (sensory) trigeminal nucleus

VENTRICLES (v)

vas	aqueduct, superior tectal
v4a	fourth ventricle, anterior
v4i	fourth ventricle, isthmal
v4ir	fourth ventricle, isthmal recess
v4l	fourth ventricle, lateral recess

PONTINE AREA (X)

X3a	anterior pons (or differentiating field)
X3p	posterior pons (or differentiating field)
X9	dorsal tegmental nucleus
X10	ventral tegmental nucleus
X15	lateral vestibular nucleus
X16	superior vestibular nucleus
X17	medial vestibular nucleus
X23	cochlear nuclei (or differentiating field)
X33	isthmus (or differentiating field)
X34	locus coeruleus

MEDULLA (Y)

Y3	medulla (or differentiating field)
Y17	ambiguus nucleus

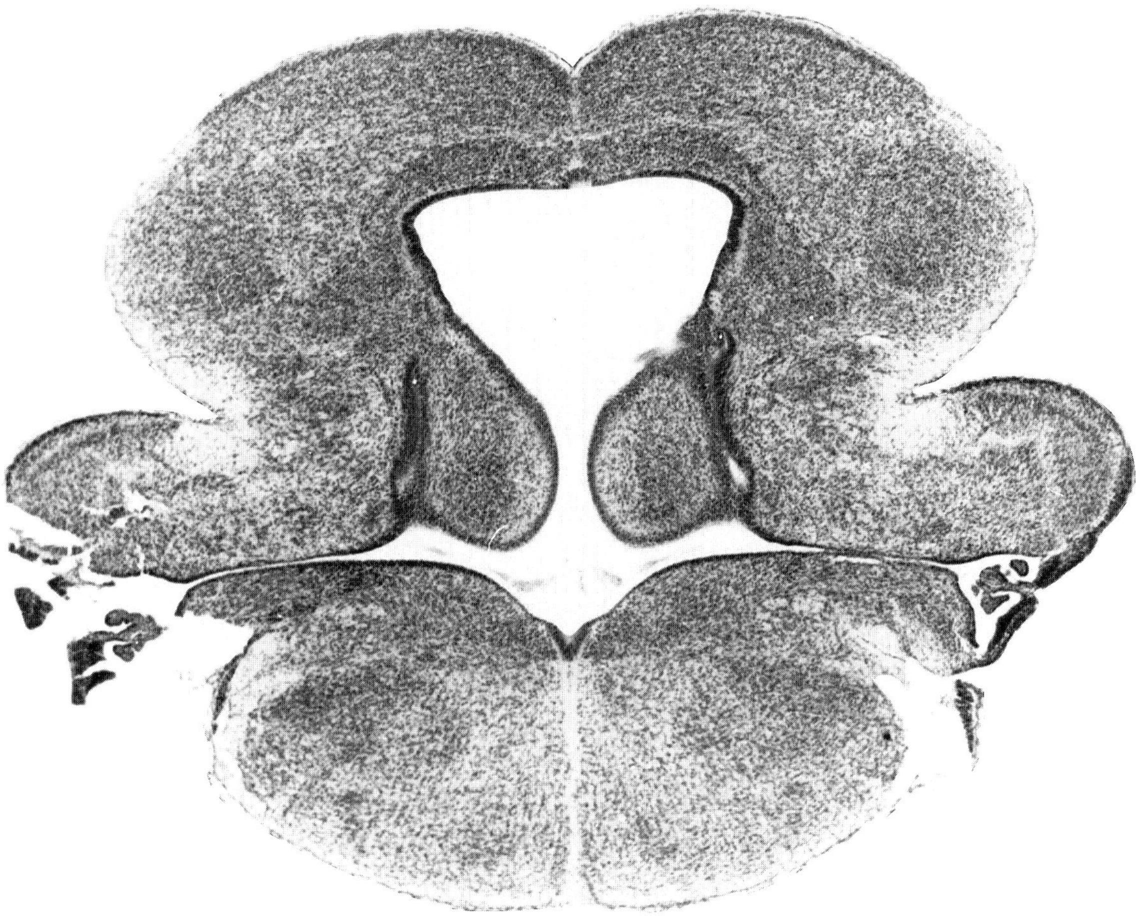

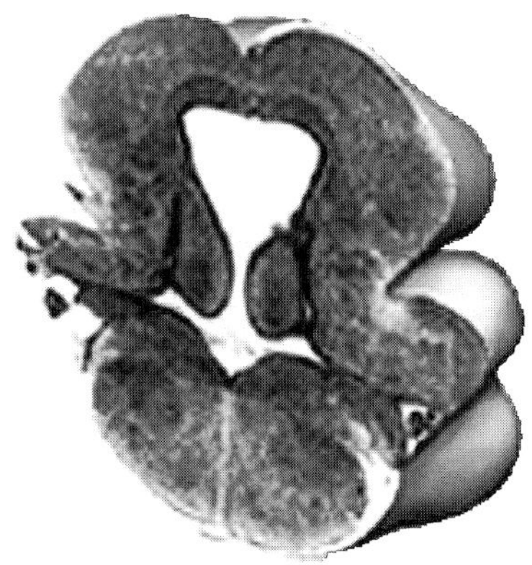

CORONAL FIGURE 20

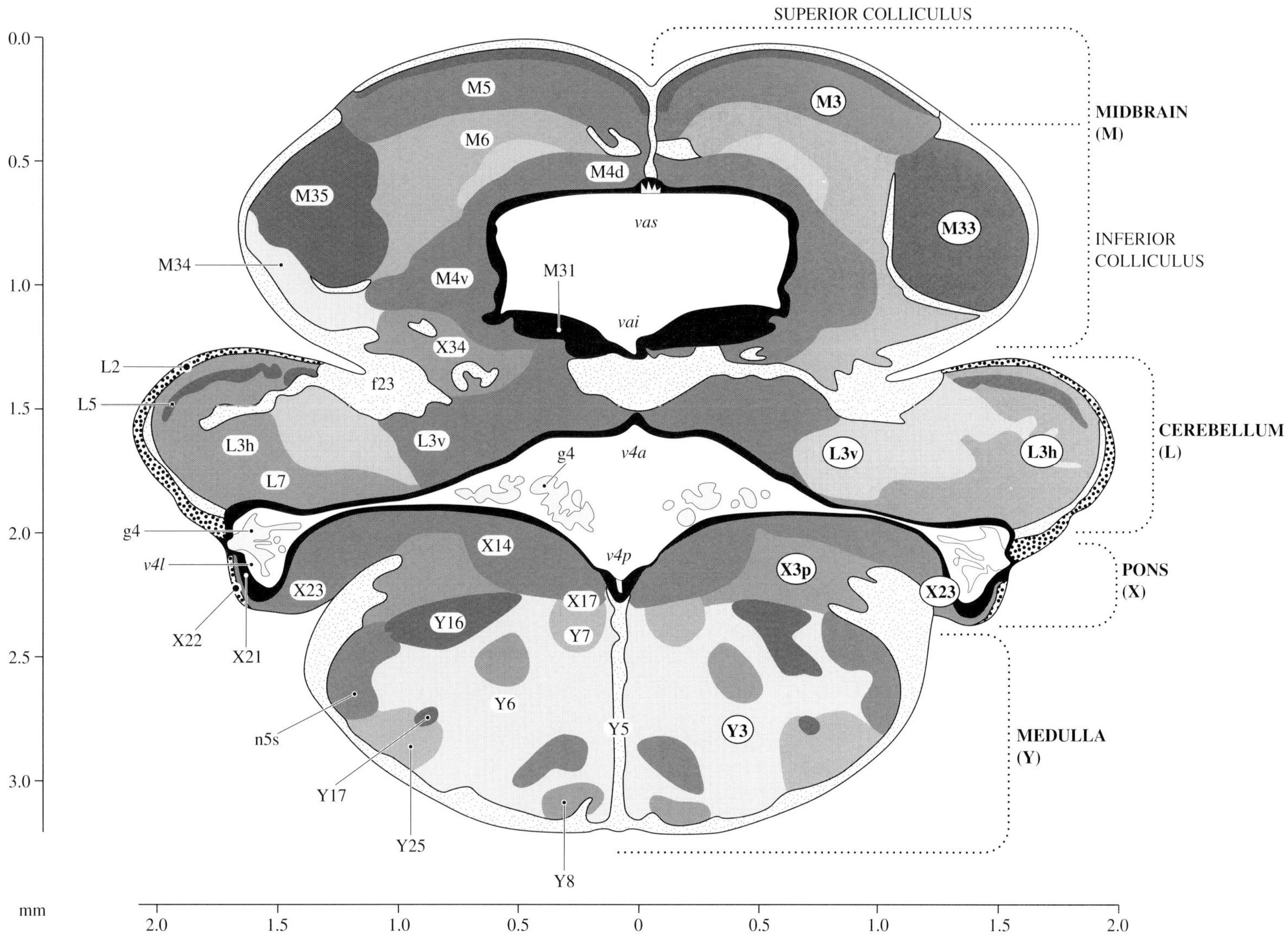

FIBER TRACTS (f)

f23	superior cerebellar peduncle

GLANDS AND PARANEURAL STRUCTURES (g)

g4	choroid plexus of fourth ventricle

CEREBELLUM (L)

L2	external germinal layer
L3h	cerebellar hemisphere (or differentiating field)
L3v	cerebellar vermis (or differentiating field)
L5	Purkinje cell layer
L7	lateral deep nucleus

MIDBRAIN (M)

M3	superior colliculus (or differentiating field)
M4d	dorsal periaqueductal gray
M4v	ventral periaqueductal gray
M5	superior colliculus, superficial layers
M6	superior colliculus, deep layers
M31	inferior tectal neuroepithelium
M33	inferior colliculus (or differentiating field)
M34	external nucleus of inferior colliculus
M35	central nucleus of inferior colliculus

NERVES AND NUCLEI, CRANIAL (n)

n5s	spinal trigeminal nucleus

VENTRICLES (v)

vai	aqueduct, inferior tectal
vas	aqueduct, superior tectal
v4a	fourth ventricle, anterior
v4l	fourth ventricle, lateral recess
v4p	fourth ventricle, posterior horn

PONTINE AREA (X)

X3p	posterior pons (or differentiating field)
X14	spinal vestibular nucleus
X17	medial vestibular nucleus
X21	cochlear nuclei neuroepithelium
X22	cochlear external germinal layer
X23	cochlear nuclei (or differentiating field)
X34	locus coeruleus

MEDULLA (Y)

Y3	medulla (or differentiating field)
Y5	medullary raphe
Y6	medullary reticular formation
Y7	prepositus nucleus
Y8	inferior olive
Y16	solitary nucleus
Y17	ambiguus nucleus
Y25	lateral reticular nucleus

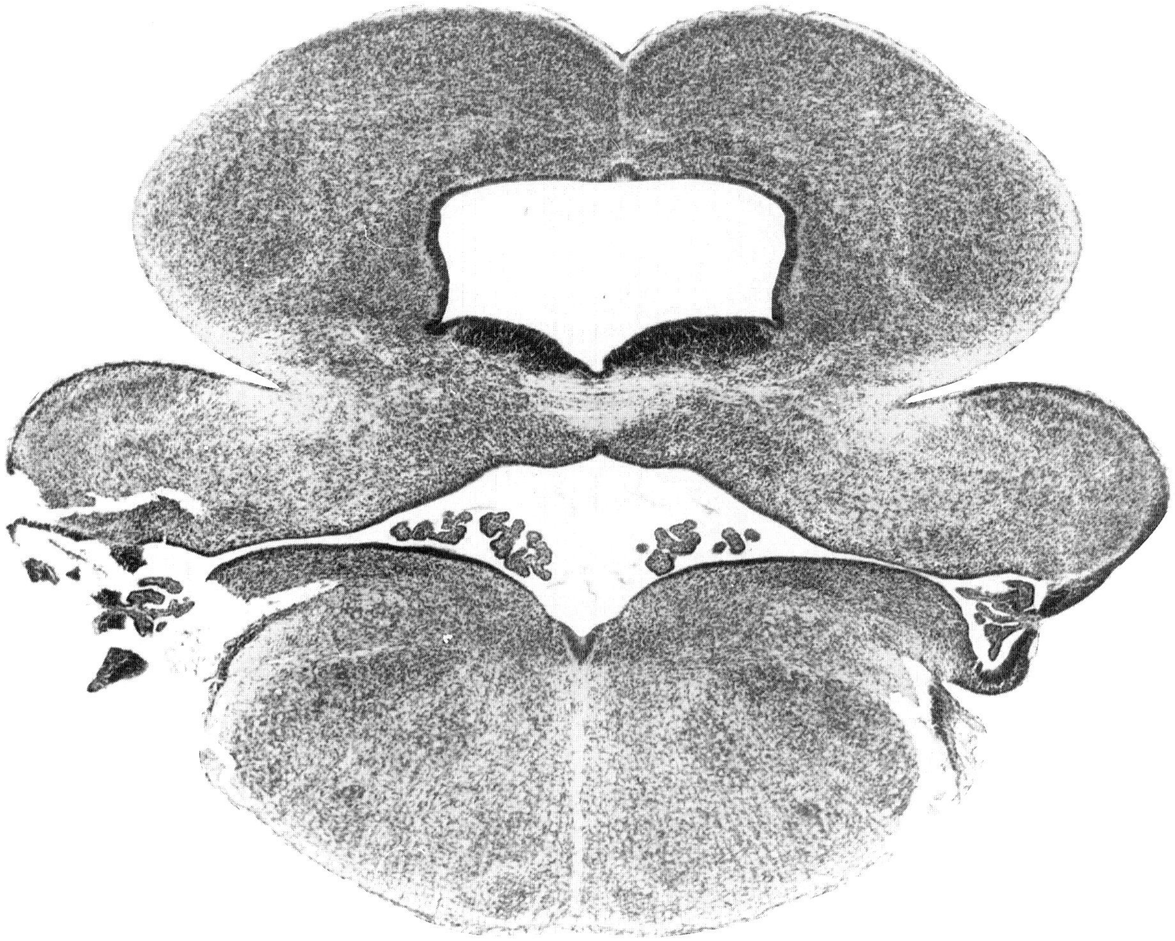

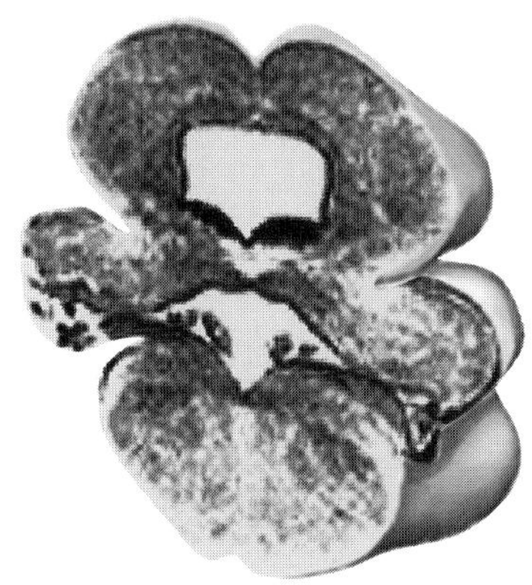

CORONAL FIGURE 21

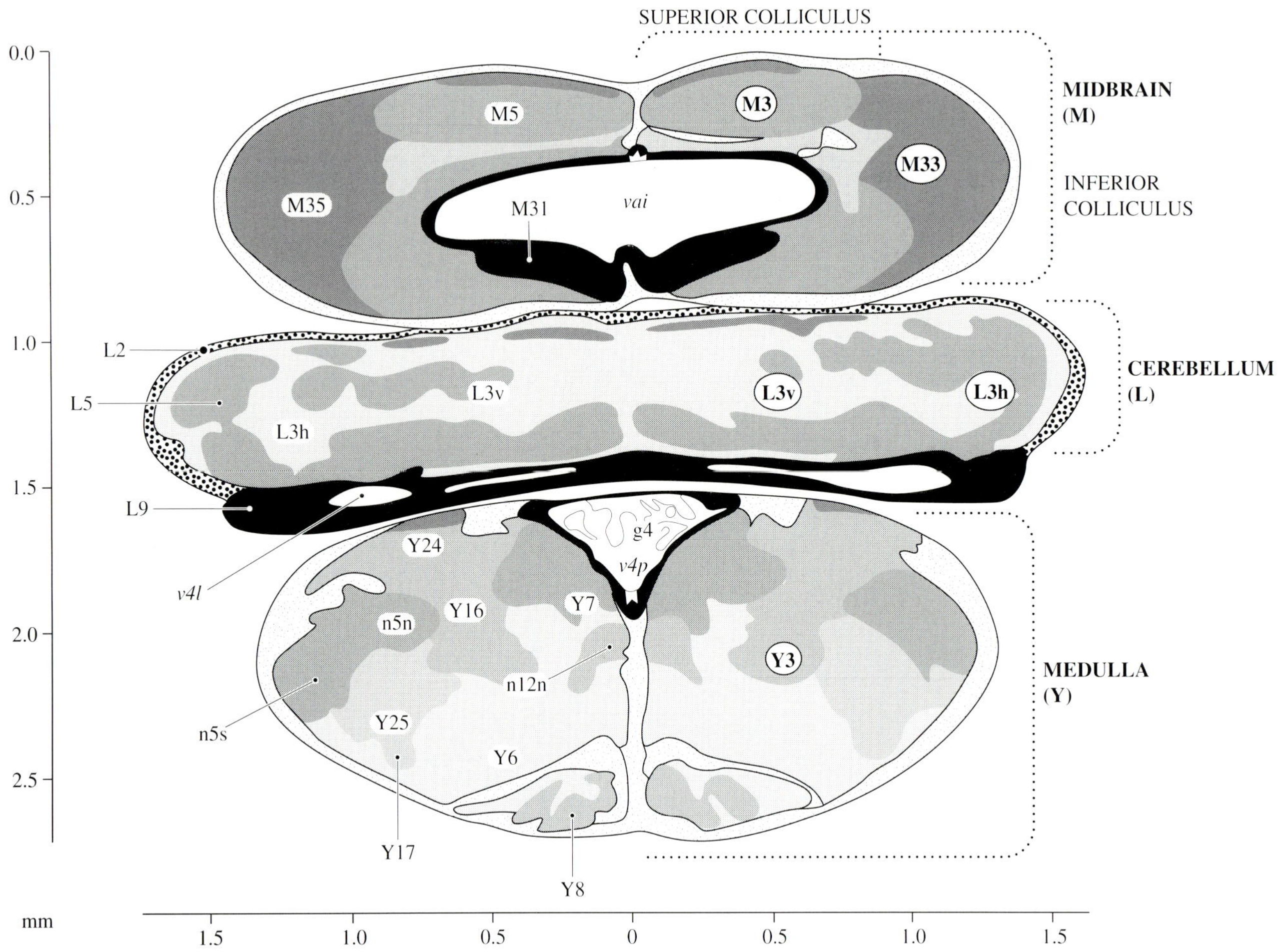

GLANDS AND PARANEURAL STRUCTURES (g)

g4 choroid plexus of fourth ventricle

CEREBELLUM (L)

L2 external germinal layer
L3h cerebellar hemisphere (or differentiating field)
L3v cerebellar vermis (or differentiating field)
L5 Purkinje cell layer
L9 cerebellar germinal trigone

MIDBRAIN (M)

M3 superior colliculus (or differentiating field)
M5 superior colliculus, superficial layers
M31 inferior tectal neuroepithelium
M33 inferior colliculus (or differentiating field)
M35 central nucleus of inferior colliculus

NERVES AND NUCLEI, CRANIAL (n)

n5n trigeminal motor nucleus
n5s spinal trigeminal nucleus
n12n hypoglossal nucleus

VENTRICLES (v)

vai aqueduct, inferior tectal
v4l fourth ventricle, lateral recess
v4p fourth ventricle, posterior horn

MEDULLA (Y)

Y3 medulla (or differentiating field)
Y6 medullary reticular formation
Y7 prepositus nucleus
Y8 inferior olive
Y16 solitary nucleus
Y17 ambiguus nucleus
Y24 external cuneate nucleus
Y25 lateral reticular nucleus

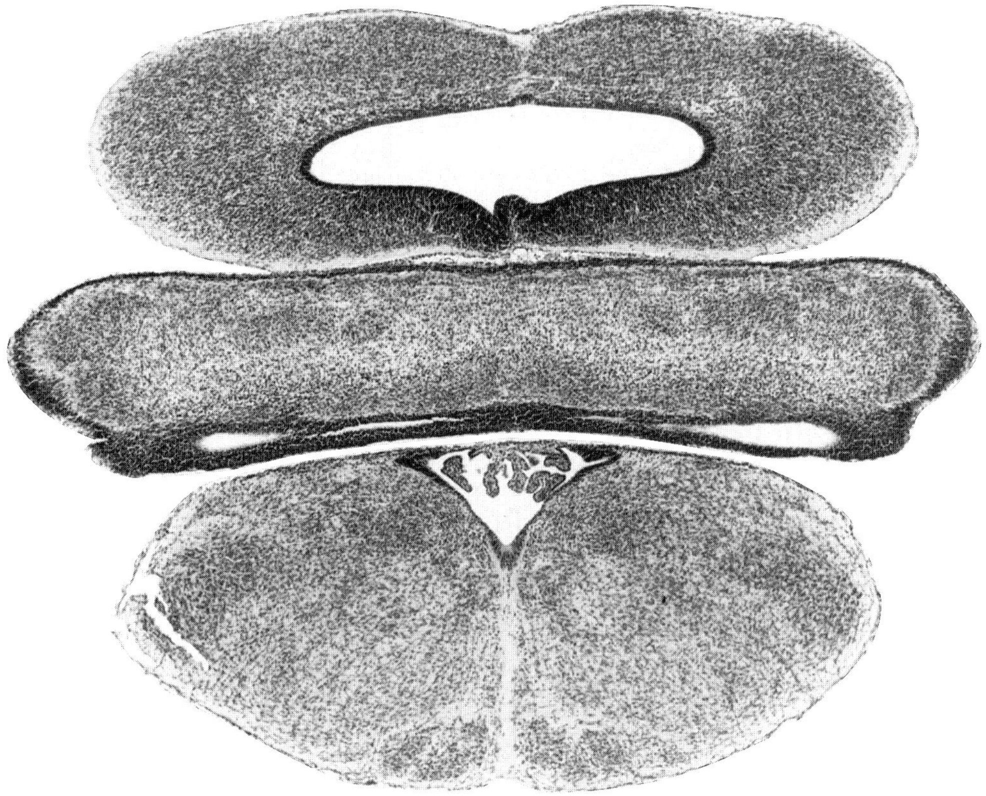

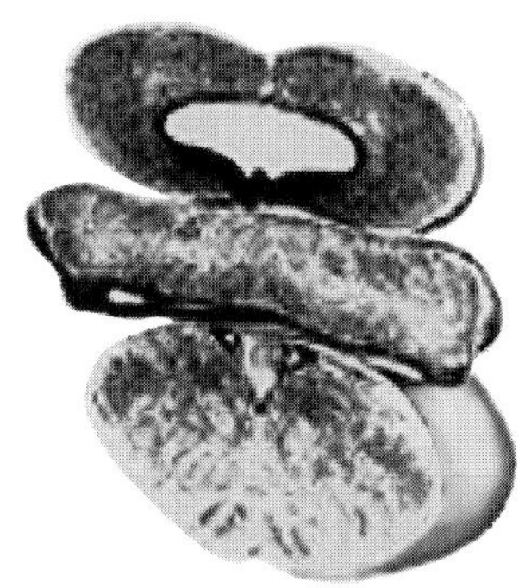

CORONAL FIGURE 22

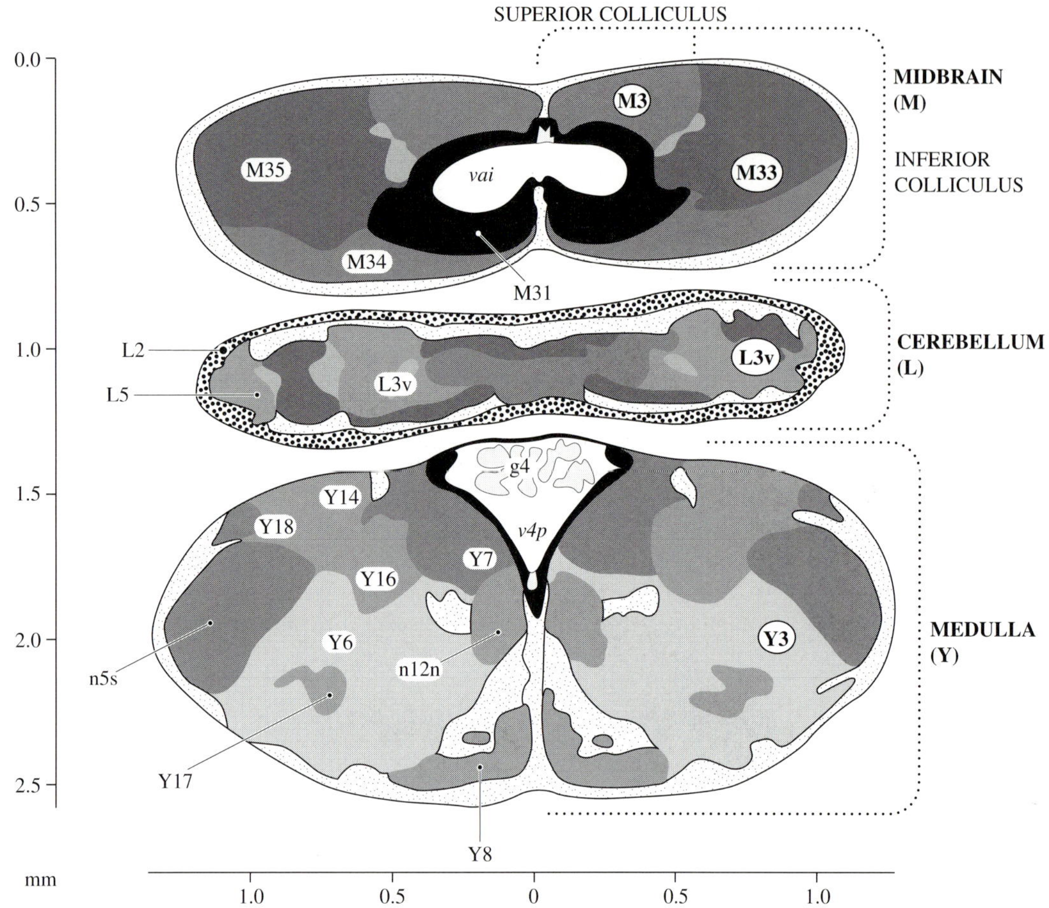

GLANDS AND PARANEURAL STRUCTURES (g)

g4 choroid plexus of fourth ventricle

CEREBELLUM (L)

L2 external germinal layer
L3v cerebellar vermis (or differentiating field)
L5 Purkinje cell layer

MIDBRAIN (M)

M3 superior colliculus (or differentiating field)
M31 inferior tectal neuroepithelium
M33 inferior colliculus (or differentiating field)
M34 external nucleus of inferior colliculus
M35 central nucleus of inferior colliculus

NERVES AND NUCLEI, CRANIAL (n)

n5s spinal trigeminal nucleus
n12n hypoglossal nucleus

VENTRICLES (v)

vai aqueduct, inferior tectal
v4p fourth ventricle, posterior horn

MEDULLA (Y)

Y3 medulla (or differentiating field)
Y6 medullary reticular formation
Y7 prepositus nucleus
Y8 inferior olive
Y14 gracile nucleus
Y16 solitary nucleus
Y17 ambiguus nucleus
Y18 magnocellular reticular nucleus

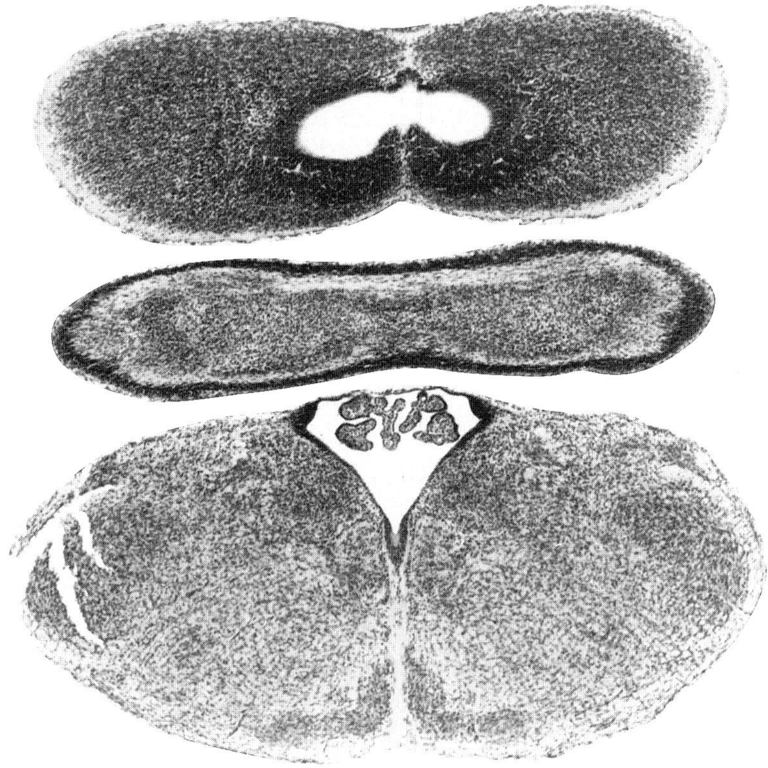

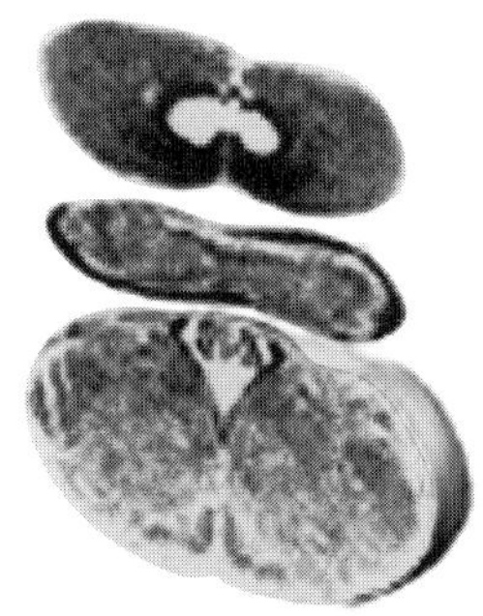

THE E22 BRAIN

LEFT, ANTERIOR VIEW

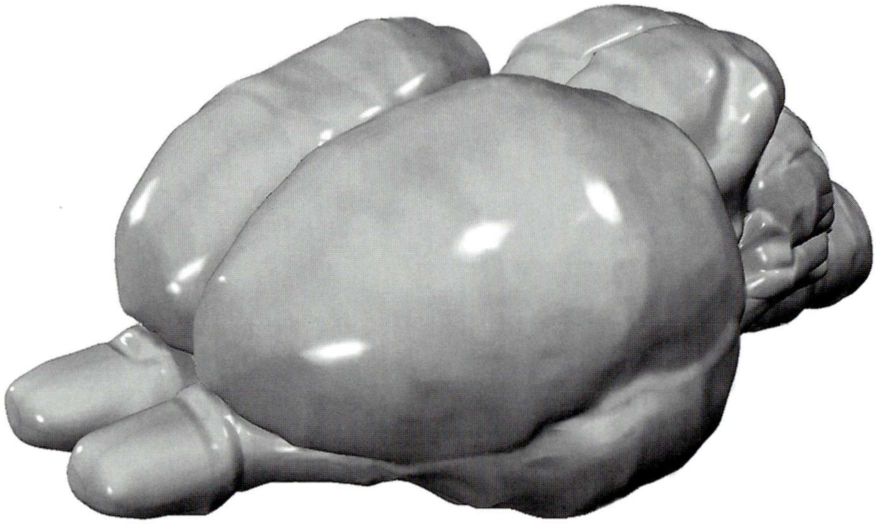

RIGHT, POSTERIOR VIEW

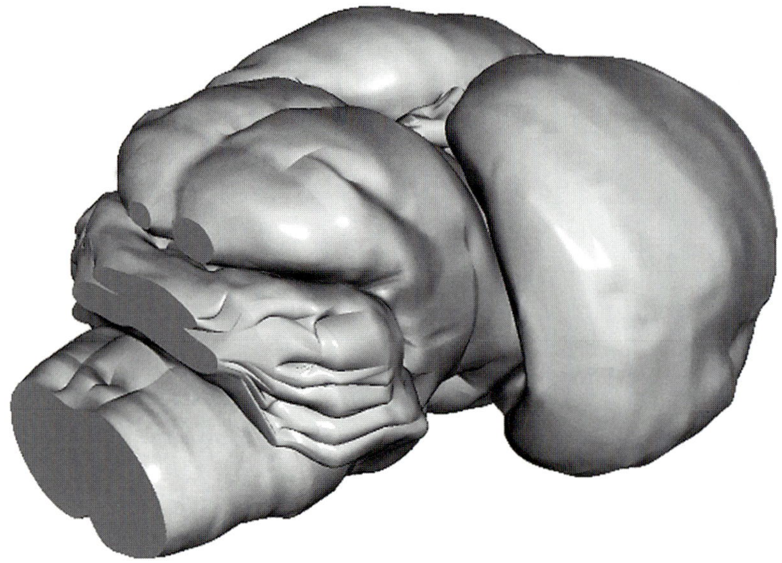

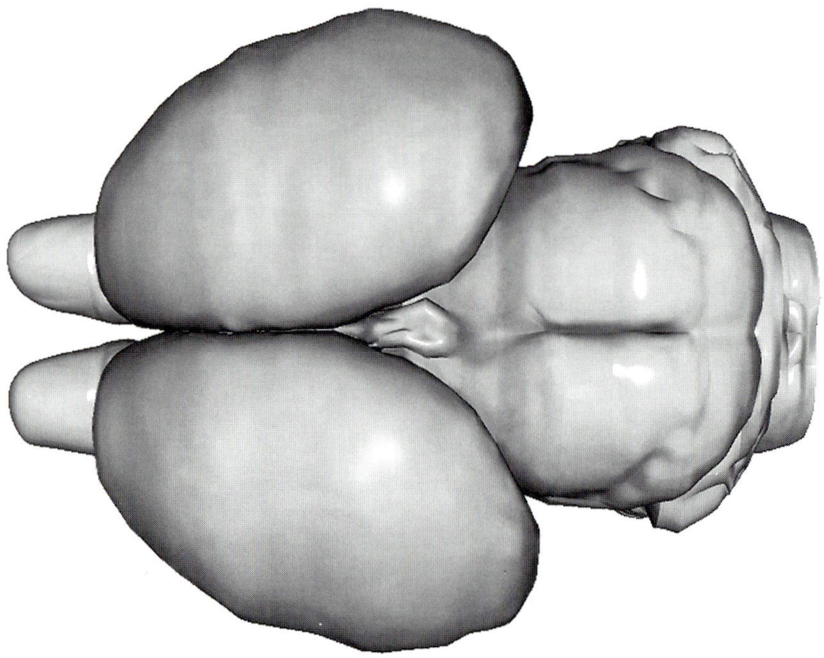

DORSAL VIEW

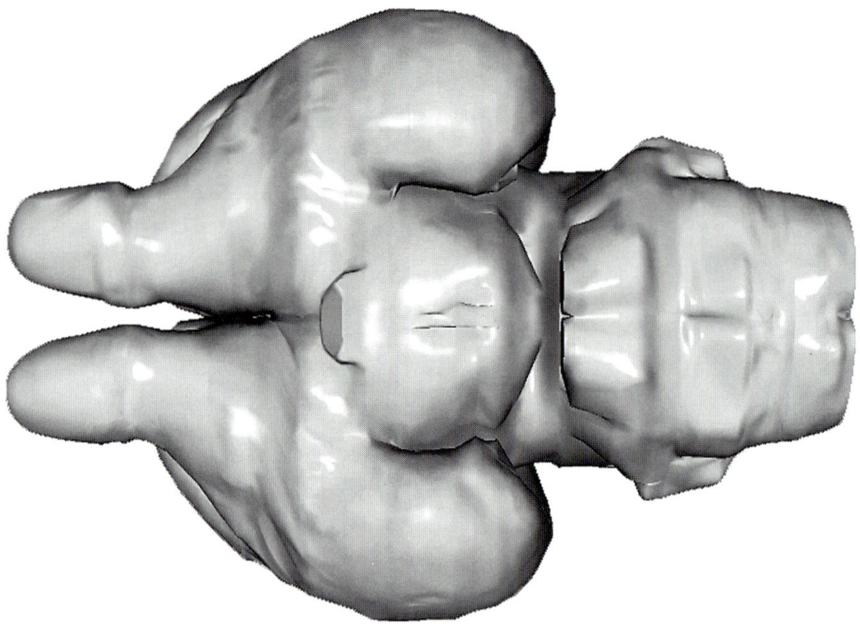

VENTRAL VIEW

SAGITTAL

FIGURES and PLATES 1-8

SAGITTAL FIGURE 1

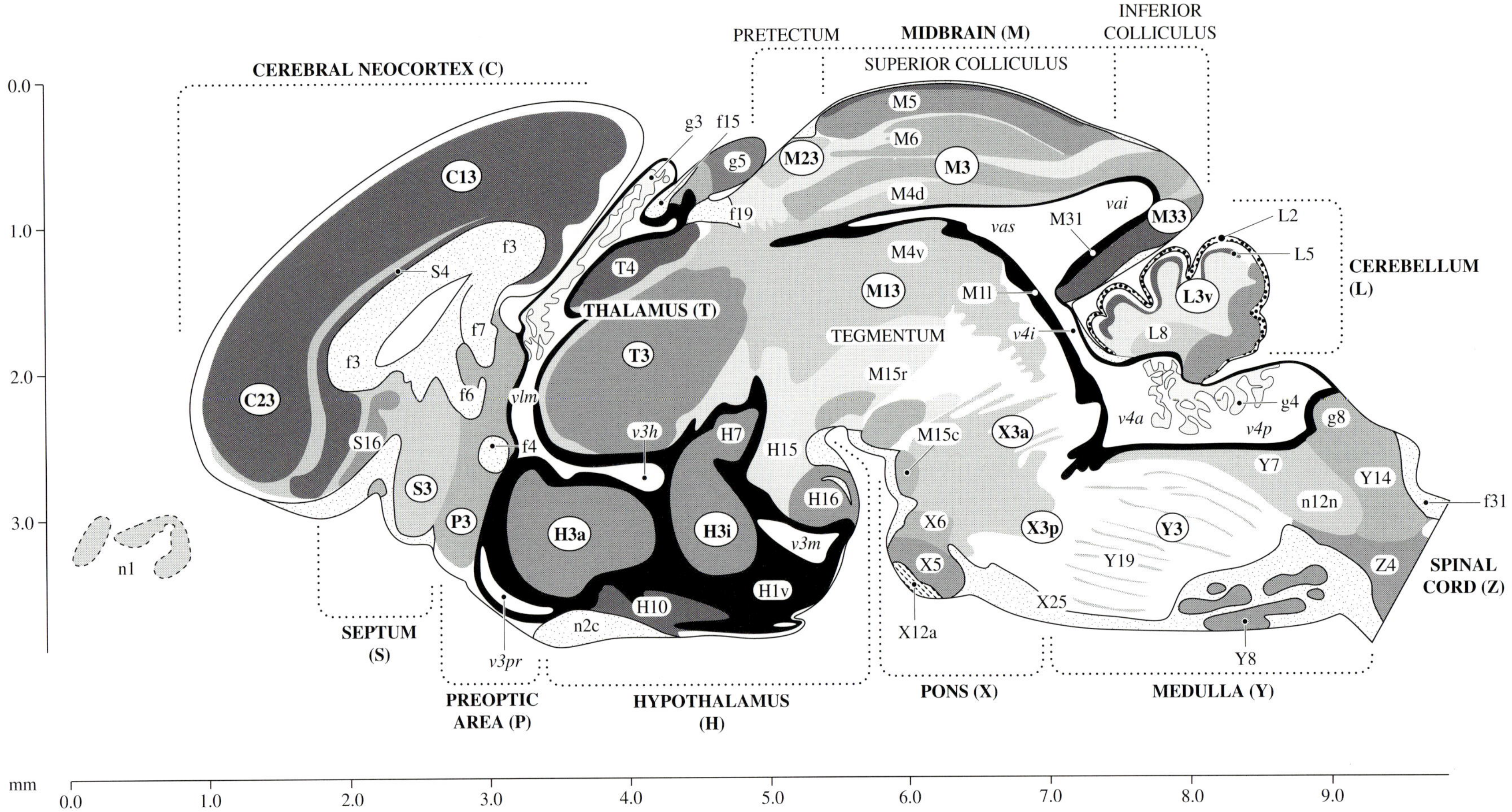

CEREBRAL CORTEX (C)

C13 cingulate cortex (or differentiating field)

C23 orbital and insular area (or differentiating field)

FIBER TRACTS (f)

f3 corpus callosum
f4 anterior commissure
f6 fornix
f7 hippocampal commissure
f15 stria medullaris
f19 posterior commissure
f31 cuneate fasciculus

GLANDS AND PARANEURAL STRUCTURES (g)

g3 choroid plexus of third ventricle
g4 choroid plexus of fourth ventricle
g5 pineal gland (primordium)
g8 area postrema

HYPOTHALAMUS (H)

H1v ventral hypothalamic neuroepithelium,
H3a anterior hypothalamus (or differentiating field)
H3i intermediate hypothalamus (or differentiating field)
H7 zona incerta
H10 anterobasal nucleus
H15 premammillary nucleus
H16 medial mammillary nucleus

CEREBELLUM (L)

L2 external germinal layer
L3v cerebellar vermis (or differentiating field)
L5 Purkinje cell layer
L8 medial deep nucleus (fastigial)

MIDBRAIN (M)

M3 superior colliculus (or differentiating field)
M4d dorsal periaqueductal gray
M4v ventral periaqueductal gray
M5 superior colliculus, superficial layers
M6 superior colliculus, deep layers
M11 tegmental neuroepithelium
M13 tegmentum (or differentiating field)
M15c interpeduncular nucleus, caudal
M15r interpeduncular nucleus, rostral
M23 pretectum (or differentiating field)
M31 inferior tectal neuroepithelium
M33 inferior colliculus (or differentiating field)

NERVES AND NUCLEI, CRANIAL (n)

n1 olfactory nerve
n2c optic chiasma
n12n hypoglossal nucleus

PREOPTIC AREA AND OPTIC EPITHELIA (P)

P3 preoptic area (or differentiating field)

SEPTAL AREA (S)

S3 septum (or differentiating field)
S4 indesium griseum
S16 taenia tecta

THALAMUS (T)

T3 thalamus (or differentiating field)
T4 medial habenular nucleus

VENTRICLES (v)

vai aqueduct, inferior tectal
vas aqueduct, superior tectal
vlm foramen of Monro
v3h third ventricle, hypothalamic
v3m third ventricle, mammillary recess
v3pr third ventricle, preoptic area
v4a fourth ventricle, anterior
v4i fourth ventricle, isthmal
v4p fourth ventricle, posterior

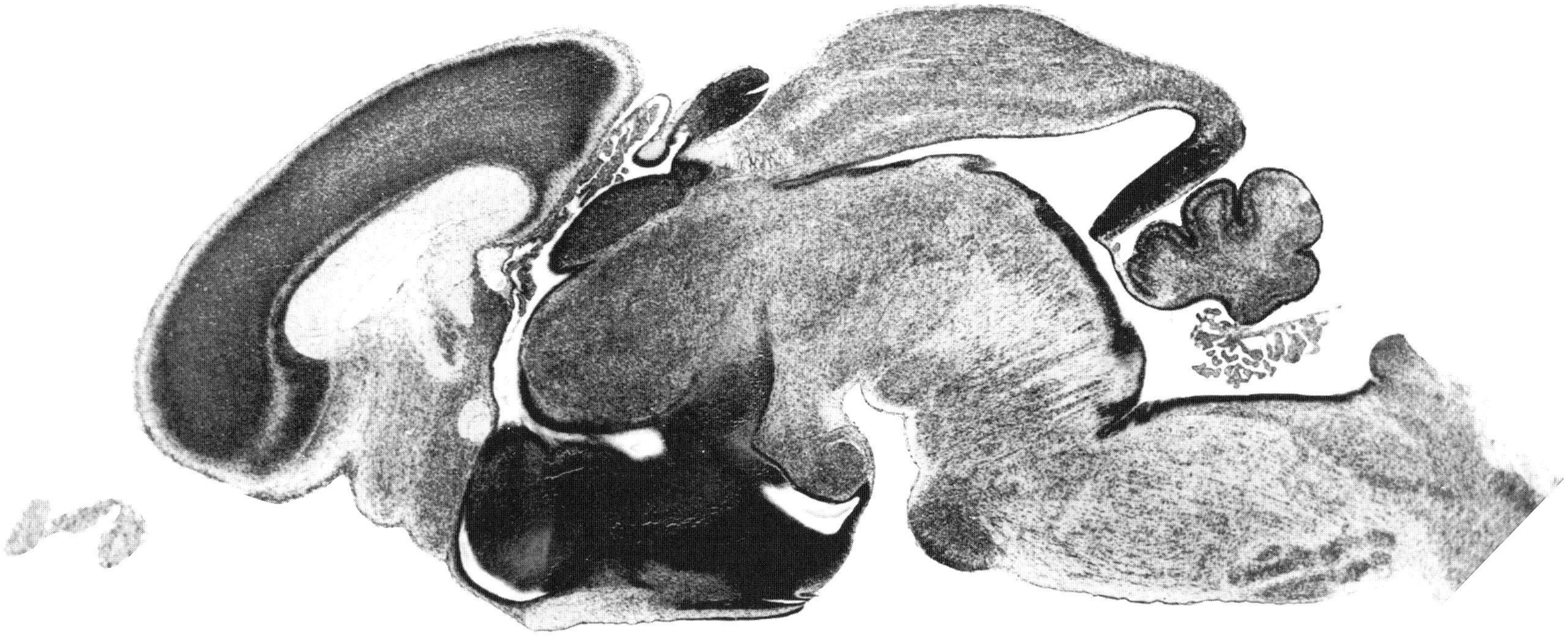

PONTINE AREA (X)

X3a anterior pons (or differentiating field)
X3p posterior pons (or differentiating field)
X5 pontine gray nucleus
X6 pontine reticulotegmental nucleus
X12a anterior precerebellar extramural migration
X25 trapezoid body

MEDULLA (Y)

Y3 medulla (or differentiating field)
Y7 prepositus nucleus
Y8 inferior olive
Y14 gracile nucleus
Y19 gigantocellular reticular nucleus

SPINAL CORD (Z)

Z4 ventral horn

SAGITTAL FIGURE 2

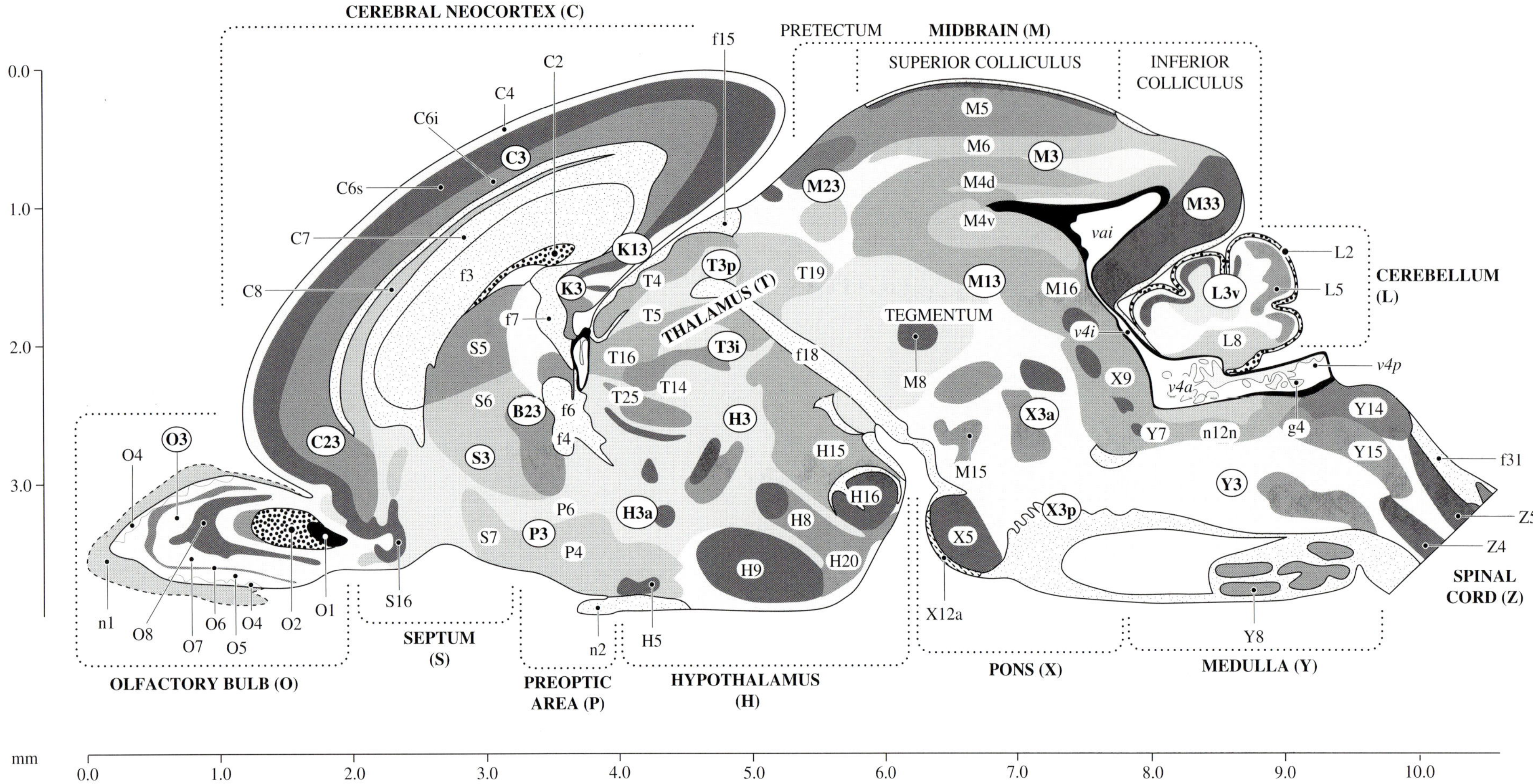

BASAL GANGLIA (B)

B23 bed nucleus of stria terminalis (or differentiating field)

CEREBRAL CORTEX (C)

C2 neocortical subventricular zone
C3 neocortex (or differentiating field)
C4 molecular layer (or primordial plexiform layer)
C6i cortical plate, infragranular part
C6s cortical plate, supragranular part
C7 white matter
C8 subplate
C23 orbital and insular area (or differentiating field)

FIBER TRACTS (f)

f3 corpus callosum
f4 anterior commissure
f6 fornix
f7 hippocampal commissure
f15 stria medullaris
f18 habenulopeduncular tract
f31 cuneate fasciculus

GLANDS AND PARANEURAL STRUCTURES (g)

g4 choroid plexus of fourth ventricle

HYPOTHALAMUS (H)

H3 hypothalamus (or differentiating field)
H3a anterior hypothalamus (or differentiating field)
H5 suprachiasmatic nucleus
H8 dorsomedial nucleus
H9 ventromedial nucleus
H15 premammillary nucleus
H16 medial mammillary nucleus
H20 arcuate nucleus

HIPPOCAMPAL FORMATION (K)

K3 hippocampus (or differentiating field)
K13 subicular area (or differentiating field)

CEREBELLUM (L)

L2 external germinal layer
L3v cerebellar vermis (or differentiating field)
L5 Purkinje cell layer
L8 medial deep nucleus (fastigial)

MIDBRAIN (M)

M3 superior colliculus (or differentiating field)
M4d dorsal periaqueductal gray
M4v ventral periaqueductal gray
M5 superior colliculus, superficial layers
M6 superior colliculus, deep layers
M8 red nucleus
M13 tegmentum (or differentiating field)
M15 interpeduncular nucleus
M16 dorsal raphe
M23 pretectum (or differentiating field)
M33 inferior colliculus (or differentiating field)

NERVES AND NUCLEI, CRANIAL (n)

n1 olfactory nerve
n2 optic nerve
n12n hypoglossal nucleus

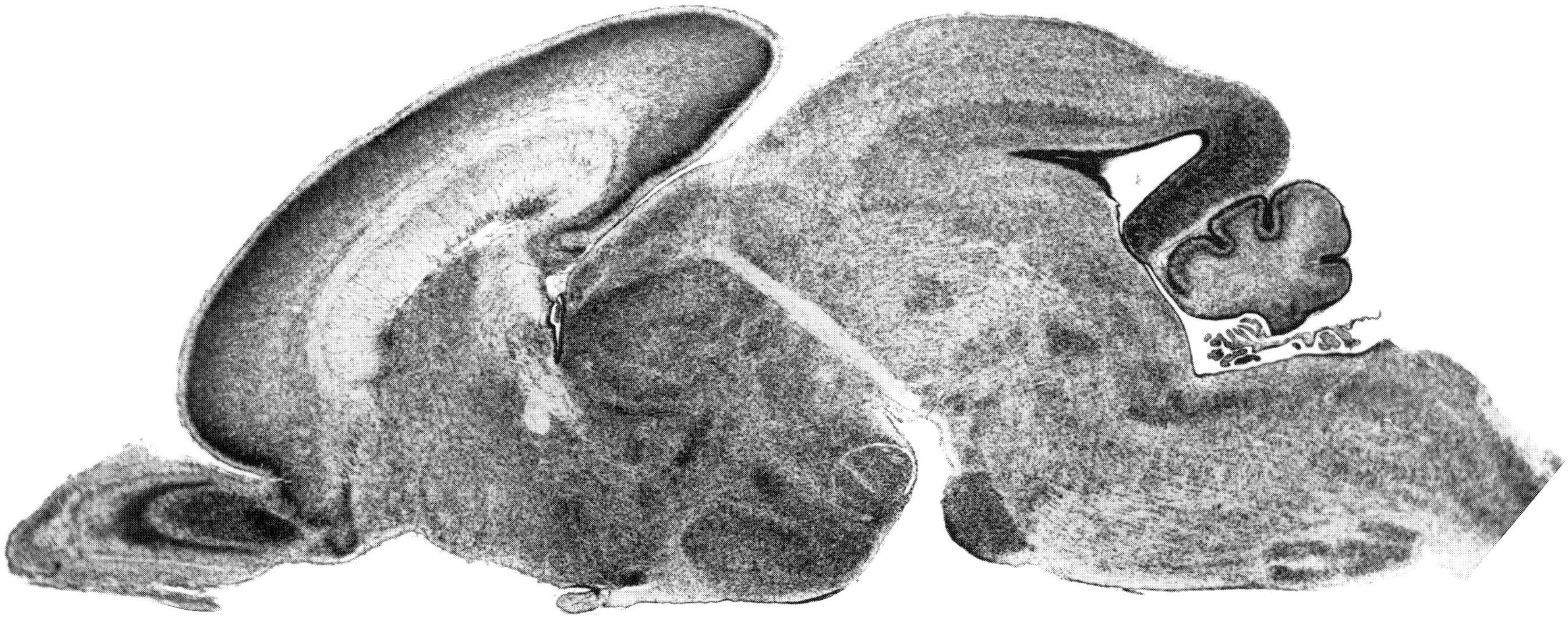

OLFACTORY BULB (O)

O1 olfactory bulb neuroepithelium
O2 olfactory bulb subventricular zone
O3 olfactory bulb (or differentiating field)
O4 glomerular layer
O5 external plexiform layer
O6 mitral cell layer
O7 internal plexiform layer
O8 granular layer

PREOPTIC AREA AND OPTIC EPITHELIA (P)

P3 preoptic area (or differentiating field)
P4 medial preoptic region
P6 superior preoptic region

SEPTAL AREA (S)

S3 septum (or differentiating field)
S5 lateral septum
S6 medial septum
S7 diagonal band (or differentiating field)
S16 taenia tecta

THALAMUS (T)

T3i intermediate thalamus (or differentiating field)
T3p posterior thalamus (or differentiating field)
T4 medial habenular nucleus
T5 lateral habenular nucleus
T14 centromedian nucleus
T16 mediodorsal nucleus
T19 parafascicular nucleus
T25 reticular nucleus

VENTRICLES (v)

vai aqueduct, inferior tectal
v4a fourth ventricle, anterior
v4i fourth ventricle, isthmal
v4p fourth ventricle, posterior

PONTINE AREA (X)

X3a anterior pons (or differentiating field)
X3p posterior pons (or differentiating field)
X5 pontine gray nucleus
X9 dorsal tegmental nucleus
X12a anterior precerebellar extramural migration

MEDULLA (Y)

Y3 medulla (or differentiating field)
Y7 prepositus nucleus
Y8 inferior olive
Y14 gracile nucleus
Y15 cuneate nucleus

SPINAL CORD (Z)

Z4 ventral horn
Z5 dorsal horn

SAGITTAL FIGURE 3

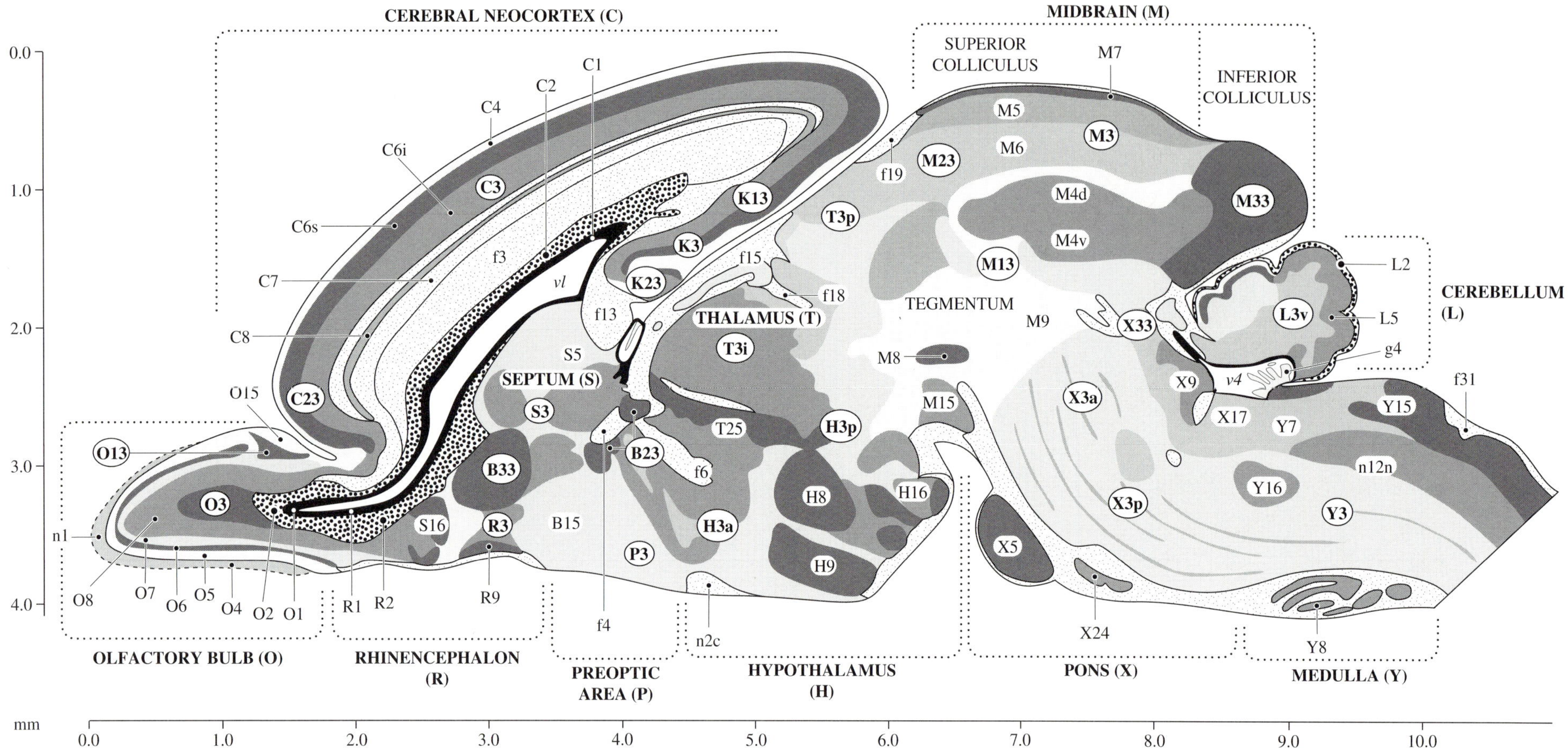

BASAL GANGLIA (B)

B15	substantia innominata
B23	bed nucleus of stria terminalis (or differentiating field)
B33	nucleus accumbens (or differentiating field)

CEREBRAL CORTEX (C)

C1	neocortical neuroepithelium
C2	neocortical subventricular zone
C3	neocortex (or differentiating field)
C4	molecular layer (or primordial plexiform layer)
C6i	cortical plate, infragranular part
C6s	cortical plate, supragranular part
C7	white matter
C8	subplate
C23	orbital and insular area (or differentiating field)

FIBER TRACTS (f)

f3	corpus callosum
f4	anterior commissure
f6	fornix
f13	fimbria
f15	stria medullaris
f18	habenulopeduncular tract
f19	posterior commissure
f31	cuneate fasciculus

GLANDS AND PARANEURAL STRUCTURES (g)

g4	choroid plexus of fourth ventricle

HYPOTHALAMUS (H)

H3a	anterior hypothalamus (or differentiating field)
H3p	posterior hypothalamus (or differentiating field)
H8	dorsomedial nucleus
H9	ventromedial nucleus
H16	medial mammillary nucleus

HIPPOCAMPAL FORMATION (K)

K3	hippocampus (or differentiating field)
K13	subicular area (or differentiating field)
K23	dentate gyrus (or differentiating field)

CEREBELLUM (L)

L2	external germinal layer
L3v	cerebellar vermis (or differentiating field)
L5	Purkinje cell layer

MIDBRAIN (M)

M3	superior colliculus (or differentiating field)
M4d	dorsal periaqueductal gray
M4v	ventral periaqueductal gray
M5	superior colliculus, superficial layers
M6	superior colliculus, deep layers
M7	superior colliculus, stratum opticum
M8	red nucleus
M9	midbrain reticular formation
M13	tegmentum (or differentiating field)
M15	interpeduncular nucleus
M23	pretectum (or differentiating field)
M33	inferior colliculus (or differentiating field)

NERVES AND NUCLEI, CRANIAL (n)

n1	olfactory nerve
n2c	optic chiasma
n12n	hypoglossal nucleus

OLFACTORY BULB (O)

O1	olfactory bulb neuroepithelium
O2	olfactory bulb subventricular zone
O3	olfactory bulb (or differentiating field)
O4	glomerular layer
O5	external plexiform layer
O6	mitral cell layer
O7	internal plexiform layer
O8	granular layer
O13	accessory olfactory bulb (or differentiating field)
O15	external plexiform layer (accessory olfactory bulb)

PREOPTIC AREA AND OPTIC EPITHELIA (P)

P3	preoptic area (or differentiating field)

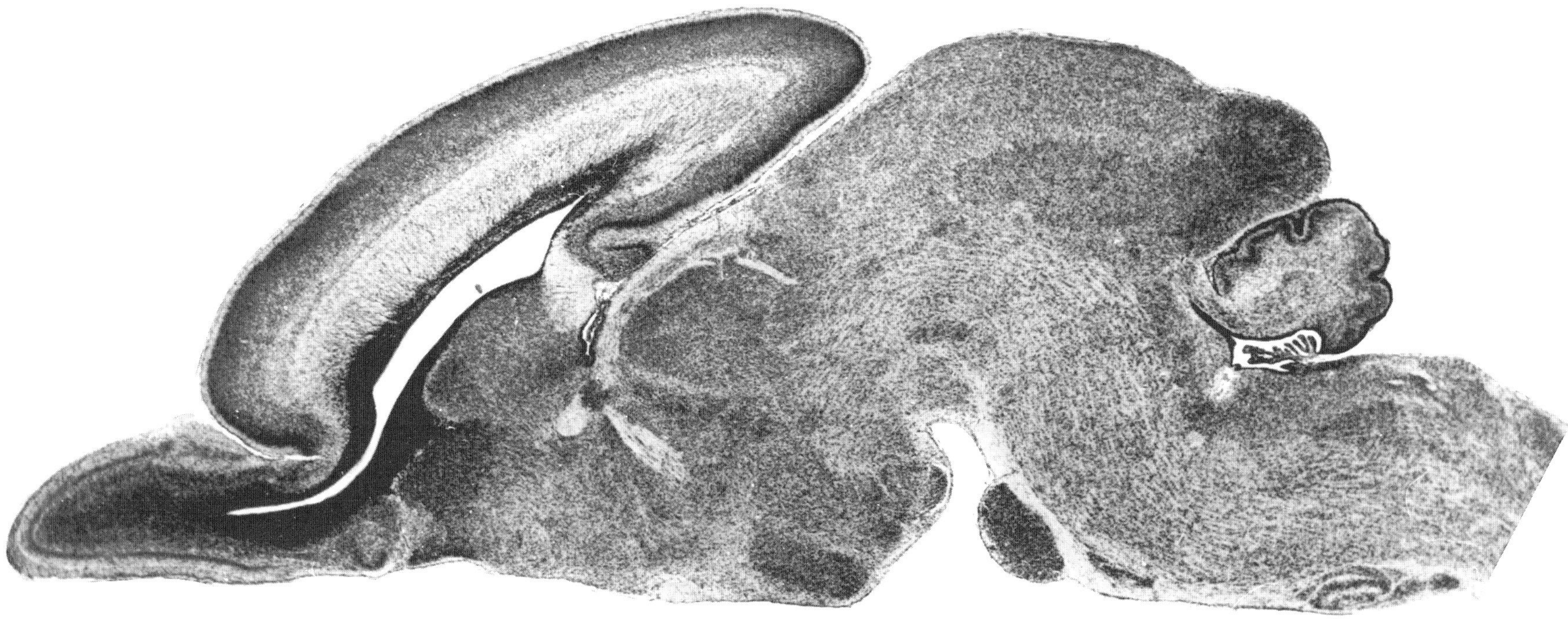

RHINENCEPHALON (R)

R1 rhinencephalic neuroepithelium
R2 rhinencephalic subventricular zone
R3 rhinencephalon (or differentiating field)
R9 olfactory tubercle

SEPTAL AREA (S)

S3 septum (or differentiating field)
S5 lateral septum
S16 taenia tecta

THALAMUS (T)

T3i intermediate thalamus (or differentiating field)
T3p posterior thalamus (or differentiating field)
T25 reticular nucleus

VENTRICLES (v)

vl lateral ventricle
v4 fourth ventricle

PONTINE AREA (X)

X3a anterior pons (or differentiating field)
X3p posterior pons (or differentiating field)
X5 pontine gray nucleus
X9 dorsal tegmental nucleus
X17 medial vestibular nucleus
X24 superior olivary nucleus
X33 isthmus (or differentiating field)

MEDULLA (Y)

Y3 medulla (or differentiating field)
Y7 prepositus nucleus
Y8 inferior olive
Y15 cuneate nucleus
Y16 solitary nucleus

SAGITTAL FIGURE 4

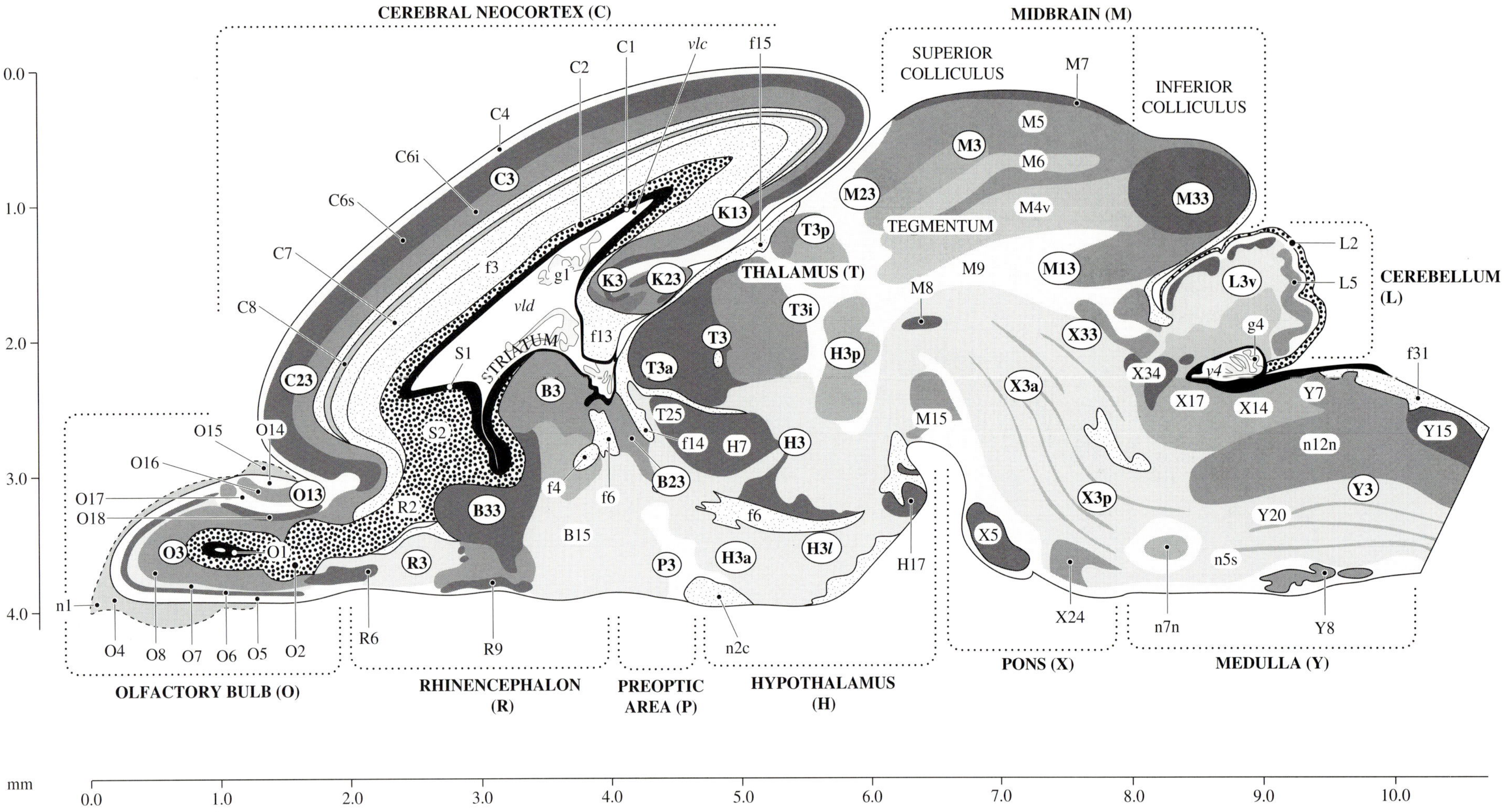

BASAL GANGLIA (B)

- **B3** striatum (or differentiating field)
- B15 substantia innominata
- **B23** bed nucleus of stria terminalis (or differentiating field)
- **B33** nucleus accumbens (or differentiating field)

CEREBRAL CORTEX (C)

- C1 neocortical neuroepithelium
- C2 neocortical subventricular zone
- **C3** neocortex (or differentiating field)
- C4 molecular layer (or primordial plexiform layer)
- C6i cortical plate, infragranular part
- C6s cortical plate, supragranular part
- C7 white matter
- C8 subplate
- **C23** orbital and insular area (or differentiating field)

FIBER TRACTS (f)

- f3 corpus callosum
- f4 anterior commissure
- f6 fornix
- f13 fimbria
- f14 stria terminalis
- f15 stria medullaris
- f31 cuneate fasciculus

GLANDS AND PARANEURAL STRUCTURES (g)

- g1 choroid plexus of lateral ventricle
- g4 choroid plexus of fourth ventricle

HYPOTHALAMUS (H)

- **H3** hypothalamus (or differentiating field)
- **H3a** anterior hypothalamus (or differentiating field)
- **H3*l*** lateral hypothalamus (or differentiating field)
- **H3p** posterior hypothalamus (or differentiating field)
- H7 zona incerta
- H17 lateral mammillary nucleus

HIPPOCAMPAL FORMATION (K)

- **K3** hippocampus (or differentiating field)
- **K13** subicular area (or differentiating field)
- **K23** dentate gyrus (or differentiating field)

CEREBELLUM (L)

- L2 external germinal layer
- **L3v** cerebellar vermis (or differentiating field)
- L5 Purkinje cell layer

MIDBRAIN (M)

- **M3** superior colliculus (or differentiating field)
- M4v ventral periaqueductal gray
- M5 superior colliculus, superficial layers
- M6 superior colliculus, deep layers
- M7 superior colliculus, stratum opticum
- M8 red nucleus
- M9 midbrain reticular formation
- **M13** tegmentum (or differentiating field)
- M15 interpeduncular nucleus
- **M23** pretectum (or differentiating field)
- **M33** inferior colliculus (or differentiating field)

NERVES AND NUCLEI, CRANIAL (n)

- n1 olfactory nerve
- n2c optic chiasma
- n5s spinal trigeminal nucleus
- n7n facial motor nucleus
- n12n hypoglossal nucleus

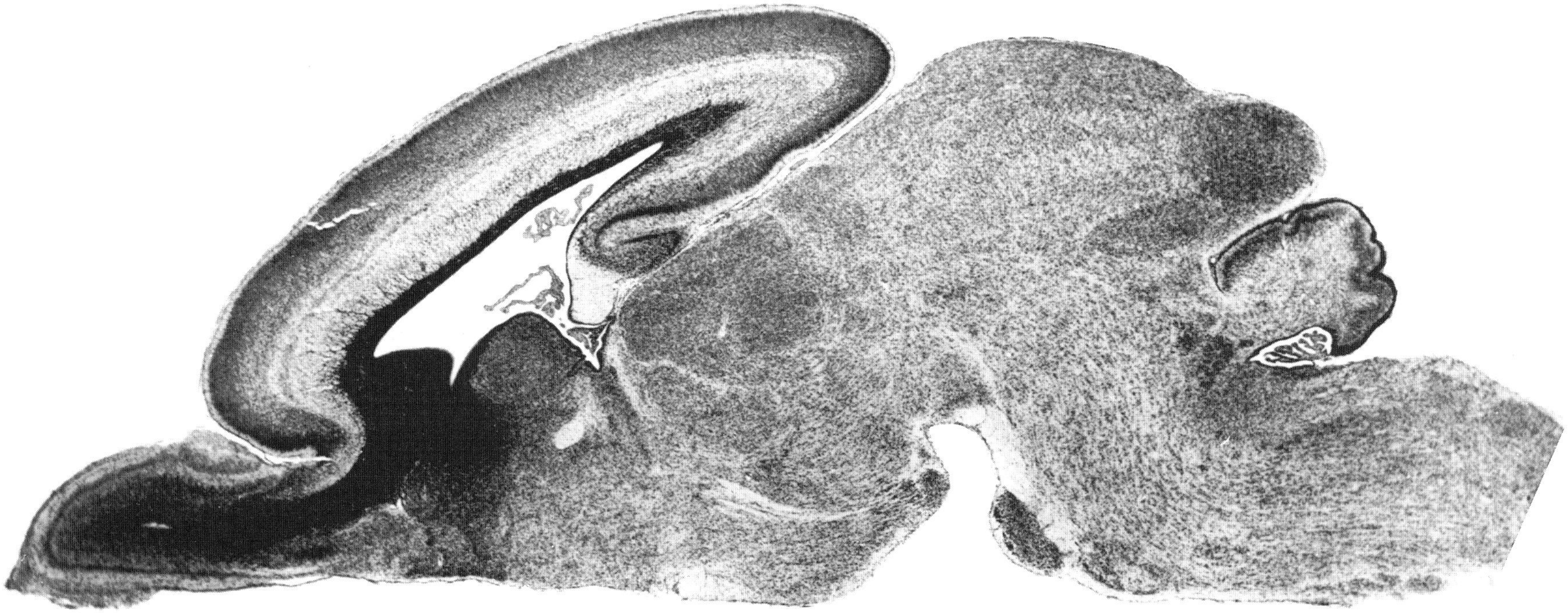

OLFACTORY BULB (O)

O1 olfactory bulb neuroepithelium
O2 olfactory bulb subventricular zone
O3 olfactory bulb (or differentiating field)
O4 glomerular layer
O5 external plexiform layer
O6 mitral cell layer
O7 internal plexiform layer
O8 granular layer
O13 accessory olfactory bulb (or differentiating field)
O14 glomerular layer (accessory olfactory bulb)
O15 external plexiform layer (accessory olfactory bulb)
O16 mitral cell layer (accessory olfactory bulb)
O17 internal plexiform layer (accessory olfactory bulb)
O18 granular layer (accessory olfactory bulb)

PREOPTIC AREA AND OPTIC EPITHELIA (P)

P3 preoptic area (or differentiating field)

RHINENCEPHALON (R)

R2 rhinencephalic subventricular zone
R3 rhinencephalon (or differentiating field)
R6 anterior olfactory nucleus
R9 olfactory tubercle

SEPTAL AREA (S)

S1 septal neuroepithelium
S2 septal subventricular zone

THALAMUS (T)

T3 thalamus (or differentiating field)
T3a anterior thalamus (or differentiating field)
T3i intermediate thalamus (or differentiating field)
T3p posterior thalamus (or differentiating field)
T25 reticular nucleus

VENTRICLES (v)

vlc lateral ventricle, central
vld lateral ventricle, dorsal
v4 fourth ventricle

PONTINE AREA (X)

X3a anterior pons (or differentiating field)
X3p posterior pons (or differentiating field)
X5 pontine gray nucleus
X14 spinal vestibular nucleus
X17 medial vestibular nucleus
X24 superior olivary nucleus
X33 isthmus (or differentiating field)
X34 locus coeruleus

MEDULLA (Y)

Y3 medulla (or differentiating field)
Y7 prepositus nucleus
Y8 inferior olive
Y15 cuneate nucleus
Y20 parvicellular reticular nucleus

SAGITTAL FIGURE 5

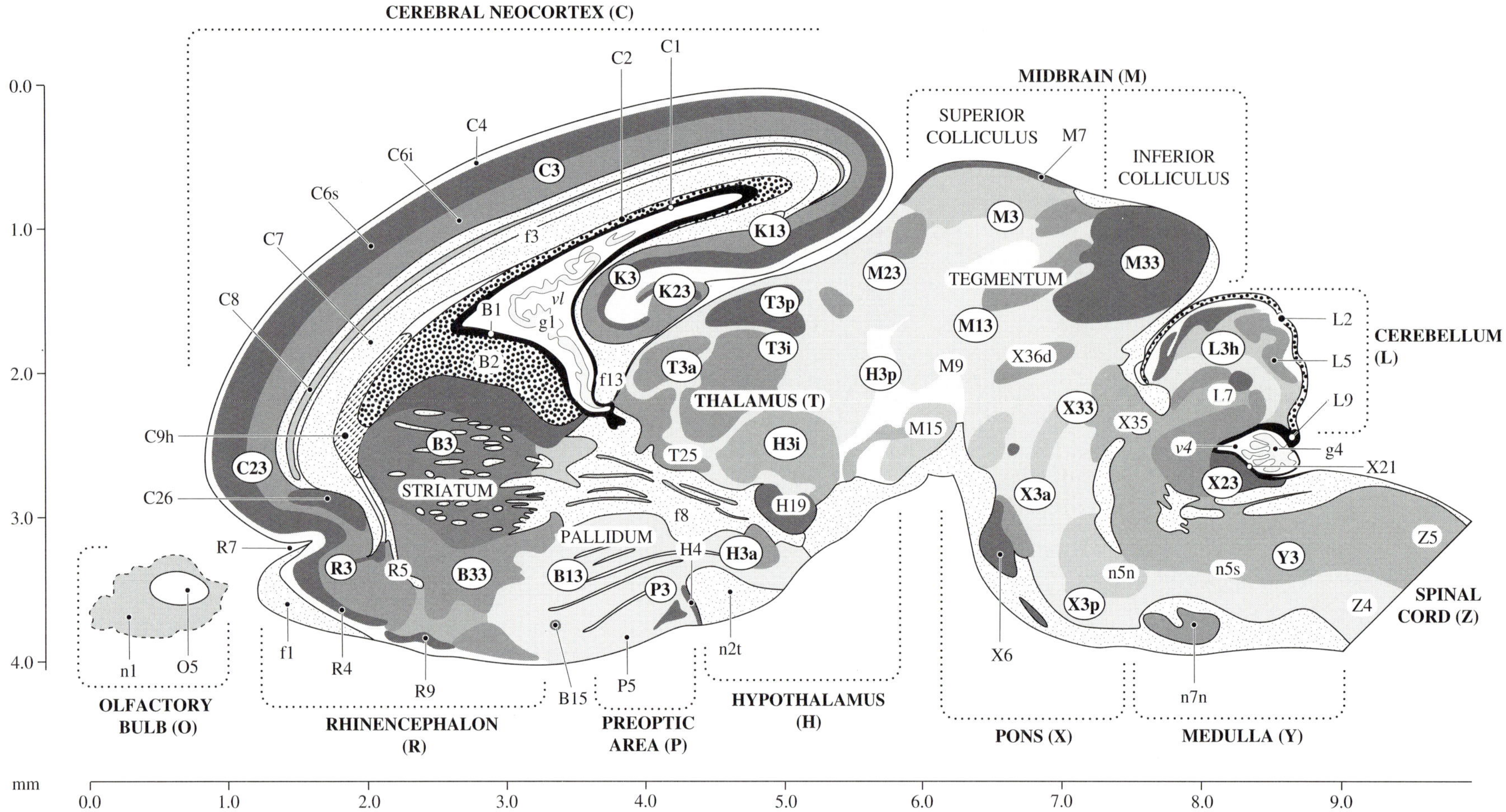

BASAL GANGLIA (B)

B1	striatal neuroepithelium
B2	striatal subventricular zone
B3	striatum (or differentiating field)
B13	pallidum (or differentiating field)
B15	substantia innominata
B33	nucleus accumbens (or differentiating field)

CEREBRAL CORTEX (C)

C1	neocortical neuroepithelium
C2	neocortical subventricular zone
C3	neocortex (or differentiating field)
C4	molecular layer (or primordial plexiform layer)
C6i	cortical plate, infragranular part
C6s	cortical plate, supragranular part
C7	white matter
C8	subplate
C9h	lateral migratory stream, head
C23	orbital and insular area (or differentiating field)
C26	claustrum

FIBER TRACTS (f)

f1	lateral olfactory tract
f3	corpus callosum
f8	internal capsule
f13	fimbria

GLANDS AND PARANEURAL STRUCTURES (g)

g1	choroid plexus of lateral ventricle
g4	choroid plexus of fourth ventricle

HYPOTHALAMUS (H)

H3a	anterior hypothalamus (or differentiating field)
H3i	intermediate hypothalamus (or differentiating field)
H3p	posterior hypothalamus (or differentiating field)
H4	supraoptic nucleus
H19	subthalamic nucleus

HIPPOCAMPAL FORMATION (K)

K3	hippocampus (or differentiating field)
K13	subicular area (or differentiating field)
K23	dentate gyrus (or differentiating field)

CEREBELLUM (L)

L2	external germinal layer
L3h	cerebellar hemisphere (or differentiating field)
L5	Purkinje cell layer
L7	lateral deep nucleus (dentate)
L9	cerebellar germinal trigone

MIDBRAIN (M)

M3	superior colliculus (or differentiating field)
M7	superior colliculus, stratum opticum
M9	midbrain reticular formation
M13	tegmentum (or differentiating field)
M15	interpeduncular nucleus
M23	pretectum (or differentiating field)
M33	inferior colliculus (or differentiating field)

NERVES AND NUCLEI, CRANIAL (n)

n1	olfactory nerve
n2t	optic tract
n5n	trigeminal motor nucleus
n5s	spinal trigeminal nucleus
n7n	facial motor nucleus

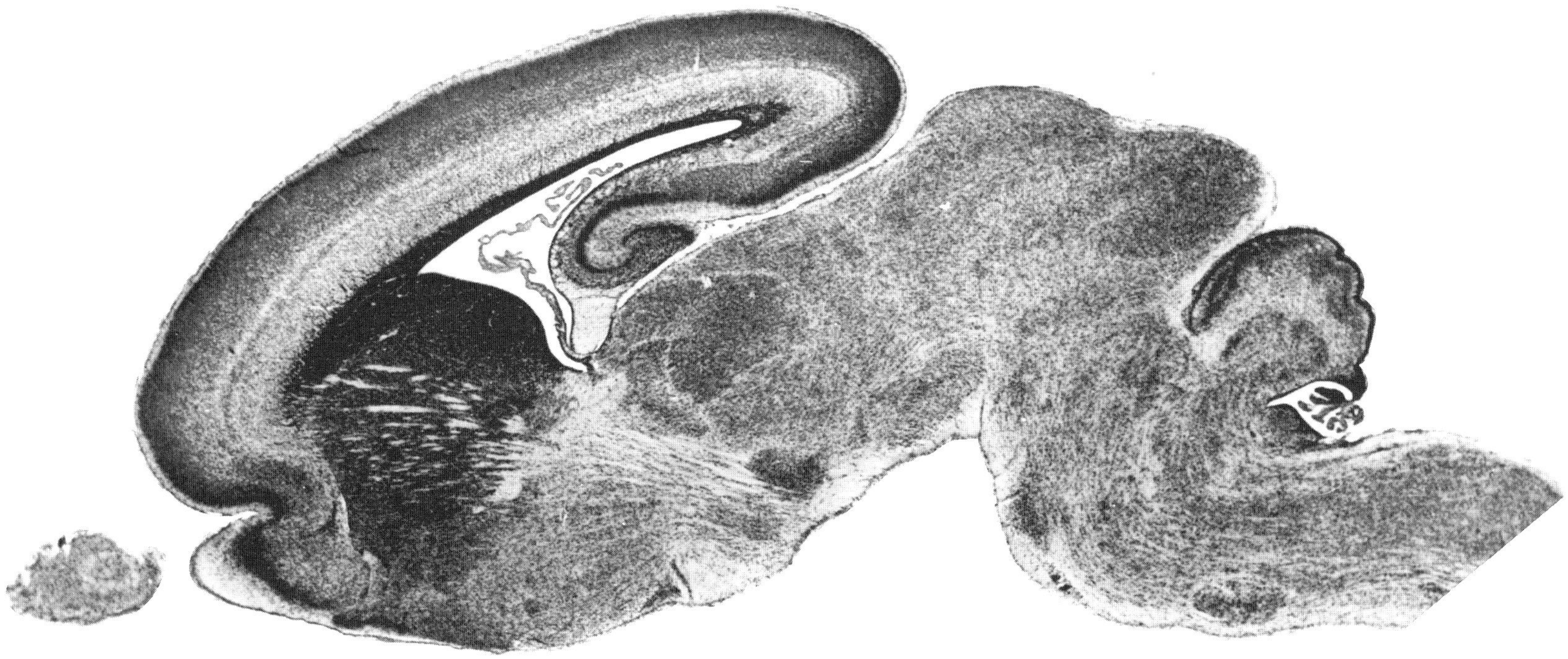

OLFACTORY BULB (O)

O5 external plexiform layer

PREOPTIC AREA AND OPTIC EPITHELIA (P)

P3 preoptic area (or differentiating field)
P5 lateral preoptic region

RHINENCEPHALON (R)

R3 rhinencephalon (or differentiating field)
R4 piriform cortex
R5 endopiriform nucleus
R7 rhinal fissure
R9 olfactory tubercle

THALAMUS (T)

T3a anterior thalamus (or differentiating field)
T3i intermediate thalamus (or differentiating field)
T3p posterior thalamus (or differentiating field)
T25 reticular nucleus

VENTRICLES (v)

vl lateral ventricle
v4 fourth ventricle

PONTINE AREA (X)

X3a anterior pons (or differentiating field)
X3p posterior pons (or differentiating field)
X6 pontine reticulotegmental nucleus
X21 cochlear nuclei neuroepithelium
X23 cochlear nuclei (or differentiating field)
X33 isthmus (or differentiating field)
X35 parabrachial nucleus
X36d nucleus of the lateral lemniscus, dorsal

MEDULLA (Y)

Y3 medulla (or differentiating field)

SPINAL CORD (Z)

Z4 ventral horn
Z5 dorsal horn

SAGITTAL FIGURE 6

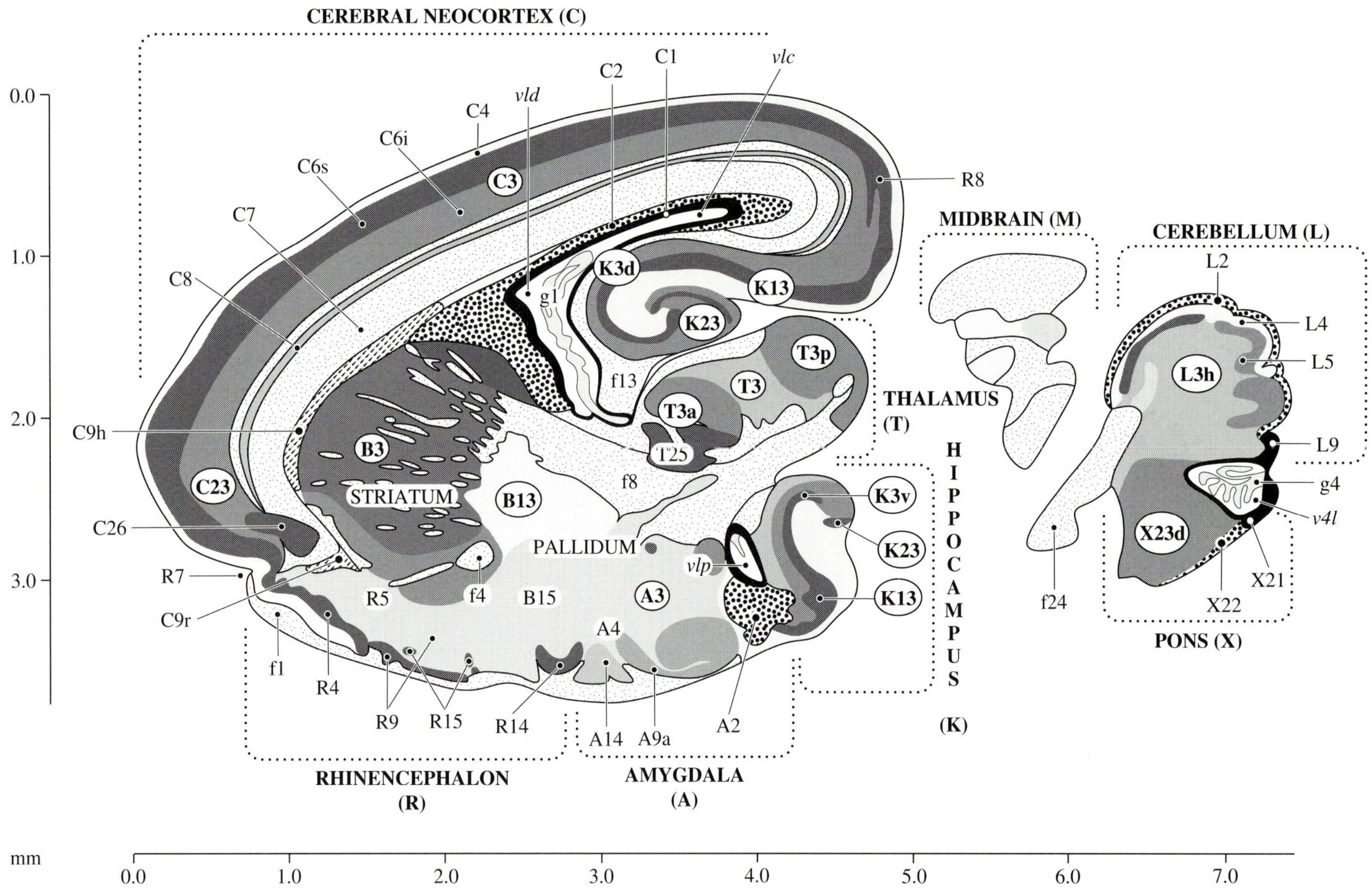

AMYGDALA (A)

A2 amygdaloid subventricular zone
A3 amygdala (or differentiating field)
A4 anterior amygdaloid area
A9a cortical amygdaloid nucleus, anterior
A14 medial amygdaloid nucleus

BASAL GANGLIA (B)

B3 striatum (or differentiating field)
B13 pallidum (or differentiating field)
B15 substantia innominata

CEREBRAL CORTEX (C)

C1 neocortical neuroepithelium
C2 neocortical subventricular zone
C3 neocortex (or differentiating field)
C4 molecular layer (or primordial plexiform layer)
C6i cortical plate, infragranular part
C6s cortical plate, supragranular part
C7 white matter
C8 subplate
C9h lateral migratory stream, head
C9r lateral migratory stream, reservoir
C23 orbital and insular area (or differentiating field)
C26 claustrum

FIBER TRACTS (f)

f1 lateral olfactory tract
f4 anterior commissure
f8 internal capsule
f13 fimbria
f24 middle cerebellar peduncle

GLANDS AND PARANEURAL STRUCTURES (g)

g1 choroid plexus of lateral ventricle
g4 choroid plexus of fourth ventricle

HIPPOCAMPAL FORMATION (K)

K3d dorsal hippocampus (or differentiating field)
K3v ventral hippocampus (or differentiating field)
K13 subicular area (or differentiating field)
K23 dentate gyrus (or differentiating field)

CEREBELLUM (L)

L2 external germinal layer
L3h cerebellar hemisphere (or differentiating field)
L4 molecular layer
L5 Purkinje cell layer
L9 cerebellar germinal trigone

RHINENCEPHALON (R)

R4 piriform cortex
R5 endopiriform nucleus
R7 rhinal fissure
R8 entorhinal cortex
R9 olfactory tubercle
R14 nucleus of the lateral olfactory tract
R15 islands of Calleja

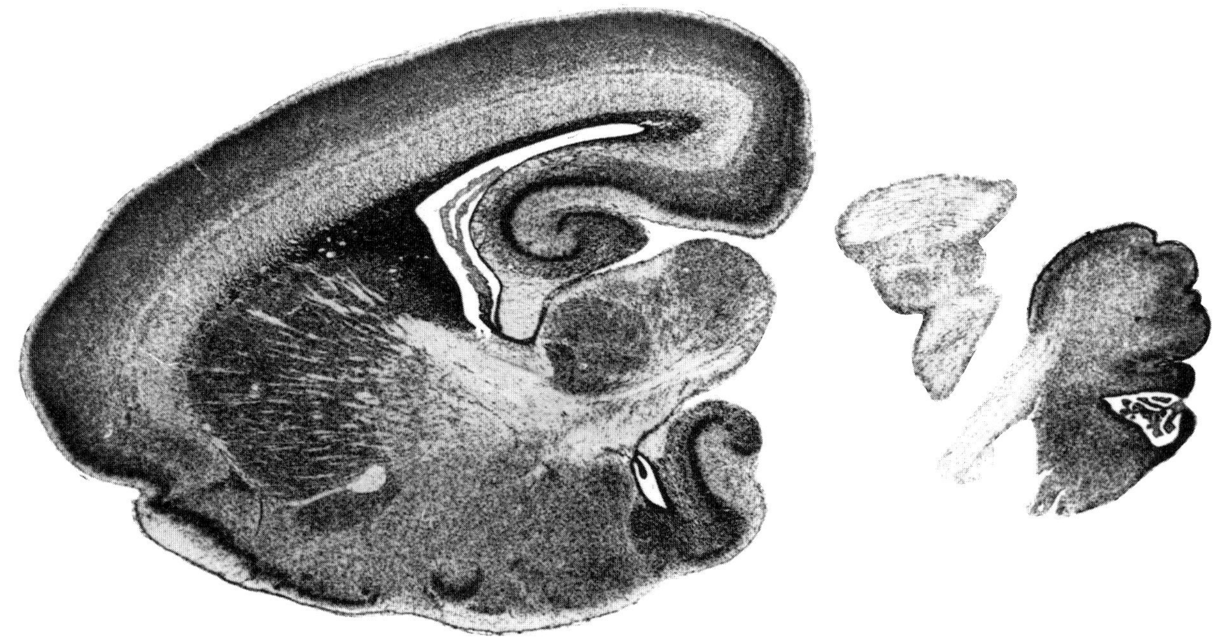

THALAMUS (T)

T3 thalamus (or differentiating field)
T3a anterior thalamus (or differentiating field)
T3p posterior thalamus (or differentiating field)
T25 reticular nucleus

VENTRICLES (v)

vlc lateral ventricle, central
vld lateral ventricle, dorsal
vlp lateral ventricle, posterior horn
v4l fourth ventricle, lateral recess

PONTINE AREA (X)

X21 cochlear nuclei neuroepithelium
X22 cochlear external germinal layer
X23d dorsal cochlear nucleus (or differentiating field)

SAGITTAL FIGURE 7

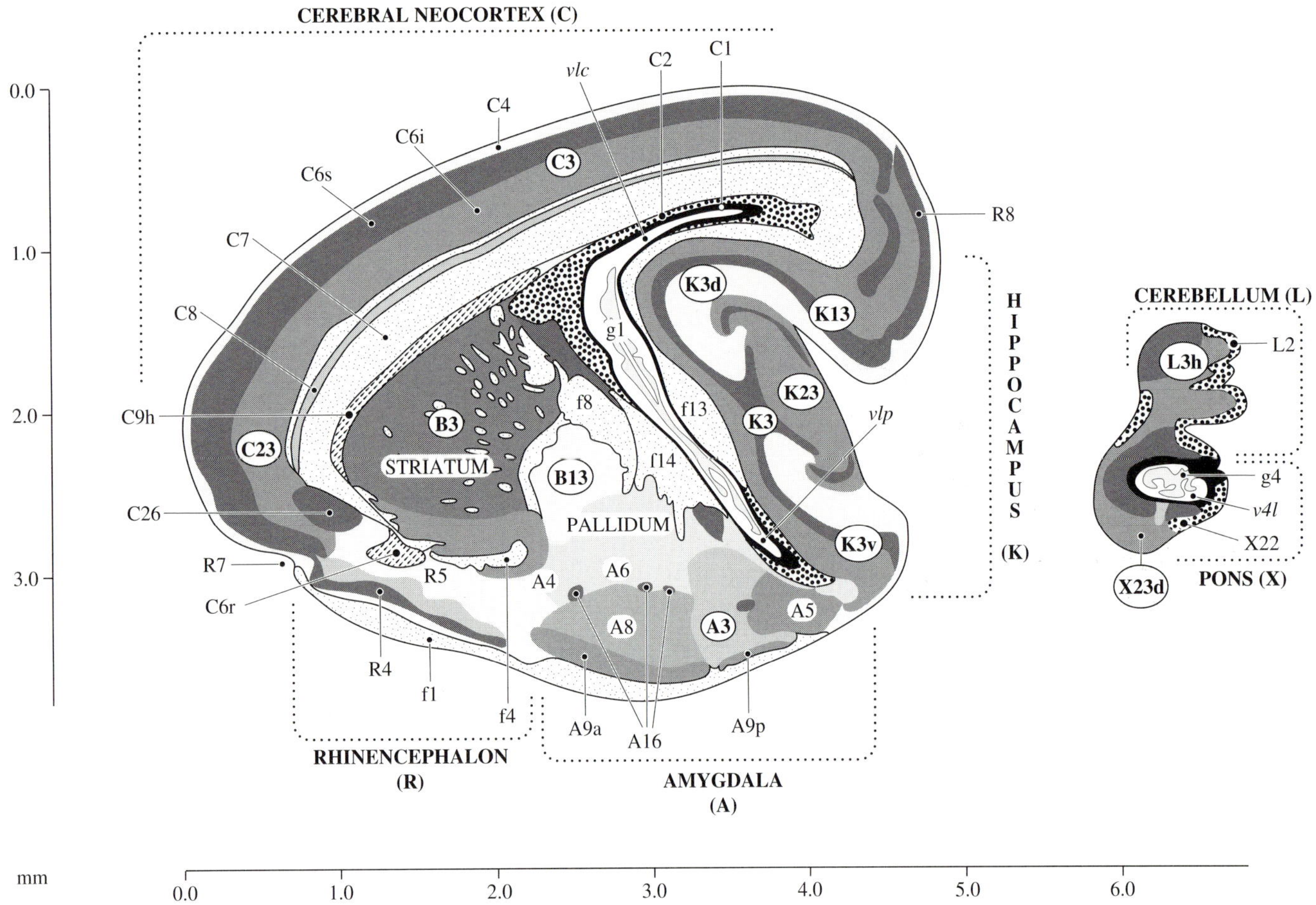

AMYGDALA (A)

A3 amygdala (or differentiating field)
A4 anterior amygdaloid area
A5 amygdala-hippocampal area
A6 central amygdaloid nucleus
A8 basolateral amygdaloid nucleus
A9a cortical amygdaloid nucleus, anterior
A9p cortical amygdaloid nucleus, posterior
A16 intercalated masses

BASAL GANGLIA (B)

B3 striatum (or differentiating field)
B13 pallidum (or differentiating field)

CEREBRAL CORTEX (C)

C1 neocortical neuroepithelium
C2 neocortical subventricular zone
C3 neocortex (or differentiating field)
C4 molecular layer (or primordial plexiform layer)
C6i cortical plate, infragranular part
C6s cortical plate, supragranular part
C7 white matter
C8 subplate
C9h lateral migratory stream, head
C9r lateral migratory stream, reservoir
C23 orbital and insular area (or differentiating field)
C26 claustrum

FIBER TRACTS (f)

f1 lateral olfactory tract
f4 anterior commissure
f8 internal capsule
f13 fimbria
f14 stria terminalis

GLANDS AND PARANEURAL STRUCTURES (g)

g1 choroid plexus of lateral ventricle
g4 choroid plexus of fourth ventricle

HIPPOCAMPAL FORMATION (K)

K3 hippocampus (or differentiating field)
K3d dorsal hippocampus (or differentiating field)
K3v ventral hippocampus (or differentiating field)
K13 subicular area (or differentiating field)
K23 dentate gyrus (or differentiating field)

CEREBELLUM (L)

L2 external germinal layer
L3h cerebellar hemisphere (or differentiating field)

RHINENCEPHALON (R)

R4 piriform cortex
R5 endopiriform nucleus
R8 entorhinal cortex
R7 rhinal fissure

VENTRICLES (v)

vlc lateral ventricle, central
vlp lateral ventricle, posterior horn
v4l fourth ventricle, lateral recess

PONTINE AREA (X)

X22 cochlear external germinal layer
X23d dorsal cochlear nucleus

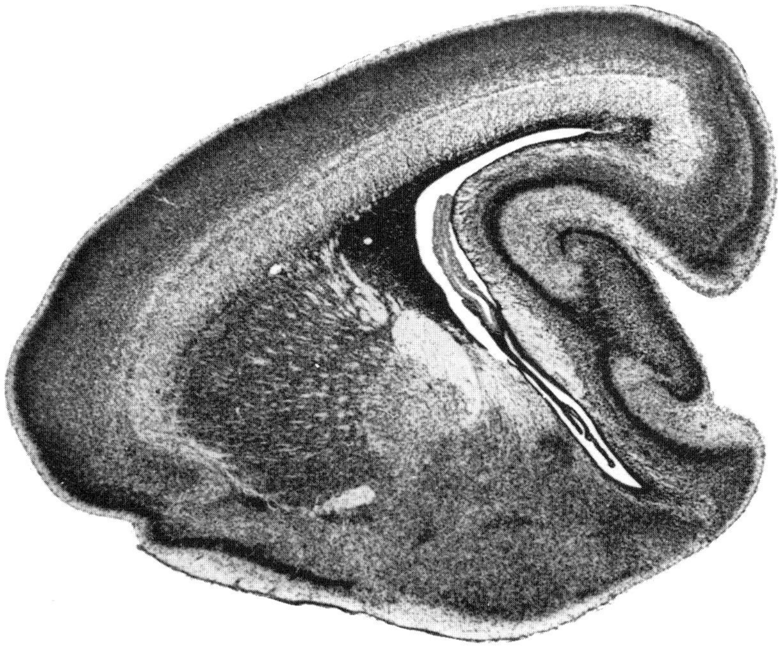

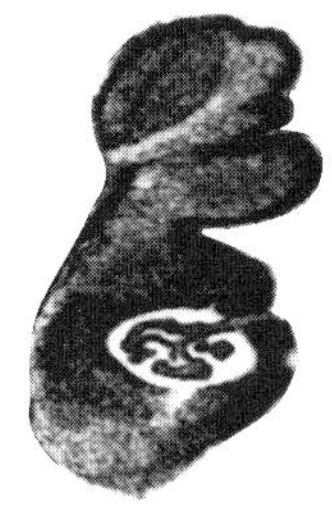

SAGITTAL FIGURE 8

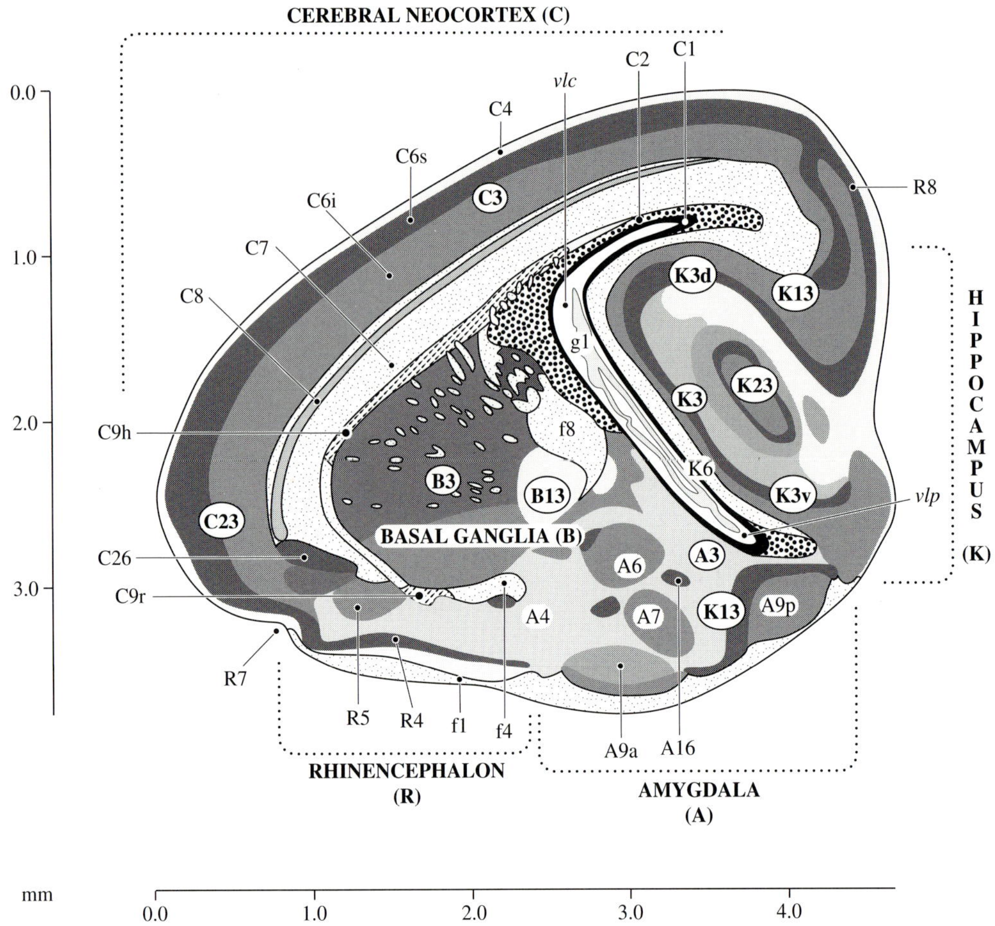

AMYGDALA (A)

A3 amygdala (or differentiating field)
A4 anterior amygdaloid area
A6 central amygdaloid nucleus
A7 basomedial amygdaloid nucleus
A9a cortical amygdaloid nucleus, anterior
A9p cortical amygdaloid nucleus, posterior
A16 intercalated masses

BASAL GANGLIA (B)

B3 striatum (or differentiating field)
B13 pallidum (or differentiating field)

CEREBRAL CORTEX (C)

C1 neocortical neuroepithelium
C2 neocortical subventricular zone
C3 neocortex (or differentiating field)
C4 molecular layer (or primordial plexiform layer)
C6i cortical plate, infragranular part
C6s cortical plate, supragranular part
C7 white matter
C8 subplate
C9h lateral migratory stream, head
C9r lateral migratory stream, reservoir
C23 orbital and insular area (or differentiating field)
C26 claustrum

FIBER TRACTS (f)

f1 lateral olfactory tract
f4 anterior commissure
f8 internal capsule

GLANDS AND PARANEURAL STRUCTURES (g)

g1 choroid plexus of lateral ventricle

HIPPOCAMPAL FORMATION (K)

K3 hippocampus (or differentiating field)
K3d dorsal hippocampus (or differentiating field)
K3v ventral hippocampus (or differentiating field)
K6 alveus
K13 subicular area (or differentiating field)
K23 dentate gyrus (or differentiating field)

RHINENCEPHALON (R)

R4 piriform cortex
R5 endopiriform nucleus
R7 rhinal fissure
R8 entorhinal cortex

VENTRICLES (v)

vlc lateral ventricle, central
vlp lateral ventricle, posterior horn

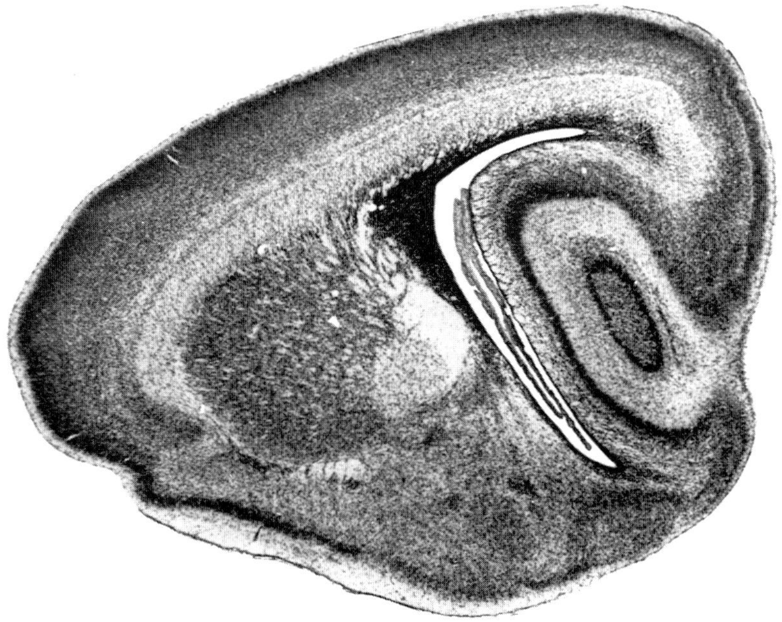

CORONAL

FIGURES and PLATES 1-27

CORONAL FIGURE 1

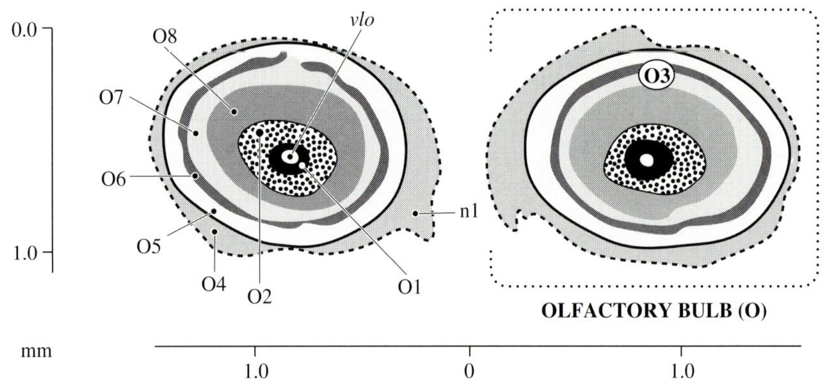

NERVES AND NUCLEI, CRANIAL (n)

n1 olfactory nerve

OLFACTORY BULB (O)

O1 olfactory bulb neuroepithelium
O2 olfactory bulb subventricular zone
O3 olfactory bulb (or differentiating field)
O4 glomerular layer
O5 external plexiform layer
O6 mitral cell layer
O7 internal plexiform layer
O8 granular layer

VENTRICLES (v)

vlo olfactory ventricle

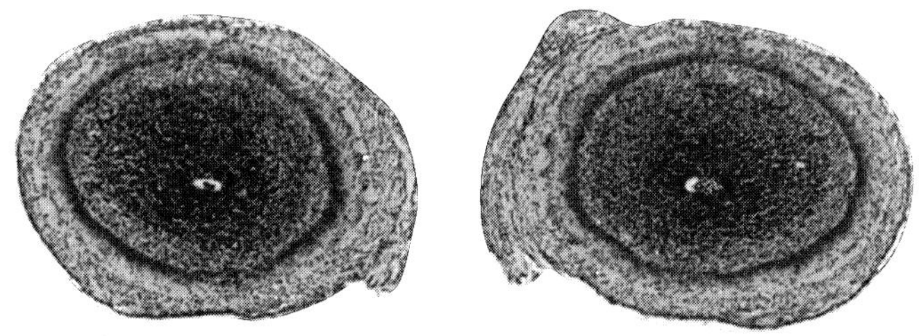

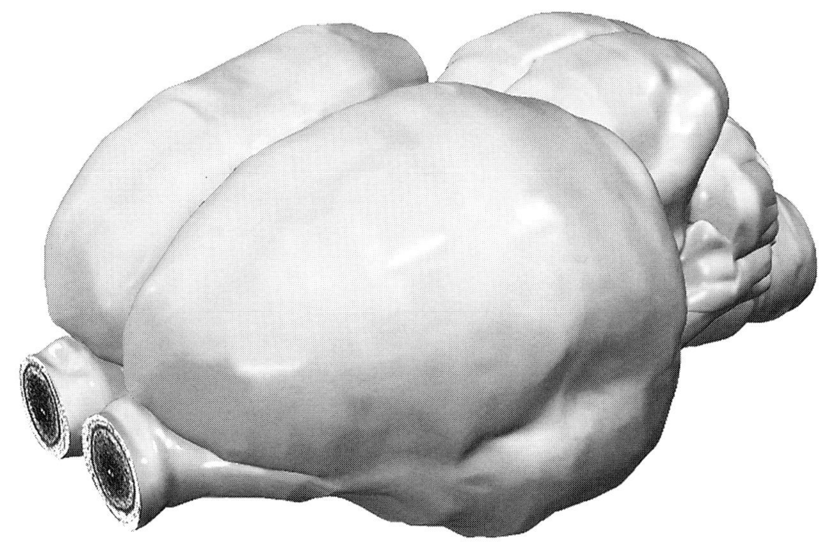

CORONAL FIGURE 2

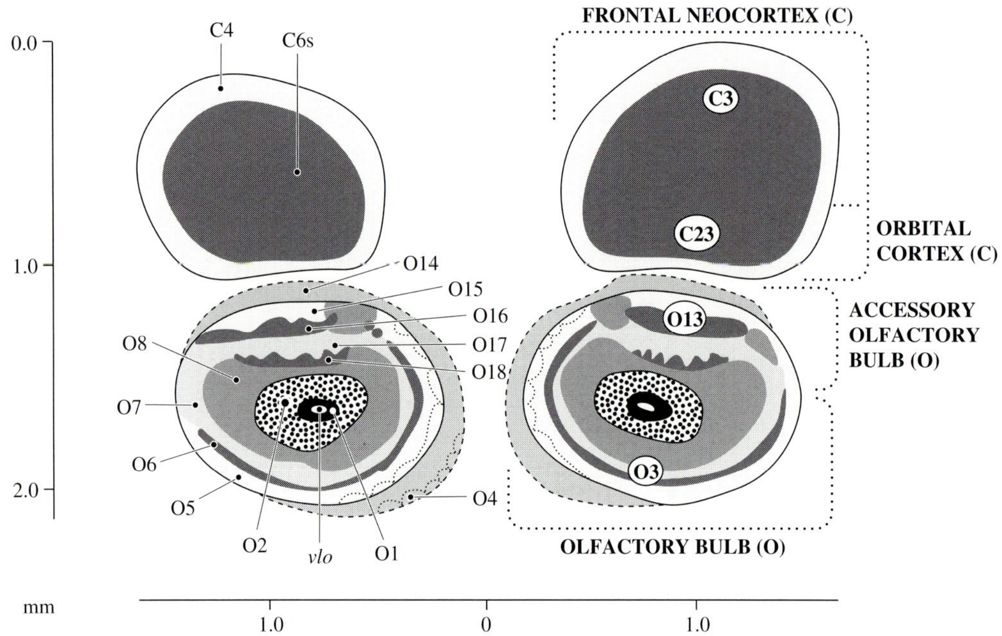

CEREBRAL CORTEX (C)

C3 neocortex (or differentiating field)
C4 molecular layer (or primordial plexiform layer)
C6s cortical plate, supragranular part
C23 orbital and insular area (or differentiating field)

OLFACTORY BULB (0)

O1 olfactory bulb neuroepithelium
O2 olfactory bulb subventricular zone
O3 olfactory bulb (or differentiating field)
O4 glomerular layer
O5 external plexiform layer
O6 mitral cell layer
O7 internal plexiform layer
O8 granular layer
O13 accessory olfactory bulb (or differentiating field)
O14 glomerular layer (accessory olfactory bulb)
O15 external plexiform layer (accessory olfactory bulb)
O16 mitral cell layer (accessory olfactory bulb)
O17 internal plexiform layer (accessory olfactory bulb)
O18 granular layer (accessory olfactory bulb)

VENTRICLES (v)

vlo olfactory ventricle

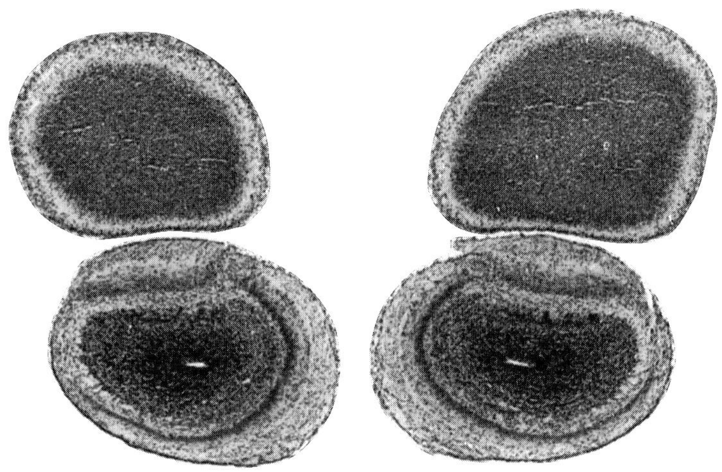

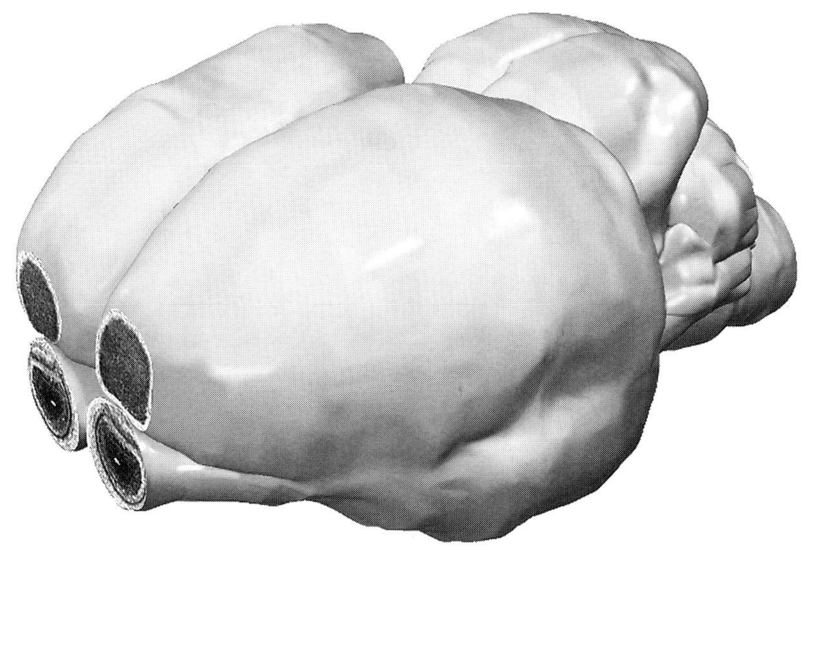

CORONAL FIGURE 3

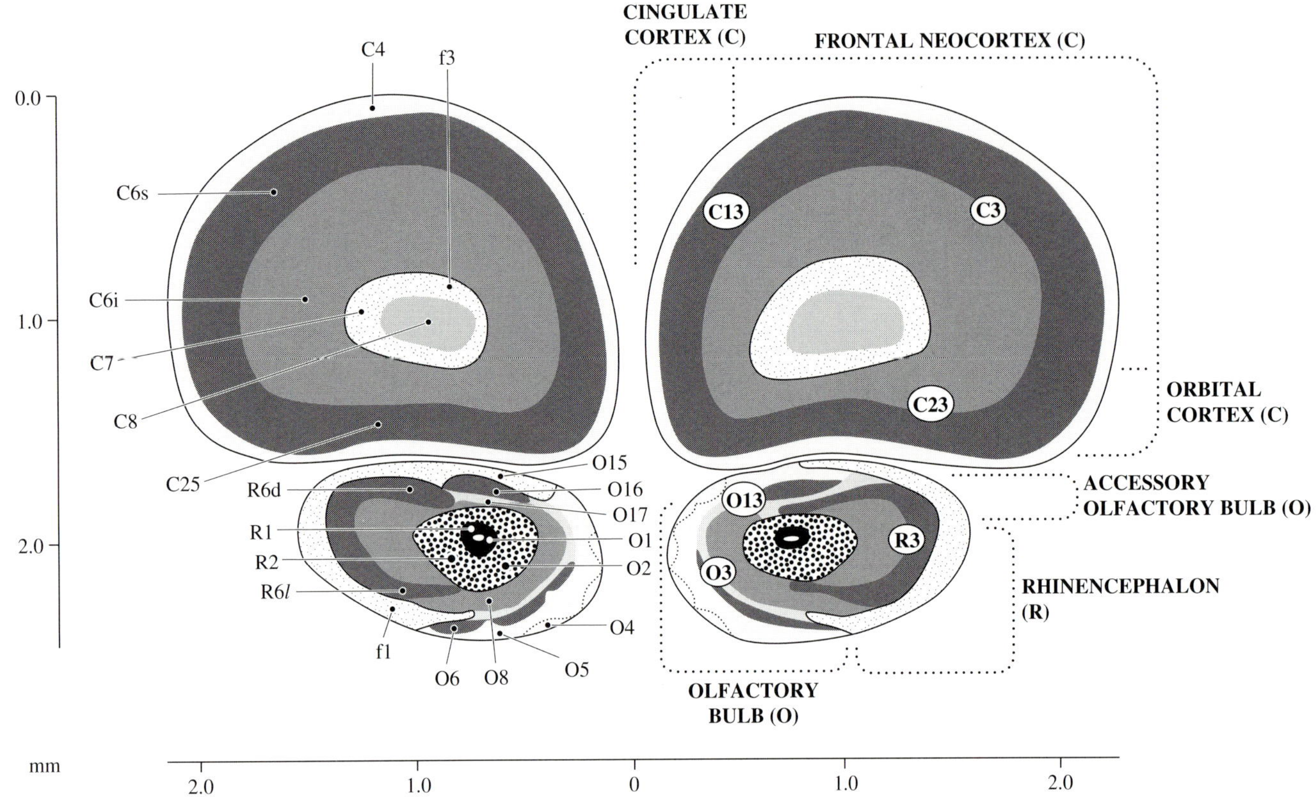

CEREBRAL CORTEX (C)

- **C3** neocortex (or differentiating field)
- C4 molecular layer (or primordial plexiform layer)
- C6i cortical plate, infragranular part
- C6s cortical plate, supragranular part
- C7 white matter
- C8 subplate
- **C13** cingulate cortex (or differentiating field)
- **C23** orbital and insular area (or differentiating field)
- C25 orbital cortex

OLFACTORY BULB (O)

- O1 olfactory bulb neuroepithelium
- O2 olfactory bulb subventricular zone
- **O3** olfactory bulb (or differentiating field)
- O4 glomerular layer
- O5 external plexiform layer
- O6 mitral cell layer
- O8 granular layer
- **O13** accessory olfactory bulb (or differentiating field)
- O15 external plexiform layer (accessory olfactory bulb)
- O16 mitral cell layer (accessory olfactory bulb)
- O17 internal plexiform layer (accessory olfactory bulb)

FIBER TRACTS (f)

- f1 lateral olfactory tract
- f3 corpus callosum

RHINENCEPHALON (R)

- R1 rhinencephalic neuroepithelium
- R2 rhinencephalic subventricular zone
- **R3** rhinencephalon (or differentiating field)
- R6d anterior olfactory nucleus, dorsal
- R6*l* anterior olfactory nucleus, lateral

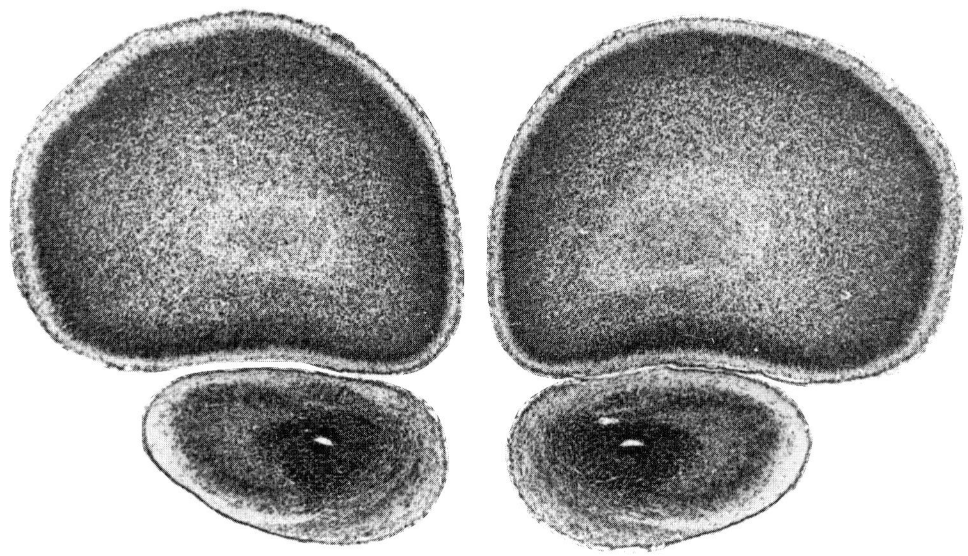

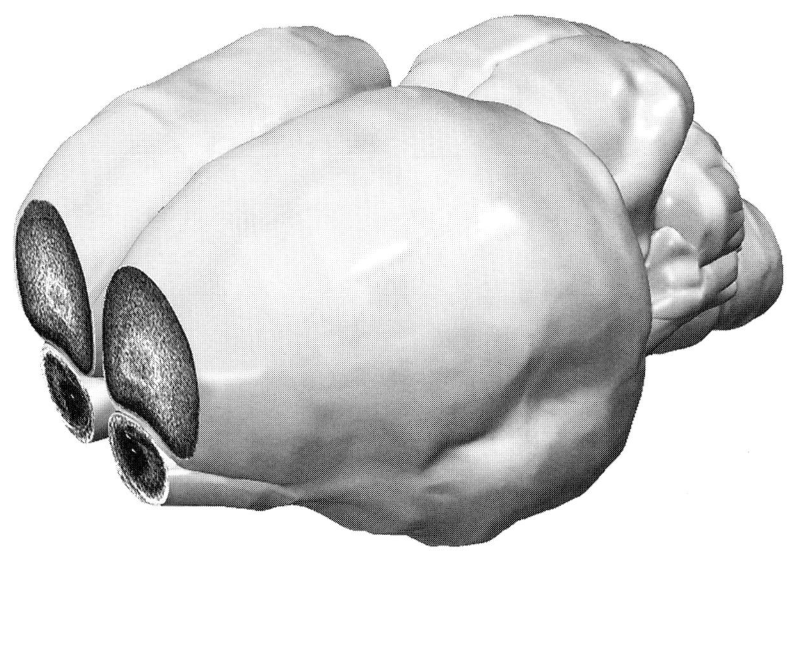

CORONAL FIGURE 4

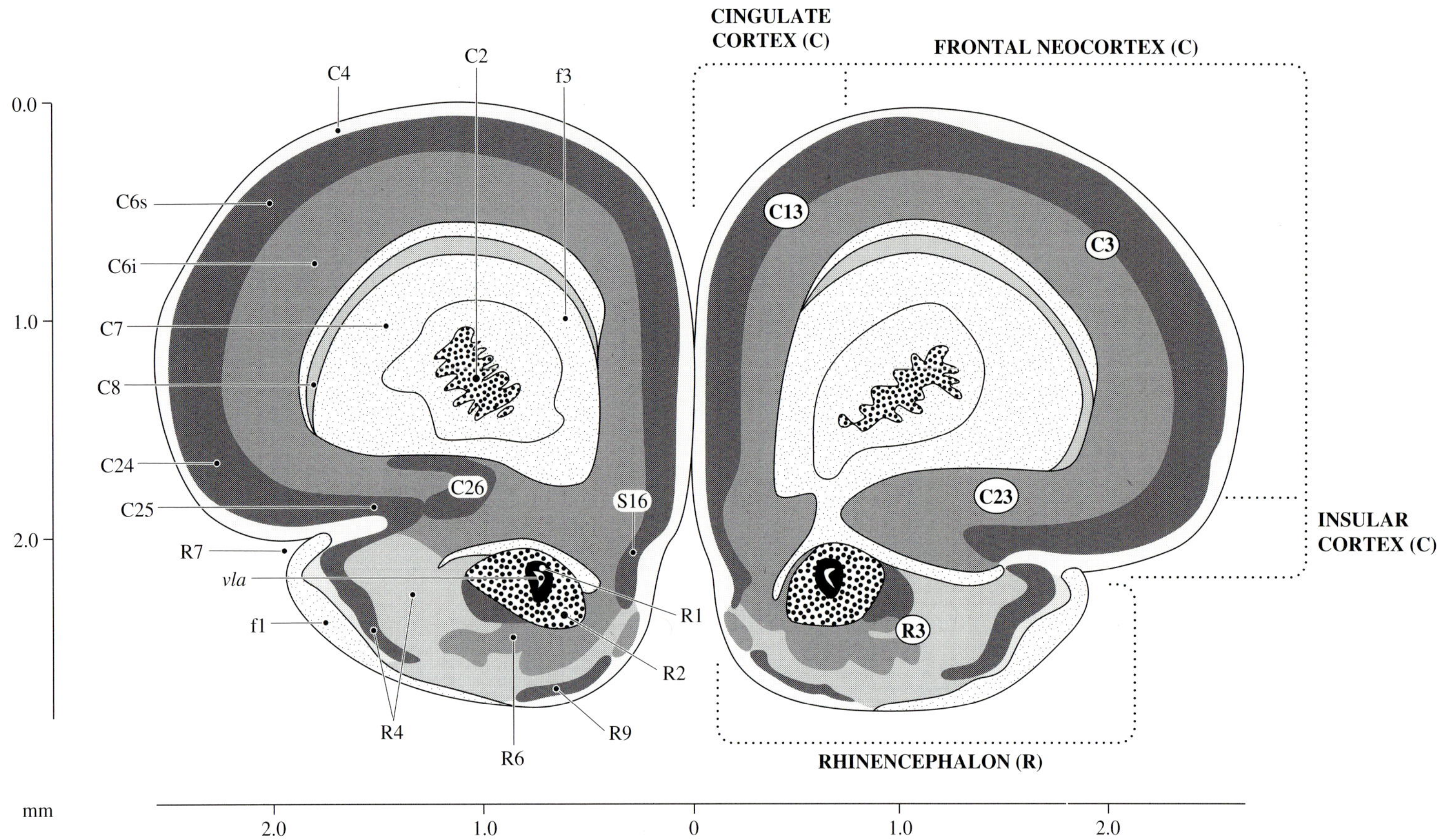

CEREBRAL CORTEX (C)

C2 neocortical subventricular zone
C3 neocortex (or differentiating field)
C4 molecular layer (or primordial plexiform layer)
C6i cortical plate, infragranular part
C6s cortical plate, supragranular part
C7 white matter
C8 subplate
C13 cingulate cortex (or differentiating field)
C23 orbital and insular area (or differentiating field)
C24 insular cortex
C25 orbital cortex
C26 claustrum

FIBER TRACTS (f)

f1 lateral olfactory tract
f3 corpus callosum

RHINENCEPHALON (R)

R1 rhinencephalic neuroepithelium
R2 rhinencephalic subventricular zone
R3 rhinencephalon (or differentiating field)
R4 piriform cortex
R6 anterior olfactory nucleus
R7 rhinal fissure
R9 olfactory tubercle

SEPTAL AREA (S)

S16 taenia tecta

VENTRICLES (v)

vla lateral ventricle, anterior horn

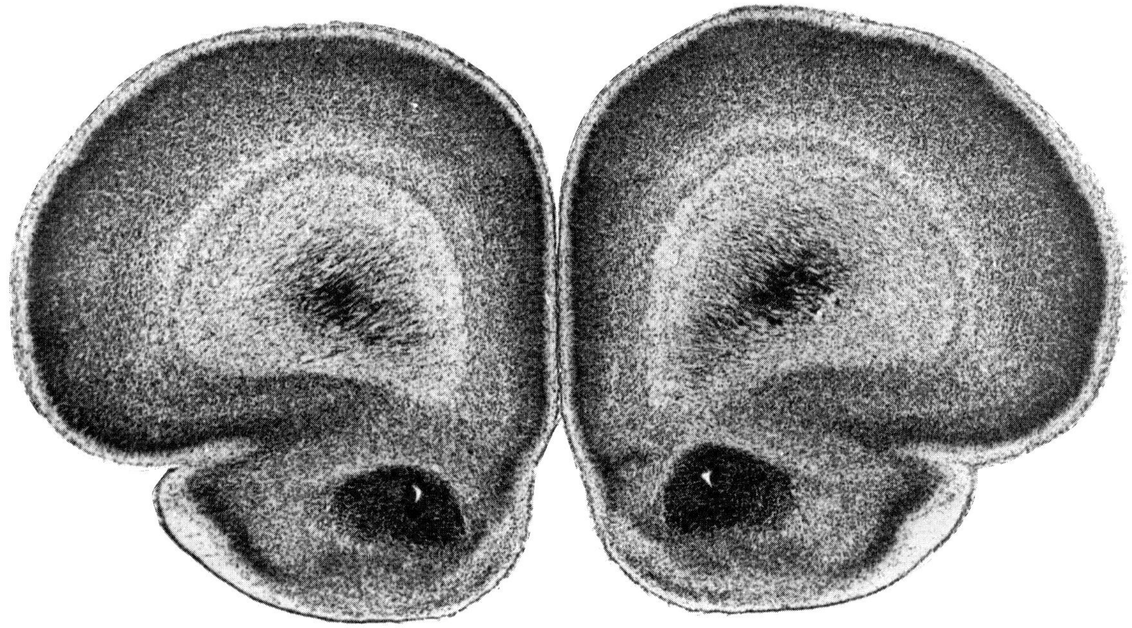

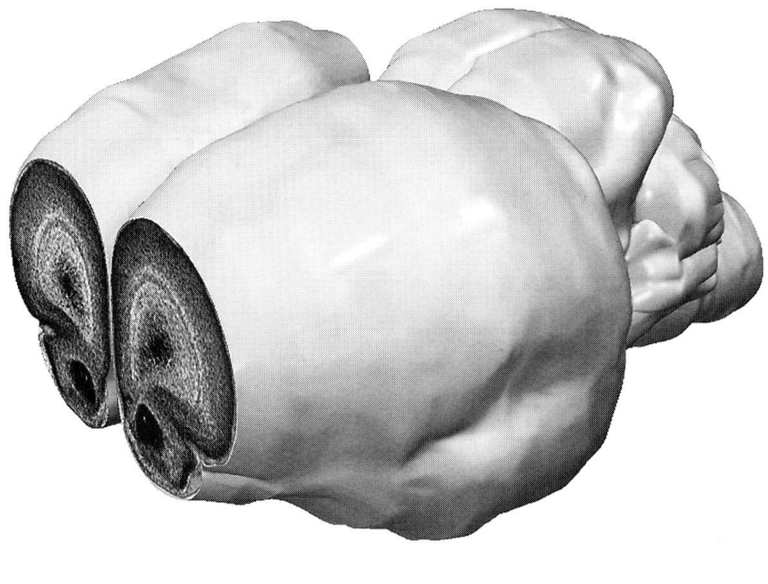

CORONAL FIGURE 5

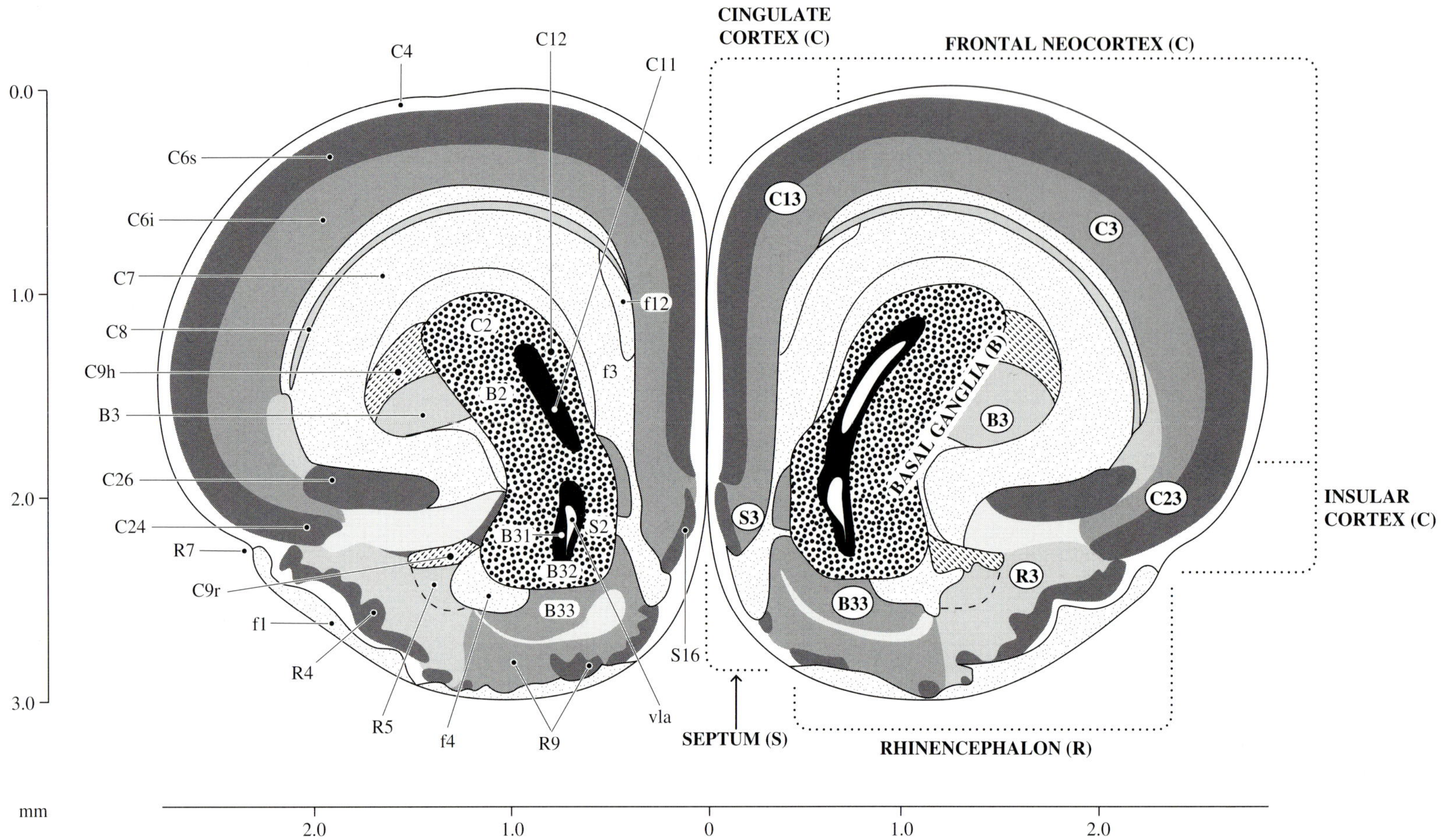

BASAL GANGLIA (B)

B2 striatal subventricular zone
B3 striatum (or differentiating field)
B31 accumbens neuroepithelium
B32 accumbens subventricular zone
B33 nucleus accumbens (or differentiating field)

CEREBRAL CORTEX (C)

C2 neocortical subventricular zone
C3 neocortex (or differentiating field)
C4 molecular layer (or primordial plexiform layer)
C6i cortical plate, infragranular part
C6s cortical plate, supragranular part
C7 white matter
C8 subplate
C9h lateral migratory stream, head
C9r lateral migratory stream, reservoir
C11 cingulate cortex neuroepithelium
C12 cingulate cortex, subventricular zone
C13 cingulate cortex (or differentiating field)
C23 orbital and insular area (or differentiating field)
C24 insular cortex
C26 claustrum

FIBER TRACTS (f)

f1 lateral olfactory tract
f3 corpus callosum
f4 anterior commissure
f12 cingulum bundle

RHINENCEPHALON (R)

R3 rhinencephalon (or differentiating field)
R4 piriform cortex
R5 endopiriform nucleus
R7 rhinal fissure
R9 olfactory tubercle

SEPTAL AREA (S)

S2 septal subventricular zone
S3 septum (or differentiating field)
S16 taenia tecta

VENTRICLES (v)

vla lateral ventricle, anterior horn

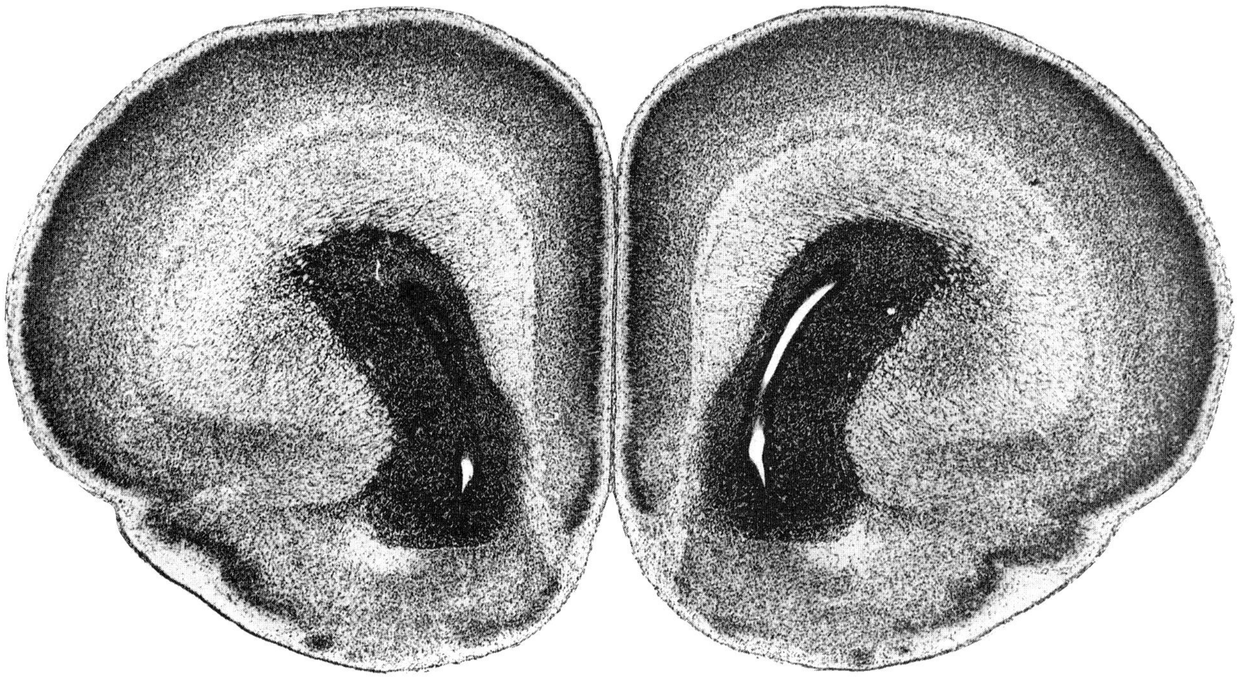

CORONAL FIGURE 6

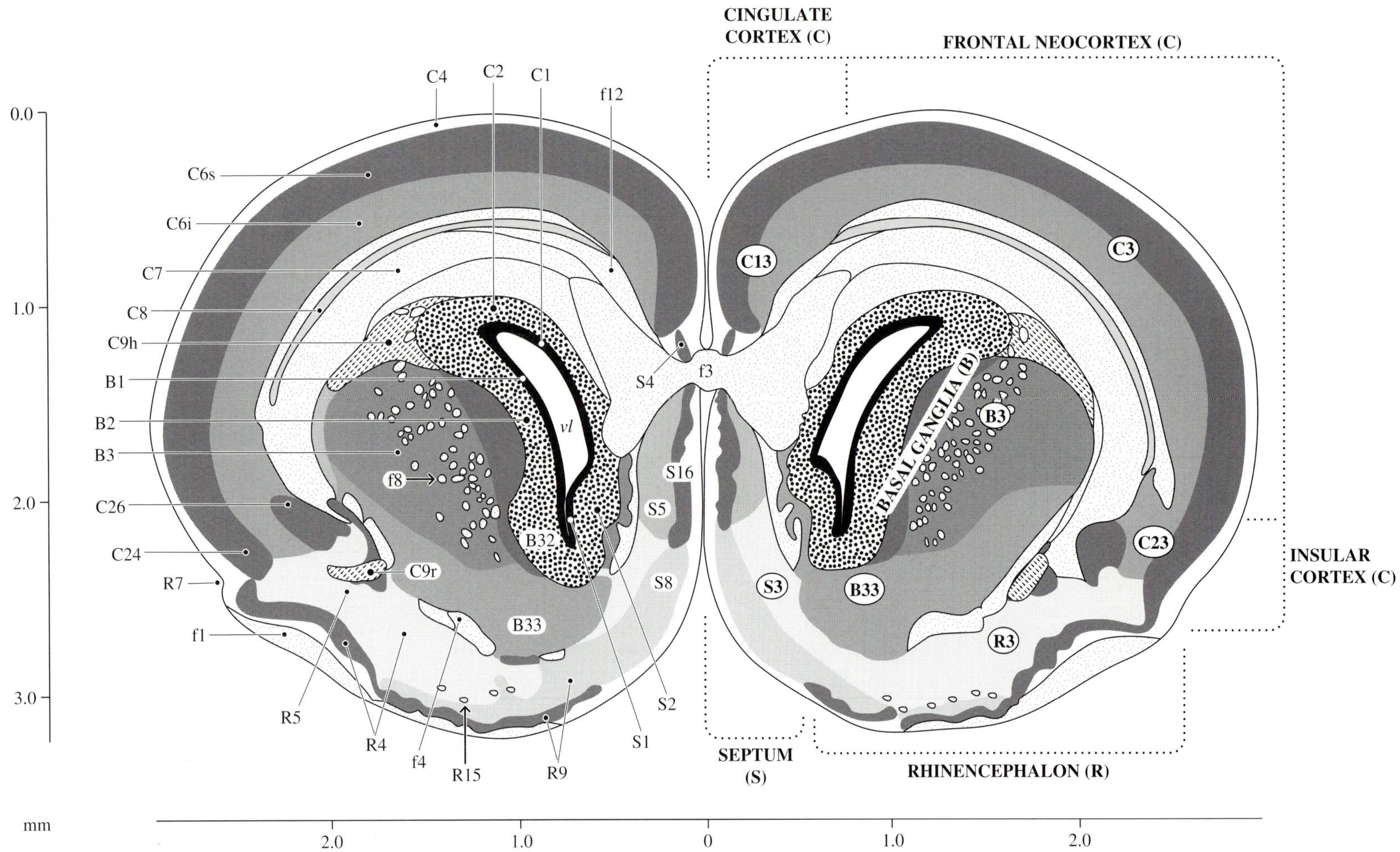

BASAL GANGLIA (B)

B1	striatal neuroepithelium
B2	striatal subventricular zone
B3	striatum (or differentiating field)
B32	accumbens subventricular zone
B33	nucleus accumbens

CEREBRAL CORTEX (C)

C1	neocortical neuroepithelium
C2	neocortical subventricular zone
C3	neocortex (or differentiating field)
C4	molecular layer (or primordial plexiform layer)
C6i	cortical plate, infragranular part
C6s	cortical plate, supragranular part
C7	white matter
C8	subplate
C9h	lateral migratory stream, head
C9r	lateral migratory stream, reservoir
C13	cingulate cortex (or differentiating field)
C23	orbital and insular area (or differentiating field)
C24	insular cortex
C26	claustrum

FIBER TRACTS (f)

f1	lateral olfactory tract
f3	corpus callosum
f4	anterior commissure
f8	internal capsule
f12	cingulum bundle

RHINENCEPHALON (R)

R3	rhinencephalon (or differentiating field)
R4	piriform cortex
R5	endopiriform nucleus
R7	rhinal fissure
R9	olfactory tubercle
R15	islands of Calleja

SEPTAL AREA (S)

S1	septal neuroepithelium
S2	septal subventricular zone
S3	septum (or differentiating field)
S4	indesium griseum
S5	lateral septum
S8	diagonal band (vertical limb)
S16	taenia tecta

VENTRICLES (v)

vl	lateral ventricle

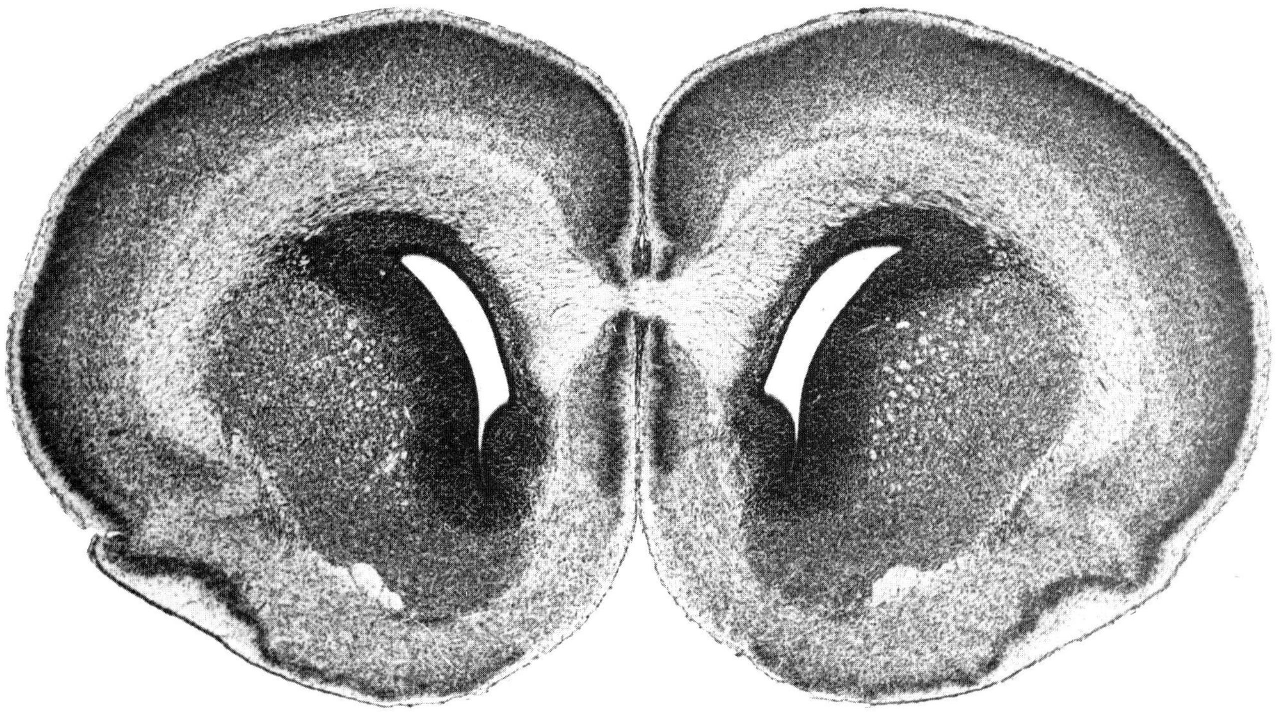

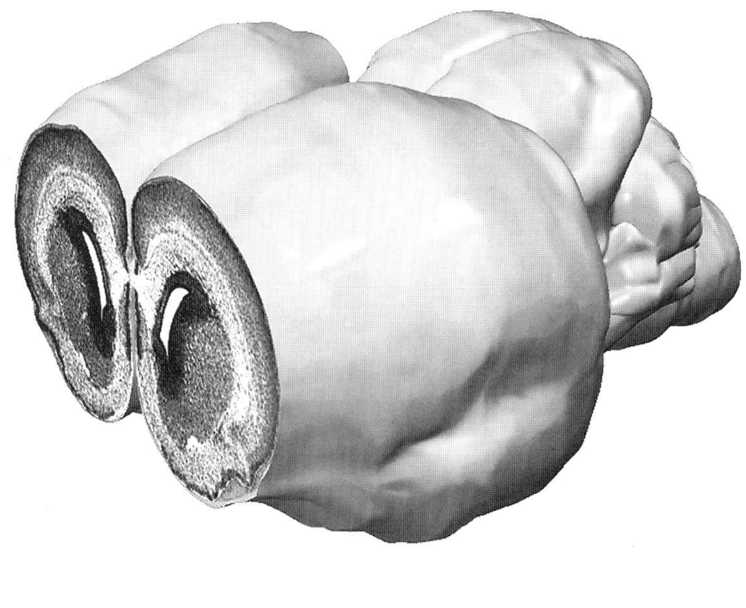

CORONAL FIGURE 7

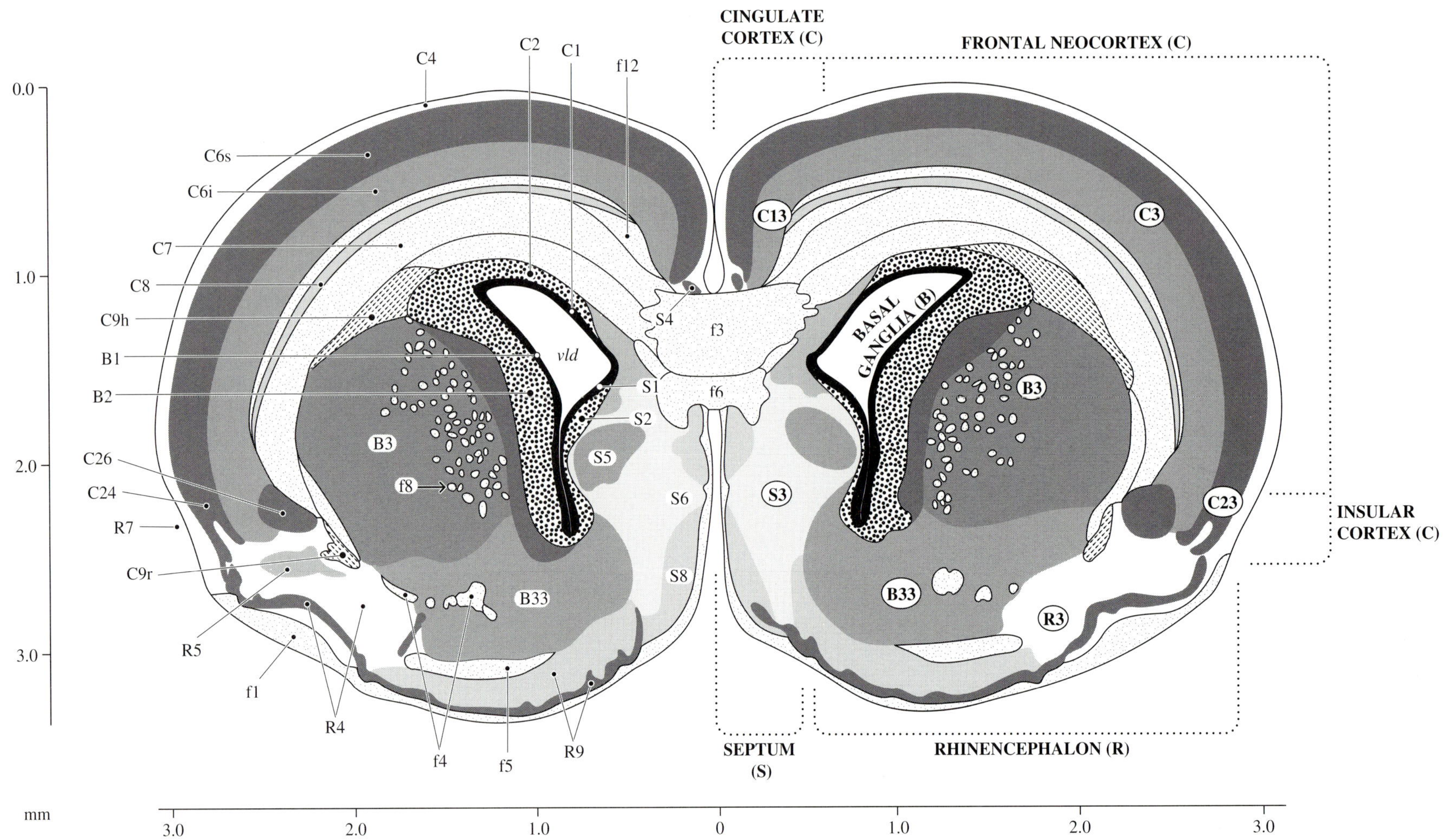

BASAL GANGLIA (B)

B1	striatal neuroepithelium
B2	striatal subventricular zone
B3	striatum (or differentiating field)
B33	nucleus accumbens (or differentiating field)

CEREBRAL CORTEX (C)

C1	neocortical neuroepithelium
C2	neocortical subventricular zone
C3	neocortex (or differentiating field)
C4	molecular layer (or primordial plexiform layer)
C6i	cortical plate, infragranular part
C6s	cortical plate, supragranular part
C7	white matter
C8	subplate
C9h	lateral migratory stream, head
C9r	lateral migratory stream, reservoir
C13	cingulate cortex (or differentiating field)
C23	orbital and insular area (or differentiating field)
C24	insular cortex
C26	claustrum

FIBER TRACTS (f)

f1	lateral olfactory tract
f3	corpus callosum
f4	anterior commissure
f5	medial forebrain bundle
f6	fornix
f8	internal capsule
f12	cingulum bundle

RHINENCEPHALON (R)

R3	rhinencephalon (or differentiating field)
R4	piriform cortex
R5	endopiriform nucleus
R7	rhinal fissure
R9	olfactory tubercle

SEPTAL AREA (S)

S1	septal neuroepithelium
S2	septal subventricular zone
S3	septum (or differentiating field)
S4	indesium griseum
S5	lateral septum
S6	medial septum
S8	diagonal band (vertical limb)

VENTRICLES (v)

vld	lateral ventricle, dorsal

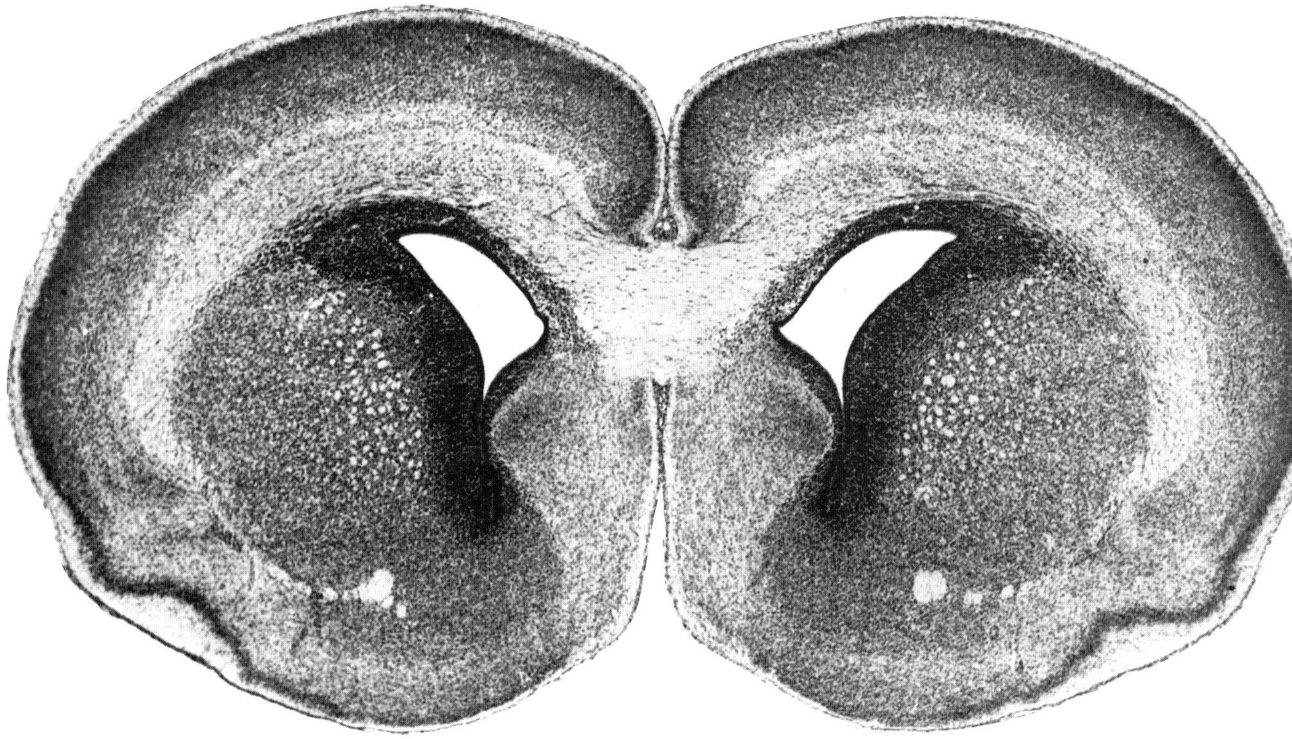

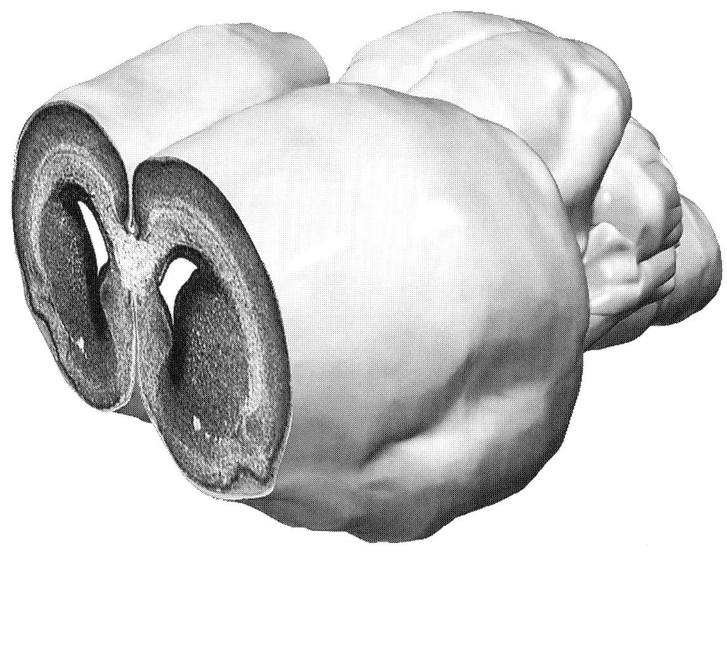

CORONAL FIGURE 8

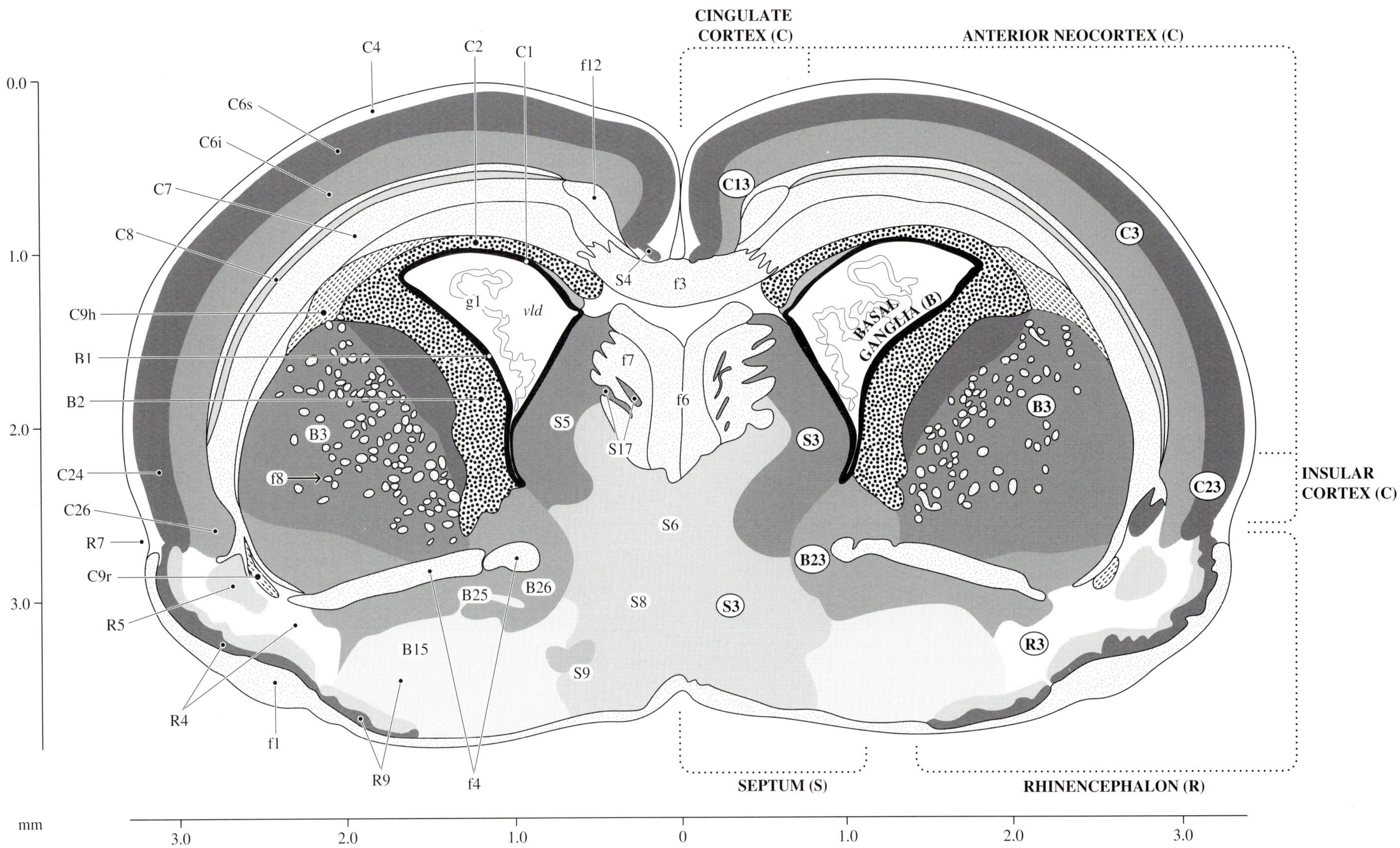

BASAL GANGLIA (B)

B1 striatal neuroepithelium
B2 striatal subventricular zone
B3 striatum (or differentiating field)
B15 substantia innominata
B23 bed nucleus of stria terminalis (or differentiating field)
B25 bed nucleus of stria terminalis, ventral division
B26 bed nucleus of stria terminalis, medial division

CEREBRAL CORTEX (C)

C1 neocortical neuroepithelium
C2 neocortical subventricular zone
C3 neocortex (or differentiating field)
C4 molecular layer (or primordial plexiform layer)
C6i cortical plate, infragranular part
C6s cortical plate, supragranular part
C7 white matter
C8 subplate
C9h lateral migratory stream, head
C9r lateral migratory stream, reservoir
C13 cingulate cortex (or differentiating field)
C23 orbital and insular area (or differentiating field)
C24 insular cortex
C26 claustrum

FIBER TRACTS (f)

f1 lateral olfactory tract
f3 corpus callosum
f4 anterior commissure
f6 fornix
f7 hippocampal commissure
f8 internal capsule
f12 cingulum bundle

GLANDS AND PARANEURAL STRUCTURES (g)

g1 choroid plexus of lateral ventricle

RHINENCEPHALON (R)

R3 rhinencephalon (or differentiating field)
R4 piriform cortex
R5 endopiriform nucleus
R7 rhinal fissure
R9 olfactory tubercle

SEPTAL AREA (S)

S3 septum (or differentiating field)
S4 indesium griseum
S5 lateral septum
S6 medial septum
S8 diagonal band (vertical limb)
S9 diagonal band (horizontal limb)
S17 septofimbrial nucleus

VENTRICLES (v)

vld lateral ventricle, dorsal

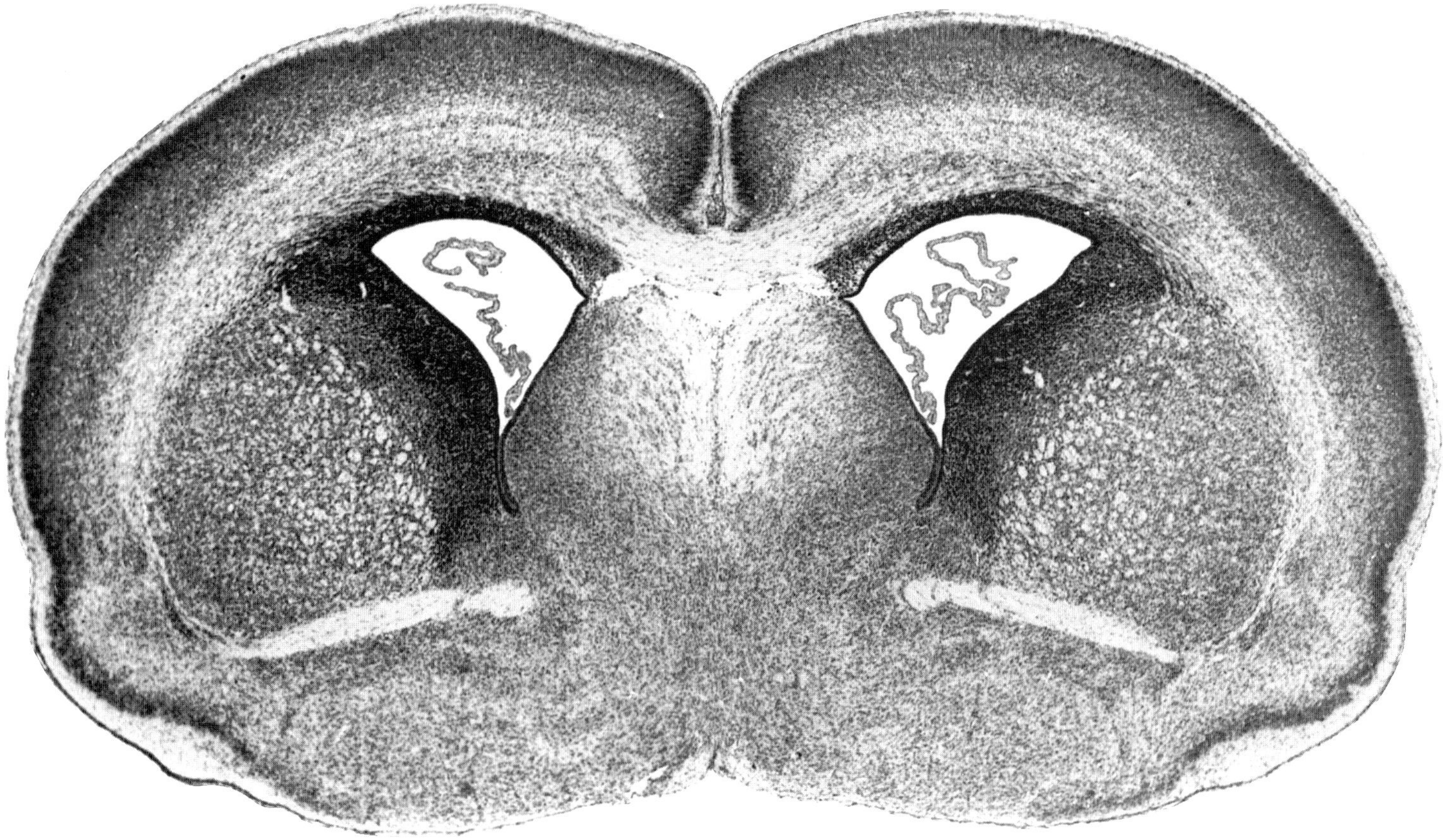

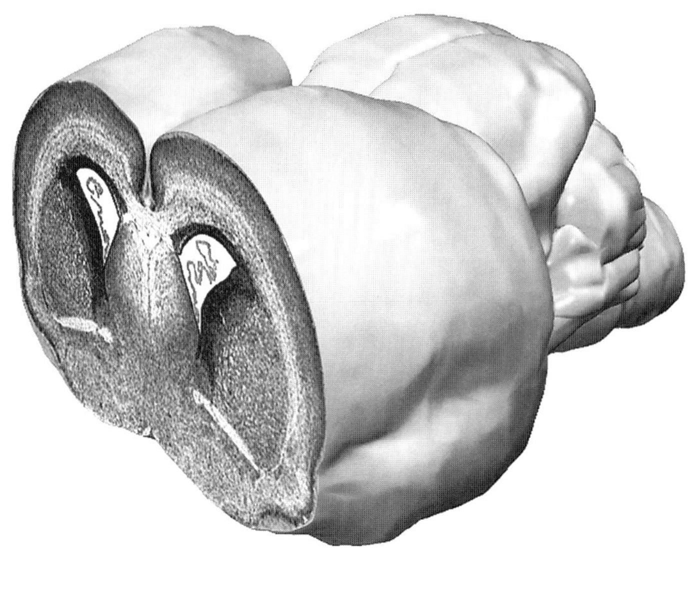

CORONAL FIGURE 9

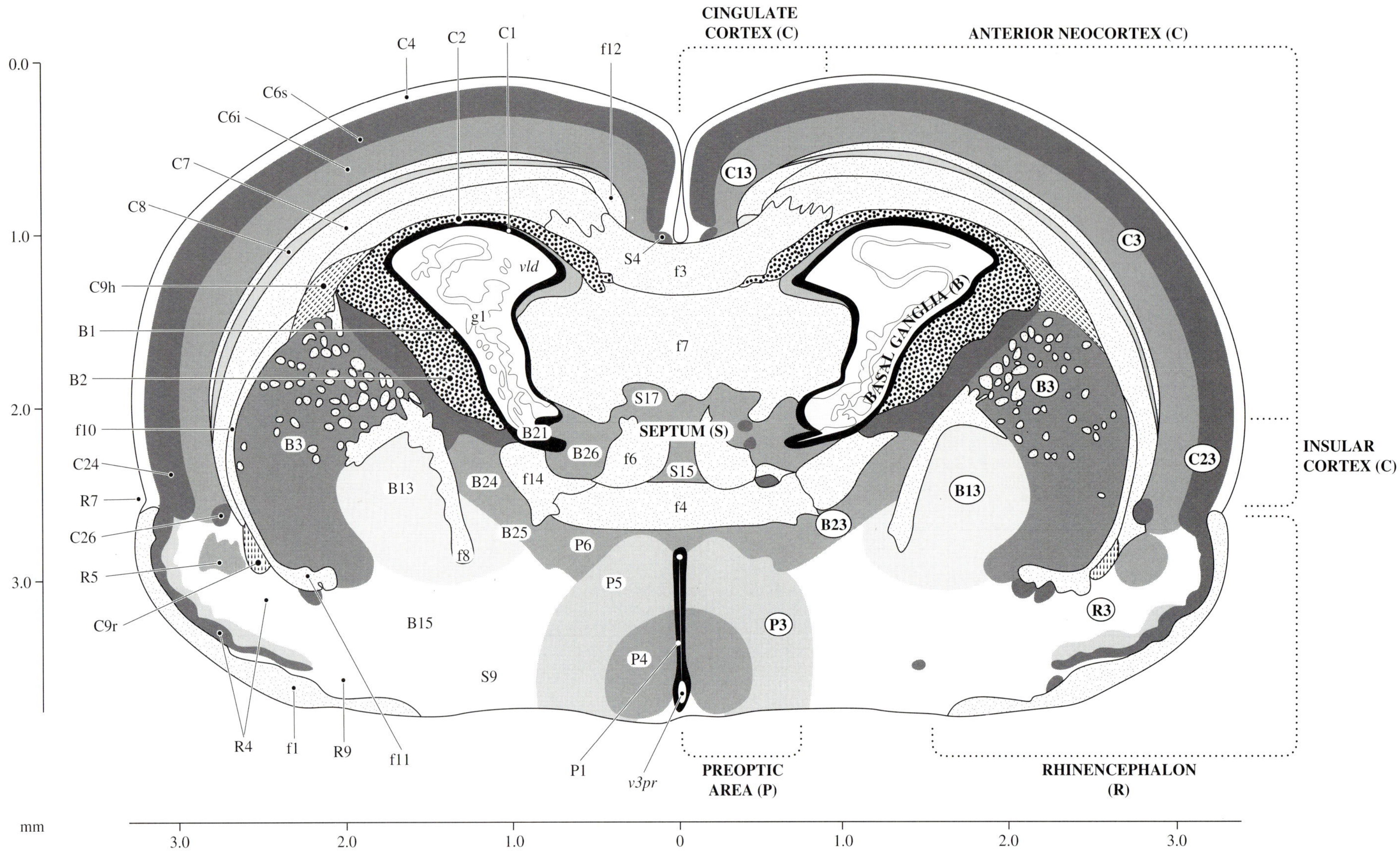

BASAL GANGLIA (B)

- B1 striatal neuroepithelium
- B2 striatal subventricular zone
- **B3** striatum (or differentiating field)
- **B13** pallidum (or differentiating field)
- B15 substantia innominata
- B21 strionuclear neuroepithelium
- **B23** bed nucleus of stria terminalis (or differentiating field)
- B24 bed nucleus of stria terminalis, dorsal division
- B25 bed nucleus of stria terminalis, ventral division
- B26 bed nucleus of stria terminalis, medial division

CEREBRAL CORTEX (C)

- C1 neocortical neuroepithelium
- C2 neocortical subventricular zone
- **C3** neocortex (or differentiating field)
- C4 molecular layer (or primordial plexiform layer)
- C6i cortical plate, infragranular part
- C6s cortical plate, supragranular part
- C7 white matter
- C8 subplate
- C9h lateral migratory stream, head
- C9r lateral migratory stream, reservoir
- **C13** cingulate cortex (or differentiating field)
- **C23** orbital and insular area (or differentiating field)
- C24 insular cortex
- C26 claustrum

FIBER TRACTS (f)

- f1 lateral olfactory tract
- f3 corpus callosum
- f4 anterior commissure
- f6 fornix
- f7 hippocampal commissure
- f8 internal capsule
- f11 external capsule
- f12 cingulum bundle
- f14 stria terminalis

GLANDS AND PARANEURAL STRUCTURES (g)

- g1 choroid plexus of lateral ventricle

PREOPTIC AREA AND OPTIC EPITHELIA (P)

- P1 preoptic neuroepithelium
- **P3** preoptic area (or differentiating field)
- P4 medial preoptic region
- P5 lateral preoptic region
- P6 superior preoptic region

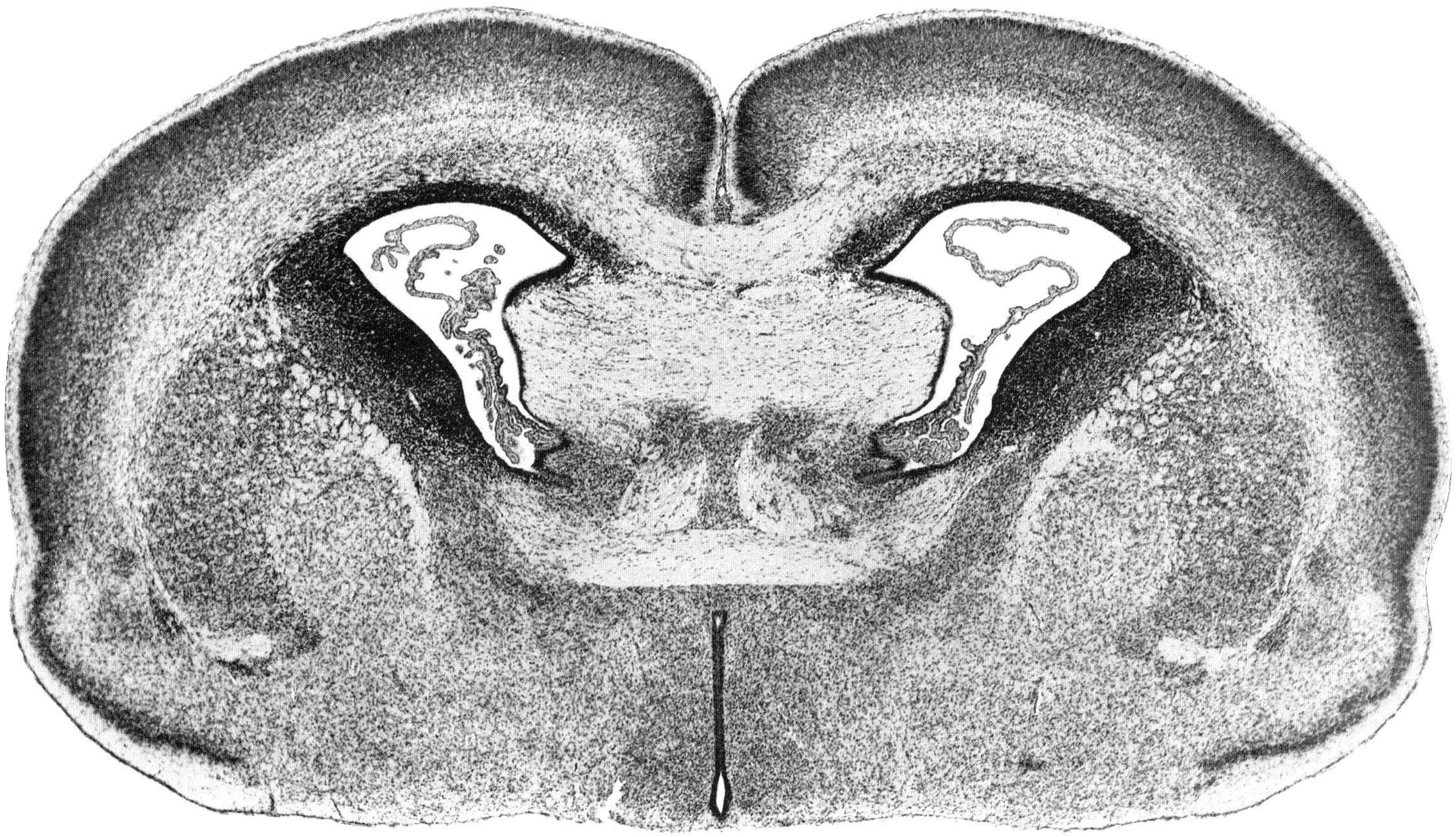

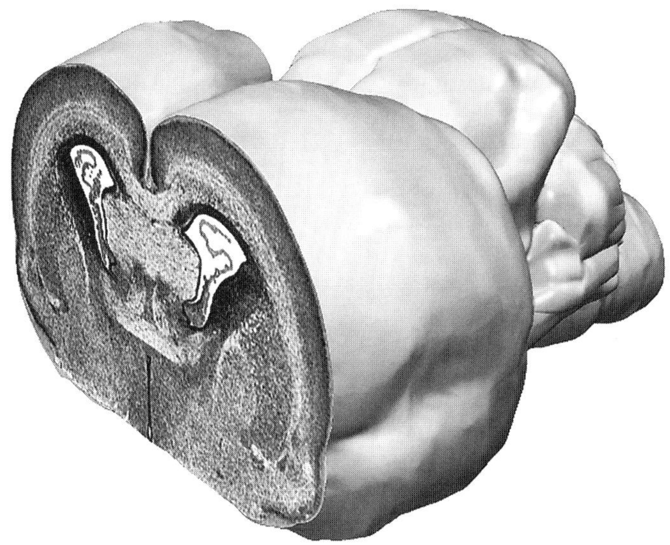

RHINENCEPHALON (R)

R3 rhinencephalon (or differentiating field)
R4 piriform cortex
R5 endopiriform nucleus
R7 rhinal fissure
R9 olfactory tubercle

SEPTAL AREA (S)

S4 indesium griseum
S9 diagonal band (horizontal limb)
S15 triangular septal nucleus
S17 septofimbrial nucleus

VENTRICLES (v)

vld lateral ventricle, dorsal
v3pr third ventricle, preoptic area

CORONAL FIGURE 10

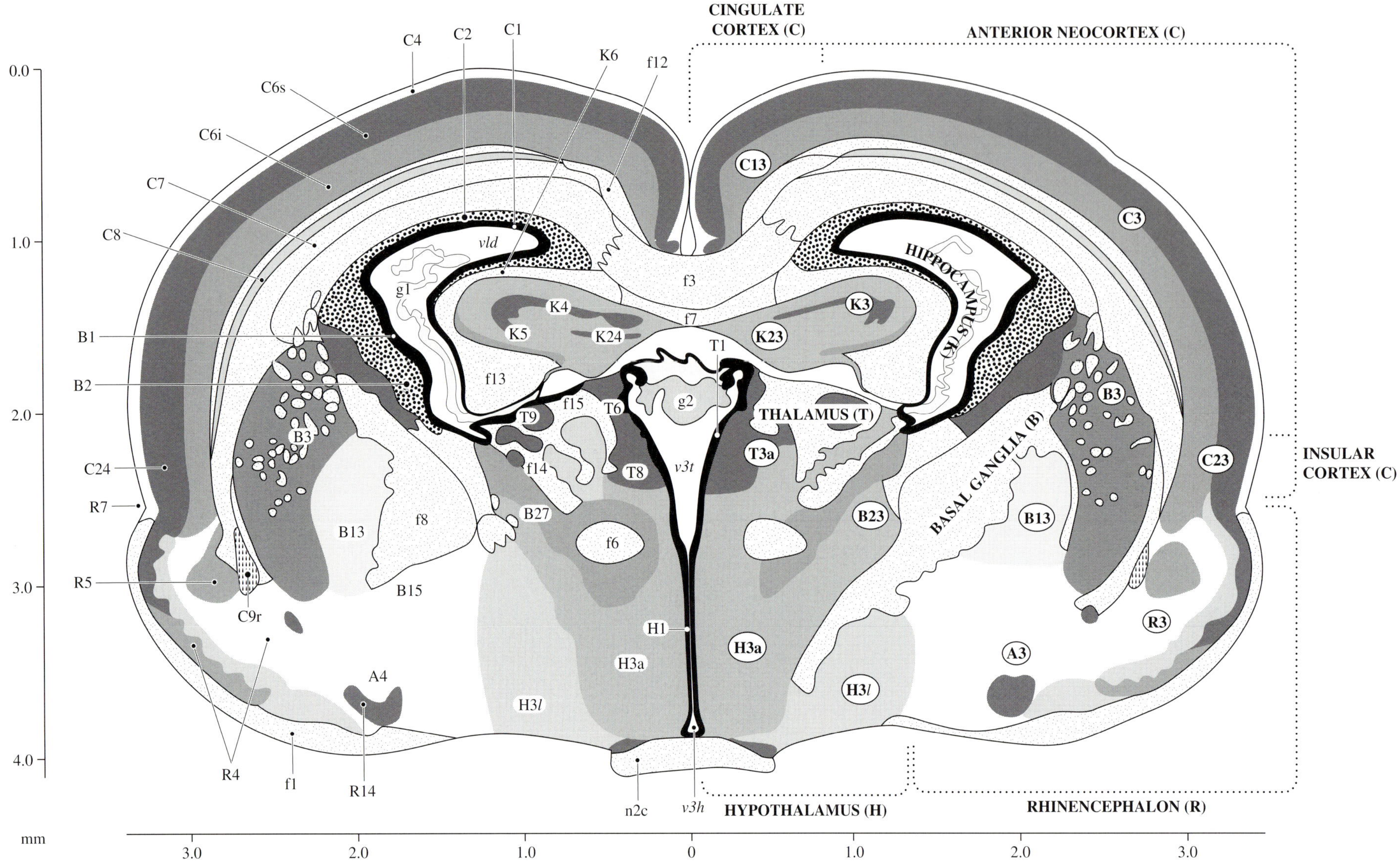

AMYGDALA (A)

A3 amygdala (or differentiating field)
A4 anterior amygdaloid area

BASAL GANGLIA (B)

B1 striatal neuroepithelium
B2 striatal subventricular zone
B3 striatum (or differentiating field)
B13 pallidum (or differentiating field)
B15 substantia innominata
B23 bed nucleus of stria terminalis (or differentiating field)
B27 bed nucleus of stria terminalis, intermediate division

CEREBRAL CORTEX (C)

C1 neocortical neuroepithelium
C2 neocortical subventricular zone
C3 neocortex (or differentiating field)
C4 molecular layer (or primordial plexiform layer)
C6i cortical plate, infragranular part
C6s cortical plate, supragranular part
C7 white matter
C8 subplate
C9r lateral migratory stream, reservoir
C13 cingulate cortex (or differentiating field)
C23 orbital and insular area (or differentiating field)
C24 insular cortex

FIBER TRACTS (f)

f1 lateral olfactory tract
f3 corpus callosum
f6 fornix
f7 hippocampal commissure
f8 internal capsule
f12 cingulum bundle
f13 fimbria
f14 stria terminalis
f15 stria medullaris

GLANDS AND PARANEURAL STRUCTURES (g)

g1 choroid plexus of lateral ventricle
g2 subfornical organ

HYPOTHALAMUS (H)

H1 hypothalamic neuroepithelium
H3a anterior hypothalamus (or differentiating field)
H3*l* lateral hypothalamus (or differentiating field)

HIPPOCAMPAL FORMATION (K)

K3 hippocampus (or differentiating field)
K4 CA1 field
K5 CA3 field
K6 alveus
K23 dentate gyrus (or differentiating field)
K24 dentate granular layer

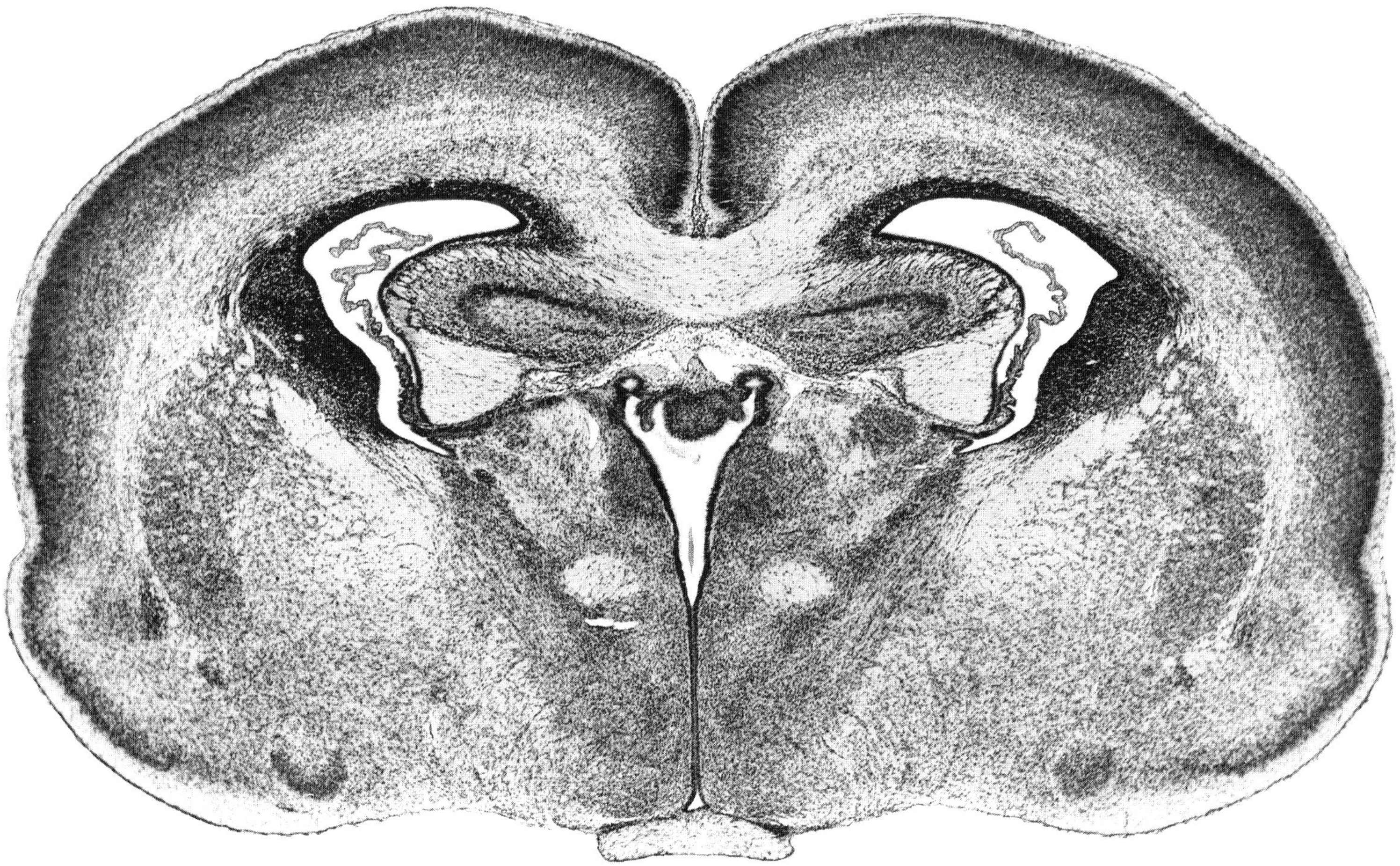

NERVES AND NUCLEI, CRANIAL (n)

n2c optic chiasma

RHINENCEPHALON (R)

R3 rhinencephalon (or differentiating field)
R4 piriform cortex
R5 endopiriform nucleus
R7 rhinal fissure
R14 nucleus of the lateral olfactory tract

THALAMUS (T)

T1 thalamic neuroepithelium
T3a anterior thalamus (or differentiating field)
T6 paraventricular nucleus
T8 anteromedial nucleus
T9 anteroventral nucleus

VENTRICLES (v)

vld lateral ventricle, dorsal
v3h third ventricle, hypothalamic
v3t third ventricle, thalamic

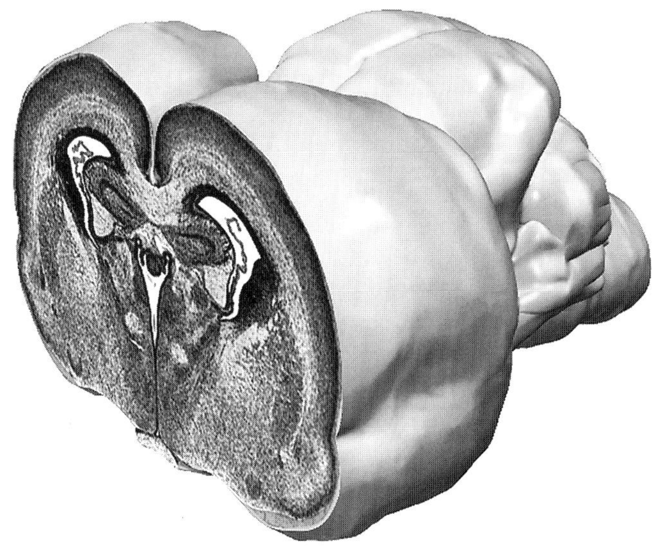

CORONAL FIGURE 11

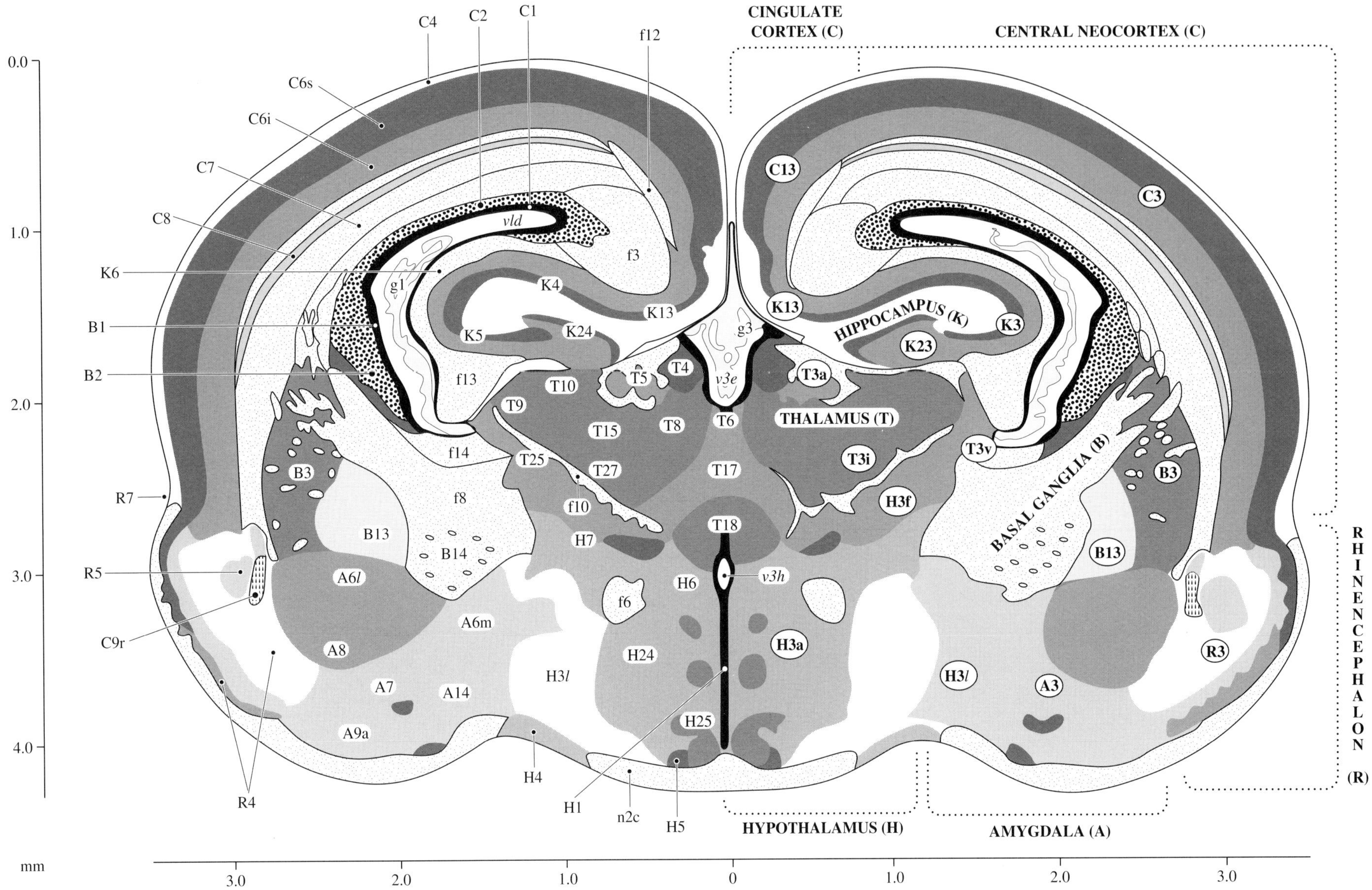

AMYGDALA (A)

A3 amygdala (or differentiating field)
A6*l* central amygdaloid nucleus, lateral
A6m central amygdaloid nucleus, medial
A7 basomedial amygdaloid nucleus
A8 basolateral amygdaloid nucleus
A9a cortical amygdaloid nucleus, anterior
A14 medial amygdaloid nucleus

BASAL GANGLIA (B)

B1 striatal neuroepithelium
B2 striatal subventricular zone
B3 striatum (or differentiating field)
B13 pallidum (or differentiating field)
B14 entopeduncular nucleus

CEREBRAL CORTEX (C)

C1 neocortical neuroepithelium
C2 neocortical subventricular zone
C3 neocortex (or differentiating field)
C4 molecular layer (or primordial plexiform layer)
C6i cortical plate, infragranular part
C6s cortical plate, supragranular part
C7 white matter
C8 subplate
C9r lateral migratory stream, reservoir
C13 cingulate cortex (or differentiating field)

FIBER TRACTS (f)

f3 corpus callosum
f6 fornix
f8 internal capsule
f10 external medullary lamina
f12 cingulum bundle
f13 fimbria
f14 stria terminalis

GLANDS AND PARANEURAL STRUCTURES (g)

g1 choroid plexus of lateral ventricle
g3 choroid plexus of third ventricle

HYPOTHALAMUS (H)

H1 hypothalamic neuroepithelium
H3a anterior hypothalamus (or differentiating field)
H3f fields of Forel (or differentiating field)
H3*l* lateral hypothalamus (or differentiating field)
H4 supraoptic nucleus
H5 suprachiasmatic nucleus
H6 paraventricular nucleus
H7 zona incerta
H24 anterodorsal nucleus
H25 anteroventral nucleus

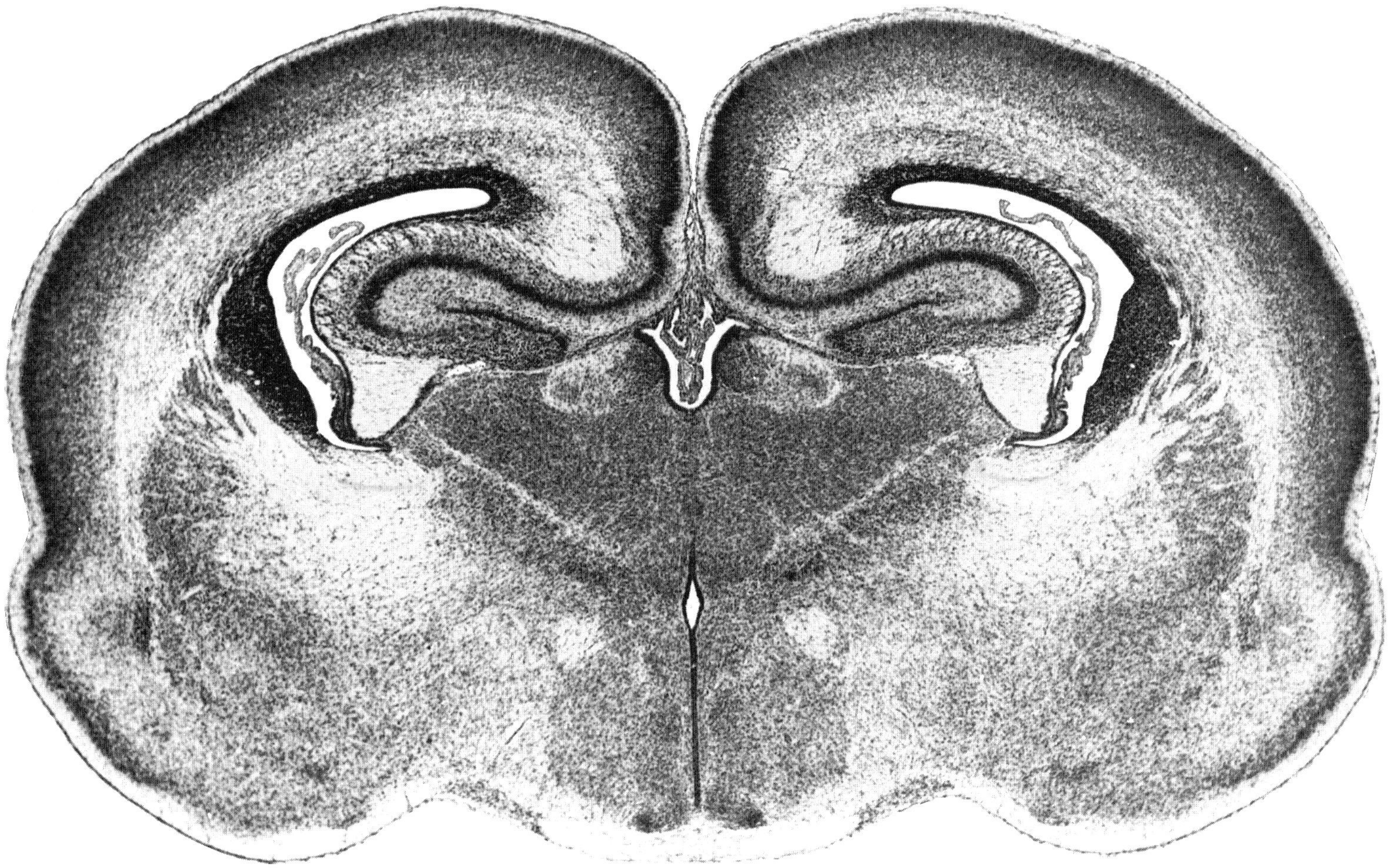

HIPPOCAMPAL FORMATION (K)

K3 hippocampus (or differentiating field)
K4 CA1 field
K5 CA3 field
K6 alveus
K13 subicular area (or differentiating field)
K23 dentate gyrus (or differentiating field)
K24 dentate granular layer

NERVES AND NUCLEI, CRANIAL (n)

n2c optic chiasma

RHINENCEPHALON (R)

R3 rhinencephalon (or differentiating field)
R4 piriform cortex
R5 endopiriform nucleus
R7 rhinal fissure

THALAMUS (T)

T3a anterior thalamus (or differentiating field)
T3i intermediate thalamus (or differentiating field)
T3v ventral thalamus (or differentiating field)
T4 medial habenular nucleus
T5 lateral habenular nucleus
T6 paraventricular nucleus
T8 anteromedial nucleus
T9 anteroventral nucleus
T10 anterodorsal nucleus
T15 centrolateral nucleus
T17 rhomboid nucleus
T18 reuniens nucleus
T25 reticular nucleus
T27 ventrolateral nuclear complex

VENTRICLES (v)

vld lateral ventricle, dorsal
v3e third ventricle, epithalamic
v3h third ventricle, hypothalamic

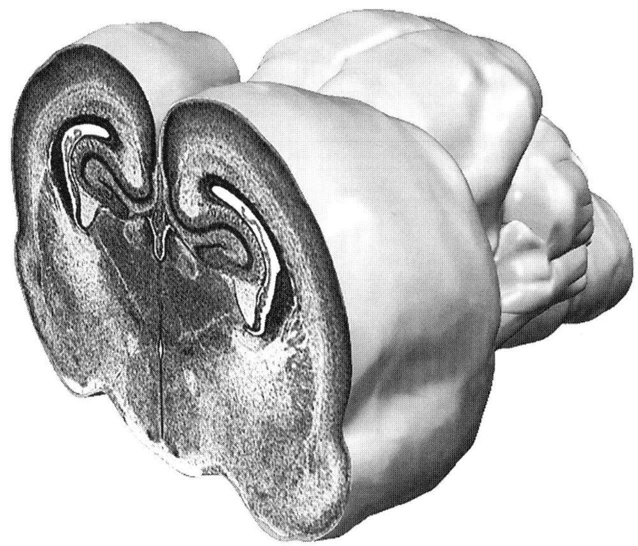

CORONAL FIGURE 12

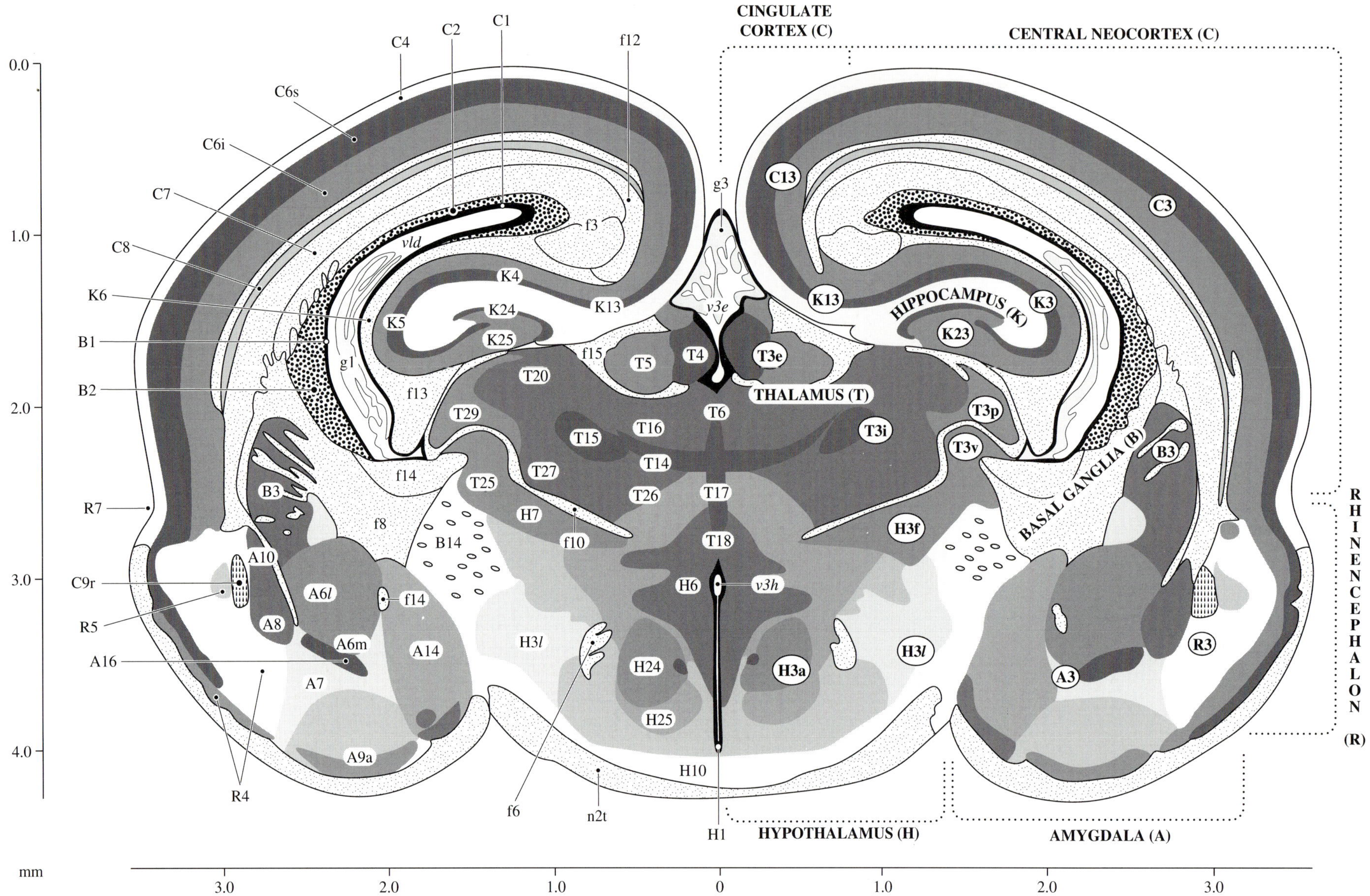

AMYGDALA (A)

A3 amygdala (or differentiating field)
A6*l* central amygdaloid nucleus, lateral
A6m central amygdaloid nucleus, medial
A7 basomedial amygdaloid nucleus
A8 basolateral amygdaloid nucleus
A9a cortical amygdaloid nucleus, anterior
A10 lateral amygdaloid nucleus
A14 medial amygdaloid nucleus
A16 intercalated masses

BASAL GANGLIA (B)

B1 striatal neuroepithelium
B2 striatal subventricular zone
B3 striatum (or differentiating field)
B14 entopeduncular nucleus

CEREBRAL CORTEX (C)

C1 neocortical neuroepithelium
C2 neocortical subventricular zone
C3 neocortex (or differentiating field)
C4 molecular layer (or primordial plexiform layer)
C6i cortical plate, infragranular part
C6s cortical plate, supragranular part
C7 white matter
C8 subplate
C9r lateral migratory stream, reservoir
C13 cingulate cortex (or differentiating field)

FIBER TRACTS (f)

f3 corpus callosum
f6 fornix
f8 internal capsule
f10 external medullary lamina
f12 cingulum bundle
f13 fimbria
f14 stria terminalis
f15 stria medullaris

GLANDS AND PARANEURAL STRUCTURES (g)

g1 choroid plexus of lateral ventricle
g3 choroid plexus of third ventricle

HYPOTHALAMUS (H)

H1 hypothalamic neuroepithelium
H3a anterior hypothalamus (or differentiating field)
H3f fields of Forel (or differentiating field)
H3*l* lateral hypothalamus (or differentiating field)
H6 paraventricular nucleus
H7 zona incerta
H10 anterobasal nucleus
H24 anterodorsal nucleus
H25 anteroventral nucleus

HIPPOCAMPAL FORMATION (K)

K3 hippocampus (or differentiating field)
K4 CA1 field
K5 CA3 field
K6 alveus

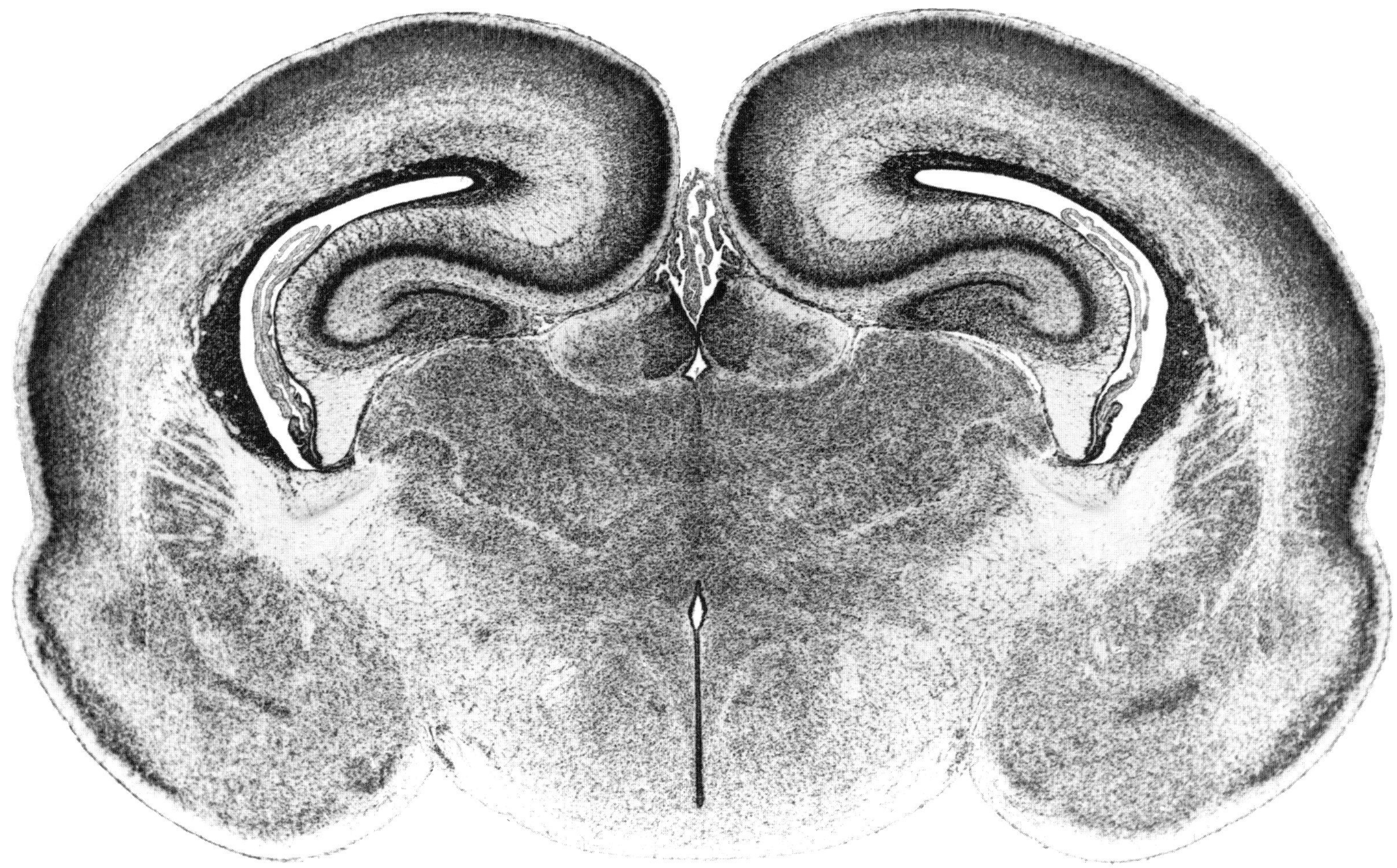

K13 subicular area (or differentiating field)
K23 dentate gyrus (or differentiating field)
K24 dentate granular layer
K25 dentate polymorph cell layer

NERVES AND NUCLEI, CRANIAL (n)

n2t optic tract

RHINENCEPHALON (R)

R3 rhinencephalon (or differentiating field)
R4 piriform cortex
R5 endopiriform nucleus
R7 rhinal fissure

THALAMUS (T)

T3e epithalamus (or differentiating field)
T3i intermediate thalamus (or differentiating field)
T3p posterior thalamus (or differentiating field)
T3v ventral thalamus (or differentiating field)
T4 medial habenular nucleus
T5 lateral habenular nucleus
T6 paraventricular nucleus
T14 centromedian nucleus
T15 centrolateral nucleus
T16 mediodorsal nucleus
T17 rhomboid nucleus
T18 reuniens nucleus
T20 laterodorsal nucleus
T25 reticular nucleus
T26 ventral medial nucleus
T27 ventrolateral nuclear complex
T29 dorsal lateral geniculate nucleus

VENTRICLES (v)

vld lateral ventricle, dorsal
v3e third ventricle, epithalamic
v3h third ventricle, hypothalamic

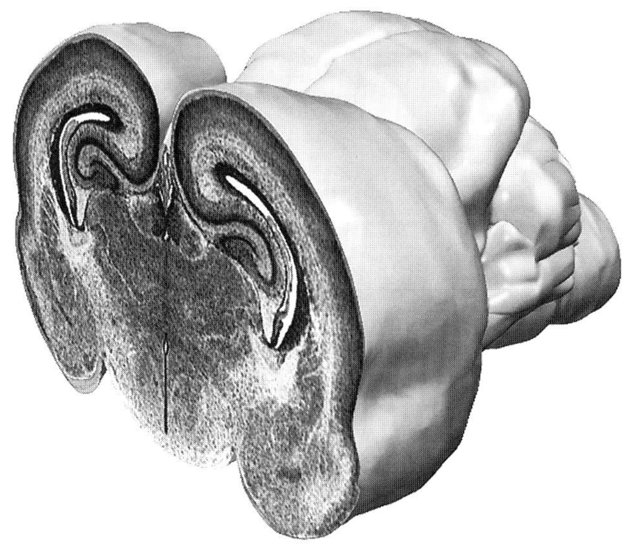

CORONAL FIGURE 13

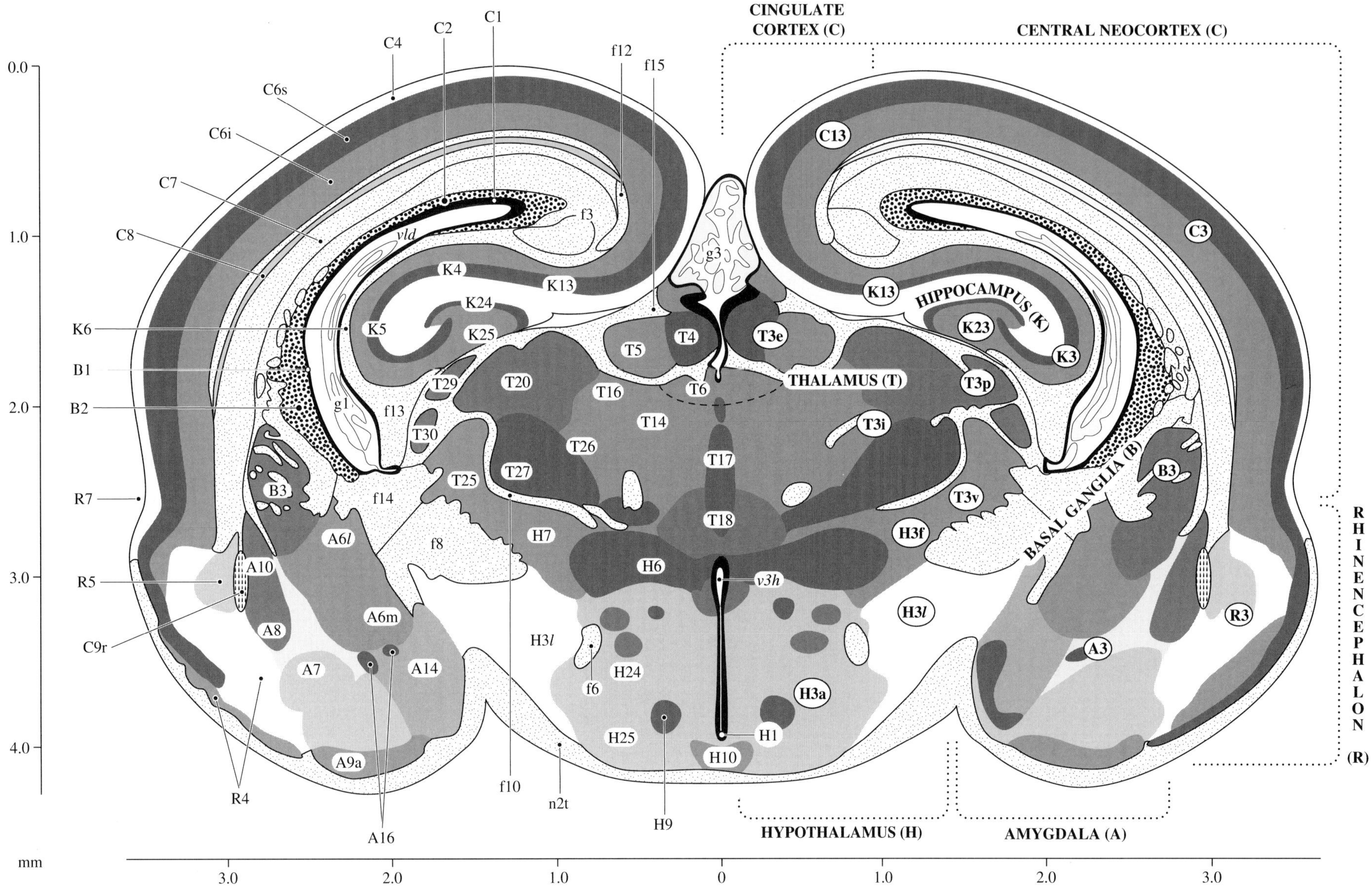

AMYGDALA (A)

- **A3** amygdala (or differentiating field)
- A6*l* central amygdaloid nucleus, lateral
- A6m central amygdaloid nucleus, medial
- A7 basomedial amygdaloid nucleus
- A8 basolateral amygdaloid nucleus
- A9a cortical amygdaloid nucleus, anterior
- A10 lateral amygdaloid nucleus
- A14 medial amygdaloid nucleus
- A16 intercalated masses

BASAL GANGLIA (B)

- B1 striatal neuroepithelium
- B2 striatal subventricular zone
- **B3** striatum (or differentiating field)

CEREBRAL CORTEX (C)

- C1 neocortical neuroepithelium
- C2 neocortical subventricular zone
- **C3** neocortex (or differentiating field)
- C4 molecular layer (or primordial plexiform layer)
- C6i cortical plate, infragranular part
- C6s cortical plate, supragranular part
- C7 white matter
- C8 subplate
- C9r lateral migratory stream, reservoir
- **C13** cingulate cortex (or differentiating field)

FIBER TRACTS (f)

- f3 corpus callosum
- f6 fornix
- f8 internal capsule
- f10 external medullary lamina
- f12 cingulum bundle
- f13 fimbria
- f14 stria terminalis
- f15 stria medullaris
- f16 medial lemniscus

GLANDS AND PARANEURAL STRUCTURES (g)

- g1 choroid plexus of lateral ventricle
- g3 choroid plexus of third ventricle

HYPOTHALAMUS (H)

- H1 hypothalamic neuroepithelium
- **H3a** anterior hypothalamus (or differentiating field)
- **H3f** fields of Forel (or differentiating field)
- **H3*l*** lateral hypothalamus (or differentiating field)
- H6 paraventricular nucleus
- H7 zona incerta
- H9 ventromedial nucleus
- H10 anterobasal nucleus
- H24 anterodorsal nucleus
- H25 anteroventral nucleus

HIPPOCAMPAL FORMATION (K)

- **K3** hippocampus (or differentiating field)
- K4 CA1 field
- K5 CA3 field
- K6 alveus
- **K13** subicular area (or differentiating field)
- **K23** dentate gyrus (or differentiating field)

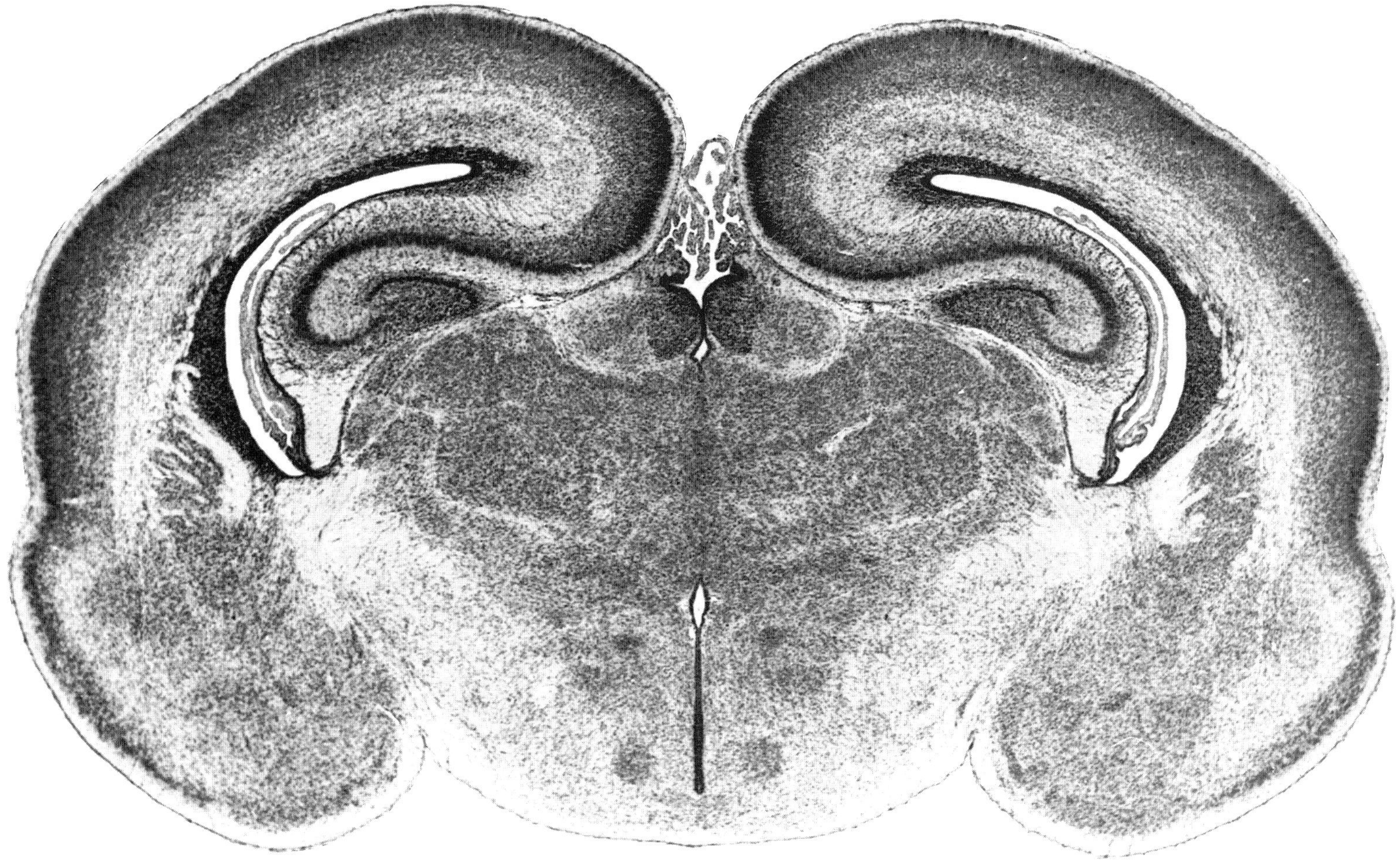

K24 dentate granular layer
K25 dentate polymorph layer

NERVES AND NUCLEI, CRANIAL (n)

n2t optic tract

RHINENCEPHALON (R)

R3 rhinencephalon (or differentiating field)
R4 piriform cortex
R5 endopiriform nucleus
R7 rhinal fissure

THALAMUS (T)

T3e epithalamus (or differentiating field)
T3i intermediate thalamus (or differentiating field)
T3p posterior thalamus (or differentiating field)
T3v ventral thalamus (or differentiating field)
T4 medial habenular nucleus
T5 lateral habenular nucleus
T6 paraventricular nucleus
T14 centromedian nucleus
T16 mediodorsal nucleus
T17 rhomboid nucleus
T18 reuniens nucleus
T20 laterodorsal nucleus
T25 reticular nucleus
T26 ventral medial nucleus
T27 ventrolateral nuclear complex
T29 dorsal lateral geniculate nucleus
T30 ventral lateral geniculate nucleus

VENTRICLES (v)

vld lateral ventricle, dorsal
v3h third ventricle, hypothalamic

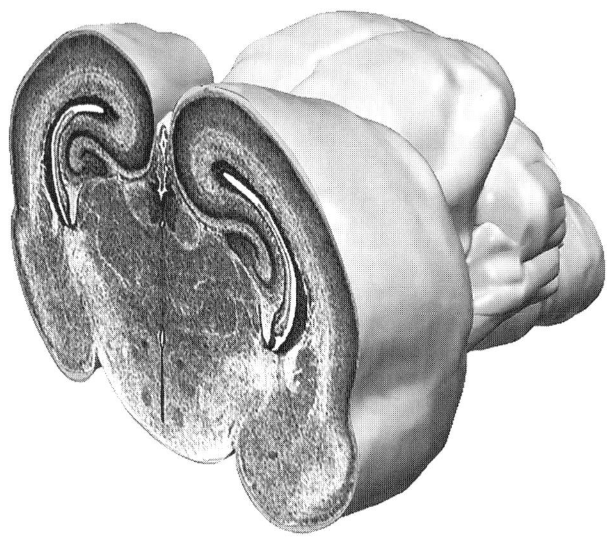

CORONAL FIGURE 14

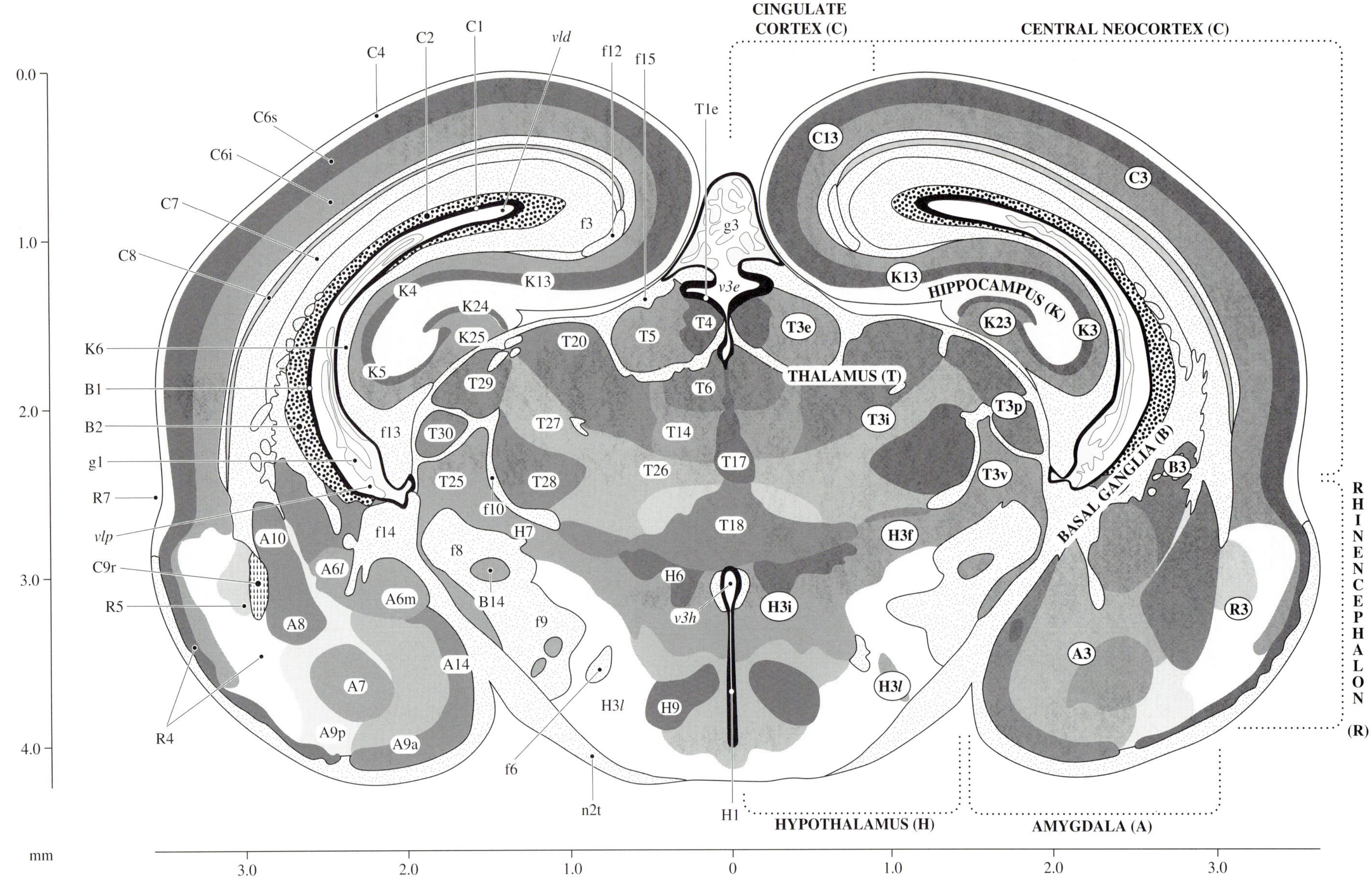

AMYGDALA (A)

A3 amygdala (or differentiating field)
A6*l* central amygdaloid nucleus, lateral
A6m central amygdaloid nucleus, medial
A7 basomedial amygdaloid nucleus
A8 basolateral amygdaloid nucleus
A9a cortical amygdaloid nucleus, anterior
A9p cortical amygdaloid nucleus, posterior
A10 lateral amygdaloid nucleus
A14 medial amygdaloid nucleus

BASAL GANGLIA (B)

B1 striatal neuroepithelium
B2 striatal subventricular zone
B3 striatum (or differentiating field)
B14 entopeduncular nucleus

CEREBRAL CORTEX (C)

C1 neocortical neuroepithelium
C2 neocortical subventricular zone
C3 neocortex (or differentiating field)
C4 molecular layer (or primordial plexiform layer)
C6i cortical plate, infragranular part
C6s cortical plate, supragranular part
C7 white matter
C8 subplate
C9r lateral migratory stream, reservoir
C13 cingulate cortex (or differentiating field)

FIBER TRACTS (f)

f3 corpus callosum
f6 fornix
f8 internal capsule
f9 cerebral peduncle
f10 external medullary lamina
f12 cingulum bundle
f13 fimbria
f14 stria terminalis
f15 stria medullaris

GLANDS AND PARANEURAL STRUCTURES (g)

g1 choroid plexus of lateral ventricle
g3 choroid plexus of third ventricle

HYPOTHALAMUS (H)

H1 hypothalamic neuroepithelium
H3f fields of Forel (or differentiating field)
H3i intermediate hypothalamus (or differentiating field)
H3*l* lateral hypothalamus (or differentiating field)
H6 paraventricular nucleus
H7 zona incerta
H9 ventromedial nucleus

HIPPOCAMPAL FORMATION (K)

K3 hippocampus (or differentiating field)
K4 CA1 field
K5 CA3 field
K6 alveus
K13 subicular area (or differentiating field)
K23 dentate gyrus (or differentiating field)
K24 dentate granular layer
K25 dentate polymorph cell layer

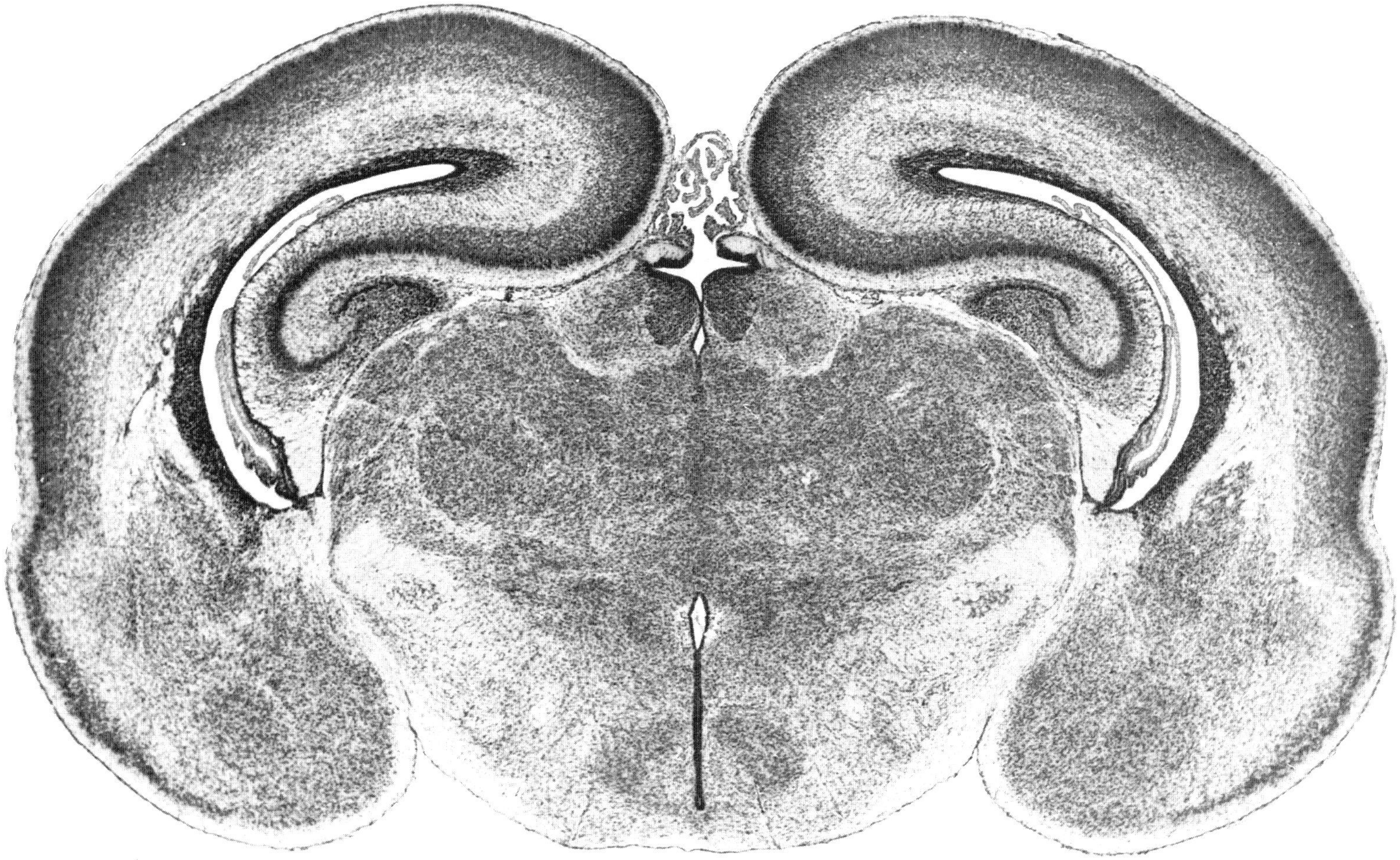

NERVES AND NUCLEI, CRANIAL (n)

n2t optic tract

RHINENCEPHALON (R)

R3 rhinencephalon (or differentiating field)
R4 piriform cortex
R5 endopiriform nucleus
R7 rhinal fissure

THALAMUS (T)

T1e epithalamic neuroepithelium
T3e epithalamus (or differentiating field)
T3i intermediate thalamus (or differentiating field)
T3p posterior thalamus (or differentiating field)
T3v ventral thalamus (or differentiating field)
T4 medial habenular nucleus
T5 lateral habenular nucleus
T6 paraventricular nucleus
T14 centromedian nucleus
T17 rhomboid nucleus
T18 reuniens nucleus
T20 laterodorsal nucleus
T25 reticular nucleus
T26 ventral medial nucleus
T27 ventrolateral nuclear complex
T28 ventrobasal nuclear complex
T29 dorsal lateral geniculate nucleus
T30 ventral lateral geniculate nucleus

VENTRICLES (v)

vld lateral ventricle, dorsal
vlp lateral ventricle, posterior horn
v3e third ventricle, epithalamic
v3h third ventricle, hypothalamic

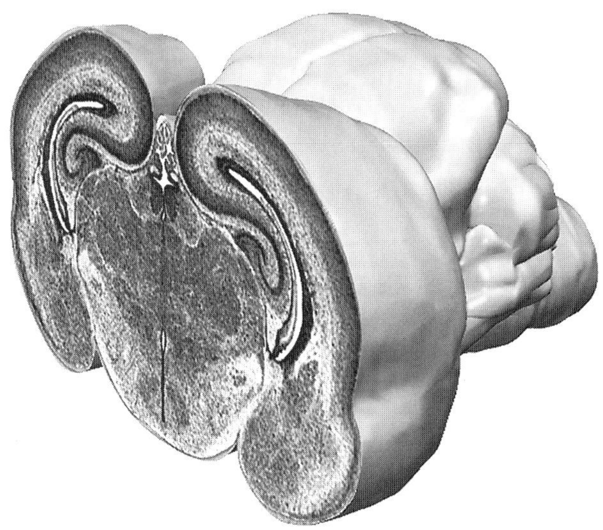

CORONAL FIGURE 15

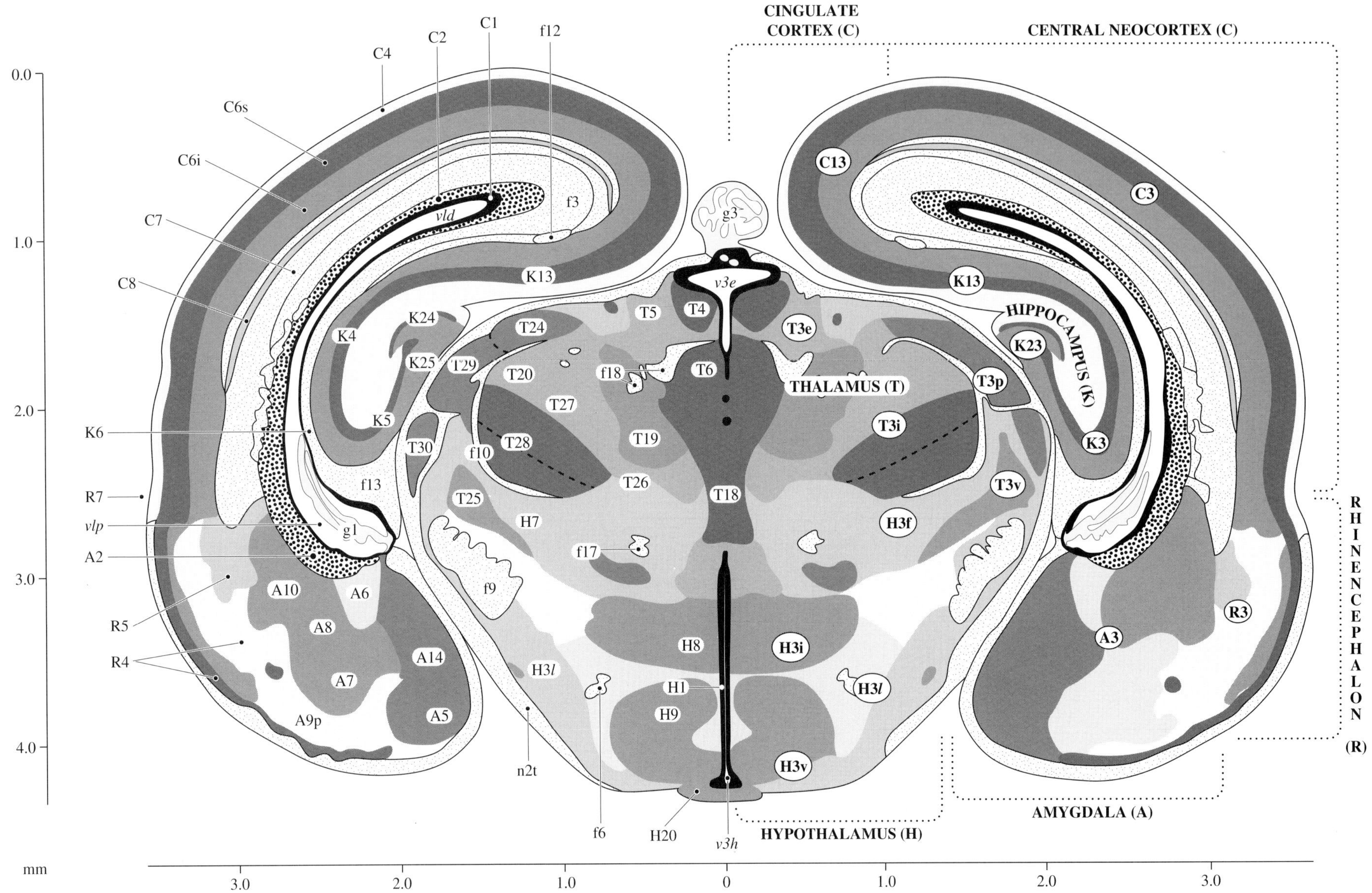

AMYGDALA (A)

A2 amygdaloid subventricular zone
A3 amygdala (or differentiating field)
A5 amygdala-hippocampal area
A6 central amygdaloid nucleus
A7 basomedial amygdaloid nucleus
A8 basolateral amygdaloid nucleus
A9p cortical amygdaloid nucleus, posterior
A10 lateral amygdaloid nucleus
A14 medial amygdaloid nucleus

CEREBRAL CORTEX (C)

C1 neocortical neuroepithelium
C2 neocortical subventricular zone
C3 neocortex (or differentiating field)
C4 molecular layer (or primordial plexiform layer)
C6i cortical plate, infragranular part
C6s cortical plate, supragranular part
C7 white matter
C8 subplate
C13 cingulate cortex (or differentiating field)

FIBER TRACTS (f)

f3 corpus callosum
f6 fornix
f9 cerebral peduncle
f10 external medullary lamina
f12 cingulum bundle
f13 fimbria
f17 mammillothalamic tract
f18 habenulopeduncular tract

GLANDS AND PARANEURAL STRUCTURES (g)

g1 choroid plexus of lateral ventricle
g3 choroid plexus of third ventricle

HYPOTHALAMUS (H)

H1 hypothalamic neuroepithelium
H3f fields of Forel (or differentiating field)
H3i intermediate hypothalamus (or differentiating field)
H3*l* lateral hypothalamus (or differentiating field)
H3v ventral hypothalamus (or differentiating field)
H7 zona incerta
H8 dorsomedial nucleus
H9 ventromedial nucleus
H20 arcuate nucleus

HIPPOCAMPAL FORMATION (K)

K3 hippocampus (or differentiating field)
K4 CA1 field
K5 CA3 field
K6 alveus
K13 subicular area (or differentiating field)
K23 dentate gyrus (or differentiating field)
K24 dentate granular layer
K25 dentate polymorph cell layer

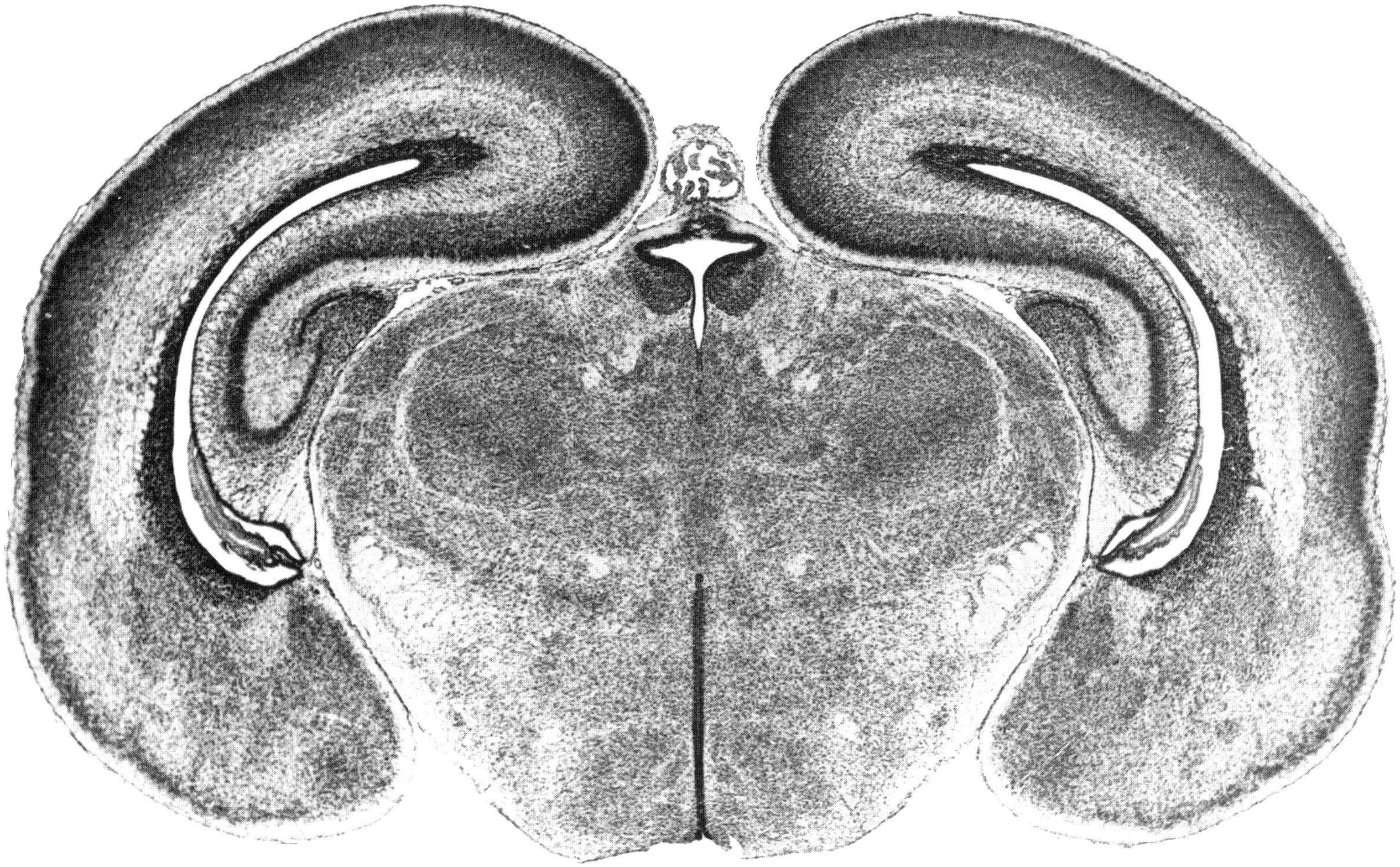

NERVES AND NUCLEI, CRANIAL (n)

n2t optic tract

RHINENCEPHALON (R)

R3 rhinencephalon (or differentiating field)
R4 piriform cortex
R5 endopiriform nucleus
R7 rhinal fissure

THALAMUS (T)

T3e epithalamus (or differentiating field)
T3i intermediate thalamus (or differentiating field)
T3p posterior thalamus (or differentiating field)
T3v ventral thalamus (or differentiating field)
T4 medial habenular nucleus
T5 lateral habenular nucleus
T6 paraventricular nucleus
T18 reuniens nucleus
T19 parafascicular nucleus
T20 laterodorsal nucleus
T24 lateral posterior nucleus
T25 reticular nucleus
T26 ventral medial nucleus
T27 ventrolateral nuclear complex
T28 ventrobasal nuclear complex
T29 dorsal lateral geniculate nucleus
T30 ventral lateral geniculate nucleus

VENTRICLES (v)

vld lateral ventricle, dorsal
vlp lateral ventricle, posterior horn
v3e third ventricle, epithalamic
v3h third ventricle, hypothalamic

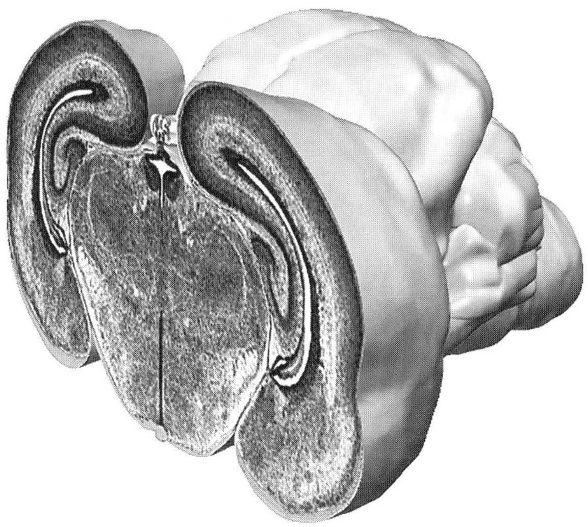

CORONAL FIGURE 16

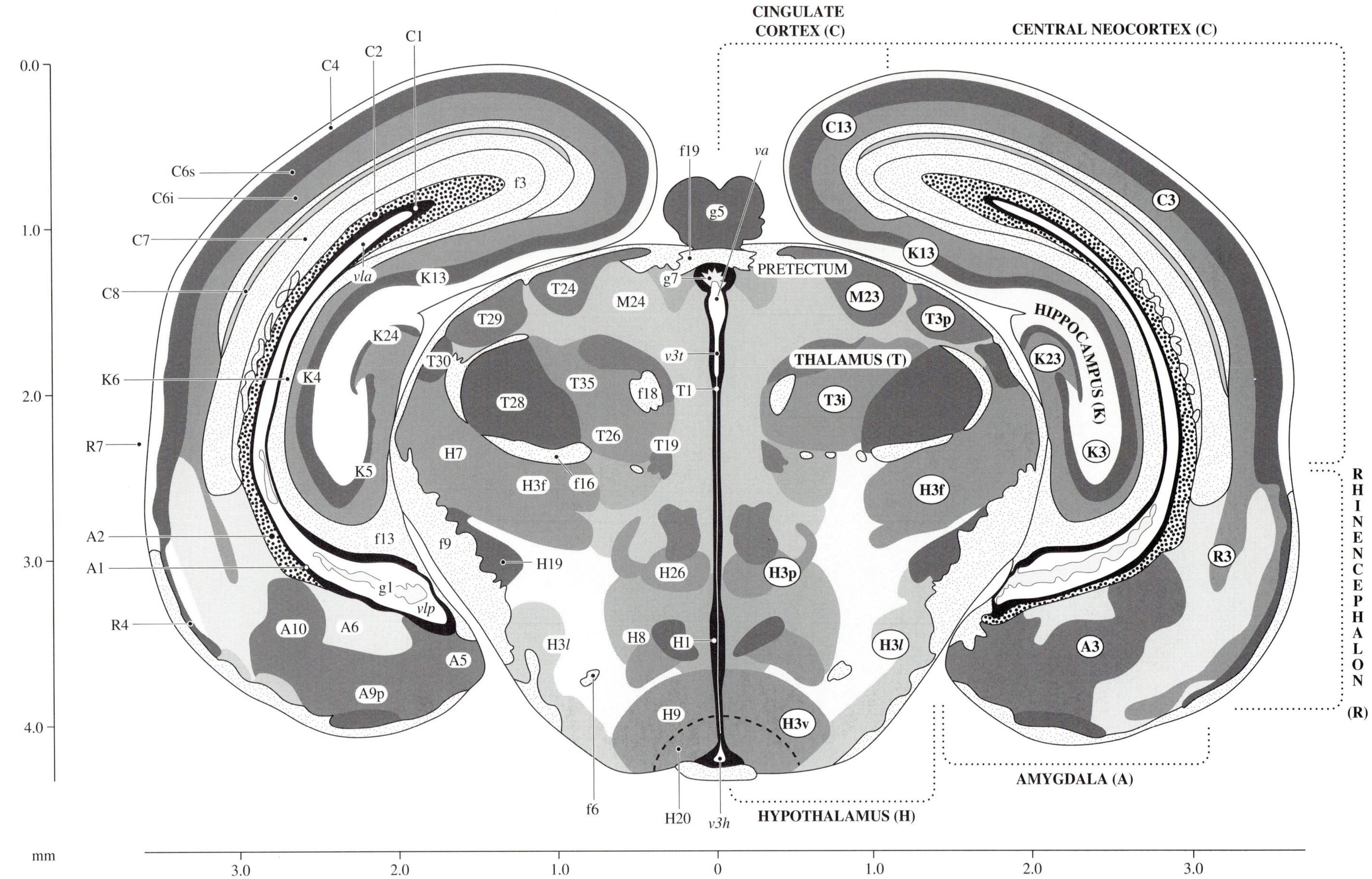

AMYGDALA (A)

A1	amygdaloid neuroepithelium
A2	amygdaloid subventricular zone
A3	amygdala (or differentiating field)
A5	amygdala-hippocampal area
A6	central amygdaloid nucleus
A9p	cortical amygdaloid nucleus, posterior
A10	lateral amygdaloid nucleus

CEREBRAL CORTEX (C)

C1	neocortical neuroepithelium
C2	neocortical subventricular zone
C3	neocortex (or differentiating field)
C4	molecular layer (or primordial plexiform layer)
C6i	cortical plate, infragranular part
C6s	cortical plate, supragranular part
C7	white matter
C8	subplate
C13	cingulate cortex (or differentiating field)

FIBER TRACTS (f)

f3	corpus callosum
f6	fornix
f9	cerebral peduncle
f13	fimbria
f16	medial lemniscus
f18	habenulopeduncular tract
f19	posterior commissure

GLANDS AND PARANEURAL STRUCTURES (g)

g1	choroid plexus of lateral ventricle
g5	pineal gland
g7	subcommissural organ

HYPOTHALAMUS (H)

H1	hypothalamic neuroepithelium
H3f	fields of Forel (or differentiating field)
H3*l*	lateral hypothalamus (or differentiating field)
H3p	posterior hypothalamus (or differentiating field)
H3v	ventral hypothalamus (or differentiating field)
H7	zona incerta
H8	dorsomedial nucleus
H9	ventromedial nucleus
H19	subthalamic nucleus
H20	arcuate nucleus
H26	posterior hypothalamic area

HIPPOCAMPAL FORMATION (K)

K3	hippocampus (or differentiating field)
K4	CA1 field
K5	CA3 field
K6	alveus
K13	subicular area (or differentiating field)
K23	dentate gyrus (or differentiating field)
K24	dentate granular layer

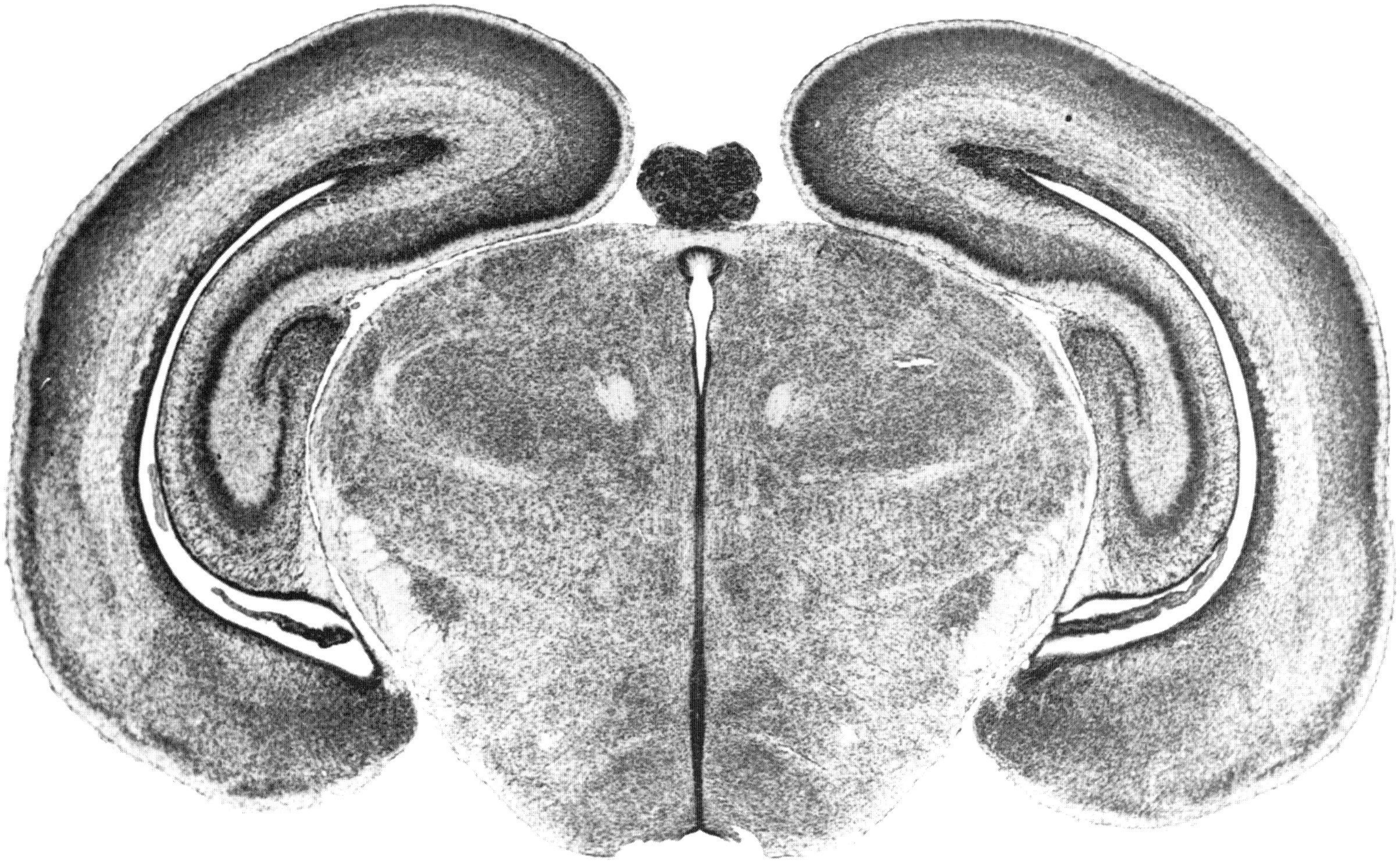

MIDBRAIN (M)

M23 pretectum (or differentiating field)
M24 anterior pretectal nucleus

RHINENCEPHALON (R)

R3 rhinencephalon (or differentiating field)
R4 piriform cortex
R7 rhinal fissure

THALAMUS (T)

T1 thalamic neuroepithelium
T3i intermediate thalamus (or differentiating field)
T3p posterior thalamus (or differentiating field)
T19 parafascicular nucleus
T24 lateral posterior nucleus
T26 ventral medial nucleus
T28 ventrobasal nuclear complex
T29 dorsal lateral geniculate nucleus
T30 ventral lateral geniculate nucleus
T35 posterior nuclear complex

VENTRICLES (v)

va aqueduct
vla lateral ventricle, anterior horn
vlp lateral ventricle, posterior horn
v3h third ventricle, hypothalamic
v3t third ventricle, thalamic

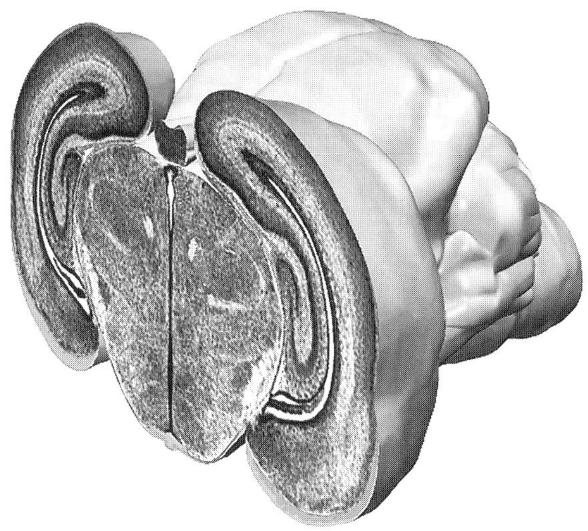

CORONAL FIGURE 17

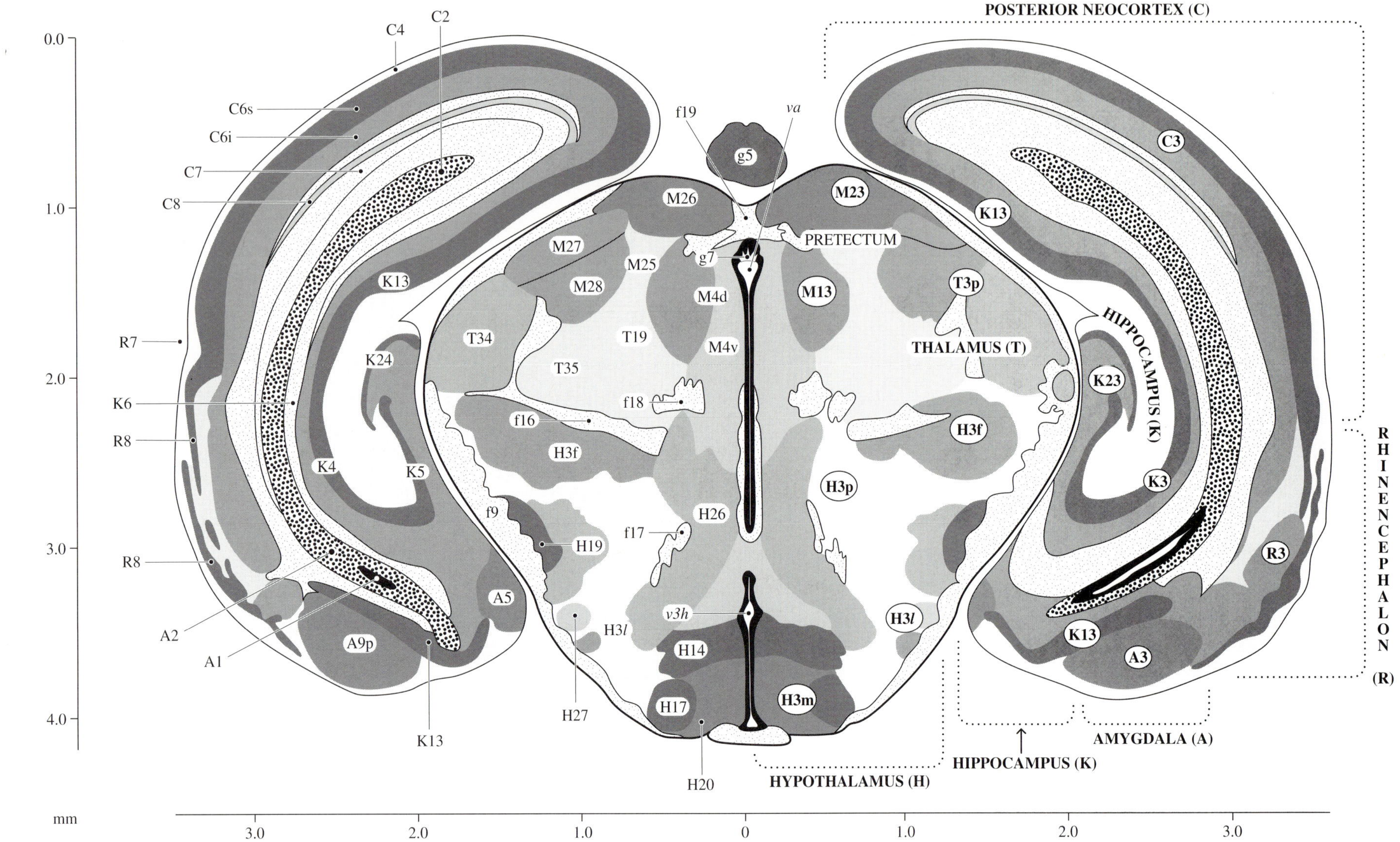

AMYGDALA (A)

A1 amygdaloid neuroepithelium
A2 amygdaloid subventricular zone
A3 amygdala (or differentiating field)
A5 amygdala-hippocampal area
A9p cortical amygdaloid nucleus, posterior

CEREBRAL CORTEX (C)

C2 neocortical subventricular zone
C3 neocortex (or differentiating field)
C4 molecular layer (or primordial plexiform layer)
C6i cortical plate, infragranular part
C6s cortical plate, supragranular part
C7 white matter
C8 subplate

FIBER TRACTS (f)

f9 cerebral peduncle
f16 medial lemniscus
f17 mammillothalamic tract
f18 habenulopeduncular tract
f19 posterior commissure

GLANDS AND PARANEURAL STRUCTURES (g)

g5 pineal gland
g7 subcommissural organ

HYPOTHALAMUS (H)

H3f fields of Forel (or differentiating field)
H3*l* lateral hypothalamus (or differentiating field)
H3m mammillary body (or differentiating field)
H3p posterior hypothalamus (or differentiating field)
H14 supramammillary nucleus
H17 lateral mammillary nucleus
H19 subthalamic nucleus
H20 arcuate nucleus
H26 posterior hypothalamic area
H27 tuberomammillary nucleus

HIPPOCAMPAL FORMATION (K)

K3 hippocampus (or differentiating field)
K4 CA1 field
K5 CA3 field
K6 alveus
K13 subicular area (or differentiating field)
K23 dentate gyrus (or differentiating field)
K24 dentate granular layer

MIDBRAIN (M)

M4d dorsal periaqueductal gray
M4v ventral periaqueductal gray
M13 tegmentum (or differentiating field)
M23 pretectum (or differentiating field)
M25 nucleus of the posterior commissures
M26 olivary pretectal nucleus
M27 nucleus of the optic tract
M28 posterior pretectal nucleus

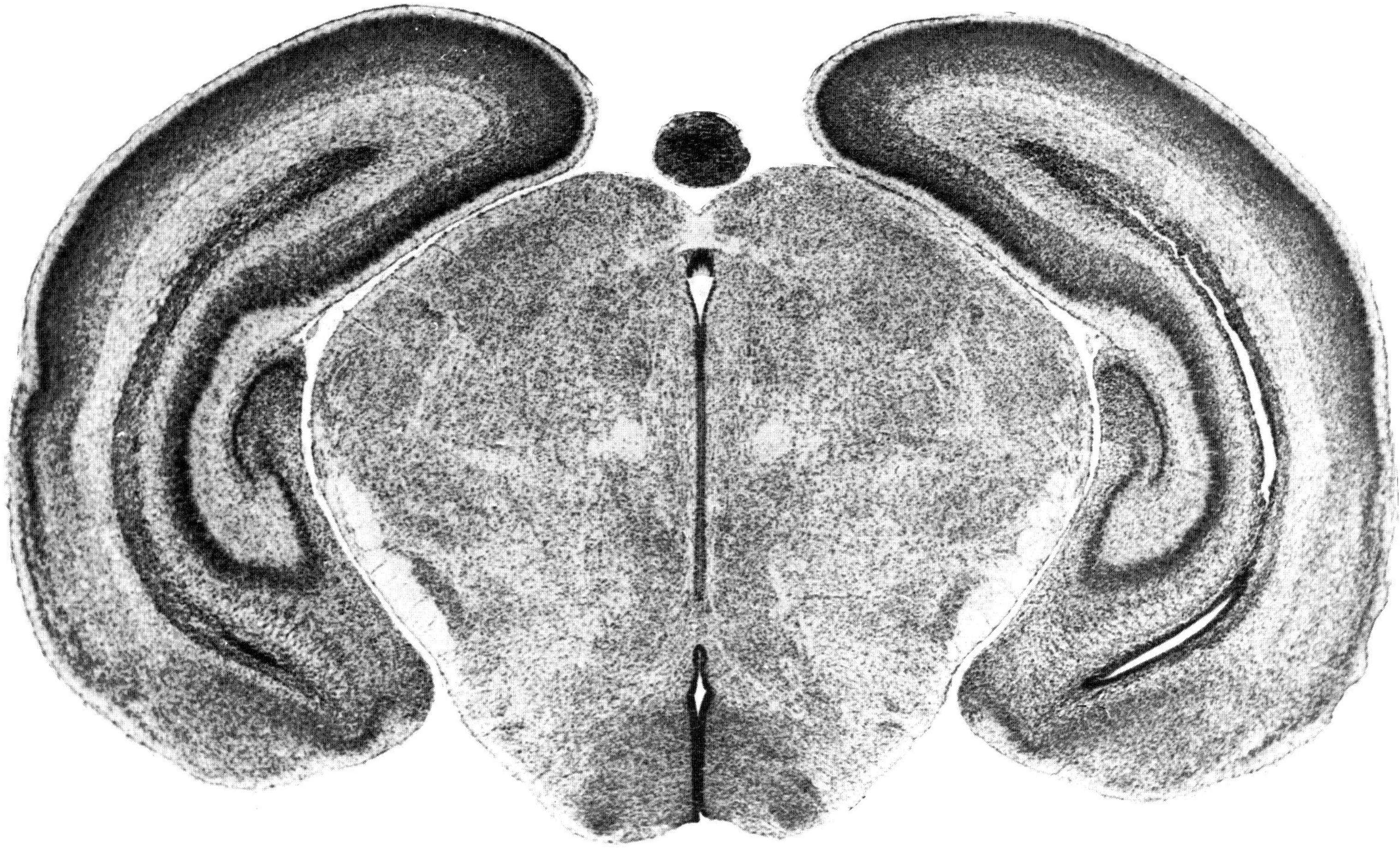

RHINENCEPHALON (R)

R3 rhinencephalon (or differentiating field)
R7 rhinal fissure
R8 entorhinal cortex

THALAMUS (T)

T3p posterior thalamus (or differentiating field)
T19 parafascicular nucleus
T34 medial geniculate nucleus
T35 posterior nuclear complex

VENTRICLES (v)

va aqueduct
v3h third ventricle, hypothalamic

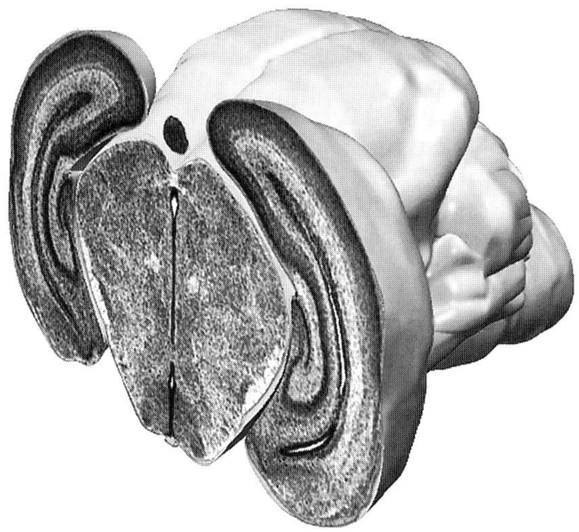

CORONAL FIGURE 18

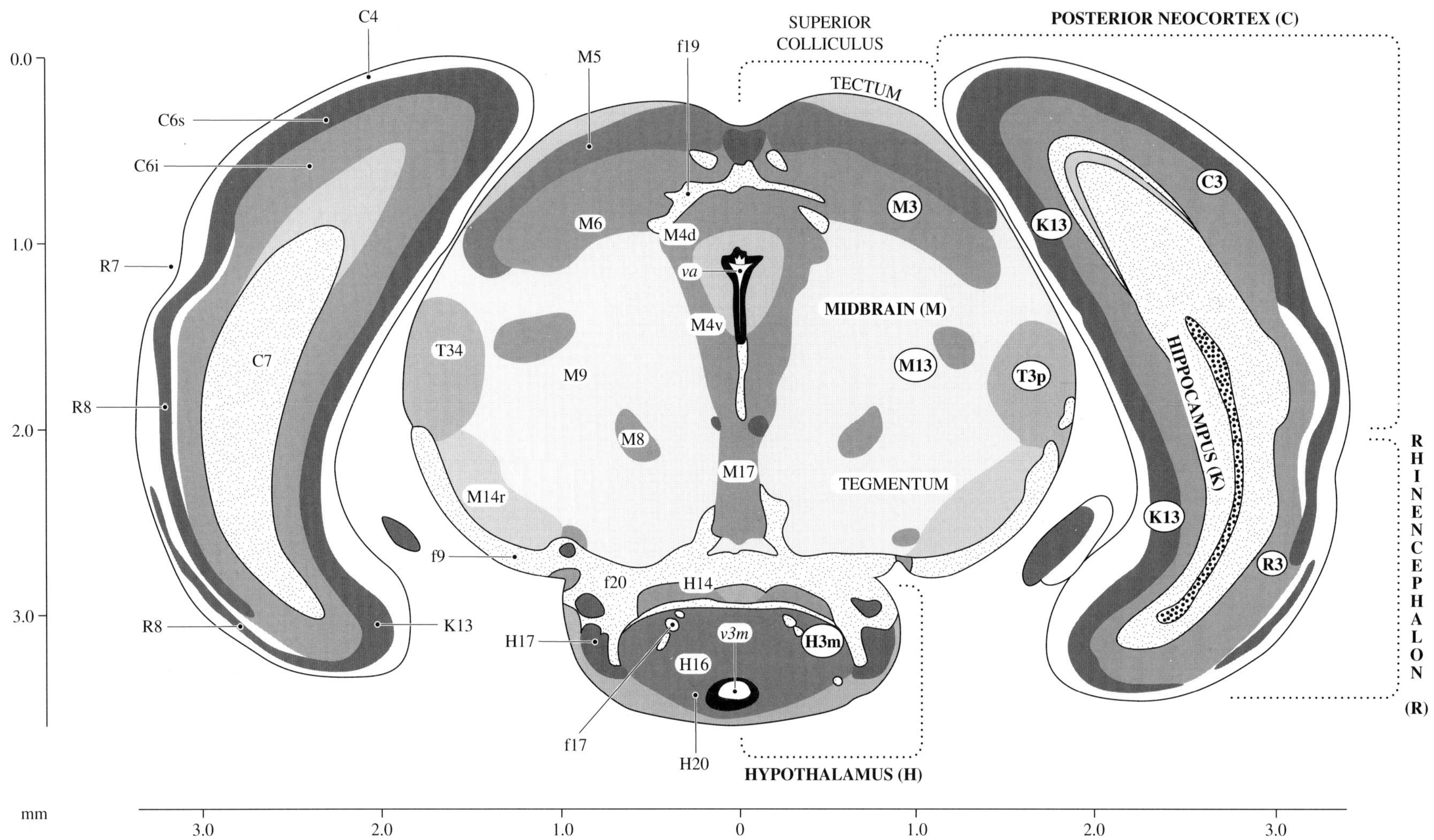

CEREBRAL CORTEX (C)

C3 neocortex (or differentiating field)
C4 molecular layer (primordial plexiform layer)
C6i cortical plate, infragranular part
C6s cortical plate, supragranular part
C7 white matter

FIBER TRACTS (f)

f9 cerebral peduncle
f17 mammillothalamic tract
f19 posterior commissure
f20 supramammillary decussation

HYPOTHALAMUS (H)

H3m mammillary body (or differentiating field)
H14 supramammillary nucleus
H16 medial mammillary nucleus
H17 lateral mammillary nucleus
H20 arcuate nucleus

HIPPOCAMPAL FORMATION (K)

K13 subicular area (or differentiating field)

MIDBRAIN (M)

M3 superior colliculus (or differentiating field)
M4d dorsal periaqueductal gray
M4v ventral periaqueductal gray
M5 superior colliculus, superficial layers
M6 superior colliculus, deep layers
M8 red nucleus
M9 midbrain reticular formation
M13 tegmentum (or differentiating field)
M14r substantia nigra, pars reticulata
M17 ventral tegmental nucleus (mesencephalic)

RHINENCEPHALON (R)

R3 rhinencephalon (or differentiating field)
R7 rhinal fissure
R8 entorhinal cortex

THALAMUS (T)

T3p posterior thalamus (or differentiating field)
T34 medial geniculate nucleus

VENTRICLES (v)

va aqueduct
v3m third ventricle, mammillary recess

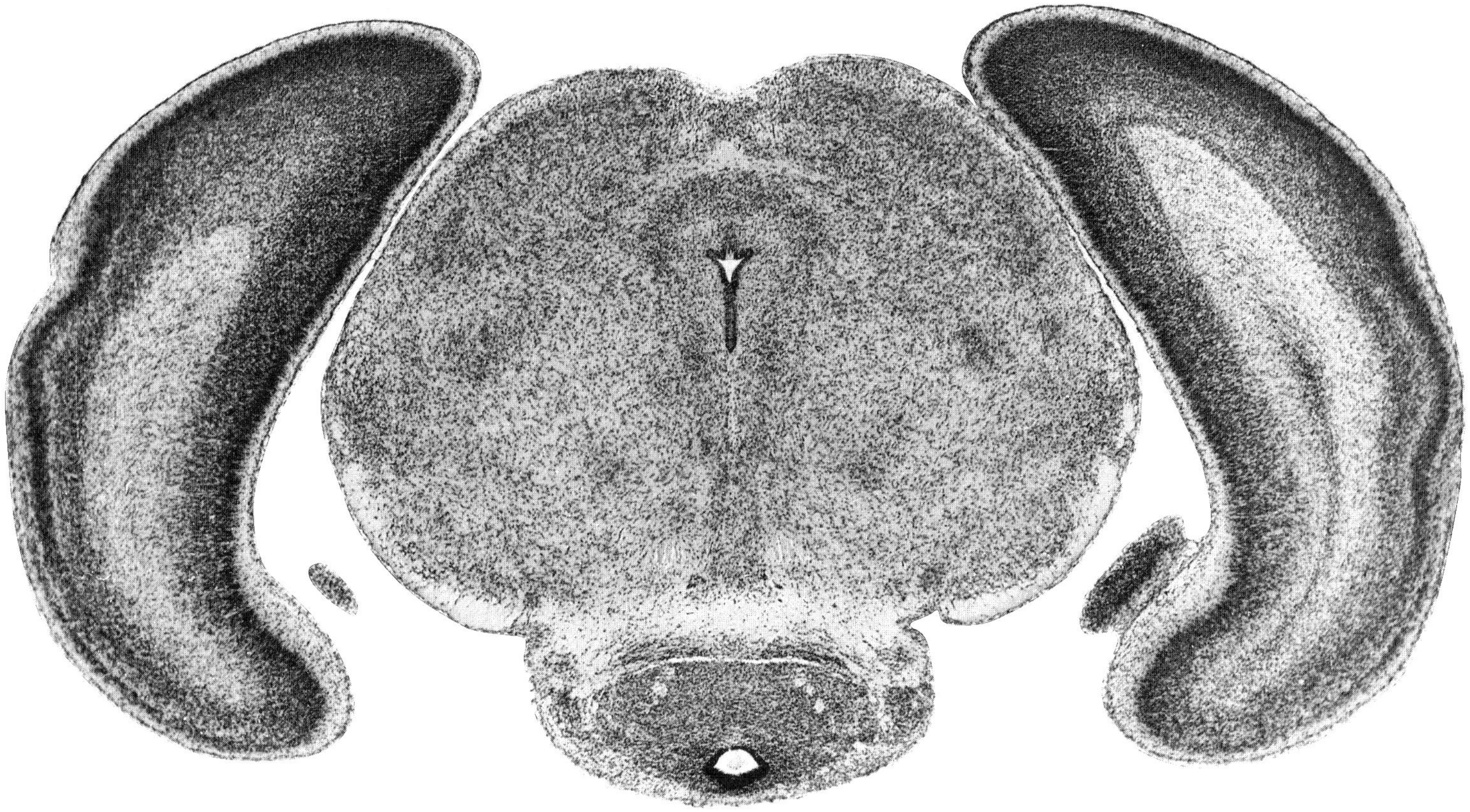

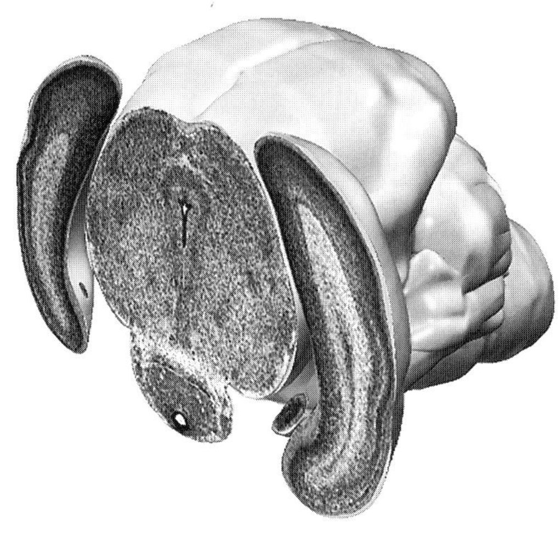

CORONAL FIGURE 19

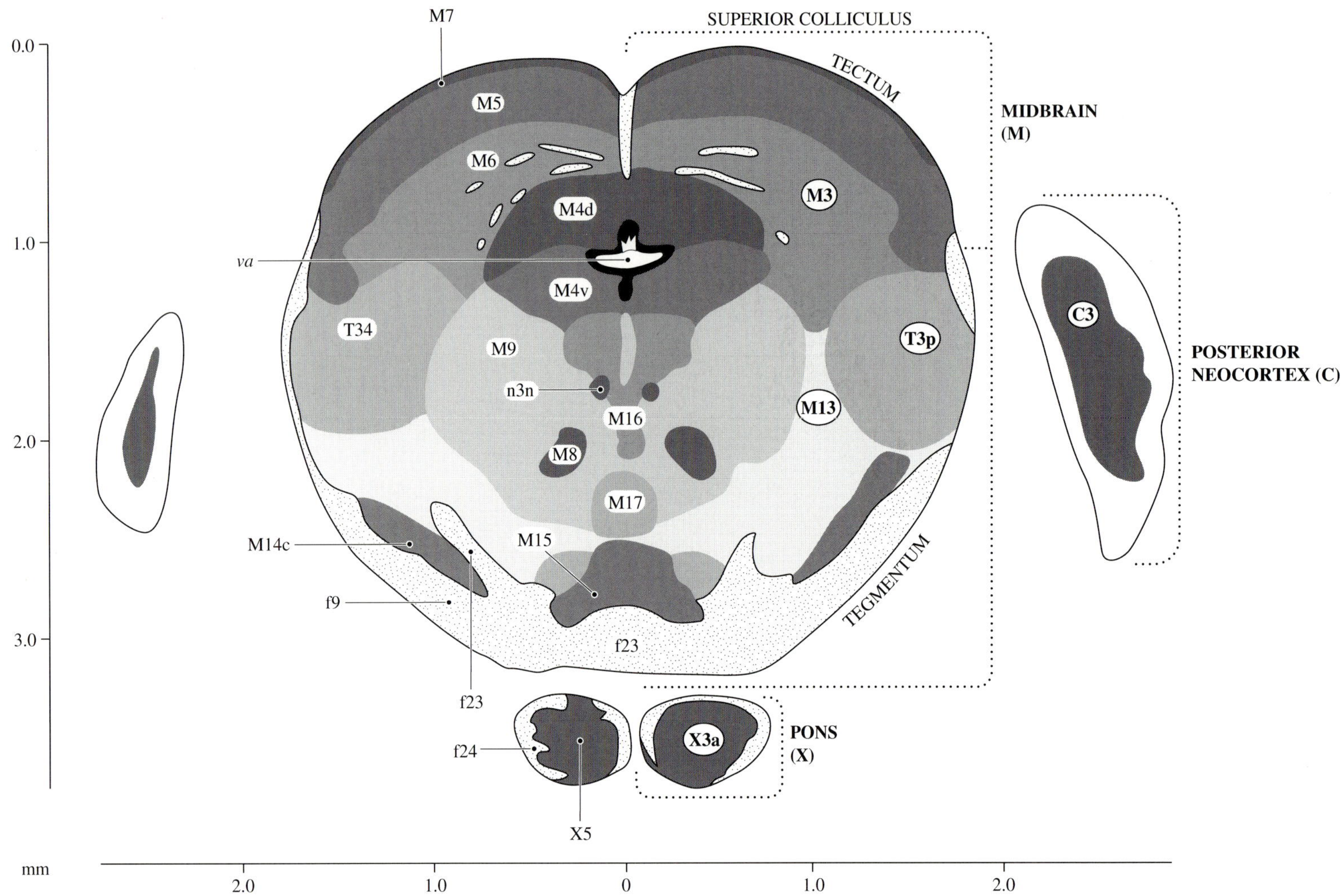

CEREBRAL CORTEX (C)

C3 neocortex (or differentiating field)

FIBER TRACTS (f)

f9 cerebral peduncle
f23 superior cerebellar peduncle
f24 middle cerebellar peduncle

MIDBRAIN (M)

M3 superior colliculus (or differentiating field)
M4d dorsal periaqueductal gray
M4v ventral periaqueductal gray
M5 superior colliculus, superficial layers
M6 superior colliculus, deep layers
M7 superior colliculus, stratum opticum
M8 red nucleus
M9 midbrain reticular formation
M13 tegmentum (or differentiating field)
M14c substantia nigra, pars compacta
M15 interpeduncular nucleus
M16 dorsal raphe (mesencephalic)
M17 ventral tegmental nucleus (mesencephalic)

NERVES AND NUCLEI, CRANIAL (n)

n3n oculomotor nucleus

THALAMUS (T)

T3p posterior thalamus (or differentiating field)
T34 medial geniculate nucleus

VENTRICLES (v)

va aqueduct

PONTINE AREA (X)

X3a anterior pons (or differentiating field)
X5 pontine gray nucleus

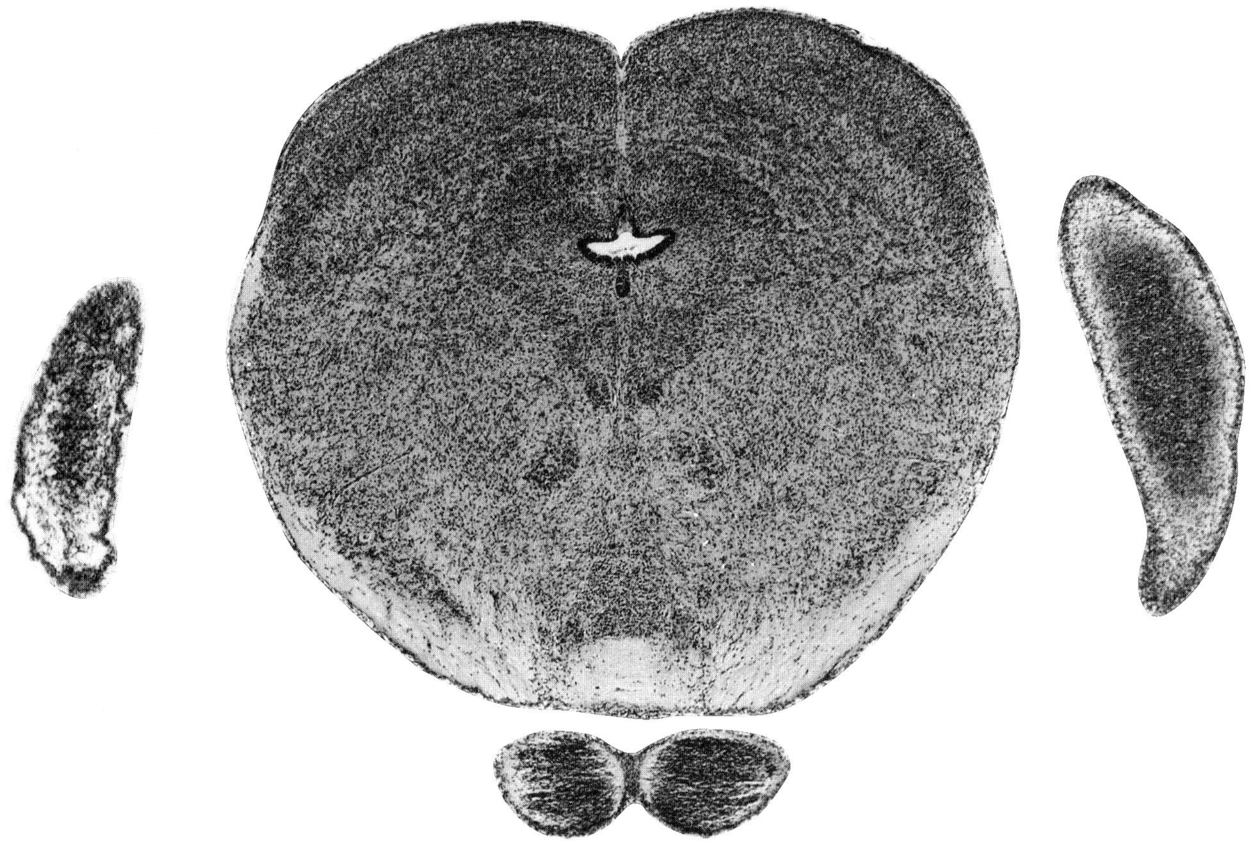

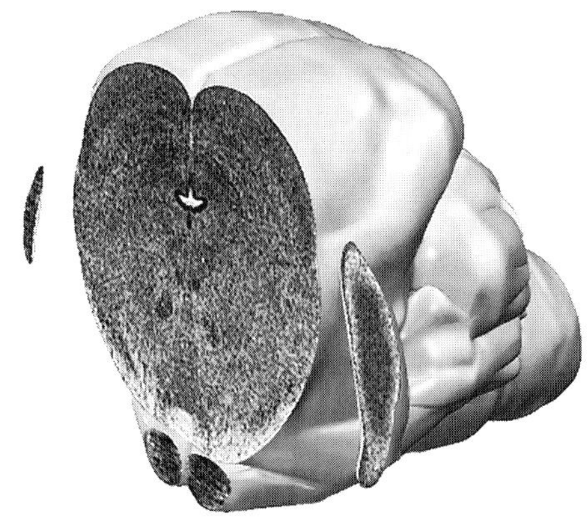

CORONAL FIGURE 20

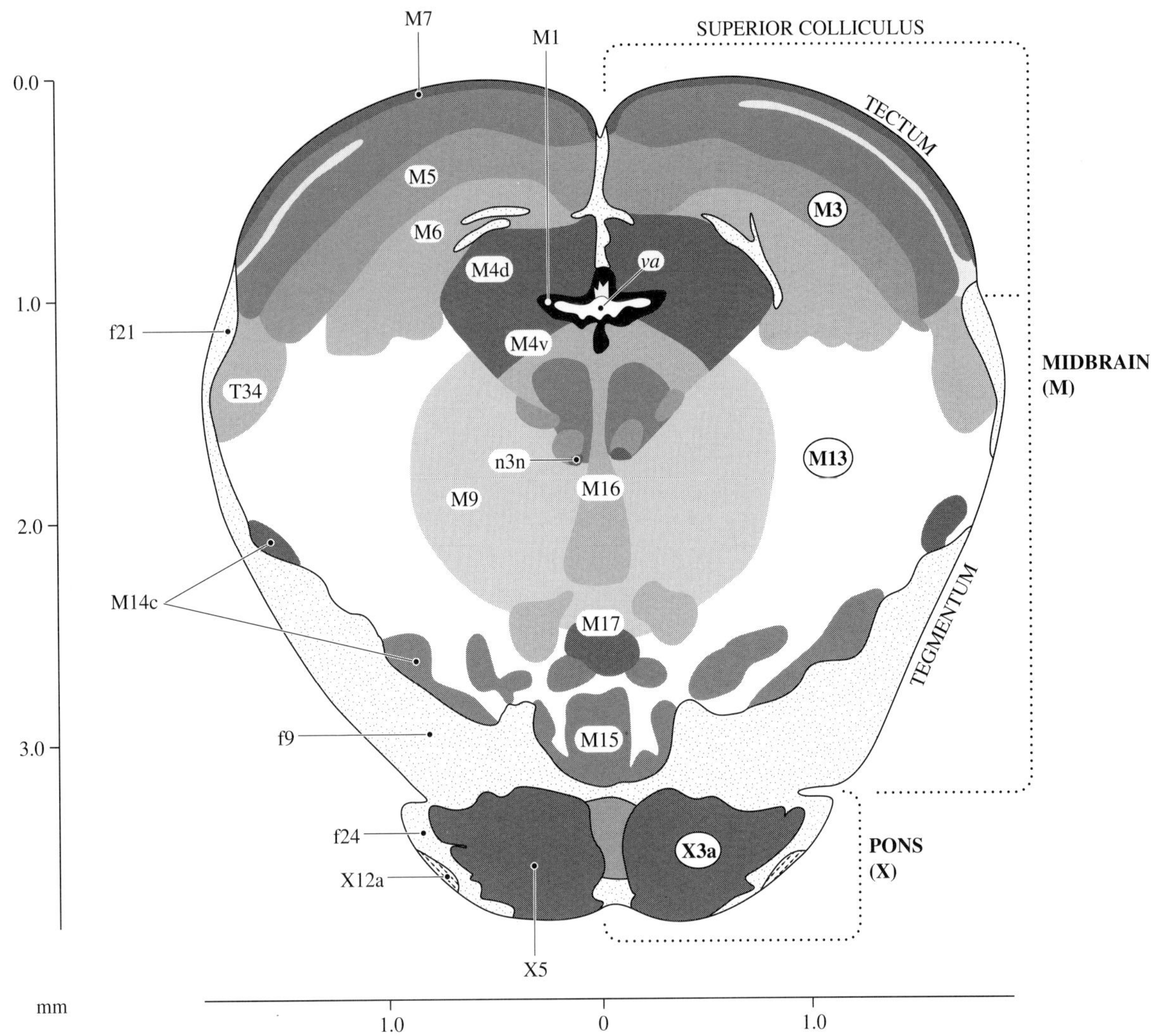

FIBER TRACTS (f)

f9 cerebral peduncle
f21 brachium of superior colliculus
f24 middle cerebellar peduncle

MIDBRAIN (M)

M1 superior tectal neuroepithelium
M3 superior colliculus (or differentiating field)
M4d dorsal periaqueductal gray
M4v ventral periaqueductal gray
M5 superior colliculus, superficial layers
M6 superior colliculus, deep layers
M7 superior colliculus, stratum opticum
M9 midbrain reticular formation
M13 tegmentum (or differentiating field)
M14c substantia nigra, part compacta
M15 interpeduncular nucleus
M16 dorsal raphe
M17 ventral tegmental nucleus (mesencephalic)

NERVES AND NUCLEI, CRANIAL (n)

n3n oculomotor nucleus

THALAMUS (T)

T34 medial geniculate nucleus

VENTRICLES (v)

va aqueduct

PONTINE AREA (X)

X3a anterior pons (or differentiating field)
X5 pontine gray nucleus
X12a anterior precerebellar extramural migration

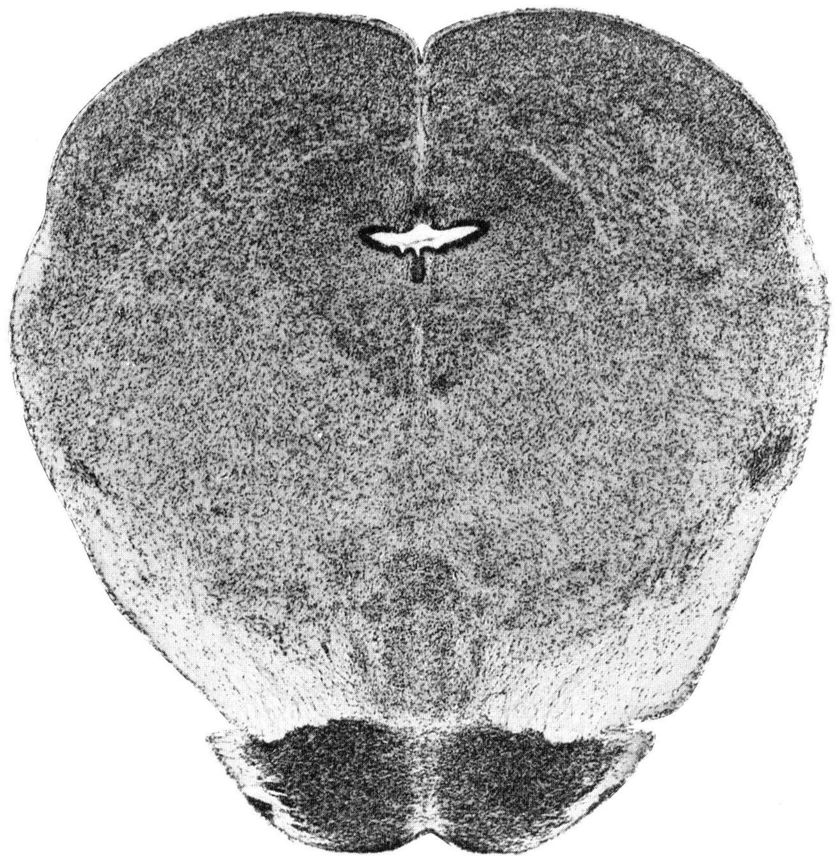

CORONAL FIGURE 21

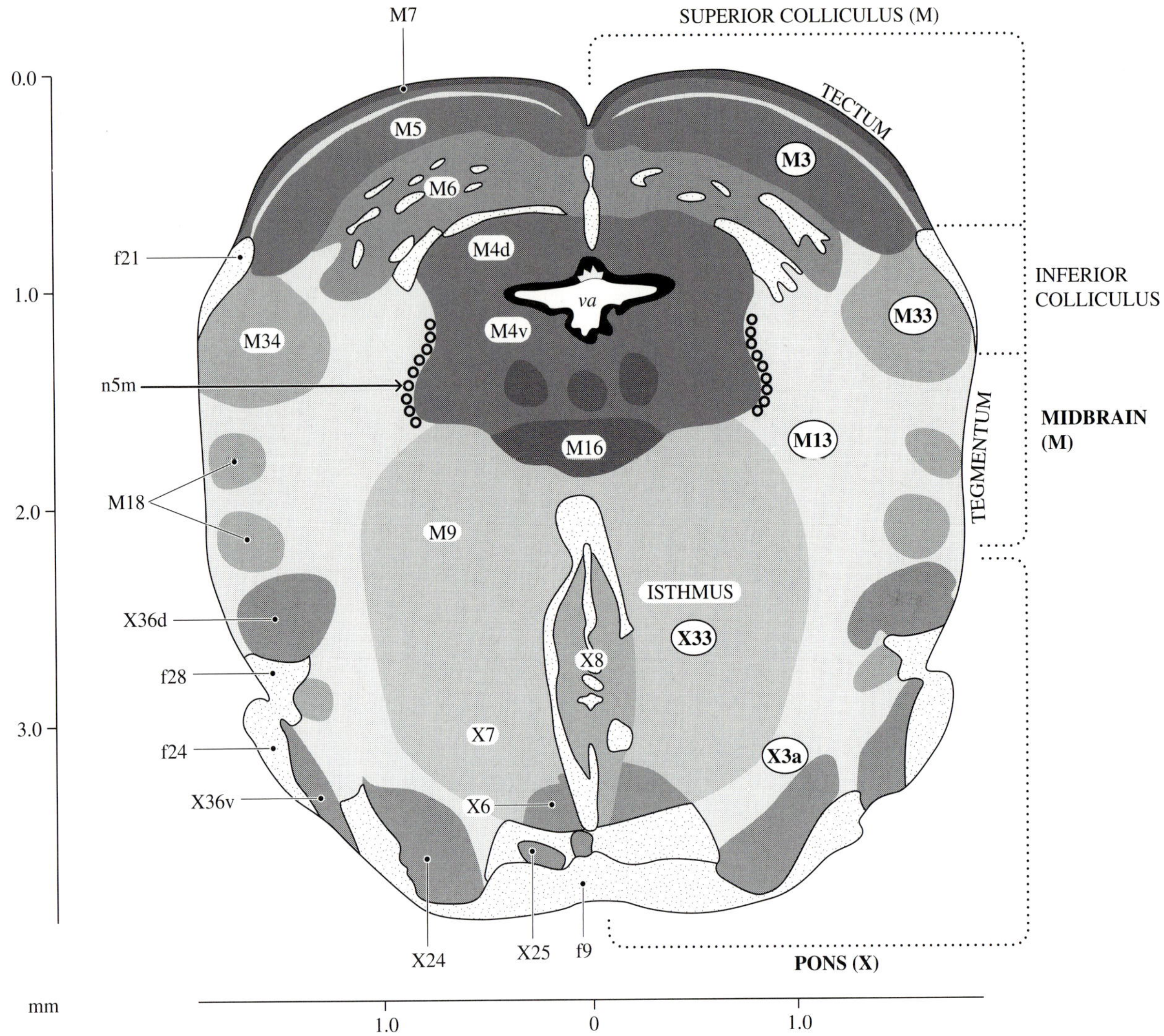

FIBER TRACTS (f)

f9	cerebral peduncle
f21	brachium of superior colliculus
f24	middle cerebellar peduncle
f28	lateral lemniscus

MIDBRAIN (M)

M3	superior colliculus (or differentiating field)
M4d	dorsal periaqueductal gray
M4v	ventral periaqueductal gray
M5	superior colliculus, superficial layers
M6	superior colliculus, deep layers
M7	superior colliculus, stratum opticum
M9	midbrain reticular formation
M13	tegmentum (or differentiating field)
M16	dorsal raphe
M18	parabigeminal nucleus
M33	inferior colliculus (or differentiating field)
M34	external nucleus of inferior colliculus

NERVES AND NUCLEI, CRANIAL

n5m	mesencephalic trigeminal nucleus

VENTRICLES (v)

va	aqueduct

PONTINE AREA (X)

X3a	anterior pons (or differentiating field)
X6	pontine reticulotegmental nucleus
X7	pontine reticular formation
X8	superior central raphe nucleus
X24	superior olivary nucleus
X25	trapezoid body
X33	isthmus (or differentiating field)
X36d	nucleus of lateral lemniscus, dorsal
X36v	nucleus of lateral lemniscus, ventral

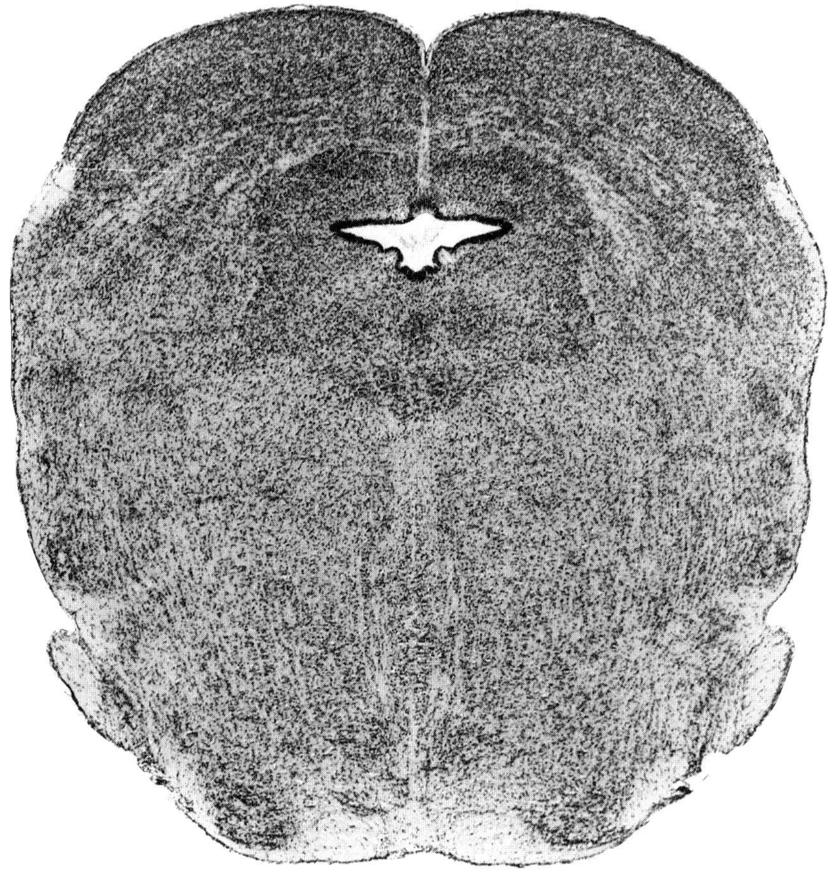

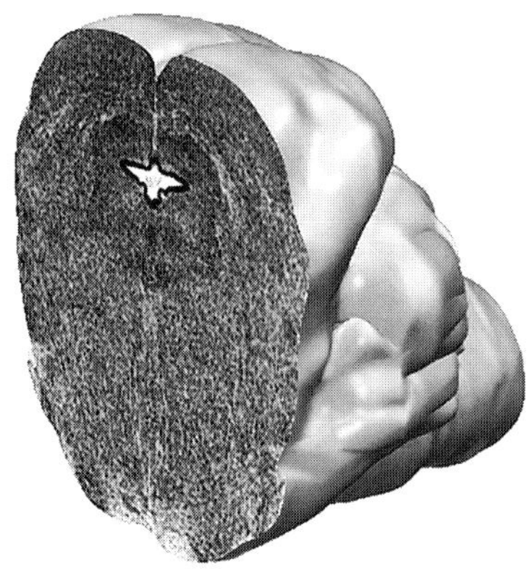

CORONAL FIGURE 22

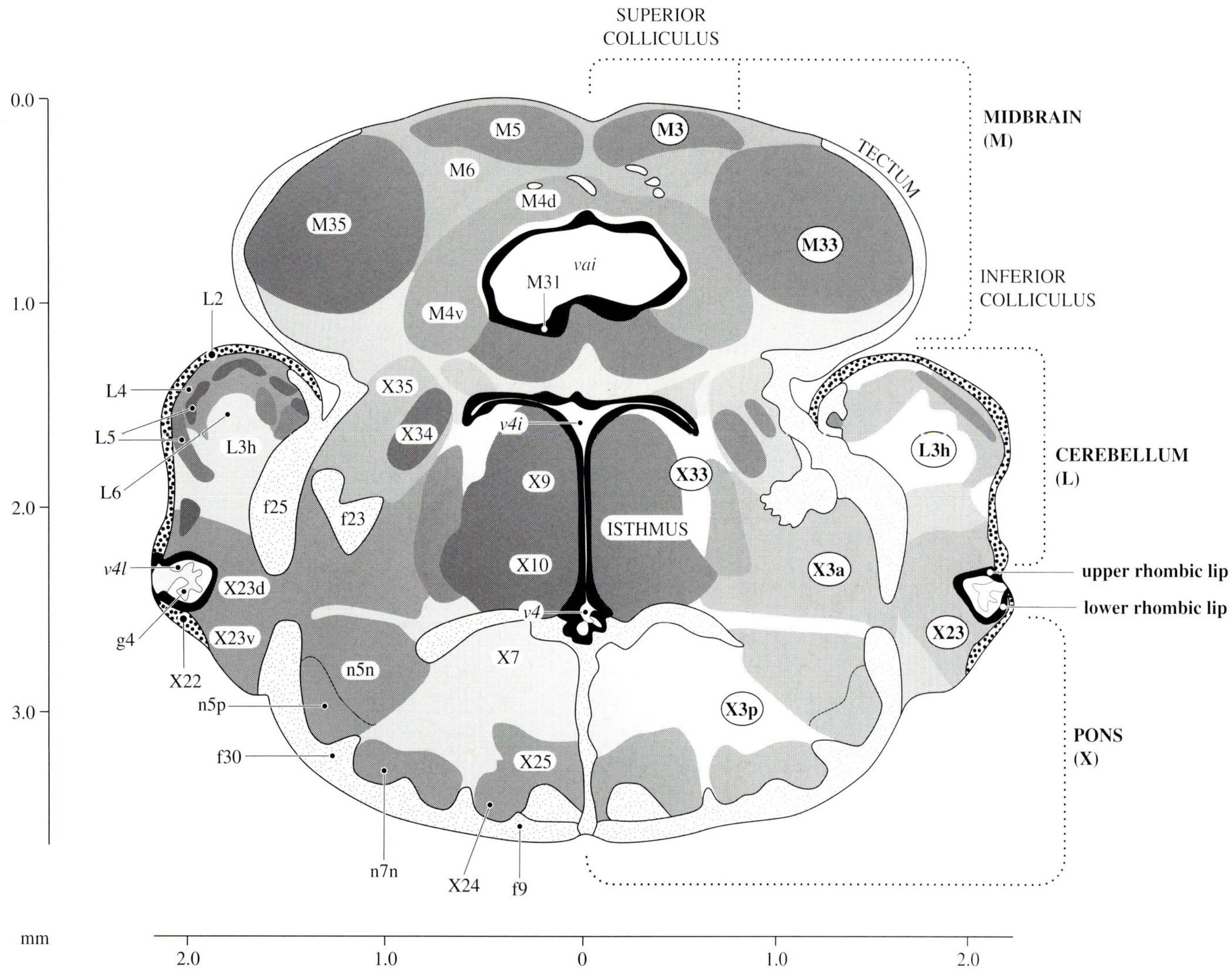

FIBER TRACTS (f)

- f9 cerebral peduncle
- f23 superior cerebellar peduncle
- f25 inferior cerebellar peduncle
- f30 spinal tract of trigeminal

GLANDS AND PARANEURAL STRUCTURES (g)

- g4 choroid plexus of fourth ventricle

CEREBELLUM (L)

- L2 external germinal layer
- **L3h** cerebellar hemisphere (or differentiating field)
- L4 molecular layer
- L5 Purkinje cell layer
- L6 medullary layer

MIDBRAIN (M)

- **M3** superior colliculus (or differentiating field)
- M4d dorsal periaqueductal gray
- M4v ventral periaqueductal gray
- M5 superior colliculus, superficial layers
- M6 superior colliculus, deep layers
- M31 inferior tectal neuroepithelium
- **M33** inferior colliculus (or differentiating field)
- M35 central nucleus of inferior colliculus

NERVES AND NUCLEI, CRANIAL (n)

- n5n trigeminal motor nucleus
- n5p principal (sensory) trigeminal nucleus
- n7n facial motor nucleus

VENTRICLES (v)

- *vai* aqueduct, inferior tectal
- *v4* fourth ventricle
- *v4i* fourth ventricle, isthmal
- *v4l* fourth ventricle, lateral recess

PONTINE AREA (X)

- **X3a** anterior pons (or differentiating field)
- **X3p** posterior pons (or differentiating field)
- X7 pontine reticular formation
- X9 dorsal tegmental nucleus
- X10 ventral tegmental nucleus
- X22 cochlear external germinal layer
- **X23** cochlear nuclei (or differentiating field)
- X23d dorsal cochlear nucleus (or differentiating field)
- X23v ventral cochlear nucleus (or differentiating field)
- X24 superior olivary nucleus
- X25 trapezoid body
- **X33** isthmus (or differentiating field)
- X34 locus coeruleus
- X35 parabrachial nucleus

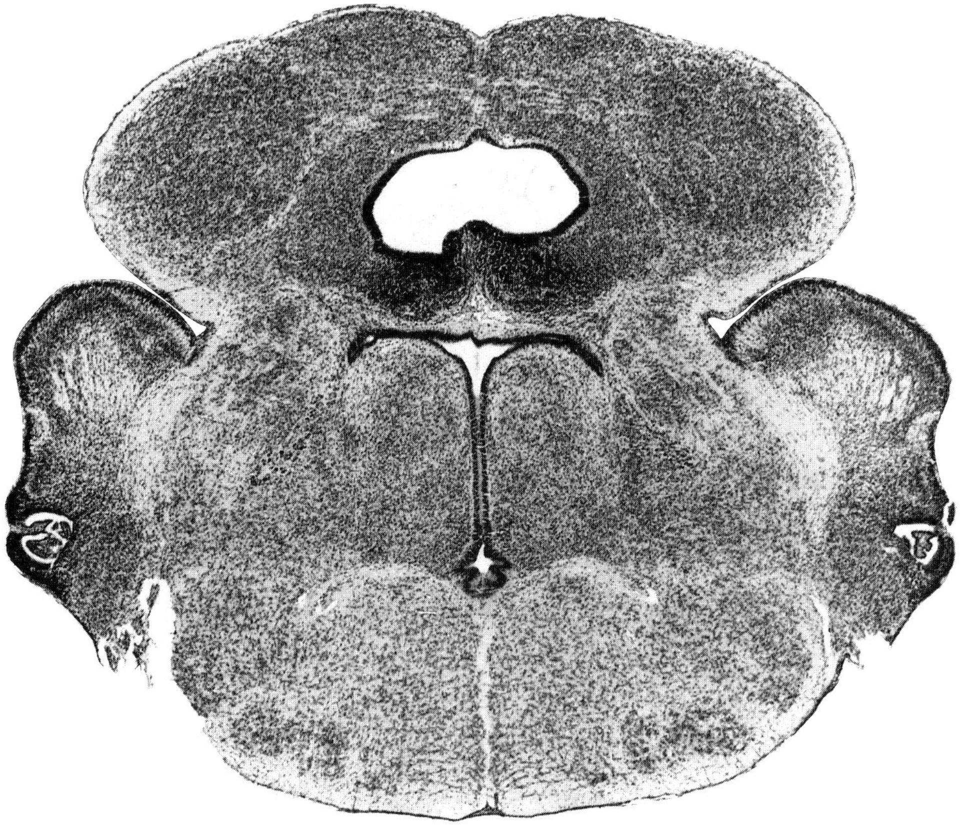

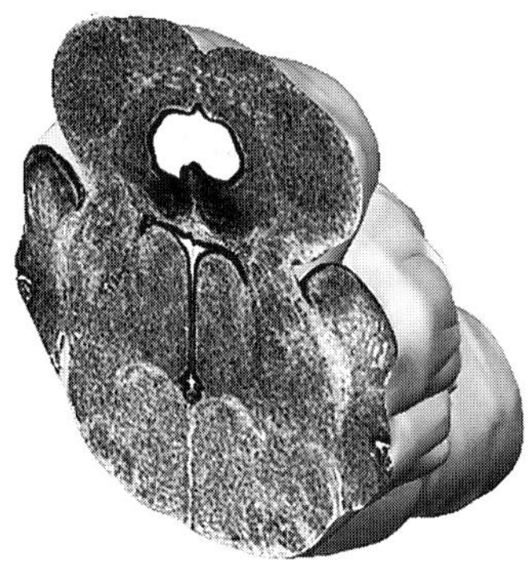

CORONAL FIGURE 23

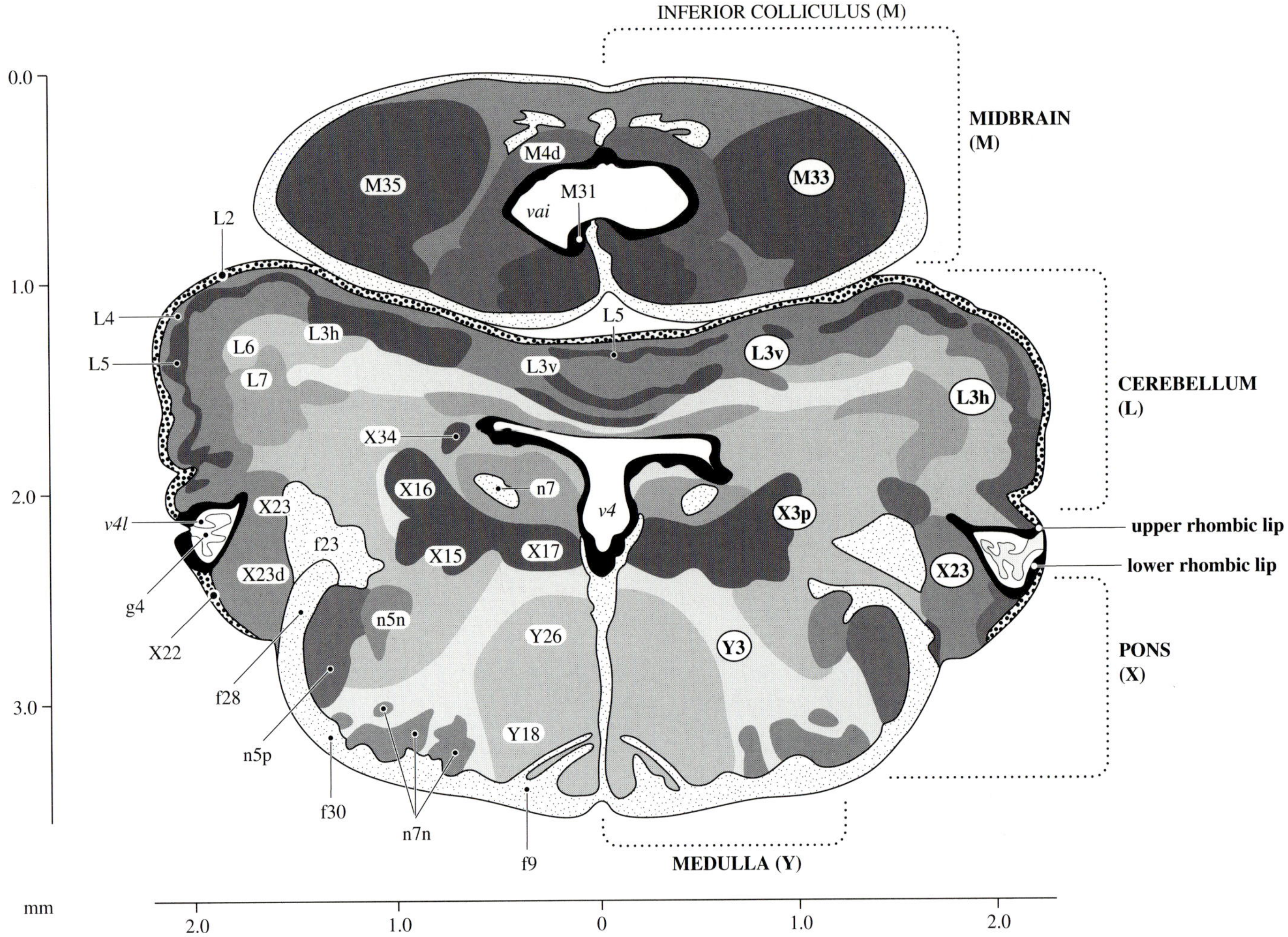

FIBER TRACTS (f)

- f9 cerebral peduncle
- f23 superior cerebellar peduncle
- f28 lateral lemniscus
- f30 spinal tract of trigeminal

GLANDS AND PARANEURAL STRUCTURES (g)

- g4 choroid plexus of fourth ventricle

CEREBELLUM (L)

- L2 external germinal layer
- **L3h** cerebellar hemisphere (or differentiating field)
- **L3v** cerebellar vermis (or differentiating field)
- L4 molecular layer
- L5 Purkinje cell layer
- L6 medullary layer
- L7 lateral deep nucleus (dentate)

MIDBRAIN (M)

- M4d dorsal periaqueductal gray
- M31 inferior tectal neuroepithelium
- **M33** inferior colliculus (or differentiating field)
- M35 central nucleus of inferior colliculus

NERVES AND NUCLEI, CRANIAL (n)

- n5n trigeminal motor nucleus
- n5p principal (sensory) trigeminal nucleus
- n7 facial nerve
- n7n facial motor nucleus

VENTRICLES (v)

- *vai* aqueduct, inferior tectal
- *v4* fourth ventricle
- *v4l* fourth ventricle, lateral recess

PONTINE AREA (X)

- **X3p** posterior pons (or differentiating field)
- X15 lateral vestibular nucleus
- X16 superior vestibular nucleus
- X17 medial vestibular nucleus
- X22 cochlear external germinal layer
- **X23** cochlear nuclei (or differentiating field)
- X23d dorsal cochlear nucleus (or differentiating field)
- X34 locus coeruleus

MEDULLA (Y)

- **Y3** medulla (or differentiating field)
- Y18 magnocellular reticular nucleus
- Y26 medial reticular nucleus

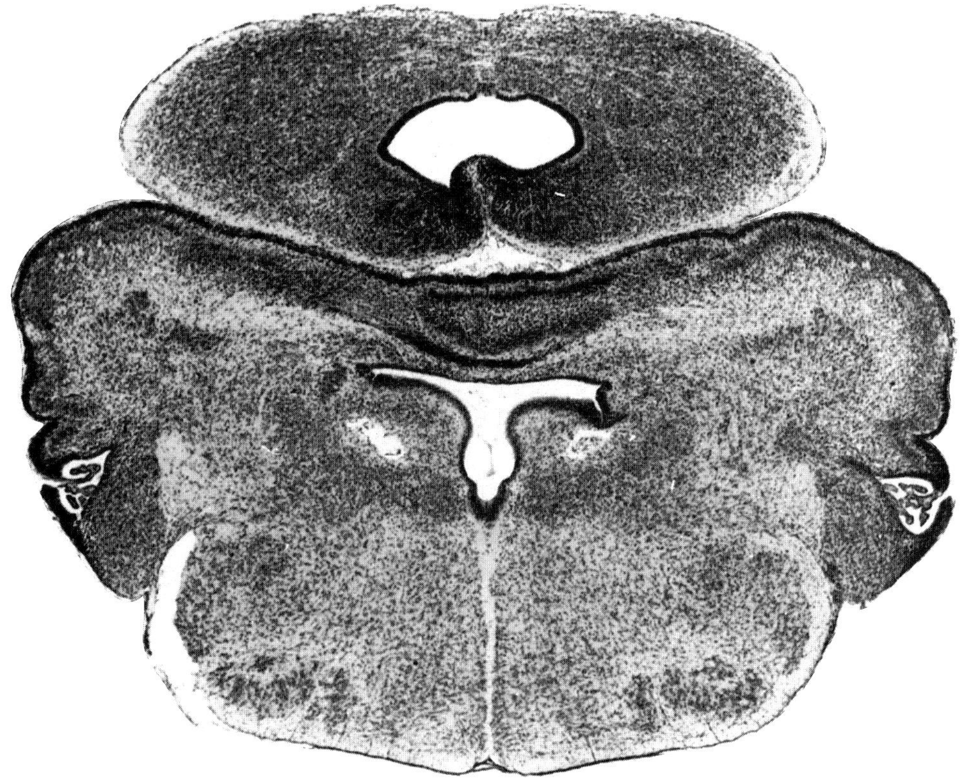

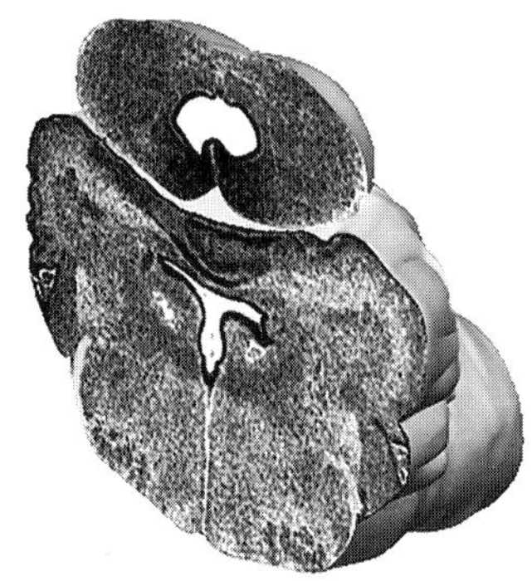

CORONAL FIGURE 24

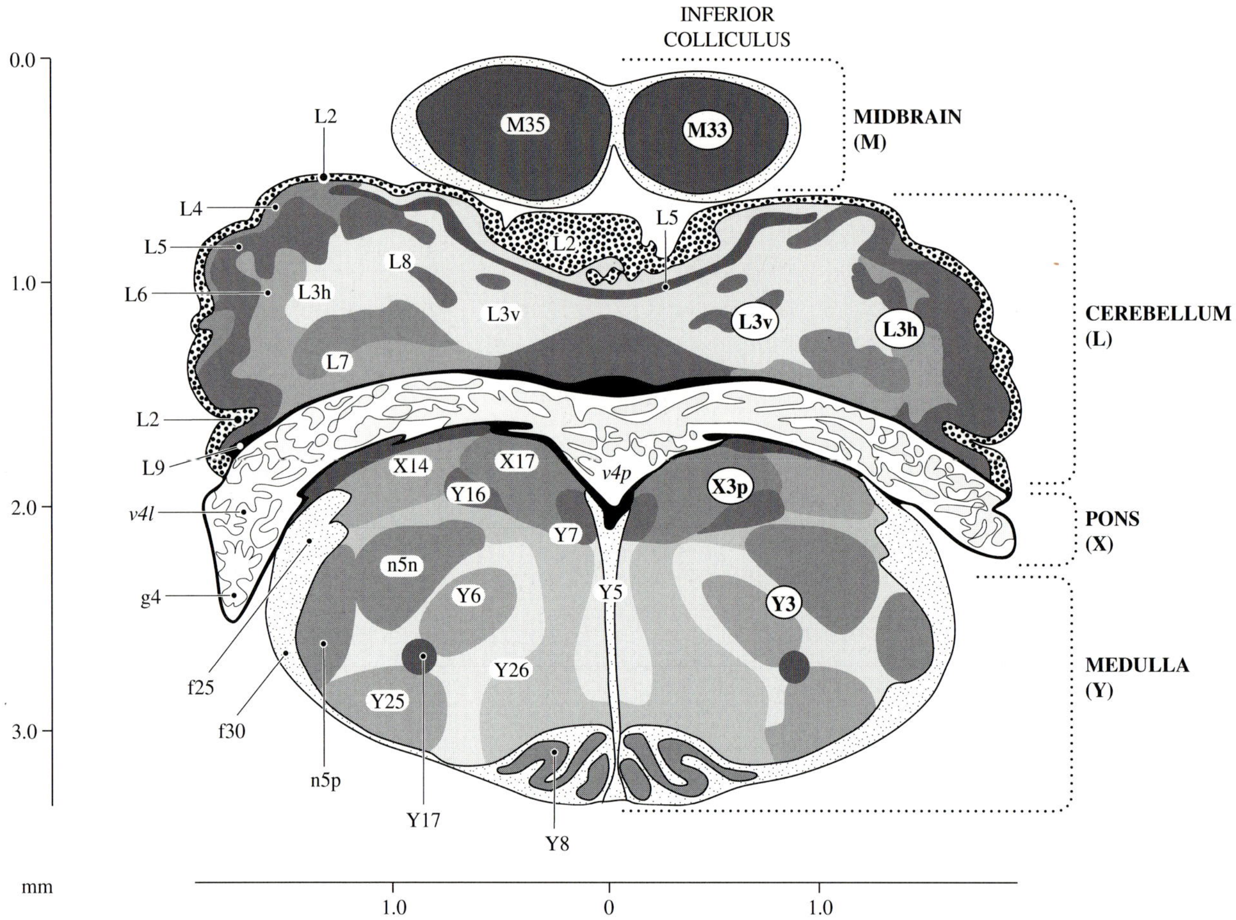

FIBER TRACTS (f)

f25 inferior cerebellar peduncle
f30 spinal tract of trigeminal

GLANDS AND PARANEURAL STRUCTURES (g)

g4 choroid plexus of fourth ventricle

CEREBELLUM (L)

L2 external germinal layer
L3h cerebellar hemisphere (or differentiating field)
L3v cerebellar vermis (or differentiating field)
L4 molecular layer
L5 Purkinje cell layer
L6 medullary layer
L7 lateral deep nucleus (dentate)
L8 medial deep nucleus (fastigial)
L9 cerebellar germinal trigone

MIDBRAIN (M)

M33 inferior colliculus (or differentiating field)
M35 central nucleus of inferior colliculus

NERVES AND NUCLEI, CRANIAL (n)

n5n trigeminal motor nucleus
n5p principal (sensory) trigeminal nucleus

VENTRICLES (v)

v4l fourth ventricle, lateral recess
v4p fourth ventricle, posterior

PONTINE AREA (X)

X3p posterior pons (or differentiating field)
X14 spinal vestibular nucleus
X17 medial vestibular nucleus

MEDULLA (Y)

Y3 medulla (or differentiating field)
Y5 medullary raphe
Y6 medullary reticular formation
Y7 prepositus nucleus
Y8 inferior olive
Y16 solitary nucleus
Y17 ambiguus nucleus
Y25 lateral reticular nucleus
Y26 medial reticular nucleus

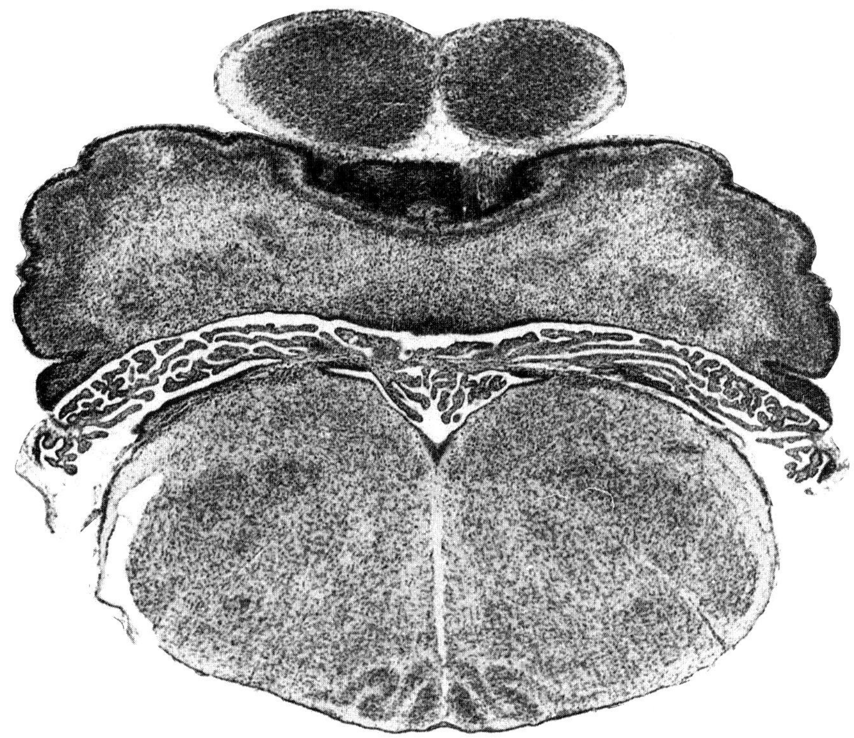

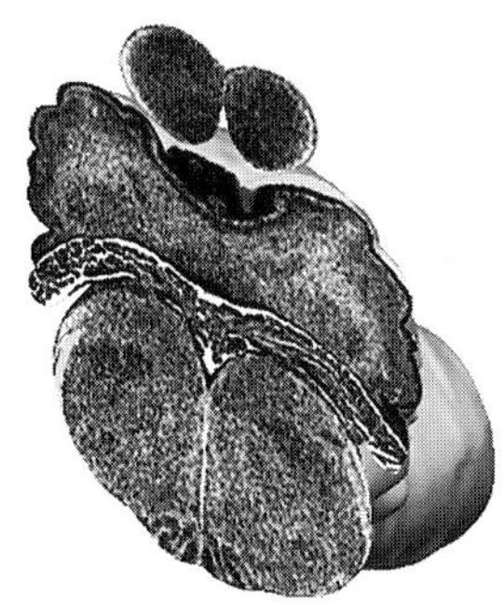

CORONAL FIGURE 25

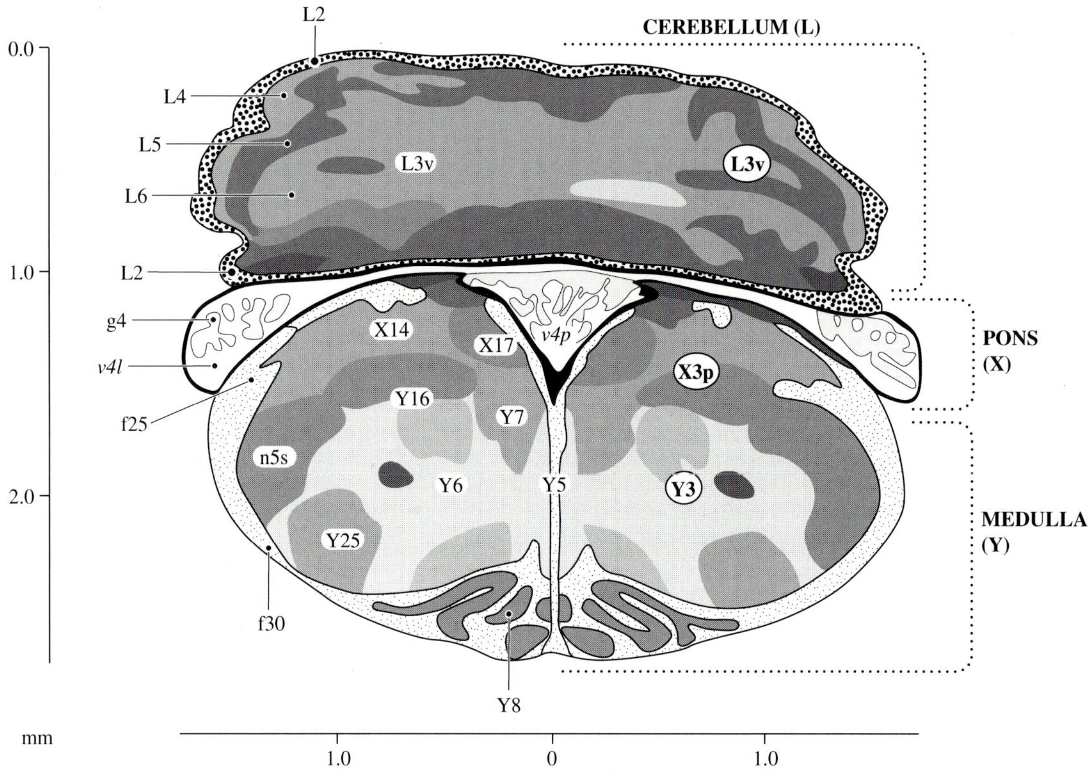

FIBER TRACTS (f)

f25 inferior cerebellar peduncle
f30 spinal tract of trigeminal

GLANDS AND PARANEURAL STRUCTURES (g)

g4 choroid plexus of fourth ventricle

CEREBELLUM (L)

L2 external germinal layer
L3v cerebellar vermis (or differentiating field)
L4 molecular layer
L5 Purkinje cell layer
L6 medullary layer

NERVES AND NUCLEI, CRANIAL (n)

n5s spinal trigeminal nucleus

VENTRICLES (v)

v4l fourth ventricle, lateral recess
v4p fourth ventricle, posterior

PONTINE AREA (X)

X3p posterior pons (or differentiating field)
X14 spinal vestibular nucleus
X17 medial vestibular nucleus

MEDULLA (Y)

Y3 medulla (or differentiating field)
Y5 medullary raphe
Y6 medullary reticular formation
Y7 prepositus nucleus
Y8 inferior olive
Y16 solitary nucleus
Y25 lateral reticular nucleus

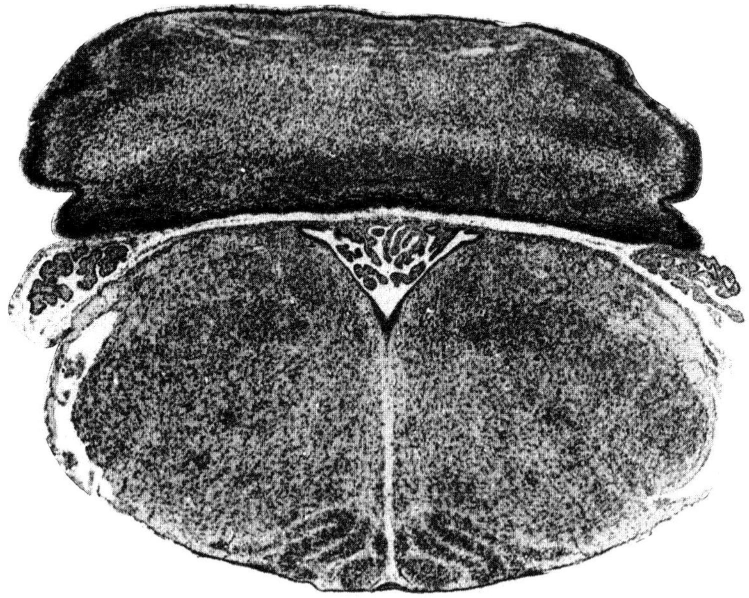

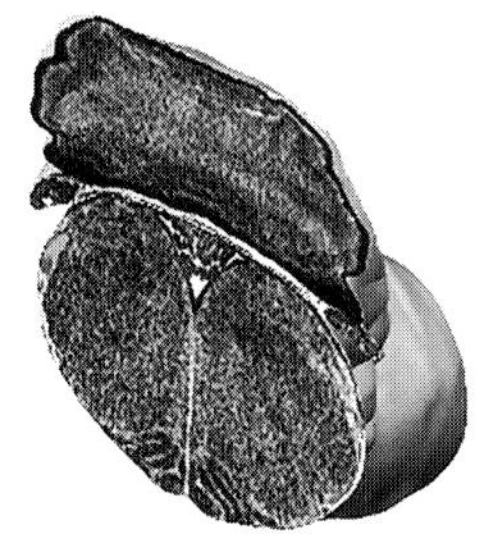

CORONAL FIGURE 26

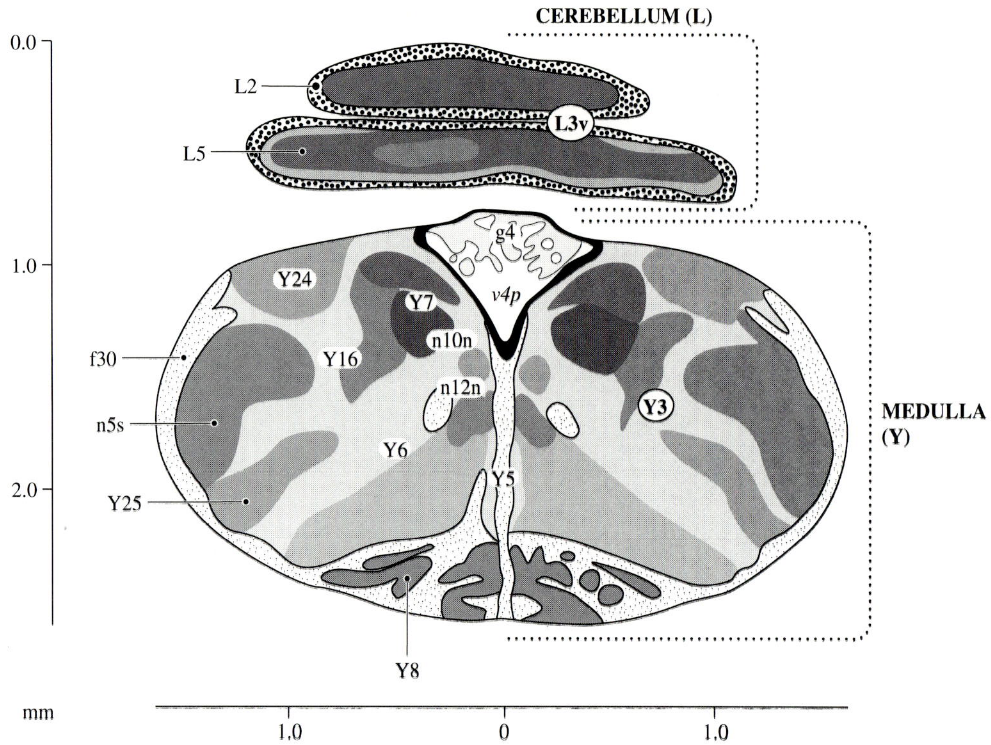

FIBER TRACTS (f)

f30 spinal tract of trigeminal

GLANDS AND PARANEURAL STRUCTURES (g)

g4 choroid plexus of fourth ventricle

CEREBELLUM (L)

L2 external germinal layer
L3v cerebellar vermis (or differentiating field)
L5 Purkinje cell layer

NERVES AND NUCLEI, CRANIAL (n)

n5s spinal trigeminal nucleus
n10n dorsal motor nucleus of vagus
n12n hypoglossal nucleus

VENTRICLES (v)

v4p fourth ventricle, posterior

MEDULLA (Y)

Y3 medulla (or differentiating field)
Y5 medullary raphe
Y6 medullary reticular formation
Y7 prepositus nucleus
Y8 inferior olive
Y16 solitary nucleus
Y24 external cuneate nucleus
Y25 lateral reticular nucleus

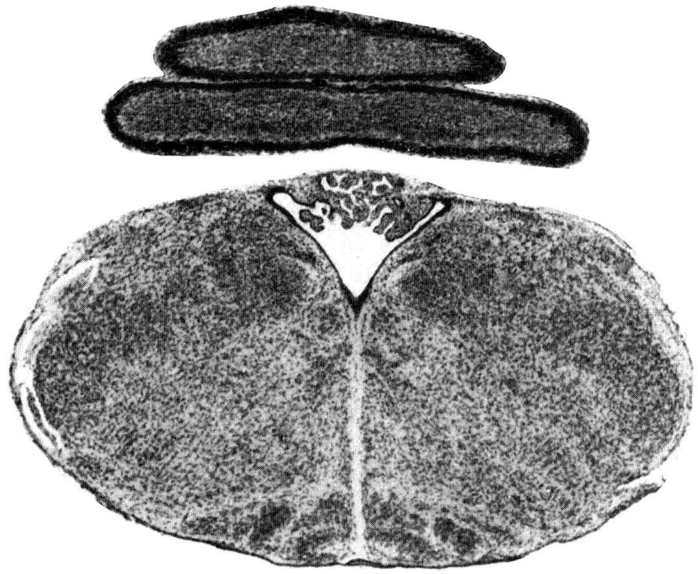

CORONAL FIGURE 27

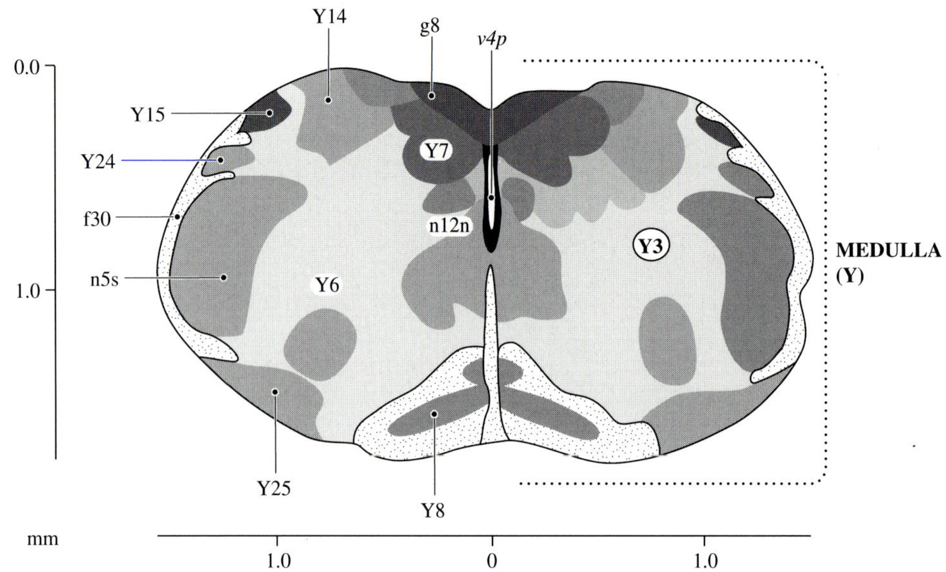

FIBER TRACTS (f)

f30 spinal tract of trigeminal

GLANDS AND PARANEURAL STRUCTURES (g)

g8 area postrema

NERVES AND NUCLEI, CRANIAL (n)

n5s spinal trigeminal nucleus
n12n hypoglossal nucleus

VENTRICLES (v)

v4p fourth ventricle, posterior

MEDULLA (Y)

Y3 medulla (or differentiating field)
Y6 medullary reticular formation
Y7 prepositus nucleus
Y8 inferior olive
Y14 gracile nucleus
Y15 cuneate nucleus
Y24 external cuneate nucleus
Y25 lateral reticular nucleus

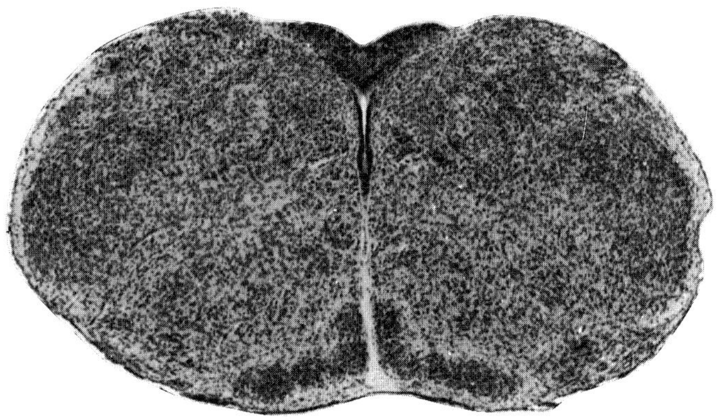

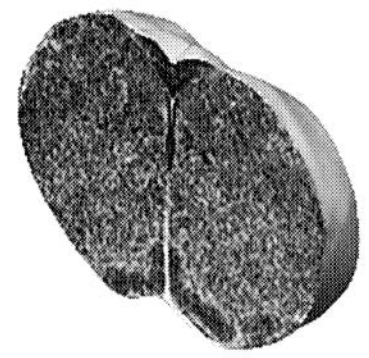